NON-GAUSSIAN STATISTICAL COMMUNICATION THEORY

IEEE Press
445 Hoes Lane
Piscataway, NJ 08854

IEEE Information Theory Society, Sponsor

NON-GAUSSIAN STATISTICAL COMMUNICATION THEORY

DAVID MIDDLETON

IEEE SERIES ON DIGITAL & MOBILE COMMUNICATION

IEEE
IEEE PRESS

WILEY
A JOHN WILEY & SONS, INC., PUBLICATION

Published simultaneously in Canada

Library of Congress Cataloging-in-Publication Data:

Print ISBN: 9780470948477

Printed in the United States of America

10 9 8 7 6 5 4 3 2 1

To:

William A. Von Winkle (1930–2007),
Inspiring Director of Research (NUSC), and loyal friend;

and to my Wife,
Joan Bartlett Middleton,
who provided the essential support and encouragement
needed in creating this book and in fulfilling my life.

CONTENTS

Foreword xv

Visualizing the Invisible xvii

Acknowledgments xxi

About the Author xxiii

Editor's Note xxv

Introduction 1

1 Reception as a Statistical Decision Problem 15

1.1 Signal Detection and Estimation, 15
1.2 Signal Detection and Estimation, 17
1.2.1 Detection, 17
1.2.2 Types of Extraction, 20
1.2.3 Other Reception Problems, 21
1.3 The Reception Situation in General Terms, 22
1.3.1 Assumptions: Space–Time Sampling, 22
1.3.2 The Decision Rule, 25
1.3.3 The Decision Problem, 26
1.3.4 The Generic Similarity of Detection and Extraction, 27
1.4 System Evaluation, 27
1.4.1 Evaluation Functions, 27
1.4.2 System Comparisons and Error Probabilities, 30
1.4.3 Optimization: Bayes Systems, 31
1.4.4 Optimization: Minimax Systems, 32
1.5 A Summary of Basic Definitions and Principal Theorems, 35
1.5.1 Some General Properties of Optimum Decision Rules, 35
1.5.2 Definitions, 36
1.5.3 Principal Theorems, 37

1.5.4 Remarks: Prior Probabilities, Cost Assignments, and System Invariants, 38
1.6 Preliminaries: Binary Bayes Detection, 40
1.6.1 Formulation I: Binary On–Off Signal Detection, 42
1.6.2 The Average Risk, 43
1.6.3 Cost Assignments, 43
1.6.4 Error Probabilities, 45
1.7 Optimum Detection: On–Off Optimum Processing Algorithms, 46
1.7.1 The Logarithmic GLRT, 48
1.7.2 Remarks on the Bayes Optimality of the GLR, 48
1.8 Special On–Off Optimum Binary Systems, 50
1.8.1 Neyman–Pearson Detection Theory, 50
1.8.2 The Ideal Observer Detection System, 51
1.8.3 Minimax Detectors, 52
1.8.4 Maximum Aposteriori (MAP) Detectors from a Bayesian Viewpoint, 53
1.8.5 Bayesian Sequential Detectors, 57
1.9 Optimum Detection: On–Off Performance Measures and System Comparisons, 57
1.9.1 Error Probabilities: Optimum Systems, 58
1.9.2 Error Probabilities: Suboptimum Systems, 65
1.9.3 Decision Curves and System Comparisons, 66
1.10 Binary Two-Signal Detection: Disjoint and Overlapping Hypothesis Classes, 69
1.10.1 Disjoint Signal Classes, 69
1.10.2 Overlapping Hypothesis Classes, 70
1.11 Concluding Remarks, 73
References, 74

2 Space–Time Covariances and Wave Number Frequency Spectra: I. Noise and Signals with Continuous and Discrete Sampling **77**
2.1 Inhomogeneous and Nonstationary Signal and Noise Fields I: Waveforms, Beam Theory, Covariances, and Intensity Spectra, 78
2.1.1 Signal Normalization, 79
2.1.2 Inhomogeneous Nonstationary (*Non-WS-HS*) Noise Covariances, 80
2.1.3 Narrowband Fields, 83
2.1.4 Noise and Signal Field Covariances: Narrowband Cases, 88
2.2 Continuous Space–Time Wiener–Khintchine Relations, 91
2.2.1 Directly Sampled Approximation of the W–Kh Relations (Hom-Stat Examples), 93
2.2.2 Extended Wiener–Khintchine Theorems: Continuous Inhomogeneous and Nonstationary Random (Scalar) Fields, 95
2.2.3 The Important Special Case of Homogeneous—Stationary Fields—Finite and Infinite Samples, 100
2.3 The W–Kh Relations for Discrete Samples in the Non-Hom-Stat Situation, 102
2.3.1 The Amplitude Spectrum for Discrete Samples, 102
2.3.2 Periodic Sampling, 107

2.4 The Wiener–Khintchine Relations for Discretely Sampled Random Fields, 108
2.4.1 Discrete Hom-Stat Wiener–Khintchine Theorem: Periodic Sampling and Finite and Infinite Samples, 110
2.4.2 Comments, 112
2.5 Aperture and Arrays—I: An Introduction, 115
2.5.1 Transmission: Apertures and Their Fourier Equivalents, 116
2.5.2 Transmission: The Propagating Field and Its Source Function, 120
2.5.3 Point Arrays: Discrete Spatial Sampling, 126
2.5.4 Reception, 129
2.5.5 Narrowband Signals and Fields, 134
2.5.6 Some General Observations, 137
2.6 Concluding Remarks, 138
References, 139

3 Optimum Detection, Space–Time Matched Filters, and Beam Forming in Gaussian Noise Fields 141
3.1 Optimum Detection I: Selected Gaussian Prototypes—Coherent Reception, 142
3.1.1 Optimum Coherent Detection. Completely Known Deterministic Signals in Gauss Noise, 142
3.1.2 Performance, 146
3.1.3 Array Processing II: Beam Forming with Linear Arrays, 150
3.2 Optimum Detection II: Selected Gaussian Prototypes—Incoherent Reception, 154
3.2.1 Incoherent Detection: I. Narrowband Deterministic Signals, 154
3.2.2 Incoherent Detection II. Deterministic Narrowband Signals with Slow Rayleigh Fading, 169
3.2.3 Incoherent Detection III: Narrowband Equivalent Envelope Inputs—Representations, 172
3.3 Optimal Detection III: Slowly Fluctuating Noise Backgrounds, 176
3.3.1 Coherent Detection, 176
3.3.2 Narrowband Incoherent Detection Algorithms, 180
3.3.3 Incoherent Detection of Broadband Signals in Normal Noise, 183
3.4 Bayes Matched Filters and Their Associated Bilinear and Quadratic Forms, I, 188
3.4.1 Coherent Reception: Causal Matched Filters (Type 1), 190
3.4.2 Incoherent Reception: Causal Matched Filters (Type 1), 192
3.4.3 Incoherent Reception-Realizable Matched Filters; Type 2, 195
3.4.4 Wiener–Kolmogoroff Filters, 198
3.4.5 Extensions: Clutter, Reverberation, and Ambient Noise, 200
3.4.6 Matched Filters and Their Separation in Space and Time I, 202
3.4.7 Solutions of the Discrete Integral Equations, 207
3.4.8 Summary Remarks, 214

3.4.9 Signal-to-Noise Ratios, Processing Gains, and Minimum Detectable Signals. I, 214
3.5 Bayes Matched Filters in the Wave Number–Frequency Domain, 219
3.5.1 Fourier Transforms of Discrete Series, 219
3.5.2 Independent Beam Forming and Temporal Processing, 230
3.6 Concluding Remarks, 235
References, 235

4 Multiple Alternative Detection **239**
4.1 Multiple-Alternative Detection: The Disjoint Cases, 239
4.1.1 Detection, 240
4.1.2 Minimization of the Average Risk, 242
4.1.3 Geometric Interpretation, 244
4.1.4 Examples, 245
4.1.5 Error Probabilities, Average Risk, and System Evaluation, 250
4.1.6 An Example, 253
4.2 Overlapping Hypothesis Classes, 254
4.2.1 Reformulation, 255
4.2.2 Minimization of the Average Risk for Overlapping Hypothesis Classes, 257
4.2.3 Simple (K + 1) - ary Detection, 259
4.2.4 Error Probabilities, Average and Bayes Risk, and System Evaluations, 260
4.3 Detection with Decisions Rejection: Nonoverlapping Signal Classes, 262
4.3.1 Optimum (K + 1) - ary Decisions with Rejection, 264
4.3.2 Optimum (K + 1) - ary Decision with Rejection, 265
4.3.3 A Simple Cost Assignment, 266
4.3.4 Remarks, 267
References, 270

5 Bayes Extraction Systems: Signal Estimation and Analysis, $p(H_1) = 1$ **271**
5.1 Decision Theory Formulation, 272
5.1.1 Nonrandomized Decision Rules and Average Risk, 272
5.1.2 Bayes Extraction With a Simple Cost Function, 274
5.1.3 Bayes Extraction With a Quadratic Cost Function, 278
5.1.4 Further Properties, 281
5.1.5 Other Cost Functions, 283
5.2 Coherent Estimation of Amplitude (Deterministic Signals and Normal Noise, $p(H_1) = 1$), 287
5.2.1 Coherent Estimation of Signal Amplitude Quadratic Cost Function, 287
5.2.2 Coherent Estimation of Signal Amplitude (Simple Cost Functions), 290

5.2.3 Estimations by (Real) θ Filters, 291
5.2.4 Biased and Unbiased Estimates, 293
5.3 Incoherent Estimation of Signal Amplitude (Deterministic Signals and Normal Noise, $p(H_1) = 1$), 294
5.3.1 Quadratic Cost Function, 294
5.3.2 "Simple" Cost Functions SCF_1 (Incoherent Estimation), 298
5.4 Waveform Estimation (Random Fields), 300
5.4.1 Normal Noise Signals in Normal Noise Fields (Quadratic Cost Function), 300
5.4.2 Normal Noise Signals in Normal Noise Fields ("Simple" Cost Functions), 301
5.5 Summary Remarks, 304
References, 305

6 Joint Detection and Estimation, $p(H_1) \leq 1$: I. Foundations **307**
6.1 Joint Detection and Estimation under Prior Uncertainty [$p(H_1) \leq 1$]: Formulation, 309
6.1.1 Case 1: No Coupling, 312
6.1.2 Case 2: Coupling, 314
6.2 Optimal Estimation [$p(H_1) \leq 1$]: No Coupling, 315
6.2.1 Quadratic Cost Function: MMSE and Bayes Risk, 316
6.2.2 Simple Cost Functions: UMLE and Bayes Risk, 319
6.3 Simultaneous Joint Detection and Estimation: General Theory, 326
6.3.1 The General Case: Strong Coupling, 326
6.3.2 Special Cases I: Bayes Detection and Estimation With Weak Coupling, 331
6.3.3 Special Cases II: Further Discussion of $\gamma^*_{p<1|QCF}$ for Weak or No Coupling, 333
6.3.4 Estimator Bias ($p \leq 1$), 336
6.3.5 Remarks on Interval Estimation, $p(H_1) \leq 1$, 338
6.3.6 Detection Probabilities, 339
6.3.7 Waveform Estimation ($p \leq 1$): Coupled and Uncoupled D and E, 341
6.3.8 Extensions and Modifications, 342
6.3.9 Summary Remarks, 345
6.4 Joint D and E: Examples–Estimation of Signal Amplitudes [$p(H_1) \leq 1$], 350
6.4.1 Amplitude Estimation, $p(H_1) = 1$, 352
6.4.2 Bayes Estimators and Bayes Error, $p(H_1) \leq 1$, 355
6.4.3 Performance Degradation, $p < 1$, 358
6.4.4 Acceptance or Rejection of the Estimator: Detection Probabilities, 367
6.4.5 Remarks on the Estimation of Signal Intensity $I_o \equiv a_o^2$, 371
6.5 Summary Remarks, $p(H)_1 \leq 1$: I—Foundations, 378
References, 379

7 Joint Detection and Estimation under Uncertainty, $p_k(H_1) < 1$. II. Multiple Hypotheses and Sequential Observations **381**

7.1 Jointly Optimum Detection and Estimation under Multiple Hypotheses, $p(H_1) \leq 1$, 382

7.1.1 Formulation, 383

7.1.2 Specific Cost Functions, 389

7.1.3 Special Cases: Binary Detection and Estimation, 396

7.2 Uncoupled Optimum Detection and Estimation, Multiple Hypotheses, and Overlapping Parameter Spaces, 400

7.2.1 A Generalized Cost Function for K-Signals with Overlapping Parameter Values, 402

7.2.2 QCF: Overlapping Hypothesis Classes, 403

7.2.3 Simple Cost Functions ($SCF_{1,2}$): Joint $D+E$ with Overlapping Hypotheses Classes, 406

7.3 Simultaneous Detection and Estimation: Sequences of Observations and Decisions, 407

7.3.1 Sequential Observations and Unsupervised Learning: I. Binary Systems with Joint Uncoupled $D+E$, 407

7.3.2 Sequential Observations and Unsupervised Learning: II. Joint $D+E$ for Binary Systems with Strong and Weak Coupling, 414

7.3.3 Sequential Observations and Unsupervised Learning: III. Joint $D+E$ Under Multiple Hypotheses with Strong and Weak Coupling and Overlapping Hypotheses Classes, 417

7.3.4 Sequential Observations and Overlapping Multiple Hypothesis Classes: Joint $D+E$ with No Coupling, 423

7.3.5 Supervised Learning (Self-Taught Mode): An Introduction, 425

7.4 Concluding Remarks, 428

References, 432

8 The Canonical Channel I: Scalar Field Propagation in a Deterministic Medium **435**

8.1 The Generic Deterministic Channel: Homogeneous Unbounded Media, 437

8.1.1 Components of the Generic Channel: Coupling, 438

8.1.2 Propagation in An Ideal Medium, 439

8.1.3 Green's Function for the Ideal Medium, 441

8.1.4 Causality, Regularity, and Reciprocity of the Green's Function, 448

8.1.5 Selected Green's Functions, 449

8.1.6 A Generalized Huygens Principle: Solution for the Homogeneous Field α_H, 453

8.1.7 The Explicit Role of the Aperture or Array, 463

8.2 The Engineering Approach: I—The Medium and Channel as Time-Varying Linear Filters (Deterministic Media), 465

8.2.1 Equivalent Temporal Filters, 466

8.2.2 Causality: Extensions of the Paley–Wiener Criterion, 471

8.3 Inhomogeneous Media and Channels—Deterministic Scatter and Operational Solutions, 473
8.3.1 Deterministic Volume and Surface Scatter: The Green's Function and Associated Field $\alpha^{(Q)}$, 475
8.3.2 The Associated Field and Equivalent Solutions for Volumes and Surfaces, 478
8.3.3 Inhomogeneous Reciprocity, 480
8.3.4 The GHP for Inhomogeneous Deterministic Media including Backscatter, 484
8.3.5 Generalizations and Remarks, 493
8.4 The Deterministic Scattered Field in Wave Number–Frequency Space: Innovations, 494
8.4.1 Transform Operator Solutions, 496
8.4.2 Commutation and Convolution, 498
8.5 Extensions and Innovations, Multimedia Interactions, 499
8.5.1 The $\hat{\eta}$-Form: Multimedia Interactions, 500
8.5.2 The Feedback Operational Representation and Solution, 503
8.5.3 An Estimation Procedure for the Deterministic Mass Operators $\hat{Q}$ and $\hat{\eta}$, 507
8.5.4 The Engineering Approach II: Inhomogeneous Deterministic Media, 508
8.6 Energy Considerations, 509
8.6.1 Outline of the Variation Method, 510
8.6.2 Preliminary Remarks, 512
8.6.3 Energy Density and Density Flux: Direct Models—A Brief Introduction, 514
8.6.4 Equal Nonviscous Elastic Media, 516
8.6.5 Energy Densities and Flux Densities in the Dissipative Media, 522
8.6.6 Extensions: Arrays and Finite Duration Sources and Summary Remarks, 527
8.7 Summary: Results and Conclusions, 535
References, 536

9 The Canonical Channel II: Scattering in Random Media; "Classical" Operator Solutions 539
9.1 Random Media: Operational Solutions—First- and Second-Order Moments, 541
9.1.1 Operator Forms: Moment Solutions and Dyson's Equation, 543
9.1.2 Dyson's Equation in Statistically Homogeneous and Stationary Media, 551
9.1.3 Example: The Statistical Structure of the Mass Operator $\hat{Q}_1^{(d)}$, with $\langle \hat{Q} \rangle = 0$, 560
9.1.4 Remarks, 564
9.2 Higher Order Moments Operational Solutions for The Langevin Equation, 565

9.2.1 The Second-Order Moments: Analysis of the Bethe–Salpeter Equation (BSE), 565
9.2.2 The Structure of $\hat{Q}_{12}^{(d)}$, 568
9.2.3 Higher-Order Moment Solutions ($m \geq 3$) and Related Topics, 570
9.2.4 Transport Equations, 572
9.2.5 The Gaussian Case, 574
9.2.6 Very Strong Scatter: Saturation $\|\hat{\eta}\| \simeq 1$, 575
9.2.7 Remarks, 579
9.3 Equivalent Representations: Elementary Feynman Diagrams, 580
9.3.1 Diagram Vocabulary, 581
9.3.2 Diagram Approximations, 586
9.3.3 A Characterization of the Random Channel: First- and Second-Order-Moments I, 594
9.3.4 Elementary Statistics of the Received Field, 596
9.4 Summary Remarks, 598
References, 599

Appendix A1 **601**

Index **617**

FOREWORD

"The statistical theory of communication has proven to be a powerful methodology for the design, analysis, and understanding of practical systems for electronic communications and related applications. From its origins in the 1940's to the present day, this theory has remained remarkably vibrant and useful, while to evolve as new modes of communication emerge. The publication in 1960 of *Introduction to Statistical Communication Theory (ISCT)* was a landmark for the field of statistical communication theory – random processes, modulation and detection, signal extraction, information theory – we combined into a single, unified treatment at a level of depth and degree of completeness which have not been matched in any subsequent comprehensive work. Moreover, *ICST* introduced a further interdisciplinary feature, in which relevant physical characteristics of communication channels were incorporated into many topics."

I wrote these words in my foreword to the 1996 IEEE Press reissue of David Middleton's classic book *Introduction to Statistical Communication Theory (ISCT)*. In the decades after he wrote *ISCT* and well into the 2000s, Dr. Middleton pursued a very active research program motivated by the above last-mentioned feature of his book, namely, the incorporation of physics into communication theory. Among other considerations, this work notably brought in two features that were not evident in *ISCT*: the spatial dimension and the fact that many of the noise phenomena arising in applications are not well modeled as Gaussian random processes. Through this research, he greatly expanded our understanding of the physical aspects of communication, and this book was conceived as a synthesis of that work.

David Middleton was a giant in the field of statistical communication theory, and the physicist's perspective he brought to the field was somewhat rare. Sadly, he did not live to complete this book, his full vision for which is described in the introductory chapter. However, the material that he did complete is nevertheless of considerable value in exposing the insights that he gained over more than six decades in the field.

H. Vincent Poor
Princeton, New Jersey

VISUALIZING THE INVISIBLE

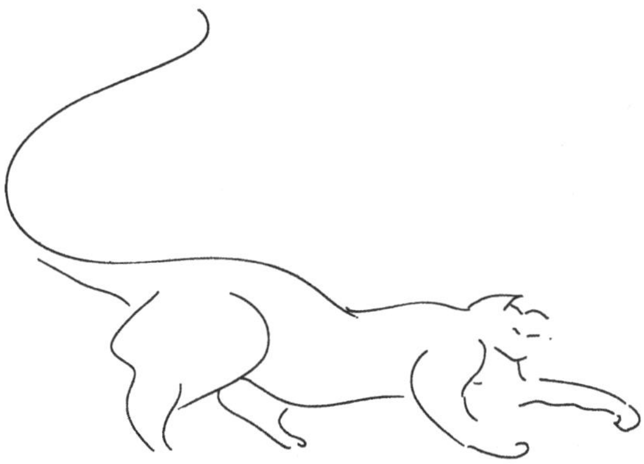

Hanging on my apartment wall is a framed line drawing entitled *Cat with Mouse* from 1959. The abstract profile of tail and rear haunches connects in swift sweeping lines across the paper to form pointed ears, whiskers, and culminates in a half hidden mouse enveloped in its paws, its fate uncertain. It has a Picasso-like quality, curvilinear strokes of the oil crayon made with verve, remarkable for the empty space surrounding the lines as for the seductive feline form. The artist was David Blakeslee Middleton, my father, and judging from the date on the drawing he probably made this sketch one evening after a day in his study proofreading the galleys for his soon to be published *An Introduction to Statistical Communication Theory*. Drawing on a big manila pad was one of his ways to unwind, perhaps to counterbalance the intense analytical left brain activity with some right brain visual and emotional release through imaginative sketches. Oftentimes this was also a clever way to entertain his young children before bedtime. I can vividly recall various creatures—cats in particular were a favorite—and other-worldly landscapes magically taking shape on the blank paper. Witnessing these forms unfolding from my father's pen, to an impressionable child it seemed that science and drawing were naturally linked thus:

$$\textit{father} = \textit{scientist},\ \textit{father} = \textit{artist},\ \textit{therefore scientist} = \textit{artist}.$$

In this Preface I'd like to share a few recollections about how the arts, and drawing in particular, played an integral part in David Blakeslee Middleton's creative process. And since I am an architect, the reader will hopefully forgive "artistic license" to speculate that his ability to visualize an otherwise invisible world of signals and noise through drawing contributed to the evolution of his research in communication theory over a sixty year period, culminating in this book.

My father was gifted with a marvelous drawing hand and a free imagination. No doubt being raised in a family of creative artists fostered this talent. His mother played violin. His father was a poet, one uncle a playwright, another uncle the Austrian émigré painter Joseph Amadeus Fleck, and one aunt was the author of well known western novels including *My Friend Flicka*. Dad was married twice, to women who were both gifted artists. On top of this arsenal of artistic influences, my father was a seasoned classical pianist. While never quite ready for Carnegie Hall, he would practice regularly and with evident pleasure into the late evening hours the sonatas of Beethoven, Schubert and Chopin. He accrued over many years a large library of history, biography, and many books on painting and sculpture.

The inclination—possibly the compulsion—to draw was, I think, a natural part of his life and heritage. It manifested itself on an endless stream of drawings on small white note or oversized manila artist pads, on gift cards for friends and family, doodles with authorial inscriptions in his books, and diagrams and marginalia surrounding his manuscripts on yellow lined writing tablets. He drew animals—phantasmagorical or real—incessantly, and when clearing out his papers after his death in 2008, my siblings and I discovered a trove of these drawings. One can discern the evolution of his drawing style from an early reliance on ruled lines and graphs, along with somewhat rigid Cubist-like animal profiles and caricatures, to some three decades later drawings full of squiggles and agitated contours, where straight lines are only rarely evident. In one image of a dragon (another favorite subject) the profile is so diffused and fuzzy it seems like the poor creature has been intercepted by intense noise signals and is being jostled to the point of being truly indeterminate. Over time, this tendency to greater abstraction of figure, of fluctuating line weight, and verve of the freehand line may also be indicative of his later research encompassing quantum theory and its attendant complexity and fluidity. My father's technical diagrams and illustrations, usually drawn by hand and directly transcribed into his first two books and a host of his published research papers, appear to become more visually dynamic and complex over time. This may due to the increasing complexity of the science involved, or to better explain the concept to the reader, or both.

I doubt a day of work went by that my father didn't draw a graph or figure while working his physics. I would stop in his study upstairs to say hi and there would be invariably a handful of colored pens, always fine point, scattered on his crowded desk or stowed in his shirt pocket (I never saw a pocket liner—but often an ink stain ruining another button-down Oxford). A typical draft manuscript would have at least several squiggles and amoeboid-like drawings, with darts of straight lines, attended by numeric or alphabetical punctuation, piercing through or oscillating around the perimeters of these figures. Parabolic arcs, waves both regular and asymmetrical, conic sections and planar intersections, vast irregular rhomboids and tori: these would all be lovingly drawn, in a fine line, exacting proportion, and precise annotation. They strike the uninitiated (at least this non-scientist) as abstract volumes and forms in a state of visual tension, yet somehow poised, not unlike figures in a Miro or early Kandinsky painting. As a young adult, struggling in calculus and marginally apprehending the finer points of Newton's Three Laws, these drawings made physics more accessible, more "real." To this day I remain baffled as to the actual science, but they gave graphic shape and substance to equations describing a part of the electromagnetic spectrum. These diagrams remain for me a rich formal vocabulary intriguing in their geometry and multi-dimensions, making visible the world of detecting signals in a cosmos of noise.

My father's working method was really like that of a visual artist or even a composer. If one looks at any of his early drafts of his abstracts or papers it resembles a score by Beethoven, full of scratches and ink blotches and re-scribed measures. One would never perceive them as "Mozartian", immaculately and precisely inscribed the first time set to paper. Indeed, some "final" drafts of his abstracts or even the chapters for this book look like collages. They are full of rub outs, white-outs (liquid or tape!), edits scribbled vertically in the margins in multiple ink colors, here and there whole sections cut out with a razor and replaced with revised equations and text. Few pages in a final draft would escape this sort of surgery. And almost everything was done long hand; typewriters just weren't capable of constructing the architecture of these dense and complex equations. While the technology of computational devices evolved rapidly in his lifetime, my father rarely if ever availed himself of a computer, even when they became a household appliance. Trained on a slide rule, for many years his only concession to modern hardware was a 1987 Hewlett Packard hand held calculator with Reverse Polish Notation only occasionally dusted off to verify a certain summation. Otherwise the entire math for his abstracts and various research papers was done in his head, augmented with a set of yellow or white lined paper pads to work through some of the calculations. I was astonished to notice that over all the decades he almost always wrote his manuscripts on ink and paper. Pencil drafts are rarely evident. And despite the time it required, I can discern an almost joyful immersion in the tactility of "cut and paste". I can not attest to the science, but the manuscript process provided one heck of show: It was "messy vitality" at its finest!

"Creation," as the architect Le Corbusier once said, "is a patient search." I grew up with my father modeling this dictum every day. Dad's efforts had all the dead ends and small victories familiar to any creative artist—or scientist: many hours of quiet, methodical sorting and testing, punctuated by intermittent *sturm und drang* of frustration, occasional late nights of revisions to remove some small theoretical imperfection, or re-drafting to get it "just right". Science is a rigorous, un-sentimental, and empirical pursuit. This effort was one of mental exertion, but it seemed to me often one of physical endurance for my father. Sometimes in the early evening when he came downstairs from his study and said "it was a hard day at the office," he surely meant it.

It wasn't until his last days in the hospital, still proofing and re-writing sections of the chapters of this book while cancer gnawed away at him, that I began to appreciate the implication of his intention to include "a space-time treatment" in the title of this monograph—and to apprehend how much visualization was a key to his particular type of research. Not unlike his drawing style, I believe his research evolved from a cool "classical" foundation in statistics to the rarefied and "romantic" world of quantum theory. He didn't say much about this compulsion to draw, but didn't have to: one can easily discern the delight and care he took to making the diagrams illuminate, punctuate, and indeed more richly describe what his summations and differentials express (to my untrained eye) in their inscrutable syntax. Only later in life did I come to realize that not all scientists had this gift, and that for many their "patient search" was less tactile and literally more cerebral.

I'm convinced my father's research relied on intensely visualizing the natural environment—imagining the shape and texture of a stochastic universe—in order to tease out the abstract equations that would accurately account for the sometimes predictable but more likely random events and features of a world full of physical uncertainty. It is a world of turbulence, under the sea or in the air, waves of all shapes and sizes and curves, chaotic and sublime. Today this world is super-saturated with signals, from radar and sonar and radio, from ELF to microwave broadcasts, cell phones and Blue-Tooth and E-Z Pass readers, all seeking a receptor of one sort or another, always under threat of being scattered. As I reflect on how David Blakeslee Middleton's work sought to find order and predictability in chaos, I am awed at how fantastically rich the physics of something that can not be seen readily has had such an impact on our modern life, and the role my father played in enhancing the connective-ness we today take for granted.

David Blakeslee Middleton

ACKNOWLEDGMENTS

Publication of David Blakeslee Middleton's *opus finale* could not have been accomplished without the wonderful support and efforts of his colleagues, collaborators, and editors. These include Dawn M. Ernst-Smith and Nancy Russell, who had the acuity of vision and the patience of saints in transcribing my father's handwriting into cogent and editable form. Those who provided my father encouragement and careful editing over the many years of gestation of this book include Dr. Leon Cohen, Dr. Joseph Goodman, and Dr. Julius Bussgang, who provided guidance and support for the purpose of this publication, and most especially Dean Vincent Poor of Princeton University's Department of Electrical Engineering, who gave valuable advice and encouragement throughout the later stages of this work, and who organized a symposium in honor of my father at Princeton in 2008. My family is especially indebted to: Dr. John Anderson of Lund University, who has painstakingly proofread the manuscript, and has gracefully provided a Editor's Note describing the context of this work; Sanchari Sil at Thomson Digital for meeting the exacting challenges of typesetting and composition; and the Editors and Publishers at John Wiley & Sons – most especially Danielle LaCourciere, Senior Production Editor and Taisuke Soda, Senior Editor – who collectively had the diligence and vision to get this work to press despite some formidable obstacles. The Middleton family expresses our heartfelt gratitude to all involved for bringing to completion a task my father had hoped to finish himself.

David Blakeslee Middleton FAIA
New York City, 2011

ABOUT THE AUTHOR

Physicist and pioneer of modern Statistical Communication Theory, David Middleton devoted his career to the study of how information is transferred from one point in space-time to another, with ground-breaking applications in radar, underwater listening devices, satellite technology, and wireless communications. Born in 1920 in New York City, he attended Deerfield Academy and graduated from Harvard College in 1942. During WW II, as a Special Research Associate at Harvard's Radio Research Laboratory, he refined the analysis and subsequent effectiveness of aluminum "chaff" used to jam enemy radar, protecting American aircraft from detection. He received his M.A in 1945 and Ph.D. from Harvard in 1947. From 1949–54 he was an Assistant Professor of Applied Physics at Harvard.

In 1960, Dr. Middleton published "An Introduction to Statistical Communication Theory". Widely translated and reprinted, it remains the seminal text for the field. In 1965 he published "Topics in Communication Theory". For over fifty years he taught, inspired and occasionally confounded his graduate and doctoral students at Harvard, RPI, Columbia, Johns Hopkins, Texas, Rice, and University of Rhode Island. From 1954 on he was a Consulting Physicist to various government agencies, advisory boards, private industry and served for seven years on the U.S. Naval Research Advisory Committee. During his lifetime Dr. Middleton published over 160 papers and abstracts.

Dr. Middleton was a Life Fellow of the Institute of Electrical and Electronic Engineers, and Fellow of the American Physical Society, the Acoustical Society of America, and American Association for the Advancement of Science. He died in the city of his birth on 16 November 2008.

EDITOR'S NOTE

When IEEE Press and its Series on Digital and Mobile Communication signed a contract with David Middleton in 1996, we could not expect that the book would take 16 years and run to 24 chapters. By the time of his passing in 2008, David had completed versions of 10 chapters, and more than one version of several. How should one prepare such a manuscript for publication? Unlike F.X. Suessmayr, the young assistant who completed Mozart's Requiem, I have not made up new text and whole chapters. The Requiem became a beloved piece of music, even with its Suessmayr chapters, but after much deliberation we have decided to limit David's book to nine chapters. Successive chapters are more rough in form and contain newer material, and it becomes steadily more risky to guess Middleton's intentions. With Chapter 10, we could find no practical way to create a reliable text. Those who would like to see for themselves–and perhaps attempt a revision–can view this part of the manuscript on the Wiley book Web site.

Thus, we present a book of nine chapters. David Middleton's original "Introduction" is reproduced as he wrote it, and the reader can see there the magnificent opus that he had in mind. In the rest of the book, we have removed all mention of the missing parts and we have aimed to make the book a coherent work in nine chapters. Exercises existed for Chapters 4 and 6 only, and these are included. There were to be many appendices, but the text for only two exists; these are included.

Middleton's style employed a multitude of equation and section references, and these presented a special problem because they often referred to early versions now lost. Usually they could be re-established, but where there was a serious risk of error, they were deleted. References to conclusions reached in Chapters 11–24 were softened or deleted, since in many cases it was doubtful that they exist. Many typed manuscript passages were not proofread by the author, and contained obvious errors and misspelled foreign names that had to be corrected. With thousands of corrections, it is certain there are errors and misguesses in this editorial process. We regret this, and ask for the indulgence of the reader.

What then is this book about? As the title suggests, it is about detection and estimation when statistics are neither Gaussian nor homogeneous. For example, in Chapter 2, Middleton extends Wiener–Khintchine theory to this case. But Middleton believes that the essence of the problem is transmission medium, and he devotes the later of the nine chapters to complicated multilayer inhomogeneous media, whose transmission is as much by diffusion as by electromagnetic waves. Another recurring theme in the book, as pointed out by Vincent Poor, is processing in four-dimensional space–time. A perhaps more subtle theme is sonar and signaling in the ocean. His hope seems to have been that later chapters would tie together these frightfully complex media with traditional detection and estimation. He had a vision for how to do this. Could he have carried it out as he wished? Can anyone carry it out, or is a practical and understandable solution beyond our comprehension? We will not find the answer this time around.

The attentive reader can find more than these technical matters. David Middleton's book is a window to a past now nearly forgotten: to mid-twentieth century pioneers in detection, estimation, and signal processing; to organizations that changed the world; to a Cold War with doomsday submarines; and to a much smaller research community, sometimes employed by opposing armies, who nonetheless knew each other's work. At the end of Chapter 3, readers can find an interesting history of the matched filter, which Middleton helped discover. An historical oddity is that Middleton was a coauthor of the first paper, Vol. 1, No. 1, p. 1, published by the prestigious *IEEE Transactions on Information Theory* (Ref. 1 in Chapter 4). It is also interesting that Middleton's day-to-day research world was almost free of computers. He mentions them from time to time, but a researcher in this field today would base his or her thinking much more on algorithms and what they could and could not do, and verify the work every step of the way with computations.

We wish you happy reading!
John B. Anderson

INTRODUCTION

In his Introduction, David Middleton refers to the parts of his book that were not completed; we have left them in place so that the reader may see the original plan for the book — Editors.

This Introduction explains my purpose of writing this book and its earlier companion [1]. It is based on the observation that *communication* is the central operation of discovery in all the sciences. In its "active mode," we use it to "interrogate" the physical world, sending appropriate "signals" and receiving nature's "reply." In the "passive mode," we receive nature's signals directly. Since we never know *a priori* what particular return signal will be forthcoming, we must necessarily adopt a probabilistic model of communication. This has developed over approximately 70 years since its beginning into a statistical communication theory (SCT). Here, it is the *set* or *ensemble* of possible results that is meaningful. From this ensemble, we attempt to construct an appropriate model format, based on our understanding of the observed physical data and on the associated statistical mechanism, analytically represented by suitable probability measures.

Since its inception in the late 1930s, and in particular subsequent to World War II, SCT has grown into a major field of study. As we have noted above, SCT is applicable to all branches of science. The latter itself is inherently and ultimately probabilistic at all levels. Moreover, in the natural world, there is always a random background "noise" as well as an inherent *a priori* uncertainty in the presentation of deterministic observations, that is, those that are specifically obtained, *a posteriori*.

THE BOOK'S TITLE

Let me now begin with a brief explanation of the title of the book.

Non-Gaussian Statistical Communication Theory, David Middleton.

Elements of Non-Gaussian Space–Time Statistical Communication Theory, Propagation, Noise, and Signal Processing in the Canonical Channel

My choice of "elements" is intended to signify a connected development of fundamental material, but with an exposition that is inevitably incomplete, with many important topics necessarily omitted, primarily for reasons of space. "Elements," however, includes the propagation physics of the channel, the role of spatial coupling (e.g., apertures and arrays), and noise models, both physically founded. The analyses also treat deterministic and random scatter, Doppler effects, and, of course, four-dimensional (i.e., space and time) signal processing, with particular attention to threshold reception in arbitrary noise environments. *Non-Gaussian noise* receives special analysis, since it is a phenomenon of increasing practical importance. Moreover, it is a topic that presents much greater complexities than the familiar Gaussian noise model, which has dominated so much of recent as well as earlier studies.

In addition, the class of signals considered here is entirely general or "canonical,"[1] so that the coding results of parallel studies in *Information Theory*[2] [2] can be readily applied in specific cases. This book (Book 2) may also be considered an extension of Book 1 (*An Introduction to Statistical Communication Theory*, [1]). Book 1 considers primarily random processes and continuously sampled noise and signals. Here, on the other hand, Book 2 deals with many earlier features of Book 1. These that require a *four-dimensional space–time formulation* now involve *random fields*. Particular attention is also given here to the physics of propagation. In this context, another portion of this book is then devoted to physical problems of signal detection and extraction in a Bayesian formulation, with particular attention to threshold (or weak signal) operation.

Finally, both homogeneous and inhomogeneous media are considered here. Such media are *linear* provided their equations of propagation are themselves linear where the requirement, of course, is that the superposition principle holds: If α is a typical field in such media, we have, symbolically for two fields $\alpha = (\alpha_1, \alpha_2) \rightarrow \alpha_{12} \equiv \alpha_1 + \alpha_2$. On the other hand, for nonlinear media $(\alpha_1, \alpha_2) \rightarrow \alpha_{12} \neq \alpha_1 + \alpha_2$: superposition is violated. Furthermore, stationarity or nonstationarity itself does not invalidate linearity or nonlinearity. Note, however, that the presence of a (finite) boundary is itself an inhomogeneity of the medium, and thus is a component of nonlinearity by the above definition. The presence of scattering elements (inhomogeneities) is also a major topic of interest as are the probability distributions generated by such scattering elements.

COMMUNICATION THEORY, THE SCIENTIFIC METHOD, AND THE DETAILED ROLE OF SCT

As I have noted above, this book is primarily an analytical presentation. For numerical results, it may be regarded as providing a set of macroalgorithms, to direct the computation of the desired numerical quantities in specific cases. Because of the availability of cheap and powerful computing today as well as the modest software costs, such numerical results should be readily and quickly available, once the needed programs (software) are obtained

[1] Here the usage of "canonical" is to indicate a form independent of a particular choice in specific applications or branch of physics.

[2] See Fig. 1.

from the aforementioned macroalgorithms. These macroalgorithms are the consequence of a well-known scientific methodology whose basic principles are stated in somewhat simplified terms below:

$$\text{Hypothesis} + \text{Experiment} = \text{Theory}^{3} \tag{1}$$

Loosely stated, "hypothesis" is a conjecture or proposition; experiment is the procedure required to verify or to disprove the proposition. Verification here implies replication by any qualified observer any number of times. "Theory" is the result of successful verification. A theory is thus potentially acceptable (or not) as long as it is empirically verified by appropriate observation. A fruitful theory is one that not only accounts for the observed phenomena but also leads to further discoveries. For science, the arena of application is the physical world, where it is the ontology "what and how" of the universe, not its "why."

Here I employ two familiar types of theory. These I call *a posteriori* and *a priori* theories. In the former, everything is already known, for example, equations of state, boundary conditions, propagation relations, and so on. It remains only to calculate a numerical result, where all of its elements collectively constitute a unique representation, that is, a deterministic result. For the latter, the same structure exists, but it is not known which representation is *a priori* present, except that a particular result will have a certain probability of actually occurring in a universe of possible outcomes. It is this fundamental uncertainty that is the key added feature, defining the field of *statistical* communication theory. It is this probabilistic nature, combined with a set of deterministic representations and associated probability measures, that in turn defines the subject. This concept that was introduced systematically in the 1930 was accelerated by the Second World War (1939–1945), and has continued into the peace time explosion of the new science and its corresponding technology.

Apart from the broad and fundamental impact, SCT is the science indicated at the beginning of this Introduction. SCT also has a microstructure that has the specifically detailed role of including the physical sciences. Figure 1 illustrates other interdisciplinary relationships.

Figure 1 represents my subjective view of where statistical communication theory fits into the scientific enterprise and specifically where the signal processing and channel physics lie in this hierarchy. The double set of arrows (↑↓, ↓↑) indicates the interrelationships of the various fields of study and emphasizes their interdependence. Intimately involved in all of this is the role of technology, which provides the instruments of discovery, investigation, and study. The direct arrow in the diagram between SCT and the physical science further emphasizes the aforementioned fact that *communication* is the link connecting the natural world with the methodology used to study it. In fact, communication in one form or another, appropriate to the scale of the phenomenon under study, is the necessary vehicle here. The progression is from the micro to the macro, that is, the very small at the quantum level (e.g., the quantum fluctuations of the vacuum) to the astronomical dimensions of the fluctuating gravity fields of galaxies [3].

As we have noted at the beginning, physical science is also based on model building. This in turn requires both the interrogation for and the reception of data from the physical

[3] This simple relation is well understood by the scientific community. It is often confused by the public, where the terms "theory" and "hypothesis" are frequently interchanged and the essential role of the "experiment" is often omitted or misinterpreted. Such confusion can have serious consequences, since it is the public that ultimately supports scientific endeavors.

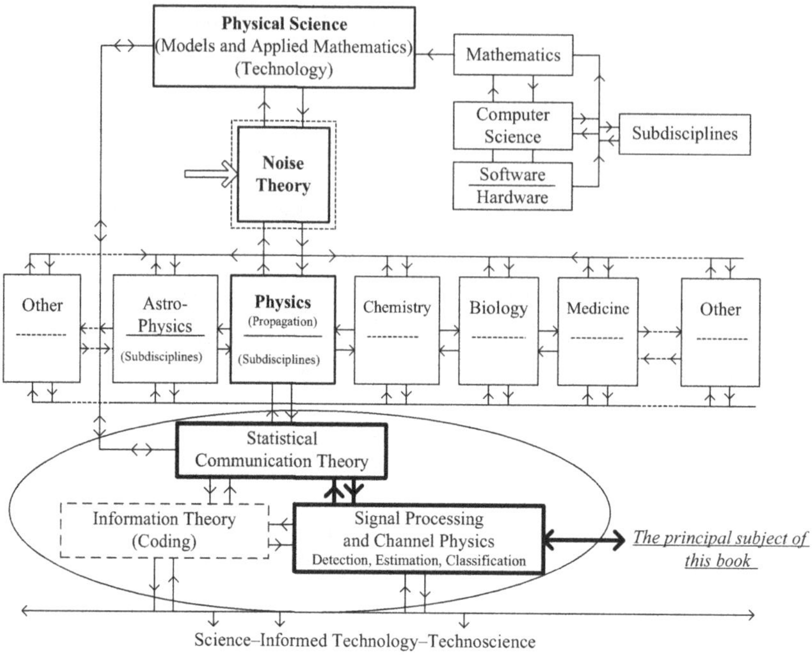

FIGURE 1 The role of statistical signal processing (SSP) in the physical sciences.

world. The probabilistic nature of the physical world is inherent according to modern understanding (as exemplified by Heisenberg's uncertainty principle and the behavior of subatomic particles), along with the ultimate uncertainty of measurement itself. The former is a fundamental property of matter and energy; the latter is independently a problem of technique. These remarks may be concisely summarized by the following three observations:

(1) Physical science is a *model building* operation.[4]
(2) Physical science is fundamentally a *probabilistic endeavor,* refer to Eq. (2).
(3) *Communication* is the process by which *hypothesis* is converted *into theory*, refer to Eq. (1).

Thus, the communication process either actively interrogates the real world or passively receives its "messages." This in either case embodies the role of experimental verification (or rejection), refer to Eq. (1). Although there is nothing really original about these remarks (1)–(3), they may serve as useful reminders of the scientific process. The quantitative language of science is, of course, mathematics.

Note the key place of noise theory in the hierarchy of Fig. 1. Although the role of noise is mostly a negative one from the point of view of treatment here in "Signal Processing" it has

[4] Whether model building is an act of discovery, an approximation of a reality independent of the observer (Platonism), or an act of invention, a Cartesian picture of reality dependent on the observer, is an unresolved philosophical question. The ("cogito ergo sum") empirical success of quantum mechanics, where the observer is part of the system, suggests at least some combination of the two, which then reduces to the not so simple physical reality; see [4].

proved to be a highly productive field in the broader context of modern physics since the beginning of the twentieth century [5]. For example, Einstein used it to prove the existence of atoms in 1905, which was a highly controversial topic up to that time. A host of other eminent scientists, among them Boltzmann and Langevin [5], also advanced its theory. See, for example, Ref. [3–6]. Since "noise," a basic random phenomenon, pervades every field of science, its study since then has yielded a host of discoveries and new methodologies. Its history has indeed proved "glorious," as Cohen has so aptly described it in his recent enlightening review article [6].

Although I have focused here on the deleterious and ubiquitous effects of noise on signal reception (particularly, in Parts 1 and 3 of the present book), its physical and analytical description, especially for *non-Gaussian noise*, is a necessary and significant major subject for discussion (cf. Part 2, Chapters 11–13). The dominant relation of noise to the physical problems inherent in signal processing is well known. Its important companion discipline in SCT, *Information Theory*,[5] is emphasized by the direct arrow in Fig. 1, as well as its connection to the physics of propagation. The close mutual relationship of the ensuing technology is also noted and is a major part of the advances discussed here and appearing in all fields of science.

THE SCOPE OF THE ANALYSIS AND NEW GENERAL RESULTS

Before summarizing the contents of this book, let me describe the physical domain of most of its contents. Topics not treated here are also noted. The principal areas of application are mainly acoustical and electromagnetic. Exceptions are quantum mechanical (the very small) and astrophysical (the very large), where elements of statistical communication theory are also specifically although briefly considered. In all cases from SCT viewpoint, we have to deal with noise, and signals in such noise, propagating in space–time. From SCT viewpoint, these are determined by the physical properties of the channel. Throughout, the appropriate language is statistical, specifically for the inherently random character of these channels and for reception, namely, detection and estimation in the face of uncertainty, that is, in an *a priori* theory.

We are also dealing here mostly with media that are regarded as continuous. These may be described by the following simple hierarchy:

(1) *Vacuum*: Empty space, no matter present at all. This is typically the usual assumption made here for most electromagnetic propagation. Such media, of course, do not support conventional acoustic propagation.

(2) *Gas*: A low-density continuum, for example, earth's atmosphere, and other low-density environments. These media clearly do not support a *shear*: $\nabla \times \boldsymbol{\alpha} = 0$, ($\alpha =$ displacement field; here α is also said to be *irrotational.*)

[5] In the Information Theory community, the space–time formulation represents the simultaneous use of separate multiple channels, which may be statistically related and possibly coupled for simultaneous new versions, of one or more signals received together. The details of the spatial environment in the immediate neighborhood of the separate receivers, however, are not directly considered from a physical point of view. The effects of the different receiver locations are subsumed in the different received waveforms. These are usually suitably combined in the receiver to take advantage of their common structure at the transmitter. The noise background at the receivers on the other hand can have noticeable statistical correlations, which can enhance the received signal(s) in reception, as is typical of MIMO (multiple input–multiple output) reception [7].

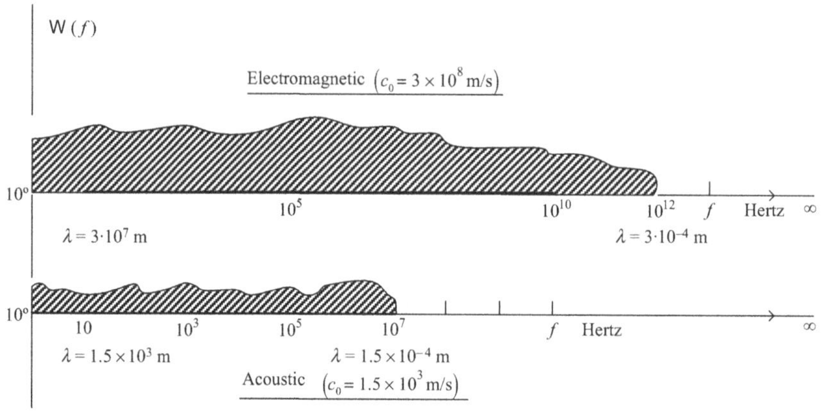

FIGURE 2 Current frequency range of signals used in general electromagnetic and acoustic communications, for example, radar, sonar, and telecommunications (from the relation $\lambda f = c_0$).

(3) *Liquid*: Usually of greater density than gas, for example, the ocean. These also do not support a shear, for example, $\nabla \times \boldsymbol{\alpha} = \mathbf{0}$. Here we consider the propagation of acoustic waves primarily.

(4) *Solids*: These are dense enough to maintain their shape or at most suffer minor distortions from a rest condition, that is, undergo and restore shape modifications. Such media are comparatively elastic, deformable, and restorable. These bodies can support a shear, for example, $\nabla \times \boldsymbol{\alpha} \neq \mathbf{0}$. When subject to stresses alone, these deformable media produce a displacement (vector) $\Delta = \boldsymbol{\alpha}$ that is *solenoidal*, that is, $\nabla - \boldsymbol{\alpha} = 0$. Electromagnetic and acoustic radiation is also possible in all the above media (except the latter in a vacuum). The magnitude of the results in the acoustic and electromagnetic cases depends, of course, on the physical properties of the media involved. For example, EM radiation in fresh and salt water is significantly weaker than acoustic radiation, but may be enhanced in certain conducting solids vis-à-vis vacuum or liquids. In all cases, we must pay attention to the appropriate conductive properties of the medium in question.

We define the domains of the analysis here in terms of the range of frequencies used for the signals and hence for the accompanying propagation whether ambient or signal generated (i.e., scattering). For acoustic applications, we have (for an average speed of sound in water $c_0 = 1.5 \times 10^3$ m/s), a frequency range is the order of (1–10^7 Hz) or in wavelengths[6] from $\lambda = 1.5 \times 10^3$ m to $\lambda = 1.5 \times 10^{-4}$ m. For the EM cases, we consider frequencies in the order of (10^0–10^{12} Hz) with the speed of light in space $c_0 = 3 \times 10^8$ m/s, which in wavelengths is in the order of (3×10^7–3×10^{-4} m). (These frequency or equivalent wavelength intervals, are, of course, somewhat loose and exceeded in any optical and quantum mechanical examples that are discussed here.)

Figure 2 illustrates the ranges of frequencies to be used in the (i) electromagnetic and (ii) acoustic applications.

[6] I use $f\lambda = c_0$, a dispersionless approximation in both acoustic and EM cases.

SOME SPECIFIC NEW RESULTS

With the above in mind, let me note now what I believe to be the major *new* material presented in this book. These are discussed in detail in subsequent chapters, some of which have also been published in recent journal papers. Note the following (mostly new) SCT topics specifically:

(1) A *space–time formulation*, largely *with discrete sampling* of continuous noise and signal fields. This includes *space–time matched filters* for optimum performance in reception, including quantification of *system degradation*, in particular for ultrawide-band (UWB) signals when optimal space–time (*ST*) processing is replaced by the usual *separate* space and time ($S \otimes T$) processing in conventional receivers.

(2) A theory of *jointly coupled detection and estimation*, as developed by the author and R. Esposito, and generalized in Chapters 6 and 7, with additional new references.

(3) *Extensions of classical noise theory* (Langevin equations, Fokker–Planck approximations, and classical scatter methods) to include random spatial phenomena (Chapter 9 of Part 2).

(4) *New methods in scattering theory* (1997–), from which first- and second-order probability densities are directly obtained (refer to Chapter 10 of Part 2 for this physical statistics (PS) approach vis-à-vis the limited classical statistical physics (SP) techniques discussed in Chapter 9).

(5) *Physically-based non-Gaussian noise models* (Class A and class B canonical noise distributions), developed since 1974 by the author and presented in Chapters 11 and 12, of Part 2.

(6) A systematic program of *threshold* (or weak signal) *detection and estimation* for general noise fields, particularly for non-Gaussian noise, and including as a special case, the Gaussian noise of earlier treatments (Part 3).

(7) The physics of propagation, in context of channel description and its space–time coupling, in transmission and reception for a variety of media, with general boundary conditions.

(8) An emphasis on the *interdisciplinary character* of the presentation throughout (see Fig. 1), with specific examples: astrophysics and computer traffic probabilities. This also leads to the development of *canonical expressions* independent of particular physical disciplines, which permits their treatment by a common methodology.

(9) An extension of the scalar field analysis of Part 2 (Chapters 8–14) to *vector fields* in Part 4, specifically illustrated by the full electromagnetic field, with the typical reception applications.

(10) A very concise treatment of quantum detection and estimation is presented in Part 4.

OVERVIEW: PARTS 1–4

The following provides a brief summary of the contents ot this book.

Note that Chapters 11–24 were not completed — Ed.

An Introduction to Non-Gaussian Statistical Communication Theory—A Bayesian Formulation

Part 1 presents most of the basic statistical apparatus needed to apply statistical decision theory from the Bayesian probabilistic point of view. These are the ultimate problems of statistical communication theory, namely, optimal and suboptimal detection (*D*) and extraction (*E*) of desired signal and noise. Chapters 1–3 introduce the subjects of space–time optimality of detection. These employ space–time Bayes matched filters and Gaussian fields, as well as coupling of the canonical channel[7] to receiver and transmitter. Inhomogeneous and nonstationary (non-Hom-Stat) conditions are considered as well as the more usual Hom-Stat situations, along with an introductory discussion of one- to three-dimensional arrays for explicit channel coupling. Discrete sampling is also introduced in these chapters, in contrast to the earlier treatment of Book 1 [1], in accordance with the usual digitalized handling of most data today. These first three chapters provide an introduction to the chapters that follow. Chapters 4 and 5 are reviews of much of the material in Chapters 21–23 of Book 1 [1] extended to space as well as time. On the other hand, Chapters 6 and 7 present mostly new material, namely, a theory of jointly coupled signal detection *and* extraction [cf. (1) above], for both binary and *M*-ary ($M > 2$) signals in noise.

Part 2. The Canonical Channel: Deterministic and Random Fields, Non-Gaussian Noise Models and Doppler

Part 2 introduces some essential elements of the classical theory of propagation. This is needed in our effort to apply these elements to the quantitative description of the channel itself, which is the origin of much of the noise which interferes and limits reception. The aim here is to go beyond the "black box" labeled "noise" and postulated *ad hoc* statistics, to the physically observed and analytically derived statistical distributions, in both time *and* space. Thus, propagation is a complex operation, involving the structure of the medium, boundaries, boundary and initial conditions. This is especially the case for inhomogeneous scattering media, and which in addition may be absorptive (i.e. dissipative). Chapter 8 discusses with deterministic cases, examples of *a posteriori* formulations. These in turn may next be regarded as "representations," which in turn form a statistical ensemble when suitable probability measures are assigned to them. This randomizing feature is characteristic of the *a priori* theory mentioned above (I), and which is our main concern here.

Two classes of problems are considered in Chapter 9: (1), where the ensemble of dynamical equations are deterministic and the driving source is itself a random field $G_T(\mathbf{R}, t)$, so that the resulting field $a(\mathbf{R}, t)$ is the probabilistic solution to $L^{(0)}\alpha_H = -G_T$ (or $\alpha_H = \hat{M}^{(0)}G_T$); $\hat{M}$ = integral Green's function. The second class of problem is the more difficult one of scattering, represented by the ensemble of equations of the form $\alpha^{(Q)} = \alpha_H + \hat{h}(Q)\alpha^{(Q)}$, which is nonlinear in the scattering elements $(\sim \hat{Q})$. The *a priori* approach is next introduced for the classical treatment of scattering. It is noted that again here scattering is a *nonlinear* property (with respect to the scattering elements) of such random media. (It is also a similar type of nonlinearity for the deterministic inhomogeneous media of Chapter 8.) The principal results here are the low-order moments of the governing probability distributions. Only in special cases (involving linear Langevin

[7] That is, in analytic forms applicable to a variety of different specific physical cases, i.e. acoustic, or electromagnetic, etc.

equations, Gaussian statistics, and Markoff assumptions) of the first class, are analytic solutions for a full treatment generally available. Chapter 10 following, however, remedies this situation with a new, purely probabilistic approach. Its advantages, and limitations with respect to the classical treatment of Chapter 9, are also discussed in Chapter 10.

Chapter 11 and 12 turn next to canonical forms,[8] as well as physical, derivations of three major classes of non-Gaussian noise, namely Poisson, *Class A, and Class B noise*,[9] which represent most physical random noise processes, including Thermal, Shot, and Impulse Noise. Chapter 11 presents the first-order pdf's of these three general classes of noise, Chapter 12 extends the treatment to the second-order cases. Physical non-Gaussian noise models are also considerably more analytically complex than Gaussian noise models are often encountered in practical receiving systems. Moreover, Poisson Class A and B noise (of interest to us here in the frequency ranges of Fig. 2, cf. Chapters 10 and 12 of [1]) are now usually a dominant component of interference in the channel. As a possible alternative to these relating complex, physical models we use Chapter 13 to present a brief treatment of various common, so-called *ad hoc* noise pdf's. These have comparatively simple analytic forms but their relationship to the underlying physical mechanisms is relatively tenuous. Chapter 14 concludes Part 2 with formulations for deterministic and random doppler, which arise in many practical situations when the transmitting and/or receiving sensor platforms are in relative nation to one another, and or to a fixed frame of reference.

Part 3: Threshold Theory: Detection and Estimation – Structure and Performance.

Part 3 is devoted to optimum and near optimum cases of threshold signal processing in general noise environments, in particular non-Gaussian noise whose explicit pdf's are obtained in Part 2. Here we are concerned not only with the structures of the detector and estimator, which are themselves generally sufficient statistics, but also with their performance. Chapter 15 develops canonical forms of such signal processors from the general arbitrary pdf's of the noise. This is done for both additive, and multiplicative signals and noise, such as these produced in inhomogeneous media with scattering elements, c.f. Chapter 16. We then apply these results to both optimum and suboptimum signal detection, while Chapter 17 considers analogous results for signal estimation. Chapters 18 and 19 are devoted respectively to examples from fluids, i.e., underwater acoustics, from elastic solids, and analogous electromagnetic vector field formulations, all in weak-signal regimes, which permit a general treatment.

Part 4: Special Topics

Part 4 concludes our general treatment and consists of a variety of special problems, based on the results of Parts 1-2 and selected papers. Chapter 20 describes acoustic problems connected with the reception of sonar signals in the ocean, where wave surface, volume, and bottom scatter are the principal interference mechanisms. Chapter 21 extends the analysis for radar in the full electromagnetic formulation. Chapter 22 considers next a variety

[8] That is, in analytic forms applicable to a variety of different specific physical cases, i.e. acoustic, or electromagnetic, etc.

[9] The author's designation.

of special problems listed below, which address briefly various additional features of the preceding analyses of Parts 1–3.

(i) Effect on performance of the separation of space (S) and time in reception, $[S \otimes T]$ vs. $[ST]$; See (xi) below;
(ii) Path integrals, for Class A and B noise;
(iii) Optical communications and quantum effects; lasers as technical enablers;
(iv) (Introductory) exposition of *Quantum Mechanics for Communications* (lasers, optics, etc.);
(v) Matched field processing (*MFP*);
(vi) Noise signals in noise – the (non-singular) Gaussian case for space–time fields;
(vii) Soliton models, spectrum; Wind-wave surface structures;
(viii) Astrophysics and computer LAN traffic applications;
(ix) Signal fading and multipath;
(x) Ocean wave surface models – Surface scatter, etc.
(xi) Ultra Wide-band systems (see (i) above)
(xii) Propagations for time-reversal or reciprocal media.

The book concludes with a series of Appendixes, mathematical relations needed above, and additional references.

LEVEL OF TREATMENT

The analytic requirements are comparable to that required in texts on theoretical physics or engineering. Reference [8] is noted in particular with the addition of the special probability methods guided by SCT. From the physical point of view here, the slogan "more vigor than rigor" is to be expected. The (ε, δ) and so on, of pure mathematics are implied, and are presented elsewhere. For example, the Dirac delta function $\delta(x - x_0)$ and its generalizations have been shown to belong to the class of "generalized functions." These are described in Lighthill's book [9], based on the concepts of Lebesque, Stieltjes, and L. Schwartz (cf. Chapters 2 and 3 of Ref. [9]), and the various extensions of the limit concept in integration. In addition, to facilitate the handling of the mathematical details in the propagation models, an operator formalism is frequently employed here. This provides a certain measure of compactness to the analytic treatment in many cases (see, in particular, Part 2).

The class of functions representing the physical models used here and in similar problems can and has been shown to give correct answers for these physical models, in addition to satisfying one's intuitive expectations. I have thus tried to avoid "cluttering up" the physical arguments represented by these macroalgorithms (refer to the first section of Introduction) by avoiding the full rigor of the associated "pure" mathematics, with its often arcane (to us) symbolisms and operations.

Finally, I have also included examples of the so-called *Engineering Approach* to these problems. Here, the (linear) canonical channel is represented by a *linear time–variable filter*. It is shown that this representation is valid, that is, it is equivalent to the general physical description of the channel in the cases treated here, *only in the far-field (Fraunhofer) regimes*. Moreover, it does not explicitly indicate range effects, namely, the attenuation of the propagating field due to "spreading." Additional conditions on the channel itself are that the signal applied to each sensor of the transmitting array must be the same and that the receiving portion of coupling to the channel must be an all-pass network (see Section 8.2). These conditions can often be met in practice, and they are usually acceptable in applications.

REFERENCING, AND SOME TOPICS NOT COVERED

In addition to the above, let me add a few remarks about the referencing, with respect to the selection of book and journal articles. These selections are based on a number of criteria and personal observations:

(1) Obviously important and pertinent books.
(2) The need for an inevitably limited number of sources, chosen now from thousands of possibilities, which are available via various search engines (Google, AOL, etc.).
(3) Recognition that there is an historical record, which is both informative and needed. This must necessarily involve a relatively small, finite number of books and papers to be manageable and thereby readily useful. Many of the references that I have used here since 1942 (including those cited in later editions of Ref. [1]) are from the formative period (1940–1970) in the development of SCI. These are still pertinent today.
(4) That any finite selection inevitably reflects the subjective choices of the author. However, I feel that these choices deserve attention, although they are now ignored in much of the current literature that appears to have a "corporate memory" of only a decade and a half [15]. The "new" is not necessarily better.
(5) During the 70 years of my activity in statistical communication theory, as it developed during World War II and subsequently (Fig. 1), I have encountered much relevant material (cf. (4) above). From these I have chosen the references used throughout. Undoubtedly, I have missed many others, for which I ask the readers' indulgence.

In addition, there are also many important topics in SCT that of necessity I have had to omit, in order to keep the sheer size of the presentation under some control. Clearly, a major field of equal importance is *information theory,* essentially the theory of coding [2, 10], to be applied to the canonical signals postulated here (Fig. 1). The purpose of such limits is (1) to preserve an acceptable combination of probabilistic methods (Part 1), (2) to present not an entirely trivial account of the relevant physics (Part 2), (3) to give the development of threshold theory (Part 3), which extends the applicability of SCT to the important (and more difficult) cases involving non-Gaussian noise, in addition to the usual Gaussian treatment, and (4) to illustrate (in Part 4) the scope of SCT through its applicability to a variety of diverse special topics.

Thus, Book 1 [1] may be regarded as a treatment of SCT involving temporal *processes* only, whereas Book (2) here provides an extension of SCT to space–time *fields*. Both books require the same level of capability; both are primarily research monographs, at the doctoral level, and both require a measure of familiarity with mathematical physics, as well as theoretical engineering (cf. References below). Of course, suitably prepared graduate students can also expect to find the book useful. In this connection, the role of the problems included here in is the same: to provide useful and special results, in addition to the text itself.

ACKNOWLEDGMENTS

I wish to thank my many colleagues who critiqued various chapters and who in some instances cast critical eyes on the entire text. They have been most helpful indeed. Their number include H. Vincent Poor, Professor and Dean of Engineering at Princeton University; Prof. Leon Cohen, Hunter College Graduate Physics Department; Prof. Steven S. Kay, Engineering & Computer Science Department, University of Rhode Island; Prof. J. W. Goodman, Optical Science, Stanford University; Prof. John Proakis, Electrical and Computer Engineering, Northeastern University; Dr. Julian J. Bussgang, founder of *Signatron*, Massachusetts; Prof. V. Bhargava, Dean of Engineering, University of British Columbia; and Prof. Eric Jakeman, Royal Society, (University of Manchester, GB).

My grateful thanks to my colleague Dr. William A. Von Winkle, Associate Director of the Naval Under Sea Warfare Center (NUWC) of the Navy (alas, now recently deceased (1930–August, 12, 2007)), who supported and encouraged my work during the period of 1960–1980. In addition, I am also particularly indebted to the U.S. Navy's Office of Naval Research, which through the years (1949–2005) has also supported much of my original research on the above topics, in addition to various private Industrial Research departments during this period (see DM, *Who's Who in America*).

For my errors of commission and omission, I take full responsibility. My reviewers have added positive comments and corrective support, for which I am most grateful.

In addition, I wish to thank the two young ladies, Ms. Dawn M. Goldstein (Centreville, VA) and Ms. Nancy S. Russell (Niantic, CT), who have been my most accomplished technical typists. They have had a very difficult (handwritten) manuscript to deal with. It is because of their efforts over quite a few years that they have produced a beautiful text.

Finally, and not by no means the least, I am deeply grateful to my wife, Joan Bartlett Middleton, for her consideration and encouragement over many years (1992–2010) during which I have been creating this book.

REFERENCES

1. D. Middleton, *An Introduction to Statistical Communication Theory*, 1st ed., International Series in Pure and Applied Physics, McGraw-Hill, New York, 1960–1972, 2nd ed., Peninsula Publishing Co., Los Altos, CA, 1987; 3rd ed. (Classic Edition), IEEE Press, Piscataway, NJ, John Wiley & Sons, Inc., 1996; also See: *Topics in Communication Theory,* McGraw-Hill, New York, 1965.
2. C. E. Shannon and W. Weaver, *The Mathematical Theory of Communication*, University of Illinois Press, Urbana, 1949.
3. S. Chandrasekhar, Stochastic Problems in Physics and Astronomy, *Rev. Mod. Phys.*, **15**(1), 1–91 (1943).
4. R. B. Griffiths, *Consistent Quantum Mechanics*, Cambridge University Press, 2003.
5. M. Lax, Fluctuations from the Non-Equilibrium Steady State, *Rev. Mod. Phys.*, **32**(1), 25–64, 1960; Classical Noise: Nonlinear Markoff Processes, *ibid.* 38 (2), 359–379, 1966;

Classical Noise: Langevin Methods, *ibid.* 38 (3), 544–566, 1966 Also see: Influence of Trapping, Diffusion, and Recombination on Carrier Concentration Fluctuations, *J. Phys. Chem. Solids*, **14**, 248–267, 1960. See also supplementary references to 3rd edition of Ref. [1]—Chapter 10, p. 1113, and Ref. [3]—Chapter 11, *ibid.* These papers contain many additional references to related work for 1966 and earlier.

6. L. Cohen, The History of Noise, *IEEE Signal Process Mag.*, 20–45, November 2005.
7. B. M. Hochwald, B. Hassik, T. L. Marzetta (Guest Eds.), The Academic and Industrial Embrace of Space–Time Methods, *IEEE Trans. Inform. Theory*, **49**, 2329, 2003.
8. P. M. Morse and H. Feshbach, *Methods of Theoretical Physics*, Vols. 1 and 2, International Series in Pure and Applied Physics, McGraw-Hill, New York, 1953 (a comprehensive nonrandom physical account of propagation, etc.).
9. M. J. Lighthill, *Fourier Series and Generalized Functions*, Cambridge University Press, 1958, Chapter 2.
10. M. Bykovsky, *Pioneers of the Information Era: History of the Development of Communication Theory*, Technosphera, Moscow, 2005. p. 375.
11. I. Tolstoy and C. S. Clay, *Ocean Acoustics, Theory and Experiment in Underwater Sound*, McGraw-Hill, New York, 1966.
12. S. M. Rytov, Yu. A. Kravtsov, and V. I. Tatarskii, *Vvedenie v. Statisticheskuyu Radiofiziku, Sluchainue Protscesui*, Nauka, Moscow, 1976; *Sluchainuie Polya*, 1978, originally in Russian; *Principles of Statistical Radiophysics, Elements of Random Process Theory*, 1987; *Correlation Theory of Random Processes*, 1988; *Elements of Random Fields*, 1989; *Wave Propagation Through Random Media,* 1989; Springer, New York (in English). Basically, classical theory: Gaussian random processes; continuous media "microstructure" and boundary conditions; finite, low-order moments with structure, boundaries, and then nonlinear media, refer to Chapter 9 of the present book; classical theory.
13. N. Wax, (Ed.) *Selected Papers on Noise and Stochastic Processes*, Dover Publications, New York, 1954.
14. D. Middleton and R. Esposito, Simultaneous Optimum Detection and Estimation of Signals in Noise. *IEEE Trans. Inform. Theory*, **IT-14**, (3) 434–444, 1968; New Results in the Theory of Simultaneous Optimum Detection and Estimation of Signals in Noise, *Problemy Peredachi Informatii*, **6** (2), 3–20, 1970.
15. D. Middleton, Reflections and Reminiscences, *Information Theory Society Newsletters*, **55**(3), 3, 5–7 (2005).
16. C. W. Helstrom, *Quantum Detection and Estimation Theory*, Vol. 123, Mathematics in Science and Engineering, Academic Press, New York, 1976.
   Some additional references are noted, which are also used in much of the book, with many others mainly specific to individual chapters. These as follows:
17. H. Margenau and G. M. Murphy, *The Mathematics of Physics and Chemistry*, D. Van Nostrand, New York, 1943.
18. G. N. Watson, *Theory of Bessel Functions*, 2nd ed., Cambridge University Press, New York, 1944.
19. G. A. Campbell and R. M. Forster, *Fourier Integrals*, D. Van Nostrand, New York, 1948.
20. J. A. Stratton, *Electromagnetic Theory*, McGraw-Hill, New York, 1941.
21. V. A. Kotelnikov, *The Theory of Optimum Noise Immunity* (translated from Russian by R. A. Silverman), McGraw-Hill, New York, 1959 (originally Kotelnikov's doctoral dissertation in 1947. published in the Soviet Union in 1956).
22. R. B. Lindsey, *Mechanical Radiation*, International Series in Pure and Applied Physics, McGraw-Hill, New York, 1960.

23. L. Brillouin, *Science and Information Theory*, 2nd ed., Academic Press, New York, 1962; also see: *Wave Propagation in Periodic Structures (Electric Filters and Crystal Lattices)*, International Series in Pure and Applied Physics, McGraw-Hill, New York, 1946.
24. A. Ishimaru, *Wave Propagation and Scattering in Random Media,* Vol. I, *Single Scatter and Transport Theory,* Vol. II, *Multiple Scatter, Turbulence, Rough Surfaces, and Remote Sensing*, Academic Press, 1978 (classical theory, refer to Refs [20, 21]).
25. I. S. Gradshteyn and I. N. Ryzhik, *Table of Integrals, Series and Products*, (corrected and enlarged edition by Alan Jeffrey), 1980.
26. J. W. Goodman, *Statistical Optics*, John Wiley & Sons, Inc., New York, 1985; also see: *Speckle Phenomenon in Optics,* Roberts and Co., Englewood, CO, 2007.
27. G. W. Gardiner, *Handbook of Stochastic Methods*, 2nd ed., Springer, New York, 1985.
28. W. C. Chew, *Waves and Fields in Inhomogeneous Media*, Series on Electromagnetic Waves, IEEE Press, New York (originally published by D. Van Nostrand-Reinhold, 1990; reprinted by IEEE Press, 1995 (classical, nonrandom, *a posteriori* theory).
29. C. W. Helstrom, *Elements of Signal Detection and Estimation*, Prentice-Hall, 1995 (SCT with Gaussian noise).
30. H. V. Poor, *An Introduction to Signal Detection and Estimation*, 2nd ed., Springer, New York, 1994.
31. N. G. VanKampen, *Stochastic Processes in Physics and Chemistry* (revised and enlarged edition), North Holland Pub., New York, 1992.
32. J. V. Candy, *Model-Based Signal Processing*, IEEE Press and John Wiley & Sons, Inc., New York, 2006.

For further references to information theory, besides Ref. [2] above, there is an extensive bibliography. We list a short introductory samples below of fundamental work (from 1953–1979):

(i) S. Goldman, *Information Theory*, Prentice-Hall, New York, 1953.
(ii) A. Feinstein, *Foundations of Information Theory*, McGraw-Hill, New York, 1958.
(iii) R. M. Fano, *Transmission of Information*, John Wiley & Sons, Inc., New York, 1961.
(iv) David Slepian (Ed.), Key Papers in the *Development of Information Theory,* IEEE Press, 1973.
(v) E. R. Berlekamp (Ed.), Key Papers in the *Development of Coding Theory,* IEEE Press, 1973.

1

RECEPTION AS A STATISTICAL DECISION PROBLEM

1.1 SIGNAL DETECTION AND ESTIMATION

As we have noted above, our aim in this chapter is to provide a concise review of Bayesian decision methods that are specifically adapted to the basic problems of signal detection (D) and estimation (E). From Fig. 1.1b we can express the reception situation concisely in a variety of equivalent ways through the following operational relations:

A. *Data Processing at the Receiver.*

$$\left(\hat{\mathbf{T}}_{\mathrm{D}} \text{ or } \hat{\mathbf{T}}_{\mathrm{E}}\right)X = Y, \quad \text{or} \quad \left(\hat{\mathbf{T}}_{\mathrm{D}}\hat{\mathbf{R}} \text{ or } \hat{\mathbf{T}}_{\mathrm{E}}\hat{\mathbf{R}}\right)\alpha = Y, \tag{1.1.1}$$

where X is the data input from the spatial processor $\hat{\mathbf{R}}$, that is, the receiving aperture $\hat{\mathbf{R}}\left(\equiv \hat{\mathbf{T}}_{\mathrm{AR}}\right)$, to the temporal data processing elements $\left(\hat{\mathbf{T}}_{\mathrm{D}}, \hat{\mathbf{T}}_{\mathrm{E}}\right)$; Y represents the output from these processors. In more detail from Fig. 1.1b we can also write the following.

B. *Data Input to Processors.*

$$X = \hat{\mathbf{R}}\alpha = \left(\hat{\mathbf{R}}\hat{\mathbf{T}}_{\mathrm{M}}^{(N)}\hat{T}_{\mathrm{AT}}\right)S_{\mathrm{in}}, \tag{1.1.1a}$$

in which α is the propagating field in the medium, which contains ambient sources and scattering elements embodied in the operator $\hat{\mathbf{T}}_{\mathrm{M}}^{(N)}$.

Non-Gaussian Statistical Communication Theory, David Middleton.

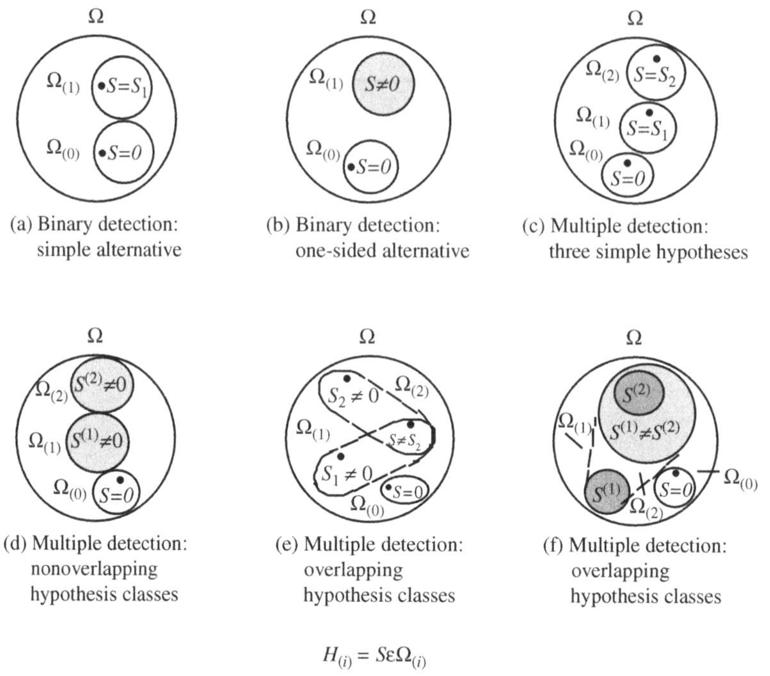

FIGURE 1.1 Signal and hypothesis classes in detection.

C. *Field in the Medium.*

$$\alpha = \hat{\mathbf{T}}_{\mathrm{M}}^{(N)}\hat{\mathbf{T}}_{\mathrm{AT}}S_{\mathrm{in}}. \tag{1.1.1b}$$

The input or injected signal S_{in} and output "decisions" $\{v\}$ are described operationally by the following.

D. *Input Signals and Decision Outputs.*

$$S_{\mathrm{in}} = \hat{\mathbf{T}}_{\mathrm{mod}}\hat{\mathbf{T}}_{\mathrm{e}}\{u\}; \quad \{v\} = \hat{\mathbf{T}}_{\mathrm{d}}Y = \hat{\mathbf{T}}_{\mathrm{d}}\big(\hat{\mathbf{T}}_{\mathrm{D}} \text{ or } \hat{\mathbf{T}}_{\mathrm{E}}\big)X, \tag{1.1.1c}$$

where now $\{u\}$ is a set of "messages" to be transmitted and the "decisions" $\{v\}$ fall into two (not necessarily disjoint) classes: ("yes"/"no") for detection (D) and a set of numbers representing measurements, namely estimates of received signal properties or parameters. Comparing Fig. 1.1a and b we see that the "compact" channel operators $\hat{\mathbf{T}}_{\mathrm{T}}^{(N)}$, and so on, in Eq. (1.1.1d) are

E. *Components of the Compact Operators.*

$$\mathbf{T}_{\mathrm{T}}^{(N)} = \hat{\mathbf{T}}_{\mathrm{AT}}\hat{\mathbf{T}}_{\mathrm{mod}}\hat{\mathbf{T}}_{\mathrm{e}}; \quad \hat{\mathbf{T}}_{\mathrm{M}}^{(N)} = \hat{\mathbf{T}}_{\mathrm{M}}^{(N)}; \quad \hat{\mathbf{T}}_{\mathrm{R}}^{(N)} = \hat{\mathbf{T}}_{\mathrm{d}}\hat{\mathbf{T}}_{\mathrm{D/E}}\hat{\mathbf{T}}_{\mathrm{AR}}. \tag{1.1.1d}$$

We emphasize here and subsequently (unless otherwise indicated) that when the signal (if any) is present in X and therefore in the received data Y, Eq. (1.1.1), it is the *received signal* $S_{\mathrm{Rec}}(= S)$. The received signal is, or course, *not* the signal S_{in} originally injected into the medium. This dichotomy occurs because the medium and canonical channel as a whole modify and generally contaminate S_{in}, with additive and signal-dependent noise (clutter and reverberation) as well as varieties of ambient noise and interference, in addition to such inherent phenomena as absorption and dispersion. All this, of course, is what makes achieving effective reception of the desired signals the challenging problem that we seek to resolve in subsequent chapters. (We remark that S_{Rec} may be generated either by the desired source, S_{in}, or by some undesired source, such as interference, or by a combination of both.)

With this in mind we see that our goals in Chapters 1–7 are first to establish explicit analytic connections between the received data X, the physical realities which affect them (via $\hat{\mathbf{T}}_{\text{M}}^{(N)}$), and the successful extraction of the desired signal, initially as $S(=S_{\text{Rec}})$, and eventually through attained knowledge of the medium (the inverse problem) to obtain acceptable reproduction of the original signal S_{in}. This chapter introduces the formal decision structure for achieving this, while Chapters 2–7 following provide the canonical algorithms (operations on the input data) and performance measures to be used subsequently for specific applications. We remark that the operations involving *coding* [$(\hat{\mathbf{T}}_{\text{d}}, \hat{\mathbf{T}}_{\text{e}})$ in (1.1.1c), and shown in Figure 1.1] belong to the domain of *Information Theory* per se [2], which is outside the scope of the present volume.[1]

1.2 SIGNAL DETECTION AND ESTIMATION

We begin our decision—theoretic formulation with a general description of the two principal reception problems, namely detection and estimation (sometimes called extraction) of signals in noise, expressed operationally above by Eq. (1.1.1). We first introduce some terminology, taken partly from the field of statistics, partly from communication engineering, and review the problems in these terms. We shall also point out some considerations that must be kept in mind concerning the given data of these problems. Later, in Sections 1.3–1.4, we generalize the reception problem, state it in mathematical language, and outline the nature of its solutions.[2]

1.2.1 Detection

The problem of the detection of a (received)[3] signal in noise is equivalent to one which, in statistical terminology, is called the problem of *testing hypotheses*: here, the hypothesis that noise alone is present is to be tested, on the basis of some received data, against the hypothesis (or hypotheses) that a signal (or one of several possible signals) is present.

Detection problems can be classified in a number of ways: by the number of possible signals that need to be distinguished, by the nature of the hypotheses, by the nature of the data and their processing, and by the characteristics of the signal and noise statistics. These will now be described in greater detail.

[1] However, through selected references we shall provide connections to these topics at appropriate points in this book.

[2] As an introduction to the methods of statistical inference, see, for example, the treatments of Kendall [3] and Cramér [4]; also Luce and Raiffa, [5].

[3] Note the comment following Eq. (1.1.1d) above.

1.2.1.1 ***The Number of Signal Classes to be Distinguished*** This is equal to the number of hypotheses to be tested but does not depend on their nature. A *binary detection system* can make but two decisions, corresponding to two hypotheses, while a *multiple alternative detection system* [6, 7] makes more than two decisions. For the time being, we deal only with the binary detection problem (the multiple alternative cases are discussed in Chapter 4 ff.).

1.2.1.2 ***The Nature of the Hypotheses*** Here the received signal is a desired system input during the interval available for observation of the mixture of signal and noise. Noise (homogeneous—*Hom-Stat* stationary or nonstationary inhomogeneous — *non-Hom-Stat*) is an undesired input, considered to enter the system independently[4] of the signal and to affect each observation according to an appropriate scheme whereby the two are combined.[5] The class of all possible (desired) system inputs is called the *signal class* and is conveniently represented as an abstract space (*signal space*) in which each point corresponds to an individual received signal.

A hypothesis, which asserts the presence of a single signal at the input is termed a simple *hypothesis*. A *class* (or *composite*) *hypothesis*, on the other hand, asserts the presence at the input of an unspecified member of a specified subclass of signals; that is, it reads "some member of subclass k (it does not matter which member) is present at the input." Such a subclass is called a *hypothesis class*. Hypothesis classes may or may not overlap (cf. Fig. 1.1e and f).

Usually, one hypothesis in detection asserts the presence of noise alone (or the absence of any signal) and is termed the *null hypothesis*. In binary detection, the other hypothesis is called the *alternative*. If the alternative is a class hypothesis and the class includes all nonzero signals involved in the problem, it is termed a *one-sided alternative*. It is a *simple alternative* if there is but one nonzero signal in the entire signal space (which signal must therefore contain nonrandom parameters only). Figure 1.1 illustrates some typical situations. In each case, the class of all possible system inputs is represented by signal space Ω.

The hypothesis classes are enclosed by dashed lines and denoted as $\Omega_{(k)}$, where the subscript refers to the hypothesis: that is, the kth hypothesis states that the signal is a member of $\Omega_{(k)}$, or, symbolically, $H_k : S\varepsilon\Omega_{(k)}$. In Fig. 1.1a and b are shown two binary cases corresponding to the simple alternative ($\Omega_{(1)}$ contains one point) and the one-sided alternative ($\Omega_{(1)}$ contains all nonzero system inputs). The latter would occur, for example, if all signals in a binary detection problem were the same except for a random amplitude scale factor, governed by, say, a Gaussian distribution. Figure 1.1c and d shows multiple hypothesis situations where the hypothesis classes do not overlap, while Fig. 1.1e and f represents situations where overlapping can occur, in (e) with single-point classes and in (f) when the classes are one - sided or composite. Many different combinations can be constructed, depending on the actual problems at hand. In the present treatment, we shall confine our attention to the nonoverlapping cases, although the general approach is in no way restricted by our so doing. [But see for example, Section 1.10.2 ff.]

[4] In most applications, but, of course, scattered radiation is signal dependent (cf. Chapters 7 and 8 ff.).

[5] There can be noiselike signals also, but these are not to be confused with the noise background. It is frequently convenient to speak of "noise alone" at the input, and this is to be interpreted as "no signal of any kind" present. In physical systems, which do not, of course, use ideal (i.e., noise-free) elements, noise may be introduced at various points in the system, so that care must be taken in accounting for the manner in which signal and noise are combined.

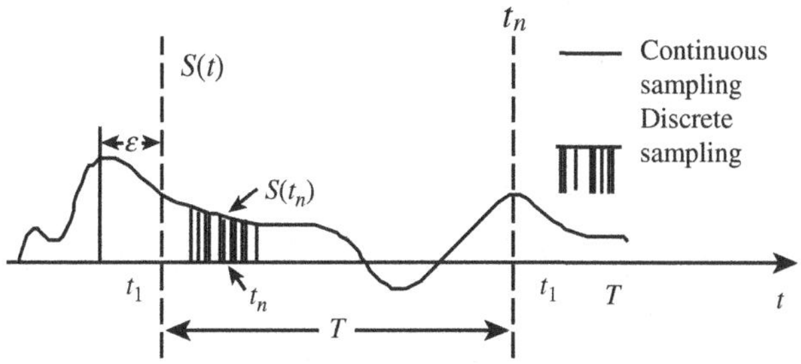

FIGURE 1.2 A temporal signal waveform, showing discrete and continuous time sampling, the epoch ε, and the observation period, T.

1.2.1.3 The Nature of the Data and Their Processing The observations made on the mixture of signal and noise during the observation period may consist of a discrete set of values (*discrete or digital sampling*) or may include a continuum of values throughout the interval (*continuous*, or *analogue sampling*) (cf. Fig. 1.2). Whether one procedure or the other is used is a datum of the problem. In radar, for example, detection may (to a first approximation) be based on a discrete set of successive observations, while, in certain communication cases, a continuous-wave signal may be sampled continuously.

Similarly, it is a datum of the problem whether or not the observation interval, that is, the interval over which the reception system can store the data for analysis, is fixed or variable. In the latter case, one can consider *sequential detection.* A sequential test proceeds in steps, deciding at each stage either to terminate the test or to postpone termination and repeat the test with additional data. In applications of decision theory, it turns out that the analysis divides conveniently at the choice between the sequential and nonsequential. The theory of each type is complete in a certain sense, and additional restrictions on the tests may not be imposed without compromising this completeness. It is, of course, true that since the class of sequential tests includes nonsequential tests as a special subclass, a higher grade of performance may be expected, on the average, under the wider assumption [8, 9].

1.2.1.4 The Signal and Noise Statistics The nature of these quantities is clearly of central importance, as it is upon them that specific calculations of performance depend. In general, individual sample values *cannot* be treated as statistically independent, and this inherent correlation between the sample values over the observation period, in both the continuous and discrete cases, is an essential feature of the problem.

We begin first with temporal waveforms, extending this signal class presently to space–time signals, in Section 1.3.1 ff. Temporal signals may be described in quite general terms involving both random and deterministic parameters. Thus, we write $S(t) = S(t, \varepsilon; a_0, \boldsymbol{\theta})$. Here, ε is an *epoch*, or time interval, measured between some selected point in the "history" of the signal S and, say, the beginning of the observation period $(t_1,\ t_1 + T)$, relating the observer's to the signal's timescale, as indicated in Fig. 1.2; a_0^2 is a scale factor, measuring (relative) intensity of the signal with respect to the noise background; and $\boldsymbol{\theta}$ denotes all other descriptive parameters, such as pulse duration, period, and so on, which may be needed to specify the signal; S itself gives the "shape," or functional *form*, here of the wave in time.

No restriction is placed on the received signal other than that it have finite energy in the observation interval. It may be entirely random, partly random (e.g., a "square wave" with random durations), or entirely causal or deterministic [e.g., a sinusoid, or a more complex structure that is nevertheless uniquely specified by $S(t)$]. Signals for which the epoch ε assumes a fixed value are said to be *coherent* (with respect to the observer), while if ε is a random variable, such signals are called *incoherent*. Coherent signals may have random parameters and thus belong to subclasses of Ω containing more than one member. Coherent signals corresponding to subclasses containing but a single member will be called *completely coherent*. From these remarks, it is clear that an incoherent signal cannot belong to such an elementary class. The description of the noise is necessarily statistical, and here we distinguish between noise belonging to *stationary* and *nonstationary* processes [10, 11]. Generalizations of the noise structure to include partially deterministic waves offer no conceptual difficulties.

1.2.2 Types of Extraction

We use the term *extraction* here to describe a reception process that calls for an estimate of the received signal itself or one or more of its descriptive parameters.

Signal extraction, like detection, is a problem that in other areas has received considerable attention from statisticians and has been known under the name of *parameter estimation*. A certain terminology has become traditional in the field, which we shall mention presently. We can classify extraction problems under three headings: the nature of the estimate, the nature of the data processing, and the statistics of signal and noise. Much of what can be said under these headings has already been mentioned above. A few more comments may be helpful.

Information about the signal may be available in either of two forms: it may be given as an elementary random process in time, defined by the usual hierarchy of multidimensional distribution functions [12] or it may be a known function of time, containing one or more random parameters with specified distributions. In the latter case, the random parameters may be time independent, or, more generally, they may be themselves random processes (e.g., a noise-modulated sine wave). Clearly there is, as in detection, a wide variety of possible situations. They may be conveniently classified as follows:

1.2.2.1 The Nature of the Estimate A *point estimate*[6] is a decision that the signal or one or more of its parameters have a definite value. An *interval estimate*[6] is a decision that such a value lies within a certain interval with a given probability. Among point estimates, it is useful to make a further distinction between *one-dimensional* and *multidimensional estimates*. An illustration of the former is the estimate of an amplitude scale factor constant throughout the interval, while an estimate of the signal itself throughout the observation period is an example of the latter.

1.2.2.2 The Nature of the Data Processing When the value of a time-varying quantity $X(t)$, (1.1.1b) at a particular instant is being estimated, the relationship between the time t_λ for which the estimate is valid and the times at which data are collected becomes important (cf. Fig. 1.3). If t_λ coincides with one of the sampling instants, the estimation process is termed *simple estimation*, or *simple extraction*. If, on the other hand, t_λ does not coincide with any sampling instant, the process is called *interpolation*, or *smoothing*, when t_λ lies within the observation interval $(t_1, t_1 + T)$ and *extrapolation*, or *prediction*, when t_λ lies outside $(t_1, t_1 + T)$. Systems of these types may estimate the value of the signal itself or

[6] See Cramér [4] *op. cit.*, for a further discussion of conventional applications.

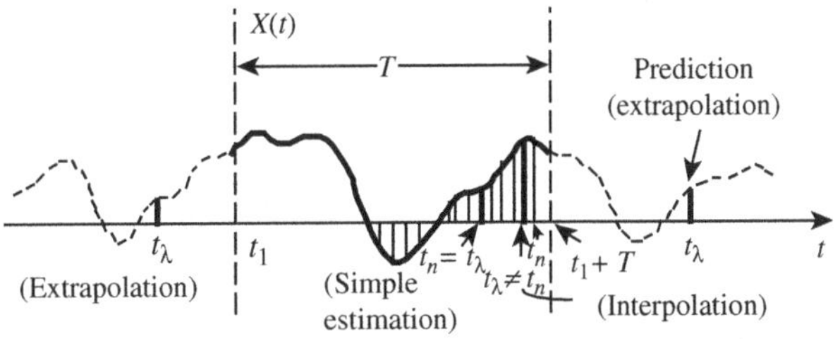

FIGURE 1.3 Simple estimation, interpolation (smoothing), and extrapolation (prediction).

alternatively that of a time-varying signal parameter or some functional of the signal, such as its derivative or integral.

Frequently, a requirement of linearity may be imposed on the optimum system (which is otherwise almost always nonlinear), so that its operations may be performed by a linear network, or sometimes certain specific classes of nonlinearity may be allowed. An important question, then, is the extent to which performance is degraded by such constraints.

1.2.2.3 The Signal and Noise Statistics As in detection, the finite sample upon which the estimate is based may be discrete or continuous, correlated or uncorrelated, and the random processes stationary or nonstationary, ergodic or nonergodic. In a similar way, we may speak of *coherent* and *incoherent extraction* according to whether the received signal's epoch ε is known exactly or is a random variablc. Thc signals thcmsclvcs may bc structurally determinate, that is, the functions S have definite analytic forms; or they may be *structurally indeterminate*, when the S are described only in terms of a probability distribution. A sinusoid is a simple example of the former, while a purely random function is typical of the latter. The case where the signal is known completely does not arise in extraction.

1.2.3 Other Reception Problems

Reception itself may require a combination of detection and extraction operations. Extraction presupposes the presence of a signal at the input, and sometimes this cannot be assumed. We may then perform detection and extraction simultaneously and judge the acceptability of the estimate according to the outcome of the detection process. The problem here is that estimation is performed under uncertainty as to the signal's presence in the received data, which in turn leads to biased estimates that must be suitably accounted for. The analytic results for this new situation are developed and illustrated in detail in Chapters 5 and 6 following. The procedure is schematically illustrated in Figure 1.4, including possible coupling between the detector and extractor.

In our reception problems here, the system designer usually has little control over the received signal, since the medium, embodied in $\hat{\mathbf{T}}_{\mathrm{T}}^{(N)}$, is specified *a priori*. The present definition of the problem states that each possible signal is prescribed, together with its probability of occurrence, and the designer cannot change these data. However, a different strategic situation confronts the designer of a system for transmitting messages from point to point through a noisy channel, since he is then permitted to control the way in which he matches the signal to thc channcl. Thc encoding process ($\hat{\mathbf{T}}_{\mathrm{e}}$), Fig. 1.1 is accordingly concerned with finding what class of signal is most effective against channel noise ($\hat{\mathbf{T}}_{\mathrm{M}}^{(N)}$) and

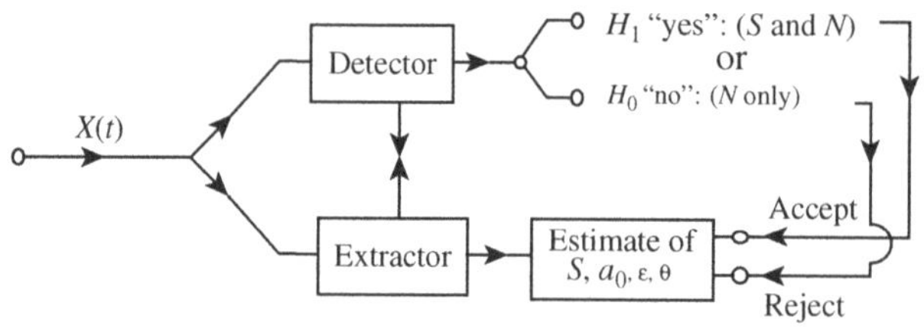

FIGURE 1.4 Reception involving joint signal detection and extraction.

how best to represent messages by such signals (Sections 6.1, 6.5.5 [1]). It is not directly a reception problem, except in the more general situation mentioned earlier (cf. also Section 23.2 [1]), where simultaneous adjustment of the transmission and reception operations $\hat{\mathbf{T}}_T^{(N)}$, $\hat{\mathbf{T}}_R^{(N)}$ is allowed. Decoding $(\hat{\mathbf{T}}_d)$, of course, is a special form of reception in which the nature of the signals and their distributions are intimately related to the noise characteristics. Moreover, for the finite samples and finite delays available in practice, this is always a nontrivial problem, since it is impossible in physical cases[7] to extract messages (in finite time) from a noisy channel without the possibility of error.

1.3 THE RECEPTION SITUATION IN GENERAL TERMS

Let us now consider the main elements of a general reception problem. We have pointed out earlier that the reception problem can be formulated as a decision problem and that consequently certain information must be available concerning the statistics of signal and noise. We have also indicated that some assumptions are necessary concerning the nature of the data and of the sampling interval and procedures. Finally, we must prescribe a criterion of excellence by which to select an optimum system and must specify the set of alternatives among the decisions to be made.

In our present formulation, we shall make certain assumptions concerning these elements. For definiteness, these assumptions will not be the most general, but they will be sufficiently unrestrictive to exhibit the generality of the approach. Later, in Section 1.4.3, we shall discuss the reasoning by which some of these restrictions are removed.

1.3.1 Assumptions: Space–Time Sampling

Concerning the statistics of signal and noise, we shall assume for the present exposition that both are known *a priori* and as well as the discretely sampled received data **X**. (In subsequent chapters we shall consider various techniques for handling the problem of unknown or unavailable priors.)

We further extend the sampling process here to space as well as time, since the array operators $\hat{\mathbf{T}}_{AR}$ and $\hat{\mathbf{T}}_{AT}$, cf. (1.1.1) et seq. sample the data *field* established in the medium by the signal and noise sources. We further assume that the sampling intervals, or sample size, in time are fixed and of finite duration T and similarly in space, that the *array* or aperture size is likewise finite. Thus, in time $n = 1, \ldots, N$ data elements can be acquired, at each of

[7] Strictly speaking, there is always some noise, although in certain limiting situations this may be a very small effect and hence to an excellent approximation ignorable vis-à-vis the signal.

$m = 1, \ldots, M$ spatial points[8]. Accordingly, we obtain a total of $J = MN$ data components in the received space–time sample.

We employ the following component designations: $j = mn =$ (space $\times$ time), so that $j = 11, 12, \ldots, 1N$ represents the N time samples at spatial point 1; $j = 21, 22, 23, \ldots, 2N$ similarly denotes the N time samples at spatial point 2, and so on. Thus, *j is a double index numeric*, obeying the convention that the first index (m) refers to the spatial point in question while the second (n) indicates the nth time sample point in T. Specifically, we write $X_{j=mn} = X(\mathbf{r}_m, t_n)$, $S_j = S(\mathbf{r}_m, t_n)$, $N_j = N(\mathbf{r}_m, t_n)$, respectively for the received data $\mathbf{X}$, the received signal $\mathbf{S}$, and noise $\mathbf{N}$, at point $\mathbf{r} = \mathbf{r}_m$ in space and at time $t = t_n$ (see 1.1.1). Furthermore, it is sometimes convenient to introduce a single index numeric, k. Thus, we write for j and k the following equivalent numbering systems:

$$\begin{pmatrix} j \\ k \end{pmatrix} \begin{matrix} = \\ = \end{matrix} \begin{pmatrix} 1,1 \\ 1 \end{pmatrix}, \begin{pmatrix} 1,2 \\ 2 \end{pmatrix}, \ldots, \begin{pmatrix} (1,N) \\ N \end{pmatrix}; \begin{pmatrix} 2,1 \\ N+1 \end{pmatrix}, \begin{pmatrix} 2,2 \\ N+2 \end{pmatrix}, \ldots, \begin{pmatrix} 2,N \\ 2N \end{pmatrix}; \ldots;$$
$$\begin{pmatrix} M,1 \\ (M-1)N+1 \end{pmatrix}, \begin{pmatrix} M,2 \\ (M-1)N+2 \end{pmatrix}, \ldots, \begin{pmatrix} M,N \\ MN \end{pmatrix}. \tag{1.3.1}$$

The double index j is convenient when we need explicitly to distinguish the spatial from the temporal portion of the received field, in processing. It is also useful when we impose the constraint of space and time separability on operations at the receiver, such as array or aperture design, independent of optimization of the temporal processing, a usual although approximate procedure in practice. Of course, j may also be treated as a single index if we order it according to the equivalent scheme (1.3.1), that is, let $j \to k$. This alternative form is often required when quantitative, that is, numerical, results are desired. The formal structure of the sampling process itself is described in detail at the beginning of Section 1.6.1, cf. Eq. (1.6.2a).

At this point we make no special assumption concerning the criterion of optimality, but we do assume, for the sake of simplicity, that the decision to be made by the system is to select among a finite number L of alternatives. Figure 1.5a and b illustrates the problem. A set of decisions γ are to be made about a received signal $\mathbf{S}$, based on data $\mathbf{X}$, in accordance

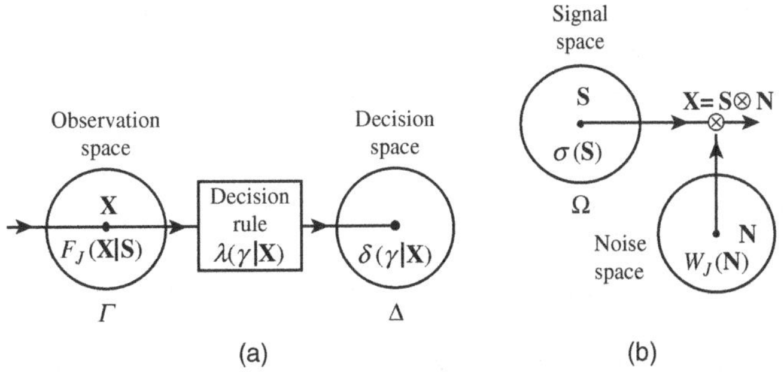

FIGURE 1.5 The reception situation. (a) Observation and decision space; (b) signal space and noise space. ⊗ indicates a combination, not necessarily additive, of received signal and noise.

[8] The spatiotemporal structure of both discrete element arrays and continuous apertures are discussed in "Chapter 8–9". See [13, 14].

with a decision rule $\delta(\gamma|\mathbf{X})$, as shown at Fig. 1.5a. Here $\boldsymbol{\gamma} = (\gamma_1, \ldots, \gamma_L)$, $\mathbf{S} = [S_j]$, $\mathbf{X} = [X_j]$, and $\mathbf{N} = [N_j]$ are vectors, and the subscripts on the components of $\mathbf{S}$ and $\mathbf{X}$ are ordered in time so that $S_{mn} = S(\mathbf{r}_m, t_n)$, $X_{mn} = X(\mathbf{r}_m, t_n)$, and so on, with $0 \leq t_1 \leq t_2 \ldots \leq t_n \leq \ldots \leq T_N \leq T$. (Ordering the spatial indexes is arbitrary, essentially a convenience suggested by the structure of the array $(\hat{\mathbf{T}}_{RT})$ sampling the input field α, cf. (1.1.1b) and Fig. 1.1.) Thus, the components of $\mathbf{X}$ form the *a posteriori* data of the sample upon which some decision γ_l is to be made.

Each of the quantities *received signal* $\mathbf{S}$, *noise* $\mathbf{N}$, *received data* $\mathbf{X}$, *decisions* $\boldsymbol{\gamma}$ can be represented by a point in an abstract space of the appropriate dimensionality. The occurrence of particular values is governed in each instance by an appropriate probability density function. These are multidimensional density functions, which are to be considered discrete or continuous depending on the discrete or continuous nature of the spaces and of corresponding dimensionality.

Here, we introduce $\sigma(\mathbf{S})$, $W_J(\mathbf{N})$, and $F_J(\mathbf{X}|\mathbf{S})$, respectively, as the probability–density functions for the received signal, for noise, and for the data $\mathbf{X}$ when $\mathbf{S}$ is given. (Note that $F_J(\mathbf{X}|\mathbf{0}) = W_J$ ($\mathbf{X}$, by definition.)) As mentioned earlier, the possible (received) signals $\mathbf{S}$ may be represented as points in a space Ω over which the *a priori* distribution $\sigma(\mathbf{S})$ is defined. Information about the signal and its distribution may be available in either of two forms: it may be given directly as an elementary random process, that is, the distribution $\sigma(\mathbf{S})$ is immediately available, as a datum of the problem (stationary, Gaussian, nonstationary, non-Gaussian, etc.). Or, as is more common, the signal $\mathbf{S}$ is a known function of one or more random parameters $\boldsymbol{\theta} = (\boldsymbol{\theta}_1, \ldots, \boldsymbol{\theta}_{\hat{M}})$, and it is the distribution[9] $\sigma(\boldsymbol{\theta})$ of these parameters which is given rather than $\sigma(\mathbf{S})$ itself. In fact, the reception problem may require decisions about the parameters instead of the signals.

We can also raise the question of what to do when $\sigma(\mathbf{S})$ is not known beforehand (or perhaps only partially known), contrary to our assumption here. Such situations are in fact encountered in practice, where it is considered risky or otherwise unreasonable to assume that complete knowledge of $\sigma(\mathbf{S})$ is available. This question is a difficult one, and a considerable portion of decision theory is devoted to providing a reasonable answer to it. [It is taken up initially in Sections 1.4.3 and 1.4.4 as well as discussed further in Section 23.4 of Ref. [1], and it is shown that even in this case the above formulation of the reception problem can be retained in its essentials.]

We can further compound the complexity of the reception problem and inquire into what to do when not only $\sigma(\mathbf{S})$, but also $W_J(\mathbf{N})$, the distribution of the noise, is partially or completely unknown *a priori*. In such a case, specification of $\mathbf{S}$ is not enough to determine $F_J(\mathbf{X}|\mathbf{S})$. In statistical terminology $F_J(\mathbf{X}|\mathbf{S})$ is then said to belong to a *nonparametric* family. Error probabilities, associated with possible incorrect decisions, cannot then be computed directly and the system can no longer be evaluated in such terms, so that the question of optimization is reopened. Nonparametric inference from the general point of view of decision theory has been discussed previously by several investigators [15–17][10] but will not be considered further here.

The discussion of possible generalizations, which has so far dealt with the decision space and with the statistics of signal and noise, can also be extended to one other topic, that is, the method of data acquisition. In the line of reasoning that led to the above formulation of the

[9] See Cramér [4] *op. cit.*

[10] See Gibson and Melsa [18] for more recent telecommunication applications.

reception problem, we assumed for convenience that the data were sampled discretely and that the sampling interval was fixed and finite. Actually, neither of these assumptions is strictly necessary. The sampling process can be continuous. Cases of this type are discussed in Chapters 19–23 of Ref. [1].

We observe also that the length of the sampling interval need not be kept fixed. In fact, the idea of a variable sampling interval leads to the notion of *sequential* decisions. A reception system that is based on sequential principles proceeds in steps, deciding, after each sample of data has been processed, whether or not to come to a conclusion or whether to extend the sampling interval and to take another reading. The class of sequential reception systems is broad and contains the nonsequential type discussed so far as a subclass.[11]

1.3.2 The Decision Rule

We begin by observing that the decision rule is represented as a probability. This may seem somewhat surprising. A reception system operating according to such a decision rule would not function like a conventional receiver, which generates a certain and definite output $\boldsymbol{\gamma}$ from a given set of inputs $\mathbf{X}$. Rather, it would contain a battery of chance mechanisms of a well-specified character. A given set of inputs would actuate the corresponding mechanism, which in turn would generate one of the L possible outputs $\boldsymbol{\gamma}$ with a certain probability, each mechanism in general with a different probability. Arrangements such as this will probably appear quite artificial but they are necessary concepts, at least in principle, for it can be shown that devices with chance mechanisms as their outputs can be superior in performance, under certain circumstances, to the conventional ones.

Accordingly, $\delta(\boldsymbol{\gamma}|\mathbf{X})$ is the conditional probability of deciding $\boldsymbol{\gamma}$ when $\mathbf{X}$ is given. More specifically, since the space Δ is here assumed to contain a finite number of decisions $\boldsymbol{\gamma}$, the decision rule $\delta(\boldsymbol{\gamma}|\mathbf{X})$ assigns a probability between (or equal to) 0 and 1 to each decision γ_l $(l = 1, \ldots, L)$, the distribution depending on $\mathbf{X}$. In most cases of practical interest, δ is either 0 or 1 for each $\mathbf{X}$ and $\boldsymbol{\gamma}$ in this case and is called a *nonrandomized decision rule*. The opposite case, a *randomized decision rule*, is not excluded from this general formulation, although, as we shall see in subsequent applications, the decision rules reduce to the nonrandom case for all the systems treated here.

We note now that the *key feature of the decision situation is that* $\delta(\boldsymbol{\gamma}|\mathbf{X})$ *is a rule for making the decisions* $\boldsymbol{\gamma}$ *from a posteriori data* $\mathbf{X}$ *alone*, that is, without knowledge of, or dependence upon, the particular $\mathbf{S}$ that results in the data $\mathbf{X}$. The *a priori* knowledge of the signal class and signal distribution, of course, is built into the optimum-decision rule, but the probability of deciding $\boldsymbol{\gamma}$, given $\mathbf{X}$, is independent of the particular $\mathbf{S}$; that is, $\boldsymbol{\gamma}$ is algebraically independent of $\mathbf{S}$, although statistically dependent upon it. This may be expressed as

$$\delta(\boldsymbol{\gamma}|\mathbf{X}) = \delta(\boldsymbol{\gamma}|\mathbf{X}, \mathbf{S}), \tag{1.3.1}$$

which states that the probability (density) of deciding $\boldsymbol{\gamma}$, given $\mathbf{X}$, is the same as the probability density of $\boldsymbol{\gamma}$, given both $\mathbf{X}$ and $\mathbf{S}$. Thus, *the decision rule* $\delta(\boldsymbol{\gamma}|\mathbf{X})$ *is the mathematical embodiment of the physical system used to process the data and yield decisions.*

[11] Earlier work on sequential detection is represented by Ref. [8] and [9] here, and more fully by Refs. [2, 32, 33, 36, 38–40] of Chapter 20 of [1]. For a recent, comprehensive treatment, see Chapter 9 of Helstrom [14].

Both fixed and sequential procedures are included in this formulation, and in both cases we deal with terminal decisions. We remark also that the Wald theory of sequential tests [8] introduces a further degree of freedom over the fixed-sample cases through the adoption of a second cost function, the "cost of experimentation" [8, 9]. In the general theory, we are free to limit the class of decision rules, in advance, to either of the above types without compromising the completeness of the theory of either type.

1.3.3 The Decision Problem

In order to give definite structure to the decision process, we must prescribe a criterion of excellence, in addition to *a priori* probabilities $\sigma(\mathbf{S})$ and $W_J(\mathbf{N})$. By this we mean the following: The decisions that are to be made by the reception system must be based on the given data $\mathbf{X}$, which, because of their contamination with noise, constitute only incomplete clues to the received signal $\mathbf{S}$. And, of course, as we have already noted at the beginning of the chapter, the received signal $\mathbf{S}$ itself is already modified by the medium through which it has been propagated, so that $\mathbf{S}_{\text{in}} \neq \mathbf{S}$, cf. Eqs. (1.1.1a–1.1.1c). Therefore, whatever the decision rule $\delta(\boldsymbol{\gamma}|\mathbf{X})$ that is finally adopted, the decisions to which it leads cannot always be correct (except possibly in the unrealizable limit $T \to \infty$). Thus, it is clear that whenever there is a nonzero probability of error some sort of value judgment is implied; in fact, the former always implies (1) a decision process and (2) a numerical cost assignment of some kind to the possible decisions. The units in which such a cost, or value, is measured are essentially irrelevant, but the relative amounts associated with the possible decisions are not.

In order to formulate the decision problem, a *loss* F $(\mathbf{S}, \boldsymbol{\gamma})$is assigned to each combination of decisions $\boldsymbol{\gamma}$ and signal $\mathbf{S}$ (the latter selecting a particular distribution function of $\mathbf{X}$, in accordance with some prior judgment of the relative importance of the various correct and incorrect decisions. Each decision rule may then be rated by adopting an *evaluation or risk function* $\mathsf{E}(\mathsf{F})$ (for example, the mathematical expectation of loss), which takes into consideration both the probabilities of correct and incorrect decisions and the losses associated with them. There are, of course, many ways of assigning loss, and hence many different risk functions. One example, which has been very common in statistics and in communication theory, is the squared-error loss. This type of loss is used in extraction problems in which the decision to be rendered is an estimation of a signal after it has been contaminated with noise. In this case, the loss is taken to be proportional to the square of the error in this estimation. Other examples are discussed in chapters 3–7.

We may now state the reception problem in the following general terms:

Given the family of distribution functions $F_J(\mathbf{X}|\mathbf{S})$, the *a priori* signal probability distribution $\sigma(\mathbf{S})$, the class of possible decisions, and the loss and evaluation functions F and $\mathsf{E}(\mathsf{F})$, the problem is to determine the best rule $\delta(\boldsymbol{\gamma}|X)$ for using the data to make decisions.

In arriving at this statement we have introduced a number of somewhat restrictive assumptions. We now give a brief heuristic discussion of what can be done to remove them. To begin with, the statement of the reception problem in these terms is actually more general than the argument that led up to it, a fact that requires some comment. A quick review of that argument shows, on the one hand, that the restriction of the decisions $\boldsymbol{\gamma}$ to a finite number L of alternatives $\boldsymbol{\gamma} = (\gamma_1, \gamma_2, \ldots, \gamma_L)$ is irrelevant and that a denumerably infinite number may equally well be used. In fact, the extension to a continuum of possible alternatives is simply a matter of reinterpretation. The decision rule $\delta(\boldsymbol{\gamma}|\mathbf{X})$ that was introduced above as a discrete probability distribution must in this case be interpreted as a probability–density functional; that is, $\delta(\boldsymbol{\gamma}|\mathbf{X})\,\mathbf{d}\boldsymbol{\gamma}$ is the probability that $\boldsymbol{\gamma}$ lies between $\boldsymbol{\gamma}$ and $\boldsymbol{\gamma}+\mathbf{d}\boldsymbol{\gamma}$, given $\mathbf{X}$. To represent a nonrandomized decision rule in this case, we interpret $\delta(\boldsymbol{\gamma}|\mathbf{X})$ as a Dirac δ-function [see Eq. (1.4.14) ff., for example]. Usually, the family of distribution functions is not given directly and must be found from a given noise distribution $W_J(\mathbf{N})$ and the mode of combining signal and noise.

1.3.4 The Generic Similarity of Detection and Extraction

Figure 1.5 emphasizes that decision rules are essentially transformations that map observation space into decision space. In detection, each point of observation space Γ (or $\mathbf{X}$) is mapped into the various points constituting the space Δ of terminal decisions. For example, the simplest form of binary detection is the same as dividing Γ into two regions, one corresponding to "no signal" and the other to "signal and noise," and carrying out the operation of decision in one step, since only a single alternative is involved. The binary detection problem is then the problem of how best to make this division. The extension to multiple alternative detection situations is made in analogous fashion: one has now three or more alternative divisions of Γ, with a corresponding set of decisions leading to a final decision [7]. Similarly, in extraction each point of Γ is mapped into a point of the space Δ of terminal decisions, which in this instance has the same structure as the signal space Ω. If the dimensionality of Δ is smaller than that of Γ (as is usually the case in estimating signal parameters), the transformation is "irreversible"; that is, many points of Γ go into a single point of Δ. In this way, extraction may also be thought of as a division of Γ into regions, so that, basically, detection and extraction have this common and generic feature and are thus not ultimately different operations. It is merely necessary to group the points of Δ corresponding to $\mathbf{S}\neq 0$ into a single class labeled "signal and noise" to transform an extractor into a detector. Conversely, detection systems may be regarded as extractors followed by a threshold device that separates, say, $\mathbf{S}=0$ from $\mathbf{S}\neq 0$. However, *a system optimized for the one function may not necessarily be optimized for the other*, and it is in this sense, that we consider detection and extraction as separate problems for analysis.

1.4 SYSTEM EVALUATION

In this section, we shall apply the concepts discussed above to a description of the problem of evaluating system performance, including that of both optimum and suboptimum types. It is necessary first to establish some reasonable method of evaluation, after which a number of criteria of excellence may be postulated, with respect to which optimization may then be specifically defined.

1.4.1 Evaluation Functions

As mentioned in Section 1.3.3, $\mathsf{F}(\mathbf{S},\boldsymbol{\gamma})$ is a *generalized loss function*, adopted in advance of any optimization procedure, which assigns a *loss*, or *cost*, to every combination of system input and decision (system output) in a way which may or may not depend on the system's operation. Actual evaluation of system performance is now made as mentioned earlier, provided that we adopt an evaluation function $\mathsf{E}(\mathsf{F})$ that takes into account all possible modes of system behavior and their relative frequencies of occurrence and assigns an overall loss rating to each system or decision rule. One obvious choice of E is the *mathematical expectation* E, or *average value*, of F, and it is on this reasonable but arbitrary choice which the present theory is based for the most part.[12]

At this point, it is convenient to define two different loss ratings for a system, one of which is used to rate performance when the signal input is fixed and the other to take account of *a priori* signal probabilities. For a given $\mathbf{S}$, we have first:

The Conditional Loss Rating.[13] $\mathsf{L}(\mathbf{S},\delta)$ *of* δ *is defined as the conditional expectation of loss:*

$$\mathsf{L}(\mathbf{S},\delta) = E_{\mathbf{X}|\mathbf{S}}\{\mathsf{F}[\mathbf{S},\gamma(\mathbf{X})]\} = \int_{\Gamma} \mathbf{dX} \int_{\Delta} \mathbf{d}\boldsymbol{\gamma} \mathsf{F}(\mathbf{S},\boldsymbol{\gamma}) F_J(\mathbf{X}|\mathbf{S})\delta(\boldsymbol{\gamma}|\mathbf{X}). \tag{1.4.1}$$

By this notation we include discrete as well as continuous spaces Δ; for the former, the integral over Δ is to be interpreted as a sum and $\delta(\boldsymbol{\gamma}|\mathbf{X})$ as a probability, rather than as a probability density. (See the remarks at the end of Section 1.3.3.)

Actually, as will be seen in Section 1.4.4, the conditional loss rating is most significant when the *a priori* probability $\sigma(\mathbf{S})$ is unknown. However, when $\sigma(\mathbf{S})$ is known, we use this information to rate the system by averaging the loss over both the sample and the signal distributions:

The Average Loss Rating. $\mathsf{L}(\sigma,\delta)$ of δ *is defined as the (unconditional) expectation of loss when the signal distribution is* $\sigma(\mathbf{S})$:

$$\mathsf{L}(\sigma,\delta) = E_{\mathbf{X}|\mathbf{S}}\{\mathsf{F}(\mathbf{S},\boldsymbol{\gamma})\} = \int_{\Omega} \mathbf{dS} \int_{\Gamma} \mathbf{dX} \int_{\Delta} \mathbf{d}\boldsymbol{\gamma}\, \mathsf{F}(\mathbf{S},\boldsymbol{\gamma})\sigma(\mathbf{S}) F_J(\mathbf{X}|\mathbf{S})\delta(\boldsymbol{\gamma}|\mathbf{X}). \tag{1.4.2}$$

Some remarks are appropriate concerning the loss function F. In the statistical literature, F is usually a function that assigns to each combination of signal and decision a certain loss, or *cost*, which is independent of δ:

$$\mathsf{F}_1 = C(\mathbf{S},\boldsymbol{\gamma}). \tag{1.4.3}$$

In the present analysis, we restrict our discussion chiefly to systems whose performance is rated according to simple loss functions[14] of this nature. There exists a substantial body of theory for this case, and certain very general statements can be made about optimum systems derived under this restriction (cf. Wald's complete class theorem, admissibility [25], and so on; see Section 1.5 ff.

[12] Other linear or nonlinear operations for E are possible and should not be overlooked in subsequent generalizations (see the comments in Section 1.5.4).

[13] This quantity is called the *a priori risk* in Wald's terminology [25].

[14] We shall use the term *risk*, henceforth, as synonymous with this simple cost, or loss.

We point out, however, that a more general type of loss function can be constructed. In fact, one such function is suggested by information theory. For, if we let

$$F_2 = -\log p(\mathbf{S}|\boldsymbol{\gamma}), \tag{1.4.4}$$

where $p(\mathbf{S}|\boldsymbol{\gamma})$ is the *a posteriori* probability of $\mathbf{S}$ given $\boldsymbol{\gamma}$, the average loss rating [Eq. (1.4.2)] becomes the well-known equivocation of information theory [2, 26] (Section 6.5.2 of Ref. [1]). This loss function can be interpreted as a measure of the "uncertainty" (or "surprisal") about $\mathbf{S}$ when $\boldsymbol{\gamma}$ is known [26], (Section 6.2.1 of Ref. [1]). It is an example of a more general type than the simple cost function [Eq. (1.4.3)]. For, unlike $C(\mathbf{S}|\boldsymbol{\gamma})$, which depends on $\mathbf{S}$ and $\boldsymbol{\gamma}$ alone, Eq (1.4.4) depends also on the decision rule in use and cannot be preassigned independently of δ. Loss functions like Eq. (1.4.4) are more difficult to deal with, and some of the general statements (Section 1.5) that can be derived for Eq. (1.4.3) clearly do not hold true for Eq. (1.4.4). In Chapter 22 of Ref. [1], however, it is shown that close connections may exist between results based on the two types of loss function.

The conditional and average loss ratings of δ may now be written, from Eqs. (1.4.1)–(1.4.4), as

I. *Conditional Risk:*

$$r(\mathbf{S},\delta) = \int_\Gamma \mathbf{dX} F_J(\mathbf{X}|\mathbf{S}) \int_\Delta \mathbf{d}\boldsymbol{\gamma} C(\mathbf{S}|\gamma)\delta(\gamma|\mathbf{X}). \tag{1.4.5}$$

II. *Average Risk:*

$$R(\sigma,\delta) = E\{r(\mathbf{S},\delta)\} = \int_\Omega r(\mathbf{S},\delta)\sigma(\mathbf{S})\mathbf{dS}, \tag{1.4.6a}$$

or

$$R(\sigma,\delta) = \int_\Omega \sigma(\mathbf{S})\mathbf{dS} \int_\Gamma \mathbf{dX} F_J(\mathbf{X}|\mathbf{S}) \int_\Delta \mathbf{d}\boldsymbol{\gamma} C(\mathbf{S}|\boldsymbol{\gamma})\delta(\boldsymbol{\gamma}|\mathbf{X}). \tag{1.4.6b}$$

III. *Conditional Information Loss:*

$$h(\mathbf{S},\delta) = -\int_\Gamma \mathbf{dX} F_J(\mathbf{X}|\mathbf{S}) \int_\Delta \mathbf{d}\boldsymbol{\gamma} [\log p(\mathbf{S}|\boldsymbol{\gamma})]\delta(\boldsymbol{\gamma}|\mathbf{X}). \tag{1.4.7}$$

IV. *Average Information Loss:*

$$H(\sigma,\delta) = E\{h(\mathbf{S},\delta)\} = \int_\Omega h(\mathbf{S},\delta)\sigma(\mathbf{S})\mathbf{dS}, \tag{1.4.8a}$$

or

$$H(\sigma,\delta) = -\int_\Omega \sigma(\mathbf{S})\mathbf{dS} \int_\Gamma \mathbf{dX} F_J(\mathbf{X}|\mathbf{S}) \int_\Delta \mathbf{d}\boldsymbol{\gamma} [\log p(\mathbf{S}|\boldsymbol{\gamma})]\delta(\boldsymbol{\gamma}|\mathbf{X}). \tag{1.4.8b}$$

The last of these is the well-known "equivocation" of information theory[15] (cf. Sections 6.5.2 and 6.5.3 of Ref. [1].

As we have already mentioned in Section 1.1, **S**, when deterministic, is a function of a set of random parameters[16] $\boldsymbol{\theta}$, and frequently it is the parameters $\boldsymbol{\theta}$ about which decisions are to be made, rather than about **S** itself (see, e.g., Section 1.4.2). Similar to Eqs. (1.4.5) and (1.4.6), the conditional and average risks for this situation may be expressed as[17]

$$r(\boldsymbol{\theta}, \delta) = \int_{\Gamma} \mathbf{dX} F_J(\mathbf{X}|\mathbf{S}(\boldsymbol{\theta})) \int_{\Delta} \mathbf{d}\boldsymbol{\gamma} C(\boldsymbol{\theta}|\gamma)\delta(\gamma|\mathbf{X}), \tag{1.4.9}$$

and

$$R(\sigma, \delta)_{\boldsymbol{\theta}} = \int_{\Omega_{\boldsymbol{\theta}}} r(\boldsymbol{\theta}, \delta)\sigma(\boldsymbol{\theta})\mathbf{d}\boldsymbol{\theta}. \tag{1.4.10}$$

Here, of course, $r(\boldsymbol{\theta}, \delta)$ and $R(\sigma, \delta)_{\boldsymbol{\theta}}$ are not necessarily the same as $r(\mathbf{s}, \delta)$,$R(\sigma, \delta)$ above, nor is the form of $\boldsymbol{\gamma}$ either. Notice that the cost function $C(\boldsymbol{\theta}|\boldsymbol{\gamma})$ is usually a different function of $\boldsymbol{\theta}$ from $C[\mathbf{S}(\boldsymbol{\theta}), \boldsymbol{\gamma}]$ also. Considerable freedom of choice as to the particular conditional and average risks is thus frequently available to the system analyst, although the appropriate choice is often dictated by the problem in question. Finally, observe that $r(\mathbf{S}, \delta)$ and $R(\sigma, \delta)$ for decisions about $\mathbf{S} = \mathbf{S}(\boldsymbol{\theta})$ are still given by Eqs. (1.4.5), (1.4.6) where $\sigma(\mathbf{S})\mathbf{dS}$ is replaced by its equivalent $\sigma(\boldsymbol{\theta})\mathbf{d}\boldsymbol{\theta}$ in Eqs. (1.4.6a) and (1.4.6b), with a corresponding change from Ω-space (for **S**) to $\Omega_{\boldsymbol{\theta}}$-space (for $\boldsymbol{\theta}$) according to the transformations implied by $\mathbf{S} = \mathbf{S}(\boldsymbol{\theta})$. Similar remarks apply for the conditional and average information losses, Eqs. (1.4.7) and (1.4.8), as well.

1.4.2 System Comparisons and Error Probabilities

The expressions (1.4.1) and (1.4.2) for the loss ratings can be put into another and often more revealing form, which exhibits directly the rôle of the error probabilities associated with the various possible decisions. Let $p(\boldsymbol{\gamma}|\mathbf{S})$ be the conditional probability[18] that the system in question makes decisions $\boldsymbol{\gamma}$ when the signal is **S** and a decision rule $\delta(\boldsymbol{\gamma}|\mathbf{X})$ is adopted, so that

$$p(\boldsymbol{\gamma}|\mathbf{S}) = \int_{\Gamma} F_J(\mathbf{X}|\mathbf{S})\delta(\boldsymbol{\gamma}|\mathbf{X})\mathbf{dX}. \tag{1.4.11}$$

Comparison with the conditional risk (1.4.5) shows that the latter may be written

$$r(\mathbf{S}, \delta) = \int_{\Delta} \mathbf{d}\boldsymbol{\gamma} p(\gamma|\mathbf{S})C(\mathbf{S}, \gamma), \tag{1.4.12}$$

[15] Note that when **S** can assume a continuum of values (as is usually the case), we must replace the probability $p(\mathbf{S}|\gamma)$ in Eq. (1.4.4) by the corresponding probability *density* $w(\mathbf{S}|\boldsymbol{\gamma})$ and include in Eqs. (1.4.7) and (1.4.8) the absolute entropy (cf. Section 6.4.1 of Ref. [1]).

[16] For simplicity, these are assumed to be time-invariant here; the generalization to include time variations $\boldsymbol{\theta} = \boldsymbol{\theta}(t)$ is straightforward.

[17] Here and henceforth, unless otherwise indicated, we adopt the notational convention that the principal argument of a function distinguishes that function from other functions: thus, $\sigma(\mathbf{S}) \neq \sigma(\boldsymbol{\theta}), p(\mathbf{X}) \neq p(\boldsymbol{\theta})$, and so on; however, $\sigma[\mathbf{S}(\boldsymbol{\theta})] = \sigma(\mathbf{S})$ and so on.

[18] Or probability density, when $\boldsymbol{\gamma}$ represents a continuum of decisions (as in extraction, cf. Chapter 5).

which is simply the sum of the costs associated with all possible decisions for the given $\mathbf{S}$, weighted according to their probability of occurrence. In a similar way, we can obtain the probability (density) of the decisions $\boldsymbol{\gamma}$ by averaging Eq. (1.4.11) with respect to $\mathbf{S}$, for example,

$$p(\boldsymbol{\gamma}) = \langle p(\gamma|\mathbf{S})\rangle_S = \int_\Omega \sigma(\mathbf{S})\mathbf{dS}\int_\Gamma \mathbf{dX}F_j(\mathbf{X}|\mathbf{S})\delta(\gamma|\mathbf{X}). \tag{1.4.13}$$

Since we shall be concerned in what follows almost exclusively with nonrandomized decision rules, particularly in applications, we see that $\delta(\gamma|\mathbf{X})$ may be expressed as

$$\delta(\boldsymbol{\gamma}|\mathbf{X}) = \delta[\boldsymbol{\gamma} - \boldsymbol{\gamma}_\sigma(\mathbf{X})], \tag{1.4.14}$$

where the δ of the right-hand member is now the Dirac δ-function. Here it is essential to distinguish between the decisions $\boldsymbol{\gamma}$ and the functional operation $\gamma_\sigma(\mathbf{X})$ performed on the data by the system. The subscript σ reminds us that this operation depends in general on signal statistics. With Eq. (1.4.14), the probability (density) of decisions $\boldsymbol{\gamma}$ on condition $\mathbf{S}$, which may represent correct or incorrect decisions, can be written

$$p(\boldsymbol{\gamma}|\mathbf{S}) - \int_\Gamma F_J(\mathbf{X}|\mathbf{S})\delta[\boldsymbol{\gamma} - \boldsymbol{\gamma}_\sigma(\mathbf{X})]\mathbf{dX} = \int_{-\infty}^{\infty}\cdots\int e^{i\tilde{\xi}\boldsymbol{\gamma}}\frac{d\xi}{(2\pi)^L}\int_\Gamma F_J(\mathbf{X}|\mathbf{S})e^{-i\tilde{\xi}\gamma_\sigma(\mathbf{X})}\mathbf{dX}, \tag{1.4.15}$$

(cf. [1], Section 17.2.1). This reveals the explicit system operation. Equation (1.4.15) in particular provides a direct way of calculating $p(\boldsymbol{\gamma}|\mathbf{S})$ for any system once its system structure $\boldsymbol{\gamma}_\sigma(\mathbf{X})$ is known. As for Eq. (1.4.13), we can also obtain the probability density of $\boldsymbol{\gamma}$ itself by averaging $p(\boldsymbol{\gamma}|\mathbf{S})$ [Eq. (1.4.15)] over $\mathbf{S}$.

Comparison of explicit decision systems now follows directly. For example, this may be done by determining which has the smallest average loss rating $L(\sigma, \delta)$, which, in terms of *average* risk Eq. (1.4.6a), involves the comparison of $R(\sigma, \delta_1)$ and $R(\sigma, \delta_2)$ for two systems with system functions $\boldsymbol{\gamma}_\sigma(\mathbf{X})_1$ and $\boldsymbol{\gamma}_\sigma(\mathbf{X})_2$ [Eq. (1.4.14)]. In a similar fashion, one can compare also $H(\sigma, \delta_1)$ and $H(\sigma, \delta_2)$ [Eq. (1.4.8)]. Note that not only optimum but suboptimum systems may be so handled once $\boldsymbol{\gamma}_\sigma(\mathbf{X})$ is specified, so that now one has a possible quantitative method of deciding in practical situations between "good," "bad," "fair," "best," and so on, where the comparisons are consistently made within a common criterion and where the available information can be incorporated in ways appropriate to each system under study. We emphasize that this consistent framework for system comparison is one of the most important practical features of the theory, along with its ability to indicate the explicit structure of optimum and suboptimum systems, embodied in the decision rule $\delta(\boldsymbol{\gamma}|\mathbf{X})$.

1.4.3 Optimization: Bayes Systems

In Section 1.4.1, we have seen how average and conditional loss ratings may be assigned to any system, once the evaluation and cost functions have been selected. We now define what we mean by an optimum decision system. We state a definition first for the case where complete knowledge of the *a priori* signal probabilities $\sigma(\mathbf{S})$ is assumed and in which, from what has been said above, evaluation from the point of view of the *average* loss $R(\sigma, \delta)$ is most appropriate. Consider, then, that *one system is "better" than another if its average loss rating is smaller for the same application (and criterion), and that the "best," or optimum, system is the one with the smallest average loss rating*. (The preassigned costs, of course, are the same.) We call this optimum system a *Bayes system*:

> *A Bayes system obeys a Bayes decision rule δ^*, where δ^* is a decision rule whose average loss rating L is smallest for a given a priori distribution σ.* (1.4.16)

For the risk and information criteria of Eqs. (1.4.6a) and (1.4.8a), this becomes

$$R^* = \min_{\delta} R(\sigma, \delta) = R(\sigma, \delta^*) : \textit{Bayes risk}, \tag{1.4.16a}$$

and

$$H^* = \min_{\delta} H(\sigma, \delta) = H(\sigma, \delta^*) : \textit{Bayes equivocation}. \tag{1.4.16b}$$

The former minimizes the average risk (or cost), while the latter minimizes the equivocation. Bayes decision rules (for the given F) form a *Bayes class*, each member of which corresponds to a different *a priori* distribution[19] $\sigma(\mathbf{S})$.

1.4.4 Optimization: Minimax Systems

When the *a priori* signal probabilities are not known or are only incompletely given, definition of the optimum system is still open. A possible criterion for optimization in such cases is provided by the *Minimax decision rule* δ_M^*, or Bayes rule associated with the conditional risk $r(\mathbf{S}, \delta)$. As indicated by our notation, there is one conditional risk figure attached to each possible signal $\mathbf{S}$. In general, these risks will be different for different signals, and there will be a minimum among them, say $r(\mathbf{S}, \delta)_{\max}$. The Minimax rule is, roughly speaking, the decision rule that reduces this maximum as far as possible. More precisely:

> *The Minimax decision rule δ_M^* is the rule for which the maximum conditional loss rating $\mathsf{L}(\mathbf{S},\delta)_{\max}$, as the signal $\mathbf{S}$ ranges over all possible values, is not greater than the maximum conditional loss rating of any other decision rule δ.* (1.4.17)

Thus, in terms of conditional risk r, or conditional information loss h, we may write

$$\begin{aligned} \max_{\mathbf{S}} r(\mathbf{S}, \delta_{\mathrm{M}}^*) &= \max_{\mathbf{S}} \min_{\delta} r(\mathbf{S}, \delta) \le \max_{\mathbf{S}} r(\mathbf{S}, \delta), \\ \max_{\mathbf{S}} h(\mathbf{S}, \delta_{\mathrm{M}}^*) &= \max_{\mathbf{S}} \min_{\delta} h(\mathbf{S}, \delta) \le \max_{\mathbf{S}} h(\mathbf{S}, \delta). \end{aligned} \tag{1.4.17b}$$

[19] Of course, it is possible that different $\sigma(\mathbf{s})$ may lead to identical decision rules, but aside from this possible ambiguity we observe that a Bayes criterion is entirely appropriate when $\sigma(\mathbf{s})$ is known, since it makes full use of all available information.

Wald has shown[20] under certain rather broad conditions (see Sections 1.5.2 and 1.5.3 ff.) that $\max_{\mathbf{S}} \min_{\delta} r(\mathbf{S}, \delta) = \min_{\delta} \max_{\mathbf{S}} r(\mathbf{S}, \delta)$ for the risk (i.e., simple cost) formulations, from which the significance of the term "Minimax" becomes apparent. Whether or not a corresponding result holds for the information-loss formulation remains to be established.

We may also express the Minimax decision process in terms of the resulting average risk. From Section 1.5.3 ff., Theorems 1, 4, 5, 9, we have the equivalent Minimax formulation

$$\left.\begin{aligned} R_{\mathrm{M}}^*(\sigma_0, \delta_{\mathrm{M}}^*) = \max_{\sigma} R^*(\sigma, \delta^*) &= \max_{\sigma} \min_{\delta} R(\sigma, \delta) \\ &= \min_{\delta} \max_{\sigma} R(\sigma, \delta) \end{aligned}\right\} = \text{Minimax average risk}^{21} \tag{1.4.18}$$

this last from Eq. (1.4.16a) and Section 1.5.2 ff., definition 7a. Thus, the Minimax average risk is the largest of all the Bayes risks, considered over the class of *a priori* signal distribution $\{\sigma(\mathbf{S})\}$. The distribution $\sigma_0 (= \sigma_{\mathrm{M}}^*)$ for which this occurs is called the *least favorable distribution*. Accordingly, the Minimax decision rule δ_{M}^* [obtained by adjusting the Bayes rule δ^* as σ is varied, cf. Eq. (1.4.18)] is one which gives us the least favorable, or "worst," of all Bayes — that is, "best" — systems. Geometrically, the Minimax situation of $\sigma \to \sigma_0$, $\delta \to \delta_{\mathrm{M}}^*$, $R(\sigma, \delta) \to R_{\mathrm{M}}^*(\sigma_0, \delta_{\mathrm{M}}^*)$ is represented by a saddle point of the average-risk surface over the (σ, δ) plane, as Fig. 1.6 indicates. The existence of $\sigma_0, \delta_{\mathrm{M}}^*$ and this saddle point follows from the appropriate theorems (cf. Section 1.5.3).

The Minimax decision rule has been the subject of much study and also of some adverse criticism. It has been argued that it is often too conservative to be very useful. However, it is also true that there are situations in which the Minimax rule is unquestionably an excellent choice. Figure 1.7a illustrates these remarks.

Here we have presented the case where the maximum conditional loss rating of all other decision rules $\delta_1, \delta_2, \ldots$ exceeds that for δ_{M}^* and where even most of the minimum loss ratings are also noticeably larger than the corresponding minimum for δ_{M}^*. Sometimes, however, we may have the situation shown in Fig. 1.7b, where δ_{M}^* leads to excessive loss ratings, except for a comparatively narrow range of values of $\mathbf{S}$. In the latter case, δ_{M}^* is perhaps too conservative, and a more acceptable decision rule might be sought.[22] The Minimax procedure does, at any rate, have the advantage of guarding against the worst case, but also may be too cautious for the more probable states of the input to the system. When the costs are preassigned and immutable, the possible conservatism of Minimax

[20] The minimax theorem was first introduced and proved by Von Neumann, in an early paper on the theory of games [27]. For a further account, see Von Neumann and Morgenstern [28], Section 17.6, p. 154; also Ref. [5].

[21] This is the average risk associated with the Minimax decision rule.

[22] These Minimax risk curves have a single distinct maximum. The least favorable *a priori* distribution σ_0 is this case consequently concentrates all of its probability mass at the signal value corresponding to the maximum conditional risk (a δ-function distribution for continuous signal space), since by definition σ_0 must maximize Bayes (average) risk. Existence of a least favorable distribution is here ensured by our assumptions *A* to *D* of Section 1.5.3, which correspond to Wald's assumptions 3.1–3.7 [25]. Roughly speaking, the Minimax conditional risk must equal its maximum value for all signals to which the least favorable distribution assigns a nonzero *a priori* probability (see Wald's theorems 3.10 and 3.11). Thus, a Minimax conditional risk curve with two distinct and equal maxima could have a corresponding σ_0 with probability concentrated at either of the two maxima or distributed between them, while if one maximum were larger than the other, the mass would have to be concentrated only at the larger, and so on. Or again, if σ_0 were nonzero over a finite interval, the corresponding Minimax conditional risk would be constant over this range (but might take on other, smaller values outside).

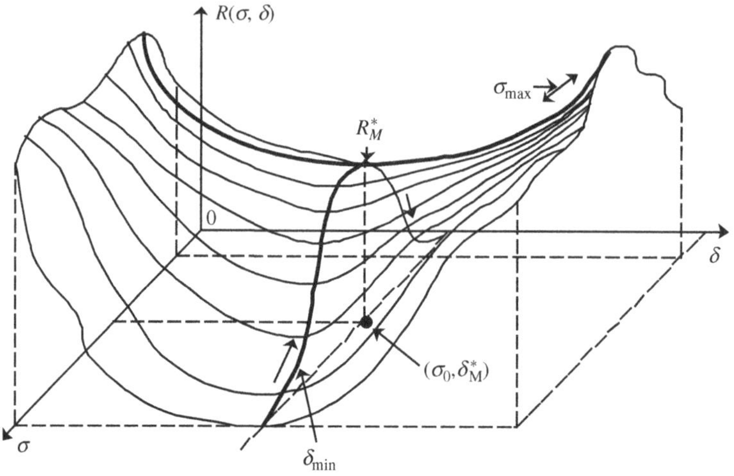

FIGURE 1.6 Average risk as a function of decision rule and *a priori* signal distribution, showing a Minimax saddle point.

cannot be avoided without choosing another criterion.[23] However, in many cases where the actual values of the preassigned costs are left open to an *a posteriori* adjustment, it may be possible by a more judicious cost assignment to modify δ_M^* more along the lines of Fig. 1.7a, where the "tails" of $\max_S r(\mathbf{S}, \delta_M^*)$ are comparable to those of $r(\mathbf{S}, \delta)$ (all $\mathbf{S}$), and thus eliminate, at least in part, the conservative nature of the decision process.[24]

The Bayes decision rule makes the fullest use of *a priori* probabilities (when these are known) and in a sense assumes the most favorable system outcome. The Minimax decision rule, on the other hand, makes no use at all of these *a priori* probabilities (for the good reason that they are not available to the observer) and in the same sense assumes the worst case [cf. Eq. (1.4.18) and Fig. 1.7]. In practical cases, an important problem is to find δ_M^*. No general simple procedure is available, although δ_M^* always exists in the risk formulation. From the definitions of δ^* and δ_M^*, however, it can be shown that a *Bayes decision rule whose*

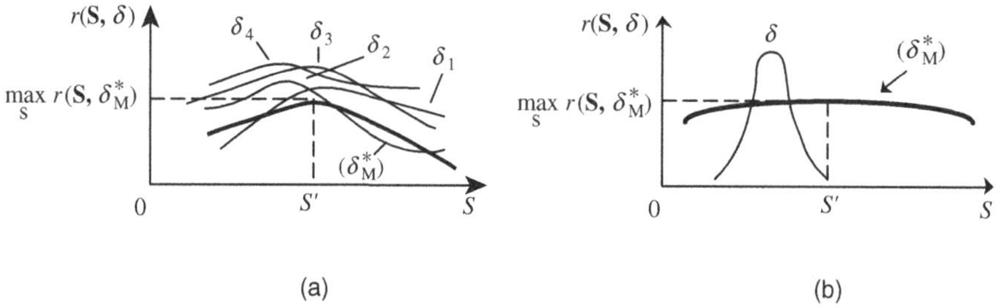

FIGURE 1.7 (a) An acceptable Minimax situation. (b) A Minimax situation that is possibly too conservative.

[23] For example, the "Minimax regret" criterion or the Hurwicz criterion, and so on [29].

[24] Hodges and Lehmann [30] have discussed some intermediate situations where $\sigma(\mathbf{s})$ is partially known on the basis of previous experience. Also, see Section 1.5 ff.

conditional loss rating L *is the same for all signals is a Minimax rule* (see Section 1.5.3 and Theorem 7). Thus, if we can find a δ^* for which this is the case, we have also determined a least favorable *a priori* distribution $\sigma_M^*(S)$ for this $\delta^* (= \delta_M^*)$, which follows from the above, that is,

$$r(\mathbf{S}, \delta^*) \left[\text{or } h(\mathbf{S}, \delta^*)\right] = \text{constant, all } \mathbf{S}. \tag{1.4.19}$$

Note that a non-Bayes rule whose conditional loss rating is constant for all $\mathbf{S}$ is not necessarily Minimax. When Eq. (1.4.19) holds, it furnishes a useful method for finding δ_M^* in each case.

1.5 A SUMMARY OF BASIC DEFINITIONS AND PRINCIPAL THEOREMS

We conclude this chapter now with a short summary of some of the principal results, theorems, and so on which give decision theory its general scope and power in our present application to communication problems. For proofs and further discussion, the reader is referred to the appropriate sections of Wald [25] and other pertinent references.

1.5.1 Some General Properties of Optimum Decision Rules

The practical utility of an optimum procedure lies to a considerable extent in its uniqueness: there is not a number of different reception systems with the same optimal properties. For the unique optimum, the problem of choosing the simplest (or least expensive) from the point of view of design is automatically resolved. For this reason and for the central one of optimization itself, it is important to know the properties of optimum decision rules and when they may be expected to apply in physical systems. We state now two of the main results on which subsequent applications are based:

1.5.1.1 Admissible Decision Rules We note, first, that the conditional loss rating of a decision rule depends, of course, on the particular signal present at the input. One decision rule may have a smaller rating than another for some signals and a larger one for others. If the conditional loss rating of δ_1 never exceeds that of δ_2 for *any* value of $\mathbf{S}$, and is actually less than that of δ_2 for at least one $\mathbf{S}$, then δ_1 is said to be *uniformly better* than δ_2. This leads, accordingly, to the notion of *admissibility*:

A decision rule is admissible if no uniformly better one exists. (1.5.1)

Observe that with this definition an admissible rule is not necessarily uniformly better than any other; other rules can have smaller ratings at particular $\mathbf{S}$ (Fig. 1.7a). However, they cannot be better for *all* $\mathbf{S}$.

It follows, then, that, *if a Bayes or Minimax rule is unique, it is admissible*. The converse is not true, since an admissible rule is not necessarily Bayes or Minimax. Accordingly, no system that does not minimize the average risk (loss rating) can be uniformly better than a Bayes system [for the same $\sigma(\mathbf{S})$], and no system that does not minimize the maximum conditional risk (loss rating) can be uniformly better than a Minimax system. Admissibility is an important additional optimum property of unique Bayes and Minimax decision systems.

1.5.1.2 The Complete Class Theorem This is Wald's fundamental theorem ([25], Theorem (3.20)) concerning complete classes of decision rules. We say first that a class D of decision rules is *complete* if, for any δ not in D, we can find a δ^* in D such that δ^* is uniformly better than δ. If D contains no subclass which is complete, D is a *minimal complete class*. Wald has shown that *for the simple loss functions* [Eqs. (1.4.3), (1.4.5), (1.4.6)] *the class of all admissible Bayes decision rules is a minimal complete class*, under a set of conditions that are certainly satisfied for most if not all physical situations (cf. Sections 1.5.2 and 1.5.3 ff.). For the same set of conditions, any Minimax decision rule can be shown to be a Bayes rule with respect to a certain *least favorable a priori* distribution $\sigma^*_M(\mathbf{S})$, and the existence of $\sigma^*_M(\mathbf{S})$, as well as of the Bayes and Minimax rules themselves, is assured. The complete class theorem thus establishes an optimum property of the Bayes class as a whole. To the author's knowledge no complete class theorem has as yet been demonstrated for the information loss ratings of Eqs. (1.4.7) and (1.4.8), nor have the general conditions for the existence of Bayes and Minimax rules for such measures been established. However, some results on the characterization of Bayes tests with this measure, for detection, are given in Ref. [1], Chapter 22.

1.5.2 Definitions[25]

It is assumed that decisions $\boldsymbol{\gamma}$ are to be made about a signal $\mathbf{S}$, based on observations $\mathbf{X}$ whose occurrence is governed by the conditional distribution–density function $F_j(\mathbf{X}|\mathbf{S})$. The decision rule $\delta(\boldsymbol{\gamma}|\mathbf{X})$ is the probability (density) that $\boldsymbol{\gamma}$ will be decided when the observation is $\mathbf{X}$, regardless of $\mathbf{S}$.

Risk theory is based on the following definitions:

(1) It is assumed that a *cost* $C(\mathbf{S},\boldsymbol{\gamma})$ is preassigned to every possible combination of signal $\mathbf{S}$ and decision γ_l, $l = 1, \ldots, L$, in the problem.

(2) The *conditional risk* $r(\mathbf{S},\delta)$ of using a decision rule δ is the expected value of the cost when the signal is S:

$$r(\mathbf{S},\delta) = \int_\Gamma \int_\Delta C(\mathbf{S},\boldsymbol{\gamma})\delta(\boldsymbol{\gamma}|\mathbf{X})F_J(\mathbf{X}|\mathbf{S})\mathbf{dX}\,\mathbf{d}\boldsymbol{\gamma}. \tag{1.5.2}$$

(3) The *average risk* $R(\sigma,\delta)$ of using δ is the expected value of $r(\mathbf{S},\delta)$ in view of the *a priori* probability (density) $\sigma(\mathbf{S})$:

$$R(\sigma,\delta) = \int_\Omega r(\mathbf{S},\delta)\sigma(\mathbf{S})\mathbf{dS}. \tag{1.5.3}$$

(4) A *Minimax* decision rule δ^*_M is one whose maximum conditional risk is not greater than the maximum conditional risk of any other δ:

$$\max_S r\left(\mathbf{S},\delta^*_M\right) \le \max_S r(\mathbf{S},\delta), \quad \text{for all } \delta. \tag{1.5.4}$$

[25] Wald's book [25] is referred to in the following as SDF. Wald's paper [25a] is recommended as an introduction to the subject.

(5) A Bayes decision rule δ^* is one whose average risk is smallest for a given *a priori* distribution $\sigma(S)$:

$$R(\sigma, \delta^*) = \min_{\delta} R(\sigma, \delta), \text{ for all } \delta. \tag{1.5.5}$$

(6) A decision rule δ_1 is *uniformly better* than a decision rule δ_2 if the conditional risk of δ_1 does not exceed that of δ_2 for any value of **S** and is actually less than that of δ_2 for some particular **S**.

(7) A decision rule is *admissible* if no uniformly better one exists.

a. An admissible rule is *not* necessarily uniformly better than any other; that is, other rules can have smaller risks at a particular value of **S**. The point is that they cannot be better at *all* values of **S**.

b. An admissible rule need not be Minimax. Clearly, δ^*_{M} could have a larger risk than δ at some values of **S** and still have a smaller maximum risk.

(8) A class D of decision rules is complete if for any δ not in D we can find a δ'^* in D such that δ'^* is uniformly better than δ. If D contains no subclass which is complete, it is a *minimal complete* class.

1.5.3 Principal Theorems

We assume that the following conditions are fulfilled:

A. $F_J(\mathbf{X}|\mathbf{S})$ is continuous in **S**.

B. $C(\mathbf{S}, \boldsymbol{\gamma})$ is bounded in **S** and $\boldsymbol{\gamma}$.

C. The class of decision rules considered is restricted to either (1) nonsequential rules or (2) sequential rules.

D. **S** and $\boldsymbol{\gamma}$ are restricted to finite closed domains.

These conditions are more restrictive in some cases than those imposed by Wald but are sufficient for our purposes. Specifically, Wald's assumptions [25]: (3.1), (3.2), (3.3) are covered by conditions A and B, (3.5) and (3.6) by condition C, and (3.4) and (3.7) by condition D.

Under these assumptions the following theorems exist:

(1) The decision problem viewed as a zero-sum two-person game is *strictly determined*:

$$\max_{\sigma} \min_{\delta} R(\sigma, \delta) = \min_{\delta} \max_{\sigma} R(\sigma, \delta), \qquad \text{(SDF Theorem 3.4)} \tag{1.5.6}$$

(2) For any *a priori* $\sigma(\mathbf{S})$, there exists a Bayes decision rule $\delta*$ relative to $\sigma(\mathbf{S})$ (SDF Theorem 3.5).

(3) A Minimax decision rule exists (SDF Theorem 3.7).

(4) A *least favorable a priori* distribution $\sigma_0(\mathbf{S})$ exists:

$$\min_{\delta} R(\sigma_0, \delta) = \max_{\sigma} \min_{\delta} R(\sigma, \delta), \qquad \text{(SDF Theorem 3.14)} \tag{1.5.7}$$

(5) Any Minimax decision rule is Bayes relative to a least favorable *a priori* distribution (SDF Theorem 3.9).

(6) The class of all Bayes decision rules is complete relative to the class of all decision rules for which the conditional risk is a bounded function of **S** (SDF Theorem 3.20). (Kiefer [31] shows that the restriction of the set of decision rules to those for which the conditional risk is a bounded function of **S** is unnecessary. He also shows that the class of all admissible decision functions is minimal complete. See in addition Wald's remarks following Theorem 3.20 in SDF.)

The facts below follow from the definitions of Section 1.5.2:

(7) A Bayes decision rule δ_{M}^* whose conditional risk is constant is a Minimax decision rule. [This follows from definitions 4, 5, and 7. For suppose δ_{M}^* were not Minimax. Then there would exist a δ' with smaller maximum risk and smaller average risk with respect to $\sigma_0(\mathbf{S}) = \sigma_{\mathrm{M}}^*(\mathbf{S})$. This contradicts the definition of δ_{M}.]

(8) If a Bayes decision rule is unique, it is admissible. [For suppose $\delta*$ were Bayes with respect to $\sigma(\mathbf{S})$ and not admissible. Then a uniformly better δ' would exist; that is, $r(\mathbf{S},\delta') \leq r(\mathbf{S},\overline{\delta^*})$ for all **S**, with equality for some **S**. But this implies that the average risk of δ' with respect to $\sigma(\mathbf{S})$ is less than that of $\delta*$ with respect to $\sigma(\mathbf{S})$. This contradicts the definition of $\delta*$.]

(9) A Minimax decision rule has a smaller *maximum* average risk than any other. [This follows from the fact that the average risk cannot exceed the maximum conditional risk. Of course, for some particular $\sigma(\mathbf{S})$ another test might have smaller average risk than the Minimax with the same $\sigma(\mathbf{S})$.]

Finally, it is of considerable importance practically to be able to avoid randomized decision rules. We have quoted one theorem due to Hodges and Lehmann [15] on this point. Others may be found in some work of Dvoretzky et al. [32].

1.5.4 Remarks: Prior Probabilities, Cost Assignments, and System Invariants

From our discussion, it is clear that *a priori* probabilities play an essential part in the formulation and application of decision theory. In a general way we may say that Bayesian methods of statistical analysis offer two main approaches to providing prior probabilities. One approach, the "subjective" approach, treats probability as the measure of confidence, or plausibility, which we are willing to assign to a proposition or event. The other approach, the so-called "objective" approach, is based on the classic "frequency of occurrence" or prior history of the event in question. Both are, and have been, open to criticism: the subjective viewpoint, of course, introduces the observer's judgment, albeit quantitatively as a probability assignment. On the other hand, the "objective" alternative is limited by its dependence on a "history," or frequency of occurrence, which may not exist. If the event has no history up to the present but can be conceived as a physical possibility, one possible way out is to create an ensemble of virtual event outcomes, and hence generate a resulting "prior" probability measure of this physically possible event. An important pragmatic justification of the subjective viewpoint is that it couples the observer's probabilistic models and decision making to the real world. And this is accomplished by providing the "plausible" priors, which the Bayesian formulation requires[26].

Both viewpoints offer the needed coupling of the observer's models to real-world applications, although the subjective approach appears to be the one more favored by scientists and engineers. In either case, statistical decision theory (SDT) can be used with various methods (Minimax among them) to provide the needed distributions. For these reasons, the quantification of *a priori* distributions is usually one of the chief problems to be faced in practical situations. In any case, the rôle of *a priori* information cannot be shrugged off or avoided [33–35].

The problem of cost assignment also must be carefully examined, since it provides another important link to the actual situation and its significance in the larger world of events. In this way, the connection between, say, the design of an optimum or near-optimum system and the operational aspects of the original problem is made with a theory of values, which seeks some over-all *raison d'être* for cost assignments in the particular case. Similar remarks apply for other loss functions. Thus, an ongoing task is to seek out other meaningful criteria (besides F_1 and F_2) and establish (if this be possible) similar optimal properties, such as admissibility and the complete class theorem for them also.

Another problem of general importance is to discover the "invariants" of various classes of detection and extraction systems. Perhaps the most important example here is *threshold* or *weak-signal reception, and in particular, reception in non-Gaussian noise.* This is because predicting the level of acceptable weak-signal reception provides limiting lower bounds on performance, expressed for instance in terms of "minimum detectable signal," or minimum acceptable estimation error. A canonical theory of threshold reception is not only generally possible, under benign constraints, but has been evolving for general noise and signals over the last four decades[27] in the Bayesian statistical decision theory (BSDT) formulation. This approach provides optimal processing algorithms, probability measures of performance, and permits evaluation and comparisons with suboptimum procedures.

Finally, it is evident that types of optimization other than Bayes and Minimax are possible: one can take as a criterion minimum average risk *with constraints*, say, on higher moments of the risk function for example, or other evaluation functions like F_2 [Eq. (1.4.4)]. However, an analytical theory for uniqueness, admissibility, and so on, comparable to Wald's for the simple cost function remains to be developed in such instances.

We turn now to the further development of Sections 1.1–1.5. Having presented the underlying theory above, let us reserve the detailed treatment of estimation to Chapter 6 and proceed to the detection problem and some of its more explicit consequences in Sections 1.6–1.10. Examples on estimation are presented in Chapter 5. Extensions are given

[26] Closely related to, and supported by, the Bayesian idea of probability measures of the plausibility assigned to an event or hypothesis, is the very old principle of *Ockham's razor*, which for our scientific purposes may be stated in contemporary terms as "choose, or favor, the simplest hypothesis over more complicated competing alternatives," which in Bayesian terms means selecting the simpler hypothesis as being more likely (i.e., having a larger probability) of being correct. These ideas are discussed more fully, with physical examples, in Refs. [33–35]; see also the Introduction here.

[27] Earlier attempts at a canonical treatment are given by the author in Ref. [19], Sections 19.4 and 21.2.5 (1960). Along similar lines we also note Section 2.7 of Ref. [36].

in Chapters 3 and 4. New material in Chapters 6 and 7 considers the related problem of joint detection and estimation.

1.6 PRELIMINARIES: BINARY BAYES DETECTION [19, 21, 36–38]

In this chapter so far we have described the main elements of space–time signal processing, namely, detection, estimation, and related applications, from a general Bayesian viewpoint. This includes employing statistical decision theoretic methods and parametric statistical models. Here we shall focus in more detail on general formulations of optimal and suboptimal detection. Specifically, in the context of the generic structures presented in Sections 1.1–1.5 above, we shall outline a general theory of single-alternative detection systems $T_{\mathrm{R}}^{(N)} = (T_{\mathrm{R}}^{(N)})_{\mathrm{det}}$, for the common and important cases where the data acquisition period (or, as we shall somewhat more loosely call it, the observation period) is fixed at the outset.[28] Since the decisions treated here have only two possible outcomes, we call them *binary decisions*, and the corresponding detection process, *binary detection*, in order to distinguish them from the multiple-alternative situations examined later in Chapter 4.

There are two types of binary detection processes, depending on whether the hypothesis classes refer to a decision between one or two possible signals. Thus, from Section 1.2 previously, we write symbolically

$$\text{I.}\quad H_1 : \mathbf{S} \otimes \mathbf{N} \text{ versus } H_0 : \mathbf{N} \tag{1.6.1a}$$

for the situation where we are asked to decide between H_1: received signal of class $\mathbf{S}$ with noise, versus H_0: noise alone. For the second situation we write

$$\text{II.}\quad H_2 : \mathbf{S}_2 \otimes \mathbf{N} \text{ versus } H_1 : \mathbf{S}_1 \otimes \mathbf{N} \tag{1.6.1b}$$

in which the decision is between the choice of a received signal of class $\mathbf{S}_2$ versus one from class $\mathbf{S}_1$, where both signals are accompanied [$\otimes$] by noise, not necessarily additive. In both instances the decision is to be made from the received data $\mathbf{X}$. We can further anatomize the structure represented by I and II above, according to their general application, as summarized in Table 1.1.

In the radar and sonar cases the received signal (S) represents a target. In a telecommunications environment S is the desired, received communication waveform and the ambient noise N_{A} embodies the (usually) similar signals or "interference," while N_{REC} is receiver noise. Here $N_{\mathrm{A+REC}} = N_{\mathrm{A}} + N_{\mathrm{REC}}$. In nonadditive situations such as envelope detection, signal and noise are combined ($S \otimes N$) nonlinearly, including the receiver noise as well as any ambient noise and interference (i.e. unwanted signals). Here and subsequently unless otherwise indicated, the term "signal" shall refer to the received (desired) signal at the output of the receiving aperture or array, $\hat{R}_0$. (see, the Introduction.)

We begin our discussion with a Bayesian formulation of the one-signal or "on–off" cases (I), Sections 1.6–1.8, including performance measures and a structure of system

[28] Variable observation periods are referred to in Section 1.8.5.

TABLE 1.1 Binary Hypothesis Classes

I.	Radar, Sonar	: H_1 : $S(S_{in}) + N(S_{in}) + N_{A+REC}$	versus H_0 : $N(S_{in}) + N_{A+REC}$	: "Addition" : $S \otimes N = S + N$
"on–off"	Telcom	: H_1 : $S(S_{in}) + N_{A+REC}$	versus H_0 : N_{A+REC}	: ″ ″
II.	Telcom	: H_2 : $S_2(S_{in}) + N_{A+REC}$	versus H_1 : $S_1(S_{in}) + N_{A+REC}$	: ″ ″
	Telcom + Scatter	: H_2 : $S_2(S_{in}) + N(S_{min}) + N_{A+REC}$	versus H_1 : $S_1(S_{in}) + N(S_{min}) + N_{A+REC}$	: ″ ″

comparisons (Section 1.9). This is followed by the formal extension of the theory to the two-signal cases (II), Section 1.10. (A summary is given in Chapter 2 of some illustrative exact results.) We observe, moreover, that exact results are the exception rather than the rule in practical applications, so that approximate methods must be employed if we are to achieve useful analytical and numerical results.

1.6.1 Formulation I: Binary On–Off Signal Detection

Binary detection problems in communication systems have been studied probabilistically in terms of tests of hypotheses since the 1940s. The original formulation in terms of statistical decision theory stems from the 1950s ([1], Chapter 18 [19]).[29] The principal objectives of the following sections here are to obtain (1) a formulation of the binary detection problem itself and (2) by so doing, to indicate how these different viewpoints can not only be reëstablished by the decision theoretical approach but extended to include situations of general practical significance. Specifically, we first derive a general class of Bayes systems and the rather well-known result that several other detection systems considered previously are special cases of this.

Before we begin to develop the elements of statistical communication theory (SCT) outlined in Sections 1.1–1.5, let us establish the effects of sampling of the input fields at the receiver. We have considered principal modes of sampling the continuous input field: (1) a continuous procedure, which essentially reproduces the original field and (2) a discrete sampling procedure that produces a series of sampled values, at the space–time points $(\mathbf{r}_m, t_n)$, where both are obtained during a finite (or infinite) interval Δ (or $\Delta = \infty$). These operations are respectively represented by

$$T_S(\alpha(\mathbf{r},t))_C = X(\mathbf{r},t), \quad \text{and} \quad T_S(\alpha(\mathbf{r},t))_D = X(\mathbf{r}_m,t_n) = X_j, \quad \text{on} \quad \Delta = |\mathbf{R}_0|T \leq \infty, \tag{1.6.2a}$$

where explicitly

$$T_S(\,)_C \equiv \int_{-\Delta/2}^{\Delta/2} \delta(\mathbf{r}' - \mathbf{r})\delta(t' - t)(\,)d\mathbf{r}'dt'; \quad T_S(\,)_D = \int_{-\Delta/2}^{\Delta/2} \delta(\mathbf{r}' - \mathbf{r}_m)\delta(t' - t_n)(\,)d\mathbf{r}'dt' \tag{1.6.2b}$$

and $\mathbf{dr}' = dx'dy'dz' = dr_{x'}dr_{y'}dr_{z'}$ or a lesser dimensionality, depending on the sampling process employed. The effects of these two sampling procedures on the input field, as we shall see later in Sections 2.3.1 and 2.3.2, are quite different when applied to ordered data streams in the discrete and continuous cases, for example, in the formation of apertures, arrays and beam patterns (Section 2.5).

[29] For the earlier studies, based for the most part on a second-moment theory (e.g., signal-to-noise ratios, etc.), see the references at the end of Chapters 19 and Reference Supplements, pp. 1103–1109 (1960); pp. 1111–1120 (1996), of Ref. [1]. Somewhat later studies, also included therein, employing a more complete statistical approach and, leading up to and in some instances coinciding with certain aspects of the present theory, are described more fully here and in Ref. [1], Part 4 and Ref. [19]. For more recent work see the references at the end of this chapter [21, 37].

1.6.2 The Average Risk

First we use F_1 [Eq. (1.4.3)] as our loss function and determine optimum systems of the Bayes class, which, as we have seen [Eq. (1.4.6b)], are defined by minimizing *the average risk*

$$R(\sigma, \delta) = \int_{\Omega} \mathbf{ds}\sigma(\mathbf{S}) \int_{\Gamma} F_J(\mathbf{X}|\mathbf{S})\mathbf{dX} \int_{\Delta} \mathbf{d}\gamma C(\mathbf{S}, \gamma)\delta(\gamma|\mathbf{X}) \tag{1.6.3}$$

We recall that in binary detection we test the hypothesis H_0 that noise alone is present against the alternative H_1 of a signal and noise, so that there are but two points $\gamma = (\gamma_0, \gamma_1)$, respectively, in decision space Δ. For the moment, allowing the possibility that the decision rule δ may be randomized, we let $\delta(\gamma_0|\mathbf{X})$ and $\delta(\gamma_1|\mathbf{X})$ be the probabilities that γ_1 and γ_0 are decided,[30] given $\mathbf{X}$. Since definite, terminal decisions are postulated here, some decision is always made and therefore

$$\delta(\gamma_0|\mathbf{X}) + \delta(\gamma_1|\mathbf{X}) = 1. \tag{1.6.3a}$$

Denoting by $\mathbf{S}$ the input signal that may occur during the observation interval, we may express the two hypotheses concisely as $H_0 : S\varepsilon\Omega_0$ and $H_1 : S\varepsilon\Omega_1$, where Ω_0 and Ω_1 are the appropriate nonoverlapping hypothesis classes, as discussed in Section 1.2.1.2. In binary detection, the null class Ω_0 usually contains only one member, corresponding to no signal. The signal class Ω_1 may consist of one or more nonzero signals. It is now convenient to describe the occurrence of signals within the nonoverlapping classes Ω_0, Ω_1 by density functions $w_0(\mathbf{S})$, $w_1(\mathbf{S})$, normalized over the corresponding spaces, for example,

$$\int_{\Omega_0} w_0(\mathbf{S})\mathbf{ds} = 1 \quad \int_{\Omega_1} w_1(\mathbf{S})\mathbf{ds} = 1. \tag{1.6.4}$$

If q and $p\ (= 1 - q)$ are respectively the *a priori* probabilities that some one signal from Ω_0 and Ω_1 will occur, the *a priori* probability distribution $\sigma(\mathbf{S})$ over the total signal space $\Omega = \Omega_0 + \Omega_1$ becomes

$$\sigma(\mathbf{S}) = qw_0(\mathbf{S}) + pw_1(\mathbf{S}) = q\delta(\mathbf{S} - 0) + pw_1(\mathbf{S}), \tag{1.6.5}$$

this last when there is but one (zero) signal in class Ω_0. Equation (1.6.5) represents the *one-sided alternative* mentioned in Section 1.2.1.2, while if there is only a single signal in class Ω_1 as well, Eq. (1.6.5) becomes $\sigma(\mathbf{S}) = q\delta(\mathbf{S} - 0) + p\delta(\mathbf{S} - \mathbf{S}_1)$, $(\mathbf{S}_1 \neq 0)$, and we have an example of the *simple alternative* situation. In both cases, $\int \sigma(\mathbf{S})\mathbf{ds} = 1$, by definition of p, q, and w.

1.6.3 Cost Assignments

The next step in our application of risk theory is to assign a set of costs to each possible combination of signal input and decision. For this we chose $\mathrm{F}_1 = C(\mathbf{S}, \boldsymbol{\gamma})$, Eq. (1.4.3), as our cost function. We illustrate the discussion with the assumption of one-sided alternatives,

[30] Since the number of alternatives is finite and discrete, the decision rule is represented by a probability (cf. the remarks following the statement of the general reception problem in Section 1.3.3).

noted above, and uniform costs, although the method is not restricted by such choices. Thus, for the binary on–off cases considered here there are four cost assignments: two for possible correct decisions and two for possible incorrect decisions. It is convenient to represent these by a (2×2) *cost matrix* $\mathbf{C}(\mathbf{S}, \boldsymbol{\gamma})$:

$$\mathbf{C}(\mathbf{S}, \boldsymbol{\gamma}) = \begin{bmatrix} C_{1-\alpha} & C_\alpha \\ C_\beta & C_{1-\beta} \end{bmatrix} \equiv \begin{bmatrix} C_0^{(0)} & C_1^{(0)} \\ C_0^{(1)} & C_1^{(1)} \end{bmatrix}, \tag{1.6.6}$$

where the rows represent costs associated with the hypothesis states H_0, H_1, and the columns costs assigned to the various decisions γ_0, γ_1. Thus, we write

$$\text{“failure”}\begin{cases} C_\alpha \equiv C_1^{(0)}, & \text{cost of deciding (incorrectly) that a signal is present,} \\ & \text{when actually only noise occurs; the decision } H_1 \text{ is false.} \\ C_\beta \equiv C_0^{(1)}, & \text{cost of deciding (incorrectly) that a signal is } \textit{not} \text{ present,} \\ & \text{when it actually is; the decision } H_0 \text{ is false.} \end{cases}$$

$$\text{“success”}\begin{cases} C_{1-\alpha} \equiv C_0^{(0)}, & \text{cost of deciding (correctly) that there is no signal,} \\ & \text{only noise, that is, the decision } H_0 \text{ is true.} \\ C_{1-\beta} \equiv C_1^{(1)}, & \text{cost of deciding (correctly) that a signal is present;} \\ & \text{the decision } H_1 \text{ is true.} \end{cases}$$

Consistent with the meaning of “correct” and “incorrect”, that is, equivalently, “success” and “failure,” with respect to the possible decisions, we require that

$$\begin{aligned} &C_{1-\alpha} < C_\alpha; C_{1-\beta} < C_\beta : \text{ “failure”costs more than “success”;} \\ &\therefore \det\mathbf{C} = C_{1-\alpha}C_{1-\beta} - C_\alpha C_\beta < 0. \end{aligned} \tag{1.6.6a}$$

Here observe that the costs are assigned vis-à-vis the possible *signal classes* (hypothesis states) and not with respect to any one signal in a signal class, which in the case of composite hypotheses, contains more than one member, (Section 1.2.1.2). Similarly, H_0 here refers to noise only, representing a specified class of noise processes where, without loss of generality, we can also postulate that $C_{1-\alpha} = 0,\ C_{1-\beta} = 0$, that is, there is no net gain or “profit” from a correct decision.[31] The best we can expect in this situation, if we are forced to adjust the costs, in that success may cost us nothing: $C_{1-\alpha} = C_{1-\beta} = 0$.

We specify next that $F_J(\mathbf{X}|\mathbf{S})$ is continuous in $\mathbf{S}$, that fixed-sample tests only are considered, and that the assumptions needed for the validity of risk theory (Section 1.5.3) are applicable to the received data $\mathbf{X}$ and signals $\mathbf{S}$ in the following, and that these are random or deterministic quantities. Thus, the average cost or risk may now be found from (1.4.6b) by integrating over the two points (γ_1, γ_2) in the decision space Δ. The result is

$$\begin{aligned} R(\sigma, \delta) = \int_\Gamma \{ &[qC_{1-\alpha}F_J(\mathbf{X}|\mathbf{0}) + pC_\beta \langle F_J(\mathbf{X}|\mathbf{S})\rangle_S]\delta(\gamma_0|\mathbf{X}) \\ &+ [qC_\alpha F_J(\mathbf{X}|\mathbf{0}) + pC_{1-\beta}\langle F_J(\mathbf{X}|\mathbf{S})\rangle_S]\delta(\gamma_1|\mathbf{X})\}\mathbf{dX}, \end{aligned} \tag{1.6.7}$$

[31] This is achieved by setting $C_\alpha \to C_{\alpha'} = C_\alpha - C_{1-\alpha}, C_\beta \to C_{\beta'} = C_\beta - C_{1-\beta}$.

and

$$p\langle F_j(\mathbf{X}|\mathbf{S})\rangle_S = \int_{\Omega_1} \sigma(\mathbf{S}) F_J(\mathbf{X}|\mathbf{S})\mathbf{ds} = p\int_{\mathbf{S}} w_1(\mathbf{S}) F_J(\mathbf{X}|\mathbf{S})\mathbf{ds}, \tag{1.6.7a}$$

from (1.6.5). When the signal processes owe their statistical natures solely to a set of random parameters $\boldsymbol{\theta}$, that is, when the signals are deterministic (a usual case is practice), then (1.6.7a) has the equivalent form

$$p\langle F_J(\mathbf{X}|\mathbf{S}(\boldsymbol{\theta}))\rangle_\theta = p\int_{\boldsymbol{\theta}} w_1(\boldsymbol{\theta}) F_J(\mathbf{X}|\mathbf{S}(\boldsymbol{\theta}))\mathbf{d}\boldsymbol{\theta}. \tag{1.6.7b}$$

In detail, we have accordingly from [(1.6.7a) and (1.6.7b)] the defining relations

$$\langle F_J(\mathbf{X}|\mathbf{S})\rangle_{\mathbf{S}} = \int_{\mathbf{S}} w_1(\mathbf{S}) F_J(\mathbf{XS}))\mathbf{ds};\ \langle F_J(\mathbf{X}|\mathbf{S}(\boldsymbol{\theta}))\rangle_{\boldsymbol{\theta}} = \int_{\boldsymbol{\theta}} w_1(\boldsymbol{\theta}) F_J(\mathbf{X}|\mathbf{S}(\boldsymbol{\theta}))\mathbf{d}\boldsymbol{\theta} \tag{1.6.7c}$$

where the dimensionality of w_1 is w_J or w_L.

1.6.4 Error Probabilities

The average cost (1.6.7) can be more compactly expressed in terms of the conditional error probabilities and conditional probabilities of correct decisions. To see this, let us begin by introducing the conditional and total error probabilities:

$\alpha \equiv \alpha(\gamma_1|H_0) =$ conditional probability of incorrectly deciding that a signal is present when only noise occurs. This is known in statistics as a Type I error probability. Here in SCT it is called the *false alarm probability*, for example, $\alpha = \alpha_F \equiv p_F$

$\beta \equiv \beta(\gamma_0|H_1) =$ conditional probability of incorrectly deciding that only noise occurs, when a signal (in class H_1) is actually present. Analogously to the above, this is often called a Type II error probability, or in SCT, the false rejection probability of the signal.

(1.6.8)

The corresponding total error probabilities are $q\alpha$ and $p\beta$, where α and β are now specified in detail by

$$\alpha = \int_\Gamma F_J(\mathbf{X}|\mathbf{0})\delta(\gamma_1|\mathbf{X})\mathbf{dX} \text{ and } \beta = \int_\Gamma \langle F_J(\mathbf{X}|\mathbf{S})\rangle_S \delta(\gamma_0|\mathbf{X})\mathbf{dX}. \tag{1.6.8a}$$

Alternatively, the conditional and total probabilities of correct decisions are

$$1-\alpha = \int_\Gamma F_J(\mathbf{X}|\mathbf{0})\delta(\gamma_0|\mathbf{X})\mathbf{dX} \equiv 1 - p_{\mathrm{F}};\quad 1-\beta = \int_\Gamma \langle F_J(\mathbf{X}|\mathbf{S})\rangle_S \delta(\gamma_1|\mathbf{X})\mathbf{dX} \equiv p_{\mathrm{D}}, \tag{1.6.9}$$

where we have used (1.6.3). The quantity $1-\beta$, (1.6.9), is then the conditional probability of (correct) signal detection p_D or in statistical terminology, the *power of the test,* while $\alpha(= p_{\mathrm{F}})$, (1.6.8a), is called the *significance level*, or *test size.*

Applying (1.6.5) to (1.6.9) for the total probability of a decision $\boldsymbol{\gamma} = \gamma_0 = H_0$: no signal, or $\boldsymbol{\gamma} = \gamma_0 = H_1$: a signal in noise, we find respectively that

$$p(\gamma_0) = q(1-\alpha) + p\beta = \int_\Gamma \left[qF_J(\mathbf{X}|\mathbf{0}) + p\langle F_J(\mathbf{X}|\mathbf{S})\rangle_S\right]\delta(\gamma_0|\mathbf{X})\mathbf{dX}, \qquad (1.6.10a)$$

$$p(\gamma_1) = q\alpha + p(1-\beta) = \int_\Gamma \left[q\langle F_J(\mathbf{X}|\mathbf{0})\rangle + p\langle F_J(\mathbf{X}|\mathbf{S})\rangle_S\right]\delta(\gamma_1|\mathbf{X})\mathbf{dX}, \qquad (1.6.10b)$$

from which we note that

$$p(\gamma_0) + p(\gamma_1) = 1, \ (p \neq q = 1), \qquad (1.6.10c)$$

as expected. Using (1.6.8)–(1.6.10), we readily obtain a more compact form for the average risk (1.6.7), namely,

$$R(\sigma, \delta) = \left\{qC_{1-\alpha} + pC_\beta\right\} - q\{C_\alpha - C_{1-\alpha}\}\alpha + p\left\{C_\beta - C_{1-\beta}\right\}\beta \qquad (1.6.11)$$

in terms of the Types 1 and 2 error probabilities, cf. Eqs. (1.6.8 and 1.6.9). In terms of the probabilities of *correct decisions* this becomes

$$R(\sigma, \delta) = \left\{qC_\alpha + pC_\beta\right\} - q\{C_\alpha - C_{1-\alpha}\} - p\left\{C_\beta - C_{1-\beta}\right\}(1-\beta) \qquad (1.6.11a)$$

$$= R_0 - q(C_\alpha - C_{1-\alpha})(1-p_{\mathrm{F}}) - p\left(C_\beta - C_{1-\beta}\right)p_{\mathrm{D}} \qquad (1.6.11b)$$

cf. (1.6.8 and 1.6.9) above. The quantity

$$R_0 \equiv qC_{1-\alpha} + pC_{1-\beta} (\geq 0), \qquad (1.6.12)$$

is called the *irreducible risk,* here a quantity that is prefixed once the costs and *a priori* probabilities ($p_1 = 1 - q$) are established. Thus, the corresponding average risk $R(\sigma, \delta)$ here deals with all signals in H_1, as well as the noise (H_0).

From Eq. (1.4.5), the *conditional risk* becomes similarly

$$r(\mathbf{S}) = (1-\alpha')C_{1-\alpha} + \alpha' C_\alpha; \mathbf{S} = \mathbf{0}; \quad [1-\beta'(\mathbf{S})]C_{1-\beta} + \beta'(\mathbf{S})C_\beta; \mathbf{S} \neq \mathbf{0}, \qquad (1.6.13)$$

where now the *simple conditional error probabilities* α' *and* β' are distinguished from the *class conditional error probabilities* α *and* β, Eq. (1.6.8a) above, according to

$$\alpha' \equiv \int_\Gamma F_J(\mathbf{X}|\mathbf{0})\delta(\gamma_1|\mathbf{X})\mathbf{dX} (= \alpha); \quad \beta'(\mathbf{S}) = \int_\Gamma F_J(\mathbf{X}|\mathbf{S})\delta(\gamma_0|\mathbf{X})\mathbf{dX} \neq \beta. \qquad (1.6.13a)$$

Finally, note that all of the above go over directly into analogous expressions in the case of random signal parameters.

1.7 OPTIMUM DETECTION: ON–OFF OPTIMUM PROCESSING ALGORITHMS

The criterion of optimization here (and subsequently) is chosen to be the minimization of average risk (1.6.11). Thus, by suitable choice of decision rule δ_0 [or δ_1, since δ_0 and δ_1 are related by Eq. (1.6.3) in these binary cases], the average risk $R(\sigma, \delta)$, Eq. (1.6.7) or (1.6.11), is minimized by making the error probabilities as small as possible, consistent with Eq. (1.6.3) and the constraints (1.6.6) et seq. on the preassigned costs.[32] We assume here, for the moment, that the cost function $C(\mathbf{S}, \gamma)$ is chosen so that overlapping hypothesis (or signal) classes are not included.[33] Here and in Sections 1.8 and 1.9 we first derive the *optimal processing* or detection *algorithms* in their various generic binary forms. In Section 1.9, we shall then consider the evaluation of performance, as measured by the Bayes risk, or its equivalent probabilities of error and correct signal detection.

Eliminating $\delta(\gamma_1|\mathbf{X})$ with the help of Eq. (1.6.3a), we may express Eq. (1.6.9) as

$$R(\sigma, \delta) = \mathsf{R}_0 + p(C_\beta - C_{1-\beta}) \int_\Gamma \delta(\gamma_0|\mathbf{X})[\Lambda(\mathbf{X}) - K] F_J(\mathbf{X}|\mathbf{0}) \mathbf{dX} \tag{1.7.1}$$

where

$$\Lambda(\mathbf{X}) \equiv \frac{p}{q} \frac{\langle F_J(\mathbf{X}|\mathbf{S}) \rangle}{F_J(\mathbf{X}|\mathbf{0})} \tag{1.7.2}$$

is a *generalized likelihood ratio* GLR[34] and K is a *threshold*:

$$K \equiv \frac{C_\alpha - C_{1-\alpha}}{C_\beta - C_{1-\beta}} (> 0), \tag{1.7.3}$$

with $\mathrm{R}_0 = qC_0^{(0)} + pC_1^{(1)}$ the *irreducible risk*(1.6.12). Since δ_0, $F(\mathbf{X}|\mathbf{0})$, $C_\beta - C_{1-\beta}$, and so on, are all positive (or zero), we see directly that R can be minimized by choosing $\delta(\gamma_0|\mathbf{X}) \to \delta^*(\gamma_0|\mathbf{X})$ to be unity when $\Lambda < 0$ and zero when $\Lambda \geq K$. Thus, we decide

$\gamma_0 : H_0$ if $\Lambda(\mathbf{X}) < K$
namely, we set $\delta^*(\gamma_0|\mathbf{X}) = 1$ for any $\mathbf{X}$ that yields this inequality. From Eq. (1.6.3) this means also that

$$\delta^*(\gamma_1|\mathbf{X}) = 0. \tag{1.7.4a}$$

The acceptance region of X for which $\delta_0^* = 1$, $\delta_1^* = 0$ is Γ_0, that is, Γ_0 contains all $\mathbf{X}$ satisfying the inequality $\Lambda(\mathbf{X}) < K$.

$\gamma_1 : H_1$ if $\Lambda_J(\mathbf{X}) \geq K$,
that is, we choose $\delta^*(\gamma_1|\mathbf{X}) = 1$ for all $\mathbf{X}$ satisfying this inequality (and equality) and consequently require that

$$\delta^*(\gamma_0|\mathbf{X}) = 0. \tag{1.7.4b}$$

Here Γ_1 denotes the acceptance region of Γ for which

$$\delta_0^* = 0, \; \delta_1^* = 1.$$

[32] Equivalently, $R(\sigma, \delta)$ is minimized by *maximizing* the probabilities of correct decisions; cf. Eq. (1.6.11a).

[33] For the generalization to include overlapping classes, including stochastic as well as deterministic signals, see Section 1.10 ff.

[34] We note that Λ is more general than the classical likelihood ratio $F_J(\mathbf{X}|\mathbf{S})/F_J(\mathbf{X}|\mathbf{0})$, that is, the ratio of the conditional probability densities of $\mathbf{X}$ with and without $\mathbf{S}$ fixed. The generalized likelihood ratio (1.7.2) reduces to this form when the *a priori* probabilities p and q are equal and the signal space contains but one point, corresponding to the very special case of a completely deterministic signal.

We remark that $\delta_{1,0}^{*}$ are nonrandomized decision rules directly deduced from the minimization process itself. From Eq. (1.6.8) for these optimum rules we may write the Bayes or minimum average risk specifically as

$$R^{*}\left(\sigma,\delta^{*}\right)=\mathbf{R}_{0}+p\left(C_{\beta}-C_{1-\beta}\right)\left[\frac{K}{\mu}\alpha^{*}+\beta^{*}\right],\ \mu\equiv p/q. \tag{1.7.5}$$

The procedures described above in (1.7.4a and 1.7.4b) present a form of *generalized likelihood-ratio test* (*GLRT*). This general Bayesian definition must be distinguished from GLRT alternatives when the *a priori* probability distributions of the signal or signal parameters are replaced by their conditional maximum likelihood estimates. [As optimal likelihood-ratios, albeit constrained, these conditional GLRTs are a special subset of the general Bayes class of likelihood-ratio detectors minimizing average risk.]

1.7.1 The Logarithmic GLRT

In actual applications it is usually much more convenient to replace the likelihood-ratio Λ by its logarithm, as we shall see presently.[35] This in no way changes the optimum character of the test, since any monotonic function of Λ may serve as test function. Thus, the optimum decision process [Eqs. (1.7.4a and 1.7.4b)] is simply reëxpressed as

Decide

$$\begin{aligned}
&\gamma_0 : H_0 \ \textit{if} \log\Lambda(\mathbf{X}) < \log K \ \textit{with} \quad \text{or} \quad \gamma_1 : H_1 \ \textit{if} \log\Lambda(\mathbf{X}) \geq \log K \ \textit{with}\\
&\qquad \delta^*(\gamma_0|\mathbf{X}) = 1 \qquad\qquad \delta^*(\gamma_0|\mathbf{X}) = 0\\
&\qquad \delta^*(\gamma_1|\mathbf{X}) = 0, \qquad\qquad \delta^*(\gamma_1|\mathbf{X}) = 1.
\end{aligned} \tag{1.7.6}$$

The likelihood ratio, and equivalently here its logarithm, embody the actual receiver structure $\mathbf{T}_{\mathrm{R}}^{(N)}$; namely, the operation the detector must perform on the received data $\mathbf{X}$ in order to reach an optimal decision as to the presence or absence of a signal (of class S) in noise. The optimum detection situation is schematically illustrated in Fig. 1.8.

Thus, choosing $\delta_0 \rightarrow \delta_0^{*}$, $\delta_1 \rightarrow \delta_1^{*}$(1.7.4a and 1.7.4b) or (1.7.6), may be stated alternatively: *Make the decision for which the a posteriori risk (or cost) is least.* It is important to observe that this is clearly a direct extension of the original *Theorem of Bayes, or Bayes' Rule*, namely, "chose that hypothesis with the greatest *a posteriori* probability, given the (data) **X**," to include now the various costs associated with the decision process.

1.7.2 Remarks on the Bayes Optimality of the GLR

The complete class theorem (see Section 1.2.1) for the risk formulation assures us that we have an optimum test and *that all such tests based on the likelihood ratio* [Eq. (1.7.2)] *are*

[35] In fact, any monotonic function of the likelihood ratio (1.7.2) is potentially suitable as an optimal (Bayes) test statistic, since (1.7.2), and $F_{\mathrm{mono}}(\Lambda)$, are *sufficient statistics* because Λ, and $F_{\mathrm{mono}}(\Lambda)$, contain all the relevant information for deciding H_1 versus H_0. [See Section 1.9.1.1. for a more detailed discussion of *sufficiency*.]

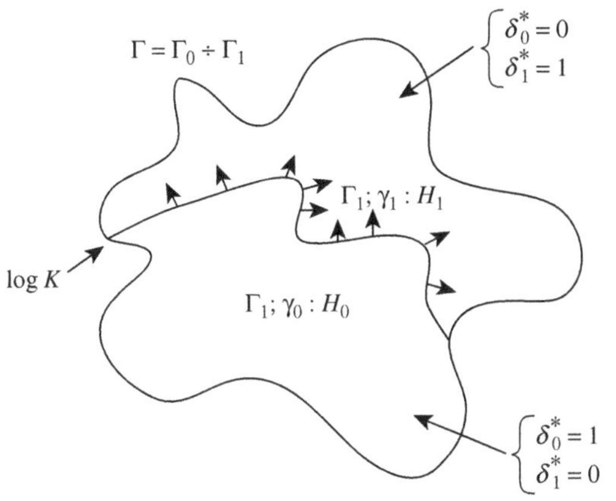

FIGURE 1.8 Optimum binary on–off detection.

Bayes tests (see Section 1.5.1). The Bayes risk R*(1.7.5) and the average risk R [Eqs. (1.6.11)] for general (not necessarily optimum) systems become, respectively,

$$R^* = \mathrm{R}_0 + p(C_\beta - C_{1-\beta})\left(\frac{K}{\mu}\alpha^* + \beta^*\right), \mu \equiv \frac{p}{q}, \tag{1.7.7a}$$

$$R = \mathrm{R}_0 + p(C_\beta - C_{1-\beta})\left(\frac{K'}{\mu}\alpha + \beta\right) \tag{1.7.7b}$$

with K given by Eq. (1.7.3), while $R_0 = qC_{1-\alpha} + pC_{1-\beta}$ in either instance [Eq. (1.6.12)]. The threshold K' for the nonideal cases[36] may or may not be equal to K.

Note, incidentally, from Eq. (1.7.7a) that if we differentiate $\left(R^{(*)} - \mathrm{R}_0\right)/p(C_\beta - C_{1-\beta})$ with respect to α^* or α we have at once

$$d\beta^*/d\alpha^* = -K/\mu;\ d\beta/d\alpha = -K'/\mu, \tag{1.7.8a}$$

relations that are of use in describing a receiver's performance characteristics. In fact, since $1 - \beta^* = p_{\mathrm{D}}^*$, (1.6.9), (1.6.11a), and since $\alpha^* = p_{\mathrm{F}}^*$, the (conditional) probability of "false alarm," use have

$$\frac{dp_{\mathrm{D}}^*}{dp_{\mathrm{F}}^*} = K/\mu, \quad \text{and} \quad \frac{d(pp_{\mathrm{D}}^*)}{d(qp_{\mathrm{F}}^*)} = \frac{dP_{\mathrm{D}}^*}{dP_{\mathrm{F}}^*} = K, \tag{1.7.8b}$$

where P_{D}^* and P_{F}^* are respectively the *un*conditional probabilities of correct detection and false alarm. In terms of the conditional probability, K/μ is the slope of the curve $p_{\mathrm{D}}^* = F^*(p_{\mathrm{F}}^*|\ldots)$, for example, $dp_{\mathrm{D}}^*/dp_{\mathrm{F}}^* = K/\mu$ when one plots $\beta_{\mathrm{D}}^* = 1 - \beta^*$ versus $p_{\mathrm{F}}^* = \alpha_{\mathrm{F}}^*$. This latter case is generally called the *receiver operating characteristic* (*ROC*) of the Bayes system here. Similarly, with $1 - \beta = p_{\mathrm{D}}, \alpha = p_{\mathrm{F}}$ for suboptimum systems, with slope $dp_{\mathrm{D}}/dp_{\mathrm{F}} = K/\mu, p_{\mathrm{D}}$ versus p_{F} is the corresponding (suboptimum) ROC curve.

[36] One always has a definite (nonzero) threshold when a definite decision is made.

Variants of the relations between $p_{\mathrm{D}}^{(*)}$ and $p_{\mathrm{F}}^{(*)}$ are noted in practice, with analogous relations involving p_{D}^{*}, P_{F}^{*} or P_{D} and P_{F}. [See Section 1.9 and Eq. (1.9.10a) et seq.]

1.8 SPECIAL ON–OFF OPTIMUM BINARY SYSTEMS

A variety of important special cases of the optimum general (fixed-sample) binary detection procedures discussed in Section 2.2 now requires our attention. These are all characterized by one or more constraints on the error probabilities in minimizing the average risk. We begin with the well known *Neyman–Pearson detector*.

1.8.1 Neyman–Pearson Detection[37] Theory

Here the constraint is on the false alarm probability $\alpha_{\mathrm{F}}(=p_{\mathrm{F}})$. We require it to remain fixed, hence the alternative designation of *constant false alarm (CFA) detector* or constant false alarm *rate* detector (CFAR), for sequences of decisions. Moreover, from the viewpoint of decision theory, we require the total Type I error probability $q\alpha_{\mathrm{F}}$ to remain fixed, while minimizing the total Type II error probability $p\beta$. This is expressed as

$$\mathsf{R}_{\mathrm{NP}}^{*} \equiv \min_{\delta}\,(p\beta + \lambda q\alpha) = p\beta_{\mathrm{NR}}^{*} + \lambda q\alpha, \tag{1.8.1}$$

where λ is an as yet undetermined multiplier. Minimization is with respect to the decision rule, in the usual way cf. (1.7.4a and 1.7.4b), (1.7.6) and subject to the fact of a definite decision. From Eq. (1.6.8a), we write explicitly

$$\begin{aligned}\mathsf{R}_{\mathrm{NP}}^{*} &= \min_{\delta}\left[p\int_{\Gamma}\langle F_J(\mathbf{X}|\mathbf{S})\rangle_G\,\delta(\gamma_0|\mathbf{X})\mathbf{dX} + \lambda q\int_{\Gamma} F_J(\mathbf{X}|\mathbf{0})\delta(\gamma_1|\mathbf{X})\mathbf{dX}\right]\\ &= \min_{\delta}\left\{\int_{\Gamma}\mathbf{dX}\delta(\gamma_0|\mathbf{X})\left[p\langle F_J(\mathbf{X}|\mathbf{S})\rangle_S - \lambda q F_J(X|0)\right]\right\} + \lambda q.\end{aligned} \tag{1.8.2}$$

For this to be a minimum it is clear that we must choose the δ's such that when

$$\begin{aligned}&p\langle F_n(\mathbf{X}|\mathbf{S})\rangle_S \geq \lambda q F_n(\mathbf{X}|\mathbf{0}),\ \textit{we set}\ \delta^*(\gamma_1|\mathbf{X}) = 1,\ \delta^*(\gamma_0|\mathbf{X}) = 0\\ &\textit{and decide signal and noise},\\ &\text{or when}\\ &p\langle F_n(\mathbf{X}|\mathbf{S})\rangle_S < \lambda q F_n(\mathbf{X}|\mathbf{0}),\ \textit{we set}\ \delta^*(\gamma_0|\mathbf{X}) = 1,\ \delta^*(\gamma_1|\mathbf{X}) = 0\end{aligned} \tag{1.8.3}$$

and decide noise alone,
cf. (1.7.4a and 1.7.4b). Thus, we have the GLRT

$$\mathbf{T}^{(N)}[\mathbf{X}]_{\mathrm{NP}} = \Delta_{\mathrm{NP}} = \frac{p\langle F_n(\mathbf{X}|\mathbf{S})\rangle_S}{qF_n(\mathbf{X}|\mathbf{0})}\begin{cases}\geq \lambda \text{ decide } \gamma_1, \text{ or}\\ < \lambda \text{ decide } \gamma_0,\end{cases} \tag{1.8.4}$$

which establishes the likelihood nature of the detection system [cf. Eq. (1.7.2)].

[37] See Section 19.2.1 and Ref. [26] therein of Ref. [1] for additional remarks regarding the classical Neyman–Pearson hypothesis test.

Comparison with Eq. (1.7.2) et seq. shows that the undetermined multiplier λ here plays the role of threshold K, and R^*_{NP} is (except for a scale factor) the corresponding Bayes risk for this threshold $\lambda \equiv K_{NP}$. (The decision regions Γ_1, Γ_0 for $\mathbf{X}$ are pictured in Fig. 1.8, for log Λ. However, λ is not arbitrary but is determined by the constraint of a preassigned value of the conditional Type I error probability α, for example,

$$\alpha_{NP} = \int_\Gamma F_J(\mathbf{X}|\mathbf{0})\delta^*(\gamma_1|\mathbf{X})\mathbf{dX} = \alpha_{NP}(\lambda = K_{NP}) \tag{1.8.5}$$

from Eq. (1.6.8a) and the nature of the optimum decision rule [Eq. (1.8.4)].

We can also write Eq. (1.8.4) in the more classical form, either by an obvious modification, or from a computation of min $_\delta(\beta + \lambda\alpha)$ by precisely the same sort of argument given above, namely,

$$\Lambda_{NP} \equiv \frac{\langle F_n(\mathbf{X}|\mathbf{S})\rangle_S}{F_n(\mathbf{X}|\mathbf{0})} = \frac{\lambda}{\mu} = K'_{NP}, \tag{1.8.6}$$

with $\mu \equiv p/q$ (1.7.7a), and where now K'_{NP} is a new threshold or significance level, into which have been absorbed the *a priori* probabilities (p, q) and the cost ratio λ. Thus, in the subclass of Neyman–Pearson tests the significance level K'_{NP} is set by choosing α_{NP}, or, equivalently α_{NP} is specified for a predetermined level K'_{NP}. In either instance, it is clear from the preceding remarks that such a formulation *implies* a specific set of *a priori* probabilities (p, $q = 1-p$) and a cost ratio K_{NP} if we are to apply this optimum detection procedure to physical situations. In practice, as has been noted in Section 1.7.1 above, the logarithmic form of the GLRT (1.8.4), with $\lambda \to \log \lambda$ now, is the usually preferred form. CFAR (i.e., Neyman–Pearson) detectors are commonly used in radar and sonar applications, where the practical constraint is keeping the false alarm rate (for sequences of decisions) suitably low, for operational reasons.

1.8.2 The Ideal Observer Detection System

Another way of designing a fixed-sample one-sided alternative test is to require that the *total* probability of error $q\alpha + p\beta$ be minimized, instead of just $p\beta$ as above. An observer who makes a decision in this way is called an *Ideal Observer* [40]. As in the Neyman–Pearson case, this may be set up as a variational problem and shown to yield a likelihood-ratio test with $K = K_I = 1$. Specifically, we want

$$R^*_I = \min_\delta (q\alpha + p\beta) = q\alpha^*_I + p\beta^*_I \tag{1.8.7a}$$

where now α and β are jointly minimized in the sum by proper choice of the decision rule. Using Eqs. (1.7.7a and 1.7.7b) again, we can write Eq. (1.8.7a) as

$$R^*_I = \min_\delta \left\{ \int_\Gamma \delta(\gamma_0|X)\left[p\langle F_J(\mathbf{X}|\mathbf{S})\rangle_S - qF_J(\mathbf{X}|\mathbf{0})\right]\mathbf{dX} \right\} + q. \tag{1.8.7b}$$

From this it follows at once that the decision procedure for the Ideal Observer is

Decide signal and noise when $\Lambda \geq 1$, that is, set *Decide noise alone when* $\Lambda < 1$, that is, set

$$\delta^*(\gamma_1|\mathbf{X}) = 1 \qquad \qquad \delta^*(\gamma_0|\mathbf{X}) = 1$$

or

$$\delta^*(\gamma_0|\mathbf{X}) = 0, \qquad \qquad \delta^*(\gamma_1|\mathbf{X}) = 0. \tag{1.8.8}$$

Accordingly, the Ideal Observer system $\mathbf{T}^{(N)}[\mathbf{X}]_1$ is a Bayes detector with threshold K_I of unity. The fact that both the Neyman–Pearson and Ideal Observer systems yield likelihood-ratio tests of the type (1.7.2) follows from the optimum performance they require. Since they *are* likelihood-ratio tests, they belong to the Bayes risk class and accordingly share the general optimum properties possessed by that class, including uniqueness and admissibility (cf. Section 1.5.1). Unlike the Neyman–Pearson detectors (1.8.1), which are particularly appropriate to radar, sonar, and similar hypothesis situations where the decision costs are unsymmetrical, that is, $K \neq 1, > 0$, (1.8.2), the Ideal Observer is the usual choice in many telecommunication applications (e.g., telephone, wireless telephony, etc.) where the costs associated with each class of decision are equal, so that now $K = 1$, cf. (1.8.8). Note, finally, that both these special classes of optimum detection system employ nonrandomized decision rules.

1.8.3 Minimax Detectors

There is yet another possible solution to the detection problem that, like the two just discussed, is optimum in a certain sense and which leads to a likelihood-ratio test. This is the *Minimax detection rule*.

When the *a priori* signal probabilities $\sigma(\mathbf{S})$ are unknown, the Minimax criterion discussed in Section 1.4.4 provides one possible definition of optimum system performance. As we have seen, a Minimax system for binary detection (and the disjoint hypothesis classes of the present chapter) can be regarded as a likelihood-ratio system for some least favorable distribution $\sigma = \sigma_0 = \sigma_{\mathrm{M}}^*$. Once this distribution is found, the Bayes system is completely determined. Now, in order to find it, we may take advantage of the fact that a likelihood-ratio system with the same conditional risks for all signals is Minimax in consequence of the definitions of Bayes and Minimax systems (Section 1.4.4). The procedure is briefly described below.

First, we require the conditional probabilities $\alpha', \beta'|\alpha' = \beta'(S)$ [cf. Eq. (1.6.13a)] to be the result of a Bayes decision rule, which here means a likelihood-ratio test, with σ as yet unspecified. Then, as different σ are tried, different α', β' result (for the same threshold K, which depends only on the preassigned costs). The conditional risks [Eq. (1.6.13)] for $\mathbf{S} = 0$ and for $\mathbf{S} \neq 0$ (all $\mathbf{S}$) will vary. As one increases, the other must decrease, in consequence of the admissibility and uniqueness of the particular Bayes rule corresponding to our choice of σ. If, then, there exists a σ for which the conditional risks [Eq. (1.6.13)] are equal[38] for all $\mathbf{S}$, we have the required Minimax rule δ_{M}^* and associated least favorable prior distribution σ_{M}^*. We remember, moreover, that, if the equation between conditional risks has no solution, this does *not* mean that a Minimax rule or least favorable distribution does not exist, only that other methods must be discovered for determining it.

[38] There can be no system for which both conditional risks together can be less than this, since this is a Bayes test.

As an example, suppose we have the *simple* alternative detection problem, when p and q are unknown, and $w_1(\mathbf{S}) = \delta(\mathbf{S} - \mathbf{S}_I)$ here. From Eq. (1.7.2), we have for the Bayes test

$$\Lambda = \frac{pF_J(\mathbf{X}|\mathbf{S}_I)}{qF_J(\mathbf{X}|\mathbf{0})} \gtrless K \tag{1.8.9}$$

where $\alpha'(=\alpha)$, $\beta'(=\beta)$ are the corresponding error probabilities [cf. Eq. (1.6.8a)]. The α, β are functions of p and q since the decision rule $\delta = \overset{*}{\delta}_{\mathrm{M}}$ depends on p and q through Λ. Thus, as p and q are varied, α, β also are changed, as the boundary between the critical and acceptance region varies. Equating the conditional risks [Eq. (1.6.7)] in order to determine the least favorable $p = \overset{*}{p}_{\mathrm{M}}$, $q = \overset{*}{q}_{\mathrm{M}}\left(= 1 - \overset{*}{p}_{\mathrm{M}}\right)$, we have

$$\left[1 - \alpha\left(\overset{*}{p}_{\mathrm{M}}, \overset{*}{q}_{\mathrm{M}}\right)\right] C_{1-\alpha} + \alpha\left(\overset{*}{p}_{\mathrm{M}}, \overset{*}{q}_{\mathrm{M}}\right) C_\alpha = \left[1 - \beta\left(\overset{*}{p}_{\mathrm{M}}, \overset{*}{q}_{\mathrm{M}}\right)\right] C_{1-\beta} + \beta\left(\overset{*}{p}_{\mathrm{M}}, \overset{*}{q}_{\mathrm{M}}\right) C_\beta, \tag{1.8.10}$$

provided that a solution exists.

Alternatively, the least favorable a priori distribution $\overset{*}{\sigma}_{\mathrm{M}}$, and therefore the Minimax decision rule, may be found in principle from the basic definitions (see Section 1.5.3, Theorems 1–5). That is, $\sigma_0 = \overset{*}{\sigma}_{\mathrm{M}}$ is the *a priori* distribution that maximizes the Bayes risk (cf. Fig. 1.6). Since every Bayes decision rule is associated with a specific *a priori* distribution, however, the Bayes rule changes as this distribution is varied for maximum risk. As a result, this method of finding the extremum may be technically difficult to implement. It is applicable, however, when the previous method (based on uniform conditional risk) fails.

In the case of the one-sided alternative where $w_1(\mathbf{S})$ is known but again p and q are not, the same procedure may be tried when now Eq. (1.7.2) is used in place of Eq. (1.8.9). Finally, if $w_1(\mathbf{S})$ is unspecified, or if neither p, q, nor $w_1(\mathbf{S})$ is given [i.e., if $\sigma(\mathbf{S})$ is completely unavailable to the observer], Eq. (1.8.10) with Eq. (1.7.2) still applies when a solution exists, although the task of finding $\overset{*}{\sigma}_{\mathrm{M}}$ may be excessively formidable. In any case, an explicit evaluation of $\overset{*}{(\alpha')}_{\mathrm{M}}$ and $\overset{*}{(\beta')}_{\mathrm{M}}$ from Eq. (1.6.13a) when $\delta = \overset{*}{\delta}_{\mathrm{M}}$ therein, may be carried out by methods outlined in Section 1.8.1 and illustrated in succeeding sections.

We observe that the Minimax error probabilities $\overset{*}{(\alpha)}_{\mathrm{M}}$, $\overset{*}{(\beta)}_{\mathrm{M}}$ are fixed quantities, independent of the actual *a priori* probabilities p, q, $w_1(\mathbf{S})$ chosen by nature. The average Minimax risk $\overset{*}{R}_{\mathrm{M}}$ is given formally by writing $\overset{*}{R}_{\mathrm{M}}$ for R^* in Eq. (1.7.7a) and replacing p, q, and so on, and α^*, β^* by $\overset{*}{p}_{\mathrm{M}}$, $\overset{*}{q}_{\mathrm{M}}, \ldots$ and $\overset{*}{\alpha}_{\mathrm{M}}$, $\overset{*}{\beta}_{\mathrm{M}}$ therein. The difference $\left(\overset{*}{R}_{\mathrm{M}} - R^* \geq 0\right)$ between the Bayes (σ known) and Minimax average risk (σ unknown) is thus one useful measure of the price we must pay for our ignorance of nature's strategy (i.e., here nature's choice of p, q, etc.). For further discussion, see Section 20.4.8 of Ref. [1].

1.8.4 Maximum Aposteriori (MAP) Detectors from a Bayesian Viewpoint

Another approach to treating unknown, or unavailable *a priori* pdfs that are exceedingly difficult to evaluate in the likelihood-ratio Λ (1.7.2) (usually of random signal parameters $\boldsymbol{\theta}$ or waveform $\mathbf{S}$), is to employ a suitably optimized estimate of $\boldsymbol{\theta}$, or $\mathbf{S}$. "Suitably optimized" means here that an appropriate likelihood ratio results and hence belongs in the family of Bayes tests, that is, one which yields a minimum average risk, $\overset{*}{R}_{\mathrm{MAP}}$ consistent with the available prior information and the constraints imposed by the receiver's ignorance and/or simplifications. For the latter reason, of course, $R^*_{\text{MAP}} - R^* \geq 0 : R^*_{\text{MAP}}$ is larger than (or at best equal to) the Bayes or minimum average risk with all prior information used, for reasons similar to the Minimax cases discussed above.

To see how such Bayes tests are obtained, we begin by considering the situation where the *a priori* pdf $w_L(\boldsymbol{\theta})$, of the L signal parameters $\boldsymbol{\theta}(=\boldsymbol{\theta}_1, \ldots, \boldsymbol{\theta}_m)$, is available but is difficult to treat in Λ, (1.7.2). Using Bayes's theorem we may write for the integrand of the numerator of Λ,

$$w_L(\boldsymbol{\theta})F_J(\mathbf{X}|\mathbf{S}(\boldsymbol{\theta})) = w_L(\boldsymbol{\theta}|\mathbf{X})W_J(\mathbf{X})(\equiv W_{J\times L}(\mathbf{X},\boldsymbol{\theta})), \tag{1.8.11}$$

$$\text{with } W_J(\mathbf{X}) = \int_{\boldsymbol{\theta}} w_L(\boldsymbol{\theta})W_J(\mathbf{X}|\boldsymbol{\theta})\mathbf{d}\boldsymbol{\theta}. \tag{1.8.11a}$$

Consequently, we have the conditional pdf of $\boldsymbol{\theta}$ given $\mathbf{X}$, namely the *a posteriori* pdf of $\boldsymbol{\theta}$ represented by

$$w_L(\boldsymbol{\theta}|\mathbf{X}) = w_L(\boldsymbol{\theta})F_J(\mathbf{X}|\mathbf{S}(\boldsymbol{\theta}))/W_J(\mathbf{X}). \tag{1.8.11b}$$

Next, we use (1.8.11b) in (1.6.9) for the (conditional) Bayes average probability of correct signal detection, namely,

$$1-\beta^* = \int_\Gamma \mathbf{dX}\,\delta(\gamma_1|\mathbf{X})\int_{\boldsymbol{\theta}} w_L(\boldsymbol{\theta}|\mathbf{X})W_J(\mathbf{X})\mathbf{d}\boldsymbol{\theta}, \tag{1.8.12}$$

which we now maximize by choosing that estimate $\hat{\boldsymbol{\theta}}^*$ which in turn maximizes the integrand (in $\boldsymbol{\theta}$). Thus, we seek that estimate $\hat{\boldsymbol{\theta}}^*$ which maximizes the average value of correct signal detection, for example, $1-\hat{\beta}^*$. The estimate $\hat{\boldsymbol{\theta}}$ is found from the $\boldsymbol{\theta} \in \Omega_0$ for which the *a posteriori probability* $w_L(\boldsymbol{\theta}|\mathbf{X})$ *is maximum*, namely, from

$$w_L\left(\hat{\boldsymbol{\theta}}^*|\mathbf{X}\right) \geq w_L(\boldsymbol{\theta}|\mathbf{X}), \tag{1.8.13a}$$

or equivalently from (1.8.11b) : $w_L\left(\hat{\boldsymbol{\theta}}^*\right)F_J\left(\mathbf{X}\middle|\mathbf{S}\left(\hat{\boldsymbol{\theta}}^*\right)\right) \geq w_L(\boldsymbol{\theta})F_J(\mathbf{X}|\mathbf{S}(\boldsymbol{\theta}))$, all $\boldsymbol{\theta} \in \Omega_0$,
(1.8.13b)

Here $W_J(\mathbf{X})$ is dropped as irrelevant to the estimation process because it does not contain $\boldsymbol{\theta}$. Accordingly, it is customary to call $\hat{\boldsymbol{\theta}}^* = \hat{\boldsymbol{\theta}}(\mathbf{X})^*$ here a *MAP*, or *maximum a posteriori probability* estimate, which depends, of course, on the received data $\mathbf{X}$. [Note that the *a posteriori* probability $w_L(\boldsymbol{\theta}|\mathbf{X})$ *depends explicitly on the prior probability* $w_L(\boldsymbol{\theta})$, as a consequence of (1.8.11b) in (1.8.12a and 1.8.12b).]

Our next step in obtaining the desired *MAP detector* Λ_{MAP} is to replace the pdf $w_L(\boldsymbol{\theta})$ in the GLR (1.7.2) by the new pdf[39]

$$\hat{w}_L(\boldsymbol{\theta})^* \equiv \delta\left(\boldsymbol{\theta} - \hat{\boldsymbol{\theta}}^*(\mathbf{X})\right), \tag{1.8.14}$$

[39] For a Bayes formulation of estimation and associated Bayes risk, see chapter 5.

which corresponds to the maximizing operation (1.8.13a and 1.8.13b) for the integrand of the GLR (1.7.2). The result is directly the classical MAP test

$$\textit{Classical MAP Test}: \quad \Lambda_{\mathrm{MAP}}\big|_{\mathrm{classical}} = \mu F_J\left(\mathbf{X}\middle|\mathbf{S}\left(\hat{\boldsymbol{\theta}}^*\right)\right)/F_J(\mathbf{X}|\mathbf{0})\begin{Bmatrix} \geq K & : \text{decide } S\otimes N \\ < & : \text{decide } N \end{Bmatrix}. \tag{1.8.15}$$

Several important points need to be emphasized regarding the maximizing condition (1.8.13a and 1.8.13b) for $\hat{\boldsymbol{\theta}}^*$. First, as we shall see in Chapter 5 ff., (1.8.13a) defines an *unconditional maximum likelihood estimate (UMLE)*, because of the presence of the *a priori* pdf $w_L(\boldsymbol{\theta})$. The conditional pdf $w_L(\boldsymbol{\theta}|X)$ must be determined from (1.8.11b), which in turn depends explicitly on $w_L(\boldsymbol{\theta})$. *Accordingly, MAP detectors cannot avoid the requirements of an explicit knowledge of* $w_L(\boldsymbol{\theta})$, which limits their use when $w_L(\boldsymbol{\theta})$, or $\sigma(\boldsymbol{\theta})$(1.6.5), is not available. A second and more serious difficulty with the classical result (1.8.15) above, and one which appears to be universally unacknowledged, is with the estimation process itself, as embodied in (1.8.13a and 1.8.13b). The point here is that $p(= p(H_1))$ is less than unity: there is a detection procedure indicated. *The data* $\mathbf{X}$ *do not always contain the desired signal* $\mathbf{S}(\boldsymbol{\theta})$: 100q% of the time the data sample $\mathbf{X}$ contains no signal, only noise. As such the result of employing (1.8.13a and 1.8.13b) which assumes $p = 1$, *yields a biased-estimate* $\boldsymbol{\theta}^* = \hat{\boldsymbol{\theta}}^*_{p=1}$, with an average positive bias *in the magnitude of the estimates,* cf. Chapter 6. However, as the analysis there shows, this situation can be remedied by using the *unbiased estimate* $\hat{\boldsymbol{\theta}}^*_{p<1} = p\hat{\boldsymbol{\theta}}^*_{p=1}$, appropriate to the UMLE process when $p < 1$ (and for $p = 1$). (It is shown in Chapters 5 and 6 that in the context of the Bayes risk formulation for detection and estimation, the UMLEs are derived by minimization of so called "simple cost" functions [cf. Sections 21.2.2 and 21.2.3 [1] and Chapter 5. ff.][40]). Accordingly, the classical result (1.8.15) needs to be replaced by the correct result:

$$\text{MAP Test } (p<1): \quad \Lambda_{\mathrm{MAP}}\big|_{p<1} = \mu F_J\left(\mathbf{X}\middle|\mathbf{S}\left(p\hat{\boldsymbol{\theta}}^*(\mathbf{X})_{p=1}\right)\right)/F_J(\mathbf{X}|\mathbf{0})\begin{Bmatrix} \geq K & : \text{decide } S\otimes N \\ < & : \text{decide } N \end{Bmatrix}. \tag{1.8.16}$$

As we shall see in Chapter 5 even $\hat{\boldsymbol{\theta}}^*(X)_{p=1}$, is itself not always easily obtained.

Finally, there is a variant of the $\mathrm{MAP}|_{p<1}$ detector that can be used when the *a priori* pdf $w_L(\boldsymbol{\theta})$ *is essentially uniform or is at least slowly varying over the range of values of* $\boldsymbol{\theta}$ *in* $F_J(X|S(\boldsymbol{\theta}))$ *where* F_J *is significant*. Thus our maximization of the (conditional) average probability of correct signal detection is accomplished by maximizing $\boldsymbol{\theta}$ according to the condition

$$F\left(\mathbf{X}\middle|\mathbf{S}\left(\hat{\boldsymbol{\theta}}\right)\right) \geq F(\mathbf{X}|\mathbf{S}(\boldsymbol{\theta})\,,) \quad \text{all } \boldsymbol{\theta} \in \Omega_{\boldsymbol{\theta}}. \tag{1.8.17}$$

Although $\hat{\boldsymbol{\theta}}$ is now *apparently* a conditional estimate (no *a priori* pdf $w_L(\boldsymbol{\theta})$), *from the unconditional Bayes viewpoint* (1.8.17) *implies that* $w_L(\boldsymbol{\theta})$ *is uniform, consistent with our requirement (1.8.17) represents the only significant variation of the integrand with* $\boldsymbol{\theta}$.

[40] For these maximum likelihood cases, including (1.8.18) below, the effective part of the maximizing estimation procedure is simply p. The relation (1.8.13a and 1.8.13b) is used when $p \leq 1$. See the analysis for Eqs. (6.3.24)–(6.3.30).

Thus, $w_L(\boldsymbol{\theta}) \doteq w_{oL}$, a constant, and now the integrand of Λ, (1.7.2), from (1.8.17), can be written

$$\int_{\boldsymbol{\theta}} w_{oL}(\boldsymbol{\theta}) F\left(\mathbf{X}\middle|\mathbf{S}\left(\hat{\boldsymbol{\theta}}(\mathbf{X})\right)\right) \mathbf{d}\boldsymbol{\theta} = F\left(\mathbf{X}\middle|\mathbf{S}\left(\hat{\boldsymbol{\theta}}(\mathbf{X})\right)\right), \tag{1.8.18}$$

since $\int_{\boldsymbol{\theta}} w_L(\boldsymbol{\theta})(= w_{0L})\mathbf{d}\boldsymbol{\theta} = 1$. Of course, the conditions on $w_L(\boldsymbol{\theta})$ leading to (1.8.17) must be obeyed for the results to be acceptably accurate. The resulting MAP Test, $p < 1$, here must also take into account the proper estimator for $p < 1$. Again, this is $p\hat{\boldsymbol{\theta}}_{p=1}(\mathbf{X})$, where $\hat{\boldsymbol{\theta}}_{p=1} = \hat{\boldsymbol{\theta}}^*_{p=1}|_{\text{uniform}}$, which is the equivalent UMLE now with a uniform *a priori* pdf $\boldsymbol{\theta} \in \Omega_{\boldsymbol{\theta}}$. The MAP test here becomes from (1.8.15) and the above

$$\text{MAP Test}|_{\text{uniform}} : \Lambda_{\text{MAP}}|_{p<1,\ \text{uniform}} = \mu F_J\left(\mathbf{X}\middle|\mathbf{S}\left(p\hat{\boldsymbol{\theta}}^*_{p=1,\ \text{uniform}}\right)\right)/F_J(\mathbf{X}|\mathbf{0})\left\{\begin{matrix}\geq \\ <\end{matrix} K\right\}\begin{matrix}\text{: decide } S \otimes N \\ \text{: decide } N.\end{matrix} \tag{1.8.19}$$

Once more, this procedure maximizes the average probability of correct signal detection, now without detailed knowledge of a parametric *a priori* pdf $w_L(\boldsymbol{\theta})$, except again that it be effectively uniform over values of $\boldsymbol{\theta}$ where F is significant *and* with recognition of the fact that $p < 1$.

The associated Bayes risk for these MAP detectors has the form of (1.7.7a), namely,

$$R^*_{\text{MAP}} = R_0 + p(C_\beta - C_{1-\beta})\left(\frac{K}{\mu}\alpha^*_{\text{MAP}} + \beta^*_{\text{MAP}}\right). \tag{1.8.20}$$

We remark once more that $R^*_{\text{MAP}} \geq R^*$, the minimum average risk for the fully known prior pdf's, including $p\,(= 1 - q)$. For quantitative results we must of course evaluate the conditional error probabilities $\alpha^*_{\text{MAP}}, \beta^*_{\text{MAP}}$, and for comparison, α^*, β^*, as well. General expansions for α^*, β^*, α^*_{MAP}, and so on are derived in Section 1.9 ff., from which, in turn, explicit results can be obtained either exactly or approximately by a variety of analytical methods. In any case, the relation $R^*_{\text{MAP}} \geq R^*$ is basically attributable to the fact that R^*_{MAP} employs only partial information regarding the prior pdfs of (here) the parameters $\boldsymbol{\theta}$, in the form of estimates, whereas the Bayes risk R^* uses the true and entire pdfs for $\boldsymbol{\theta}$.

Similar arguments for MAP detection of received signal *waveforms* $\mathbf{S}$ give at once the desired counterparts to (1.8.16) and (1.8.19):

$$\text{MAP Test } (p < 1)\text{: } \Lambda_{\text{MAP}}|_{p<1} = \mu F_J\left(X\middle|p\hat{\mathbf{S}}^*(\mathbf{X})_{p=1}\right)/F_J(\mathbf{X}|\mathbf{0})$$

$$\text{MAP Test}|_{\text{uniform}} : \Lambda_{\text{MAP}}|_{p<1;\ \text{uniform}} = \frac{\mu F_J\left(X\middle|p\hat{\mathbf{S}}^*(\mathbf{X})_{p=1|\text{uniform}}\right)}{F_J(\mathbf{X}|\mathbf{0})}\left\{\begin{matrix}\geq \\ <\end{matrix} K \begin{matrix}\text{: decide } S \otimes N \\ \text{: decide } N\end{matrix}\right\}, \tag{1.8.21}$$

where (1.8.13a and 1.8.13b) and (1.8.18), with $\boldsymbol{\theta}$ replaced by $\mathbf{S}$, now provide the maximizing condition for $\hat{\mathbf{S}}^*(\mathbf{X})_{p=1}$ in (1.8.21).

Finally, an alternative way of handling the MAP estimators in detection is discussed in Chapter 6. Here the biased nature of the estimator is handled by a strongly coupled estimator — a detector system that does not require *explicit a priori* knowledge of $p\ (=p(H_1))$ when $p<1$. This is accomplished by the application of appropriate thresholds on the estimators and detector, along with feedback of the detector's decision $(H_1 \text{ or } H_0)$ regarding acceptance or rejection of these MAP estimators as well as presence or absence of the signal. Nevertheless, it should be remembered that in the Bayesian formulation prior probabilities (among them p or $q=1-p$), are always at least implied. An extensive discussion of the pros and cons of these methods is given in Section 23.4 of Ref. [1].

1.8.5 Bayesian Sequential Detectors

Other variations in the form of the likelihood detector are also possible. In all of the above, sample size (J) is fixed and minimization of the average risk, generally, involves minimizing the appropriate error probabilities. In sequential detection, however, the false alarm and Type II error probabilities are preset and the aim is to reach a decision ($S \otimes N$ or N) in the shortest time, that is, for the smallest sample size, *on the average*. Thus, sample size J is now the random variable (as well as the data $\mathbf{X}$). Minimization of the average risk is now minimization of the "average cost of experimentation," defined as being proportional to sample size. This Bayes risk can be expressed as

$$R^*_{\text{seq}} = q\alpha C_\alpha + p\beta C_\beta + pC_o \min_{\delta \to \delta^*} \left\langle J\left(\mathbf{X}|\mathbf{S},\delta\right)^*\right\rangle_{\mathbf{X}}, \ j \to J^*(\text{termination}) \qquad (1.8.22)$$

where C_o is the cost per unit trial (per unit of $j=11, 12, \ldots, J$). In many (but not all) cases, $\delta \to \delta^*_{\text{seq}}$ yields a likelihood detector for the optimum structure. If $y_s\left(\equiv \log\Lambda_{J-\text{seq}}\right)$ is this likelihood detector, then the best procedure involves a *double* threshold, instead of the single threshold (log K) characteristic of the fixed-sample tests described above. The detection process is described by

$$\begin{aligned} &\text{Sequential (Binary) } \textit{Test}:\ \text{If } B\ (=\beta/(1-\alpha)) < y_s < A\ (=(1-\beta)/\alpha):\ \text{continue test } j \to j+1 \\ &\qquad \text{If } y_s \geq A:\ \text{test terminates } j \to J(\mathbf{X}|\mathbf{S},\delta_{\text{seq}})^*\ \text{decide } H_1:\ S \otimes N \\ &\qquad \text{If } y_s < B:\ \text{test terminates } j \to J(X|S,\delta_{\text{seq}})^*\ \text{decide } H_0:\ N. \end{aligned} \qquad (1.8.23)$$

The theory of sequential tests is due primarily to Wald [8], with its application to signal detection subsequently initiated by Bussgang and Middleton [9], with further development by Blasbalg [22] and others; see also Basseville and Nikiforov [23]. Chapter 9 of Helstrom [14] provides a comprehensive account of the subject with additional references. Further discussion here is outside the scope of this book.

1.9 OPTIMUM DETECTION: ON–OFF PERFORMANCE MEASURES AND SYSTEM COMPARISONS

A second and equally significant task of Bayes SCT, along with the determination and practical interpretation of the optimal data processing algorithms ($\Lambda(\mathbf{X})$, $\log\Lambda(\mathbf{X})$, $\Lambda_{\text{MAP}}(\mathbf{X})$, etc.), is the evaluation of optimum system performances and performance comparisons with suboptimum receivers $G(\mathbf{X})$. The latter is particularly important because practical systems are themselves never strictly optimum: optimality is an ideal, to be approached under the inevitable constraints of usually limited knowledge of the environment and bounded economic resources. Nevertheless, optimality and its explicit formulations provide a guide to the key elements of (1) effective practical system design, (2) limiting measures of performance against which the practical system can be compared and often improved, and (3) insights regarding the critical channel structures that inhibit performance. Accordingly, modeling the communication environment, that is, translating the physics of the channel $\left(\hat{\mathbf{T}}^{(N)}\right)$ into relevant mathematical relationships becomes a third major task. This will be treated in detail in Chapters 8 and 9, but it needs to be borne in mind here, because of its ultimate influence on the actual probability measures that constitute the elements of the (Bayes) risks by which performance is evaluated.

Useful measures of performance all depend in some way on the conditional error probabilities $\alpha^*, \beta^*, \alpha, \beta, \ldots,$ and so on, cf. Sections 1.6.2 and 1.7 above. Since the error probabilities are also functions of the received signal and of the parameters of the accompanying noise, comparisons of systems performance can also be made in terms of such quantities as well, under a variety of conditions, involving both optimality and suboptimality.

1.9.1 Error Probabilities: Optimum Systems

Our first problem is to provide some way of determining the error probabilities $\alpha^{(*)}$, $\beta^{(*)}$, α_{MAP}, β_{MAP}, and so on, which occur for Bayes and non-Bayes (suboptimum) systems.

We begin with the Bayes class, namely, those described in Sections 1.7 and 1.8 above, where the decision rules $\delta^*(\gamma_0|\mathbf{X}), \delta^*(\gamma_1|\mathbf{X})$ are determined according to (1.7.4a and 1.7.4b). With the help of the transformation $x = \log \Lambda(\mathbf{X})$, cf. (1.7.7a and 1.7.7b), we can write the following expressions for the conditional class probabilities of the Types I and II errors in the Bayesian cases

$$\alpha^* = \int_{\log K}^{\infty} dx \int_{\Gamma} F_J(\mathbf{X}|\mathbf{0})\delta[x - \log \Lambda(\mathbf{X})]\mathbf{dX} = \int_{\log K}^{\infty} Q_1(x)dx \tag{1.9.1a}$$

and

$$\beta^* = \int_{-\infty}^{\log K} dx \int_{\Gamma} \langle F_J(\mathbf{X}|\mathbf{S}(\boldsymbol{\theta}))\rangle_{\mathbf{S}\text{ or }\boldsymbol{\theta}}\delta[x - \log \Lambda(\mathbf{X})]\mathbf{dX} = \int_{-\infty}^{\log K} P_1(x)dx, \tag{1.9.1b}$$

Here Q_1 and P_1 are respectively given by

$$Q_1(x) = \int_{\Gamma} F_J(\mathbf{X}|\mathbf{0})\delta[x - \log \Lambda(\mathbf{X})]\mathbf{dX};\ P_1(x) = \int_{\Gamma} \langle F_J(\mathbf{X}|\mathbf{S}(\boldsymbol{\theta}))\rangle_{\mathbf{S}\text{ or }\boldsymbol{\theta}}\delta[x - \log \Lambda(\mathbf{X})]\mathbf{dX}. \tag{1.9.2}$$

From the fact that $F_J(\mathbf{X}|\mathbf{0})$ and $\langle F_J(\mathbf{X}|\mathbf{S}(\boldsymbol{\theta}))\rangle$ are themselves probability densities and that $x - \log \Lambda(\mathbf{X})$ is also a random variable when considered over the ensemble of possible values of $\mathbf{X}$, it follows that the Q_1, P_1 of (1.9.1a) and (1.9.1b) are the probability densities of x with respect to the distributions associated with the hypothesis states H_0 and H_1, respectively. We shall elaborate on this further below, cf. Eq. (1.9.4a) et. seq.

1.9.1.1 Sufficient Statistics and Monotonic Mapping The mapping for **X**- to x-space by means of the transformation $x - \log \Lambda(\mathbf{X})$ is mathematically quite arbitrary; (see Section 1.7.1). Any monotone function $x = F(\Lambda)$ can be used without altering the values of the error probabilities. This is to be expected, since $F(\Lambda)$, like Λ here, remains a *sufficient statistic*.[41] The analytic consequences of monotonicity in turn are readily demonstrated by the relations

$$\alpha^* \equiv \int_K^\infty q_1(y)^* dy = \int_K^\infty Q_1(F(y))F'(y)dy = \int_{K_{\text{new}}=F(K)}^\infty Q_1(x)dx, x = F(y). \tag{1.9.3}$$

Since $y = \Lambda$ and $x = F(y) = F(\Lambda) = \log \Lambda$ here, $F'(y)dy = dx = d\Lambda/\Lambda$ with $F(K) = \log K$, so that (1.9.1a) results. A similar procedure gives β^*(1.9.1b). In fact, for any monotonic relation $z = G(\mathbf{X})$, where $x = F(G) = F(z)$ it follows that the general (not necessarily optimum) conditional error probabilities α and β can be represented by

$$\begin{aligned} \alpha \equiv \int_K^\infty q_1(z)dz = \int_K^\infty \hat{Q}_1(F)\left(\frac{dF}{dz}\right)dz = \int_{K_{\text{new}}=F(K)}^\infty Q_1(x)dx; \\ \text{etc. for } \beta \equiv \int_{-\infty}^K \hat{p}_1(z)dz = \int_{-\infty}^{K_{\text{new}}=F(K)} \hat{P}_1(x)dx. \end{aligned} \tag{1.9.3a}$$

Although a new threshold $K_{\text{new}} = F(K)$ is established by the transformation F, *the key result is that the error probabilities $\alpha^{(*)}, \beta^{(*)}$ remain unchanged under any monotonic mapping*. Moreover, $\alpha_F^*, \beta^* \to 0$ and $P_D^* \to 1$ as the signal $S \to \infty$ vis-à-vis the accompanying noise, and likewise $(\alpha_F^*, \beta^*) \to 1$ when $S \to 0$. In the optimum cases this also means that $K_{\text{new}}[= K_{\text{new}}(S)]$ depends on the signal in such a way that these limiting results are achieved. The resulting decision process is *consistent* as $S \to \infty$. For suboptimum systems similar results will occur, depending on our choice of test statistic $z = G(\mathbf{X})$: however, not all choices lead to consistency.

Similarly, if several successive monotonic mappings are carried out, that is $x = G_1(y)$ and $z = G_2(x) = G_{21}(y)$, then (1.9.3a) becomes generally

$$\alpha = \int_K^\infty q_1(y)dy = \int_{K_1=G_1(K)}^\infty Q_1(x)dx = \int_{K_2=G_2(K_1)=G_{21}(K)} Q_2(z)dz, \text{ etc.,} \tag{1.9.3b}$$

and for the optimum cases $\alpha \to \alpha^*, y = \Lambda$, and so on. As noted above, monotonicity guarantees that $\alpha^{(*)}$ and $\beta^{(*)}$ remain unchanged in value, although their analytic forms are now different from the original expressions. The practical importance of this is that it very often allows us to evaluate performance analytically, without recourse to numerical methods, by suitable simplifying choices of monotonic transformations (e.g., $x = \log \Lambda$ instead of Λ itself). We shall see several examples employing these general results in Sections 3.2–2, 3, 5 subsequently.

[41] We recall that $\Lambda(\mathbf{X})$ is a sufficient statistic if specifying **X** in addition to $x = \Lambda(\mathbf{X})$ does not in any way increase our knowledge of the signal **S** (which is implicit in Λ, cf. $\langle F_J(\mathbf{X}|\mathbf{S})\rangle$(1.7.2)). Analytically, [Section 22.1.1 of Ref. [1] and, p. 1010] the n.+s. condition that Λ is a sufficient statistic is the requirement that the pdf $w_J(\mathbf{S}|\Lambda) = F_{\text{mono}}[f(\mathbf{S})g(\Lambda)]$, that is, that w_J is a monotonic function of the factors f, g here.

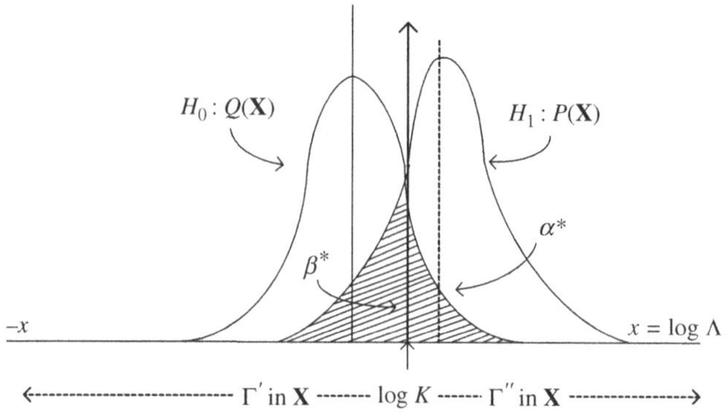

FIGURE 1.9 Probability densities and decision regions for $x = \log\Lambda$, Eq. (1.7.6).

In any case, we see that the δ-functions in (1.9.1a) and (1.9.2) pick out the region Γ^* in $\mathbf{X}$-space for which $\delta^*(\gamma_1|\mathbf{X}) = 1$ in the case of α^*, with a similar interpretation for β^* when $\delta^*(\gamma_0|\mathbf{X}) = 1$. Figure 1.9 shows the two regions in x-space when $x = \log\Lambda(\mathbf{X})$, with the dotted line at $x = \log K$ separating the decision regions for H_0 and H_1.

1.9.1.2 Error Probabilities and Contour Integration [24] Returning to Eqs. (1.9.1a and 1.9.1b) and (1.9.2) and using the integral exponential form for the δ-functions therein, we can write at once the characteristic functions (c.f.s) associated with the pdfs Q_1, P_1

$$F_1(i\xi)_Q = E_{H_0}\left\{e^{i\xi\log\Lambda(\mathbf{X})}\right\} = \int_\Gamma e^{i\xi\log\Lambda(\mathbf{X})} F_J(\mathbf{X}|\mathbf{0})\mathbf{dX} \tag{1.9.4a}$$

$$F_1(i\xi)_P = E_{H_1}\left\{e^{i\xi\log\Lambda(\mathbf{X})}\right\} = \int_\Gamma e^{i\xi\log\Lambda(\mathbf{X})} \langle F_J(\mathbf{X}|\mathbf{S}(\boldsymbol{\theta}))\rangle\mathbf{dX}, \tag{1.9.4b}$$

for which the corresponding pdfs are, from (1.9.2)

$$Q_1(x) = \mathsf{F}^{-1}\left\{F_1(i\xi)_Q\right\} = \int_{-\infty}^{\infty} e^{-i\xi x}\frac{d\xi}{2\pi}\int_\Gamma e^{i\xi\log\Lambda(\mathbf{X})} F_J(\mathbf{X}|\mathbf{0})\mathbf{dX} \tag{1.9.5a}$$

$$P_1(x) = \mathsf{F}^{-1}\{F_1(i\xi)_P\} = \int_{-\infty}^{\infty} e^{-i\xi x}\frac{d\xi}{2\pi}\int_\Gamma e^{i\xi\log\Lambda(\mathbf{X})} \langle F_J(\mathbf{X}|\mathbf{S}(\boldsymbol{\theta}))\rangle\mathbf{dX}. \tag{1.9.5b}$$

As noted earlier [[1], Eq. (19.32a) and Problem 17.8] there is a simple formal relation between P_1 and Q_1, which follows from the identity

$$\int_\Gamma e^{-i\xi x} F_J(\mathbf{X}|\mathbf{0})\mathbf{dX} \equiv \mu\int_\Gamma e^{(i\xi-1)x}\langle F_J(\mathbf{X}|\mathbf{S})\rangle\mathbf{dX};\; x = \log\Lambda = \frac{\mu\langle F_J(\mathbf{X}|\mathbf{S})\rangle_{\boldsymbol{\theta}}}{F_J(\mathbf{X}|\mathbf{0})} \tag{1.9.6}$$

(which is readily established on using $x = \log\Lambda$ explicitly in (1.9.6)). Thus, from (1.9.4a and 1.9.4b) it is seen at once that $F_1(i\zeta)_Q = \mu e^{-x}F_1(i\xi)_P$, so that using this relation in (1.9.5a and 1.9.5b) gives the desired relation

$$Q_1(X) = \mu e^{-x}P_1(x) \tag{1.9.7}$$

which is sometimes useful in the explicit evaluation of error probabilities, particularly when the (sometimes) easier to evaluate $Q_1(x)$ can be found.

Potentially useful alternative relations for the error probabilities, now in terms of the characteristic functions (1.9.4a and 1.9.4b), can be obtained as follows. Applying (1.9.4a and 1.9.4b) to (1.9.5a and 1.9.5b), and then in (1.9.1a and 1.9.1b), we first extend the domain of ξ by analytic continuation to appropriate regions of the complex ξ-plane, in order to ensure convergence of the integrals $\int_{\log K}^{\infty} e^{-i\xi x}dx$, $\int_{\infty}^{\log K} e^{-i\xi x}dx$ in the reversal of the orders of integration which we employed in the above. The results are then the inverse Fourier transforms

$$\alpha^* = \int_{-\infty-ic}^{\infty-ic} \frac{e^{-i\xi\log K}}{2\pi i\xi} F_1(i\xi)_Q d\xi = \int_{C^{(-)}} \frac{e^{-i\xi\log K}}{2\pi i\xi} F_1(i\xi)_Q d\xi, \tag{1.9.8a}$$

and

$$\beta^* = \int_{-\infty+ic}^{\infty+ic} \frac{e^{-i\xi\log K}}{-2\pi i\xi} F_1(i\xi)_P d\xi = \int_{C^{(+)}} \frac{e^{-i\xi\log K}}{-2\pi i\xi} F_1(i\xi)_P d\xi, \tag{1.9.8b}$$

where $\mathbf{C}^{(-)}, \mathbf{C}^{(+)}$ are respectively contours extending from $-\infty$ to $+\infty$ along the real axis, indented downward and upward about any singularities on this axis, usually at $\xi = 0$, as shown in Fig. 1.10. (We note the equivalence of the contours $[(-\infty \mp ic), (\infty \mp ic)]$ and $\mathbf{C}^{(-)}, \mathbf{C}^{(+)}$, since the contributions of the paths A_0A', A_1A''; B'', B_1, $B'B_0$ vanish at $\mp\infty$..) Simple poles on the ξ-axis or within the rectangular paths $\mathbf{C}^{(-)} + B_0B' + A'A_0$ and $\mathbf{C}^{(+)} + B_1B'' + A''A_1$ are handled in the usual way with the help of Cauchy's theorem,[42] extended to include any branch points by appropriate modification of the contours. For example, Fig. 1.11 shows some equivalent contours when the integrands (1.9.8a and 1.9.8b) contain a branch insert at $(= 0)$.[43] Equivalent contours are also obtained by

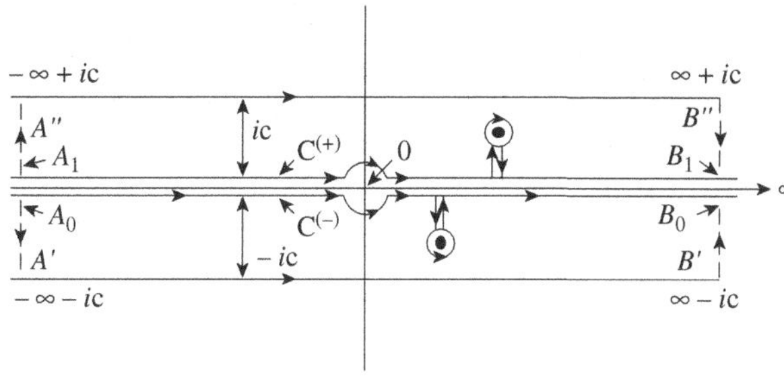

FIGURE 1.10 Equivalent contours of integration for the error probabilities (α^*, β^*), (1.9.8a and 1.9.8b).

[42] We cannot use circular arcs $\xi = \rho e^{i\phi_\xi}$ since $\rho \to \infty$ and $-\pi \le \phi_\xi \le 0$ or $0 \le \phi_\xi \le \pi$ generally, and have their contributions vanish, leaving $\mathbf{C}^{(\mp)}$. This depends on $F_1/\xi = F = \exp(-\xi^2/2)$, $A_1 > 0$, is a case in point. We can, however, eliminate A_1A'' by letting $0 \to +\infty$, etc., with $\mathbf{C}^{(+)}$ indented upward by ε, about any singularities on the Re ζ-axis, and so on.

[43] A rather extensive discussion of Fourier and Laplace transforms is available in Sections 2.2.4 and 2.2.3 of Ref. [1], including extensive references, along with applications to filters (Chapter 2), rectification, modulation (Chapters 5, 12, 13, 15), and Bayes detection results (Chapters 19, 20, 23), also in Ref. [1]. See Refs. [24] and [41] as well, along with [42, 43] for related analytical tools.

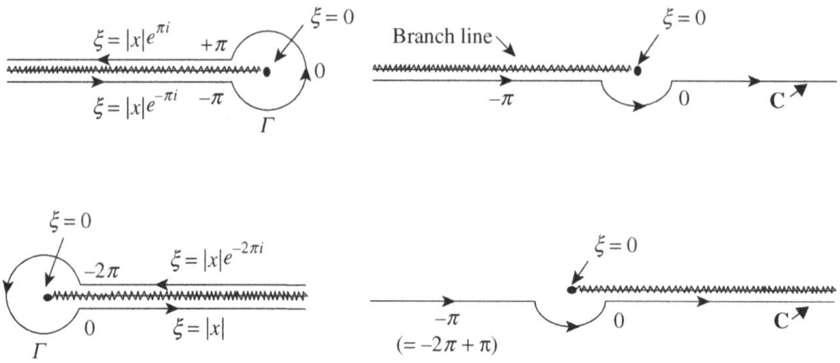

FIGURE 1.11 Some equivalent contours of integration when $F_1(i\xi)/\xi$ contains a branch point at $\xi = 0$; [24, 41].

setting $\xi = s/i$ in the above. The result is a rotation by $-i = e^{-\pi i/2}$ of the ξ-plane contours.[42]

Finally, it is useful for subsequent applications to provide explicit results for determining the (class) error probabilities (α^*, β^*) for the various types of binary Bayes, that is, optimum detectors discussed in Chapter 3, along with general expressions of their associated risks or costs. For this we shall formally employ the c.f.s (1.9.4a and 1.9.4b) to be used in (1.9.8a and 1.9.8b) directly:

$$\alpha^* = \int_{C^{(-)}} \frac{e^{-i\xi \log K}}{2\pi i\xi} F_1(i\xi)_Q d\xi \quad \beta^* = \int_{C^{(+)}} \frac{e^{-i\xi \log K}}{-2\pi i\xi} F_1(i\xi)_P d\xi, \tag{1.9.9}$$

with the general Bayes case:

(1) *General On–Off Bayes.*

$$[\text{Section 1.7}]\ F_1(i\xi)_Q = \int_\Gamma e^{i\xi \log \Lambda} F_J(\mathbf{X}|\mathbf{0})\mathbf{dx};\ \Lambda(\mathbf{X}) = \frac{\mu\langle F_J(\mathbf{X}|\mathbf{S}(\boldsymbol{\theta}))\rangle_{\boldsymbol{\theta}}}{F_J(\mathbf{X}|\mathbf{0})},\ (1.7.2),\ (1.7.6), \tag{1.9.9a}$$

$$F_1(i\xi)_P = \int_\Gamma e^{i\xi \log \Lambda} \langle F_J(\mathbf{X}|\mathbf{S}(\boldsymbol{\theta}))\rangle_{\boldsymbol{\theta}} \mathbf{dx}; \tag{1.9.9b}$$

$$R^*(\sigma, \delta^*) = \mathsf{R}_0 + p(C_\beta - C_{1-\beta})\left(\frac{K}{\mu}\alpha^* + \beta^*\right),\ (1.7.1),\ (1.7.5), \tag{1.9.9c}$$

with α^*, β^* given by (1.9.1a), via (1.9.4a and 1.9.4b) in (1.9.5a and 1.9.5b).

(2) *Neyman–Pearson.*

$$[\text{Section 1.8.1}]\ \alpha^* = \alpha_F = \int_{C^{(-)}} \frac{e^{-i\xi \log K_{NP}}}{2\pi i\xi} F_1(i\xi)_Q d\xi, \quad F_{1Q} = (\text{Eq.1.9.9a});\ K = K_{NP}(\alpha_F) \tag{1.9.10a}$$

$$\beta_{\mathrm{NP}}^{*}=\int_{C^{(+)}}\frac{e^{-i\xi\log K_{\mathrm{NP}}(\alpha_{\mathrm{F}})}}{-2\pi i\xi}F_{1}(i\xi)_{P}d\xi,\ F_{1P}=\mathrm{Eq.}(1.9.9\mathrm{b}); \tag{1.9.10b}$$

$$R_{\mathrm{NP}}^{*}(\sigma,\delta_{\mathrm{NP}}^{*})=C_{0}\left[p\beta_{\mathrm{NP}}^{*}+K_{\mathrm{NP}}(\alpha_{\mathrm{F}})q\alpha_{\mathrm{F}}\right],\ \mathrm{Eqs.}(1.7.7\mathrm{a\ and\ }1.7.7\mathrm{b}),(1.8.5). \tag{1.9.10c}$$

(3) *Ideal-Observer.*

$$[\text{Section 1.8.2}]\ \alpha_{I}^{*}=\int_{C^{(-)}}F_{1}(i\xi)_{Q}\frac{d\xi}{2\pi i\xi};\ (1.9.9\mathrm{a})\ \text{for}\ F_{1}(i\xi)_{Q};\ (K=1); \tag{1.9.11a}$$

$$\beta_{I}^{*}=\int_{C^{(+)}}F_{1}(i\xi)_{P}\frac{d\xi}{-2\pi i\xi},\ (1.9.9\mathrm{b})\ \text{for}\ F_{1}(i\xi)_{P}; \tag{1.9.11b}$$

$$R_{I}^{*}(\sigma,\delta_{I}^{*})=C_{0}(q\alpha_{I}^{*}+p\beta_{I}^{*}),\ \mathrm{Eqs.}\ (1.8.7\mathrm{a\ and\ }1.8.7\mathrm{b}). \tag{1.9.11c}$$

(4) *Minimax.*

$$[\text{Section 1.8.3}]\ F_{1}(i\xi)_{Q_{\mathrm{M}}}=\int_{\Gamma}e^{i\xi\log\Lambda_{\mathrm{M}}}F_{J}(\mathbf{X}|\mathbf{0})\mathbf{dx};\ \Lambda=\Lambda_{\mathrm{M}}=\frac{\mu_{\mathrm{M}}\langle F_{J}(\mathbf{X}|\mathbf{S}(\boldsymbol{\theta}_{\mathrm{M}}))\rangle_{\boldsymbol{\theta}_{\mathrm{M}}}}{F_{J}(\mathbf{X}|\mathbf{0})} \tag{1.9.12a}$$

$$F_{1}(i\xi)_{P_{\mathrm{M}}}=\int_{\Gamma}e^{i\xi\log\Lambda_{\mathrm{M}}}\langle F_{J}(\mathbf{X}|\mathbf{S}(\boldsymbol{\theta}_{\mathrm{M}}))\rangle_{\boldsymbol{\theta}_{\mathrm{M}}}\mathbf{dx},\ \text{cf. }(1.9.9\mathrm{a\ and\ }1.9.9\mathrm{b}) \tag{1.9.12b}$$

$$R^{*}(\sigma_{\mathrm{M}}^{*},\delta^{*})=\mathsf{R}_{0}+p(C_{\beta}-C_{1-\beta})\left(\frac{K}{\mu}\alpha_{\mathrm{M}}^{*}+\beta_{\mathrm{M}}^{*}\right); \tag{1.9.12c}$$

$\alpha^{*}\rightarrow\alpha_{\mathrm{M}}^{*}$, $\beta^{*}\rightarrow\beta_{\mathrm{M}}^{*}$ fixed and determined from (1.9.12a and 1.9.12b) in (1.9.9a and 1.9.9b).

(5) *MAP Detectors, MAP1,2.*

$$[\text{Section 1.8.4}]\ F_{1}(i\xi)_{Q-\mathrm{MAP}_{1}}=\int_{\Gamma}e^{i\xi\log\Lambda_{\mathrm{MAP}_{1}}}F_{J}(\mathbf{X}|\mathbf{0})\mathbf{dx};$$

$$\Lambda(\mathbf{X})_{\mathrm{MAP}_{1}}=\frac{\mu F_{J}\left(\mathbf{X}|\mathbf{S}\left(p\hat{\boldsymbol{\theta}}_{p=1}^{*}\right)\right)}{F_{J}(\mathbf{X}|\mathbf{0})},\ \text{cf. Eq. }(1.8.16) \tag{1.9.13a}$$

$$F_{1}(i\xi)_{P-\mathrm{MAP}_{1}}=\int_{\Gamma}e^{i\xi\log\Lambda_{\mathrm{MAP}_{1}}}\langle F_{J}(\mathbf{X}|\mathbf{S}(\boldsymbol{\theta}))\rangle_{\boldsymbol{\theta}}\mathbf{dx} \tag{1.9.13b}$$

[Here the actual or "true" pdf $w_{L}(\boldsymbol{\theta})$ is known, in order to obtain $\hat{\boldsymbol{\theta}}^{*}$ in the maximization process, cf. (1.8.13a) et. seq. Hence $\langle\ \rangle_{\boldsymbol{\theta}}$ involves $w_{L}(\boldsymbol{\theta})$ in $\langle F_{J}\rangle_{\boldsymbol{\theta}}$(1.9.13b): $p\hat{\boldsymbol{\theta}}^{*}(\mathbf{X})_{p=1}$ is the required *unbiased* estimate of $\mathbf{S}$.]

We observe that $\alpha^*_{\mathrm{MAP}_1}$, $\beta^*_{\mathrm{MAP}_1}$, follow from (1.9.13a and 1.9.13b) in (1.9.8), cf. (1.9.9c). Also, that

$$R^*_{\mathrm{MAP}_1} = R^*\left(\sigma, \delta^*_{\mathrm{MAP}_1}\right) = \mathrm{R}_0 + p(C_\beta - C_{1-\beta})\left(\frac{K}{\mu}\alpha^*_{\mathrm{MAP}_1} + \beta^*_{\mathrm{MAP}_1}\right), \text{cf. (1.8.7) and (1.9.13c);}$$

MAP2.

$$F_1(i\xi)_{Q-\mathrm{MAP}_2} = \int_\Gamma e^{i\xi\log\Lambda_{\mathrm{MAP}_2}} F_J(\mathbf{X}|\mathbf{0})\mathbf{dx};\ \begin{cases} \Lambda(\mathbf{X})_{\mathrm{MAP}_2} = \dfrac{\mu\left\langle F_J\left(\mathbf{X}|\mathbf{S}\left(p\hat{\boldsymbol{\theta}}^*_{p=1|\mathrm{uniform}}\right)\right)\right\rangle}{F_J(\mathbf{X}|\mathbf{0})} \\ \text{cf. (1.8.19) : here } w_L(\boldsymbol{\theta}) \equiv \text{uniform } \boldsymbol{\theta} \in \Omega_{\boldsymbol{\theta}} \end{cases}, \tag{1.9.14a}$$

$$F_1(i\xi)_{P-\mathrm{MAP}_2} = \int_\Gamma e^{i\xi\log\Lambda_{\mathrm{MAP}_2}} \langle F_J(\mathbf{X}|\mathbf{S}(\boldsymbol{\theta}))\rangle_{\boldsymbol{\theta}:\ \mathrm{uniform}} \boldsymbol{dX}; \tag{1.9.14b}$$

$(\alpha^*, \beta^*)_{p-\mathrm{MAP}_2}$ follow from (1.9.5a and 1.9.5b) in (1.9.8).

$$R^*_{(\mathrm{uniform}, \delta^*_{\mathrm{MAP}_2})} = R_0 + p(C_\beta - C_{1-\beta})(\frac{K}{\mu}\alpha^* + \beta^*)_{\mathrm{MAP}_2} \tag{1.9.14c}$$

(6) *Sequential Detection.* Here α, β are preset: the test statistic is

$$[\text{Section 1.8.5}]\ \Lambda = \Lambda(\mathbf{X}|J) = \mu\frac{\langle FJ(\mathbf{X}|\mathbf{S}(\boldsymbol{\theta}))\rangle_{\boldsymbol{\theta}}}{F_J(\mathbf{X}|\mathbf{0})} \tag{1.9.15a}$$

$$\left\{\begin{array}{l} \dfrac{\beta}{1-\alpha} < \Lambda(\mathbf{X}|j) < \dfrac{1-\beta}{\alpha} : \text{ continue test, } j \to J+1; \\ \Lambda(\mathbf{X}|J^*) \geq \dfrac{1-\beta}{\alpha} : \text{ decide } H_1;\ \Lambda\left(\mathbf{X}|J^*\right) < \dfrac{\beta}{1-\alpha} : \text{ decide } H_0 \\ j \to J^* : \text{ terminating sample size.} \end{array}\right\} \tag{1.9.15b}$$

$$R^*_{\mathrm{seq}}\left(\sigma, \delta^*_{\mathrm{seq}}\right) = q\alpha C_\alpha + p\beta C_\beta + pC_0 \min_{\delta \to \delta^*_{\mathrm{seq}}} \left\langle J\left(\mathbf{X}|\mathbf{S}, J\right)^*\right\rangle^*_{\mathbf{X}}. \tag{1.9.15c}$$

In all of the above (except **(6)**) we observe that the various error probabilities (α^*, β^*) are also functions of the prior probabilities (p, q), as well as the parameters $\boldsymbol{\theta}$ of the signal and of the noise, through $\log\Lambda(\ )$ of the various optimal detectors above. When it is the signal waveform **S** with which we are directly concerned, rather than its parameters $\boldsymbol{\theta}$, we simply replace $\mathbf{S}(\boldsymbol{\theta})$, and so on, with **S**, or $p\hat{\mathbf{S}}^*$ and so on, cf. **(5)** for the MAP detectors. All of the detectors are optimal within the general Bayesian framework here. However, since they all provide likelihood-ratio tests representative of the level of optimum performance which they demand, they differ in their average costs of decision (Bayes risk). This occurs primarily because of the various constraints imposed upon the signal parameters and their distributions: the more constrained and the more approximative of the actual distributions, the larger the Bayes risk. Thus, ignorance of the true distribution imposes an average risk penalty, cf. remarks in Section 1.8.3, which can be determined by comparing R^*, (1.9.9c), with the other average risks, (1.9.10c), (1.9.11c), and so on.

1.9.2 Error Probabilities: Suboptimum Systems

The approach of Section 1.9.1 is in no way restricted to optimum systems. For example, in the case of an actual preselected detection system, with a threshold K' (implying at least a cost *ratio*), and a structure represented by $G(\mathbf{X})$ [$\neq \Lambda(\mathbf{X})$ usually] the conditional probabilities of the Type I and Type II errors are now described by analogues of (1.9.1a,b), namely,

$$\alpha = \int_{\log K'}^{\infty} dz \int_{\Gamma} F_J(\mathbf{X}|\mathbf{0})\delta[z - \log G(\mathbf{X})]\mathbf{dX} = \int_{\log K'}^{\infty} q_1(z)dz \tag{1.9.16a}$$

$$\beta = \int_{-\infty}^{\log K'} dz \int_{\Gamma} \langle F_J(\mathbf{X}|\mathbf{S}(\boldsymbol{\theta}))\rangle_{\boldsymbol{\theta}\text{ or }\mathbf{S}}\delta[z - \log G(\mathbf{X})]\mathbf{dX} = \int_{-\infty}^{\log K'} p_1(z)dz. \tag{1.9.16b}$$

The conditional error probabilities α', β', Eq. (1.6.13a), are again obtained on omitting the average $\langle\,\rangle_{\boldsymbol{\theta}\text{ or }\mathbf{S}}$ over parameters or waveform. The distributions (pdfs) of $y = \log G(\mathbf{X})$ are respectively given by (1.9.2), with $\Lambda(\mathbf{X})$ replaced by $G(\mathbf{X})$ under H_0, H_1. The c.f.s of q_1, p_1 are likewise described by

$$F_1(i\xi)_{q_1} = E_{H_0}\left\{e^{i\log G(\mathbf{X})}\right\} = \int_{\Gamma} e^{i\xi\log G(\mathbf{X})}F_J(\mathbf{X}|\mathbf{0})\mathbf{dX} \tag{1.9.17a}$$

$$F_1(i\xi)_{p_1} = E_{H_1}\left\{e^{i\log G(\mathbf{X})}\right\} = \int_{\Gamma} e^{i\xi\log G(\mathbf{X})}\langle F_J(\mathbf{X}|\mathbf{S}(\boldsymbol{\theta}))\rangle_{\boldsymbol{\theta}\text{ or }\mathbf{S}}\mathbf{dX}, \tag{1.9.17b}$$

cf. (1.9.4a and 1.9.4b). [Figure 1.9 applies here also, provided that we replace K by K', α^* by α', Q_1 by q_1, etc.]

The same procedure used to obtain α^* and β^*, (1.9.8a and 1.9.8b) et. seq., applies directly here for α and β. We have directly

$$\alpha = \int_{-\infty-ic'}^{\infty-ic'} \frac{e^{-i\xi\log K'}}{2\pi i\xi} F_1(i\xi)_{q1}d\xi = \int_{C^{(-)'}} \frac{e^{-i\xi\log K'}}{2\pi i\xi} F_1(i\xi)_{q1}d\xi \tag{1.9.18a}$$

$$\beta = \int_{-\infty+ic'}^{\infty+ic'} \frac{e^{-i\xi\log K'}}{-2\pi i\xi} F_1(i\xi)_{p1}d\xi = \int_{C^{(+)'}} \frac{e^{-i\xi\log K'}}{-2\pi i\xi} F_1(i\xi)_{p1}d\xi, \tag{1.9.18b}$$

where $c', \mathbf{C}^{(\pm)'}$ are similar to $c, \mathbf{C}^{(\pm)}$ in (1.9.8a and 1.9.8b) and in Fig. 1.10. Note, however, that the relation (1.9.7) connecting the pdfs Q_1 and P_1 of $x = \log \Lambda(\mathbf{X})$ under H_0 and H_1 does *not* hold for q_1 and p_1, cf. (1.9.16a and 1.9.16b). On the other hand, it is still true if $u = F(z)$, $z = G(\mathbf{X})$, that for any monotonic function $F(z)$, α and β remain unchanged: here $F = \log z = \log G(\mathbf{X})$ specifically, (1.9.3a) above.

Accordingly, with (1.9.16a and 1.9.16b) or (1.9.18a and 1.9.18b), we are able, at least in principle, to determine the average risk $R(\sigma, \delta)$

$$[(1.6.11)]: \qquad R(\sigma,\delta) = \mathsf{R}_0 + q(C_\alpha - C_{1-\alpha})\alpha + p(C_\beta - C_{1-\beta})\beta \tag{1.9.19}$$

cf. (1.6.11a and 1.6.11b), and then compare the performance of the suboptimum system $G(\mathbf{X})$, or $\log G$, with that of the corresponding Bayes detectors (Sections 1.8 and 1.9.1), as outlined below.

1.9.3 Decision Curves and System Comparisons

The relations (1.9.1a and 1.9.1b) with the Bayes risk R^*, (1.7.7), and with the average risk R, enable us to compare the performance of actual and optimum systems for the same purpose and of course for the same input signals and noise statistics.

We note first that the error probabilities α^*, β^* and α, β, which appear in R^* and in R, are functions of a_0, the *input signal-to-noise* [(rms) amplitude] *ratio*[44], defined according to $a_0 = \left(\langle S^2\rangle/\langle N^2\rangle\right)^{1/2}$. Curves of average risk as a function of a_0 (or of any other pertinent signal parameters such as sample size J, and other structure parameters) are called *decision curves*. It is in terms of these that specific system comparisons may be made. Figure 1.12 illustrates a typical situation, involving an ideal and an actual detection system for the same purpose.[44] Thus, if we choose the same threshold ($K = K'$) and assign the same costs [Eq. (1.6.6)] to each possible decision, then the average risk R for all a_0 will exceed the corresponding Bayes risk R^*, as indicated.

One definition of *minimum detectable signal*[45] *is that input signal-to-noise ratio* $(a_0)_{\min}$ *that yields an average risk* R_0 *that is some specified fraction of the maximum average risk*, that is, *the* a_0 *for which* $R_0 = \eta R_{\max} (0 < \eta < 1)$. ($R_{\max}$, in physical situations at least, occurs for $a_0 = 0$.) System comparison can now be carried out in a variety of ways, of which the following are some examples (Fig. 1.12):

(1) $(a_0)^*_{\min\eta}$, versus $(a_0)_{\min\eta}$ (in general, $R_0 \neq R_0^*$ for the same η);
(2) $(a_0)^*_{\min\eta_1}$, versus $(a_0)_{\min\eta_1}$ (for $R_0^* = R_0''$, which determines η_1, η_2);
(3) R_0^* versus R' [for the same $(a_0)^*_{\min\eta}$].

Another definition of the *minimum detectable signal*, $\langle a_0^2\rangle^*_{\min} \left(\neq \langle a_0^2\rangle^2_{\min\eta}\right)$ which is explicitly related to detector structure and performance and easier to calculate, is obtained

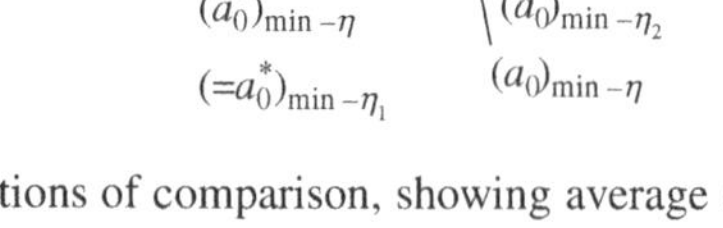

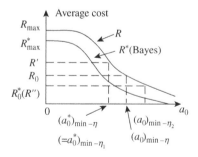

FIGURE 1.12 Typical situations of comparison, showing average and Bayes risks and minimum detectable signals.

[44] Frequently, a_0 is random over the signal class, so that the appropriate ratio is $\bar{a}_0$, or $\overline{a_0^2}$, and so on, depending on the system. See the examples in Section 3.2 ff.

[45] See Sections 19.3.3, 20.3.1, 20.4 of Ref. [1] and Refs. [1, 2] of Chapter 19, Ref. [1] for more details.

from the *detection parameter*. This quantity, in turn, explicitly determines the performance probabilities $P_D^{(*)}$, or $P_D^{(*)}$ and $P_F^{(*)}$ in the optimum and suboptimum cases. It is defined and discussed in detail in Section 3.1.2 and is employed throughout this book.[46]

1.9.3.1 Betting Curves Another decision curve, also useful for comparison, is the *betting curve*, introduced originally by Siegert [40], which relates the probability $W_1(a_0; J)$ of a correct decision (Section 19.3.3 [1]) to the input signal-to-noise ratio a_0. This is defined by

$$W_1(a_0; J) = 1 - (\alpha q + \beta p) = 1 - (P_F + P_D). \tag{1.9.20}$$

For optimum systems Eq. (1.9.20) thus becomes

$$W_1(a_0; J) = 1 - \left(\alpha^* q + \beta^* p\right) = 1 - \left[P_D^* + P_F^*\right]. \tag{1.9.20a}$$

For the Neyman–Pearson and Ideal Observer we may replace α and β by the appropriate α^* and β^* [Eqs. (19.4.1a and 19.4.1b) in [1]], since these systems were shown in Sections 1.8.1 and 1.8.2 to be Bayes with suitable assumptions on the cost ratio (Fig. 1.13).

It is often convenient to use normalized betting curves, defined by ((20.135a and 20.135b) of Ref. [1]), which for the Neyman–Pearson and Ideal Observer become specifically here

$$W_1(a_0; J)_{NP} = \left(W_1|_{NP} - q\alpha_F^*\right)/\left(1 - q\alpha_F^*\right), \tag{1.9.21a}$$

$$W_1(a_0; J)_I = \left[W_{1-I} - (p \text{ or } q)\right]/\left[1 - (p \text{ or } q)\right], \tag{1.9.21b}$$

where (p or q) means that the larger of the two is to be used, and W_1 is given by (1.9.20). The Bayes risks (1.8.2) and (1.8.7a) can be expressed more compactly in terms of the betting curve (1.9.20a) by

$$R_{NP}^* = C_0\{1 - W_{1-NP} + \alpha_{NP} q(K_{NP} - 1)\}, \tag{1.9.22a}$$

$$R_I^* = C_0\{1 - (W_{1-I})\}, \tag{1.9.22b}$$

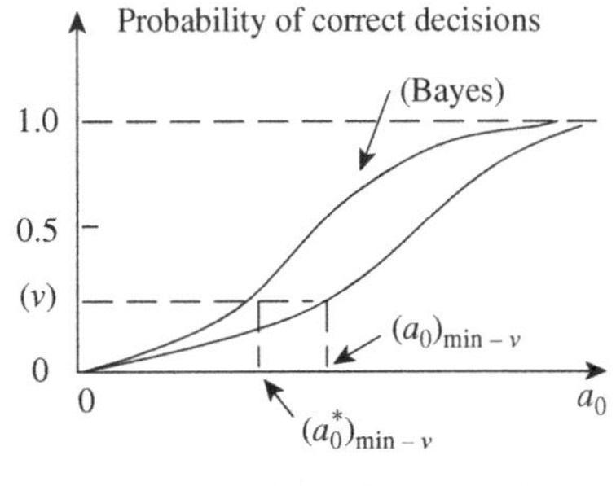

FIGURE 1.13 Betting curves and associated minimum detectable signals.

[46] Examples of $\langle a_0^2 \rangle_{min}$ are also noted in Sections 20.3 and 20.4 of Ref. [1].

with more involved forms when the W_1s are replaced by their normalized representation. The conditional probabilities may be calculated as before [cf. Eqs. (1.9.1a and 1.9.1b) and (1.9.16a and 1.9.16b)] and may be used in a similar way for system comparison, with the minimum detectable signal defined now in terms of an input which leads to a given percentage ν of successful decisions at the output (cf. Fig. 1.9). Each point on the betting curve, considered as a function of α and β, corresponds to a point on the risk curve, which is also a function of these α, β. Thus, comparisons in terms of "success" are equivalent to those on a risk basis (except for the scale that sets the absolute cost).

1.9.3.2 Performance versus Sample Size An additional description and comparison of system performance, often of considerable interest, is given by the behavior of the minimum detectable signal as a function of the acquisition, or integration time $T(\sim J$, the sample size). This relationship is found from the set of average risk curves [Eq. (1.6.11)] (cf. Fig. 1.12) or the betting curves [Eqs. (1.9.20 and 1.9.20a)], as J assumes all allowed values. The examples considered in Section 20.4, [1] provide some further illustrations.

1.9.3.3 Other Performance Measures Still another variant of the general average risk curve is given by the probability of successfully deciding that a signal is present (i.e., the alternative hypothesis H_1), as a function of input signal-to-noise ratio or other significant system parameters, for example, the decision or "deflection" parameter. Thus, one may use the conditional probability $P_D^* = 1 - \beta^*$, or the total probability $P_D^* = pp_D^* = p(1 - \beta^*)$, versus a_0, T (or J), and so on, for optimum systems. Here, usually, $\alpha\,(= \alpha_F^*)$ is fixed, so that a Neyman–Pearson system is essentially employed. For suboptimum systems, one has similarly $P_D = pp_D = p(1 - \beta)$, versus a_0, and so on, for the desired performance curve and system comparisons.

It is also sometimes convenient to use a normalized decision curve for defining minimum detectable signals and making system comparisons. However, since normalization is an arbitrary procedure, there is no unique or compelling general reason for doing it. In any case, for system comparison care must be taken that common criteria be used under identical conditions. This usually means that comparisons should be made on the basis of the unnormalized or absolute risk curves instead, since normalization may sometimes disguise or diminish significant differences.

We remark, finally, that in the construction, operation, and evaluation of these binary detection processes we have assumed that the signal parameters, *or their average values*, if they are originally random, are known or preset beforehand from a decision curve and that they are then inserted into Λ [Eq. (1.7.2)] so that the scale of Λ can be fixed and an actual test [Eq. (1.7.4a)] carried out with the same parameter values. However, it may be that these "true" parameters, that is, the values actually occurring when a signal is present or average values appropriate to the signal class in question, are not given beforehand, in which case the test of Eq. (1.7.4) can still be carried out, but we are unable to specify the error probabilities α, β uniquely and so cannot determine the Bayes or average risk uniquely. We note some examples of this in the case of the Bayes sequential detectors referred to in Section 1.8.5. For the most part, however, it is not unrealistic to assume at least a knowledge of the required moments of the signal parameters or by some such process as Minimax, to define a class of Bayes receivers for the problem at hand which guards against least favorable situations in some operationally meaningful sense. Unless otherwise indicated, we shall assume henceforth that the appropriate statistics of the signals (and noise) parameters are specified and used.

1.10 BINARY TWO-SIGNAL DETECTION: DISJOINT AND OVERLAPPING HYPOTHESIS CLASSES

The previous on–off analysis here in this chapter is readily extended to the binary two-signal detection cases, where the hypothesis situation is now $H_1 : S_1 \otimes N$ versus $H_2 : S_2 \otimes N$. The two-signal cases are important in many telecommunications applications (Chapter 3 ff.) and when S_1 (or S_2) may represent an interfering or otherwise unwanted signal in radar and sonar environments. It is therefore both useful and instructive to generalize the on–off formalism to include the presence of a signal of class S_2 (in noise) vis-à-vis a signal of Class S_1 (also in noise). Unlike the on–off cases of the preceding sections, where the hypothesis classes are required always to be disjoint [cf. Fig. 1.1b and remarks after Eq. (1.6.3)], there is the additional possibility that the two-signal classes may overlap and thus be nondisjoint. For this latter situation, however, a more sophisticated viewpoint is required [44], as will be seen below.

1.10.1 Disjoint Signal Classes

We consider first the simpler case of disjoint signal classes, where now $\Omega = \Omega_1 + \Omega_2$ and $\Omega_1 \cap \Omega_2$ is empty. In place of (1.6.5), we have

$$\sigma(\mathbf{S}) = p_1 w_1(\mathbf{S}_1) + p_2 w_2(\mathbf{S}_2), \text{ and} \int_\Omega \sigma(\mathbf{S})\mathbf{dS} = 1, \tag{1.10.1}$$

this last as before (cf. remarks after Eq. (1.6.5)), since $p_1 + p_2 = 1$, with p_1 and p_2, respectively, the *a priori* probabilities of a signal of Class 1 (or 2) occurring in the data sample $\mathbf{X}$. Equation (1.6.4) becomes

$$\int_{\Omega_1} w_1(\mathbf{S}_1)\mathbf{dS}_1 = \int_{\Omega_2} w_2(\mathbf{S}_2)\mathbf{dS}_2 = 1, \tag{1.10.2}$$

with w_1, w_2 the pdfs of $\mathbf{S}_1$ and $\mathbf{S}_2$, or their respective random parameters $\boldsymbol{\theta}_1$, $\boldsymbol{\theta}_2$ in $\mathbf{S}_{1,2}(\boldsymbol{\theta}_{1,2})$. Equation (1.6.3) and cost matrix (1.6.6) are modified in an obvious way to

$$\delta(\gamma_1|\mathbf{X}) + \delta(\gamma_2|\mathbf{X}) = 1;\ \mathbf{C}(\mathbf{S},\boldsymbol{\gamma}) = \begin{bmatrix} C_1^{(1)} & C_2^{(1)} \\ C_1^{(2)} & C_2^{(2)} \end{bmatrix},\ \boldsymbol{\gamma} = [\gamma_1, \gamma_2]. \tag{1.10.3}$$

now with $C_1^{(1)} < C_2^{(1)}$; $C_2^{(2)} < C_1^{(2)}$ to ensure again that "failure" is more expensive than "success" cf. (1.6.6a), and where the upper index as before designates the true state of affairs and the lower the associated decision. The average risk (1.6.7) is accordingly modified to

$$\begin{aligned} R(\sigma,\delta) = \int_\Gamma \Big\{ &\Big[p_1 C_1^{(1)} \langle F_J(\mathbf{X}|\mathbf{S}_1)\rangle_1 + p_2 C_1^{(2)} \langle F_J(\mathbf{X}|\mathbf{S}_2)\rangle_2\Big]\delta(\gamma_1|\mathbf{X}) \\ &+ \Big[p_2 C_2^{(1)} \langle F_J(\mathbf{X}|\mathbf{S}_1)\rangle_1 + p_2 C_2^{(2)} \langle F_J(\mathbf{X}|\mathbf{S}_2)\rangle_2\Big]\delta(\gamma_2|\mathbf{X})\Big\}\mathbf{dX} \end{aligned} \tag{1.10.4}$$

in which

$$p_i\langle F_J(\mathbf{X}|\mathbf{S}_i)\rangle_i = \int_\Gamma \sigma(\mathbf{S}_i) F_J(\mathbf{X}|\mathbf{S}_i)\mathbf{dS}_i = p_i \int_{\mathbf{S}_i \text{ or } \boldsymbol{\theta}_i} w_i(\mathbf{S}_i \text{ or } \boldsymbol{\theta}_i) F_J(\mathbf{X}|\mathbf{S}_i)\mathbf{dS}_i(\text{or } \mathbf{d}\boldsymbol{\theta}_i), i = 1, 2, \tag{1.10.4a}$$

for averages over signal waveform $\mathbf{S}_i$, or parameters $\boldsymbol{\theta}_i$ in $\mathbf{S}_i(\boldsymbol{\theta}_i)$.

The error probabilities are similarly modified:

$$\left.\begin{aligned} \alpha \to \beta_2^{(1)} \equiv \beta_2^{(1)}(\gamma_2|H_1) &= \text{conditional probability of incorrectly deciding that a Class 2} \\ &\quad\ \text{signal is present, when actually a Class 1 signal occurs;} \\ \beta \to \beta_1^{(2)} \equiv \beta_1^{(2)}(\gamma_1|H_2) &= \text{the reverse of the above.} \end{aligned}\right\} \tag{1.10.5a}$$

Similarly, the respective conditional probabilities of correct decisions are

$$\beta_2^{(2)} = \beta_2^{(2)}\left(\gamma_2|H_2\right); \quad \beta_1^{(1)} = \beta_1^{(1)}(\gamma_1|H_1), \tag{1.10.5b}$$

with $p_2\beta_2^{(2)}$, $p_1\beta_1^{(1)}$, $p_2\beta_1^{(2)}$, $p_1\beta_2^{(1)}$ the corresponding total probabilities of correct and incorrect decisions.

The average risk takes the compact forms:

$$R(\sigma, \delta) = \left(p_1 C_1^{(1)} + p_2 C_2^{(2)}\right) + p_2\left(C_2^{(1)} - C_1^{(2)}\right)\beta_2^{(1)} + p_2\left(C_1^{(2)} - C_2^{(2)}\right)\beta_1^{(2)}, \text{ or} \tag{1.10.6a}$$

$$R(\sigma, \delta) = \left(p_1 C_2^{(1)} + p_2 C_1^{(2)}\right) - p_1\left(C_2^{(1)} - C_1^{(1)}\right)\beta_1^{(1)} - p_2\left(C_1^{(2)} - C_2^{(2)}\right)\beta_2^{(2)}, \tag{1.10.6b}$$

the former in terms of the *error* probabilities, the latter in terms of the probabilities of correct decision.

1.10.2 Overlapping Hypothesis Classes (F. C. Ogg Jr. [44])

When the signal classes S_1, S_2 are not disjoint but overlap (i.e., $S_1 \cup S_2 \neq 0$), the usual definitions of correct and incorrect decisions are no longer valid, since it is no longer certain whether or not an error has been made. Let us suppose, for example, that signal class S_1 consists of deterministic signals of the type $S(\boldsymbol{\theta}_1)$ and signal class S_2 of the type $S(\boldsymbol{\theta}_2)$, where the waveforms (S) of the two classes are the same and each has the same type of random parameter(s); for example, $\boldsymbol{\theta}_1, \boldsymbol{\theta}_2 = \boldsymbol{\theta}$ in both instances represent a common set of random parameters, but with different distribution densities, $w_1(\boldsymbol{\theta}) \neq w_2(\boldsymbol{\theta})$. Any given signal S may belong to either signal class, but S will usually belong to one class with greater probability than to the other. It is reasonable to assign to the more probable decision a lesser cost. Thus, if $\boldsymbol{\theta} = a$ represents a random amplitude, for instance, and if the amplitude a of a particular S lies close to the mean value of $w_1(a)$ but well out on the "tail" of $w_2(a)$, a larger value is assigned to the loss function $\mathsf{F}(S, \gamma_2)$ than to the loss function $\mathsf{F}(S, \gamma_1)$ for the more probably correct decision.

Accordingly, it is clear that the cost assignment should be related to the probability that the signal belongs to *each* of the classes. This can be accomplished in a variety of ways, but the simplest is to require specifically that (1) $\mathsf{F}(S, \gamma)$ be continuous in the prior probabilities (p_1, p_2, w_1, w_2) and (2) that $F(S, \gamma)$ reduce to the usual cost assignments whenever the signal belongs to a disjoint signal class ($S_1 \cap S_2 = 0$). For the systems considered here, based on constant preset costs, an *extension* of the "constant" cost function F_1, Eq. (1.4.3), satisfying these conditions is [44]

$$C(S,\gamma_i) = \left[C_1^{(i)} p_1 w_1(\boldsymbol{\theta}) + C_2^{(i)} p_2 w_2(\boldsymbol{\theta})\right] / \sigma(\boldsymbol{\theta}) \qquad i = 1,\, 2 \tag{1.10.7}$$

where $\sigma(\boldsymbol{\theta}) = p_1 w_1(\boldsymbol{\theta}) + p_2 w_2(\boldsymbol{\theta})$ is the prior of $\boldsymbol{\theta}$ for deterministic signals $S(\boldsymbol{\theta})$. In general, we have for signal waveforms

$$C(S,\gamma_i) = \left[C_1^{(i)} p_1 w_1(\mathbf{S}) + C_2^{(i)} p_2 w_2(\mathbf{S})\right] / \sigma(\mathbf{S}) \tag{1.10.8}$$

with $\sigma(\mathbf{S})$ given by Eq. (1.10.1). Thus, by a similar argument Eq. (1.10.8) applies for the case of completely stochastic signals $\mathbf{S}$, where now $w_1(\mathbf{S}) \neq w_2(\mathbf{S})$. With Eq. (1.10.7) and (1.10.8) the average risk $R(\sigma,\delta)$ reduces to the original expression (1.10.4) for disjoint classes. Overlapping classes that involve the null signal ($\mathbf{S}=$noise alone) are handled in the same way, now with

$$C(\mathbf{S},\gamma_i) = \left[C_0^{(i)} q w_0(\mathbf{S}) + C_1^{(i)} p w_1(\mathbf{S})\right] / \sigma(\mathbf{S}) \qquad i = 0, 1, \tag{1.10.9}$$

where $\sigma(\mathbf{S})$ is given by (1.6.5) and $R(\sigma,\delta)$ by (1.6.7), and so on. In this way we unite the treatment of overlapping and nonoverlapping signal classes, employing the formalism of the latter as before but now including all the signal types of practical interest.

Let us next calculate the average risk, based on (1.6.7), and determine the minimum average (i.e., Bayes) risk. We observe first that the average risk (1.6.7) must first contain the component exhibited in (1.6.7), here for the two original cases obeying (1.10.7), namely,

$$\begin{aligned} R(\sigma,\delta) = \int_\Gamma &\left[\left\{p_1 C_1^{(1)} \left\langle F_J\left(\mathbf{X}|\mathbf{S}^{(1)}\right)\right\rangle + (\,)\right\} + \left\{p_2 C_2^{(2)} \left\langle F_J\left(\mathbf{X}|\mathbf{S}^{(2)}\right)\right\rangle + (\,)\right\}\right] \delta(\gamma_1|\mathbf{X})\mathbf{dX} \\ &+ \left[\left\{p_1 C_2^{(1)} \left\langle F_J\left(\mathbf{X}|\mathbf{S}^{(1)}\right)\right\rangle + (\,)\right\} + \left\{p_2 C_2^{(2)} \left\langle F_J\left(\mathbf{X}|\mathbf{S}^{(2)}\right)\right\rangle + (\,)\right\}\right] \delta(\gamma_2|\mathbf{X})\mathbf{dX}, \end{aligned} \tag{1.10.10a}$$

where the components of the disjoint (i.e., nonoverlapping) component are explicitly given. The quantities $\langle F_J(\mathbf{X}|\mathbf{S}^{(i)})\rangle = \int_\Omega w_i(\boldsymbol{\theta}) F_J(\mathbf{X}|\mathbf{S}^{(i)}(\boldsymbol{\theta}))\mathbf{d}\boldsymbol{\theta}$, $i = 1,\, 2$, here. The overlap contributions are seen to be from (1.10.7):

$$\left.\begin{aligned} &\text{coefficient of } \delta(\gamma_1|\mathbf{X}) : \; C_2^{(1)} p_2 \left\langle F_J(\mathbf{X}|\mathbf{S}^{(2)})\right\rangle, \;\; C_2^{(1)} p_1 \left\langle F_J(\mathbf{X}|\mathbf{S}^{(1)})\right\rangle \\ &\text{coefficient of } \delta(\gamma_2|\mathbf{X}) : \; C_1^{(2)} p_2 \left\langle F_J(\mathbf{X}|\mathbf{S}^{(2)})\right\rangle, \;\; C_1^{(2)} p_1 \left\langle F_J(\mathbf{X}|\mathbf{S}^{(1)})\right\rangle \end{aligned}\right\}. \tag{1.10.10b}$$

Using the relation $\delta(\gamma_1|\mathbf{X}) = 1 - \delta(\gamma_2|\mathbf{X})$: a decision is always made, and dividing (and multiplying) each term in (1.10.10a) by $qF(\mathbf{X}|\mathbf{0})$, we obtain

$$\begin{aligned} R(\sigma,\delta) = &\int_\Gamma qF_J(\mathbf{X}|\mathbf{0}) \left[C_1^{(1)}\Lambda^{(1)} + C_2^{(1)}\Lambda^{(2)} + C_1^{(2)}\Lambda^{(2)} + C_2^{(1)}\Lambda^{(1)}\right]\mathbf{dX} \\ &+ \int_\Gamma qF_J(\mathbf{X}|\mathbf{0}) \left[(C_2^{(1)}\Lambda^{(1)} + C_1^{(2)}\Lambda^{(2)} + C_2^{(2)}\Lambda^{(2)} + C_1^{(2)}\Lambda^{(1)}) \right. \\ &\left. - (C_1^{(1)}\Lambda^{(1)} + C_2^{(1)}\Lambda^{(2)} + C_1^{(2)}\Lambda^{(2)} + C_2^{(1)}\Lambda^{(1)})\right] \cdot \delta(\gamma_2|\mathbf{X})\mathbf{dX}, \end{aligned} \tag{1.10.11a}$$

where $\Lambda^{(1)} = p_1 \int_\Omega F_J(\mathbf{X}|\mathbf{S}^{(1)}) w_1(\boldsymbol{\theta}) \mathbf{d}\boldsymbol{\theta}$, $\Lambda^{(2)} = p_2 \int_\Omega F_J(\mathbf{X}|\mathbf{S}^{(2)}) w_2(\boldsymbol{\theta}) \mathbf{d}\boldsymbol{\theta}$. The first term of (1.10.11a) reduces to the irreducible risk

$$\mathsf{R}_{02} \equiv p_1 C_1^{(1)} + p_2 C_2^{(1)} + p_2 C_1^{(2)} + p_1 C_2^{(1)} = p_1\left(C_1^{(1)} + C_2^{(1)}\right) + p_2\left(C_2^{(1)} + C_1^{(2)}\right) > 0, \tag{1.10.11b}$$

Since $F_J(\mathbf{X}|\mathbf{0}) \geq 0$, the second term is clearly minimized when $\delta(\gamma_2|\mathbf{X}) = 1$, that is a (nonrandom) decision is made that signal $\mathbf{S}^{(2)}$ is present, where the expression in [ ] is set equal to zero. This latter gives us the result (collecting $\Lambda^{(2)}$s and $\Lambda^{(1)}$s)

$$\Lambda^{(2)}\left(C_1^{(2)} + C_2^{(2)}\right) + \Lambda^{(1)}\left(C_2^{(1)} + C_1^{(2)}\right) \leq \Lambda^{(1)}\left(C_1^{(1)} + C_2^{(1)}\right) + \Lambda^{(2)}\left(C_2^{(1)} + C_1^{(2)}\right),$$
$$\therefore \Lambda^{(2)}\left(C_2^{(2)} - C_2^{(1)}\right) \leq \Lambda^{(1)}\left(C_1^{(1)} - C_1^{(2)}\right),$$

and since "failure" is more expensive than "success," that is, $C_2^{(1)} - C_2^{(2)}, C_1^{(2)} - C_1^{(1)} > 0$, we have finally

$$\delta(\gamma_2|\mathbf{X}) = 1, \ \therefore \delta(\gamma_1|\mathbf{X}) = 0, \text{ if } \quad \Lambda^{(2)} > \left(\frac{C_1^{(2)} - C_1^{(1)}}{C_2^{(1)} - C_2^{(2)}}\right)\Lambda^{(1)} \text{or } \Lambda^{(2)} \geq K_{12}\Lambda^{(1)}; \ K_{12} > 0 \tag{1.10.12}$$

For the decisions $\delta(\gamma_1|\mathbf{X}) = 1$, we have $\delta(\gamma_2|0) = 0$ and $\Lambda^{(2)} \leq K_{12}\Lambda^{(1)}$. It is to be noted that the effects of overlap Eq. (1.10.10b) leave the decision process unchanged: they are the same as for the nonoverlapping cases. This is a direct consequence of the choice of cost function (1.10.7), which now unites the treatment of both types of signal class (overlapping and nonoverlapping), as stated above. We observe, however, that the irreducible risk R_{02} (1.10.11b) contains four terms rather than two $\left(p_1 C_1^{(1)} + p_2 C_1^{(2)}\right)$. Figure 1.14 shows the decision regions for (1.10.12), in logarithm forms, that is, $\log\Lambda_2$ versus $\log\Lambda_1 + \log K_{12}$.

The case of overlapping classes involving the null signal (1.10.9) follows at once. Setting 2 equal to 1, and 1 equal to 0 in the above gives the result

$$\left.\begin{aligned} &\text{decide } \delta(\gamma_1|\mathbf{X}) = 1 : \Lambda^{(1)} \geq \left(\frac{C_0^{(1)} - C_0^{(0)}}{C_1^{(0)} - C_1^{(1)}}\right), \text{ or } \Lambda^{(2)} \geq K_{01}, K_{01} > 0 \\ &\quad \left(\text{and } \delta(\gamma_0|\mathbf{X}) = 0\right) \\ &\text{decide } \delta(\gamma_0|\mathbf{X}) = 1 : \ \Lambda^{(1)} < K_{01} \\ &\quad \left(\text{and } \delta(\gamma_1|\mathbf{X}) = 0\right) \end{aligned}\right\}. \tag{1.10.13}$$

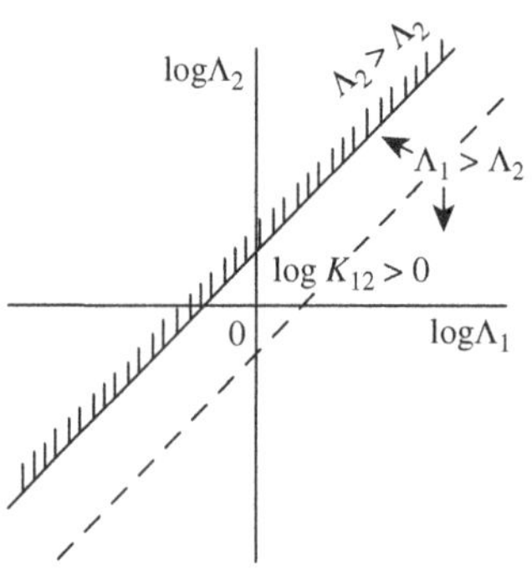

FIGURE 1.14 Decision regions for Λ_1 and Λ_2, for $K_{12} > 1$; for $0 < K_{12} \leq 1$ the boundary lies below the dotted line.

The irreducible risk, however, remains unchanged, namely, (1.10.11b). For stochastic signals (**S**), **S** replaces the deterministic **S(θ)** in (1.10.10a)–(1.10.13). Finally, we observe from Section 4.2 following that these results are extendable to the (K+1)-ary or K-ary signal classes: the decision process remains the disjoint result, but the irreducible risks and Bayes risks are themselves different.

The error probabilities, average, and Bayes risks, are obtained here from (1.6.8) and Section 1.6.2 generally. The results are

$$\beta^{(1)} = \int_{\Gamma} \left\langle F(\mathbf{X}|\mathbf{S}^{(1)})\right\rangle \delta(\gamma_2|\mathbf{X})\mathbf{dX}; \quad \beta^{(2)} = \int_{\Gamma} \left\langle F(\mathbf{X}|\mathbf{S}^{(2)})\right\rangle \delta(\gamma_1|\mathbf{X})\mathbf{dX} \tag{1.10.14}$$

$$\begin{aligned} R(\sigma,\delta) &= p_1\left(C_1^{(1)}+C_2^{(1)}\right)\left(1-\beta^{(1)}\right)+p_2\left(C_2^{(1)}+C_1^{(2)}\right)\beta^{(2)} \\ &\quad + p_1\left(C_2^{(1)}+C_1^{(2)}\right)\beta^{(1)}+p_2\left(C_1^{(2)}+C_2^{(2)}\right)\left(1-\beta^{(2)}\right) \\ &= p_1\left(C_1^{(2)}-C_1^{(1)}\right)\beta^{(1)}+p_2\left(C_2^{(1)}-C_2^{(2)}\right)\beta^{(2)}+R_{02} \end{aligned} \tag{1.10.15a}$$

and[47] for optimality, we have

$$R(\sigma,\delta)^* = p_1\left(C_1^{(2)}-C_1^{(1)}\right)\beta^{(1*)}+p_2\left(C_2^{(1)}-C_2^{(2)}\right)\beta^{(2*)}+R_{02}. \tag{1.10.15b}$$

The evaluation of $\beta^{(1^*)}$, $\beta^{(2^*)}$ associated with the Bayes risk, and in the suboptimum cases, is formally accomplished in Section 1.9. In the case of the null signal, we have

$$R(\sigma,\delta)^* = p_0\left(C_0^{(1)}-C_0^{(0)}\right)\alpha^*+p_1\left(C_1^{(0)}-C_1^{(1)}\right)\beta^*+R_{01} \tag{1.10.16}$$

with $(\alpha^*,\beta^*) \to (\alpha,\beta)$ in the average (nonoptimal) risk.

1.11 CONCLUDING REMARKS

In this first chapter we have obtained some of the principal concepts and techniques used in SCT, based on the fundamental viewpoint of a Bayesian statistical decision theory developed mainly in the mid-twentieth century. A concise topical description of the major elements of both may be gleaned from Sections 1.1–1.10 above, which in turn contain the guiding principles and definitions as well as generic examples. Their implementation is one of the principal aims of the present book, along with the extension to random space-time *fields*. Another is the general use of discrete sampling methods, in conjunction with the physical world of four-dimension, namely space and time, as distinct from earlier analyses devoted to stochastic time processes alone. Thus, to summarize briefly, we employ Sections 1.1–1.5 to provide the formal structure of SDT, basically, a concise description of the fundamental concepts involved. Sections 1.6–1.8 are an illustrative introduction to binary detection and a variety of optimization procedures, with the extension in Section 1.10 to a two-signal binary formulation, in which disjoint and overlapping signal classes are treated.

Optimality, and its approximation, is another goal of the analysis, in conjunction with the Bayesian philosophy used here, with possible constraints imposed by system demands and always subject to the specifics of the physical environment. As will be seen (in Chapters 8, 9

[47] Here we require "failure" always to be more expensive than "success," so that $C_1^{(2)} > C_1^{(1)}, C_2^{(1)} > C_2^{(2)}$.

particularly), the propagation physics needs to be specifically introduced, as it is in many ways a major controlling factor in successful operation. Thus, the physics of the channel in pertinent detail is required. The convenient "black box" approach of additive Gaussian and (often) deterministic interference, with *ad hoc* statistics, loosely based on a postulated random process, in many cases does not represent a full or realistic model of the environment. In Chapter 9, we shall quantitatively describe *non-Gaussain noise* (fields and processes) based on the underlying noise mechanisms, with attention to their spatial as well as their temporal properties. Here our aim is to provide probability distributions (or densities), not just the lower order moments.

In Chapter 2 following, we shall begin this journey from generality to statistical detail by considering first the space–time covariance of a noise field and various conditions that determine its properties. Other relations then follow, in particular the four-dimensional Wiener–Khintchin (W–Kh) theorem.

REFERENCES

1. D. Middleton, *An Introduction to Statistical Communication Theory*, McGraw-Hill, New York, 1960; IEEE Press, Classic Reissue Piscataway, NJ. Chapter 18, 1996. (See also Ref. [19] for other editions.).
2. C. E. Shannon, A Mathematical Theory of Communication, *Bell System Tech. J.*, **27**, 379, 623, 1948.
3. M. G. Kendall, *The Advanced Theory of Statistics*, 5th ed., Vol. II, Griffin, London, 1952.
4. H. Cramér, *Mathematical Methods of Statistics*, Princeton University Press, Princeton, NJ, 1946.
5. R. D. Luce and H. Raiffa, *Games and Decisions*, John Wiley & Sons, Inc., New York, 1957.
6. D. Middleton and D. Van Meter, Detection and Extraction of Signals in Noise from the Point of View of Statistical Decision Theory, *J. Soc. Ind. Appl. Math.*, **3**, 192–253, 1955; **4**, 86–149, 1956.
7. D. Middleton and D. Van Meter, On Optimum Multiple Alternative Detection of Signals in Noise, *IRE Trans. Inform. Theory*, **IT-1**, 1–9, 1955.
8. A. Wald, *Sequential Analysis*, John Wiley & Sons, Inc., New York, 1947.
9. J. Bussgang and D. Middleton, Optimum Sequential Detection of Signals in Noise, *IRE Trans. Inform. Theory*, **IT-1**, 5, 1955.
10. J. L. Doob, *Stochastic Processes*, Chapters 10 and 11, John Wiley & Sons, Inc., New York, 1953.
11. A. Blanc-Lapierre and R. Fortet, *Théorie des Fonctions Aléatoires*, Masson et Cie, Paris, 1953.
12. M. C. Wang and G. E. Uhlenbeck, On the Theory of the Brownian Motion. II, *Rev. Modern Phys.*, **17**, 323, 1945.
13. D. Middleton, Threshold Detection in Correlated Non-Gaussian Noise Fields, *IEEE Trans. Inform. Theory*, **41**(4), 976–1000, 1995.
14. C. W. Helstrom, *Elements of Signal Detection and Estimation*, Prentice-Hall, Englewood Cliffs, NJ, 1995.
15. J. L. Hodges and E. L. Lehmann, Some Problems in Minimax Point Estimation, *Am. Math. Stat.*, **21**, 182, 1950.
16. W. Hoeffding, Optimum Non-parametric Tests, *Proceedings of the 2d Berkeley Symposium on Mathematical Statistics and Probability*, 1950, p. 83.
17. E. L. Lehmann and C. Stein, On the Theory of Some Nonparametric Hypotheses, *Am. Math. Stat.*, **20**, 28, 1949.

18. J. D. Gibson and J. L. Melsa, *Introduction to Nonparametric Detection with Applications*, Academic Press, New York, 1975; reprinted, IEEE Press, Piscataway, NJ, 1996.
19. D. Middleton, *An Introduction to Statistical Communication Theory*, 1st ed., McGraw-Hill, New York, 1960–1972, 2nd Ed., Peninsula Publishing Co., Los Altos, CA, 1987–1996.
20. D. Middleton, *Topics in Communication Theory*, McGraw-Hill, New York, 1965–1972. Peninsula Publishing, Los Altos, CA, 1987.
21. H. V. Poor, *An Introduction to Signal Processing and Estimation*, Springer-Verlag, New York, 1988.
22. H. Blasbalg, The Relationship of Sequential Filter Theory to Information Theory and Its Application to the Detection of Signal in Noise by Bernoulli Trials, *IRE Trans. Inform. Theory* **IT-3**, 122–131, 1957.
23. M. Basseville and I. V. Nikiferov, *Detection of Abrupt Changes*, Prentice Hall, Englewood Cliffs, NJ, 1993.
24. N. W. McLechdaln, *Complex Variable and Operational Calculus*, Cambridge University Press, London and New York, 1939.
25. (a) A. Wald, *Statistical Decision Functions*, John Wiley & Sons, Inc., New York, 1950; (b) A. Wald, Basic Ideas of a General Theory of Statistical Decision Rules, *Proceedings of the International Congress of Mathmaticians*, 1950, p. 231.
26. P. M. Woodward, *Probability and Information Theory with Applications to Radar*, Chapters 4 and 5, McGraw-Hill, New York, 1953.
27. J. Von Neumann, Zur Theorie der Gesellschaftsspiele, *Math. Ann.*, **100**, 295, 1928.
28. J. Von Neumann and O. Morgenstern, *Theory of Games and Economic Behavior*, 3rd ed., Princeton University Press, Princeton, NJ, 1953.
29. R. Radner and J. Marschak, Note on Some Proposed Decision Criteria, in R. M. Thrall, C. H. Coombs,and R. L. Davis (Eds.), *Decision Processes,* Section V, John Wiley & Sons Inc., New York, 1954.
30. J. L. Hodges and E. L. Lehmann, The Use of Previous Experience in Reaching Statistical Decisions, *Ann. Math. Stat.*, **23**, 296, 1952.
31. J. Kiefer, On Wald's Complete Class Theorem," *Ann. Math. Stat.*, **24**, 70, 1953.
32. A. Dvoretzky, A. Wald, and J. Wolfowitz, Elimination of Randomization in Certain Statistical Decision Procedures and Zero-sum Two Person Games, *Ann. Math. Stat.*, **22**, 1, 1951.
33. E. T. Jaynes, in R. D. Rosenkranz (Ed.) *Papers on Probability, Statistics and Statistical Physics, D. Reidel*, Dordrecht, Holland, 1983.
34. J. O. Berger, *Statistical Decision Theory and Bayesian Analysis*, 2nd ed., Springer-Verlag, New York, 1985.
35. W. H. Jeffreys and J. O. Berger, Ockham's Razor and Bayesian Analysis, *Am. Scientist*, **80**, 64–73, 1992.
36. D. Middleton, *Topics in Communication Theory*, McGraw-Hill, New York, 1965.
37. C. W. Helstrom, *Statistical Theory of Signal Detection*, 2nd ed. Pergamon Press, New York, 1968 (also Ref. [14] above).
38. W. B. Davenport and W. L. Root, *An Introduction to the Theory of Random Signals and Noise,* McGraw-Hill, New York, 1958.
39. S. S. Wilks, *Mathematical Statistics*, John Wiley & Sons, Inc., New York, 1962.
40. A. J. F. Siegert, Preface and Section 7.5 of J. L. Lawson and G. E. Uhtenbech, *Threshold Signals*, MIT Radiation Laboratory Series, Vol. **24**, McGraw-Hill, New York, 1950.
41. G. A. Campbell and R. M. Foster, *Fourier Integrals for Practical Applications*, 3rd ed., D. Van Nostrand, New York, 1951.

42. J. T. Cushing, *Applied Analytical Mathematics for Physical Scientists,* John Wiley & Sons, New York, 1975.

43. I. S. Gradshteyn and I. H. Rizhik, *Tables of Integrals*, Series and Products, Corrected and Enlarged Edition (Ed. Alan Jeffrey), Academic Press, New York, 1980.

44. F. C. Ogg, Jr., A Note on Bayes Detection of Signals, *IRE Trans. Inform. Theory*, **IT-10**, 57–60, 1964.

45. V. A. Kotelnikov, *The Theory of Optimum Noise Immunity* (Translated from Russian by R.A. Silverman), McGraw-Hill Book Co, NY, 1959.

46. M. Bykhovskiy, *Pioneers of the Information Era*, Technosphera, Moscow, 2006, p. 19.

2

SPACE–TIME COVARIANCES AND WAVE NUMBER FREQUENCY SPECTRA: I. NOISE AND SIGNALS WITH CONTINUOUS AND DISCRETE SAMPLING

The purpose of Chapter 2 is to introduce directly and in more detail the various *statistical elements* of the generic communication process. This is principally a channel phenomenon, determined by the physical properties of the medium through which propagation of signals and noise is effected. In dealing with both noise and signal, we must account for their statistical elements in signals with random parameters, for interference and noise fields, which are principally random. For our initial discussion, we shall assume that the noise process and noise fields are purely random (i.e., have no deterministic component). More complex environments can then be readily introduced, with "memory," as we shall see in Chapters 8–9.

We begin with the random space–time field and specifically with its fundamental statistic, the space–time covariance, which includes its wave number frequency (WNF) transform, the intensity spectrum. This dual relationship is well known as the Wiener–Khintchine (W–Kh) theorem (Section 3.2 [1a] of Ref. [1]) in its familiar temporal frequency form for random processes. Since we are dealing with fields throughout, we must comply with its extensions to four dimensions, for example, space–time and its transmission to wave number frequency. (Yaglom [2] gives an extended rigorous mathematical treatments in small doses; we shall consider the discrete four-dimensional case (Sections 2.1 and 2.2).)

In Section 2.1, inhomogeneous and nonstationary covariances are specifically considered. In Section 2.2, we provide an analysis of the intensity spectrum and its associated

Non-Gaussian Statistical Communication Theory, David Middleton.

covariance function for inhomogeneous, nonstationary (i.e., non-Hom-Stat) random fields, as well as for the more commonly assumed Hom-Stat and isotropic Hom-Stat cases. In particular, this includes the Wiener–Khintchine relations and their extensions to the more general non-Hom-Stat situations ([1] Section 3.2.3,[3], [4]), which are important for detection and estimation involving both Gaussian and non-Gaussian noise. We conclude in Section 2.5 with an introductory discussion of apertures and arrays, followed by a brief summary of the main results of the chapter in Section 2.6.

We remind the reader that the space–time formulation presented here (and throughout) involves for the most part discrete sampling, with sample numbers as described in Section 1.3.1.

2.1 INHOMOGENEOUS AND NONSTATIONARY SIGNAL AND NOISE FIELDS I: WAVEFORMS, BEAM THEORY, COVARIANCES, AND INTENSITY SPECTRA

Before we can obtain explicit results, namely, optimum detection and extraction algorithms and their associated performance, we must construct relevant models of the received noise fields involved, structures that necessarily embody their physical content. For example, signals can be deterministic or random, broadband or narrowband. Noise fields are usually nonstationary and inhomogeneous, signal fields the same. Both have space–time structures, although in reception the spatial characteristics are often implicitly designated simply by an index (m), designating a point in the received field. In general, of course, many noise fields are non-Gaussian as are most of the often accompanying signal fields. For the present discussion, we consider scalar fields $X(\mathbf{R},t)$ only, unless otherwise indicated.

In addition, considerable space here is devoted to the *covariance*, and subsequently in Section 2.2 to its associated intensity spectrum, because of its importance in detection and estimation. Not only is this the case for many systems where Gaussian noise is present but also for the often more important situations where non-Gaussian noise is the dominant interference, particularly in critical threshold signal regimes. For the most part, we shall consider the received (and transmitted) fields, and their statistics, consisting of discretely sampled quantities, as a consequence of the space–time sampling procedure described in the beginning of Section 1.3.1.

We begin accordingly (cf. Section 1.3.1, Eq. (1.3.1), and Section 1.6.1) by treating the (real) scalar fields in question as values at specific discrete "sample points" $X(\mathbf{r}_m, t_n) = X_j$. Thus, for the resulting space–time samples, we use the double index $j = mn$ to designate these sample points: here $m = 1, 2, \ldots, M$ and $n = 1, 2, \ldots, N$ are respectively space and time index numbers. Accordingly, $J = MN$ represents the totality of these (nonoverlapping) points in a physical continuum, that is, an acoustic field, or a component, usually the dominant one, of an electromagnetic field, for example; $\mathbf{X} = [X_j]$ denotes a sample data vector of such a received field.[1] Similarly, $\mathbf{X}\tilde{\mathbf{X}} = [X_j X_{j'}] = [X_{j'} X_j]$ is a (real) symmetric (square) matrix of such sampled field data, and $\langle \mathbf{X}\tilde{\mathbf{X}} \rangle = [K_X(\mathbf{R}_m, t_n; \mathbf{R}_{m'}, t_{n'})]$ with $\langle \mathbf{X} \rangle = 0$, is the covariance (matrix) of this field.. When the (sampled) field is *homogeneous and stationary* — in more complete terminology, *wide sense homogeneous*

[1] This is the spatially sampled output from the receiving (point) sensor at $P(\mathbf{R}_m, t_n)$, namely, $X = \hat{\mathbf{R}}\alpha$, where $\alpha(\mathbf{R}, t)$ is the external field in which the receiver is embedded. For more detail, see (1.6.2a)–(1.6.2b).

and wide sense stationary (WS-HS), applicable only to this covariance or second moment state generally — K_X depends only on the difference of the coordinates, that is, $\mathbf{R}_{m'} - \mathbf{R}_m$ and $t_{n'} - t_n$, so that we can accordingly write

$$\langle X_j X_{j'} \rangle = K_X(\mathbf{R}_m, t_n; \mathbf{R}_{m'}, t_{n'}) = K_X(\mathbf{R}_{m'} - \mathbf{R}_m, t_{n'} - t_n) = K_X(\Delta\mathbf{R}, \tau), \ \langle X_j \rangle = 0, \tag{2.1.1}$$

where $\Delta\mathbf{R} = \mathbf{R}_{m'} - \mathbf{R}_m$ and $\tau = t_{n'} - t_n$, with the added properties

$$K_X(\Delta\mathbf{R}, \tau) = K_X(-\Delta\mathbf{R}, -\tau), \quad (\text{but } K_X(\pm\Delta\mathbf{R}, \mp\tau) \neq K_X(\Delta\mathbf{R}, \tau)), \tag{2.1.1a}$$

as a consequence of its WS-HS nature. It is also easily shown that $K_X(\mathbf{0}, \mathbf{0}) \geq |K_X(\Delta\mathbf{R}, \tau)|$.

For an *isotropic* random field X, a stricter condition on the coordinates applies, expressing the fact that now the (statistical) properties of the random field are independent of spatial and temporal *directions*. The covariance here becomes

$$K_X(|\Delta\mathbf{R}|, |\tau|) = K_X(|\mathbf{R}_{m'} - \mathbf{R}_m|, |t_2 - t_1|). \tag{2.1.1b}$$

Not any function of the type (2.1.1, 2.1.1b) can be a covariance function, except those K_X that have a unique Fourier transform that is everywhere positive and vanishes sufficiently rapidly at infinity, with further special conditions including the reality of K_X (from the assumed reality of X). (For further details, see the beginning pages of Sections 21 and 22 of Yaglom [2], which also includes the general theory of multidimensional scalar and vector field covariances and their associated spectra, that is, generalizations of the Wiener–Khintchine theorem in one dimension (see Section 3.2.2 of Ref. [1], especially Ref. [1a].)

2.1.1 Signal Normalization

Let us consider a real signal field $\alpha(\mathbf{R}, t)_S$, sampled at various (point) sensors of the receiving array, located at $\mathbf{R}_m, \mathbf{R}_{m'}$, $m, m' = 1, \ldots, M$ (Fig. 2.1) and at time t_n in the interval $(t_0 \leq t_n \leq t_0 + T)$, $n = 1, \ldots, N$. Thus, $\mathbf{S} = [S_{j=mn}]$ is the *received* space–time signal vector after sampling. Accordingly, the received signal is canonically represented by

$$\mathbf{S} = \hat{\mathbf{R}}\alpha(\mathbf{R}, t_n)_S = [\hat{R}_m \alpha(\mathbf{R}, t_n)_S] = [S_{j=mn}] \equiv \left[A_{mn}^{(m)} s_n^{(m)} / \sqrt{2}\right], \tag{2.1.2}$$

where in more detail

$$\text{Scale}: A_{0n}^{(m)} = A_0^{(m)}\left(t_n - \varepsilon; \boldsymbol{\theta}^{(m)}\right); \quad \text{waveform}: s_n^{(m)} = s^{(m)}\left(t_n - \varepsilon; \boldsymbol{\theta}^{(m)}\right); \quad \left\langle s^{(m)^2} \right\rangle = 1. \tag{2.1.2a}$$

The $\boldsymbol{\theta}^{(m)}$ are possible (random) parameters associated with the sensors' outputs; ε is an epoch, relating the received data interval (T) to some point in time on the input signal (Fig. 2.1). Here, $s_n^{(m)}$ is constrained by the ensemble average condition $\langle s^{(m)^2} \rangle \equiv 1$, where now we define $s_n^{(m)}$ by $s^{(m)} \equiv \hat{\alpha}_n^{(m)} f^{(m)}(t_n)$, on t_n in $t_0 \leq t_n \leq T + t_0$. Thus, the scale $\hat{\alpha}_n^{(m)}$ of $f^{(m)}(t_n)$ is determined by the constraining condition $\langle s_n^{(m)^2} \rangle$, namely,

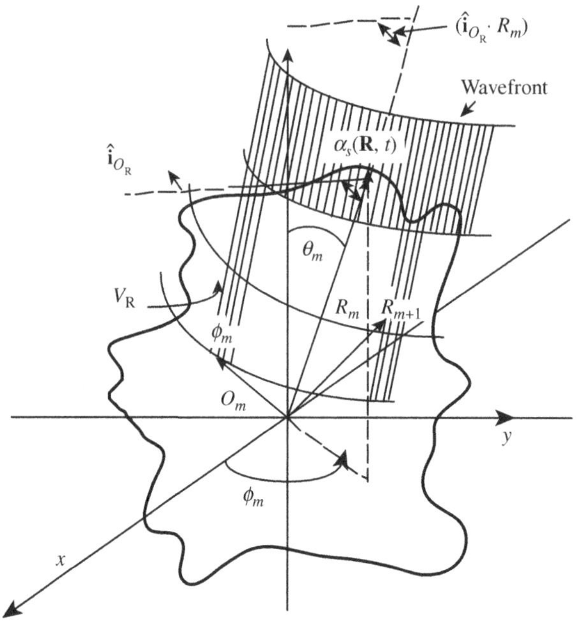

FIGURE 2.1 Received signal wave front sampled at points $P(\mathbf{R}_m, t_n)$ in space–time by an (point) array in V_R.

$\hat{\alpha}_n^{(m)} = \left(\left\langle f^{(m)}(t_n)^2\right\rangle\right)^{-1/2}$, all (m, n). In this way, the element of "scale" is $s_n^{(m)}$ transferred to $A_n^{(m)}$. Moreover, $A_{0n}^{(m)}$ and $s_n^{(m)}$ may be statistically connected, for example, $\left\langle A_{0n}^{(m)} s_n^{(m)}\right\rangle \neq \left\langle A_n^{(m)}\right\rangle\left\langle s_n^{(m)}\right\rangle$, although we can usually safely require that $A_n^{(m)}$ and $s_n^{(m)}$ be statistically independent.

Equation (2.1.2) is readily extended to *narrowband signals* by writing

$$s_j|_{\text{nb}} = s_n^{(m)}|_{\text{nb}} = \sqrt{2}\cos\left[\omega_0(t_n - \varepsilon) - \phi_n^{(m)}\right], \tag{2.1.3a}$$

and the full signal waveform is accordingly

$$\mathbf{S}|_{\text{nb}} = \left[a_{0n}^{(m)} s_n^{(m)} \sqrt{\psi_j}\right] = \left[\frac{A_0^{(m)}\left(t_n - \varepsilon; \boldsymbol{\theta}^{(m)}\right)}{\sqrt{2}} \cdot \sqrt{2}\left[\cos\omega_0(t_n - \varepsilon) - \phi_n^{(m)}\right]\right]. \tag{2.1.3b}$$

When the signal field is homogeneous and the sensor gains are equal, then $A_{0n}^{(m)} = A_{0n}$, $\phi_n^{(m)} = \phi_n$.

2.1.2 Inhomogeneous Nonstationary (*Non-WS-HS*) Noise Covariances

We consider next the case where the (real) noise field is both nonstationary and inhomogeneous (*non-WS-HS*). We have from the definition of the (now) space–time covariance the following $(J \times J)$ matrix for the *outputs* of the m, m' sensors at times $t_n, t_{n'}$:

$$\mathbf{K}_X \equiv \left[\left\langle\left(X_j - \langle X_j\rangle\right)\left(X_{j'} - \langle X_{j'}\rangle\right)\right\rangle\right] = \left[\langle Y_j Y_{j'}\rangle\right],\ Y_{j,j'} = X_j - \langle X_j\rangle \text{ and so on}, \quad (2.1.4)$$

with the variances

$$\sigma_{X_j}^2 = \overline{\left(X_j - \bar{X}_j\right)^2}, \quad \sigma_{X_{j'}}^2 = \overline{\left(X_{j'} - \bar{X}_{j'}\right)^2}, \quad \text{or} \quad \sigma_{Y_{j,j'}}^2 = \overline{Y_j Y_{j'}}. \quad (2.1.4a)$$

Here, $\mathbf{X} = [X_j] = [X(\mathbf{r}_m, t_n)]$ represents the sensor outputs at time t_n and position n, the result of sampling the input field $\alpha(\mathbf{r}, t)$. The *scale-normalized form* of $\mathbf{K}_X$ is given by

$$\mathbf{k}_x = \left[\overline{x_j x_{j'}} - \bar{x}_j \bar{x}_{j'}\right] = \left[\left\langle\left(X_j - \langle X_j\rangle\right)\left(X_{j'} - \langle X_{j'}\rangle\right)\right\rangle \Big/ \sigma_{X_j}\sigma_{X_{j'}}\right]. \quad (2.1.5)$$

When $\overline{x_j} = \overline{x_{j'}} = 0$, the more frequent situation in most applications, particularly for narrowband systems (refer to Sections 2.1.3 and 2.1.4), Eq. (2.1.5) reduces to

$$\mathbf{k}_x = \left[\overline{x_j x_{j'}}\right] = \left[\overline{X_j X_{j'}} \Big/ \sqrt{\psi_j \psi_{j'}}\right]; \quad \overline{x_j} = \overline{x_{j'}} = 0 \quad \text{with} \quad x_j = X_j \Big/ \sqrt{\psi_j} \text{ and so on.} \quad (2.1.6)$$

Even when the covariance $\mathbf{k}_x$ is scale normalized, it is not necessarily homogeneous and stationary, that is, $[\mathbf{k}_x(\mathbf{r}_m, t_n; \mathbf{r}_{m'}, t_{n'})] \neq [\mathbf{k}_x(\mathbf{r}_{m'} - \mathbf{r}_m, t_{n'} - t_n)]$, although all main diagonal terms are unity because of the normalization (2.1.6). We also observe that

$$\frac{K_X(\mathbf{r}_m, t_n; \mathbf{r}_{m'}, t_{n'})}{\left(\overline{X_{mn}^2}\,\overline{X_{m'n'}^2}\right)} = k_x(\mathbf{r}_m, t_n; \mathbf{r}_{m'}, t_{n'}) \quad \text{and} \quad \therefore \quad k(\mathbf{r}_m, t_n; \mathbf{r}'_m, t'_n) = 1. \quad (2.1.6a)$$

If $\mathbf{K}_X$ is Hom-Stat, it should also be noted that $\left|\mathbf{K}_X\right| \leq K_X(\mathbf{0}, \mathbf{0}) = \sigma_X^2\left[\sigma_{X_j}^2 \delta_{jj'}\right]$, and in particular, $\left|\mathbf{k}_x\right| \leq \mathbf{k}_x(\mathbf{0}, \mathbf{0}) = \left[\sigma_{x_j}^2 \delta_{jj'}\right] = \left[1 \cdot \delta_{jj'}\right]$ with $\delta_{jj'} = \delta_{nn'} \cdot \delta_{mm'}$, the familiar Kronecker delta.

If inhomogeneity and nonstationarity (*non*-WS-HS) *are due to scale alone* and the *normalized* sensor outputs are otherwise at least locally homogeneous and stationary, then (2.1.4) reduces to the equivalent form $\left(\bar{X}_j = \bar{x}_j = 0 \text{ and so on}\right)$ [2]

$$\mathbf{K}_X = \Psi_J \hat{\mathbf{k}}_x(\Delta\mathbf{R}_{m'm}, \Delta t_{n'n}) = \Psi_J\left[\left(\hat{\eta}_j \hat{\eta}_{j'}\right)^{1/2} \mathbf{k}_x(\Delta\mathbf{R}, \Delta t)_{jj'}\right];$$

$$\Delta\mathbf{R}_{mm} \equiv \mathbf{r}_{m'} - \mathbf{r}_m; \quad \Delta t_{n'n} \equiv t_{n'} - t_n, \quad (2.1.7)$$

where ψ_J and $\hat{\eta}_j$ are scaling factors:

$$\psi_J \equiv J^{-1}\sum\nolimits_j \psi_j \quad \text{and} \quad \psi_j = \left(\psi_j/\psi_J\right)\psi_J = \hat{\eta}_j \psi_J \quad \therefore \quad \hat{\eta}_j \equiv \psi_j/\psi_J, \quad (2.1.8a)$$

[2] We shall use **R** and **r** equivalently and interchangeably throughout, unless there is a distinction to be made in particular applications.

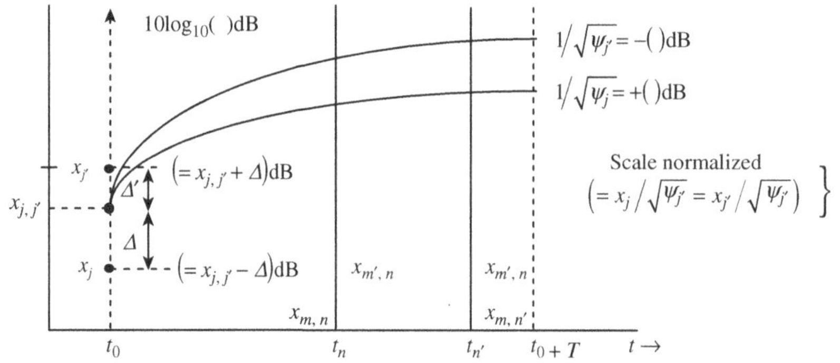

FIGURE 2.2 Normalizing (intensities) $\sqrt{\psi_j}, \sqrt{\psi_{j'}}$ for sensor outputs of a sampled noise field (refer to Eqs. (2.1.6–2.1.8a).

and $\mathbf{k}_x$ (2.1.6) is now $\mathbf{k}_x(\Delta\mathbf{R}, \Delta t)_{jj'}$. In many cases, stationarity (but not homogeneity) is well approximated, so that $\psi_j = \psi_m$ and ψ_J is modified to

$$\psi_J \to \psi_M = \frac{1}{M}\sum\nolimits_m \psi_m, \quad \text{with} \quad \hat{\eta}_j \doteq \hat{\eta}_m \psi_m / \psi_M. \tag{2.1.8b}$$

A *practical approximation* here to the ensemble averages ψ_j (and therefore to ψ_J, ψ_M) employs a *time average of each sensor's output noise intensity* under the Hom-Stat conditions assumed here, namely:

$$\psi_j \doteq \int_T \psi_j dt/T = \bar{\psi}_j^t = (\bar{\psi})_m, \quad \text{now with } \psi_j \doteq \bar{\psi}_m^t \equiv \psi'_m, \tag{2.1.9}$$

so that (approximately) $\psi_J\left(\hat{\eta}_j \hat{\eta}_{j'}\right)^{1/2} \doteq \psi_j \doteq \psi'_m$ and

$$\therefore \quad \mathbf{k}_x = \left[\overline{x_j x_{j'}}\right] \doteq \left[\overline{X_j X_{j'}} \Big/ \sqrt{\psi'_m \psi'_{m'}}\right], \tag{2.1.10}$$

which preserves the WS-HS property of the scale-normalized data ($\sim \{x_j\}$).

In the general case of temporal nonstationarity as well, the normalization procedure of (2.1.6) is indicated, the practical problem being the lack of the ensemble averages required for $\psi_j (= \psi_{m,n})$. However, if several local channel interrogations are empirically available, often assisted by an average, that is, deterministic propagation model, sample ensemble values of ψ_j can be estimated. This can be done with suitable small-sample statistical tests (e.g., Kolmogrov–Smirnov [5]) to assess the stability of the medium in time and space and, therefore, the local WS-HS of the data from which each $\psi_j (= \psi_{mn})$ is estimated. Figure 2.2 illustrates the normalization process involved. Note once more, however, that this scale-normalized sampled field may still not be WS-HS (refer to remarks following Eq. (2.1.6)). Also, we remark that "time" here is proportional to range, that is, $t = |\mathbf{R}|/c_0$, in such applications as radar and sonar.

In practice, some sensors (m') may be inoperative. For such cases, $\psi_{m'',n} = 0$, all n.

Consequently, the number of effective sensors in M is m'', to be used in constructing the scale-normalizing factors ψ_J, ψ_M, ψ_m. In addition, we may expect distortion in the resulting beam pattern designed for M sensor outputs.[3] Although the original input field may be Hom-Stat, beam forming by various combinations of the sensor outputs renders the field apparently inhomogeneous, as does any variation in sensor output levels. The latter effect, of course, can be largely removed in practice by the scale renormalization $\psi'_j(\doteq \psi_j)$ applied to (2.1.6), as illustrated schematically in Fig. 2.2. In essence, beam forming produces a concentrated, nonuniform wave number spectrum, which modifies the energy distribution of the received field.

2.1.3 Narrowband Fields

In many common applications, we must deal with narrowband noise and signal transmissions and reception, particularly in the frequency domain and often in wave number space as well. It is well known that such narrowband waves can be expressed as a linear combination of slowly varying components about some high-frequency "carrier" reference frequency. Here, we briefly exploit the structure of such narrowband phenomena to gain additional insight into how such narrowband representations are constructed.

2.1.3.1 Narrowband Conditions We begin with the Fourier representation of the real space–time field $\alpha(\mathbf{R}, t)$ in the neighborhood V_R of a receiving array or aperture:

$$\alpha(\mathbf{r}, t) = \int_{-\infty[\nu]}^{\infty} \mathbf{d}^3\boldsymbol{\nu} \to \mathbf{d}^{(3)}\boldsymbol{\nu}\, e^{-2\pi i \boldsymbol{\nu}\cdot\mathbf{r}} \int_{-\infty}^{\infty} S(\boldsymbol{\nu}, f)_\alpha e^{i\omega t} df, \quad \omega = 2\pi f. \tag{2.1.11}$$

The real nature of $X(\mathbf{r}, t)$ requires that

$$S(\boldsymbol{\nu}, f)_\alpha = S(-\boldsymbol{\nu}, -f)^*_\alpha, \tag{2.1.12}$$

where (*) as usual denotes the complex conjugate. (As can be easily seen, substituting the right-hand side of (2.1.12) in (2.1.11) leaves $X(\mathbf{r}, t)$ unchanged, as expected.) Rewriting (2.1.11) and using (2.1.12), we obtain

$$\alpha(\mathbf{r}, t) = \int_{-\infty[\boldsymbol{\nu}]}^{\infty} e^{-2\pi i \boldsymbol{\nu}\cdot\mathbf{r}} \mathbf{d}^3\boldsymbol{\nu} \left[\int_{-\infty}^{0} S(-\boldsymbol{\nu}, -f)^*_\alpha e^{i\omega t} df + \int_{0}^{\infty} S(\boldsymbol{\nu}, f)_\alpha e^{i\omega t} df \right]. \tag{2.1.13}$$

Letting $\int^{\infty} S_\alpha e^{i\omega t} df \equiv b(\boldsymbol{\nu}, t)$, we see that the first term of (2.1.13) becomes $b(-\boldsymbol{\nu}, t)^*$, so that (2.1?13) reduces to

$$\alpha(\mathbf{r}, t) = \int_{-\infty(\nu)}^{\infty} e^{-2\pi i \boldsymbol{\nu}\cdot\mathbf{r}} \left\{ b(-\boldsymbol{\nu}, t)^* + b(\boldsymbol{\nu}, t) \right\} \mathbf{d}^3\boldsymbol{\nu}. \tag{2.1.14}$$

[3] See Sections 3.2.1.4 and 3.2.2 ff.

Setting $\boldsymbol{\nu} = -\boldsymbol{\nu}'$, we find that the first term of (2.1.14) becomes

$$\int_{-\infty}^{\infty} e^{2\pi i \boldsymbol{\nu}' \cdot \mathbf{r}} b(\boldsymbol{\nu}', t)^* \mathbf{d}^3\boldsymbol{\nu} = \left(\int_{-\infty(\boldsymbol{\nu})}^{\infty} e^{-2\pi i \boldsymbol{\nu}' \cdot \mathbf{r}} b(\boldsymbol{\nu}', t) \mathbf{d}^3\boldsymbol{\nu}' \right)^* \equiv B(\mathbf{r}, t)^*, \tag{2.1.15}$$

so that now the space–time field can be expressed as

$$\left. \begin{aligned} \alpha(\mathbf{r}, t) &= B(\mathbf{r}, t)^* + B(\mathbf{r}, t) = 2\,Re\,B(\mathbf{r}, t) \\ &= 2Re \int_{-\infty(\boldsymbol{\nu})}^{\infty} e^{-2\pi i \boldsymbol{\nu} \cdot \mathbf{r}} b(\boldsymbol{\nu}, t) \mathbf{d}^3\boldsymbol{\nu} = 2Re \int_{-\infty(\boldsymbol{\nu})}^{\infty} e^{-2\pi i \boldsymbol{\nu} \cdot \mathbf{r}} \mathbf{d}^3\boldsymbol{\nu} \int_{-\infty}^{\infty} S(\boldsymbol{\nu}, f)_\alpha e^{i\omega t} df \end{aligned} \right\}, \tag{2.1.16}$$

which is an exact relation.

Next, we observe that since the energy $\left(\sim |S(\boldsymbol{\nu}, f)_\alpha|^2\right)$ is concentrated here in the volume ΔV about $(\boldsymbol{\nu}_0, f_0)$ (Fig. 2.3), $|S_\alpha|^2$ exceeds zero significantly only when $|\boldsymbol{\nu}_0| - \Delta\nu < |\boldsymbol{\nu}_0| < |\boldsymbol{\nu}_0| + \Delta\nu$ and $f_0 - \Delta f < f_0 < f_0 + \Delta f$. This dual condition also clearly holds for $|S_\alpha| > 0$ only when $|\boldsymbol{\nu}| \sim |\boldsymbol{\nu}_0| (> 0)$, $f \sim f_0 (> 0)$ in these same intervals. We now let $f \equiv f_0 + f'$ $\boldsymbol{\nu} \equiv \boldsymbol{\nu}_0 + \boldsymbol{\nu}'$ in (2.6.16) to get

$$\alpha(\mathbf{r}, t) = 2Re \left\{ e^{i\omega_0 t - 2\pi i \boldsymbol{\nu}_0 \cdot \mathbf{r}} \cdot \int_{[-\nu_0]}^{\infty} e^{-2\pi i \boldsymbol{\nu}' \cdot \mathbf{r}} \mathbf{d}^3\boldsymbol{\nu}' \cdot \int_{-f_0}^{\infty} S(\boldsymbol{\nu}_0 + \boldsymbol{\nu}', f_0 + f')_\alpha e^{i\omega' t} df' \right\}, \tag{2.1.17}$$

which is still exact, where $\nu' > -\nu_0$ and $f' > -f_0$. But the *narrowband condition* postulates that $|\boldsymbol{\nu}_{\min}| \simeq \Delta\nu \gg -\nu_0$ and $f_{\min} \simeq \Delta f \gg -f_0$, so that *to an excellent approximation we can*

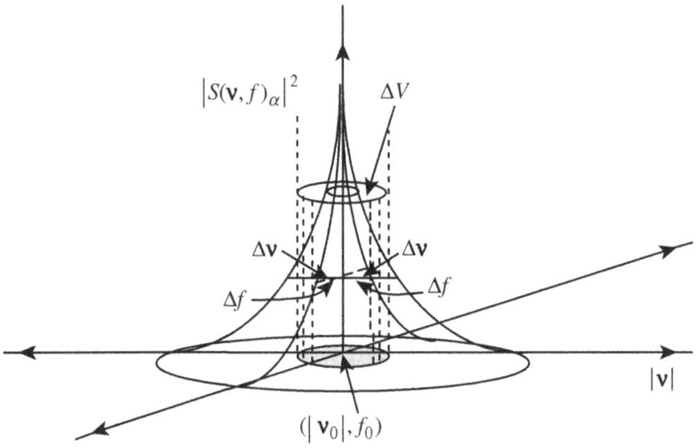

FIGURE 2.3 Intensity spectrum of a space–time "narrowband" noise or signal field.

replace $-|\nu_0|$ and $-f_0$ by $-\infty$ in (2.1.16), since $|S_X| > 0$ only for ΔV about $\int(\nu_0, f_0)$. The (now approximate) desired result is from (2.1.17):

$$\alpha(\mathbf{r},t) \doteq Re\left\{ e^{-i\omega_0 t - 2\pi i \boldsymbol{\nu}_0 \cdot \mathbf{r}} \int_{-\infty}^{\infty} e^{-2\pi i \boldsymbol{\nu}' \cdot \mathbf{r}} \mathbf{d}^3\boldsymbol{\nu}' \int_{-\infty}^{\infty} S_0(\boldsymbol{\nu}', f')_\alpha e^{i\omega' t} df' \right\} \tag{2.1.18}$$

with

$$\left.\begin{aligned} S_0(\boldsymbol{\nu}', f')_\alpha &\equiv 2S(\boldsymbol{\nu}_0 + \boldsymbol{\nu}', f_0 + f')_\alpha|_{\text{nb}} \text{ in and about } (\nu_0, f_0) \in \Delta V \\ &= 0,\ \boldsymbol{\nu}' < -\boldsymbol{\nu}_0,\ f' < -f_0; \quad (\nu', \nu_0 = |\boldsymbol{\nu}'|, |\boldsymbol{\nu}_0|) \end{aligned}\right\}, \tag{2.1.19}$$

where the last relation of (2.1.19) is exact.

Since (2.1.19) applies, $S_0(-\boldsymbol{\nu}', -f')^*_\alpha = 2S_0(\boldsymbol{\nu}_0 - \boldsymbol{\nu}', f_0 - f')^*_\alpha$ and thus $S_0(-\boldsymbol{\nu}', -f')^*_\alpha \neq S_0(\boldsymbol{\nu}', f')_\alpha$, with the integrals in (2.1.17) generally complex. This in turn leads to the concept of the *space–time complex envelope*, which is an extension of the more familiar purely temporal complex envelope of earlier communication treatments (e.g., [1]). Accordingly, we define the *complex envelope of the nb field* $\alpha(\mathbf{r}, t)$ *by*

$$E_0(\mathbf{r},t)_\alpha \equiv \int_{-\infty[\boldsymbol{\nu}]}^{\infty} e^{-2\pi i \boldsymbol{\nu} \cdot \mathbf{r}} \mathbf{d}^3\boldsymbol{\nu} \int_{-\infty}^{\infty} S_0(\boldsymbol{\nu}, f)_\alpha e^{i\omega t} df \ (= \text{complex}) \equiv E_{0\alpha} e^{-i\phi_{0\alpha}}. \tag{2.1.20}$$

Here, $\phi_{0\alpha} = \phi_0(\mathbf{r}, t)_\alpha$ is a slowly varying real phase, like the envelope $|E_{0\alpha}|$, vis-à-vis (cos/sin)$(\omega_0 t - 2\pi\boldsymbol{\nu}_0 \cdot \mathbf{r})$. Applying (2.1.20) to (2.1.18) allows to write the real result for the narrowband field:

$$\alpha(\mathbf{r},t) \doteq Re\left\{ |E_0(\mathbf{r},t)_\alpha| e^{i[\omega_0 t - \mathbf{k}_0 \cdot \mathbf{r} - \phi_0(\mathbf{r},t)_\alpha]} \right\} = |E_{0\alpha}| \cos(\omega_0 t - \mathbf{k}_0 \cdot \mathbf{r} - \phi_0 \alpha), \tag{2.1.21}$$

with $\mathbf{k}_0 \equiv 2\pi\boldsymbol{\nu}_0$ a vector wave number specifying the direction $\hat{\mathbf{i}}_0$ of the propagating field's wave fronts in the vicinity of the domain V_R occupied by the receiving sensors. Thus, $\mathbf{k}_0 = \hat{\mathbf{i}}_0 k_0$, where the explicit structure of k_0 itself depends on the physical character of the medium supporting the propagation of the fields.[4] Equivalent forms of (2.1.21) are

$$\begin{aligned} \alpha(\mathbf{r},t) &\doteq \alpha_c(\mathbf{r},t)\cos\hat{\Phi}_\alpha(\Delta\phi_\alpha) + \alpha_s(\mathbf{r},t)\sin\hat{\Phi}_\alpha(\Delta\phi_\alpha); \quad \hat{\Phi}_\alpha(\Delta\phi_\alpha) = \omega_0 t - \mathbf{k}_0 \cdot \mathbf{r} \\ &= (|E_{0\alpha}|\cos\phi_{0\alpha})\cos\hat{\Phi}_\alpha + ||E_{0\alpha}|\sin\phi_{0\alpha})\sin\hat{\Phi}_\alpha \\ &= \sqrt{\alpha_c^2 + \alpha_s^2}\cos\left[\hat{\Phi}_\alpha - \tan^{-1}(\alpha_s/\alpha_c)\right] \text{ and so on.} \end{aligned} \tag{2.1.21a}$$

2.1.3.2 Narrowband Sensor Outputs Since we are primarily interested here in the sensor *outputs* $X(\mathbf{r}_m, t)$, $m = 1, \ldots, M$, which now represent the sampled field at the discrete

[4] This point is discussed and illustrated by the results of Chapters 8 and 9.

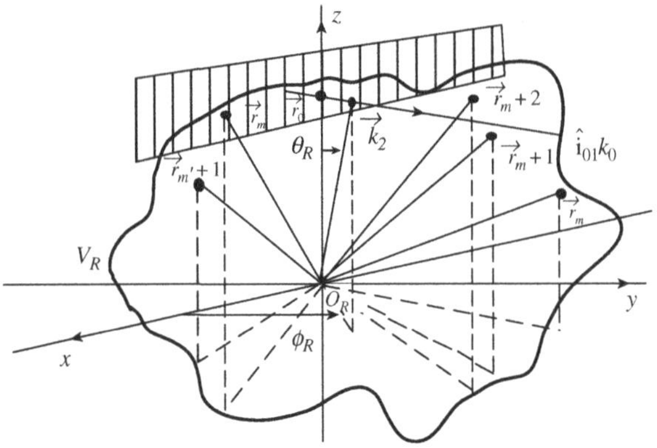

FIGURE 2.4 Unconnected sensor elements at the points $\mathbf{r}_m, \mathbf{r}_{m+1}, \ldots, \mathbf{r}_k$ in a receiving domain V_R with "center" at O_R where $\{X(\mathbf{r}_m, t)\}$ are the element outputs [Eqs. (2.1.22 and 2.1.23)] and $\hat{\mathbf{i}}_0$ is the unit vector in the direction of the wavefront, that is, $\mathbf{k}_0 = \hat{\mathbf{i}}_0 k_0$(2.1.21).

spatial points $\{\mathbf{r}_m\}$ but continuous in time (Fig. 2.4). These are subsequently to be suitably combined and processed, ultimately for signal extraction. Let us consider first the vector $\hat{\mathbf{R}}\alpha = [X(\mathbf{r}_m, t)]$, which is the column vector set of sensor outputs, namely.

$$\hat{\mathbf{R}}\alpha = \left[\hat{R}_m\alpha\right] = \left[\left[\int_{V_R} A_m\delta(\mathbf{r}-\mathbf{r}_m)\cdot\left\{\int_{-\infty}^{\infty} h_R^{(m)}\left(\tau, t\middle|\mathbf{r}\right)\alpha(\mathbf{r}, t-\tau)d\tau\right\}\mathbf{d}^3V_R(\mathbf{r})\right]\right]$$
$$= [X(\mathbf{r}_m, t)], \tag{2.1.22}$$

with α in more detail equal to $\alpha\left(\hat{\mathbf{i}}_k\cdot\hat{\mathbf{i}}_m r_m/c_0, t-\tau\right)$ where $\hat{\mathbf{i}}_k$ is the direction of the wave front of this incoming field and $\mathbf{k}_0 = \hat{\mathbf{i}}_k k_0$. In (2.1.22), $h_R^{(m)}$ is the time-varying (linear) Green's function of the mth sensor's filter.

For narrowband fields, narrowband filters are appropriate. From earlier work, (pp. 98–100 of Ref. [2], [1]), we find that the filter outputs are explicitly

$$X(\mathbf{r}_m, t) \doteq \int_{-\infty}^{\infty} h_0^{(m)}(\tau, t|\mathbf{r}_m)E_0(\mathbf{r}_m, t-\tau)_\alpha \cos\left[\hat{\omega}_0 t - \mathbf{k}_0\cdot\mathbf{r}_m - \phi_0(\mathbf{r}_m, t-\tau)_\alpha\right.$$
$$\left. +\omega_D(t-\tau) - \gamma_0(\tau|\mathbf{r}_m)\right]d\tau, \tag{2.1.23}$$

with

$$\left.\begin{aligned} &\omega_D \equiv \omega_0 - \hat{\omega}_0(\ll \hat{\omega}_0, \omega_0);\; h_0^{(m)}(t|\mathbf{r}_m)e^{-i\gamma_0(t|\mathbf{r}_m)} = \int_{-\infty}^{\infty} Y_0(f'|\mathbf{r}_m)e^{i\omega' t}dt,\; f' = f - \hat{f}_0 \\ &\text{where} \\ &\qquad Y_0(f'|\mathbf{r}_m) \equiv Y(f' + \hat{f}_0|\mathbf{r}) = \mathsf{F}\left\{h_0(t|\mathbf{r}_m)e^{-i\gamma_0(t|\mathbf{r}_m)}\right\} \end{aligned}\right\}, \tag{2.1.23a}$$

where $\omega_D(=2\pi f_D)$ measures the amount of "detuning" between the "center frequency" f_0 of the input field and the center frequency $(\hat{f}_0)$ of the narrowband sensor outputs.[5]

Often it is reasonable to assume that without noticeable error, the array sensors are tuned to the center frequency of the incoming field, so that $\omega_D = 0$, that is, $\hat{f}_0 = f_0$, as we shall postulate henceforth unless otherwise required. As a result, Eq. (2.1.23) can be written in the following equivalent forms:

$$X(\mathbf{r}_m, t) \doteq Re\left\{ e^{-i\omega_0 t - i\mathbf{k}_0\cdot\mathbf{r}_m} \int_{-\infty}^{\infty} h_0^{(m)}(\tau, t|\mathbf{r}_m) E_0(\mathbf{r}_m, t-\tau)_\alpha e^{-i\phi_0(\mathbf{r}_m, t-\tau)_\alpha - i\gamma_0(\tau|\mathbf{r}_m)} d\tau \right\} \tag{2.1.24a}$$

$$= Re\left\{\left[2\hat{\alpha}(\mathbf{r}_m, t) - i\hat{\beta}(\mathbf{r}_m, t)\right] e^{i\omega_0 t - i\mathbf{k}_0\cdot\mathbf{r}_m}\right\}, \quad \text{with } \omega_D = 0 \tag{2.1.24b}$$

where specifically now

$$\left.\begin{aligned} \hat{\alpha}(\mathbf{r}_m, t) &= Re \int_{-\infty}^{\infty} h_0^{(m)}(t-\tau, t|\mathbf{r}_m) E_0(\mathbf{r}_m, \tau)_\alpha e^{-i\phi_0(\mathbf{r}_m,\tau)_\alpha - i\gamma_0(t-\tau, t|\mathbf{r}_m)} d\tau \\ \hat{\beta}(\mathbf{r}_m, t) &= Im \int_{-\infty}^{\infty} h_0^{(m)}(t-\tau, t|\mathbf{r}_m) E_0(\mathbf{r}_m, \tau)_\alpha e^{-i\phi_0(\mathbf{r}_m,\tau)_\alpha - i\gamma_0(t-\tau|\mathbf{r}_m)} d\tau \end{aligned}\right\}. \tag{2.1.24c}$$

Accordingly, we can write alternatively for the output of the mth sensor into the space–time processing portion of the receiver:

$$X(\mathbf{r}_m, t) \doteq \hat{\alpha}(\mathbf{r}_m, t)\cos(\omega_0 t - \mathbf{k}_0\cdot\mathbf{r}_m) + \hat{\beta}(\mathbf{r}_m, t)\sin(\omega_0 t - \mathbf{k}_0\cdot\mathbf{r}_m) \tag{2.1.25a}$$

$$= X_c(\mathbf{r}_m, t)\cos\hat{\Phi}_m(t) + X_s(\mathbf{r}_m, t)\sin\hat{\Phi}_m(t), \quad \hat{\Phi}_m(t) \equiv \omega_0 t - \mathbf{k}_0\cdot\mathbf{r}_m, \tag{2.1.25b}$$

$$= \left(X_c^2 + X_s^2\right)^{1/2}\cos\left[\hat{\Phi}_m(t) - \tan^{-1}(X_s/X_c)\right] = E(\mathbf{r}_m, t)_X \cos\left(\hat{\Phi}_m(t) - \psi_{s/c}^{(m)}\right), \tag{2.1.25c}$$

with $X_{c_m} = \hat{\alpha}(\mathbf{r}_m, t), X_{s_m} = \hat{\beta}(\mathbf{r}_m, t)$ for the "in-phase" (cosine) and "out-of-phase" or quadrature (sine) components, respectively, and $\psi_{s/c}^{(m)} \equiv \tan^{-1}(X_s/X_c)$.

[5] The approximation arises from neglecting the essentially zero contribution of the additive high-frequency terms that lie outside the main portion of the narrow-band of the sensors. Thus, (refer to 2.63 (a,b) of Ref. [1]),

$$h_R^{(m)}(\tau, t|\mathbf{r}_m)) \doteq 2h_0(\tau, t|\mathbf{r}_m)\cos[\omega_0\tau - \gamma_0(\tau|\mathbf{r}_m)], \tag{2.1.23b}$$

and $E_{0\alpha}$, h_0, γ_0, $\phi_{0\alpha}$ are real, slowly varying quantities compared to $\cos\hat{\omega}_0\tau$, $\cos\omega_0\tau$ and so on. The narrowband filter responses h_0, γ_0 may be obtained from $h_R^{(m)}$ by applying the "high-Q" (≡narrowband) approximation for the filter in question or they may be obtained from $\mathsf{F}^{-1}\{Y_0\}$, (2.1.23a). See the discussion in (**5**), Section 2.2 of Ref. [1].

It should be emphasized that our results above apply to both noise and signal fields, α_N and α_S, where $X(\mathbf{r}_m, t)$ can represent either class of sampled input field. However, for the most part, $X(\mathbf{r}_m, t)$ in previous (and subsequent) discussions designates the received output of a sensor array or aperture. This can consist of noise alone or a mixture of signal and noise to be subsequently processed by the receiver elements, whereas $S(\mathbf{r}_m, t)$ is explicitly a signal output of a sensor element. Thus, we can write explicitly for the latter in these narrowband cases (before temporal sampling) the direct analogues of (2.1.25a):

$$S(r_m, t) \doteq \hat{\alpha}_S(\mathbf{r}_m, t)\cos(\omega_0 t - \mathbf{k}_0 \cdot \mathbf{r}_m) + \hat{\beta}_S(\mathbf{r}_m, t)\sin(\omega_0 t - \mathbf{k}_0 \cdot \mathbf{r}_m) \tag{2.1.26a}$$

$$= S_c(\mathbf{r}_m, t)\cos\hat{\Phi}_m(t) + S_c(\mathbf{r}_m, t)\sin\hat{\Phi}_m(t) = \left(S_c^2 + S_s^2\right)^{1/2}\cos\left[\hat{\Phi}_m(\mathbf{t}) - \tan^{-1}(S_s/S_c)\right], \tag{2.1.26b}$$

or as with Eqs. (2.1.2 and 2.1.3):

$$S(r_m, t) \equiv A_0^{(m)} s^{(m)}(t)\Big/\sqrt{2}, \quad \text{with} \quad A_0^{(m)}(t) = \sqrt{S_c^2 + S_s^2},$$
$$s^{(m)}(t) = \sqrt{2}\cos\left[\hat{\Phi}_m(t) - \psi_{s/c}^{(m)}(t)\right], \tag{2.1.27}$$

where, again, $\hat{\Phi}_m(t) \equiv \omega_0 t - \mathbf{k}_0 \cdot \mathbf{r}_m$ (2.1.25b), and $A_0^{(m)}(t)$, $\psi_{s/c}^{(m)}(t)$ are slowly varying vis-à-vis $\cos\hat{\Phi}_m$, $\sin\hat{\Phi}_m$. Note that the slowly varying amplitude and phase variables are contained in $\hat{\alpha}_s$, $\hat{\beta}_s$ here. Moreover, if we include an epoch ε, in the manner of Eqs. (2.1.2–2.1.3), then $t \to t - \varepsilon$, with $\hat{\Phi}_m(t) \to \omega_0(t - \varepsilon) - \mathbf{k}_0 \cdot \mathbf{r}_m$, in Eqs. (2.1.25) and (2.1.26).

2.1.4 Noise and Signal Field Covariances: Narrowband Cases

Here, we develop the covariance structures of both noise and signal fields, when these are narrowband, in the sense of Section 2.1.3.

2.1.4.1 Narrowband Noise In Section 2.1.2, we have considered the general situation of inhomogeneous nonstationary (non-WS-HS) noise fields from the viewpoint of their covariance functions after space–time sampling. Here, we extend our treatment to the important cases of narrowband fields α_{noise}, in the sense of (2.1.15–2.1.27) and as described by (2.1.21) before space–time sampling, and by (2.1.25b) after such sampling, now with $t \to t_n$ therein. Normalizing by $x_j = X_j/\sqrt{\psi_j}$ (2.1.7) and setting $\bar{x}_j = 0$, since these noise fields are narrowband, we can write this normalized field in the vector form[6]

$$\mathbf{x}(\mathbf{r}, \mathbf{t}) = [x(\mathbf{r}_m, t_n)] = [x_j] = \left[x_{c_j}\cos\hat{\Phi}_j + x_{s_j}\sin\hat{\Phi}_j\right], \quad j = mn, \tag{2.1.28}$$

[6] It is always possible formally to represent any field, broad- or narrowband, by (2.1.28). In the broadband cases however, x_{c_j}, x_{s_j} are not slowly varying. The concepts of envelope and phase here lose their usual physical interpretation, although they can be redefined in terms of appropriate Hilbert transforms; see (**3**) of Sections 7.5.3 of Ref. [1] and (**6**) of Section 2.2.5 of Ref. [1].

with $x_{c,s} = (X_c, X_s)/\sqrt{\psi_j}$, where $\hat{\Phi}_j \equiv \omega_0(t_n - \mathbf{k}_0 \cdot \mathbf{r}_m/\omega_0)$ (2.1.25b). As before (2.1.25a) x_{c_j}, x_{s_j} are the slowly varying "in-phase" and "out-of-phase" (quadrature) components of $\mathbf{x} = [x_j]$ and $\hat{\Phi}_j = \omega_0 t - \mathbf{k}_0 \cdot \mathbf{r}_m$ (without "steering").

We next proceed to the calculation of the narrowband[7] noise fields covariance of $\mathbf{x}(\mathbf{r}, t)$, (2.1.28) from (2.1.5 et seq.), remembering that now $\bar{\mathbf{x}}_j = 0$ here. We obtain the following (square) matrix:

$$\mathbf{k}_N = [k_N(\mathbf{r}_m, t_n; \mathbf{r}_{m'}, t_{n'})] = \overline{\mathbf{x}\tilde{\mathbf{x}}} = \Big[\overline{x_{c_j} x_{c_{j'}}} \cos\hat{\Phi}_j \cos\hat{\Phi}_{j'} + \overline{x_{s_j} x_{s_{j'}}} \sin\hat{\Phi}_j \sin\hat{\Phi}_{j'} + \overline{x_{c_j} x_{s_{j'}}} \cos\hat{\Phi}_j \sin\hat{\Phi}_{j'} + \overline{x_{s_j} x_{c_{j'}}} \sin\hat{\Phi}_j \cos\hat{\Phi}_{j'}\Big] \tag{2.1.29a}$$

or

$$\mathbf{k}_N = \Big[(\boldsymbol{\rho}_c)_{jj'} \cos\hat{\Phi}_j \cos\hat{\Phi}_{j'} + (\boldsymbol{\rho}_s)_{jj'} \sin\hat{\Phi}_j \sin\hat{\Phi}_{j'} + (\boldsymbol{\rho}_{cs})_{jj'} \cos\hat{\Phi}_j \sin\hat{\Phi}_{j'} + (\boldsymbol{\rho}_{sc})_{jj'} \sin\hat{\Phi}_j \cos\hat{\Phi}_{j'}\Big], \tag{2.1.29b}$$

which defines the covariances $(\boldsymbol{\rho}_c, \boldsymbol{\rho}_s, \boldsymbol{\rho}_{cs}, \boldsymbol{\rho}_{sc})$. Writing for the time delay from the mth sensor to some reference point in the receiving array

$$\Delta t_j \equiv t_n - (\mathbf{k}_0 \cdot \mathbf{r}_m)/\omega_0; \quad \Delta t_{j'} \equiv t_{n'} - (\mathbf{k}_0 \cdot \mathbf{r}_{m'})/\omega_0, \tag{2.1.30}$$

we have explicitly for these generally non-Hom-Stat, slowly varying covariances

$$\boldsymbol{\rho}_c = [\boldsymbol{\rho}_c(\Delta t_j, \Delta t_{j'})] = (\boldsymbol{\rho}_c)_{jj'}, \text{ and so on.} \tag{2.1.31}$$

When a *steering vector* $(\mathbf{k}_{0R})$ is introduced, which is the usual case in practical applications, the time delay Δt_j is modified to

$$\Delta\tau_j \equiv \Delta t_j - \mathbf{k}_{0R} \cdot \mathbf{r}_m/\omega_0 = t_n - (\mathbf{k}_0 - \mathbf{k}_{0R}) \cdot \mathbf{r}_m/\omega_0. \tag{2.1.32}$$

These covariances reduce for at least homogeneity and wide sense stationarity to the expected difference relations $\boldsymbol{\rho}_c = [\boldsymbol{\rho}_c(\Delta\tau_j - \Delta\tau_{j'})]$ and so on. (Note that although in general the covariance $\mathbf{k}_N$, (2.1.29), is still scale normalized (2.1.7–2.1.10), it is not necessarily both wide sense homogeneous and wide sense stationarity.) Nevertheless, by applying the "narrowband" conditions (2.1.19), we obtain the (approximate) equivalent forms when local wide sense homogeneity and stationarity are applicable, which allow to write (2.1.29b) now as

[7] We assume that the narrowband noise field is generated by narrowband, nonscatter ambient sources. If the noise is produced by scattering in the medium of propagation of the original signal, there are nonvanishing (statistical) mean values, that is, coherent components ($\bar{x} \neq 0$) are possible. If $\mathbf{y}$ is such a scatter field, letting $\mathbf{x} = \mathbf{y} - \bar{\mathbf{y}}$ in (2.1.28) allows at once to include the contributions of any nonvanishing mean value components. Accordingly, we have $\boldsymbol{\rho}_c = \overline{x_{c'} \tilde{x}_c}^{(x)} = \langle (y_{c'} - \bar{y}_c)(y_{c'} - \bar{y}_c) \rangle^{(x)}$ and so on.

$$\mathbf{k}_{\mathrm{N}} = \overline{\mathbf{x}\tilde{\mathbf{x}}} \doteq \left[\left(\frac{\boldsymbol{\rho}_{\mathrm{c}} + \boldsymbol{\rho}_{\mathrm{s}}}{2}\right)_{jj'} \cos(\hat{\Phi}_{j'} - \hat{\Phi}_j) + \left(\frac{\boldsymbol{\rho}_{\mathrm{cs}} + \boldsymbol{\rho}_{\mathrm{sc}}}{2}\right)_{jj'} \sin(\hat{\Phi}_{j'} - \hat{\Phi}_j)\right]$$

$$= \left[(\boldsymbol{\rho}_0)_{jj'} \cos(\hat{\Phi}_{j'} - \hat{\Phi}_j) + (\boldsymbol{\lambda}_0)_{jj'} \sin(\hat{\Phi}_{j'} - \hat{\Phi}_j)\right], \quad (2.1.33a)$$

with

$$\boldsymbol{\rho}_0 \equiv (\boldsymbol{\rho}_{\mathrm{c}} + \boldsymbol{\rho}_{\mathrm{s}})/2; \quad \boldsymbol{\lambda}_0 \equiv (\boldsymbol{\rho}_{\mathrm{cs}} + \boldsymbol{\rho}_{\mathrm{sc}})/2. \quad (2.1.33b)$$

This shows at once that $\boldsymbol{\rho}_0 = \tilde{\boldsymbol{\rho}}_0$ and $\boldsymbol{\lambda}_0 = -\tilde{\boldsymbol{\lambda}}_0$. Moreover, since $(\mathbf{k}_{\mathrm{N}})_{jj} = 1$, all j, it follows that $(\boldsymbol{\rho}_0)_{jj} = 1$, $(\boldsymbol{\lambda}_0)_{jj} = 0$, the latter directly from the relation $\boldsymbol{\lambda}_0 = -\tilde{\boldsymbol{\lambda}}_0$ above. Equation (2.6.33) is *exact* when $\boldsymbol{\rho}_{\mathrm{s}} = \boldsymbol{\rho}_{\mathrm{c}}$ and $\boldsymbol{\rho}_{\mathrm{cs}} = -\boldsymbol{\rho}_{\mathrm{sc}}$: this means that $\boldsymbol{\rho}_{\mathrm{c}} = \left[\boldsymbol{\rho}_{\mathrm{c}}(\Delta\tau_j - \Delta\tau_{j'})\right]$ and $\boldsymbol{\rho}_0 = \left[\boldsymbol{\rho}_0(\Delta\tau_j - \Delta\tau_{j'})\right]$, with $\left[\lambda_0(\Delta\tau_j - \Delta\tau_{j'})\right] = -\left[\lambda_0(\Delta\tau_{j'} - \Delta\tau_j)\right]$. Then, the scale-normalized covariance $\mathbf{k}_{\mathrm{N}}$ and its slowly varying components are now widesense *Hom-Stat*.[8]

2.1.4.2 ***Narrowband Signals*** In case of signal, refer to Eqs. (2.1.26a, 2.1.26b and 2.1.27), we distinguish two different classes: (i) the more common one where the signal is deterministic and only the epoch phase $\phi_\varepsilon = \omega_0\varepsilon$ is high frequency (and random for incoherent reception (Section 1.2.1.4); and (ii) the occasional one in which the signal itself is essentially a random field, without significant deterministic structure, and is thus analogous to the accompanying noise.

For (i), based on Eqs. (2.1.33a and 2.1.33b), we now define

$$\Phi_j \equiv \hat{\Phi}_j(\Delta\tau_j) - \phi_j; \quad \phi_j \left(= \phi_n^{(m)}(\Delta\tau_j)\right) \text{a slowly varying signal phase vis-à-vis } \hat{\Phi}_j. \quad (2.1.34)$$

For the normalized signal in general and explicitly for the narrowband cases here,

$$\hat{s}_j \equiv a_{0n}^{(m)} s^{(m)} = A_{0n}^{(m)} s_n^{(m)} \Big/ \sqrt{2\psi_j}; \quad \hat{s}_j|_{\mathrm{nb}} = \frac{A_0^{(m)}(\Delta t_j - \varepsilon)}{\sqrt{2\psi_j}} \cdot \sqrt{2}\cos\left[\omega_0(\Delta\tau_j - \varepsilon) - \phi_n^{(m)}\right], \quad (2.1.35)$$

with

$$s_n^{(m)}|_{\mathrm{nb}} \equiv \sqrt{2}\cos\left[\omega_0(\Delta\tau_j - \varepsilon) - \phi^{(m)}(\Delta\tau_j - \varepsilon)\right], \text{ refer to (2.1.1)–(2.1.3) and (2.1.27).}$$

Accordingly, we may proceed formally as above for the purely random noise cases to write their covariances in the usual matrix form for these sampled fields, subject once more to the

[8] In general, particularly for applications where detection is reverberation or clutter dominate, we may expect the field to be *non-Hom.-Stat.*, so that $\boldsymbol{\rho}_{\mathrm{c}} \neq \boldsymbol{\rho}_{\mathrm{s}}$, $\boldsymbol{\rho}_{\mathrm{cs}} \neq -\tilde{\boldsymbol{\rho}}_{\mathrm{sc}}$: these variances are unequal.

narrowband conditions of Section 2.1.4.1, in the locally *Hom-Stat* situations. Thus, for the deterministic signal cases of (i), we may define the signal covariance by

$$\mathbf{K}_{\hat{s}} \equiv \hat{s}\tilde{\hat{s}} \doteq \left[(\hat{\boldsymbol{\rho}}_0)_{jj'} \cos\left(\hat{\Phi}_{j'} - \hat{\Phi}_j\right) + \left(\hat{\boldsymbol{\lambda}}_0\right)_{jj'} \sin\left(\hat{\Phi}_{j'} - \hat{\Phi}_j\right) \right], \text{ refer to Eqs. (2.1.33a and 2.1.33b).} \tag{2.1.36}$$

Now specifically from (2.1.33a and 2.1.33b), we have

$$(\hat{\boldsymbol{\rho}}_0)_{jj'} \equiv \left(\boldsymbol{\rho}_0^{(s)}\right)_{jj'} \cos\Delta\phi_{jj'} - \left(\boldsymbol{\lambda}_0^{(s)}\right)_{jj'} \sin\Delta\phi_{jj'}; \; \left(\hat{\boldsymbol{\lambda}}_0\right)_{jj'} \equiv \left(\boldsymbol{\rho}_0^{(s)}\right)_{jj'} \sin\Delta\phi_{jj'} + \left(\boldsymbol{\lambda}_0^{(s)}\right)_{jj'} \cos\Delta\phi_{jj'}, \tag{2.1.37a}$$

with $\Delta\phi_{jj'} \equiv \phi_{j'} - \phi_j$, where $\boldsymbol{\rho}_0^{(s)}$ and $\boldsymbol{\lambda}_0^{(s)}$ obey

$$\boldsymbol{\rho}_0^{(s)} \equiv \left(\boldsymbol{\rho}_c^{(s)} + \boldsymbol{\rho}_s^{(s)}\right)/2; \; \left(\boldsymbol{\rho}_{cs}^{(s)} + \boldsymbol{\rho}_{sc}^{(s)}\right)/2 \equiv \boldsymbol{\lambda}_0^{(s)} = -\tilde{\boldsymbol{\lambda}}_0^{(s)}; \; \left(\boldsymbol{\rho}_0^{(s)}\right)_{jj} = 1; \; \left(\boldsymbol{\lambda}_0^{(s)}\right)_{jj} = 0, \tag{2.1.37b}$$

analogous to (2.1.33a et seq.). Again, like the case of the narrowband noise field $\mathbf{x}$, Eqs. (2.1.28, 2.1.37a and 2.1.37b) are exact (at least) in the locally *Hom-Stat* condition, where $\boldsymbol{\rho}_c^{(s)} = \boldsymbol{\rho}_s^{(s)}, \boldsymbol{\rho}_{cs}^{(s)} = -\boldsymbol{\rho}_{sc}^{(s)}$.

Finally, for the purely noiselike signals of (ii) above, we have $\Phi_j \to \hat{\Phi}_j$, since ϕ_j are essentially absorbed into $\boldsymbol{\rho}_c^{(s)}, \boldsymbol{\rho}_s^{(s)}$, and so on, and therefore ultimately into $\hat{\boldsymbol{\rho}}_0$ and $\hat{\mathbf{x}}_0$ in (2.1.36). However, because the signal level is normalized here in terms of the noise outputs of the sensors $(\sim \psi_j, \psi_{j'})$ at $\mathbf{r}_m$, $\mathbf{r}_{m'}$, refer to (2.1.7) et seq., $\mathbf{K}_{\hat{s}}(\mathbf{r}_m, t_n; \mathbf{r}_m, t_n)$ is different from unity, unlike the scale-normalized covariance of $\mathbf{k}_N(\mathbf{r}_m, t_n; \mathbf{r}_m, t_n)$.

2.2 CONTINUOUS SPACE–TIME WIENER–KHINTCHINE RELATIONS[9,10]

The classical result for the Wiener–Khintchine relations, namely, that the Fourier transform of the temporal covariance is proportional to the intensity of the frequency spectrum, refer to Section 3.2.2 of Ref. [1] et seq., is directly extendable to the covariance and its transform of the space–time field $X(\mathbf{r},t)$ by a generalization of Loève's approach [6] (Refs. [1], pp 142 and 143, and [7], Chapter 4, and Section 2.2.2 following). In Section 2.2, we shall consider both the Hom-Stat and non-Hom-Stat situations, with particular attention to the latter, because of their frequent occurrence in practice. In addition, we shall devote considerable space to discrete sampling of the received field itself, required by the digital processing employed in many applications (Section 2.3 ff.)

[9] We write the abbreviation *W–Kh* to distinguish the Wiener–Khintchine relations from the Wiener–Kolmogoroff filters, which are abbreviated *W–K*.

[10] $\mathbf{R}$, $\mathbf{r}$, $\boldsymbol{\nu}$ are in rectangular coordinates, unless otherwise indicated (for example, spherical, polar coordinates).

We consider first the case of continuous sampled fields $X(\mathbf{r},t) = \alpha(\mathbf{r},t)$ that are wide sense homogeneous and stationary (Hom-Stat). The intensity spectrum of $X(\mathbf{r},t)$ is defined by an obvious extension of the purely temporal case[11] to

$$\mathsf{W}_X(\boldsymbol{\nu},f)_{\Delta\to\infty} = \lim_{\Delta\to\infty} \mathbf{E}\left\{\frac{2}{\Delta^{(4)}}\left|S_X(\boldsymbol{\nu},f)_\Delta\right|^2\right\} = \mathsf{W}_X(\boldsymbol{\nu},f) \tag{2.2.1}$$

with the covariance of the sampled field as (one-half) the Fourier transform of (2.2.1). The two Wiener–Khintchine relations extended to space–time for this case are given explicitly by (2.2.1a and 2.2.1b):

$$\begin{aligned}\mathsf{W}_X(\boldsymbol{\nu},f) &= 2\mathsf{F}_{\Delta\mathbf{R}}\mathsf{F}_\tau\{K_X(\Delta\mathbf{R},\tau)\}\\ &= 2\iint\limits_{-\infty}^{\infty} K_X(\Delta\mathbf{R},\tau)e^{2\pi i\boldsymbol{\nu}\cdot\Delta\mathbf{r}-2\pi if\tau}\mathbf{d}^3(\Delta\mathbf{R})d\tau\end{aligned} \tag{2.2.1a}$$

where

$$\left\{\begin{array}{l}\Delta\mathbf{R} = \mathbf{R}_2 - \mathbf{R}_1 = \hat{\mathbf{i}}_1(\mathbf{R}_{2x} - \mathbf{R}_{1x}) + \hat{\mathbf{i}}_2(\mathbf{R}_{2y} - \mathbf{R}_{1y}) + \hat{\mathbf{i}}_3(\mathbf{R}_{2z} - \mathbf{R}_{1z}),\quad \tau = \tau_2 - \tau_1\\ d^3(\Delta\mathbf{R}) \equiv d(\Delta R_x)d(\Delta R_y)d(\Delta R_z),\quad \mathbf{r} = \hat{\mathbf{i}}_x x + \hat{\mathbf{i}}_y y + \hat{\mathbf{i}}_z z\end{array}\right\}.$$

Here, $\boldsymbol{\nu} = \hat{\mathbf{i}}_x\nu_x + \hat{\mathbf{i}}_y\nu_y + \hat{\mathbf{i}}_z\nu_z$ (and $f = \omega/2\pi$) is a *vector wave number* with $\nu_x = 1/\lambda_x$ whose reciprocals are wavelengths. The wave numbers are also specified in (x, y, z) — rectangular coordinate system frequency is given by $f = \omega/2\pi$, with $\omega(= 2\pi f)$ being an angular frequency. Similarly, $\mathbf{k} = 2\pi\boldsymbol{\nu} = \left(\hat{\mathbf{i}}_x k_x + \hat{\mathbf{i}}_y k_y + \hat{\mathbf{i}}_z k_z\right)$ are vector *angular wave numbers*. The inverse of (2.2.1a) is easily shown to be

$$\begin{aligned}K_X(\Delta\mathbf{R},\tau) &= \frac{1}{2}\mathsf{F}_{\mathbf{k}}^{-1}\mathsf{F}_\omega^{-1}\{\mathsf{W}(\mathbf{k}/2\pi,\omega/2\pi)\}\\ &= \frac{1}{2}\iint\limits_{-\infty}^{\infty}\mathsf{W}_X(\mathbf{k}/2\pi,\omega/2\pi)e^{-i\mathbf{k}\cdot\Delta\mathbf{R}+i\omega\tau}\frac{\mathbf{d}^3\mathbf{k}}{(2\pi)^3}\frac{d\omega}{2\pi}.\end{aligned} \tag{2.2.1b}$$

Equations (2.2.1a and 2.2.1b) are the space–time extensions of the Wiener–Khintchine theorem, refer to Section 3.2.2 of Ref. [1].[12] These reduce to the well-known forms for the purely temporal covariance and intensity spectrum:

$$\mathsf{W}_X(f) \equiv 2\mathsf{F}_\tau\{K_X\} = 2\int_{-\infty}^{\infty} K_X(\tau)e^{-i\omega\tau}d\tau;\quad K_X(\tau) = \frac{1}{2}\mathsf{F}^{-1}\{\mathsf{W}_X\} = \frac{1}{2}\int_{-\infty}^{\infty}\mathsf{W}_X(f)e^{-i\omega\tau}df. \tag{2.2.2}$$

[11] See Section 3.2.2 of Ref. [1], pp. 141–143, for an outline of the proof due to Loève [8] in the temporal cases, which can serve as a basis for an extension to suitably defined random functions of space and time.

[12] Note that the dimensions of K_X are [amplitude]^2 while that of corresponding intensity spectrum W_X are [amplitude]^2 interval [length]^3 [time] = $[A^2][L^3][T]$ or $[A^2]$/[wave number^3][frequency].

Even if the noise field is *scale* inhomogeneous and scale nonstationary (in various combinations), Eqs. (2.2.1a, 2.2.1b, 2.2.2) still apply, now to the normalized covariance with W_X replaced by W_x and K_X by k_x, with the same arguments $(\Delta\mathbf{R}, \tau)$. Again, we remind the reader of the well-known necessary and sufficient conditions on the covariance that its Fourier transform be nonnegative with the fall-off in frequency and wave number sufficiently rapid at infinity. The limiting cases $K_X(\Delta\mathbf{R}, \Delta t) = \delta(\Delta\mathbf{R} - \Delta\mathbf{R}_0)\delta(\tau - t_0)$ and so on are employed when the concept of periodic $(\Delta R_0 \neq 0,\ \tau_0 \neq 0)$ and "white" spectra and "dc" phenomena are included $(\Delta\mathbf{R}_0 = 0,\ \tau_0 = 0)$. See Ref. [7], Sections 21 and 22 for details.

2.2.1 Directly Sampled Approximation of the W–Kh Relations (Hom-Stat Examples)

The Wiener–Khintchine relations (2.2.1a and 2.2.1b) for random noise (and signal) fields can also be expressed (approximately) in a discrete or sampled formulation. The resulting series expressions can be useful for sampling procedures employed in many practical applications. Among these are the design and implementation of digital matched filters, which in turn have desirable optimal properties, for example, in the reception process (Section 3.5 ff). However, we note here that discrete samples of the covariance of signal and noise fields are *not* the same as the covariance of these sampled fields. In fact, the latter required a considerably more indirect and analytically involved procedures than that described in the present section. (See Eqs. 2.2.3 et seq.)

Physical fields are not naturally factorable into solely spatial and temporal portions, that is, $X(\mathbf{R}_m, t_n) \neq A(\mathbf{R}_m)B(t_n)$, and neither are their associated statistics, for example, $K_X(\Delta\mathbf{R}, \tau) \neq K_X^{(S)}(\Delta\mathbf{R})\, K_X^{(T)}(\tau)$.[13] However, from the point of view of operational convenience, as well as design, separation of space and time is often *imposed* on the receiver. The result is constrained operation, where the temporal and spatial parts can be separately optimized, which is in principle suboptimum to the unconstrained case that includes the interactions between them. Quantitatively, the degree of suboptimality of the former vis-à-vis the latter is usually assumed to be small. Because of the difficulty in measuring this effect, it is usually ignored. The convenience of the imposed separability outweighs its assumed small magnitude. The technique discussed here and subsequently offer the possibility of quantifying this restriction (See Section 3.4.7).

The imposed condition of *separating space and time* now allows to express the covariance K_X and the associated spectrum of W–Kh relations (2.2.1a and 2.2.1b) for these fields as

$$K_X(\Delta\mathbf{R}, \tau) = K_X^{(S)}(\Delta\mathbf{R})K_X^{(T)}(\tau) = K_X^{(S\otimes T)}; \quad \mathsf{W}_X(\boldsymbol{\nu}, f) = \mathsf{W}_X^{(S)}(\boldsymbol{\nu})\mathsf{W}_X^{(T)}(f). \tag{2.2.3}$$

In matrix form, these become the Kronecker product of space and time components:

$$\mathbf{K}_X = \left[(K_X)_{jj'}\right] = [K_X(\mathbf{r}_{m'} - \mathbf{r}_m; t_{n'} - t_n)] = \left[K_X^{(S)}(\mathbf{r}_{m'} - \mathbf{r}_m) \otimes \left[K_X^{(T)}(t_{n'} - t_n)\right]\right] = [a_{mm'}\mathbf{B}],$$

[13] An exception occurs when either $A(\Delta\mathbf{R})$ or $B(\tau)$, or both, are monofrequentic, that is, $A(\mathbf{R})$ is described by a single wave number $(\boldsymbol{\nu} - \boldsymbol{\nu}_0)$ and $B(\tau)$ by $B(\tau) = (1/2)\cos\omega_0\tau$, and $A(\Delta\mathbf{R}) = K_X(\mathbf{R}_2 - \mathbf{R}_1)$. Even if all space–time samples are independent that is, the matrix $K_X = \left[K_{mn,m'n'}\delta_{mm'}\delta_{nn'}\right] = \left[K_{mn,mn}\right] \neq \left[K_{mm',nn'}\right]$ and so on, $\mathbf{K}_X$ does not factor.

that is,

$$\mathbf{K}_X = \mathbf{K}_X^{(S)} \otimes \mathbf{K}_X^{(T)} = \mathbf{A} \otimes \mathbf{B} = [a_{mm'} b_{nn'}], \quad \mathbf{A} = [a_{mm'}]; \quad \mathbf{B} = [b_{nn'}], \tag{2.2.4}$$

where $\mathbf{A}$ and $\mathbf{B}$, respectively, are functions of space and time. Thus, $\mathbf{A} = M \times M$ matrix and $\mathbf{B} = N \times N$ matrix and the product is an $(MN \times MN) = (J \times J)$ square matrix, as expected. Furthermore, the elements of $\mathbf{A}$ and $\mathbf{B}$ can be written $a_{mm'} = a_{m'-m}, b_{nn'} = b_{n-n'}$ from (2.2.3). From this result (2.2.3), we obtain the expression corresponding to the continuous forms (2.2.1a and 2.2.1b).

Next, we can obtain (approximate) discrete W–Kh relations by observing first that the elements of $\mathbf{A}$ and $\mathbf{B}$ can also be written $a_{mm'} = a_k, k = m' - m$, and $b_{nn'} = b_l$ with $l = n' - n$ because of the assumed WS-HS property of the field $X(\mathbf{r}, t)$. Then, the elements of integration in W–Kh relation (2.2.1a) can be written $\Delta^3\mathbf{K} = a_k b_l \Delta^3\mathbf{R}\Delta t$, where $\Delta^3\mathbf{R}\Delta t = \Delta x \Delta y \Delta z \Delta t$ is the four-dimensional volume element. From the continuous forms (2.2.1a and 2.2.1b) we find directly that the corresponding discrete form of these integral relations is approximately

$$\left.\begin{aligned} \mathsf{W}_X^{(S\otimes T)}(\boldsymbol{\nu}_q, f_p) &\doteq 2\sum_{k=-M}^{M}\sum_{l=-N}^{N} a_k b_l e^{2\pi i \boldsymbol{\nu}_q \cdot \mathbf{R}_k - 2\pi i f_p \tau_l}\Delta^3\mathbf{R}\Delta t \\ &\doteq 2\sum_{k=-M}^{M}\sum_{l=-N}^{N} K_X^{(S)}(\mathbf{R}_k) K_X^{(T)}(\tau_l) e^{2\pi i \boldsymbol{\nu}_q \cdot \mathbf{R}_k - 2\pi i f_p \tau_l}\Delta^3\mathbf{R}\Delta t \\ &\doteq \mathsf{W}_X^{(S)}(\boldsymbol{\nu}_q)\mathsf{W}_X^{(T)}(f_p) \end{aligned}\right\}, \tag{2.2.5a}$$

where $q = (-M, \ldots, +M)$ and $p = (-N, \ldots, +N)$. Here specifically $\tau_l = (n' - n)\Delta t = l\Delta t$, $l = n' - n$, and $\mathbf{R}_k = \mathbf{r}_{m+k} - \mathbf{r}_m = \hat{\mathbf{i}}_1 R_{1k} + \hat{\mathbf{i}}_2 R_{2k} + \hat{\mathbf{i}}_3 R_{3k} = \hat{\mathbf{i}}_1(x_{m+k} - x_m) + \hat{\mathbf{i}}_2(y_{m+k} - y_m) + \hat{\mathbf{i}}_3(z_{m+k} - z_m) \equiv k\Delta R$, where l, and k are integers. Similarly, we have[14] $\boldsymbol{\nu}_q = \boldsymbol{\nu}_{s+q} - \boldsymbol{\nu}_s \equiv q\Delta\boldsymbol{\nu}$ and $f_p = (r' - r)\Delta f = p\Delta f$, (q, r) integers. The inverse relations, namely, the (approximate) discrete form of (2.7.1b), are similarly obtained, with the volume element now $\Delta \mathsf{W} = \alpha_q \beta_l \Delta\boldsymbol{\nu}\Delta f$, where $\Delta^3\boldsymbol{\nu} = \Delta\nu_x \Delta\nu_y \Delta\nu_z$. We have

$$\left.\begin{aligned} K_X^{(S\otimes T)}(\mathbf{R}_k, \tau_l) &\doteq \frac{1}{2}\sum_{q=-M}^{M}\sum_{p=-N}^{N} \alpha_q \beta_p e^{-2\pi i \boldsymbol{\nu}_q \cdot \mathbf{R}_k + 2\pi i f_p \tau_l}\Delta^3\boldsymbol{\nu}\Delta f \\ &\doteq \frac{1}{2}\sum_{q=-M}^{M}\sum_{p=-N}^{N} \mathsf{W}_X^{(S)}(\boldsymbol{\nu}_q)\mathsf{W}^{(T)}(f_p) e^{-2\pi i \nu_q \cdot \mathbf{R}_k + 2\pi i f_p \tau_l}\Delta^3\boldsymbol{\nu}\Delta f \\ &\doteq K_X^{(S)}(\mathbf{R}_k) K_X^{(T)}(\tau_l) \end{aligned}\right\}, \tag{2.2.5b}$$

with the imposed separability again evident. A physical demonstration of WS-HS by direct calculation of the noise (and signal) fields, including the specific physical conditions that must be satisfied therein, can be made. In any case, we emphasize that separability of space

[14] $\Delta\boldsymbol{\nu} \neq \Delta^3\boldsymbol{\nu}$; see remarks following Eq. (2.2.1a).

and time here is an *imposed* ensemble property. (Note that when the separability constraint is removed, $K_X^{S\otimes T} \to K_X^{(ST)}(\mathbf{R}_k, \tau_l)$ and $\mathsf{W}_X^{(S\otimes T)} \to \mathsf{W}_X^{(ST)}(\boldsymbol{\nu}_q, f_p)$ in (2.2.5a and 2.2.5b).

Since Gaussian random processes are essentially defined by their covariance functions K_X (from which associated means and second-order moments are immediately obtained), the resulting pdf values (of discrete, sampled values) can be constructed at once, including various types of inhomogeneities, nonstationary, and so on, exhibited by the direct sampling of K_X and W_X in the discussion above. For non-Gaussian processes and fields, however, higher order covariances, which are *not* functions of the first-order K_X, appear explicitly as well as implicitly in their statistical description. Therefore, it is usually *not* enough to describe inhomogeneity or nonstationarity in second-order, that is, wide sense terms alone (see Chapter 8 following).

Finally, we must distinguish here between the sampling procedure above and those in Section 2.2.3. The former is an analytic approximation applied directly to the W–Kh relations themselves [Eq. (2.2.5a and 2.2.5b)], whereas the latter is the actual physical operation applied to the received field, in the manner of Eqs. (1.6.2a). Thus, the former is a method of approximation, while the latter represents discrete sampling of the continuous field data.

2.2.2 Extended Wiener–Khintchine Theorems: Continuous Inhomogeneous and Nonstationary Random (Scalar) Fields

As we have already noted (Section 2.1.2), real nonstationary processes and inhomogeneous fields are the rule rather than the exception in practice. Frequently, we can assume with some justification from experiment that a received field is effectively, or at least to a good approximation homogeneous and stationary (Hom-Stat) in the vicinity of the receiver. However, in many cases, we must contend with the fact that the field in question is not so obliging. We must accordingly take this bothersome fact into account if we are to achieve effective or even near-maximal detection and estimation and the correct decisions and measurements that follow. We have already seen [Eqs. (2.2.1 and 2.2.2)] that the covariance and the associated intensity spectrum are among the important statistics that describe such random fields whose analytical representation must now be extended to account for their non-Hom-Stat condition. We therefore seek appropriate generalizations of the usual Hom-Stat Wiener–Khintchine cases discussed in Section 2.2.1. Sections 2.2.2–2.2.3 and the following present some of these generalizations.

2.2.2.1 Continuous Non-Hom-Stat Fields Here, the (real) field in the neighborhood of the receiver is continuously sampled for a finite period and over a finite spatial interval (1.6.2) to yield a continuous data stream $X(\mathbf{r},t)$ after sampling,[15] of space–time "size" $\Delta^{(4)} = |\mathbf{R}_0 = |X_0Y_0Z_0||T$. During this interval, $X(\mathbf{r},t)$ is assumed to be non-Hom-Stat. Let us now consider the amplitude wave number frequency spectrum of the sample $X(\mathbf{r},t)_\Delta$ in Δ, which is the Fourier transform of $X(\mathbf{r},t)$:

[15] We assume the simplest sampling situation, where $\alpha(\mathbf{r},t) = \kappa_0\alpha_{\text{input}}(\mathbf{r},t)$, with κ_0 representing the role of the transducer in changing from the input field dimensions to other dimensionalities, such as, pressure to current displacement. In most cases, we shall absorb the conversion parameter κ into $\propto$, that is, $\alpha(\mathbf{r},t) = \alpha_{\text{input}}$, and so on to electric potential (volts).

$$S_X(\boldsymbol{\nu},f)_\Delta = \int_{-\Delta/2}^{\Delta/2} X(\mathbf{r},t)_\Delta e^{2\pi i\mathbf{r}\cdot\boldsymbol{\nu}-2\pi itf}\mathbf{d}^3\mathbf{r}dt, \quad \text{with} \quad \Delta^{(4)} = \left(R_{0_x}R_{0_y}R_{0_z}\right)T = \Delta,\ \langle X\rangle = 0, \tag{2.2.6a}$$

and $\therefore \langle S_X\rangle = 0$, where we have applied (1.6.2b) to the input field to obtain $X(\mathbf{r},t)_\Delta = X(\mathbf{r},t), (\mathbf{r},t) \in \Delta$; and 0 elsewhere. Here, as usual, $\mathbf{d}^3\mathbf{r} = dxdydz$ and $\boldsymbol{\nu} = \hat{\mathbf{i}}_1\nu_x + \hat{\mathbf{i}}_2\nu_y + \hat{\mathbf{i}}_3\nu_z$, where now $X(\mathbf{r},t)$ is understood to represent the *ensemble* of representation of X. The inverse transform of $S_X(\boldsymbol{\nu},f)_\Delta$ is directly

$$\int_{-\infty}^{\infty} \cdots \int \mathbf{d}^3\boldsymbol{\nu}df\, S_X(\boldsymbol{\nu},f)_\Delta e^{-2\pi i\boldsymbol{\nu}\cdot\mathbf{r}+2\pi ift} = X(\mathbf{r},t)_\Delta, \quad \in \Delta\left(=\Delta^{(4)}\right),$$
$$= 0, \text{ elsewhere}, \langle X\rangle = 0, \tag{2.2.6b}$$

with $\mathbf{d}^3\boldsymbol{\nu} = d\nu_x d\nu_y d\nu_z$ in rectangular coordinates. The (continuously) sampled field $X(\mathbf{r},t)$ may also be the result of a two- or one-dimensional sampling operation, that is, $\mathbf{r} = \hat{\mathbf{i}}_1 r_x + \hat{\mathbf{i}}_2 r_y$ or $\mathbf{r} = \hat{\mathbf{i}}_1 r_x$, for example, in which case only $\boldsymbol{\nu} = \hat{\mathbf{i}}_1\nu_x + \hat{\mathbf{i}}_2\nu_y$ or $\boldsymbol{\nu} = \hat{\mathbf{i}}_1\nu_x$ respectively appear in the spectrum, and $\mathbf{d}^3\mathbf{r}dt = dr_x dr_y \delta(z-0)dz$ or $dr_x\delta(y-0)\delta(z-0)dr_y dr_z$ represents the differential elements of integration.[16] Also, it may be more convenient in some cases to use spherical or polar coordinates in (2.2.6a and 2.2.6b), (for which the Jacobian of the transformation is $\nu^2 d\nu \sin\theta d\theta d\phi$ or $\nu d\nu d\phi$, with $\boldsymbol{\nu}\cdot\mathbf{r} = \left(\hat{\mathbf{i}}_\nu\cdot\hat{\mathbf{i}}_r\right)\nu|\mathbf{r}|$) (2.2.1a above.).

We next use the definition $\mathsf{W}_{X|\Delta,\,\Delta<\infty}$ in (2.2.1) to obtain this intensity spectrum, based on the finite sample Δ in (2.2.6a), namely,

$$\mathsf{W}_X(\boldsymbol{\nu},f)_\Delta \equiv \mathbf{E}\left\{\frac{2}{\Delta^{(4)}}\left|S_X(\boldsymbol{\nu},f)_\Delta\right|^2\right\} = \frac{2}{\Delta^{(4)}}\overline{|\mathsf{F}\{X_\Delta\}|^2}, \tag{2.2.7}$$

which is assumed to exist from (2.2.6a and 2.2.6b). Thus, recalling from (2.1.1) that the (auto) covariance is symmetric in ($\mathbf{r}$,t), obtain

$$\therefore \mathsf{W}_X(\boldsymbol{\nu},f)_\Delta = \frac{2}{\Delta}\int_{-\Delta^{(3)}/2,-\Delta/2}^{\Delta^{(3)}/2,\Delta/2} \overline{X(\mathbf{r}_1,t_1)X(\mathbf{r}_2,t_2)}\, e^{2\pi i\boldsymbol{\nu}\cdot(\mathbf{r}_1-\mathbf{r}_2)-2\pi if(t_1-t_2)}\mathbf{d}^3\mathbf{r}_1\mathbf{d}^3\mathbf{r}_2 dt_1 dt_2$$
$$= \frac{2}{\Delta}\int_{-\Delta^{(3)}/2,-\Delta/2}^{\Delta^{(3)}/2,\Delta/2} K_X(\mathbf{r}_1,t_1;\mathbf{r}_2,t_2)\, e^{2\pi i\boldsymbol{\nu}\cdot(\mathbf{r}_1-\mathbf{r}_2)-2\pi if(t_1-t_2)}\mathbf{d}^3\mathbf{r}_1\mathbf{d}^3\mathbf{r}_2 dt_1 dt_2. \tag{2.2.8}$$

This, in turn, is finite, positive (or zero), when K_X is the (untruncated) covariance, with $\langle X\rangle = 0$ here. Since

[16] We shall use r_x for x, and so on and vice versa, including $\mathbf{r}$ for $\hat{\mathbf{i}}_1 x + \hat{\mathbf{i}}_2 y + \hat{\mathbf{i}}_3 z$, interchangeably.

$$\int_{-\infty}^{\infty} e^{2\pi i \boldsymbol{\nu}\cdot(\mathbf{r}_2-\mathbf{r}_1+\mathbf{r})}\mathbf{d}^3\boldsymbol{\nu}\int_{-\infty}^{\infty} e^{2\pi i f(t_2-t_1+t)}df = \boldsymbol{\delta}(\mathbf{r}_2-\mathbf{r}_1+\mathbf{r})\delta(t_2-t_1+t), \quad (2.2.9a)$$

the double, that is, wave number–frequency Fourier transform of $\mathsf{W}_X(\boldsymbol{\nu},f)_\Delta$, defined over the ensemble $X(\mathbf{r},t)_\Delta$, is

$$\int_{-\infty}^{\infty} \mathsf{W}_X(\boldsymbol{\nu},f)_\Delta e^{-2\pi i \boldsymbol{\nu}\cdot\mathbf{r}+2\pi i f t}\mathbf{d}^3\boldsymbol{\nu}df$$

$$= \frac{2}{\Delta}\int_{-\Delta^{(3)}/2,-\Delta/2}^{\Delta^{(3)}/2,\Delta/2} K_X(\mathbf{r}_1,t_1;\mathbf{r}_2,t_2)\mathbf{d}^3\mathbf{r}_1\mathbf{d}^3\mathbf{r}_2 dt_1 dt_2\cdot\int_{-\infty}^{\infty} e^{2\pi i\boldsymbol{\nu}\cdot(\mathbf{r}_2-\mathbf{r}_1+\mathbf{r})+2\pi i f(t_2-t_1+t)}\mathbf{d}^3\boldsymbol{\nu}df, \quad (2.2.9b)$$

which gives for (2.2.8) the result

$$\int_{-\infty}^{\infty} \mathsf{W}_X(\boldsymbol{\nu},f)_\Delta e^{-2\pi i \boldsymbol{\nu}\cdot\mathbf{r}+2\pi i f t}\mathbf{d}^3\boldsymbol{\nu}df = \frac{2}{\Delta}\int_{-\Delta/2}^{\Delta/2} K_X(\mathbf{r}_1,t_1;\mathbf{r}_1-\mathbf{r},t_1-t)\mathbf{d}^3\mathbf{r}_1 dt_1. \quad (2.2.10)$$

Equations (2.2.8 and 2.2.10) are the *non-Hom-Stat versions* of the (wide sense) homogeneous–stationary Wiener–Khintchine relations (2.2.1a and 2.2.1b), and the extension of Eqs. (3.37–3.39) of Ref. [1] to now include space as well as time.

At this point, it is convenient to introduce the more compact notation[17]

$$\mathbf{p}_1 \equiv (\mathbf{r}_1,t_1),\ \mathbf{p}_2 \equiv (\mathbf{r}_2,t_2),\ \text{or in vector form explicitly,}$$

$$\left\{\mathbf{p}_{(1,2)} \equiv \mathbf{r}_{(1,2)} - \hat{\mathbf{i}}_4 t_{(1,2)},\ \mathbf{q} \equiv \boldsymbol{\nu} + \hat{\mathbf{i}}_4 f,\right. \quad (2.2.11)$$

with

$$\mathbf{r} = \hat{\mathbf{i}}_x r_x + \hat{\mathbf{i}}_y r_y + \hat{\mathbf{i}}_z r_z \quad \text{and} \quad \mathbf{q} = \hat{\mathbf{i}}_x \nu_x + \hat{\mathbf{i}}_y \nu_y + \hat{\mathbf{i}}_z \nu_z + \hat{\mathbf{i}}_4 f = \hat{\mathbf{i}}_1 R_x^{-1} + \hat{\mathbf{i}}_2 R_y^{-1} + \hat{\mathbf{i}}_3 R_z^{-1} + \hat{\mathbf{i}}_4 f \quad (2.2.11a)$$

and where

$$\mathbf{d}^4\mathbf{p} \equiv \mathbf{d}^3\mathbf{r}dt = dxdydz \text{ or } dr_x dr_y dr_z dt \quad \text{and} \quad \mathbf{d}^4\mathbf{q} = \mathbf{d}^3\boldsymbol{\nu}df = d\nu_x d\nu_y d\nu_z df \quad (2.2.11b)$$

where notationally $F(\mathbf{p})$ is a function of $\mathbf{p}$, that is, $\mathbf{p}$, $|\mathbf{p}|$, p_1, ..., p_4, and so on. In fact, $F(\mathbf{p})$ is usually a functional of $\mathbf{p}$ in $F(g(\mathbf{p}))$, where $g(\mathbf{p}) = p_x$ and so on.

[17] Thus, we have $K(\mathbf{p}) = K(\mathbf{r}_m - \hat{i}_4\mathbf{t}_n)$ or $K(\mathbf{r}_m, t_n)$, $K(\mathbf{p}_1,\mathbf{p}_2) = K(\mathbf{r}_m - \hat{i}_4 t_n, \mathbf{r}_{m'} - \hat{i}_4 t_{n'})$ or $K(\mathbf{r}_m, \mathbf{t}_n;\ \mathbf{r}_{m'}, \mathbf{t}_{n'})$: the notation is self-explanatory.

We, therefore, write (2.2.8 and 2.2.10) as the following:

$$I.\ \mathit{Spectrum}:\ \mathsf{W}_X(\boldsymbol{\nu},f)_\Delta = \frac{2}{\Delta}\int_{-\Delta^{(3)}/2,-\Delta/2}^{\Delta^{(3)}/2,\Delta/2} K_X(\mathbf{p}_1,\mathbf{p}_2)e^{2\pi i\mathbf{q}\cdot(\mathbf{p}_1-\mathbf{p}_2)}\mathbf{d}^4\mathbf{p}_1\mathbf{d}^4\mathbf{p}_2$$

$$= \mathsf{W}_X(\mathbf{q})_\Delta \qquad (2.2.12a) \qquad (2.2.12a)$$

$$II.\ \mathit{Covariance}:\ \frac{2}{\Delta}\int_{-\Delta^{(4)}/2}^{\Delta^{(4)}/2} K_X(\mathbf{p}_1,\mathbf{p}_1-\mathbf{p})\mathbf{d}^4\mathbf{p}_1 = \int_{-\infty}^{\infty}\mathsf{W}_X(\mathbf{q})_\Delta e^{-2\pi i\mathbf{q}\cdot\mathbf{p}}\mathbf{d}^4\mathbf{q} \qquad (2.2.12b)$$

(These results, Eqs. (2.2.7–2.2.12b), are the generalizations of the purely temporal results of Eqs. (3.37–3.39) of Ref. [1] to space–time.)

For the real non-Hom-Stat fields typical of physical communication processes, where $\Delta \to \infty$, it can be demonstrated that the right-hand number (of 2.2.12b) exists, provided

$$\left.\begin{array}{ll} \text{(i)} & K_X \text{ is suitably continuous at } t_1 = t_2 = 0 \text{ and } \mathbf{r}_1 = \mathbf{r}_2 = 0; \\ \text{(ii)} & X(\mathbf{r},t) \text{ possess a } \textit{finite} \text{ space and time average } \langle X(\mathbf{r},t)\rangle \\ & \text{for almost all members of the ensemble } X; \text{ and} \\ \text{(iii)} & X(\mathbf{r},t) \text{ is postulated to be an (real) entirely random field.} \end{array}\right\} \qquad (2.2.13)$$

Accordingly, as $\Delta \to \infty$, we finally obtain[18]

$$\langle K_X(\mathbf{p}_1,\mathbf{p}_1-\mathbf{p})\rangle = \frac{1}{2}\int_{-\infty}^{\infty}\mathsf{W}_X(\boldsymbol{\nu},f)_\infty e^{-2\pi i\mathbf{q}\cdot\mathbf{p}}\mathbf{d}^4\mathbf{q} = \frac{1}{2}\mathsf{F}^{-1}_{\nu,f}\{\mathsf{W}_{X,\infty}\} \qquad (2.2.13a)$$

with

$$2\int_{-\infty}^{\infty}\langle K_X(\mathbf{p}_1,\mathbf{p}_1-\mathbf{p})\rangle e^{2\pi i\mathbf{q}\cdot\mathbf{p}}\mathbf{d}^4\mathbf{p} = \mathsf{W}_X(\boldsymbol{\nu},f)_\infty = 2\mathsf{F}_{\mathbf{r},t}\{\langle K_X\rangle\}, \qquad (2.2.13b)$$

where $\lim_{\Delta\to\infty}\mathsf{W}_X(\boldsymbol{\nu},f)_\Delta \to \mathsf{W}_X(q)_\infty$. Equations (2.2.13a and 2.2.13b) constitute an extended form of the more familiar W–Kh theorem (2.2.1a and 2.2.1b).

A hybrid expression for (2.2.13a and 2.2.13b) follows directly when the spatial domain $|\Delta\mathbf{R}|$ remains finite, while $T \to \infty$, a situation often approached in practice where the spatial portion of the data samples necessarily remain bounded. Then, $\Delta^{(4)} = |\Delta^3\mathbf{R}|T \to |\Delta^3\mathbf{R}|$ $(T \to \infty)$, $|\Delta\mathbf{R}| \to \infty$, symbolically, and $\mathsf{W}_X(\boldsymbol{\nu},f)_\infty \to \mathsf{W}_X(\boldsymbol{\nu},f)_{|\Delta\mathbf{R}|,\infty}$ in (2.2.13a and

[18] A proof of the existence of (2.2.13a and 2.2.13b) when $\Delta \to \infty$ may be found as an extension of the Hom-Stat cases from the temporal proof outlined in Section 3.2.2 of Ref. [1], refer to footnote, Section 3.4.5 of Ref. [1], p. 194.

2.2.13b). In both cases, when the non-Hom-Stat, nonstationary conditions result in $|\langle X(\mathbf{r},t)\rangle^2| \to \infty$ (for almost all members of the ensemble), the limits on W_X do not exist as $\Delta \to \infty$ or $T \to \infty$, and thus Eqs. (2.2.13a and 2.2.13b) are meaningless. The case where the intensity level for the field $X(\mathbf{r},t)$ increases with time, or space and time, is one such example. On the other hand, for suitably bounded nonstationarities, the relations (2.2.13a and 2.2.13b) can be applied.

Since $K_X(\mathbf{r}_1,t_1;\mathbf{r}_2,t_2) = K_X(\mathbf{r}_2,t_2;\mathbf{r}_1,t_1)$, that is, is symmetric for these real fields, we see that $\langle K_X(\mathbf{p}_1,\mathbf{p}_1-\mathbf{p})\rangle = \langle K_X(\mathbf{p}_1,\mathbf{p}_1+\mathbf{p})\rangle$, from which it follows that $\mathsf{W}_X(\nu,f) = \mathsf{W}_X(-\nu,-\mathbf{f})$ (2.1.12)a); $\langle K_X(\mathbf{p}_1,\mathbf{p}_1-\mathbf{p})\rangle$ is an even function $(\boldsymbol{\nu}, f)$. The critical analytic complexity resulting from the non-Hom-Stat nature of the field is that the covariance K_X now depends on when the field is initially observed, so that it is no longer possible to take advantage of the inherent simplification of the convolutional form when Hom-Stat occurs. Ensemble averages are now always required, since we cannot use the (wide sense) ergodic theorem (Section 1.7.1 of Ref. [1]) to replace the average by observations of a single representation $X^{(j)}(\mathbf{r},t)$ in the space–time domain.

As a further example of the need for the statistical average (< >), we find for the case of *cross-covariance* of two possibly related field ensembles $X(\mathbf{r},t)$ and $Y(\mathbf{r},t)$, the resulting extension of the Wiener–Khintchine theorem (2.2.12a and 2.2.12b) in the non-Hom-Stat cases:

$$I.\ \textit{Spectrum}:\ \mathsf{W}_{XY}(\mathbf{q})_\Delta = \frac{2}{\Delta}\int_{-\Delta/2}^{\Delta/2} K_{XY}(\mathbf{p}_1,\mathbf{p}_2)e^{2\pi i\mathbf{q}\cdot(\mathbf{p}_1-\mathbf{p}_2)}\mathbf{d}^4\mathbf{p}_1\mathbf{d}^4\mathbf{p}_2 \qquad (2.2.14a)$$

$$II.\ \textit{Covariance}:\ \frac{2}{\Delta}\int_{-\Delta^4/2}^{\Delta^4/2} K_{XY}(\mathbf{p}_1,\mathbf{p}_1-\mathbf{p})\mathbf{d}^4\mathbf{p}_1 = \int_{-\infty}^{\infty} \mathsf{W}_{XY}(\mathbf{q})_\Delta e^{-2\pi i\mathbf{q}\cdot\mathbf{p}}\mathbf{d}^4\mathbf{q}. \qquad (2.2.14b)$$

The corresponding expressions as $\Delta \to \infty$ and $\mathsf{W}_{XY}(\mathbf{q})_\Delta \to \mathsf{W}_{XY}(\mathbf{q})$, with $\mathsf{W}_{XY}(\mathbf{q})$ (almost) everywhere finite, and subject to (i–iii) (2.2.13), are found at once paralleling the procedure for (2.2.13a and 2.2.13b), namely,

$$II.\ \textit{Covariance}:\ \langle K_{XY}(\mathbf{p}_1,\mathbf{p}_1-\mathbf{p})\rangle = \frac{1}{2}\int_{-\infty}^{\infty} \mathsf{W}_{XY}(\mathbf{q})e^{-2\pi i\mathbf{q}\cdot\mathbf{p}}\mathbf{d}^4\mathbf{q} = \frac{1}{2}\mathsf{F}_{\mathbf{q}}^{-1}\{\mathsf{W}_{XY}\} \qquad (2.2.15a)$$

$$I.\ \textit{Spectrum}:\ \mathsf{W}_{XY}(\mathbf{q}) = 2\int_{-\infty}^{\infty} \langle K_{XY}(\mathbf{p}_1,\mathbf{p}_1-\mathbf{p})\rangle e^{2\pi i\mathbf{q}\cdot\mathbf{p}}\mathbf{d}^4\mathbf{p} = 2\mathsf{F}_{\mathbf{r},t}^{-1}\{\langle K_{XY}\rangle\}, \qquad (2.2.15b)$$

where

$$K_{XY}(\mathbf{p}_1,\mathbf{p}_1-\mathbf{p}) = K_{YX}(\mathbf{p}_1,\mathbf{p}_1+\mathbf{p}), \quad \text{and} \quad \mathsf{W}_{XY}(\mathbf{q}) = \mathsf{W}_{YX}(\mathbf{q})^*. \qquad (2.2.15c)$$

The concept of the intensity spectrum of a nonhomogeneous, nonstationary random field is accordingly extended from the more familiar Hom-Stat cases by the procedure above

(2.2.7–2.2.15c). As can be seen, considerably more sample data are required to approximate the defining covariances, because it is the ensemble mathematically regarded as having an infinite number of members that is invoked here. In high-frequency radar applications, a large number of relevant samples can be obtained in the short time available for the decision process. Where sufficient computational power is available, it should be possible to obtain good approximations for K_X (and K_{XY}) in real time and thus to replace the theoretical ideal $K_X(p_1, p_1 - p)$ with effective estimates. In sonar applications, on the other hand, effective estimates of K_X may not be so precise. An exception occurs when the non-Hom-Stat condition is comparatively weak, allowing one to assume essentially. Hom-Stat conditions. In any case, it is the stability (or instability) of the medium that is usually the controlling factor here.

2.2.3 The Important Special Case of Homogeneous—Stationary Fields—Finite and Infinite Samples

When the Hom-Stat condition is applicable, then $K_X(\mathbf{p}_1, \mathbf{p}_2) = K_X(\mathbf{p}_2 - \mathbf{p}_1) = K_X(\mathbf{p}_1 - \mathbf{p}_2)$ on the infinite sample interval. For *finite samples* ($\Delta < \infty$), we easily see the following on applying Eq. (2.2.12a):

$$I.\ Spectrum: \left.\begin{aligned} \mathsf{W}_X(\mathbf{q})_\Delta &= \frac{2}{\Delta}\int_{-\Delta^{(8)}/2}^{\Delta^{(8)}/2} K_X(\mathbf{p}_1 - \mathbf{p}_2) e^{2\pi i \mathbf{q}(\mathbf{p}_1-\mathbf{p}_2)} \mathbf{d}^4\mathbf{p}_1 \mathbf{d}^4\mathbf{p}_2 \\ &= 2\int_{-\infty}^{\infty} K_X(\mathbf{p})_\Delta e^{2\pi i(\mathbf{q}\cdot\mathbf{p})} \mathbf{d}^4\mathbf{p} \geq 0, \quad K_X(\mathbf{p}) = K_X(-\mathbf{p}) \end{aligned}\right\} \tag{2.2.16}$$

Here, $\mathbf{p} = \mathbf{p}_1 - \mathbf{p}_2$, that is, $\mathbf{p} = \mathbf{r}_1 - \mathbf{r}_2 - \hat{i}_4(t_2 - t_1) = \mathbf{r} - \hat{i}_4\tau$, with $\mathbf{r} = \mathbf{r}_1 - \mathbf{r}_2$, $\tau = t_2 - t_1$. In addition, we have compactly

$$K_X(\mathbf{p})_\Delta = K_X(\mathbf{p})_\infty(1 - |\mathbf{p}^3|/\Delta^4); \ = 0, \ |\Delta^4| < |\mathbf{p}|, \tag{2.2.16a}$$

or equivalently,

$$K_X(\mathbf{p})_\Delta = K_X(\mathbf{p})\left(1 - \left|\mathbf{R}^{(3)}\right|/\Delta^4\right)\left(1 - |\tau|/\Delta^4\right); \ = 0, \ \left|\mathbf{R}^{(3)}\right| < \left|\mathbf{r}^{(3)}\right|; \ T < |\tau|, \tag{2.2.16b}$$

refer to (2.2.6a), where $\Delta^4 = \left|\mathbf{R}^3\right|T$, $\left(\left|\mathbf{R}^3\right| = R_x R_y R_z\right)$ is the domain of the (ensemble) of finite samples.

Taking the Fourier transform of both sides of (2.2.16) gives the following in these Hom-Stat cases.

$$II.\ Covariance: \quad K_X(\mathbf{p})_\Delta = \frac{1}{2}\int_{-\infty}^{\infty} \mathsf{W}_X(\mathbf{q})_\Delta e^{-2\pi i \mathbf{q}\cdot\mathbf{p}} \mathbf{d}^4\mathbf{q}. \tag{2.2.16c}$$

To establish the W–Kh theorem in case of infinite sample intervals is not difficult.[19] We need to show that $\lim_{\Delta\to\infty} \mathsf{W}_{X-\Delta}$ approaches a definite limit, that is,

$$\lim_{\Delta\to\infty} \mathsf{W}_X(\mathbf{q})_\Delta = 2\int_{-\infty}^{\infty} K_X(\mathbf{p}) e^{2\pi i \mathbf{q}\cdot\mathbf{p}} \mathbf{d}^4\mathbf{p} = 2\mathsf{F}\{K_X\} = \mathsf{W}_X(\mathbf{q}), \tag{2.2.17a}$$

so that as required

$$K_X(\mathbf{p}) = \frac{1}{2}\int_{-\infty}^{\infty} \mathsf{W}_X(\mathbf{q}) e^{-2\pi i (\mathbf{q}\cdot\mathbf{p})} \mathbf{d}^4\mathbf{q} = \frac{1}{2}\mathsf{F}^{-1}\{\mathsf{W}_X(\mathbf{q})\}. \tag{2.2.17b}$$

We accordingly outline the steps showing $\lim_{\Delta\to\infty} \mathsf{W}_X(\mathbf{q})_\Delta = \mathsf{W}_X(\mathbf{q})$ [Eq. (2.2.17a)]. First, we require $K_X(\mathbf{p})$ in (2.2.16a) to be continuous and then begin with the nonnegative form Eq. (2.2.17a). We next have the following:

(1) Equation (2.2.16a) multiplied[20] by $(1 - |\mathbf{q}|/\mathbf{q}_0)e^{-2\pi i \mathbf{p}\cdot\mathbf{q}}$, where $|\mathbf{q}| = |v_x v_y v_z \cdot f|$ and $|\Delta v| f = v_{0x} v_{0y} v_{0z} f_0 \equiv |\mathbf{q}_0|$; (2.2.18)

(2) Then, integrate it with respect to $\mathbf{q}$ over the region $(-|\mathbf{q}_0|/2, |\mathbf{q}_0|/2)$ to obtain

$$\int_{-\mathbf{q}_0}^{\mathbf{q}_0} (1 - |\mathbf{q}|/\mathbf{q}_0) \mathsf{W}_X(\mathbf{q})_\Delta e^{2\pi i \mathbf{q}\cdot\mathbf{p}} \mathbf{d}^3\mathbf{q} = 2\int_{-\infty}^{\infty} K_X(\mathbf{p})_\Delta \prod_{l=1}^{4} \left\{ \frac{\sin[q_{0l}(p'_l - p_l)/2]}{q_l(p'_l - p_l)/2} \right\}^2 |\mathbf{q}_0| \mathbf{d}^3\mathbf{p}', \tag{2.2.19}$$

where $\mathbf{dp} = dp_1 dp_2 dp_3 dp_4 = dr_x dr_y dr_z dt$; $q_{0l} = |v_{0l}|$, $l = 1,2,3$; $q_{04} = f_0$, and $p_l = r_l$, $l = 1,2,3$, $p_4 = -x$.

(3) The coefficient of $e^{-2\pi i \mathbf{q}\cdot\mathbf{p}}$ in the left integrand is nonnegative. Therefore, this integrand is proportional to a characteristic function or, equivalently, to a proper pdf.

(4) According to the continuity theorem of Lévy, in Section 10.4 of Ref. [1], the right-hand side of (2.2.19) converges uniformly in every finite interval to $K_X(p)_\Delta$ as $|\mathbf{q}_0| \to \infty$, under the above requirement that $K_X(\mathbf{p})_\Delta$ be continuous.

(5) Consequently, $K_X(\mathbf{p})_\Delta$ for which $K_X(0)_\Delta = K_X(0) > 0$ is interpretable as a characteristic function, except for the scale factor $K_X(0)_\Delta$.

(6) Also, it follows from the Lévy continuity theorem that as $\Delta \to \infty$ in every finite interval $\mathbf{p}$, $K_X(\mathbf{p})_\Delta$ is also a characteristic function when divided by $K_X(0)$. Thus, since $\lim_{\Delta\to\infty} K_X(\mathbf{p})_\Delta = K_X(\mathbf{p})$, the result (2.2.17a) follows, proving the W–Kh theorem (2.2.17a and 2.2.17b) for these Hom-Stat fields $X(\mathbf{r},t)$. Moreover, since $K_X(\mathbf{p}) = K_X(-\mathbf{p})$ and $\mathsf{W}_X(\mathbf{q}) = \mathsf{W}_X(-\mathbf{q})$, we see that $K_X(\mathbf{p})$ and $\mathsf{W}_X(\mathbf{q})$ are cosine

[19] This demonstration (2.2.18–2.2.20) is a simple extension of the proof of Loève [4]. See the footnote, p. 142 of Ref. [1].

[20] This is to be interpreted like Eqs. (2.2.16a and 2.2.16b).

Fourier transforms of one another, for example,

$$\mathsf{W}_X(\mathbf{q}) = 4\int_0^{\infty} K_X(\mathbf{p})e^{\cos\, 2\pi i\mathbf{q}\cdot\mathbf{p}}\mathbf{d}^4\mathbf{p};\quad K_X(\mathbf{p}) = \int_0^{\infty} \mathsf{W}_X(\mathbf{q})e^{\cos 2\pi i\mathbf{q}\cdot\mathbf{p}}\mathbf{d}^4\mathbf{q} \qquad (2.2.20)$$

refere to (2.2.17a and 2.2.17b). (See the discussion in Section 3.2.2, and following it, of Ref. [1] for the purely temporal cases. See also Section 2.2.1 of this chapter for an introductory treatment of the space–time cases.

2.3 THE W–Kh RELATIONS FOR DISCRETE SAMPLES IN THE NON-HOM-STAT SITUATION

In Section 2.2.1, we have presented a direct approximation of the continuous Hom-Stat W-Kh form, followed in Section 2.2.2 by the continuous extensions of W–Kh relations to the non-Hom-Stat cases. Now, as a preliminary to the treatment of apertures and arrays in Section 2.3, we need to examine in more detail the role of discrete space–time sampling of the received input field $\alpha(\mathbf{r},t)$ itself and its sampled output $X(\mathbf{r}_m, t)$, since the covariance of such samples is not the same as the sampled covariance. As we shall see below, this requires a somewhat more intricate analysis than the continuous treatment of Sections 2.2.1 and 2.2.2 because we are replacing a continuum with a series of discrete values of the input field, taken at *dimensionless* points. Accordingly, we begin by applying the discrete sampling operator of Eq. (2.1.2) to the continuous input field $\alpha(\mathbf{r},t)$, to consider first the following.

2.3.1 The Amplitude Spectrum for Discrete Samples

Using (1.6.2a and 1.6.2b), we find that for the finite intervals considered here, the individual samples of received input are

$$X(\mathbf{r}_m, t_n) = \mathbf{T}_S(\alpha)_d = \kappa_0 \int_{-\mathbf{R}^{(3)}/2}^{\mathbf{R}^{(3)}/2} \mathbf{d}^3\mathbf{r} \int_{-T/2}^{T/2} \alpha(\mathbf{r}, t)_{\text{input}}\boldsymbol{\delta}(\mathbf{r}-\mathbf{r}_m)\delta(t-t_n)dt = \kappa_0\alpha(\mathbf{r}_m, t_n)_{\text{input}}$$

$$= \alpha(\mathbf{r}_m, t_n),\quad -\frac{\Delta}{2} \le \mathbf{r}_m, t_n \le \frac{\Delta}{2};\ = 0 \quad \text{elsewhere.} \qquad (2.3.1)$$

Here, Δ is the finite space–time sample interval $\Delta\left(=\left|\mathbf{R}^{(3)}\right|T\right)$ applied to α_{input} and containing the sample $\alpha(\mathbf{r}_m, t_n)$.[21]

Discrete sampling in space–time is seen to be straightforward here, directly yielding the expected result $X_j = \alpha_j$, refer to Eq. (2.3.1). But for the resulting spectrum, it is a more complex affair, requiring a modification of the usual Fourier transform definition. We begin by considering the analogy with Eq. (4.2) of Ref. [1], namely,

[21] The conversion factor κ_0 represents the effect of the physical transducer employed by the sensor at $(\mathbf{r}_m, t_n)$ to convert the external field (usually) to a suitable voltage or current.

$$S_y(f) = \int_{-T/2}^{T/2} y(t)e^{-2\pi\omega t}dt \Rightarrow \sum_{-N/2}^{N/2} \Delta_n y_n e^{-2\pi i\omega n T_0} = S_{y_n}(f), -f_0 \le f \le f_0 \\ = 0, \text{elsewhere} \Bigg\}, \quad (2.3.2)$$

With $y_n = y(t_n)$, $\Delta_n = (t_{n+1} - t_n) = \Delta T_n$, and for periodic samplings, $t_n = nT_0$ and $\Delta_n = \Delta T_n = t_0$ (refer also to Eq. (71) versus Eq. (48) of Ref. [9]). We can now define a preliminary relation for a space–time spectrum, namely,[22]

$$S_{X_j}(\boldsymbol{\nu},f) = S_{X_j}(\mathbf{q}) \equiv X_j \cdot \Delta_{oj}^{(4)} \cdot e^{2\pi i \mathbf{p}_j \cdot \mathbf{q}},$$
$$-\Delta_{o_j}/2 \le \mathbf{r}_m, t_n \le \Delta_{o_j}/2; \quad = 0, \text{elsewhere in space-time}, \quad (2.3.3)$$

for which $\Delta_{0_j}^{(4)} \equiv \Delta_{0_m}^{(3)} \Delta t_n$ in the extension to space as well as time, like the interval $\Delta t_n \equiv t_{n+1} - t_n$ between time samples in earlier work [10], with X_j everywhere finite on the sample interval $\Delta_{0_j}^{(4)}$. For $\mathbf{p}_j$ and $\mathbf{q}$, we have

$$\mathbf{p}_j = \mathbf{r}_m - \hat{\mathbf{i}}_4 t_n \quad \text{and} \quad \mathbf{q} = \boldsymbol{\nu} + \hat{\mathbf{i}}_4 f, \quad \Delta\mathbf{r}_m = \mathbf{r}_{mn} - \mathbf{r}_m. \quad (2.3.4)$$

(We observe that the dimensionality of S_{X_j} are $[A][L]^3[T]$ or $[A][f]^{-1}\left[|\boldsymbol{\nu}|^3\right]^3$ as required since $S_{X_j}(\mathbf{q} = \boldsymbol{\nu},f)$ is a four-dimensional spectrum, three in wave number space and one in frequency space.) Thus, the components of $\mathbf{p}_j$ are the *intervals* in space $\Delta\mathbf{r}_m \equiv \mathbf{r}_{m+1} - \mathbf{r}_m = \hat{\mathbf{i}}_{\Delta m}|\Delta\mathbf{r}_m|$ and in time $\Delta t_n \equiv t_{n+1} - t_n$ of a point array. The spheres in Fig. 2.7 represent the spatial domains of the vectors $\mathbf{r}_m$, $\mathbf{r}_{m+1}$, and so on. In addition, the quantities α_j, X_j, S_{X_j} can represent *ensemble* values in the overall sample interval that is taken from the continuous field α_{input} in the manner of Eq. (2.3.1). See Figures 2.5–2.7

Note that the spectrum in (2.3.3) for the point datum X_j is apparently unbounded.[23] In fact, (2.3.3) is periodic in wave number and frequency (for each $j = mn$), with period $\Delta_{0_j}^{(4)}$ and is completely specified in the primary interval $\left(-1/2\Delta_{0_j}^{(4)}, 1/2\Delta_{0_j}^{(4)}\right)$ for each j. By suitable choice of the scalar factor $\Delta_{0_j}^{(4)}$ in (2.3.3), we can remedy the spectral indeterminacy. This is achieved by choosing a primary interval to limit the domain of ($\boldsymbol{\nu}$ and f) for each jth interval.[24] Accordingly, we see from this, and the earlier example (2.3.2), that $\Delta_{0_j}^{(4)}$ must represent the spatial and temporal *sampling interval*, here extending from $\mathbf{r}_m$ and t_n at the jth sample by the amounts $|\Delta\mathbf{r}_m|$ and $\Delta t_n = T_{0_n}$, respectively. Thus, in (2.3.3), we have for $\Delta\mathbf{r}_m^{(3)}\Delta t_n$

$$\Delta_{0_j} = |\mathbf{r}_{m+1} - \mathbf{r}_m|(t_{n+1} - t_n) = |\Delta\mathbf{r}_m| T_{0_n}; \quad = 0 \text{ elsewhere}, \quad (2.3.5)$$

where $j, j+1$ represent consecutive space and time points. We shall verify that this is indeed the case for Δ_{0_j} when we evaluate the inverse transform $X_J = \mathsf{F}^{-1}(S_J)$, refer to Eq. (2.3.3) ff.

[22] See the text following Eq. (2.3.1).

[23] The nonlinear character of the discrete sampling operator T_d in (2.3.1) produces the infinite "white" spectrum over $(\boldsymbol{\nu},f)$ periodically, which represents the additional frequencies generated by T_d.

[24] By requiring the wave number–frequency spectrum to vanish outside this primary interval, we avoid the resulting infinite energy catastrophe, which is clearly nonphysical and which is not now implicit in our model of space–time samples on bounded intervals. The spectrum is still continuous, in the primary interval, but is null bounded everywhere else.

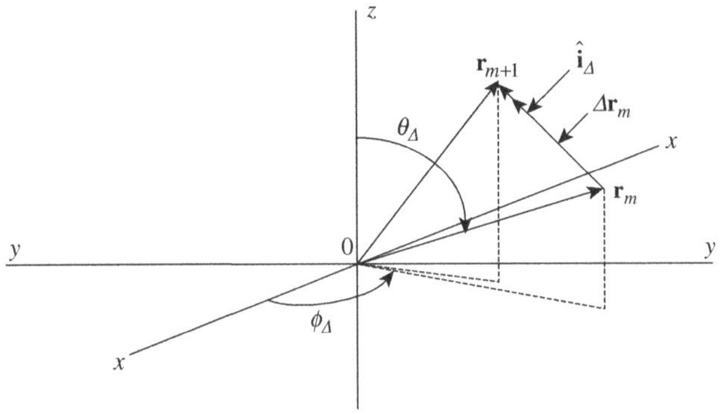

FIGURE 2.5 Geometry of $\Delta\mathbf{r}_m\left(=\hat{\mathbf{i}}_\Delta|\Delta\mathbf{r}_m|\right)$, the interval defined by $(\mathbf{r}_m, \mathbf{r}_{m+1})$ [Eq. (2.3.4)].

For the *j*th interval, the corresponding domains of $\boldsymbol{\nu}$ and f in wave number–frequency space are given by the reciprocal $\boldsymbol{\nu}_{0_m}$ and f_{0_n} of the domains of $\Delta\mathbf{r}_m$ and T_{0_n}, namely,

$$\left.\begin{array}{l} -\boldsymbol{\nu}_{0_m} \le \boldsymbol{\nu} \le \boldsymbol{\nu}_{0_m} \;\text{ or }\; 0 \le \nu \le (\nu_0)_m,\; 0 \le \phi_\nu \le 2\pi,\; 0 \le \theta_\nu \le \pi, \text{ and } 0 \le f \le f_{0_n} \\ \text{where} \\ -\mathbf{R}_{0_m} \le \Delta\mathbf{r}_m \le \mathbf{R}_{0_m} \;\text{ or }\; 0 \le |\Delta\mathbf{r}_m| \le R_{0_m},\; 0 \le \phi_m \le 2\pi,\; 0 \le \theta_m \le \pi \end{array}\right\}, \tag{2.3.6}$$

expressed in spherical coordinates, by which we reveal, as expected, the linear dimensionality of $\Delta\mathbf{r}_m$ and hence the reciprocal linearity of $|\boldsymbol{\nu}| = \nu = [L^{-1}]$. The entire region occupied by the set of data discrete points is represented symbolically by $\mathbf{R}_0$, where we also have the bounding conditions $(-\mathbf{R}_0 \le \mathbf{R}_{0_m} \le \mathbf{R}_0)$ and $(0 \le T_{0_n} \le T_n)$. Accordingly, we find now that the upper limits of $|\boldsymbol{\nu}|$ and t are respectively for each component of $\boldsymbol{\nu}$

$$(\nu_0)_m = 1/\mathbf{R}_{0_m}, \text{ with } (\phi_{\nu m}, \theta_m) \text{ in } (0, 2\pi; 0, \pi), \text{ and } f_{0_n} = 1/T_{0_n}. \tag{2.3.7a}$$

Thus, it is the finite spatial intervals between consecutive points that determine the individual finite continuous domains of the corresponding wave numbers and frequencies for these generally *aperiodic* samples.

The transform variables $\mathbf{p}$, $\mathbf{q}$, (2.3.4), representing the location of successive points in space–time and wave number–frequency space are explicitly in spherical coordinates $\left(\hat{\mathbf{i}}_1 = \hat{\mathbf{i}}_x, \text{ and so on}\right)$

$$\left.\begin{array}{l} \mathbf{q} = \hat{\mathbf{i}}_\nu \nu + \hat{\mathbf{i}}_4 f, \quad \text{with} \quad \hat{\mathbf{i}}_\nu = \hat{\mathbf{i}}_1 \cos\phi_\nu \sin\theta_\nu + \hat{\mathbf{i}}_2 \sin\phi_\nu \sin\theta_\nu + \hat{\mathbf{i}}_3 \cos\theta_\nu \\ \mathbf{r}_m = \hat{\mathbf{i}}_m r_m, \quad \text{and} \quad \hat{\mathbf{i}}_m = \hat{\mathbf{i}}_1 \cos\phi_m \sin\theta_m + \hat{\mathbf{i}}_2 \sin\phi_m \sin\theta_m + \hat{\mathbf{i}}_3 \cos\theta_m \end{array}\right\}. \tag{2.3.7b}$$

From (2.3.7b), we easily see that

$$\hat{\mathbf{i}}_\nu \cdot \hat{\mathbf{i}}_m = \cos\phi_\nu \sin\theta_\nu \cos\phi_m \sin\theta_m + \sin\phi_\nu \sin\theta_\nu \sin\phi_m \sin\theta_m + \cos\phi_\nu \cos\theta_m = \cos\psi_{\nu m}, \tag{2.3.8}$$

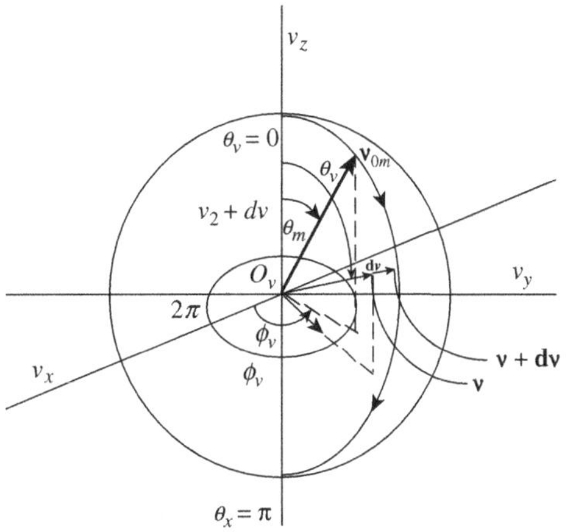

FIGURE 2.6 The domain of $\boldsymbol{\nu}$ and ν_{0_m} in wave number space for the line interval $\Delta\mathbf{r}_m = \mathbf{r}_{m+1} - \mathbf{r}_m$.

which is cosine of the angle between $\hat{\mathbf{i}}_\nu$ and $\hat{\mathbf{i}}_m$, and correspondingly, (x, y, z) - components of $\boldsymbol{\nu}$ and $\mathbf{r}_m$ are

$$\left.\begin{aligned} &\nu_x = |\boldsymbol{\nu}|\cos\phi_\nu \sin\theta_\nu; \;\; \nu_y = |\boldsymbol{\nu}|\sin\phi_\nu \sin\theta_\nu; \;\; \nu_z = |\boldsymbol{\nu}|\cos\phi_\nu, \text{ with } |\boldsymbol{\nu}| = \left(\nu_x^2 + \nu_y^2 + \nu_z^2\right)^{1/2} \\ &(\mathbf{r}_m)_x = |\mathbf{r}_m|\cos\phi_m \sin\theta_m; \;\; (\mathbf{r}_m)_y = |\mathbf{r}_m|\sin\phi_m \sin\theta_m; \;\; (\mathbf{r}_m)_z = |\mathbf{r}_m|\cos\theta_m \end{aligned}\right\}. \tag{2.3.9}$$

From (2.3.4) and (2.3.7b–2.3.9), we readily obtain the appropriate form of the transform variables in (2.3.3) for a straight line in both **r**-space and $\boldsymbol{\nu}$-space (Fig. 2.6):

$$\mathbf{p}_j \cdot \mathbf{q} = \hat{\mathbf{i}}_m \cdot \hat{\mathbf{i}}_\nu |\mathbf{r}_m||\boldsymbol{\nu}| - \hat{\mathbf{i}}_4 f T_n; \;\; \mathbf{dq} = d\nu d\phi_\nu d\theta_\nu df, \tag{2.3.10}$$

From (2.3.10), we observe that it is $\mathbf{p}_j$ that reflects the orientation in space of the interval $|\Delta\mathbf{r}_m|$. For example, if $\mathbf{r}_m$ lie only on the xy-plane, then $\theta_m = \pi/2$ and $\mathbf{p}_j = |\mathbf{r}_m|\left(\hat{\mathbf{i}}_1 \cos\phi_m + \hat{\mathbf{i}}_2 \sin\phi_m\right) - \hat{\mathbf{i}}_4 T_n$. Hence, only the wave numbers $\nu_x = |\nu|\cos\phi_\nu \sin\theta_\nu$, $\nu_y = |\boldsymbol{\nu}|\sin\phi_\nu \sin\theta_\nu$ associated with the sampled field appear in the spatial spectrum. Similarly, for point (or equivalently sensor) data samples aligned along an axis, say the x-axis, we have $(\theta_m = \pi/2, \phi_m = 0)$, so that $\mathbf{p}_j = \hat{\mathbf{i}}_1 (r_m)_x - \hat{\mathbf{i}}_4 T_n$. Then, as seen in Section 2.2.2.1, it is the components of $\mathbf{p}_j$ that determine the components of $\mathbf{q}$ in the exponent of (2.3.3) and hence in $S_{X_j}(\mathbf{q})$ and its inverse. Also, as a consequence of this, the differential element in (2.3.10) becomes for these examples $\mathbf{dq} = d\nu d\phi_\nu d\theta_\nu df$ and $\mathbf{dq} = d\nu \delta(\phi_\nu - 0) d\phi \delta(\theta_\nu - \pi/2) d\theta$, with $\Delta_{0_j} = 2\pi^2 \nu_{0_m} f_{0_n}$ and ν_{0_m}, respectively, in (2.3.3) and (2.3.7a) ff. In all cases as expected, the dimensionalities of $|\mathbf{r}_m|$ and $|\boldsymbol{\nu}|$ are $[L]$ and $[L]^{-1}$, respectively, so that the dimensionality of $|\mathbf{p}_j|$ and $|\mathbf{q}|$ are $[LT]$ and $[LT]^{-1}$, as required of the amplitude spectrum (2.3.3) here (in addition to the dimensionality of the field samples $[X_j]$). Figures 2.7 and 2.8 illustrates the space–time geometry of a typical jth = *interval.*

We are now ready to determine the amplitude wave number–frequency spectrum of the entire sample of data points $X_j, j = (0, 0), \ldots, MN$. For convenience in comparison with the results of Section 2.2.2, we consider the case of $M + 1, N + 1$ sample points, M, N even, in the interval $|\mathbf{R}_0|T$, so that we have MN intervals in space and time each bounded by

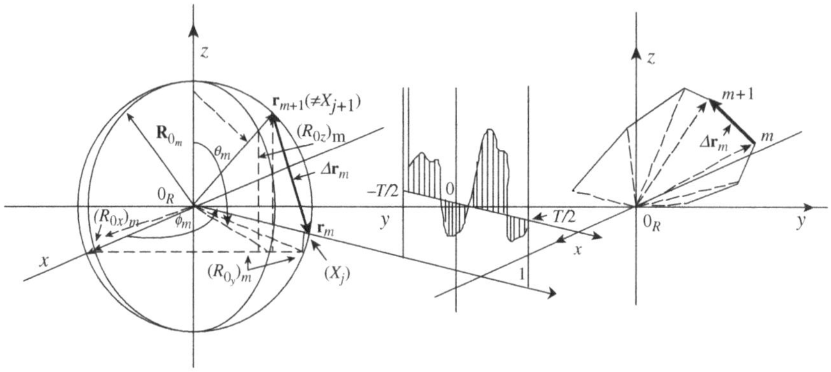

FIGURE 2.7 Geometry of the vector interval $\Delta\mathbf{r}_m$ in spherical coordinates, cf. Eq. 2.3.7b, and an array of points $m = 1, \ldots, M$ in space, with a typical interval $\Delta\mathbf{r}_m$, etc., and time samples.

consecutive sample points, selected under the principle of the "nearest neighbor." The intervals are equally divided about $M/2$, that is, extending from $-M/2$ to $M/2$ in each half of the sample space $|\mathbf{R}_0|T$, as shown in the figures. From 2.2.7b) and (2.3.10), the scalar length of a typical interval in space–time is $|\Delta\mathbf{r}_m|T_{0_n}$ and the entire length of the array of points is the sum of the intervals as defined above, namely,

$$\Delta_J = \sum_j^J \Delta_{0_j} = \sum_{m,n}^J |\Delta\mathbf{r}_m|^3 T_{0_n}. \tag{2.3.11}$$

For the connected intervals, we have (for the ensemble)

$$X_J = \sum_{-J/2}^{J/2} X_j; \;\; \overline{X_j} = 0; \quad \text{and} \quad S_J(\boldsymbol{\nu}, f)_\Delta = \sum_{-J/2}^{J/2} S_{X_j}(\mathbf{q})$$
$$= \sum_{-J/2}^{J/2} X_j \Delta_{oj} e^{2\pi i \mathbf{p}_j \cdot \mathbf{q}} \left(= \sum_{-J/2}^{J/2} \alpha_j \Delta_j e^{2\pi i \mathbf{p}_j \cdot \mathbf{q}} \right) \tag{2.3.12}$$

from (2.3.3). It is evident on taking the inverse Fourier transform of S_Δ with the help of (2.3.7b) and (2.3.12) that

$$X_J = \sum_{-J/2}^{J/2} X_j = \mathsf{F}^{-1}\{S_{J,\Delta}\} = \sum_{-J/2}^{J/2} \int_{-\nu_{0m}/2}^{\nu_{0m}/2} d\nu \int_0^{2\pi} d\phi \int_0^{\pi} d\theta \int_{-f_{0n}/2}^{f_{0n}/2} S_{X_j}(\mathbf{q}) e^{-2\pi i \mathbf{q} \cdot \mathbf{p}_j} df$$
$$2\pi^2 \nu_{0_m} f_{0_n} = \Delta_{0_j}^{-1}; \quad \therefore \nu_{0_m} = \left(2\pi^2 |\Delta\mathbf{r}_m|\right)^{-1}, \tag{2.3.13}$$

as required, with ν replaced by $|\boldsymbol{\nu}| = \nu$ and $\mathbf{q}$ given by (2.3.7b). From (2.3.11), we see that in terms of wave number–frequency bounds (ν_{0_m}, f_{0_n}), $\Delta_{0_j} = (2\pi^2 \text{ or } 4\pi)^{-1}$. Thus, the wave number–frequency domain of $S_J(\mathbf{q})_\Delta$ consisting of the sum of the individual domains of $S_j(\mathbf{q})_{\Delta j}$ is also finite. Like them, it also vanishes outside its own Fourier dimensional $(\boldsymbol{\nu}, f)$ intervals, being continuous within that interval.

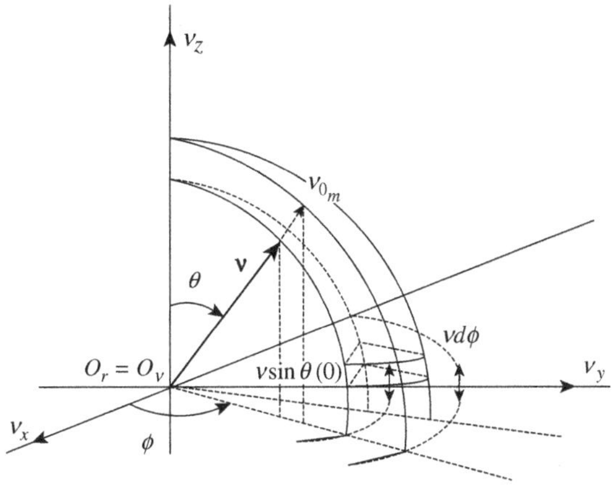

FIGURE 2.8 Geometry of the differential elements $d\nu$, $d\theta$, $d\phi$.

2.3.2 Periodic Sampling

Instead of the deterministically irregular sampling process of the above, both in space with respect to interval length and in direction, as illustrated in Figs. 2.5 and 2.6, our results above may be considered simplified if we use periodic sampling in space and time. With the same interval distance and temporal separations, we have, for $f_{0_n} = f_0 = T_0^{-1}$,

$$\left.\begin{array}{l} \text{(i) } \Delta\mathbf{r}_m = \hat{\mathbf{i}}_m r_0 = \hat{\mathbf{i}}_m (r_{m+1} - r_m) \text{ and } \begin{pmatrix} T_{0_n} = T_0 \\ T = nT_0 \end{pmatrix} \\ \quad \text{(different direction of } r_m) \\ \text{(ii)} \begin{pmatrix} \Delta\mathbf{r}_m = \hat{\mathbf{i}}_0 r_0 \\ \mathbf{r}_m = m\hat{\mathbf{i}}_0 r_0 \end{pmatrix} \text{ and } \begin{pmatrix} T_{0_n} = T_0 \\ T = nT_0 \end{pmatrix} \text{ (straight line in space)} \end{array}\right\}; \boldsymbol{\nu} = \hat{\mathbf{i}}_\nu \nu = \hat{\mathbf{i}}_\nu / r_0. \tag{2.3.14a}$$

In any case, with periodic sampling, we have the following for the overall array length Δ_J:

$$\Delta_J^{(4)} = J\Delta_{0_j} = J\Delta_0 = Jr_0T_0, \quad \text{with } \Delta_{0_j} \equiv \Delta_0 = r_0^3 T_0. \tag{2.3.14b}$$

Note that if in (ii) r_0 is not parallel to an axis, it is then possible to obtain the direction of a (localized) source producing X_j. (We shall discuss this in more detail in Section 2.5 when considering actual arrays of sensors in space.) Figure 2.9a and b shows examples of aperiodic and periodic sampling in time.

Finally, Eq. (2.3.3) becomes for the entire sample

$$\mathit{Case}(i) : (2.3.14a) : \quad S_J(\boldsymbol{\nu}, f)_\Delta = \sum_{-M,N/2}^{M,N/2} X\left(\hat{\mathbf{i}}_m r_0, nT_0\right)\left(r_0^3 T_0\right) e^{2\pi i\left[\left(\hat{\mathbf{i}}_\nu \cdot \hat{\mathbf{i}}_m\right) r_0 \nu - nfT_0\right]}, \quad \nu = |\boldsymbol{\nu}|, \tag{2.3.15a}$$

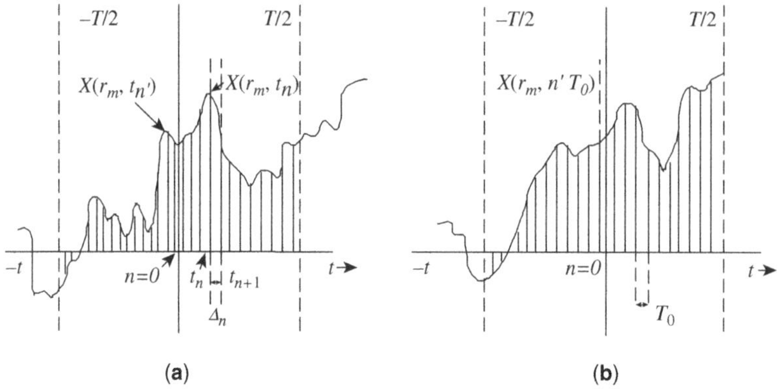

FIGURE 2.9 (a) Aperiodic sampling of $X(\mathbf{r}, t)$: $X(\mathbf{r}_m, t_n)$ and so on (2.3.3), in $(-T/2, T/2)$. (b) Periodic sampling of $X = X(\mathbf{r}_m, nT_0)$ in $(-T/2, T/2)$ [Eq. (2.3.14a)].

$$Case(ii) : (2.3.14a) : \quad S_J(\boldsymbol{\nu}, f)_\Delta = \sum_{-M,N/2}^{M,N/2} X\left(m, \hat{\mathbf{i}}_0 r_0, nT_0\right)\left(r_0^3 T_0\right) e^{2\pi i\left[\left(\hat{\mathbf{i}}_\nu \cdot \hat{\mathbf{i}}_0\right) m r_0 \nu - nfT_0\right]} \tag{2.3.15b}$$

with the basic interval now $\Delta_{0j} = r_0^3 T_0 \equiv \Delta_0$, independent of $j\,(=mn)$, and the total interval is $\Delta_J = MN\Delta_0 = Jr_0^3 T_0$, refer to Eq. (2.3.14b). The single data sample is still $X_j = \alpha_j\left(=K_0\alpha_{\text{input}}\right)$. From (2.3.4 and 2.3.7b), we have again for the basic space–time coordinates,

$$\mathbf{p}_j = m\hat{\mathbf{i}}_m r_0 - \hat{\mathbf{i}}_4 nT_0; \quad \mathbf{q} = \hat{\mathbf{i}}_\nu \nu + \hat{\mathbf{i}}_4 f, \quad \text{and} \quad \gamma_0 d\nu d\theta d\phi df, \tag{2.3.15c}$$

where the limit on ν in the periodic version of the integral over $S_X(\mathbf{q})$ is given by (2.3.13). For these three-dimensional cases, the normalizing factor is specifically $\gamma_0 \equiv \sigma/\pi\nu_0^3 f_0 = (\sigma/\pi)R_{0_m}T_0$ again, refer to Eq. (2.3.13a). Equations (2.3.8–2.3.10) provide some of the general relations modified here by the assumption of periodicity.

2.4 THE WIENER–KHINTCHINE RELATIONS FOR DISCRETELY SAMPLED RANDOM FIELDS

With the results (2.3.10–2.3.13) above, we are now ready to calculate W–Kh relations for finite samples of *discrete data*, that is, for $\{X_j\}$, under the quite general conditions of *non-Hom-Stat inputs and aperiodic sampling*. We begin by modifying the definition of spectral intensity (2.2.7) used in the continuous cases of Section 2.2.2 to

$$\mathsf{W}_{X_J}(\mathbf{q})_\Delta \equiv \mathrm{E}\left\{\frac{2}{\Delta_J}|S_\Delta(\mathbf{q})|^2\right\} = \frac{2}{\Delta_J}\overline{|\mathsf{F}(X_J)|^2}, \tag{2.4.1}$$

cf. (2.3.12). With the help of (2.3.13), this becomes

$$\mathsf{W}_{X_J}(\mathbf{q})_\Delta = \frac{2}{\Delta_J}\sum_{-J/2}^{J/2}\sum_{-J'/2}^{J'/2}\overline{X_j X_{j'}}\,\Delta_{0_j}\Delta_{0_{j'}}e^{2\pi i\left(\mathbf{p}_j - \mathbf{p}_{j'}\right)\cdot\mathbf{q}} \tag{2.4.2a}$$

$$= \frac{2}{\Delta_J} \sum_{-J/2,-J'/2}^{J/2,J'/2} K_X(\mathbf{p}_j, \mathbf{p}_{j'}) \Delta_{0_j} \Delta_{0_{j'}} e^{2\pi i (\mathbf{p}_j - \mathbf{p}_{j'}) \cdot \mathbf{q}} \tag{2.4.2b}$$

with Δ_J, Eq. (2.3.11), the effective overall sample size in space–time, for aperiodic (nonoverlapping) individual samples, in the finite domain $\Delta = |\mathbf{R}_0|T$, with K_X being the covariance of the sample values X_j at points j, j' in Δ. For periodic samples, $\Delta_J = J\Delta_0 = Jr_0^3 T_0$ and $\mathbf{p}_{j,j'} = \hat{\mathbf{i}}_m r_0 - \hat{\mathbf{i}}_4 T_0,\ \hat{\mathbf{i}}_{m'} r_0 - \hat{\mathbf{i}}_4 T_0$, refer to Eq. (2.3.14a). Equations (2.4.2a and 2.4.2b) are the discrete analogues of the continuous case (2.2.8) above.

Let us now obtain the other member of W–Kh transform pair, paralleling Eqs. (2.2.9–2.2.12b). For this, we need the following result relating the delta functions $\delta(\mathbf{x}_j - \mathbf{x}_{j'})$ and the Kronecker delta $\delta^j_{j'} (= \delta_{jj'})$:

$$\lim_{\Delta \to 0} \Delta \cdot \delta(\mathbf{x}_{j'} - \mathbf{x}_j) = \delta^j_{j'} = \delta_{jj'} \quad \text{or} \quad \lim_{\Delta \to 0} \delta^j_{j'}/\Delta = \delta(\mathbf{x}_{j'} - \mathbf{x}_j). \tag{2.4.3}$$

Then, for any value of $\mathbf{p}_k$ in $\Delta (= |\mathbf{R}_0|T)$, we have with respect to the Fourier transform of the intensity spectrum (2.4.2a)

$$\mathsf{F}^{-1}\{\mathsf{W}_{X_J}(\mathbf{q})_\Delta\}_{\mathbf{p}_k} = 2 \sum_{-J/2,-J'/2}^{J/2,J'/2} \frac{\Delta_{0_j}}{\Delta_J} K_X(\mathbf{p}_j, \mathbf{p}_{j'}) \left[\lim_{\Delta_{0_j} \to \Delta'_0 \to 0} \Delta'_0 \int_{-\infty}^{\infty} e^{2\pi i (\mathbf{p}_j - \mathbf{p}_{j'} - \mathbf{p}_k) \cdot \mathbf{SQ}} \mathbf{dq} \right]. \tag{2.4.4}$$

The expression in the brackets [ ] is given by (2.4.3) $\Delta'_0 \delta(\mathbf{p}_{j'} - [\mathbf{p}_j - \mathbf{p}_k]) = \delta^{j-k}_{j'}$, the Kronecker delta. Summing over j' gives $K_X(\mathbf{p}_{j'}, \mathbf{p}_j - \mathbf{p}_k)_\Delta$, with $\Delta_{0_j}/\Delta_J = \Delta_{0_j}/\sum_j \Delta_{0_j}$, from (2.3.11). The result is finally

$$\left(2\Big/\sum_j \Delta_{0_j}\right) \sum_{-J/2}^{J/2} \Delta_{0_j} K_X(\mathbf{p}_j, \mathbf{p}_j - \mathbf{p}_k) = \int_{-\infty}^{\infty} \mathsf{W}_{X_J}(\mathbf{q})_\Delta e^{-2\pi i \mathbf{q} \cdot \mathbf{p}_k} \mathbf{dq} < \infty. \tag{2.4.5}$$

We observe that the intensity spectrum (2.4.2) is composed of separate primary intervals that are individually bounded in wave number–frequency–space according to $\Delta_{0_j}^{-1}$, refer to Eq. (2.3.13).

For *periodic sampling* in space and time, we use (2.3.14 and 2.3.15) to obtain directly the simpler result:

$$\textit{Periodic Sampling}: \quad \frac{2}{J} \sum_{-J/2}^{J/2} K_X(\mathbf{p}_j, \mathbf{p}_j - \mathbf{p}_k)_\Delta = \int_{-\infty}^{\infty} \mathsf{W}_X(\mathbf{q})_\Delta e^{-2\pi i \mathbf{q} \cdot \mathbf{p}_k} \mathbf{dq} = \mathsf{F}_d^{-1}\{\mathsf{W}_X(\mathbf{q})_\Delta\} \tag{2.4.5a}$$

where $\mathbf{p}_j$, $\mathbf{q}$, and so on are given explicitly now by (2.3.15c). The intensity spectrum (2.4.2b) is modified to the following:

$$\text{(I)}\textit{Periodic Sampling}: \quad \mathsf{W}_{X_J}(\mathbf{q})_\Delta = \frac{2\Delta_0}{J} \sum_{-J/2}^{J/2} \sum_{-J'/2}^{J'/2} K_X(\mathbf{p}_j, \mathbf{p}_{j'}) \Delta_0 e^{2\pi i (\mathbf{p}_j - \mathbf{p}_{j'}) \cdot \mathbf{q}} \tag{2.4.5b}$$

These are the expected discrete analogues of (2.2.10b) for the continuous cases, now for periodically sampled data $\{X_j\}$ in the finite domain Δ, when the received field is non-Hom-Stat. (Note also that because $\mathbf{K}_X$ is real and symmetric, $K_X(\mathbf{p}_j, \mathbf{p}_j - \mathbf{p}_k) = K_X(\mathbf{p}_j, \mathbf{p}_j + \mathbf{p}_k)$ and hence $\mathsf{W}_X(\mathbf{q}) = \mathsf{W}_X(-\mathbf{q})$, that is, $\mathsf{W}_X(\boldsymbol{\nu}, f) = \mathsf{W}_X(-\boldsymbol{\nu}, -f)$, refer to Eq. (2.2.12).

Provided we can apply conditions (ii) and (iii) of (2.3.13), we can extend (2.4.5) to the infinite sample interval, that is, $|\Delta| \to \infty$ for non-Hom-Stat fields obeying (i)–(iii). The results are again analogous to (2.3.8a and 2.3.8b).

$$\mathsf{W}_{\{X_j\}}(\mathbf{q})_\infty = 2\int_{-\infty}^{\infty} \langle K_X(\mathbf{p}_j, \mathbf{p}_j - \mathbf{p}_k)\rangle e^{2\pi i \mathbf{p}_k \cdot \mathbf{q}} \mathbf{dp}_k, \tag{2.4.6a}$$

with

$$\langle K_X(\mathbf{p}_j, \mathbf{p}_j - \mathbf{p}_k)\rangle = \frac{1}{2}\int_{-\infty}^{\infty} \mathsf{W}_{\{X_j\}}(\mathbf{q})_\infty e^{-2\pi i \mathbf{q} \cdot \mathbf{p}_k} \mathbf{dq}, \tag{2.4.6b}$$

once more provided $\langle |X_j|^2 \rangle < \infty$ in the entire region $-\infty \le \Delta \le \infty$. Here, the wave number–frequency region now extends over the infinite domain $\Delta \to \pm\infty$. For the periodic cases, $\Delta_{0_j} = \Delta_0 = r_0 T_0$, (2.3.14b), and one uses (3.3.35b) for the coordinates, in (2.4.1–2.4.5) and subsequently, for example, in (2.4.6a–2.4.12b).

In the hybrid cases, where the spatial sample is bounded ($|\Delta \mathbf{R}| < \infty$) but the temporal portion is unbounded, that is, $T \to \infty$, we have

$$\textit{II. Covariance:} \quad 2\sum_{-M/2}^{M/2}\sum_{-\infty}^{\infty} K_X(\mathbf{r}_m, t_n; \mathbf{r}_m - \mathbf{r}_k, t_n - t_k) = \int_{-\infty}^{\infty} \mathsf{W}_{\{X\}}(\mathbf{q})_{|\mathbf{R}_0|,\infty} e^{2\pi i \mathbf{q} \cdot \mathbf{p}_k} \mathbf{dq} \tag{2.4.7a}$$

with the transform pair member (2.4.2b) in detail, namely:

$$\textit{I. Spectrum:} \quad \left\{ \begin{aligned} \mathsf{W}_{\{X_j\}}(\mathbf{q})_{|\mathbf{R}_0|,\infty} &= \frac{2}{M}\sum_{-M/2}^{M/2}\sum_{-M'/2}^{M'/2} |\Delta \mathbf{r}_m| \lim_{N\to\infty}\frac{1}{N}\sum_{-(n,n')x-\infty}^{n,n'=\infty} T_{0_n} K_X(\mathbf{r}_m, t_n; \mathbf{r}_{m'}, t_{n'}) \\ &\cdot e^{2\pi i[(\mathbf{r}_m - \mathbf{r}_{m'})\cdot \mathbf{v} - f(n-n')]T}. \end{aligned} \right. \tag{2.4.7b}$$

The complexity of these results, and in particular for the intensity spectrum, refer to Eqs. (2.4.2b and 2.4.2b), arises largely from the non-Hom-Stat character of the fields, as well as technically from the explicit addition of the spatial contributions.

2.4.1 Discrete Hom-Stat Wiener–Khintchine Theorem: Periodic Sampling and Finite and Infinite Samples

Similar to the continuous cases of Section 2.2.2 above, when the discrete samples $\{X_j\}$ of finite duration ($\Delta < \infty$) are extracted from the *Hom-Stat fields of infinite duration* ($\Delta \to \infty$), we have directly from (2.4.2b), using (2.3.14b) for Δ_0,

$$I.\ \textit{Spectrum}:\qquad \mathsf{W}_{\{X_j\}}(\mathbf{q})_\Delta = \frac{2}{\Delta_J}\sum_{-J/2}^{J/2}\sum_{-J'/2}^{J'/2} K_X(\mathbf{p}_j - \mathbf{p}_{j'})\Delta_0^2 e^{2\pi i(\mathbf{p}_j-\mathbf{p}_{j'})\cdot\mathbf{q}}$$
$$= \frac{1}{K}\sum_{-K}^{K} K_X^{(\Delta)}(\mathbf{p}_k)\cdot\Delta_0 e^{2\pi i\mathbf{p}_k\cdot\mathbf{q}}(\geq 0) = 2\mathsf{F}\{K_X(\mathbf{p}_k)_\Delta\}, \tag{2.4.8}$$

Here, $\mathbf{p}_k = \mathbf{p}_j - \mathbf{p}_{j'} = \mathbf{p}_{j'} - \mathbf{p}_j$, and so on, because of the Hom-Stat condition, and

$$K_X^{(\Delta)}(\mathbf{p}_k) = K_X(\mathbf{p}_k)_\infty(1 - |\mathbf{p}_k|/\Delta_J); \quad = 0, \Delta_J < |\mathbf{p}_k|, \quad \text{with } K = MN, \tag{2.4.8a}$$

$$1 - |\mathbf{p}_k|/\Delta_J = (1 - |\mathbf{r}_R|/\Delta|\mathbf{R}|)(1 - |t_k|/T), \quad |\mathbf{p}_k| < \Delta, \tag{2.4.8b}$$

where we have used the discrete form of the continuous identity[25]

$$\frac{1}{\Delta}\int_{-\Delta/2,-\Delta/2}^{\Delta/2,\Delta/2} F(\mathbf{x}_1 - \mathbf{x}_2)e^{2\pi i\mathbf{q}\cdot(\mathbf{x}_1-\mathbf{x}_2)}\mathbf{d}^3\mathbf{x}_1\mathbf{d}^3\mathbf{x}_2 = \int_{-\Delta}^{\Delta} F^{(\Delta)}(\mathbf{x})e^{2\pi i\mathbf{q}\cdot\mathbf{x}}\mathbf{d}^3\mathbf{x}, \tag{2.4.9}$$

where $F^{(\Delta)}(\mathbf{x}) = \left(1 - \left|\mathbf{x}^{(3)}\right|/\Delta\right),\ \left|\mathbf{x}^{(3)}\right| \leq 0; = 0,\ \left|\mathbf{x}^{(3)}\right| \geq \Delta$. Specifically, this discrete form of (2.4.8a) is given by

$$\sum_{-K}^{K} F^{\Delta}(\mathbf{x}_k)e^{2\pi i\mathbf{q}\cdot\mathbf{p}_k} = \sum_{-J/2}^{J/2}\sum_{-J'/2}^{J'/2}\frac{\Delta_0^{(4)}}{\Delta}F(\mathbf{x}_j - \mathbf{x}_{j'})e^{2\pi i\mathbf{q}\cdot(\mathbf{x}_j-\mathbf{x}_{j'})}. \tag{2.4.10}$$

For covariance, we obtain the other member of the transform pair $\mathsf{F}^{-1}\left\{\mathsf{W}_{(X_j)}\right\}$ as follows. We write from (2.4.8),

$$\frac{1}{2}\int_{-\infty}^{\infty}\mathsf{W}_{\{X_j\}}(\mathbf{q})_\Delta e^{-2\pi i\mathbf{q}\cdot\mathbf{p}_l}\mathbf{d}^3\mathbf{q} = \frac{1}{K}\sum_{-K}^{K} K_X^{(\Delta)}(\mathbf{p}_k)\left\{\Delta_0^{(4)}\int_{-\infty}^{\infty} e^{2\pi i(\mathbf{p}_k-\mathbf{p}_l)\cdot\mathbf{q}}\mathbf{d}^3\mathbf{q}\right\}. \tag{2.4.11a}$$

Now, from (2.4.3), the expression in the braces is just $|\delta_l^k|$, unchanged in the limit $\Delta_0^{(4)} \to 0$, so that (2.4.11a) becomes the desired result:

$$II.\ \textit{Covariance}:\qquad K_X(\mathbf{p}_k)_\Delta = \frac{1}{2}\int_{-\infty}^{\infty}\mathsf{W}_{\{X_j\}}(\mathbf{q})_\Delta e^{-2\pi i\mathbf{q}\cdot\mathbf{p}_k}\mathbf{dq}\Big|_{\mathbf{p}_l\to\mathbf{p}_k} \tag{2.4.11b}$$

with $\mathsf{F}\{K_{X-\Delta}\}$, (2.4.8), the associated intensity spectrum for the finite samples ($\Delta < \infty$) of these Hom-Stat fields on the infinite interval.

[25] Note that $K^{(\Delta)}$, $F^{(\Delta)}$ are not the same as K_Δ, and so on in (2.4.5)–(2.4.7).

Again, when the infinite sampling interval is considered, the discrete situation here presents similar difficulties to those encountered in the continuous cases above, when $\Delta \to \infty$, refer to Section 2.2.3, Eqs. (2.2.17a) et seq. However, the steps (1)–(6) of Eqs. (2.2.18)–(2.2.20), when modified for discrete sampling, may also be used to establish Loève's results when $\Delta \to \infty$. We then obtain

$$\text{I. Spectrum:} \qquad \mathsf{W}_{\{X_j\}}(\mathbf{q})_{\infty} = 2\Delta_0^{(4)} \sum_{-\infty}^{\infty} K_X(\mathbf{p}_k)_{\infty} e^{2\pi i \mathbf{p}_k \cdot \mathbf{q}} = 2\mathsf{F}\{K_X(\mathbf{p}_k)_{\infty}\}. \tag{2.4.12a}$$

and

$$\text{II. Covariance:} \qquad K_X(\mathbf{p}_k)_{\infty} = \frac{1}{2} \int_{-\infty}^{\infty} \mathsf{W}_{\{X_j\}}(\mathbf{q})_{\infty} e^{-2\pi i \mathbf{q} \cdot \mathbf{p}_k} \mathbf{dq} = \frac{1}{2} F^{-1}\left\{\mathsf{W}_{\{X_j\}}(\mathbf{q})_{\infty}\right\}. \tag{2.4.12b}$$

These are the discrete counterparts of (2.2.17a, 2.2.17b, and 2.2.20) for the cases of continuous sampling on the infinite sampling interval ($\Delta \to \infty$).

2.4.2 Comments

The discrete sampling procedure of Section 2.2.3 vis-à-vis the continuous procedures of Section 2.2.2 needs some further remarks. The latter is a straightforward extension of the continuous field $\alpha(\mathbf{r}', t')$ sampled in the interval Δ to become the continuous data stream $X(\mathbf{r},t)$: from (1.6.2), $X = T_s(\alpha)_c = \alpha = \kappa_0 \alpha_{\text{input}}$, where κ_0 as before[26] is the only effect of this transducer in the (linear) all-pass filter whose weighting function (i.e., Green's function) is $\kappa_0 \delta(\mathbf{r}' - \mathbf{r})\delta(t' - t)$. In case of discrete samples of the input field, however, the sampling operation $X(\mathbf{r}_m, t_n) = T_s(\alpha)_d = \kappa_0 \alpha_{\text{input}}(\mathbf{r}_m, t_n)$ transforms the continuous input into a series of ordered *numbers* $\{X_j\} = \{X(\mathbf{r}_m, t_n)\}$ in the interval Δ, at a series of discontinuous, dimensionless *points*. To construct a function of X_j, and $\sum_j X_j$, such as the amplitude spectrum, for example, we must define an *interval* between consecutive points. This in turn reestablishes a continuous function on Δ, which employs the sampled data at the discrete points and allows to perform the desired evaluation. Since the points are dimensionless, we can, in principle, use any interval measure at these points. For our purposes (refer to Section 2.3.1), we have chosen the one-dimensional straight line connections, illustrated in Fig. 2.7. This yields direct analogues of the corresponding continuous forms of Section 2.2.2. The establishment of the interval enables us to define a corresponding region for transform variable $\mathbf{q}$ [Eqs. (2.3.4) and (2.3.7b)], and thus for the resulting wave number–frequency spectra.

At this point, it is helpful to summarize the principal results of Sections 2.2–2.4, in particular for generalizations of the Wiener–Khintchine relations. For this purpose, we include Table 2.1. The important role of the covariance and its Fourier transform, the intensity spectrum of the field $\{X_j(\mathbf{r}_m, t_n)\}$, arises in a variety of situations: (i) It is the governing statistics in the detection and extraction of signals in Gaussian noise (refer to

[26] See footnote 21.

TABLE 2.1 Intensity Spectra and (Auto) Covariance Functions of Received Space–Time Fields $X(\mathbf{r}_m, t')$: General Wiener–Khintchine Forms

Type: Remarks	Intensity Spectra $\mathsf{W}_{(\,)}(\mathbf{q})_{(\,)}$	Covariance $K_{X\|_{(\,)}}$
1. *Non-Hom-Stat.*; $\langle X\rangle = 0$ *continuous sampling* on $\Delta < \infty$; Section 2.2.2.1	$\frac{2}{\Delta}\int_{-\Delta/2}^{\Delta/2} K_X(\mathbf{p}_1, \mathbf{p}_2) e^{2\pi i \mathbf{q}\cdot(\mathbf{p}_1-\mathbf{p}_2)} \mathbf{dp}_1 \mathbf{dp}_2$; Eq. (2.2.12a)	$\frac{2}{\Delta}\int_{-\Delta/2}^{\Delta/2} K_X(\mathbf{p}_1, \mathbf{p}_1 - \mathbf{p})\mathbf{dp}_1 = \mathsf{F}^{-1}\{\mathsf{W}_\Delta\}$ Eq.(2.2.12b)
2. Same as (1), but $\Delta \to \infty$	$2\int_{-\infty}^{\infty} \langle K_X(\mathbf{p}_1, \mathbf{p}_1 - \mathbf{p})\rangle e^{2\pi i \mathbf{q}\cdot\mathbf{p}} \mathbf{dp}$; Eq. (2.2.13b)	$\langle K_X(\mathbf{p}_1, \mathbf{p}_1 - \mathbf{p})\rangle = {}^1\!/\!_2 \mathsf{F}^{-1}\{\mathsf{W}_\infty\}$ $\left(K_{x\|\Delta\to\infty = K_X}\right) (\mathsf{W}_\infty = \mathsf{W})$ Eq. (2.2.13a)
3. Hom-Stat.; *continuous sampling* on $\Delta < \infty$; $\mathbf{p} = \mathbf{p}_1 - \mathbf{p}_2$, or $\mathbf{p}_2 - \mathbf{p}_1 = \mathbf{p}$; Section 2.2.3	$2\int_{-\infty}^{\infty} K_X(\mathbf{p})_\Delta e^{2\pi i \mathbf{q}\cdot\mathbf{p}} \mathbf{dp}$; Eq. (2.2.16)	$K_X(\mathbf{p})_\Delta = \frac{1}{2}\mathsf{F}^{-1}\{\mathsf{W}_\Delta\}$ [Eqs. (2.2.16)–(2.2.16c)
4. Same as (3), but $\Delta \to \infty$ $\langle \|X^2\|\rangle < \infty$	$2\int_{-\infty}^{\infty} K_X(p) e^{2\pi i \mathbf{q}\cdot\mathbf{p}} \mathbf{dp}$; Eq. (2.2.17a)	$K_X(\mathbf{p}) = \frac{1}{2}\mathsf{F}^{-1}\{\mathsf{W}\}$ [Eq. (2.2.17b)] $\left(K_X^{(\Delta)} \to K_X\right)$ $(W_\Delta \to W_\infty)$
5.[a] *Non-Hom-Stat.*; $\langle X\rangle = 0$ *discrete aperiodic sampling* on $\Delta < \infty$; $J = MN$ (No. of intervals) $X \to \{X_j\}$	$\frac{2}{\Delta_J}\sum_{-J/2}^{J/2}\sum_{-J'/2}^{J'/2} \Delta_{0_j}\Delta_{0_{j'}}\left[\left(K_X(\mathbf{p}_j, \mathbf{p}_{j'})_d \cdot e^{2\pi i(\mathbf{p}_j - \mathbf{p}_{j'})\cdot\mathbf{q}}\right)\right.$; (2.4.2b)	$\frac{2}{\Delta_J}\sum_{-J/2}^{J/2} \Delta_{0_j} K_X(\mathbf{p}_j, \mathbf{p}_j - \mathbf{p}_k)_d = \mathsf{F}^{-1}\{\mathsf{W}_\Delta\}$ [Eq. (2.4.5)] $\Delta_J = \sum_J \Delta_{0_j}$

(*continued*)

TABLE 2.1 *(Continued)*

Type: Remarks	Intensity Spectra $\mathsf{W}_{(\,)}(\mathbf{q})_{(\,)}$	Covariance $K_{X\|_{(\,)}}$
6. *Hom-Stat; discrete periodic sampling* $\Delta < \infty;\ X \to \{X_j\}$	$\frac{1}{K}\sum_{K}^{K} K_X^{(\Delta)}(\mathbf{p}_k)_d \Delta_0 e^{2\pi i \mathbf{p}_k \cdot \mathbf{q}}$; (2.4.8)	$K_X(\mathbf{p}_k)_{d\Delta} = \frac{1}{2}\mathsf{F}^{-1}\{\mathsf{W}_\Delta\}$ [Eq. (2.4.11b)]
7. Same as (6), but $\Delta \to \infty \ \left\|\left\langle X_j^2 \right\rangle\right\| < \infty$	$2\sum_{-\infty}^{\infty} K_X(\mathbf{p}_k)_d \Delta_0 e^{2\pi i \mathbf{p}_k \cdot \mathbf{q}}$; (2.4.12a)	$K_X(\mathbf{p}_k) = \frac{1}{2}\mathsf{F}^{-1}\{\mathsf{W}_\infty = \mathsf{W}\}$ $(K_{X,\Delta} \to K_{X,\infty})$ [Eq. (2.4.12b)]

(i) Continuous sampling in space and time } $\mathbf{p} = \mathbf{r} - \hat{\mathbf{i}}_4 t$

(ii) Discrete aperiodic sampling in space and time; } $\mathbf{p} \to \mathbf{p}_j = \mathbf{r}_m - \hat{\mathbf{i}}_4 n T_0$

Discrete periodic sampling : [Eqs. (2.3.14a) and (2.3.14b) et seq.} $\Delta_{0_j}^{(4)} = r_0 T_0 = \Delta_0^{(4)}$

$\mathbf{p}_j = m\hat{\mathbf{i}}_m r_0 - \hat{\mathbf{i}}_4 n T_0$

(iii) Hybrid: discrete sampling in space, continous in time } $\mathbf{p} = \mathbf{r}_m - \hat{\mathbf{i}}_4 t$

[a] For periodic sampling in this case, see Eqs. (2.4.5a,b).

Chapters 3–6 ff.]. (ii) It is a critical statistic in the extended theory of threshold detection and estimation of signals in *non-Gaussian* noise when correlated noise samples are encountered. (iii) It is a key element in determining the structure of the matched filters, employed in both optimum and suboptimum detection and estimation (Sections 3.3–3.5, and Chapters 4–6 ff.). As is evident from Table 2.1, it is the non-Hom-Stat character of these sampled fields that increases considerably the complexity of the results. This is primarily the consequence of employing ensembles of data as opposed to the single representation for a Hom-Stat sample, that is, being able to assume that (at least locally) the ergodic condition applies practically (see [1]).

Besides the results for continuous sampling on the space–time interval Δ, discussed in Section 2.2.2 for the non-Hom-Stat cases, we have presented in Section 2.3.1 a corresponding treatment for discrete sampling. Finally, examples of the hybrid cases involving discrete sampling in space, for application to arrays (cf. Section 2.5), with continuous sensor outputs in time, are also given in Section 2.4. These are also good models of many actual array systems in practice. Generalization of the continuous random field concept is available in Chapter 4, Part I, of Yaglom [2], and references therein (pp. 500–505), including books on statistical inference, that is, measurement. For the latter, from the point of view of communication engineering, see, for example, Ref. [11].

2.5 APERTURE AND ARRAYS—I: AN INTRODUCTION[27]

In transmission and reception, coupling to the medium is represented by the operators $\mathbf{T}_A, \mathbf{T}_R$, cf. Fig. 1.1a and b and Eqs. (1.1.1a) and (1.1.1d). This is accomplished physically with the help of *apertures* and *arrays* of sensor elements. By aperture, we mean a continuous sensor body to each element $\mathbf{d\xi}$ to which is applied a transmitted signal or a received field. Arrays, on the other hand, consist of separate sensor bodies. These are well approximated in practice by continuous discrete "point" elements in space, connected physically to a common reference. Furthermore, with each sensor is associated a delay that in concert with the other sensors acts to direct the energy of the signal or field, that is, *to form a beam*. The larger the array or aperture, the narrower or more focused the beam, for transmission or reception. In this way, space as well as time is used to increase the effectiveness of transmission and reception. The role of apertures and arrays is of course well known and well documented: see, for example, Refs. [12–14], and more recently Refs. [6, 10, 15]. The important engineering treatise of Van Trees [16] on this subject is particularly noted.

Here, we begin by considering apertures and arrays alone, without an accompanying noise background, to illustrate the elements of their operation. Later, in Chapters 2–5, we shall examine their role as essential elements of Bayes matched filters for the optimum test statistics $\left(\sim \Psi_x^*\right)$ and performance parameters $\left(\Psi_s^*,\ \text{and so on}\right)$, where background noise is now present. In previous studies, such noise has generally been considered to be Gaussian, Hom-Stat, and usually spectrally white in wave number–frequency space. In the later treatment, we have removed this restriction and also consider non-Hom-Stat noise fields, along with similar signal fields. This includes emphasis on a more physical description of aperture and array operation and its relation to the generation of the associated fields.

[27] The present section is an extension of earlier work of the author [17] Sections 4.1 and 4.2, and mostly subsequent studies [7, 18] involving the application of apertures and arrays [3, 4, 7, 18].

In particular, besides the requisite definitions, conditions (i.e., causality, stability, etc.), the various Fourier transform relations are presented in Section 2.5.1. In Sections 2.5.2 and 2.5.3 following, we illustrate the role of the aperture in the generation of the field, specifically for an isotropic Helmholtz medium [19]. Section 2.5.4 thus treats the reciprocal case of reception. Approximations of the general results for narrowband outputs are included in Section 2.5.5.

2.5.1 Transmission: Apertures and Their Fourier Equivalents

We begin with continuous functions, continuous sampling, and linear time invariant filters, the latter distributed in space in a finite interval $|\mathbf{R}_0|$. The four-dimensional Green's function, that is, space–time *aperture weighting function* associated with these *transmitting apertures* (Tr) is $h(\mathbf{r}, \mathbf{t})_{\mathrm{Tr}}$. Here, $h(\mathbf{r}, t)_{\mathrm{Tr}}$ is *physically realizable* or *causal*,[28] that is, operates only on the past (and immediate present) of its input. For the spatial portion of these filters, the related condition of *spatial causality* is the *radiation condition*. These space–time causality conditions are described analytically in chapter 8. (See Section 2.2.5(3) of Ref. [1].) The various Fourier transforms of the *continuous* function $h(\mathbf{r}, t)_{\mathrm{Tr}}$, which has the dimensions $[L^3T]^{-1}$, are as follows:

(1) *The Aperture Transfer or System Function:*

$$Y(\mathbf{r},f)_{\mathrm{Tr}} = \mathsf{F}_t\{h(\mathbf{r}, t)_{\mathrm{Tr}}\} = \int_{-\infty}^{\infty} h(\mathbf{r}, t)_{\mathrm{Tr}} e^{-i\omega\, t} dt, \; \omega = 2\pi f, \tag{2.5.1a}$$

with its inverse.

(2) *The Aperture Weighting Function (Green's Function):*

$$h(\mathbf{r}, t)_{\mathrm{Tr}} = \mathsf{F}_f^{-1}\{Y_{\mathrm{Tr}}\} = \int_{-\infty}^{\infty} Y(\mathbf{r}, t)_{\mathrm{Tr}} e^{i\omega\, t} df. \tag{2.5.1b}$$

(3) *The Aperture Beam Function:*

$$\mathsf{A}(\boldsymbol{\nu},f)_{\mathrm{Tr}} = \mathsf{F}_{\mathbf{r}}\{Y_{\mathrm{Tr}}\} = \int_{-\infty}^{\infty} Y(\mathbf{r},f)_{\mathrm{Tr}} e^{2\pi i \boldsymbol{\nu}\cdot\mathbf{r}} \mathbf{d}^3\mathbf{r} \tag{2.5.2a}$$

$$= \mathsf{F}_{\mathbf{r}}\mathsf{F}_t\{h_{\mathrm{Tr}}\} = \int_{-\infty,(\Delta)}^{\infty}\int h(\mathbf{r}, t)_{\mathrm{Tr}} e^{2\pi i \boldsymbol{\nu}\cdot\mathbf{r} - i\omega\, t} dt\mathbf{d}^3\mathbf{r}. \tag{2.5.2b}$$

[28] Noncausal filters can, of course, be employed, but they require the whole set of data within the data interval before they can process any of it.

Here, $Y_{TR} \neq 0$ when the spatial domain is finite, that is, $|\mathbf{R}_0|$, and vanished outside this interval, that is, $Y_{Tr,|\mathbf{R}_0|} \neq 0,\ \mathbf{r} \in |\mathbf{R}_0|; = 0,\ r \notin |R_0|$. Correspondingly, the domain of the space–time filter where $h_{Tr} \neq 0$ is $|R_0|T = \Delta$, $h_{Tr} = 0$, outside Δ, expressed by $|R_0| = T = 0$ together. h_{Tr} is realizable or causal for $t > 0$ — and obeys the radiation conditions (cf. chapter 8), properties that govern their behavior when applied to signals and their generation as fields. Similarly, we have the partial inverse.

(4) *The Aperture System Function:*

$$\mathsf{F}_\nu^{-1}\{\mathsf{Y}_{Tr}\} = \int_{-\infty}^{\infty}\!\!\int \mathsf{Y}(\boldsymbol{\nu},f)_{Tr} e^{-2\pi i \mathbf{r}\cdot\boldsymbol{\nu}} \mathbf{d}^3\boldsymbol{\nu} = Y(\mathbf{r},f)_{Tr}, \tag{2.5.3a}$$

with the total inverse

(5) *The Aperture Weighting Function:*

$$\mathsf{F}_{\boldsymbol{\nu}}^{-1}\mathsf{F}_f^{-1}\{\mathsf{A}_{Tr}\} = \int_{-\infty}^{\infty}\!\!\int \mathsf{A}(\boldsymbol{\nu},f)_{Tr} e^{-2\pi i \mathbf{r}\cdot\boldsymbol{\nu} + i\omega t} \mathbf{d}^3\boldsymbol{\nu} df = h(\mathbf{r},t)_{Tr}. \tag{2.5.3b}$$

Here, $\boldsymbol{\nu}$ as before (cf. Section 2.2.2), is a vector wave number, that is, $\boldsymbol{\nu} = \hat{\mathbf{i}}_\nu \nu = \hat{\mathbf{i}}_1/\lambda_x + \hat{\mathbf{i}}_2/\lambda_y + \hat{\mathbf{i}}_3/\lambda_z$, with $\lambda = \left(\lambda_x^2 + \lambda_y^2 + \lambda_z^2\right)^{1/2}$ in rectangular coordinates, and $2\pi\,\boldsymbol{\nu} = \mathbf{k}$ is the corresponding angular vector wave number. Since the spatial domain of the aperture is necessarily finite ($|\mathbf{R}_0| < \infty$), we have in more detail the representations

$$\int_{|\mathbf{R}_0|} (\,)\mathbf{d}^3\mathbf{r} \equiv \int_{-\infty}^{\infty} (\,)_{|R_0|}\mathbf{d}^3\mathbf{r}, \quad \text{and} \quad \int_{0-}^{\infty} (\,)_{T=+\infty} dt = \int_{-\infty}^{\infty} (\,)_T dt \tag{2.5.4a}$$

with $\mathbf{dr} = dxdydz = dr_x dr_y dr_z$, $\mathbf{d}\boldsymbol{\nu} = d\nu_x d\nu_y d\nu_z$ in general, with the dimensionality of $\mathbf{v}$ depending on that of $\mathbf{r}$, that is, if $\mathbf{r} = \hat{\mathbf{i}}_1 r_x + \hat{\mathbf{i}}_2 r_2$, then $\boldsymbol{\nu} = \hat{i}_1\nu_x + \hat{i}_2\nu_y$, and $\mathbf{dr} = dr_x dr_y$, $\mathbf{d}\boldsymbol{\nu} = d\nu_x d\nu_y$, and so on. In the more condensed notation of Section 2.2.2 et seq., we may write

$$\mathbf{p} = \mathbf{r} - \hat{\mathbf{i}}_4 t = \hat{\mathbf{i}}_1 r_x + \hat{\mathbf{i}}_2 r_y + \hat{\mathbf{i}}_3 r_z - \hat{\mathbf{i}}_4 t \quad \text{and} \quad \mathbf{q} = \boldsymbol{\nu} + \hat{\mathbf{i}}_4 f = \hat{\mathbf{i}}_1 \boldsymbol{\nu}_x + \hat{\mathbf{i}}_2 \boldsymbol{\nu}_y + \hat{\mathbf{i}}_3 \boldsymbol{\nu}_z + \hat{\mathbf{i}}_4 f, \quad \text{with}$$
$$\mathbf{dp} = \mathbf{dr}dt,\ \mathbf{dq} = \mathbf{d}\boldsymbol{\nu}df \tag{2.5.4b}$$

in the rectangular coordinate system used here for the apertures, and where now $\mathsf{A}(\boldsymbol{\nu},f)_{Tr} = \mathsf{A}(\mathbf{q})_{Tr}$ and $h(\mathbf{r},\mathbf{t})_{Tr} = h(\mathbf{p})_{Tr}$ are the abbreviated forms for the relations above. Finally, since the Green's function h_{Tr} is real, we have

$$Y(\mathbf{r},f)_{Tr} = Y(\mathbf{r},-f)^*_{Tr} \quad \text{and} \quad \mathsf{A}(\boldsymbol{\nu},f)_{Tr} = \mathsf{A}(-\nu,-f)^*_T \quad \left(\text{or } \mathsf{A}(\mathbf{p})_T = \mathsf{A}(-\mathbf{p})^*_T\right), \tag{2.5.5}$$

from (2.1.12) applied to (2.5.1a) and (2.5.2b).

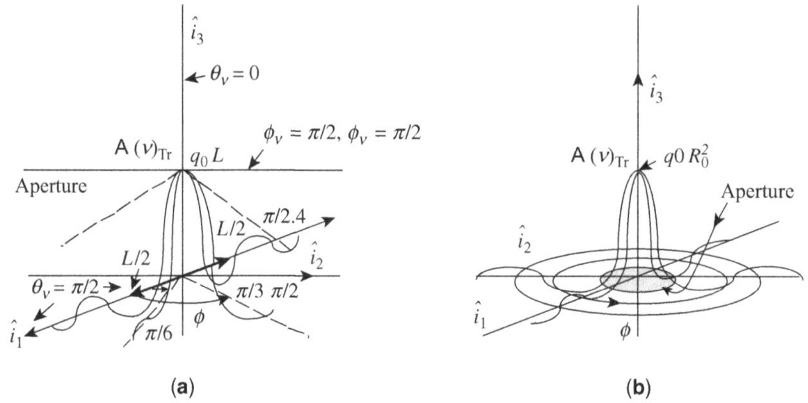

FIGURE 2.10 (a) Sections of aperture beam Eq. (2.5.6) (finite line aperture on x-axis); beam in three dimensions. (b) Symmetrical aperture in xy-plane; beam in three dimensions (Eq. 2.5.8).

2.5.1.1 Two Beam Pattern Examples It is instructive to determine the beam pattern $\mathsf{A}_{\rm Tr}$, Eq. (2.5.2a), in some special cases. Here, we consider the spatial portion of the beam pattern as is often done independent of the frequency dependent part: we regard $\boldsymbol{\nu}$ as the independent variable, ignoring the fact that $\boldsymbol{\nu} = \boldsymbol{\nu}(f)$ (cf. (2.5.16b) ff.) Thus, we let $h(\mathbf{r},t) = H(\mathbf{r})$ and absorb the time-dependent feature, so that in effect $h(\mathbf{r},t)$ is "frozen" at some value t', that is, $h(\mathbf{r},t) = H(\mathbf{r})$. Next, we let the weighting function $H(\mathbf{r}) = q_0(r_x)$ along the x-axis, between $(-L/2, L/2)$, so that the beam pattern is from (2.5.2a)

$$\mathsf{A}^{(1)}(\boldsymbol{\nu})_{\rm Tr} = \int_{-L/2}^{L/2} q_0(r_x) e^{2\pi i \nu_x r_x} dr_x \big|_{q(r_x)=q_0} = (q_0 L) \frac{\sin(\pi L \nu \cos\phi_\nu \sin\theta_\nu)}{\pi L \nu \cos\phi_\nu \sin\theta_\nu} \tag{2.5.6}$$

from $\nu_x = \nu \cos\phi_\nu \sin\theta_\nu$, (2.3.9), where $L\nu = L/\lambda$, $\lambda = \left(\lambda_x^2 + \lambda_y^2 + \lambda_z^2\right)^{1/2}$, with ν having some nonzero value in the three-space occupied by $\boldsymbol{\nu}$. Figure 2.10a shows some sections of $\mathsf{A}(\boldsymbol{\nu})_{\rm Tr}$ along different angles $(\phi_\nu = 0, \pi/6, \pi/3, \pi/2;\ \theta_\nu = \pi/2)$. Note that the dimension of the resulting beam pattern is $\sim[L]$, as expected.

Our second example is displayed in Fig. 2.10b. Here, we again postulate a uniform weighting (q_0), now over a circular disk of finite radius R_0 in the xy-plane, that is, $\theta_r = \pi/2$. The beam pattern is again distributed in three dimensions, like the example above. In this case, it is convenient to use spherical coordinates, although the aperture itself is two dimensional. We have for a source density now given by $q_0 = [L]^{-2}$ the pattern

$$\mathsf{A}^{(2)}(\boldsymbol{\nu})_{\rm Tr} = \int_0^{2\pi} q_0 d\phi_r \int_0^{R_0} r dr e^{2\pi i \nu r (\hat{\mathbf{i}}_\nu \cdot \hat{\mathbf{i}}_r)} = q_0 \int_0^{2\pi} d\phi_r \int_0^{R_0} r dr \exp\{2\pi i \nu r \cos(\phi_r - \phi_\nu) \sin\phi_\nu\} \tag{2.5.7}$$

where $\hat{\mathbf{i}}_\nu \cdot \hat{\mathbf{i}}_r = \sin\phi_\nu \cdot \cos(\phi_r - \phi_\nu)$, from (2.5.7)–(2.5.8). Since $\int_0^x x J_0(x) dx = J_1(x)$, as is readily seen from the expansion of $J_0(x)$, we obtain

$$\mathsf{A}^{(2)}(\boldsymbol{\nu})_{\rm Tr} = q_0 \int_0^{R_0} r J_0(ar) dr = q_0 \frac{R_0}{a} J_1(aR_0), \quad a \equiv 2\pi\nu \sin\theta_\nu. \tag{2.5.8}$$

This beam pattern is also dimensionless, as expected, that is, $\sim [L^2]$, and $\boldsymbol{\nu}$ again occupies all of ν-space, vanishing $O|\boldsymbol{\nu}|^{-3/2}$ as $|\boldsymbol{\nu}| \to \infty$. Figure 2.10b sketches a beam from a three-dimensional aperture, whose beam pattern is obtained from

$$\mathsf{A}^{(3)}(\boldsymbol{\nu})_{\mathrm{Tr}} = q_0 \int_0^{R_0} r^2 dr \int_0^{2\pi} d\phi \int_0^{\pi} e^{2\pi i(\nu r)(\hat{\mathbf{i}}_\nu \cdot \hat{\mathbf{i}}_r)} \sin\theta \, d\theta, \tag{2.5.8a}$$

where $\hat{\mathbf{i}}_\nu \cdot \hat{\mathbf{i}}_r$ is given by the appropriately modified form of (2.3.8), namely,

$$\hat{\mathbf{i}}_\nu \cdot \hat{\mathbf{i}}_r = (\cos\phi_\nu \sin\theta_\nu)\cos\beta_r \sin\theta_r + (\sin\phi_\nu \sin\theta_\nu)\sin\phi_\nu \sin\phi_r + (\cos\theta_\nu)\cos\theta_r. \tag{2.5.8b}$$

2.5.1.2 The Aperture as Energy "Lens" Let us now apply a signal sample of space–time extent Δ to the aperture filter $h(\mathbf{r}, t)_T$, the (volume) element $\mathbf{d}^{(3)}\boldsymbol{\xi}\ (= d\xi_x d\xi_y d\xi_z)$ in the region $\left(\boldsymbol{\xi}^{(3)} - \mathbf{d}^{(3)}\boldsymbol{\xi}/2,\ \boldsymbol{\xi}^{(3)} + \mathbf{d}\boldsymbol{\xi}^{(3)}/2\right)$, for all positions[29] $\boldsymbol{\xi}$ in the aperture occupying the space $V_T(\boldsymbol{\xi})$, that is, $\boldsymbol{\xi} \in V_T(\boldsymbol{\xi})$. The output of this element is

$$S_{\mathrm{Tr}}(\boldsymbol{\xi}, t)_\Delta \mathbf{d}^3\boldsymbol{\xi} = \int_{-\infty}^{\infty} S_{\mathrm{in}}(\boldsymbol{\xi}, \tau)_\Delta h(\boldsymbol{\xi}, t-\tau)_{\mathrm{Tr}} d\tau \mathbf{d}^3\boldsymbol{\xi}; \quad S_{\mathrm{in}} \in \Delta; \quad S_{\mathrm{in}} = 0, \quad S_{\mathrm{in}} \notin \Delta \tag{2.5.9a}$$

Replacing h_T by its equivalent (2.1.1a) and $S_{\mathrm{in},\Delta}$ by its Fourier transform, we have for (2.5.9a)

$$S_{\mathrm{Tr}}(\boldsymbol{\xi}, t)_\Delta = \int_{-\infty}^{\infty} Y(\boldsymbol{\xi}, f)_{\mathrm{Tr}} S_{\mathrm{in}}(\boldsymbol{\xi}, f)_\Delta e^{i\omega t} df, \quad \omega = 2\pi f, \tag{2.5.9b}$$

where, of course, $S_{\mathrm{in}}(\boldsymbol{\xi}, f)_\Delta$ (nonuniformly) occupies the entire *frequency* domain $(-\infty < f < \infty)$ generated by the truncated signal sample $S_{\mathrm{in}}(\boldsymbol{\xi}, t)_\Delta$. The filtered output $S_T(\boldsymbol{\xi}, t)_\Delta$ (2.5.9b) is *the signal source per unit volume* $\mathbf{d}\boldsymbol{\xi}$, which plays a key role in the generation of the transmitted field, as we shall see in Section 2.5.2 (Figure 2.13b shows the beam.)

To illustrate the beam forming or "focusing" character of the aperture, let us consider the total energy from all the emitting elements in $V_T(\boldsymbol{\xi})$ before their propagation into space over a time period T_0, that is, $E(T_0) = \int_0^{T_0} S_T(t)_\Delta^2 dt$. We begin with (2.5.9b):

$$\begin{aligned} S_T(t)_\Delta &= \int_{-\infty}^{\infty} S_T(\boldsymbol{\xi}, t)_\Delta \mathbf{d}^3\boldsymbol{\xi} \\ &= \int_{-\infty}^{\infty} \mathbf{d}^3\boldsymbol{\xi} \int_{-\infty}^{\infty} e^{i\omega t} \left(\int_{-\infty}^{\infty} \mathsf{A}(\boldsymbol{\nu}, f)_{\mathrm{Tr}} e^{-2\pi i \boldsymbol{\xi}\cdot\boldsymbol{\nu}} \mathbf{d}^3\boldsymbol{\nu} \int_{-\infty}^{\infty} S_{\mathrm{in}}(\boldsymbol{\nu}', f) e^{-2\pi i \boldsymbol{\xi}\cdot\boldsymbol{\nu}'} \mathbf{d}^3\boldsymbol{\nu}' \right) df, \end{aligned} \tag{2.5.10a}$$

[29] For transmission, we use the convention $\mathbf{r} \to \boldsymbol{\xi}$ here (cf. Eqs. (2.5.12) et seq. For reception, $\mathbf{r} \to \boldsymbol{\eta}$, refer to Section 2.5.3.

from (2.5.3a) and $\mathsf{F}_{\boldsymbol{\nu}}^{-1}\{S_{\text{in}}(\boldsymbol{\nu},f)\} = S_{\text{in}}(\boldsymbol{\xi},f)_{\Delta}$. Since

$$\int_{-\infty}^{\infty} e^{-2\pi i\boldsymbol{\xi}\cdot(\boldsymbol{\nu}+\boldsymbol{\nu}')}\mathbf{d}^3\boldsymbol{\xi} = \delta(\boldsymbol{\nu}'+\boldsymbol{\nu}) = \delta(\nu_x'+\nu_x)\delta(\nu_y'+\nu_y)\delta(\nu_z'+\nu_z), \tag{2.5.10b}$$

we see that (2.5.10a) becomes

$$S_T(t)_{\Delta} = \int_{-\infty}^{\infty} e^{i\omega t}df \int_{-\infty}^{\infty} \mathsf{A}(\boldsymbol{\nu},f)_{\text{Tr}} S_{\text{in}}(-\boldsymbol{\nu},f)\mathbf{d}^3\boldsymbol{\nu}. \tag{2.5.10c}$$

From this, we can illustrate the directed energy role of the aperture beam former A_T by calculating

$$\begin{aligned} E(T_0) &= \int_0^{T_0} S_T(t)_{\Delta}^2 dt = \int_{-\infty}^{\infty} S(t)_{\Delta} S(t)_{\Delta}^{*} dt = \int_{-\infty}^{\infty} df \left| \int_{-\infty}^{\infty} A(\boldsymbol{\nu},f)_{\text{Tr}} S_{\text{in}}(-\boldsymbol{\nu},f)\mathbf{d}^3\boldsymbol{\nu} \right|^2 \\ &= \int_{-\infty}^{\infty} df \left| \int_{-\infty}^{\infty} B_{\text{in}}(\boldsymbol{\nu},f)_{\text{Tr}}\mathbf{d}^3\boldsymbol{\nu} \right|^2 = \int_{-\infty}^{\infty} E_{T_0}(f)df \equiv E(T_0), \end{aligned} \tag{2.5.11}$$

where $E_{\text{Tr}}(f)$ is the energy density (in Hertz), or intensity, of the projectable radiation in the beam. The larger the aperture in space, the narrower or the more focused the beam. [This is easily inferred from (2.5.2a) on letting $\mathbf{r}$ be large in $Y(\mathbf{r},f)_{\text{Tr}}$, with Y_T such that ($\mathbf{r}$-large) Y_{Tr} is nonvanishing, then $\mathsf{F}_{\mathbf{r}}\{Y_{\text{Tr}}\} = \mathsf{A}(\nu,f)_{\text{Tr}} \to \delta(\boldsymbol{\nu}-\mathbf{0})F(f)$.]

2.5.2 Transmission: The Propagating Field and Its Source Function

However, our physical picture is still incomplete: we need to examine the explicit function of the beam $\sim(A_T S_{\text{in}})$ as it is projected into space. For this, we must specify a *propagation equation* appropriate to the medium that supports it. Accordingly, we choose the common generic model of a homogeneous isotropic medium governed by the well-known time-dependent Helmholtz equation, or wave equation,

$$\nabla^2\alpha(\mathbf{R},t) - \frac{1}{c_0^2}\frac{\partial^2}{\partial t^2}\alpha(\mathbf{R},t) = G_T(\boldsymbol{\xi},t); \quad = 0, \quad G_T \notin V_T(\boldsymbol{\xi}), \tag{2.5.12}$$

where c_0 is the speed of propagation of a wave front.[30] In addition, (2.5.9a) obeys the initial condition $\alpha(\mathbf{R},t) = 0,\ t < t'$, and the boundary condition that the medium is *homogeneous*, that is, here c_0 is constant and the medium contains neither scatterers nor boundaries (other than that of the source, V_T). Furthermore, it is *isotropic*, that is, has the same properties independent of direction. Additional conditions are *spatial causality* or the *radiation condition*, which ensures that only outgoing waves from the source are propagated and only time-related solutions are possible. (Other conditions are specified in Section 8.1.3 ff.)

[30] Note that the dimensions of the field α are $[A][L^{-1}][T^{-1}]$ and $\therefore [G_T] = [A]/[L^3][T]$ = amplitude pair (unit volume $\times$ unit time), where amplitude $[A]$ is determined by the medium, including a vacuum.

Here, G_T is a *source function*, associated with the volume element $\mathbf{d\xi}$, that is, (2.5.9a) and (2.5.9b), in the aperture $V_T(\boldsymbol{\xi})$, give, $G_T = S_{\mathrm{Tr}}(\boldsymbol{\xi}, t)_\Delta$.

The solution of (2.5.12) is well known to be (Section 7.3 of Ref. [19]; and here see chapter 8).

$$\alpha(\mathbf{R}, t) = \int_{V_T} G_T(\boldsymbol{\xi}, t - r(\boldsymbol{\xi})/c_0)\mathbf{d}^3\boldsymbol{\xi}/4\pi r(\boldsymbol{\xi}), \quad \mathbf{r}(\boldsymbol{\xi}) = \mathbf{R} - \boldsymbol{\xi} \tag{2.5.12a}$$

or

$$\alpha(\mathbf{R}, t) = \int_{V_T} \frac{\mathbf{d}^3\boldsymbol{\xi}}{4\pi|\mathbf{R} - \boldsymbol{\xi}|} \int_{-\infty}^{\infty} Y(\boldsymbol{\xi}, f)_{\mathrm{Tr}} S_{\mathrm{in}}(\boldsymbol{\xi}, f)_\Delta e^{i\omega(t - |\mathbf{R}-\boldsymbol{\xi}|/c_0)} df \tag{2.5.12b}$$

by (2.5.9b). This can be expressed alternatively and more conveniently to accommodate a unity of initial conditions here with the help of the contour integral:

$$\alpha(\mathbf{R}, t) = \int_{V_T} \frac{\mathbf{d}^3\boldsymbol{\xi}}{4\pi|\mathbf{R} - \boldsymbol{\xi}|} \int_{-i\infty+d}^{i\infty+d} Y(\boldsymbol{\xi}, s/2\pi i)_{\mathrm{Tr}} S_{\mathrm{in}}(\boldsymbol{\xi}, s/2\pi i)_\Delta e^{s(t - |\mathbf{R}-\boldsymbol{\xi}|/c_0)} \frac{ds}{2\pi i}, \tag{2.5.12c}$$

where $Re(s) < d$; $0 < d$, and all singularities of the integrand lie to the left of d. The (usually) straight line contour $(-i\infty + d, +i\infty + d)$, that is, $s = i\omega + d$ here is called a *Bromwich contour* and is represented by $\int_{\mathrm{Br}}$, (Chapter 4 of Ref. [20]), and s is a complex variable, that is, a complex angular frequency (see also Ref. [16] where $s \to p$ therein, and chapter 8 for a more detailed discussion). The geometry of (2.5.12a)–(2.5.12c) is shown in Fig. 2.11. Equation (2.5.12) is valid for both broad- and narrowband signals.

2.5.2.1 The Far-Field Approximation For many applications, we are interested in fields at considerable distances from the aperture $V_T(\boldsymbol{\xi})$, namely, fields in the Fraunhofer or *far-field* region, where the maximum dimension of the aperture is much smaller than $\mathbf{R}$, that is,

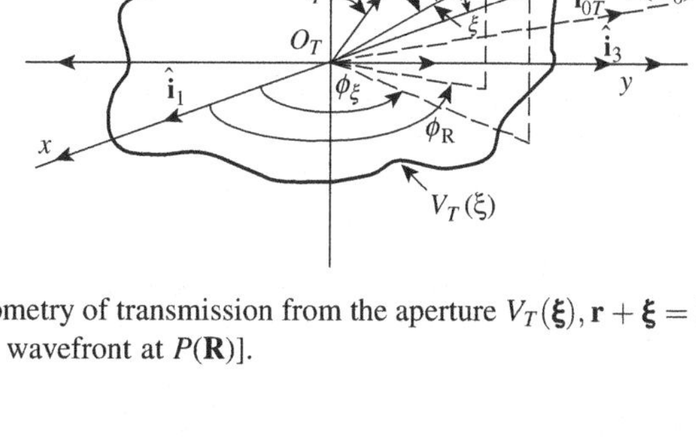

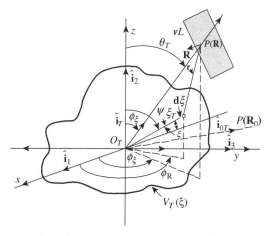

FIGURE 2.11 Geometry of transmission from the aperture $V_T(\boldsymbol{\xi})$, $\mathbf{r} + \boldsymbol{\xi} = \mathbf{R}$: [▥: Wave front at $P(\mathbf{R})$; $\hat{\mathbf{i}}$ = normal, to wavefront at $P(\mathbf{R})$].

$|\mathbf{R}| \gg |\boldsymbol{\xi}|_{max}$. The condition enables us to simplify (2.5.12) considerably, enabling us to display the role of the aperture as beam former explicitly in space. We begin with the Fraunhofer condition and write

$$\begin{aligned} r = |\mathbf{R}-\boldsymbol{\xi}| &= \{(\mathbf{R}-\boldsymbol{\xi})\cdot(\mathbf{R}-\boldsymbol{\xi})\}^{1/2} = R\left(1-[2\xi\cdot\mathbf{R}/R^2-\boldsymbol{\xi}\cdot\boldsymbol{\xi}/R^2]\right)^{1/2} \\ &= R-\boldsymbol{\xi}\cdot\hat{\mathbf{i}}_T+\frac{1}{2R}\left(\boldsymbol{\xi}\cdot\boldsymbol{\xi}-\boldsymbol{\xi}\cdot\hat{\mathbf{I}}_T\boldsymbol{\xi}\right)+O\left(|\boldsymbol{\xi}|^3/R^2\right) \end{aligned} \tag{2.5.13}$$

where $\hat{\mathbf{i}}_T \equiv \mathbf{R}/|\mathbf{R}|$. We have explicitly

$$\hat{\mathbf{i}}_T = \hat{\mathbf{i}}_1 \cos\phi_T \sin\theta_T + \hat{\mathbf{i}}_2 \sin\phi_T \sin\theta_T + \hat{\mathbf{i}}_3 \cos\phi_T;$$

$$\hat{\mathbf{i}}_\xi = \hat{\mathbf{i}}_1 \cos\phi_\xi \sin\theta_\xi + \hat{\mathbf{i}}_2 \sin\phi_\xi \sin\theta_\xi + \hat{\mathbf{i}}_3 \cos\theta_\xi \tag{2.5.13a}$$

$$\boldsymbol{\xi} = \hat{\mathbf{i}}_\xi \xi;\quad \boldsymbol{\xi}\cdot\boldsymbol{\xi} = |\xi|^2;\quad \boldsymbol{\xi}\cdot\mathbf{I}_T\cdot\boldsymbol{\xi} = \left(\boldsymbol{\xi}\cdot\hat{\mathbf{i}}_T\right)^2 = \left(\hat{\mathbf{i}}_\xi\cdot\hat{\mathbf{i}}_T\right)^2|\boldsymbol{\xi}|^2 = |\boldsymbol{\xi}|^2\cos^2\psi_{\xi T}, \tag{2.5.13b}$$

where $\psi_{\xi T}$ is the angle between $\boldsymbol{\xi}$ and $\hat{\mathbf{i}}_T$ (or $\mathbf{R}$) with the subscripts ν replaced by ξ and m by T). The *dyadic* (in rectangular coordinates) is

$$\hat{\mathbf{I}}_T = \hat{\mathbf{i}}_T\hat{\mathbf{i}}_T = \text{diag.}\left[\hat{\mathbf{i}}_{1T}\hat{\mathbf{i}}_{1T}, \hat{\mathbf{i}}_{2T}\hat{\mathbf{i}}_{2T}, \hat{\mathbf{i}}_{3T}\hat{\mathbf{i}}_{3T}\right],\quad \text{with } \hat{\mathbf{i}}_{1T} = \hat{\mathbf{i}}_1\cos\phi_T\sin\theta_T, \text{ and so on,} \tag{2.5.13c}$$

and $\hat{\mathbf{I}}_T$ itself is a dyadic or second-rank tensor, or equivalently, a square matrix with only diagonal elements [7]. Using this in (2.5.11), we obtain

$$\frac{1}{2R}\left(\boldsymbol{\xi}\cdot\boldsymbol{\xi}-\boldsymbol{\xi}\cdot\hat{\mathbf{I}}_T\cdot\boldsymbol{\xi}\right) = \frac{|\boldsymbol{\xi}|^2}{2R}\left[1-\left(\hat{\mathbf{i}}_\xi\cdot\hat{\mathbf{i}}_T\right)^2\right] \tag{2.5.13d}$$

for the third term of (2.5.13a), so that we can write

$$r = |\mathbf{r}| \doteq R-\boldsymbol{\xi}\cdot\hat{\mathbf{i}}_T+\frac{|\boldsymbol{\xi}|^2}{2R}\left[1-\left(\hat{\mathbf{i}}_\xi\cdot\hat{\mathbf{i}}_T\right)^2\right] \doteq R-\boldsymbol{\xi}\cdot\hat{\mathbf{i}}_T+O\left(|\boldsymbol{\xi}|^2/2R\right) \tag{2.5.13e}$$

for the far-field or Fraunhofer approximation.[31] A condition here for (2.5.13) to be valid is that

$$\frac{|\boldsymbol{\xi}|^2_{max}}{2R}\left\langle 1-\left[\hat{\mathbf{i}}_\xi\cdot\hat{\mathbf{i}}_T\right]^2\right\rangle_{\psi_{\xi T}} = \frac{|\boldsymbol{\xi}|^2_{max}}{2R}\cdot\frac{1}{2\pi}\int_0^{2\pi}\left(1-\cos^2\psi_{\xi T}\right)d\psi_{\xi T} = \frac{|\boldsymbol{\xi}_{max}|^2}{4R} \ll |\boldsymbol{\xi}|_{max} \text{ or } |\boldsymbol{\xi}_{max}| \ll 4R, \tag{2.5.14}$$

[31] For points in space closer to the aperture, such that (2.5.13) must be considered in the exponent, the point $\mathbf{R}$ is said to be in the *Fresnel region*, where now the aperture beam function, A_T, and hence the beam generated by it, as well as the field itself, is no longer independent of aperture structure ($\boldsymbol{\xi}$). The amplitude of the field ($\sim R^{-1}$), however, is still not noticeably affected.

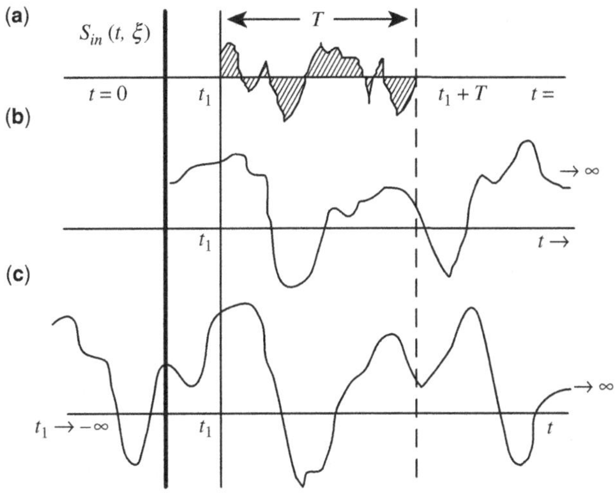

FIGURE 2.12 A (broadband) signal (S_{in}) applied to the aperture. (a) Finite duration. (b) Semi-infinite duration. (c) Steady state $(-\infty \leq t \leq \infty)$.

namely, the condition that the point of observation **R** in space be sufficiently far away from the aperture's average maximum dimension $|\boldsymbol{\xi}_{\max}| = |\boldsymbol{\xi}|_{\max}\left\langle 1 - \left[\hat{\mathbf{i}}_{\xi} \cdot \hat{\mathbf{i}}_{T}\right]^2\right\rangle_{\psi_{\xi_T}}^{1/2}$ for the far-field condition (2.5.14) to be satisfied.

Next, we consider the fact that the signal to the aperture is applied only for a time T (Fig. 2.12a), starting at $t = t_1$ and ending at $t = t_1 + T$, as in Fig. 2.12a. The integrand over frequency is modified by inserting the factor $(1 - e^{-sT})$, which cancels the contribution of the first term when $t > T$. Furthermore, the far-field approximation (2.5.13c) allows us to write (2.5.12c) as

$$\alpha(\mathbf{R}, t)_{\text{FF}} \doteq \int_{\text{Br}_1} \frac{e^{s(t-R/c_0)}}{4\pi R}\left(1 - e^{-sT}\right)\frac{ds}{2\pi i} \cdot \int_{-\infty}^{\infty} Y(\boldsymbol{\xi}, s/2\pi i)_{\text{Tr}} e^{i\boldsymbol{\xi}\cdot\hat{\mathbf{i}}_T s/ic_0} S_{\text{in}}\left(\boldsymbol{\xi}, \frac{s}{2\pi i}\right)_{\Delta} \mathbf{d}^3\boldsymbol{\xi}, \tag{2.5.15}$$

From this and Eqs. (2.5.2a) and (2.5.3b), we can get

$$\hat{\mathbf{i}}_T s/2\pi i c_0 = \boldsymbol{\nu}_T, \tag{2.5.15a}$$

a (complex) wave number (which in turn becomes the real wave number of Section 2.2.2 when $s \rightarrow 2\pi f$). Following the procedure of (2.5.10a)–(2.5.10c), we readily find that (2.5.15) can be represented in turn by

$$\alpha(\mathbf{R}, t)_{\text{FF}} \doteq \alpha(\mathbf{R}, t - R/c_0) = \int_{\text{Br}_1} \frac{e^{s(t-R/c_0)}}{4\pi R}\left(1 - e^{-sT}\right)\frac{ds}{2\pi i} \cdot \int_{-\infty}^{\infty} \mathsf{A}\left(\boldsymbol{\nu}_T - \hat{\boldsymbol{\nu}}, \frac{s}{2\pi i}\right)_{\text{Tr}} S_{\text{in}}\left(\hat{\boldsymbol{\nu}}, \frac{s}{2\pi i}\right)\mathbf{d}^3\hat{\boldsymbol{\nu}}, \tag{2.5.15b}$$

with $|\boldsymbol{\xi}_{\max}| \ll 4R$ (2.5.14). Equation (2.5.15b) is the desired far-field expression for the field at **R** in the direction indicated by the unit vector $\hat{\mathbf{i}}_T$. Inserting a *steering vector* $\boldsymbol{\nu}_{O_T}$ into the aperture, where each element at $\mathbf{d\xi}$ has an appropriate delay with respect to the common reference point O_T (Fig. 2.15), now enables us to direct the energy of the source (2.5.11) to an arbitrary location, say $P(\mathbf{R}_0)$ of Fig. 2.11. For this more general situation, (2.5.15b) becomes

$$\alpha(\mathbf{R},t)_{\mathrm{FF}} \doteq \int_{\mathrm{Br}_1} \frac{e^{s(t-R/c_0)}}{4\pi R}\left(1-e^{-sT}\right)\frac{ds}{2\pi i}\cdot\int_{-\infty}^{\infty} \mathsf{A}(\boldsymbol{\nu}_T-\boldsymbol{\nu}_{O_T}-\hat{\boldsymbol{\nu}}, s/2\pi i)_{\mathrm{Tr}} S_{\mathrm{in}}(\hat{\boldsymbol{\nu}},f)_{\Delta}\mathbf{d}^3\hat{\boldsymbol{\nu}} \tag{2.5.16a}$$

for the truncated input $S_{\mathrm{in}}(\boldsymbol{\xi},t)_{\Delta}=S_{\mathrm{in}}(\boldsymbol{\xi},t)$, $S_{\mathrm{in}}\in\Delta:(V_T(\boldsymbol{\xi}),-t_1\le t\le t_1+T), t_1 = R/c_0; S_{\mathrm{in}}=0,\ S_{\mathrm{in}}\notin\Delta$, as shown in Figs. 2.6 and 2.7. Equation (2.5.16a) applies for a continuous sample in $(t_1,\ t_1+T)$ (Fig. 2.12a). When $T\to\infty$, the factor $1-e^{-sT}$ reduces to 1, that is, in Fig. 2.12b. If $t_1=-\infty$, namely, the source has been "on" since $t_1=-\infty$, then the integrand of the Br_1 contour is exp st, and the contour itself is the imaginary axis, with a possible indentation about poles, and so on, on this line, that is, one has a Fourier transform. When $\boldsymbol{\nu}_{O_T}=\boldsymbol{\nu}_T$, the maximum energy in the beam ($\sim\mathsf{A}S_{\mathrm{in}}$) (2.5.1.2) is focused on the point $\mathbf{R}=\mathbf{R}_0$. For these ideal media, which among other benign properties are nondispersive, the speed of propagation c_0 is a constant independent of frequency, ([19], pp. 477–479, and Chapter 9 generally). This means that the defining wavelength–frequency relation $\lambda f=c_0=$ constant is obeyed, namely,

$$c_0(f)=c_0=\lambda f \quad \text{and} \quad \therefore\ |\boldsymbol{\nu}|=\nu=f/c_0, \quad \text{since} \quad \nu=1/\lambda; \quad \mathbf{d}^3\boldsymbol{\nu}=\nu^2\sin\theta d\theta d\phi, \tag{2.5.16b}$$

From this we see *that*[32] $\boldsymbol{\nu}_T,\boldsymbol{\nu}_{O_T}$ *are functions of frequency*, allowing us to write (2.5.16), finally, with the help of (2.5.16b):

$$\left.\begin{aligned}\alpha(\mathbf{R},t)_{\mathrm{FF}} &\doteq \int_{\mathrm{Br}_1}\frac{e^{s(t-R/c_0)}}{4\pi R}\left(1-e^{-sT}\right)\frac{ds}{2\pi i}\\ &\cdot\int_{-\infty}^{\infty}\hat{\boldsymbol{\nu}}^2\mathbf{d}\hat{\boldsymbol{\nu}}\int_0^{\pi}\sin\hat{\theta}d\hat{\theta}\int_0^{2\pi}\mathsf{A}\left(\Delta\hat{\mathbf{i}}_T s/2\pi i c_0-\hat{\boldsymbol{\nu}},s/2\pi i\right)_{\mathrm{Tr}} S_{\mathrm{in}}(\hat{\boldsymbol{\nu}},s/2\pi i)d\hat{\phi}\end{aligned}\right\} \tag{2.5.17a}$$

where $\alpha=0,\ t<R_0/c,\ t>R_0/c+T$ and where

$$\hat{\mathbf{i}}_T=\text{Eq. (2.5.13a)};\quad \Delta\hat{\mathbf{i}}_T\equiv\hat{\mathbf{i}}_T-\hat{\mathbf{i}}_{O_T};\quad \left(\boldsymbol{\nu}_T-\boldsymbol{\nu}_{O_T}=\Delta\boldsymbol{\nu}_T=\Delta\hat{\mathbf{i}}_T f/c_0=\Delta\hat{\mathbf{i}}_T(s/2\pi i c_0)\right);$$
$$(s=\text{complex angular frequency}). \tag{2.5.17b}$$

As noted earlier (2.5.12c), the singularities of the integrand all lie to the left of d, that is, *Re* $s<d, d>0$. In the Fresnel region, closer to the source, the condition (2.5.14) is not satisfied and this comparatively simple result (2.5.17a) for a homogeneous, in fact isotropic, medium ($\mathbf{R}\to$ R, and c_0 constant) no longer holds because of the influence of the additional

[32] Here, $\hat{\nu}$ is a variable of integration, independent of the physical quantities $\boldsymbol{\nu}_{O_T},\boldsymbol{\nu}_T$, and so on.

$\boldsymbol{\xi}$-dependent term (2.5.14) in the phase. This "nearer far-field" region, however, can be important in applications such as synthetic aperture radar (SAR) and sonar (SAS), and in telecommunications, where the far-field conditions (2.5.14) are not satisfied. Physically, it is apparent from (2.5.16a)–(2.5.17b) that the spatial and temporal portions of the space–time field $\alpha(\mathbf{R},t)$ are *not* generally separable: $ST \neq S \otimes T$ symbolically. Separability can be imposed artificially, here at the transmitter, by designing the aperture or array structures independent of the temporal portions, which is a useful convenience for the system designer. However, separability is a constraint and as such reduces performance, at least in principle, with respect to the optimum, unconstrained joint space–time design. The latter properly accounts for the physical fact of wave number dependence on frequency (2.5.16b), particularly in the broadband cases, which can be described by (2.5.17b), as well as for many narrowband signals. This dependence, however, increases the difficulty of implementation, since the steering vector $\boldsymbol{\nu}_{O_T}\left(= \hat{\mathbf{i}}_{O_T} f/c_0\right)$ itself is a function of all the frequencies in the driving signal S_{in}.

Only for *monofrequentic signals*, that is, sinusoids, *is strict separation possible.* If the bandwidths of the input signal, namely, the wave number and temporal frequency bandwidths, are both sufficiently narrow, then separability $(S \otimes T)$ is, in practice, possible with ignorable error. (For details, see Section 3.5 ff.)

Besides the far-field condition (2.5.14), there is another condition on the aperture in its capacity as a space–time filter (2.5.1a), namely, that it does not significantly distort the projected signal S_{in}. Ideally, no distortion would be the goal. This requires that the aperture have the weighting function $h(\boldsymbol{\xi}', t')_{\text{Tr}} = H_0 \delta(\boldsymbol{\xi}' - \boldsymbol{\xi})\delta(t' - t)$ of an "all-pass" filter, which is equivalent to $\mathsf{A}_{\text{Tr}} = \mathsf{A}_{\text{Tr}}|_\delta = H_0 \delta(\hat{\boldsymbol{\nu}} - \Delta\boldsymbol{\nu}_T)\delta\left(\frac{s'-s}{2\pi i}\right)$. The resulting field is

$$\left.\begin{aligned}\alpha(\mathbf{R}, t)_{\text{FF}} = \alpha(\mathbf{R}, t - R/c_0)\Big|_{\substack{\text{FF}\\ \delta}} &\doteq \int_{\text{Br}_1} \frac{e^{s(t-R/c_0)}(1 - e^{-sT})}{4\pi R} S_{\text{in}}(\Delta\boldsymbol{\nu}_T(s/2\pi i), s/2\pi i)\boldsymbol{\nu}\frac{ds}{2\pi i} \\ &= 0, \quad R/c_0 \leq t \leq R/c + T,\end{aligned}\right\} \tag{2.5.18}$$

where $|_\delta$ refers to the all-pass nature of h_{Tr}, the aperture weighting function (2.5.1b). When $\Delta\boldsymbol{\nu}_T = 0$, that is, when $\hat{\mathbf{i}}_T = \hat{\mathbf{i}}_{O_T}$ the beam formed by the aperture is pointed in the direction of a potential target at $\mathbf{R}_0$ (Fig. 2.11). Then, (2.5.18) becomes

$$\alpha(\mathbf{R}_0, t)\Big|_{\substack{\text{FF}\\ \delta}} \doteq S_{\text{in}}(0, t - R_0/c_0)_\Delta / 4\pi R_0, \quad \begin{cases} t \leq R_0/c_0 \leq t + T; & = 0, \text{elsewhere}, \\ \Delta\hat{\mathbf{i}}_T = 0 \end{cases} \tag{2.5.18a}$$

for the undistorted, "on-target" signal, "painted" by this transmitter, as shown in Fig. 2.13a. One criterion of performance is to measure the ratio of the intensity of the field at a point $\mathbf{R}_0$ when the beam pattern A_T is not "all-pass," that is, (2.4.17a) and (2.4.17b), to the ideal case (2.5.18a) where it is,

$$\left|\alpha(\mathbf{R}_0, t)_{\text{FF}}\right|^2 = \text{Eqs. (2.5.17a) and (2.5.17b)}| = S_{\text{in}}(0, t - R_0/c_0)|^2 / 4\pi R_0 \equiv \gamma_0 \leq 1. \tag{2.5.19}$$

Other, related criteria are also clearly possible.

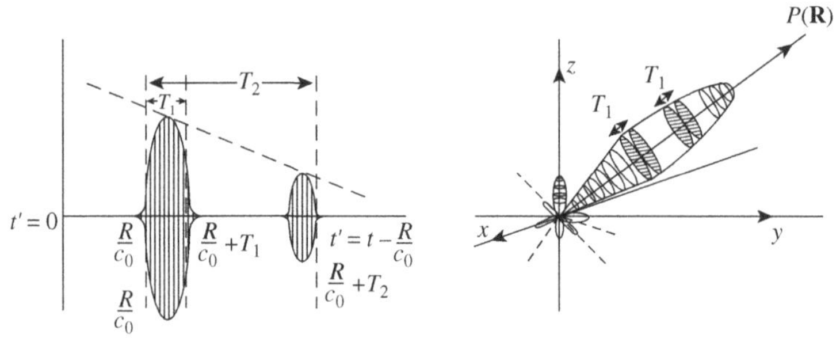

FIGURE 2.13 (a) Field $\alpha(\mathbf{R},t)$ after transmission in three dimensions. (b) A beam pattern.

2.5.3 Point Arrays: Discrete Spatial Sampling

In contrast to the aperture, whose elements are effectively connected to form a continuum, arrays are regarded as essentially discrete elements in space. An excellent practical approximation is to consider the discrete elements as "points," defined physically as being smaller than the shortest wavelength employed at the input of the array. At each of these points, there is a linear filter (time-invariant at the moment), which produces a delayed, filtered version of the input S_{in}. These outputs are then added, again forming a beam of energy from the transmitter, which is then propagated into the medium. The beam is steered in some appropriate direction, in a precisely similar fashion to the operation discussed above in Section 2.5.2.

However, for the spatial Fourier transform $\mathsf{F}_{r_m}^{(d)}\{\ \}$ of this sampled Green's function, we must now employ the analysis of Section 2.3.1, where $\mathsf{F}\{\ \}$ is replaced by $\mathsf{F}^{(d)}\{\ \}$:

$$\mathsf{F}\{X(\boldsymbol{r})\} = \int_{-\infty}^{\infty} X(r)e^{2\pi i\boldsymbol{\nu}\cdot\mathbf{r}}\boldsymbol{dr} \rightarrow \sum_{m=1}^{M} \mathsf{F}_{r_m}^{(d)}\{X_m\} \equiv \sum_{m=1}^{M} X_m\Delta_m e^{2\pi i\boldsymbol{\nu}\cdot\mathbf{r}_m}, \tag{2.5.20}$$

with $\Delta_m = |\mathbf{r}_{m+1} - \mathbf{r}_m|$ the interval between sample points, as described in Section 2.3.2.1.[33]

The aperture weighting (or Green's) function, (2.5.1b), is now replaced by $h(\mathbf{r}_m,\ t),\ t > 0$, at each point $\mathbf{r} = \mathbf{r}_m,\ m = 1, \ldots, M$. Accordingly, the array beam function (2.8.2b), when considered apart from propagation, becomes[33]

$$\begin{aligned}\mathsf{A}_M^{(\mathrm{dc})}(\boldsymbol{\nu},f)_{\mathrm{Tr}} &= \sum_{m=1}^{M}\Delta_m \int_{-\infty}^{\infty} h^{(m)}(\mathbf{r}_m,t)_{\mathrm{Tr}}e^{2\pi i\boldsymbol{\nu}\cdot\mathbf{r}_m - i\omega t}\Delta dt \\ &= \sum_{m=1}^{M}\Delta_m \cdot Y_m(\mathbf{r}_m,f)_{\mathrm{Tr}}e^{2\pi i\boldsymbol{\nu}\cdot r_m - i\omega t}, \quad \Delta_m = |\mathbf{r}_{m+1} - \mathbf{r}_m|\end{aligned} \tag{2.5.20a}$$

The spatial portions of the other aperture relations (2.5.1)–(2.5.3) are similarly modified.

[33] It is only a matter of convenience here to use the positive octant and time axis for our discrete samples instead of the symmetrical sampling plans of Section 2.3.3.1.

In the special case of the propagated *field* (Section 2.5.2) in the specific instance of a Helmholtz medium (*H*), we see at once that the source function G_T becomes $G_T(\xi_m, t)$ and that the field at a point $P(\mathbf{R})$ is now represented by[34]

$$\text{Eq. } (2.5.12a)^{34} : \alpha^{(dc)}(\mathbf{R}, t)_H = \sum_{m=1}^{M} \frac{G_T^{(dc)}(\xi_m, t - r(\xi_m)/c_0)_m}{4\pi r(\xi_m)}, \quad r(\xi_m) = |\mathbf{R} - \xi_m|, \tag{2.5.21a}$$

$$\text{Eq. } (2.5.12b)^{34} : \alpha^{(dc)}(\mathbf{R}, t)_H = \sum_{m=1}^{M} \frac{1}{4\pi|\mathbf{R} - \xi_m|} \int_{Br_1} Y_m{}^{(dc)}(\xi_m, s/2\pi i)_{\mathrm{Tr}} \, S_{\text{in}-m}^{(dc)}(\xi_m, s/2\pi i)_\Delta e^{s(t-|\mathbf{R}-\xi_m|/c_0)} \frac{ds}{2\pi i}, \tag{2.5.21b}$$

and for the far-field:

$$\text{Eq. } (2.5.12c)^{34} : \alpha^{(dc)}(\mathbf{R}, t)\Big|_{\mathrm{FF}}^{\mathrm{H}} \doteq \int_{\mathrm{Br}_1} \frac{e^{s(t-R/c_0)}(1 - e^{-sT})}{4\pi R} \frac{ds}{2\pi i} \cdot \sum_{m=1}^{M} Y_m^{(dc)}(\xi_m, s/2\pi i)_{\mathrm{Tr}} S_{\text{in}-m}^{(dc)}(\xi_m, s/2\pi i)_\Delta e^{2\pi i \xi_m \cdot (\boldsymbol{\nu}_T - \boldsymbol{\nu}_{\mathrm{OT}})}, \tag{2.5.21c}$$

where we have introduced a steering vector $\boldsymbol{\nu}_{O_T}$ into the phase component $2\pi i \xi_m \cdot (\boldsymbol{\nu}_T - \boldsymbol{\nu}_{O_T})$. (We note again (2.5.16b), that $\boldsymbol{\nu}_T$ and $\therefore \boldsymbol{\nu}_{O_T}$ are functions of frequency, where now f is extended to the complex quantity $s/2\pi i$.)

Now from (2.3.3), (2.3.15), and (2.5.20), we see that the *discrete* Fourier transforms of $Y_m^{(dc)}$ and $S_{\text{in}-m}^{(dc)}$ are given by

$$\begin{aligned} Y_m^{(dc)} \Delta_m e^{2\pi i \boldsymbol{\nu} \cdot \xi_m} &= \therefore \; Y_m^{(dc)}(\boldsymbol{\nu}, s/2\pi i), \\ S_{\text{in}-m}^{(dc)} \Delta_m e^{2\pi i \boldsymbol{\nu} \cdot \xi_m} &= \therefore \; S_{\text{in}-m}^{(dc)}(\boldsymbol{\nu}, s/2\pi i), \end{aligned} \tag{2.5.22}$$

with $\Delta_{\mathrm{om}} = |r_{m+1} - r_m|$, for the mth interval between sensors (2.5.28). Substitution into (2.5.21c) and noting because of the finite interval Δ for $S_{\text{in}-m}^{(dc)}$, and $Y_m^{(dc)}$ that we can replace $\sum_{m=1}^{M}$ by the infinite sum $\sum_{m=-\infty}^{\infty}$, however, does nothing to simplify (2.5.21c)), even when we restrict the array to a straight line in space, that is, $\xi_m = \left(\hat{\mathbf{i}}_0, r_0\right) m$. The reason, of course, is that $A_m^{(dc)}, S_{(\Delta, m)}^{(dc)}$ remain functions of $m(= 1, 2, \ldots, M)$, because of which it is generally not possible to effectuate the sum $\sum_m$. Thus, it is much more convenient to deal

[34] The superscript (dc) is to remind the reader that the quantities so designated have been discretely sampled, here in space (*d*), but are continuously sampled (observed) in time (*c*). Then, in the same fashion (dd) represents discrete sampling in both space and time, while (cc), or no superscript, indicates that observation, that is, sampling in space and time are both continuous. Thus, (dc) here implies a type of sampling in space–time (ST).

with the result (2.5.21c) itself than to use its equivalent Fourier transform $F^{(d)}\left(Y^{(\mathrm{dc})}S_{\mathrm{in}}^{(\mathrm{dc})}\right)$ and have to contend with the sum of each of m-different beam patterns, that is,

$$\sum_{m=1}^{M}\int_{[\nu]}\int_{[\nu']} \mathsf{A}_m^{(\mathrm{dc})} S_{\mathrm{in}-m}^{(\mathrm{dc})}\Big|_{(\Delta)} e^{2\pi i \xi_m\cdot(\Delta\boldsymbol{\nu}_T-\boldsymbol{\nu}-\boldsymbol{\nu}')}\mathbf{d\nu}\,\mathbf{d\nu}', \quad \text{with } \Delta\boldsymbol{\nu}_T=\boldsymbol{\nu}_T-\boldsymbol{\nu}_{O_T}, \tag{2.5.22a}$$

where the integrations over $(\boldsymbol{\nu}, \boldsymbol{\nu}')$ are along the line of Eq. (2.3.13), here restricted to space only.

2.5.3.1 Same Signal Inputs to the Array However, there is one important and frequently employed situation where comparatively simple results are obtained, namely, the case where *the same input signal is applied to each of the M sensors*, with Green's functions $h^{(m)}(\mathbf{r}, t)$. Then, $S_{\mathrm{in}-m} = S_{\mathrm{in}}(t)_T$ is independent of sensor location (ξ_m), and (2.5.21c) for the transmitted field can be written

$$\alpha(\mathbf{R}, t)\Bigg|_{\substack{\mathrm{H}\\ \mathrm{FF}\\ S_{in}}} \doteq \int_{Br_1} S_{\mathrm{in}}(s/2\pi i)_\Delta \frac{(1-e^{-st})e^{s(t-R/c_0)}}{4\pi R} A_M^{(\mathrm{dc})}(\Delta\boldsymbol{\nu}_T, s/2\pi i)_{\mathrm{Tr}}\frac{ds}{2\pi i}, \tag{2.5.23}$$

where

$$\mathsf{A}_M^{(\mathrm{dc})}(\Delta\boldsymbol{\nu}_T, s/2\pi i)_{\mathrm{Tr}} = \sum_{m=1}^{M}\int_{[\boldsymbol{\nu}]} \mathsf{A}_m^{(\mathrm{dc})}(\boldsymbol{\nu}, s/2\pi i)_{\mathrm{Tr}} e^{-2\pi i(\boldsymbol{\nu}-\Delta\boldsymbol{\nu}_T)\cdot\xi_m}\mathbf{d\nu}. \tag{2.5.23a}$$

From (2.3.13) and the discussion following the integral over $\mathbf{d\nu}$ $(=\nu^2\sin\theta d\nu d\phi d\phi)_m$ for the interval Δ_m comes

$$\int_{-(\nu_0-\Delta\nu_T/2)}^{(\nu_0-\Delta\nu_T)/2} \sin\theta\,\nu^2 d\nu \int_0^{2\pi} d\phi \int_0^{\pi} A_M^{(\mathrm{dc})}|_{\mathrm{Tr}} e^{-2\pi i(\nu-\Delta\nu_T)\cdot\xi_m}d\theta = Y_m^{(\mathrm{dc})}(\xi_m, s/2\pi i)_{\mathrm{Tr}},\ e^{-2\pi i\Delta\boldsymbol{\nu}_T\cdot\xi_m} \tag{2.5.24}$$

which reduces (2.5.23a) to the result of (2.5.22):

$$\mathbf{A}_M^{(\mathrm{dc})}(\Delta\boldsymbol{\nu}_T, s/2\pi i) = \sum_{m=1}^{M} Y_m^{(\mathrm{dc})}(\xi_m, s/2\pi i)e^{-2\pi i\Delta\boldsymbol{\nu}_T\cdot\xi_m}. \tag{2.5.24a}$$

Now the resultant beam pattern $\mathsf{A}_M^{(\mathrm{dc})}$, (2.5.23a), of the transmitting array is the sum of the beam patterns associated with m sensors at ξ_m, $m = 1, \ldots, M$, when the *same signal* is applied to each of them, even through the weighting functions may be different for different m (2.5.20).

Again, from (2.5.16b) we observe that for general input signals, space and time are not naturally separable, since $\Delta_T\boldsymbol{\nu} = \left(\hat{\mathbf{i}}_T - \hat{\mathbf{i}}_{0_T}\right)s/2\pi i c_0$: separation of the two is *an imposed constraint*. For example, let us assume that the local beam patterns $\mathsf{A}^{(\mathrm{dc})}$ are proportional to a constant A_m multiplied by a phase shift, namely,

$$\mathsf{A}_m^{(\mathrm{dc})}|_{\mathrm{Tr}} = Y(\boldsymbol{\xi}_m, s/2\pi i)_{\mathrm{Tr}} e^{2\pi i(\boldsymbol{\nu}-\Delta\boldsymbol{\nu}_T)\cdot\xi_m} = \left(A_m \cdot e^{-2\pi i \Delta\boldsymbol{\nu}_T \cdot \boldsymbol{\xi}_m}\right) e^{2\pi i \boldsymbol{\nu}\cdot\boldsymbol{\xi}_m} \tag{2.5.24a}$$

and thus

$$Y_m^{(\mathrm{dc})} = A_m e^{-2\pi i \Delta\boldsymbol{\nu}_T\cdot\boldsymbol{\xi}_m}, \text{ refer to Eq. (2.5.22)}, \tag{2.5.24b}$$

is similarly modified. Accordingly, the overall beam pattern $\mathsf{A}_M^{(\mathrm{dc})}$ of the array of sensors becomes

$$\mathsf{A}_M^{(\mathrm{dc})}(\Delta\boldsymbol{\nu}_T,\ s/2\pi i)_{\mathrm{Tr}} = \sum_{m=1}^{M} A_m e^{-2\pi i \Delta\boldsymbol{\nu}_T\cdot\boldsymbol{\xi}_m} = \sum_{m=1}^{M} A_m e^{-s\Delta\hat{\mathbf{i}}_T\cdot\boldsymbol{\xi}_m/c_0},$$
$$\Delta\hat{\mathbf{i}}_T \equiv \left(\hat{\mathbf{i}}_T - \hat{\mathbf{i}}_{OT}\right), \tag{2.5.25}$$

which is even simpler if the weightings A_m are independent of m, that is, $A_m = A$.

When the array is a straight line in space with equal spacing (r_0) between sensors, that is, then it is possible to evaluate the sum in (2.5.25) exactly by a finite geometric series $\Sigma_M x^m = (1 - x^{M+1})/(1 - x)$, $|x| < 1$. (An example of this situation is given in Eqs. (3.1.22)–(3.1.24) following.) Note that the beam patterns presented here and earlier are generally complex but under the requirement that the filter's weighting function $h^{(m)}(\mathbf{r},\ t)$ is real, vide (2.5.5). The frequency dependence (i.e., on s) of the beam pattern is explicitly shown here in (2.5.25) again, revealing once more the general nonseparability of the spatial and temporal parts of the propagating field.

The formal effect of discrete sampling in space is twofold: (1) to change the domain of integration of the wave numbers from $(-\infty \leq \nu \leq \infty)$ to the finite interval $-[\boldsymbol{\nu}_0]_m/2 \leq \boldsymbol{\nu} \leq [\boldsymbol{\nu}_0]_m/2$, refer to Eq. (2.5.24), and (2) to modify the explicit forms of the beam pattern and input signal spectrum according to (2.5.22). Equations (2.5.21a)–(2.5.25) apply for the general case of arbitrary lengths and directions of the sampling intervals Δ_m, (Fig. 2.6). If we include discrete sampling in time, as well as space, then the results above are extended with the help of Section 2.3 directly, for example, $\Delta_{0_m} \to \Delta_{0_j}$ $(j = mn)$, $t \to t_n$. However, since it is not possible physically to generate δ-function pulses, we will always have signals of finite, nonzero duration in transmission. Accordingly, we do not need to consider the fully discrete cases further here. Reception is another matter when we are required to digitalize the received input signal (and noise).

2.5.4 Reception

The reception process is largely the reverse of transmission. That reception is not completely the inverse operation, as is well known, is due to the effects of the channel, which exhibits ambient noise, signal-generated noise, that is, interference and scattering, and the often inherent inhomogeneity of the medium itself. However, the methodology of Sections 2.5.1–2.5.3 can be applied here with appropriate modifications. The relevant geometry is shown in Fig. 2.14.

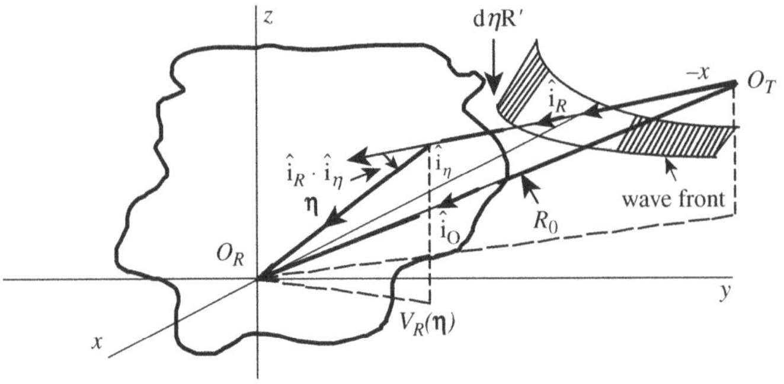

FIGURE 2.14 Geometry of a receiving aperture $V_R(\boldsymbol{\eta})$, where $\mathbf{r}+\boldsymbol{\eta}=\mathbf{R}$ with sensor element at $\mathbf{d}\boldsymbol{\eta}$, source point at $P(R)$. Here $\hat{\mathbf{i}}_\eta \cdot \hat{\mathbf{i}}_R = \cos\Phi_{\eta R}$.

The continuous input field to the sampler $\hat{\mathbf{T}}_s(\;)_{c \text{ or } d}$ (2.5.26), which selects a finite segment of the received fields is represented by

$$X(\mathbf{r},\, t) = \hat{T}_s^{(cc)}\alpha(\mathbf{r},\, t) = \alpha(\mathbf{r},\, t)_\Delta,\; (r \in |\mathbf{R}|,\, t \in T); \quad = 0 \quad \text{elsewhere}, \qquad (2.5.26a)$$

or the (column) vector

$$[X(r_m,\, t_n)] = \left[\hat{T}_s^{(dd)}\alpha(\mathbf{r},\, t)\right] = \left[\alpha(r_m, t_n)_\Delta\right], \quad \Delta = |\mathbf{R}|T; \; = 0, \quad \text{otherwise}. \quad (2.5.26b)$$

For much of analysis throughout we are concerned with Eq. (2.5.26b), namely, the set of discrete space–time samples of data (Chapters 3,4, and so on following), which constitutes the input to the detectors and estimates of our processing systems. We next take advantage of the results of Sections 2.5.1 and 2.5.2 and consider some specific cases involving reception of Helmholtz fields (Eq. 2.5.12) in an ideal medium, processed by apertures and arrays. As noted in Section 2.5.2, these are respectively continuous and discrete operators $\hat{R}^{(c)}$ and $\hat{\mathbf{R}}^{(d)}$, respectively, on the input $X(\mathbf{r},\, t)$ or $X(\mathbf{r}_m,\, t_n)$ [(2.5.26a) and (2.5.26b)]. As expected, each produces a single but different result, since they stem from different operations. Accordingly, we begin with the following.

2.5.4.1 Continuous Sampling (Helmholtz Medium) The Helmholtz medium here is assumed to be ideal, that is, nondispersive, infinite, and contains no other elements than the receiver and transmitter. Accordingly, for reception, we begin by writing the results of Section 2.5.2 for the transmitted field $\alpha_T(\mathbf{R},\, t)$, particularly Eqs. (2.5.12a)–(2.5.16a), along with Eq. (2.5.9b), as a guide for the corresponding received field $\alpha(\boldsymbol{\eta},\, t)_R$ at a typical sensor element ($\sim \mathbf{d}\boldsymbol{\eta}$), when transmitter (T) and receiver (R) are in each other's far-field (FF) or Fraunhofer region. The relationships of the respective component terms are readily seen to be in the present case of continuous sampling:

Transmission (T|FF) Reception (R|FF)

$$\text{Eq. (2.5.9b)} \quad G_T = S_{in}^{(cc)}(\boldsymbol{\xi},\, t)_{\Delta|T} \quad \rightarrow \quad \alpha^{(cc)}(\boldsymbol{\eta},\, t)_{\Delta|T}|_{FF} \quad : \text{received field input} \qquad (2.5.27a)$$

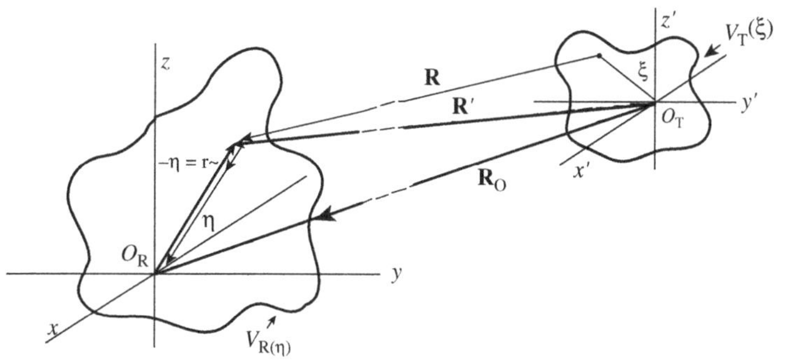

FIGURE 2.15 Schema of transmitter and receiver in each other's far-field.

$$\text{Eq. (2.5.1a)} \qquad Y^{(cc)}(\boldsymbol{\xi}, s/2\pi i)_T \;\rightarrow\; Y^{(cc)}(\boldsymbol{\eta}, s/2\pi i)_R \;:\; \text{receiver aperture system function} \tag{2.5.27b}$$

$$\left.\begin{array}{llll} \text{Eq. (2.5.13c) : phase : } |\mathbf{R}-\boldsymbol{\xi}| & \doteq R - \mathbf{i}\boldsymbol{\xi}\cdot\hat{\mathbf{i}}_T & \rightarrow |R_0-\boldsymbol{\eta}| : & \text{phase,} \\ \text{amplitude :} & \doteq R, |\boldsymbol{\xi}| \ll R & \rightarrow R_0 \quad : & \text{amplitude} \end{array}\right\}, |R_0| \gg |\eta| \tag{2.5.27c}$$

For (2.5.27c) we see from Fig. 2.15 that

$$\boldsymbol{\eta} + \mathbf{R}' + \xi = \mathbf{R}_0 \;\text{ or }\; \boldsymbol{\eta} = \mathbf{R}_0 - (\mathbf{R}' + \boldsymbol{\xi}), \text{ where } \mathbf{R}' \approx \mathbf{R}_0 \text{ and } |\boldsymbol{\xi}| \ll R', R_0, \;\therefore\; \mathbf{R}' = \mathbf{R}_0 - \boldsymbol{\eta}, \tag{2.5.27d}$$

where $\mathbf{R}'$ is the distance from $\mathbf{P}(\boldsymbol{\xi})$ in the transmitter's aperture to the sensing element at $\mathbf{P}(\mathbf{R}')$ in the receiver. The input signal from the receiver (R) to its temporal processing element is accordingly, refer to Eqs. (2.5.12c) and (2.5.15)[35]

$$\left.\begin{aligned} Z^{(cc)}(t|R_0)_R = \hat{R}\alpha_{\Delta|R} &= \int_{V_R} \alpha^{(cc)}(\boldsymbol{\eta}|\mathbf{R}_0, t)_{\Delta|T} d\eta \doteq \int_{V_R} \frac{d\boldsymbol{\eta}}{4\pi R_0} \int_{Br_1} Y^{(cc)}\left(\boldsymbol{\eta}, \frac{s}{2\pi i}\right)_R \alpha^{(cc)}(\eta, t)_{\Delta|T|FF} e^{s(t-|\mathbf{R}_0-\boldsymbol{\eta}|c_0)} \frac{ds}{2\pi i} \\ &\doteq \int_{Br_1} \frac{e^{s(t-\mathbf{R}_0/c_0)}(1-e^{-sT})}{4\pi R_0} \frac{ds}{2\pi i} \int_{V_R-(\boldsymbol{\eta})} Y^{(cc)}\left(\boldsymbol{\eta}, \frac{s}{2\pi i}\right)_R e^{i\eta\cdot\hat{i}_R s/c_0} \alpha^{(cc)}\left(\eta, \frac{s}{2\pi i}\right)_{\Delta|T|FF} d\boldsymbol{\eta} \end{aligned}\right\} \tag{2.5.28a}$$

where we have included the factor $(1-e^{-sT})$ to account for the input, as well as the received signal, truncated in time to the interval (O, T) (2.5.15). Adding a steering vector $\exp\left(-\boldsymbol{\eta}\cdot\hat{\mathbf{i}}_{O_R} s/c_0\right)$ to the aperture element $Y_R^{(cc)}$ associated with $d\boldsymbol{\eta}$ and using (2.5.10b), with the Fourier transform $\mathsf{F}_{\boldsymbol{\eta}}$ of $Y_R^{(cc)}$ and $\alpha_T^{(cc)}$, gives us directly the analogies of (2.5.15b) and (2.5.16a), namely,[35]

[35] Refer to Fig. 2.14.

$$Z^{(cc)}(t|R_0) = \int_{V_R(\boldsymbol{\eta})} \alpha^{(cc)}(\boldsymbol{\eta}, t)_\Delta \Big|_{\substack{R\\ FF}} \mathbf{d}\boldsymbol{\eta} \doteq \int_{Br_1} \frac{e^{s(t-\mathbf{R}_0/c_0)}}{4\pi R_0}\left(1-e^{-sT}\right)\frac{ds}{2\pi i} \int_{-\infty(\boldsymbol{\nu})}^{\infty}$$

$$\cdot \mathsf{A}^{(cc)}\left(\Delta\boldsymbol{\nu}_R - \boldsymbol{\nu}, \frac{s}{2\pi i}\right)_R \alpha^{(cc)}\left(\boldsymbol{\nu}, \frac{s}{2\pi i}\right)_{\Delta|T}\Big|_{FF} \mathbf{d}\boldsymbol{\nu} \qquad (2.5.28b)$$

where $\Delta\boldsymbol{\nu}_R \equiv \boldsymbol{\nu}_R - \boldsymbol{\nu}_{O_R} = \Delta\hat{\mathbf{i}}_R f/c_0$.

Here, from (2.5.16a), we have explicitly for the transmitted field from the source at O_T to the receiver element $\mathbf{d}\boldsymbol{\eta}$:[36]

$$\alpha^{(cc)}(\mathbf{R}_0, t)\Big|_{\substack{\Delta|T\\ FF}} = \int_{Br_1} \frac{e^{s'(t-|\mathbf{R}_0|/c)}}{4\pi R_0}\left(1-e^{-s'T}\right)\frac{ds'}{2\pi i} \int_{-\infty(\nu)}^{\infty} \mathsf{A}^{(cc)}\left(\Delta\boldsymbol{\nu}_T - \boldsymbol{\nu}', \frac{s'}{2\pi i}\right)_T S_{\text{in}}\left(\boldsymbol{\nu}, \frac{s'}{2\pi i}\right)_\Delta \mathbf{d}\boldsymbol{\nu}'$$

$$(2.5.28c)$$

where, as in Section 2.5.2, $\Delta\boldsymbol{\nu}_T \equiv \boldsymbol{\nu}_T - \boldsymbol{\nu}_{O_T} = \Delta\hat{\mathbf{i}}_T f/c_0$.

The double transform of the input field in (2.5.27) is $\mathsf{F}_{\eta,\tau}\left(\alpha^{(cc)}(\boldsymbol{\eta}, t)\right) = S_\alpha(\hat{\boldsymbol{\nu}}, s/2\pi i)_\Delta$, where $\alpha^{(cc)} = 0$ if $\tau < R/c_0$, $\tau < R/c_0 + T$, and the spatial domain $V_T(\boldsymbol{\xi})_{\text{Rec}}$ of the transmitting aperture is finite, that is, $|\mathbf{R}_0| < \infty$, as is the region $V_R(\boldsymbol{\eta})$ occupied by the receiving aperture $\mathsf{A}^{(cc)}_{\text{Rec}}$ in (2.5.27). The vector distance $\mathbf{R}$ locates the (point) source precisely in the far-field of the receiver, when $\Delta\boldsymbol{\nu}_R = 0$. Note that *unlike the case of transmission, where knowledge of the type of medium must be explicit in addition to its statistical properties, only the latter can be specified for* α_{Rec} *in reception.*

When the *same* signal is applied to the aperture elements ($\mathbf{d}\boldsymbol{\xi}$) of the transmitter. Eqs. (2.5.28b) and (2.5.28c) reduce to the simpler relations:

$$Z^{(cc)}(t|\mathbf{R}_0) = \int_{V_R(\boldsymbol{\eta})} \alpha^{(cc)}(\boldsymbol{\eta}, t)_\Delta \Big|_{\substack{R\\ FF}} \doteq \int_{Br_1} \frac{e^{s(t-R_0/c_0)}}{4\pi R_0}$$

$$\times\left(1-e^{-sT}\right)\mathsf{A}^{(cc)}\left(\Delta\boldsymbol{\nu}_R, \frac{s}{2\pi i}\right)_R \alpha^{(cc)}\left(\frac{s}{2\pi i}\right)\Big|_{\substack{\Delta|T\\ FF}} ds/2\pi i, \qquad (2.5.29)$$

where now the input signal to the receiver is

$$\alpha^{(cc)}(\mathbf{R}_0, t)\Big|_{\substack{\Delta|T\\ FF}} \doteq \int_{Br_1} \frac{e^{s'(t-\mathbf{R}_0/c)}}{4\pi R_0}\left(1-e^{-s'T}\right)\mathsf{A}^{(cc)}\left(\Delta\boldsymbol{\nu}_T, \frac{s'}{2\pi i}\right)_T S_{\text{in}}\left(\frac{s'}{2\pi i}\right)_\Delta \frac{ds'}{2\pi i},$$

$$(2.5.29a)$$

refer to Eq. (2.5.17a).

[36] Refer to Fig. 2.14.

2.5.4.2 Discrete Sampling Let us next consider the following two cases in reception: (I) discrete sampling in space, appropriate to arrays of "point" sensors, and (II) discrete sampling in both space and time, that is, the complete discretization of the received field and its subsequent processing by the receiving array. For case (I), we obtain for the receiver version of (2.5.21):

$$\text{I.}\quad Z^{(\mathrm{dc})}(\Delta\boldsymbol{\nu}_{\mathrm{R}}) = \hat{\mathbf{R}}^{(\mathrm{dc})} Z(\mathbf{r}, t')_{\mathrm{Rec}}\Big|_{\substack{H\\ \mathrm{FF}}} \doteq \int_{Br_1} e^{st}\left(1 - e^{-sT}\right)$$

$$\cdot \sum_{m=1}^{M} Y_m^{(\mathrm{dc})}(\eta_m, s/2\pi i)_{\mathrm{Rec}} \alpha^{(\mathrm{dc})}(\boldsymbol{\eta}_m, s/2\pi i)\Big|_{\Delta\,\substack{H\\ \mathrm{FF}}} e^{2\pi i \boldsymbol{\eta}_m \cdot \Delta\boldsymbol{\nu}_{\mathrm{R}}} \frac{ds}{2\pi i}, \tag{2.5.30}$$

where $\hat{\mathbf{R}}^{(\mathrm{dc})} = \tilde{\mathbf{1}}\mathbf{R}^{(\mathrm{dc})}$ and $\alpha(\boldsymbol{\eta}_m, t) \in \Delta;\ = 0$, otherwise. Here, now $Y_{\mathrm{Rec}}^{(\mathrm{dc})}$ embodies the receiving array, refer to Eq. (2.5.2a). With discrete spatial sampling of the array, we encounter the same problem of complexity of the result as we did in applying (2.5.22) for transmission, refer to Eqs. (2.5.22) and (2.5.23). Here, however, we cannot assume that the input signal, that is the field, is independent of sensor position, so that (2.5.30) remains the simpler alternative to (2.5.22a) when applied to reception.

For case (II), the results (2.3.10), (2.3.12), and (2.3.13) with (2.3.4) can be applied to give for the output of the receiver, before subsequent processing,

$$\begin{aligned} \text{II.}\quad Z_j^{(\mathrm{dd})}(\Delta\boldsymbol{\nu}_R) &= \hat{R}^{(\mathrm{dd})}\alpha(\boldsymbol{\eta}, t)_{\mathrm{Rec}} = \sum_{n=1}^{N}\sum_{m=1}^{M} \alpha^{(\mathrm{dd})}(\boldsymbol{\eta}_m, t_n)_{\mathrm{Rec}} e^{2\pi i \Delta\boldsymbol{\nu}_R \cdot \boldsymbol{\eta}_m} \\ &= \sum_{n=1}^{N}\sum_{m=1}^{M} \int_{[f_{0_n}]} df \int_{[\boldsymbol{\nu}_{0m}]} d\hat{\boldsymbol{\nu}} \mathsf{A}_m^{(\mathrm{dd})}(\hat{\boldsymbol{\nu}}, f)_{\mathrm{Rec}} e^{-2\pi i(\hat{\boldsymbol{\nu}} - \Delta\boldsymbol{\nu}_R)\cdot \boldsymbol{\eta}_m + 2\pi i f t_n} \end{aligned}, \tag{2.5.30a}$$

where $t_n = nT_{0_n} = nT_0, f_{0_n} = 1/T_0$ for periodic sampling, (2.3.14a) et seq., and

$$\mathsf{A}_m^{(\mathrm{dd})}(\boldsymbol{\nu}, f)_{\mathrm{Rec}} = \Delta_{o_j}\alpha_j^{(\mathrm{dd})} e^{2\pi i \mathbf{q}\cdot\mathbf{p}_j} \quad \text{and} \quad \Delta\boldsymbol{\nu}_{\mathrm{R}} = \boldsymbol{\nu}_{\mathrm{R}} - \boldsymbol{\nu}_{O_R}; \quad j = m, n; \quad q = \boldsymbol{\nu} + \hat{i}_4 f, \tag{2.5.30b}$$

where $(m, n) \in M, N$ and 0 otherwise. Here $\Delta_{o_j} = |\boldsymbol{\eta}_{m+1} - \boldsymbol{\eta}_m| T_n \equiv \Delta\boldsymbol{\eta}_m T_n$, which is the space–time interval between samples, refer to Eq. (2.3.5) et seq. The quantity $\Delta\boldsymbol{\nu}_{\mathrm{R}}$ represents the wave numbers associated with the beam direction $\hat{\mathbf{i}}_{\mathrm{R}}$ and a steering vector pointed at a position $\left(\hat{\mathbf{i}}_{O_{\mathrm{R}}}\right)$. We observe that $Z^{(\mathrm{dc})}$ and $Z^{(\mathrm{dd})}$ are largest when $\Delta\boldsymbol{\nu}_{\mathrm{R}} = 0$, that is, when the receiver is pointed at the transmitter, as expected. An alternative to these cases is to use continuous sampling to obtain $Z(\boldsymbol{\eta}, t)$, $(\boldsymbol{\eta}, t) \in \Delta$, in the manner of (2.5.26) and then to apply the sampling procedure, $\mathbf{T}_{\mathrm{s}}^{(\mathrm{dd})}$, Eq. (1.6.2b). The result is the (vector) output of the receiver after the sampling:

$$\left.\begin{aligned}\left[Z^{(dd)}(\boldsymbol{\eta}_m, t_n)\right] &= \mathbf{T}_S^{(dd)}\{Z(\boldsymbol{\eta}, t)\} = \mathbf{T}_S^{(dd)}\hat{R}^{(cc)}\alpha(\boldsymbol{\eta}, t)_{\Delta_R}, \quad (1, 1 < m, n < M, N); \quad = 0, \quad \text{elsewhere} \\ &= \mathbf{T}_S^{(dd)}\int_{Br_1} e^{st}(1 - e^{-sT})\frac{ds}{2\pi i}\int_{-\infty(\hat{\boldsymbol{\nu}})}^{\infty} A_0^{(cc)}(\Delta_R\boldsymbol{\nu} - \hat{\boldsymbol{\nu}}, s/2\pi i)_{Rec} S_{\alpha_{\Delta R}}(\hat{\boldsymbol{\nu}}, s/2\pi i)\mathbf{d}\hat{\boldsymbol{\nu}}\end{aligned}\right\}, \tag{2.5.31}$$

where $T_S^{(cc)}\alpha = \alpha_{\Delta_R}$, $X(\boldsymbol{\eta}, t) \in \Delta$; $= 0$, elsewhere. In general, because of the relation $c_0 = f\lambda$, (2.5.16b), for ideal media and for more complex media where $c_0 \to c(f) = f\lambda(f)$, $\nu = \nu(f)(= \lambda^{-1}(f))$, the wave number $\nu(f)$ is a function of frequency, that is, $\Delta\boldsymbol{\nu}_R = \boldsymbol{\nu}_T(f) - \boldsymbol{\nu}_{O_T}(f)$. As we have noted earlier, space and time are not generally separable.

2.5.5 Narrowband Signals and Fields

When the transmitted signal S_{in} or the received field α_{Rec} is *sufficiently* narrowband, considerable simplification is possible in our preceding results and their implementation. From (2.1.18) and (2.1.19), we see at once that the aperture functions $A^{(cc)}|_{Tr,Rec}$ are now represented by

$$\mathsf{A(cc)}(\boldsymbol{\nu}, f)\Big|_{\substack{Tr\\Rec}} \left\{\begin{aligned}&= 2\mathsf{A}(\boldsymbol{\nu}_0 + \boldsymbol{\nu}', f_0 + f')_\alpha|_{nb} \equiv \mathsf{A0}(\boldsymbol{\nu}', f'), \quad f_0 - \Delta f < f < f_0 + \Delta f \\ &= 0 \text{ elsewhere}\end{aligned}\right\}, \tag{2.5.32a}$$

and similarly for the various inputs

$$S_{in,\alpha}(\boldsymbol{\nu}, f)\left\{\begin{aligned}&= 2S_{in,\alpha}(\boldsymbol{\nu}_0 + \boldsymbol{\nu}', f_0 + f')_{nb} \equiv S_0(\boldsymbol{\nu}', f')_{in,\alpha}, \quad f_0 - \Delta f < f < f_0 + \Delta f \\ &= 0 \text{ elsewhere.}\end{aligned}\right\}. \tag{2.5.32b}$$

Since $f = f_0 + f'$, $\boldsymbol{\nu} = \boldsymbol{\nu}_0 + \boldsymbol{\nu}'$ the bounds on f', $\boldsymbol{\nu}'$ are equivalently

$$-\Delta f < f' < \Delta f; \quad -\Delta\boldsymbol{\nu} < \boldsymbol{\nu}' < \Delta\boldsymbol{\nu}. \tag{2.5.32c}$$

2.5.5.1 *Transmission*

Consequently, for transmission (see Section 2.5.2.1) the continuous narrowband version of (2.5.17a), with the help of (2.1.18)–(2.1.19) and (2.5.32a), becomes specifically for these Helmholtz media[37]

$$\alpha^{(cc)}(\mathbf{R}, t)\Big|_{\substack{H\\FF\\nb}} = Re\left\{e^{i\omega_0(t-R/c_0)}\int_{Br_1}\frac{e^{s'(t-R/c_0)}\left(1 - e^{-s'T}\right)}{4\pi R}\left(\frac{ds}{2\pi i}\right)\right.$$
$$\left.\cdot\int_{-\infty[\hat{\boldsymbol{\nu}}']}^{\infty} \mathsf{A}_0^{(cc)}\left(\Delta\hat{\mathbf{i}}_T s'/2\pi i c_0 - \hat{\nu}', s'/2\pi i c_0\right)_{Tr} S_0(\boldsymbol{\nu}', s'/2\pi i c_0)_{Tr}\mathbf{d}\hat{\boldsymbol{\nu}}'\right\} \tag{2.5.33a}$$

[37] Note from (2.1.18)–(2.1.21ab), the appearance of (Re) in (2.5.33a), since $\alpha(\mathbf{R}, t)$, being physical, is always a real quantity.

$$S_{\text{in},\Delta} \neq 0, \quad R/c_0 < t < R/c_0 + T; \quad S_{\text{in}} \in \Delta; \quad = 0,\ S_{\text{in}} \notin \Delta,$$

with $\Delta\hat{\mathbf{i}}_{\text{T}} = \hat{\mathbf{i}}_{\text{T}} - \hat{\mathbf{i}}_{O_{\text{T}}}$, $(\boldsymbol{\nu}'_T - \boldsymbol{\nu}'_{O_{\text{T}}}) = \Delta\hat{\mathbf{i}}_T s'/2\pi i c_0$, from (2.5.16b) and (2.5.17b), where $\boldsymbol{\nu}_{\text{T}}, \boldsymbol{\nu}_{O_{\text{T}}}$ are now functions of f', for the ideal medium here. For sufficiently narrowband signals, we can choose a corresponding narrowband aperture such that $\mathsf{A}_0^{(\text{cc})} \to \mathsf{A}_0^{(\text{cc})}(\boldsymbol{\nu}_0, f_0)$, that is, $(\nu_0 + \nu', f_0 + f' \doteq \nu_0, f_0)$, with $\mathsf{A}_0^{(\text{cc})}$ now effectively a constant at and immediately about $(\boldsymbol{\nu}_0, f_0)$. Accordingly, we may replace s' by s_0 and $\hat{\boldsymbol{\nu}} \neq 0$ in $\mathsf{A}_0^{(\text{cc})}$, which then reduces (2.5.33a) to the much simpler result[37]

$$\alpha(\mathbf{R}, t)\Big|_{\substack{H\\ \text{FF}\\ \text{nb}}} \doteq Re\left\{\frac{\mathsf{A}_0^{(\text{cc})}\left(\Delta\hat{\mathbf{i}}_{\text{T}} f_0/c_0, f_0\right)_{\text{Tr}}}{4\pi R} e^{i\omega_0(t-R/c_0)} \int_{Br_1} e^{s'(t-R/c_0)}\left(1 - e^{-s'T}\right) S_0(s'/2\pi i)_{\text{in}} \frac{ds'}{2\pi i}\right\}$$

$$\doteq Re\left\{\mathsf{A}_0^{(\text{cc})}\left(\Delta\hat{\mathbf{i}}_{\text{T}} f_0/c_0, f_0\right)_{\text{Tr}} S_{\text{in}}(t - R/c_0)_{\text{nb},\Delta}/4\pi R\right\};$$

$$S_{\text{in}}(t - R/c_0)_{\Delta,\text{nb}} \neq 0,\ R/c_0 < t < R/c_0 + T; \quad = 0 \text{ elsewhere.} \tag{2.5.33b}$$

Note that in this case of sufficiently narrowband signals, and therefore, sufficiently narrowband apertures, space and time processing are *effectively* separable, that is, $ST \doteq S \otimes T$. Thus, the design and implementation of the array can be carried out independent of the signal, which is customarily done in practice. The keywords here are "sufficiently narrowband," particularly at very high frequencies (ν_0, f_0), where (ν', f'), although relatively small vis-à-vis the central frequencies (ν_0, f_0), may be actually sufficiently large for the aperture to seriously modify the input signal. In this situation, of course, (2.5.33b) fails and (2.5.33a) must be used, unless the more general "broadband" formulation (2.5.17b) is required. (We observe here that the narrowband condition applies to both wave numbers $(\boldsymbol{\nu}')$ and temporal frequencies (f').) For transmission arrays (Section 2.5.3), which employ *discrete spatial sampling* of the input signal, similar results for the field of narrowband signals are obtained when s is replaced by s', $\Delta\boldsymbol{\nu}_{\text{T}}$ by $\boldsymbol{\nu}'_{\text{T}} - \boldsymbol{\nu}'_{O_{\text{T}}}$, $\mathsf{A}_M^{(\text{dc})}$ by $\mathsf{A}_{0_m}^{(\text{dc})}$ and so on, in (2.5.22) for $\alpha(\mathbf{R}, t) = \alpha^{(\text{dc})}(\mathbf{R}, t)$. This can be expressed directly in terms of the aperture function $A_{0_m}^{(\text{dc})}$ in the important case of the *same now narrowband input signal* applied to each array element at $\boldsymbol{\xi}_m$, $m = 1, \ldots, M$. The result is specifically

$$\alpha^{(\text{dc})}(\mathbf{R}, t)\Big|_{\substack{H\\ \text{FF}\\ \text{nb}\\ S_{\text{in}} = S_0}} \doteq Re\left\{e^{i\omega_0(t-R/c_0)} \int_{Br_1} S_0(s'/2\pi i)_{(\text{in},\Delta)}\left(1 - e^{-s'T}\right)\frac{e^{s'(t-R/c_0)}}{4\pi R}\mathsf{A}_{0_m}^{(\text{dc})}(\Delta\boldsymbol{\nu}'_{\text{T}}, s'/2\pi i)_{\text{Tr}}\frac{ds'}{2\pi i}\right\},$$

$$S_{\text{in},\Delta} \neq 0, \quad R/c_0 < t < R/c_0 + T; \quad S_{\text{in},\Delta} \in \Delta; \quad = 0, S_{in,\Delta} \notin \Delta, \tag{2.5.34}$$

with

$$\mathsf{A}_{0_M}^{(\text{dc})}(\Delta_{\text{T}}\boldsymbol{\nu}'_T, s'/2\pi i) = \sum_{m=1}^{M} \int_{-\infty(\nu)}^{\infty} \mathsf{A}_{0_m}^{(\text{dc})}(\boldsymbol{\nu}', s'/2\pi i)_{\text{Tr}} e^{-2\pi i(\boldsymbol{\nu}' - \Delta\boldsymbol{\nu}_{\text{T}})\cdot\boldsymbol{\xi}_m} \mathbf{d}\boldsymbol{\nu}' \Bigg\}. \tag{2.5.34a}$$

A special case of some importance occurs when the local beam patterns $\mathsf{A}_{0_m}^{(\mathrm{dc})}$, that is, the pattern per sensor for this narrowband situation, are represented by

$$\mathsf{A}_{0_m}^{(\mathrm{dc})} = \mathsf{A}_{0_m}^{(\mathrm{dc})} e^{2\pi i(\boldsymbol{\nu}' \cdot \boldsymbol{\xi}_m)}, \quad \Delta\hat{\mathbf{i}}_{\mathrm{T}} = \hat{\mathbf{i}}_{\mathrm{T}} - \hat{\mathbf{i}}_{O_{\mathrm{T}}}, \text{ and so on,} \tag{2.5.35}$$

refer to Eq. (2.5.16b). The full beam pattern of the sensor array in this special case is

$$\mathsf{A}_{0_M}^{(\mathrm{dc})} = \sum_{m=1}^{M} \mathsf{A}_{0_m}^{(\mathrm{dc})} e^{2\pi i \Delta\boldsymbol{\nu}'_{\mathrm{T}} \cdot \boldsymbol{\xi}_m} = \sum_{m=1}^{M} \mathsf{A}_{0_m}^{(\mathrm{dc})} e^{s'\left(\hat{\mathbf{i}}_T - \hat{\mathbf{i}}_{O_\mathrm{T}}\right) \cdot \boldsymbol{\xi}_m / c_0}, \tag{2.5.36}$$

which is a "shaded" beam pattern (via $\mathsf{A}_{0_m}^{(\mathrm{dc})}$), reducing to the still simpler and familiar form with uniform shading in these narrowband cases, namely:

$$\mathsf{A}_{0_M}^{(\mathrm{dc})} = \mathsf{A}_{0_m}^{(\mathrm{dc})} \sum_{m=1}^{M} e^{i\left(\hat{\mathbf{i}}_{\mathrm{T}} - \hat{\mathbf{i}}_{O_{\mathrm{T}}}\right) \cdot \boldsymbol{\xi}_m s' / c_0}. \tag{2.5.36a}$$

[See (2.5.24a)–(2.5.25) for the general case of more complex beam patterns, including broadband as well as narrowband width (in wave number–frequency).]

2.5.5.2 Reception When the field in reception is narrowband, we proceed in a similar manner to that above for transmission, refer to Section 2.5.4, adapting the general results for (2.5.27) and (2.5.31), again with the help of (2.5.32a)–(2.5.32c) and (2.1.18)–(2.1.21a). We obtain for the case of continuous sampling (2.5.27) in this instance,

$$Z^{(\mathrm{cc})}(t)_{\mathrm{nb}} = \left[\hat{R}\alpha(\boldsymbol{\eta}, t)_{nb}\right]^{(\mathrm{cc})} = Re\left\{ e^{i\omega_0 t} \int_{\mathrm{Br}_1} e^{st}\left(1 - e^{-s'T}\right) \frac{ds'}{2\pi i} \int_{-\infty(\hat{\boldsymbol{\nu}}')}^{\infty} \mathsf{A}_0^{(\mathrm{cc})}\left(\Delta\hat{\mathbf{i}}_{\mathrm{R}} s'/2\pi i c_0 - \hat{\boldsymbol{\nu}}', s'/2\pi i\right)_{\mathrm{Rec}} \cdot S_0(\hat{\boldsymbol{\nu}}', s'/2\pi i)_{\alpha,\Delta} \mathbf{d}\hat{\boldsymbol{\nu}}' \right\}, \tag{2.5.37}$$

where $Z^{(\mathrm{cc})}(t)_{\mathrm{nb}} \neq 0$, $Z^{(\mathrm{cc})}(t)_{\mathrm{nb}} \in \Delta$; $= 0$ otherwise, with $\Delta\boldsymbol{\nu}'_{\mathrm{R}} = \boldsymbol{\nu}'_{\mathrm{R}} - \boldsymbol{\nu}'_{0_R}$, and so on, and $\mathsf{A}_0^{(\mathrm{cc})}$, $S_0|_{\alpha,\Delta}$ given by (2.5.32a)–(2.5.32c). At the other extreme of *discrete sampling,* we sample the continuous output of the array when the incoming field is narrowband, getting for the *output* the vector of data

$$\left[Z^{(\mathrm{dd})}(\mathbf{r}_m, t_n)\right]_{\mathrm{nb}} = T_{\mathrm{S}}^{(\mathrm{dd})} Z^{(\mathrm{cc})}(\mathbf{r}, t)_{\mathrm{nb}} = T_{\mathrm{S}}^{(\mathrm{dd})}\left[\hat{R}\alpha(\mathbf{r}, t)_{\mathrm{nb}}\right]^{(\mathrm{cc})}, \tag{2.5.38}$$

where the discrete sampling operator $T_{\mathrm{S}}^{(\mathrm{dd})}$ is given explicitly by (1.6.2b), and $\left(\hat{R}\alpha_{\mathrm{nb}}\right)^{(\mathrm{cc})}$ by (2.5.37).

When the field is sufficiently narrowband in its space–time bandwidth, as for (2.5.33b) in the case of transmission $\left(S_{\mathrm{in}} = S_{\mathrm{in,\,nb}}\right)$, we obtain by the same argument the simple result for reception:

$$\left[Z^{(\mathrm{dd})}(\mathbf{r}_m, t_n)_{\mathrm{nb}}\right] = T_{\mathrm{S}}^{(\mathrm{dd})} Re\left\{e^{-i\omega_0 t} \mathsf{A}_0^{(\mathrm{cc})}\left(\Delta\hat{\mathbf{i}}_{\mathrm{R}} f_0/c_0, f_0\right)_{\mathrm{Rec}} \alpha(\mathbf{r}, t)_{\mathrm{nb}}^{(\mathrm{cc})}\right\}. \tag{2.5.39a}$$

This becomes explicitly on using (1.6.2b)

$$\left[Z^{(\mathrm{dd})}(\mathbf{r}_m, t_n)_{\mathrm{nb}}\right] = Re\left\{e^{-i\omega_0 t_n} \mathsf{A}_0^{(\mathrm{cc})}\left(\Delta\hat{\mathbf{i}}_{\mathrm{R}} f_0/c_0, f_0\right)_{\mathrm{Rec}} \alpha^{(\mathrm{cc})}(\mathbf{r}_m, t)_{\mathrm{nb}}\right\} \neq 0, (\mathbf{r}_m, t) \in \Delta$$
$$= 0, \text{ elsewhere} \tag{2.5.39b}$$

where

$$\left.\begin{array}{l} \alpha(\mathbf{r}_m, t_n)_{\mathrm{nb}} = \left|E(\mathbf{r}_m, t_n)_\alpha\right| \cos\left[\omega_0 t_n - 2\pi \boldsymbol{\nu}_{OL} \cdot \mathbf{r}_m - \boldsymbol{\phi}_0(\mathbf{r}_m, t_n)_\alpha\right] \neq 0, (\mathbf{r}_m, t_n) \in \Delta, \\ \qquad = 0, \text{ elsewhere.} \end{array}\right\}. \tag{2.5.40}$$

From (2.1.25a)–(2.1.25c) $|E_\alpha|$ is the (absolute) value of the complex envelope and $\phi_{0\alpha}$ is its phase. Both are narrowband. We remark that $\left[Z_{m,\,n}^{(\mathrm{dd})}\right]$ is the (vector) of sampled values of the narrowband field *after reception* by the array and before subsequent processing.

2.5.6 Some General Observations

In Section 2.5, we have examined in an introductory fashion some of the definitions and elements of apertures and arrays, in conjunction with a simple propagation model, that is, a Helmholtz medium represented by the well-known partial different equation (2.5.12) in an infinite, ideal medium, excluding the source. This also includes a more general field generated by a source and a field (not necessarily a Helmholtz one) that itself is a source for a typical receiver with an aperture or array. For the former, we have the conversion of a purely temporal source into a space–time field ($\hat{\mathbf{T}}_{\mathrm{AT}}$, Fig. 1.1b). For the latter, we have the reverse procedure of conversion of a space–time field into a temporal process in the receiver ($\hat{\mathbf{T}}_{\mathrm{AR}}$, Fig. 1.1b).

From the analysis presented in Section 2.5, we may make certain general as well as particular observations regarding the transmission and reception operations. Although apparently well known, these do not appear to have been discussed much previously. They are as follows:

(1) Generally, beam patterns are functions of frequency as is the associated wave number. This also means that the required steering vectors must be a similar function of frequency as is the associated wave number in order to achieve effective steering and to maintain the shape of the beam, refer to Eqs. (2.5.16a)–(2.5.17b).

(2) *Physically, space and time are coupled together in the propagation of a field* by the finite speed or the wave velocity c_0 (which is a constant in our Helmholtz example, refer to Eq. (2.5.12), characteristic of a nondispersive medium). *It is thus not possible that space and time phenomena are naturally separated, that is* $ST \neq S \otimes T$. Requiring $S \otimes T$, that is, separation of spatial and temporal operations, in processing these *ST* fields while often chosen is accordingly a constraint, namely, a reduction in optimality (Sections 2.5.3, 3.4.6, 3.4.7 following).

(3) Discrete sampling (i.e., discretization leading to digitalization) modifies the incoming field after processing by the receiving aperture or array, refer to Eqs. (2.5.31)

versus (2.5.27). The modification depends on which stage(s) in reception the sampling occurs. (See the papers referred to in Refs. [21–33] for further discussion and analysis.)

(4) The wave number–frequency apertures or aperture beam functions (2.5.2a) and (2.5.2b), that is, the Fourier transform of the aperture weighting function $h(\mathbf{r}, t)$, are *not* generally Bayes-matched filters as defined and discussed in Sections 3.4 and 3.5 following. This is because they do not consider the signal received in the accompanying noise. They are, however, proportional (within a constant) to Bayes-matched filters when the noise fields are "white," that is, have a constant wave number–frequency intensity spectrum.

For specific details, consider again Sections 2.5.1–2.5.3, which provide an introduction to the role of aperture and arrays in the transmission and reception of these (scalar) fields.

2.6 CONCLUDING REMARKS

As noted in Chapter 1, we have provided a more detailed formulation of the binary detection problem. Particular attention has been given to various classes of spatiotemporal Bayes detectors, their associated costs of decision, and the formal procedures for evaluating performance, measured by the relevant probabilities of correct detection and false alarms. The generalized likelihood ratio (GLR) $\Lambda(\mathbf{X})$, of course, embodies the algorithm whereby signal and noise are processed, and the GLRT is the associated statistical test of the hypotheses $H_1 : S \oplus N$ versus $H_0 : N$, namely, signal and noise versus noise alone. In most of this development, space and time play equally significant roles, since we must deal with channels that have spatial as well as temporal features. A brief summary of the principal topics considered so far in this chapter is noted below:

(i) The first half of this chapter, Section 2.1–2.4, is devoted to the Wiener–Khintchine (W–Kh) relations and in particular, to their extensions to the non-Hom-Stat noise (and signal) situations, where both continuous and discrete sampling of the input field $\alpha(\mathbf{r}, t)$ are specifically required.

(ii) In Section 2.1, broadband and narrowband noise, signals, and their various covariance structures in space–time are considered.

(iii) Section 2.2.1 treats the case of homogeneous–stationary fields and their associated W–Kh results.

(iv) Section 2.2.2 provides extensions of the W–Kh relations to the non-Hom-Stat cases for continuous sampling.

(v) Discrete sampling of random fields is treated in Section 2.3, which requires a modified approach because of the point nature of the sampled field.

(vi) In Section 2.4, the Wiener–Khintchine relations for discretely sampled random fields are considered.

(vii) Section 2.5 provides an introductory treatment of apertures and arrays (with continuous and discrete spatial sampling) for transmission and reception, illustrated by propagation in an ideal Helmholtz medium, with specific attention to beam formation.

The results of this chapter provide some of the formal structure for the specific applications to follow, where actual realizations of these theoretical results are obtained, sometimes in their optimal form and more frequently in near-optimum approximations.

REFERENCES[38]

1. (a) D. Middleton, *An Introduction to Statistical Communication Theory, McGraw-Hill, New York, 1960–1972; IEEE Press, Classic Reprint, Piscataway, NJ, 1996; (b) D. Middleton, Topics in Communication Theory*, McGraw-Hill, New York, 1965.
2. A. M. Yaglom, *Correlation Theory of Stationary and Related Random Functions. I. Basic Results; II. Supplementary Notes and References*, Springer-Verlag, New York, 1987.
3. D. Middleton, Acoustic Scattering for Composite Wind-Wave Surfaces in "Bubble-Free" Régimes, *IEEE J. Oceanic Eng.*, **OE-14**(1), pp. 17–75 1989.
4. D. Middleton, Threshold Detection in Correlated Non-Gaussian Noise Fields, *IEEE Trans. Inform. Theory*, **41**(4), 976–1000, 1995.
5. S. Siegal, *Non-parameters Statistics*, Kolmogoff–Smirnov Test, McGraw-Hill, New York, 1956. pp. 47–52, 127–136.
6. R. T. Compton, *Adaptive Antennas*, Prentice-Hall, Englewood Cliffs, NJ, 1988.
7. D. Middleton, Channel Modeling and Threshold Signal Processing in Underwater Acoustics: An Analytical Overview, *IEEE J. Oceanic Eng.*, **OE-12**(1) 4–28, 1987.
8. M. Loève, Sur les Fonctions Aliatoires Stationaires de Seconde Ordre, *Rev. Scientifiques*, **83**, 207 1945.
9. R. S. Phillips, in H. M. James, N. B. Nichols,and R. S. Phillips (Eds.) *Theory of Servomechanisms*, MIT Radiation Laboratory Series, Vol. 25, McGraw Hill, New York, 1947. Chapter 6.
10. S. U. Pillai, *Array Signal Processing*, Springer-Verlag, New York, 1989; see also the references therein on pp. 103–105, 179–182, 212, 218.
11. S. M. Kay, *Modern Spectral Estimation* (Theory and Applications), Prentice Hall, Englewood Cliffs, NJ, 1988.
12. B. D. Steinberg, *Principles of Aperture and Array System Design*, John Wiley & Sons, Inc., New York, 1976.
13. C. S. Clay and H. Medwin, *Acoustical Oceanography*, John Wiley, & Sons Inc., New York, 1977.
14. S. Haykin (Ed.), *Array Processing-Applications to Radar*, Dowden and Hutchinson and Ross, Stroudsburg, PA, 1989.
15. S. Haykin (Ed.), *Array Signal Processing*, Prentice-Hall, Englewood Cliffs, NJ, 1985.
16. H. L. Van Trees, *Optimum Array Processing*, John Wiley & Sons Inc., New York, 2002.
17. D. Middleton, A Statistical Theory of Reverberation and Similar First-Order Scattered Fields. I. Waveforms and the General Process, *IEEE Trans. Inform. Theory*, **IT-13**,(3) 372–392, 1967.
18. D. Middleton and H. L. Groginsky, Detection of Random Acoustic Signals by Receivers with Distributed Elements—Optimum Receiver Structures for Normal Signals and Noise Fields, *J. Acous. Soc. Am.*, **38**(5), 727–727, 1963.
19. P. M. Morse and H. Feshbach, *Methods of Theoretical Physics*, Vol. 1, Section 7.3 and Eq. 7.3.13, McGraw Hill, New York, 1953. see also Chapter 8 of Ref. [1].
20. N. W. McLachlan, *Complex Variables and Operational Calculus*, Cambridge University Press, London, NY, 1939.

[38] Many early ($\leq$1958) applications to specific detection problems based on the general methods and described here are given in Chapter 19 and supplements to Ref. [1].

21. N. K. Bose, Editor, Multidimensional Signal Processing, *Proc. IEEE (Special Issue)*, **78** (4), 1990. See pp. 590–597, and subsequent material. Also, earlier work, for example, in Refs. [22–23], along these lines.
22. H. Miyakawa, Sampling Theorem of Stationary Stochastic Variables in Multi-Dimensional Space, *J. Inst. Elec. Commun. Eng.*, **42**, 421–427, 1959.
23. D. P. Petersen and D. Middleton, Sampling and Reconstruction of Wave-Number-Limited Functions in *N*-Dimensional Euclidean Spaces, *Inform. Control*, **6**(4), 1962 and references. See also, D. P. Petersen and D. M Middleton, On Representative Observations, *Tellers*, **15**(4), 1963; *Inform. Control*, **7**, 445–476, 1964; *IEEE Trans. Inform. Theory*, **IT-11** (6), pp. 18–30, 1963.
24. H. Cramér, *Mathematical Methods of Statistics*, Section 10.4, Princeton University Press, Princeton, NJ, 1946.
25. A. V. Oppenheim and R. W. Schafer, *Digital Signal Processing*, Prentice-Hall, Englewood Cliffs, NJ, 1975.
26. M. I. Skolnik, *Introduction to Radar Systems*, McGraw-Hill, New York, 1962. Chapter 7.
27. P. M. Morse and K. U Ingard, *Theoretical Acoustics*, McGraw-Hill, New York, 1968.
28. H. M. James, N. B. Nichols and R.S. Phillips, *Theory of Servomechanisms,* MIT Radiation Laboratory Services, Vol. 25, McGraw-Hill, New York, 1947.

3

OPTIMUM DETECTION, SPACE–TIME MATCHED FILTERS, AND BEAM FORMING IN GAUSSIAN NOISE FIELDS

An exact reduction of the generalized likelihood ratio, or its monotonic equivalents, and the exact evaluation of the error probabilities in detection are not usually possible. In fact, canonical but approximate analytical results can be obtained only in the critical limiting situations of weak signal operation. However, there are some results that are exact for all signal levels. These owe their tractability in part to the originally Gaussian nature of the noise fields in combination with the mode of reception, that is, whether or not reception is coherent or incoherent, and to the particular type of doppler distortions and amplitude fading (if any). Here, we shall consider several classes of such problems, where a major difference from earlier work (Chapters 19–23 of Ref. [1]) is the extension to spatially distributed noise and signal fields that are inhomogeneous and nonstationary. Among the new results is our exact treatment of *broadband* incoherently received signals and noise, and the cases where the *background noise* is slowly fading (Section 3.3). Also, among the new results is a rather extended examination of space–time discrete matched filters and the inherent beam-forming capabilities of their spatial parts (Sections 3.4 and 3.5).

The organization of this chapter is as follows: Section 3.1 presents a treatment of coherent detection when everything is known about the signal except its presence or absence. Section 3.2 provides actual optimum results in selected cases, representing typical *exact* examples of broad- and narrowband reception. These are discussed for coherent detection, including detector algorithms and optimal performance, as well as some typical examples of array performance. The important cases of incoherent reception are next, including, besides the optimum algorithms, performance given in terms of the well-known Q-function, for both

Non-Gaussian Statistical Communication Theory, David Middleton.

Neyman–Pearson (NP) and Ideal Observers (IO). A section specifically on array processing in conjunction with the temporal aspects of reception provides some further insight into the role of spatial processing.

These sections are then followed by several sections involving variations on incoherent reception: A discussion of the effects of Rayleigh fading on detection and treatment of optimum narrowband reception in terms alternative to the fully narrowband treatment given earlier in Section 3.2. Section 3.3 treats coherent and incoherent detection when the background noise is slowly varying. Section 3.3 continues with a treatment of broadband incoherent reception, involving the eigenvalues and eigenfunctions of the space–time covariance. Section 3.3 then concludes with the definition of the signal-to-noise ratio at the output in terms of the minimum detectable signal and the processing gain. Section 3.4 contains an extended discussion of Bayes space–time matched filters, including the Wiener–Kolmogoroff cases and space–time variable examples. Section 3.4 also treats the imposed separability of space and time. Section 3.5 then concludes our study with a variety of approximate results, including matched filters as optimal beam forming systems. Again, the important special case of the imposed separation of space–time fields on these filters is included here.

Finally, we note that our general results can easily be specialized directly to the more familiar and previously considered cases where the noise is locally homogeneous and stationary.

3.1 OPTIMUM DETECTION I: SELECTED GAUSSIAN PROTOTYPES—COHERENT RECEPTION

Having presented the fundamentals of the Gaussian field, determined practically by its covariance function, we are now in a position to evaluate a number of examples of coherent and incoherent detection in such Gaussian noise. These examples range from the coherent cases of everything known *a priori* about the signal except its presence or absence to the more complex cases of incoherent reception involving not only ignorance about the signal epoch but also the presence of fading and variability of the accompanying noise level. The list is, of course, not complete, but it represents typical prototypes illustrating some of the techniques used to accommodate the more involved situations encountered in practice. These prototypes are exact examples and have interest in their own right, including as they do the spatial as well as the temporal aspects of the received data. It is assumed that the signal is "on" and sampled at $t_n, 0 \leq t_n \leq N$, during the observation period $(0, T)$. Moreover, the original field itself is received over a spatial domain determined by an array containing $m = 1, \ldots, M$ (point) sensors. We note that in the general context of detection (and estimation, refer to Chapter 5) here and throughout the book, such arrays are defined by their connection to a common reference point and combination $(= \Sigma_m)$ of their outputs at this point to form a "beam," as discussed in Section 2.3.

3.1.1 Optimum Coherent Detection.[1] Completely Known Deterministic Signals in Gauss Noise

Let us begin with the simple but nontrivial example of optimum *On–Off* detection in Gauss noise of an otherwise completely known signal at the receiver. The normalized, canonical

[1] For the definition of *coherent reception*, see Section 1.2–1.4.

signal is represented by Eqs. (3.1.1a) and (3.1.1b). The epoch ε and parameters $\boldsymbol{\theta}^{(m)}$ are also known *a priori*, that is, $\varepsilon = \varepsilon_0, \boldsymbol{\theta}^{(m)} = \boldsymbol{\theta}_0^{(m)}$. What is not known is whether or not the signal is present in the accompanying (additive) noise. Our problem here is the familiar On-Off test of the simple hypotheses H_1: $S+N$ versus H_0: N, with the signal class in H_1 consisting of one member, refer to Fig. 1.1a and Section 1.2.1. In particular, we wish to determine:

(1) Whether or not the completely deterministic signal S, whose structures and strength are fully specified at the receiver, is present in the received data $\mathbf{X} = \mathbf{X}(\mathbf{r}_m, t_n) = [X_{j=mn}], 1, 1 \leq m, n \leq J$.
(2) The (Bayes) optimum receiver $\hat{\mathbf{T}}_{\text{R-OPT}}$ (Fig. I.1) for this purpose, namely, the decision algorithm; and
(3) The expected performance, by the methods of Section 1.7, obtaining the various probabilities α^*, β^* of decision error and of correct decisions, that is, $1-\alpha^*$, $1-\beta^*$, including the Bayes risk or cost.

Reception is coherent and if the signal is present, its structure is completely known at the receiver. Our results apply generally for any signal, independent of narrow- or broad-bandedness.

3.1.1.1 The Detection Algorithm The noise pdf's are here under H_0 and H_1 in normalized form and with $\bar{\mathbf{x}} = 0$:

$$H_0: w_J(\mathbf{x}) = \frac{e^{-(1/2)\tilde{\mathbf{x}}\mathbf{k}_N^{-1}\mathbf{x}}}{(2\pi)^{J/2}\sqrt{\det \mathbf{k}_N}}; \; H_1: w_J(\mathbf{x}-\mathbf{as}) = \frac{\exp\left[-(\tilde{\mathbf{x}}-\tilde{\mathbf{a}}\mathbf{s})\mathbf{k}_N^{-1}(\mathbf{x}-\mathbf{as})/2\right]^-}{(2\pi)^{J/2}\sqrt{\det \mathbf{k}_N}}; \; \mathbf{x} = \left[\frac{X_j}{\sqrt{\psi_j}}\right] \tag{3.1.1}$$

where the normalized signal $\hat{\mathbf{s}}$ has a variety of equivalent forms:

$$\hat{\mathbf{s}} = \mathbf{as} = \left[S_j/\sqrt{\psi_j}\right] = \left[A_{oj}/\sqrt{\psi_j}\right] = \left[a_{0j}s_j\right] = \left[\frac{A_{0j}}{\sqrt{\psi_j}}s_n^{(m)}\right]; \; a_0 = \frac{A_0}{\sqrt{\psi}} \; \therefore \; a_{0j} = a_0\sqrt{\psi/\psi_j}, \tag{3.1.1a}$$

for broadband waves. For narrowband signals, we have the vector

$$\hat{\mathbf{s}} = \mathbf{as} = \left[\frac{A_{0j}}{\sqrt{2\psi_j}}\sqrt{2}\cos(\omega_0 t_n - \phi_j)\right] = \left[\frac{A_{0j}}{\sqrt{2\psi_j}}\sqrt{2}s_n^{(m)}\right]; \; a_0 = \frac{A_0}{\sqrt{2\psi}} \therefore \; a_{0j} = a_0\sqrt{2\psi/2\psi_j} \tag{3.1.1b}$$

with $A_{0j} = A_{on}^{(m)}(t_n)$ now slowly varying compared to $\sqrt{2}\cos(\omega_0 t - \phi_j)$. The intensities of the various signals are ΣA_{0j}^2 and $(1/2)\Sigma A_{0j}^2$, since $S_j|_{nb} = \left({}^{A_{oj}}/\sqrt{2}\right)\sqrt{2}\cos(\omega_0 t_n - \phi_j)$ gives $A_{0j}^2/2$ for its intensity. The normalized intensities are $\Sigma_j A_{0j}^2/\psi_j$ and $\Sigma_j A_{0j}^2/2\psi_j$, respectively, for broad- and narrowband signals. Here we let $\hat{\mathbf{s}}$ represent either (normalized) signal, as the case may be.

Forming the likelihood ratio and taking its logarithm give directly

$$z = \log\boldsymbol{\Lambda}(\mathbf{x}) = \left(\log\mu - \frac{1}{2}\tilde{\mathbf{a}}\tilde{\mathbf{s}}\,\mathbf{k}_{\mathrm{N}}^{-1}\mathbf{a}\mathbf{s}\right) + \tilde{\mathbf{a}}\tilde{\mathbf{s}}\,\mathbf{k}_{\mathrm{N}}^{-1}\mathbf{x}, \tag{3.1.2}$$

with the normalized signal in the two different cases—broadband and narrowband—given now by $\hat{\mathbf{s}}$. The terms in parentheses are the *bias* and the term involving $\mathbf{x}$ contains the data. Thus, using (3.1.3a) and 3.1.3b), we have compactly for both cases,[2]

$$\Psi^{*}_{s-\mathrm{coh}} = \tilde{\mathbf{a}}\tilde{\mathbf{s}}\,\mathbf{k}_{\mathrm{N}}^{-1}\mathbf{a}\mathbf{s} = \tilde{\hat{\mathbf{s}}}\mathbf{k}_{\mathrm{N}}^{-1}\hat{\mathbf{s}} = \sum_{jj'} A_{0j}A_{0j'}\left(\mathbf{k}_{N}^{-1}\right)_{jj'}\Big/\sqrt{\psi_j\psi_{j'}}\ (\mathrm{bb}), \tag{3.1.3a}$$

or

$$= \sum_{jj'} \frac{A_{0j}A_{0j'}\left(\mathbf{k}_{N}^{-1}\right)_{jj'}}{2\sqrt{\psi_j\psi_{j'}}}\left\{\sqrt{2}\cos\left(\omega_0 t_n - \phi_j\right)\cdot\sqrt{2}\cos\left(\omega_0 t_{n'} - \phi_{j'}\right)\right\},\ \mathrm{nb} \tag{3.1.3b}$$

In case of preformed beams ($M = 1$), time enters only in the temporal structure of the algorithm (3.1.2). The result then reduces to the simpler forms $\left(\text{with } \psi_j = \psi\right)$

$$\left.\begin{aligned} \log\boldsymbol{\Lambda}(\mathbf{x}) = \log\mu - \frac{1}{2}\tilde{\hat{\mathbf{s}}}\mathbf{k}_{\mathrm{N}}^{-1}\hat{\mathbf{s}} + \tilde{\hat{\mathbf{s}}}\mathbf{k}_{\mathrm{N}}^{-1}\mathbf{x};\ \left(\hat{\mathbf{s}}_{\mathrm{bb}}\right)_j = A_{0_n}/\sqrt{\psi};\ \left(\hat{\mathbf{s}}_{\mathrm{nb}}\right)_n &= \frac{A_{0n}}{\sqrt{2\psi}}\sqrt{2}\cos(\omega_0 t_n - \phi_n) \\ = \sqrt{2a_j};\quad a_j = a_{0n}s_n;\quad \overline{s_n^2} &= 1; \\ a_{0_n} = A_{0_n}/\sqrt{2\psi}&. \end{aligned}\right\} \tag{3.1.4}$$

The *test statistic* is

$$\Psi^{*}_{x-\mathrm{coh}} = \tilde{\mathbf{x}}\mathbf{k}_{\mathrm{N}}^{-1}\hat{\mathbf{s}}. \tag{3.1.5a}$$

Generally, the *decision process* is specifically

$$\begin{matrix} \text{Decide}\, H_1 : S+N \ \text{ if} \\ \text{Decide}\, H_0 : N \ \text{ if} \end{matrix} \begin{bmatrix} \Psi^{*}_{x-\mathrm{coh}} \geq \log(K_{\mathrm{coh}}/\mu) + \Psi^{*}_{s}/2 \\ \Psi^{*}_{x-\mathrm{coh}} < \log(K_{\mathrm{coh}}/\mu) + \Psi^{*}_{s}/2 \end{bmatrix}_{\mathrm{coh}}. \tag{3.1.5b}$$

Where preformed beams are used and only temporal processing is carried out, the decision process reduces to the simpler form,

$$H_1 : a_0\Psi^{*}_{x-\mathrm{coh}} \geq \log(K_{\mathrm{coh}}/\mu) + \frac{a_0^2}{2}\Psi^{*}_{s-\mathrm{coh}},\quad \text{or}\quad H_0 : a_0\Psi^{*}_{x-\mathrm{coh}} < \log(K_{\mathrm{coh}}/\mu) + a_0^2\Psi^{*}_{s-\mathrm{coh}}/2. \tag{3.1.5c}$$

For the Neyman–Pearson detector, the threshold $K_{\mathrm{coh}} > 0$, while for the Ideal Observer $K_{\mathrm{coh}} = 1$.

[2] $\Psi^{*}_{s-\mathrm{coh}}$ may also be regarded as a "generalized" signal-to-noise (power) ratio since $\Psi^{*}_{s-\mathrm{coh}} = \tilde{\hat{\mathbf{s}}}\mathbf{k}_{\mathrm{N}}^{-1}\hat{\mathbf{s}} = \Sigma s_i s_j\left(\mathbf{k}_{\mathrm{N}}^{-1}\right)_{ij} = \sum_{ij}\left(\mathbf{k}_{\mathrm{S}}\right)_{ij}\left(\mathbf{k}_{\mathrm{N}}^{-1}\right)_{ij}\left(\sim (S/N)^2\right)$, where $\mathbf{k}_{\mathrm{N}}^{-1} = \tilde{\mathbf{k}}_{\mathrm{N}}^{-1}$. See also $\Psi^{*}_{\mathrm{S\text{-}inc}}$, Section 3.2.1 following.

3.1.1.2 Space–Time Matched Filter The test statistic $\Psi^*_{x-\mathrm{coh}} = \tilde{\mathbf{a}\mathbf{s}}\,\mathbf{k}_{\mathrm{N}}^{-1}\boldsymbol{x}$ (3.1.5) represents the weighted *space–time cross-correlation of the signal replica* ($\sim \tilde{\mathbf{a}\mathbf{s}}\,\mathbf{k}_{\mathrm{N}}^{-1}$)*with the received data* $\mathbf{x}$. (Here, we equate the received signal with the signal replica, although generally the two are not strictly equivalent in practical cases.) All signal parameters are also known *a priori*. Note, however, the comparative generality of the noise covariance, which can include the effects of inhomogeneity and nonstationarity here. Note also that the optimum processor $\Psi_{x-\mathrm{coh}}$ is linear in the data $\mathbf{x}$ $[= [x(\mathbf{r}_m, t_n)]]$ as a direct consequence of coherent reception and of the Gaussian character of the additive noise. However, coherent reception can be highly nonlinear in the received data when the noise itself is non-Gaussian. Finally, we observe that $\tilde{\hat{\mathbf{s}}}\mathbf{k}_{\mathrm{N}}^{-1}$ $(= \mathbf{k}_{\mathrm{N}}^{-1}\hat{\mathbf{s}})$ is a (discrete Bayes) *space–time matched filter,*[3] that is,

$$\mathbf{k}_{\mathrm{N}}^{-1}\hat{\mathbf{s}} = \mathbf{H}, \quad \text{or} \quad \left[\sum\nolimits_{j'} (\mathbf{k}_{\mathrm{N}})_{jj'}\mathbf{H}_{j'}\right] = \left[\hat{s}_j\right] = \mathbf{as}. \tag{3.1.6}$$

Here, specifically, $\hat{\mathbf{s}} = \hat{\mathbf{s}}_{\mathrm{nb}}$ or $\hat{\mathbf{s}}_{\mathrm{bb}}$, as given by (3.1.1a) and (3.1.1b), and the filter is optimum in the Bayesian sense. Note that if we use nonnormalized forms, we have the equivalent relations $\mathbf{K}_{\mathrm{N}}\hat{\mathbf{H}} = \mathbf{S}$, where $\hat{\mathbf{H}} = \left[H_j/\psi_j^{1/2}\right]$.

In more detail, the matched filter $\mathbf{H} = \mathbf{k}_{\mathrm{N}}\hat{\mathbf{s}}$ obeys the discrete *set* of equations, which we call *discrete integral equations*, by analogy to their integral counterparts:

$$\sum_{m',\,n'} k_{\mathrm{N}}(\mathbf{r}_m, t_n; \mathbf{r}'_m, t'_n) H(\mathbf{r}_{m'}, t_{n'}) = \hat{s}_{m,n} = \hat{s}(\mathbf{r}_m, t_n),$$

with

$$0 \le (\mathbf{r}_m, t_n) \le \left([\mathbf{R}]_M, T_N\right); \quad 0 \le (\mathbf{r}_{m'}, t_n) \le \left([\mathbf{R}]_{M'}, T_{N'}\right); \quad j = m, n = 1, 1, \ldots, M, N, \tag{3.1.6a}$$

where $[R_M]$ is the value of $\mathbf{R}$ at $\mathbf{r}_M$. Now the matched filters in time have a direction, from a past to the "now" of operation. Thus, we desire to accumulate the past until the present. Their responses, that is, their "memory," must have the form $H(\mathbf{r}_m, t_n) = H(\mathbf{r}_m, T_N - t_n)$, $0 \le t_n \le T_N$, and is zero outside this interval. For the spatial filter, there is no preferred direction, backward or forward in space, so that $H(\mathbf{r}_m, \ldots) = H(-\mathbf{r}_m, \ldots)$ and the direction is optional. Thus, the complete (discrete) space–time filter is, of course, linear and is written $H(\mathbf{r}_m, T_N - t_n)$, $0 \le \mathbf{r}_m \le [\mathbf{R}]_M$, $0 \le t_n \le T_N$, and is zero outside this space–time interval. The spatial constraint is determined by the "size" of the aperture or array, and a realizable time filter (operating only on the past) is obtained, for example, with a switch at $t = T_N$, or practically with a transversal or delay-line filter. The size of the time window is determined by the observation period. For independent but inhomogeneous and nonstationary noise samples, the normalized covariance is $\mathbf{k}_{\mathrm{N}} = \left[\delta_{jj'}\right]$. The discrete set of equations (3.1.6a) is at once solved. The result is for narrowband waves:

$$[H_j] = \left[\delta_{jj'}H_{j'}\right] = [H(\mathbf{r}_m, T_N - t_n)] = \left[\hat{s}_j\right]. \tag{3.1.6b}$$

(For the full covariance, it is $\hat{\mathbf{H}} = \left[S_j/\psi_j^{1/2}\right]$.)

[3] Matched filters and the matched filter concept are discussed further in Sections 3.4 and 3.5 ff., including extensions to space as well as time. As shown in Section 3.4.4, the Bayes matched filter $\mathbf{H} = \mathbf{k}_{\mathrm{N}}^{-1}\hat{\mathbf{s}}$ here is also a Wiener–Kolmogoroff (W–kh) filter.

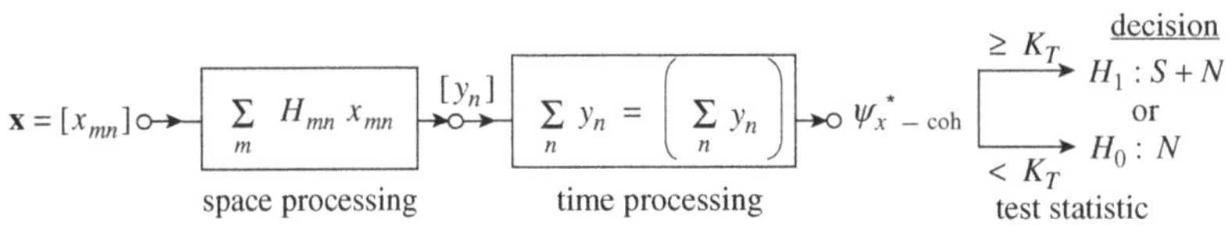

FIGURE 3.1 Spatial and time processing for optimum broad- or narrowband coherent detection of (completely) deterministic signals in Gaussian noise fields with linear arrays (refer to Section 3.1.3.)

Figure 3.1 shows a schematic of the matched filter $[H_j]$, that is, the solution of (3.1.6a), for operation in the general case, for both broad- and narrowband inputs, depending on the signal $\hat{s}_j$.

3.1.2 Performance

Our next step now is to obtain the various probability measures of performance. For this we need the following well-known relation (e.g., Eq. 7.26 of Ref. [1]).

$$I(\boldsymbol{\xi}) = \int_{-\infty}^{\infty}\!\cdots\!\int e^{i\tilde{\boldsymbol{\xi}}\mathbf{y} - \frac{1}{2}\tilde{\mathbf{y}}A\mathbf{y}}\, d\mathbf{y} = \frac{(2\pi)^{J/2}}{(\det \mathbf{A})^{1/2}}\, e^{-\tilde{\boldsymbol{\xi}}\mathbf{A}^{-1}\boldsymbol{\xi}/2}, \qquad (3.1.7)$$

where $\boldsymbol{\xi}$, $\mathbf{y}$ are column vectors, $\boldsymbol{\xi} = [\xi_j]$, $\mathbf{y} = [y_j]$, and $\mathbf{A} = \tilde{\mathbf{A}}$ is a symmetric $J \times J$ matrix, such that $\tilde{\mathbf{y}}\mathbf{A}\mathbf{y}$ and $\tilde{\boldsymbol{\xi}}\,\mathbf{A}^{-1}\boldsymbol{\xi}$ are positive definite (see Section 7.3 of Ref. [1] and references therein). With the help of (3.1.7), we can obtain the characteristic functions (cf's) of $z = \log \boldsymbol{\Lambda}(\mathbf{x})$, (3.1.2), under H_0 and H_1. This is carried out by using the relations in (3.1.1), respectively, for $z = \log \boldsymbol{\Lambda}(\mathbf{x})$. The result is

$$\left\{\begin{matrix} H_0: \\ H_1: \end{matrix}\right. \langle F_1(i\xi)\rangle\Big|_{P_1}^{Q_1} = \left\langle \exp\left(i\xi\left[\log\mu - \Psi_s^*/2\right] - \xi^2\Psi_s^*/2\right)_{\text{coh}}\right\rangle_{Q_1,P_1}, \quad \mu = p/q, \qquad (3.1.8)$$

where the upper sign applies to the pdf Q_1 and the lower to P_1. Taking the indicated transform $\mathsf{F}^{-1}\{z\} = \langle z\rangle\Big|_{Q_1}^{P_1}$ gives for the pdf values under H_0, H_1:

$$\left.\begin{matrix} Q_1(z) \\ P_1(z) \end{matrix}\right\} = \left(2\pi\sigma_0^{*2}\right)^{-1/2} \exp\left(-\left[z - \log\mu \pm \sigma_0^{*2}/2\right]^2 / 2\sigma_0^{*2}\right)_{\text{coh}}, \qquad (3.1.9)$$

where

$$\sigma_{0-\text{coh}}^{*2} \equiv \Psi_{s-\text{coh}}^{*} \quad \text{Eq.(3.1.3)}. \qquad (3.1.9a)$$

Here, $\sigma_{0-\text{coh}}^{*2}$ is often called *the detection parameter* and is more precisely *the detection performance parameter (DPP). This is a form of output signal-to-noise (intensity) ratio.*

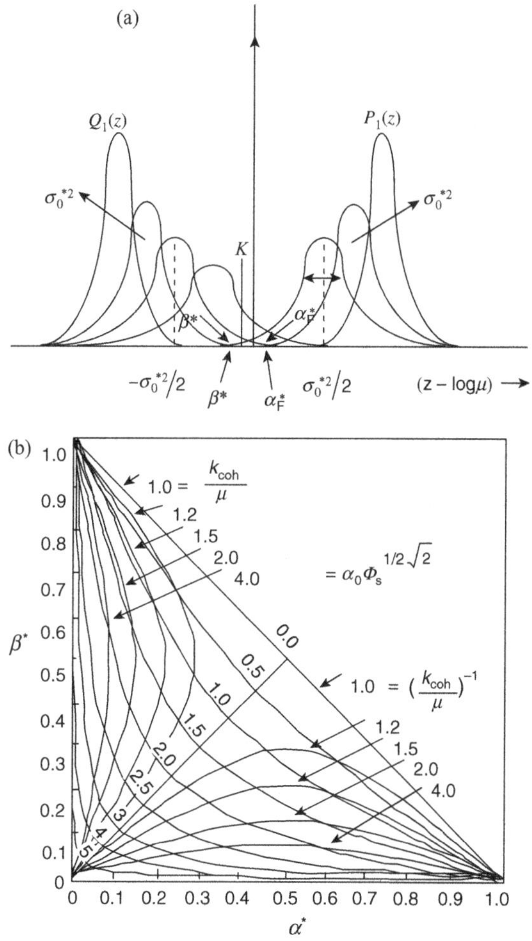

FIGURE 3.2 (a) Probability densities of $z = \log\boldsymbol{\Lambda}(\mathbf{x})$Eq. (3.1.9), under H_0: $Q_1(z)$ and H_1: $P_1(z)$. Error probabilities $\beta_0^{(1)*}\,(=\beta^*)$ versus $\beta_1^{(0)*}\,(=\alpha_F^*)$ for optimum coherent binary detection of a known signal in additive normal noise (a form of ROC diagram), Eq. (3.1.10).

This is interpreted more fully in the remainder of Chapter 3. Figure 3.2 shows a typical set of curves of the pdf values $Q_1(z)$ and $P_1(z)$ for the test statistic $z = \log\boldsymbol{\Lambda}$, (3.1.9) and (3.1.9a). Basically, Ψ_s^*, both for coherent and incoherent detection, may also be defined as an *(output) signal-to-noise ratio* from the detector. Thus, the conditional error probabilities may be obtained directly. The exact results here are

$$\left.\begin{matrix}\alpha_F^* \\ \beta^*\end{matrix}\right\} = \frac{1}{2}\left\{1 - \textcircled{H}\left[\frac{\sigma^*_{0-\mathrm{coh}}}{2\sqrt{2}} \pm \frac{\log(K_{\mathrm{coh}}/\mu)}{\sqrt{2}\,\sigma^*_{0-\mathrm{coh}}}\right]\right\}, \quad \text{with } \textcircled{H}(x) = (2/\sqrt{\pi})\int_0^x e^{-t^2}dt \equiv \operatorname{erf} x. \tag{3.1.10}$$

The optimum probability of correctly deciding that the signal is present in the received data $\mathbf{x}$ is then

$$P_{\mathrm{D}}^{*}=pp_{\mathrm{D}}^{*}=p\cdot(1-\beta^{*})=\frac{p}{2}\left\{1+\textcircled{H}\left[\frac{\sigma_{0-\mathrm{coh}}^{*}}{2\sqrt{2}}-\frac{\log(K_{\mathrm{coh}}/\mu)}{\sqrt{2}\sigma_{0-\mathrm{coh}}^{*}}\right]\right\}=\frac{p}{2}\left\{1+\textcircled{H}\left[\frac{\sigma_{0-\mathrm{coh}}^{*}}{\sqrt{2}}-\textcircled{H}^{-1}(1-2\alpha_{\mathrm{F}}^{*})\right]\right\} \tag{3.1.10a}$$

from (3.1.10). Here, $\alpha^{*}\left(=\alpha_{\mathrm{F}}^{*}\right)$ is the false alarm probability, which when preselected is equivalent to determining the threshold (or cost ratio) K, from

$$\log K_{\mathrm{coh}}=\log\mu-\sigma_{0-\mathrm{coh}}^{*2}/2+\sigma_{0-\mathrm{coh}}^{*}\sqrt{2}\,\textcircled{H}^{-1}(1-2\alpha_{\mathrm{F}}^{*}). \tag{3.1.11}$$

Figure 3.2 shows β^{*} versus α^{*} from (3.1.10). This is a form of *receiver operating characteristic* (*ROC*).[4]

Finally, the various Bayes risks [see Section (1.6.4)] may be obtained as linear combinations of the conditional error probabilities.

3.1.2.1 Neyman–Pearson and Ideal Observers These detectors are special forms of the Bayes detector. For the NP detector (Section 1.8.1), we have

$$\left.\begin{array}{l}\mathrm{NP}:\ \alpha^{*}=\alpha_{\mathrm{F}}^{*}\ \text{fixed (or equivalently } K_{\mathrm{coh}}(>0\ \text{fixed})),\\ \text{then}\\ \qquad\beta^{*}\rightarrow\beta_{\mathrm{NP}}^{*}=\dfrac{1}{2}\left\{1-\textcircled{H}\left[\dfrac{\sigma_{0-\mathrm{coh}}^{*}}{\sqrt{2}}-\textcircled{H}[1-2\alpha_{\mathrm{F}}^{*}]\right]\right\},\\ \text{with}\\ \qquad P_{\mathrm{D}}^{*}|_{\mathrm{NP}}=pp_{\mathrm{D}}^{*}|_{\mathrm{NP}}=p\cdot(1-\beta_{\mathrm{NP}}^{*}),\quad\alpha^{*}=\alpha_{\mathrm{fixed}}^{*}.\end{array}\right) \tag{3.1.12}$$

When $\sigma_{0-\mathrm{coh}}^{*}\rightarrow\infty$, $\beta_{\mathrm{NP}}^{*}\rightarrow 0$ as expected. The NP detector is an example of a *constant false alarm rate* (CFAR) detector, where "rate" refers to a succession of decisions in time, with fixed threshold or equivalently a constant false alarm rate.

For the *Ideal Observer*, on the other hand (cf. Section 1.8.2), α^* and β^* are linearly combined and then optimized. Equations (3.1.10) still apply, now with $K_{\mathrm{coh}}=1$. With a "symmetric channel," defined here by the *a priori* probabilities $p=q=1/2$, or $\mu=1$, we have a decision error probability

$$\mathrm{IO}:\qquad \mu=1=K_{\mathrm{coh}}:\quad P_{e}^{*}=\frac{1}{2}(\alpha^{*}+\beta^{*})=\frac{1}{2}\left\{1-\textcircled{H}\left[\sigma_{0-\mathrm{coh}}^{*}/2\sqrt{2}\right]\right\} \tag{3.1.13a}$$

and a Bayes risk $R^{*}|_{\mathrm{IO}}=C_{0}P_{\mathrm{e}}^{*}$, refer to Section (1.8.2) and (3.1.13a). The probability of correctly detecting a signal is

$$P_{\mathrm{c}}^{*}=1-P_{\mathrm{e}}^{*}=\frac{1}{2}\left\{1+\textcircled{H}\left(\sigma_{0}^{*}/2\sqrt{2}\right)\right\}. \tag{3.1.13b}$$

[4] ROC curves usually plot $p_{\mathrm{D}}^{(*)}(=1-\beta^{(*)})$ versus $\alpha^{(*)}$.

The decision process itself is now from (3.1.5), with $K_{\text{coh}} = 1$:

$$\text{IO}: \quad \left.\begin{array}{ll} \text{decide } H_1\text{: } S+N, \text{ if} & \Psi^*_{x-\text{coh}} \geq \Psi^*_{s-\text{coh}}/2 \\ \text{decide } H\text{: } N, \text{ namely, } P^*_e, \text{ if} & \Psi^*_{x-\text{coh}} < \Psi^*_{s-\text{coh}}/2 \end{array}\right\}, \quad \mu = K = 1. \tag{3.1.13c}$$

When $\sigma^2_{0-\text{coh}} \to \infty$, $P^*_e = 0$ and $P^*_c = 1$, no error, and when $\sigma^2_{0-\text{coh}} \to 0$, $P^*_c = 1/2 = P^*_e$, the expected *a priori* probability is $p = q = 1/2$.

If should be noted that when the "channel" is unsymmetric, that is, $p \neq q$, we must determine P^*_e from P_c, Eq. (3.1.10), namely:[5]

$$P^*_c = 1 - P^*_e = \left\{\frac{1}{2} + \frac{q}{2}\,\textcircled{H}\left[\frac{\sigma^*_{0-\text{coh}}}{2\sqrt{2}} + \frac{\log\mu}{\sqrt{2}\sigma^*_{0-\text{coh}}}\right] + \frac{p}{2}\,\textcircled{H}\left[\frac{\sigma^*_{0-\text{coh}}}{2\sqrt{2}} - \frac{\log\mu}{\sqrt{2}\sigma^*_{0-\text{coh}}}\right]\right\}, \tag{3.1.14a}$$

and

$$\therefore\ P^*_e = q\alpha^* + p\beta^* = \frac{1}{2}\left\{1 - q\,\textcircled{H}\left[\frac{\sigma^*_{0-\text{coh}}}{2\sqrt{2}} - \frac{\log\mu}{\sqrt{2}\sigma^*_{0-\text{coh}}}\right] - p\,\textcircled{H}\left[\frac{\sigma^*_{0-\text{coh}}}{2\sqrt{2}} + \frac{\log\mu}{\sqrt{2}\sigma^*_{0-\text{coh}}}\right]\right\}. \tag{3.1.14b}$$

The decision process depends on $\sigma^*_{0-\text{coh}}$, but the individual error probabilities $q\alpha^*$, $p\beta^*$ are no longer equal. Equation (3.1.14b) reduces to (3.1.13b) for the symmetrical channel ($\mu = 1$), which is the familiar result (3.1.13c), with an acceptable or unacceptable error probability according to (3.1.13a). When $\sigma^*_{0-\text{coh}} \to \infty$, then $P^*_c = 1$ and $P^*_e = 0$. On the other hand, when $\sigma^*_{0-\text{coh}} \to 0$, we find that

$$\left.\begin{array}{rll} P^*_c = & \frac{1}{2}(1+q-p) = q, & \log\mu > 0 \\ = & \frac{1}{2}(1-q+p) = p, & \log\mu < 0 \end{array}\right\}, \tag{3.1.14c}$$

with the respective error probabilities

$$\left.\begin{array}{rll} P^*_e = & \frac{1}{2}(1-q+p) = p, & \log\mu > 0 \\ = & \frac{1}{2}(1+q-p) = q, & \log\mu < 0 \end{array}\right\}. \tag{3.1.14d}$$

As we have remarked above, the NP (or CFAR) detector is normally employed when the cost consequences of missing a signal are relatively large, as in radar and sonar applications. On the other hand, when the probability of decision error is not costly (per decision), as in telephony using error correcting codes (and usually when *a priori* signal and "no signal"

[5] The threshold K_{coh} is always unity in these IO cases.

probabilities are equal), then the Ideal Observer approach is appropriate. Accordingly, P_e^*[(3.1.13a) and (3.1.13b)] is now the measure of performance error P_e^* (Eq. 3.1.14b) in the more general case of the unsymmetric channel. Other binary detectors, of Section 1.6, may similarly employ NP or IO procedures, as appropriate. Finally, we note that $\hat{\mathbf{s}}$ can be replaced by $\langle \hat{s} \rangle \neq 0$: the statistical average $< >$ does not result in a vanishing value, so the coherent operation is still possible. For *incoherent* reception, $\langle \hat{\mathbf{s}} \rangle = 0$ and more complex detector structures are needed (Section 3.2.1).

3.1.3 Array Processing II: Beam Forming with Linear Arrays

As we have previously noted (Section 2.5), forming a beam consists of combining sensor outputs with appropriate delays (spatial processing) in order to maximize the (received) signal energy observed in the direction of the source in question (when, or course, there is a source to be observed).

Accordingly, let us apply the results of Section 2.5.4 for reception and illustrate the concept here with the following simple example. We begin with the case of a received sampled field $\{X(\mathbf{r}_m, t)\}$ that contains an additive noise of specified covariance, which in turn consists of *independent* (normal) *noise samples*, that is, $\mathbf{k}_N = [\delta_{jj'}]$ and $\mathbf{K}_N = [\psi_j \delta_{jj'}] = [\psi_{mn} \delta_{mm',nn'}]$.[6] This gives for the test statistic (3.1.5a) the simple result

$$\Psi^*_{x-\text{coh}} = \tilde{\mathbf{x}} \mathbf{k}_N^{-1} \hat{\mathbf{s}} = \psi_j^{-1} \sum_j X_j S_j. \tag{3.1.15}$$

Here, specifically, the signal and data samples are represented by their Fourier transforms:

$$S_T^{(m)}(t_n - \varepsilon_0 - \Delta\tau_m) = \int_{-\infty}^{\infty} S_T^{(m)}(f)^* e^{-i\omega(t_n - \varepsilon_0 - \Delta\tau_m)} df \neq 0, 0 \le t_n \le T; \quad = 0, \text{ elsewhere}, \tag{3.1.15a}$$

$$X_j = X_T^{(m)}(t_n - \varepsilon_0) = \int_{-\infty}^{\infty} S_X^{(m)}(f) e^{i\omega(t_n - \varepsilon_0)} df, \quad \text{real} \neq 0, 0 \le t_n \le T; \quad = 0, \text{ elsewhere}, \tag{3.1.15b}$$

both at time $t_n - \varepsilon_0$. Summing over $-\infty \le n \le \infty$, and since $S_T^{(m)}$, $X_T^{(m)}$ vanish outside $(0, T)$, gives

$$\Psi^*_{x-\text{coh}} = \sum_m^M \sum_{n=-\infty}^{\infty} \frac{1}{\psi_j} \iint_{-\infty}^{\infty} S_T^{(m)}(f)^* S_X^{(m)}(f') e^{-i(\omega - \omega') n \Delta t} \cdot e^{i(\omega - \omega')(\varepsilon_0) + i\omega \Delta\tau_m} df df'. \tag{3.1.16}$$

[6] Usually, K_N does not consist of independent samples, that is, $K_N \neq [\psi_j \delta_{jj'}]$. A full treatment is given in Section 3.2 et seq. and in the chapters following.

But we can extend the limits on the summation, since the truncated series vanishes for $|n| > N$. Thus, we have (for $\Delta t = N/T$)

$$\sum_n e^{i(\omega-\omega')n\Delta t} = \sum_{n=-\infty}^{\infty} e^{i(\omega-\omega')n\Delta t} = \frac{1}{\Delta t}\delta(f'-f), \quad -\frac{f_0}{2} < f, f' < f_0/2, \quad \text{or} \quad \frac{-1}{2\Delta t} < f, f' < \frac{1}{2\Delta t}, \tag{3.1.17}$$

refer to Problem 4.2, Eq. (4.8) of [1], where $\Delta t = T_0 = N/T = 1/f_0$ and $2B (= 1/N\Delta t = 1/T)$, the bandwidth of $S_T^{(m)}(f)$, $S_X^{(m)}(f)$. Consequently, (3.1.16) reduces to

$$\Psi^*_{x\text{-coh}} \doteq \sum_m^M \frac{N}{T\psi_m} \int_{-\infty}^{\infty} S_T^{(m)}(f)^* S_X^{(m)}(f) e^{i\omega\Delta\tau_m} df = \sum_m^M \frac{N}{T\psi_m} \int_{0-}^{T+} S_T^{(m)}(t) X_T^{(m)}(t+\Delta\tau_m) dt \Bigg\}, \tag{3.1.18a}$$

which becomes

$$= \sum_m^M \frac{N}{T\psi_m} \int_{\Delta\tau_m}^{T+\Delta\tau_m} S_T^{(m)}(t-\Delta\tau_m) X^{(m)}(t) dt = \sum_m^M \frac{NA}{T\psi_m} \int_{\Delta\tau_m}^{T+\Delta\tau_m} h_T^{(m)}(T-t+\Delta\tau_m) X^{(m)}(t) dt. \tag{3.1.18b}$$

We note that the term "line array" used here is distinct from "linear array." The former refers to the geometrical arrangement of the sensors in space, while the latter describes the connectivity of the sensors in forming a beam. This last term is the matched filter form, for each sensor, $(m = 1, \ldots, M)$, that is, $h_T^{(m)} = h_T^{(m)}(T-t+\Delta t_n) = h_T^{(m)}(T)$, $t = \Delta\tau_m$. As noted at the beginning of this section, beam forming is the process of adding the various appropriately delayed sensor outputs, as indicated by the summation over the M sensors.

We can further simplify the results (3.1.18a) and (3.1.18b) by using a variant of ψ_J. This requires replacing the often unavailable ensemble values ψ_j by the practical *time averages*:

$$\psi_m \doteq \psi'_m = T^{-1} \int_0^T X^{(m)}(t)^2 dt, \quad \text{so that } \psi = \psi'_m = M^{-1} \sum_m \psi'_m \tag{3.1.18c}$$

obtained from the input when there is no signal accompanying the noise. Renormalizing each sensor output, that is, those m for which $X^{(m)} \neq 0$ in (0, T), namely, the sensors for which there is an output in (0,T), is achieved by multiplying it with each time average ratio $\psi_m/\psi_M \doteq \psi'_m/\psi'_M$ where $\psi'_m = M^{-1}\sum_m \psi_m$. We thus eliminate the dependence of sensor output on the sensor number (m) to get the still simpler result:

$$\Psi^*_{x\text{-coh}} \doteq \frac{NA}{T\psi_m} \sum_m \int_{\Delta\tau_m}^{T+\Delta\tau_m} h_T^{(m)}(T-t+\Delta\tau_m) X^{(m)}(t) dt. \tag{3.1.18d}$$

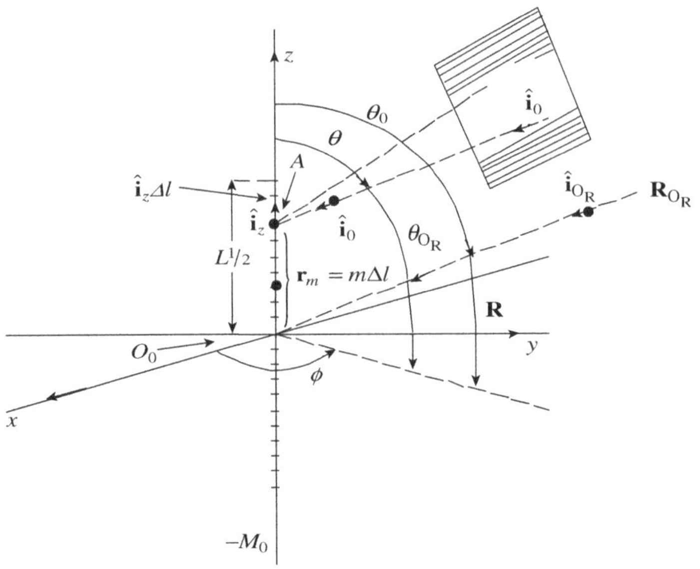

FIGURE 3.3 Schema of a section of a wave front of direction $\hat{\mathbf{i}}_0$ impinging on the *mth* sensor in a vertical line array $(-\hat{\mathbf{i}}_z L/2,\ \hat{\mathbf{i}}_z L/2)$.

(The matched filter form (3.1.18b) and (3.1.18d) now has ψ_m replaced by ψ'_m.) Of course, care must be taken to normalize X for only those sensors (m) for which there is an output $(\psi'_m > 0)$.

To proceed further, we must next introduce some explicit sensor geometry relative to the incoming signal field. We do this with the additional assumption here that the medium is "ideal," in that in the vicinity of the receiving array the medium is essentially homogeneous and stationary. Accordingly, a typical sensor at $\mathbf{r}_m$ is shown in Fig. 3.3, with wave front delay $\hat{\mathbf{i}}_0 \cdot \mathbf{r}_m / c_0$ at A and a *steering vector* $-\hat{i}_{\mathrm{O_R}} \cdot \mathbf{r}_m / c_0$, directed from the reference point $\mathrm{O_R}$ to an element of the received *steered beam, for which a delay (referred to* $\mathrm{O_R}$*) is*

$$\Delta\tau_m \equiv \left(\hat{\mathbf{i}}_0 - \hat{\mathbf{i}}_{\mathrm{O_R}}\right) \cdot \mathbf{r}_m / c_0 = \Delta\hat{\mathbf{i}}_{\mathrm{O_R}} \cdot \mathbf{r}_m / c_0. \tag{3.1.19}$$

Here, c_0 is the speed of the incoming wave front and $\hat{\mathbf{i}}_0$, $\hat{\mathrm{i}}_{\mathrm{O_R}}$ are unit vectors, as shown in Fig. 3.3.

As an example, let us consider a symmetrical vertical line array[7] about $\mathrm{O_R}$, where now $\Delta\tau_m = m\Delta l(\cos\theta_0 - \cos\theta_{\mathrm{O_R}})$, with directional steering, in which $-M_0, \ldots, M_0$ $(=2M_0 + 1 = M)$ is the total number of sensors. Using (3.1.19) in (3.1.18a), we obtain explicitly

$$\Psi^*_{x\text{–coh}} \doteq \frac{N}{T\psi'_M} \sum_{-M_0}^{M_0} \int_{-\infty}^{\infty} S_T^{(|m|)}(f)^* S_X^{(|m|)}(f) e^{i\omega m \Delta l \Delta\theta_{\mathrm{O_R}}/c_0} df \quad (\Delta\theta_{\mathrm{O_R}} \equiv \cos\theta_0 - \cos\theta_{\mathrm{O_R}}), \tag{3.1.20}$$

[7] We repeat that the term "line array" used here is distinct from "linear array."

where $\Delta\tau_m = m\Delta l\,\Delta\theta_{O_R}/c_0$. In terms of wave numbers, $\Psi_{x-\mathrm{coh}}$ is equivalently (since $kc_0 = 2\pi f$)

$$\Psi^*_{x-\mathrm{coh}} \doteq \frac{Nc_0}{2\pi\psi'_M T}\sum_{-M_0}^{M_0}\int_{-\infty}^{\infty} S_T^{(|m|)}\left(\frac{kc_0}{2\pi}\right)^* S_X^{(|m|)}\left(\frac{kc_0}{2\pi}\right)e^{ikm\Delta l\,\Delta\theta_{O_R}}dk. \tag{3.1.21}$$

When the signal replica is *narrowband and S(t) and X(t) are the same from each sensor*, Eq. (3.1.21) becomes

$$\Psi^*_{x-\mathrm{coh}} \doteq \frac{Nc_0}{2\pi\psi'^T_M}\left\{1+2\sum_{m=0}^{M}\cos[mk_0\Delta l\Delta\theta_R]\right\}\int_{-\infty}^{\infty} S_T\left(\frac{k'c_0}{2\pi}\right)^* S_X\left(\frac{k'c_0}{2\pi}\right)dk', \quad k_0 = \omega_0/c_0. \tag{3.1.22}$$

On summing the cosine series $(\sum_m)$ with the aid of the geometric series $\sum_{m=1}^{M} r^m = (1-r^{M+1})/(1-\mathrm{r})$, $\Psi^*_{x-\mathrm{coh}}$ reduces to

$$\Psi^*_{x-\mathrm{coh}} \doteq \frac{Nc_0}{2\pi\psi'^T_M}\left\{\cos M\Delta_0 + \sin M\Delta_0\cdot\left(\frac{\sin\Delta_0}{1-\cos\Delta_0}\right)\right\}\int_{-\infty}^{\infty} S_T\left(\frac{k'c_0}{2\pi}\right)^* S_X\left(\frac{k'c_0}{2\pi}\right)dk', \tag{3.1.23}$$

where now

$$\cos M\Delta_0 + \left(\frac{\sin\Delta_0}{1-\cos\Delta_0}\right)\sin M\Delta_0 \equiv \mathsf{A}^*_R(\Delta\theta, f_0) \tag{3.1.24}$$

is the *beam pattern*, with $\Delta_0 \equiv k_0\Delta l\Delta\theta_0 = k_0\Delta l(\cos\theta_0 - \cos\theta_{O_R})$. When the beam is directed toward the source of the incoming signal, then $\Delta_0 = 0$ and the beam pattern $\mathsf{A}_R \sim (2M_0+1)$ is maximal, proportional to the number of sensors. Accordingly, the maximum available aperture is[8] $\mathbf{A}_R = (2M_0+1)\Delta l$.

Finally, similar to the results above for $\Psi_{x-\mathrm{coh}}$, Eqs. (3.1.15) et seq., we see that the *detection parameter* $\sigma^{*2}_{0-\mathrm{coh}} = \Psi^*_{s-\mathrm{coh}}$, (3.1.9a), can be represented in more detail by

$$\left.\begin{aligned}\sigma^{*2}_{0-\mathrm{coh}} = \Psi^*_{s-\mathrm{coh}} &= \frac{N}{T\psi'_M}\sum_m\int_{-\infty}^{\infty} s_T^{(m)}(t)s_T^{(m)}(t+\Delta\tau_m)dt\\ &= \frac{N}{\psi'_M}\sum_m\int_{\Delta\tau_m}^{T} s^{(m)}(t+\Delta\tau_m)s^{(m)}(t)dt/T \doteq \frac{N}{\psi'_M}\sum_m\int_{\Delta\tau_m}^{T} A_0^{(m)}(t)^2\cos\omega_0\Delta\tau_n dt/2T\Bigg|_{\mathrm{nb}}\end{aligned}\right\}, \tag{3.1.25}$$

[8] For a more extensive discussion of arrays, apertures, beam forming, and so on, see Section 3.4.

this last in the narrowband cases. When the beam is directed at a signal source, that is, $\hat{\mathrm{i}}_{0_R} = \hat{\mathrm{i}}_0$, then $\Delta\tau_m = 0$, and we see that[9]

$$\sigma^{*2}_{0-\mathrm{coh}}\Big|_{\max} = \frac{N}{\psi'_M}\sum_m \int_0^T S^{(m)}(t)^2 dt/T = \frac{N}{\psi'_M}\sum_m E^{(m)}_{s-\max},$$

where (3.1.25a)

$$E^{(m)}_{s-\max} = \int_0^T S^{(m)}(t)^2 dt/T = \int_0^T A_0^{(m)}(t)^2/2T\Big|_{\mathrm{nb}}$$

is the maximum average signal power in the output of the mth sensor in the receiving array. Writing $\bar{E}_{s-\max} = M^{-1}\sum_m E^{(m)}_{\max}$, we can put (3.2.25a) into an equivalent form more useful for later use, namely,

$$\sigma^{*2}_{0-\mathrm{coh}}\Big|_{\max} = MN\bar{E}_{s-\max}/\psi'_M \tag{3.1.26}$$

3.2 OPTIMUM DETECTION II: SELECTED GAUSSIAN PROTOTYPES—INCOHERENT RECEPTION

When reception is incoherent (in the sense of Section 1.2), we have generally a more difficult situation than in the case of coherent reception of completely known signals (Section 3.1), even with an additive normal noise background. For broadband signals, where signal bandwidth (in frequency and wave number) is a large fraction of the mean wave number–frequency, exact solutions are not usually possible, although canonical approximations can be obtained in threshold reception. This is also the case for narrowband signals. However, for the latter, if *RF* epoch (ε) is the only random signal parameter, exact results can be obtained with normal noise, as the relations below demonstrate. (These are also important in the more general cases of *threshold reception* in non-Gaussian noise.) As in Section 3.1, we can carry out the analysis in the general situation of normal, inhomogeneous–nonstationary noise.

3.2.1 Incoherent Detection: I. Narrowband Deterministic Signals[10]

Here, we start again with Gaussian statistics where $\bar{\mathbf{x}} = 0$, but since there is usually no *a priori* preferred distribution of epoch ε we choose (a^1/a *Occam's* razor) a uniform distribution of signal waveform over one cycle of the high-frequency center frequency f_0 of the narrowband signal: $w(\varepsilon) = 1/\Delta$, $\Delta = f_0^{-1}$ (and zero outside this interval). Accordingly, we now consider the following results.

[9] Here, $S_j = S(\mathbf{r}_m, t_n) = S^{(m)}(t_n) = A^{(m)}_{on} s^{(m)}_n/\sqrt{2} \to S^{(m)}\ (t = n\Delta t) = \left[A_0^{(m)}(t) s^{(m)}(t)/\sqrt{2}\right]_{c\tau = t_n}$.

[10] For a definition of incoherent reception, see Section 1.2.1.4.

3.2.1.1 Incoherent Detector Structures, Statistical Tests, and Matched Filters We now specifically have for the likelihood ratio

$$\Lambda(\mathbf{x})\Big|_{\substack{\text{Gauss}\\ \text{inc}\\ \text{nb}}} = \mu\left\langle e^{-\frac{1}{2}\hat{\mathbf{a}}\tilde{\mathbf{s}}\mathbf{k}_N^{-1}\mathbf{a}\mathbf{s} + \tilde{\mathbf{x}}\mathbf{k}_N^{-1}\mathbf{a}\mathbf{s}}\right\rangle_{\varepsilon} = \mu\left\langle e^{-\frac{1}{2}\Psi^*_{s-\mathrm{inc}} + \Psi^*_{x-\mathrm{inc}}}\right\rangle_{\varepsilon} \qquad (3.2.1)$$

where $\mathbf{x} = [x(\mathbf{r}_m, t_n)] = \left[X_j/\sqrt{\psi_j}\right]$. We next define the following narrowband signal vectors by

$$\mathbf{a} = \left[\left(A_{on}^{(m)}/\sqrt{\psi_j}\right)\cos\omega_0\left(t_n - \phi_n^{(m)}\right)\right]; \quad \mathbf{b} = \left[\left(A_{on}^{(m)}/\sqrt{\psi_j}\right)\sin\left(\omega_0 t_n - \phi_n^{(m)}\right)\right], \quad j = mn, \qquad (3.2.2)$$

where, as required for these space–time samples, $j\,(=mn)$ is a double index. The phase $\phi_n^{(m)}$ contains spatial information for both propagation and location, which in more detail can be written

$$\phi_n^{(m)} \equiv (\mathbf{k}_0 - \mathbf{k}_{0_R})\cdot\mathbf{r}_m + \psi_n^{(m)}, \quad \text{and} \quad \Phi_j \equiv \omega_0 t_n - \phi_n^{(m)}. \qquad (3.2.2a)$$

Then, **a** and **b** become[11]

$$\mathbf{a} = [a_j] = [A_j\cos\Phi_j]; \quad \mathbf{b} = [b_j] = [A_j\sin\Phi_j]. \qquad (3.2.2b)$$

Accordingly, we have for the input signal (**as**)

$$\left.\begin{aligned}\hat{\mathbf{a}} = \left[A_j\cos\Phi_j\cos\omega_0\varepsilon + A_j\sin\Phi_j\sin\omega_0\varepsilon\right] &= \left[A_j\cos(\Phi_j - \omega_0\varepsilon)\right]\\ &= \left[\frac{A_{0_n}^{(m)}}{\sqrt{2\psi_j}}\sqrt{2}\cos\left(\omega_0(t_n-\varepsilon) - \phi_n^{(m)}\right)\right]\\ &= \mathbf{a}\cos\omega_0\varepsilon + \mathbf{b}\sin\omega_0\varepsilon\end{aligned}\right\}, \qquad (3.2.3)$$

$$\text{with } A_j = A_{0_n}^{(m)}/\sqrt{\psi_j}. \qquad (3.2.3a)$$

We note that Φ_j is independent of ε and that $A_{0_n}^{(m)}$ and $\phi_n^{(m)}$ are slowly varying vis-à-vis $\omega_0(t_n - \varepsilon)$.

[11] Note that $A_j = A_{o_n}^{(m)}/\sqrt{2\psi_j} \neq A_{0j}\left(= A_{o_n}^{(m)}\right)$, without the normalization used in Section 3.1.1; **a**, **â**, **b** are of course functions of $\mathbf{r}_m$ and t_n here and throughout, as are the covariances.

Since $\mathbf{as} = \hat{\mathbf{a}}$, we have for the detection performance parameter (DPP) (3.1.9a)

$$\Psi^*_{s-\text{inc}} \equiv \widetilde{\mathbf{as}}\,\mathbf{k}_N^{-1}\mathbf{as} = \tilde{\hat{\mathbf{a}}}\mathbf{k}_N^{-1}\hat{\mathbf{a}} \tag{3.2.4}$$

$$= \sum_{jj'} \left(\mathbf{k}_N^{-1}\right)_{jj'} \hat{a}_j \hat{a}_{j'} = \sum_{jj'} \left(\mathbf{k}_N^{-1}\right)_{jj'} \left(a_j \cos\omega_0\varepsilon + b_j \sin\omega_0\varepsilon\right)\left(a_{j'}\cos\omega_0\varepsilon + b_{j'}\sin\omega_0\varepsilon\right); \tag{3.2.4a}$$

with

$$\left.\begin{aligned} \hat{a}_j\hat{a}_{j'} &= a_j a_{j'}\cos^2\omega_0\varepsilon + b_j b_{j'}\sin^2\omega_0\varepsilon + \left(a_j b_{j'} + a_{j'} b_j\right)\sin\omega_0\varepsilon\cos\omega_0\varepsilon \\ &= A_j A_{j'}\left[\cos\Phi_j\cos\Phi_{j'}\cos^2\omega_0\varepsilon + \sin\Phi_j\sin\Phi_{j'}\sin^2\omega_0\varepsilon\right] \\ &\quad + \frac{1}{2}\left(A_j A_{j'} + A_{j'} A_j\right)\left(\cos\Phi_j\sin\Phi_{j'} + \sin\Phi_j\cos\Phi_{j'}\right)\sin^2\omega_0\varepsilon \end{aligned}\right\}. \tag{3.2.4b}$$

Now we see that since

$$\left\{\begin{aligned} \cos\Phi_j\cos\Phi_{j'} &= \frac{1}{2}\cos\left(\Phi_j + \Phi_{j'}\right) + \frac{1}{2}\cos\left(\Phi_j - \Phi_{j'}\right) \doteq \frac{1}{2}\cos\left(\Phi_j - \Phi_{j'}\right) \\ \sin\Phi_j\sin\Phi_{j'} &= \frac{1}{2}\cos\left(\Phi_j - \Phi_{j'}\right) - \frac{1}{2}\cos\left(\Phi_j + \Phi_{j'}\right) \doteq \frac{1}{2}\cos\left(\Phi_j - \Phi_{j'}\right) \end{aligned}\right.$$

$$\left\{\begin{aligned} \cos\Phi_j\sin\Phi_{j'} &= \frac{1}{2}\sin\left(\Phi_{j'} + \Phi_j\right) + \frac{1}{2}\sin\left(\Phi_{j'} - \Phi_j\right) \doteq \frac{1}{2}\sin\left(\Phi_{j'} - \Phi_j\right) \\ \sin\Phi_j\cos\Phi_{j'} &= \frac{1}{2}\sin\left(\Phi_{j'} + \Phi_j\right) - \frac{1}{2}\sin\left(\Phi_{j'} - \Phi_j\right) \doteq -\frac{1}{2}\sin\left(\Phi_{j'} - \Phi_j\right) \end{aligned}\right\} = 0, \tag{3.2.4c}$$

which *exhibits the narrowband condition*, wherein the rapidly oscillating terms $\cos\left(\Phi_j + \Phi_{j'}\right)$, $\sin\left(\Phi_j + \Phi_{j'}\right)$, involving the argument $\omega_0(t_n + t_{n'})$, are clearly negligible. Thus, (3.2.4) and (3.2.4a) reduce to

$$\left.\begin{aligned} \hat{a}_j\hat{a}_{j'} = a_j a_{j'} \doteq \frac{1}{2}A_j A_{j'}\cos\left(\Phi_j - \Phi_{j'}\right) \text{ and } \therefore \tilde{\hat{\mathbf{a}}}\mathbf{k}_N^{-1}\hat{\mathbf{a}} &\doteq \left[\frac{1}{2}\sum_{jj'} A_j A_{j'}\left(\mathbf{k}_N^{-1}\right)_{jj'}\cos\left(\Phi_j - \Phi_{j'}\right)\right] \\ &\doteq \frac{1}{2}\left(\tilde{\mathbf{a}}\mathbf{k}_N^{-1}\mathbf{a} + \tilde{\mathbf{b}}\mathbf{k}_N^{-1}\mathbf{b}\right) \end{aligned}\right\}. \tag{3.2.4d}$$

Thus, $\Psi^*_{s-\text{inc}}(>0)$ and the DPP (3.2.4) finally becomes

$$\Psi^*_{s-\text{inc}} = \widetilde{\mathbf{as}}\mathbf{k}_N^{-1}\mathbf{as} \doteq \frac{1}{2}\left(\tilde{\mathbf{a}}\mathbf{k}_N^{-1}\mathbf{a} + \tilde{\mathbf{b}}\mathbf{k}_N^{-1}\mathbf{b}\right) = \tilde{\mathbf{a}}\mathbf{k}_N^{-1}\mathbf{a} = \tilde{\mathbf{b}}\mathbf{k}_N^{-1}\mathbf{b}, \quad \text{refer to Eq. (3.2.4)}, \tag{3.2.5}$$

which no longer depends on ε in the narrowband approximation here. Then,

$$\left\langle \exp\left(-\Psi^*_{s-\text{inc}}/2\right)\right\rangle_\varepsilon = \exp\left(-\Psi^*_{s-\text{inc}}/2\right) \tag{3.2.5a}$$

since $\Psi^*_{s-\text{inc}}$ is now independent of ε. (Note that $\Psi^*_{s-\text{inc}}$ may be expressed either as $\tilde{\mathbf{a}}\mathbf{k}_\mathrm{N}^{-1}\mathbf{a}$ or as $\tilde{\mathbf{b}}\mathbf{k}_\mathrm{N}^{-1}\mathbf{b}$, as well as by the symmetric form $(\tilde{\mathbf{a}}\mathbf{k}_\mathrm{N}\mathbf{a}+\tilde{\mathbf{b}}\mathbf{k}_\mathrm{N}\mathbf{b})/2$, that since $k_{jj}=\tilde{k}_{jj}$, $\Psi^*_{s-\text{inc}}$ can be regarded here as a kind of generalized signal-to-noise (intensity) ratio, since (3.2.5) can be written $\Psi^*_s=\sum_{ij}(\mathbf{k}_\mathrm{S})_{ij}(\cdot\mathbf{k}_\mathrm{N}^{-1})_{ij}$, Note also the *formal* similarity of Eqs. (3.2.4) and (3.2.5), that is, $\Psi^*_{s-\text{inc}}|_{\text{nb}}(=\hat{\mathbf{a}}\mathbf{k}_\mathrm{N}^{-1}\hat{\mathbf{a}})$, to the earlier detection parameter $\Psi^*_{s-\text{coh}}=\hat{\mathbf{s}}\mathbf{k}_\mathrm{N}^{-1}\hat{\mathbf{s}}$ [Eq. (3.1.3a)], which applies for both broad- and narrowband signals.

In a similar fashion, using the various narrowband approximations in (3.2.4b), we can also show that

$$\begin{pmatrix}\tilde{\mathbf{a}}\mathbf{k}_\mathrm{N}^{-1}\mathbf{a}\\ \tilde{\mathbf{b}}\mathbf{k}_\mathrm{N}^{-1}\mathbf{b}\end{pmatrix}=\sum_{jj'}A_jA_{j'}\left(\mathbf{k}_\mathrm{N}^{-1}\right)_{jj'}\begin{pmatrix}\cos\Phi_j\cos\Phi_{j'}\\ \sin\Phi_j\sin\Phi_{j'}\end{pmatrix}\doteq\frac{1}{2}\cos\left(\Phi_j-\Phi_{j'}\right)\Bigg) \tag{3.2.6a}$$

so that

$$\therefore\ \tilde{\mathbf{a}}\mathbf{k}_\mathrm{N}^{-1}\mathbf{a}=\tilde{\mathbf{b}}\mathbf{k}_\mathrm{N}^{-1}\mathbf{b}=\Psi^*_{s-\text{inc}}. \tag{3.2.6b}$$

Furthermore, remembering $\mathbf{k}_\mathrm{N}$ (and $\therefore\ \mathbf{k}_\mathrm{N}^{-1}$) are symmetric, on using (3.2.4c) again one can write for the "cross-terms"

$$\left.\begin{aligned}C&\equiv\tilde{\mathbf{a}}\mathbf{k}_\mathrm{N}^{-1}\mathbf{b}=\textstyle\sum_{jj'}A_jA_{j'}\left(\cos\Phi_j\sin\Phi_{j'}\doteq\frac{1}{2}\sin\left(\Phi_{j'}-\Phi_j\right)\right)\\ D&\equiv\tilde{\mathbf{b}}\mathbf{k}_\mathrm{N}^{-1}\mathbf{a}=\textstyle\sum_{jj'}A_jA_{j'}\left(\sin\Phi_j\cos\Phi_{j'}\doteq-\frac{1}{2}\sin\left(\Phi_{j'}-\Phi_j\right)\right)\end{aligned}\right\}, \tag{3.2.7}$$

and interchanging j and j' in D gives at once $C\doteq D$. Since $C+D\doteq0$, or $C\doteq-D$, it follows that $\tilde{\mathbf{a}}\mathbf{k}_\mathrm{N}^{-1}\mathbf{b}\doteq\tilde{\mathbf{b}}\mathbf{k}_\mathrm{N}^{-1}\mathbf{a}\doteq0$ in these narrowband cases. Finally, a number of alternative, compact forms for $\Psi^*_{s-\text{inc}}$, (3.2.5) are available. Again, letting $\mathbf{c}=\mathbf{k}_\mathrm{N}^{-1}\mathbf{a}$, $\mathbf{d}=\mathbf{k}_\mathrm{N}^{-1}\mathbf{b}$ so that $\tilde{\mathbf{a}}\mathbf{k}_\mathrm{N}^{-1}\mathbf{a}=\tilde{\mathbf{a}}\mathbf{c}=\tilde{\mathbf{b}}\mathbf{k}_\mathrm{N}^{-1}\mathbf{b}=\tilde{\mathbf{b}}\mathbf{d}$, we can therefore write

$$\Psi^*_{s-\text{inc}}\doteq\tilde{\mathbf{a}}\mathbf{c}=\tilde{\mathbf{b}}\mathbf{d}=\frac{1}{2}\left(\tilde{\mathbf{a}}\mathbf{c}+\tilde{\mathbf{b}}\mathbf{d}\right)=\frac{1}{2}Re\,\tilde{\mathbf{A}}\hat{\mathbf{h}}^*\mathbf{e}^{i\Phi}=\frac{1}{2}\sum_j Re\,A_j\hat{H}^*_je^{i\Phi_j},\quad \hat{h}^*_j=c_j-id_j, \tag{3.2.7a}$$

where $\hat{H}$ represents a complex *matched filter response*, refer to discussion following Eq. (3.2.13a).

Next, for the test statistic $\Psi^*_{x-\text{inc}}(x,\varepsilon)\equiv\tilde{\mathbf{x}}\mathbf{k}_\mathrm{N}^{-1}\mathbf{as}=\tilde{\mathbf{x}}\mathbf{k}_\mathrm{N}^{-1}(\mathbf{a}\cos\omega_0\varepsilon+\mathbf{b}\sin\omega_0\varepsilon)$ in (3.2.1), we can write equivalently

$$\Psi^*_{x-\text{inc}}(x,\varepsilon)_{\text{inc}}=\left[\left(\tilde{\mathbf{x}}\mathbf{k}_\mathrm{N}^{-1}\mathbf{a}\right)^2+\left(\tilde{\mathbf{x}}\mathbf{k}_\mathrm{N}^{-1}\mathbf{b}\right)^2\right]^{1/2}\cos(\omega_0\varepsilon-\Phi(\mathbf{x}));\quad \Phi(\mathbf{x})=\tan^{-1}\left(\tilde{\mathbf{x}}\mathbf{k}_\mathrm{N}^{-1}\mathbf{b}/\tilde{\mathbf{x}}\mathbf{k}_\mathrm{N}^{-1}\mathbf{a}\right), \tag{3.2.8}$$

so that in (3.2.1) we get

$$\left\langle\exp\left(\Psi^*_x(\mathbf{x},\varepsilon)_{\text{inc}}\right)\right\rangle_\varepsilon=\langle\exp\{F(\mathbf{x})\cos[\omega_0\varepsilon-\Phi(\mathbf{x})]\}\rangle_\varepsilon=\sum_{n=0}^{\infty}\varepsilon_nI_n(F(\mathbf{x}))\langle\cos n(\omega_0\varepsilon-\Phi)\rangle_\varepsilon, \tag{3.2.8a}$$

$F(\mathbf{x})$ is also independent of ε in the present narrowband approximation, refer to Eq. (3.2.2, 2a), for example,

$$F(\mathbf{x}) \equiv \Psi^*_{x-\text{inc}} = \left(\tilde{\mathbf{x}}\mathbf{k}_N^{-1}\mathbf{a}\right)^2 + \left(\tilde{\mathbf{x}}\mathbf{k}_N^{-1}\mathbf{b}\right)^2 = \tilde{\mathbf{x}}\mathbf{k}_N^{-1}\left(\mathbf{a}\tilde{\mathbf{a}} + \mathbf{b}\tilde{\mathbf{b}}\right)\mathbf{k}_N^{-1}\mathbf{x}. \tag{3.2.8b}$$

Here in (3.2.8a), I_n is a modified Bessel function of order n, and ε_n is the Kronecker symbol, with $\varepsilon_n = 1,\ \varepsilon_n = 2, n \geq 2$ in the usual way. For the uniform pdf $w_1(\varepsilon) = 1/\Delta,\ \Delta = f_0^{-1}$ discussed above, only the term $n=0$ is nonvanishing, with the result that

$$\left\langle \exp\left(\Psi^*_x(\mathbf{x},\varepsilon)_{\text{inc}}\right)\right\rangle_\varepsilon = I_0\left(\left[\left(\tilde{\mathbf{x}}\mathbf{k}_N^{-1}\mathbf{a}\right)^2 + \left(\tilde{\mathbf{x}}\mathbf{k}_N^{-1}\mathbf{b}\right)^2\right]^{1/2}\right). \tag{3.2.9}$$

(Unlike the DPP $\Psi^*_{s-\text{inc}}$, the test statistic $\Psi^*_{x-\text{inc}}$ must employ *both terms* **a** and **b** involving the signal, refer to Eq. (3.2.8))

Accordingly, for the narrowband (Bayes) optimum *incoherent* detector, the complete test statistic[12] is

$$U(\mathbf{x})^* \equiv \log\mathbf{\Lambda}(\mathbf{x}) = \log\mu - \frac{1}{2}\Psi^*_{s-\text{inc}} + \log I_0\left(\sqrt{\Psi^*_{x-\text{inc}}}\right), \tag{3.2.10}$$

where I_0 is a modified Bessel function of the first kind and $\Psi^*_{s-\text{inc}}$ is generally given by (3.2.5). The resulting decision process is

$$\text{Decide } H_1\text{: } S+N \quad \text{if} \quad U(\mathbf{x})^* \geq \log K_{\text{inc}} \quad \text{or} \quad \text{Decide } H_0\text{: } N \quad \text{if} \quad U(\mathbf{x})^* < \log K_{\text{inc}}. \tag{3.2.11}$$

However, for determining detection performance, the complete test statistic (3.2.10) does not provide analytically tractable pdf values under H_0 and H_1. Consequently, at this point, it is necessary to replace the original test statistic $U(\mathbf{x})$, (3.2.10), by a much simpler optimum monotonic function of the data[13] than $I_0\left(F(\mathbf{x})^{1/2}\right)$ and by the corresponding new threshold $K_{T-\text{inc}}$. This is possible, and optimality is preserved because the new, simpler monotonic functional (of x) is also a sufficient statistic.[14] The optimum test statistic and decision process are accordingly chosen to be

$$Z \equiv \Psi^*_{x-\text{inc}} \begin{cases} = \left(\tilde{\mathbf{x}}\mathbf{k}_N^{-1}\mathbf{a}\right)^2 + \left(\tilde{\mathbf{x}}\mathbf{k}_N^{-1}\mathbf{b}\right)^2, \\ = \mathbf{x}\mathbf{k}_N^{-1}\left(\mathbf{a}\tilde{\mathbf{a}} + \mathbf{b}\tilde{\mathbf{b}}\right)\mathbf{k}_N^{-1}\mathbf{x} \end{cases}$$

$$\left\{ \begin{matrix} \text{where we Decide } H_1\text{: } S+N & \text{if} & \Psi^*_{x-\text{inc}} \geq K_{T-\text{inc}} \\ \text{Decide } H_0\text{: } \quad N & \text{if} & \Psi^*_{x-\text{inc}} < K_{T-\text{inc}} \end{matrix} \right\}. \tag{3.2.12}$$

[12] Henceforth we shall use the term "test statistic" loosely to mean Ψ^*_x or some function of Ψ^*_x. The meaning should be evident from the analytical context.

[13] See the remarks in Section 1.6.

[14] See Section 1.9 for details.

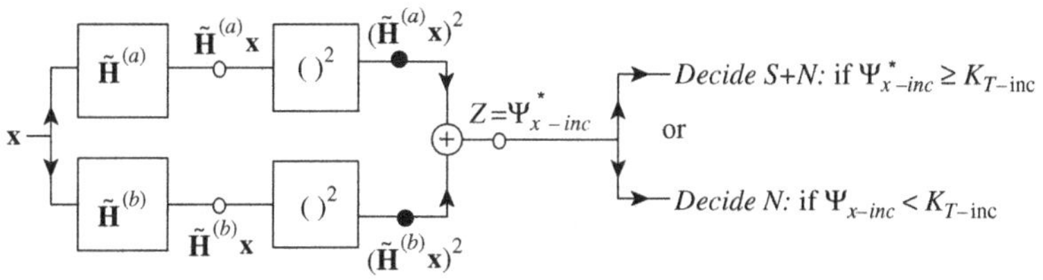

FIGURE 3.4 The generic detection algorithm, Eqs. (3.2.12) and (3.2.12b), for the incoherent reception of narrowband signals in additive normal noise, with quadratic array processing, which are part of $(\tilde{\mathbf{H}}\mathbf{x})^2$, cf. Fig. 3.7 ff.

Here, the new threshold $K_{T-\text{inc}}$ is related to the original threshold K_{inc} specifically by

$$K_{T-\text{inc}}(K_{\text{inc}}) = I_0^{(-1)}\left[\left(\frac{K_{\text{inc}}}{\mu} e^{\Psi^*_{s-\text{inc}}/2}\right)^2\right] > 0, \tag{3.2.12a}$$

where I_0^{-1} is the inverse Bessel function.[15] In terms of generic *matched filter* [(3.1.6) et seq. and Sections 3.4, 3.4.4, and 3.4.7], $\mathbf{H}^{(a)} = \mathbf{k}_\text{N}^{-1}\mathbf{a}$, $\mathbf{H}^{(b)} = \mathbf{k}_\text{N}^{-1}\mathbf{b}$ here, Eq. (3.2.12) is alternatively represented by

$$Z = \Psi^*_{x-\text{inc}} = \left(\tilde{\mathbf{x}}\mathbf{H}^{(a)}\right)^2 + \left(\tilde{\mathbf{x}}\mathbf{H}^{(b)}\right)^2. \tag{3.2.12b}$$

Figure 3.4 illustrates the detection algorithm.

The new test statistic has a variety of equivalent representations besides (3.2.9) and (3.2.12b):

$$Z = \Psi^*_{x-\text{inc}} = \tilde{\mathbf{x}}\mathbf{k}_\text{N}^{-1}(\mathbf{a}\tilde{\mathbf{a}} + \mathbf{b}\tilde{\mathbf{b}})\mathbf{k}_\text{N}^{-1}\mathbf{x} = \tilde{\mathbf{x}}\mathbf{k}_\text{N}^{-1}\Psi^*_{\text{ss}}\mathbf{k}_\text{N}^{-1}\mathbf{x}, \quad \text{with} \quad \Psi^*_{\text{ss}} \equiv \mathbf{a}\tilde{\mathbf{a}} + \mathbf{b}\tilde{\mathbf{b}} = [A_j A_{j'} \cos(\Phi_j - \Phi_{j'})]. \tag{3.2.13}$$

(Note that both $\Psi^*_{s-\text{inc}}$ and $\Psi^*_{x-\text{inc}}$ are $(J \times J)$ positive definite, symmetrical quadratic forms, a fact that will prove useful in the analysis of array processing in Section 3.2.1.4. If we let $\tilde{\mathbf{y}} = \tilde{\mathbf{x}}\mathbf{k}_\text{N}^{-1}$ (or $\mathbf{y} = \mathbf{k}_\text{N}^{-1}\mathbf{x}$) where $\mathbf{k}_\text{N}$ (and $\therefore \mathbf{k}_\text{N}^{-1}$) are symmetric as well as positive definite, we can express the test statistic even more compactly as

$$Z = \Psi^*_{x-\text{inc}} = \tilde{\mathbf{y}}\Psi^*_{\text{ss}}\mathbf{y}, \quad \text{with} \quad \mathbf{y} = \mathbf{k}_\text{N}^{-1}\mathbf{x} \quad \text{and} \quad \therefore\ \mathbf{x} = \mathbf{k}_\text{N}\mathbf{y} = \sum\nolimits_{j'} (\mathbf{k}_\text{N})_{jj'} y_{j'}. \tag{3.2.13a}$$

[15] It is not even necessary to use the original threshold. The final one (here $K_{T-\text{inc}}$) will suffice. (However, (3.2.12a) is needed to determine the limiting forms β^* when $\Psi^*_{s-\text{inc}} \to \infty$, or 0 [(3.2.20a) ff.)].

Thus, $\mathbf{y}$ is a new data vector, obtained linearly from the original received data vector $\mathbf{x} = [x_j]$. Letting $\mathbf{c} = \mathbf{k}_N^{-1}\mathbf{a}$, $\mathbf{d} = \mathbf{k}_N^{-1}\mathbf{b}$, with $\hat{\mathbf{H}} = \mathbf{c} + i\mathbf{d}$, we see that the test statistic $\Psi_{x-\text{inc}}$ can also be compactly expressed as

$$\Psi^*_{x-\text{inc}} = (\tilde{\mathbf{x}}\mathbf{c})^2 + (\tilde{\mathbf{x}}\mathbf{d})^2 = |\tilde{\mathbf{x}}(\mathbf{c}+i\mathbf{d})|^2 = \left|\tilde{\mathbf{x}}\hat{\mathbf{H}}\right|^2, \quad \text{with} \quad \hat{\mathbf{H}} = \mathbf{c}+i\mathbf{d}, \quad \text{and} \quad \hat{\mathbf{H}} = \left[\sum_{j'} \left(\mathbf{k}_N^{-1}\right)_{jj'} A_{j'} e^{i\Phi_{j'}}\right]; \tag{3.2.13b}$$

($\mathbf{c}$, $\mathbf{d}$, and $\hat{\mathbf{H}}$ are all column vectors of J elements.) Also, in terms of the real component parts of $\hat{\mathbf{H}}(= \mathbf{c} + i\mathbf{d})$, we can equally well write

$$\Psi^*_{x-\text{inc}} = (\tilde{\mathbf{c}}\mathbf{x})^2 + \left(\tilde{\mathbf{d}}\mathbf{x}\right)^2 = \tilde{\mathbf{x}}\hat{\mathbf{H}}\tilde{\hat{\mathbf{H}}}^*\mathbf{x} = \left(\hat{\mathbf{H}}^{(a)}\mathbf{x}\right)^2 + \left(\hat{\mathbf{H}}^{(b)}\mathbf{x}\right)^2, \quad \text{with} \quad \hat{\mathbf{H}} = \hat{\mathbf{H}}^{(a)} + i\hat{\mathbf{H}}^{(b)}, \tag{3.2.13c}$$

so that the test statistic $\Psi^*_{x-\text{inc}}$ can be represented by structures like those shown in Figs. 3.2–3.7 ff. In (3.2.13) or (3.2.13c), $\Psi^*_{x-\text{inc}}$ embodies a generalized or matrix *autocorrelation* of the received data $\mathbf{x}$ (or $\mathbf{y}$) with the (matrix) *autocorrelation* of the (sum of the) in-phase and out-of-phase autocorrelation components of the signal replica matrix $\Psi^*_{\text{ss}}\left(\doteq \Psi^*_{\text{ss-Rec}}\right)$, as shown in (3.2.13a). (Note also that $\Psi^*_{x-\text{inc}}$, Eq. (3.2.8b), and $\Psi^*_{s-\text{inc}}$, Eq. (3.2.5), are approximate under the present narrowband condition.) The quantity $\Psi^*_{s-\text{inc}}$ (3.2.5), appearing in the threshold $K_{T-\text{inc}}$, (3.2.12a), has a number of important and related interpretations, which we shall discuss in Section 3.2.1.2.

3.2.1.2 Performance: Probability Distributions and Detection Probabilities In order to evaluate performance, we must first obtain the pdf values $w_1(Z|H_0)$, $w_1(Z|H_1)$, where Z is the test statistic (3.2.12) and (3.2.13), namely, $Z = \Psi^*_{x-\text{inc}}$. We must then apply these pdf values accordingly to the procedures of Section 1.6 to find the various error probabilities and probabilities of correct decisions for these On–Off binary, incoherent, narrowband, Gaussian detection scenarios.

Accordingly, for (3.2.12), we seek the respective pdf values of the *normalized square* envelope $E^2 = Z$, where $\mathsf{E}^2 = \left(\tilde{\mathbf{x}}\mathbf{k}_N^{-1}\mathbf{a}\right)^2 + \left(\tilde{\mathbf{x}}\mathbf{k}_N^{-1}\mathbf{b}\right)^2 = Z = u^2 + v^2$. Here, $u \equiv \tilde{\mathbf{x}}\mathbf{k}_N^{-1}\mathbf{a}$, $v \equiv \tilde{\mathbf{x}}\mathbf{k}_N^{-1}\mathbf{b}$, and we are, of course, dealing with narrowband fields, so that it is meaningful to consider the existence of the envelope. We begin by observing that u and v are Gaussian random variables, with the second-order pdf ($\bar{u} = \bar{v} = 0$)

$$w_2(u,v) = \frac{e^{-\left(u^2+v^2-2uv\rho\right)/2\left(1-\rho^2\right)\psi_0}}{2\pi(1-\rho^2)\psi_0}, \quad \psi_0 = \overline{u^2} = \overline{v^2}, \quad \rho_0 = \overline{uv}/\psi_0, \tag{3.2.14}$$

and characteristic function (cf) here

$$\mathsf{F}\{w_2\} = F_2(i\xi_1, i\xi_2)_{u,v} = \exp\left[-\frac{1}{2}\left(\xi_1^2 + \xi_2^2 - 2\xi_1\xi_2\rho\right)\psi_0\right]. \tag{3.2.14a}$$

It is readily seen that (since $\mathbf{k}_N = \tilde{\mathbf{k}}_N$)

$$\bar{u}^2 = \tilde{\mathbf{a}}\mathbf{k}_N^{-1}\bar{\mathbf{x}}\tilde{\mathbf{x}}\mathbf{k}_N^{-1}\mathbf{a} = \tilde{\mathbf{a}}\mathbf{k}_N^{-1}\mathbf{a} = \psi_0\mathbf{a}; \quad \bar{v}^2 = \tilde{\mathbf{b}}\mathbf{k}_N^{-1}\mathbf{b} = \psi_0\mathbf{b}; \quad \overline{uv} = \tilde{\mathbf{a}}\mathbf{k}_N^{-1}\mathbf{b} = \tilde{\mathbf{b}}\mathbf{k}_N^{-1}\mathbf{a} \doteq 0 \tag{3.2.14b}$$

from (3.2.6a)–(3.2.7), with $\bar{u}^2 = \bar{v}^2 = \psi_0$, cf. (3.2.14b). Accordingly, u and v are uncorrelated, so that now

$$w_2(u, v) = w_1(u)w_1(v) = \frac{e^{-u^2/2\psi_0}}{\sqrt{2\pi\psi_0}} \cdot \frac{e^{-v^2/2\psi_0}}{\sqrt{2\pi\psi_0}}; \quad F_2 = e^{-(1/2)\xi_1^2\psi_0} \cdot e^{-(1/2)\xi_2^2\psi_0}. \quad (3.2.15)$$

Also, we note from (3.2.6a) and (3.2.6b) that

$$\psi_0 = \left(\tilde{\mathbf{a}}\mathbf{k}_N^{-1}\mathbf{a} + \tilde{\mathbf{b}}\mathbf{k}_N^{-1}\mathbf{b}\right)/2 = \sum\nolimits_{jj'} A_j A_{j'} \left(\mathbf{k}_N^{-1}\right)_{jj'} \cos\left(\Phi_{j'} - \Phi_j\right) = \Psi^*_{s-\text{inc}}. \quad (3.2.16)$$

Next, we employ the well-known results for the pdf values of the (normalized) envelope E under H_0 and H_1 (see Sections 9.2 and 9.2.1 of Ref. [1]), namely,

$$w_1(\mathsf{E}|H_0) = \frac{\mathsf{E}}{\psi_0} e^{-\mathsf{E}^2/2\psi_0}; \quad w_1(\mathsf{E}|H_1) = \frac{\mathsf{E}}{\psi_0} e^{-(\mathsf{E}^2 + b_0^2)/2\psi_0} I_0(\mathsf{E}b_0/\psi_0); \quad \mathsf{E} \geq 0, \quad (3.2.17)$$

where b_0 is a (normalized) signal envelope. In the present case, $b_0 = \Psi^*_{s-\text{inc}}$, which is now quadratic in $\mathbf{a}$ (or $\mathbf{b}$), refer to Eq. (3.2.16). Thus, we see that (3.2.17) becomes

$$w_1(\mathsf{E}|H_0) = \frac{\mathsf{E}}{\Psi^*_{s-\text{inc}}} e^{-\mathsf{E}^2/2\Psi^*_{\text{inc}}}; \quad w_1(\mathsf{E}|H_1) = \frac{\mathsf{E}}{\Psi^*_{s-\text{inc}}} e^{-\mathsf{E}^2/2\Psi^*_{s-\text{inc}} - \Psi^*_{s-\text{inc}}/2} I_0(\mathsf{E}); \quad \mathsf{E} \geq 0. \quad (3.2.17a)$$

Finally, since we need these pdf values for $Z = \mathsf{E}^2$, we readily observe that (3.2.17a) reduces directly to (see Fig. 3.5)

$$w_1(Z|H_0) = \left(2\Psi^*_{s-\text{inc}}\right)^{-1} e^{-Z/2\Psi^*_{\text{inc}}};$$

$$w_1(Z|H_1) = \left(2\Psi^*_{s-\text{inc}}\right)^{-1} e^{-Z/2\Psi^*_{s-\text{inc}} - \Psi^*_{s-\text{inc}}/2} I_0\left(\sqrt{Z}\right); \quad Z \geq 0. \quad (3.2.18)$$

The associated cfs are easily shown to be ([4], p. 718, 4, $\nu = 0$, $\beta = 1$, $\alpha = (1/2)\Psi^*_{s-\text{inc}} - i\xi$):

$$F_1(i\xi|H_0)_Z = \int_0^\infty e^{i\xi Z} w_1(Z|H_0)dZ = \left(1 - 2i\xi\Psi^*_{s-\text{inc}}\right)^{-1} \quad (3.2.18a)$$

$$F_1(i\xi|H_1)_Z = \int_0^\infty e^{i\xi Z} w_1(Z|H_1)dZ = \frac{e^{i\xi\Psi^*_{s-\text{inc}}/\left(1-2i\xi\Psi^*_{s-\text{inc}}\right)}}{1 - 2i\xi\Psi^*_{s-\text{inc}}}. \quad (3.2.18b)$$

The inversion of the cfs yields as expected the corresponding pdf values (3.2.18) (as can be verified from Ref. [2], No. 655.1, p. 79.)

The false alarm probability is now readily determined with the help of (3.2.18), namely,

$$\alpha_F^* = \int_{K_T(K)_{\text{inc}}}^{\infty} w_1(Z|H_0)dZ = \int_{K_{T-\text{inc}}}^{\infty} e^{-Z/2\Psi_{s-\text{inc}}^*} \frac{dZ}{2\Psi_{s-\text{inc}}^*} = e^{-K_{T-\text{inc}}/2\Psi_{s-\text{inc}}^*}. \tag{3.2.19a}$$

The false "rest" or rejection probability β^* (of an actual signal) is similarly obtained with the help of (3.2.8):

$$\beta^* = \int_0^{K_{T-\text{inc}}} w_1(Z|H_1)dZ = 1-Q\left(\sqrt{\Psi_{s-\text{inc}}^*}, \sqrt{K_{T-\text{inc}}/\Psi_{s-\text{inc}}^*}\right) = 1-Q\left(\sqrt{\Psi_{s-\text{inc}}^*}, \sqrt{-2\log\alpha_F^*}\right), \tag{3.2.19b}$$

with $K_{T-\text{inc}}\left(= -2\Psi_{s-\text{inc}}^*\log\alpha_F^*\right)$ given by (3.2.12a) in terms of the original threshold K_{inc}, cf. (3.2.11).[16]

The (tabulated) Q-function is defined by Ref. [3]:

$$Q(\alpha,\beta) \equiv \int_{\beta\geq 0}^{\infty} e^{-\alpha^2/2-\lambda^2/2}\lambda I_0(\alpha\lambda)d\lambda, \tag{3.2.20}$$

with

$$Q(\alpha,0) = Q(0,0) = Q(\infty,0) = 1; \quad Q(0,\infty) = Q(\alpha,\infty) = 0; \quad Q(0,\beta) = e^{-\beta^2/2}. \tag{3.2.20a}$$

Expanding the Bessel function in (3.2.20) and integrating gives

$$Q(\alpha,\beta) = e^{-\alpha^2/2}\sum_{n=0}^{\infty}\left(\frac{\alpha^2}{2}\right)^n \frac{1}{n!^2}\Gamma(n+1;\beta^2/2), \quad \text{where} \quad \Gamma(\mu+1;z) \equiv \int_z^{\infty} t^{\mu}e^{-t}dt \tag{3.2.21a}$$

is the incomplete Γ-function, with $\Gamma(\mu+1;z) = \Gamma(\mu+1)-\Gamma(\mu+1,z)$ ([4], Section 8.35 and [5], Section 6.5, including Tables). Of particular interest in establishing the result $p_D^* = 1$ below, for sufficiently large input signals, we have

$$\lim_{\alpha\to\infty} Q(\alpha,\alpha^{\varepsilon}) = 1, [0 \leq \varepsilon < 1], \tag{3.2.21b}$$

which is simply a statement that for $\alpha \to \infty$, the concentration of the integrand grows more rapidly to the larger values of λ than does the lower limit ($\beta = \alpha^{\varepsilon}$). The reverse is the case when $\alpha \to 0$, which accounts for $Q = 0$ values in this limit. For the probability p_D^* of correct

[16] Henceforth, we omit the designation $|_Z$, refer to Section 1.7, for example, since it should be clear from the context which sufficient statistic is being used for the GLR.

detection, we have

$$p_D^* = 1-\beta^* = \int_{K_T}^{\infty} w_1(Z|H_1)dZ = Q\left(\sqrt{\Psi^*_{s-\mathrm{inc}}}, \sqrt{K_{T-\mathrm{inc}}/\Psi^*_{s-\mathrm{inc}}}\right)$$

$$= e^{-\Psi^*_{s-\mathrm{inc}}/2}\sum_{n=0}^{\infty}\left(\Psi^*_{s-\mathrm{inc}}/2\right)^n \frac{1}{n!^2}\Gamma\left(n+1; K_{T-\mathrm{inc}}/2\Psi^*_{s-\mathrm{inc}}\right) \tag{3.2.22}$$

or

$$= e^{-\Psi^*_{s-\mathrm{inc}}/2}\sum_{n=0}^{\infty}\left(\frac{\Psi^*_{s-\mathrm{inc}}}{2}\right)^n \frac{1}{n!^2}\Gamma\left(n+1; -\log\alpha_F^*\right). \tag{3.2.22a}$$

Note that

$$Q(0,\beta) = Q\left(\left[K_T/\Psi^*_{s-\mathrm{inc}}\right]^{1/2}\right) = \exp\left(-K_{T-\mathrm{inc}}/2\Psi^*_{s-\mathrm{inc}}\right) = \alpha_F^*. \tag{3.2.22b}$$

The detection probabilities $\alpha_F^*, \beta^*, p_D^*$, etc. are "exact" for all levels of input signal as long as the narrowband conditions (3.2.4a), (3.2.6a,b, 7) hold. *As we shall see presently,* $\Psi^*_{s-\mathrm{inc}}$, (cf. Eqs. (3.2.5) and (3.2.22c), and subsequently through Chapter 3), *like* $\Psi^*_{s-\mathrm{coh}}$ (3.1.9a)[17], *is the relevant detection parameter,* here designated by $\sigma^{*2}_{0-\mathrm{inc}}$:

$$\sigma^{*2}_{0-\mathrm{inc}} = \Psi^*_{s-\mathrm{inc}} = \left(\tilde{\mathbf{a}}\mathbf{k}_N^{-1}\mathbf{a} + \tilde{\mathbf{b}}\mathbf{k}_N^{-1}\mathbf{b}\right)/2 = \tilde{\mathbf{a}}\mathbf{k}_N^{-1}\mathbf{a} = \tilde{\mathbf{b}}\mathbf{k}_N^{-1}\mathbf{b}, \tag{3.2.22c}$$

for those cases where an "exact" treatment for all signal levels (in Gauss noise) is possible. Figure 3.6 shows α_F^* vs. β^* with $\Psi^*_{s-\mathrm{inc}}(=\hat{\sigma}^{*2}_{s-\mathrm{inc}})$ as parameter. Note that as $\Psi^*_{s-\mathrm{inc}}$ becomes larger, both α_F^* and β^* become smaller, consistent with $\alpha_F^* \to 0$ and $p_D^* \to 1$ as $\Psi^*_{s-\mathrm{inc}} \to \infty$ with this increasing signal input and consequent increasing output signal-to-noise ratios. When considering *threshold reception*, however, we shall see that the detection parameter $(\sigma^{*2}_{0-\mathrm{inc}})$ for incoherent reception takes a different form. This is a consequence of the canonical form of the weak signal development of the likelihood ratio ($\log\Lambda$).

Finally, it is important to observe that although we can use (3.2.12a) to establish the quantitative connection here between the original threshold K_{inc} and the new threshold $K_{T-\mathrm{inc}}$ for the new, statistically equivalent test statistic Z, (3.2.12), practically it is simpler to use $K_{T-\mathrm{inc}}$ itself. Its value can be readily found from our choice of false alarm probability α_F^*, conditioned on the minimum probability of detection p_D^* that we demand for the task at hand. Again, this is possible because of the sufficiency and monotonicity of the new test statistic $\Psi^*_{x-\mathrm{inc}}$, as explained in Section 3.2.1.1. The new threshold $K_{T-\mathrm{inc}}$ then also guarantees the

[17] See the comment following Eq. (3.2.5a) above.

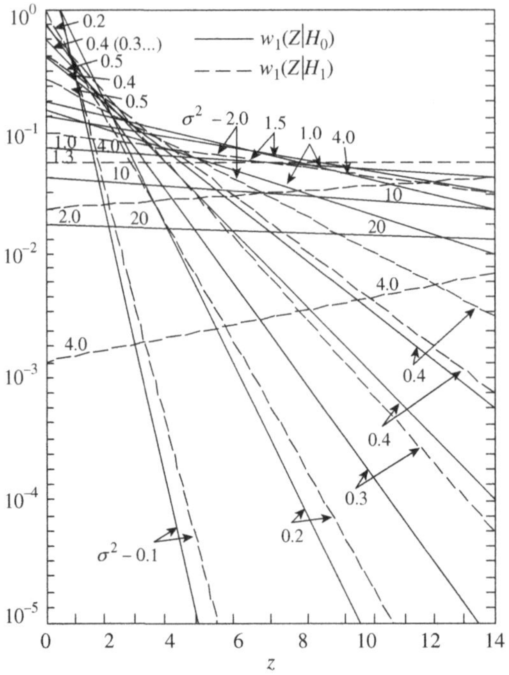

FIGURE 3.5 The pdf's (3.2.18) of Z (3.2.12), the optimum detector structure incoherent detection.

desired limiting behavior as the signal becomes very large or vanishes. However, we observe that $K_{\text{T-inc}} = 0\left(\Psi_{\text{s-inc}}^{1+\varepsilon}\right), 0 < \varepsilon < 1$, *for all* $0 \leq \Psi_{\text{s}} \leq \infty$. (This can be readily established by inverting (3.2.12a).)

3.2.1.3 Neyman–Pearson and Ideal Observers Here, we use the previous results of Section 1.8.1 where the threshold K ($\neq 1$ generally) is given by (3.2.12a), to describe first the *Neyman–Pearson* or *CFAR detector* (prechosen $\alpha_{\text{F}}^{*} > 0$) explicitly. We have for correct decisions

$$\text{NP Detector}: \quad \text{Decide}\, H_1 : S+N, \ \text{if} \ \Psi_{x\text{-inc}}^{*} \geq K_{T\text{-inc}}, \quad \text{or Decide} \ H_0 : N, \text{if} \ \Psi_{x\text{-inc}}^{*} < K_{T\text{-inc}}, \tag{3.2.23}$$

where $Z = \Psi_{x\text{-inc}}^{*}$ is described by (3.2.12) above. The performance $\left(p_{\text{D}}^{*} = 1 - \beta^{*}, \alpha_{\text{F}}^{*}\right)$ [Eq. (3.2.19a) and (3.2.19b)], is given by (3.2.22b), with $P_{\text{D}}^{*} = pp_{\text{D}}^{*}$ as before. Figure 3.4 also shows the procedure, where $K_{T\text{-inc}}$ is determined from (3.2.19a) by a suitable choice of false alarm probability. The threshold $K_{T\text{-inc}}$ is related to the original threshold by (3.2.12a). Note that β^{*} vanishes as $\Psi_{s\text{-inc}}^{*} \to \infty$, as required, with $\alpha_{\text{F}}^{*} > 0$ and fixed (cf. 3.2-21b).

For the *Ideal Observer*, refer to (1.8.2), we can use the new threshold $K_T = 0\left(\Psi_{s\text{-inc}}^{*(1+\varepsilon)}\right)$, $0 < \varepsilon < 1$, and $\mu\, (= p/q)$ can be different from unity, (i.e., the "unsymmetrical" channel). In any case, the (minimum or) Bayes probability of error P_{e}^{*} is $q\alpha_{I}^{*} + p\beta_{I}^{*}$, where α_{I}^{*} and β_{I}^{*} are jointly minimized, so that for these narrowband Bayesian incoherent systems in additive normal noise, one has directly from (3.2.19a) and (3.2.19b),

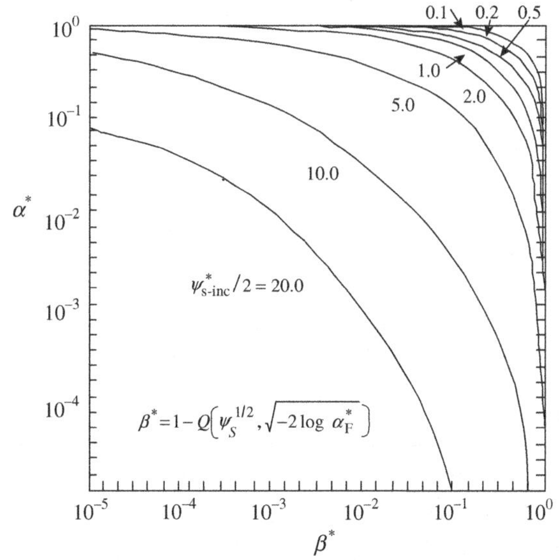

FIGURE 3.6 Error probabilities for incoherent detection of narrowband deterministic signals of known amplitude, $\alpha_F^* = \alpha_{F_{NP}}^*$ fixed, and given by (3.2.19a), $\beta^* = \beta_{NP}^*$ (3.2.19b) for the *Neyman–Pearson or CFAR Detectors*. For the Ideal Observer, P_e is obtained from (3.2.24a) and (3.2.24b), $\alpha^* = \alpha_I^*$, $\beta^* = \beta_I^*$.

$$\text{IO}: \quad P^*_{e-\text{inc}} = qe^{-K_{I-\text{inc}}/2\Psi^*_{s-\text{inc}}} + p\left[1 - Q\left(\sqrt{\Psi^*_{s-\text{inc}}}, \sqrt{K_{I-\text{inc}}/\Psi^*_{s-\text{inc}}}\right)\right], \tag{3.2.24a}$$

$$= q\alpha_I^* + p\left[1 - Q\left(\sqrt{\Psi^*_{s-\text{inc}}}, \sqrt{-2\log\alpha_I^*}\right)\right] = q\alpha_I^* + p\beta_I^* \tag{3.2.24b}$$

and comments following Eq. (3.2.22b) above. Thus, we decide

$$\text{or} \quad \left.\begin{array}{ll} S+N: & \text{if} \quad \Psi^*_{x-\text{inc}} \geq 1 \\ N: & \text{if} \quad \Psi^*_{x-\text{inc}} < 1 \end{array}\right\}, \tag{3.2.25}$$

and in each case P_e is given by (3.2.24). (See Sections 1.8.1, and 1.8.2 for the derivations and discussions of the *NP* and *IO* classes of optimum binary (here On–Off) detectors.) In all instances, the error probability P_e^* becomes smaller as $\Psi_{s-\text{inc}}$ becomes larger. For Q in the above, we may also use (3.2.22a). Again, we have, as expected from (3.2.21b), $P_e \to 0$ as $\Psi^*_{s-\text{inc}} \to \infty$, while $P_e \to p$ when $\Psi^*_{s-\text{inc}} \to 0$. Figure 3.4 also illustrates the decision process, however with $K_{T-\text{inc}}$ replaced by $K^*_{I-\text{inc}}$.

3.2.1.4 Array Processing (Quadratic Arrays) Unlike the linear arrays of the optimum coherent detector in the Gaussian regimes exemplified in Section 3.1.3, the space–time optimum Gaussian incoherent narrowband detectors of Section 3.2.1 embody *quadratic array structures*. This is shown in Fig. 3.7, along with the linear arrays of the coherent cases above.

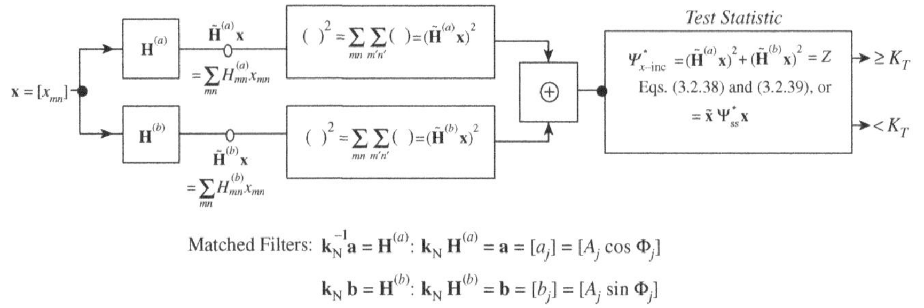

FIGURE 3.7 Spatial (and temporal) processing for optimum narrowband incoherent detection of deterministic signals in Gaussian noise: quadratic (product) arrays.

The latter employ a linear combination $\left(\sim \sum_m ( )_m\right)$ of suitably delayed sensor outputs, while the former use *a product* of such linear combinations $\left(\sim \sum_{mm'} ( )_{mm'}\right)$. The linear character in the received data $\mathbf{x}$ of coherent reception dictates the linear array structure. Correspondingly, it is the quadratic dependence of incoherent reception here that ensures the press of quadratic array processing, however dynamically weighted $\left[\sim \sum_{mm'} ( )_{mm'} \times \sum_{nn'} ( )_{nn'}\right]$ each sensor output may be.

To see this analytically, let us consider the optimum space–time processor $Z = \Psi^*_{x\text{-inc}}$ [(3.2.12) and (3.2.13)], *first renormalizing the output noises* $\psi_j, \psi_{j'}$ *of the* (m, m') *sensors by the procedure described above*[18] in (3.1.18c). This entails multiplying the various elements $x_j x_{j'} C_{jj'}$ of $\Psi_{x\text{-inc}}$, (3.2.13), by $\gamma_{jj'kl}$, where

$$\gamma_{jj'kl} \equiv \frac{\sqrt{\psi_j \psi_{j'}}}{\psi'_M} \cdot \frac{\sqrt{\psi_k \psi_l}}{\psi'_M} \tag{3.2.26}$$

and where, of course, ensemble averages have been replaced by time averages, in the manner of (3.1.18c), which also defines ψ'_M. The result is explicitly for (3.2.12) and (3.2.13) now

$$\Psi^*_{x\text{-inc}} \to \Psi^{*\prime}_{x\text{-inc}} \equiv \tilde{\hat{\mathbf{x}}}\hat{\mathbf{C}}\hat{\mathbf{x}} = \sum_{jj'} \hat{x}_j \hat{x}_{j'} \hat{C}_{jj'}, \tag{3.2.27}$$

with $\hat{x}_j = X_j / \sqrt{\psi'_M} = x_j \hat{\eta}_j$, and so on. The symmetric matrix $\hat{\mathbf{C}}$ is represented in detail by

$$\hat{\mathbf{C}} = \tilde{\hat{\mathbf{C}}} = \left[C_{jj'} = \sum_{ke} \left(\mathbf{k}_N^{-1}\right)_{jk} \left(\hat{a}_{0_k}\hat{a}_{0_l} + \hat{b}_{0_k}\hat{b}_{0_l}\right) \left(\mathbf{k}_N^{-1}\right)_{lj'} \right], \tag{3.2.27a}$$

and the renormalized signal components in (3.2.27) are now typically

$$\left.\begin{array}{c} \hat{a}_{0_k} \\ \hat{b}_{0_k} \end{array}\right\} \equiv \left.\begin{array}{c} A_{0_k}/\sqrt{\psi'_M} \\ B_{0_k}/\sqrt{\psi'_M} \end{array}\right\} \equiv \left\{\begin{array}{c} \left(A^{(k')}_{0_{k''}}/\sqrt{\psi'_M}\right) \cos\left[\omega_0 t_{k''} - \phi^{(k')}_{k''}\right] \\ \left(B^{(k')}_{0_{k''}}/\sqrt{\psi'_M}\right) \sin\left[\omega_0 t_{k''} - \phi^{(k')}_{k''}\right] \end{array}\right\}, \tag{3.2.27b}$$

[18] We introduce this alternative renormalization for (most) practical applications, usually because of the noncomputability of noise power ensemble averages. However, renormalization itself is not necessary to demonstrate the presence of the quadratic operations.

with $k \to l$ for $\hat{a}_{0_l}, \hat{b}_{0_l}$, refer to (3.2.27a) and (3.2.27b). (In the above, $(j \cdots j')$ are the usual double indexes, that is, $j = mn$, $j' = m'n'$, and similarly, $k = k'k''$, $l = l'l''$, where the first index represents "space" (m, m', k', l'), that is, the sensor "number," associated with sensor location in space, while the second index (n, n', k'', l'') indicates the "time" at which the received field at a particular point is sampled.) As before, refer to (3.1.18c), only those (m) sensors producing an output (noise) $(\psi_m > 0)$ are to be used in normalization. Also as before, $\hat{\mathbf{H}}^{(a)}$, $\hat{\mathbf{H}}^{(b)}$ obtained from $\mathbf{k}_N \hat{\mathbf{H}}^{(a)} = \hat{\mathbf{a}}_0$ and $\mathbf{k}_N \hat{\mathbf{H}}^{(b)} = \hat{\mathbf{b}}_0$ are *discrete space–time matched filters*, refer to (3.2.13b) and (3.2.13c) above, consisting both slowly varying and rapidly varying components. Finally, we observe from (3.2.8b) and (3.2.12) and more directly from (3.2.29), as expected, that the test statistic $\Psi^*_{x-\text{inc}}$ is represented by the *autocorrelation of the received data with themselves*, a characteristic of incoherent reception. This is to be compared with the coherent case (Section 3.1.1), where the signal and data appear as a cross-correlation (Section 3.1.3).

Example 3.1 In many applications, considerable simplification of processing, often with small loss in performance, can be achieved by using effectively independent noise samples, that is, $\mathbf{k}_N = [\delta_{jj'}]$.[19] Then the renormalized test statistic (3.2.27), $\Psi^{*\prime}_{x-\text{inc}}$ reduces to

$$\Psi^{*\prime}_{x-\text{inc}} \doteq \sum_{jj'} \hat{x}_j \hat{x}_{j'} \left\{ \frac{A_{0_n}^{(m)} A_{0_{n'}}^{(m')}}{\psi'_M} Re\left(\exp\left[i\omega_0 (t_n - t'_n) - i\omega_0 \cdot \Delta\tau_{mm'}/c_0 - i\Delta\phi_{nn'}^{(m,m')} \right] \right) \right\} \tag{3.2.28}$$

with $\hat{x}_j = X_j / \sqrt{\psi'_M}$ and so on. Now for these quadratic arrays, the steered beam (using $\hat{i}_{O_R}$) involves the difference of path delays along $\mathbf{r}_m$ and $r_{m'}$, namely, $\Delta\tau_m - \Delta\tau_{m'} \equiv \Delta\tau_{mm'}$, or in more detail

$$\Delta\tau_{mm'} \equiv \left(\hat{\mathbf{i}}_0 - \hat{\mathbf{i}}_{O_R}\right) \cdot \mathbf{r}_m / c_0 - \left(\hat{\mathbf{i}}_0 - \hat{\mathbf{i}}_{O_R}\right) \cdot \mathbf{r}_{m'} / c_0 = \left(\hat{\mathbf{i}}_0 - \hat{\mathbf{i}}_{O_R}\right) \cdot (\mathbf{r}_m - \mathbf{r}_{m'}) / c_0 \equiv \Delta\hat{\mathbf{i}}_0 \cdot \Delta\mathbf{r}_{mm'} / c_0, \tag{3.2.28a}$$

vide Fig. 3.3 and (3.1.19) above, on inserting the required path delays $\Delta\tau_m, \Delta\tau_{m'}$; in addition, we use the abbreviations $\Delta\hat{\phi}_{jj'} = \Delta\phi_{n,n'}^{(m,m')} \equiv \phi_n^{(m)} - \phi_{n'}^{(m')}$, $t_{nn'} \equiv t_n - t_{n'}$.

Example 3.2 A still simpler and often useful result here arises when $\Delta\hat{\phi} = 0$ and $A_{0_n}^{(m)} = A_{0_{n'}}^{(m')} = A_{0_n}$, with $\psi_j = \psi$, all j for cw uniform signal and noise fields over the receiving array. Then, (3.2.28) reduces directly to an expression that clearly exhibits the product nature of this simplified test statistic, namely, the rather elegant result

$$\Psi^{*\prime}_{x-\text{inc}}\Big|_{\text{cw}} \doteq \frac{A_0^2}{\psi} \left| \sum_{mn}^{J=MN} \hat{x}_{mn} e^{-i\omega_0 t_n} e^{-i\omega_0 \left(\hat{\mathbf{i}}_0 - \hat{\mathbf{i}}_{O_R}\right) \cdot \mathbf{r}_m / c_0} \right|^2, \quad \mathbf{k}_N = [\delta_{jj'}], \tag{3.2.29}$$

for (truncated) uniform space–time cw fields. Equations (3.2.28) and (3.2.29) are examples of *adaptive beam forming* since the components of the array elements depend on the

[19] $\delta_{jj'} = \delta_{mm'} \delta_{nn'}$: from physical consideration, since "space" is compared to "space," and "time" to "time."

particular space–time received data $\hat{\mathbf{x}}$. Moreover, the resulting beam formation is the *product* of two array structure outputs $\left(\sim \left|\sum_m(\,)\right|^2\right)$. Note that this test statistic is maximized when $\hat{\mathbf{i}}_{O_R} = \hat{\mathbf{i}}_0$, that is, when (under H_1) each linear array output is itself maximized. Thus, we have

$$\Psi^{*\prime}_{x-\text{inc}}\Big|_{\substack{cw \\ \hat{\mathbf{i}}_{O_R}=\hat{\mathbf{i}}_0}} \doteq \frac{A_0^2}{\psi}\left(\sum_{mn}^{J}\hat{x}_{mn}\right)^2 = \frac{A_0^2}{\psi}\sum_{jj'}\hat{x}_j\hat{x}_{j'};\quad \bar{\hat{x}}_j = 0. \tag{3.2.30}$$

This result is seen to be a discrete autocorrelation of the received data with themselves. (Equations (3.2.29) and (3.2.30), in addition, must obey the simplifying condition cited above at the beginning of this paragraph, where, in addition, the discrete matched filters are now simply the vectors $\hat{H}^{(a)} = \hat{\mathbf{a}}_0 = [(A_0/\sqrt{\psi})\cos(\omega_0 t_n - \phi_0)]$ and $\hat{H}^{(b)} = \hat{\mathbf{b}}_0 = [(A_0/\sqrt{\psi})\sin(\omega_0 t_n - \phi_0)]$.

Example 3.3 Similar observations apply for the key parameter $\Psi^*_{s-\text{inc}}$ of the pdf values $W_1(Z|H_0, H_1)$ governing the error probabilities of detection. Renormalizing the sensor outputs according to (3.1.18c) et seq. from (3.2.5) to (3.2.7a), we can write explicitly, showing the array structure:

$$\Psi^*_{s-\text{inc}} = \frac{1}{2}\left(\tilde{\mathbf{a}}\mathbf{k}_N^{-1}\mathbf{a} + \tilde{\mathbf{b}}\mathbf{k}_N^{-1}\mathbf{b}\right) = \frac{1}{2}\sum_{jj'}\left(\mathbf{k}_N^{-1}\right)_{jj'}\frac{A_{0_n}^{(m)}A_{0_{n'}}^{(m')}}{\psi'_M}Re\left\{e^{i\omega_0\left[t_{nn'} - \Delta\hat{\mathbf{i}}_0\,\cdot\,\Delta\mathbf{r}_{mm'}/c_0\right] - i\Delta\phi_{jj'}}\right\}. \tag{3.2.31}$$

Since $\mathbf{k}_N = \tilde{\mathbf{k}}_N$, $(\det\mathbf{k}_N > 0)$, it is possible to find an orthogonal $(J \times J)$ matrix $\mathbf{Q}$, $\mathbf{Q}\tilde{\mathbf{Q}} = \tilde{\mathbf{Q}}\mathbf{Q} = \mathbf{I}$ to diagonalize $\mathbf{k}_N$ in terms of its eigenvalues[20] and so express $\Psi^*_{s-\text{inc}}$

$$\Psi^*_{s-\text{inc}} = \left.\begin{aligned}&\frac{1}{2}\sum_j\left(u_j^2 + v_j^2\right)\lambda_j \doteq (1/2)\sum_j\frac{A_{0_n}^{(m)^2}}{\psi'_M}\\ &+\frac{1}{2}\sum_{jj'}{}'\frac{A_{0_n}^{(m)}A_{0_{n'}}^{(m')}}{\psi'_M}Re\left\{e^{i\omega_0\hat{\Phi}_{jj'}}\right\}\end{aligned}\right\}, \tag{3.2.32}$$

where $\hat{\Phi}_{jj'} \equiv \left(\Delta t_{nn'} - \Delta\hat{\mathbf{i}}_0\cdot\Delta\mathbf{r}_{mm'}/c_0 - \Delta\phi_{jj'}\right)$ and where $\Sigma'_{jj'}$ denotes a sum in which the terms $(j = j')$ are omitted. The first terms of (3.2.7) are always positive here, since $\mathbf{k}_N$ is positive definite, and thus the eigenvalue λ (or diagonal terms) of $\tilde{\mathbf{Q}}^{-1}\mathbf{k}_N\mathbf{Q} = [\lambda_j\delta_{jj'}] = \mathbf{\Lambda}$ are also positive. The second set of terms $(\sim \Sigma'_{jj'})$ yields an oscillating component that perforce, if negative, must always be smaller than the (positive) diagonal $(j = j')$ contribution in the right number of the second equation. In fact, because of the oscillating nature of the off-diagonal terms $(\sim \Sigma'_{jj'})$, we expect them to be small and even negligible

[20] Thus, let $(\mathbf{u}, \mathbf{v})$ be vectors, such that $\mathbf{Qu} = a$, $\mathbf{Qv} = b$, $\therefore$ $\tilde{\mathbf{a}}\mathbf{k}_N^{-1}\mathbf{a} = \tilde{\mathbf{u}}\tilde{\mathbf{Q}}^{-1}\mathbf{k}_N^{-1}\mathbf{Qu}$, with $\mathbf{Q} = \tilde{\mathbf{Q}}^{-1}$ an orthogonal matrix such that $\tilde{\mathbf{Q}}^{-1}\mathbf{k}_N^{-1}\mathbf{Q} = \mathbf{\lambda}_a^{-1}$ and $\therefore$ $\tilde{\mathbf{a}}\mathbf{k}_N^{-1}\mathbf{a} = \sum_j u_j^2\lambda_j^{-1}(>0)$ since all eigenvalues (λ_j) of $\mathbf{k}_N^{-1}$ are positive as is $\tilde{\mathbf{a}}\mathbf{k}_N^{-1}\mathbf{a}$, which is also positive definite, with $\mathbf{Q} = \tilde{\mathbf{Q}}$, and $\therefore$ $\tilde{\mathbf{Q}}^{-1}\mathbf{k}_N\mathbf{Q} = [\lambda_j\delta_{jj'}]$. The same result applies for $\tilde{\mathbf{b}}\mathbf{k}_N^{-1}\mathbf{b} = \sum_j v_j^2\lambda_j^{-1}\left((\geq 0)_{jj'}\right)$. Here, $(\mathbf{u}, \mathbf{v})$ are the eigenvectors associated with $(\mathbf{a}, \mathbf{b})$, with respect to the kernel $\mathbf{k}_N$; $[\lambda_j\delta_{jj'}]$ is the (square) matrix of eigenvalues. For further discussion see Section 7.3.1 of Ref. [1] and Appendix A1 of this book.

vis-à-vis the diagonal term. When $\Delta\hat{\mathbf{i}}_0 = 0$, that is, when the (quadratic) array is directed toward the signal source, we see that

$$\Psi^*_{s-\text{inc}} \doteq \frac{1}{2}\sum_{mn} A_{on}^{(m)^2} \Big/ \psi_M \approx \frac{1}{2}\Psi^*_{s-\text{inc}}\Big|_{\max}, \tag{3.2.32a}$$

as expected.

Again, from (3.2.19)–(3.2.22), it is also evident that $\Psi^*_{s-\text{inc}}$ plays the rôle of *detection parameter* $\sigma^{*2}_{0-\text{inc}}(=\Psi^*_{s-\text{inc}})$ in this now incoherent detection situation, similar to the coherent case of Section 3.1 above. To emphasize this point, we rewrite (3.2.19a) as

$$\alpha^*_{\text{F}} = e^{-K_T/2\sigma^{*2}_{0-\text{inc}}}; \quad p^*_{\text{D}} = Q\left(\sigma^*_{0-\text{inc}}, \left[K_T/\sigma^{*2}_{0-\text{inc}}\right]^{1/2}\right) = 1-\beta^*. \tag{3.2.33}$$

Accordingly, the larger $\sigma^*_{0-\text{inc}}$, the smaller the false alarm probability α^*_{F} and the larger the (conditional) probability of detection, as expected. From (3.2.31), it also follows as expected that $\Psi^*_{s-\text{inc}}$ and $\therefore \sigma^{*2}_{0-\text{inc}}$ are maximized for any fixed size of space–time sample $J = MN$ when in the process of beam formation one sets $\hat{\mathbf{i}}_{0_{\text{R}}} = \hat{\mathbf{i}}_0$: the receiving beam is pointed in the direction of the signal source.

3.2.2 Incoherent Detection II. Deterministic Narrowband Signals with Slow Rayleigh Fading[21]

Here, we introduce a dimensionless amplitude factor $\hat{a}_0$ in $\Psi^*_{x-\text{inc}}$ (3.2.5) and in $\Psi^*_{x-\text{inc}}$ (3.2.12) and (3.2.13), so that $\Psi^*_{s-\text{inc}} \to \hat{a}_0^2\Psi^*_{s-\text{inc}}$, $\Psi^*_{x-\text{inc}} \to \hat{a}_0^2\Psi^*_{x-\text{inc}}$ with the result that the likelihood ratio (3.2.1) can be written explicitly

$$\Lambda(x) = \mu\left\langle \exp\left(-\frac{1}{2}\hat{a}_0^2\Psi^*_{s-\text{inc}} + \hat{a}_0\Psi^*_{x-\text{inc}}\right)\right\rangle_{\varepsilon,\hat{a}_0}, \quad \mu \equiv p/q. \tag{3.2.34}$$

Again, ε is uniformly distributed over a cycle of the high-frequency "carrier" $f_0(=\omega_0/2\pi)$.[22] The scale or amplitude factor $\hat{a}_0$, whose value does not change during the reception interval (i.e., "slow fading"), is assumed to obey a Rayleigh pdf, namely,

$$w_1(\hat{a}_0) = \frac{\hat{a}_0 e^{-\hat{a}_0^2/2\sigma_{\text{R}}^2}}{\sigma_{\text{R}}^2}, \quad \hat{a}_0 \geq 0, \quad \text{with} \quad \sigma_{\text{R}}^2 = \langle\hat{a}_0^2\rangle - \langle\hat{a}_0\rangle^2 > 0. \tag{3.2.35}$$

Since physically fading may usually be regarded as a sum of a large number of independent unresolvable multipath components of propagation in the medium in question, by a Central Limit Theorem (CLT) argument its envelope statistics have the Rayleigh form (3.2.35). Then, using the result $\exp U(\mathbf{x})^*$, (3.2.10), which is the result of the average over the epochs

[21] See Ref. [6] for an extensive treatment of fading; see also the earlier work [7].

[22] See the remarks on p. 154 and Eq. (3.2.5a) et seq.

$\{\varepsilon\}$, for example, $\langle\,\rangle_\varepsilon$, we next obtain the average $\langle\,\rangle_{\hat{a}_0}$ over the amplitude scale factor $\hat{a}_0$:

$$\boldsymbol{\Lambda}(\mathbf{x})_{\mathrm{R}} = \langle \boldsymbol{\Lambda}(\mathbf{x}, \hat{a}_0)\rangle_{\hat{a}_0} = \frac{\mu}{\sigma_{\mathrm{R}}^2}\int_0^\infty e^{-\frac{1}{2}\left(\Psi^*_{s-\mathrm{inc}}+\sigma_{\mathrm{R}}^{-2}\right)y^2}\, y\, I_0\left(y\sqrt{\Psi^*_{x-\mathrm{inc}}}\right)dy \tag{3.2.36a}$$

$$= \frac{\mu}{1+\sigma_{\mathrm{R}}^2\Psi^*_{s-\mathrm{inc}}}\, e^{\sigma_{\mathrm{R}}^2\Psi^*_{x-\mathrm{inc}}/2\left(1+\sigma_{\mathrm{R}}^2\Psi^*_{s-\mathrm{inc}}\right)} \tag{3.2.36b}$$

with the help of Hankel's first exponential integral ((A1.49), [1])

$$\int_0^\infty J_\nu(az)z^{\mu-1}e^{-b^2z^2}dz = \frac{\Gamma\left(\dfrac{\nu+\mu}{2}\right)}{2b^\mu\Gamma(\nu+1)}\left(\frac{a}{2b}\right)^\nu {}_1F_1\left(\frac{\nu+\mu}{2};\nu+1;-\mathrm{a}^2/4\mathrm{b}^2\right),$$

$$Re(\mu+\nu)>0;\ |\arg b| < \pi/4. \tag{3.2.36c}$$

Thus, the detector structure now becomes explicitly the simple relation

$$U(\mathbf{x})^*_{\mathrm{R}} \equiv Z_{\mathrm{R}} = \log \boldsymbol{\Lambda}(\mathbf{x})_{\mathrm{R}} = B_{\mathrm{R}} + C_{\mathrm{R}}\Psi^*_{x-\mathrm{inc}}, \tag{3.2.37}$$

where

$$B_{\mathrm{R}} \equiv \log\mu - \log\left(1+\sigma_{\mathrm{R}}^2\Psi^*_{s-\mathrm{inc}}\right) = \log\left[\mu/\left(1+\sigma_{\mathrm{R}}^2\Psi^*_{s-\mathrm{inc}}\right)\right];\quad C_{\mathrm{R}} \equiv \sigma_{\mathrm{R}}^2/2\left(1+\sigma_{\mathrm{R}}^2\Psi^*_{s-\mathrm{inc}}\right). \tag{3.2.38}$$

Since the original decision process uses the threshold $\log K$, refer to Eq. (3.2.11), we see for $U(\mathbf{x})^*_{\mathrm{R}}$ that the decision process (for correct decisions) here is

$$\text{or}\quad \left.\begin{array}{l} \text{Decide } H_1\text{: } S+N \text{ if: } U(x)^*_{\mathrm{R}} \geq \log K_{\mathrm{inc}}, \quad \text{or} \quad \Psi^*_{x-\mathrm{inc}} \geq (\log K_{\mathrm{inc}} - B_{\mathrm{R}})/C_{\mathrm{R}} \equiv K_R \\ \text{Decide } H_0\text{: } N \text{ if: } U(x)^*_{\mathrm{R}} < \log K_{\mathrm{inc}}, \quad \text{or} \quad \Psi^*_{x-\mathrm{inc}} < (\log K_{\mathrm{inc}} - B_{\mathrm{R}})/C_{\mathrm{R}} \equiv K_{\mathrm{R}} \end{array}\right\}, \tag{3.2.39}$$

where the new threshold K_{R} is given by

$$K_{\mathrm{R}} = \frac{2\lambda_{\mathrm{R}}^2}{\sigma_{\mathrm{R}}^2}\left|\log\left(\frac{K_{\mathrm{inc}}}{\mu}\lambda_{\mathrm{R}}^2\right)\right| > 0;\quad \lambda_{\mathrm{R}}^2 \equiv 1+\sigma_{\mathrm{R}}^2\Psi^*_{s-\mathrm{inc}}. \tag{3.2.39a}$$

This is to be contrasted with $K_{T-\mathrm{inc}} = \left\{I_0^{(-1)}\left((K_{\mathrm{inc}}/\mu)\exp\left[\Psi^*_{s-\mathrm{inc}}/2\right]\right)\right\}^2 (>0)$, Eq. (3.2.12a), when there is no Rayleigh fading. In any case, as mentioned earlier (refer to Section 3.2.1.2 et seq.), it is much easier to obtain the new threshold K_{R} directly from our choice of false alarm probability, $\alpha^*_{\mathrm{F}}|_{\mathrm{R}}$, (3.2.40) ff.

In addition, the test statistic $\Psi^*_{x-\mathrm{inc}}$ remains unchanged (cf. (3.2.13)) from the original narrowband optimal incoherent cases of Section 3.2.1. This also means that $\Psi^*_{x-\mathrm{inc}}$ has the same representation in terms of the matched filters $\mathbf{H}^{(a)}$, $\mathrm{H}^{(b)}$, namely, Eq. (3.2.12b). The results of this section accordingly apply directly here, provided that the old threshold $\mathbf{K}_{T-\text{inc}}$ is replaced by K_R above (3.2.39a). Thus, the pdf values $w_1(Z_\text{R}|H_0, H_1)$ are given by (3.2.18) with Z replaced by Z_R here, with the cf's $F_1(i\xi|Z \to Z_\text{R})_{H_0,H_1}$ (3.2.18a) and (3.2.18b). The error probabilities (a_F^*, β^*) and the probability of correct detection p_D^* are also given by (3.2.19a) (3.2.19b), and (3.2.22) with (3.2.39a) now, namely[23] $K_{T-\text{inc}} \to K_\text{R}$.

$$\alpha_\text{F}^*|_\text{R} = e^{-K_\text{R}/2\Psi_{s-\text{inc}}^*};\ \beta_\text{R}^* = 1 - Q\left(\sqrt{\Psi_{s-\text{inc}}^*}, \sqrt{K_\text{R}/\Psi_{s-\text{inc}}^*}\right) = 1 - Q\left(\sqrt{\Psi_{s-\text{inc}}^*}, \sqrt{-2\log(\alpha_\text{F}^*|_\text{R})}\right), \tag{3.2.40}$$

and

$$p_\text{D}^*|_\text{R} = Q\left(\sqrt{\Psi_{s-\text{inc}}^*}, \sqrt{K_\text{R}/\Psi_{s-\text{inc}}^*}\right);\quad K_\text{R}/\Psi_{s-\text{inc}} = -2\log(\alpha_\text{F}^*|_\text{R}), \tag{3.2.41}$$

Furthermore, Figure 3.6 applies here also for the *detection parameter*, and $\Psi_{s-\text{inc}}^* = \sigma_{0-\text{inc}}^{*2}$ as before. The discussion of *array processing* in Section 3.2.2 carries over in a similar way and is directly applicable to these "slow" Rayleigh amplitude cases, including Fig. 3.7 and Eqs. (3.2.23)–(3.2.32a).

The decision process (3.2.39) above also represents a *Neyman–Pearson* test of the hypotheses. For *the Ideal Observer,* Eq. (3.2.24) now applies, with $K_{\text{R}-\text{I}}/\Psi_{s-\text{inc}}^*$ given by $-2\log \alpha_{\text{R}-\text{I}}^*$:

$$\text{IO}:\quad P_{e-\text{R}}^* = qe^{-K_{\text{R}-I}/2\Psi_{s-\text{inc}}^*} + p\left[1 - Q\left(\sqrt{\Psi_{s-\text{inc}}^*},\ \sqrt{K_{\text{R}-I}/\Psi_{s-\text{inc}}^*}\right)\right] \tag{3.2.42a}$$

$$= q\alpha_{\text{R}-I}^* + p\left(1 - Q\left(\sqrt{\psi_{s-\text{inc}}^*}, -2\log\alpha_{\text{R}-I}^*\right)\right) \tag{3.2.42b}$$

(Equation (3.2.39a) ensures that $P_{e-\text{R}}^*$ is properly bounded, that is, $0 \le P_e^* \le 1$.) Again, we have for the decision process from (3.2.39):

$$\left\{\begin{array}{ll} \text{Decide } H_1\text{: } S+N, & \text{if } \Psi_{x-\text{inc}}^* \ge K_\text{R}|_{K_\text{inc}=1} (= -B_\text{R}/C_\text{R}) \\ \text{Decide } H_0\text{: } N, & \text{if } \Psi_{x-\text{inc}}^* < K_\text{R}|_{K_\text{inc}=1} \end{array}\right\}. \tag{3.2.43}$$

Figures 3.6 and 3.7 also illustrate this decision process, with the threshold now given by $K_{\text{R}-I}|_{K_\text{inc}} = 1$. We observe once more as $\Psi_{s-\text{inc}}^* \to \infty$, with the help of (3.2.21b) that $P_{e-\text{R}}^* \to 0$. Thus, when indefinitely large output signals occur, perfect detection is theoretically possible: there is no error on the average, and the signal is truly received. At the other extreme of $\Psi_{s-\text{inc}}^* \to 0$, we have $P_{e-\text{R}} = q$, also as before, refer to Section 3.2.1.3, which $\to$ 0, refer to Eqs. (3.2.20) and (3.2.20a). (As before, we may use (3.2.19b) and (3.2.20a) in (3.2.40) and (3.2.42), appropriately modified for slow Rayleigh fading.)

[23] Unfortunately, the results for H_1 in case (2) of Ref. [8], pp. 41–45, are incorrect.

3.2.3 Incoherent Detection III: Narrowband Equivalent Envelope Inputs—Representations

In the analyses of Sections (3.2.1) and (3.2.1.2), the signal and noise field components are expressed in terms of their narrowband, high-frequency elements, namely, **a**, **b**, $\cos\Phi_j$, $\sin\Phi_{j'}$, vide (3.2.2) and (3.2.3). In particular, this is also the case for the detection parameter $\Psi^*_{s-\mathrm{inc}}$, (3.2.5) and (3.2.7b), and the test statistic $\Psi^*_{x-\mathrm{inc}}$ (3.2.8b) and (3.2.12). Since it is also sometimes required to employ explicitly the slowly varying or *envelope* components of these narrowband fields, we shall also derive them here and include them alternatively in $\Psi^*_{s-\mathrm{inc}}$ and $\Psi^*_{x-\mathrm{inc}}$. Equation (3.2.31) suggests that we can write in the form[24] (with $\Phi_j = \hat{\Phi}_j - \phi_j$ as before):

$$\Psi^*_{s-\mathrm{inc}} = \frac{\tilde{\mathbf{a}}\mathbf{k}_N^{-1}\mathbf{a} + \tilde{\mathbf{b}}\mathbf{k}_N^{-1}\mathbf{b}}{2} = \left\{\widetilde{\cos\boldsymbol{\Phi}}\,\boldsymbol{\Psi}_s^{(c)}\cos\boldsymbol{\Phi} + \widetilde{\sin\boldsymbol{\Phi}}\,\boldsymbol{\Psi}_s^{(s)}\sin\boldsymbol{\Phi}\right\}_{\mathrm{inc}} \quad \text{or}$$
$$= \sum_{jj'}\left\{\left(\boldsymbol{\Psi}_s^{(c)}\right)_{jj'}\cos\Phi_j\cos\Phi_{j'} + \left(\boldsymbol{\Psi}_s^{(s)}\right)_{jj'}\sin\Phi_j\sin\Phi_{j'}\right\}_{\mathrm{inc}}, \tag{3.2.44}$$

where $\Psi_s^{(c)}$, $\Psi_c^{(s)}$ are the *envelope* or slowly varying components of the "carrier" or rapidly varying terms. Similarly, Eqs. (3.2.27a) and (3.2.27b) also suggest that the test statistic can be represented by[24]

$$\Psi^*_{x-\mathrm{inc}} = \left(\tilde{\mathbf{x}}\mathbf{k}_N^{-1}\mathbf{a}\right)^2 + \left(\tilde{\mathbf{x}}\mathbf{k}_N^{-1}\mathbf{b}\right)^2 = \left(\widetilde{\cos\boldsymbol{\Phi}}\,\boldsymbol{\Psi}_x^{(c)}\cos\boldsymbol{\Phi} + \widetilde{\sin\boldsymbol{\Phi}}\,\boldsymbol{\Psi}_x^{(s)}\sin\boldsymbol{\Phi}\right)_{\mathrm{inc}}$$
$$= \sum_{jj'}\left\{\left(\boldsymbol{\Psi}_x^{(c)}\right)_{jj'}\cos\Phi_j\cos\Phi_{j'} + \left(\boldsymbol{\Psi}_x^{(s)}\right)_{jj'}\sin\Phi_j\sin\Phi_{j'}\right\}_{\mathrm{inc}}. \tag{3.2.45}$$

These coefficients of $\cos\Phi_j\cos\Phi_{j'}$, and so on will be given explicitly in Sections 3.2.3.1 and 3.2.3.3. It is important to remember that the arguments of all the components, both slowly and rapidly varying, contain the appropriate time delays $\Delta\tau_j, \Delta t_j$, and so on. However, the slowly varying elements are usually independent of the spatial delays, unless the receiving array aperture is a sizable fraction of the (average) wavelength.

3.2.3.1 ***Evaluation of*** $\boldsymbol{\Psi}^{(c)}_{s-\mathrm{inc}}, \boldsymbol{\Psi}^{(s)}_{s-\mathrm{inc}}$ We proceed as follows, considering first $\mathbf{a} + \mathbf{b}$ (3.2.2b), namely,

$$\mathbf{a} + \mathbf{b} = \left[A_j\cos\Phi_j + A_j\sin\Phi_j\right], \quad \text{with} \quad A_j = A_{0_n}^{(m)}/\psi_j, \tag{3.2.46}$$

[24] Here we have replaced $\hat{\Phi}_j$ by $\hat{\Phi}_j - \phi_j = \Phi_j$ of (3.2.2) and so on, in $\mathbf{k}_N$, where now $\Phi_j = \omega_0\Delta t_j - \phi = \omega_0\Delta t_j$, including the slowly varying phase component $\phi_j(\Delta t_j)$ in the argument Φ_j. When a *steering vector*, $\mathbf{k}_{O_R}$, is included, $\Delta\tau_j = t_n - (k_0 - k_{O_R})\cdot r_m/\omega_0 - \phi_j(\Delta t_j)/\omega_0$. Practically, ϕ_j/ω_0 can be introduced as an additional time delay (or equivalently a phase shift) in the output path from the mth sensor to the chosen reference point O_R in the receiving array, refer to Fig. 3.3. By this device we are able to simplify the analysis somewhat, where now $\Phi_j = \omega_0\Delta\tau_j - \phi_j$ in place of $\hat{\Phi}_j = \omega_0\Delta\tau_j$.

since $(\tilde{\mathbf{a}}+\tilde{\mathbf{b}})\mathbf{k}_N^{-1}(\mathbf{a}+\mathbf{b}) \doteq \tilde{\mathbf{a}}\mathbf{k}_N^{-1}\mathbf{a}+\tilde{\mathbf{b}}\mathbf{k}_N^{-1}\mathbf{b} = 2\Psi^*_{s-\text{inc}}$ under the narrowband approximation (3.2.3a). Next, we introduce a new vector:

$$\mathbf{c} = \mathbf{k}_N^{-1}(\mathbf{a}+\mathbf{b}) = \mathbf{c}_a + \mathbf{c}_b, \quad \text{with} \quad \mathbf{c}_a = \mathbf{k}_N^{-1}\mathbf{a}, \quad \mathbf{c}_b = \mathbf{k}_N^{-1}\mathbf{b} \tag{3.2.47a}$$

and write

$$\mathbf{c} = \left[c_{cj}\cos\Phi_j + c_{sj}\sin\Phi_j\right]; \quad \therefore \mathbf{c}_a = \left[c_{cj}\cos\Phi_j\right], \quad \text{and} \quad \mathbf{c}_b = \left[c_{sj}\sin\Phi_j\right] \tag{3.2.47b}$$

$$\therefore \ \mathbf{a}+\mathbf{b} = \mathbf{k}_N\mathbf{c} = \mathbf{k}_N\left[c_{cj}\cos\Phi_j + c_{sj}\sin\Phi_j\right] = \left[A_j\cos\Phi_j + A_j\sin\Phi_j\right], \tag{3.2.47c}$$

where c_{cj} and c_{sj} are slowly varying vis-à-vis $(\cos\Phi_j, \sin\Phi_j)$ in the usual way. Using (2.6.33a) and (2.6.33b) for $\mathbf{k}_N$, we can write (3.2.47c) in more detail as

$$A_j(\cos\Phi_j+\sin\Phi_j) = \sum\nolimits_{j'}\left\{(\rho_0)_{jj'}\cos(\Phi_{j'}-\Phi_j)+(\lambda_0)_{jj'}\sin(\Phi_{j'}-\Phi_j)\right\}\left\{c_{cj'}\cos\Phi_{j'}+c_{sj'}\sin\Phi_{j'}\right\}. \tag{3.2.48}$$

Expanding the right-hand side of (3.2.48), invoking the narrowband conditions $\cos^2\Phi_{j'} = \sin^2\Phi_{j'} \doteq 1/2$, $\sin\Phi_{j'}\cos\Phi_{j'} \doteq 0$, and comparing coefficient of $\cos\Phi_j$, $\sin\Phi_j$ give directly the *slowly varying relations*

$$2\mathbf{A} = \rho_0\mathbf{c}_c + \lambda_0\mathbf{c}_s : 2\mathbf{A} = \rho_0\mathbf{c}_s - \lambda_0\mathbf{c}_c, \quad \mathbf{A} = \left[A_{0_j}/\sqrt{\psi}\right] = \left[A_j\right](\neq A_{0_j}). \tag{3.2.49}$$

Solving for $\mathbf{c}_c$ and $\mathbf{c}_s$ yields the desired results[25]

$$\mathbf{c}_c = 2\left[\mathbf{I}+\boldsymbol{\lambda}_0(\boldsymbol{\rho}_0-\boldsymbol{\lambda}_0)^{-1}(\boldsymbol{\rho}_0+\boldsymbol{\lambda})\right]^{-1}\boldsymbol{\rho}_0^{-1}\mathbf{A}; \quad \mathbf{c}_s = 2\left[\mathbf{I}-\boldsymbol{\rho}_0^{-1}\boldsymbol{\lambda}_0(\boldsymbol{\rho}_0+\boldsymbol{\lambda}_0)^{-1}(\boldsymbol{\rho}_0+\boldsymbol{\lambda}_0)\right]^{-1}\boldsymbol{\rho}_0^{-1}\mathbf{A}. \tag{3.2.50}$$

With the help of (3.2.47b), with $\mathbf{c}_a$ and $\mathbf{c}_b$, and with $\mathbf{c}_c, \mathbf{c}_s$ given by (3.2.50), we obtain the slowly varying components of (3.2.44) on applying the narrowband conditions to $\tilde{\mathbf{a}}\mathbf{c}_a/2$ and $\tilde{\mathbf{b}}\mathbf{c}_b/2$, namely:

$$\sum_{j'}\left(\Psi^{(c)}_{s-\text{inc}}\right)_{jj'} = \frac{1}{2}(\tilde{\mathbf{a}}\mathbf{c}_a)_j \doteq \frac{1}{4}A_jc_{cj}; \quad \sum_{j'}\left(\Psi^{(s)}_{s-\text{inc}}\right)_{jj'} = \frac{1}{2}(\tilde{\mathbf{b}}\mathbf{c}_s)_j \doteq \frac{1}{4}A_jc_{sj}. \tag{3.2.51}$$

[25] This result is achieved by equating the two right members of (3.2.49), namely, $\boldsymbol{\rho}_0\mathbf{c}_c + \boldsymbol{\lambda}_0\mathbf{c}_s = \boldsymbol{\rho}_0\mathbf{c}_s - \boldsymbol{\lambda}_0\mathbf{c}_c$, with the consequence that $\mathbf{c}_c = (\boldsymbol{\rho}_0+\boldsymbol{\lambda}_0)^{-1}(\boldsymbol{\rho}_0-\boldsymbol{\lambda}_0)\mathbf{c}_s$, and so on in the second relation followed by solving for $\mathbf{c}_s$. Different, equivalent forms are obtained, for example, by starting with the first equation or the second relation, in (3.2.49).

From (3.2.51) and (3.2.44) above, we have finally the first of the desired general results in terms of the slowly varying components A_j, c_{cj}, and c_{sj}:

$$\begin{aligned}\Psi^*_{s-\text{inc}}\big|_{\text{nb}} &= \frac{1}{4}\sum_{jj'}\{A_j c_{cj}\cos\Phi_j\cos\Phi_{j'} + A_j c_{sj}\sin\Phi_j\sin\Phi_{j'}\}\\ &= \frac{1}{2}\left(\tilde{\mathbf{a}}\mathbf{k}_N^{-1}\mathbf{a} + \tilde{\mathbf{b}}\mathbf{k}_N^{-1}\mathbf{b}\right) = \sum_{jj'}\left(\Psi^{(c)}_{s-jj'}\cos\Phi_j\cos\Phi_{j'} + \Psi^{(s)}_{jj'}\sin\Phi_j\sin\Phi_{j'}\right) \doteq \frac{1}{2}\left(\Psi^{(c)^*}_{s-\text{inc}} + \Psi^{(s)^*}_{s-\text{inc}}\right).\end{aligned} \tag{3.2.52}$$

3.2.3.2 Spectral Symmetry In the important and usual cases of spectral symmetry, where $\boldsymbol{\lambda}_0 = \mathbf{0}$ (refer to Ref. [1], and Section 7.5.3), we see from (3.2.49) that

$$\mathbf{c}_c = 2\boldsymbol{\rho}_0^{-1}\mathbf{A} = \mathbf{c}_s \quad (\boldsymbol{\lambda}_0 = \mathbf{0}), \tag{3.2.53}$$

and consequently (3.2.52) simplifies to

$$\boldsymbol{\lambda}_0 = \mathbf{0} : \Psi^*_{s-\text{inc}}\big|_{\text{nb}} = \frac{1}{2}\mathbf{A}\boldsymbol{\rho}_0^{-1}\mathbf{A} = \frac{1}{2}\sum_{jj'} A_j A_{j'} (\boldsymbol{\rho}_0)^{-1}_{jj'} = \sum_{jj'} \frac{A^{(m)}_{0_n} A^{(m)}_{0_{n'}}}{2\psi_{m,n}} (\boldsymbol{\rho}_0)_{jj'}, \tag{3.2.54}$$

which is the "slowly varying" equivalent of (3.2.31), where again $A_{0_j} = A^{(m)}_{0_n}(\Delta\tau_j)$, and so on. Note that in the special case of independent noise field sampling, that is, $\boldsymbol{\rho}_0 = [\delta_{jj'}] = [\delta_{mm'}\delta_{nn'}]$, Eq. (3.2.54) reduces further to

$$\boldsymbol{\rho}_0 = [\delta_{jj'}] : \Psi^*_{s-\text{inc}} = \frac{1}{2}\sum_j A_j^2 = \frac{1}{2}\sum_j A^{(m)}_{0_n}(\Delta\tau_j)^2_{2\psi_j}; \quad \Delta\tau_j \doteq t_n - (\mathbf{k}_0 - \mathbf{k}_{0_R})\cdot\mathbf{r}_m/\omega_0, \tag{3.2.55}$$

which is precisely Eq. (3.2.32a) in the original narrowband or "rapidly varying" cases in Section 3.2.1. (Generally, ϕ_j/ω_0 is negligible vis-à-vis the rest of $\Delta\tau_j$.)

3.2.3.3 Evaluation of the Test Statistics $\Psi^*_{x-\text{inc}}$, $\Psi^{(c)}_{x-\text{inc}}$, $\Psi^{(s)}_{x-\text{inc}}$ Determining the slowly varying components of $\Psi^*_{x-\text{inc}}$ is a somewhat more involved process than for $\Psi^*_{s-\text{inc}}$ in Section 3.2.3.1. We still proceed as in Section 3.2.3.1, starting with $\left(\tilde{\mathbf{x}}\mathbf{k}_N^{-1}\mathbf{a}\right)^2$ and $\left(\tilde{\mathbf{x}}\mathbf{k}_N^{-1}\mathbf{b}\right)^2$ and again making the appropriate narrowband reductions of the various trigonometric terms, which remove the rapidly varying contributions as before. The result is found to be

$$\left(\Psi^{(c)}_{x-\text{inc}}\right)_{jj'} \doteq \frac{1}{4}\left(x_{cj}x_{cj'}c_{cj}c_{cj'}\right); \quad \left(\Psi^{(s)}_{x-\text{inc}}\right)_{jj'} \doteq \frac{1}{4}\left(x_{sj}x_{sj'}c_{sj}c_{sj'}\right), \tag{3.2.56}$$

which when inserted into (3.2.45) yields the desired "slowly varying" forms for $\Psi^*_{x-\text{inc}}$, along with the usual rapidly-varying components $[\sim \cos(\Phi_{j'} - \Phi_j)]$. We proceed with the reduction:

$$\Psi^*_{x-\text{inc}} \doteq \frac{1}{8}\sum_{jj'}\left\{(\tilde{\mathbf{x}}_c\mathbf{c}_c)_j(\tilde{\mathbf{x}}_c\mathbf{c}_c)_{j'} + (\tilde{\mathbf{x}}_s\mathbf{c}_s)_j(\tilde{\mathbf{x}}_s\mathbf{c}_s)_{j'}\right\}_{\text{inc}}\cos(\Phi_{j'}-\Phi_j)$$

$$\doteq \frac{1}{8}\sum_j\left[(\mathbf{x}_c\mathbf{c}_c)_j^2 + (\mathbf{x}_s\mathbf{c}_s)_j^2\right] = \frac{1}{2}\left(\Psi^{(c)^*}_{x-\text{inc}} + \Psi^{(s)^*}_{x-\text{inc}}\right);\quad \Psi^{(c),(s)^*}_{x-\text{inc}} = \sum_{jj'}\left(\Psi_x^{(c),(s)}\right)_{jj'}. \tag{3.2.57}$$

The general vectors $\mathbf{c}_c$, $\mathbf{c}_s$ are given as before by (3.2.50).

For the *symmetric cases* ($\boldsymbol{\lambda}_0 = \mathbf{0}$), (3.2.53) applies. Then, (3.2.57) reduces to the useful result for most applications:

$$(\boldsymbol{\lambda}_0 = \mathbf{0}) : \Psi^*_{x-\text{inc}} = \frac{1}{2}\sum_{jj'}\left\{(\tilde{\mathbf{x}}_c\boldsymbol{\rho}_0^{-1}\mathbf{A})^2_{jj'} + (\tilde{\mathbf{x}}_s\boldsymbol{\rho}_0^{-1}\mathbf{A})^2_{jj'}\right\}_{\text{inc}}\cos(\Phi_{j'}-\Phi_j) \doteq \frac{1}{2}\sum_j\left[(\tilde{x}_{cj}A_j)^2 + (\tilde{x}_{sj}A_j)^2\right]$$

$$= \frac{1}{2}\left(\Psi^{(c)^*}_{x-\text{inc}} + \Psi^{(c)^*}_{s-\text{inc}}\right) \tag{3.2.58}$$

which is just the equivalent form of (3.2.27) when $\mathbf{k}_N = \tilde{\mathbf{k}}_N$. When the noise samples are statistically independent (which for the Gaussian fields also implies strict stationarity and homogeneity), then $\boldsymbol{\rho}_0 = [\delta_{jj'}]$ and (3.2.58) reduces to[26]

$$\Psi^*_{x-\text{inc}} = \frac{1}{2}\sum_{jj'}\left\{(\tilde{\mathbf{x}}_c\mathbf{A}\tilde{\mathbf{A}}\mathbf{x}_c)_{jj'} + (\tilde{\mathbf{x}}_s\mathbf{A}\tilde{\mathbf{A}}\mathbf{x}_s)_{jj'}\right\}_{\text{inc}}\cos(\Phi_{j'}-\Phi_j) \doteq \frac{1}{2}\sum_j A_j^2\left(x_{cj}^2 + x_{sj}^2\right). \tag{3.2.58a}$$

Thus,

$$x_j x_{j'} A_j A_{j'} = \frac{1}{2}(x_{cj}x_{cj'} + x_{sj}x_{sj'})A_j A_{j'} \tag{3.2.58b}$$

in terms of the narrowband received data $x_{j,j'}$. The results (3.2.57) and (3.2.58) are again maximized when the receiving array is directed toward the signal source, that is, when $\Phi_{j'}-\Phi_j = \omega_0[\Delta t_{nn'} - (\mathbf{k}_0 - \mathbf{k}_{OR})\cdot(\mathbf{r}_{m'}-\mathbf{r}_m)] = \omega_0\Delta t_{nn'} - \Delta\phi_{jj'}\mathbf{k}_{OR} = \mathbf{k}_0$; and $\Delta\phi_{jj'} \equiv \phi_{j'} - \phi_j \doteq \phi_{j'}(\Delta t'_j) - \phi_j(\Delta t'_j) = \phi^{(m')}(t_{n'}) - \phi^{(m)}(t_n)$, refer to Eq. (3.2.28). The amplitude factors $A_j, A_{j'}$ likewise reduce to $A_0^{(m)}(t_n - \phi_j/\omega_0) \doteq A_0^{(m)}(t_n)$, since $|\phi_j/\omega_0| << t_n$, and so on for these slowly varying components. Similar remarks apply to x_{cj}, x_{sj}, namely, $x_{cj}(\Delta\tau'_j) \doteq x_{cj}(\Delta_j\tau) = x_{cj}(t_n)$ (see footnote 24).

[26] Note that the diagonal ($j' = j$) terms, refer to Eq. (3.2.58a), are $\sum_j\left(x_{cj}^2 + x_{sj}^2\right)A_j^2/2$, where $\left(x_{cj}^2 + x_{sj}^2\right)/2 = x_j^2$, this last from the diagonal terms of $\mathbf{x}\tilde{\mathbf{x}}$. If $A_j = A_0/\sqrt{\psi}$, a constant all j, then

$$\text{diag}\,\Psi_{x-\text{inc}} = \frac{A_0^2}{\psi}\sum\left(x_{cj}^2 + x_{sj}^2\right)/2 = \frac{A_0^2}{\psi}\sum_j x_j^2, \tag{i}$$

$$\therefore \langle\text{diag}\,\Psi_{x-\text{inc}}\rangle_x = \frac{A_0^2}{\psi}\sum_j\overline{x_j^2} = \frac{A_0^2}{\psi}J,\quad \text{since}\quad \overline{x_j^2} = 1 = \left(\overline{x_{cj}^2} + \overline{x_{sj}^2}\right)/2,\quad \overline{x_{cj}^2} = \overline{x_{sj}^2} = 1, \tag{ii}$$

as required for these normalized noise components.

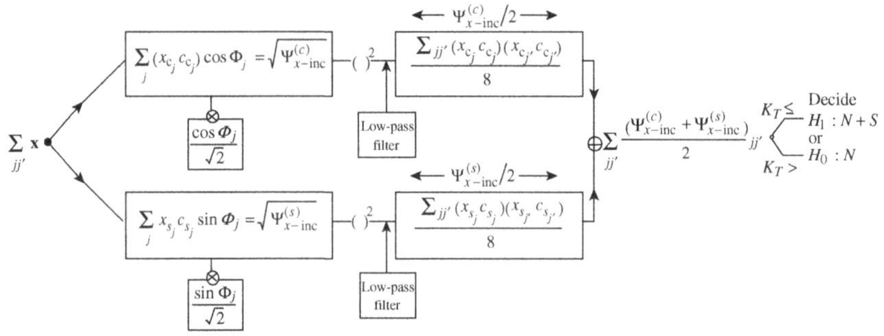

FIGURE 3.8 The slowly varying equivalent of (3.2.12), where $\Psi_{x-\text{inc}} \doteq (\Psi^{(c)^*}_{x-\text{inc}} \Psi^{(s)^*}_{x-\text{inc}})/2$ (see Figs. 3.4 and 3.7).

3.2.3.4 Performance Probabilities Figure 3.8 shows how the low-frequency components may be used directly in determining the performance probabilities. From (3.2.57), we have the following: Similarly, $\Psi^*_{s-\text{inc}} \doteq (1/2)\left(\Psi^{(c)^*}_{s-\text{inc}} + \Psi^{(s)^*}_{s-\text{inc}}\right)$ [(3.2.44) and (3.2.52). One simply substitutes the right-hand expression for $\Psi^*_{s-\text{inc}}$, and for $\Psi_{x-\text{inc}} \doteq \left(\Psi^{(c)^*}_{x-\text{inc}} + \Psi^{(s)^*}_{x-\text{inc}}\right)/2$.

3.3 OPTIMAL DETECTION III: SLOWLY FLUCTUATING NOISE BACKGROUNDS

In Section 3.2, we presented an example of the (Bayes) optimum detection of slowly fading narrowband signals, namely, slow Rayleigh fading, refer to Eq. (3.2.35) in Gaussian noise, a typical situation in practice.[27] Here, we consider a different class of fading phenomena, in which it is the accompanying (additive) noise background that exhibits slow fading over the observation interval (0, *T*), that is, negligible fading during any one interval (0, *T*), but fading from interval to interval. The fading mechanism in this instance is usually a form of unresolvable multipath, from a variety of interfering sources that slowly and irregularly reinforce or cancel each other or from changes in the geometry of scattering surfaces vis-à-vis the receiver from interval to interval, which can be common events in radar and sonar applications.

3.3.1 Coherent Detection

For such fluctuating Gaussian (and often non-Gaussian) noise fields, an excellent statistical description is given by the gamma (Γ-) pdf of the fluctuating noise intensity ψ:

$$w_1(\psi|\beta)_\Gamma = \frac{(\psi/a)^\beta}{a\Gamma(\beta+1)} e^{-\psi/a}, \beta > -1; \quad = 0, \ \psi < 0, \quad \text{with } a = \bar{\psi}/(\beta+1). \quad (3.3.1)$$

For the present, we also assume that the field intensity ψ is locally wide sense homogeneous and stationary, that is, $\psi_j = \psi_{mn} = \psi$ over any chosen finite interval (0, *T*). A detailed

[27] For further development of fading analysis and results, see the references, especially [6], [7].

discussion, and justification on physical grounds, of the Γ pdf (3.3.1) can be given. It can also be shown that $\beta + 1$ is usually numerically small Ref. [9] and Appendix [$<$O(10)] in many applications. However, at the other extreme of many essentially independent fluctuation cells in (0, T), that is, $\beta \to \infty$, one has

$$\lim_{\beta\to\infty} w_1(\psi|\beta)_\Gamma = \lim_{\beta\to\infty}\left[\frac{e^{-(\psi-\bar{\psi})^2/2\bar{\psi}^2(\beta+1)}}{\left[2\pi\bar{\psi}^2(\beta+1)\right]^{1/2}}\right]_{\beta>>1} = \delta(\psi-\bar{\psi}). \tag{3.3.1a}$$

This in effect replaces a possible representative value ψ of noise intensity in (0, T) by its ensemble average $\bar{\psi}$. [The result when applied to Gauss or Rayleigh pdf is just the original Gauss or Rayleigh pdf, now with ($\psi = \bar{\psi}$).]

Here, as in Section 3.2.2, our effort is directed to obtaining the pdf values of the (log-) likelihood ratio $\log \Lambda(\mathbf{x})$ under H_0 and H_1, rather than limiting our attention to the pdf values of the noise and signal and noise fields themselves individually, as is the aim in Chapter 10. The analytical procedures are quite different: the present approach not only requires (1) a derivation of different pdf values subject to (3.3.1) but also (2) determining the monotonic nature of H_1 and H_0 pdf values as respective test statistics, for all $\beta > -1$. As a result of (2), it turns out that it is greatly possible to simplify the final form for a still statistically sufficient test statistic and to use the earlier results of Sections 3.1.1 and 3.2.2 to obtain the needed performance probabilities $\alpha_F^*|_\Gamma$ and $p_D^*|_\Gamma = 1-\beta_\Gamma^*$. This, in turn, enables us to obtain optimum test statistics and performance for the important class of K-noise distributions [9–12], which commonly occur in radar and sonar applications.

Accordingly, we begin first with (fully) coherent reception, followed by the familiar narrowband incoherent case.

3.3.1.1 The Detection Algorithm Since the background noise intensity ψ is fluctuating *slowly* from interval to interval, we must "denormalize" Eq. (3.1.1) to exhibit ψ explicitly. The result is

$$w_J(\mathbf{X}-\mathbf{S}) = \frac{\exp[-Q(\mathbf{X}-\mathbf{S})/2\psi]}{(2\pi\psi)^{J/2}(\det \mathbf{k}_\mathrm{N})^{1/2}}, \quad \text{with } Q(\mathbf{X}-\mathbf{S}) \equiv Q_1 = (\tilde{\mathbf{X}}-\tilde{\mathbf{S}})\mathbf{k}_\mathrm{N}^{-1}(\mathbf{X}-\mathbf{S}) > 0, \tag{3.3.2a}$$

and[28]

$$\therefore w_J(\mathbf{X}) = \frac{\exp[-Q(\mathbf{X})/2\psi]}{(2\pi\psi)^{J/2}(\det \mathbf{k}_\mathrm{N})^{1/2}}; \quad Q(\mathbf{X}) \equiv Q_0 = \tilde{\mathbf{X}}\mathbf{k}_\mathrm{N}^{-1}\mathbf{X} > 0, \tag{3.3.2b}$$

so that the test statistic is explicitly

$$\log \mathbf{\Lambda}(\mathbf{X})_{\mathrm{coh}} = \log\mu + \log\left\langle \left[2\pi\psi(\det \mathbf{k}_\mathrm{N})^{1/J}\right]^{-J/2} e^{-Q_1/2\psi}\right\rangle_\psi - \log\left\langle \left[2\pi\psi(\det \mathbf{k}_\mathrm{N})^{1/J}\right]^{-J/2} e^{-Q_0/2\psi}\right\rangle_\psi \tag{3.3.3}$$

[28] The quantity Q used throughout Section 3.3.5 is not to be confused with the Q-function $Q(\alpha,\beta)$, (3.2.20) et seq.

where the average $\langle\,\rangle_{\psi}$ is to be carried out using (3.3.1).[29] With the help of the relation (Ref. [4], 4th ed., pp. 340–349)

$$I_1 \equiv \int_0^{\infty} y^{\alpha} e^{-y-B/y} dy = 2B^{(\alpha+1)/2} K_{|\alpha+1|}\left(2\sqrt{B}\right) = 2^{-\alpha}\left(2\sqrt{B}\right)^{\alpha+1} K_{|\alpha+1|}\left(2\sqrt{B}\right) \tag{3.3.4}$$

where $K_{|\alpha+1|}\left(2\sqrt{B}\right)$, $B > 0$, is a modified Bessel function of the second kind,[30] order $(\alpha + 1)$, now applied to (3.3.3). The result is

$$\log \mathbf{\Lambda}_{\text{coh}} = \log\{\mu F_1(\mathbf{X}, \mathbf{S})/F_0(\mathbf{X}, 0)\}, \tag{3.3.5}$$

where specifically

$$F_1(\mathbf{X}, \mathbf{S}) = C_{J/2} Z_1^{\beta-J/2+1} K_{|\beta-J/2+1|}(Z_1); \quad Z_1 \equiv \sqrt{2(\beta+1)Q_1/\bar{\psi}}, \quad \beta > -1 \tag{3.3.5a}$$

$$\left.\begin{aligned} F_0(\mathbf{X}, \mathbf{0}) &= C_{J/2} Z_0^{\beta-J/2+1} K_{|\beta-J/2+1|}(Z_0); \quad Z_0 \equiv \sqrt{2(\beta+1)Q_0/\bar{\psi}} \\ C_{J/2} &= \left(\frac{\beta+1}{4\pi\bar{\psi}}\right)^{J/2} \left\{2^{\beta}\Gamma(\beta+1)(\det \mathbf{k}_{\mathrm{N}})^{1/2}\right\}^{-1} \end{aligned}\right\}. \tag{3.3.5b}$$

We note again that since Q_0 and Q_1 are positive definite (Z_0, $Z_1 > 0$), F_1, F_0 are real. Moreover, we also realize that $F(\mathbf{X}, \mathbf{0})$ and $F(\mathbf{X}, \mathbf{S})$, (3.3.5a) and (3.3.5b), are the (J-dimensional) *amplitude* pdf values, (3.3.2a) and (3.3.2b), of the well-known K-distribution of importance in radar and sonar [9–12]. Accordingly, the decision process is

$$\left.\begin{aligned} &\text{Decide } H_1\text{: } S+N \text{ if:} \\ &\text{Decide } H_0\text{: } N \text{ if:} \end{aligned}\right\} \log \mathbf{\Lambda}_{\text{coh}} = \log\left\{\mu Z_1^{\gamma} K_{|\gamma|}(Z_1)/Z_0^{\gamma} K_{|\gamma|}(Z_0)\right\} \begin{matrix} \geq \\ < \end{matrix} \Big\} \log K_{T-\text{coh}}, \tag{3.3.6}$$

where $\gamma = \beta - J/2 + 1$, $\beta > -1$, is real. A further condition on γ is that $J > 2(\beta + 1)$, that is $\gamma < 0$, which is needed to ensure that both numerator and denominator of $\log \mathbf{\Lambda}_{\text{coh}}$ are monotonic in their arguments.

To reduce (3.3.6) to a more manageable form, we employ the monotonic nature[31] of Q_0, Q_1, and so of $Q_0 - Q_1$ in the received data $\mathbf{X}$ and use once again (Section 3.2.1 and Eqs. (3.2.10) and (3.2.12) et seq.) the fact that Q_0 and Q_1 are *sufficient statistics*. The monotonic character of Q_0 and Q_1 may be established as follows: First, we observe that when $-\gamma = n + 1/2\ (> 0)$, refer to (12), p. 80 of Ref. [13],

[29] Here, as in most of the preceding examples, it is assumed that the signal, if present, is *a priori* known at the receiver.

[30] See Section 3.61 of Ref. [13] Eqs. (5) and (6) in conjunction with (8), p. 78 therein.

[31] Q_0, Q_1 are each monotonic in $\mathbf{X}$, being symmetric positive definite forms. The difference $Q_0 - Q_1$ here is likewise monotonic in $\mathbf{X}$, for which the n. + s. condition is seen to be $\sum_i x_i(s_i \lambda_i) > 0$ (or < 0), where $\lambda_i > 0$ since k_{β}^{-1} is real symmetric and the quadratic forms Q_0, Q_1 are positive definite and s_i can be specified positive *a priori*.

$$Z^{-(n+1/2)}K_{n+1/2}(Z) = Z^{-n}e^{-Z}\sqrt{\frac{\pi}{2}}\sum_{l=0}^{n}\frac{(n+l)!}{(n-l)!l!}(2Z)^{-l}, \quad Z > 0 \tag{3.3.6a}$$

is monotonically decreasing, since $(d/dZ)\sum_l(\,)_l < 0$ for all $Z \geq 0$. Moreover, $K_{|\gamma|}$ is a continuous exponentially decreasing function for all $|\gamma|$, with magnitudes increasing with the order $|\gamma|$. The simple candidates here are suggested by the structure itself of Z_0 and Z_1, in $K_{|\gamma|}$. Thus, the numerator and denominator of (3.3.6) are each monotonically decreasing functions of Z_0, Z_1 as they increase. This suggests that the simplest candidate for monotonic substitution of the original likelihood ratio (3.3.5) is given by Z_0, Z_1 directly. Accordingly, our choice of test statistic is the equivalent sufficient statistic[32]

$$\log \boldsymbol{\Lambda}'_{\text{coh}} = Z_0^2 - Z_1^2 = \frac{4(\beta+1)}{\bar{\psi}}\left(\tilde{\mathbf{X}}\mathbf{k}_N^{-1}\mathbf{S} - \frac{1}{2}\tilde{\mathbf{S}}\mathbf{k}_N^{-1}\mathbf{S}\right), \tag{3.3.6b}$$

refer to Eqs. (3.3.2a) and (3.3.2b), which is clearly monotonic in $\mathbf{X}$, $\tilde{\mathbf{S}}\mathbf{k}_N^{-1}\mathbf{S}$ being prespecified.[33] The optimal decision process (3.3.6) becomes the still simpler equivalent, where $\boldsymbol{\Lambda}'_{\text{coh}}$ is replaced by $\boldsymbol{\Lambda}^*_{T-\text{coh}}$, that is, the common factor $4(\beta+1)/\bar{\psi}$ on the right hand side of (3.3.6b) is eliminated, $\log \boldsymbol{\Lambda}^*_{T-\text{coh}} = [4(\beta+1)/\bar{\psi}]^{-1}$. Thus, this $\log \boldsymbol{\Lambda}'_{\text{coh}}$ decision process is now represented by

$$\text{Decide}: \left\{\begin{matrix} H_1 \\ \text{or} \\ H_0 \end{matrix}\right\} : Z^*_{\text{coh}} \equiv \log \boldsymbol{\Lambda}^*_{T-\text{coh}}\left\{\begin{matrix} \geq \\ < \end{matrix}\right\}\log K^*_{T-\text{coh}}, \text{ or } \bar{\Psi}^*_{xs-\text{coh}}\left\{\begin{matrix} \geq \\ < \end{matrix}\right\}\log\left(\frac{K^*_{T-\text{coh}}}{\mu}\right) + \frac{1}{2}\bar{\Psi}^*_{s-\text{coh}}, \tag{3.3.7a}$$

refer to Eq. (3.1.5b), where in detail we define

$$\bar{\Psi}^*_{xs-\text{coh}} \equiv \tilde{\mathbf{X}}\mathbf{k}_N^{-1}\mathbf{S}/\bar{\psi} = \left(\tilde{\mathbf{x}}\mathbf{k}_N^{-1}\mathbf{a_0 s}\right)\bar{\psi}; \quad \bar{\Psi}^*_{s-\text{coh}} \equiv \tilde{\mathbf{S}}\mathbf{k}_N^{-1}\mathbf{S}/\bar{\psi} = \left(\mathbf{a}_0\tilde{\mathbf{s}}\mathbf{k}_N^{-1}\mathbf{a_0 s}\right)\bar{\psi} \equiv \overline{\sigma_0^{2,}}^*_{-\text{coh}}. \tag{3.3.7b}$$

It is relevant to note once more that the equivalent threshold $K^*_{T-\text{coh}}$ in (3.3.7b) depends on $\bar{\Psi}^*_{s-\text{coh}}$, which is also the *detection parameter* $\overline{\sigma_0^{2,}}^*_{-\text{coh}}$, refer to Eq. (3.1.9a). In addition, $K^*_{T-\text{coh}}$ also depends on $\bar{\Psi}^*_{s-\text{coh}}$, but logarithmically here, so that (3.3.7a) is obeyed and $\bar{\Psi}^*_{s-\text{coh}}/2$ is then the dominant term as $\bar{\Psi}^*_{s-\text{coh}} \to \infty$.

Finally, as expected, $\bar{\Psi}^*_{xs-\text{coh}}$ represents the output of a (Wiener–Kolmogoroff) *matched filter* refer to Section 3.4.4 ff., for

$$\bar{\Psi}_{xs-\text{coh}} = \tilde{\mathbf{x}}\left(\mathbf{k}_N^{-1}\mathbf{S}/\bar{\psi}\right) = \left(\tilde{\mathbf{x}}\mathbf{k}_N^{-1}\mathbf{s}\right)_{\psi} = \left(\tilde{\mathbf{x}}\mathbf{W}\right)_{\psi} \tag{3.3.7c}$$

where $(\mathbf{H})_{\bar{\psi}} \equiv (\mathbf{W})_{\bar{\psi}} \equiv \mathbf{k}_N^{-1}\mathbf{S}/\bar{\psi}^{1/2}$ and, therefore, $\mathbf{k}_N(\mathbf{W})_{\bar{\psi}} = \mathbf{S}/\bar{\psi}^{1/2}$ in the observation space (and zero elsewhere).

[32] Q_0, Q_1 are each monotonic in $\mathbf{X}$, being symmetric positive definite forms. The difference of their squares here is likewise monotonic in $\mathbf{X}$, provided $\sum_i x_i(s_i\lambda_i) > 0$, refer to footnote 31.

[33] Alternatively, $\mathbf{k}_N^{-1}\mathbf{S} \sim \mathbf{k}^{-1}\mathbf{s} = \mathbf{W}$, where $\mathbf{W}$ is a Wiener-Kolmogoroff filter (i.e. a Bayes matched filter (Type 1), cf. Section 3.4.4 ff., which maximizes $(S/N)^2_{\text{out}}$, incidentally exemplifying the monotonicity of (3.3.6b) in $\mathbf{X}$.

3.3.1.2 Coherent Detector Performance We see that for (3.3.7c) we may appropriately modify our previous results (3.1.8)–(3.1.11). Thus, the pdf values of $\bar{\Psi}^*_{s-\text{coh}}$ are given here by (3.1.9) where $\sigma^{*2}_{0-\text{coh}}$ is replaced by $\bar{\sigma}^{*2}_{0-\text{coh}}$, (3.3.7b). The desired performance probabilities follow directly from (3.1.10) and (3.1.11) with the obvious modifications engendered by the comparisons of $K^*_{T-\text{coh}}$, (3.3.7a) and (3.1.5b). The results are as follows:

$$\left.\begin{matrix}\alpha^*_{\text{F-coh}}\\ \beta^*_{\text{F-coh}}\end{matrix}\right\} = \frac{1}{2}\left\{1 - \textcircled{H}\left[\frac{\sqrt{\bar{\sigma}^{2*}_{0-\text{coh}}}}{2\sqrt{2}} \pm \frac{\log\left(K^*_{T-\text{coh}}/\mu\right)}{\sqrt{2}\sqrt{\bar{\sigma}^{2*}_{0-\text{coh}}}}\right]\right\} \tag{3.3.8a}$$

$$\begin{aligned} p^*_{D-\text{coh}} = 1-\beta^*_{\text{coh}} &= \frac{1}{2}\left\{1 + \textcircled{H}\left[\frac{\sqrt{\bar{\sigma}^{2*}_{0-\text{coh}}}}{2\sqrt{2}} - \frac{\log\left(K^*_{T-\text{coh}}/\mu\right)}{\sqrt{2}\left(\bar{\sigma}^{2,*}_{0-\text{coh}}\right)^{1/2}}\right]\right\} \\ &= \frac{1}{2}\left\{1 + \textcircled{H}\left[\frac{\sqrt{\bar{\sigma}^{2*}_{0-\text{coh}}}}{\sqrt{2}} - \textcircled{H}^{-1}\left(1-2\alpha^*_{\text{F-coh}}\right)\right]\right\}, \end{aligned} \tag{3.3.8b}$$

with

$$\log K^*_{T-\text{coh}} = \log\mu - \sqrt{\bar{\sigma}^{*2}_{0-\text{coh}}/2} + \sqrt{\bar{\sigma}^{*2}_{0-\text{coh}}}\sqrt{2}\,\textcircled{H}^{-1}\left(1-2\alpha^*_{\text{F-coh}}\right). \tag{3.3.8c}$$

The "ROC" diagrams of Figure 3.2 again apply now with K/μ replaced therein by $K^*_{T-\text{coh}}/\mu$ and $a_0\Phi_s^{1/2}\sqrt{2}$ by $\bar{\Phi}^*_{s-\text{coh}}$. Similarly, the results of Section 3.1.3, regarding the array processing embodied in $\bar{\sigma}^{*2}_{0-\text{coh}}$ apply as well, as do the formulations in Section 3.1.2.1 for the Neyman–Pearson and Ideal Observers.

3.3.2 Narrowband Incoherent Detection Algorithms

Here, the test statistic is modified with the help of (3.2.10), after the average over the uniformly distributed RF epoch (ε) to give[34]

$$\begin{aligned} \log \boldsymbol{\Lambda}_{\text{inc}} = \log\mu + \log&\left\langle \psi^{-J/2} e^{-\left(\Psi^*_{xx}+\Psi^*_{ss}\right)_{\text{inc}}/2\psi} I_0\left(\sqrt{\Psi^*_{xs-\text{inc}}/\psi}\right)\right\rangle_\psi \\ &-\log\left\langle \psi^{-J/2}\exp\left[-\Psi^*_{xx-\text{inc}}/2\psi\right]\right\rangle_\psi, \end{aligned} \tag{3.3.9}$$

which remains monotonic in the test statistic $\Psi^*_{xs-\text{inc}}$. It is defined here by

$$\Psi^*_{xs-\text{inc}} \equiv \left(\tilde{\mathbf{X}}\mathbf{k}_N^{-1}\mathbf{A}\right)^2 + \left(\tilde{\mathbf{X}}\mathbf{k}_N^{-1}\mathbf{B}\right)^2 = \tilde{\mathbf{X}}\mathbf{k}_N^{-1}\left(\mathbf{A}\tilde{\mathbf{A}} + \mathbf{B}\tilde{\mathbf{B}}\right)\mathbf{k}_N^{-1}\mathbf{X}. \tag{3.3.9a}$$

[34] Since $\langle\,\rangle_\varepsilon$ is uniform over an RF cycle, refer to (3.2.8a) et seq., it may equally well be regarded as being uniform over all (integral) RF cycles $(0\to\infty)$, which still preserves the value and monotonicity of the resulting average (3.3.9). One obtains the same result for $\langle\,\rangle_{\varepsilon,\psi}$ as for $\langle\,\rangle_{\psi,\varepsilon}$, where the monotonicity of I_0 is again preserved.

Similarly, we define

$$\left.\begin{aligned} &\Psi^*_{\text{ss-inc}} \equiv \left(\tilde{\mathbf{A}}\mathbf{k}_{\text{N}}^{-1}\mathbf{A}+\tilde{\mathbf{B}}\mathbf{k}_{\text{N}}^{-1}\mathbf{B}\right)/2, \quad \left.\begin{aligned}\text{with } \mathbf{A} &\equiv \left[\sqrt{\psi}A_j\cos\Phi_j\right]\\ \mathbf{B} &\equiv \left[\sqrt{\psi}A_j\sin\Phi_j\right]\end{aligned}\right\}\text{Eq.}(3.2.5)\\ &\Psi^*_{xx\text{-inc}} \equiv \tilde{\mathbf{X}}\mathbf{k}_{\text{N}}^{-1}\mathbf{X}, \quad \text{and } \Phi_j \equiv \Delta t_j - \phi^{(m)}(t_n), \quad A_j = A_{0_n}^{(m)}(t_n), \text{Eqs. (3.2.2a) and (3.2.3)} \end{aligned}\right\}, \tag{3.3.9b}$$

and in summary again we have

$$\begin{aligned} \Delta t_j &\equiv t_n - \mathbf{k}_0\cdot\mathbf{r}_m/\omega_0; \quad \Delta t'_j \equiv t_n - \mathbf{k}_0\cdot\mathbf{r}_m/\omega_0 - \phi_j/\omega_0\\ \Delta\tau_j &\equiv t_n - (\mathbf{k}_0 - \mathbf{k}_{0_{\text{R}}})\cdot\mathbf{r}_m/\omega_0; \quad \Delta\tau'_j \equiv t_n - (\mathbf{k}_0 - \mathbf{k}_{0_{\text{R}}})\cdot\mathbf{r}_m/\omega_0 - \phi_j/\omega_0, \end{aligned} \tag{3.3.9c}$$

these last relations including a steering vector $\mathbf{k}_{0_{\text{R}}}$, refer to Eq. (3.2.28a) et seq. The matched filters here are now given by the W–K filters $\mathbf{H}^{(a)} = \mathbf{k}_{\text{N}}^{-1}\mathbf{A}$, $\mathbf{H}^{(b)} = \mathbf{k}_{\text{N}}^{-1}\mathbf{B}$, (Section 3.4.4) with

$$\left(\bar{\Psi}^*_{xs\text{-inc}}\right)^{1/2} = \left[\left(\tilde{\mathbf{X}}\mathbf{H}^{(a)}\right)^2 + \left(\tilde{\mathbf{X}}\mathbf{H}^{(b)}\right)^2\right]^{1/2}. \tag{3.3.9d}$$

Since I_0 is monotonic in its argument, the form of the second term of (3.3.9) suggests a simpler monotonic function, namely, $I_0 \to \exp(\Psi_{\text{ss}}/2\psi)$. Then, carrying out the averages over ψ in (3.3.9) with the help of (3.3.1) and (3.3.4), and following the procedures of (3.3.5)–(3.3.6b), we obtain finally the result:

$$\text{Decide}: \left\{\begin{matrix} H_1\\ \text{or}\\ H_0\end{matrix}\right\} : Y^*_{\text{inc}} \equiv \log\mathbf{\Lambda}^*_{T\text{-inc}} \left\{\begin{matrix}\geq\\ <\end{matrix}\right\} \log K^*_{T\text{-inc}} \quad\text{or}\quad \bar{\Psi}^*_{xs\text{-inc}} \left\{\begin{matrix}\geq\\ <\end{matrix}\right\} \log\left(K^*_{T\text{-inc}}/\mu\right) + \bar{\Psi}^*_{\text{ss}} \tag{3.3.10}$$

where $Y^*_{\text{inc}} = Y^*_0 - Y^*_1$ and

$$\begin{gathered} Y^*_0 = \left(\bar{\Psi}^*_{xx}\right)_{\text{inc}}; \quad Y^*_1 = \left(\bar{\Psi}^*_{xx} + \bar{\Psi}_{\text{ss}} - \bar{\Psi}^*_{xs}\right)_{\text{inc}}, \quad \text{with } \bar{\Psi}^*_{xs\text{-inc}} = \left\langle \bar{\Psi}^*_{xx\text{-inc}}\right\rangle_{\psi}\\ \text{and } \bar{\Psi}^*_{\text{ss-inc}} = \left\langle\bar{\Psi}_{\text{ss-inc}}\right\rangle = \left\langle\overline{\sigma^{*2}_{0\text{-inc}}}\right\rangle \end{gathered} \tag{3.3.11}$$

(with the averages of the test statistic obtained from (3.3.9a) on dividing by $\bar{\psi}$). This result is similar to (3.3.7a) and (3.3.7b), except that the factor 2 is missing in $\bar{\Psi}^*_{\text{ss-inc}}$, although the *form* of $\bar{\Psi}^*_{\text{s-inc}}$ is similar to $\bar{\Psi}^*_{\text{s-coh}}$, and $\bar{\Psi}^*_{xs\text{-inc}}$ is quite different from $\bar{\Psi}^*_{xs\text{-coh}}$, refer to Eq. (3.3.9a) versus (3.3.7b).

3.3.2.1 Incoherent Detector Performance We note that the test statistic is now formally $\bar{\Psi}^*_{xs\text{-inc}}$. Accordingly, we can use the previously determined performance probabilities

(3.2.19a) and (3.2.19b), with $\Psi^*_{s-\text{inc}}$ replaced by $\bar{\Psi}^*_{ss-\text{inc}}$ and $K_{T-\text{inc}}$ by $K^*_{T-\text{inc}}$. The explicit results are

$$\alpha^*_{\text{F}} = e^{-K^*_{T-\text{inc}}/2\bar{\Psi}^*_{ss-\text{inc}}} \quad 1-\beta^* = Q\left(\sqrt{\bar{\Psi}^*_{ss-\text{inc}}}, \sqrt{-2\log\alpha^*_{\text{F}}}\right) \tag{3.3.12}$$

which also gives the false rejection probability β^*. When $\bar{\Psi}^*_{ss-\text{inc}} \to \infty$, $K^*_{T-\text{inc}} \to \bar{\Psi}^*_{ss-\text{inc}}/2 \to \infty$, and thus $p^*_{\text{D}} = Q(\alpha, c\log\alpha) \to 1$; $0 < c < 1$, as $\alpha \to \infty$ (3.2.20a). Also, $\alpha^*_{\text{F}} \to 0$ as required, since $\left(K^*_{T-\text{inc}}\right)^2/b\bar{\Psi}^{**}_{ss-\text{inc}}/2 \to \infty$ in (3.3.12). Figures 3.6 and 3.8 are appropriate here with the relevant changes of parameters, that is, $\Psi^*_{s-\text{inc}} \to \bar{\Psi}^*_{ss-\text{inc}}$; α^*_{F}, β^* are now represented by (3.3.12). Finally, observe that in the generic narrowband incoherent formulations of Section 3.2.2, the relevant threshold $K_{T-\text{inc}}$ appears as the square root for β^* in the Q-function (3.2.19b) et seq. and in the first power in the exponent of α^*_{F}, (3.2.19a).

3.3.2.2 Neyman–Pearson (i.e., CFAR) and Ideal Observers Here (3.3.11) represents the decision process when $\bar{\Psi}^*_{xs-\text{inc}}$ is the test statistic, with threshold $K^*_{T-\text{inc}}\left(\bar{\Psi}^*_{ss-\text{inc}}\right)$, for Neyman-Pearson detectors for which α^*_{F} (3.3.12) is *a priori* chosen (refer to remarks preceding Section 3.2.1.3). Again, Figs. 3.6 and 3.8, with suitably modified parameters, are representative of performance.

The *Ideal Observer* (Section 1.8.2) requires the joint minimization of $q\alpha_I + p\beta_I \to q\alpha^*_I + p\beta^*_I$. The result for these incoherent cases is from (3.3.12) the signal (symbol) error probability:

$$P^*_{e-\text{inc}} = qe^{-K^*_{I-\text{inc}}/2\bar{\Psi}^*_{ss-inc}} + p\left[1 - Q\left(\sqrt{\bar{\Psi}^*_{ss-\text{inc}}}, \sqrt{-2\log\alpha^*_{\text{F}}}\right)\right]. \tag{3.3.13}$$

The threshold $K^*_{I-\text{inc}}$ in Eq. (3.12) is most easily found by choosing *a priori* an acceptable symbol error probability $P^*_{e-\text{inc}}$, given the maximum value of $\bar{\Psi}^*_{ss-\text{inc}}$ available. With a symmetrical channel, that is, $p = q = 1/2$, and $\bar{\Psi}^*_{ss-\text{inc}} \to \infty$, $Q \to 1$ (refer to remarks subsequent to (3.3.12) and footnote 34), we have $\left(P^*_{e-\text{inc}}\right)_{\text{sym}} = 0$ as well as $P^*_{e-\text{inc}}|_{p\neq 1/2} = 0$, as expected. With sufficiently strong signals vis-à-vis the accompanying (Gaussian) noise the symbol error probability vanishes. On the other hand, when $\bar{\Psi}^*_{ss-\text{inc}} \to 0$, $P^*_{e-\text{inc}} = p$, since[35] $Q(0,\infty) = 0$: all decisions are in error by an amount p. In general, the decision process here is once more the following (3.2.25):

$$\text{IO}: \quad \text{or} \quad \left.\begin{array}{ll} \text{Decide } H_1 : S+N, & \text{if } \left(\bar{\Psi}^*_{sx-\text{inc}}\right) \geq 1 \\ \text{Decide } H_0 : N, & \text{if } \left(\bar{\Psi}^*_{sx-\text{inc}}\right) < 1 \end{array}\right\}, \tag{3.3.14}$$

with P^{**}_{e} given by (3.3.13). (The series form of Q, [Eqs. (3.2.22a) and (3.2.22b)], may also be used here to calculate $P^*_{e-\text{inc}}$ from (3.3.13).).

[35] Also see Section 3.2.1.

3.3.3 Incoherent Detection of Broadband Signals in Normal Noise

In Sections 3.2.1–3.2.3, it was shown that optimal narrowband incoherent detection has closed form solutions under a variety of *a priori* conditions. However, for broadband signals[36] in Gaussian noise, the incoherent case presents difficulties. This is mainly because the (time) epoch is no longer over an RF cycle, but may extend over a sizable function of the signal's duration, because the explicit form of the signal $\left\{\hat{s}_n^{(m)}\right\}$ does not separate into slowly varying and rapidly varying components, such as $A_0^{(m)}(t_n)\sqrt{2}\cos\left[\omega_0(t-\varepsilon)-\phi_n^{(m)}\right]$, refer to Eq. (3.2.3). This makes the evaluation of the generic form of the likelihood ratio impossible to evaluate in any useful exact form:

$$\log \boldsymbol{\Lambda}(\mathbf{x}) = \mu\left\langle \exp\left[-\frac{1}{2}\widetilde{\mathbf{as}}\,\mathbf{k}_{\mathrm{N}}^{-1}\mathbf{as} + \widetilde{\mathbf{as}}\,\mathbf{k}_{\mathrm{N}}^{-1}\mathbf{x}\right]\right\rangle_{\varepsilon,\mathbf{s}} = \mu\left\langle \exp\left[-\frac{1}{2}\Psi^{*}_{\mathrm{s-inc}} + \Psi^{*}_{x-\mathrm{inc}}\right]\right\rangle_{\varepsilon,\mathbf{s}}. \tag{3.3.15}$$

However, in the threshold signal cases, a complete canonical theory is possible asymptotically, when the effective number of independent samples is large. This, in turn, can serve as a starting point for a class of generally suboptimum approximation, $\log G(\mathbf{x}) = g(\mathbf{x})$, at stronger signals, levels that are the canonical forms for the threshold signals. In Section 3.3.3.1, below we describe how this is accomplished.

3.3.3.1 Optimum Threshold Detection in Gaussian Noise[37] These structures, extended from Sections 20.1.1 and 20.3.1 and (2) of Ref. [1] to include spatial processing, are found with some obvious modifications to be

$$\log \boldsymbol{\Lambda}(x) \doteq g(\mathbf{x})^{*}_{\mathrm{inc}} = \log \mu + \left[B^{(2)} + B^{(4)}\right] + \bar{\Psi}^{*}_{x-\mathrm{inc}}(\mathbf{x}), \quad \bar{\mathbf{x}} = 0, \tag{3.3.16}$$

where $\left[B^{(2)} + B^{(4)}\right]$ are "bias" terms, necessary to ensure the local asymptotic normality (LAN) of the test statistic $\bar{\Psi}^{*}_{x-\mathrm{inc}}$. Specifically, refer to Eqs. (20.11a)–(20.11c) of Ref. [1], the components of $g(\mathbf{x})^{*}$ are as follows, since $\mathbf{k}_{\mathrm{s}} = \tilde{\mathbf{k}}_{\mathrm{s}}$; $\mathbf{k}_{\mathrm{N}} = \tilde{\mathbf{k}}_{\mathrm{N}}$, and where $\mathbf{J} = \mathbf{J}(\mathbf{S})$:

$$\left.\begin{aligned} B^{(2)} &\equiv -\frac{1}{2}\overline{\tilde{\hat{\mathbf{s}}}\mathbf{k}_{\mathrm{N}}^{-1}\hat{\mathbf{s}}} = -\frac{1}{2}\operatorname{trace}\mathbf{k}_{\mathrm{N}}\mathbf{J} = -\frac{1}{2}\operatorname{trace}\mathbf{k}_{\hat{s}}\mathbf{k}_{\mathrm{N}}^{-1}; \quad \mathbf{J} \equiv \mathbf{k}_{\mathrm{N}}^{-1}\left\langle\hat{\mathbf{s}}\tilde{\hat{\mathbf{s}}}\right\rangle\mathbf{k}_{\mathrm{N}}^{-1} = \mathbf{k}_{\mathrm{N}}^{-1}\mathbf{k}_{\hat{s}}\mathbf{k}_{\mathrm{N}}^{-1} \\ B^{(4)} &\equiv -\frac{1}{4}\operatorname{trace}(\mathbf{k}_{\mathrm{N}}\mathbf{J})^{2}; \quad \operatorname{trace}(\mathbf{k}_{\mathrm{N}}\mathbf{J})^{2} = \operatorname{trace}\left(\overline{\hat{\mathbf{s}}\tilde{\hat{\mathbf{s}}}\mathbf{k}_{\mathrm{N}}^{-1}}\right) = \operatorname{trace}\left(\mathbf{k}_{\hat{s}}\mathbf{k}_{\mathrm{N}}^{-1}\right)^{2} = \operatorname{trace}\left(\left\langle\tilde{\hat{\mathbf{s}}}\mathbf{k}_{\mathrm{N}}^{-1}\hat{\mathbf{s}}\right\rangle\right)^{2} \end{aligned}\right\} \tag{3.3.16a}$$

and

$$\bar{\Psi}^{*}_{x-\mathrm{inc}}(\mathbf{x}) = \tilde{\mathbf{x}}\left(\mathbf{k}_{\mathrm{N}}^{-1}\left\langle\hat{\mathbf{s}}\tilde{\hat{\mathbf{s}}}\right\rangle\mathbf{k}_{\mathrm{N}}^{-1}\right)\mathbf{x} = \tilde{\mathbf{x}}\mathbf{J}\mathbf{x}; \quad \mathbf{x} = \left[x_j\right]; \quad \hat{\mathbf{s}} = A_{on}^{(m)}(t_n)s^{(m)}(t_n)/\sqrt{2\psi_j} \tag{3.3.16b}$$

[36] These signals may be deterministic or random.

[37] The full threshold treatment of signal detection and extraction was to be given in Part 3—*Ed.*

refer to Eq. (3.1.3a), with $\mathbf{k}_{\hat{s}} = \overline{a_0^2}\mathbf{k}_S = \overline{a_0^2}\mathbf{k}_N^{-1}\mathbf{k}_S\mathbf{k}_N^{-1}$. In condensed form, (3.3.16) is finally

$$\log \Lambda(\mathbf{x}) \doteq g(\mathbf{x})^*_{\text{inc}} = \log \mu + A_J + \tilde{\mathbf{x}}\mathbf{J}\mathbf{x}; \quad A_J = -B^{(2)} - B^{(4)}; \quad \mathbf{x} = 0 \tag{3.3.16c}$$

This suggests for *detection of general broadband signals in normal noise*, which are threshold optimum (*and thus suboptimum in the stronger signal regimes*) that we can employ:

$$g(\mathbf{x})^{(*)}_{\text{inc}} = \log \mu + A_J + \tilde{\mathbf{x}}\mathbf{J}\mathbf{x}; \quad \text{and generally}, g(\mathbf{x}) = A'_J + \tilde{\mathbf{x}}\mathbf{J}_0\mathbf{x}, \quad \mathbf{J}_0 = \text{arbitrary}. \tag{3.3.17}$$

The quantity A'_J may be adjusted to provide at least *consistency* of detector performance, that is, $\lim_{J\to\infty} \alpha_F \to 0, \lim_{J\to\infty} p_D \to 1$. Accordingly, we shall briefly examine the threshold results for (3.3.17) and consider *en passant* the results for the incoherent prototype ($\sim \tilde{\mathbf{x}}\mathbf{J}_0\mathbf{x}$) at all signal levels.

Because the test statistic is also asymptotically normal for large samples (when they also contain a sufficient number of effectively independent samples), we can write the false alarm probability α_F^* and probability of detection $p_D^* = 1-\beta^*$ in the broadband (and narrowband) threshold régimes as

$$\alpha_F^* \cong \frac{1}{2}\left\{1 - Ⓗ\left[\frac{\left(\text{trace}\sqrt{\Psi^*_{\hat{\tilde{s}}-\text{inc}}}\right)^{1/2}}{4} + \frac{\log(K/\mu)}{\left(\text{trace}\sqrt{\Psi^*_{\hat{\tilde{s}}-\text{inc}}}\right)^{1/2}}\right]\right\}, \tag{3.3.18a}$$

where trace $\left(\mathbf{k}_{\hat{s}}\mathbf{k}_N^{-1}\right)^2 \equiv \Psi^*_{\hat{\tilde{s}}-\text{inc}} = \left\langle \tilde{\hat{\mathbf{s}}}\mathbf{k}_N^{-1}\mathbf{k}_{\hat{s}}\mathbf{k}_N^{-1}\hat{\mathbf{s}}\right\rangle = \left(\mathbf{k}_{\hat{s}}\mathbf{k}_N^{-1}\right)^2$, and

$$p_D^* = 1-\beta^* \cong \frac{1}{2}\left\{1 + Ⓗ\left[\frac{\left(\sqrt{\Psi^*_{\hat{\tilde{s}}-\text{inc}}}\right)^{1/2}}{4} - \frac{\log(K/\mu)}{\left(\sqrt{\Psi^*_{\hat{\tilde{s}}-\text{inc}}}\right)^{1/2}}\right]\right\}, \tag{3.3.18b}$$

refer to Eq. (20.9.1), Section 20.3.1.2, [1], where $\mathbf{k}_{\hat{s}} = \left[\left\langle A_{on}^{(m)}A_{on'}^{(m)}\hat{s}_n^{(m)}\hat{s}_n^{(m')}\right\rangle \big/ 2\sqrt{\psi_j\psi_{j'}}\right]$, and $\mathbf{k}_s = \left[\left\langle s_n^{(m)}s_{n'}^{(m')}\right\rangle\right], j, j' = mn, m'n'$, refer to Eq. (3.1.3a), and $\mathbf{k}_N = \left[\overline{x_jx_{j'}}\right]$, $\overline{x_j} = 0$ as usual. Not only is *consistency* obeyed, but also α_F^* and p_D^* approach these limits optimally: this is an example of a LAN detector Eq. (3.3.16) that is AO (asymptotically optimum). Note, also, that $\mathbf{k}_{\hat{s}}\mathbf{k}_N^{-1}$ is, in effect, a generalized *signal-to-noise* (power) *ratio*, spread out over the correlation history of the signal and the noise, namely, $\Sigma_j(\mathbf{k}_S)_{ij}\left(\mathbf{k}_N^{-1}\right)_{jk} = (S/N)^2_{ik}$, all i, k.

3.3.3.2 Incoherent Detection (All Signal Levels) Next, let us use (3.3.16c) at all broadband signal levels for an *incoherent* detector model. Although (3.3.16c) remains optimum for weak signals, it becomes suboptimum at stronger signals. The characteristic function for the first-order pdf under H_1 (and by inclusion, H_0) is given by Eq. (17.32b) [1], where the test statistic is given by $g(\mathbf{x})^* = A_J + \tilde{\mathbf{x}}\mathbf{J}\mathbf{x}$, $\mathbf{J} = \mathbf{k}_N^{-1}\overline{\mathbf{k}_{\hat{s}}\tilde{\hat{\mathbf{s}}}}\mathbf{k}_N^{-1}$, in keeping with the threshold optimal behavior of $g(\mathbf{x})^*$. The result is explicitly

$$F_1(i\xi|H_1) = \frac{\left\langle \exp\left\{ i\xi A_J - \frac{1}{2}\tilde{\hat{\mathbf{s}}}\mathbf{k}_N^{-1}\left[\mathbf{I}-(\mathbf{I}-2i\xi\mathbf{k}_N\mathbf{J})^{-1}\right]\hat{\mathbf{s}}\right\}\right\rangle_{\hat{\mathbf{s}}}}{[\det(\mathbf{I}-2i\xi\mathbf{k}_N\mathbf{J})]^{1/2}} \tag{3.3.19a}$$

with

$$F_1\left(i\xi|H_0\right) = [\det(\mathbf{I}-2i\xi\mathbf{k}_N\mathbf{J})]^{-1/2}, \quad \mathbf{k}_N\mathbf{J} = \mathbf{k}_{\hat{s}}\mathbf{k}_N^{-1}, \tag{3.3.19b}$$

obtained on setting $\hat{\mathbf{s}} = \mathbf{0}$ for the null hypothesis H_0. Here, $\mathbf{k}_N\mathbf{J} = \mathbf{k}_{\hat{s}}\mathbf{k}_N^{-1}$, from (3.3.16a), with $A_J = B^{(2)} + B^{(4)}$ specifically. (Note the average over $\hat{\mathbf{s}}$ in (3.3.19a), which also usually implies over the parameters of $\hat{\mathbf{s}}$.)

From Appendix A1, we may obtain explicit evaluations of the transforms of the cf values (3.3.19a) and (3.3.19b). As an example, let us assume that the eigenvalues of $\mathbf{k}_N\mathbf{J}$ are all positive and discrete, and are arranged in order of descending magnitude $\lambda_1^{(+)} > \lambda_2^{(+)} > \cdots > \lambda_k^{(+)} > \cdots > \lambda_{k=MN}^{(+)}$. (The matrix $\mathbf{k}_N\mathbf{J}$ is real and not necessarily symmetric.) The integrals in question are then the (first-order) pdf values of $g(\mathbf{x})^*$ obtained from the transformation $p = -i\xi$, $z = y - A_J = \tilde{\mathbf{x}}\mathbf{J}\mathbf{x}$ and are found to be

$$w_1(z|H_1) = \left\langle e^{-(1/2)\hat{\mathbf{s}}\mathbf{k}_N^{-1}\hat{\mathbf{s}}} \int_{-\infty:}^{\infty} e^{pz} \prod_{k=1}^{J} \frac{e^{(1/2)c_k(\hat{\mathbf{s}})^2/\left(1+a\lambda_k^{(+)}p\right)}}{\left(1+a\lambda_k^{(+)}p\right)^{1/2}} \frac{dp}{2\pi i}\right\rangle_{\hat{\mathbf{s}}}, \quad a = \text{real} > 0, \tag{3.3.20}$$

where $c_k(\hat{s})^2 = \sum_l \hat{s}_k \left(\mathbf{k}_N^{-1}\right)_{kl}\hat{s}_l$. In addition to the multiple (single) branch points at $p_k = -1/a\lambda_k^{(+)}$, there are multiple (single) essential singularities at these same points, as exhibited by the exponential terms in (3.2.20). The result for H_0 simplifies to

$$w_1(z|H_0) = \int_{-\infty i}^{\infty i} e^{pz} \prod_{k=1}^{J}\left(1+a\lambda_k^{(+)}p\right)^{-1/2} \frac{dp}{2\pi i}, \quad a = \text{real} > 0. \tag{3.3.20a}$$

Note that the eigenvalue $\left\{\lambda_k^{(+)}\right\}$ depends only on $\mathbf{k}_N\mathbf{J}$ and not on the signal. (Henceforth, since it is assumed that all the eigenvalues are positive, we drop the superscript $(+)$. Care must be taken here because of the multiple branch points, since $\gamma(=1/2)$ is nonintegral. One must stay on the same branch, so that the integrated function always remains on that branch and returns to its original value after the complex variable p makes one complete circuit of the contour. This introduces a *branch factor* $B_{\gamma=1/2} = \exp\left[-2\pi i(1-1/2)\cdot {}_kC_2\right]$, where ${}_kC_2 = k(k-1)/2$, namely, $B_{1/2} = \exp[(-\pi i/2)k(k-1)]$, which must be included. The result is

$$w_1(z|H_1) = \left\langle e^{-(1/2)\hat{\mathbf{s}}\mathbf{k}_N^{-1}\hat{\mathbf{s}}} \sum_{k=1}^{J} e^{-\pi i_k C_2} w_{1k}(z|H_1) {\prod_{l=1}^{J}}' \frac{e^{c_l(\hat{s})^2/2(1-\lambda_l/\lambda_k)}}{|1-\lambda_l/\lambda_k|^{1/2}}\right\rangle_{\hat{\mathbf{s}}} \tag{3.3.21}$$

where the prime (′) on the product indicates that $l \neq k$ and where explicitly

$$w_{1k}(z|H_1) = \frac{[2\lambda_k]^{1/4}}{z^{1/4}[2\lambda_k]} e^{-c_k(\hat{s})^2/2 - z/2\lambda_k} I_{-1/2}\left(\sqrt{\frac{zc_k^2(\hat{s})}{\lambda_k}}\right), z > 0; \quad = 0, z < 0. \quad (3.3.21a)$$

(For c_k, see line following (3.2.20).)

For $w_1(z|H_0)$, Equation (3.3.21a) specializes directly to

$$w_1(z|H_0) = \sum_{k=1}^{J} e^{-\pi i_k C_2} \cdot w_{1k}(z|H_0) \prod_{l=1}^{J}{}' |1 - \lambda_l/\lambda_k|^{-1/2} \quad (3.3.22)$$

with

$$w_{1k}(z|H_0) = \frac{[z/2\lambda_k]^{-1/2} e^{-z/2\lambda_k}}{2\lambda_k \sqrt{\pi}}, z > 0; \quad = 0, z = 0. \quad (3.3.22a)$$

(Note that $w_{1k}(z|H_0, H_1)$ are probability densities, that is, $\int_0^\infty w_{1k} dz = 1$ and $w_{1k} \geq 0; z > 0$.)

The performance probabilities are facilitated by noting that the Bessel function can be represented by

$$I_{-1/2}\left(\sqrt{zc_k(\hat{s})^2/\lambda_k}\right) = \sqrt{\frac{2\lambda_k}{\pi c_k(\hat{s})^2}} \cosh \sqrt{\frac{zc_k(\hat{s})^2}{\lambda_k}}, \quad (3.3.22b)$$

so that finally, on setting $z = u^2 = z_k/\lambda_k$, we obtain

$$p_{kD}^{(*)} = 1 - \beta_k^{(*)} = \int_{K_T}^{\infty} w_{1k}(z|H_1) dz = \frac{2^{3/4}\sqrt{\hat{c}_k}}{\sqrt{\pi}} \int_{(K_T/\lambda_k)^{1/2}}^{\infty} \frac{\cosh \hat{c}_k u}{(\hat{c}_k u)^{3/2}} e^{-c_k^2/2 - u^2/2} du, \quad (3.3.23)$$

which when substituted into (3.3.21) gives us the formal solution for our problem, at *all* signal levels. In this case, it remains threshold optimum, in which the alternative method of Section 3.3.3.1 is to be preferred. The corresponding false alarm probability is found to be (3.3.22a), with

$$\left(\alpha_F^{(*)}\right)_k = \int_{K_T}^{\infty} w_{1k}(z|H_0) dz = 1 - \Gamma(1/2, K_T/2\lambda_k); \quad \Gamma(\alpha, x) \equiv \int_0^x t^{\alpha-1} e^{-t} dt / \Gamma(\alpha), \quad (3.3.24)$$

where $\Gamma(\alpha, x)$ is one form of definition of the incomplete Γ-function ([4], p. 940 et seq.). Evaluation of (3.3.21) and (3.3.23), however, must be by numerical methods. Thus, we have for the false alarm probability, substituting (3.3.24) into (3.3.22),

$$\alpha_F^* = \sum_{k=1}^{J} e^{-\pi i_k C_2} \prod_{l=1}^{J}{}' |1 - \lambda_l/\lambda_k|^{-1/2} \{1 - \Gamma(1/2, K_T/2\lambda_k)\}. \quad (3.3.25)$$

Similarly, we obtain for the (conditional) probability of detection the formal (and formidable) result:

$$p^{(*)} = \left\langle e^{-\tilde{\hat{\mathbf{s}}}\mathbf{k}_N^{-1}\hat{\mathbf{s}}/2} \sum_{k=1}^{J} e^{\pi i_k C_2} \prod_{l=1}^{J}{}' \frac{e^{c_l(\hat{s})^2/2(1-\lambda_l/\lambda_k)}}{|1-\lambda_l/\lambda_k|^{1/2}} \cdot \frac{2^{3/4} c_k^{1/2}}{\sqrt{\pi}} \int_{[K_T/\lambda_k]^{1/2}}^{\infty} \frac{\cosh c_k u}{(c_k u)^{1/2}} e^{-c_k^2/2-u^2/2} du \right\rangle_{\hat{\mathbf{s}}}. \tag{3.3.26}$$

Again, we remark that these results are *suboptimum* for the larger signal inputs, since the test statistic $A_J + \tilde{\mathbf{x}}(\mathbf{k}_N^{-1}\mathbf{k}_{\hat{s}}\mathbf{k}_N^{-1})\mathbf{x}$, for *incoherent broadband* reception, is only optimal for threshold signals, refer to Section 3.3.3.1. Finally, revisiting (3.3.26) in the form

$$p^{(*)} = 1 - \left\langle e^{-\tilde{\hat{\mathbf{s}}}\mathbf{k}_N^{-1}\hat{\mathbf{s}}/2} \sum_{k=1} (\,) \cdot \int_0^{(K_T/\lambda_k)^{1/2}} \frac{\cosh c_k u}{(c_k u)^{1/2}} e^{-c_k^2/2-u^2/2} du \right\rangle_{\hat{\mathbf{s}}} \tag{3.3.26a}$$

shows that as $\tilde{\hat{\mathbf{s}}}\mathbf{k}_N^{-1}\mathbf{s} \to \infty$, $p^{(*)} \to 1$ as expected.

3.3.3.3 A Coherent Detection Problem A formally closely related problem to the preceding is one in which the test statistic is

$$y = \log G_1(\mathbf{x}) = A_J + \tilde{\mathbf{x}}\mathbf{J}\mathbf{x}, \quad \mathbf{J} = \mathbf{J}(\mathbf{s}), \tag{3.3.27}$$

where $\mathbf{J}$ is a symmetric matrix $\mathbf{J} = \tilde{\mathbf{J}}$ and is otherwise arbitrary. Here, $\mathbf{x} = \mathbf{n} + \mathbf{s}$ is the input, which consists of normal, zero-mean noise and an arbitrary signal. The observation process is *coherent* and the signal and data ($\mathbf{x}$) are broadband, but with a quadratic test statistic, unlike the coherent cases discussed in Section 3.1.1. However, we drop the average (of the epoch) over $\hat{s}$, where now the epoch is assumed known. The result is precisely (3.3.25) and (3.3.26) for (3.3.27), but without the averages $\langle\,\rangle_{\hat{\mathbf{s}}}$, but ultimately still requiring a numerical integration for H_1. The eigenvalues are determined from $\mathbf{k}_N\mathbf{J}$, and, as before, are independent of signal $\hat{\mathbf{s}}$. The characteristic functions $F_1(i\xi|H_1)$, $F(i\xi|H_0)$ of the associated pdf values are given by Eqs. (3.3.19a) and (3.3.19b), without the average $\langle\,\rangle_{\hat{\mathbf{s}}}$. Equations (3.3.25) and (3.3.26) are exact, although the test statistic (3.3.27) is generally suboptimum. (Equations (3.3.25) and (3.3.26) are also the *exact* solutions to Example 2, Section 17.2.1 of Ref. [1], with or without the averages over $\hat{\mathbf{s}}$.)

3.3.3.4 Narrowband Incoherent Detection with Asymmetrical Intensity Spectrum When $\hat{\mathbf{s}}$ is narrowband *and* the accompanying narrowband noise has an asymmetrical intensity spectrum, the calculation of the pdf values under H_1, H_0 for the test statistic (3.3.27) when reception is *incoherent* may be accomplished with the help of (3.2.51) and (3.2.52) and the analysis of Section 3.2.3. We observe that

$$\tilde{\hat{\mathbf{s}}}\mathbf{k}_N^{-1}\hat{\mathbf{s}} \doteq \frac{1}{2}\left[\Psi_{s-\mathrm{inc}}^{(c)*} + \Psi_{s-\mathrm{inc}}^{(s)*}\right]_{\mathrm{nb}} = \frac{1}{4}\sum_j \left(A_j c_{cj} + A_j c_{sj}\right) = \Psi_{s-\mathrm{inc}}^{*}\big|_{\mathrm{nb}} \tag{3.3.28}$$

Accordingly, we have

$$c_j\hat{s}^2 \equiv \sum_{j'} \tilde{\hat{s}}_j \left(\mathbf{k}_N^{-1}\right)_{jj'} \hat{s}_{j'} = \left(A_j c_{cj} + A_j c_{sj}\right)/4, \tag{3.3.28a}$$

and

$$\therefore \sum_j c_j(\hat{s})^2 = \sum_{jj'} \tilde{\hat{s}}_j \left(\mathbf{k}_N^{-1}\right)_{jj'} \hat{s}_{j'} = \tilde{\hat{\mathbf{s}}}\mathbf{k}_N^{-1}\hat{\mathbf{s}} = \Psi^*_{s-\text{inc}}\big|_{\text{nb.}} \tag{3.3.28b}$$

In the usual case of *symmetrical* spectra, where (3.2.53) applies, that is, $\mathbf{c}_c = \mathbf{c}_s = 2\,\boldsymbol{\rho}_0^{-1}\mathbf{A}$, $(\boldsymbol{\lambda}_0 = \mathbf{0})$, the above reduces to

$$\tilde{\hat{\mathbf{s}}}\mathbf{k}_N^{-1}\hat{\mathbf{s}} = \frac{1}{2}\sum_{jj'} A_j A_{j'} (\boldsymbol{\rho}_0)^{-1}_{jj'} = \sum_{jj'} \frac{A_{0_n}^{(m)} A_{0_{n'}}^{(m')}}{2\sqrt{\psi_j \psi_{j'}}} (\boldsymbol{\rho}_0)^{-1}_{jj'} = \sum_{jj'} \hat{A}_j \hat{A}_{j'} (\boldsymbol{\rho}_0)^{-1}_{jj'}; \quad \hat{A}_j = A_{0_n}^{(m)} / \sqrt{2\psi_j}, \text{ and so on} \tag{3.3.29}$$

Thus, for the test statistic (3.3.27), where now $\hat{s}$ and the noise are narrowband and reception is incoherent, the results (3.3.25) and (3.3.26), *without*$\langle\,\rangle_{\hat{S}:\varepsilon}$, apply for the false alarm probability and probability of detection.

Other related results are discussed in Section 17.2 of Ref. [1], in particular, Examples 2 and 3, and in Section 17.2.3 of Ref. [1], the distribution of the intensity spectrum of a normal process. Section 17.3 of Ref. [1] gives some additional examples of first-order probability densities following a (full-wave) quadratic detector, broad- and narrowband inputs, also for originally normal noise inputs. See, in addition, Appendix A1 for the extension to space–time fields.

3.4 BAYES MATCHED FILTERS AND THEIR ASSOCIATED BILINEAR AND QUADRATIC FORMS, I [14–16][38]

In the preceding sections of this chapter, we have introduced the generic discrete space–time matched filter $\mathbf{H} = \mathbf{k}_N^{-1}\mathbf{a}$ (refer to Section 3.1.1.2). This is an essential component of the optimum detector and estimator (cf. Chapters 5 and 6). Such filters are implicitly optimized as part of the Bayes formulation (Chapter 1), which in turn provides the general criteria of optimality employed throughout this book. Because of its importance in signal processing, let us examine the general concept of the matched filter[39] in more detail.

We begin by first considering the general concept of the matched filter itself, extended to space as well as time. This is needed in the analysis of system structure, which is typified by the linear and nonlinear operations on the input data $\mathbf{x}$. These are represented by the

[38] See the Historical Note at the end of Section 3.5. For other, later work see Ref. [17] and Ref. 3-11 therein. See also Sections 16.2 and 16.3 of Ref. [1].

[39] The terms "matching," "matched," and so on always represent a form of optimization here and throughout with respect to Bayes criteria, refer to Section 1.3.3; unless otherwise indicated, it also implies the presence of (point) sensors at each designated point in space, which in turn have path delays to a common reference point, usually O_T or O_R, for instance, in Fig. 3.16 and also Fig. 2.14.

following quadratic forms for the test statistic, characteristic of signals in Gaussian noise and of threshold signals in non-Gaussian noise as well, specifically

$$\Psi_{J-\mathrm{coh}}^{(1)} = \tilde{\mathbf{x}}\mathbf{A}^{(1)}\mathbf{s}; \quad \Psi_{J-\mathrm{inc}}^{(2)} = \tilde{\mathbf{x}}\mathbf{A}^{(2)}s\mathbf{x}, \quad \text{where } \mathbf{A}^{(1)} = \tilde{\mathbf{A}}^{(1)}; \quad \mathbf{A}^{(2)}(s) = \tilde{\mathbf{A}}^{(2)}(s), \tag{3.4.1}$$

and $\mathbf{x} = [X/\sqrt{\psi_j}\,]$, as before, refer to (3.2.1) et seq. Here, $\Psi_{J-\mathrm{coh}}^{(1)}$ and $\Psi_{J-\mathrm{inc}}^{(2)}$ are the result of Bayes optimality procedures (Section 3.1.3), where $\mathbf{x}$, $\mathbf{s}$ are J-component vectors and $\mathbf{A}^{(1)}$, $\mathbf{A}^{(2)}(s)$ in Eq. (3.4.1a) are real symmetric ($J \times J$) space–time, positive finite matrixes. In particular we have

$$\mathbf{A}^{(1)} = \mathbf{k}_{\mathrm{N}}^{-1} \quad \text{and } \mathbf{A}^{(2)}(s) = \mathbf{k}_{\mathrm{N}}^{-1}\mathbf{s}\tilde{\mathbf{s}}\mathbf{k}^{-1}, \quad \mathbf{k}_{\mathrm{N}}^{-1}(\mathbf{a}\tilde{\mathbf{a}} + \mathbf{b}\tilde{\mathbf{b}})\mathbf{k}_{\mathrm{N}}^{-1}, \quad \mathbf{k}_{\mathrm{N}}^{-1}\overline{\mathbf{s}\tilde{\mathbf{s}}}\,\mathbf{k}_{\mathrm{N}}^{-1} = \mathbf{k}_{\mathrm{N}}^{-1}\hat{\mathbf{k}}_{\mathrm{S}}\mathbf{k}_{\mathrm{N}}^{-1}, \tag{3.4.1a}$$

the basic bilinear and quadratic forms for coherent and incoherent reception. The noise covariances embodied in $\mathbf{A}^{(1)}$ and $\mathbf{A}^{(2)}$ are generally not homogeneous and stationary unless otherwise indicated. The signals (as before, Chapter 1) for which the matched filters are to be used are themselves deterministic or may even be purely random.

Broadly stated, matching is a form of optimization that attempts to enhance in some appropriate sense the reception of a desired signal *in the undesired noise background.* Matched filters here, as in earlier work, are required to be linear (but not necessarily time-invariant) and may be used in conjunction with other, subsequent zero-memory nonlinear elements and linear integrating devices [14–17]. However, in contrast to earlier definitions, our present definition of a matched filter is considerably more general and includes the previous examples in Chapters 2 and 3 as special cases, in addition to the extension to space–time fields, as we shall note below.

The structure of matched filters depends on the following:

(1) Nature of the signal
(2) Statistics of the accompanying noise, and the way in which it combines with the signal
(3) Role of space as well as time
(4) Criterion of optimality that is chosen

Since (1) and (2) are essentially *a priori* data, it is (3) and (4), the extension to space and the choice of criterion, that permit the generalization of the earlier definitions. Most of (4) has been based on energy calculations, that is, on some form of maximization of signal energy vis-à-vis that of the noise without direct reference to the actual decision process implied in reception. Usually, these matched filters have been obtained by maximizing a signal-to-noise ratio and may for thus be called *S/N*-matched filters. By recognizing that reception here implies a definite decision process, we extend this matched filter concept to space and base it on the Bayes decision rules considered previously in Chapters 1 and 2. When this is done, such optimum (linear) filters are called *Bayes matched filters* [16] or more precisely *space–time* Bayes matched filters. Their precise structure, of course, depends on the properties of the signal and noise, as well as on the decision criterion that generates the characteristic structural components of the types in Eq. (3.4.1).

3.4.1 Coherent Reception: Causal[40] Matched Filters (Type 1)

We consider first $\Psi^{(1)}_{J-\text{coh}}$ in Eq. (3.4.1), where the operations on the normalized data $\mathbf{x}\left(=\left[X_j/\sqrt{\psi_j}\right]\right)$ are required to be both linear and realizable (or equivalently "causal").[40] The normalized signal vector $\mathbf{s}(=[S/\sqrt{2\psi}])$ is assumed to be nonvanishing (on all finite subintervals) in the observation interval ($\mathbf{r}_m \in [0, \mathbf{R}]$, $t_n \in [0, T]$) so that $\Psi^{(1)}_{J-\text{coh}}$ is correspondingly nonvanishing. Then, let $\mathbf{H}$ be a (column) vector[41] such that

$$\mathbf{A}^{(1)}\mathbf{s} = \mathbf{H}^{(1)}, \quad \text{and} \quad \therefore \Psi^{(1)}_{J-\text{coh}} = \tilde{\mathbf{x}}\mathbf{H}^{(1)} = \tilde{\mathbf{H}}^{(1)}\mathbf{x} = \sum_{j=1}^{J=} x(r_m, t_n) H^{(1)}(\mathbf{r}_m, t_n), \quad (3.4.2)$$

which includes the spatial as well as the temporal nature of the input data $\mathbf{x}$. Here $\mathbf{A}^{(1)}$ is positive definite and symmetrical. Next, we set the vector

$$\mathbf{H}^{(1)} = \left[H_j^{(1)}\right] = \left[H^{(1)}(\mathbf{r}_m, T - t_n)\right], \quad (m, n = 1, \ldots, M; 1, \ldots, N), \quad j = mn, \quad (3.4.3)$$

where H_j is the weighting function of a time-invariant (i.e., nontime varying), realizable, discrete space–time filter, with a readout at time T $(=N\Delta t)$ and at positions $\mathbf{r}_m (m = 1, 2, \ldots, M)$. For the quadratic form $\Psi^{(1)}_{J-\text{coh}}$, (3.4.2), $\mathbf{H}^{(1)}$ is the solution of the discrete integral equation $\mathbf{A}^{(1)}\mathbf{s} = \mathbf{H}^{(1)}$, which can also be written in terms of the signal $\mathbf{s}$ as $[\mathbf{A}^{(1)}]^{-1}\mathbf{H}^{(1)} = \mathbf{s}$, refer to Eq. (3.4.4). Thus, $\Psi^{(1)}_{J-\text{coh}}$ from (3.4.2) may be considered to be the output of this discrete linear filter, which is a *Bayes matched filter* when $\Psi^{(1)}_{J-\text{coh}}$ is the result of a Bayesian optimization process. The output of this filter is a cumulative maximum at the readout $t = T$. Filters of this type are realized practically by a delay line with suitable weighting and readout. These Bayes matched filters are closely related to, and in some cases are identical, to the space–time matched filters [14–18]. We call them *Bayes matched filters of the first kind, Type 1*. Figure 3.9 illustrates this filter, which is a realization of data function $\Psi^{(1)}_{J-\text{coh}}$, Eq. (3.4.1).

When $\mathbf{A}^{(1)} = \mathbf{k}_N^{-1}$, as is the case with the noise fields discussed in this chapter, we have $\mathbf{H}^{(1)} = \mathbf{k}_N^{-1}\mathbf{s}$, where now $\mathbf{H}^{(1)}$ is obtained from the discrete integral equation.

$$\mathbf{k}_N\mathbf{H}^{(1)} = \mathbf{s} \ (\neq 0, \in (\mathbf{R}, T); \quad = 0, \quad \notin (\mathbf{R}, T)). \quad (3.4.4)$$

[40] "Realizability" or "causality" refers to operations only on the past (and present) of the data, in particular, within the data time $(0, T)$. For these filters, it also requires for the spatial part of the data at the receiver to obey a radiation condition, namely, the condition that only outgoing radiation from the source be propagated. Mathematically, causality for these filters is expressed by an extension of the Paley–Wiener criterion to include these propagation effects. Noncausal or "nonrealizable" systems are still usable, but they require the entire data sample in order to obtain a result at any intermediate or final instant.

[41] See Eq. (1.3.1) for indexing. Here, we use the time index $n = 1, \ldots, N$ at each spatial point (m), repeating for different points $m = 1, \ldots, M$. The result is a sample vector of length $J = MN$ elements.

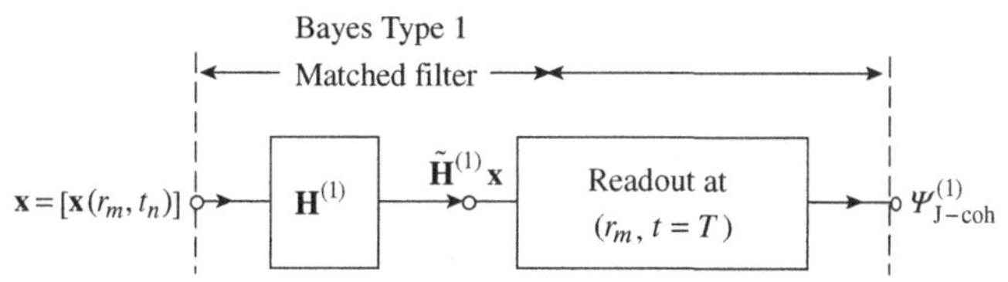

FIGURE 3.9 Representation of $\Psi^{(1)}_{J-\text{coh}}$ by a Bayes matched filter $\left(\mathbf{H}^{(1)}\right)$ of the first kind, Type 1 (realizable).

3.4.1.1 The "White Noise" Case In the important special case where the accompanying noise is "white" in space and time, we have for the normalized covariance $\mathbf{k}_\text{N}$,

$$\int_{-\infty}^{\infty} \cdots \int e^{i(\mathbf{k}-\mathbf{k}_0)\cdot\mathbf{r}} \mathbf{dr} = \delta(\mathbf{k}-\mathbf{k}_0) \quad \text{and} \quad \int_{-\infty}^{\infty} e^{-i(\omega-\omega_0)t} dt = \delta(f-f_0), \quad \omega = 2\pi f, \text{and so on.} \tag{3.4.5}$$

In the discrete cases correspondingly, we have $\mathbf{k}_\text{N} = \left[\delta_{jj'}\right] = \left[\delta_{mm'}\delta_{nn'}\right]$. Then, $\mathbf{H}^{(1)} = \langle\mathbf{s}\rangle$ and $\Psi^{(1)}_{J-\text{coh}}$ become (considered as an ensemble)

$$\Psi^{(1)}_{J-\text{coh}}\Big|_{\text{white}} = \sum_{mn} x(\mathbf{r}_m, t_n)\langle s(\mathbf{r}_m, t_n)\rangle = \sum_{mn} x(\mathbf{r}_m, t_n) H^{(1)}(\mathbf{r}_m, T-t_n). \tag{3.4.6}$$

When $\mathbf{x} = \langle\mathbf{s}\rangle + \mathbf{n}, \mathbf{n} = \mathbf{0}$ and the noise samples n_j are at least wide sense stationary and homogeneous, (3.4.6) can be expressed as

$$\left\langle \Psi^{(1)}_{J-\text{coh}} \right\rangle_\mathbf{n} = \sum_{mn} \langle s(\mathbf{r}_m, t_n)\rangle^2. \tag{3.4.6a}$$

Note that if we write $\mathbf{H}^{(1)}$ as (3.4.3) and use (3.4.5), we obtain Eq. (3.4.6b) in the continuous form:

$$\begin{aligned} \mathbf{H}^{(1)} = \left[h^{(1)}(\mathbf{r}_m, T-t_n)\Delta\mathbf{r}\Delta t\right], \;\therefore\; \left\langle \Psi^{(1)}_{J-\text{coh}} \right\rangle_\mathbf{n} &= 2\sum_{\text{mn}} \langle s(r_m, t_n)\rangle^2 \Delta\mathbf{r}\Delta t / W_v W_0 \\ &\to \int_\mathbf{R} \frac{\mathbf{dr}}{W_v} \int_{-\infty}^{\infty} \frac{2}{W_0} \langle s(r,t)\rangle^2 dt \end{aligned} \tag{3.4.6b}$$

Now let us consider a *cross-correlation receiver* (detector) for coherent reception for a typical member of the ensemble, where the observation period is large compared to the correlation time of the noise. We have

$$\begin{aligned} R_{xs}(\Delta\mathbf{r}, \Delta t) = \tilde{x}_1 s_2 &= \sum_j (\langle s\rangle + n)_j \langle s_j\rangle + \Delta_j \doteq \sum_j \left[\langle s\rangle\langle s_j\rangle + \Delta_j\right]; \;\; \Delta_j = \Delta\mathbf{r}\Delta t \\ &\doteq \sum_j \langle s(\mathbf{r}_m, t_n)\rangle\langle s(\mathbf{r}_m + \Delta\mathbf{r}, t_n + \Delta t)\rangle = \sum_j \langle s(\mathbf{r}_m, t_n)\rangle^2 = \Psi^{(1)}_{J-\text{coh}}\Big|_{\text{white}} \end{aligned} \tag{3.4.7}$$

when we set $\Delta\mathbf{r} = \mathbf{0}$, that is, $\mathbf{r}_M = \mathbf{R}$, and $\Delta t = 0$, that is, $t_\text{N} = T$, namely, the desired maximum at the readout time $t_\text{N} = T$ and place $M = \mathbf{R}$ at the end of the data sample.

Thus, in the particular case of normal noise, the correlation detector is optimum, refer to (3.4.6a), (3.4.6b), and (3.4.7). It also maximizes the (S/N) ratio generally in non-Gaussian noise, although this correlation receiver is now no longer optimum.

3.4.2 Incoherent Reception: Causal Matched Filters (Type 1)

In this case, the general quadratic form in question is $\Psi_{J-\mathrm{inc}}^{(2)} = \tilde{\mathbf{x}}\mathbf{A}^{(2)}\mathbf{x}$ where the additive noise is Gaussian or non-Gaussian. Because this form is nonlinear in the data, we may expect that its structure in terms of a realizable (or causal) discrete linear matched filter and other processing elements is more complex than that for $\Psi_{J-\mathrm{coh}}^{(1)}$ in Section 3.4.1. In fact, the matched filters here are not uniquely determined—a variety of such filters is possible, as we shall see in Sections 3.4.2–3.5.3. The filters themselves are now for the most part represented by square or triangular *matrices*, embodying *time-variable operations*, as distinct from the vector forms of the time-invariant cases (apart from truncation) characteristic of coherent reception described in Section 3.4.1.

Here, Type 1 matched filter for incoherent reception involves such nonlinear devices as zero-memory square-law detectors or multipliers, as well as ideal "integrators," and is not unique.

To see this, we first introduce a (real) *time-varying linear, discrete filter* represented by $(J \times J)$*triangular* square matrix[42]

$$\mathbf{H}^{(2,1)} = \left[H_{jj'}^{(2,1)'}\right],$$

with truncation, as a result of finite sample size (finite J), which as before is determined by the physical bounds $(0, |\mathbf{R}|); (0, T)$ of the acquisition interval. We let the output of this time-variable filter be $\mathbf{x}_{\mathrm{F}} = \mathbf{H}^{(2,1)}\mathbf{x}$ and consequently require that $\Psi_J^{(2)}$ be reduced to

$$\Psi_{J-\mathrm{inc}}^{(2)}\left(\equiv \tilde{\mathbf{x}}\mathbf{A}^{(2)}\mathbf{x}\right) = \tilde{\mathbf{x}}_{\mathrm{F}}\tilde{\mathbf{H}}^{(2,1)-1}\mathbf{A}^{(2)}\mathbf{H}^{(2,1)-1}\mathbf{x}_{\mathrm{F}} = \tilde{\mathbf{x}}_{\mathrm{F}}\mathbf{x}_{\mathrm{F}} = \Sigma_j(\mathbf{x}_{\mathrm{F}})_j^2, \quad \text{with} \quad \mathbf{x}_{\mathrm{F}} = \mathbf{H}^{(2,1)}\mathbf{x}. \tag{3.4.8}$$

To see this, we first choose $\mathbf{Q}$, a *congruent matrix*, with the property that a symmetric, positive definite quadratic form, here $\mathbf{A}^{(2)}$, is (nonuniquely) reducible to the sum of squares, that is, $\tilde{\mathbf{Q}}\mathbf{A}^{(2)}\mathbf{Q} = \mathbf{\Lambda}_\beta = \mathrm{diag}(\ldots, \beta_j, \ldots)$. Then we make the transformation $\mathbf{x} = \mathbf{Q}\hat{\mathbf{x}}$ and $\therefore \tilde{\mathbf{x}}\mathbf{A}^{(2)}\mathbf{x} = \tilde{\hat{\mathbf{x}}}(\tilde{\mathbf{Q}}\mathbf{A}^{(2)}\mathbf{Q})\hat{\mathbf{x}} = \tilde{\hat{\mathbf{x}}}\mathbf{\Lambda}_\beta\hat{\mathbf{x}}$ $(= \sum_j \hat{x}_j^2 \beta_j)$. To reduce this further to the form (3.4.8), that is, to obtain a pure sum of squares in the data $\mathbf{x}_{\mathrm{F}}$, we make the further transformation $\mathbf{x}_{\mathrm{F}} = [\sqrt{\beta_j}\hat{x}_j] = \mathbf{\Lambda}_\beta\hat{\mathbf{x}}$, so that

$$\tilde{\mathbf{x}}\mathbf{A}^{(2)}\mathbf{x} = \tilde{\hat{\mathbf{x}}}\mathbf{\Lambda}_\beta\hat{\mathbf{x}} = \tilde{\mathbf{x}}_{\mathrm{F}}\left(\mathbf{\Lambda}_\beta^{-1/2}\tilde{\mathbf{Q}}\mathbf{A}^{(2)}\mathbf{Q}\mathbf{\Lambda}_\beta^{-1/2}\right)\mathbf{x}_{\mathrm{F}} = \tilde{\mathbf{x}}_{\mathrm{F}}\left(\tilde{\mathbf{P}}\mathbf{A}^{(2)}\mathbf{P}\right)\mathbf{x}_{\mathrm{F}} = \tilde{\mathbf{x}}_{\mathrm{F}}\mathbf{I}\mathbf{x}_{\mathrm{F}}, \tag{3.4.8a}$$

[42] See Eq. (1.3.1) for alternative indexing. Here, a typical element is $j = mn = k$ (single digits) and $A_{jj'} = A_{mn,m'n'} = A_{kk'}$, where k and k' are respectively mn and $\acute{m}\acute{n}$. The result is a square matrix of $K \times K = MN \times MN = J \times J$ elements. The time elements associated with each, $k, \acute{k}$ or each $m, \acute{m}$ can be easily identified from the alternative description of Eq. (1.3.1).

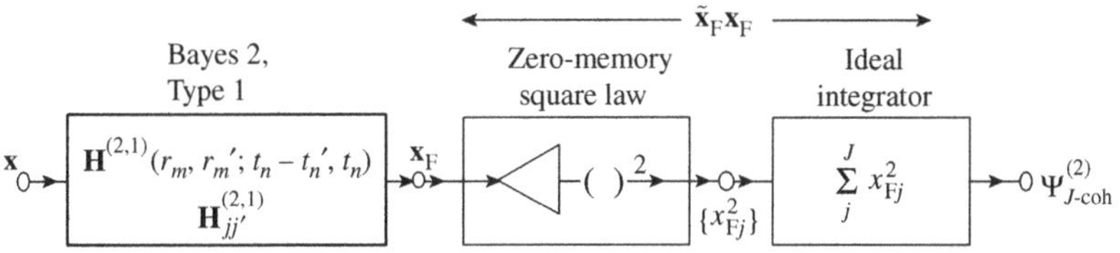

FIGURE 3.10 Resolution of $\Psi_{J-\mathrm{coh}}^{(2)}(\mathbf{x})$ by a Bayes matched filter of the *Second Kind, Type 1*, Eq. (3.4.8) (time-varying and realizable).

where $\mathbf{P} = \boldsymbol{\Lambda}_\beta^{-1/2}\mathbf{Q}$. The reducing congruent matrix $\mathbf{Q}$ is well known to be *upper* triangular (Section 10.16 of Ref. [19]). Comparing (3.4.8a) with (3.4.8b), we see for this discrete matched filter here that

$$\mathbf{H}^{(2,1)} = \left(\boldsymbol{\Lambda}^{-1/2}\mathbf{Q}\right)^{-1} = \mathbf{P}^{-1}, \quad \text{and} \quad \mathbf{P}^{-1} \neq \tilde{\mathbf{P}}^{-1}, \quad \text{since } \mathbf{Q} \neq \tilde{\mathbf{Q}}, \tag{3.4.8b}$$

with $\mathbf{Q}$ having the form

$$\mathbf{Q} = \begin{bmatrix} 1 & & Q_{jj'} \quad j < j' \\ & 1 & \\ & \mathbf{0} & \ddots & \\ & j > j' & & 1 \end{bmatrix}. \tag{3.4.8c}$$

The matrix $\mathbf{P}$ here is also upper triangular. See Section 10.16 of Ref. [19] for further details. (Note that even though $\tilde{\mathbf{Q}}\mathbf{A}^{(2)}\mathbf{Q} = \boldsymbol{\Lambda}_\beta$ reduces $\mathbf{A}^{(2)}$ to diagonal form, $\boldsymbol{\Lambda}_\beta$ does not represent $[\lambda]$, the diagonal matrix of the eigenvalues, since $\boldsymbol{\Lambda}_j = \beta_j$ are *not* the solutions of $\det(\mathbf{A}-\lambda\mathbf{I}) = 0$.)

The structure of $\Psi_{J-\mathrm{inc}}^{(2)}$ accordingly consists of the time-varying filter $\mathbf{H}^{(2,1)}$, followed by a zero-memory square-law operation and ideal "integrator," as shown in Fig. 3.10. Because $\mathbf{A}^{(2)}$ is required to be symmetrical and positive definite, like $\mathbf{A}^{(1)}$ in (3.4.1) and (3.4.2), it has an inverse $\mathbf{A}^{(2)-1}$. Because of the bounded input data sample, that is, the truncation condition, the matrices $\mathbf{A}^{(2)} = \mathbf{0}$ and $\mathbf{H}^{(2,1)} = \mathbf{0}$ when $(j, j' > J, < 11)$, or equivalently $(m, m') > M, < 1$; $(n, n') > N, < 1$.

Next, we use Eqs. (3.4.8a)–(3.4.8c) and obtain

$$\tilde{\mathbf{H}}^{(2,1)-1}\mathbf{A}^{(2)}\mathbf{H}^{(2,1)-1} = \mathbf{I}, \quad \therefore \ \mathbf{A}^{(2)} = \tilde{\mathbf{H}}^{(2,1)}\mathbf{H}^{(2,1)} \equiv \hat{\boldsymbol{\rho}}^{(2)} = \tilde{\hat{\boldsymbol{\rho}}}^{(2)}, \tag{3.4.9a}$$

where $\hat{\boldsymbol{\rho}}^{(2)}$, of course, is symmetric, positive definite, and is specifically here

$$\hat{\boldsymbol{\rho}}^{(2)} = \mathbf{k}_\mathrm{N}^{-1}\hat{\mathbf{k}}_\mathrm{S}\mathbf{k}_\mathrm{N}^{-1}; \quad \hat{\mathbf{k}}_\mathrm{S} = \langle \mathbf{s}\tilde{\mathbf{s}} \rangle = \frac{\langle \mathbf{a}\tilde{\mathbf{a}} \rangle + \langle \mathbf{b}\tilde{\mathbf{b}} \rangle}{2}. \tag{3.4.9b}$$

We add ($^\wedge$) to those covariances $\hat{\mathbf{k}}_\mathrm{S} \neq \mathbf{k}_\mathrm{N}$ that are normalized by $\psi_{\mathrm{N}_j}, \psi_{\mathrm{N}_{j'}}$, that is, the normalizing factors are $\left(\psi_{\mathrm{S}_j}\psi_{\mathrm{S}_j}/\psi_{\mathrm{N}_j}\psi_{\mathrm{N}_{j'}}\right)^{1/2} \neq 1$. See Section 3.4.5, Eq. (3.4.28a).

In keeping with the realizability condition, we see from the above that **H** is a *lower* triangular matrix from Eqs. (3.4.8b) and (3.4.8c). Noting that $H(r_m, t_n; r_{m'}, t_{n'})$ can be written equivalently as $H(r_m, r_{m'}; t_n - t_{n'}, t_n)$ we have

$$\mathbf{H}^{(2,1)} = \begin{bmatrix} \nwarrow & & \mathbf{0} \\ & H_{jj} & \\ H_{jj'} & j > j' & \searrow \end{bmatrix}, \quad \text{that is,} \quad \left[H_{jj'}^{(2,1)}\right] = H^{(2,1)}(\mathbf{r}_m, t_n; \mathbf{r}_{m'}, t_{n'}) = \left[H^{(2,1)}(\mathbf{r}_m, \mathbf{r}_{m'}; t_n - t_{n'}, t_n)\right]$$
$$= 0, \begin{pmatrix} j' > j, \text{ and } j', j > J, \text{ or} \\ m, m' > M;\ n' > n;\ \text{and } n, n' > N \end{pmatrix}, \tag{3.4.9c}$$

which also vanishes for $j, j' < 1$, that is, $m, m' < 1$ and $n, n' < 1$, as defined by the size (J) of the matrix, which in turn expresses the physical fact that the filter does not yet have an input. (The diagonal in Eq. (3.4.9c) is recognized as part of a Cholesky matrix, whose diagonal terms are eigenvalues of the Cholesky matrix, (c_{jj}), but are *not* the eigenvalues of $\mathbf{A}^{(2)}$—Section 5.2, p. 88, of Ref. [20].)

We call the discrete filter $\mathbf{H}^{(2,1)}$ a (discrete) Bayes matched filter of the *second kind, Type 1*. Its properties are readily summarized:

(i) $\mathbf{H}^{(2,1)}$ is *linear* and *time-variable*, that is,

$$\begin{aligned} \mathbf{H}^{(2,1)} &= [H^{2,1}(\mathbf{r}_m, t_n; \mathbf{r}_{m'}, t_{n'})] \\ &= \left[H^{(2,1)}(\mathbf{r}_m, \mathbf{r}_{m'}; t_n - t_{n'}, t_n)\right], \end{aligned}$$

refer to Eq. (3.5.9c)

(ii) $H_{jj'}^{(2,1)} = 0, j' > j$, indicating that $\mathbf{A}^{(2)}$ may be reduced to the unit matrix **I** (refer to remarks preceding Eq. (3.4.9a)), with a congruent matrix **P** that is upper diagonal, so that $\mathbf{H}^{(2,1)}$ is lower diagonal [Eq. (3.4.9c)]. This congruent matrix $\left[H_{jj'}^{(2)}\right]$ has the following (refer to form p. 308 of Ref. [19]):

$$\mathbf{H}^{(2,1)} = \begin{bmatrix} H_{11}^{(2,1)} & & & & & \\ H_{21}^{(2,1)} & H_{22}^{(2,1)} & & & \mathbf{0} & \\ H_{31}^{(2,1)} & H_{32}^{(2,1)} & H_{33}^{(2,1)} & & j' > j & \\ \vdots & \vdots & & \ddots & & \\ \vdots & \vdots & & & H_{jj}^{(2,1)} & \\ \vdots & \vdots & & & & \ddots \\ H_{J1}^{(2,1)} & \cdots & \cdots & \cdots & \cdots & \cdots\ H_{JJ}^{(2,1)} \end{bmatrix} = \begin{bmatrix} \Lambda_1^{(A)1/2} & & & & & \\ X_{21} & \Lambda_2^{(A)1/2} & & & \mathbf{0} & \\ X_{31} & X_{32} & \ddots & & & \\ \vdots & \vdots & & \ddots & & \\ \vdots & \vdots & & & \Lambda_j^{(A)1/2} & \\ \vdots & \vdots & & & & \ddots \\ X_{J1} & X_{J2} & \cdots & \cdots & \cdots & \cdots\ \Lambda_J^{(A)} \end{bmatrix}^{-1} = (\Lambda_\beta \mathbf{Q})^{-1} = \mathbf{P}^{-1} \tag{3.4.10}$$

(iii) Accordingly, $\mathbf{H}^{(2,1)}$ represents a realizable or causal filter, since it operates only on the past of the data ($t_{t'}^{n} > t_n$, (3.4.9c)), and is required also to obey the radiation condition.

(iv) Note that $\mathbf{H}^{(2,1)} \neq \tilde{\mathbf{H}}^{(2,1)}$ (3.4.10).

The set of discrete nonlinear integral equations (3.4.9c) determining the elements of $\mathbf{H}^{(2,1)}$, namely, $\tilde{\mathbf{H}}^{(2,1)}\mathbf{H}^{(2,1)} = \hat{\boldsymbol{\rho}}^{(2)}$, can be equivalently expressed in detail as

$$\left.\begin{aligned}(1 \le j,j' \le J): \hat{\rho}_{jj'}^{(2)} &= \sum_k H_{kj}^{(2,1)} H_{kj'}^{(2,1)} = \sum_{i=1}^{M}\sum_{l=1}^{N} H^{(2,1)}(\mathbf{r}_i, \mathbf{r}_m; t_l - t_n, t_l) H^{(2,1)}(\mathbf{r}_i, \mathbf{r}_{m'}; t_l - t_{n'}, t_l)\\ &\qquad 1 \le m, m' \le M; 1 \le n, n' \le N\\ &= 0, n > l, n' > l; \mathbf{r}_i, \mathbf{r}_m \text{ outside } (0, |\mathbf{R}|).\end{aligned}\right\} \tag{3.4.11}$$

The quadratic form $\Psi_{J-\text{inc}}^{(2)}$ (3.4.8) becomes accordingly

$$\Psi_{J-\text{inc}}^{(2)}(\mathbf{x}) = \tilde{\mathbf{x}}\hat{\boldsymbol{\rho}}^{(2)}\mathbf{x} = \sum_{j=1}^{J}\left(\sum_{j'=1}^{\infty=j} H_{jj'}^{(2,1)} x_{j'}\right)^2$$

$$= \sum_{m,n=1}^{J}\left(\sum_{(m',n'=j')=1}^{(m,n=j)} H_{jj'}^{(2,1)}(r_m, r_{m'}; t_n - t_{n'}, t_n) x_{j'}\right)^2 = \sum_{j=1}^{J} x_{Fj}^2, \tag{3.4.12}$$

which is one interpretation of the quadratic form $\Psi_{J-\text{inc}}^{(2)}$, shown in Fig. 3.10.

A variant of the linear causal filter, Eqs. (3.4.8)–(3.4.9b), is to replace it with a causal *invariant* filter represented here by the vector $\mathbf{H}^{(2,1a)}$, and a time-variable switch or readout at $t = t_n \le 0$, that is, $\mathbf{H}^{(2,1a)} = [H_j^{(2,1a)}(\mathbf{r}_m, t_{n'} - t_n)]$, with $H_j^{(2,1a)}(\mathbf{r}_m, t'_n - t_n) = 0, t_n < 0$. Equations (3.4.8)–(3.4.12) remain unchanged except that $\mathbf{H}^{(2,1)}$ is replaced by $\mathbf{H}^{(2,1a)} = [H_j^{(2,1a)}]$. Figure 3.10 remains essentially the same, except that the triangular matrix $\mathbf{H}^{(2,1)}$ is replaced by the *vector* $\mathbf{H}^{(2,1a)}$ plus the time-varying switch (refer to Fig. 5 of Ref. [16]).

3.4.3 Incoherent Reception-Realizable Matched Filters; Type 2

A second equivalent resolution of $\Psi_{J-inc}^{(2)}$ yields a Bayes matched filter $\mathbf{H}^{(2,2)}$ of the *second kind, Type 2*. To see this, let us use Eq. (3.4.9a) and observe first that we can also write for $\boldsymbol{\rho}^{(2)}$

$$\boldsymbol{\rho}^{(2)} \equiv \left[\rho_{jj'}^{(2)}\right] \equiv \left[H_{jj'}^{(2,2)}\right]; \quad \mathbf{H}^{(2,2)} \ne \mathbf{H}^{(2,1)}, \quad \text{since } \boldsymbol{\rho}^{(2)} = \tilde{\boldsymbol{\rho}}^{(2)},$$

$$\text{refer to Eqs. (3.4.9a) and (3.4.9b).} \tag{3.4.13}$$

The matrix $\mathbf{H}^{(2,2)}$ is, so far, not a causal time-varying filter, since $\mathbf{H}^{(2,2)}$ is symmetrical. However, the constraint of noncausality can be removed by the artifice of deleting all terms above the diagonal and doubling those below it, that is, keeping $j > j'$ for operations on the "past," where $\mathbf{H}^{(2,2)} \ne \mathbf{0}$ and setting $\mathbf{H}^{(2,2)} = 0$ for all $j' > j$ for the "future" of the data stream. We take advantage of the fact that for any symmetric matrix $\mathbf{A}^{(2)} = \tilde{\mathbf{A}}^{(2)}$, the quadratic form $\tilde{\mathbf{x}}\mathbf{A}^{(2)}\mathbf{x}$ remains unchanged if we write it as

$$\sum_{jj'}^{J} x_j A_{jj'}^{(2)} x_{j'} = 2\sum_{j'=1}^{J}\sum_{j'=1}^{j'\le j} x_j x_{j'} A_{jj'}^{(2)} \varepsilon_{jj'}^{-1} \tag{3.4.14}$$

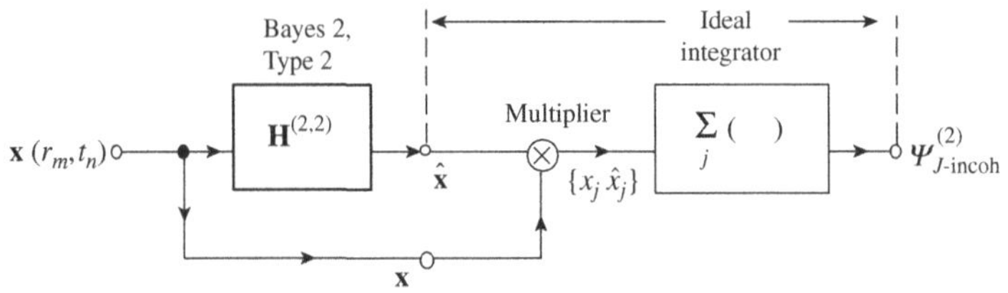

FIGURE 3.11 Resolution of $\Psi^{(2)}_{J-\text{inc}}$ by a Bayes (time-varying and causal) matched filter of the *second kind, Type 2*, Eqs. (3.4.13) and (3.4.15).

and then set $A^{(2)}_{jj'} = 0, j' > j$. Thus, with (3.4.13) in (3.4.14), we have

$$\Psi^{(2)}_{J-inc} = \sum_{jj'}^{J} x_j x_{j'} H^{(2,2)}_{jj'} = \sum_{j=1}^{J} x_j \left(\sum_{j'=1}^{j \geq j'} x_{j'} H^{(2,2)}\left(r_m, r'_m; t_{n'} - t_n, t'_n\right) \right) = \sum_{j=1}^{J} x_j \hat{x}_j,$$

$$\text{where } \hat{x}_j = \sum_{j'=1,1}^{j' \leq j} x_{j'} H^{(2,2)}_{jj'}. \tag{3.4.15}$$

This result is interpreted as a causal time-varying filter, followed by a simple (i.e., zero-memory) multiplier of its output with the input data, which is then integrated, as shown in Fig. 3.11. This matched filter is given formally at once by (3.4.13). For example, in the case, $\mathbf{A}^{(2)} = \mathbf{k}_N^{-1}\hat{\mathbf{k}}_s\mathbf{k}_N^{-1}$, which arises in incoherent reception of signals ($\hat{\mathbf{k}}_s = \langle \mathbf{s}_1 \hat{\mathbf{s}}_2 \rangle$ or $\hat{\mathbf{k}}_s$ is normal noise), is simply $\mathbf{H}^{(2,2)} = \mathbf{k}_N^{-1}\hat{\mathbf{k}}_S\mathbf{k}_N^{-1}$. On the other hand, solutions for $\mathbf{H}^{(2,1)}$ are determined from (3.4.11) and are much more difficult to achieve, because of the nonsymmetric nature of $\mathbf{H}^{(2,1)}$.

Another variant in the resolution of $\Psi^{(2)}_{J-\text{inc}}$ can occur when $\mathbf{A}^{(2)}$ factors into the matrix product of two vectors, that is, $\mathbf{A}^{(2)} = \mathbf{y}\tilde{\mathbf{y}}$. This factorization, when possible, is closely related and in fact identical to (S/N) matched filters of the earlier theory ([14, 15] and Section 16.3 of Ref. [1]). This class of filter we shall call a Bayes matched filter of the *second kind, Type 2a*, which yields a time-invariant and realizable weighting function:

$$\mathbf{y} \equiv \mathbf{H}^{(2,2a)} = \left[H^{(2,2a)}(\mathbf{r}_m, T - t_n) \right]. \tag{3.4.16}$$

The quantity $\Psi^{(2)}_{J-\text{incoh}}$ is now constructed according to

$$\Psi^{(2)}_{J-\text{incoh}} = \tilde{\mathbf{x}}(\mathbf{y}\tilde{\mathbf{y}})\mathbf{x} = (\tilde{\mathbf{x}}\mathbf{y})^2 = \left(\sum_{j=1}^{J} x_j H^{(2,2a)}_j \right)^2 = \left(\sum_{j=mn} x_j H^{(2,2a)}(\mathbf{r}_m, T - t_n) \right)^2, \tag{3.4.17}$$

where $\mathbf{y} = \mathbf{H}^{(2,2a)}$ is a *vector*. Here, $\Psi^{(2)}_{J-\text{inc}}$ is interpreted as a time-invariant, causal filter followed by a ideal square-law rectifier, in the manner of Fig. 3.12. Note that unlike the previous two cases, there is no final integration.

Once $\mathbf{A}^{(2)}$ is factored, whenever this can be done, $\mathbf{H}^{(2,2a)}$ is determined from (3.4.16), and except for possible scale factors, is unique.

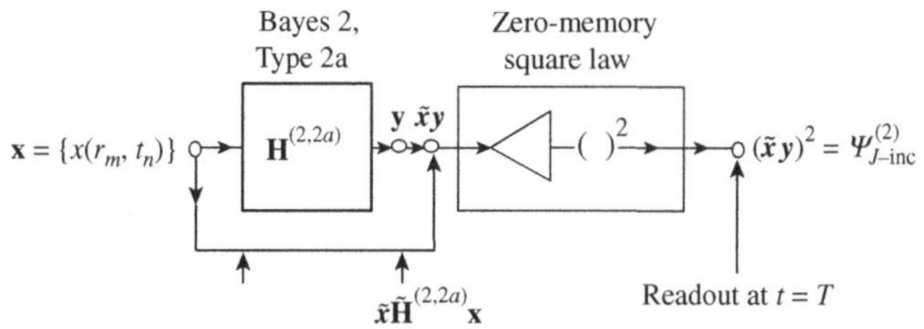

FIGURE 3.12 Resolution of $\Psi^{(2)}_{J\text{-inc}}(\mathbf{x})$ by a Bayes matched filter of the second kind, Type 2a, Eq. (3.5.16) (invariant and causal).

Analytically, factoring $\mathbf{A}^{(2)}$ into two components $\mathbf{A}^{(2)^{1/2}} \cdot \mathbf{A}^{(2)^{1/2}}$ is also possible when $\mathbf{A}^{(2)}$ is composed of one or more symmetrical positive definite covariance matrices of random processes and fields [21]. A simple example is provided by the case where the signal is deterministic and its covariance takes the form $\hat{\mathbf{k}}_S = \langle \mathbf{s} \rangle \langle \tilde{\mathbf{s}} \rangle$. This occurs in the situation described is Section 3.2.1 where the signal is narrowband and depends only on a uniformly distributed RF epoch ε, Eq. (3.2.12). Then $\mathbf{A}^{(2)}$ is typically

$$\mathbf{A}^{(2)} = \mathbf{k}_N^{-1}\hat{\mathbf{k}}_S\mathbf{k}_N^{-1}, \quad \text{with} \quad \hat{\mathbf{k}}_S = \mathbf{s}\tilde{\mathbf{s}} \text{ or } \left(\mathbf{a}\tilde{\mathbf{a}} + \mathbf{b}\tilde{\mathbf{b}}\right)/2 \tag{3.4.18}$$

which allows us to write (since $\mathbf{k}_N = \tilde{\mathbf{k}}_N$)

$$\mathbf{A}^{(2)} = \left(\mathbf{k}_N^{-1}\mathbf{s}\right)\left(\tilde{\mathbf{s}}\mathbf{k}_N^{-1}\right) = \left(\mathbf{k}_N^{-1}\mathbf{s}\right)^2 = \mathbf{y}\tilde{\mathbf{y}}, \text{ and so on,} \tag{3.4.19a}$$

with

$$\mathbf{y} = \mathbf{H}^{(2,2a)}, \text{a vector,} \quad \text{or} \quad \mathbf{y}^{(a)} = \mathbf{H}_a^{(2,2a)}; \quad \mathbf{y}^{(b)} = \mathbf{H}_b^{(2,2b)}, \quad \text{and } \mathbf{H}^{(2,2a)} = \mathbf{k}_N^{-1}\langle \mathbf{s} \rangle, \tag{3.4.19b}$$

and thus (3.4.16) follows. Here, $\mathbf{H}^{(2,2a,b)}$ are *vectors.*

For deterministic signals *with other random parameters*, and purely random signals, we have $\mathbf{k}_S = \langle \mathbf{s}\tilde{\mathbf{s}} \rangle$, which is different from zero, so that

$$\mathbf{A}^{(2)} = \mathbf{k}_N^{-1}\overline{\mathbf{s}\tilde{\mathbf{s}}}\,\mathbf{k}_N^{-1} = \mathbf{k}_N^{-1}\hat{\mathbf{k}}_S\mathbf{k}_N^{-1}, \quad \text{and} \quad \therefore \quad \mathbf{A}^{(2)} = \left(\mathbf{k}_N^{-1}\hat{\mathbf{k}}_S^{1/2}\right)^2 = \mathbf{u}\tilde{\mathbf{u}} = \tilde{\mathbf{u}}\mathbf{u}. \tag{3.4.20}$$

Then, we can write

$$\mathbf{H}^{(2,2b)} = \mathbf{u} = \text{a square (symmetric) matrix,}$$

$$\text{with} \quad \mathbf{H}^{(2,2b)} = \mathbf{k}_N^{-1}\hat{\mathbf{k}}_S^{1/2}, \neq \left(H^{(2,2a)}_{(a,b)}, \mathbf{H}^{(2,2a)}\right), \text{refer to Eq. (3.4.20)} \tag{3.4.21}$$

Note that if $\mathbf{H}^{(2,2a)} = \mathbf{k}_N^{-1}\langle \mathbf{s} \rangle = \mathbf{H}^{(1)}$, then Eq. (3.4.4) is a *vector.* This also occurs if $\mathbf{A}^{(1)} = \mathbf{k}_N^{-1}$ and $\mathbf{A}^{(2)} = \mathbf{k}_N^{-1}\mathbf{s}\tilde{\mathbf{s}}\,\mathbf{k}_N^{-1}$, that is, $\mathbf{s}\tilde{\mathbf{s}} = \left(\mathbf{a}\tilde{\mathbf{a}} + \mathbf{b}\tilde{\mathbf{b}}\right)/2$, which is now the product of two or more vectors. This is a consequence of the role of the additive Gaussian noise for the completely deterministic narrowband signal in coherent detection, and the same signal in incoherent detection with only a (uniformly) random RF epoch. (For non-Gaussian noise, these matched filters are still useful as a first approximation to optimality, particularly in threshold regimes.)

Finally, from Figs. 3.7–3.12, it is seen that the matched filters here are, of course, linear filters as required. This is the case without exception for coherent reception (Fig. 3.9). However, for incoherent reception, one has a choice, provided a suitable zero-memory *nonlinear* device is also included, refer to Figs. 3.10–3.12. Here, the matched filter can be time-invariant (Fig. 3.12) or time-variable (Fig. 3.10), unlike the coherent cases, which are always time-invariable adjustment always required to aid the matching process.

3.4.4 Wiener–Kolmogoroff Filters [22, 23]

As we have seen in the preceding Sections (3.4.1–3.4.3), the matched filter occurs in a variety of forms and is an essential component of the optimum detection process. Also as we have noted, such filters are inherently optimized as part of the Bayes or minimum average risk procedure. They are, of course, linear and deterministic and appear in a variety of forms: time-invariant, time-variable, causal, and noncausal, and all embody in some way the minimization of the effects of the noise accompanying the desired signal at the end of the observation interval (in both space and time), subject to various constraints upon the signal.

Matched filters that have the structure $\mathbf{H}^{(\;)} = \mathbf{k}_\mathrm{N}^{-1}\langle \mathbf{s}\rangle$ are known as Wiener–Kolmogoroff filters [22, 23], after the men who (independently) first studied them. (Refer to Section 16.3.1 of Ref. [1]; also see Ref. [24] and Chapter 16, Notes, pp. 1114 and 1115 of Ref. [1]). These same filters also maximize the output signal-to-noise ratios in linear systems [14, 15] and in the nonlinear systems of threshold detection. In addition, they minimize the mean square error between input signal and output signal corrupted by the noise, and as such are also known as MMSE filters. Finally, they also belong to the Bayes class of matched filters, which are part of optimum detection itself, as demonstrated below (Section 16.3.2 of Ref. [1] and Chapter 4 of [8]).

We show now that the W–K filters are also equivalent to the above time-invariant Bayes matched filters of detection, specifically to $\mathbf{H}^{(1)}$, (3.4.1), and so on, for coherent detection (see Table 3.1); and for $\mathbf{H}^{(2,2)}$, refer to (3.4.13), etc. For coherent detection, let us begin with the case of a deterministic signal with random parameters. Specifically, we consider the general class for which $\langle \mathbf{s}\rangle_{\boldsymbol{\theta}}$ is different from zero, that is, $\langle \mathbf{s}\rangle_{\boldsymbol{\theta}} \neq 0$. The noise field is additive (it may or may not be Gaussian, homogeneous, and (stationary), with a symmetric (real) covariance matrix $\mathbf{k}_\mathrm{N}(=\tilde{\mathbf{k}}_\mathrm{N})$. We wish now to minimize its variance, that is, its average intensity $\bar{I}_N$ at the output while keeping a constant signal output intensity—the condition for a W–K filter. If the (deterministic) vector $\mathbf{W}$ represents the W–K filter, we have initially

$$\tilde{\mathbf{W}}\langle \mathbf{s}\rangle = C_0(>0) : \text{constant output}\langle \mathbf{s}\rangle = \langle \mathbf{s}\rangle_{\boldsymbol{\theta}}, \text{and}\, \bar{I}_N = \left\langle \left(\tilde{\mathbf{W}}\mathbf{n}\right)^2\right\rangle = \left\langle \tilde{\mathbf{W}}\overline{\mathbf{n}\tilde{\mathbf{n}}}\mathbf{W}\right\rangle = \tilde{\mathbf{W}}\mathbf{k}_\mathrm{N}\mathbf{W}. \tag{3.4.22}$$

Next, we wish to minimize $\bar{I}_N$, subject to the constraint $\tilde{\mathbf{W}}\langle \mathbf{s}\rangle = C_0$, or equivalently,

$$\mathsf{L} \equiv \bar{I}_N + \lambda\tilde{\mathbf{W}}\langle \mathbf{s}\rangle, \quad \therefore \quad \delta\mathsf{L} = \delta\tilde{W}(2\mathbf{k}_\mathrm{N}\mathbf{W} + \lambda\langle \mathbf{s}\rangle) = 0, \tag{3.4.23}$$

where λ is a Lagrange multiplier (a constant) and the vector variation is $\delta\mathbf{W}\left(=\delta\tilde{\mathbf{W}}\right)$, since $\mathbf{k}_\mathrm{N} = \tilde{\mathbf{k}}_\mathrm{N}$, in the usual fashion. Setting $\delta\tilde{\mathbf{W}} = 0$, we obtain the extremum from (3.4.23), namely,

$$\mathbf{W} = (-\lambda/2)\mathbf{k}_{\mathrm{N}}^{-1}\langle\mathbf{s}\rangle = \langle\tilde{\mathbf{s}}\rangle\mathbf{W} = C_0; \quad \therefore \quad (-\lambda/2) = C_0/\langle\tilde{\mathbf{s}}\rangle\mathbf{k}_{\mathrm{N}}^{-1}\langle\mathbf{s}\rangle = C_0/\Phi_{\langle s\rangle} \equiv C_1, \tag{3.4.24}$$

which last determines the factor $(-\lambda/2)$. Accordingly, the weighting function of the W–K filter is

$$\mathbf{W} = C_0\mathbf{k}_{\mathrm{N}}^{-1}\langle\mathbf{s}\rangle/\Phi_{\langle s\rangle} = C_1\mathbf{k}_{\mathrm{N}}^{-1}\langle\mathbf{s}\rangle, \quad \text{where } \Phi_{\langle s\rangle} = \langle\tilde{\mathbf{s}}\rangle\mathbf{k}_{\mathrm{N}}^{-1}\langle\mathbf{s}\rangle = \sum_{jj'} s_j\left(\mathbf{k}_{\mathrm{N}}^{-1}\right)_{jj'}s_{j'}(>0). \tag{3.4.25}$$

To establish the minimizing nature of $\mathbf{W}$, that is, $\bar{I}_N$, (3.4.22), is a minimum, we must show that $\delta^2\mathsf{L} > 0$. This is easily done, since from (3.4.23) and (3.4.24) we obtain from $\delta\tilde{\mathbf{W}} = \left[2\mathbf{k}_{\mathrm{N}}\mathbf{W} - 2C_0 \cdot \langle\mathbf{s}\rangle/\Phi_{\langle s\rangle}\right]$

$$\delta^2\mathsf{L} = \delta^2\tilde{\mathbf{W}} \cdot \mathbf{0} + \delta\tilde{\mathbf{W}}\mathbf{k}_{\mathrm{N}}\boldsymbol{\delta}\mathbf{W} = \boldsymbol{\delta}\tilde{\mathbf{W}}\,\mathbf{k}_{\mathrm{N}}\boldsymbol{\delta}\mathbf{W} = \delta^2\bar{I}_{\mathrm{N}} > 0, \tag{3.4.26}$$

since $\bar{I}_N > 0$ and $\boldsymbol{\delta}\tilde{\mathbf{W}}\,\mathbf{k}_{\mathrm{N}}\,\boldsymbol{\delta}\mathbf{W}$ is positive definite. We obtain at once the discrete integral equation that $\mathbf{W}$ obeys, namely,

$$\mathbf{k}_{\mathrm{N}}\mathbf{W} = C_0\langle\mathbf{s}\rangle \quad \text{or} \quad \sum_{m'n'} k_{\mathrm{N}}(\mathbf{r}_m, t_n; \mathbf{r}_{m'}, t_{n'})W(\mathbf{r}_{m'}, t_{n'}) = C_0\langle s(\mathbf{r}_m, t_n)\rangle \tag{3.4.26a}$$

for all $(\mathbf{r}_m, t_n)$ in $(\mathbf{0}, |\mathbf{R}|; 0, T)$.

Comparisons with the matched filter $\mathbf{H}^{(1)} = \mathbf{k}_{\mathrm{N}}^{-1}\mathbf{s}$, derived from the optimum detector analysis above (Section 3.4.1) gives $\mathbf{W} = C_1\mathbf{H}_1$. Thus, except for the (positive) constant C_1 and the fact that now $\hat{\mathbf{s}} = \langle\hat{\mathbf{s}}\rangle$, *the W–K filter* (3.4.25) *is identical to the generic matched filter, Type I,* $\mathbf{H}^{(1)}$, *of the Bayes detection analysis of Sections 3.1 and 3.2*, here specifically in the case of originally Gaussian noise. (By adjusting C_1 to unity, that is, choosing $C_0 = \Phi_{\langle s\rangle}$, we establish the identity.) The equivalence of the W–K filter to $\mathbf{H}^{(2,2a)}$, $\mathbf{H}^{(4,1a)}$, and $\mathbf{H}^{(4,2b)}$ for incoherent reception in detection in Gaussian noise is established from the fact that these matched filters have the same form as $\mathbf{W}$. Accordingly, if we now require the W–K filter to be identical to the generic Bayes matched filter $\mathbf{H}^{(1)}$, *and thereby incorporate the W–K filter in the general Bayesian framework of statistical decision theory* (Chapter 1), we get $C_1 = 1$, so that $\mathbf{W} = \mathbf{H}^{(1)} = \mathbf{k}_{\mathrm{N}}^{-1}\langle\mathbf{s}\rangle$ here. Note, incidentally, during the observation interval or data acquisition period $(0 \le \mathbf{r}_m \le \mathbf{R}, 0 \le t_n \le T)$, the average signal intensity $\bar{I}_S$ at the output of the filter is held constant, refer to Eq. (3.4.22). We have here

$$I_S = (\bar{I}_S) = \left(\tilde{\mathbf{W}}\langle\mathbf{s}\rangle\right)^2 = \tilde{\mathbf{W}}\langle\mathbf{s}\rangle\langle\tilde{\mathbf{s}}\rangle\mathbf{W} = \tilde{\mathbf{W}}\mathbf{k}_{\langle s\rangle}\mathbf{W} = C_0^2 = \Phi_{\langle s\rangle}^2; \quad \mathbf{k}_{\langle s\rangle} = \langle\mathbf{s}\rangle\langle\tilde{\mathbf{s}}\rangle. \tag{3.4.27}$$

On the other hand, *when* $\langle\mathbf{s}\rangle = 0$, as is the case for incoherent detection and in some cases coherent detection as well, *it is not possible to employ a Wiener filter*. One must use a causal *time-varying filter*, of the second kind, Types 1 and 2, for example, $\mathbf{H}^{(2,1),(2,2)}$, $\mathbf{H}^{(4,1)}$, and $\mathbf{H}^{(4,2)}$, Sections 3.4.3, which are now described by matrices, not vectors like $\mathbf{W}$.

Finally, however, we must observe that the operations here are fundamentally different from the original MMSE tasks. The matched filter is a component of the detection process, whose ultimate outcome is expressed in terms of probabilities of correct and incorrect decisions. The W–K filter on the other hand is used for *linear estimation* (E), producing a minimum mean square estimator, in particular, with the signal known to be present. It depends statistically on the covariance of the noise only, which can be non-Gaussian. Detection (D), however, is sensitive to the full statistical nature of the noise, that is, to the joint pdf values, not only to the covariance. Moreover, in the general framework of decision theory, the cost functions of detection are different, basically constant costs versus quadratic error constraints in the W–K situation. That the matched filter in the cases where $\langle s\rangle \neq 0$ and the W–K filter are formally the same (to within a constant scale factor) in these different D and E operations is largely a statistical coincidence, attributable to the additive *Gaussian nature of the noise* (in detection) and to the required *linear* character of the two filter types.

3.4.5 Extensions: Clutter, Reverberation, and Ambient Noise

When the additive noise contains such signal-dependent noise as clutter and reverberation, (i.e., scatter noise) as well as ambient noise (which later can contain intentional or unintentional jamming), we can easily extend the matched filters to include these additional components. In place of (normalized) covariance of the usual added noise $\mathbf{k}_\mathrm{N}$, we have now[43]

$$\mathbf{k}_\mathrm{N} \rightarrow \mathbf{k}_\mathrm{N} + \hat{\mathbf{k}}_\mathrm{C}(S_\mathrm{RCC}) + \hat{\mathbf{k}}_\mathrm{I}, \quad \text{where } \hat{\mathbf{k}}_\mathrm{C} = \left[(b_\mathrm{C})^2_{jj'}(\mathbf{k}_\mathrm{C})_{jj'}\right];\ \hat{\mathbf{k}}_\mathrm{I} = \left[(b_\mathrm{I})^2_{jj'}(\mathbf{k}_\mathrm{I})_{jj'}\right]. \tag{3.4.28}$$

Thus,

$$(b_\mathrm{C})^2_{jj'} = \sqrt{\psi_{\mathrm{C}_j}\psi_{\mathrm{C}_{j'}}/\psi_{\mathrm{N}_j}\psi_{\mathrm{N}_{j'}}};\quad (b_\mathrm{I})^2_{jj'} = \sqrt{\psi_{\mathrm{I}_j}\psi_{\mathrm{I}_{j'}}/\psi_{\mathrm{N}_j}\psi_{\mathrm{N}_{j'}}}. \tag{3.4.28a}$$

Here $\hat{\mathbf{k}}_\mathrm{C}$ and $\hat{\mathbf{k}}_\mathrm{I}$ represent the covariance of the scatter noise and the jamming, respectively, normalized to the noise ψ_{N_j}. These latter may be highly non-Gaussian, especially when produced by rough terrain or wave surfaces, or such other sources as communications, lightning, and vehicular emissions (cf. Chapter 8). In addition, scatter or signal-generated sources can include "coherent," that is, resolvable multipath. Not only are these noise types encountered in electromagnetic propagation, their counterparts also occur in acoustic environments, both in the ocean and in the atmosphere. However, since matched filters are by definition linear and at most time-variable, it is sufficient to consider the first-order covariance (3.4.28) in their design and implementation, where the different relative levels of $\hat{\mathbf{k}}_\mathrm{C}, \hat{\mathbf{k}}_\mathrm{I}$ vis-à-vis $\mathbf{k}_\mathrm{N}$ are accounted for by their intensity scaling factors $(b_\mathrm{C})^2_{jj'}, (b_\mathrm{I})^2_{jj'}$, (3.4.28a).

Let us proceed formally and begin with the case of ambient and scatter noise, $\mathbf{k}_\mathrm{N} + \hat{\mathbf{k}}_\mathrm{C}(S_\mathrm{inc})$ in the quadratic forms $\Psi^{(3)}_{J\text{-coh}}$ and $\Psi^{(4)}_{J\text{-inc}}$:

$$\begin{aligned} \Psi^{(3)}_{J\text{-coh}} &= \tilde{\mathbf{x}}\mathbf{A}^{(3)}\langle\mathbf{s}\rangle, \quad \text{where} \quad \mathbf{A}^{(3)} = \left(\mathbf{k}_\mathrm{N} + \hat{\mathbf{k}}_\mathrm{C}\right)^{-1}; \\ \Psi^{(4)}_{J\text{-inc}} &= \tilde{\mathbf{x}}\mathbf{A}^{(4)}\mathbf{x}, \quad \text{with} \quad \mathbf{A}^{(4)} = \left(\mathbf{k}_\mathrm{N} + \hat{\mathbf{k}}_\mathrm{C}\right)^{-1}\hat{\mathbf{k}}_\mathrm{S}(\mathbf{k}_\mathrm{N} + \mathbf{k}_\mathrm{C})^{-1}, \end{aligned} \tag{3.4.29}$$

where $\hat{\mathbf{k}}_\mathrm{S} = \overline{a_0^2}\mathbf{k}_\mathrm{S}$ with $\overline{a_0^2}$ normalized to ψ_M, so that $|(k_\mathrm{S})_{jj'}| \leq 1$. From Eqs. (3.4.28) and (3.4.29), we can write

[43] Most of the covariances in this book are symmetric and scale-normalized, although the unnormalized representations are not necessarily Hom-Stat.

$$\Psi^{(3)}_{J-\text{coh}} = \tilde{\mathbf{x}}\mathbf{H}^{(3)} = \tilde{\mathbf{x}}\left(\mathbf{k}_{\text{N}} + \hat{\mathbf{k}}_{\text{C}}\right)^{-1}\langle\mathbf{s}\rangle, \quad \text{and} \quad \therefore \quad \mathbf{H}^{(3)} = \left(\mathbf{k}_{\text{N}} + \hat{\mathbf{k}}_{\text{C}}\right)^{-1}\langle\mathbf{s}\rangle, \text{ a vector,} \tag{3.4.30}$$

from which we can calculate $\mathbf{H}^{(3)}$, either from the following discrete integral equation or directly from (3.4.30), namely,

$$\left(\mathbf{k}_{\text{N}} + \hat{\mathbf{k}}_{\text{C}}\right)\mathbf{H}^{(3)} = \langle\mathbf{s}\rangle \quad \text{or} \quad \mathbf{H}^{(3)} = \left(\mathbf{k}_{\text{N}} + \hat{\mathbf{k}}_{\text{C}}\right)^{-1}\langle\mathbf{s}\rangle, (O; |\mathbf{R}|, T); \quad = 0 \text{ elsewhere.} \tag{3.4.31}$$

Like $\mathbf{H}^{(1)}$ in (3.4.3) and (3.4.4), $\mathbf{H}^{(3)}$ is causal, and is illustrated in Fig. 3.9 with obvious change in notation. A correlation receiver, Eq. (3.4.7), in "white" noise is a simple example, maximizing its output at $t_n = T$.

For incoherent reception, the generic quadratic form provides the obvious extensions of (3.4.8). We have

$$\Psi^{(4)}_{J-\text{inc}} = \tilde{\mathbf{x}}\mathbf{A}^{(4)}\mathbf{x}, \quad \text{with } \mathbf{A}^{(4)} \equiv \boldsymbol{\rho}^{(4)} \equiv \left(\mathbf{k}_{\text{N}} + \hat{\mathbf{k}}_{\text{C}}\right)^{-1}\hat{\mathbf{k}}_{\text{S}}\left(\mathbf{k}_{\text{N}} + \hat{\mathbf{k}}_{\text{C}}\right)^{-1} = \tilde{\mathbf{H}}^{(4,1)}\mathbf{H}^{(4,1)} = \tilde{\boldsymbol{\rho}}^{(4)}. \tag{3.4.32}$$

Paralleling the analysis of (3.4.9a)–(3.4.12), we obtain directly

$$\Psi^{(4)}_{J-\text{inc}} = \tilde{\mathbf{x}}\boldsymbol{\rho}^{(4)}\mathbf{x} = \tilde{\mathbf{x}}\tilde{\mathbf{H}}^{(4,1)}\mathbf{H}^{(4,1)}\mathbf{x}, \quad \text{with} \quad \tilde{\mathbf{H}}^{(4,1)}\mathbf{H}^{(4,1)} = \boldsymbol{\rho}^{(4)} = \tilde{\boldsymbol{\rho}}^{(4)} = \mathbf{u}\tilde{\mathbf{u}}. \tag{3.4.33}$$

Here $\mathbf{H}^{(4,1)}$, like $\mathbf{H}^{(2,1)}$, is a *lower triangular* matrix, refer to Eq. (3.4.9c) et seq. so that $\mathbf{H}^{(4,1)} \neq \tilde{\mathbf{H}}^{(4,1)}$ and is determined by $\boldsymbol{\rho}^{(4)}$, Eq. (3.4.32), as a result of the reduction of $\boldsymbol{\rho}^{(4)}$ by a suitable *congruent* transformation, refer to Eq. (3.4.8). $\mathbf{H}^{(4,1)}$, like $\mathbf{H}^{(2,1)}$, is a time-variable causal filter. Figure 3.10 also applies here, with obvious notational changes, and Eq. (3.4.10) shows its structure. Equation (3.4.12) gives $\Psi^{(4)}_{J-\text{inc}}$ in more detail, with $\mathbf{H}^{(4,1)}$ replacing $\mathbf{H}^{(2,1)}$, and so on.

A similar result applies to the decomposition of $\Psi^{(4)}_{J-\text{inc}}$ when we express it in the alternative form:

$$\Psi^{(4)}_{J-\text{inc}} = \tilde{\mathbf{x}}\mathbf{H}^{(4,2)}\mathbf{x}; \quad \boldsymbol{\rho}^{(4)} = \mathbf{H}^{(4,2)} = \tilde{\boldsymbol{\rho}}^{(4)}, \quad \text{with } \mathbf{H}^{(4,2)} \neq \tilde{\mathbf{H}}^{(4,1)}, \tag{3.4.34}$$

since $\mathbf{H}^{(4,2)}$ is a symmetric matrix, whereas $\mathbf{H}^{(4,1)}$ is a triangular one, refer to Eq. (3.4.10). However, the same technique used to make $\mathbf{H}^{(2,2)}$ causal (refer to Eq. (3.4.13) et seq.) may be employed here, with $\mathbf{A}^{(4)}$ given by (3.4.32) et seq., so that we can write alternatively to (3.4.13):

$$\Psi^{(4)}_{J-\text{inc}} = \tilde{\mathbf{x}}\tilde{\mathbf{H}}^{(4,2)}\mathbf{x} = \sum_{j}^{J} x_j \hat{x}_j; \quad \hat{\mathbf{x}} = \mathbf{H}^{(4,2)}\mathbf{x}, \quad \text{where } \mathbf{H}^{(4,2)} = \boldsymbol{\rho}^{(4)}. \tag{3.4.35}$$

In this case, Fig. 3.11 applies again.

Finally, we have cases $\mathbf{H}^{(4,1a)}, \mathbf{H}^{(4,1b)}$ as a variant of case $\mathbf{H}^{(4,1)}$, when $\boldsymbol{\rho}^{(4)}$ factors according to (3.4.18) and (3.4.20)[44], that is,

$$\boldsymbol{\rho}^{(4,1a)} = \left(\mathbf{k}_\mathrm{N} + \hat{\mathbf{k}}_\mathrm{C}\right)^{-1}\mathbf{a}\left[\tilde{\mathbf{a}}\left(\mathbf{k}_\mathrm{N} + \hat{\mathbf{k}}_\mathrm{C}\right)^{-1}\right] + \left[\left(\mathbf{k}_\mathrm{N} + \hat{\mathbf{k}}_\mathrm{C}\right)^{-1}\mathbf{b}\right]\left[\tilde{\mathbf{b}}\left(\mathbf{k}_\mathrm{N} + \hat{\mathbf{k}}_\mathrm{C}\right)^{-1}\right] \\ \text{or } \left(\mathbf{k}_\mathrm{N} + \hat{\mathbf{k}}_\mathrm{C}\right)^{-1}\langle\mathbf{s}\rangle\langle\tilde{\mathbf{s}}\rangle\left(\mathbf{k}_\mathrm{N} + \hat{\mathbf{k}}_\mathrm{C}\right)^{-1} \tag{3.4.36a}$$

and

$$\boldsymbol{\rho}^{(4,2b)} = \left(\mathbf{k}_\mathrm{N} + \hat{\mathbf{k}}_\mathrm{C}\right)^{-1}\hat{\mathbf{k}}_\mathrm{S}^{1/2} \cdot \hat{\mathbf{k}}_\mathrm{S}^{1/2}\left(\mathbf{k}_\mathrm{N} + \hat{\mathbf{k}}_\mathrm{C}\right)^{-1}. \tag{3.4.36b}$$

The first relation arises in the case of purely incoherent reception involving narrowband noise and signals, where only the RF epoch of the latter ε is uniformly distributed. The second occurs when $\hat{\mathbf{k}}_\mathrm{S} = \langle\mathbf{s}\tilde{\mathbf{s}}\rangle$. Then, we have formally,

$$\Psi_{J-\mathrm{inc}}^{(4a)} = \tilde{\mathbf{x}}\boldsymbol{\rho}^{(4,2a)}\mathbf{x} = \tilde{\mathbf{z}}\mathbf{z}, \quad \text{with} \quad \mathbf{z} = \tilde{\mathbf{H}}^{(4,2a)}\mathbf{x} \quad \text{and} \quad \mathbf{H}^{(4,2a)} = \left(\mathbf{k}_\mathrm{N} + \hat{\mathbf{k}}_\mathrm{C}\right)^{-1}\langle\mathbf{s}\rangle, \tag{3.4.37a}$$

$$\Psi_{J-inc}^{(4b)} = \tilde{\mathbf{x}}\boldsymbol{\rho}^{(4,2b)}\mathbf{x} = \tilde{\mathbf{v}}\mathbf{v}, \quad \text{with} \quad \mathbf{v} = \tilde{\mathbf{H}}^{(4,2b)}\mathbf{x} \quad \text{and} \quad \mathbf{H}^{(4,2b)} = \left(\mathbf{k}_\mathrm{N} + \hat{\mathbf{k}}_\mathrm{C}\right)^{-1}\hat{\mathbf{k}}_\mathrm{S}^{1/2}. \tag{3.4.37b}$$

Here $\mathbf{H}^{(4,2a)}$ is a *vector*, representing a causal time-invariant filter, and $\mathbf{H}^{(4,2b)}$ is a square symmetric *matrix*, which can be rendered causal by the methods of Section 3.4.3, refer to Eqs. (3.4.13) et seq. Figure 3.9 is applicable to $H^{(4,2a)}$, while Fig. 3.11 represents $H^{(4,2b)}$, which is a time-variable filter.

3.4.6 Matched Filters and Their Separation in Space and Time I

Here we examine the role of the matched filter on the *test statistic* Ψ_{x-J} (Eq. (3.4.1)) and on the detection parameter Ψ_{x-J}, when the separation of space and time processing is imposed at the receiver. This is a constraint on optimality, since for the received fields space and time are not generally physically separable (Section 2.5). The result can be a degradation of performance, especially for broadband signals. The trade-off is usually a simpler processing procedure when this condition is imposed, as it permits separate optimizations of the space and time portions of reception.

3.4.6.1 Separation of $\Psi_J^{(1)} = \tilde{\mathbf{x}}\mathbf{k}_\mathrm{N}^{-1}\mathbf{a}$ We begin by considering the input to the receiver of the general type $\tilde{\mathbf{x}}\mathbf{k}_\mathrm{N}^{-1}\mathbf{a}$ (Eq. (3.4.1)). First, we introduce the matched filter $\mathbf{H} = \mathbf{k}_\mathrm{N}^{-1}\mathbf{a}$, expressed in more detail as the solutions to the set of equations.

$$\mathbf{k}_\mathrm{N}\mathbf{H} = \mathbf{a} \quad \text{or} \quad \sum_{m',n'} k_\mathrm{N}(\mathbf{r}_m, t_n; \mathbf{r}_{m'}, t_{n'} \leftarrow t_n) H^{(a)}(\mathbf{r}_{m'}, t'_n) = a(\mathbf{r}_m, t_n), \quad \text{with}$$
$$0 \leq (\mathbf{r}_m, t_n) \leq (\mathbf{R}, T); \quad 0, \text{elsewhere}; \quad j = mn = 1, \ldots, J. \tag{3.4.38}$$

[44] Recall that $\mathbf{k}_\mathrm{N}, \hat{\mathbf{k}}_\mathrm{C}, \hat{\mathbf{k}}_\mathrm{S}$, and so on are all symmetric and positive definite.

Then, $\tilde{\mathbf{x}}\mathbf{k}_N^{-1}\mathbf{a} = \tilde{\mathbf{x}}\mathbf{H}^{(a)} = \sum_j x_j H_j^{(a)} = \sum_{m,n} H_{m,n}^{(a)} x_{m,n}$ and so on, which is the familiar result in the general case, refer to Eqs. (3.1.6) and (3.1.6a). (A similar result holds for $\tilde{\mathbf{x}}\mathbf{k}_N^{-1}\mathbf{b}$, namely, $\tilde{\mathbf{x}}\mathbf{k}_N^{-1}\mathbf{b} = \sum_{j=m,n} H_{m,n}^{(b)} x_{m,n}$ in the narrowband cases of Section 3.2.)

Next, we introduce the separation of space and time. This gives us (also for the broadband cases when $\mathbf{a} \to \hat{\mathbf{s}}$)

$$\tilde{\mathbf{x}}\mathbf{k}_N^{-1}\mathbf{a} = \tilde{\mathbf{x}}\mathbf{H}^{(a)} = \sum_{mn} x_m H_m^{(a)} u_n \hat{H}_n^{(a)} = \sum_m x_m H_m^{(a)} \cdot \sum_n u_n \hat{H}_n^{(a)}, \quad \text{where now } x_j = x(r_m)u(t_n)$$
$$\tilde{\mathbf{x}}\mathbf{k}_N^{-1}\mathbf{b} = \tilde{\mathbf{x}}\mathbf{H}^{(b)} = \sum_{j=mn} x_m u_n H_m^{(b)} \hat{H}_n^{(b)} \tag{3.4.39}$$

The discrete integral equations (3.4.38) now become (for nb noise)

$$\sum_{m'} k_N(\mathbf{r}_m, \mathbf{r}_{m'}) \begin{pmatrix} H^{(a)}(\mathbf{r}_{m'}) \\ H^{(b)}(\mathbf{r}_{m'}) \end{pmatrix} \begin{matrix} = \\ = \end{matrix} \begin{pmatrix} a(\mathbf{r}_m) \\ b(\mathbf{r}_m) \end{pmatrix}; \quad m = 1, \ldots, M$$
$$0 \leq \mathbf{r}_1, \mathbf{r}_2, \ldots, \mathbf{r}_M \leq \mathbf{R} \tag{3.4.40a}$$

$$\sum_{n'} k_N(t_n, t'_n) \begin{pmatrix} \hat{H}^{(a)}(t'_n) \\ \hat{H}^{(b)}(t'_n) \end{pmatrix} \begin{matrix} = \\ = \end{matrix} \left.\begin{pmatrix} \hat{a}(t_n) \\ \hat{b}(t_n) \end{pmatrix}\right\}; \quad n = 1, \ldots, N$$
$$0 \leq t_1, t_2, \ldots, t_N \leq T \tag{3.4.40b}$$

Equation (3.4.39) can also be written

$$\left(\tilde{\mathbf{x}}\mathbf{k}_N^{-1}\mathbf{a}\right)_{\text{nb}} = \sum_m A_m \sum_n B_n = \left(\tilde{\mathbf{1}}_M \cdot \mathbf{A}\right)\left(\tilde{\mathbf{1}}_N \cdot \mathbf{B}\right), \quad \begin{matrix} \mathbf{1}_M = [1, 1, \ldots, 1_M] \\ \mathbf{1}_N = [1, 1, \ldots, 1_N]. \end{matrix} \tag{3.4.41}$$

To achieve $\left(\tilde{\mathbf{x}}\mathbf{k}_N^{-1}\mathbf{a}\right)_{\text{nb}}^2$, we proceed as follows:

$$\left.\begin{aligned} \left(\tilde{\mathbf{x}}\mathbf{k}_N^{-1}\mathbf{a}\right)_{\text{nb}}^2 &= (\tilde{\mathbf{x}}\mathbf{H})^2 \Rightarrow \left|\sum_{m,n} \left(x_m H_m u_n \hat{H}_n\right)\right|^2 = \sum_{mn}\sum_{m'n'} \left(x_m H_m x_{m'} H_{m'}\right)\left(u_n \hat{H}_n u_{n'} \hat{H}_{n'}\right) \\ &= \sum_{mm'}\sum_{nn'} \left(\mathbf{A}^{(a)} \otimes \mathbf{B}^{(a)}\right); \quad \text{where} \quad \mathbf{A}^{(a)} = [A_{mm'}] = [x_m H_m x_{m'} H_{m'}] \\ &\qquad \mathbf{B}^{(a)} = [B_{nn'}] = \left[u_n \hat{H}_n u_{n'} \hat{H}_{n'}\right] \end{aligned}\right\}. \tag{3.4.42}$$

To diagram this, we write as in Fig. 3.1.3.

3.4.6.2 *Separated Structure of the Test Statistic* $\Psi^*_{x-\text{inc}}$

We are now ready to give the complete structure of $\Psi^*_{x-\text{inc}} = \left(\tilde{\mathbf{x}}\mathbf{k}_N^{-1}\mathbf{a}\right)^2 + \left(\tilde{\mathbf{x}}\mathbf{k}_N^{-1}\mathbf{b}\right)^2$ in the separable cases. Referring to Eqs. (3.2.5)–(3.2.9),

$$\Psi^*_{x-\text{inc}}\big|_{\text{nb}} = \left(\tilde{\mathbf{x}}\mathbf{k}_N^{-1}\mathbf{a}\right)^2 + \left(\tilde{\mathbf{x}}\mathbf{k}_N^{-1}\mathbf{b}\right)^2 = \tilde{\mathbf{x}}\left[\mathbf{k}_N^{-1}\left(\mathbf{a}\tilde{\mathbf{a}} + \mathbf{b}\tilde{\mathbf{b}}\right)\mathbf{k}_N^{-1}\right]\mathbf{x} \equiv \tilde{\mathbf{x}}\mathbf{C}\mathbf{x}. \tag{3.4.43}$$

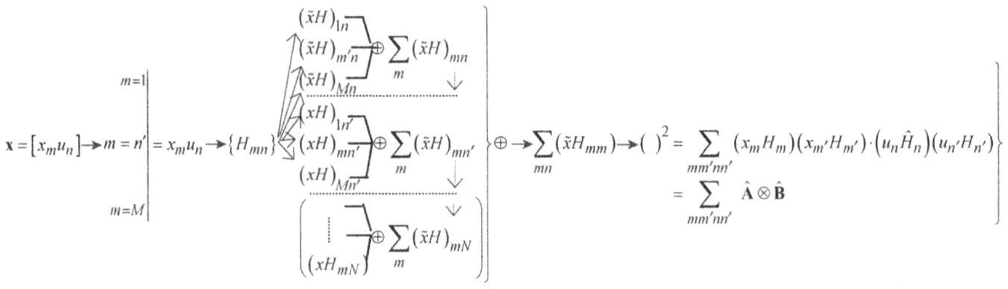

FIGURE 3.13 Schematic of $\tilde{\mathbf{x}}\mathbf{k}_{\mathrm{N}}^{-1}\mathbf{a}$, when $\mathbf{x} = [x(r_m)u(t_n)]$ is separated into space and time components.

For this we need a parallel branch for $\left(\tilde{\mathbf{x}}\mathbf{k}_{\mathrm{N}}^{-1}\mathbf{b}\right)^2$ and as above a designation to distinguish the matched filters $\mathbf{H}^{(a)}$ from $\mathbf{H}^{(b)}$. For $\left(\tilde{\mathbf{x}}\mathbf{k}_{\mathrm{N}}^{-1}\mathbf{b}\right)^2$, following the analysis for $\left(\tilde{\mathbf{x}}\mathbf{k}^{-1}\mathbf{a}\right)$, we have

$$\left(\tilde{\mathbf{x}}\mathbf{k}_{\mathrm{N}}^{-1}\mathbf{b}\right)^2 = \sum_{mm'}\sum_{nn'} \mathbf{A}^{(b)} \otimes \mathbf{B}^{(b)} = \sum_{mm'}\sum_{nn'} \left(x_m H_m^{(b)} x_m H_{m'}^{(b)}\right)\left(u_n^{(b)} \hat{H}_n^{(b)} u_{n'}^{(b)} \hat{H}_{n'}^{(b)}\right)_{\mathrm{nb}}. \tag{3.4.44}$$

Thus, the complete reduction of the separable space–time data is

$$\Psi^*_{x-\mathrm{inc}} \to \Psi^*_{x-\mathrm{inc}}\big|_{\mathrm{sep.}} = \sum_{mm'}\sum_{nn'} \Big[\left(x_m H_m^{(a)} x_{m'} H_{m'}^{(a)}\right)\left(u_n^{(a)} \hat{H}_n^{(a)} u_{n'}^{(a)} \hat{H}_{n'}^{(a)}\right) + \left(x_m H_m^{(b)} x_{m'} H_{m'}^{(b)}\right)\left(u_n^{(b)} \hat{H}_n^{(b)} u_{n'}^{(b)} \hat{H}_{n'}^{(b)}\right)\Big]_{\mathrm{nb}} \tag{3.4.45}$$

with a second branch added at $\mathbf{H} \to H^{(b)}_{j=mn}$ in Fig. 3.13.

3.4.6.3 Separated Structure of the Detection Parameter Ψ^*_{s} Performance is measured for coherent detection using the detection parameter $\Psi^*_{s-\mathrm{coh}} = (\hat{\tilde{\mathbf{s}}}\mathbf{k}_{\mathrm{N}}^{-1}\hat{\mathbf{s}})_{S\otimes T}$, refer to Eqs. (3.1.3a) and (3.1.9a) for the coherent cases treated here, and for the incoherent cases, with the help of $\Psi^*_{s-\mathrm{inc}} = \left[\left(\tilde{\mathbf{a}}\mathbf{k}_{\mathrm{N}}^{-1}\mathbf{a} + \tilde{\mathbf{b}}\mathbf{k}_{\mathrm{N}}^{-1}\mathbf{b}\right)/2\right]_{S\otimes T:\mathrm{nb}}$, refer to Eqs. (3.2.5), (3.2.19a), and (3.2.22). We have

$$\Psi^*_{s-\mathrm{coh}}\big|_{\mathrm{sep.}} = \sum_{mm'} \underbrace{\hat{s}_m \left(\mathbf{k}_{\mathrm{N}}^{-1}\right)^{(r)}_{mm'} \hat{s}_{m'}} \cdot \sum_{nn'} \underbrace{u_n \left(\mathbf{k}_{\mathrm{N}}^{-1}\right)^{(r)}_{nn'} u_{n'}}\ ; \mathbf{k}_{\mathrm{N}}^{(x)} = k_{\mathrm{N}}(\mathbf{r}_m, \mathbf{r}_{m'}),\ \mathbf{k}_{\mathrm{N}}^{(t)} = k_{\mathrm{N}}(t_n, t_{n'})$$

$$= \sum \hat{s}_m H_m \cdot \sum_n u_n \hat{H}_n \tag{3.4.46}$$

for the coherent cases, where $H_m, \hat{H}_n$ are the respective matched filters, each respectively determined by the set $m = 1, \ldots, M$, and $n - 1, \ldots, N$ equations, in the manner of (3.4.40a) and (3.4.40b). For the incoherent cases, $\Psi^*_{s-\mathrm{inc}|\mathrm{sep.}}$ becomes

$$\Psi^*_{s-\text{inc}}\big|_{\text{sep}} = \frac{1}{2}\left(\sum_m a_m H_m^{(a)} \sum_n u_n \hat{H}_n^{(a)} + \sum_m b_m H_m^{(b)} \sum_n v_n \hat{H}_n^{(b)}\right)_{\text{nb}}, \quad a_n = u(t_n), v_n = v(t_n), \tag{3.4.47}$$

which is a simpler result than $\Psi_{x-\text{inc}|\text{sep.}}$. Alternatively, solving for $H_m^{(a)(b)}, \hat{H}_n^{(a)(b)}$ gives the pair of discrete (sets of) equations (3.4.41) in the general cases of imposed separability. Performance in the two situations is compared by contrasting the effects of $\Psi^*_{s-\text{coh}}, \Psi^*_{s-\text{inc}}$ with $\Psi^*_{s-\text{coh}}\big|_{\text{sep}}, \Psi^*_{s-\text{inc}}\big|_{\text{sep}}$, respectively, in the expressions for the error probabilities under H_0 and H_1. We have from (3.1.3a) and Eqs. (3.4.46) and (3.4.47)

$$\Psi^*_{s-\text{coh}} = \sum_{j=mn} \hat{s}(r_m, t_n) H(\mathbf{r}_m, t_n) \geq \sum_{j=mn} \hat{s}(\mathbf{r}_m) H(\mathbf{r}_m) u(t_n) \hat{H}(t_n) \equiv \Psi^*_{s-\text{coh}}\big|_{\text{sep}}, \tag{3.4.48a}$$

$$\Psi^*_{s-\text{inc}} = \frac{1}{2}\sum_{j=mn}\left[a_j H_j^{(a)} + b_j H_j^{(b)}\right] \geq \frac{1}{2}\sum_{mn}\left(a_m u_n H_m^{(a)} \hat{H}_n^{(a)} + b_m v_n H_m^{(b)} \hat{H}_n^{(b)}\right) \equiv \Psi^*_{s-\text{inc}}\big|_{\text{sep}-\text{nb}}, \tag{3.4.48b}$$

with the unconstrained case giving a higher (or equal) value of Ψ^*_s than the constrained case, since a constraint generally limits optimality.

The matched filters in time have a direction, from a point in the past to the "now" of operation. Thus, we wish the filter to "accumulate" the past until the present, and so we set their response (i.e., memory) to have the form $H_n = H(T_N - t_n), 0 \leq t_n \leq T_N$; 0 otherwise, where $T_N = N\Delta t = t_N$. (A delay-line filter with cutoff at $t = T$, the end of the observation period (T_N) will serve.) For the spatial filter, there is not a preferred direction: forward or backward in space in arbitrary, and hence $H(\mathbf{r}_m) = H(-\mathbf{r}_m)$. We choose one or the other and accumulate, the direction being optimal. Thus, we represent the matched filters of the unconstrained cases by $H_{mn} = H(\mathbf{r}_m, T_N - t_n), 0 \leq r_m \leq |\mathbf{R}|, 0 < t_n < T_N$; and $= 0$ outside this space–time interval, with the spatial constraint determined by the size of the aperture (or array) and time constraints specified by a setting on the delay line, namely, by the observation period $(0, T_N)$. In the common occurrence of (imposed) separability of space and time operations, the matched filters response is $H_{mn} = H_m \hat{H}_n = H(\mathbf{r}_m) H(T_N - t_n)$, with the aforementioned limits.

3.4.6.4 "White" Noise in Space and Time A situation of frequent occurrence is one where the noise is "white" in space and time, so that $\delta_{jj'} = \delta_{mm'}\delta_{nn'}$, or at least is approximated by this condition. Then, the discrete integral equations are immediately soluble, since $\mathbf{k}_\text{N} = \delta_{jj'} = \delta_{mm'}\delta_{nn'}$ and therefore

$$\text{(unconstrained):} \quad H(\mathbf{r}_m, t_n) = a(\mathbf{r}_m, t_n)$$

$$\text{(constrained):} \quad H_m^{(a)} = a(\mathbf{r}_m); \quad H_m^{(b)} = b(\mathbf{r}_m); \quad \hat{H}_n^{(a)} = \hat{a}(t_n); \quad \hat{H}_n^{(b)} = \hat{b}(t_n). \tag{3.4.49}$$

Specifically, for the coherent and narrowband incoherent cases, we have

$$\hat{s}_j = a_{on}^{(m)} s_0^{(m)}; \quad (\hat{s}_j)_{\text{sep}} = \frac{A_m \hat{s}_m \cdot A_n \hat{s}_n}{\sqrt{\psi_j}} = \frac{A(\mathbf{r}_m)\hat{s}(\mathbf{r}_m) \cdot A_0(t_n)\hat{s}(t_n)}{\sqrt{\psi_j}}, \tag{3.4.50a}$$

Unseparated: *Separated:*

$$\left.\begin{matrix} a_j \\ b_j \end{matrix}\right]_{\text{nb}} \begin{matrix} = A_j \cos\Phi_j = \dfrac{A_{0_n}^{(m)}}{\sqrt{\psi_j}} \cos((\omega_0 t_n - \phi_n) - (\mathbf{k}_0 - \mathbf{k}_{O_R}) \cdot \mathbf{r}_m) \\ = A_j \sin\Phi_j = \dfrac{A_{0_n}^{(m)}}{\sqrt{\psi_j}} \sin(\omega_0 t_n - \phi_n - (\mathbf{k}_0 - \mathbf{k}_{O_R}) \cdot \mathbf{r}_m) \end{matrix} \right\}; \quad \left.\begin{matrix} a_j \\ b_j \end{matrix}\right]_{\text{sep}} \begin{matrix} = \dfrac{A_m(\mathbf{r}_m) A_0(t_n)}{\sqrt{\psi_j}} [\cos()_n \cos()_m + \sin()_n \sin()_m] \\ = \dfrac{A(\mathbf{r}_m) A_0(t_n)}{\sqrt{\psi_j}} [\sin()_n \cos()_n + \cos()_m \sin()_m]. \end{matrix} \tag{3.4.50b}$$

The performance parameters (3.4.46) and (3.4.47) become with the help of (3.4.49).

$$\Psi^*_{s\text{-coh(sep)}} = \sum_m \hat{s}_m^2 \sum_n u_n^2; \quad \Psi^*_{s\text{-inc}}\Big|^{\text{sep}}_{\text{nb}} = \frac{1}{2} \sum_{mn} \left(a_m^2 u_n^2 + b_m^2 v_n^2\right)_{\text{nb}} \tag{3.4.51}$$

$$= \psi^{-1} \sum_{mn} A_m^2 s_m^2 A_n^2 s_n^2 \doteq \frac{1}{2\psi} \sum_{mn} A_m^2 A_n^2 = \frac{1}{2\psi} \sum_{mn} A_0^{(m)}(r_m)^2 A_0(t_n)^2. \tag{3.4.52}$$

For the *nonseparated* cases, we have again using

$$\begin{aligned} \Psi^*_{s\text{-coh}} &= \sum_{mn} \hat{s}_n^{(a)^2} = \left(2\psi \sum_{mn} A_0^2(\mathbf{r}_m, t_n) s_n^{(m)}\right); \Psi^*_{s\text{-inc}}\Big|_{\text{nb}} \\ &\doteq \frac{1}{2} \sum_{mn} \left[a(\mathbf{r}_m, t_n)^2 + b(\mathbf{r}_m, t_n)^2\right] = \frac{1}{2} \sum_{mn} A_0^{(m)}(\mathbf{r}_m, t_n)^2_{\text{nb}}, \end{aligned} \tag{3.4.53}$$

which shows that except for special cases (e.g., $A_0^{(m)}(\mathbf{r}_m, t_n)^2 = A_0^{(n)}(\mathbf{r}_m)^2 A_0(t_n)^2$), they are *not* equal.

In addition, the test functions Ψ^*_x in the two situations are also not equal. We have

$$\Psi^*_{x\text{-coh}} = \tilde{\mathbf{H}}\mathbf{x} = \tilde{\hat{\mathbf{s}}}\mathbf{x} = \sum\nolimits_j \hat{s}_j x_j$$

$$\rightarrow \sum_{mn} H_m x_m H_n v_n = \sum_{mn} x(\mathbf{r}_m) a(\mathbf{r}_m) u(t_n) \hat{a}(t_n) = \left[\sum_m x(\mathbf{r}_m) a(\mathbf{r}_m)\right] \sum_n u(t) \hat{a}(t_n)$$

$$= \mathsf{A}_\text{R} \sum u(t) \hat{a}(t) \tag{3.4.54a}$$

$$\left.\begin{aligned}\Psi^*_{x-\text{inc}} &= \left(\tilde{\mathbf{H}}^{(a)}\mathbf{x}\right)^2 + \left(\tilde{\mathbf{H}}^{(b)}\mathbf{x}\right)^2 = \tilde{\mathbf{x}}\left(\mathbf{H}^{(a)}\tilde{\mathbf{H}}^{(a)} + \mathbf{H}^{(b)}\tilde{\mathbf{H}}^{(b)}\right)\mathbf{x}\\ &\rightarrow \sum_{mn}\sum_{m'n'}\left[\left(x_m x_{m'} H_m^{(a)} H_{m'}^{(a)}\right)\left(u_n u_{n'} \hat{H}_n \hat{H}_{n'}\right) + (\,)^{(b)}_{mm'}(\,)^{(b)}_{nn'}\right]\\ &= \sum_{mn} x_m^2\left(a_m^2 \hat{a}_n^4 + b_m^2 \hat{b}_n^4\right) = \sum_{mn} x(\mathbf{r}_m)^2\left[a(\mathbf{r}_m)^2 \hat{a}(t_n)^4 + b(\mathbf{r}_m)^2 \hat{b}(t_n)^4\right]\\ &= \mathsf{A}_{\mathrm{R}}^{(a)}\sum_n \hat{a}_n^4 + \mathsf{A}_{\mathrm{R}}^{(b)}\sum_n \hat{b}_n^4\end{aligned}\right\}\ \text{separated} \tag{3.4.54b}$$

where the differences are directly seen. (The different indexes (m, n) here and above denote different functions.) The A_{R}, $\mathsf{A}_{\mathrm{R}}^{(a)}$, $\mathsf{A}_{\mathrm{R}}^{(b)}$ are the array contributions, and exhibit equivalencies for the case $(M = 1)$, where the array is separately optimized or not. Since the unconstrained cases give a higher (or equal) value of Ψ^*_x and Ψ^*_s, we may expect that these test functions will provide a higher (or equal) value against the prechosen threshold. Note that if we use for the signal **a** or **b** (a choice that is arbitrary, refer to Eq. (3.2.31)), we need to consider only the terms involving (**a**) or (**b**) in $\Psi^*_{s-\text{inc}}$ above. For the test statistic, however, both terms **a** and **b** are required. In case of $\Psi^*_{s-\text{inc}}$, this simplifies the relation in Sections 3.4.6.3 and 3.4.6.4. In the more general situation involving averages over the signal parameters, we replace **s** by $\langle\mathbf{s}\rangle$, **a** by $\langle\mathbf{a}\rangle$, and so on in the quadratic forms (3.4.1) defining these matched filters (represented by $H_n^{(a)}$, $\hat{H}_n^{(A)}$ et seq.). The averages over the space and time factors are likewise separated, as they are for the solutions to the discrete integral equations for the matched filters.

3.4.7 Solutions of the Discrete Integral Equations

Finally, there remains the task of evaluating the various discrete integral equations that define the matched filters discussed in Sections 3.4.1–3.4.6. These matched filters are basically described by three types of disparate relations, with a number of variants, summarized in Section 3.4.8 following, refer to Table 3.1. The generic types are

(i) $\mathbf{k}_{\mathrm{N}}\begin{pmatrix}\mathbf{H}^{(1)}sW\\ \mathbf{H}^{(2,2a)}\end{pmatrix} = \langle\mathbf{s}\rangle$

(ii) Eqs.(3.4.21) $\mathbf{k}_{\mathrm{N}}\begin{pmatrix}\mathbf{H}^{(2,2b)}\\ \mathbf{H}^{(2,1)}\end{pmatrix} = \hat{\mathbf{k}}_{\mathrm{S}}^{1/2}$

(iii) $\mathbf{k}_{\mathrm{N}}\mathbf{H}^{(2,2)} = \hat{\mathbf{k}}_{\mathrm{S}}\hat{\mathbf{k}}_{\mathrm{N}}^{-1}$ (3.4.55)

in the space–time sampling interval $\Delta = [S] \otimes [T]$, and is 0 outside this interval, with the normalized covariance of noise and signal (where appropriate).[45] These noise and signal covariances are furthermore postulated to be real, symmetric, and positive definite, and unless otherwise indicated, represent *non*-Hom-Stat fields.

[45] Note that for the unnormalized forms, one has $\mathbf{K}_{\mathrm{N}} = \left[\sqrt{\psi_j\psi_{j'}}(k_{\mathrm{N}})_{jj'}\right]$ and $\mathbf{k}_{\mathrm{N}}$ in (i) is replaced by $\mathbf{K}_{\mathrm{N}} = \hat{\mathbf{H}}^{(\,)} = \langle\mathbf{S}\rangle$, where $\hat{\mathbf{H}}^{(\,)} = \left[\mathbf{H}^{(\,)}/\psi_j^{1/2}\right]$, vide (3.1.6) et seq., and $\hat{\mathbf{H}}^{(\,)}$ now has the dimensions of $[S]^{-1}$, $[S]$ denoting signal amplitude. Alternatively, all the normalized quantities in (3.4.55) are dimensionless. See p. 142 and the remarks in Section 3.4.7.2.

The solutions to these equations essentially require at least finding the eigenvalues of the various covariances.[46] This may be accomplished in a variety of efficient ways, depending on the size of the covariance matrixes involved. Before referring to their specific numerical techniques [19, 20], let us obtain the formal solutions for these (real) matched filters by employing orthogonal matrices $(\mathbf{R}_1, \mathbf{R}_2, \ldots)$ to reduce them to diagonalized form, that is, in terms of their eigenvalues. The following examples illustrate the procedures for the Bayes matched filter types summarized in Eq. (3.4.55) and derived in Sections 3.4.1–3.4.5.

Here, the real normalized covariance $\mathbf{k}_\mathrm{N}$, which represents the more general situation of non-Hom-Stat fields, is

(i) symmetric and positive definite, that is, $\mathbf{k}_\mathrm{N} = \tilde{\mathbf{k}}_\mathrm{N}$;

(ii) has a defined inverse, also symmetric and positive definite, that is, det $\mathbf{k}_\mathrm{N} \neq 0, \mathbf{k}_\mathrm{N}^{-1} = \left(\tilde{\mathbf{k}}_\mathrm{N}\right)^{-1}$;

(iii) has real positive eigenvalues, none of which are zero, which is the necessary and sufficient condition for (i) and for which a real orthogonal matrix $\mathbf{R}$ is obtainable.[47] The orthogonal matrix $\mathbf{R}$ has the defining property that

$$\tilde{\mathbf{R}}\mathbf{R} = \mathbf{R}\tilde{\mathbf{R}} = \mathbf{I} \quad \text{or} \quad \mathbf{R} = \tilde{\mathbf{R}}^{-1}, \quad \tilde{\mathbf{R}} = \mathbf{R}^{-1}, \tag{3.4.56}$$

where $\mathbf{I}$ as usual is the identity matrix $\mathbf{I} = \left[\delta_{jj'}\right] = \mathrm{diag}(1, 1, \ldots, 1)$. The orthogonal matrix $\mathbf{R}_1$ diagonalizes $\mathbf{k}_\mathrm{N}$, that is, $\tilde{\mathbf{R}}\mathbf{k}_\mathrm{N}\mathbf{R}_1 = \boldsymbol{\Lambda}_\mathrm{N}\left(\lambda_{\mathrm{N}_j}\right)$. Let us consider the following examples:

Example 1: $\quad \mathbf{k}_n\mathbf{H}^{(1)} = \mathbf{s} : \therefore \quad \mathbf{H}^{(1)} = \mathbf{k}_\mathrm{N}^{-1}\mathbf{s}$

$$
\begin{aligned}
&\text{or}\\
&\mathbf{H}^1\mathbf{R}_1 = \mathbf{k}_\mathrm{N}^{-1}\mathbf{s}\mathbf{R}_1 = \mathbf{k}_\mathrm{N}^{-1}\tilde{\mathbf{R}}_1\tilde{\mathbf{s}}\\
&\therefore \mathbf{R}_1\tilde{\mathbf{R}}_1\tilde{\mathbf{H}}^{(1)} = \mathbf{R}_1\mathbf{k}_\mathrm{N}^{-1}\tilde{\mathbf{R}}\tilde{\mathbf{s}}\\
&\tilde{\mathbf{H}}^{(1)} = \boldsymbol{\Lambda}_\mathrm{N}^{-1}\tilde{s} \quad \text{or} \quad \mathbf{H}^{(1)} = \mathbf{s}\boldsymbol{\Lambda}_\mathrm{N}^{-1},\\
&\text{or}\\
&\mathbf{H}^{(1)}\left[\sum_{j'} s_j\delta_{jj'} \Big/ \lambda_{Nj'}\right] = \left[s_j/\lambda_{Nj}\right]
\end{aligned}
\tag{3.4.57}
$$

$$
\left.\begin{aligned}
\text{where } \boldsymbol{\Lambda}_\mathrm{N} &= \left[\lambda_{Nj}\cdot\delta_{jj'}\right] = \tilde{\boldsymbol{\Lambda}}_\mathrm{N}\\
\text{or} \quad \boldsymbol{\Lambda}_\mathrm{N}^{-1} &= \left[\delta_{jj'}/\lambda_{Nj}\right] = \tilde{\boldsymbol{\Lambda}}_\mathrm{N}^{-1}
\end{aligned}\right\}
$$

where we have used the diagonalizing properties of $\mathbf{R}_1$, refer to the various steps leading to (3.4.57). The result is the eigenvalue reduction of the normalized covariance $\mathbf{k}_\mathrm{N}$ and the

[46] The square root of a real matrix requires that the matrix be symmetric and at least positive semidefinite (Chapter 6, Section 5, pp. 92 and 93 of Ref. [21].

[47] See pp. 54–58 of Ref. [21] for these and for other properties of general symmetric matrices, refer to Chapter 4, ibid.

desired solution for the matched filter $\mathbf{H}^{(1)}$, which can be made identical to $\mathbf{W}$, the W–K filter (3.4.26a), namely:

$$\mathbf{H}^{(1)} = \left[s_j/\lambda_{Nj}\right] = W, (\text{Eq.}(3.4.26\text{a}) \text{ with } C_1 = 1, \text{p. } 185). \tag{3.4.58}$$

In the situation where the coherent average of the signal $\langle \mathbf{s} \rangle$ is different from zero, we see at (3.4.57)

$$\textit{Example 1a}: \qquad \mathbf{k}_N\mathbf{H}^{(1)} = \langle \mathbf{s} \rangle : \mathbf{H}^{(1)} = \left[\langle s_j \rangle/\lambda_{Nj}\right] = \langle \mathbf{W}_s \rangle \tag{3.4.58a}$$

For the unnormalized cases, $\mathbf{K}_N$, we obtain

$$\textit{Example 1b}: \mathbf{K}_N\hat{\mathbf{H}}^{(1)} = \langle \mathbf{S} \rangle : \hat{\mathbf{H}}^{(1)} = \left[\langle S_j \rangle \Big/ \hat{\lambda}_{Nj}\right] = \left[\langle s_j \rangle \Big/ \sqrt{\psi_j \lambda_{Nj}}\right], \quad \text{with } \hat{\lambda}_{Nj} = \psi_j \lambda_{Nj}. \tag{3.4.58b}$$

When we consider the cases of signal samples represented by symmetric (positive definite) matrices like $\hat{\mathbf{k}}_s^{1/2}$, refer to (3.4.21) we must use another (necessary and sufficient) property of orthogonal matrices, namely, that *one orthogonal matrix can simultaneously diagonalize two symmetrical matrices* (like $\mathbf{k}_N$ and $\hat{\mathbf{k}}_s^{1/2}$), *provided they commute*, which they clearly do here (see Chapter 4, Theorem 5 of Section 11 of Ref. [21]) The Bayes matched filters in this example are as follows:

$$\textit{Example 2}: \mathbf{H}^{(2)} = \mathbf{k}_N^{-1}\hat{\mathbf{k}}_s^{1/2} : \tilde{\mathbf{R}}_2\mathbf{H}^{(2)}\mathbf{R}_2 = \tilde{\mathbf{R}}_2\mathbf{k}_N^{-1}\mathbf{R}_2\tilde{\mathbf{R}}_2\hat{\mathbf{k}}_s^{1/2}\mathbf{R}_2 = \mathbf{\Lambda}_N^{-1}\mathbf{\Lambda}_s^{1/2}$$
$$\text{by Theorem 5, Section 11, Chapter 4 of Ref. [19]}, \tag{3.4.59a}$$

Multiply the right-hand side by $\mathbf{R}_2^{-1}$, followed by multiplication of the left-hand side by $\mathbf{R}_2$, to get

$$\mathbf{R}_2\tilde{\mathbf{R}}_2\mathbf{H}^{(2)} = \mathbf{R}_2\mathbf{R}_2^{-1} = \mathbf{H}^{(2)} = \mathbf{R}_2\mathbf{\Lambda}_N^{-1}\mathbf{\Lambda}_s^{1/2}\mathbf{R}_2^{-1} = \mathbf{R}_2\mathbf{R}_2^{-1}\hat{\mathbf{\Lambda}}_s^{1/2}\mathbf{\Lambda}_N^{-1} \tag{3.4.59b}$$

or

$$\text{Eq. (3.4.21)}: \quad \mathbf{H}^{(2,1)(2,2b)}\mathbf{\Lambda}_s^{1/2}\mathbf{\Lambda}_N^{-1} = \left[\sum_{j'} \lambda_{N_j}^{-1}\delta_{jj'}\lambda_{sj'k}^{1/2}\right] = \left[\lambda_{N_j}^{-1}\delta_{jj'}\lambda_s^{1/2}\right]. \tag{3.4.60}$$

By the same method, we get $\mathbf{H}^{(2,2)} = \mathbf{k}_N^{-1}\hat{\mathbf{k}}_s\mathbf{k}_N^{-1}$

$$\text{Eq.(3.4.13)}: \quad \mathbf{H}^{(2,2)} = \left[\hat{\lambda}_{sj}\delta_{jj'}/\lambda_{Nj}^2\right], \tag{3.4.60b}$$

Note that $\mathbf{R}_1, \mathbf{R}_2$, and so on are different orthogonal matrices, depending on the different covariances involved, that is, $\mathbf{k}_N^{-1}$ and $\mathbf{k}_s^{1/2}$.

A somewhat extended version of the above technique can be used in case of two or more added noise covariances, say, involving clutter, interference, or reverberation, as well as general receiver or ambient noise in the channel, that is, Section 3.4.5. We obtain, for example, in case of clutter and ambient noise fields,

$$\text{Eqs. (3.4.31), (3.4.36a), and (3.4.37b)}: \left(\mathbf{k}_{\mathrm{N}}+\hat{\mathbf{k}}_{\mathrm{C}}\right)\mathbf{H}^{(4,2a)}=\langle\mathbf{s}\rangle;$$

$$\text{Eq. (3.4.37b)}: \left(\mathbf{k}_{\mathrm{N}}+\hat{\mathbf{k}}_{\mathrm{C}}\right)\mathbf{H}^{(4,2b)}=\hat{\mathbf{k}}_{\mathrm{S}}^{1/2} \tag{3.4.61}$$

where now we must include a *relative normalizing factor* for the additional noise component $\hat{\mathbf{k}}_{\mathrm{C}}$ ((3.4.74a) following). Again, using an orthonormal matrix to effect the reduction of the sum $(\mathbf{k}_{\mathrm{N}}+\mathbf{k}_{\mathrm{C}})$, for example,

$$\left(\tilde{\mathbf{R}}_3\mathbf{k}_{\mathrm{N}}\mathbf{R}_3+\tilde{\mathbf{R}}_3\hat{\mathbf{k}}_{\mathrm{C}}\mathbf{R}_3\right)\tilde{\mathbf{R}}_3\mathbf{H}^{(4,2a)}\mathbf{R}_3=\tilde{\mathbf{R}}_3\langle\mathbf{s}\rangle\mathbf{R}_3, \quad \text{or} \quad (\mathbf{\Lambda}_{\mathrm{N}}+\mathbf{\Lambda}_{\mathrm{C}})\mathbf{H}^{(4,2a)}=\langle\mathbf{s}\rangle \tag{3.4.61a}$$

we find, as expected, the (vector) matched filter response

$$\mathbf{H}^{(4,1a)(4,2a)}=\left[\left\{(\lambda_{\mathrm{N}})_j+b_j^2(\lambda_{\mathrm{C}})_j\right\}^{-1}\langle s_j\rangle\right]. \tag{3.4.61b}$$

Similarly, we obtain for $\mathbf{H}^{(4,2b)}$ the matrix

$$\mathbf{H}^{(4,2b)}=\left[\left(\hat{\lambda}_S^{1/2}\right)_j\delta_{jj'}\Big/\left\{(\lambda_{\mathrm{N}})_j+b_j^2(\lambda_{\mathrm{C}})_j\right\}\right]. \tag{3.4.62}$$

Here and wherever $\left(\hat{\lambda}_{\mathrm{S}}\right)_j$ appears, for the normalization with respect to $(\psi_{\mathrm{N}})_j$ we have

$$\left(\hat{\lambda}_S\right)_j=\left(\psi_{Sj}/\psi_{Nj}\right)\lambda_{Sj}. \tag{3.4.62a}$$

In case of "white" noise in the receiver, that is, $\mathbf{k}_{\mathrm{N}}=\left[\delta_{jj'}\right]=\left[\delta_{mm'}\delta_{nn'}\right]$, we have $(\lambda_{\mathrm{N}})_j=1$ all $(m=m', n=n')$, that is, $\mathbf{\Lambda}_{\mathrm{N}}=\mathbf{I}$, so that (3.4.57)–(3.4.59) all reduce to the simpler forms:

$$\mathbf{H}^{(1)}=\langle\mathbf{s}_j\rangle;\quad \mathbf{H}^{(2,2b)}=\left[\left(\hat{\lambda}_S^{1/2}\right)_j\delta_{jj'}\right];\quad \mathbf{H}^{(2,1)}=\left[\left(\hat{\lambda}_S\right)_j^{1/2}\delta_{jj'}\right];\quad \mathbf{H}^{(2,2)}=\left[\hat{\lambda}_{Sj}\delta_{jj'}\right], \tag{3.4.63}$$

Equations (3.4.61b) and (3.4.62) reduce to

$$\mathbf{H}^{(4,2a)}=\left[\langle s_j\rangle/\left(1+b_j^2(\lambda_{\mathrm{C}})_j\right)\right];\quad \mathbf{H}^{(4,2b)}=\left[\left(\hat{\lambda}_S^{1/2}\right)_j\delta_{jj'}/\left(1+b_j^2\lambda_{C_j}\right)\right], \tag{3.4.64}$$

respectively a vector and a square matrix.

3.4.7.1 Separation of Space and Time[48] Physical fields are functions of space and time, that is, $\alpha=\alpha(\mathbf{r},t|ST)$, which are not naturally separable into functions of space alone

[48] Again, we have non-Hom-Stat covariances, both unnormalized $(\mathbf{K}_{\mathrm{N}})$ and normalized $(\mathbf{k}_{\mathrm{N}})$, and so on.

and time alone, that is, $\alpha(\mathbf{r}, t|ST) \neq \alpha(\mathbf{r}|S)\alpha(t|T)$. Nevertheless, it is often an important convenience to be able to treat them as separable. In narrowband situations, it is a good first-order approximation to do so. However, for the broadband cases, separability is definitely a suboptimizing constraint, which becomes serious for signals and the accompanying noise that have significant bandwidths. For such signals and noise, the spatial problems of coupling to the medium, embodied in the aperture or array, become significant. The physical elements are frequency sensitive so that the aperture or array has different electrical or acoustic "sizes," and hence produces beam patterns of different resolution over the range of frequencies employed. At the same time, in the temporal domain, the necessarily fixed sample-size, that is, processing time, is differently affective for different portions of the band. The matched filters in this scenario are then much more complex functions of frequency than in the narrowband cases, which latter can be designed for a fixed central frequency f_0, and is then independent of frequency variations. In Section 3.5, we briefly explore the general broadband case in wave number—frequency space, that is, Fourier four-space. This is done from the viewpoint of the matched filter as *system function*, $\mathsf{F}_d^{-1}\{\mathbf{H}\} = Y(\mathbf{u}, f)$, rather than as *weighting function* $(\mathbf{H})$. Here, the spatial and temporal roles, joint and separated, are more revealing of their functions than the space–time structures considered in Section 3.4.

Let us now consider these separated cases in which the single space–time matched filter is separated into two distinct matched filters, one for space and the other for time processing. Analytically, we represent this by the Kronecker products.

$$\mathbf{k}_N\mathbf{H} = \langle \mathbf{s} \rangle : \quad \mathbf{k}_N = \mathbf{k}_N^{(S)} \otimes \mathbf{k}_N^{(T)}; \quad \mathbf{H} = \mathbf{H}^{(S)} \otimes \mathbf{H}^{(T)}; \quad \langle \mathbf{s} \rangle = \langle \mathbf{s} \rangle^{(S)} \otimes \langle \mathbf{s} \rangle^{(T)}. \tag{3.4.65}$$

We then use the product relation $(\mathbf{A} \otimes \mathbf{B})(\mathbf{C} \otimes \mathbf{D}) = (\mathbf{AC}) \otimes (\mathbf{BD})$, where $\mathbf{A}$, $\mathbf{C}$ are $M_1 \times M_1$ matrices and $\mathbf{B}$, $\mathbf{D}$ are $N \times N$ matrices, (Chapter 12, Sections 5–9, 11, and 12 of Ref. [21]), to write for $\mathbf{k}_N\mathbf{H}^{(1)} = \langle \mathbf{s} \rangle$ in the separated condition the discrete integral equations

$$\left(\mathbf{k}_N^{(S)}\mathbf{H}^{(S)}\right) \otimes \left(\mathbf{k}_N^{(T)}\mathbf{H}^{(T)}\right) = \left\langle \mathbf{s}^{(S)} \right\rangle \otimes \left\langle \mathbf{s}^{(T)} \right\rangle; \quad = 0 \text{ outside } \Delta = [S] \otimes [T]. \tag{3.4.66a}$$

(Observe that for the complete separation of space and time postulated here, separate ensemble averages over the signal are required, refer to Eq. (3.4.65).) Equation (3.4.66a) is equivalent to

$$\mathbf{k}_N^{(S)}\mathbf{H}^{(S)} = \left\langle \mathbf{s}^{(S)} \right\rangle; \quad \mathbf{k}^{(T)}\mathbf{H}^{(T)} = \left\langle \mathbf{s}^{(T)} \right\rangle; \quad = 0 \text{ outside } \Delta, \tag{3.4.66b}$$

with similar relations for the other examples of (3.4.55) and Table 3.1 following. Now the solutions corresponding to (3.4.66b) and to separated versions of (3.4.57)–(3.4.62) become.

$$\mathbf{H}^{(1)} = \mathbf{H}^{(1)(S \otimes T)} = \mathbf{H}^{(1),(S)} \otimes \mathbf{H}^{(1),(T)} = \left[\left(\lambda_N^{(S)}\right)_m \left\langle s_m^{(S)} \right\rangle\right] \otimes \left[\left(\lambda_N^{(T)}\right)_n \left\langle s_n^{(T)} \right\rangle\right] \tag{3.4.67}$$

$$\mathbf{H}^{(2,1)} = \mathbf{H}^{(2,2b)} = \mathbf{H}^{(2,2b)(S)} \otimes \mathbf{H}^{(2,2b),(T)} = \left[\left(\hat{\lambda}_S^{(S)}\right)_m^{1/2} \delta_{mm'} \Big/ \left(\lambda_N^{(S)}\right)_m\right] \otimes \left[\left(\hat{\lambda}_S^{(T)}\right)_n^{1/2} \delta_{nn'} \Big/ \left(\lambda_N^{(T)}\right)_n\right], \tag{3.4.68}$$

$$\mathbf{H}^{(2,2)} = \left[\left(\hat{\lambda}_S^{(S)}\right)_m \delta_{mm'} / \left(\lambda_N^{(S)}\right)_m^2\right] \otimes \left[\left(\hat{\lambda}_S^{(T)}\right)_n \delta_{nn'} / \left(\lambda_{\mathrm{N}}^{(T)}\right)_n^2\right]. \tag{3.4.69}$$

In a similar way, we readily find that the separated counterparts to (3.4.61b) and (3.4.62) are

$$\begin{aligned}\mathbf{H}^{(4,2a)} = \mathbf{H}^{(4,2a)(S)} \otimes \mathbf{H}^{(4,2a),(T)} &= \left[\left\{\left(\lambda_{\mathrm{N}}^{(S)}\right)_m + b_m^2\left(\lambda_{\mathrm{C}}^{(S)}\right)_m\right\}^{-1}\left\langle s_m^{(S)}\right\rangle\right] \\ &\otimes \left[\left\{\left(\lambda_{\mathrm{N}}^{(T)}\right)_n + b_n^2\left(\lambda_{\mathrm{C}}^{(T)}\right)_n\right\}^{-1}\left\langle s_n^{(T)}\right\rangle\right],\end{aligned} \tag{3.4.70}$$

the Kronecker product of two vectors, and the Kronecker product of two matrices is for (3.4.62):

$$\begin{aligned}\mathbf{H}^{(4,2b)} = \mathbf{H}^{(4,2b)(S)} \otimes \mathbf{H}^{(4,2b),(T)} &= \left[\left(\hat{\lambda}_S^{(S)}\right)_m^{1/2} \delta_{mm'} \Big/ \left\{\left(\lambda_{\mathrm{N}}^{(S)}\right)_m + b_m^2\left(\lambda_{\mathrm{C}}^{(S)}\right)_m\right\}\right] \\ &\otimes \left[\left(\hat{\lambda}_S^{(T)}\right)_n^{1/2} \delta_{nn'} \Big/ \left\{\left(\lambda_{\mathrm{N}}^{(T)}\right)_n + b_n^2\left(\lambda_{\mathrm{C}}^{(T)}\right)_n\right\}\right].\end{aligned} \tag{3.4.71}$$

(Of course, $\lambda_m^{(S)} \neq \lambda_n^{(T)}$ generally, since they are solutions of different discrete equations, (3.4.66b)) For "white" noise in the receiver, that is, $\mathbf{k}_{\mathrm{N}}^{(S\otimes T)} = [\delta_{mm'}] \otimes [\delta_{nn'}]$, we have $\left(\lambda_{\mathrm{N}}^{(S)}\right)_m = 1$, $\left(\lambda_{\mathrm{N}}^{(T)}\right)_n = 1$, which simplifies the results (3.4.67)–(3.4.71) considerably, refer to Eqs. (3.4.63) and (3.4.64).

3.4.7.2 Unnormalized Covariances The results above apply for *normalized* covariances, for example, (3.4.55). Moreover, these results can be readily related to the unnormalized cases, where $\mathbf{K}_{\mathrm{N}} = \left[(\mathbf{K}_{\mathrm{N}})_{jj'}\right] = \left[\sqrt{\psi_{Nj}\,\psi_{Nj'}}(\mathbf{k}_{\mathrm{N}})_{jj'}\right]$ and $\mathbf{K}_{\mathrm{C}} = \left[(\mathbf{K}_{\mathrm{C}})_{jj'}\right] = \left[\sqrt{\psi_{Cj}\psi_{Cj'}}(\mathbf{k}_{\mathrm{N}})_{jj'}\right]$. Similarly, we have for the unnormalized signals and W–K filter, $\hat{\mathbf{H}}^{(1)}$. $\langle\mathbf{S}\rangle = \left[s_j\psi_j^{1/2}\right]$, $\hat{\mathbf{H}}^{(1)} = \left[H_j^{(1)}\psi_j^{-1/2}\right]$. To see this in the additive noise case, we simply write

$$(\mathbf{K}_{\mathrm{N}} + \mathbf{K}_{\mathrm{C}})\hat{\mathbf{H}}^{(3)} = \langle\mathbf{S}\rangle \quad \text{or} \quad \hat{\mathbf{H}}^{(3)} = \left[\sqrt{\psi_{Nj}\,\psi_{Nj'}}(\mathbf{k}_{\mathrm{N}})_{jj'} + \sqrt{\psi_{Cj}\psi_{Cj'}}(\mathbf{k}_{\mathrm{C}})_{jj'}\right]^{-1}\left[\langle s_{j'}\rangle\sqrt{\psi_{Nj'}}\right], \tag{3.4.72}$$

which in diagonalization by the orthogonal matrix $\mathbf{R}$ becomes, since $\hat{\mathbf{H}}^{(3)} = \left[H_j^{(3)}\psi_{Nj}^{-1/2}\right]$,

$$\begin{aligned}\hat{\mathbf{H}}^{(3)} = \left[\hat{H}_j^{(3)}\right] &= \left[\delta_{jj'}\left\{\left(\hat{\lambda}_{\mathrm{N}}\right)_j + \left(\hat{\lambda}_{\mathrm{C}}\right)_j\right\}^{-1}\right]\left[\langle s_{j'}\rangle\sqrt{\psi_{Nj'}}\right] \\ &= \left[\delta_{jj'}\sqrt{\psi_{Nj'}}\left\{\left(\hat{\lambda}_{\mathrm{N}}\right)_j + \left(\hat{\lambda}_{\mathrm{C}}\right)_j\right\}^{-1}\right]\left[\langle s_{j'}\rangle\right].\end{aligned} \tag{3.4.73a}$$

The eigenvalues of $\mathbf{K}_{\mathrm{N}}$ and $\mathbf{K}_{\mathrm{C}}$ are respectively $\left(\hat{\lambda}_{\mathrm{N}}\right)_j$ and $\left(\hat{\lambda}_{\mathrm{C}}\right)_j$, so that

$$\left(\hat{\lambda}_{\mathrm{N}}\right)_j = \psi_{Nj}\,\lambda_{Nj};\quad \left(\hat{\lambda}_{\mathrm{C}}\right)_j = \psi_{Cj}\lambda_{Cj};\quad \left(\hat{\lambda}_{\mathrm{N}}\right)_j + \left(\hat{\lambda}_{\mathrm{C}}\right)_j = \psi_{Nj}\left(\lambda_j + (\psi_{Cj}/\psi_{Nj})\lambda_{\mathrm{C}}\right). \tag{3.4.73b}$$

Accordingly, (3.4.73a) becomes, in normalized form:

$$\mathbf{H}^{(3)} = \mathbf{\Lambda}_{\mathrm{NC}}^{-1}\langle\mathbf{s}\rangle = \left[\delta_{jj'}\left\{\lambda_{Nj} + (\psi_{Cj}/\psi_{Nj})\lambda_{Cj}\right\}^{-1}\right]\left[\langle s_j\rangle\right] = \left[\frac{\delta_{jj'}}{\lambda_{Nj} + b_j^2\lambda_{Cj}}\right]\langle\mathbf{s}\rangle,\ \text{a vector;} \tag{3.4.74}$$

$$\left\{\begin{array}{l} b_j^2 \equiv \psi_{Cj}/\psi_{Nj} \geq 0 \\ \therefore \hat{\mathbf{k}}_{\mathrm{C}} = \left[\left(\psi_{Cj}\psi_{Cj'}/\psi_{Nj}\psi_{Nj'}\right)^{1/2}(k_{\mathrm{C}})_{jj'}\right] \equiv \left[b_{jj'}^2 (k_{\mathrm{N}})_{jj'}\right] \end{array}\right\}. \tag{3.4.74a}$$

The ratio $b_j^2(\equiv \psi_{Cj}/\psi_{Nj})$, of course, appears in the appropriate place in all diagonalized expressions involving an additional noise component, such as that for clutter and so on. In (3.4.60) et seq., we have written $\hat{\mathbf{k}}_{\mathrm{C}}$ for this covariance, normalized to ψ_{N_j} in its reduced (i.e., diagonal) form, where $(\mathbf{k}_{\mathrm{N}})_{jj} = (k_{\mathrm{C}})_{jj} = 1$. (This normalizing factor, including the off-diagonal terms of $\hat{\mathbf{k}}_{\mathrm{C}}$, is in general $b_{jj'}^2$, (3.4.74a).) Equation (3.4.72) can be expressed equivalently as

$$\hat{\mathbf{H}}^{(3)} = \hat{\mathbf{H}}^{(4,2a)} = (\mathbf{K}_{\mathrm{N}} + \mathbf{K}_{\mathrm{C}})^{-1}\langle\mathbf{S}\rangle = \left[\left\{(\mathbf{k}_{\mathrm{N}})_{jj'} + b_{jj'}^2(\mathbf{k}_{\mathrm{C}})_{jj'}\right\}(\psi_{Nj}\psi_{Nj'})^{1/2}\right]^{-1}\langle\mathbf{S}\rangle = \left[\hat{H}_j\psi_{Nj}^{-1/2}\right], \tag{3.4.75}$$

which also shows the structure of the *unnormalized* covariances in terms of those normalized with respect to $\psi_{Nj}\psi_{Nj'}$.

In the case of separated space and time operations considered in Section 3.4.7.1, we write

$$\left\{\hat{\mathbf{H}}^{(3)}\left(=\hat{\mathbf{H}}^{(4,2a)}\right)\right\}_{S\otimes T} = \left[\left(\mathbf{K}_{\mathrm{N}}^{(S)} + \mathbf{K}_{\mathrm{C}}^{(S)}\right)^{-1}\right]\left\langle\mathbf{S}^{(S)}\right\rangle \otimes \left[\left(\mathbf{K}_{\mathrm{N}}^{(T)} + \mathbf{K}_{\mathrm{C}}^{(T)}\right)^{-1}\right]\left\langle\mathbf{S}^{(T)}\right\rangle \tag{3.4.76a}$$

$$= \left[\frac{\delta_{mm}}{(\lambda_{\mathrm{N}})_m + b_m^2(\lambda_{\mathrm{C}})_m}\right]\left\langle\mathbf{s}^{(S)}\right\rangle \otimes \left[\frac{\delta_{nn}}{(\lambda_{\mathrm{N}})_n + b_n^2(\lambda_{\mathrm{C}})_n}\right]\left\langle\mathbf{s}^{(T)}\right\rangle, \tag{3.4.76b}$$

with similar relations for $\hat{\mathbf{H}}^{(4,1a)}$. For $\hat{\mathbf{H}}^{(4,2b)}$, the average $\left\langle\mathbf{S}^{(S)}\right\rangle$ is replaced by $\left(\hat{\lambda}_{\mathrm{S}}^{(S)}\right)_m$ and $\left\langle s^{(T)}\right\rangle \to \left(\hat{\lambda}_{\mathrm{S}}^{(T)}\right)_n$, with $\left(\hat{\lambda}_{\mathrm{S}}\right)_{m,n}^{(S),(T)} = \left(\psi_{\mathrm{S}}^{(S)}/\psi_{\mathrm{N}}^{(S)}\right)_m\left(\lambda_{\mathrm{S}}^{(S)}\right)_m, \left(\psi_{\mathrm{S}}^{(T)}/\psi_{\mathrm{N}}^{(T)}\right)_n\left(\lambda_{\mathrm{S}}^{(T)}\right)_n$. The eigenvalues in these separated cases are $\hat{\lambda}_{j|S\otimes T} = \hat{\lambda}_m^{(S)}\hat{\lambda}_n^{(T)}$, refer to Appendix A.

Finally, when the *beam is preformed*, that is, $M = 1$, and only the temporal processing can be optimized, that is, $\mathbf{H} = \left[H_{n,n'}\right] = [H(t_n, t_{n'})]$, the space–time solutions above for these matched filters reduce to the simpler forms where the eigenvalues Eqs. (3.4.57) and (3.4.58)) et seq. depend only on the index n and $j, j' \rightarrow n, n'$ in $\langle \mathbf{s}_j \rangle, \delta_{jj'}$, and so on. One adds a gain factor $(g_0 > 0)$ to $\mathbf{H}$ here, and replaces the operations $(\;)^{(S)}\otimes$ by $\mathbf{1}\otimes$, or unity.

3.4.7.3 Matrix Reduction and Evaluation Finally, methods of reducing the matrices in these discrete integral equations to eigenvalue (i.e., diagonal) form and obtaining specific numerical results are QR decomposition, SVD (singular value decomposition), reduction by Cholesky matrices, among others (Appendix A.5, A.6, A.63 of Ref. [31]). These, among others, are also discussed in Ref. [20, 22, 23, 25], (refer to Bibliography, Appendix A of Ref. [31]). Chapter 10 of Ref. [21] describes some of these concisely in a physical context. All these techniques are to be implemented by appropriate computer programs for the required numerical results (Appendix A.6 of Ref. [31]).

3.4.8 Summary Remarks

We have seen from the results of Section 3.4 that there are two classes of matched filter: Class I is a *J (=MN) column vector*, and Class II is a *(J × J) square or triangular matrix*, where either class may apply to coherent or incoherent reception and deterministic or random signals. Table 3.1 provides a short summary of their classification, type, weighting function form (vector or matrix), illustrated in the text above.

As we can see from the table 3.1, the Wiener–Kolmogoroff filter $\mathbf{W}$ plays a prominent role as the optimizing component of these quadratic forms $\Psi^{(1)}, \Psi^{(2)}$, Eq. (3.4.1). The W–K filter is distinguished by its linear vector character and its space–time invariance. It can also appear nonlinearly (i.e., quadratically) in the anatomization of the quadratic form from which it is derived, for example, Nos. 3, 4, 10–12 of Table 3.4.1. The reduction of the generic quadratic forms (3.4.1) and (3.4.1a) is clearly not unique: there are many possible matched filters, but all are causal, that is, realizable, however complex in structure, and all are capable of handling non-Hom-Stat covariances. Matched filters for the latter are of course space–time variable.

Finally, in the special case when the accompanying additive noise is spectrally "white" in the wave number–frequency domain, all our preceding results simplify greatly, since then $\mathbf{k}_\mathrm{N} = \left[1_{Nj}\delta_{ij'}\right]$ and the matched filters are now proportional to the replica signals used in the receiver.

3.4.9 Signal-to-Noise Ratios, Processing Gains, and Minimum Detectable Signals. I

For the detection problems discussed in Sections 3.1–3.3 and in particular for the various matched filters treated in Sections 3.4.1–3.4.7 and illustrated in Figs. 3.9–3.12, the key parameter appearing in the probability measures of performance (e.g., $1-\beta^*, \alpha_\mathrm{F}^*$, etc.) is the quadratic form $\Psi^*_{s-\mathrm{coh}}$, Eq. (3.1.3a) and (3.1.3b), $\Psi^*_{s-\mathrm{inc}}$, Eqs. (3.2.5) and (3.2.13). This includes extensions involving more complex forms of (additive) noise backgrounds, refer to Eq. (3.4.55), and is true for suboptimum detectors as well, represented by $\Psi_{s-\mathrm{coh}}, \Psi_{s-\mathrm{inc}}$, where the resulting decisions are *consistent*, that is, $p_\mathrm{D} \rightarrow 1$ as sample size becomes infinite.

3.4.9.1 Coherent Detection For broad- or narrowband signals (3.1.1a) and (3.1.1b), we see from (3.1.3) and (3.1.9) that

$$\Psi^*_{s-\mathrm{coh}} \equiv \sigma_0^{*2} = \tilde{\hat{\mathbf{s}}}\mathbf{k}_\mathrm{N}\hat{\mathbf{s}} = \tilde{\hat{\mathbf{s}}}\mathbf{H}^{(1)} = \tilde{\mathbf{H}}^{(1)}\hat{\mathbf{s}}^{(1)} = 2\left(\frac{J\tilde{\mathbf{H}}^{(1)}\hat{\mathbf{s}}}{2\tilde{\hat{\mathbf{s}}}\hat{\mathbf{s}}}\right)\left(\frac{\tilde{\hat{\mathbf{s}}}\hat{\mathbf{s}}}{J}\right) = 2\Pi^*_\mathrm{coh} \cdot \left(a_{0\ \mathrm{coh}}^2\right)^*_\mathrm{min} \tag{3.4.77}$$

TABLE 3.1 Discrete Matched Filters: Class I and II

	Bayes No., Type	Equation No.	Fig. No.	Discrete Green's Function[a]: Section 3.4.1–3.4.5	Remarks: (All Filters Linear + Causal)	$A^{(2)}$
1	1, Type 1	(3.4.4)	(3.9)	$\mathbf{H}^{(1)} = \mathbf{k}_N^{-1}\mathbf{s} = \mathbf{W}$	Vector; ST invariant, coherent	
2	1, Type 1, 1a	(3.4.26a)	(3.9)	$\mathbf{H}^{(1a)} = \mathbf{k}_N^{-1}\langle\mathbf{s}\rangle = \langle\mathbf{W}\rangle$	Vector; ST invariant, coherent	
3	2, Type 2a	(3.4.19b)	(3.11)	$\mathbf{H}^{(2,2a)} = \mathbf{k}_N^{-1}\langle\mathbf{s}\rangle = \mathbf{H}^{(1,a)} = \mathbf{y} = \langle\mathbf{W}\rangle$	Vector; ST invariant, incoherent	$y\tilde{y}$
4	2, Type 2a, 2b	(3.4.19b)	(3.12)	$\mathbf{H}_{a,b}^{(2,2a;2b)} = \mathbf{k}_N^{-1}(\mathbf{a} \text{ or } \mathbf{b}) = \mathbf{H}^{(1)} = \mathbf{W}$	Vector; ST invariant, incoherent	$y\tilde{y}$
5	2, Type 1	(3.4.9b)	(3.10)	$\tilde{\mathbf{H}}^{(2,1)}\mathbf{H}^{(2,1)} = \boldsymbol{\rho}^{(2)}$	Triangle; ST variable; incoherent	$\tilde{u}u$
6	2, Type 2	(3.4.13)	(3.11)	$\mathbf{H}^{(2,2)} = \hat{\boldsymbol{\rho}}^{(2)}$	Square; ST variable; incoherent	
7	2, Type 2b	(3.4.21)	(3.10)	$\mathbf{H}^{(2,2b)} = \mathbf{k}_N^{-1}\hat{\mathbf{k}}_S^{1/2} = \mathbf{u}$	Square; ST variable; incoherent	
8	4, Type 1	(3.4.32)	(3.10)	$\tilde{\mathbf{H}}^{(4,1)}\mathbf{H}^{(4,1)} = \hat{\boldsymbol{\rho}}^{(4)}$	Triangle; ST variable; incoherent	$\tilde{u}u$
9	4, Type 2	(3.4.34)	(3.11)	$\mathbf{H}^{(4,2)} = \hat{\boldsymbol{\rho}}^{(4)}$	Square; ST variable; incoherent	
10	3, Type 1	(3.4.31)	(3.9)	$\mathbf{H}^{(3)} = \left(\mathbf{k}_N + \hat{\mathbf{k}}_C\right)^{-1}\langle\mathbf{s}\rangle = \mathbf{H}^{(2,2a)} = \mathbf{W}$	vector; ST invariant, coherent	
11	4, Type 2a	(3.4.37a) and (3.4.37b)	(3.9)	$\mathbf{H}^{(4,2a)} = \mathbf{H}^{(3)} = \quad '' \quad = \mathbf{W}; \mathbf{z}$	vector; ST invariant, coherent	$z\tilde{z}$
12	4, Type 1a	(3.4.36a) and (3.4.36b)	(3.11)	$\mathbf{H}^{(4,1a)} = \left(\mathbf{k}_N + \hat{\mathbf{k}}_C\right)^{-1}\langle\mathbf{s}\rangle = \mathbf{H}^{(1)} = \mathbf{W}; \mathbf{z}$	vector, incoherent	
13	4, Type 2b	(3.4.37a) and (3.4.37b)	(3.11)	$\mathbf{H}^{(4,2b)} = \left(\mathbf{k}_N + \hat{\mathbf{k}}_C\right)^{-1}\hat{\mathbf{k}}_S^{1/2} = \mathbf{H}^{(2,2b)}; \mathbf{u}$	square; ST variable, incoherent	$\tilde{u}u$

[a] For eigenfunction solutions, Section 3.4.7.

(i) Noise covariances are generally non-Hom-Stat, but may be Hom-Stat; specifically, they appear in $\Psi_J^{(1)}, \Psi_J^{(2)}$ as $\mathbf{k}_N^{-1}, \mathbf{k}_N^{-1}\langle s\rangle\langle\tilde{s}\rangle\mathbf{k}_N^{-1}, \mathbf{k}_N^{-1}\hat{\mathbf{k}}_S\mathbf{k}_N^{-1}$, refer to Eq. (3.4.1a).
(ii) Signals may be deterministic or purely random.
(iii) All matrices (except triangular) are square, symmetric, and positive definite (the triangular is positive definite).
(iv) All quantities are normalized to ψ_{Nj} and are thus dimensionless; $\hat{\mathbf{k}}_C = \left[b_{jj'}^2(\mathbf{k}_C)_{jj'}\right]$, while $\hat{\mathbf{k}}_C$ is normalized to $\left[(\psi_{Nj})(\psi_N)_{j'}\right]^{1/2}$ as is $\mathbf{k}_N$; Section 3.4.7.2, Eq. (3.4.74a).

where now $J = MN$ as before and the *processing gain* Π^*_{coh} and *associated minimum detectable signal* $\left(a^2_{0-\text{coh}}\right)^*_{\min}$ are defined and represented here by[49]

$$\Pi^*_{\text{coh}} \equiv J\tilde{\mathbf{H}}^{(1)}\hat{\mathbf{s}}/2\tilde{\hat{\mathbf{s}}}\hat{\mathbf{s}} \quad \text{and} \quad \left(a^2_{0-\text{coh}}\right)^*_{\min} \equiv \tilde{\hat{\mathbf{s}}}\hat{\mathbf{s}}/J \tag{3.4.78}$$

for these coherently received signals. Broadly speaking, we may say that the processing gain measures how much the minimum detectable signal is increased to produce the output signal, subject to the decision H_1 or H_0. The minimum detectable signal is the smallest input signal that can be detected under H_1. The quantities are, of course, to be understood probabilistically, as part of the Bayes criterion of detection postulated in this book, as discussed generally in Chapter 1.

As specific examples, from (3.1.3a) and (3.1.3b), we obtain for broadband signals

$$\Pi^*_{\text{coh-bb}} = J\tilde{\hat{\mathbf{s}}}\mathbf{k}_N^{-1}\hat{\mathbf{s}}/2\tilde{\hat{\mathbf{s}}}\hat{\mathbf{s}}\big|_{\text{bb}} = J\sum_{jj'}^{J}\frac{A_{0j}\cdot A_{0j'}\left(\mathbf{k}_N^{-1}\right)_{jj'}}{2\sqrt{\psi_j\psi_{j'}}}\Bigg/\sum_{jj'}^{J}\frac{A^2_{0j}}{\psi_j}\Bigg]_{\text{bb}} \tag{3.4.79a}$$

and

$$\left(a^2_{0-\text{coh}}\right)^*_{\min-\text{bb}} = \frac{\tilde{\hat{\mathbf{s}}}\hat{\mathbf{s}}}{J}\Bigg|_{\text{bb}} = \frac{1}{J}\sum_{j}^{J}\frac{A^2_{0j}}{\psi_j}\Bigg|_{\text{bb}} \tag{3.4.79b}$$

In case of narrowband signals, we readily see that from (3.1.3b),

$$\Pi^*_{\text{coh-nb}} = \text{Eq. (3.4.79a), with factor 2 inserted in the denominators of each sum} \tag{3.4.80a}$$

$$\left(a^2_{0-\text{coh}}\right)^*_{\text{nb}} = \text{Eq.(3.4.79b), with a factor 2 in the denominator.} \tag{3.4.80b}$$

Finally, observing that $\mathbf{H}^{(1)}$ is a Bayes matched filter of the first kind, refer to Eq. (3.4.2) and Fig. 3.9, we can express the processing gain Π^*_{coh} more compactly in terms of eigenvalues of $\mathbf{k}_N$, namely, Eq. (3.4.57):

$$\Pi_{\text{coh}|\text{bb,nb}} = J\sum_{j}^{J}\left(\hat{s}^2_j/\lambda_{Nj}\right)/2\sum_{j}^{J}\hat{s}^2_j, \tag{3.4.81}$$

with the minimum detectable signals represented by (3.4.78), and in detail by (3.4.79b) and (3.4.80b).

3.4.9.2 Narrowband Incoherent Detection When an exact treatment is possible for the narrowband cases (Sections 3.2 and 3.3), we have the more complex relations

$$\Psi^*_{s-\text{inc}}\big|_{\text{nb}} \equiv \sigma^{*2}_{0-\text{inc}} \doteq \left(\tilde{\mathbf{a}}\mathbf{k}^{-1}\mathbf{a} + \tilde{\mathbf{b}}\mathbf{k}^{-1}\mathbf{b}\right)/2 = \left(\tilde{\mathbf{H}}^{(a)}\mathbf{a} + \tilde{\mathbf{H}}^{(b)}\mathbf{b}\right)\Big/2 \tag{3.4.82a}$$

[49] The factor 2 in (3.4.77) and (3.4.83) is used in the definition of processing gains here to make it conform to the more general definition that must be used in the threshold non-Gaussian noise cases.

from (3.2.22c), where now the matched filters are $\mathbf{H}^{(a)}, \mathbf{H}^{(b)}$, with $\mathbf{k}_N\mathbf{H}^{(a)} = \mathbf{a}, \mathbf{k}_N\mathbf{H}^{(b)} = \mathbf{b}$, where $\mathbf{a}$ and $\mathbf{b}$ are given by (3.2.2). This is equivalent to

$$\Psi^*_{s-\mathrm{inc}}\Big|_{\mathrm{nb}} = 2\left\{\frac{J\left(\tilde{\mathbf{H}}^{(a)}\mathbf{a} + \tilde{\mathbf{H}}^{(b)}\mathbf{b}\right)}{2\left(\tilde{\mathbf{a}}\mathbf{a} + \tilde{\mathbf{b}}\mathbf{b}\right)}\right\} \cdot \left\{\frac{\tilde{\mathbf{a}}\mathbf{a} + \tilde{\mathbf{b}}\mathbf{b}}{2J}\right\} \equiv 2\Pi^*_{\mathrm{inc}} \cdot \left(a^2_{0-\mathrm{inc}}\right)^*_{\min}\Big|_{\mathrm{nb}}, \quad (3.4.83)$$

where specifically

$$\Pi^*_{\mathrm{inc-nb}} \equiv J\left(\tilde{\mathbf{H}}^{(a)}\mathbf{a} + \tilde{\mathbf{H}}^{(b)}\mathbf{b}\right)/2\left(\tilde{\mathbf{a}}\mathbf{a} + \tilde{\mathbf{b}}\mathbf{b}\right); \quad \left(a_0^2\right)^*_{\mathrm{min-nb}} \equiv \left(\tilde{\mathbf{a}}\mathbf{a} + \tilde{\mathbf{b}}\mathbf{b}\right)\Big/2J. \quad (3.4.84)$$

In terms of eigenvalues of $\mathbf{k}_N$, refer to Eq. (3.4.57), the processing gain finally becomes

$$\Pi^*_{\mathrm{inc-nb}} = J\sum_j^J\left(\left(a_j^2 + b_j^2\right)/\lambda_{Nj}\right)\Big/2\sum_j\left(a_j^2+b_j^2\right)\Big|_{\mathrm{nb}} = J\sum_{j=1}^J\left(A_j^2/\lambda_{Nj}\right)\Big/2\sum_{j=1}^J A_j^2, \quad (3.4.85)$$

with

$$\mathbf{a} = \left[A_j\cos\Phi_j\right]; \quad \mathbf{b} = \left[A_j\sin\Phi_j\right]; \quad A_j \equiv A_{on}^{(m)}\Big/\sqrt{\psi_j}; \quad \Phi_j \equiv w_0t_n - \phi_m^{(m)}, \quad j = mn, \quad (3.4.85a)$$

and where the minimum detectable signal is now from (3.4.84).

$$\left(a_0^2\right)^*_{\mathrm{min-inc}}\Big|_{\mathrm{nb}} = \sum_{j=1}^J A_j^2\Big/2J. \quad (3.4.86)$$

From Section 3.4.4, we observe that the Bayes matched filters $\mathbf{H}^{(1)}, \mathbf{H}^{(2)}, \mathbf{H}^{(b)}$ are all W–K filters here, as a consequence of the particular structure of the detection Ψ^*_s. This is clearly *not* generally the case for the test statistic Ψ^*_x, refer to Sections 3.2.1, 3.2.2, and Section 3.3, where more complex matched filters are often required, in the manner of Sections 3.4.2, 3.4.3, and 3.4.5.

3.4.9.3 Signal-to-Noise Intensity Ratios The generalized signal-to-noise (intensity) ratios in the above optimum cases may be directly defined by the relations

$$(S/N)^{2*}_{\mathrm{out-}\binom{\mathrm{coh}}{\mathrm{inc}}} \equiv 2\Pi^*_{\binom{\mathrm{coh}}{\mathrm{inc}}}(S/N)^{2*}_{\binom{\mathrm{coh}}{\mathrm{inc}}} = \Psi^*_{\mathrm{s-}\binom{\mathrm{coh}}{\mathrm{inc}}}; \quad (S/N)^{2*}_{\mathrm{in-}\binom{\mathrm{coh}}{\mathrm{inc}}} \equiv a^2_{0-\binom{\mathrm{coh}}{\mathrm{inc}}}. \quad (3.4.87)$$

In terms of eigenvalues $\{\lambda_{Nj}\}$ of $\mathbf{k}_N$, from Eqs. (3.4.81), (3.4.85) and (3.4.86) these generalized signal-to-noise ratios become specifically

$$\left(\frac{S}{N}\right)^{2*}_{\mathrm{out-coh}} = \frac{1}{(2)J}\sum_{j=1}^J \hat{s}_j^2/\lambda_{Nj} = \Psi^*_{\mathrm{s-coh}} = \sigma^{2*}_{0-\mathrm{coh}}; \quad \left(\frac{S}{N}\right)^{2*}_{\mathrm{out-inc}} = \frac{1}{J}\sum_{j=1}^J\left(A_j^2/2\lambda_{Nj}\right) = \Psi^*_{\mathrm{s-inc}} = \sigma^{2*}_{0-\mathrm{inc}} \quad (3.4.88)$$

with $\hat{s}_j = A_j = A_{0_n}^{(m)}/\sqrt{\psi_j}$, refer to Eq. (3.1.1a), and $\left(a_{0-\text{coh}}^2\right)_{\min}^*$, $\left(a_{\text{inc}}^2\right)_{\min}^*$ given respectively by (3.4.79b) and (3.4.86). (Note the factor 2 in the denominator of the first set of equations, which applies when $\hat{s}_j$ is narrowband, refer to Eq. (3.1.1b).)

Similarly, relations to (3.4.87) also apply for suboptimum systems belonging to the class that has Ψ_s^* and so on as its optimum, namely,

$$\left[\Psi_s = \sigma_0^2 = (S/N)_{\text{out}}^2 = 2\Pi(S/N)_{\text{in}}^2 = 2\Pi\left(a_0^2\right)_{\min}\right]_{\substack{\text{coh}\\ \text{inc}}} \tag{3.4.89}$$

This enables us to compare a suboptimum with the corresponding optimum system. For example, with the same sample size and minimum detectable signals (the more usual situation in practice), we can write

$$\sigma_0^2 = \Phi_d^* \sigma_0^{2*} \quad \text{or} \quad \sigma_0^2/\sigma_0^{2*} = \Phi_d^* \le 1, \quad \text{where } \Phi_d^* = \Pi/\Pi^*; \quad \left(a_0^2\right)_{\min} = \left(a_0^{2*}\right)_{\min} \tag{3.4.90}$$

and Φ_d^* is a *degradation factor*. Comparisons between two suboptimum systems are also possible:

$$\sigma_{01}^2 = (\Phi_d)_{12}\sigma_{02}^2, \quad (\Phi_d)_{12} \le 1, \quad \sigma_{02}^2 \ge \sigma_{01}^2, \quad \text{where}(\Phi_d)_{12} = \Pi_{01}/\Pi_{02}. \tag{3.4.91}$$

Similarly, comparisons can be made between three or more systems (for the same purpose). We have $\sigma_{0l}^2 = (\Phi_d)_{lm}\sigma_{0m}^2, \sigma_{0m}^2 > \sigma_{0l}^2$ and so on, which we can place say, in descending order, for example, $\sigma_{0l}^2 < \sigma_{0(lm)}^2 < \cdots$. In addition, we have the option to compare systems with the same probabilities of correct detection and the same sample size (J), where the minimum detectable signals must be different. Thus, we find that

$$\left(a_0^2\right)_{\min-1}^* = \Phi_d^*\left(a_0^2\right)_{\min-2}, \tag{3.4.92}$$

in case of an optimum system, which has a smaller minimum detectable signal than the suboptimum one. A third variation on comparisons is to choose equal probabilities of detection and minimum detectable signals, and determine the increase in *sample size*, $J > J^*$, required for this result, thus, $\Pi \cdot (J) = \Pi^* \cdot (J^*)$ gives us the desired relation between J and J^*.

3.4.9.4 Remarks Since $\Psi_s^{(*)}$ is the only signal-dependent parameter of the decision probabilities and does not depend on the relevant threshold (K), used in the decision process, the results of this section apply equally to CFAR and IO systems. From a more general point of view, the concepts of minimum detectable signals and processing gains are very useful in the comparison process, but they are incomplete. They, by themselves, do not provide the desired probabilities of performance. They are necessary but not sufficient, as can also be seen by the fact that they involve signal-to-noise ratios, which say nothing about false alarm probabilities and probabilities of detection. They are a form of second-moment criteria of performance, refer to Sections 5.3.4 and 5.3.5 of Ref. [1]. They are related to the relative efficiencies, and asymptotic relative efficiencies, of performance (Section 6.3.3, pp. 95–102), [28], which likewise are useful but incomplete descriptors of the associated detection probabilities.

Finally, we must emphasize that these specific definitions and results apply in the cases of reception *only when an exact treatment*,[50] such as that described in Sections 3.1–3.3 here, *is possible*, so that an appropriately simple sufficient statistic for the signal intensity can be used. In fact, the natural choice of detection parameter in these exact cases $\left[\sigma^{*2}_{0-\mathrm{coh,inc}} = \Psi^{*}_{s-\mathrm{coh,inc}}\right]$ is determined by the pdf values (under H_0, H_1) of a simplified but equivalent sufficient statistic (i.e., proportional to a likelihood ratio of Section 1.6), vide the examples in Sections 3.1, 3.2, and 3.3.1. When, as is usually the case, such is not possible and only a (canonical) threshold analysis can be constructed, the definitions of processing gain and minimum detectable signal must be suitably extended, following; see also Section 3.3.1.1.

3.5 BAYES MATCHED FILTERS IN THE WAVE NUMBER–FREQUENCY DOMAIN

Although the discrete matched filters of Sections 3.4.1–3.4.5 operate on the normalized input sampled data $\mathbf{x}$ in space–time, their wave number–frequency equivalents, that is, their discrete Fourier transforms (indicated by the subscript d) can be even more revealing of their properties. Accordingly, we employ the sampling procedures of Chapter 2, physically accomplished by an array of $M+1$ sensors distributed in space. Let us begin first by considering the matched filter $H^{(1)}(\mathbf{r}_m, t_n)$, sampled at $(\mathbf{r}_m, t_n)$, where $\mathbf{H}^{(1)}$ is a function jointly of space ($\mathbf{r}$) and time (t). Later, we shall examine the useful special case where we *impose* the separability of space and time on $H^{(1)}$ by representing it as $\mathbf{H}^{(1)}(\mathbf{r}_m, t_n) = \mathbf{H}^{(S)}(\mathbf{r}_m)\mathbf{H}^{(T)}(t_n)$.

3.5.1 Fourier Transforms of Discrete Series

We start by introducing the following 4-vectors and assume ordered, that is, *periodic* sampling of the continuous function $H(\mathbf{r}, t)$ in space and time, which means that each component of the 4-vector is subject to periodicity. Furthermore, the interval between sample points is $|\mathbf{r}_0|$, along a straight line of directionality $\hat{\mathbf{i}}_0$. Thus, for the vector of position ($\mathbf{r}_m$) and the component of time (t_n), we have

$$\mathbf{r}_m = m\hat{\mathbf{i}}_0 r_0, \quad \hat{\mathbf{i}}_4 t_n = \hat{\mathbf{i}}_4 n T_0; \quad \mathbf{p}_j \equiv \mathbf{r}_m - \hat{\mathbf{i}}_4 t_n = \hat{\mathbf{i}}_0 m|\mathbf{r}_0| - \hat{\mathbf{i}}_4 n T_0; \quad \mathbf{p}_m = \hat{\mathbf{i}}_0 m r_0, \tag{3.5.1}$$

with $|\mathbf{r}_m| = \left(r_x^2 + r_y^2 + r_z^2\right)_m^{1/2} = m|\mathbf{r}_0|$. Here, $\mathbf{r}_0$ is a unit of scalar length, m is the number of such units, and T_0 is the sampling period (Fig. 3.14). In more detail, we see that

$$\left.\begin{aligned} &\hat{\mathbf{i}}_0 = \hat{\mathbf{i}}_1 \cos\phi_0 \sin\theta_0 + \hat{\mathbf{i}}_2 \sin\phi_0 \sin\theta_0 + \hat{\mathbf{i}}_3 \cos\theta_0 \\ &r_{xm} = m r_0 \cos\phi_0 \sin\theta_0; \quad r_{ym} = m r_0 \sin\phi_0 \sin\theta_0; \quad r_{zm} = m r_0 \cos\theta_0 \end{aligned}\right\}. \tag{3.5.2}$$

[50] It is necessary to point out that here there are no additional averages *over* signal waveform: not in the completely specified signal case (Section 3.1.1), nor in the incoherent cases of Sections 3.2.1 and 3.2.2. In Sections 3.3.1–3.3.3 however, we do have additive averages $\langle\,\rangle_{\hat{s}}$ over signal. These can be accommodated by replacing $(a_0^{2,*})_{\min}$ by $\langle a_0^{2,*}\rangle_{\min}$, as in $\bar{\Psi}^{*}_{ss-\mathrm{inc}} = \bar{\sigma}^{2,*}_{0-\mathrm{inc}}$, Eq. (3.3.10a) et seq. Also, we note that the definition of the detection parameter in Ref. [28], Sec. 6.2, and in Ref. [8], is $(S/N)^2_{\mathrm{out}} = \sigma_0^{(*)2}/2$, which differs from the above by a factor 2. Similar modifications occur in the incoherent cases.

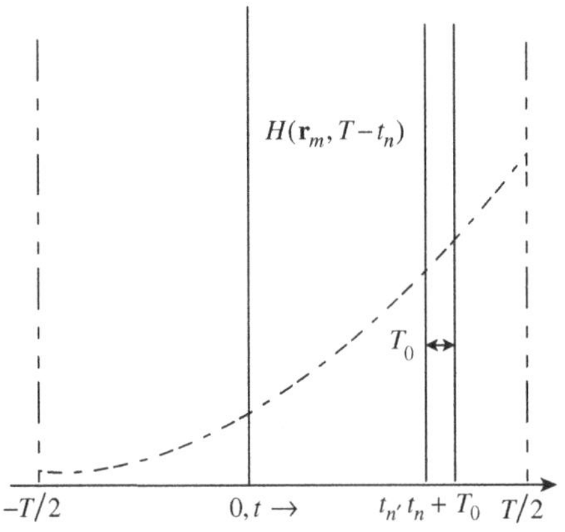

FIGURE 3.14 Discrete periodic temporal sampling of the matched filter response (weighting function) $H(\mathbf{r}_m, t_n) = H(\mathbf{r}_m, T-t_n)$ in the sample interval $(-T/2, T/2)$. (We consider only the primary intervals of the transform pair (3.5.4a) and (3.5.4b).)

In the Fourier transform space, the complementary vectors are accordingly

$$\mathbf{q} \equiv \boldsymbol{\nu} + \hat{\mathbf{i}}_4 f = \hat{\mathbf{i}}_\nu \nu + \hat{\mathbf{i}}_4 f = \hat{\mathbf{i}}_1 \nu_x + \hat{\mathbf{i}}_2 \nu_y + \hat{\mathbf{i}}_3 \nu_z + \hat{\mathbf{i}}_4 f \equiv \mathbf{q}_\nu + \hat{\mathbf{i}}_4 f;\ |\boldsymbol{\nu}| = \left(\nu_x^2 + \nu_y^2 + \nu_z^2\right)^{1/2} \tag{3.5.3}$$

where $\mathbf{q}_\nu = \mathbf{v}$ is a (vector) wave number and f is a frequency. The generalization of the temporal sampling procedure of Eq. 4.2 of [1], for the discrete *ordered* series of periodically sampled values of $H_j^{(1)}$ forming the vector $\mathbf{H}^{(1)}$ here, is the discrete Fourier transform of $\mathbf{H}^{(1)}$,

$$Y^{(1)}(\boldsymbol{\nu}, f)_d = \mathsf{F}_d\left\{\sum_j H_j^{(1)}\right\} \equiv \sum_{-J/2}^{J/2} \Delta_0 H^{(1)}(\mathbf{r}_m, nT_0) e^{2\pi i \mathbf{p}_j \cdot \mathbf{q}}, \tag{3.5.4a}$$

The inverse transform is similarly

$$\begin{aligned} \mathsf{F}_d^{-1}\{Y^{(1)}(\boldsymbol{\nu}, f)_d\} &= \int_{-[\boldsymbol{\nu}_0/2]}^{[\boldsymbol{\nu}_0/2]} \mathbf{d}\boldsymbol{\nu} \int_{-f_0/2}^{f_0/2} df\, Y^{(1)}(\boldsymbol{\nu}, f)_d e^{-2\pi i \mathbf{q} \cdot \mathbf{p}_j} \\ &= \int_{-\infty}^{\infty} \mathbf{d}\boldsymbol{\nu} \int_{-\infty}^{\infty} df\, Y^{(1)}(\boldsymbol{\nu}, f)_d e^{-2\pi i \mathbf{p} \cdot \mathbf{q}} \sum_{-J/2}^{J/2} H^{(1)}(\mathbf{r}_m, nT_0) \end{aligned} \tag{3.5.4b}$$

where we note that $\mathsf{F}_d^{(J)-1} = \Sigma_{-J/2}^{J/2} \mathsf{F}_d^{(j)-1}$ and that

$$\mathbf{d}\boldsymbol{\nu} = \nu^2 d\nu\, d\phi \sin\theta\, d\theta;\quad [\boldsymbol{\nu}_0] = 2\pi^2 \nu_0;\quad \Delta_{0j} = \Delta_0 = r_0 T, \text{refer to Eq. (2.7.34a)};\quad \mathbf{r}_m = m\hat{\mathbf{i}}_0 r_0 \tag{3.5.4c}$$

since the now equally spaced sample points of $\mathbf{H}^{(1)}$ are along a straight line in space. For the general case where m intervals have different lengths and directions, see 2.5 and

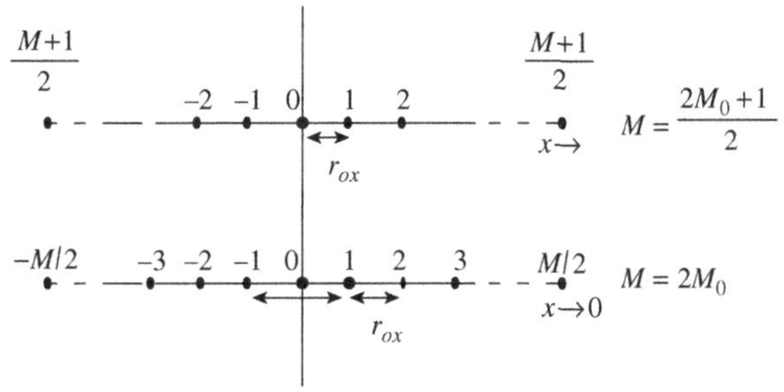

FIGURE 3.15 Symmetrical linear arrays of sensors (spatial periodic sampled values): $M = (2M_0 + 1)/2$, odd number; $M = 2M_0$ = even number, along the x-axis.

Figs. 2.11–2.13. The sums in (3.5.4a) and (3.5.4b) are over $j = m,n$ as usual on the basic, space–time sampling interval is ${}_0\Delta_0 = \mathrm{r}_0\mathrm{T}$. Figure 3.14 shows a typical time response for the matched filter. The principle domains of $\mathbf{v}$ and f are respectively $\neq \nu_0/2, (\phi = 0, 2\pi), (\theta = 0, \pi)\hat{i}_0[\nu_0]/2$, (3.5.4b) $(-f_0/2, f_0/2)$, where $f_0 = 1/T_0$. The spectral density (3.5.4a) and (3.5.4b) is periodic in wave number and frequency and is completely specified in the primary interval $(-1/2|\mathbf{R}|, 1/2|\mathbf{R}|), (-1/2T_0, 1/2T_0))$. Thus, *the spectrum is zero outside these intervals* indicated by the bounds on the integrals (3.5.4b), *as the subscript (d) also reminds us.* Both the space–time samples and the resulting wave number–frequency spectrum are limited to finite domains.

These results (3.5.4a–3.5.4c) are directly applicable to those (matched) filters that are represented previously as vectors, namely, $\mathbf{H}^{(1)}, \mathbf{H}^{(2,2a)}, \mathbf{W}, \mathbf{H}^{(3)}, \mathbf{H}^{(4,2b)}$, refer to Table 3.1. Figure 3.15 shows a typical sampling plan when M is odd or even, for linear arrays.

For the discrete space and time-variable matched filters that have the generic form $\left[H^{(2,-)}(\mathbf{r}_m, t_n; \mathbf{r}_{m'}, t'_n)\right]$, that is, are square $(J \times J)$ matrices and are generally space and time-variable, we find by similar extensions of (3.5.4a)–(3.5.4c) that the corresponding amplitude spectra here are given by

$$\begin{aligned}\hat{Y}_{\mathrm{d}}^{(2,-)}(\mathbf{v}_1,f_1;\mathbf{v}_2,f_2)_{\mathrm{d}} &= \mathbf{F}_{\mathrm{d}}^{(J\times J)}\left\{\sum_{j,j'} \hat{H}_{jj'}\right\} \\ &= \sum_{\substack{j,j' \\ -(J/2,J/2)}}^{J/2,J/2} \Delta_0^2 \cdot \hat{H}^{(2,-)}(\mathbf{r}_m, nT_0; \mathbf{r}_{m'}, n'T_0)e^{2\pi i(\mathbf{q}_1\cdot\mathbf{p}_j + \mathbf{q}_2\cdot\mathbf{p}_{j'})}\end{aligned} \tag{3.5.5a}$$

with the other member of the transform pair

$$\begin{aligned}\sum_{j\to J/2,J/2\leftarrow j'}^{J/2,J/2} \hat{H}^{(2,-)}(\mathbf{p}_j,\mathbf{p}_{j'}) &= (\mathbf{F}_d^{(J,J')})^{(-1)}\left\{\hat{Y}_{\mathrm{d}}^{(2,-)}(\mathbf{v}_1,f_1;\mathbf{v}_2,f_2)\right\} \\ &= \iint_{-[\mathbf{v}_0]^2/4}^{[\mathbf{v}_0]^2/4} d\mathbf{v}_1 d\mathbf{v}_2 \iint_{-f_0/2}^{f_0/2} \Delta_0^{-2}\hat{Y}_{\mathrm{d}}^{(2,-)}(\mathbf{v}_1,f_1;\mathbf{v}_2,f_2)_{\mathrm{d}}e^{-2\pi i(\mathbf{q}_1\cdot\mathbf{p}_j + \mathbf{q}_2\cdot\mathbf{p}_{j'})}df_1 df_2\,.\end{aligned} \tag{3.5.5b}$$

Again, the spectral density is (now doubly) periodic in wave number and frequency, where the primary intervals on $Y_{\mathrm{d}}^{(2,-)}$ are given by $(-1/2|\mathbf{R}|, 1/2|\mathbf{R}|)_{1,2}$ and $(-(1/2)T_0, (1/2)T_0)_{1,2}$ and we consider only this primary interval (see the comments following Eq. (3.5.4c)).

In addition to the discrete Fourier transforms of the periodically sampled forms of the matched filters $[H_j]$ and so on, we shall also need the discrete representation of the Hom-Stat Wiener–Khintchine relations (Section 2.2). These (unnormalized) W–Kh relations are applicable to the frequent but less usual cases of fields that are (wide sense) homogeneous and stationary, namely, here for periodically sampled fields in the primary interval $(-|\mathbf{R}|/2, |\mathbf{R}|/2; -T/2, T/2)$ represented by the $((J \times J)$ square) matrices

$$[K_N(\mathbf{r}_{m'} - \mathbf{r}_m; t_{n'} - t_n)] = [K_N(\Delta\mathbf{r}_{mm'}; (n'-n)T_0)] = [K_N(\Delta\mathbf{p}_{\hat{j}})], \quad \hat{j} = j' - j = (m'-m, n'-n), \tag{3.5.6}$$

with $\Delta\mathbf{r}_{mm'} = (m'-m)\hat{\mathbf{i}}_0 r_0$. Thus, we have specifically

$$\Delta\mathbf{p}_{\hat{j}} = \Delta\mathbf{r}_{mm'} - \hat{\mathbf{i}}_4(n'-n)T_0 = (m'-m)\hat{\mathbf{i}}_0 r_0 - \hat{\mathbf{i}}_4(n'-n)T_0 = \Delta\mathbf{p}_k, \quad k = (m'-m, n'-n). \tag{3.5.6a}$$

Here $\mathbf{q} = \mathbf{v} + \hat{\mathbf{i}}_4 f$, as before, refer to (3.5.3), as well as $f_0 = 1/T_0$ and $[\mathbf{v}_0] = 1/|\mathbf{R}_0|$, refer to (3.5.4b) and (3.5.4c). The discrete Fourier transforms of the periodic series $\hat{j} = (\hat{m}, \hat{n})$ formed by the elements of $\left[(\mathbf{k}_N)_{j,j'}\right] = [(\mathbf{k}_N)_k]$ represent the extension of the W–Kh theorem in the Hom-Stat cases to discrete periodic sampling. It is found from Section 2.2.1 to be[51]

$$\mathsf{W}_N(\mathbf{v}, f)_d = \mathsf{F}_d\left\{K_N\left(\Delta\mathbf{p}_{\hat{j}}\right)\right\} = \Delta_0 \cdot \sum_{-K}^{K} K_N(\Delta\mathbf{p}_k) e^{2\pi i \Delta\mathbf{p}_k \cdot \mathbf{q}}, \tag{3.5.7}$$

cf. (3.5.6a) above, with the corresponding member of the transform pair

$$K_N(\Delta\mathbf{p}_k)_d \doteq \mathsf{F}_d^{-1}\{\mathsf{W}_N\} = \frac{1}{2} \int_{-[\mathbf{v}_0]/2}^{[\mathbf{v}_0]/2} d\mathbf{v} \int_{-f_0/2}^{f_0/2} df\, \mathsf{W}_N(\mathbf{v}, f)_d e^{2\pi i \mathbf{q} \cdot \Delta\mathbf{p}_k} \tag{3.5.8}$$

where, as usual, $d\boldsymbol{\nu}$ is given by (3.5.4c), and for all values in the new primary interval $(-K, K)$. As before, we confine our attention to the principal interval above (setting the others

[51] Henceforth, we drop the designation on the Fourier transform operators $\mathsf{F}_d^{(J)}, \mathsf{F}_d^{(J,J')}$, and so on, as being evident from the text as to its order $(J), (J, J')$.

equal to zero, to avoid ambiguities and the possibility of attributing spurious energy to the finite data sample).

Returning now to the general situation of inhomogeneous, nonstationary random fields, we may obtain further insight into the structure of the above discrete matched filters and their processing of received signals and noise, by examining their representations in the resulting finite wave number–frequency domain.[52] This is made possible by the results of Section 3.5.1 where periodic sampling is used. To this end, let us consider the Fourier domain solution to the unnormalized discrete integral equation[53] for the general W–K filter, described by the weighting function $\hat{\mathbf{H}}_j^{(1)}$, all j for the finite interval, $(-|\mathbf{R}|/2, |\mathbf{R}|/2;\ -T/2, T/2)$, namely,

$$\sum_{j'} (\mathbf{K}_N)_{jj'} \hat{\mathbf{H}}_{j'}^{(1)} = \langle \hat{S}_{\mathrm{sig}}(\mathbf{r}_m, T-t_n) \rangle = \langle \hat{S}_j \rangle,\ (1,1) \leq j \leq J; = 0,\ \text{elsewhere}, \quad (3.5.9)$$

where periodic sampling is employed (Section 3.5.1). The deterministic signal $\langle \hat{S}_j \rangle$ embodies the time-reverse of the temporal portion of the matched filter represented by the (ordered elements of the) vector $\hat{\mathbf{H}}^{(1)}$. Equation (3.5.9) is exact and may be solved in a variety of ways, all ultimately requiring numerical methods for specific matrices in practical applications. But such solutions lack the direct physical insights that the solution in the wave number–frequency space can provide.

Accordingly, let us consider the following Fourier transform solution, which is nonvanishing only in the finite (ν, f) domain. Again, we consider the general *non-Hom-Stat cases* for the covariance function $\mathbf{K}_\mathrm{N}$ (Section 2.4.3.3) Eq. (2.4.40b); for the Hom-Stat case, see Eq. (3.5.7)). We begin by expressing the exact relation (3.5.9) in terms of its Fourier transforms $\mathsf{F}_\mathrm{d}^{-1}$:

$$\left.\begin{aligned} \mathsf{F}_\mathrm{d}^{-1}(\mathsf{W}_N(\boldsymbol{\nu},f)_d)\mathsf{F}_\mathrm{d}^{-1}\left\{\hat{Y}_H^{(1)}(\boldsymbol{\nu}',f')_d\right\} &= \mathsf{F}_\mathrm{d}^{-1}\left\{\hat{\mathsf{S}}_{\mathrm{sig}}(\boldsymbol{\nu},-f)_d\right\},(-|\mathbf{R}|/2,|\mathbf{R}|/2;\ -T/2,T/2) \\ &= 0,\ \text{elsewhere} \end{aligned}\right\} \quad (3.5.10)$$

From (3.5.4b), we now extend the limits on f_0, $[\mathbf{v}_0]$ to $(\pm\infty)$ without changing the integrals (viz, Eq. 3.5.10), since only the primary interval of the periodic spectra contributes. (This includes the wave number–frequency equivalent) of the covariance in space–time, multiplied by $J/2$, or when $\mathsf{W}_N|_d$ in (3.5.10) is set equal to $J/2\mathsf{W}_X|_\Delta$.) In either instance, these two intensity spectra have the required dimension $[S^2][L][T]$. Then, (3.5.10) can be

[52] See the pertinent remarks following Eq. (3.5.4c).

[53] Here in Section 3.5.2, we use the *unnormalized* covariance K_N, so that $\hat{\mathbf{H}}_j^{(1)}$ in (3.5.9) has the dimensions $[S]^{-1}$, where $[S]$ denotes the dimension of amplitude, that is, $\mathbf{K}_N = [S]^2$, $\langle S_{\mathrm{rec}} \rangle = [S]$. In the normalized cases of Section. 3.5.1, $\mathbf{H}^{(1)}$ and so on, and in Table 3.1 ff., are dimensionless, as is $\mathbf{k}_\mathrm{N}$.

written *exactly*

$$\sum_{j'=-\infty}^{\infty} \int\!\!\int_{-\infty}^{\infty} \mathbf{d}\boldsymbol{\nu}\, df \mathsf{W}_N(\boldsymbol{\nu},f)_d e^{-2\pi i \mathbf{q}\cdot(\mathbf{p}_j-\mathbf{p}_{j'})} \hat{Y}_H^{(1)}(\boldsymbol{\nu}',f')_d e^{-2\pi i \mathbf{q}\cdot\mathbf{p}_{j'}} \mathbf{d}\boldsymbol{\nu}' df'$$

$$= \iint_{-\infty}^{\infty} \mathbf{d}\boldsymbol{\nu}\, df \langle \mathsf{S}_{\text{sig}}(\boldsymbol{\nu},-f)_d\rangle e^{-2\pi i \mathbf{q}\cdot\mathbf{p}_j} \tag{3.5.10a}$$

With the help of the extended version of Eq. (4.8) [1], namely,

$$\sum_{j'=-\infty}^{\infty} e^{-2\pi i[(\mathbf{p}_j-\mathbf{p}_{j'})\cdot\mathbf{q}+\mathbf{p}_{j'}\cdot\mathbf{q}']} = e^{-2\pi i \mathbf{q}\cdot\mathbf{p}_j}\delta(\mathbf{q}'-\mathbf{q}),\ \frac{-|\boldsymbol{\nu}_0|}{2} < \nu' < \frac{|\boldsymbol{\nu}_0|}{2}, -f_0/2 < f' < f_0/2, \tag{3.5.11}$$

we see that (3.5.10) finally becomes for any p_j,

$$\iint_{-\infty}^{\infty} \mathbf{d}\boldsymbol{\nu}\, df \left\{ \mathsf{W}_{\mathrm{N}}(\boldsymbol{\nu},f)_d \hat{Y}_H^{(1)}(\boldsymbol{\nu},f_d) - \left\langle \hat{\mathsf{S}}_{\text{sig}}(\boldsymbol{\nu},-f)_d \right\rangle \right\} e^{-2\pi i \mathbf{q}\cdot\mathbf{p}_j} = 0. \tag{3.5.12}$$

For arbitrary, and hence all (ν, f), this gives the desired result[54]

$$\hat{Y}_H^{(1)}(\boldsymbol{\nu},f)_d = \left\langle \hat{\mathsf{S}}_{\text{sig}}(\boldsymbol{\nu},-f)_d \right\rangle \Big/ \mathsf{W}_{\mathrm{N}}(\boldsymbol{\nu},f)_d, \tag{3.5.12a}$$

with

$$\left\langle \hat{\mathsf{S}}_{\text{sig}}(\boldsymbol{\nu},-f) \right\rangle = \langle \mathsf{S}_{\text{sig}}(\boldsymbol{\nu},-f)\rangle e^{-i\omega T} \tag{3.5.12b}$$

to account for the time-reversal signal in (3.5.9). Equation (3.5.12a) is the extension to space–time (or equivalently here to wave number–frequency space) of the familiar continuous temporal matched filter[55] (cf. Section 16.2 of Ref. [1]), represented now by the discrete series of sampled values $\hat{\mathbf{H}}_{j'}$, Eq. (3.5.9). When $K_N \to K_N + K_I + K_C$, we simply replace W_{N} by $\mathsf{W}_{\mathrm{NIC}}$, the combined intensity spectra corresponding to the sum of the separate components of ambient, interference (or jamming), and scatter noise.

When the ambient (and system) noise is "white," so that Eq. (3.5.12b) for this matched filter becomes

$$\hat{Y}_H^{(1)}(\boldsymbol{\nu},-f)_d = \left\langle \hat{\mathsf{S}}_{\text{sig}}(\boldsymbol{\nu},-f)_d \right\rangle / \left[\mathsf{W}^{(ST)} + \mathsf{W}_I(\boldsymbol{\nu},f) + \mathsf{W}_C(\boldsymbol{\nu},f) \right]_d \tag{3.5.13}$$

[54] Incidentally, here the dimensions of $Y_H^{(1)}, \hat{S}_{\text{sig}}$, and W_N are respectively $[\alpha]^{-1}, [\alpha][LT]$, and $\mathsf{W}_N = \alpha^2[LT]$ in (3.5.13). In the space–time domain $\mathbf{K}_{\mathrm{N}}, \dot{H}^{(1)}$ and $\hat{S}_{\text{sig}}$ are correspondingly $[\alpha]^2, [\alpha]^{-1}, \hat{S}_{\text{sig}} = [\alpha]$ for Eq. (3.5.9) above, where $[\alpha] = [s]$ is used interchangeably for the dimension of amplitude.

[55] See the historical note at the end of Section 3.5.

where $\mathsf{W}_I, \mathsf{W}_C$ are (unnormalized) intensity spectra and

$$\mathsf{W}_{\mathrm{N}}(\boldsymbol{\nu},f) = \mathsf{W}_{\mathrm{N}}^{(ST)} = \frac{\mathsf{W}_{\mathrm{N}}^{(S)}}{2}\cdot\frac{\mathsf{W}_{\mathrm{N}}^{(T)}}{2}, \quad \text{and} \quad \mathbf{K}_{\mathrm{N}}(\mathbf{r}'-\mathbf{r}, t'-t) = \frac{\mathsf{W}_{\mathrm{N}}^{(ST)}}{2}\delta(\mathbf{r}'-\mathbf{r})\delta(t'-t), \tag{3.5.13a}$$

where $\mathsf{W}^{(ST)}$ has the dimensions $[S^2]LT$ for "white noise" in the wave number–frequency domain, corresponding to the large number of emitting independent point sources in space–time now with $\delta(\mathbf{r}'-\mathbf{r})\delta(t'-t) \to \delta\left(\hat{\mathbf{i}}_0(m'-m)r_0\right)\delta((n'-n)T_0)$ here.[56] For further optimization by "signal design," we can also use as our signal the *Green's function of the medium* in our original matched filter $\hat{\mathbf{H}}^{(1)}$, refer to Eqs. (3.5.9) and (3.5.12).

3.5.1.1 Space–Time Matched Filter as Optimum Beam Former The important and inherent feature of the spatial dimension here is that the ordered *sum* (over (m, m')) of samples $\hat{H}^{(1)}(\mathbf{p}_j)$ in (3.5.4a) now represent for these matched space–time filters *optimum beam forming by the resultant array of connected discrete point sensors, at* $\mathbf{r}_m$ $(m = 1, \dots, M)$. Thus, the spatial part of the *matched filter response function* $Y_H^{(1)}(\boldsymbol{\nu}, -f)_d$, that is, the so-called system function ([1], Section 2.2.5) here *for space–time processing embodies the formation of an optimum "beam."* With an appropriate added phase shift ($\sim\boldsymbol{\nu}_{OR}$ = a steering vector) to each element of the array, it is then possible to steer this beam by varying $\boldsymbol{\nu}_{OR}(=\hat{\mathbf{i}}_{OR}\nu_{OR})$ and then to locate potential signal sources, refer to Fig. 3.16; (this is discussed further in Section 3.4). Equation (3.5.13) is accordingly

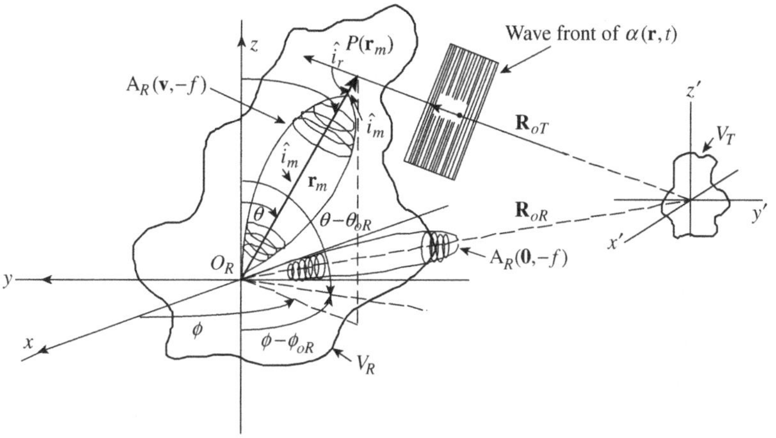

FIGURE 3.16 Receiving array element at $\mathbf{r}_m$ and (far-field) incident wave front, and steering vector $\mathbf{v}_{OR}$ (cf. Eq. 3.5.14); V_R, V_T = receiving and transmitting apertures; path delay to $O_R = \left(1-\hat{\mathbf{i}}_m\cdot\hat{\mathbf{i}}_{OR}\right)|\mathbf{R}_{OR}/c_0|$.

[56] In all cases, it is assumed that the noise and signal spectra are suitably defined to avoid singular results for the quadratic forms of Eq. (3.5.1), and that the noise spectra vanish at frequencies $0 < f < 0+$, to avoid technical problems, that is, Its integrals in the Gaussian cases.

modified to

$$Y_H^{(1)}(\boldsymbol{\nu}-\boldsymbol{\nu}_{OR},-f)_d \doteq \left\langle \hat{\mathsf{S}}_{\text{sig}}(\boldsymbol{\nu}-\boldsymbol{\nu}_{OR},-f)_d \right\rangle \Big/ \left[\mathsf{W}_{\text{N}}^{(ST)}(\boldsymbol{\nu},f)+\mathsf{W}_I(\boldsymbol{\nu}-\boldsymbol{\nu}_{OR},f)+\mathsf{W}_{\text{C}}(\boldsymbol{\nu}-\boldsymbol{\nu}_{OR},f) \right]_d \tag{3.5.14}$$

when the noise in the vicinity of the receiver can be regarded as *isotropic*. (When the noise is homogeneous only, then $\mathbf{v}$ is replaced by $\mathbf{v}-\mathbf{v}_{OR}$.) We remark that the detailed structure of $\mathbf{v}$, and hence $\mathbf{v}_{OR}$, depends on the medium of propagation, that is, whether it is "ideal," absorptive, or generally dissipative, and so on (Chapters 8–10). In the general case when the noise is non-Hom-Stat, $Y_H^{(1)}$ is also recognized as the *general aperture beam function* (cf. (2.5.2a) and (2.5.2b)) or *beam pattern* (here for discrete arrays) of the receiver:

$$\hat{Y}_H^{(1)}(\mathbf{v}-\mathbf{v}_{OR},-f_d) = \hat{\mathsf{A}}_{\text{R}}\left(\mathbf{v}-\mathbf{v}_{OR},f \middle| \hat{S}_{\text{sig}}\right). \tag{3.5.14a}$$ [57]

(An important case in practice, discussed in Section 3.5.3 presently, arises when we separate the space and time operations.) Accordingly, we may say, generally (from the result (3.5.14a) that

Beam formation in reception in noise is the Fourier transform of the aperture weighting function $\left(\hat{Y}_N^{(1)}\right)$, *and that optimal beam forming*(in the Bayes framework) *is the space – time Fourier transform of the receiver matched filter (represented here by the discrete W – K weighting function* $\hat{\mathbf{H}}^{(1)}$).

3.5.1.2 Wave Number as Functions of Frequency We next need to show that wave number (space) and frequency (time) are not physically independent. This, of course, can be practically important in the task of effective steering of the optimal (and near optimal) beams, refer to Eqs. (3.5.14) and (3.5.14a). We begin by noting that *these space–time fields, whether signals or noise, can be expressed as equivalent time-variable quantities*: instead of $\mathbf{r}$ and t being the independent variables, t' and t can replace them. Thus, for example, the *signal field* can be represented alternatively by

$$S(\mathbf{r},t) = S(\mathbf{r}/c_0, t/c_0) = S\left(t'\hat{\mathbf{i}}_r, t|c_0\right) = S\left(t', t\middle|c_0, \hat{\mathbf{i}}_r\right) = S\left(t'-t, t'\middle|c_0, \hat{\mathbf{i}}_r\right), \tag{3.5.15}$$

which emphasizes the time-variable nature of the signal field here when treated solely as a function of time at the receiver.[58] This is of course possible, from the viewpoint of the field,

[57] The dimension of $\hat{Y}_H^{(1)} = \hat{\mathsf{A}}_{\text{R}}$ is $[S]^{-1} = [\alpha]^{-1}$. In the normalized cases, $\mathbf{k}_{\text{N}}\mathbf{H}^{(1)} = \mathbf{s}$, $\hat{Y}_H^{(1)} \to Y^{(1)} = A_{\text{R}}$ is dimensionless.

[58] Of course, here there is no relative motion of receiver, transmitter, and the medium supporting the signal field, and consequently from the customary perspective of these entities, $\mathbf{r}$ is simply a fixed position in space. Accordingly, the filter elements representing the aperture or array coupling to the medium are time invariant here. When there is (relative) motion of one or more of these components of the channel, then there will be Doppler components, for example, in the received (or transmitted) field, requiring in turn a time-variable matching filter, for the shifted frequencies or more generally for a modified amplitude and scaled phase [$\sim \omega\,(at\text{-}b)$] for broadband cases.

because $S(\mathbf{r},t)$ represents a *propagating field, where space and time are related by* $|\mathbf{r}| = c_0 t$, and where c_0 is the phase (or group) velocity of propagation in the medium[59] supporting S(**r**,t). Furthermore, if we in turn represent the field by its Fourier transform, refer to Eq. (3.5.4b), in the sample interval Δ, that is,

$$S(\mathbf{r},t)_\Delta = \int_{-[\nu_0]/2}^{[\nu_0]/2} \mathbf{d}^3\boldsymbol{\nu} \int_{-f_0/2}^{f_0/2} df \mathsf{S}_{\text{sig}}(\boldsymbol{\nu},f)_d \exp(2\pi i[ft - \mathbf{r}\bullet\boldsymbol{\nu}]) \tag{3.5.15a}$$

for a simple nondispersive medium where $f\lambda = c_0$ or $f = c_0/\lambda = c_0\nu$, so that $\mathbf{r}\cdot\boldsymbol{\nu} = r\nu(\hat{\mathbf{i}}_r \cdot \hat{\mathbf{i}}_\nu) = (c_0 t)(\hat{\mathbf{i}}_r \cdot \hat{\mathbf{i}}_\nu)\nu$, the exponent in (3.5.15a) is alternatively

$$ft = c_0\nu\hat{\mathbf{i}}_\nu \cdot \hat{\mathbf{t}}, \quad \therefore \quad \boldsymbol{\nu} = (f/c_0)\hat{\mathbf{i}}_\nu \quad \text{and} \quad \hat{\mathbf{t}} \equiv \hat{\mathbf{i}}_r t', \quad \text{with } \nu = |\boldsymbol{\nu}| = f/c_0. \tag{3.5.15b}$$

Accordingly, |ν| and ν are specifically here linear functions of frequency,[60] which shows *that space and time are not naturally separable in a propagating field*, as noted earlier, (Section 2.5). In fact, this holds for any Hom-Stat (and non-Hom-Stat) field where the field variable is a function of space and time. As we shall see in Chapter 10, the specific argument above can be extended to ambient and scattered noise fields, and correspondingly to the matched filter for more complex, that is, *dispersive*, media (Section 8.2.1).

Thus in general, the dependence of wave number (ν) on frequency (f) illustrates the connectivity between space and time in the generation of field structure and must be taken into account if the matched filter is to achieve its full optimality. Because of this dependence of wave number on frequency, Eq. (3.5.15b), we can write (3.5.12b), and more generally (3.5.14) as

$$(3.5.12\text{b}): Y_H^{(1)}(\boldsymbol{\nu}(f) - \boldsymbol{\nu}_{OR}(f), f)_d = \left\langle \mathsf{S}_{\text{sig}}(\boldsymbol{\nu}(f) - \boldsymbol{\nu}(f)_{OR}, -f)_d \right\rangle \Big/ \mathsf{W}_{\text{N}}(\boldsymbol{\nu}(f) - \boldsymbol{\nu}(f)_{OR}, f), \tag{3.5.16a}$$

$$(3.5.14): Y_H^{(1)}(\boldsymbol{\nu}(f) - \boldsymbol{\nu}_{OR}(f), f)_d = \left\langle \mathsf{S}_{\text{sig}}(\boldsymbol{\nu}(f) - \boldsymbol{\nu}(f)_{OR}, -f_d)\right\rangle \Big/ \left\{ \mathsf{W}_{\text{N}}^{(ST)} + (\mathsf{W}_I + \mathsf{W}_C)_{[\boldsymbol{\nu}(f) - \boldsymbol{\nu}(f)_{OR}, f]} \right\}, \tag{3.5.16b}$$

where for optimization, the steering vector $\boldsymbol{\nu}_{OR}(f)$ *must be a similar function of frequency to* $\nu(f)$, when directed at a signal source. Again, for these specific results, it is assumed that the fields here are not necessarily homogeneous and stationary.

3.5.1.3 Clutter and Reverberation: The Inverse (Urkowitz) and (Eckart) Matched Filters Several important special cases of the Bayes matched filters discussed in Section 3.5 are also obtained as approximations of the general cases considered in

[59] When considered solely as a function of time, as in (3.5.15), a Hom-Stat field $S(\mathbf{r},t)$ becomes *nonstationary*: the delay $t'_m = (r_m/c_0)$ from the mth-sensor to the reference point (O_R) (Fig. 3.16) occurs at a time t_m independent of the "memory" time (t) of the filter at $\mathbf{r}_m$.

[60] For dispersive media, the wave numbers are *nonlinear* functions of frequency (Section 8.1.4 and Table 8.2).

Sections (3.4.1)–(3.4.7). These are the so-called *Inverse* or Urkowitz [29] and Eckart [30] *matched filters*, where the former is used in the detection of deterministic signals against signal-generated noise, that is, strong clutter or reverberation, and the latter is employed in the detection of stochastic signals generally [31]. For the *inverse filter*, we have $\mathbf{K}_N + \hat{\mathbf{K}}_C \doteq \hat{\mathbf{K}}_C$, that is, the clutter is dominant. The structure of the resultant filter is more fully revealed by the corresponding system function $Y_H^{(1)}$, refer to Eq. (3.5.13), which reduces to

$$Y_H^{(1)}(\mathbf{v}, -f)_{\text{Inverse}} \doteq \left\langle \hat{\mathsf{S}}_{\text{sig}}(\mathbf{v}, -f)_d \right\rangle \Big/ \left\{ \mathsf{W}_C(\mathbf{v}, f)_d \quad \text{or} \quad \mathsf{W}_C(\mathbf{v}-\mathbf{v}_{OR}, f) \right\}, \qquad (3.5.17)$$

respectively, for isotropic or the much less restricted case of inhomogeneous nonstationary clutter, respectively. In case of Eckart filter, the now stochastic signal may be regarded as a sequence of overlapping pulses that occur randomly and are themselves random in amplitude and duration. Here, the observation interval in space and time is $\Delta = (-|\mathbf{R}|/2, |\mathbf{R}|/2; -T/2, T/2)$ and is much larger than the duration of a signal pulse, which is now described by $\hat{S}_{\text{sig}} = \mathsf{W}_{\text{sig}}(\mathbf{v}-\mathbf{v}_{OR}, -f)^{1/2} \exp i\gamma(\mathbf{v}-\mathbf{v}_{OR}, f)$, where γ is a phase factor inserted to make the filter causal. In practice, it is the time-duration of the interval that is usually governing.

Thus, in case of inverse filter, we first assume pure clutter or reverberation, free of resolvable, that is, deterministic multipath. Here, the spectral model is a generalized version of Campbell's theorem (Section 4.6.2 and Eqs. (4.79) and (4.80), p. 236, of Ref. [1]) for at least locally Hom-Stat scatter, namely,

$$\mathsf{W}_C(\mathbf{v}, f)_d = \gamma_0 \frac{2\overline{y^2}}{\bar{\tau}} \left\langle \left| \mathsf{S}_u\left(\mathbf{v}, f|\tau\right)^2 \right| \right\rangle_\tau, \qquad (3.5.18)$$

with the corresponding covariance

$$K_C(\Delta\mathbf{r}, \Delta t) = \gamma_0 \overline{y^2} \langle \rho_u(\Delta\mathbf{r}, \Delta t|\tau) \rangle_\tau \qquad (3.5.18a)$$

$$\rho_u = \langle u(\mathbf{r}_m, t_0|\tau) u(\mathbf{r}_{m'}, t_0 + \Delta t|\tau) \rangle_\tau \equiv \rho_u(\Delta\mathbf{r}, \Delta t)$$

$$\Delta\mathbf{r} = \mathbf{r}_{m'} - \mathbf{r}_m; \Delta t = (n'-n)T_0. \qquad (3.5.18b)$$

Here, $\gamma_0 = \bar{n}_0 \bar{\tau} =$ the total number of the individual overlapping scatter events $u(\mathbf{r}_m, t_n)$, with $0 \le |u| \le \infty$, $\bar{\tau} =$ their average duration (s), and n_0 their expected number per second. (As an example, we have $u = \exp(-\beta\tau)$, $0 < \tau \le \infty$, and $\beta = 1/\bar{\tau}$ here.) Next, we make the usually reasonable assumption that the scatter returned to the receiver is proportional to the *received signal*, that is, the scatter intensity is

$$\mathsf{W}_C(\boldsymbol{\nu}, f)_d = \gamma_0 \left| \hat{\mathsf{S}}_{\text{sig}}(\boldsymbol{\nu}, -f) \right|^2 = \gamma \hat{\mathsf{S}}_{\text{sig}}(\boldsymbol{\nu}, -f) \hat{\mathsf{S}}_{\text{sig}}(\boldsymbol{\nu}, -f)^*. \qquad (3.5.19)$$

Then, from (3.5.17), we have directly for the inverse filter, the well-known approximate result

$$Y_H^{(1)}(\boldsymbol{\nu}, f)_{\text{inverse}} \doteq \gamma_0^{-1} \hat{\mathsf{S}}_{\text{sig}}(-\boldsymbol{\nu}, f)^{-1}. \qquad (3.5.20)$$

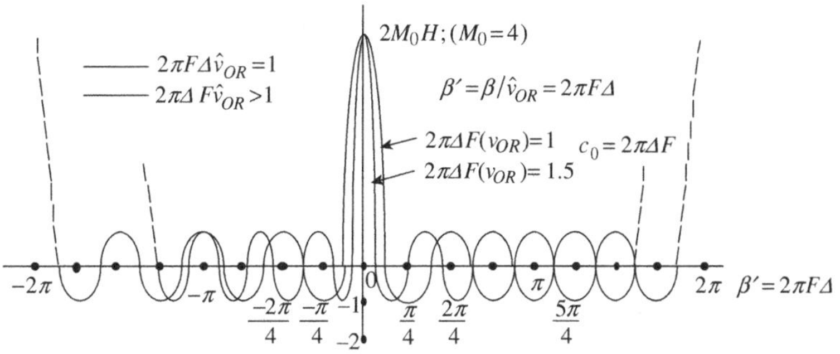

FIGURE 3.17 Beam pattern of Eq. (3.5.31a) for selected values of $\beta'(=\beta/\hat{v}_{OR})$, Eq. (3.5.31b), $\hat{v}_{OR}$ as parameter.

For beam steering an adjustable directional component, $\mathbf{v}_{OR}$, can be added (Fig. 3.16). Note that this form of matched filter does *not* employ the time-reversal signal (i.e., $Y_H^{(1)}(\boldsymbol{\nu}, -f)_{\text{inverse}}$). Furthermore, it is not realizable, without adding an appropriate phase factor [18, 31]. A variety of additional approximations are implicit in Eqs. (3.5.19) and (3.5.20), as can be seen on comparison of (3.5.18) with (3.5.19): the average $\left(2\overline{y^2}/\bar{\tau}\langle|S_u|^2\rangle\right) \doteq |S_{\text{sig}}|^2$, that is, replacing the average of the square by the unaveraged, deterministic quantity $|S_{\text{sig}}|^2$ and neglecting multiple scatter, as well as ignoring range dependence for the scatter.

With a strong signal from the transmitter and consequently strong clutter or reverberation in the medium, the appropriate matched filter in the receiver is the *inverse filter* (Eq. (3.5.20)). Its temporal behavior (at a given location on the wave front) is explained by observing that with strong signals of fixed power and not too large bandwidth, this inverse filter has the desired optimum weighting function $\mathbf{H}_C^{(a)} = \mathsf{F}_d^{-1}\{Y_N^{(1)}\}$ (Eq. (3.5.20)). As the frequency spectrum of the signal is broadened, its intensity decreases and the scatter return exhibits a progressively finer time structure. Thus, when the signal consists of a very short pulse or train of such pulses, as in the case of radar and some sonar applications, the resulting clutter or reverberation may be resolved into individual pulses with little or no time overlap. The result is a consequent improvement in detection and resolution of the desired targets. (The frequency response of the array (or aperture), as well as that of the temporal processing, must be broad enough to accommodate these spectrally broadened signals.)

However, although the resolution is improved through the progressive shortening of the emitted pulses, a point is eventually reached where the clutter component is dominated by the white noise background (cf. $\mathsf{W}_N^{(ST)}$ in (3.5.13)). Subclutter visibility of the target is lost, and the matched filter is once more the filter of the correlation detector (3.4.6b). More generally, it is the Bayes matched filter of the first kind, Type 1, embodied in the discrete (normalized) form $\mathbf{H}^{(1)}$ of Section 3.4. (The Bayes filters $\mathbf{H}^{(2,2a)}$, $\mathbf{H}^{(4,1a)}$, and $\mathbf{H}^{(4,2b)}$, respectively given by (3.4.19b), (3.4.36a), and (3.4.37b), are also of type $\mathbf{H}^{(1)}$, along with the Wiener–Kolmogoroff filters $\mathbf{W}$ of Section 3.4.4, except that the former are defined for more complex noise processes.)

For *the Eckart filter*, on the other hand, instead of a structured (i.e., deterministic) signal, we have a purely random space–time signal, which is represented approximately by

$$\hat{\mathsf{S}}_{\text{sig}}(\boldsymbol{\nu}-\boldsymbol{\nu}_{OR}, -f)_d = \mathsf{W}_{\text{sig}}(\boldsymbol{\nu}-\boldsymbol{\nu}_{OR}, f)_d^{1/2} e^{-i\gamma_d(\boldsymbol{\nu}-\boldsymbol{\nu}_O, f)}, \tag{3.5.21}$$

so that the system function of this matched filter is from (3.5.12b):

$$Y_H^{(1)}(\boldsymbol{\nu}-\boldsymbol{\nu}_{OR},-f)_d \doteq \mathsf{W}_{\text{sig}}(\boldsymbol{\nu}-\boldsymbol{\nu}_{OR},f)_d^{1/2} e^{-i\gamma_d(\boldsymbol{\nu}-\boldsymbol{\nu}_{OR},-f)} / \mathsf{W}(\boldsymbol{\nu},f)_d. \tag{3.5.22}$$

Here, a suitable phase term γ_d is also added to ensure causality. A steering vector $\mathbf{v}_{OR}$ is included to allow beam directivity (in the signal), refer to Section 3.5.2.1 above. (Because the noise here is assumed to be isotropic, steering does not alter the wave number portion of the space–time intensity spectrum. However, for non-Hom-Stat noise, the noise is directional and hence $\mathsf{W}_\text{N} = \mathsf{W}_\text{N}(\mathbf{v}-\mathbf{v}_{OR},f)_d$.

3.5.2 Independent Beam Forming and Temporal Processing

Although joint optimization of space and time processing is required for strict or unconstrained performance, it is often necessary to optimize the two separately, as a matter of practical convenience in the design of arrays (and aperture) and temporal processors (Sections 3.4.6 and 3.4.7.1). This scenario, of course, is suboptimum vis-à-vis the joint procedure, which is presented in the various forms of matched filter discussed in Section 3.5. With separate or disjoint space and time processing, the joint matched filter factors into two separate matched filters in the following way, refer to Eq. (3.4.65) et seq.:

$$\left(\mathbf{k}_\text{N}^{(S)}\mathbf{H}^{(S)}\right) \otimes \left(\mathbf{k}_\text{N}^{(T)}\mathbf{H}^{(T)}\right) = \langle\mathbf{s}\rangle^{(S)} \otimes \langle\mathbf{s}\rangle^{(T)} \tag{3.5.23a}$$

and accordingly,

$$\therefore\ \mathbf{k}_\text{N}^{(S)}\mathbf{H}^{(S)} = \left\langle \mathbf{s}^{(S)} \right\rangle \quad \text{and} \quad \mathbf{k}_\text{N}^{(T)}\mathbf{H}^{(T)} = \left\langle \mathbf{s}^{(T)} \right\rangle. \tag{3.5.23b}$$

From Eq. (3.5.10), this is equivalent to

$$\mathsf{F}_d\left\{\mathsf{W}_\text{N}^{(S)}\right\}\mathsf{F}_\text{d}^{-1}\left\{Y_H^{(1)(S)}\right\} = \mathsf{F}_d\left\{\left\langle \mathsf{S}_\text{sig}^{(S)}\right\rangle\right\};\quad \mathsf{F}_\text{d}^{-1}\left\{\mathsf{W}_\text{N}^{(T)}\right\}\mathsf{F}_\text{d}^{-1}\left\{Y_H^{(1)(T)}\right\} = \mathsf{F}_d\left\{\left\langle \mathsf{S}_\text{sig}^{(T)}\right\rangle\right\}. \tag{3.5.23c}$$

Note that this imposed separation implies that the wave number or spatial factors do *not* depend on the same frequency as the time factors (Eqs. (3.5.15), (3.5.15a), and (3.5.15b)). Accordingly, we must replace $\mathbf{v}(f)$ by $\hat{\mathbf{v}}(\hat{f})$ in the former to distinguish it from the frequency dependence of the latter. Thus, $|\hat{\mathbf{v}}(\hat{f})| = \hat{f}/c_0 \neq |\mathbf{v}(f)| = f/c_0$ in the following results for $Y_H^{(S\otimes T)}$.

Proceeding next as in (3.5.10a) and (3.5.11)–(3.5.12b), we obtain again the pair of relations:

$$\mathsf{W}_\text{N}^{(S)}(\hat{\boldsymbol{\nu}})_d Y_H^{(S)}(\hat{\boldsymbol{\nu}})_d = \left\langle \mathsf{S}_\text{sig}^{(S)}(\hat{\boldsymbol{\nu}})_d\right\rangle,\quad \hat{\boldsymbol{\nu}} = (\hat{f}/c_0)\left(\hat{\mathbf{i}}_m\cdot\hat{\mathbf{i}}_{\hat{\nu}}\right)\mathsf{W}_\text{N}^{(T)}(f)_d Y_H^{(T)}(f)_d = \left\langle \mathsf{S}_\text{sig}^{(T)}(-f)_d\right\rangle. \tag{3.5.24}$$

Combining, we have the desired disjoint form, after inserting a steering vector $(\boldsymbol{\nu}_{OR})$ (Eq. (3.5.14)),

$$Y_{H_1}^{(S\otimes T)} = Y_{H_1}^{(S)}(\hat{\boldsymbol{\nu}}-\hat{\boldsymbol{\nu}}_{OR})Y_{H_1}^{(T)}(-f) = \left\langle \mathsf{S}_{\text{sig}}^{(S)}(\hat{\boldsymbol{\nu}}-\hat{\boldsymbol{\nu}}_{OR})_d\right\rangle\left\langle \hat{\mathsf{S}}_{\text{sig}}(-f)\right\rangle/\mathsf{W}_{\text{N}}^{(S)}(\hat{\boldsymbol{\nu}}-\hat{\boldsymbol{\nu}}_{OR})\mathsf{W}_{\text{N}}^{(T)}(f) \tag{3.5.24a}$$

$$Y_{H_1}^{(S\otimes T)} = \mathsf{A}_R^{(S)}(\hat{\boldsymbol{\nu}}-\hat{\boldsymbol{\nu}}_{OR})_d\left\langle \mathsf{S}_{\text{sig}}^{(T)}(-f)_d\right\rangle\Big/\mathsf{W}_{\text{N}}^{(T)}(f), \tag{3.5.24b}$$

where

$$\mathsf{A}_{\text{R}}^{(S)}(\hat{\boldsymbol{\nu}}-\hat{\boldsymbol{\nu}}_{OR})_d \equiv \left\langle \mathsf{S}_{\text{sig}}^{(S)}(\hat{\boldsymbol{\nu}}-\hat{\boldsymbol{\nu}}_{OR})_d\right\rangle\Big/\mathsf{W}_{\text{N}}^{(S)}(\hat{\boldsymbol{\nu}}-\hat{\boldsymbol{\nu}}_{OR})_d \tag{3.5.25}$$

represents the *separately optimized receiving array or aperture*, embodied in the system function $Y_H^{(S\otimes T)}$ of the matched filter. The characteristic time-reversal of the signal, refer to Section 3.4.1, is given here by

$$\left\langle \mathsf{S}_{\text{sig}}^{(T)}(-f)_d\right\rangle = \left\langle \mathsf{S}_{\text{sig}}^{(T)}(-f)_d\right\rangle e^{-i\omega T} = \left\langle \mathsf{S}_{\text{sig}}^{(T)}(f)_d^*\right\rangle e^{-i\omega T}. \tag{3.5.26}$$

If the noise field is isotropic, in addition to being homogeneous and stationary (Hom-Stat), then $\mathsf{W}_{\text{N}}^{(S)}(\hat{\boldsymbol{\nu}}-\hat{\boldsymbol{\nu}}_{OR}) = \mathsf{W}_{\text{N}}^{(S)}(\hat{\boldsymbol{\nu}})$. The factor in the braces $\langle\,\rangle$ of (3.5.24b) is the familiar temporal matched filter.

In the situation of two or more different noise components, for example, $\mathbf{K}_{\text{N}} + \hat{\mathbf{K}}_I + \hat{\mathbf{K}}_{\text{C}}$, we find that (3.5.13), for the unseparated case, is extended to

$$Y_{H_1}^{(S\otimes T)} = \left\langle \mathsf{S}_{\text{sig}}^{(S)}\right\rangle\left\langle \mathsf{S}_{\text{sig}}^{(T)}\right\rangle/\left(\mathsf{W}_{\text{N}}^{(S)}+\mathsf{W}_I^{(S)}+\mathsf{W}_{\text{C}}^{(S)}\right)\left(\mathsf{W}_{\text{N}}^{(T)}+\mathsf{W}_I^{(T)}+\mathsf{W}_{\text{C}}^{(T)}\right). \tag{3.5.27}$$

The added terms, $\left(\mathsf{W}_{\text{N}}^{(S)}\mathsf{W}_I^{(T)}+\mathsf{W}_{\text{N}}^{(S)}\mathsf{W}_{\text{C}}^{(T)}+\mathsf{W}_I^{(S)}\mathsf{W}_{\text{N}}^{(T)}+\mathsf{W}_I^{(S)}\mathsf{W}_{\text{C}}^{(T)}+\mathsf{W}_{\text{C}}^{(S)}\mathsf{W}_{\text{N}}^{(T)}+\mathsf{W}_{\text{C}}^{(S)} +\mathsf{W}_I^{(T)}\right) > 0$ vis-à-vis $\mathsf{W}_{\text{N}}^{(S)}\mathsf{W}_{\text{N}}^{(T)}+\mathsf{W}_I^{(S)}\mathsf{W}_I^{(T)}+\mathsf{W}_{\text{C}}^{(S)}\mathsf{W}_{\text{C}}^{(T)}$ clearly reduce the magnitude of Y_{H_1} here and negatively impact the separated case $Y^{(S\otimes T)}$ with respect to the unconstrained matched filter (3.5.13). Accordingly, we may expect a possibly significant reduction in optimality here when two or more different noise fields are present in reception, as well as from the simpler separability cases involving only a single interfering noise component. Thus, the optimality properties of the separate space and time matched are individually retained, but their joint effect is less (i.e., gives a smaller detection parameter $\Psi_{s-\text{coh}}^{(*)}, \Psi_{s-\text{inc}}^{(*)}$) than the unconstrained cases (Eqs. (3.1.9a) and (3.2.22a).

3.5.2.1 An Example We can use the array (or aperture) function A_R, (3.5.25), for the space–time separable matched filter (3.5.24a) and (3.5.24b) to illustrate the versatility of its structure. From (3.5.4c), now restricted to space alone, we obtain the *beam pattern*

$$\mathsf{A}^{(S)}(\hat{\boldsymbol{\nu}}-\hat{\boldsymbol{\nu}}_{OR})_d = \sum_{-M/2}^{M/2}|\mathbf{r}_0|\left\langle H_1^{(S)}(m\hat{i}_0 r_0)\right\rangle e^{2\pi i m r_0 \hat{\mathbf{i}}_0\cdot\mathbf{q}_\nu}\Big/\mathsf{W}_{\text{N}}^{(S)}(\hat{\boldsymbol{\nu}}-\hat{\boldsymbol{\nu}}_{OR}), \tag{3.5.28}$$

where (3.5.4c) gives us $[\boldsymbol{\nu}_0]$, and $\mathbf{p}_m = \mathbf{r}_m = m\hat{\mathbf{i}}_0 r_0$ (Eq. 3.5.1) and $\mathbf{q}'\hat{\nu} = \hat{\boldsymbol{\nu}} - \hat{\boldsymbol{\nu}}_{OR}$ (Eq. 3.5.3). The simplest class of array structure occurs for $H_1^{(S)}(\mathbf{r}_m) = A_m$, a constant "shading" if A_m varies only with m. The simplest version of this in turn is $A_m = A, (> 0)$, for example, $A_m = 1$, and in both instances for "white" spatial noise, that is, $\mathsf{W}^{(S)} = \mathsf{W}_{\mathrm{N}}^{(S)}/2$, refer Eq. (3.5.13a). When the array is linear (and of course with sensors at equally spaced intervals r_0) so that $\mathbf{r}_m = \hat{\mathbf{i}} m r_0$, for example, we get the special but not the uncommon result ($M =$ odd) for $H_1^{(S)}(\mathbf{r}_m)$, on setting $|\hat{\boldsymbol{\nu}}| = |\hat{\boldsymbol{\nu}}_{OR}|$ and remembering that $\hat{\boldsymbol{\nu}} = |\hat{\boldsymbol{\nu}}|\hat{\mathbf{i}}_{\hat{\nu}}$ and $\hat{\boldsymbol{\nu}}_{OR} = |\hat{\nu}_{OR}|\hat{\mathbf{i}}_{\hat{\nu}_{OR}}$, namely,

$$\left.\begin{aligned} \hat{\mathbf{i}}_{\hat{\nu}} &= \hat{\mathbf{i}}_x \sin\theta_\nu \cos\phi_\nu + \hat{\mathbf{i}}_y \sin\theta_\nu \sin\phi_\nu + \hat{\mathbf{i}}_z \cos\theta_\nu \\ \hat{\mathbf{i}}_{\nu_{OR}} &= \left(\hat{\mathbf{i}}_x \sin\theta \cos\phi + \hat{\mathbf{i}}_y \sin\theta \sin\phi + \hat{\mathbf{i}}_z \cos\theta\right)_{(OR)} \end{aligned}\right\} \tag{3.5.29a}$$

and

$$\therefore\ H_1^{(S)}(\mathbf{r}_m) e^{2\pi i \mathbf{r}_m \cdot \mathbf{q}'_{\hat{\nu}}} = A e^{2\pi i m r_0 |\boldsymbol{\nu}_{OR}| F_R(\phi,\theta;\phi_0,\theta_0)}; \quad F_0 = (\sin\theta\cos\phi - \sin\theta_{OR}\cos\phi_{OR}). \tag{3.5.30a}$$

With $\beta = 2\pi F|\hat{\mathbf{v}}_{OR}|r_0$, the series in (3.5.28) becomes

$$\sum_{-M/2}^{M/2} e^{im\beta} = \cos M_0\beta + \left(\frac{\sin\beta}{1-\cos\beta}\right)\sin M_0\beta. \tag{3.5.30b}$$

Thus, in this example where $M = 2M_0 + 1$ and with "white" noise in space, we obtain finally for (3.5.28):

$$\mathsf{A}_{\mathrm{R}}^{(S)}(\hat{\boldsymbol{\nu}} - \hat{\boldsymbol{\nu}}_{OR})_d = \left[\cos M_0\beta + \left(\frac{\sin\beta}{1-\cos\beta}\right)\sin M_0 P\right] B_0,$$

$$\text{where } B_0 = 2A/W_{\mathrm{N}}^{(S)}[\nu_0] = A/\pi^2\nu_0 W_{\mathrm{N}}^{(S)}, \tag{3.5.31a}$$

from (3.5.4c), where now

$$\beta = 2\pi r_0\,\hat{\nu}_{OR}(\sin\theta\cos\phi - \sin\theta_{OR}\cos\phi_{OR}) = 2\pi\Delta F\hat{\nu}_{OR}. \tag{3.5.31b}$$

When the beam produced by the array is pointed at a source at (θ, ϕ), that is $\theta_{OR} = \theta$; $\phi_{OR} = \phi$, then $F = 0$ and $\therefore\ \beta = 0$, with a resulting beam maximum $\left(\mathsf{A}_{\mathrm{R}}^{(S)}\right)_{\max} = (2M_0 + 1)B_0 = MB_0$, proportional to the total number of array elements, an expected result for this type of configuration.

Since $\hat{\nu}_{OR} = \hat{f}_{OR}/c_0 = 1/\hat{\lambda}_{OR}$, where $\hat{\lambda}_{OR}\left(= \hat{\lambda}\right)$ is the wavelength of a frequency in the received field, we observe in addition that β increases with frequency, that is, at shorter wavelengths, so that the beam pattern is narrowed, although it maintains its relative shape, with a maximum again at the source (θ, ϕ). (In the neighborhood of the source, where $\beta = 0+$, the pattern is $\mathrm{A_R} \doteq (2M_0 + 1 - M_0^3\beta^2/2)B_0$, where $M_0^3\beta^2/2 \ll 1$, with $\beta = 2\pi\Delta(\pm\varepsilon)\cos(\theta_{OR} + \phi_{OR})/\hat{\lambda}$, that is, $M_0^3(2\pi\Delta)^2\cos^2(\theta_{OR} + \phi_{OR})\left(\pm\varepsilon/\hat{\lambda}\right)^2 \ll 2$. Thus, $|\varepsilon|$ must be sufficiently small vis-à-vis $\hat{\lambda}$ to satisfy this inequality. Here ε represents the angular departure of θ_{OR} from θ, and ϕ_{OR} from ϕ.) Figure 3.17 shows a typical beam pattern $(M_0 = 4)$, $\hat{v}_{OR} = 1, \hat{v}_{OR} > 1\left(\text{i.e., } 1/\hat{\lambda}_{OR} = 1, < 1\right)$ for the linear array of the example.

3.5.2.2 Narrowband Signals Finally, in the important practical cases of *narrowband signals,* the beam pattern or aperture function, that is, the wave number–frequency response function $Y_H^{(1)}(\mathbf{v}-\mathbf{v}_{OR},f)$ (Eqs. (3.5.14a), (3.5.16a), and (3.5.16b)), in both the nonseparable and separated cases (Sections 3.5.2 and 3.5.3) *is insensitive to frequency variations* in case of spatial "white" noise. In fact, this insensitivity is a measure of what one means by narrowbandedness, (Section 2.5.5). Accordingly, for such narrowband cases, one has

$$Y_H^{(S)}(\boldsymbol{\nu}(f)-\boldsymbol{\nu}_{OR}(f))_d \rightarrow Y_H^{(S)}(\boldsymbol{\nu}(f_0)-\boldsymbol{\nu}_{OR}(f_0))_d\text{:}\, f \rightarrow f_0, \text{ a constant.} \tag{3.5.32}$$

Here, f_0 is usually the central frequency of the narrowband signal and of the resulting narrowband noise accompanying it in the matched filter. This insensitivity to frequency greatly simplifies the beam structure and source localization, since $\boldsymbol{\nu}_{OR}(f_0)$ is constant and does not require the frequency scanning of $\boldsymbol{\nu}_{OR}(f)$ required in "broadband" systems. Of course, Eq. (3.5.32) is an approximation, that becomes exact only for monofrequentic signals. A condition for its acceptability may be obtained from the expansion

$$Y_1^{(S)}(\Delta\boldsymbol{\nu}) \doteq Y_1^{(S)}(\Delta\boldsymbol{\nu}(f_0)) + \Delta\boldsymbol{\nu}\cdot(\nabla_{\Delta\nu}Y)_{f=f_0} + \cdots; \quad \nabla_{\Delta\nu} = \hat{\mathbf{i}}_x\frac{\partial}{\partial(\Delta\nu_x)} + \hat{\mathbf{i}}_y\frac{\partial}{\partial(\Delta\nu_y)} + \hat{\mathbf{i}}_z\frac{\partial}{\partial(\Delta\nu_z)},$$

$$\left.\begin{aligned} &\doteq Y_1^{(S)}|_{f_0}\left(1 + \Delta\boldsymbol{\nu}\cdot(\nabla_{\Delta\nu}Y)_{f=f_0}/Y_1^{(S)}|_{f_0}\right) \doteq Y_1^{(S)}|_{f_0}\exp(\Delta\boldsymbol{\nu}\cdot\mathbf{A})_{f_0} \\ &\qquad\qquad \text{where } A_{f_0} \equiv \nabla_{\Delta\nu}Y_1^{(S)}|_{f=f_0}/Y_1^{(S)}|_{f_0} \end{aligned}\right\}$$

$$Y_1^{(S)}(\Delta\boldsymbol{\nu}) \doteq Y_1^{(S)}|_{f_0}|\Delta\boldsymbol{\nu}\cdot\mathbf{A}_{f_0}| \ll 1,\ \Delta\boldsymbol{\nu} \equiv \boldsymbol{\nu}(f) - \boldsymbol{\nu}_{OR}(f). \tag{3.5.33}$$

The condition for the approximation is thus Eq. (3.5.33), $|\Delta\boldsymbol{\nu}\cdot\mathbf{A}|_{f_0}| \ll 1$.

When the spatial noise is not white, that is, is "colored," the more general relation (3.5.28) must be used, where, however, $\hat{\boldsymbol{\nu}} = \hat{\boldsymbol{\nu}}(f_0)$ again and the steering vector $\hat{\boldsymbol{\nu}}_{OR} = \hat{\boldsymbol{\nu}}_{OR}(f_0)$. The temporal part of the matched filter in these cases of imposed separation of space and time is given by the temporal factor in (3.5.24b), with $\hat{\mathrm{S}}_{\mathrm{sig}}^{(T)}(-f)_d$ in turn given by (3.5.26). For the more general coupled cases of Section 3.5.2, the dependence on frequency of the spatial part is also removed, so that space and time are de facto separated. Thus, *with the narrowband constraint, it is also possible to treat the array or aperture independent of the temporal processing and still retain* (approximate) *space–time optimality,* where the optimum aperture function[61] is now specified by (3.5.25) and the optimum temporal processing by the time-factor in (3.5.24b).

3.5.2.3 Summary Remarks Whether or not the noise fields are nonhomogeneous and nonstationary, we can readily obtain transform, that is, wave number–frequency solutions, to the discrete integral equations defining these matched filters. This is demonstrated in Section 3.5.2. The principal results are as follows:

(1) The system function $Y_H^{(1)}$ of the matched filter consists of a *space–time beam former* or *aperture function* $\mathsf{A}_R(\boldsymbol{\nu}-\boldsymbol{\nu}_{OR}, f|S_{sig})$, (3.5.14a) (Section 3.5.2.1).
(2) The spatial part of this aperture is frequency dependent, along with the time-variable portion of the aperture. This occurs because of the physical nature of the propagating fields, (Eqs. (3.5.15a)–(3.5.16b), which couple space and time together.
(3) For the general situation (2), the separation of space and time operations is suboptimum, but is often a constraint imposed on the design of receivers for convenience in implementation.
(4) However, *in the narrowband case,* the optimal aperture is insensitive to frequency, depending only on the constant center frequency f_0 of the suitably narrowband (Eq. (3.5.33)). In fact, this is what is meant here by "narrowband." Thus, the temporal processing can be separably optimized without overall system degradation (Section 3.5.3.1).

These results (1-4) apply for the general situation of non-stationary, non-homogeneous noise fields, and include the Hom-Stat. situation as a special case.

Historical Note:
The concept of the "matched filter" was independently discovered by D.O. North [15a], in June 1943, and by J. H. Van Vleck and D. Middleton at about the same time [14a]. It was after their analysis that they learned of North's RCA Report [15a], which was duly referenced then and subsequently [14]. The methods employed in the two investigations, however, were different: North [15a] used a calculus of variations technique; JHVV and DM used the Schwartz inequality [14, 14a]. Moreover, the term "matched filter" was originally introduced by the latter authors, along with applications to other detection problems [14, 14a,14b], as the titles indicate. Reference [14] appeared in the open literature in 1946. North's important work [15] was not published until 1963. (Some of this history has been noted in *Threshold Signals*, Vol. 24 of the MIT Radiation Laboratory Series, McGraw-Hill, 1950; see also unpublished correspondence (12/7/98) of DM with Prof. Jerry Gibson, Southern Methodist University (12/29/98).)

[14a] J. H. Van Vleck and D. Middleton, "Theory of the Visual vs. Aural or Meter Reception of Radar Signals in the Presence of Noise," Harvard Radio Research Laboratory (RRL), Report No. 411-86, May 1944. See also Ref. [13], p. 218 of Vol. 24, just cited.

[14b] D. Middleton, "The Effect of a Video Filter on the Detection of Pulsed Signals in Noise," also p. 218, of Vol. 24, just cited, as well as *J. Appl. Phys.*, **21** (8), 1950.

[61] Optimality in the sense of a linear filter maximizing a quadratic form derived under a Bayes criteria for detection or estimation: see the initial remarks of Section 3.5.

[15a] D. O. North, Analysis of Factors Which Determine Signal-Noise Discrimination in Pulsed Carrier Systems, RCA Technical Report, PTR-6C, June 1943.

3.6 CONCLUDING REMARKS

Let us briefly review the results of Chapter 3. In Section 3.1, we have discussed the prototypical case of coherent detection in additive normal noise, where everything is known about the signal except its presence or absence. In Section 3.2, a number of examples of incoherent detection involving narrowband signals, also in additive normal noise, have been examined. These examples include the narrowband situation where only the signal epoch is unknown and the extension to include Rayeigh fading of the amplitude. Section 3.3 has been devoted to the case of a slowly fluctuating *noise* background and to the incoherent detection of broadband signals, in normal noise. In Section 3.4, Bayes matched filters and their many space–time formulations have been defined, discussed, and their eigenvalue solutions formally obtained. Both space–time invariant and variable filters are described, especially for dealing with incoherent reception. In Section 3.5, these matched filters are examined in terms of their Fourier transforms. In the Fourier transform domain of wave number–frequency, their equivalent representation is shown to be *the aperture function*, which is also an optimum frequency-dependent *beam former in space–time*. However, in case of sufficiently narrowband signals, this aperture function is shown to be essentially *frequency independent*. The treatment in each section of Chapter 3 has also included a variety of important special cases.

Here, in Chapter 3, we have provided an introduction to explicit results, following upon the general analysis outlined in Chapter 2, which is not limited to particular classes of noise fields and signals, and to the general methods presented therein for solutions. Chapter 3 and subsequent Chapters 4–7, are devoted to realizing the details of these general procedures, ultimately in non-Gaussian environments. It is, of course, one thing to describe a general theory and quite another to realize its potential with specific results. Chapter 3 is one such beginning. All the above is prelude to the difficult problems in detection and estimation, and elsewhere encountered when the noise fields are *non-Gaussian*, a situation that can and does occur in all applications of signal processing. Chapter 3 and likewise Chapters 4–7, accordingly, are preliminaries to the more complex and often more realistic real-world environments, where normal noise is by no means the rule and where the non-Gaussian world can be dominant.

REFERENCES

1. D. Middleton, *An Introduction to Statistical Communication Theory*, McGraw-Hill New York, 1960; IEEE Press, *Classic*, Reissue, Piscataway, NJ. 1996.
2. G. A. Campbell and R. M. Foster, *Fourier Integrals for Practical Applications*, D. Van Nostrand, New York, 3rd Printing, 1948, pp. 32–33d.
3. J. I. Marcum, Tables of Q-Functions, RAND Corp. RM-339; ASIT Doc: AS-116551, Jan. 1950.
4. I. S. Gradshteyn and I. M. Ryzhik, in *Tables of Integrals, Series, and Products*, Corrected and Enlarged Edition Alan Jeffrey, Ed. 4th Edition, Academic Press, New York, 1980, pp. 32 and 33a.

5. M. Abramowitz and I. A. Stegun (Eds.), *Handbook of Mathematical Functions,* Section 6.5 and Tables, National Bureau of Standards, Applied Math Series 55, U.S. Government Printing Office, Washington D.C., June 1964.
6. M. K. Simon and M.-S. Alouini, *Digital Communication over Fading Channels*, John Wiley & Sons, Inc., New York, 2000.
7. M. Schwartz, W. R. Bennett, and S. Stein, *Communication Systems and Techniques*, McGraw-Hill, New York, 1966.
8. D. Middleton, *Topics in Communication Theory*, McGraw-Hill, New York, 1965, and Peninsula Publishing, Los Altos, CA, 1987.
9. D. Middleton, New Physical-Statistical Methods and Models for Clutter and Reverberation: The KA-Distribution and Related Probability Structures, *OEET Ocean. Eng.*, **24**, 261–284, 1999.
10. E. Jakeman, On the Statistics of K-Distributed Noise, *J. Phys. A*, **13**, 31, 1980.
11. E. Jakeman and R. J. A. Tough, Non-Gaussian Models for the Statistics of Scattered Waves, *Adv. Phys.* **37**, 471–529, 1987, with many additional references.
12. D. Middleton, New Results in Applied Scattering Theory: The Physical-Statistical Approach, Including Strong Multiple Scatter vs. Classical Statistical-Physical Methods and the Born and Rytov Approximations vs. Exact Strong Scatter Probability Distribution," *Wave Random Media*, **12**, 99–145, 2002.
13. G. N. Watson, *Theory of Bessel Functions*, 2nd ed., Macmillan, New York, 1944.
14. J. H. Van Vleck and D. Middleton, A Theoretical Comparison of the Visual, Aural, or Meter Reception of Pulsed Signals in the Presence of Noise, *J. Appl. Phys*, **17**, 940–971, 1946. (See *Historical Note,* Section 3.5.).
15. D. O. North, An Analysis of the Factors Which Determine Signal-Noise Discrimination in Pulsed Carrier Systems, *Proceedings IEEE*, **51**, 1016–1027, 1963. (see *Historical Note,* Section 3.5.).
16. D. Middleton, On New Classes of Matched Filters and Generalizations of the Matched Filter Concept, *IRE Transactions on Information Theory*, **IT-6**, 349–360, 1960.
17. E. J. Kelly, I. S. Reed, and W. L. Root, The Detection of Radar Echoes in Noise, I., *J. Soc. Indus. Appl. Math.*, **8**(2),309–341, 1960; **8**(3), 481–507, for extraction.
18. L. A. Weinstein and V. D. Zubakov, *Extraction of Signals from Noise*, Chapter 6 (trans. by R. A. Silverman), Prentice-Hall, Englewood Cliffs, NJ, 1962.
19. H. Margenau and G. M. Murphy, *The Mathematics of Physics and Chemistry*, Van Nostrand, New York, 1943; see Section 10.15.
20. G. H. Golub and C. F. Van Loan, *Matrix Computations*, John Hopkins University Press, Baltimore, MD, 1989.
21. R. Bellman. *Introduction to Matrix Analysis*, McGraw-Hill, New York, 1960.
22. N. Wiener, Extrapolation, Interpolation, and Smoothing of Stationary Time Series, Technology Press, Cambridge, MA, 1949. (Wiener's analysis was originally contained in Report No. DIC-6037, Sec. D2, Feb. 1, 1942, MIT).
23. A. N. Kolmogoroff, "Interpolation and Extrapolation," *Bull. Acad. Sci., USSR*, **5**, 3–14, 1944.
24. A. M. Yaglom, *Correlation Theory of Stationary Random Functions: I. Basic Results; II. Supplementary Notes and References*, Springer, New York, 1987.
25. H. L. Van Trees, *Optimum Array Processing*, John Wiley & Sons, Inc., New York, 2002, Appendix.
26. S. M. Kay, *Modern Spectral Estimation Theory and Application*, Prentice-Hall, Englewood Cliffs, 1988.
27. S. Haykin, *Adaptive Filter Theory*, 3rd ed., Prentice-Hall, Upper Saddle River, 1996.

28. D. Middleton and A. D. Spaulding, Optimum Reception in Non-Gaussian Environment: II. Optimum and Suboptimum Threshold Detection in Class A and B Noise, NTIA Report 83-120, Ins. Telcom. Sci., U.S. Department of Commerce, Boulder, CO., NTIS Catalogue. PB83-241141.
29. H. Urkowitz, "Filters for Detection of Small Radar Signals in Clutter," *J. Appl. Phys.*, **24**, pp. 1024–1031, 1953.
30. C. Eckart,"The Theory of Noise Suppression by Linear Filters," Scripps Institute of Oceanography, Report S1051-44 (Oct. 8, 1951).
31. C. W. Helstrom, *Elements of Detection and Estimation*, Prentice-Hall, Englewood Cliffs, NJ, 1995, Section 10.2, and references.

4

MULTIPLE ALTERNATIVE DETECTION[1]

In Chapters 2 and 3, we have outlined a theory of optimum binary detection of space–time fields where the final output of our optimum receiver is a definite decision as to the presence or absence of a signal. For the detection situation, so far, only at most two hypothesis states have been considered: signal and noise versus noise, or a signal of one class and noise versus a signal of another class and noise (see Chapter 1). Here we shall begin by extending the analysis first to multiple-alternative situations consisting of an arbitrary number of disjoint hypothesis states, including the case of decision rejection (Section 4.1). This is followed by Section 4.2 on overlapping signal classes. Section 4.2 is based on the formulation of Section 1.10, and on decision *rejection* of signals (Section 4.3) that are equivocal as to their presence. The extension here to space–time data and operation is also formally included by introducing the sampling process denoted by the index $j = mn$. Here, again, m denotes the spatial location and n the temporal instant of the received data X_j, or field $\alpha_j = \alpha(\mathbf{r}_m, t_n)$, as first noted in Section 1.3.1 and developed in more detail in Chapters 2 and 3. A short discussion of the results concludes the chapter.

4.1 MULTIPLE-ALTERNATIVE DETECTION[2]: THE DISJOINT CASES

As we have noted above, the detection and estimation processes considered in earlier chapters are capable of generalization within the framework of decision theory. We illustrate this by considering important extensions of the binary detection process, first, for disjoint or nonoverlapping signal classes.

[1] This chapter is mostly adapted from Section 23.1 of Ref. [2], with added material (Section 4.2), and extended formally to include spatial as well as temporal sampling.

[2] This section is based in part on the original work of Middleton and Van Meter [1].

4.1.1 Detection

Although the binary systems described in Chapters 2 and 3 are quite common in practice, it is frequently necessary to consider situations involving more than two alternatives. Instead of having to distinguish one signal (and noise) out of a given class of signals from noise alone, we may be required to determine which one of several possible signals is present. For example, in communication applications a variety of different signal waveforms may be used as a signal alphabet in the course of transmission, and the receiver is asked to determine which particular waveform is actually sent in any given signal period $(0, T)$. A second example arises in radar and sonar, where it is desired to distinguish between several targets that may or may not appear simultaneously during an observation interval. In fact, whenever we are required to discriminate between more than two hypothesis classes, we have an example of *multiple-alternative detection*. Here we extend the binary theory of Chapters 2 and 3 to those multiple-alternative situations where only a single signal can appear (with noise) on any one observation, that is, to the case of *disjoint*, or nonoverlapping, hypothesis classes. The more involved situations of joint, or overlapping, hypothesis classes (an extension of Section 1.10) are next briefly considered in Section 4.2.

4.1.1.1 Formulation Let us begin with a brief formulation of the decision model. As before, the criterion of excellence (cf. Chapter 1) is the minimization of average risk, or cost. The resulting system (i.e., the indicated operations on the received data) which achieves this is the corresponding optimum detection system. Accordingly, our procedure is first to construct the average risk function [(1.4.6a) and (1.4.6b)] and then to minimize it by a suitable choice of decision rule (e.g., system structure). From Eq. 1.4.6b), the average risk can be written

$$R(\sigma, \delta) = \int_{\Omega} \sigma(\boldsymbol{S})\boldsymbol{dS} \int_{\Gamma} F_J(\boldsymbol{X}|\boldsymbol{S})d\boldsymbol{X} \int_{\Delta} \boldsymbol{d}\gamma C(\boldsymbol{S}, \gamma)\delta(\gamma|\boldsymbol{X}), \tag{4.1.1}$$

where now specifically

(1) $\sigma(\boldsymbol{S})$ is the *a priori* probability (density) governing all possible signals $\boldsymbol{S}$, in which explicitly

$$\sigma(\boldsymbol{S}) = \sum_{k=0}^{K} p_k w_k(\boldsymbol{S})\delta^{(k)}(\boldsymbol{S}), \qquad \sum_{k=0}^{K} p_k = 1, \tag{4.1.2}$$

with $p_k (k = 0, \ldots, K)$ the *a priori* probabilities that a signal of class (or type) k is present on any one observation. Here, w_k is the pdf of $\boldsymbol{S}^{(k)}$ [or of the parameters[3] θ of $\boldsymbol{S}^{(k)}(\theta)$], where $\boldsymbol{S}^{(k)}$ represents the signals of class k. The class $k = 0$ is a class of possible null signals, or "noise alone," and $\delta^{(k)}(\boldsymbol{S}) = 1$, when $\boldsymbol{S} = \boldsymbol{S}^{(k)}$, with $\delta^{(k)}(\boldsymbol{S}) = 0,\ \boldsymbol{S} \neq \boldsymbol{S}^{(k)}$.

(2) $\left[S^{(k)}(\boldsymbol{r}_1, t_1), \ldots, S^{(k)}(\boldsymbol{r}_M, t_N)\right] = \left[S_{11}^{(k)}, S_{12}^{(k)}, \ldots, S_{j=mn}^{(k)}, \ldots, S_{M}\right]$ is the kth signal vector, whose components are the sampled values of $S^{(k)}(\boldsymbol{R}, t)$ at the points $(\boldsymbol{R}_m, t_n)$

[3] In such cases we replace $\sigma(\boldsymbol{S})$ by $\sigma(\theta) = \sum_{k=0}^{K} p_k w_k(\theta)\delta^{(k)}(\boldsymbol{S})$ and Ω_θ for the region of integration.

in the observation interval ($\Delta\boldsymbol{R}$, T). Similarly, we have $\boldsymbol{X} = [X_j] = [X_{11}, \ldots, X_{MN}]$, the received-data vector, with $X_j = X_{mn}$, and so on, as before.

(3) $F_J(\boldsymbol{X}|\boldsymbol{S})$, $F_J[\boldsymbol{X}|\boldsymbol{S}(\theta)]$ are the conditional pdfs of $\boldsymbol{X}$, given $\boldsymbol{S}$.

(4) $\gamma = [\gamma_0, \ldots, \gamma_K]$ is a set of $k+1$ decisions; γ_ℓ is a decision that a signal of class ℓ is present versus all other possibilities.

(5) As before, $j(= mn)$ is a double index symbol, where $m = 1, \ldots, M$ and $n = 1, \ldots, N (J = MN)$, denoting, respectively, space and time (see Section 1.3.1 and Eq. (1.3.1).)

As before, $C(\boldsymbol{S}, \gamma)$ is a cost function [cf. Eq. (1.4.3)] that assigns to the various $\boldsymbol{S}$, for one or more possible decisions γ_ℓ, $(L = 0, \ldots, k)$, about the $\boldsymbol{S}$, some appropriate preassigned constant costs. These costs are assigned to signal classes: no cost distinction is made between different signals of the same class or type. Also, as before (Section 1.3.1), $\delta(\gamma_\ell|\boldsymbol{X})$ is a decision rule, or probability.[4] The decisions k are governed by the further condition

$$\sum_{\ell=0}^{K} \delta(\gamma_\ell|\boldsymbol{X}) = 1, \tag{4.1.3}$$

which is simply a statement of the fact that a definite decision must be made.

Constant costs are next assigned to the possible outcomes, according to our usual procedure [Eqs. (1.6.6) in the binary case]. Here we set $C_\ell^{(k)}$ = cost of deciding that a signal of class ℓ is present when actually a signal of class k occurs.[5] Thus, if $\ell \neq k$, $C_\ell^{(k)}$ is the cost of an incorrect decision, while if $\ell = k$, $C_\ell^{(\ell)}$ represents the cost of a correct decision. In all cases, we have

$$C_\ell^{(k)}\big|_{\ell \neq k} > C_\ell^{(\ell)}, \tag{4.1.4}$$

since by definition an "error" must be more "expensive" than a correct choice.

Let us consider the costs $C_\ell^{(k)}$ in more detail. For "successful" or correct decisions we have specifically

$$\begin{aligned} &C\left(\boldsymbol{S}^{(0)};\ \gamma_0\right) = C_0^{(0)};\ \text{noise alone is correctly detected;} \\ &C\left(\boldsymbol{S}^{(k)};\ \gamma_k\right) = C_k^{(k)};\ \text{a signal of class } k \text{ is correctly detected } (k = 1, \ldots, K). \end{aligned} \tag{4.1.5a}$$

The costs preassigned to "failures," or incorrect decisions, are represented by $C\left(\boldsymbol{S}^{(0)};\ \gamma_\ell\right) = C_\ell^{(0)}$; a signal of class $\ell (= 1, \ldots, K)$ is incorrectly decided when noise alone occurs, and

$$\begin{aligned} C\left(S^{(k)}; \gamma_\ell\right) = C_\ell^{(k)};\ &\text{a signal of class } \ell\ (\ell \neq k; \ell = 0, \ldots, K, \text{out of } k = 1, \ldots, K) \\ &\text{is incorrectly detected, when a signal of class } k \text{ occurs.} \end{aligned} \tag{4.1.5b}$$

[4] With $0 \leq \delta(\gamma_\ell|\boldsymbol{X}) \leq 1$, $(\ell = 0, \ldots, K)$, in the case of detection. For estimation, $\delta(\gamma|\boldsymbol{X})$ is a probability density (Sections 1.3.2, 1.3.3).

[5] We adopt the convention that the superscript on the cost $C_\ell^{(k)}$ refers to the true or actual state, while the subscript refers to the decision made.

Setting $p_0 \equiv q$, we readily obtain the average risk [Eq. (4.1.1)] after integrating over decision space Δ. The result is

$$R(\sigma,\delta) = \int_\Gamma \left(\begin{array}{l} \left[C_0^{(0)}\delta(\gamma_0|\boldsymbol{X}) + \sum_{k=1}^{K} C_k^{(0)}\delta(\gamma_k|\boldsymbol{X}) \right] qF_J(\boldsymbol{X}|\boldsymbol{0})d\boldsymbol{X} \\ + \left\{ \sum_{k=1}^{K} p_k \left[C_k^{(k)}\delta(\gamma_k|\boldsymbol{X}) + \sum_{\ell=0}^{M}{}' C_\ell^{(k)}\delta(\gamma_\ell|\boldsymbol{X}) \right] \left\langle F_J\left(\boldsymbol{X}|\boldsymbol{S}^{(k)}\right)\right\rangle_k d\boldsymbol{X} \right\} \end{array} \right), \tag{4.1.6}$$

subject to Eqs. (4.1.3) and (4.1.4), where $\langle\ \rangle_k$ denotes the statistical average over $\boldsymbol{S}^{(k)}$ (or over the random parameters of $\boldsymbol{S}^{(k)}$), and the prime on the summation signifies that $\ell \neq k$. Note that when $M = 1$ (the binary case), Eq. (4.1.6) reduces at once to Eq. (1.6.7), with obvious changes of notation.

4.1.2 Minimization of the Average Risk

At this point, it is convenient to rearrange Eq. (4.1.6) with the help of Eq. (4.1.3) by collecting coefficients of $\delta(\gamma_\ell|\boldsymbol{X})$. First, let us introduce the expressions

$$\begin{aligned} &\lambda_i^{(0)} \equiv C_i^{(0)} - C_0^{(0)}, && i = 1,\ldots,K \\ &\lambda_i^{(k)} \equiv C_k^{(k)} - C_0^{(k)}, && \\ &\lambda_k^{(i)} \equiv C_k^{(i)} - C_0^{(i)}, && i \neq k\ (k,i = 1,\ldots,K) \\ &\lambda_k^{(i)} \gtreqless 0, && k \neq i\left[\text{with } \lambda_k^{(0)} > 0(k>0),\ \lambda_k^{(k)} < 0(k \neq 0)\right] \end{aligned} \tag{4.1.7}$$

and write[6]

$$A_k(\boldsymbol{X}) \equiv \lambda_k^{(0)} + \sum_{i=1}^{M} \lambda_k^{(i)}\Lambda_i(\boldsymbol{X}), \tag{4.1.8}$$

where, as before [cf. Eq. (1.7.2), the $\Lambda_i(\boldsymbol{X})$ are generalized likelihood ratios

$$\Lambda_i(\boldsymbol{X}) \equiv p_i\left\langle F_J\left(\boldsymbol{X}|\boldsymbol{S}^{(i)}\right)\right\rangle i/qF_J(\boldsymbol{X}|\boldsymbol{0}). \tag{4.1.9}$$

For additive signals and noise, this reduces to the simpler relation

$$\Lambda_i(\boldsymbol{X}) \equiv p_i\left\langle W_J\left(\boldsymbol{X} - \boldsymbol{S}^{(i)}\right)_N\right\rangle_i / qW_J(\boldsymbol{X})_N, \tag{4.1.9a}$$

in which as before, $W_J(\boldsymbol{X})_N$ is the joint Jth-order pdf of the background noise.

After some algebra, we find that the average risk [Eq. (4.1.6)] may now be rewritten

$$R_K(\sigma,\delta) = R_{0K} + R_K \tag{4.1.10a}$$

[6] In this notation $\Lambda_i(\boldsymbol{X})$ is the abbreviation for $\Lambda\left(\boldsymbol{X}|\boldsymbol{S}^{(i)}\right)$, throughout this chapter.

$$R_{0K} \equiv qC_0^{(0)} + \sum_{k=1}^{K} p_k C_0^{(k)} (> 0) \tag{4.1.10b}$$

$$R_K \equiv \int_\Gamma \left[\sum_{k=1}^{K} \delta(\gamma_k|\boldsymbol{X}) A_k(\boldsymbol{X}) \right] qF_J(\boldsymbol{X}|\boldsymbol{0}) d\boldsymbol{X}. \tag{4.1.10c}$$

Here R_{0K} is simply the expected cost of calling every signal (including noise alone) "noise," while R_K is the portion of the average risk that can be adjusted by choice of the decision rules. Again, the precise form of the system is embodied in the likelihood ratios, while the process of detection is determined by our choice of the δs and the corresponding regions in data space Γ, which are nonoverlapping since the various hypotheses here are mutually exclusive, that is, the signals are disjoint.

Now, in order to optimize the detection operation we minimize the average risk by proper selection of the $\delta(\gamma_k|\boldsymbol{X})$. In essence, this is the problem of finding the boundaries of the critical regions for the $\Lambda_k(\boldsymbol{X})$. The argument for minimization is readily given. Since $qF_J(\boldsymbol{X}|0) \geq 0$ everywhere in Γ, the average risk [Eq. (4.1.10c)] is least where for each value of $\boldsymbol{X}$ we choose δ to minimize $\sum \delta A_k$. The procedure, accordingly, is to examine all As for the given $\boldsymbol{X}$, selecting the one (A_k) that is algebraically least[7] *and then for this same* $\boldsymbol{X}$ *choosing* $\delta(\gamma_k|\boldsymbol{X}) = 1$, $\delta(\gamma_\ell|\boldsymbol{X}) = 0$ (all $\ell \neq k$). We repeat this for all $\boldsymbol{X}$ in Γ, to obtain finally a set of conditions on the A_ks$(k = 1, \ldots, K)$. Observe from the form of R_K, where the δs appear linearly, and from the method of minimization itself, that δ is automatically a nonrandomized decision rule, that is, $\delta = 1$ or 0 only [Eq. (1.7.4a) et seq.]. Since the $A_k(\boldsymbol{X})$ may contain negative parts [Eq. (4.1.7)], we find that to minimize the average risk of Eq. (4.1.10a) and make a decision γ_k (signal of type k present in noise) on the basis of received data $\boldsymbol{X}$, the explicit conditions on the A_k are that the data $\boldsymbol{X}$ satisfy the following linear *inequalities*,

$$\begin{aligned} &A_k(\boldsymbol{X}) \leq A_\ell(\boldsymbol{X}) \quad \text{or} \quad \lambda_k^{(0)} + \sum_{i=1}^{K} \lambda_k^{(i)} \Lambda_i(\boldsymbol{X}) \leq \lambda_\ell^{(0)} + \sum_{i=1}^{K} \lambda_\ell^{(i)} \Lambda_i(\boldsymbol{X}) \\ &\text{and} \\ &A_k(\boldsymbol{X}) \leq 0 \quad \text{or} \quad \lambda_k^{(0)} + \sum_{i=1}^{K} \lambda_k^{(i)} \Lambda_i(\boldsymbol{X}) \leq 0, \\ &\qquad \text{all } \ell \neq k, \ (\ell = 1, \ \ldots, \ K) \end{aligned} \tag{4.1.11a}$$

for each decision $\gamma_k(k = 1, \ldots, K)$ in turn, with

$$\delta(\gamma_k|\boldsymbol{X}) = 1, \ \delta(\gamma_\ell|\boldsymbol{X}) = 0, \quad \text{all} \quad \ell \neq k, \ = 0, \ \ldots, \ K. \tag{4.1.11b}$$

For the remaining case of noise alone $(k = 0)$ we have the conditions

$$A_k(\boldsymbol{X}) \geq 0 \quad \text{or} \quad \lambda_k^{(0)} + \sum_{i=1}^{K} \lambda_k^{(i)} \Lambda_i(\boldsymbol{X}) \geq 0, \quad \text{each} \quad k = 1, \ldots, K \tag{4.1.12a}$$

[7] Subject (for the moment) to the assumption that for this given $\boldsymbol{X}$ there actually exists an $A_k(\boldsymbol{X})$ that is algebraically less than all other $A_\ell(\boldsymbol{X}), (\ell \neq k)$.

and

$$\delta(\gamma_0|\boldsymbol{X}) = 1, \quad \delta(\gamma_\ell|\boldsymbol{X}) = 0, \quad \ell = 1, \ldots, K. \tag{4.1.12b}$$

4.1.3 Geometric Interpretation

If now we regard the Λs as independent variables, we can at once give a direct geometric interpretation of the mutually exclusive sets of conditions (4.1.11), (4.2.12). Writing (L_k) as the value of the quantity $\lambda_k^{(0)} + \sum_{i=1}^{K} \lambda_k^{(i)} \Lambda_i(\boldsymbol{X})$ and L_k for the hypersurface[8] (L_k), we observe that, in conjunction with the hypersurfaces forming the boundaries of the first "2^{K+1}-tant,"[9] the *equalities* in the conditions (4.2.11) or (4.2.12) give the boundaries of a *closed* region within which lie all values of $\Lambda_k(\boldsymbol{X})$ associated with the decision γ_k. Each closed region is distinct from every other, and the K planar hypersurfaces that form its boundaries are then from Eqs. (4.1.11), (4.1.12) specified by

$$L_k = 0; \quad L_k - L_\ell = 0 \quad \text{all } \ell = 1, \ldots, K(\ell \neq k)(\text{for each } \gamma_k,\ k = 1,\ \ldots,\ K) \tag{4.1.13a}$$

$$L_k = 0; \quad \text{all } k = 1, \ldots, K(\ell \neq k) \quad (\text{for } \gamma_0)\,. \tag{4.1.13b}$$

Solving the K linear equations (4.1.11), (4.1.12), or Eqs. (4.1.13a), (4.1.13b), we can show in straightforward fashion that the various (distinct) K planar hypersurfaces determining the boundaries of each region all intersect at a point $\boldsymbol{K} = \left(K_0^{(1)}, \ldots, K_0^{(K)}\right)$, where now the Ks represent a set of K thresholds $A_k(\boldsymbol{X}') = K_o^{(k)} (k = 1, \ldots, K)$, in which the $\boldsymbol{X}'$ are all values of $\boldsymbol{X}$ satisfying this relation. These thresholds depend explicitly only on the preassigned costs, that is, only on the λs of Eqs. (4.1.11a), and so on. The requirement (4.1.4) ensures that the point $\boldsymbol{K}$ lies in the first 2^{k+1}-tant, that is, all $K^{(k)} \geq 0$.

A simpler variant of Eqs. (4.2.11), also of practical interest, arises when the problem becomes that of testing for the presence of one signal of class k in noise against (any one of the) other possible nonzero signals in *noise*. The case of noise alone is here eliminated.[10] Under these circumstances, the costs $C_0^{(i)}, C_0^{(0)}, C_i^{(0)}$ drop out, and the λs of Eq. (4.1.7) et seq. are simply $\lambda_k^{(i)} = C_k^{(i)} (k, i = 1, \ldots, K) \geq 0$. Consequently, if $C_k^{(i)} \geq 0$, $Ci_k^{(i)} \geq 0$, the $A_k(\boldsymbol{X})$ can now never be negative. Minimization of the average risk then gives only the first set of inequalities in Eq. (4.1.11a), with $\lambda_k^{(0)} = \lambda_\ell^{(0)} = 0$, so that we may write for the decision γ_k the modified conditions

$$\sum_{i=1}^{K} \Lambda_i(\boldsymbol{X}) C_k^{(i)} \leq \sum_{i=1}^{K} \Lambda_i(\boldsymbol{X}) C_\ell^{(i)} \quad \ell \neq k; \text{ all } \ell = 1, \ldots, M, \tag{4.1.14}$$

[8] For $K \geq 4$, $L_k = 0$ is a plane hypersurface (in Λ_1-, ..., Λ_2-space). For $K = 1, 2$, $L_k = 0$ represents a straight line in two dimensions (of Λ_1-, Λ_2-space), while, for $K = 3$, $L_k = 0$ represents a plane surface (in Λ_1-, Λ_2, Λ_3-space).

[9] For example, if $K = 1$, the first "2^2-tant" first "quadrant;" for $K = 2$, the first "2^3-tant" first "octant," and so on. Since the likelihood ratios Λ_k can never be negative, values of $A_k(\boldsymbol{X})$ must always lie in the first "2^{K+1}-tant."

[10] This is equivalent to setting $\delta(\gamma_0|\boldsymbol{X}) = 0 = 1 - \sum_{k=1}^{K} \delta(\gamma_k|\boldsymbol{X})$ [cf. Eq. (4.1.3)] in Eq. (4.1.6) and proceeding as above. The quantity $qF_J(\boldsymbol{X}|\boldsymbol{0})$ in Λ_i is only a normalizing factor here.

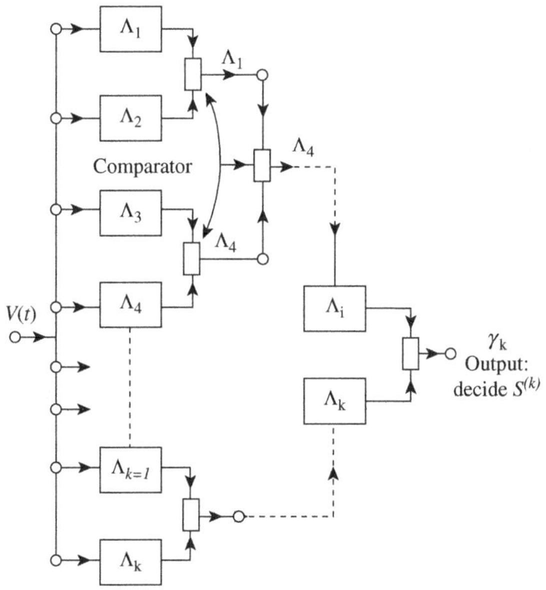

FIGURE 4.1 Schematic diagram of the decision process γ_k, for Eqs. (4.2.11), (4.2.12), or (4.1.14).

repeated for each $k = 1, \ldots, K$ in turn. The point $\boldsymbol{K}$ is now zero, and all decision regions (for the Λs) have their apexes at the origin. The bounding hyperplanes all intersect at $\boldsymbol{K} = 0$, and the equations of the boundaries of the kth region are simply

$$L_k - L_\ell = 0, \quad \text{all } \ell = 1, \ldots, K(\ell \neq k), \quad (\text{for each } \gamma_k, k = 1, \ldots, K). \tag{4.1.15}$$

To summarize, we see that the optimum $(K+1)$-ary (or K-ary) detector consists of a computer that evaluates the $\Lambda_k(\boldsymbol{X})$ for a given set of data $\boldsymbol{X}$ over the observation interval $(\boldsymbol{R}, T)$, computes the various $A_k(\boldsymbol{X})$, and then inserts the results into the inequalities (4.2.11), (4.2.12), or (4.1.14), finally making the decision γ_k associated with the one set of inequalities that is satisfied.[11] One possibility is sketched in Fig. 4.1, where a succession of intermediate binary decision is employed to yield ultimately γ_k.

4.1.4 Examples

4.1.4.1 Binary Detection The simplest and most familiar case of Eqs. (4.1.11), (4.1.12) arises when we have to distinguish $\boldsymbol{S}^{(1)} + \boldsymbol{N}$ versus $\boldsymbol{N}$ alone, so that $K = 1$. With the help of Eq. (4.1.7) in Eqs. (4.1.11), (4.1.12), we easily find that we

$$\left.\begin{array}{ll} \text{Decide } \gamma_0 : \boldsymbol{N}, & \text{if } \Lambda_1(\boldsymbol{X}) < \\ \text{Decide } \gamma_1 : \boldsymbol{S}^{(1)} + \boldsymbol{N}, & \text{if } \Lambda_1(\boldsymbol{X}) \geq \end{array}\right\} K_0^{(1)} \tag{4.1.16a}$$

[11] Of course, this is not a unique way of setting up an actual computing scheme.

where

$$K_0^{(1)} \equiv -\frac{\lambda_1^{(0)}}{\lambda_1^{(1)}} = \frac{C_1^{(0)} - C_0^{(0)}}{C_0^{(1)} - C_1^{(1)}} = \frac{C_\alpha - C_{1-\alpha}}{C_\beta - C_{1-\beta}} \tag{4.1.16b}$$

with the threshold $K_0^{(1)}$ a function of the costs only. [$C_\alpha, C_{1-\alpha}$, etc., are expressed in the earlier notation of Eq. (1.6.6a).]

The somewhat more general binary problem of distinguishing $\boldsymbol{S}^{(a)} + \boldsymbol{N}$ against $\boldsymbol{S}^{(b)} + \boldsymbol{N}$, with a, b any single integers in the range $(1 \leq a, b \leq K)$, $(a \neq b,\ K = 2)$, is readily treated. From Eq. (4.1.14), we write for the decision process

$$\begin{aligned} &\text{Decide } \gamma_a : \boldsymbol{S}^{(a)} + \boldsymbol{N}, \text{ if } \Lambda_a(\boldsymbol{X}) > \Lambda_b(\boldsymbol{X}) K_a^{(b)} \\ &\text{Decide } \gamma_b : \boldsymbol{S}^{(b)} + \boldsymbol{N}, \text{ if } \Lambda_a(\boldsymbol{X}) < \Lambda_b(\boldsymbol{X}) K_a^{(b)}, \\ &\text{where} \quad K_a^{(b)} = \frac{C_a^{(b)} - C_b^{(b)}}{C_b^{(a)} - C_a^{(a)}} > 0, \end{aligned} \tag{4.1.17}$$

which can also be expressed alternatively in terms of a single likelihood ratio

$$\Lambda_a^{(b)}(\boldsymbol{X}) \equiv p_b \left\langle F_J\left(\boldsymbol{X}|\boldsymbol{S}^{(b)}\right)\right\rangle_b \Big/ p_a \left\langle F_J\left(\boldsymbol{X}|\boldsymbol{S}^{(a)}\right)\right\rangle_a \tag{4.1.18}$$

or its reciprocal. Typical regions are shown in Fig. 4.2 (or in Fig. 4.3 if we replace Λ_1 by Λ_a, Λ_2 by Λ_b and set the point $K = 0$.

4.1.4.2 Ternary Detection In this second example, we assume that noise alone is one of the three possible alternatives, so that $K = 2$. Then, from Eqs. (4.1.11), (4.1.12), the decision process is found at once (with $k = 1, 2$; $\ell = 1, 2$). The two thresholds $K_0^{(1)}, K_0^{(2)}$ are

$$\begin{aligned} K_0^{(1)} &= \frac{\lambda_2^{(0)}\lambda_1^{(2)} - \lambda_1^{(0)}\lambda_2^{(2)}}{\Delta} \\ K_0^{(2)} &= \frac{\lambda_1^{(0)}\lambda_2^{(1)} - \lambda_1^{(1)}\lambda_2^{(0)}}{\Delta} \\ \Delta &= \lambda_1^{(1)}\lambda_2^{(2)} - \lambda_1^{(2)}\lambda_2^{(1)}, \end{aligned} \tag{4.1.19}$$

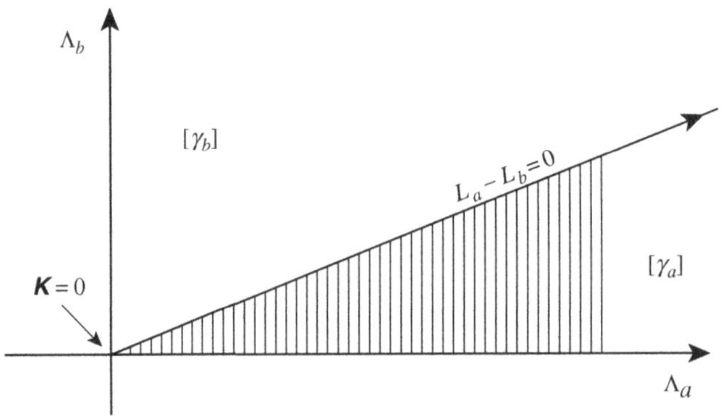

FIGURE 4.2 Region of decision for the binary case $(K = 2) : S^{(a)} + N$ vs. $S^{(b)} + N$.

and we decide

$$\gamma_0 : \boldsymbol{N}, \quad \text{when } \lambda_1^{(0)} + \lambda_1^{(1)}\Lambda_1 + \lambda_1^{(2)}\Lambda_2 > 0 \qquad \lambda_2^{(0)} + \lambda_2^{(1)}\Lambda_1 + \lambda_2^{(2)}\Lambda_2 > 0, \tag{4.1.20a}$$

$$\gamma_1 : \boldsymbol{S}^{(1)} + \boldsymbol{N}, \quad \text{when } \lambda_1^{(0)} + \lambda_1^{(1)}\Lambda_1 + \lambda_1^{(2)}\Lambda_2 < \lambda_2^{(0)} + \lambda_2^{(1)}\Lambda_1 + \lambda_2^{(2)}\Lambda_2 \qquad \lambda_1^{(0)} + \lambda_1^{(1)}\Lambda_1 + \lambda_1^{(2)}\Lambda_2 < 0, \tag{4.1.20b}$$

$$\gamma_2 : \boldsymbol{S}^{(2)} + \boldsymbol{N}, \quad \text{when } \lambda_2^{(0)} + \lambda_2^{(1)}\Lambda_1 + \lambda_2^{(2)}\Lambda_2 < \lambda_1^{(0)} + \lambda_1^{(1)}\Lambda_1 + \lambda_1^{(2)}\Lambda_2 \qquad \lambda_2^{(0)} + \lambda_2^{(1)}\Lambda_1 + \lambda_2^{(2)}\Lambda_2 < 0. \tag{4.1.20c}$$

The boundaries of the various decision regions are easily determined from Eqs. (4.1.13a) and the Λ_1, Λ_2 axes bounding the first quadrant. A typical case is illustrated in Fig. 4.3.

The decision process is particularly simple when the case of noise alone is removed and one of three possible combinations of signal and noise can now occur, for example $\boldsymbol{S}^{(a)} + \boldsymbol{N}$, $\boldsymbol{S}^{(b)} + \boldsymbol{N}$, $\boldsymbol{S}^{(c)} + \boldsymbol{N} (a \neq b \neq c; 1 \leq a, b, c \leq K; K = 3)$. From Eq. (4.1.14), we find the decision process to be

$$\text{Decide } \gamma_a : \boldsymbol{S}^{(a)} + \boldsymbol{N}, \quad \text{when } \Lambda_a > \Lambda_b K_{ab}^{(b)} + \Lambda_c K_{ab}^{(c)} \qquad \Lambda_a > \Lambda_b K_{ac}^{(b)} + \Lambda_c K_{ac}^{(c)}, \tag{4.1.21a}$$

$$\text{Decide } \gamma_b : \boldsymbol{S}^{(b)} + \boldsymbol{N}, \quad \text{according to Eq. (4.1.21a), replacing } a \text{ by } b \text{ and } b \text{ by } a \text{ therein,} \tag{4.1.21b}$$

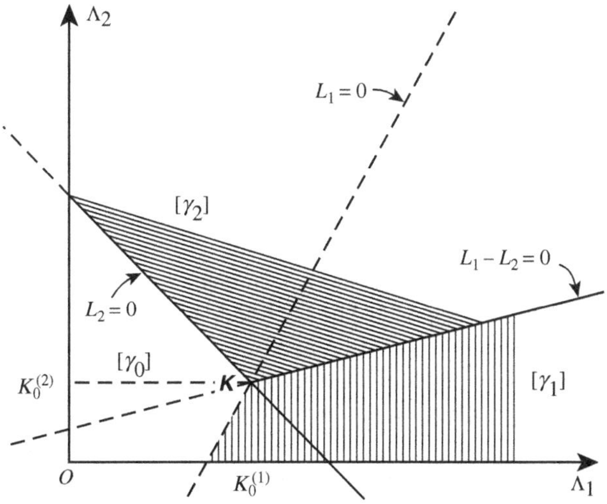

FIGURE 4.3 Regions of decision for the ternary detection $(K = 2)$:$\boldsymbol{N}$ versus $\boldsymbol{S}^{(1)} + \boldsymbol{N}$ versus $\boldsymbol{S}^{(2)} + \boldsymbol{N}$; $\lambda_1^{(2)} > 0$; $\lambda_2^{(1)} < 0$.

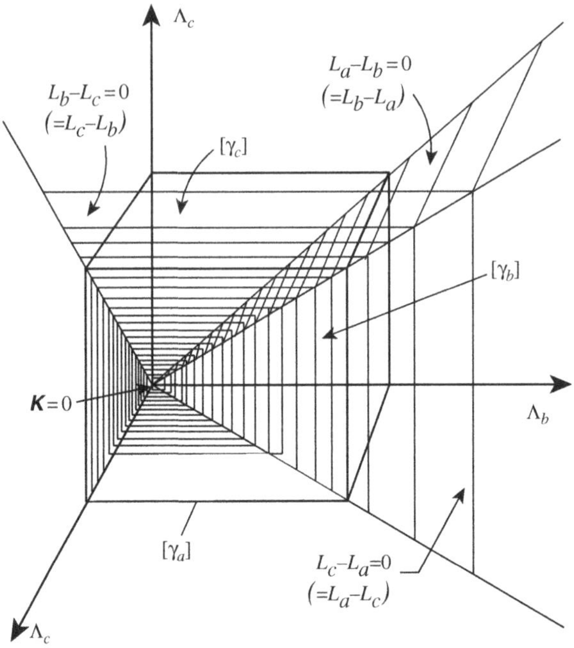

FIGURE 4.4 Regions of decision for the ternary detection $(K = 3)$:$S^{(a)} + N$ versus $S^{(b)} + N$ versus $S^{(c)} + N$; constant cost C_0 of failure, zero cost of success.

$$\text{Decide } \gamma_c : S^{(c)} + N, \quad \begin{array}{l}\text{according to Eq. (4.1.21a),}\\ \text{letting } a \to c,\ b \to a,\ c \to b \text{ therein,}\end{array} \tag{4.1.21c}$$

where the thresholds $K_{ab}^{(b)}$, and so on, are specifically

$$\begin{aligned} K_{ab}^{(b)} &\equiv \frac{C_a^{(b)} - C_b^{(b)}}{C_b^{(a)} - C_a^{(a)}} \quad K_{ab}^{(c)} \equiv \frac{C_a^{(c)} - C_b^{(c)}}{C_b^{(a)} - C_a^{(a)}} \\ K_{ac}^{(b)} &\equiv \frac{C_a^{(b)} - C_c^{(b)}}{C_c^{(a)} - C_a^{(a)}} \quad K_{ac}^{(c)} \equiv \frac{C_a^{(c)} - C_c^{(c)}}{C_c^{(a)} - C_a^{(a)}}. \end{aligned} \tag{4.1.22}$$

[The thresholds for γ_b, γ_c are found by interchanging a, b, c according to Eqs. (4.1.21b), (4.1.21c), respectively.] The bounding surfaces follow from Eq. (4.1.15), and the planes defining the first *octant* (in Λ_a-, Λ_b-, Λ_c-space) are shown for a typical case [Section 4.1.4.3] in Fig. 4.4. A variety of different regions are clearly possible in these ternary cases, depending on our choice of the preassigned costs, subject to Eq. (4.1.4), of course.

4.1.4.3 Simple $(K+1)$-ary Detection A particular case of considerable interest and simplicity [as far as the structure of the detector, e.g., Eqs. (4.1.11a), (4.1.11b), or (4.1.11a)c), is concerned] occurs when we assign the *same* constant costs $C_0(> 0)$ to all types of "failure" and zero costs to all types of "success." Then Eqs. (4.1.7) become

$\lambda_k^{(0)} = C_0$, $\lambda_k^{(\ell)} = 0 (\ell \neq k)$, $\lambda_k^{(k)} = -C_0$, and the decision process Eqs. (4.11) is governed simply by the inequalities:

$$\text{Decide } \lambda_k, \text{ if } \begin{array}{l} \Lambda_k(\boldsymbol{X}) \geq \Lambda_\ell(\boldsymbol{X}) \quad \text{all } \ell \neq k \; (\ell = 1, \ldots, K) \\ \Lambda_k(\boldsymbol{X}) \geq 1 \end{array} \tag{4.1.23}$$

for each $k(= 1, \ldots, K)$ in turn. For noise alone, Eq. (4.1.12a) is simply

$$\text{Decide } \gamma_0, \text{ if } \Lambda_k(\boldsymbol{X}) \leq 1 \quad \text{all } \ell = 1, \ldots, K. \tag{4.1.24}$$

The boundaries of the decision regions are obtained as before from Eqs. (4.1.13a) and (4.1.13b) in terms of the equalities above. For K-ary detection, where the case "noise alone" is removed, the first line of Eq. (4.1.23) provides the desired inequalities, and the boundaries of the decision regions again follow from the indicated equalities [see Eq. (4.1.15) and Fig. 4.4].

The actual structure of these optimum multiple-alternative systems depends in each case on the explicit form of $\left\langle F_J\left(\boldsymbol{X}|\boldsymbol{S}^{(k)}\right)\right\rangle_k$, as in the binary theory. Also as in the binary theory, we may resolve the threshold structure for additive signals and (normal) noise into a sequence of realizable linear and nonlinear elements: for example, when there is sample uncertainty, into the Bayes match filters, zero-memory nonlinear rectifiers, and ideal integrators (Sections 3.4 and 3.5). This yields $\log \Lambda_\ell (\ell = 1, \ldots, K)$, from which in turn the Λ_ℓ may be found to be combined according to Eqs. (4.1.11), (4.1.12), or (4.1.14) for an ultimate decision γ_k (Fig. 4.1). Similar remarks apply in the instances of sample certainty with the structure appropriate to these cases (Sections 20.1.1 and 20.2.1 of Ref. [2]). The logarithmic form of the characteristic function may also be used to advantage here in the special situation of the example in Section 4.1.4.3 [and, of course, for the binary cases, cf. the example in Section 4.1.4.1]. Instead of Eqs. (4.1.23) and (4.1.24), we can write for this particular cost assignment

$$\text{Decide } \lambda_k, \text{ if } \log \begin{array}{l} \Lambda_k(\boldsymbol{X}) \geq \log \Lambda_\ell(\boldsymbol{X}), \quad \text{all } \ell \neq k \; (\ell = 1, \ldots, K) \\ \Lambda_k(\boldsymbol{X}) \geq 0 \end{array} \tag{4.1.25}$$

for each $k(= 1, \ldots, K)$ in turn. For noise alone, we have

$$\text{Decide } \gamma_0, \text{ if } \log \Lambda_k(\boldsymbol{X}) \leq 0, \quad \text{all } k = 1, \ldots, K. \tag{4.1.26}$$

The boundaries of the decision regions follow from Eqs. (4.1.13), subject now to the additional logarithmic transformation. When the alternative "noise alone" is eliminated, Eq. (4.1.25) becomes

$$\text{Decide } \gamma_0, \text{ if } \log \Lambda_k(\boldsymbol{X}) \geq \log \Lambda_\ell(\boldsymbol{X}), \quad \text{all } \ell = 1, \ldots, K. \tag{4.1.27}$$

The optimum system consists of a sequence of operations, $\log \Lambda_k(\boldsymbol{X})$, as in Fig. 4.1, whose ultimate output is once more a decision γ_k and whose structure (each ℓ) is in threshold cases given by threshold developments of the type for normal noise backgrounds.

4.1.5 Error Probabilities, Average Risk, and System Evaluation

In order to evaluate the performance of these optimum decision systems, it is necessary to determine their Bayes risks, and for this in turn we need the error probabilities associated with the various possible decisions. Similarly, to evaluate the performance of suboptimum systems and compare them with the corresponding optimum cases, we must also determine the appropriate error probabilities. This is a conceptually straightforward generalization of Chapters 2 and 3 for the earlier binary theory, although, as we shall note presently, there are certain technical problems here not present in the simpler case, problems which make explicit calculations considerably more difficult.

Let us now consider the various conditional probabilities of error. We define

$$\begin{aligned}\alpha_k^{(0)} &\equiv \int_\Gamma \delta(\gamma_k|\boldsymbol{X})F_J\left(\boldsymbol{X}|\boldsymbol{S}^{(0)}\right)d\boldsymbol{X} \\ &\equiv \text{conditional probability of calling a null signal any} \\ &\quad\ \text{one member of } k\text{th signal class } (k=1,\ldots,K)\end{aligned} \tag{4.1.28}$$

$$\begin{aligned}\left\langle\beta_\ell^{(k)}\right\rangle_k &\equiv \int_\Gamma \delta(\gamma_\ell|\boldsymbol{X})\left\langle F_J\left(\boldsymbol{X}|\boldsymbol{S}^{(k)}\right)\right\rangle_{\boldsymbol{k}} d\boldsymbol{X} \\ &\equiv \text{conditional probability of calling any one member} \\ &\quad\ \text{the } k\text{th signal class a member of } \ell\text{th signal class (in noise)} \\ &\quad\ (\ell \neq k;\ \ell=0,\ldots,K;\ k=1,\ldots,K).\end{aligned} \tag{4.1.29}$$

The conditional probability of *correctly deciding* that any one signal member of class $k(k=0,\ldots,K)$ is present in noise is

$$\left\langle\eta_k^{(k)}\right\rangle_k \equiv \int_\Gamma \delta(\gamma_k|\boldsymbol{X})\left\langle F_J\left(\boldsymbol{X}|\boldsymbol{S}^{(k)}\right)\right\rangle_{\boldsymbol{k}} d\boldsymbol{X}, \tag{4.1.30}$$

$$\left\langle\eta_k^{(k)}\right\rangle_k = 1 - \sum_{\ell=0}^{K}{}' \left\langle\beta_\ell^{(k)}\right\rangle_k. \tag{4.1.30a}$$

In the situation where the case "noise alone" is removed, we have only $\left\langle\beta_\ell^{(k)}\right\rangle_k$ $(\ell \neq k;\ k=1,\ldots,K)$.

For optimum systems, where Eqs. (4.1.11), (4.1.12), (4.1.14) apply, it is convenient to make a change of variable and consider some monotonic function of the Λs as our new independent variables, since it is in terms of the Λs that the (optimum) decision regions for $\gamma_0,\ldots,\gamma_K$ are explicitly given. As before, we let $x_k = \log\Lambda_k (k=1,\ldots,K)$, so that Eqs. (4.1.28) and (4.1.29) may now be written[12]

$$\alpha_k^{(0)*} = \int_{[x_1]}\cdots\int_{[x_k]}\cdots\int_{[x_K]} Q_J(x_1,\ldots,x_K)dx_1,\ldots,dx_K \tag{4.1.31}$$

$$\left\langle\beta_\ell^{(k)*}\right\rangle_k = \int_{[x_1]}\cdots\int_{[x_\ell]}\cdots\int_{[x_K]} P_J^{(k)}(x_1,\ldots,x_K)dx_1,\ldots,dx_K;\ P_J^{(0)} = Q_J, \tag{4.1.32}$$

[12] See 1.9 for a discussion in the binary cases.

where $Q_J, P_J^{(k)}$ are the *joint probability densities* for the random variables $\log \Lambda_1, \ldots, \log \Lambda_K$ in the first instance with respect to the null hypothesis H_0 and in the second with respect to the alternatives H_k. Here $[x_k]$ signifies the decision region for x_k. (Some typical cases are illustrated in Figs. 4.2 and 4.3 for the transformation $x'_k = \Lambda_k$, rather than $x_k = \log \Lambda_k$). These probability densities may be written in terms of the original data $\boldsymbol{X}$ as [Section 1.9]

$$Q_J(x_1, \ldots, x_K) = \int_\Gamma F_J(\boldsymbol{X}|\boldsymbol{0}) \prod_{\ell=1}^{K} \delta(x_\ell - \log \Lambda_\ell) d\boldsymbol{X}, \tag{4.1.33}$$

$$P_J^{(k)}(x_1, \ldots, x_K) = \int_\Gamma \left\langle F_J\left(\boldsymbol{X}|\boldsymbol{S}^{(k)}\right)\right\rangle_k \prod_{\ell=1}^{K} \delta(x_\ell - \log \Lambda_\ell) d\boldsymbol{X}. \tag{4.1.34}$$

The corresponding characteristic functions are

$$F_K(i\xi)_Q = \int_\Gamma e^{i\tilde{\xi}x} F_J(\boldsymbol{X}|\boldsymbol{0}) d\boldsymbol{X}, \tag{4.1.35}$$

$$F_K^{(k)}(i\xi)_P = \int_\Gamma e^{i\tilde{\xi}x} \left\langle F_J\left(\boldsymbol{X}|\boldsymbol{S}^{(k)}\right)\right\rangle_k d\boldsymbol{X}, \tag{4.1.36}$$

A specific example with additive Gaussian noise is considered in Problem 4.3.

With Eqs. (4.1.28)–(4.1.32), we can now write the average risk [Eq. (4.1.6) or (4.1.10a)], minimized or not, as

$$\begin{aligned} R_K(\sigma, \delta)\langle * \rangle = R_{0K} + q \sum_1^K \lambda \alpha_k^{(0)(*)} \ + & \sum_{\substack{\ell = 0; k \geq 1, \\ \ell \geq 0 (k \neq \ell)}}^{K}{}' \quad p_k \lambda_\ell^{(k)} \left\langle \beta_\ell^{(k)(*)} \right\rangle_k \\ + & \sum_{k=1}^{K}{}' p_k \lambda_k^{(k)} \left\langle \eta_k^{(k)(*)} \right\rangle_k, \end{aligned} \tag{4.1.37}$$

which for the case of the excluded null signal simply omits the terms in $\lambda_k^{(0)}$ and sets $R_{0K} = 0$, with $\lambda_k^{(\ell)}(k, \ell \geq 1)$ equal to the costs $C_k^{(\ell)} \geq 0$.

The expressions (4.1.31), (4.1.32) for the error probabilities appearing in Eqs. (4.1.35) and (4.1.36) have particularly simple limits for $[x_k]$ when we make the cost assumptions $(C_0, 0)$ of the example in Section 4.1.4.3 above. We easily find that now (the primes indicate terms $\ell = k$, or ℓ, omitted in the products)

$$\alpha_k^{(0)*} = \int_0^\infty dx_k \left(\prod_{\ell=1}^{K}{}' \int_{-\infty}^{x_k} dx_\ell \right) Q_J(x_1, \ldots, x_K), \qquad k = 1, \ldots, K, \tag{4.1.38}$$

$$\left\langle \beta_\ell^{(k)*} \right\rangle_k = \int_0^\infty dx_\ell \left(\prod_{i=1}^{K}{}' \int_{-\infty}^{x_\ell} dx_i \right) P_J^{(k)}(x_1, \ldots, x_K), \quad \ell \neq k; k \geq 1, \tag{4.1.39}$$

and $$\left\langle \beta_0^{(k)*} \right\rangle_k = \int_{-\infty}^{0} dx_1 \cdots \int_{-\infty}^{0} dx_K P_J^{(k)}(x_1, \ldots, x_K). \tag{4.1.40}$$

When the noise class is omitted, Eqs. (4.1.38) and (4.1.39) do not apply and the lower limit on the first integral of Eq. (4.1.39) becomes $-\infty$ instead of 0. For the general cost assumptions [Eq. (4.1.7)], we must use Eqs. (4.1.13) or (4.1.15) (all k) to specify the boundaries on Λ_k and hence on $\log \Lambda_k$. General results for the ternary case [the example in Section 4.1.4.2] follow at once from Eqs. (4.1.19), (4.1.20) or (4.1.21) with aid of Figs. 4.3 and 4.4.

4.1.5.1 Suboptimum Systems For suboptimum systems, instead of $x_\ell = \log \Lambda_\ell(\boldsymbol{X})$ we have $y_\ell = \log G_\ell(\boldsymbol{X})$, where $G_\ell(\boldsymbol{X})$ represents the ℓth component of the actual suboptimum system in use (Section 1.9.2). Then, analogous to Eqs. (2.5.16a), (2.5.16b), and so on, Eqs. (4.1.31), (4.1.32) are modified to

$$\alpha_k^{(0)} = \int \underset{[y]}{\cdots} \int q_J(y_1, \ldots, y_K) dy_1 \cdots dy_K > \alpha_k^{(0)*}, \tag{4.1.41a}$$

$$\left\langle \beta_\ell^{(k)} \right\rangle_k = \int \underset{[y]}{\cdots} \int p_J(y_1, \ldots, y_K) dy_1 \cdots dy_K > \left\langle \beta_\ell^{(k)*} \right\rangle. \tag{4.1.41b}$$

where $q_J, p_J^{(k)}$ are now the joint pdfs of the random variables $\log G_\ell(\boldsymbol{X})$, $(\ell = 1, \ldots, K)$, and $[\boldsymbol{y}]$ denotes the decision regions for the $y_1, \ldots, y_K$, which may or may not be the same as $[\boldsymbol{x}]$ in Eqs. (4.1.31) and (4.1.32). These probability densities are expressed in terms of the data process $\boldsymbol{X}$, analogous to Eqs. (4.1.33) and (4.1.34), as

$$q_J(y_1, \ldots, y_K) = \int_\Gamma F_J(\boldsymbol{X}|\boldsymbol{0}) \prod_{i=1}^{K} \delta[y_i - \log G_i(\boldsymbol{X})] d\boldsymbol{X} \tag{4.1.42a}$$

$$p_J^{(k)}(y_1, \ldots, y_K) = \int_\Gamma \left\langle F_J\left(\boldsymbol{X}|\boldsymbol{S}^{(k)}\right)\right\rangle_k \prod_{i=1}^{K} \delta[y_i - \log G_i(\boldsymbol{X})] d\boldsymbol{X} \tag{4.1.42b}$$

with the associated characteristic functions [cf. Eqs. (1.9.17a), (1.9.17b)]

$$F_K(i\xi)_q = \int_\Gamma e^{i\xi y} F_J(\boldsymbol{X}|\boldsymbol{0}) d\boldsymbol{X} \quad F_K^{(k)}(i\xi)_p = \int_\Gamma e^{i\xi y} \left\langle F_J\left(\boldsymbol{X}|\boldsymbol{S}^{(k)}\right)\right\rangle_k d\boldsymbol{X}. \tag{4.1.43}$$

System comparisons are then made on the basis of the respective Bayes and average risks, by an obvious extension of the binary methods to these multialternative cases. However, even when the background noise is normal, additive, and independent, it is difficult to evaluate these error probabilities, since they are, in effect, K dimensional error functions whose arguments depend on the successive variables of integration. The technical problem is analogous to that of evaluating the volume cutoff from a hyperellipsoid by a series of hyperplanes. One special example of some interest, where the multiple integrals "factor" conveniently, is considered the example in Section 4.1.6. Generally, however, our expressions are not so analytically tractable, so that, while structure can be obtained, the error probabilities require a more formidable computational program.

4.1.6 An Example

Consider the multiple-alternative detection situation in which the null signal or any one of K arbitrary nonzero signals (and noise), identical in structure but differing in amplitude, may be present at the input. Detection is assumed to be *coherent*. The signals and noise are additive. Each signal set contains only one member, and the costs of correct and incorrect decisions are taken to be zero and unity, respectively. The noise itself is assumed Gaussian, with a known variance matrix. The kth hypothesis class contains but one signal, of amplitude a_{0k}, and the amplitudes are ordered: $0 < a_{01} < a_{02} \cdots < a_{0k} < \cdots a_{0K}$.

(1) We first obtain the characteristic functions and pdfs for the vector $\boldsymbol{y} = \boldsymbol{x} - \left[\log \mu_k + a_{0k}^2 \Phi_{kk}/2\right]$:

$$F_K(i\xi)_Q = e^{-\frac{1}{2}\tilde{\xi}\boldsymbol{S}_K\xi} \tag{4.1.44a}$$

$$F_K^{(k)}(i\xi)_P = e^{-i\tilde{\xi}(\boldsymbol{S}_K)_k - \frac{1}{2}\tilde{\xi}\boldsymbol{S}_K\xi} \tag{4.1.44b}$$

$$Q_J(\boldsymbol{y}) = (2\pi)^{-K/2}(det\,\boldsymbol{S}_K)^{-1/2} e^{-\frac{1}{2}\tilde{\boldsymbol{y}}\boldsymbol{S}_M^{-1}\boldsymbol{y}} \tag{4.1.45a}$$

$$P_J^{(k)}(\boldsymbol{y}) = (2\pi)^{-K/2}(det\,\boldsymbol{S}_K)^{-1/2} e^{-\frac{1}{2}\left[\tilde{y} - \left(\tilde{S}^K\right)_k\right]\boldsymbol{S}_K^{-1}\left[y - \left(S^K\right)_k\right]} \tag{4.1.45b}$$

where $\boldsymbol{S}_K \cong [a_{0i}a_{0k}\Phi_{ik}]$, $\Phi(s_i, s_k) = \tilde{\boldsymbol{s}}_i \boldsymbol{k}_N^{-1} s_k$, and $\mu_i = p/q$.

(2) The error probabilities for the optimum system here are found to be

$$\begin{aligned} \alpha_i^{(0)*} &= \int_{-A_i}^{\infty} dy_k \left(\prod_{\ell=1}^{K}{}' \int_{-\infty}^{y_i + A_i - A_\ell} dy \right) Q_J(\boldsymbol{y}) \\ &= a_{0i}^{-1}(2\pi\Phi)^{-1/2} \int_{(C_{i-} \text{ or } -A_i)}^{C_{i+}} e^{-y^2/2a_{0i}2\Phi} dy, \quad i = 1, \ldots, K, \end{aligned} \tag{4.1.46a}$$

$$\begin{aligned} \beta_0^{(k)*} &= \int_{-\infty}^{-A_i} dy_1 \cdots \int_{-\infty}^{-A_k} dy_K P_j^{(k)}(y) \\ &= a_{0i}^{-1}(2\pi\Phi)^{-1/2} \int_{-\infty}^{C_{0+}} e^{-\left(y - a_{0k}^2\Phi\right)^2/2a_{0k}2\Phi} dy, \quad k = 1, \ldots, K, \end{aligned} \tag{4.1.46b}$$

$$\begin{aligned} \beta_0^{(k)*} &= \int_{-A_i}^{\infty} dy_i \left(\prod_{\ell=1}^{K}{}' \int_{-\infty}^{y_i + A_i - A_\ell} dy \right) P_j^{(k)}(y) \\ &= a_{0i}^{-1}(2\pi)^{-1/2} \int_{C_{i-} \text{ or } A_i}^{C_{i+}} e^{-(y - a_{0i} - a_{0k}\Phi)^2/2a_{0i}2\Phi} dy \quad \ell, k = 1, \ldots, K; \ell \neq k, \end{aligned} \tag{4.1.46c}$$

where

$$A_i = \log \mu_i - a_0^2 - a_0^2\Phi_{ii}/2;\ \Phi_{ii} \equiv \Phi;\ C_{i-} \equiv \max_{\ell} C_{i\ell}(\ell < i),\ C_{i+} \equiv \min_{\ell} C_{i\ell}(\ell > i),$$

with (4.1.46d)

$$C_{i\ell} \equiv \tfrac{1}{2} a_{0i}(a_{0i} + a_{0\ell})\Phi + a_{0i}/(a_{0\ell} - a_{0i}) - \log(\mu_i/\mu_\ell).$$

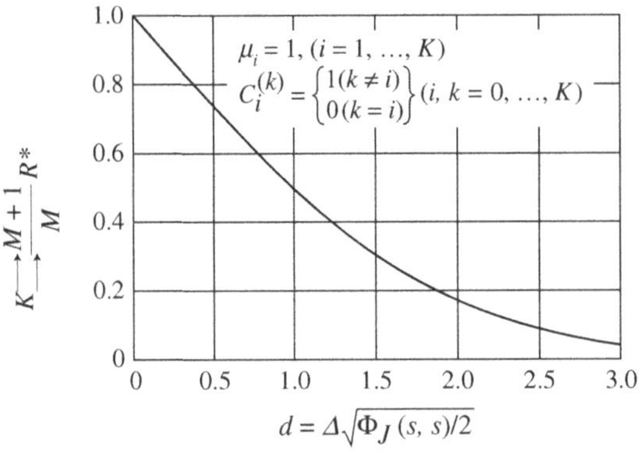

FIGURE 4.5 Bayes risk for the coherent multiple-alternative detection of K signals in Gaussian noise; simple cost assignments.

(3) Assume next that the amplitudes are uniformly spaced, so that $a_{0i} = i\Delta$ with $\Delta \equiv (\Delta A_0)/(2\psi_J)^{1/2}$, and let $u_i = 1(\text{all } i = 1, \ldots, K)$. Then the Bayes risk for these simple cost assignments becomes

$$R_K^* = \frac{K}{K+1}\left[1 - \Theta\left(\frac{d}{2}\right)\right]$$
$$d \equiv \left[\frac{\Phi_J(s,s)}{2}\right]^{1/2} \Delta \tag{4.1.47}$$

(cf. Fig. 4.5). Here Θ is the *erf* function defined on p. 369. We next verify that this result also holds for the K-ary case as well. Observe that as $K \to \infty$, the Bayes risk increases to

$$1 - \Theta(d/2), \tag{4.1.48}$$

independent of the number of alternatives, and depending only on the autocorrelation of the normalized signal and the size Δ of the increment between signals. The larger the latter the easier it is to distinguish between signals, and hence the smaller the Bayes risk (i.e., $d \gg 1$). Conversely, for large K there is negligible distinction between large and small signals. For details, see Middleton and Van Meter [1], pp. 6–9.

4.2 OVERLAPPING HYPOTHESIS CLASSES

When the signal classes overlap, as discussed in Section 1.10, the usual definitions, which depend on the classes being disjoint, do not apply. Any signal may now belong to one or more signal classes out of a total of K classes. This is the case if each contains the same type of parameter $(\theta_k, \theta_\ell), k \neq \ell$, for example, with similar waveforms, but with different pdfs for the parameters, such that the signal classes overlap. Thus, for signals of class k the parameter space may coincide with that of class ℓ by a certain amount, as does class ℓ by another amount on class k, in the manner suggested by Fig. 4.6a. Here any given signal may belong to two or

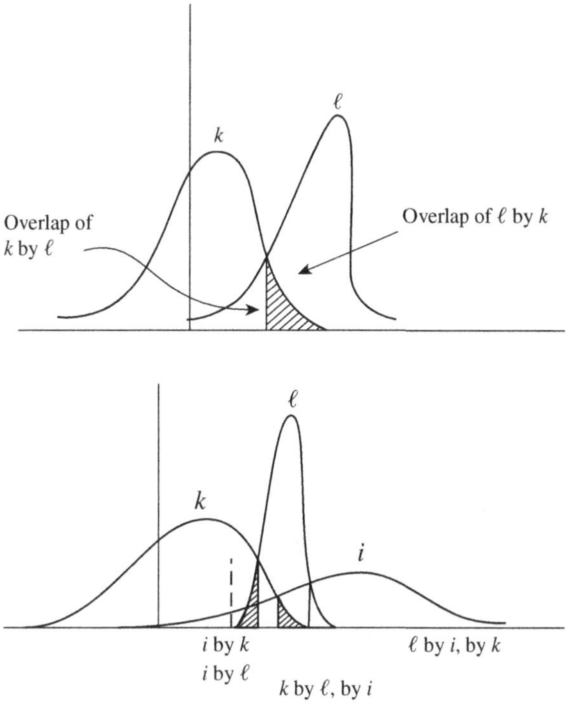

FIGURE 4.6 (a) Mutual overlap of signals of class k and class ℓ (through their parameters θ, Eq. (4.2.1)). (b) Same as part (a) for mutual overlap of signals of class i, k, and ℓ.

more classes, but will usually belong to one class with the largest probability. Accordingly, it is reasonable, as in the binary cases (Section 1.10.2), to assign the least cost to this most probable case. The overlap may not be limited to two neighboring classes (in a probabilistic sense), but may in principle involve all K of them (Fig. 4.6b). [In the degenerate cases where two or more distributions are equal, it is still possible to distinguish the signals, if the prior probabilities (p_k) are all different.] Accordingly, the cost assignments depend on the probability that each signal belong to all K classes, with varying degrees of overlap, including the fact that some distributions may *not* overlap, that is, are disjoint. It may be "nearest neighbors," if, say, the parameter in question is amplitude (i.e., scale), or it may depend on several separate signal classes, and so on.

4.2.1 Reformulation

In our present examples we limit our cost assignments to the preset constant costs previously used, and extend the "constant" cost function F_1, to the more general case for the kth decision γ_k, where "no signal" is included:

$$\begin{aligned} C_\ell^{(k)}(S(\theta),\gamma_k) &= \left[C_{\ell|0}^{(k)}p_0w_0(\theta) + C_{\ell/1}^{(k)}p_1w_1(\theta) + \ldots + C_{\ell|r}^{(k)}p_rw_r(\theta)\right]\Big/\sigma(\theta) \\ &= \sum_{r=0}^{K} C_{\ell|r}^{(k)}p_rw_r(\theta)/\sigma(\theta), \end{aligned} \tag{4.2.1}$$

where $r = 0, \ldots, r_0 \leq K$, and $\ell = 0, \ldots, K$ are decided, with $k = 0, 1, \ldots, K$. From (4.1.2), now modified for the parameters θ rather than waveforms $\boldsymbol{S}$, we have

$$\sigma(\theta) = \sum_{k=0}^{K} p_k w_k(\theta) \delta^{(k)} \left(\theta - \theta^{(k)}\right). \tag{4.2.1a}$$

In the case of stochastic signals (4.2.1) is given by

$$C_\ell^{(k)} \left(\boldsymbol{S}^{(k)}, \gamma_k\right) = \sum_{r=0}^{K} C_{\ell|r}^{(k)} p_r w_r \left(\boldsymbol{S}^{(k)}\right) / \sigma(\boldsymbol{S}), \tag{4.2.2}$$

again with $(k, \ell) = 0, 1, \ldots, K$, and $r = 0, 1, 2, \ldots, r_0 \leq K$, where $\sigma(\boldsymbol{S})$ is given by Eq. (4.1.2). The Bayes risk in once more expressed generally by (4.1.1), with $\sigma(\theta)$, or $\sigma(\boldsymbol{S})$, the *a priori* probability density governing the θ, or all possible signals $\boldsymbol{S}$. The p_r are positive numbers between (and including) $(0, 1)$, such that (4.1.2) is obeyed, in short, the p_r are again the prior probabilities that a signal of class r *can be present* (including $r = k$) on any one observation, and the class $r = 0$ represents the null signal class. Eq. (4.1.3) applies to the decision rules γ_k: a definite decision is always made, with probability 1. Failure is more expensive than success, that is, (4.1.4) applies here, where

$$\begin{aligned} C_0^{(0)}(0, \gamma_0) &= \text{Eq. (4.2.1)}: \ C_{0|0}^{(0)} p_0 w_0(\theta) \\ &= C_{0|0}^{(0)} \delta(\boldsymbol{S} - \boldsymbol{0}) : \text{noise alone is correctly detected.} \end{aligned} \tag{4.2.3}$$

Here there is no signal (not to be confused with a signal with $\theta = \boldsymbol{0}$). Similarly, we have

$$\left.\begin{aligned} &C_k^{(k)}\left(\boldsymbol{S}^k(\theta), \gamma_k\right) = C_{k|r}^{(k)} : && \text{a signal of class } k \text{ is correctly detected, with parameters } \theta, \\ & && \text{although contaminated by different parameter distributions} \\ & && r \neq k, r = 0, \ldots, K. \\ &C_\ell^{(0)}(0, \gamma_\ell) = C_{\ell|r}^{(0)} : && \text{a signal of class } \ell, (\ell = 1, \ldots, K), \text{ is incorrectly decided,} \\ & && \text{when noise alone occurs, with parameters} (\theta_\sigma = \boldsymbol{0}). \\ & && \text{Again, all parameters distributions } (> 0) \text{ contribute.} \\ &C_\ell^{(k)}\left(\boldsymbol{S}^k(\theta), \gamma_\ell\right) = C_{\ell|r}^{(k)} : && \ell \neq \text{k}: \text{a signal of class } \ell \text{ is incorrectly detected, when a} \\ & && \text{signal of class } k \text{ occurs, with contamination by all possible} \\ & && \text{parameter distributions, } r = 0, 1, \ldots, K. \end{aligned}\right\} \tag{4.2.4}$$

The average risk is from (4.1.1) and (4.2.1)–(4.2.4), now expressed as

$$R(\sigma, \delta) = \int_\Gamma \left[\begin{aligned} &\left\{ C_{0|0}^{(0)} \delta(\gamma_0|\boldsymbol{X}) + \sum_{k=1}^{K} \sum_{r=1}^{K} C_{k|r}^{(0)} \delta(\gamma_k|\boldsymbol{X}) \right\} q F_J(\boldsymbol{X}|\boldsymbol{0}) d\boldsymbol{X} \\ &+ \left\{ \begin{aligned} &\sum_{k=1}^{K} \sum_{r=1}^{K} C_{k|r}^{(k)} p_r \left\langle F_J\left(\boldsymbol{X}|\boldsymbol{S}^{(k)}(\theta)\right)\right\rangle_{\mathrm{W}_r} \delta(\gamma_k|\boldsymbol{X}) \\ &+ \sum_{k=1}^{K}{}' \sum_{r=1}^{K} C_{\ell|r}^{(k)} p_r \left\langle F_J\left(\boldsymbol{X}|\boldsymbol{S}^{(k)}(\theta)\right)\right\rangle_{\mathrm{W}_r} \delta(p_\ell|\boldsymbol{X}) \end{aligned} \right\} d\boldsymbol{X} \end{aligned} \right], \tag{4.2.5}$$

since $w_0(\boldsymbol{\theta}) = \delta(\boldsymbol{S} - \boldsymbol{0})$, where the prime over the summation (in ℓ) signifies $\ell \neq k$. [Note when the signal classes are disjoint (Section 4.1), that is, that only $r = k$ (or ℓ) in all the terms of (4.2.5), our extended result reduces to the disjoint result (4.1.6) of Section 4.1, that is, $C_{k|k}^{(0)} = C_k^{(0)}, C_{k|k}^{(k)} = C_k^{(k)}, C_{\ell|\ell}^{(k)} = C_\ell^{(k)}$, all k, ℓ.] Here the different distributions $w_r(\theta)$ of the common parameters θ influence the decisions, as well as the associated cost functions $C_{k|r}^{(k)}$, $C_{k|r}^{(\ell)}$, and the *a priori* probabilities p_r. For individual distributions that are disjoint of the other, $C_{k|r}^{(k)} = C_k^{(k)}$, $C_{\ell|r}^{(k)} = C_\ell^{(k)}$, for any particular k (or ℓ), or several k (or ℓ).

4.2.2 Minimization of the Average Risk for Overlapping Hypothesis Classes

Following Section 4.1.2 we rearrange Eq. (4.1.6), extending it to the set of costs:

$$\lambda_i^{(0)} \to \lambda_{i|r}^{(0)} = C_{i|r}^{(0)} - C_0^{(0)}; \quad i = 1, \ldots, K; \tag{4.2.6a}$$

$$\lambda_k^{(k)} \to \lambda_{k|r}^{(k)} = C_{k|r}^{(k)} - C_{0|r}^{(k)}; \quad r = 1, \ldots, K; \tag{4.2.6b}$$

$$\left.\begin{array}{ll} \lambda_k^{(i)} \to \lambda_{k|r}^{(i)} = C_{k|r}^{(i)} - C_{0|r}^{(i)}; \quad r = 0, \ldots, K; & i \neq k, (k, i) = 1, \ldots, K; \\ \lambda_{k|r}^{(i)} \lessgtr 0. & k \neq i, \text{ with } \lambda_{k|r}^{(0)} > 0, (k > 0); \lambda_{k|r}^{(k)} < 0, k \neq 0 \end{array}\right\} \tag{4.2.6c}$$

Next, following (4.1.8), we write

$$B_k(\boldsymbol{X}) \equiv \lambda_{k|0}^{(0)} + \sum_{i=1}^{K} \sum_{r=1}^{K} \lambda_{k|r}^{(i)} \Lambda(\boldsymbol{X})_{w_r}, \text{ where} \tag{4.2.7}$$

$$\Lambda_{i|r}(\boldsymbol{X}) = p_i \left\langle F_J\left(\boldsymbol{X}|\boldsymbol{S}^{(i)}(\theta)\right)\right\rangle_{w_r} \Big/ qF_J(\boldsymbol{X}|0),\ p_0 = q, \tag{4.2.7a}$$

which becomes for additive signal and noise

$$\Lambda_{i|r}(\boldsymbol{X}) = p_i \left\langle W_J\left(\boldsymbol{X} - \boldsymbol{S}^{(i)}(\theta)\right)\right\rangle_{w_r} \Big/ qW_J(\boldsymbol{X})_N, \tag{4.2.7b}$$

cf. (4.1.9a). Again, after some algebra we find that the average risk (4.2.5) can now be expressed as

$$R(\sigma, \delta) = R_{0K} + R_K, \text{ where } R_{0K} \equiv qC_0^{(0)} + \sum_{i=1}^{K} \sum_{r=1}^{K} p_r C_{0|r}^{(0)}, \text{ and} \tag{4.2.8a}$$

$$R_k = \int_\Gamma qF_J(\boldsymbol{X}|\boldsymbol{0}) \sum_{k=1}^{K} \delta(\gamma_k|\boldsymbol{X}) B_k(\boldsymbol{X}) d\boldsymbol{X}, \tag{4.2.8b}$$

cf. (4.1.10a,b,c). The quantity R_{0K} is once more the average cost of calling every signal, including noise alone, "noise," and R_K is that portion of the average risk which is adjusted by choice of the decision rules $\delta(\gamma_k|X)$. Once more, the form of the decision system is represented by the (sum of) likelihood ratios $B_k(X)$. The detection process, in turn, is determined by the choice of the δs and the corresponding regions in data space Γ as before (Section 4.1).

Our next task is to minimize the average risk,[13] that is, the Bayes risk $R^*(\sigma, \delta)$ is that risk for which R_K is a minimum, or $\sum_{k=1} B_k(X)$ is minimum, since $qF_J(X|\mathbf{0}) \geq 0$, all X, in Eq. (4.2.8b). Accordingly, we proceed to examine all the B_ks for a given X, selecting the one B_k which is algebraically least.[14] For this same data set X we then set $\delta(\gamma_k|X) = 1$, and $\delta(\gamma_\ell|X) = 0$, $\ell \neq k$. We next repeat the process for all X in Γ, to obtain at last a set of conditions on the $k = 1, \ldots, K$. (We note from the form of R_K, where the decision rules appear linearly, and from the mode of minimization, that the δ is automatically a nonrandom decision risk, that is, $\delta = 1$ or 0 only, cf. Eq. (1.7.4a) et seq.)

Since the optimum $B_k(X)^*$, like $A_k(X)$, (4.1.8), may contain negative parts, cf. (4.1.7), we see that minimizing the average risk (4.2.8b) and making a decision γ_k (signal of class k present in noise) on receipt of the data X, the specific conditions on the B_k are that the data X satisfy the following linear inequalities:

$$\left.\begin{array}{lll} B_k(X)^* \leq B_\ell(X)^*, & \text{or} & \lambda_{k|0}^{(0)} + \displaystyle\sum_{i=1}^{K}\sum_{r=1}^{K} \lambda_{k|r}^{(i)} \Lambda_{i|r}(X)_{w_r} \leq \lambda_{\ell|0}^{(0)} + \displaystyle\sum_{i=1}^{K}\sum_{r=1}^{K} \lambda_{\ell|r}^{(i)} \Lambda_{i|r}(X)_{w_r} \\ \text{and} & & \\ B_k(X)^* \leq 0, & \text{or} & \lambda_{k|0}^{(0)} + \displaystyle\sum_{i=1}^{K}\sum_{r=1}^{K} \lambda_{k|r}^{(i)} \Lambda_{i|r}(X)_{w_r} \leq 0, \text{ all } \ell \neq k, (\ell, k = 1, \ldots, K). \end{array}\right\} \tag{4.2.9a}$$

Here $\Lambda_{i|r}(X)_{w_r}$ is specifically given by (4.2.7) or (4.2.7a). For each decision $\gamma_k (k = 1, \ldots, K)$ in turn, we have

$$\delta(\gamma_k|X) = 1,\ \delta(\gamma_\ell|X) = 0,\ \text{all } \ell \neq k, (\ell = 0, \ldots, K). \tag{4.2.9b}$$

For the case of noise alone ($k = 0$), we find that the condition is

$$B_k(X)^* \leq 0, \quad \text{or} \quad \lambda_{k|0}^{(0)} + \sum_{i=1}^{K}\sum_{r=1}^{K} \lambda_{k|r}^{(i)} \Lambda_{i|r}(X)_{w_r} \geq 0, k = 1, \ldots, K, \tag{4.2.10a}$$

with

$$\delta(\gamma_0|X) = 1, \delta(\gamma_\ell|X) = 0, \quad \ell = 1, \ldots, K. \tag{4.2.10b}$$

Even though the signal classes overlap, they do so in the parameter space Ω_θ. This does not affect their behavior in data space Γ: the Λ_{i-w_r} may still be regarded as independent variables as before, Section 4.1, and treated accordingly. The difference is that the (nonoverlapping)

[13] This imposes conditions in the choice of costs functions, the *a priori* probabilities $p_i (i = 0, \ldots, K)$ being given. It is assumed that these conditions are met in practical applications, that is, the greatest costs are assigned to the least probable events, and no two (or more) events are equally probable.

[14] Again, this is subject to the condition that for a given X there exists a $B_k(X)$ that is algebraically less all the other $B_\ell(X), \ell \neq k$.

boundaries of the decision regions are more complex, exceeding the dimensionality K of the decision process.

For the simpler cases encountered in testing for the presence of one signal of class k in noise against any one of the other possible nonzero signals in noise, the case of noise alone is eliminated (cf. Section 4.1.3).[15] Here the costs $C_{0|r}^{(\ell)}, C_0^{(0)}, C_{\ell|r}^{(0)}$ are omitted in $B_k(X)$ above, and $\lambda_{k|r}^{(i)} = C_{k|r}^{(i)}, (k, i = 1, \ldots, K) \geq 0$. Thus, if $C_{k\{r}^{(i)} \geq 0, C_{i\{r}^{(i)} \geq 0$, the $B_k(X)$ can never be negative. Minimization of the average risk (4.2.8a) then yields only the first set of inequalities in (4.2.9a), with $\lambda_{k|r}^{(0)} = \lambda_{\ell|r}^{(i)} = 0$. The result is that we can write for the decision γ_k, the modified condition

$$\sum_{i=1}^{K}\sum_{r=1}^{K} C_{k|r}^{(i)}\Lambda_{i|r}(X)_{w_r} \leq \sum_{i=1}^{K}\sum_{r=1}^{K} C_{\ell|r}^{(i)}\Lambda_{i|r}(X)_{w_r}, \ \ell \neq k; \text{ all } \ell = 1, \ldots, K, \tag{4.2.11}$$

repeated for each $k (= 1, \ldots, K)$ in turn.

In summary, we observe that the optimum $(K+1)$-ary, or K-ary, detector is in effect a (more complex) computer which evaluates the $\Lambda_{i|r}(X)_{w_r}$ for a specified data set X over the observation interval $(\Delta \boldsymbol{R}, T)$. It then computes the various $B_k(X)$ and inserts the results into the series of inequalities (4.2.9a, 10a) or (4.2.11), finally making the decision γ_k for the single set of inequalities which is satisfied. This is, of course, not a unique process. The procedure sketched in Fig. 4.1, involving a sequence of intervening binary decisions to yield ultimately γ_k, is again one possibility, equally applicable to the case of overlapping signal classes.

4.2.3 Simple (K + 1) - ary Detection

As in Section 4.1.4.3, a simplified case of some interest and simplicity (though not so simple as the disjoint signal classes (Section 4.1)) occurs if we assign the *same* constant costs $C_0 (\geq 0)$ to all types of "failure" and zero costs to all types of "success." Equations (4.2.6a) and (4.2.6b) become $\lambda_{k|r}^{(0)} = C_0$, $\lambda_{k|r}^{(\ell)} = 0 (\ell \neq k)$, $\lambda_{k|r}^{(k)} = -C_0$, and the decision process (4.2.9) is governed now by the simpler inequalities

$$\text{Decide } \gamma_k : \text{if } \left.\begin{aligned} &\sum_{r=1}^{K}\Lambda_{k|r}(X)_{w_r} \leq \sum_{r=1}^{K}\Lambda_{\ell|r}(X)_{w_r} \\ &\sum_{r=1}^{K}\Lambda_{k|r}(X)_{w_r} > 1, \end{aligned}\right\} \text{all } \ell \neq k; (\ell = 1, \ldots, K), \tag{4.2.12}$$

for each $k = (1, \ldots, K)$ in turn. For noise alone $(k = 0)$, Eq. (4.2.10a) is the simple result

$$\text{Decide } \gamma_0 : \text{if } \sum_{r=1}^{K}\Lambda_{k|r}(X)_{w_r} \leq 1, \text{ all } k = 1, \ldots, K. \tag{4.2.13}$$

Again, the boundaries of the decision regions for each k overlap, and in general are the entire (hyperplane) $\Gamma \geq 0$. In the case of *K-ary detection* (i.e., "noise alone" is removed), Eq. (4.2.12) supplies the needed inequalities. (Because of the sum $\sum_{r=1}^{K}$ there is no

[15] This is again equivalent to choosing $\delta(\gamma_0|X) = 0 = 1 - \sum_{k=1}^{K}\delta(\gamma_k|X)$, cf. Eq. (4.1.3) in (4.1.6).

advantage *per se* in using the logarithm of the likelihood ratios: the result involves $\log\left(\sum_r^K\right)\Lambda_{k|r}(\boldsymbol{X})_{w_r}$, i.e., the logarithm of the sum rather than the sum of the logarithms [(4.1.25), (4.1.26), (4.1.27)].)

4.2.4 Error Probabilities, Average and Bayes Risk, and System Evaluations

To obtain the error probabilities we extend [(4.1.28) and (4.1.29) in a straightforward way:

$$\begin{aligned}\alpha_k^{(0)} \to \alpha_{k|0}^{(0)} &= \int_\Gamma \delta(\gamma_k|\boldsymbol{X})F_J(\boldsymbol{X}|\boldsymbol{0})d\boldsymbol{X} \\ &= \text{conditional probability of calling a null signal} \\ &\quad\ \text{any one member of the } k\text{th signal class, } k=(1,\ldots,K);\end{aligned} \tag{4.2.14a}$$

$$\begin{aligned}\left\langle\beta_\ell^{(k)}\right\rangle \to \left\langle\beta_\ell^{(k)}\right\rangle_{k|r} &= \int_\Gamma \delta(\gamma_\ell|\boldsymbol{X})\left\langle F_J\left(\boldsymbol{X}|\boldsymbol{S}^{(k)}(\theta)\right)\right\rangle_{w_r} d\boldsymbol{X} \\ &= \text{conditional probability of calling any one member of the} \\ &\quad\ k\text{th signal class a member of the } \ell\text{th class, subject to the} \\ &\quad\ r\text{th pdf, } w_r(\theta)\text{, of the parameters, } k=(1,\ldots,K)\text{, where} \\ &\quad\ \ell\neq k;\ \ell=0,1,\ldots,K;\ k=1,\ldots,K.\end{aligned} \tag{4.2.14b}$$

Similarly, the conditional probability of *correctly deciding* that any one signal member (including the null signal) of class k, $(k=0,1,\ldots,K)$, is present in noise, is

$$\left\langle\eta_k^{(k)}\right\rangle_k \to \left\langle\eta_k^{(k)}\right\rangle_{k|r} = \int_\Gamma \delta(\gamma_k|\boldsymbol{X})\left\langle F_J\left(\boldsymbol{X}|\boldsymbol{S}^{(k)}\right)\right\rangle_{w_r} d\boldsymbol{X} \tag{4.2.15}$$

and

$$\left\langle\eta_k^{(k)}\right\rangle_k = 1 - \sum_{\ell=0}^{K}\left\langle\beta_\ell^{(k)}\right\rangle_k \to \sum_{\ell=0}^{K}\sum_{r=0}^{K}\left\langle\beta_\ell^{(k)}\right\rangle_{k|r} \tag{4.2.16}$$

and $k=(1,\ldots,K)$. In the case where "noise alone" is removed, (4.2.14b) becomes $\left\langle\beta_\ell^{(k)}\right\rangle_{k|r}$, $(\ell\neq k; r,\ell,k=1,\ldots,K)$. With (4.2.14a)–(4.2.16), we can write the average risk (4.2.5), minimized or not minimized, as

$$\begin{aligned}R_K(\sigma,\delta)^{(*)} &= R_{0K} + q\sum_{k=1}^{K}\lambda_{k|0}^{(0)}\alpha_{k|0}^{(0)(*)} \\ &\quad + \sum_{k=1}^{K}{}' \sum_{\ell=0,k\neq\ell;r=0}^{K}\lambda_{\ell|r}^{(k)}p_r\left\langle\beta_\ell^{(k)}\right\rangle_{k|r}^{(*)} + \sum_{k=1}^{K}\lambda_{k|r}^{(k)}\sum_{r=1}^{K}p_r\left\langle\eta_k^{(k)}\right\rangle_{k|r}^{(*)}\end{aligned} \tag{4.2.17}$$

where the $(*)$ denotes the minimum or Bayes conditional probabilities and the prime ($'$) on the second sum indicates that $k\neq\ell$. For the excluded null signal, the terms in $\lambda_{k|r}^{(0)}$ are omitted, and $R_{0K}=0$, with $\lambda_{k|r}^{(\ell)}$, $(k,\ell\geq 1)$, equal to the costs $C_{k|r}^{(\ell)}$.

Finally, the error probabilities here are calculated in much the same way as for the disjoint classes [(4.1.31)–(4.1.36)], except that the averages (over the parameters θ) are computed with respect to $w_r = w_r(\theta)$. Thus, $Q_{J|0}^{(0)}, P_{J|r}^{(k)}$ are now the joint probability densities of the likelihood ratios $y_i = \Lambda_i(\boldsymbol{X})_{w_r}, (i = 1, \ldots, K)$, which are found from the fact that $\delta(\gamma_k|\boldsymbol{X}) = \delta\big(y_k - \Lambda_{k|r}(\boldsymbol{X})\big)$. We have

$$H_0 : Q_J^{(0)}(\boldsymbol{y})_{r=0}^{*} = \int_{\Gamma} \cdots \int F_J(\boldsymbol{X}|\boldsymbol{0}) \cdot \prod_{\ell=1}^{K} \delta\big(y_\ell - \Lambda_{\ell|0}(\boldsymbol{X})\big) d\boldsymbol{X};\ \boldsymbol{y} = [y_1, \ldots y_\ell, \ldots y_K]; \tag{4.2.18a}$$

$$H_k : P_J^{(k)}(\boldsymbol{y})_r^{*} = \int_{\Gamma} \cdots \int \Big\langle F_J\Big(\boldsymbol{X}|\boldsymbol{S}^{(k)}(\theta)\Big)\Big\rangle_{w_r} \prod_{\ell=1}^{K} \delta\big(y_\ell - \Lambda_{\ell|r}(\boldsymbol{X})\big) d\boldsymbol{X}, \tag{4.2.18b}$$

where

$$\begin{aligned} \mathsf{F}_K(i\xi)_{Q|0} &= \int_{\Gamma} \cdots \int e^{i\tilde{y}\xi} F_J(\boldsymbol{X}|\boldsymbol{0}) d\boldsymbol{X}; \\ \mathsf{F}_K(i\xi)_{P^{(k)}|r} &= \int_{\Gamma} \cdots \int e^{i\tilde{y}\xi} \Big\langle F_J\Big(\boldsymbol{X}|\boldsymbol{S}^{(k)}\Big)\Big\rangle_{w_r} d\boldsymbol{X} \end{aligned} \tag{4.2.18c}$$

are the characteristic functions of $\boldsymbol{y}$, where $\langle\ \rangle_{w_r}$ is the average over the parameter space appropriate to $w_r(\theta)$. Consequently, we have

$$\alpha_{k|0}^{(0)*} = \int_{[y_1]\cdots[y_K]} \cdots \int Q_J^{(0)}(y_1, \ldots, y_K)_{r=0}^{*} \, dy_1 \cdots dy_K, \tag{4.2.19a}$$

and

$$\Big\langle \beta_\ell^{(k)} \Big\rangle_{k|r}^{(0)*} = \int_{[y_1]\cdots[y_K]} \cdots \int P_J^{(k)}(y_1, \ldots, y_K)_{r>0}^{*} \, dy_1 \cdots dy_K, \tag{4.2.19b}$$

where $[y_K]$ represents the decision regions for the $y_k, k = 1, \ldots, K$.

For suboptimum systems $z_{\ell|r} = G_{\ell|r}(\boldsymbol{X})$ instead of $\mathsf{F}_{\ell|r} = \Lambda_{\ell|r}(\boldsymbol{X})$, and the results of [(4.2.18) and (4.2.19)] are modified to

$$\beta_{k|0}^{(0)} = \int_{[z_1]\cdots[z_K]} \cdots \int q_J^{(0)}(z_1, \ldots, z_K)_{r=0} \, dz_1 \cdots dz_K \quad \left(> \alpha_{k|0}^{(0)*}\right), \tag{4.2.20a}$$

$$\Big\langle \beta_\ell^{(k)} \Big\rangle_{k|r} = \int_{[z_1]\cdots[z_K]} \cdots \int p_J^{(k)}(z_1, \ldots, z_K)_{r>0} dz_1 \cdots dz_K \quad \left(> \Big\langle \beta_\ell^{(k)} \Big\rangle_{k|r}^{*}\right), \tag{4.2.20b}$$

and $[z]$ in Eqs. (4.2.20a) and (4.2.20b) represents the decision regions for the $z_1, \ldots, z_K$. The $q_{J|r}^{(0)}$ and $p_{J|r}^{(k)}$ are,[16] as before, the joint pdfs of the random variables $G_{\ell|r}(\boldsymbol{X}), \ell = 1, \ldots, K$.

[16] See, for example, early paper of Chow [3].

These pdfs are likewise given in terms of the received data processes $\boldsymbol{X}$, analogous to (4.2.18), by

$$q_J^{(0)}(z_1, \ldots, z_K)_{r=0} = \int_\Gamma F_J(\boldsymbol{X}|\boldsymbol{0}) \prod_{i=1}^K \delta\big(z_1 - G_{i|r}(\boldsymbol{X})\big) \boldsymbol{dX} \neq Q_{J|r=0}^{(0)*}, \tag{4.2.21a}$$

$$p_J^{(k)}(z_1, \ldots, z_K)_r = \int_\Gamma \left\langle F_J\left(\boldsymbol{X}|\boldsymbol{S}^{(k)}\right)\right\rangle_{W_r} \prod_{i=1}^K \delta\big(z_1 - G_{i|r}(\boldsymbol{X})\big) \boldsymbol{dX} \neq P_{J|r}^{(k)*}, \tag{4.2.22a}$$

with the associated c.f.s (4.2.18c) $F_k(i\boldsymbol{\xi})_{q|0}$, $F_k(i\xi)_{P^{(k)}|r}$, where now the exponential is $\exp(i\tilde{z}\xi)$. The average risk (and the Bayes or minimum average risk) is provided by (4.2.17), respectively with ( ) and (*).

In principle, comparing the error probabilities with various constraints, that is, Neyman–Pearson and Ideal Observer restrictions, extended to particular signals for example, permits system comparisons: nonideal systems [(4.2.20a) and (4.2.20b)] versus ideal [(4.2.19a) and (4.2.19b)]. However, even in the cases of normal noise processes and simplified costs (4.2.11), (4.2.12), and (4.2.13), the actual calculations are very difficult if not impossible analytically, with the result that numerical computation is required. An exception of some interest is given in the example of Section 4.1.6. Another exception, of much greater generality, is the important case of threshold operation, with independent sampling.

4.3 DETECTION WITH DECISIONS REJECTION: NONOVERLAPPING SIGNAL CLASSES

In the usual procedure described above, if a data sample $\boldsymbol{X}'$ is such that the inequalities (4.2.11) or (4.2.12) are satisfied for a particular k, that is, if the data "point" $\boldsymbol{X}'$ falls in the decision region associated with γ_k, we make the decision that a signal of Class k is present. However, it may be that the point $\boldsymbol{X}'$ falls too close to the boundaries of the acceptance region for γ_k, that is, for the given cost assignments the probability of decision error is too large, so that the resulting decision is actually incorrect. Because of the background noise, the point $\boldsymbol{X}'$, which really belongs in the region for $\gamma_\ell(\ell \neq k)$, say, has landed in γ_k. Of course, we can realign the boundaries to diminish this effect—that is, decrease the probabilities of error—by a readjustment of the various costs involved, when these are at our disposal. But frequently we do not have this option: not only are the costs preassigned (in any case), but these preassigned values are inflexible for other reasons. In such circumstances, we may then decide to reject all excessively doubtful decisions. Thus, we introduce a *new set of decisions*, namely, decisions to reject the decisions that particular signals are present. The purpose of this rejection procedure is to guard further against the penalties of wrong decisions, by substituting "blanks" or other indications of rejection in the noticeably doubtful cases. Accordingly, a typical rejection of a signal of Class k might be represented by $S_{\mathrm{R}}^{(k)}$, where both the signal rejected and the fact of rejection are available at the output of the receiver. Rejection procedures have application in coding for communication and computation whenever it is advantageous to take additional precautions against decision errors. The expense of incorrect decisions is, of

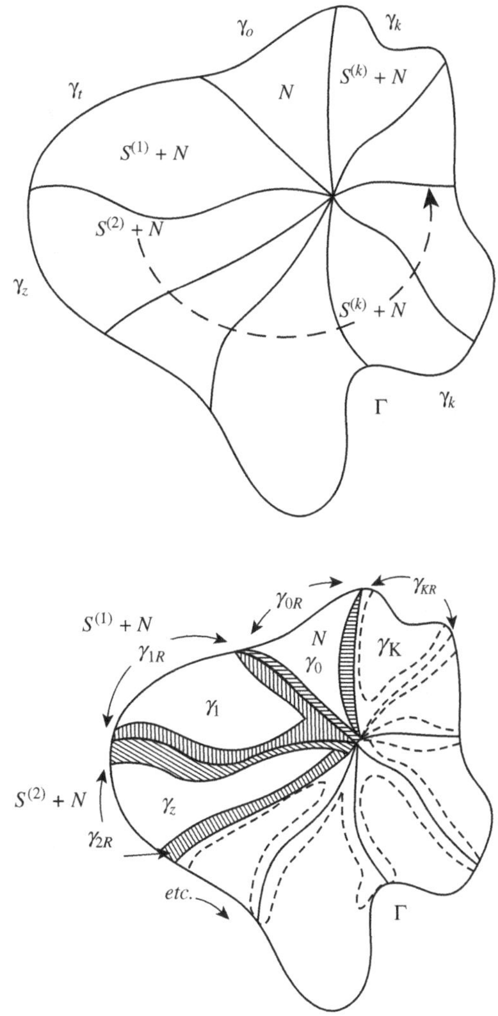

FIGURE 4.7 (a) Data space Γ, showing decision regions without rejection. (b) Same as part (a), but with rejection regions.

course, reduced at the lesser expense[17] of a lower positive decision rate. Rejection also occurs in estimation when $p(H_1) < 1$, that is, the desired signal is not known surely to be present $(p(= H_1))$; see Chapters 6 and 7 ff.

We can illustrate this schematically with the simple diagrams of Fig. 4.7a and b. These show the data space Γ divided into various regions. For the nonrejection case, there are $K+1$ distinct, nonoverlapping zones, corresponding to the decisions to accept noise or $S^{(k)}$ in noise, for example, N, $S^{(1)}+N$, $S^{(2)}+N, \ldots, S^{(K)}+N$, while in the rejection case $K+1$ additional zones have been introduced which represent the regions where acceptance of a

[17] This depends on the cost assignments of rejection and acceptance; for meaningful operation, it is clear that rejection is to be assigned a smaller cost than a decision error.

particular $S^{(k)}$ in noise (or noise alone) is excessively doubtful. Thus, if a particular data point $\boldsymbol{X}'$ falls too close to an original acceptance boundary (solid lines in Fig. 4.7a and b), our decision is to reject the original decision, whether or not it is correct or false. With this in mind, we can readily extend the analysis of Section 4.1.1 to include the additional $K+1$ decisions with rejection.

4.3.1 Optimum (K + 1) - ary Decisions with Rejection

We begin first with the cost assignments. Let us write

$C_k^{(\ell)} =$ cost of *incorrectly* deciding $\boldsymbol{S}^{(k)}$ is present when $\boldsymbol{S}^{(\ell)}$ actually occurs and of accepting the decision [$\ell \neq k$; $\ell, k = 0, \ldots, K$ ($\ell = 0, k = 0$ are "noise alone" states)];

$C_\ell^{(\ell)} =$ cost of *correctly* deciding a signal of Class ℓ is present when such a signal actually occurs and of accepting the decision ($\ell = 0, \ldots, K$);

$C_{kR}^{(\ell)} =$ cost of deciding $\boldsymbol{S}^{(k)}$ is present when $\boldsymbol{S}^{(\ell)}$ really occurs and of *rejecting* this decision (whether correct or incorrect).

A meaningful interpretation of these costs and of the subsequent decision procedures requires that

$$C_\ell^{(\ell)} < C_{\ell R}^{(\ell)} \leq C_{kR}^{(\ell)} < C_{k(\neq \ell)}^{(\ell)}, \quad \ell, k = 0, \ldots, K, \tag{4.3.1}$$

that is, the cost of a correct decision is less that the cost of a rejection, which in turn is less than the cost of an error.

Modifying Eqs. (4.1.7) to

$$\lambda_k^{(\ell)} = C_k^{(\ell)} - C_o^{(\ell)} \quad \lambda_{kR}^{(\ell)} = C_{kR}^{(\ell)} - C_o^{(\ell)}, \quad \ell, k = 0, \ldots, K, \tag{4.3.2a}$$

and setting

$$\gamma_k' = \begin{cases} \gamma_k & , \quad k = 1, \ldots, K \\ \gamma_{(k-K)\mathrm{R}} & , \quad k = K+1, \ldots, 2K+1, \end{cases} \tag{4.3.2b}$$

where $\gamma_{(K+1)\mathrm{R}} =$ decision to reject the choice of "noise alone," we extend the condition (4.1.3) to

$$\sum_{k=0}^{K} \delta(\gamma_k|\boldsymbol{X}) + \sum_{K+1}^{2K+1} \delta\left(\gamma_{(k-K)R}|\boldsymbol{X}\right) = 1. \tag{4.3.2c}$$

Then, following the steps leading to Eq. (4.1.10c), we obtain the rearranged version of the average risk,

$$R(\sigma, \delta)_{2K+1} = (R_0)_{2K+1} + \int_\Gamma \left[\sum_{k=1}^{2K+1} \delta(\gamma_k'|\boldsymbol{X}) A_k(\boldsymbol{X})\right] qF_J(\boldsymbol{X}|\boldsymbol{0})d\boldsymbol{X}, \tag{4.3.3}$$

where now

$$
\begin{aligned}
A_k(\boldsymbol{X}) &= \lambda_k^{(0)} + \sum_{\ell=1}^{K} \lambda_k^{(\ell)} \Lambda_\ell(\boldsymbol{X}), & k = 1, \ldots, K, \\
&= \lambda_{(k-K)\mathrm{R}}^{(0)} + \sum_{\ell=K+1}^{2K+1} \lambda_{(k-K)\mathrm{R}}^{(\ell-K)} \Lambda_{(\ell-K)}(\boldsymbol{X}), & k = K+1, \ldots, 2K+1.
\end{aligned}
\tag{4.3.3a}
$$

The generalized likelihood ratios Λ_ℓ, $\Lambda_{(\ell-K)}$ are given as before by Eqs. (4.1.9), (4.1.9a) with

$$
q + \sum_{\ell=1}^{K} p_\ell = 1 \text{ [cf. Eq. (4.1.2)].} \tag{4.3.3b}
$$

For optimum detection, we minimize the average risk [Eq. (4.3.3)], again by suitable choices of the decision rules, as in the nonrejection case above. The result is the following set of decision procedures:

$$
\begin{aligned}
&\text{Decide } \gamma'_k (k = 1, \ldots, 2K+1), \text{ if} \\
&\lambda_k^{(0)} + \sum_{\ell=1}^{2K+1} \lambda_k^{(\ell)} \Lambda_\ell(\boldsymbol{X}) \le \lambda_i^{(0)} + \sum_{i=1}^{2K+1} \lambda_i^{(\ell)} \Lambda_\ell(\boldsymbol{X}), \\
&\lambda_k^{(0)} + \sum_{\ell=1}^{2K+1} \lambda_k^{(\ell)} \Lambda_\ell(\boldsymbol{X}) \le 0, \quad (\text{all } i \ne k (i = 1, \ldots, 2K+1)).
\end{aligned}
\tag{4.3.4a}
$$

Then set $\delta(\gamma'_k|\boldsymbol{X}) = 1$ and $\delta(\gamma'_i|\boldsymbol{X}) = 0 (i \ne k)$. For noise alone

$$
\text{Decide } \gamma'_0, \text{ if } \lambda_k^{(0)} + \sum_{\ell=1}^{2K+1} \lambda_k^{(\ell)} \Lambda_\ell(\boldsymbol{X}) \ge 0, \text{ all } k = 1, \ldots, 2K+1. \tag{4.3.4b}
$$

Similarly, $\delta(\gamma'_0|\boldsymbol{X}) = 1$, $\delta(\gamma'_k|\boldsymbol{X}) = 0 (k \ge 1)$. Here we adopt the convention that

$$
\Lambda_\ell|_{\ell > K} = \Lambda_{(\ell-K)}; \quad \Lambda_k^{(\ell)}|_{k \text{ and/or } \ell > K} = \lambda_{(k-K)\mathrm{R}}^{(\ell) \text{ or } (\ell-K)}. \tag{4.3.4c}
$$

Accordingly, all decisions for $k \ge K+1$ represent *rejections* of the decision $k = 1, \ldots, K, 0$, respectively. As in the nonrejection case, the boundaries of the decision regions are obtained from the equalities in Eqs. (4.3.4a) and (4.3.4b).

4.3.2 Optimum $(K+1)$ - ary Decision with Rejection

Here we remove the null signal class: a signal is always present in noise. From Eq. (4.1.14), by an obvious extension of the analysis we can write the decision procedures as

$$
\text{Decide } \gamma'_k, \text{ if } \sum_{\ell=1}^{2K+1} \lambda_k^{(\ell)} \Lambda_\ell(\boldsymbol{X}) \le \lambda_i^{(0)} + \sum_{\ell=1}^{2K+1} \hat{\lambda}_i^{(\ell)} \Lambda_\ell(\boldsymbol{X}), \; i \ne k, \text{ all } i = 1, \ldots, 2K, \tag{4.3.5}
$$

where now

$$\begin{aligned}
\hat{\lambda}_k^{(\ell)} &= C_k^{(\ell)} && \ell, k = 1, \ldots, K \\
&= C_k^{(k)} && k = 1, \ldots, K \\
&= C_{(k-K)\mathrm{R}}^{(\ell)} && k = K+1, \ldots, 2K \\
&= C_{(k-K)\mathrm{R}}^{(\ell-K)} && \ell, k = K+1, \ldots, 2K \\
&= C_{k\mathrm{R}}^{(\ell-K)} && \ell \neq k \neq K, \text{ with } C_{(k-K)\mathrm{R}}^{(\ell)} = C_{k\mathrm{R}}^{(\ell-K)} = C_{(k-K)\mathrm{R}}^{(\ell-K)}.
\end{aligned} \tag{4.3.5a}$$

The likelihood ratios Λ_ℓ still obey Eqs. (4.1.9), (4.1.9a), except that $q(>0)$ is arbitrary, and the condition (4.1.2) becomes $\sum_1^K p_k = 1$. The convention of Eq. (4.3.4c) also applies, while $A_k = \sum_{\ell=1}^{2K} \hat{\lambda}_k^{(\ell)} \Lambda_\ell(\boldsymbol{X})$ [Eq. (4.1.8)]. The boundaries of the various decision regions are determined by the equalities in Eq. (4.3.5).

4.3.3 A Simple Cost Assignment

Let us consider as an example the situation where the null signal class is omitted, so that Eq. (4.3.5) applies. We set C_R = cost of rejection, C_E = cost of an error, and zero the cost of a correction decision that is accepted. Accordingly, Eq. (4.3.5a) becomes

$$\begin{aligned}
\hat{\lambda}_k^{(k)} &= 0 \\
\hat{\lambda}_k^{(\ell)} &= C_\mathrm{E} \qquad \ell \neq k, \ell = 1, \ldots, 2K; k = 1, \ldots, K \\
\hat{\lambda}_k^{(\ell)} &= C_\mathrm{R} \quad k = K+1, \ldots, 2K, \text{ with } 0 < C_\mathrm{R} < C_\mathrm{E}.
\end{aligned} \tag{4.3.6}$$

The decision rules [Eq. (4.3.5)] reduce explicitly to

Acceptance

Decide that some one $\boldsymbol{S}^{(k)} (k = 1, \ldots, K)$ is present and accept the decision, if

$$\begin{aligned}
&\Lambda_i \leq \Lambda_k && \text{(for some } k \text{(each } i = 1, \ldots, K; i \neq k)); \\
&0 \leq \frac{C_\mathrm{R} - C_\mathrm{E}}{C_\mathrm{E}} \sum_{\ell=1}^{K} \Lambda_\ell + \Lambda_k && \text{for this } k.
\end{aligned} \tag{4.3.7a}$$

Rejection

Reject a signal $\boldsymbol{S}^{(k)} (k = 1, \ldots, K)$, if

$$0 \geq \frac{C_\mathrm{R} - C_\mathrm{E}}{C_\mathrm{E}} \sum_{\ell=1}^{K} \Lambda_\ell + \Lambda_k. \tag{4.3.7b}$$

for the k above for which $\Lambda_i \leq \Lambda_k$ applies $(i \neq k)$. Once more, the equalities in Eqs. (4.3.5a) establish the boundaries of the various decision regions.

4.3.4 Remarks

As in the nonrejection situations of Section 4.1.1, the elements of the system are the likelihood ratios $\Lambda_\ell(\ell = 1, \ldots, K, 0)$ as shown in Fig. 4.1, albeit they are combined in a somewhat different fashion because of the rejection operations. In the limiting case of threshold signals, the structure once again involves a generalized (averaged) cross- or autocorrelation of the received data with itself. Moreover, if the background noise is normal, we may use the results of Chapter 3 to obtain the specific structures of the Λ_ℓ in such cases. The error probabilities and Bayes risk may be found in principle from Eqs. (4.1.31) and (4.1.32), although explicit calculations are usually difficult because of the nature of the integrands and the limits. For the simple cost assignments of Sections 4.1.3 and 4.3.3, when the *a priori* probabilities q, p_i are not known to the observer, the corresponding Minimax system can be shown to have the same structures as the above, where now $q = p_i = 1/(K+1)(\text{all } i)$: the p_is are "uniformly" distributed.

Finally, we may extend the results of Section 4.3 involving decision rejection to the case where the signal classes overlap, generalizing the analysis of Section 4.2. These results, though somewhat tedious, are straightforward, and are left to the reader (Problem P4.7).

PROBLEMS

P4.1 **(a)** Carry out the details leading from Eq. (4.1.6) to Eq. (4.1.10c).

(b) Obtain Eqs. (4.1.38)–(4.1.40) for the simple cost assignment.

(c) Establish Eqs. (4.3.7a) and (4.3.7b) for the simple cost assignment, where there is decision rejection.

P4.2 Show that in the multiple-alternative detection case of Section 4.1.1, the actual amount of information conveyed, on the average, is

$$H_{\mathrm{T}} = -H(\sigma, \delta) - q \log q - \sum_{i=1}^{K} p_i \log p_i. \tag{1}$$

Verify that the equivocation $H(\sigma, \delta)$ is

$$H(\sigma, \delta) = -\sum_{i=0}^{K} \sum_{\ell=0}^{K} P(\gamma_i, S_\ell) \log[P(\gamma_i, S_\ell)/P(\gamma_i)]. \tag{2}$$

where

$$P(\gamma_i, S_\ell) = \begin{cases} p_\ell \beta_i^{(\ell)} & \ell \neq 0 \\ \text{or} & \\ q\alpha_i^{(0)} & \ell = 0 \end{cases} \tag{3}$$

and

$$P(\gamma_i) = \begin{cases} q\alpha_i^{(0)} + p_i\left(1 - \sum_{\ell \neq i} \beta_k^{(i)}\right) + \sum_{\substack{\ell \neq i \\ \ell \neq 0}} p_{\ell i}^{(\ell)} & i \neq 0 \\ \text{or} & \\ \sum_{\ell \neq 0} p_\ell \beta_0^{(\ell)} + q\left(1 - \sum_{k \neq 0} \alpha_k^{(0)}\right) & i = 0. \end{cases} \tag{4}$$

P4.3 **(a)** A signal is present additively with noise in K channels, or there are only the various noises in these K channels. Reception is coherent, and the signal class has only one member per channel. If S_k is the version of the signal in the k channel, and if the noises are all independent normal processes, show that the optimum detector now uses the data $\boldsymbol{X}_1, \ldots, \boldsymbol{X}_K$ from the K channels according to

$$\log \Lambda_J(X_1,\ldots,X_K) = \log \mu - {}^1\!/\!_2 \sum_{k=1}^{K} a_{0k}{}^2 \Phi_k(s) + \sum_{k=1}^{K} a_{0k} \Phi_k(x,s) \gtreqless \log K, \quad (1)$$

$$\text{where } \Phi_k(s) \equiv \tilde{\boldsymbol{s}}_k \left(\boldsymbol{k}_{\mathrm{N}}^{-1}\right)_k \boldsymbol{s}_k \quad \Phi_k(x,s) \equiv \tilde{\boldsymbol{x}}_k \left(\boldsymbol{k}_{\mathrm{N}}^{-1}\right)_k \boldsymbol{s}_k \quad (1\mathrm{a})$$

in which s_k, x_k are, respectively, the (normalized) signal and data vectors in the kth channel and $(\boldsymbol{k}_{\mathrm{N}})_k = (\boldsymbol{K}_{\mathrm{N}})_k / \psi_k$ is the (normalized) covariance matrix of the noise in this channel. As before, K is a threshold, and $a_{0k}^2 = A_{0k}^2/2\psi_k$, with A_{0k} the amplitude of $S_k(t)$. Thus, one simply takes the outputs $\log \Lambda_J(\boldsymbol{X}_\ell) - (1/K)\log \mu + {}^1\!/\!_2 a_{0k}{}^2 \Phi_\ell(s) = a_{0\ell}\Phi_\ell(x,s)$, which are the weighted cross-correlations of [Eq. (1a)] of s_k with x_k, and combines them additively to get Eq. (1), which is then compared with the threshold in the usual way.

(b) Show, next, that the error probabilities of this Bayes system are for all signal levels

$$\begin{Bmatrix} \alpha^* \\ \beta^* \end{Bmatrix} = \frac{1}{2}\left(1 - \Theta\left\{\frac{\left[\sum a_{0k}{}^2 \Phi_k(s)\right]^{1/2}}{2\sqrt{2}} \pm \frac{\log(K/\mu)}{\sqrt{2}\left[\sum a_{0k}{}^2 \Phi_k(s)\right]^{1/2}}\right\}\right). \quad (2)$$

In the limiting situation where $s_k = s$, $a_{0k} = a_0$, and $(\boldsymbol{k})_k = \boldsymbol{k}_{\mathrm{N}}$, (all k), we accordingly obtain

$$\begin{Bmatrix} \alpha^* \\ \beta^* \end{Bmatrix} = \frac{1}{2}\left(1 - \Theta\left\{\frac{a_0 \Phi_s(s)^{1/2}\sqrt{K}}{2\sqrt{2}} \pm \frac{\log(K/\mu)}{\sqrt{2}a_0\Phi_s(s)^{1/2}\sqrt{K}}\right\}\right) \quad (3)$$

showing that in effect we have increased the "size" of our statistical sample: $a_0\Phi_s^{1/2}$ of the single-channel case is made larger by the factor $\sqrt{K}$. Thus, the minimum detectable signal (on an *rms* basis) *is decreased by* $\sqrt{K}$. Here Θ is the *erf* defined on p. 369.

P4.4 **(a)** Repeat Problem 4.3, but now for incoherent reception (where $\langle s \rangle_\varepsilon = 0$), and where s_k, s_J are essentially statistically independent (like the noises) from channel to channel. Hence, show that the optimum detector structure in the threshold case is

$$\log \Lambda_J(\boldsymbol{X}_1, \ldots, \boldsymbol{X}_K) \doteq \log \mu + \sum_{k=1}^{K} \left(B_{Jk}^{(2)} + B_{Jk}^{(4)}\right) + {}^1\!/\!_2 \sum_{k=1}^{K} \overline{a_{0k}{}^2} \tilde{\boldsymbol{x}}_k \bar{\boldsymbol{G}}_k \boldsymbol{x}_k \quad (1)$$

with $\bar{G}_k = \left(\boldsymbol{k}_{\mathrm{N}}^{-1}\right)_k \boldsymbol{s}_k \tilde{\boldsymbol{s}}_k \left(\boldsymbol{k}_{\mathrm{N}}^{-1}\right)_k$.

(b) Obtain the Bayes error probabilities

$$\begin{Bmatrix} \alpha^* \\ \beta^* \end{Bmatrix} \cong \frac{1}{2}\left(1 - \Theta\left\{\frac{\left(\sum \overline{a_{0k}^2}^2 \langle \Phi_{G_k} \rangle\right)^{1/2}}{4} \pm \frac{\log(K/\mu)}{\left(\sum \overline{a_{0k}^2}^2 \langle \Phi_{G_k} \rangle\right)^{1/2}}\right\}\right) \qquad (2)$$

where $\langle \Phi_{G_k} \rangle = \langle \tilde{s}_k \bar{G}_k s \rangle$. Hence, in the special case $s_k = s$, $a_{0k} = a_0$, $(k_N)_k = k_N$, and so on, we get

$$\begin{Bmatrix} \alpha^* \\ \beta^* \end{Bmatrix} \cong \frac{1}{2}\left(1 - \Theta\left\{\frac{\overline{a_0^2} \langle \Phi_{G_k} \rangle^{1/2} \sqrt{K}}{4} \pm \frac{\log(K/\mu)}{\overline{a_0^2} \langle \Phi_{G_k} \rangle^{1/2} \sqrt{K}}\right\}\right). \qquad (3)$$

Thus, the improvement (on an *rms* basis) of the minimum detectable signal with K independent channels (all containing the same signal), over the single channel case is $\sqrt[4]{K}$*for incoherent threshold reception.*

P4.5 Given the two sets of data $X_1 = a_0 S_1 + N_1$, $X_2 = a_0 S_2 + N_2$, where N_1, N_2 are normal processes that may be statistically related, if a_0 is normally distributed, show that with a quadratic cost function the Bayes estimator of a_0 is (for sampled data)

$$a_0^*(X_1, X_2) = \frac{\bar{a}_0/\sigma^2 + \tilde{x}_1 k_1^{-1} s_1 + \tilde{x}_1 G_1 s_1 + \tilde{x}_2 G_2 s_2 - \tilde{x}_1 G_3 s_2 - \tilde{x}_{21} G_3 s_1}{1/\sigma^2 + \tilde{s}_1 k_1^{-1} s_1 + \tilde{s}_1 G_4 s_1 - 2\tilde{s}_1 G_3 s_2 + \tilde{s}_2 G_5 s_1} \qquad (1)$$

where

$$\begin{aligned}
G_1 &= \tilde{k}_{12} k_1^{-1} \left(k_2 - k_{21} k_1^{-1} k_{12}\right)^{-1} k_1^{-1} k_{12} \qquad \eta^2 = \psi_2/\psi_1 \\
G_2 &= \eta^{-2} \left(k_2 - k_{21} k_1^{-1} k_{12}\right)^{-1} \\
G_3 &= \tilde{k}_{12} k_1^{-1} \left(k_2 - k_{21} k_1^{-1} k_{12}\right)^{-1} \\
G_4 &= \tilde{k}_{21} k_1^{-1} \left(k_2 - k_{21} k_1^{-1} k_{12}\right)^{-1} k_1^{-1} k_{12} \\
G_5 &= \eta^{-2} \left(k_2 - k_{21} k_1^{-1} k_{12}\right)^{-1}
\end{aligned} \qquad (2)$$

and the ks are normalized auto- and cross-covariance matrices between N_1 and N_2. (Compare these results with those of Section 4.3.1.)

P4.6 **(a)** Show that for multiple-estimation procedures the Bayes risk $R_\ell^{*(k)}$ is always equal to or less than R_ℓ^* using only a single data set X_ℓ.

(b) Extend the treatment of Section 4.1.3 to the simple cost function.

P4.7 Carry out the details of the evaluation in Section 4.2 for overlapping signal classes.

REFERENCES

1. D. Middleton and D. Van Meter, On Optimum Multiple-Alternative Detection of Signals in Noise, *IRE Trans. Inform. Theory*, **IT-1**(1), 1–9, 1955.
2. D. Middleton, *An Introduction to Statistical Communication Theory*, 2nd Reprint ed. [IEEE Press *Classic Reissue*], IEEE Press, Piscataway, NJ, 1996.
3. G. K. Chow, An Optimum Character Recognition System Using Decision Functions, *IRE Trans. Electron. Comput.*, **EC-6**, 267, 1957.
4. C. W. Helstrom, *Elements of Signal Detection and Estimation*, Sections 3.5.4, 10.1.3 10.4, Prentice-Hall Englewood Cliff, NJ, 1995.
5. M. Schwartz, W. R. Bennett, and S. Stein, *Communication Systems and Techniques*, Section 2.7, Chapters 10, 11, McGraw-Hill, New York, 1966.

5

BAYES EXTRACTION SYSTEMS: SIGNAL ESTIMATION AND ANALYSIS, $p(H_1) = 1$

In Chapters 2 and 3, we have considered binary detection systems, and in Chapter 4, multiple alternative (K-ary) detection procedures, where the central problem is to determine the presence or absence of a signal in noise. Here, on the other hand, the desired signal is known *a priori* to be present at the receiver, and we are concerned with a somewhat more general problem: that of determining the explicit structure or the descriptive parameters of a signal in noise when these are unknown to the observer. Typical examples of communication interest are the measurement of the presence, the location, and the velocity (Doppler) of a moving target by a radar or a sonar; the calculation of signal amplitude, frequency, delay, and other information-bearing features of the signal process in telephony. This also includes the extraction of waveform data in the analysis and identification of signal structures in various other communication operations. In all such cases we have to deal with a *measurement situation*: our decisions now are not the simple "yes" or "no" of binary or K-ary detection but are instead specific numbers, associated with the signal process in question. Furthermore, in practical cases we do not have "pure" data upon which to operate, but data corrupted by noise. As in detection, since observation time is limited, that is, since the available samples from which we obtain our measurement are finite, we can expect only imperfect estimates of the desired quantities. Some will be better than others, depending on our methods for reducing the effects of the interfering noise. The principal aim of the present chapter, accordingly, is to find and examine systems $\boldsymbol{T}_R$ that have this desirable property of yielding optimum estimates.

A receiving system $\boldsymbol{T}_R$ that measures the waveform or one or more of the parameters of an incoming signal ensemble we call a *signal extraction system* and the process itself *signal extraction*. As we noticed in Section 1.2.2, signal extraction is a form of statistical estimation (e.g., point estimation) analogous to signal detection as a form of hypothesis testing. Both, of

---

Non-Gaussian Statistical Communication Theory, David Middleton.

course, fall within the domain and techniques of decision theory (Section 1.2). If we define *signal analysis* to include methods of estimation as well as representation, we see, then, that signal extraction plays an important part in this theory. Here, as in detection, the goals of an adequate theory remain unchanged: to obtain optimum systems for estimation, to interpret the resulting structures $\boldsymbol{T}_R$ in terms of specific, realizable elements, and to evaluate and compare the performance of optimum and suboptimum systems. In Section 5.1, we begin with a decision theoretic formulation of signal estimation where the signal is known to be present.[1] These general methods are then applied in Sections 5.2–5.4 to various problems of signal analysis. Because of the many possible situations and criteria available, our treatment is specifically selective and thus less compact and comprehensive than our preceding treatment of detection in Chapters 2 and 3. Chapter 5 is based largely on the earlier work of Middleton and Van Meter [1] and Middleton, Chapter 21 of Ref. [2], extended to include spatial as well as temporal data.

5.1 DECISION THEORY FORMULATION

A distinguishing feature of the decision theoretic formulation for communication is the introduction of cost, or value judgments associated with the decisions. These costs play a different role in estimation than they do in detection. In the latter, they are assigned to the various correct and incorrect discrete outcomes of the detector: "yes, a signal is present (H_1)" or "no, a signal is not present (H_0)," and are represented as components of a decision threshold, as explained in Chapter 1 and illustrated with examples in Chapter 3. In signal extraction, however, the decisions are now numbers or magnitudes, since estimation is basically a measurement process.

The fact that we have a definite decision implies some sort of cost assignment. We may accordingly expect that estimation systems, like detection systems, can be naturally incorporated within the framework of decision theory on choice of an appropriate cost function. That this is the case will be demonstrated here and in succeeding sections. For example, we shall show that classical estimation methods, like maximum likelihood under rather broad conditions, may be regarded as decision systems (minimizing average risk) with respect to certain cost assignments. Optimum systems here are again Bayes systems (Section 1.4.3), as in the analogous detection theory. However, the cost functions themselves and the resulting optimal structures are now quite different. An important feature of estimation from the decision-theoric viewpoint is the variety of possible cost functions and corresponding optimum systems that can be generated. The present section is accordingly devoted to signal extraction (or estimation) from this standpoint. Specifically, it is devoted to a discussion of some cost functions of practical interest, the associated Bayes systems, and their various properties, such as the distribution of estimators and the probabilities of correct decisions.

5.1.1 Nonrandomized Decision Rules and Average Risk

Extraction in the theory of signal reception is the counterpart of parameter estimation in statistics. Here the signal parameters $\boldsymbol{\theta} = (\theta_1, \theta_2, \ldots, \theta_M)$ are the parameters of the distribution density $F_n[\boldsymbol{X}|\boldsymbol{S}(\boldsymbol{\theta})]$ governing the occurrence of the received data $\boldsymbol{X}$. Frequently, the signal parameters $\boldsymbol{\theta}$ are taken to be the signal components $\boldsymbol{S}$ themselves, as in the case of

[1] The reader is referred to results of classical estimation, as background to the Bayesian approach of this chapter.

stochastic signals and waveform estimation generally. Point estimation (Section 1.2.2), namely the direct estimation of each component of $\boldsymbol{\theta}$ or $\boldsymbol{S}$, is considered here, rather than estimation by confidence intervals (Section 6.3.5). As before (Section 1.3.1), we let $\boldsymbol{\gamma}$ represent the decision to be made about $\boldsymbol{\theta}$, where equivalently $\boldsymbol{\gamma}[=\boldsymbol{\gamma}_\theta$ or $\boldsymbol{\gamma}_\sigma]$ is the required estimator of $\boldsymbol{\theta}$ (Appendix A5.1). Observe also that, when a particular $\boldsymbol{\gamma}$ is to be an estimate of $\boldsymbol{\theta}$, the parameter and decision spaces $\boldsymbol{\Omega}_\theta$, $\boldsymbol{\Delta}$ have the same structure, that is, a one-to-one mapping exists between the two. We assume that each space contains a continuum of points and is a finite, closed region, which may, however, be taken large enough to be essentially infinite for practical purposes.[2] Similarly, when waveform estimation is desired, we set $\boldsymbol{\theta}=\boldsymbol{S}$ and the signal and decision spaces $\boldsymbol{\Omega}$, $\boldsymbol{\Delta}$ of Fig. 1.5 likewise have the same structure. The cost functions $\mathsf{F}_1 = C(\boldsymbol{\theta}, \boldsymbol{\gamma})$, $\mathsf{F}_1 = C(\boldsymbol{S}, \boldsymbol{\gamma})$ [Eq. (1.4.3)], to be used in the risk analysis, are as in detection (Section 1.6) preassigned in accordance with the external constraints of the problem and are critical in determining the specific structure of the resulting system.

An important theorem, due to Hodges and Lehmann [3], enables us to avoid randomized decision rules (Section 1.3.2) in most applications. This theorem states:

If $\boldsymbol{\Delta}$ is the real line and if $C(\boldsymbol{\theta}, \boldsymbol{\gamma})$ $C(\boldsymbol{S}, \boldsymbol{\gamma})$ is a convex function[3] of $\boldsymbol{\gamma}$ for every $\boldsymbol{\theta}$ (or $\boldsymbol{S}$), then for any decision rule $\boldsymbol{\delta}$ there exists a nonrandomized decision rule whose average risk is not greater than that of $\boldsymbol{\delta}$ for all $\boldsymbol{\theta}$ (or $\boldsymbol{S}$) in $\boldsymbol{\Omega}_\theta$ (or $\boldsymbol{\Omega}$).

This applies to one-dimensional $\boldsymbol{\theta}(=\theta_1)$, or $\boldsymbol{S}(=S_1)$, and $\boldsymbol{\gamma}(=\gamma_1)$ but can usually be extended to include multidimensional vectors [4]. Thus, when the decision rule is nonrandomized, we have from Eq. (1.4.14)

$$\delta(\boldsymbol{\gamma}|\boldsymbol{X}) = \delta[\boldsymbol{\gamma} - \boldsymbol{\gamma}_\sigma(\boldsymbol{X})] \tag{5.1.1}$$

where $\boldsymbol{\gamma}_\sigma(\boldsymbol{X}) = T_{\mathrm{R}}\{\boldsymbol{X}\}$ is the functional operation performed on the data $\boldsymbol{X}$ by the system $\boldsymbol{T}_{\mathrm{R}}$ and $\boldsymbol{\gamma}$ is, for particular $\boldsymbol{X}$, the estimate produced by the system. Since we are interested primarily in Bayes systems, the unconditional estimator $\boldsymbol{\gamma}_\sigma(\boldsymbol{X})$, rather than the conditional estimator $\boldsymbol{\gamma}_\theta(\boldsymbol{X})$, is the quantity to be optimized, by minimization of the average risk. Accordingly, for waveform extraction, $\theta = \boldsymbol{S}$, where $\boldsymbol{\gamma}(\boldsymbol{X})$ is an estimator of $\boldsymbol{S}$ based on $\boldsymbol{X}$ and on the *a priori* distribution $\sigma(\boldsymbol{S})$, while, for parameter extraction, the estimator $\boldsymbol{\gamma}_\theta(\boldsymbol{X})$ is based instead[4] on $\boldsymbol{X}$ and $\sigma(\boldsymbol{\theta})$.

For waveform estimation, in contrast to that for space and time-independent parameters, the points $\boldsymbol{P}(\boldsymbol{\gamma}, t)$ at which estimates are desired provide a convenient subdivision of the extraction process into three principal types of procedure[5]. Let us write for the waveform value under estimation $S_\lambda = S(\boldsymbol{\gamma}_\lambda, t_\lambda)$, where λ may assume a discrete (and, later, a continuous) set of values. Depending on how $(\boldsymbol{\gamma}_\lambda, t_\lambda)$ are chosen with respect to the spatial points $\boldsymbol{\gamma}_1, \ldots, \boldsymbol{\gamma}_m, \ldots, \boldsymbol{\gamma}_M$ and the times $t_1, t_2, t_n, \ldots, t_N$ in the space–time interval $(0, \boldsymbol{R}; 0, t)$ at which the received data $\boldsymbol{X}(\boldsymbol{\gamma}, t)$ are sampled (to give $\boldsymbol{X}$), we distinguish as follows.

[2] See also Section 1.5.3.

[3] A real-valued function $g(x)$ is convex in an interval (a, b) if for any x and y in (a, b) and any number $0 < \rho < 1$ one has $\rho g(x) + (1-\rho)g(y) \geq g[\rho x + 1(1-\rho)y]$.

[4] Again, we employ the convention that (unless otherwise indicated) functions of the same form and different arguments are different; that is, $\sigma(\boldsymbol{\theta}) \neq \sigma(\boldsymbol{S})$ unless $\theta = \boldsymbol{S}$; $w(\boldsymbol{X}) \neq w(\boldsymbol{S}|\boldsymbol{X})$, and so on.

[5] For space–time indexing we employ the double index convention $j = mn$ (or the single index, k) described in Section 1.3.1.

5.1.1.1 Smoothing or Interpolation Here $\gamma = j$ lies within the observation period $X(\boldsymbol{\gamma}, t)$, but it does not coincide with any of the space–time points $\gamma_i \neq (\gamma_m, t_n)$, of the data and signal elements $\left(\boldsymbol{X}_{11}, \ldots, \boldsymbol{X}_{m,n}, \ldots, \boldsymbol{X}_{M,N}\right), \left(\boldsymbol{S}_{11}, \ldots, \boldsymbol{S}_{m,n}, \ldots, \boldsymbol{S}_{M,N}\right), J = MN$, on which the estimation of $\boldsymbol{S}$ is based. [λ_j may take a single value or many values in $(0, \boldsymbol{R}; 0, t)$. Usually, however, only a single $\boldsymbol{S}_\lambda$ is to be determined, based on $\boldsymbol{X}$ and $\boldsymbol{S}$.]

5.1.1.2 Simple or Coincidental Extrapolation Here $\lambda_j = (\boldsymbol{\gamma}_m, t_n)$ is one (or more) of the space–time points at which data are obtained and for which *a priori* information concerning $\boldsymbol{S}$ is also given. Thus, $\lambda_j = (\boldsymbol{\gamma}_m, t_n)$ again lies in the observation interval, and extraction consists in obtaining suitable estimates of $\boldsymbol{S}$ based on $\boldsymbol{X}$.

5.1.1.3 Prediction or Extrapolation In this case $\lambda_j = (\boldsymbol{\gamma}_m, t_n)$ lies outside the data interval $(0, \boldsymbol{R}; 0, t)$, where no samples are taken, and we are asked now to make an estimate of $\boldsymbol{S}$ on the basis of $\boldsymbol{S}$ and $\boldsymbol{X}$ in $(0, \boldsymbol{R}; 0, t)$ (see Fig. 1.3 extended to include space–time sampling).

From Eqs. [(1.4.5) and (1.4.6)] and (1.4.14), we accordingly write for the average risk

$$R(\sigma, \delta)_S = \int_\Omega \sigma(\boldsymbol{S}, S_\lambda) dS_\lambda d\boldsymbol{S} \int_\Omega C(\boldsymbol{S}, S_\lambda; \boldsymbol{\gamma}_\sigma) F_J(\boldsymbol{X}|\boldsymbol{S}) d\boldsymbol{X}, \tag{5.1.2}$$

where $F_J(\boldsymbol{X}|\boldsymbol{S})$ does not contain $\boldsymbol{S}_\lambda$ if $\lambda = \lambda_j \neq j\,(mn)$ or if $X_j, \lambda_j \lesseqgtr (0, \boldsymbol{R}; 0, t)$, as in prediction[6]. Since $\boldsymbol{S}$ is usually a function of one or more statistical parameters $\boldsymbol{\theta}$, we can (in the case of deterministic signals) make use of the fact that $\sigma(\boldsymbol{S}, S_\lambda) d\boldsymbol{S}\, dS_\lambda$ when integrated over all Ω-space is equivalent to $\sigma(\boldsymbol{\theta}) d\boldsymbol{\theta}$, similarly integrated over all Ω-space [Ω and Ω_θ mapping into each other as a result of the transformation $\boldsymbol{S} = \boldsymbol{S}(\boldsymbol{\theta})$]. Then Eq. (5.1.2) can be put into the alternative and more convenient form

$$R(\sigma, \delta)_S = \int_{\Omega_\theta} \sigma(\boldsymbol{\theta}) d\boldsymbol{\theta} \int_\Gamma C[S(\boldsymbol{\theta}), S_\lambda(\boldsymbol{\theta}); \boldsymbol{\gamma}_\sigma] F_J[\boldsymbol{X}|\boldsymbol{S}(\boldsymbol{\theta})] d\boldsymbol{X}. \tag{5.1.2a}$$

Similarly, the average risk associated with the parameter estimation of $\boldsymbol{\theta}$ is

$$R(\sigma, \delta)_\theta = \int_{\Omega_\theta} d\boldsymbol{\theta}\, \sigma(\boldsymbol{\theta}) \int_\Gamma C(\boldsymbol{\theta}, \boldsymbol{\gamma}_\sigma) W_J(\boldsymbol{X}|\boldsymbol{\theta}) d\boldsymbol{\theta} \tag{5.1.3}$$

where $W_J(\boldsymbol{X}|\boldsymbol{\theta}) = \langle F_J[\boldsymbol{X}|\boldsymbol{S}(\boldsymbol{\theta}, \boldsymbol{\theta}')] \rangle_{\theta'}$ and where $\boldsymbol{\theta}'$ represents all other parameters, apart from $\boldsymbol{\theta}$. Of course, if there are no other parameters $\boldsymbol{\theta}'$, then $W_J(\boldsymbol{X}|\boldsymbol{\theta}) = \langle F_J[\boldsymbol{X}|\boldsymbol{S}(\boldsymbol{\theta})] \rangle$. Observe here that the $\boldsymbol{\gamma}_\sigma$ for $\boldsymbol{\theta}$ and for $\boldsymbol{S}, S_\lambda$ are different estimators in general, while the average risks R_θ, R_S likewise assume different values. As in detection, Bayes systems are obtained by the choice of decision rule, here the estimator $\boldsymbol{\gamma}_\sigma$ [Eq. (5.1.1)] which minimizes the average risk. Examples for special cost functions are considered presently in Sections 5.1.2 and 5.1.3.

5.1.2 Bayes Extraction With a Simple Cost Function

Let us now determine the optimum extraction procedure for minimizing the average risk [Eq. (5.1.2a) or (5.1.3)] when a simple, or "constant," cost function is chosen. By a *simple*, or

[6] When $\lambda = j(= mn)$, the notation is pleonastic: S_λ is absorbed into $\boldsymbol{S}$; this is illustrated in subsequent examples.

constant, cost function is meant one for which the cost of correct estimates is set equal to C_C, while the costs of incorrect estimates all have the same value $C_E(> C_C)$, regardless of how much in error they may be.

Assuming simple, or coincident, estimation $\lambda = j(mn)$ of the signal parameters $\boldsymbol{\theta} = (\theta_1, \ldots, \theta_M)$ (Section 5.1.1.2) for the moment, as well as discrete parameter and data spaces, so that $\boldsymbol{\theta}$ and $\boldsymbol{X}$ can take only discrete values, we may write the constant cost function specifically as

$$C(\boldsymbol{\theta}, \boldsymbol{\gamma}_\sigma) = \sum_{k=1}^{K} \left[C_E - (C_E - C_C)\delta_{\gamma_k\theta_k}\right] \tag{5.1.4}$$

where $\delta_{\gamma_k\theta_k}$ is the Kronecker delta ($\delta = 1,\ \gamma_k = \theta_k;\ \gamma = 0,\ \gamma_k \neq \theta_k$). Thus, we are penalized an amount C_E for each parameter ($k = 1, \ldots, K$) when our estimate γ_k is incorrect, that is, when $\gamma_k \neq \theta_k$, and, correspondingly, are assessed the smaller amount C_C when a correct decision $\gamma_k = \theta_k$ is made. From Eqs. (1.4.6) and (5.1.4), we can write the discrete analogue of the average risk [Eq. (5.2.3)] here as

$$R(\sigma, \delta)_\theta = KC_E - (C_E - C_C)\sum_{k=1}^{K}\sum_{\boldsymbol{\Omega}_\theta}\sum_{\boldsymbol{\Gamma}}\sum_{\boldsymbol{\Delta}} \delta_{\gamma_k\theta_k}\, p_\sigma(\boldsymbol{\theta})p_J(\boldsymbol{X}|\boldsymbol{\theta})\delta(\boldsymbol{\gamma}|\boldsymbol{X}), \tag{5.1.5}$$

where p_σ and p_J are respectively the *a priori probabilities* of $\boldsymbol{\theta}$ and the conditional probability of $\boldsymbol{X}$, given $\boldsymbol{\theta}$. The symbols $\sum_{\boldsymbol{\Omega}_\theta}$, and so on, indicate summation over all L_k allowed values of each $\theta_k (k = 1, \ldots, K)$, and so on. Thus we write

$$\sum_{\boldsymbol{\Omega}_\theta} \equiv \sum_{\ell_1=1}^{L_1} \cdots \sum_{\ell_k=1}^{L_K};\ \sum_{\boldsymbol{\Gamma}} \equiv \sum_{\ell_1=1}^{N_1} \cdots \sum_{\ell_J=1}^{N_J},\ \text{and so on.} \tag{5.1.6}$$

Next, let us define

$$D_k(\boldsymbol{X}, p_\sigma, \delta) \equiv \sum_{\Delta}\sum_{\boldsymbol{\Omega}_\theta} \delta_{\gamma_k\theta_k}\, p_J(\boldsymbol{X}|\boldsymbol{\theta})p_\sigma(\boldsymbol{\theta})\delta(\boldsymbol{\gamma}|\boldsymbol{X}) \tag{5.1.6a}$$

$$= \sum_{\Delta k} p_J(\boldsymbol{X}|\gamma_k)p_\sigma(\gamma_k)\delta(\gamma_k|\boldsymbol{X}) \tag{5.1.6b}$$

in which $p_J(\boldsymbol{X}|\gamma_k)$, $p_\sigma(\gamma_k)$ are the *marginal* probabilities

$$p_J(\boldsymbol{X}|\boldsymbol{\theta}_k) \equiv \sum_{\boldsymbol{\Omega}_\theta - \boldsymbol{\Omega}_{\theta_k}} p_J(\boldsymbol{X}|\boldsymbol{\theta});\quad p_\sigma(\boldsymbol{\theta}_k) = \sum_{\boldsymbol{\Omega}_\theta - \boldsymbol{\Omega}_{\theta_k}} p_\sigma(\boldsymbol{\theta}). \tag{5.1.7a}$$

The decision rule $\delta(\gamma_k|\boldsymbol{X})$ is similarly the marginal probability

$$\delta(\gamma_k|\boldsymbol{X}) = \sum_{\Delta - \Delta_k} \delta(\boldsymbol{\gamma}|\boldsymbol{X}). \tag{5.1.7b}$$

(In our present notation, $\sum_{\Delta} = \sum_{\Delta_k} \cdot \sum_{\Delta - \Delta_k}$, etc.) The average risk [Eq. (5.1.5)] can now be expressed more compactly as

$$R(\sigma, \delta)_\theta = KC_{\text{E}} - (C_{\text{E}} - C_{\text{C}}) \sum_{\boldsymbol{\Gamma}} \sum_{k=1}^{K} D_k(\boldsymbol{X}, p_\sigma, \delta). \tag{5.1.8}$$

Optimization is next achieved by a suitable choice of the M decision rules $\delta(\gamma_k \boldsymbol{X})$ $(k = 1, \ldots, K)$. It is clear that R_θ is smallest when each D_k is largest, since $D_k \geq 0\,[1 \geq p_J,\ p_\sigma,\ \delta(\gamma_k|\boldsymbol{X}) \geq 0 \text{ also}]$. Accordingly, we select $\delta(\gamma_k|\boldsymbol{X})$ to *maximize* D_{k_θ}. This is accomplished by setting

$$\delta(\gamma_k|\boldsymbol{X}) = \delta_{\gamma_k \hat{\gamma}_k} = 0, \quad \gamma_k \neq \hat{\gamma}_k|\boldsymbol{X}, \quad k = 1, \ldots, K, \tag{5.1.9}$$

where $\hat{\gamma}_k (= \hat{\theta}_k)$ is the *unconditional* maximum likelihood estimator (UMLE) of θ_k, defined by

$$p_\sigma(\hat{\theta}_k) p_J(\boldsymbol{X}|\hat{\theta}_k) \geq p_\sigma(\theta_k) p_J(\boldsymbol{X}|\theta_k), \quad \text{all } \theta_k \text{ in } \Omega_{\theta_k}. \tag{5.1.10a}$$

This UMLE, $\hat{\theta}_k$ may be obtained from (5.1.10b) ff., where, of course, L_J is now $L(\boldsymbol{X},\ \theta_k) = p_\sigma(\theta_k) p_J(\boldsymbol{X},\ \theta_k)$, with probability densities replaced by the appropriate probabilities.[7]

$$\frac{\partial}{\partial \boldsymbol{\theta}} \log L(\boldsymbol{X},\ \boldsymbol{\theta}) = \left[\frac{\partial}{\partial \theta_k} \log p_\sigma(\theta_k) p_J(\boldsymbol{X}|\theta_k) \right]_{\theta_k = \hat{\theta}_k = \hat{\gamma}_k} \Bigg\} = 0. \tag{5.1.10b}$$

The Bayes risk becomes

$$R_\theta^* \equiv \min_{\hat{\sigma}} R(\sigma, \delta)_\theta = KC_{\text{E}} - (C_{\text{E}} - C_{\text{C}}) \sum_{k=1}^{K} \sum_{\boldsymbol{\Gamma}} p_\sigma(\hat{\theta}_k) p_J(X|\hat{\theta}_k). \tag{5.1.11}$$

Note that because of the particular structure of the cost function [Eq. (5.1.4)], the Bayes estimators of $\theta_1, \ldots, \theta_K$ here are the individual UML estimators $\hat{\gamma}_k = \hat{\theta}_k$ of each θ_k. The parameters may be statistically related, as $p_J(\boldsymbol{X}|\boldsymbol{\theta})$ indicates. Each $p_\sigma(\theta_k)$, $p_J(\boldsymbol{X}|\hat{\theta}_k)$ embodies such interrelationships, but the UMLEs are determined independently from (5.1.10b), as indicated above.

When the data, parameter, and decision spaces $\boldsymbol{\Gamma}$, $\boldsymbol{\Omega}_\theta$, $\boldsymbol{\Delta}$ are continuous instead of discrete, the continuous analogue of the discrete constant cost function (5.1.4) becomes

$$C(\boldsymbol{\theta},\ \boldsymbol{\gamma}_\sigma) = \sum_{k=1}^{K} [C_{\text{E}} A'_k - (C_{\text{E}} - C_{\text{C}}) \delta(\gamma_k - \theta_k)], \tag{5.1.12}$$

where the A'_k are (positive) constants, with the dimensions of the delta functions (that is, $|\theta_k|^{-1}$) chosen so that the average risk for each θ_k is also positive (or zero).[8] This can be

[7] Note that generally $p_J(\boldsymbol{X}|\theta_k) \neq p_J(\boldsymbol{X}|\theta_l)$ $(k \neq 1$, etc.).

[8] The choice of the A'_k, while to an extent arbitrary, is closely related to the fineness with which our extraction system can distinguish between observed magnitudes (Section 5.1.5).

expressed more compactly as

$$C(\boldsymbol{\theta},\ \boldsymbol{\gamma}_\sigma) = C_0 \sum_{k=1}^{K} [A_k - \delta(\gamma_k - \theta_k)] \quad C_0 \equiv C_\mathrm{E} - C_\mathrm{C},\ A_k \equiv \frac{C_\mathrm{E}}{C_\mathrm{E} - C_\mathrm{C}} A'_k. \tag{5.1.12a}$$

Paralleling Eqs. (5.1.5) to (5.1.8), we get for the average risk

$$R(\sigma,\ \delta)_\theta = C_0 \sum_{k=1}^{K} \left[A_k - \int_\Gamma D_k(\boldsymbol{X};\ \sigma,\ \delta) d\boldsymbol{X} \right], \tag{5.1.13a}$$

with

$$D_k(\boldsymbol{X};\ \sigma,\ \delta) = \int_\Delta \sigma(\gamma_k) W_n(\boldsymbol{X}|\gamma_k) \delta(\gamma_k|\boldsymbol{X}) d\gamma_k, \tag{5.1.13b}$$

[Eq. (5.1.6b)], where now probabilities have been replaced by the corresponding *probability densities* $\sigma,\ W_n,\ \delta$.

Optimization again is achieved by selecting the decision rules $\delta(\gamma_k|\boldsymbol{X})$ so as to maximize D_k. The analogue of Eq. (5.1.9) for this purpose is

$$\delta(\gamma_k|\boldsymbol{X}) = \delta[\gamma_k - \hat{\gamma}_k(\boldsymbol{X})], \quad k = 1, \ldots, K, \tag{5.1.14}$$

where $\hat{\gamma}_k(\boldsymbol{X}) = \hat{\theta}_k(\boldsymbol{X})$ is the UML estimator of θ_k, defined by

$$\sigma(\hat{\theta}_k) W_J(\boldsymbol{X}|\hat{\theta}_k) \geq \sigma(\theta_k) W_J(\boldsymbol{X}|\theta_k), \quad \text{all } \theta_k \text{ in } \Omega_{\theta_k}. \tag{5.1.14a}$$

The required UMLE's $\hat{\theta}_k$, are once more obtained from Eq. (5.1.10b), with $L_{J_Z} = \sigma(\theta_k) W_J(\boldsymbol{X},\ \theta_k)$ now. Thus, we have equivalently

$$\left. \frac{\partial}{\partial \boldsymbol{\theta}} \left[\log \sigma(\theta_k) W_J(\boldsymbol{X}|\theta_k) \right]_{\theta_k = \hat{\theta}_k = \hat{\gamma}_{(\theta_k)^*}} \right\} = 0,\ k = 1, \ldots, K. \tag{5.1.14b}$$

The corresponding Bayes risk for this constant cost function [Eqs. (5.1.12) and (5.1.14)] is

$$R^*(\sigma,\ \delta)_\theta = C_0 \sum_{k=1}^{K} \left[A_k - \int_\Gamma \sigma(\hat{\theta}_k) W_J(\boldsymbol{X}|\hat{\theta}_k) d\boldsymbol{X} \right], \tag{5.1.15}$$

where we remember, of course, that $\hat{\theta}_k = \gamma_k(\boldsymbol{X})^*$ is a function of $\boldsymbol{X}$.

For *waveform estimation*, the constant cost function [Eq. (5.1.12a)] becomes for the coincident estimation $x = j \left(j = 11,\ 12, \ldots, \dfrac{M}{MN} \right)$ employed here,

$$C(\boldsymbol{S},\ S_\lambda;\ \boldsymbol{\gamma}_\sigma) = C(\boldsymbol{S},\ \boldsymbol{\gamma}_\sigma) = C_0 \left[A_J - \sum_{j=1}^{J} \sigma(\gamma_j - S_j) \right], \tag{5.1.16}$$

in which it is assumed that $A_k = A'_n/J$ for each point $(r_m,\ t_n)$ at which estimates of S_k are desired. Following the argument of Eqs. (5.1.13a)–(5.1.15), we see that the Bayes estimators of $S_{11}, \ldots, S_{MN}$ are again the various unconditional maximum likelihood estimators

$\hat{\gamma}_k(\boldsymbol{X}) = \hat{S}_k(\boldsymbol{X})$, while the Bayes risk (5.1.15) is modified to[9]

$$R^*(\sigma,\delta)_S = C_0\left[A_n - \sum_{j=1}^{J=MN}\int_{\Gamma}\sigma(\hat{S}_j)F_J(\boldsymbol{X}|\hat{S}_k)\,d\boldsymbol{X}\right] \quad (>0). \tag{5.1.17}$$

(Other types of "constant" cost function can be constructed (Section 5.1.5), but they are not particularly well suited to many applications, their chief defect being an excessive strictness with regard to accuracy and hence too great an average risk.)

In the above instances, the optimum estimators are all UML estimators. The estimates themselves are the parameter or signal values which maximize the joint probabilities (for discrete magnitudes) or probability densities of θ_k (or S_k) and $\boldsymbol{X}$, regarded as functions of θ_k (or S_k) with $\boldsymbol{X}$ fixed. For each $\boldsymbol{X}$, these estimates differ. If all signals S_k or parameters θ_k are *a priori* equally likely, these estimates are equivalent to the corresponding *conditional* likelihood estimates. However, regardless of the precise form of $\sigma(\theta_k)$ or $\sigma(S_k)$, these Bayes estimates are the parameters (or signals) most likely *a posteriori*, that is, for given $\boldsymbol{X}$. In general, *we can regard unconditional maximum likelihood estimators as Bayes estimators relative to an appropriate constant, or simple, cost function*. Since these optimum estimators are all UMLEs, classical maximum likelihood theory as extended to the unconditional cases may be applied in detail here. We can accordingly interpret the UMLEs of amplitude a_0 and *shape factor* θ considered in the examples of as optimum decision systems that minimize average risk with respect to a constant cost function of the type (5.1.12). In a similar way, the estimator of Eq. (5.1.12) for the example in Section 5.1.2 can be shown to be Bayes relative to the cost function of Eq. (5.1.11) (see Section 5.4.1). Finally, observe that the extension of Eqs. (5.1.16) and (5.1.17) to the interpolation and extrapolation of waveforms $\lambda'_j \neq \lambda_i$ as distinct from simple estimation $\lambda'_j = \lambda_i$ may be carried out in the same way.

5.1.3 Bayes Extraction With a Quadratic Cost Function

We begin again with the case of parameter estimation. Now, however, the cost function to be used in the estimation of the K signal parameters $\boldsymbol{\theta} = (\theta_1, \ldots, \theta_K)$ is the quadratic cost function

$$C(\boldsymbol{\theta}, \boldsymbol{\gamma}_\sigma) = C_0(\tilde{\boldsymbol{\theta}} - \tilde{\boldsymbol{\gamma}}_\sigma)(\boldsymbol{\theta} - \boldsymbol{\gamma}_\sigma) = C_0\|\boldsymbol{\theta} - \boldsymbol{\gamma}_\sigma(\boldsymbol{X})\|^2 = C_0\sum_{k=1}^{K}(\theta_k - \gamma_k)^2, \tag{5.1.18}$$

where of course $C_0 > 0$. Here C itself is convex so that the inconvenience of considering both randomized and nonrandomized decision rules may be avoided.

This cost function is the square of the "distance" between the true value and the estimate. The chief virtues of this "squared-error" or quadratic cost function for extraction are

(1) that it is convenient mathematically in the cases here where the signal is known to be present;

[9] Note that in Eq. (5.1.17), $F_J(\boldsymbol{X}|S_k) \neq F_J(\boldsymbol{X}|S)$. For example, we write $F_J(\boldsymbol{X}|S_j) = \int\sigma(\boldsymbol{S})F_J(\boldsymbol{X}|\boldsymbol{S})\,d\boldsymbol{S}'$, where $\boldsymbol{S}' = S_{11}, \ldots, S_J$ with S_j omitted.

(2) that it takes reasonable account of the fact that usually large errors are more serious than small ones; and
(3) that under certain conditions (described in Section 5.1.4) it is also an optimum or Bayes estimator for a wider class of cost functions than the quadratic [5]. Moreover, as we shall see presently, it also leads in certain cases to the earlier extraction procedures [4, 6] based on least-mean-squared-error criteria for linear systems, which we wish to include from the more general viewpoint of decision theory.

For Bayes extractors, we minimize the average risk R as before. The conditional risk [Eq. (1.4.5)] for Eq. (5.1.18) and the decision rule $\boldsymbol{\delta}[\boldsymbol{\gamma} - \boldsymbol{\gamma}_\sigma(\boldsymbol{X})]$ become

$$r(\boldsymbol{\theta}, \boldsymbol{\gamma}_\sigma) = C_0 \int_{\boldsymbol{\Gamma}} \left\|\boldsymbol{\theta} - \boldsymbol{\gamma}_\sigma(\boldsymbol{X})_\theta\right\|^2 W_n(\boldsymbol{X}|\boldsymbol{\theta}) d\boldsymbol{X}, \tag{5.1.19}$$

where care is taken to distinguish between the estimator $\boldsymbol{\gamma}_\sigma(\boldsymbol{X})$, defined for all possible $\boldsymbol{X}$, and the estimate $\boldsymbol{\gamma}_\sigma(\boldsymbol{X}')$, for a particular $\boldsymbol{X} = \boldsymbol{X}'$. The average risk follows at once from Eqs. (1.4.6a) and (1.4.6b) and is here

$$\begin{aligned} R(\sigma, \delta)_\theta = C_0 E_{X,\theta}\left\{\left\|\boldsymbol{\theta} - \boldsymbol{\gamma}_\sigma(\boldsymbol{X})_\theta\right\|^2\right\} &= C_0 \int_{\boldsymbol{\Gamma}} d\boldsymbol{X}\, w_J(\boldsymbol{X}) \\ &\times \int_{\boldsymbol{\Omega}_\theta} \left\|\boldsymbol{\theta} - \boldsymbol{\gamma}_\sigma(\boldsymbol{X})_\theta\right\|^2 w_K(\boldsymbol{\theta}|\boldsymbol{X}) d\boldsymbol{\theta}, \end{aligned} \tag{5.1.20a}$$

where we have written the probability densities

$$w_J(\boldsymbol{X}) = \int_{\boldsymbol{\Omega}_\theta} \sigma(\boldsymbol{\theta}) W_J(\boldsymbol{X}|\boldsymbol{\theta})\, d\boldsymbol{\theta}; \quad w_K(\boldsymbol{\theta}|\boldsymbol{X}) = \sigma(\boldsymbol{\theta}) W_J(\boldsymbol{X}|\boldsymbol{\theta}) / w_J(\boldsymbol{X}). \tag{5.1.20b}$$

When the cost function is differentiable (with respect to the estimator), we obtain the following general condition for an extremum of R (actually a minimum for these concave cost functions):

$$\boldsymbol{\delta} R = \int_{\boldsymbol{\Gamma}} d\boldsymbol{X} \int_{\boldsymbol{\Omega}_\theta} \sigma(\boldsymbol{\theta}) W_J(\boldsymbol{X}|\boldsymbol{\theta}) \frac{\partial C(\boldsymbol{\theta}, \boldsymbol{\gamma})}{\partial \boldsymbol{\gamma}} d\boldsymbol{\theta}\, \boldsymbol{\delta\gamma} = 0, \tag{5.1.21a}$$

or

$$\int_{\boldsymbol{\Omega}_\theta} \sigma(\boldsymbol{\theta}) W_J(\boldsymbol{X}|\boldsymbol{\theta}) \frac{\partial C(\boldsymbol{\theta}, \boldsymbol{\gamma})}{\partial \boldsymbol{\gamma}}\Big|_{\boldsymbol{\gamma}=\boldsymbol{\gamma}^*} d\boldsymbol{\theta} = 0. \tag{5.1.21b}$$

With the quadratic cost function (5.1.18), this becomes simply

$$\begin{aligned} \boldsymbol{\gamma}_\sigma^*(\boldsymbol{X})_\theta = \int_{\boldsymbol{\Omega}_\theta} \boldsymbol{\theta}\, w_K(\boldsymbol{\theta}|\boldsymbol{X})\, d\boldsymbol{\theta} &= \frac{\int_{\boldsymbol{\Omega}_\theta} \boldsymbol{\theta}\, \sigma(\boldsymbol{\theta}) W_J(\boldsymbol{X}|\boldsymbol{\theta})\, d\boldsymbol{\theta}}{\int_{\boldsymbol{\Omega}_\theta} \sigma(\boldsymbol{\theta}) W_J(\boldsymbol{X}|\boldsymbol{\theta})\, d\boldsymbol{\theta}} \\ &= \frac{\int_{\boldsymbol{\Omega}_\theta} \theta\, w_{JK}(\boldsymbol{X}, \boldsymbol{\theta})\, d\boldsymbol{\theta}}{\int_{\boldsymbol{\Omega}_\theta} w_{JK}(\boldsymbol{X}, \boldsymbol{\theta})\, d\boldsymbol{\theta}}, \end{aligned} \tag{5.1.22}$$

in which $w_{JK}(\boldsymbol{X}, \boldsymbol{\theta})$ is the joint d.d. of $\boldsymbol{X}$ and $\boldsymbol{\theta}$. Thus, the Bayes estimator for the squared-error cost function is the conditional expectation [7] of $\boldsymbol{\theta}$, given $\boldsymbol{X}$. Since both members of

Eq. (5.1.22) are vectors, Eq. (5.1.22) represents K equations between the K components of $\boldsymbol{\gamma}_\sigma^*$ and $\boldsymbol{\theta}$. Accordingly, the kth component of the estimator is

$$\begin{aligned}&\left[\boldsymbol{\gamma}_\sigma^*(X_n, \ldots, X_{MN})\right]_k \\ &\quad = \int \ldots \int \theta_k w_K(\theta_1, \ldots, \theta_M | X_{11}, X_{12}, \ldots, X_{MN})\, d\theta_1 \cdots d\theta_K \\ &\qquad\qquad\qquad\qquad k = 1, \ldots, K. \end{aligned} \tag{5.1.22a}$$

The same result follows, of course, if instead of Eq. (5.1.18) we use for each component $\boldsymbol{\theta}$, that is, if the Bayes estimate of a single component of $C = (\theta_k - \gamma_k)^2$ is required, rather than the more general vector estimate. This is a consequence of the quadratic nature of the preassigned cost function and clearly does not hold in general. (Note again that $W_J(\boldsymbol{X}|\boldsymbol{\theta}) = \langle F_J[\boldsymbol{X}|S(\boldsymbol{\theta}, \boldsymbol{\theta}')]\rangle_{\theta'}$, and if $\boldsymbol{S}$ is a function of $\boldsymbol{\theta}$ only, then $W_J(\boldsymbol{X}|\boldsymbol{\theta}) = F_J[\boldsymbol{X}|\boldsymbol{S}(\boldsymbol{\theta})]$.)

The Bayes risk is specifically, from (5.1.22) in (5.1.20a),

$$\begin{aligned} R^*(\sigma, \delta)_\theta &= C_0 E_{X,\theta}\left\{\left\|\boldsymbol{\theta} - \boldsymbol{\gamma}_\sigma^*(\boldsymbol{X})_\theta\right\|^2\right\} \\ &= C_0 E_{\theta}\left\{\tilde{\boldsymbol{\theta}}\boldsymbol{\theta}\right\} - 2C_0 E_{X,\theta}\left\{\tilde{\boldsymbol{\theta}}\boldsymbol{\gamma}_\sigma^*\right\} + C_0 E_{\theta}\left\{\tilde{\boldsymbol{\gamma}}_\sigma^*\boldsymbol{\gamma}_\sigma^*\right\}, \end{aligned} \tag{5.1.23}$$

since $\boldsymbol{\gamma}_\sigma^*$ is a function of $\boldsymbol{X}$ only, while $\boldsymbol{\theta}$ is governed solely by the d.d. $\sigma(\boldsymbol{\theta})$. Note, however, that since $\boldsymbol{X}$ is a function of $\boldsymbol{\theta}$, inasmuch as $\boldsymbol{X} = S(\boldsymbol{\theta}) + N$ here, the cross-term in Eq. (5.1.23) does *not* generally factor into the product of the individual averages $E_X\{\boldsymbol{\gamma}_\sigma^*\} \times E_{\theta}\{\tilde{\boldsymbol{\theta}}\}$.

In the case of waveform estimation, the preceding treatment is easily modified for simple estimation ($\lambda = j$). Instead of Eq. (5.1.18) one has now the cost function

$$C(\boldsymbol{S}, \boldsymbol{\gamma}_\sigma) = C_0(\tilde{\boldsymbol{S}} - \tilde{\boldsymbol{\gamma}}_\sigma)(\boldsymbol{S} - \boldsymbol{\gamma}_\sigma) \tag{5.1.24}$$

where $\boldsymbol{\gamma}_\sigma = \boldsymbol{\gamma}_\sigma(\boldsymbol{X})$ is the estimator of $\boldsymbol{S}$ itself, rather than of $\boldsymbol{\theta}$ in $\boldsymbol{S} = \boldsymbol{S}(\boldsymbol{\theta})$. Replacing $\boldsymbol{\theta}$ by $\boldsymbol{S}$ and $W_J(\boldsymbol{X}|\boldsymbol{\theta})$ by $F_J(\boldsymbol{X}|\boldsymbol{\theta})$, and so on, in (5.1.20a)–(5.1.22), we obtain the corresponding average risk, optimum estimators, and Bayes risk

$$R(\sigma, \delta)_S = C_0 / \int_{\Omega}\int_{\Gamma} \sigma(\boldsymbol{S})\|\boldsymbol{S} - \boldsymbol{\gamma}_\sigma(\boldsymbol{X})\|^2 F_J(\boldsymbol{X}|\boldsymbol{S})\, d\boldsymbol{S}\, d\boldsymbol{X}, \tag{5.1.25}$$

$$\boldsymbol{\gamma}_\sigma^*(\boldsymbol{X})_S = \frac{\int_{\Omega} \boldsymbol{S}\sigma(\boldsymbol{S}) F_J(\boldsymbol{X}|\boldsymbol{S})\, d\boldsymbol{S}}{\int_{\Omega}\sigma(\boldsymbol{S}) F_J(\boldsymbol{X}|\boldsymbol{S})\, d\boldsymbol{S}} = \frac{\int_{\Omega} \boldsymbol{S} W_J(\boldsymbol{X}, \boldsymbol{S})\, d\boldsymbol{S}}{\int_{\Omega} W_J(\boldsymbol{X}, \boldsymbol{S})\, d\boldsymbol{S}}, \tag{5.1.26a}$$

and

$$R^*(\sigma, \delta)_S = C_0 E_{X,S}\left\{\left\|\boldsymbol{S} - \boldsymbol{\gamma}_\sigma^*(\boldsymbol{X})_S\right\|^2\right\}. \tag{5.1.26b}$$

Like Eq. (5.1.2a), useful alternative forms of (5.1.25)–(5.1.26a) in the case of deterministic signal processes are obtained on replacing $\boldsymbol{S}$ by $\boldsymbol{S}(\boldsymbol{\theta})$ explicitly and $\sigma(\boldsymbol{S})$ by $\sigma(\boldsymbol{\theta})$, where now the integration is over all $\boldsymbol{\Omega}$-space. The result from [(5.1.26a) and (5.1.26b)] is

$$\boldsymbol{\gamma}_\sigma^*(\boldsymbol{X})_S = \frac{\int_{\Omega_\theta} \boldsymbol{S}(\boldsymbol{\theta})\sigma(\boldsymbol{\theta}) F_J\left(\boldsymbol{X}|\boldsymbol{S}(\boldsymbol{\theta})\right) d\boldsymbol{\theta}}{\int_{\Omega_\theta} \sigma(\boldsymbol{\theta}) F_J\left(\boldsymbol{X}|\boldsymbol{S}(\boldsymbol{\theta})\right) d\boldsymbol{\theta}}, \tag{5.1.27a}$$

with the Bayes risk

$$R^*(\sigma, \delta)_S = C_0 E_{X,\boldsymbol{\theta}}\left\{\left\|\boldsymbol{S}(\boldsymbol{\theta}) - \boldsymbol{\gamma}_\sigma^*(\boldsymbol{X})_S\right\|^2\right\}. \tag{5.1.27b}$$

Even when $\bar{\boldsymbol{S}}$ or $\bar{\gamma}_\sigma^*$ vanishes, the Bayes risk [Eqs. (5.1.26a) and (5.1.26b)] still contains the cross-term, namely,

$$R(\sigma, \delta)_S^* = C_0 E_{\boldsymbol{S}\text{ or }\boldsymbol{\theta}}\{\tilde{\boldsymbol{S}}\boldsymbol{S}\} - 2C_0 E_{X,\boldsymbol{S}\text{ or }\boldsymbol{\theta}}\{\tilde{\boldsymbol{S}}\boldsymbol{\gamma}_\sigma^*\} + C_0 E_X\{\tilde{\boldsymbol{\gamma}}_\sigma^*\boldsymbol{\gamma}_\sigma^*\} \tag{5.1.27c}$$

again, since $\boldsymbol{X}(\text{in } \boldsymbol{\gamma}_\sigma^*)$ is a function of $\boldsymbol{S}$ or $\boldsymbol{\theta}$.

5.1.4 Further Properties

Some further properties of the Bayes estimator with the quadratic cost function (5.1.18) are easily shown.

Case 1: For example, in the case of simple estimation if a fixed signal $\boldsymbol{S}_1$ is applied at the input, so that $\sigma(\boldsymbol{S}) = \delta(\boldsymbol{S} - \boldsymbol{S}_1)$, the Bayes estimator is $\boldsymbol{\gamma}_\sigma^*(\boldsymbol{X})_S = \boldsymbol{S}_1$ and is therefore *conditionally unbiased*, that is,

$$\int_{\boldsymbol{\Gamma}} \boldsymbol{\gamma}_\sigma^*(\boldsymbol{X})_S F_J(\boldsymbol{X}|\boldsymbol{S})\, d\boldsymbol{X} = \boldsymbol{S}_1. \tag{5.1.28}$$

The conditional risk for any unbiased estimator $\boldsymbol{\gamma}_\sigma(\boldsymbol{X})_S$ is actually its *conditional variance*, indicated by var $\boldsymbol{\gamma}_\sigma(\boldsymbol{X})_S$. The average risk [Eq. (5.1.25)a may be written

$$R(\sigma, \boldsymbol{\gamma}_\sigma) = C_0 \int_{\boldsymbol{\Omega}} \left[\operatorname{var} \boldsymbol{\gamma}_\sigma(\boldsymbol{X})_S\right]\sigma(\boldsymbol{S})\, d\boldsymbol{S}. \tag{5.1.29}$$

Since $R(\sigma, \boldsymbol{\gamma}_\sigma)$ is least for the Bayes estimator, we see from Eq. (5.1.29) that

The Bayes estimator for the quadratic cost function has the smallest average variance among all unbiased estimators.

Similar remarks apply for the parameter estimators $\boldsymbol{\gamma}_\sigma(\boldsymbol{X})_\theta$ above.

Case 2: When signal and noise are additive and independent and waveform estimation is required, several more important properties of Bayes estimators with quadratic cost functions and simple estimation can be demonstrated. The first of these is the so-called *translation property* [8, 9], which appears when the *a priori* signal distribution is uniform. The Bayes estimator of $\boldsymbol{S}$ is now given by Eq. (5.1.26a), where $F_J(\boldsymbol{X}|\boldsymbol{S})$ is replaced by $W_J(\boldsymbol{X} - \boldsymbol{S})_N$. With $\sigma(\boldsymbol{S})$ uniform, this estimator becomes

$$\boldsymbol{\gamma}_\sigma^*(\boldsymbol{X})_S = \frac{\int \overset{\infty}{\underset{-\infty}{\cdots}} \int \boldsymbol{S} W_J(\boldsymbol{X} - \boldsymbol{S})_N\, d\boldsymbol{S}}{\int \overset{\infty}{\underset{-\infty}{\cdots}} \int W_J(\boldsymbol{X} - \boldsymbol{S})_N\, d\boldsymbol{S}}. \tag{5.1.30}$$

Now, letting $\boldsymbol{\lambda}$ be an arbitrary fixed vector and introducing a new variable $\boldsymbol{U}$ such that $\boldsymbol{U} = \boldsymbol{S} + \boldsymbol{\lambda}$, we see that (5.1.30) gives at once

$$\boldsymbol{\gamma}_\sigma^*(\boldsymbol{X} \pm \boldsymbol{\lambda})_S = \boldsymbol{\gamma}_\sigma^*(\boldsymbol{X})_S \pm \boldsymbol{\lambda}. \tag{5.1.31}$$

For example, if a fixed signal $\boldsymbol{S} = \boldsymbol{X} - \boldsymbol{N}$ is applied to a system designed to be Bayes with respect to a uniform *a priori* distribution $\sigma(\boldsymbol{S})$, then, if the same signal is changed by an amount $\boldsymbol{\lambda}$, the system output is altered by the same amount.

Case 3: A third property is that of *Minimax*. We recall from Section 1.4.4 that if a Bayes system designed for a certain signal distribution $\sigma_0(\boldsymbol{S})$ has a conditional risk that is independent of $\boldsymbol{S}$, then it is a Minimax system and $\sigma_0(\boldsymbol{S})$ is called the *least favorable distribution*. The above Bayes system $\boldsymbol{\gamma}_\sigma^*(\boldsymbol{X})_S$ has also the conditional risk

$$r(\boldsymbol{S}, \boldsymbol{\gamma}^*) = C_0 \int_{\Gamma} \left\| \boldsymbol{S} - \boldsymbol{\gamma}^*(\boldsymbol{X})_S \right\|^2 W_J(\boldsymbol{X} - \boldsymbol{S})_N \, d\boldsymbol{X}. \tag{5.1.32}$$

Letting the domain be infinite for each component and introducing new variables $\boldsymbol{Z} = \boldsymbol{X} - \boldsymbol{S}$, we see that (5.1.21) becomes

$$r(\boldsymbol{S}, \boldsymbol{\gamma}^*) = C_0 \int_{-\infty}^{\infty} \cdots \int \left\| \boldsymbol{S} - \boldsymbol{\gamma}^*(\boldsymbol{S} - \boldsymbol{Z})_S \right\| W_J(\boldsymbol{Z})_N \, d\boldsymbol{Z} \tag{5.1.33}$$

but since the Bayes system $\boldsymbol{\gamma}^*$ here has the translation property (5.1.31), we have $\boldsymbol{\gamma}^*(\boldsymbol{S} + \boldsymbol{Z})_S = \boldsymbol{\gamma}^*(\boldsymbol{Z})_S + \boldsymbol{S}$. When this is used in Eq. (5.1.31), the conditional risk becomes independent of $\boldsymbol{S}$, showing, therefore, that

> *When signal and nose are additive and independent, the least favorable a priori distribution $\sigma_0(\boldsymbol{S})$ is a uniform one and the Bayes system $\boldsymbol{\gamma}_\sigma^*$ for this distribution is Minimax $(= \boldsymbol{\gamma}_M^*)$.*

Observe that, with deterministic signals, the signal parameters (except for amplitude $\sim a_0$) do not appear *linearly* in $\boldsymbol{S}$, so that the translation property above does not hold any more for the Bayes estimator $\boldsymbol{\gamma}_\sigma^*(\boldsymbol{X})_\theta$ even if $\sigma(\boldsymbol{\theta})$ is uniform. Moreover, this Bayes system is no longer Minimax, while the least favorable distribution $\sigma_0(\boldsymbol{S})$ is not uniform. The exception to this occurs for the signal amplitude, because of its linear relation vis-à-vis the accompanying noise.

Sometimes the added constraint of linearity is imposed upon the estimator $\boldsymbol{\gamma}(\boldsymbol{X})$, so that the estimate $\boldsymbol{\gamma}$ is required to be the output of a linear (usually physically realizable) filter with $X(r_m, t)$ as its input. Then it is not generally true that optimum extractors under this constraint are Bayes, since from Eq. (5.1.26a) the Bayes extractor is usually a nonlinear operator upon $X(r_m, t)$, even for the simple cost assignment of Section 5.1.4. In fact, as we can see from the above, $F_J(\boldsymbol{X}|\boldsymbol{S})$ must have rather special properties if Bayes and linearity are to be concomitant features of the optimum extractor. We remark, however, that some of the notions of risk theory are useful for such restricted classes of decision rules, even though the main theorems do not apply [10]. That is, if we agree that only the class of linear estimators is to be considered, we may speak of the one with the smallest average risk, the one for which the maximum conditional risk is smallest, the one with the property that no other in the class is uniformly better, and so on, settling the questions of existence and uniqueness in specific situations by construction.

Finally, recall again that these estimators $\boldsymbol{\gamma}_\sigma(\boldsymbol{X})_S$, $\boldsymbol{\gamma}_\sigma(\boldsymbol{X})_\theta$ embody the actual structure of our receiving and data processing systems, for example, $\boldsymbol{T}_R\{\boldsymbol{X}\} = \boldsymbol{\gamma}_\sigma(\boldsymbol{X})_S$ or $\boldsymbol{\gamma}_\sigma(\boldsymbol{X})_\theta$. Optimal systems from the risk point of view have now been defined with respect to at least two classes of cost function, the simple and the quadratic[10] [(5.1.4) et seq., (5.1.18) et seq.]. Some additional cost functions are considered in the next section, while the questions of structure and the distributions of the estimators themselves are examined in more detail in Section 5.2.

5.1.5 Other Cost Functions[11]

The number of possible cost functions, while theoretically infinite, is in practice limited by the dual requirements of reasonableness and computability. On the one hand, the cost of "errors" must depend on the magnitude of the "error" in some fashion such that correct decisions are penalized less than incorrect decisions. On the other hand, the cost function should be such that optimum systems (in the sense of minimum average risk, for example) can be found, at least approximately, from such conditions as Eqs. ((5.1.21a) and (5.1.21b)) and their corresponding Bayes risks in turn determined. Perhaps the most natural choice of cost function is the symmetric "distance" function

$$C(\boldsymbol{\theta}, \boldsymbol{\gamma}_\sigma) = C(|\boldsymbol{\theta} - \boldsymbol{\gamma}_\sigma|), \text{ or } C(\boldsymbol{S}, \boldsymbol{\gamma}_\sigma) = C(|\boldsymbol{S} - \boldsymbol{\gamma}_\sigma|), \tag{5.1.34}$$

where correct decisions cost a fixed amount $C^{(0)}(\geq 0)$ and incorrect decisions a greater amount; consequently, if C possess a Taylor expansion, Eq. (5.1.34) can be alternatively represented by series of the type

$$C(\boldsymbol{\theta}, \boldsymbol{\gamma}_\sigma) = C^{(0)} \sum_{ij} \left(C_\theta^{(0)}\right)_{ij} (\theta_i - \gamma_{\sigma_i})(\theta_j - \gamma_{\sigma_i}) \tag{5.1.34a}$$

with an analogous development for $C(\boldsymbol{S}, \boldsymbol{\gamma}_\sigma)$ in terms of $C_S^{(0)}$, $C_{Sij}^{(2)}$, $(S_i - \gamma_{\sigma_i})$, and so on.

Let us consider the K parameters $\boldsymbol{\theta}$ a number of K-dimensional cost functions of the type (5.1.34), which, however, are not necessarily developable in a Taylor series like Eq. (5.1.34a). The quadratic cost function of Eq. (5.1.18) is one example that we have already examined in some detail (Section 5.1.3). Another is the *exponential cost function*

$$C_2(\boldsymbol{\theta}, \boldsymbol{\gamma}_\sigma) = C_0\left\{1 - \exp\left[-{}^1\!/\!_2(\tilde{\boldsymbol{\theta}} - \boldsymbol{\gamma}_\sigma)\boldsymbol{n}^{-1}(\boldsymbol{\theta} - \boldsymbol{\gamma}_\sigma)\right]\right\}, \tag{5.1.35}$$

where $\boldsymbol{n}$ is the diagonal matrix $[\eta_k \delta_{kj}]$ and the η_k are scale factors with the dimensions of the respective parameters θ_k. The η_k are directly related to the fineness with which the estimator (or receiver $\boldsymbol{T}_\mathrm{R}\{\boldsymbol{X}\} = \boldsymbol{\gamma}_\sigma$) is able to distinguish between correct and incorrect values. According to Eq. (5.1.35), the receiver is penalized comparatively little if the estimates $\boldsymbol{\gamma}_\sigma(\boldsymbol{X})$ are close to the actual values $\boldsymbol{\theta}$ and almost the maximum amount C_0 when one or more of these estimates depart noticeably from the correct values. Unlike the expandable cost functions of Eqs. (5.1.18) and (5.1.35), we cite two additional examples involving the nondevelopable "rectangular" cost functions:

$$C_3(\boldsymbol{\theta}, \boldsymbol{\gamma}_\sigma) = C_0\left(1 - \sum_{k=1}^{K} \Delta\left|\frac{\theta_k - \gamma_k}{\eta_k}\right|\right) \tag{5.1.36}$$

[10] We remark that the Bayes estimators for quadratic cost functions also sometimes possess optimal properties with respect to a wider class of cost functions (Section 5.1.5).

[11] See the report by Ashby [11].

and
$$C_4(\boldsymbol{\theta}, \boldsymbol{\gamma}_\sigma) = C_0\left(1 - \prod_{k=1}^{K} \Delta\left|\frac{\theta_k - \gamma_k}{\eta_k}\right|\right), \tag{5.1.37}$$

where $\boldsymbol{\Delta}|x| = 1$ if $|x| < 1$ and $\boldsymbol{\Delta}|x| = 0$ when $|x| > 1$. The former assesses the observer an amount $C_0 l/K$ if $l(\geq 1)$ estimates out of the K fall outside the tolerance limit $(\theta_i \pm \eta)$, while the latter invokes the maximum penalty C_0 if any *one* estimate lies outside these limits. For many applications, the latter is too strict. It produces too high an average risk (for given η), rapidly approaching the maximum C_0 as $K \to \infty$.

Although, strictly speaking, the derivatives of the rectangular cost functions C_3, C_4 [Eqs. (5.1.38) and (5.1.39a)] do not exist, we can still obtain the solution of the extremum condition for the Bayes estimators $\boldsymbol{\gamma}^*$, in terms of delta functions. Thus, for C_3 we write

$$\frac{\partial}{\partial\gamma} C_0\left(1 - \sum_{k=1}^{K} \Delta\left|\frac{\theta_k - \gamma_k}{\eta_k}\right|\right) = C_0[\delta(\theta_k - \gamma_k + \eta_k) - \delta(\theta_k - \gamma_k - \eta_k)];\ k = 1, \ldots, K. \tag{5.1.38}$$

Substituting this into Eq. (5.1.21b) yields the set of K equations $(k = 1, \ldots, K)$

$$\sigma(\gamma_k^* - \eta_k) W_J(\boldsymbol{X}|\gamma_k^* - \eta_k) = \sigma(\gamma_k^* + \eta_k) W_J(\boldsymbol{X}|\gamma_k^* + \eta_k). \tag{5.1.39a}$$

The solutions $\boldsymbol{\gamma}_k^*$ of these relations are accordingly the optimum estimators of $\boldsymbol{\theta}$. In terms of the joint d.d. $w_k(\boldsymbol{X}, \theta_k)$ of $\boldsymbol{X}$ and θ_k we have the equivalent relations

$$w_k(\boldsymbol{X}, \gamma_k^* - \eta_k) = w_k(\boldsymbol{X}, \gamma_k^* + \eta_k); \quad k = 1, \ldots, K. \tag{5.1.39b}$$

Note that when η_k is small, we may develop both members of Eq. (5.1.39b) after first taking their logarithms, to get approximately (for each $k = 1, \ldots, K$)

$$\log w_k(\boldsymbol{X}, \gamma_k^*) - \eta_k \frac{\partial \log w_k}{\partial\theta_k}\bigg|_{\theta_k=\gamma_k^*} = \log w_k(\boldsymbol{X}, \gamma_k^*) + \eta_k \frac{\partial \log w_k}{\partial\theta_k}\bigg|_{\theta_k=\gamma_k^*}$$
$$\frac{\partial \log w_k(\boldsymbol{X}, \theta_k)}{\partial\theta_k}\bigg|_{\theta_k=\gamma_k^*} = 0;\ k = 1, \ldots, K \tag{5.1.40}$$

But this is precisely the relation specifying the unconditional maximum likelihood estimator (UMLE) of θ_k [cf. Eqs. (5.1.9) and (5.1.10a)] or, equivalently, obeying the condition (5.1.14a), since $w_k(\boldsymbol{X}, \theta_k) = \sigma(\theta_k) W_J(\boldsymbol{X}, \theta_k)$. Thus, $\boldsymbol{\gamma}^*$ is Bayes with respect to the continuous version (5.1.12a) of the constant cost function discussed earlier, in Section 5.1.2, which in turn is the limiting form as $\eta_k \to 0$ of the rectangular cost function of Eq. (5.1.36).[12] The (vector) Bayes estimator $\boldsymbol{\gamma}_\sigma^*$ is again the UML estimator $\hat{\boldsymbol{\gamma}}_\sigma$, each component of which is determined from Eq. (5.1.40), when we are able to distinguish between magnitudes arbitrarily close together in value (i.e., $\eta_k \to 0$). While such distinctions in physical cases cannot be pushed to this theoretical limit, often η_k can still be made so small that Eq. (5.1.40)

[12] After appropriate adjustment of the constants C_0, and so on, in the two cases.

is quite acceptable in practice. This limiting form, leading to the equivalent result [Eq. (5.1.40)], is called the *simple cost function of Type One, (SCF_1)*.

Similar results may be derived for the stricter rectangular cost function (5.1.37), which in the limit $\boldsymbol{n} \to 0$ leads to the unconditional maximum likelihood estimator $\hat{\boldsymbol{\gamma}}_\sigma(\boldsymbol{X}) = \gamma_\sigma^*(\boldsymbol{X})$, defined now by[13]

$$\sigma(\hat{\boldsymbol{\gamma}}_\sigma)W_J(\boldsymbol{X}|\hat{\boldsymbol{\gamma}}_\sigma) \geq \sigma(\theta)W_J(\boldsymbol{X\theta}), \text{ simultaneously all } \boldsymbol{\theta} \text{ in } \boldsymbol{\Omega}_\theta \tag{5.1.41}$$

or, equivalently, determined by the set of joint likelihood equations. The constant cost function in this instance is given by

$$C = C_0[A_K - \delta(\boldsymbol{\gamma}_\sigma - \boldsymbol{\theta})], \tag{5.1.42}$$

which is called equivalently the *strict simple cost function(SCF_2)*, [cf. Eq. (5.1.12a)]. The corresponding Bayes risk is

$$R^*(\sigma, \delta)_\theta = C_0\left[A_K - \int_\Gamma \sigma(\hat{\boldsymbol{\gamma}}_\sigma)W_J(\boldsymbol{X}|\hat{\boldsymbol{\gamma}}_\sigma)\boldsymbol{dX}\right] > 0. \tag{5.1.43}$$

A theorem [11] for determining Bayes estimators $\boldsymbol{\gamma}_\sigma^*$ can sometimes be applied when the cost function is a symmetric differentiable distance function of the type of Eq. (5.1.34). The theorem states:

Theorem I. If the joint d.d. $w(\boldsymbol{X}, \boldsymbol{\theta})$ *can be factored into the form* $w(\boldsymbol{X}, \boldsymbol{\theta}) = f_1(\boldsymbol{X})f_2[\boldsymbol{\theta} - \boldsymbol{g}(\boldsymbol{X})]$, *where* f_1 *and* g *are any functions of* X, *and if* $f_2[\boldsymbol{\theta} - \boldsymbol{g}(\boldsymbol{X})] = f_2[\boldsymbol{g}(\boldsymbol{X}) - \boldsymbol{\theta}]$ *and is unimodal about* $\boldsymbol{\theta} = \boldsymbol{g}(\boldsymbol{X})$, *where the ranges of* $\boldsymbol{\theta}$ *are* $(-\infty, \infty)$, *then, for differentiable*[14] *cost criteria [Eq. (5.1.34)], the Bayes estimator is*

$$\boldsymbol{\gamma}_\sigma^*(\boldsymbol{X}) = \boldsymbol{g}(\boldsymbol{X}). \tag{5.1.44}$$

To establish this result, observe that Eq. (5.1.21b) for determining $\boldsymbol{\gamma}_\sigma^*$ is equivalent to

$$\int_{\boldsymbol{\Omega}_\theta} f_1(\boldsymbol{X})f_2[\boldsymbol{\theta} - \boldsymbol{g}(\boldsymbol{X})]\frac{\partial C(\boldsymbol{\theta} - \boldsymbol{\gamma}_\sigma^*)}{\partial \boldsymbol{\theta}}\, \boldsymbol{d\theta} = 0. \tag{5.1.44a}$$

Since $C(\boldsymbol{\theta} - \boldsymbol{\gamma}_\sigma^*)$ is even, its derivatives are odd. With $z = \boldsymbol{\theta} - \boldsymbol{g}(\boldsymbol{X})$, and after the substitution $\boldsymbol{g}(\boldsymbol{X}) = \boldsymbol{\gamma}_\sigma^*$, Eq. (5.1.44) is now equivalent to

$$\int_{-\infty}^{\infty} \cdots \int f_2(z)\frac{\partial C(z)}{\partial z}\, \boldsymbol{dz} = 0, \tag{5.1.44b}$$

which is identically satisfied, since the integrand is odd. This establishes the result (5.1.44), where we note that (5.1.44) yields a minimum average risk because of the convex nature of the cost function and the unimodality of f_2.

[13] This is the extraction procedure discussed originally in Middleton and Van Meter [1], Section 4.1.

[14] The theorem is easily extended to include the rectangular cost functions [Eqs. (5.1.36) and (5.1.37)] in the limit of arbitrarily small η_k, and so on, the constant or "simple" cost functions [Eqs. (5.1.12a) and (5.1.42)].

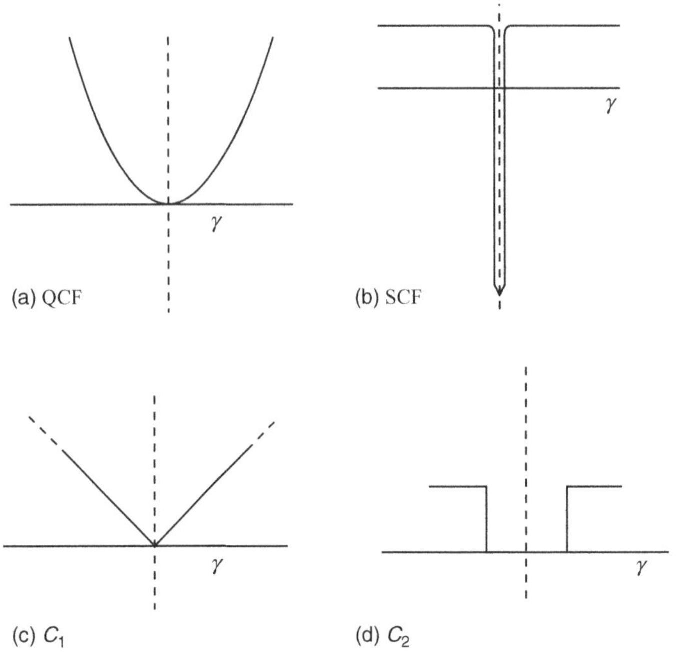

FIGURE 5.1 Some typical cost functions for signal.

A second theorem[15] is also often useful when we attempt to obtain Bayes estimation and Bayes risks for cost functions other than the quadratic:

Theorem II. For all cost functions of the type $C(\theta - \gamma_\sigma)$ where

$$\left.\begin{array}{l} C(\theta - \gamma_\sigma) = C(\gamma_\sigma - \theta) \geq 0 \\ C\left[(\theta - \gamma_\sigma)_1\right] < C\left[(\theta - \gamma_\sigma)_2\right] \quad \text{if}\left|(\theta - \gamma_\sigma)_1\right| < \left|(\theta - \gamma_\sigma)_2\right| \end{array}\right\}, \tag{5.1.45}$$

the Bayes estimator $\gamma_\sigma^*\left[= \left(\gamma_\sigma^*\right)_{\text{QCF}}\right]$ for a quadratic cost function is also optimum, that is, minimizes the average risk, for these other cost functions, provided the conditional d.d. of the parameter (θ), given $\mathbf{X}$, that is, $w(\theta|\mathbf{X})$, is unimodal and symmetric about the mode[16] [of $w(\theta|\mathbf{X})$].

Some representative cost functions of practical interest shown in Fig. 5.1 are [with $\gamma_\sigma = \gamma^* \equiv \left(\gamma_\sigma^*\right)_{\text{QCF}}$ here, for optimality]

$$C_1\left(\theta - \gamma_\sigma^*\right) = \left|\theta - \gamma_\sigma^*\right|; \tag{5.1.46a}$$

$$C_2\left(\theta - \gamma_\sigma^*\right) = \begin{cases} 0 & \left|\theta - \gamma_\sigma^*\right| < A; \\ 1 & \left|\theta - \gamma_\sigma^*\right| > A > 0; \end{cases} \tag{5.1.46b}$$

[15] This is a modified version of Sherman's results [5].

[16] Or, equivalently, that $w(\theta - \gamma_\sigma^*|\mathbf{X})$ is unimodal and symmetric about $\theta = \gamma_\sigma^*$.

$$C_3(\theta - \gamma_\sigma^*) = \begin{cases} 0 & |\theta - \gamma_\sigma^*| < A; \\ \dfrac{|\theta - \gamma_\sigma^*| - |A|}{|B| - |A|} & |A| \le |\theta - \gamma_\sigma^*| \le |B|; \\ 1 & |B| < |\theta - \gamma_\sigma^*|. \end{cases} \tag{5.1.46c}$$

Note again that $\gamma_\sigma^* = (\gamma_\sigma^*)_{QCF}$ is not generally linear in $\boldsymbol{X}$ and that, of course, the Bayes risks[17] $E_{X,\theta}\{C_1\}$, and so on, are not the same as for the quadratic cost function (5.1.23).

Finally, the results of this section may be equally well applied to the important cases of waveform estimation. For simple estimation $(\lambda = j)$, $j = 11, \ldots, MN$, one simply replaces θ by $\boldsymbol{S}$ and $W_J(\boldsymbol{X}|\boldsymbol{\theta})$ by $F_J(\boldsymbol{X}|\boldsymbol{S})$, and so on, as before. For interpolation and extrapolation, similar results are readily found, with an appropriate modification of the cost function $C(\boldsymbol{S};\ \boldsymbol{\gamma}_\sigma)$ to $C(\boldsymbol{S},\ S;\ \boldsymbol{\gamma}_\sigma)$ (cf. the beginning of Section 5.1.1). While the cost functions of the distance type lead generally to the simplest results for $\boldsymbol{\gamma}_\sigma^*$ and R^*, we are in no way restricted to their use. The informational cost function $F_2 = -\log p(\boldsymbol{S}|\boldsymbol{\gamma})$ is one example of a more complicated type, where the cost assignment depends not only on $\boldsymbol{S}$ and $\boldsymbol{\gamma}$ but on the decision rule δ as well.

5.2 COHERENT ESTIMATION OF AMPLITUDE (DETERMINISTIC SIGNALS AND NORMAL NOISE, $(p(H_1) = 1)$

A fairly comprehensive theory for amplitude estimation in normal noise can be constructed when the signal $(\boldsymbol{s})$ is deterministic, signal and noise $(\boldsymbol{n})$ are additive and independent, and when the quadratic or the simple cost functions are chosen as measures of the risk. It is further assumed that $\bar{\boldsymbol{n}}$ vanishes, the usual situation in most applications, and that observation is coherent, that is, the amplitude is the only unknown quantity. The following four examples illustrate the calculation of Bayes extractors and their associated Bayes risks: (1) coherent extraction of signal amplitude in normal noise with a quadratic cost function and normal d.d. of amplitudes; (2) the same as (1), but with the simple cost function SCF_1 $(= \text{SCF}_2$ here), which is the limit of C_3 as $\eta_k \to 0$, cf. (5.1.36) et seq.; (3) incoherent estimation of amplitude, again in normal noise with a quadratic cost function; and (4) similar to (3), but with the simple cost function SCF_1, resulting in an appropriate UML estimator. In this section, we assume (1) and Section 5.3, on the other hand, is devoted to (3) and (4), as a consequence of the fact, once more, that the amplitude parameter appears linearly in the signal representations.

5.2.1 Coherent Estimation of Signal Amplitude[18] Quadratic Cost Function

We examine first the estimation of the amplitude factor a_0 of a deterministic signal $S(t - \varepsilon_0,\ \boldsymbol{\theta}_0)$ which has a *normal* distribution of amplitudes but is otherwise completely specified at the receiver:

$$\sigma(\boldsymbol{\theta}) = \sigma(a_0) = \left(2\pi\sigma^2\right)^{-1/2} e^{-(a_0 - \bar{a}_0)^2/2\sigma^2}; \quad \sigma^2 = \overline{a_0^2} - \bar{a}_0^2. \tag{5.2.1}$$

[17] To calculate these other Bayes risks we may profitably employ the Laplace-transform techniques of Section _______ and carry out the required averages over the resulting exponential terms, as in the calculation of output covariance and spectra (cf. Chapter __ also).

[18] We assume throughout that unless otherwise indicated, signal and noise are normalized by $\psi_j^{1/2}$.

Again we assume that signal $\boldsymbol{s}$ and noise $\boldsymbol{n}$ are additive and normalized,

$$a_0\boldsymbol{s} = \left[S_j/\psi_j^{1/2}\right]; \quad \boldsymbol{x} = a_0\boldsymbol{s} + \boldsymbol{n}; \quad \boldsymbol{n} = \left[N_j/\psi_j^{1/2}\right]. \tag{5.2.1a}$$

The quadratic cost function here is $C_0(\gamma_\sigma - a_0)^2$, where γ_σ is the estimator, so that from Eq. (5.1.22) or (5.1.30) the Bayes estimator $\gamma_\sigma^* = a_0^*$ becomes

$$\boldsymbol{T}_{\mathrm{R}}^{[N]}(\boldsymbol{X}) = a_0^*(\boldsymbol{X}) = \int_{-\infty}^{\infty} a_0 w(\boldsymbol{x}, a_0)\, da_0 \Big/ \int_{-\infty}^{\infty} w(\boldsymbol{x}, a_0)\, da_0 \tag{5.2.2}$$

with the joint distribution density $w(\boldsymbol{x}, a_0)$ here specifically equal to

$$\frac{\exp\left\{-1/2\left[(\boldsymbol{x} - a_0\tilde{\boldsymbol{s}})\boldsymbol{k}_N^{-1}(\boldsymbol{x} - a_0\boldsymbol{s}) + (a_0 - a_0 s)^2/\sigma^2\right]\right\}}{(2\pi)^{J/2}(\det \boldsymbol{k}_N)^{1/2}(2\pi\sigma^2)^{1/2}}; \quad \bar{N} = 0 \tag{5.2.2a}$$

and with $\boldsymbol{k}_N$, as before, the (normalized) covariance matrix of the normal noise process for sample size J. Noting the identity

$$a_0^* = \frac{\int_{-\infty}^{\infty} a_0 e^{-A\left(a_0 - a_0^*\right)^2 + B(x)} da_0}{\int_{-\infty}^{\infty} e^{-A\left(a_0 - a_0^*\right)^2 + B(x)} da_0}, \tag{5.2.2b}$$

we can evaluate Eq. (5.2.2) very simply by completing the square in Eq. (5.2.2a). The result is the desired Bayes estimator, or optimum receiver,

$$\boldsymbol{T}_{\mathrm{R}}^{[N]}(\boldsymbol{X}) = a_0^*(\boldsymbol{X}) = \frac{\Phi_J(\boldsymbol{x}, \boldsymbol{s}) + \overline{a_0}/\sigma^2}{\Phi_J(\boldsymbol{s}, \boldsymbol{s}) + 1/\sigma^2} = \frac{\sigma^2\Phi_x + 1/\overline{a_0}}{\sigma^2\Phi_s + 1}, \tag{5.2.3}$$

where $$\Phi_J(\boldsymbol{x}, \boldsymbol{s}) = \tilde{\boldsymbol{x}}\boldsymbol{k}_N^{-1}\boldsymbol{s} \equiv \Phi_J(\boldsymbol{s}, \boldsymbol{s}) = \tilde{\boldsymbol{s}}\boldsymbol{k}_N^{-1}\boldsymbol{s} \equiv \Phi_S. \tag{5.2.3a}$$

Because of sample certainty $[p(H_1) = 1]$, the normality of the noise and the linearity of the parameter in the signal representation, this estimator is *linear* in the received data $\boldsymbol{X}$. (When the noise is non-Gaussian, the optimum estimator of amplitude is *nonlinear* in the received data.)

We observe first that as $\sigma^2 \to 0$, that is, as the *a priori* signal density becomes a delta function at $a_0 = \hat{a}_0$, the Bayes estimator reduces to $a_0^* = \bar{a}_0 = a_0$, as expected. When $\sigma^2 \to \infty$ [so that the *a priori* distribution density becomes uniform] the Bayes estimates become equal to $a_0^* = \hat{\gamma}_{a_0} = \hat{a}_0$, agreeing with the maximum (conditional and unconditional) likelihood estimates, as it should [for $\bar{N} = 0$, cf. the examples in 5.2.2]. The structure of these estimators takes the form indicated in Fig. 21.2 of Ref. [2].[19]

Because of the normal character of the noise, the normal distribution of amplitudes, and the coherent observation of the signal, the Bayes estimator of amplitude is here a linear function of the received sample data $\boldsymbol{X}$. The distribution density of this estimator is accordingly normal. To determine the explicit form of the d.d., let us begin first with the characteristic function

[19] Here, for $\Phi_J(\boldsymbol{x}, \boldsymbol{s})$ one has essentially the same elements, for example, matched filters and ideal integrators used in the coherent detection of such signals, followed, however, by a simple computer which adds $\bar{a}_0/\sigma^2$ and then divides by $\Phi_T + \sigma^{-2}$.

$$F_1(i\xi)_{a_0^*} = E_{X|H}\left\{e^{i\xi a_0^*(X)}\right\} = \int_{\Gamma} e^{i\xi\left(Axk_N^{-1}s+B\right)} w_J(X)dX, \tag{5.2.4}$$

where $w_J(X)$ is found from Eq. (5.2.2a) to be specifically

$$w_J(X) = \left[(\sigma^2\Phi_s + 1)\,(2\pi)^J\,(\det k_N)\right]^{-1/2} \exp\left(c_0 - \bar{a}_0^2/2\sigma^2\right), \\ \cdot \exp\left[-1/2\,x\left(k_N^{-1} - 2c_2 G\right)x + c_1\left(xk^{-1}s\right)\right] \tag{5.2.5}$$

with $G = k_N^{-1} s\tilde{s} k_N^{-1}$ and

$$c_0 = a_0^2/\left(2\sigma^4\Phi_S + 2\sigma^2\right); \quad c_1 = \bar{a}_0/\left(1 + \sigma^2\Phi_S\right); \quad c_2 = 2\Phi_S + 2/\sigma^2 \tag{5.2.5a}$$

and $c_0 > 0$, c_1, $c_2 \geq 0$. Carrying out the indicated operations[20] in Eq. (5.2.4) using Eq. (5.2.5), we get finally

$$F_1(i\xi)_{a_0^*} = e^{i\xi a_0 - 1/2 \sim \sigma^4\Phi_S\xi^2/\left(1+\sigma^2\Phi_S\right)}, \tag{5.2.6}$$

so that a_0^* is normally distributed with mean $\bar{a}_0$ and variance $\sigma^4\Phi_s/(1 + \sigma^2\Phi_s)$. Observe that when $\sigma^2 \to 0$, the Bayes estimator a_0^* equals $\bar{a}_0 = a_0$, as expected, since the only value of amplitude is $\bar{a}_0$ itself, that is, $w_1\left(a_0^*\right) = \delta\left(a_0^* - \bar{a}_0\right)$. On the other hand, if σ^2 is sufficiently great, then $\sigma^2\Phi_s \gg 1$ and then a_0^* is normally distributed with a variance that is independent of waveform and equal to the variance of a_0 itself. This is reasonable, since now the large spread in the possible values of signal amplitude, apart from the effects of the background noise, dominates the distribution of $a_0^*(X)$.

We may use Eq. (5.2.6) directly to obtain the first and second moments of the Bayes estimator. These are

$$\overline{a_0^*} = \bar{a}_0; \quad \overline{a_0^{*2}} = \bar{a}_0^2 + \frac{\sigma^4\Phi_S}{1 + \sigma^2\Phi_S}. \tag{5.2.7}$$

For the Bayes risk [Eq. (5.1.23)], we need the d.d. of $a_0^* - a_0$, which like that of a_0^* is also normal, since a_0^* is linear in X [cf. Eq. (5.2.3)] and a_0 is normally distributed. Using the fact that

$$E_{x|H_1}\{\Phi_x\} = \bar{a}_0\Phi_S, \quad E_{x|H_1}\left\{\Phi_x^2\right\} = \overline{a_0^2}\Phi_S^2 + \Phi_S,$$

we find after a little manipulation that

$$E_{x,a_0|H_1}\left\{\left(a_0^* - a_0\right)^2\right\} = \frac{\sigma^2}{\sigma^2\Phi_S^* + 1}. \tag{5.2.8}$$

The d.d. of $a_0^* - a_0$ is normal, c.f. (6.4.9) ff., specifically with zero mean (since $\overline{a_0^*} - \bar{a}_0$) and variance $\sigma^2/\sigma^2\Phi_S^* + 1$ above. The c.f. of $a_0^* - a_0$ is accordingly

$$F_1(i\xi)_{a_0^* - a_0} = e^{-\xi^2\sigma^2/2\left(\sigma^2\Phi_S+1\right)}. \tag{5.2.8a}$$

[20] An alternative and simpler method is to compute the c.f. of a_0^* directly from $E_{X,\,a_0|H_1}\left\{e^{i\xi a_0^*}\right\}$ with respect to the joint d.d. $w(X,\,a_0)$ [Eq. (5.2.2a)].

The Bayes risk is proportional to Eq. (5.2.8), or to $-d^2F/d\xi^2|_{\xi=0}$, from Eq. (5.2.8a), so that we can write it directly as

$$R^*(\sigma, \delta)_{a_0} = C_0\left(\overline{a_0^{*2}} - \overline{2a_0a_0^*} + \overline{a_0^2}\right) = \frac{C_0\sigma^2}{1+\sigma^2\Phi_S}. \tag{5.2.9}$$

A more detailed development of these results is given in Section 6.4.1 ff., including $w_1(y, a_0|H_1), w_1(y|a_0)_{H_1}, w_1(y|H_1)$ where $y = a_0^*(\boldsymbol{X})$, (5.2.3), preliminary to the treatment of the more general situation, $a_0^*|_{p<1}$.

As expected when $\sigma^2 \to 0$, the average risk vanishes, sine extraction is exact on the average, that is, $a_0^* = \bar{a}_0 = a_0$. For $\sigma^2 \to \infty$ there is necessarily always a finite average cost $C_0\Phi_J(s, s)^{-1}$. The Bayes risk, of course, depends on the structure of the signal and on the spectrum of the accompanying noise. This example is of some practical interest in the case where signals subject to fading are received in noise, when $\bar{a}_0$ is large and σ^2 reasonably small, so that a_0 is essentially always greater than zero. In the present case, the noise arises primarily in the receiver (and hence is unaffected by the propagation mechanism producing the fading). Observe, moreover, that a_0^* (5.2.3) or (5.2.3a) with $\sigma^2 \to \infty$ is also a *Minimax estimate*, as can be seen from the translation property [Eq. (5.1.31)], which becomes here $a_0^*(x+bs) = a_0^*(x) + bs$, with b a scalar quantity. One then uses an argument precisely parallel to that given above [Eq. (5.1.32) et seq.], to show that the conditional risk is independent of a_0. For discrete sampling, the Minimax average risk is specifically

$$R_{\mathrm{M}}^* = C_0\Phi_S(s, s)^{-1} \qquad (>R^*). \tag{5.2.9a}$$

5.2.2 Coherent Estimation of Signal Amplitude (Simple Cost Functions)

From the results of Section 5.1.2 and in particular (5.1.14b) or (5.1.40) where the limit $\eta_k \to 0$ is employed on the rectangular cost function C_J, we obtain the following equivalent relation for the resulting UMLE in the simple parameter case, $\theta = a_0$:

$$\begin{aligned}\left[\frac{\partial}{\partial\theta}\log w_1(\boldsymbol{X}, \theta)\right]_{\theta=a_0=a_0^*} &= \left[\frac{\partial}{\partial a_0}\log \sigma(a_0) = w_1(\boldsymbol{X}|a_0)\right]_{a_0=a_0^*},\\ &= \left[\frac{\partial}{\partial a_0}\log \text{Eq. (5.2.2a)}\right]_{a_0\to a_0^*}\end{aligned} \tag{5.2.10}$$

applied here for amplitude estimation, where $\sigma(a_0)$ is the pdf of a_0, given by (5.2.2a). Equation (5.2.10) becomes specifically

$$\frac{\partial}{\partial a_0}\left[\Phi_J(\boldsymbol{x},\boldsymbol{x})/2 - a_0\Phi_J(\boldsymbol{x},\boldsymbol{s}) + a_0\Phi_J(\boldsymbol{s},\boldsymbol{s}) + (a_0-\bar{a}_0)/2\sigma^2\right]\Big|_{a_0\to a_0^*} = 0, \tag{5.2.11a}$$

which yields at once

$$a_0^* = a_0^*(\boldsymbol{x})\Big|_{\mathrm{SCF}} = \frac{\Phi_J(\boldsymbol{x}, \boldsymbol{s}) + \bar{a}_0/\sigma^2}{\Phi_J(\boldsymbol{s}, \boldsymbol{s}) + 1/\sigma^2} = \frac{\sigma^2\Phi_x + \bar{a}_0}{\sigma^2\Phi_s + 1} = a_0^*(\boldsymbol{x})\Big|_{\mathrm{QCF}}, \tag{5.2.11b}$$

from (5.2.3). This shows that the UMLE for signal amplitude under the conditions of (1), (5.2.1) is equal to that derived from the QCF. This, of course, is not generally true: the estimators for the QCF and $SCF_{1,2}$ are generally different, and the associated Bayes risks are always different, as indicated, for example, by (5.2.9) versus (5.2.12) ff.

Let us now calculate the Bayes risk here for the SCF_1 (= SCF_2), from (5.1.43). We have

$$R^*(\sigma, \delta)_\theta = C_0\left[A_1 - \int w_1(\boldsymbol{x}, a_0^*)\,d\boldsymbol{x}\right], \tag{5.2.12a}$$

for which we use (5.2.2a) or the simpler result (6.2.37) in (6.4.43) ff., to obtain

$$R^*(\sigma, \delta)_\theta = C_0\left[A_1 - \left(\frac{1+\sigma^2\Phi_s}{2\pi\sigma^3\Phi_s^{1/2}}\right)e^{-a_0^2\left(1+\sigma^2\Phi_s\right)/2\sigma^4\Phi_s}\right]. \tag{5.2.12b}$$

Although these estimators are the same (5.2.11b), clearly the Bayes risks in the two cases (QCF versus $SCF_{1,2}$) are not: vide (5.2.9) versus (5.2.12b).

5.2.3 Estimations by (Real) θ Filters[21]

It is instructive to show that the estimator a_0^*, (5.2.11b), is equivalent to a W-K filter, that is, a mean square error (MMSE) filter, operating on the input data $\boldsymbol{x}(=\boldsymbol{s}+\boldsymbol{n})$ under the conditions (1) at the beginning of Section 5.2, that is, deterministic signals, additive Gaussian noise (with zero mean, $\bar{n}=0$), which is independent of the signal. Instead of minimizing the variance of the random part of the received data $\boldsymbol{x}$ for a specified received signal or signal-to-noise ratio, here at some point in space–time the usual condition (cf. Sections 16.2 and 16.3 of Ref. [2] for matched filters) of operation, we require equivalently that the noise power output of the Wiener filter be minimized under the constraint of maintaining a specified constant signal value at the end of the space–time sampling period. Accordingly, we have for this constraint

$$\tilde{\boldsymbol{w}}_s = \tilde{\boldsymbol{s}}\boldsymbol{w} = C\ \ (>0) = \text{output value at } j = J, \tag{5.2.13a}$$

where $\boldsymbol{w}$ is the vector space–time response (i.e., discrete weighting function) of the Wiener–Kolmogoroff filter. The noise output variance to be minimized is

$$\left\langle (\tilde{\boldsymbol{w}}\boldsymbol{n})^2 \right\rangle = \boldsymbol{w}\overline{\boldsymbol{n}\tilde{\boldsymbol{n}}}\boldsymbol{w} = \boldsymbol{w}\boldsymbol{k}_N\boldsymbol{w}. \tag{5.2.13b}$$

The extremum process to be carried out is therefore, from $L = \boldsymbol{w}\boldsymbol{k}_N\boldsymbol{w} + \boldsymbol{\lambda}\tilde{\boldsymbol{s}}\boldsymbol{w}$,

$$\delta(\boldsymbol{w}\boldsymbol{k}_N\boldsymbol{w} + \boldsymbol{\lambda}\tilde{\boldsymbol{s}}\boldsymbol{w}) = \delta L = 0, \tag{5.2.14}$$

and for a minimum we require $\delta L = 0,\ \delta^2 L > 0$.

Carrying out the variation indicated in (5.2.14), we have, since $\tilde{\boldsymbol{k}}_N = \boldsymbol{k}_N$:

$$\delta\, \boldsymbol{w}\left(\tilde{\boldsymbol{w}}\tilde{\boldsymbol{k}}_N + \tilde{\boldsymbol{w}}\boldsymbol{k}_N + \boldsymbol{\lambda}\tilde{\boldsymbol{s}}\right) = 0;\ \text{or}\ 2\tilde{\boldsymbol{w}}\boldsymbol{k}_N = -+\boldsymbol{\lambda}\tilde{\boldsymbol{s}},\ \text{which yields}\ \boldsymbol{w} = -(\lambda/2)\boldsymbol{k}^{-1}\boldsymbol{s}. \tag{5.2.15a}$$

[21] For a detailed discussion, including prediction and interpolation, see Section 16.2 of Ref. [2]. See also Section 3.3 in this book.

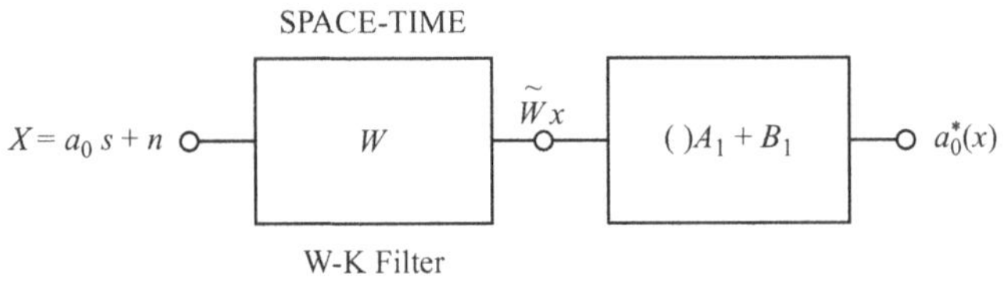

FIGURE 5.2 Coherent Bayes estimator of signal amplitude, a_0^*, for QCF = SCF$_{1,2}$.

Since by (5.2.13a) $\tilde{s}w = -\lambda/2\tilde{s}k_N^{-1}s = -\lambda/2\,\Phi_s = C$, then $-\lambda/2 = C/\Phi_s$. From the last relation in (5.2.15a) and the result for $-\lambda/2$, we get the desired expression for this Wiener filter, which as we have seen from (5.2.11b) is also Bayes optimum with respect to the simple cost function (SCF$_{1,2}$) and the QCF. It is the desired minimum[22]:

$$w = Ck_N^{-1}/\Phi_s = Ck_N^{-1}s/\tilde{s}k_N^{-1}s, \quad \tilde{s}k_N^{-1}s = \Phi_s; \quad s = s_{\text{coh}}. \tag{5.2.15b}$$

Accordingly, we see at once from (5.2.11b) that since the output of the Wiener Filter is $\tilde{w}x(=\tilde{x}w)$, with $C = 1$:

$$a_0^*(x) = (\tilde{w}x)A_1 + A_2, \text{ where } A_1 = \sigma^2/\left(\sigma^2\Phi_s + 1\right); \quad \bar{a}_0/\left(\sigma^2\Phi_s + 1\right). \tag{5.2.16}$$

If $\bar{a}_0 = 0$, then $A_2 = 0$, and if the *a priori* distribution of a_0 is uniform, that is, $\sigma^2 \to \infty$, then $A_1 = 1, A_2 = 0$ and the estimator of amplitude is simply $a_0^*(x) = \tilde{w}x$, the unscaled output of the Wiener filter. See Figure 5.2.

In the case of *suboptimum systems*, suggested by the optimum structure [(5.2.11b), (5.2.16)], we can write

$$\begin{aligned} \gamma_\sigma(x) &= a_0(x) = (\tilde{w}'x)B_1 + B_2, \\ \text{with} \qquad w' &= Gs/(\tilde{s}Gs), \quad G \neq K_N, \end{aligned} \tag{5.2.17}$$

where $G = \tilde{G}$ and B_1, B_2 are positive constants (and $\tilde{x}Gs(\sim w'x)$ has the properties described in Problem 10.8 [2]). The associated average risk, moments, and pdf of this suboptimum estimator may be obtained, as above, when the noise is Gaussian, zero mean, and independent of the (deterministic) signal (Problem 20.8 [2]). Comparisons with the optimum a_0^* then follow.

The W-K filter takes the familiar form of a (discrete) space–time matched filter (5.2.15b), on choosing for the arbitrary constant C, the (known) signal dependent quantity $\Phi_s\left(=\tilde{s}k_N^{-1}s\right)$. We have

$$w = \left[w_j\right] = \left[\sum_{k=1}^{J}\left(k_N^{-1}\right)_{jk}s_k\right] = k_N^{-1}s, \tag{5.2.18}$$

which requires the inversion of the matrix k_N to establish the components of w. From this in turn we can obtain an alternative set of relations for determining w:

$$\sum_{j'=1}^{J} w_j(k_N)_{j'k} = s_k, \text{ or } \sum_{j'=m'n'}^{J} w(r_{m'}, t_{n'})k_N(m\Delta r, m'\Delta r; t_n, t_{n'}) = s(r_{m'}, t_{n'}), \tag{5.2.18a}$$

[22] This follows from $\delta^2 L = 2(\delta\tilde{w})^2[k_N w - s/\Phi_s] + 2(\delta\tilde{w})k_N(\delta w)$. Now, $k_N w = k_N^{-1}(k_N s/\Phi_s) = s/\Phi_s$, and since k_N is positive definite, $\therefore \delta^2 L = 2(\delta\tilde{w})k_N\delta\, w > 0$, as required.

where

$$(m,\, m') = 1, \ldots, M;\ (n,\, n') = 1, \ldots, N \text{ or } (\mathbf{O} \leq \mathbf{r}_{m'},\, \mathbf{r}_{n'} < \mathbf{\Delta R}+);\ (0 \leq t_n,\, t_{n'} < T+), \tag{5.2.18b}$$

specify the domains of space and time for which $\mathbf{w}$ is defined. Note that we equivalently write $(\mathbf{r}_{m'},\, T - t_n)$ with $s(\mathbf{r}_{m'},\, t_n)$, in which the limits are as indicated in (5.2.18b). This reminds us that the filter has its maximum buildup at $t_N = T$, at which point an appropriate switch terminates the output with a readout. As before, the typical sample points in $(\mathbf{O},\, \mathbf{\Delta R}\,;\, O\,,\, T)$ are $\mathbf{r}_m = m\mathbf{\Delta r}$ and $t_n = n\Delta t$. Inverting $\mathbf{k}_N$ may be computationally simpler than solving the J-fold set of equations [(5.2.18a) and (5.2.18b)].

5.2.4 Biased and Unbiased Estimates

By definition an unbiased estimator is one whose average value is the same as the expected or true value $(\boldsymbol{\theta})$ of the quantity being estimated. Accordingly, for *conditional unbiased estimators* we can write

$$E_{X|\theta}\{\gamma_\theta(X)\} = \int_{\boldsymbol{\Gamma}} W_J(X|\theta)\gamma_\theta(X)d\mathbf{x} = \theta. \tag{5.2.19}$$

The corresponding *unconditional unbiased estimators* are

$$E_X\{\gamma_\theta(X)\} = \int_{\boldsymbol{\Gamma}} W_J(X)\gamma_\theta(X)d\mathbf{x} = \bar{\theta} \quad \left[= \int_{\Omega_\theta} \theta\sigma(\theta)d\theta\right] \tag{5.2.20}$$

from which we have alternatively

$$\begin{aligned} E_{X,\theta}\{\gamma_\theta(X)\} &= \textstyle\int_{\Omega_\theta} d\theta \int_{\boldsymbol{\Gamma}} W_J(X|\theta)\sigma(\theta)\gamma_\theta(X)d\mathbf{x} \\ &= \textstyle\int_{\Omega_\theta}\int_{\boldsymbol{\Gamma}} W_J(X,\theta)\gamma_\theta(X)d\mathbf{x}\, d\theta = \bar{\theta}. \end{aligned} \tag{5.2.21}$$

Here, as before, $W_J(X|\theta)$ is the conditional d.d. of X, given θ, while $W_J(X,\ \theta)$ is the *joint* d.d. of X and θ. *Biased estimators*, on the other hand, do not possess this desirable feature. Their expected values contain an additional function, $b(\theta)$, of the parameter in question. Accordingly, for biased estimators we have

$$E_{X|\theta}\{\gamma_\theta(X)\} = \theta + b(\theta) \quad E_X\{\gamma_\theta(X)\} = E_\theta\{\theta\} + E_\theta\{b(\theta)\} = \bar{\theta} + \overline{b(\theta)}, \tag{5.2.22}$$

respectively, for the conditional and unconditional estimators. Applying these results to (5.2.11b), remembering that $\mathbf{x} = \mathbf{n} + a_0\mathbf{s}$, we obtain at once the following conditional and unconditional estimators:

$$\text{Conditional: } \langle a^*(\mathbf{x})\rangle_{\mathbf{x}|a_0} = \frac{a_0\Phi_s + \bar{a}_0/\sigma^2}{\Phi_s + 1/\sigma^2}; \quad \text{Unconditional: } \langle a_0^*(\mathbf{x})\rangle_{\mathbf{x},\,a_0} = \bar{a}_0 \tag{5.2.23}$$

of which the last is unconditionally unbiased, whereas the first is not conditionally unbiased. When $\sigma^2 \to \infty$, the Minimax condition, then $\langle a^*(\mathbf{x})\rangle_{\mathbf{x}|a_0} = a_0$, which is conditional unbiased as well.

5.3 INCOHERENT ESTIMATION OF SIGNAL AMPLITUDE (DETERMINISTIC SIGNALS AND NORMAL NOISE, $p(H_1) = 1$)

As our second example, let us consider the previous situation again, where now the deterministic signal is required to be narrowband and reception involves sample uncertainty, here with a uniform d.d. of RF epochs ε_c. All other signal parameters are assumed known, except the amplitudes $\sim a_0$, which are random now, with distributions unknown to the observer, unlike the example above.

5.3.1 Quadratic Cost Function

As in detection, where Minimax (Section 1.8.3) or MAP procedures (Section 1.8.4) can be used, we shall adopt the subcriterion of selecting an amplitude distribution that maximizes the average uncertainty, subject to (1) a *maximum value constraint* $(a_0)_{\max} = P_{\mathrm{M}}^{1/2}$ or to (2) an *average intensity constraint* $\overline{a_0^2} = 2P_0$. The former leads to a uniform d.d. for $a_0(>0)$, with $\overline{a_0^m} = P_{\mathrm{M}}^{m/2}/(m+1)$, while the latter yields a Rayleigh d. d., with $\overline{a_0^m} = (2P_0)^{m/2}\Gamma(m/2+1)$. For the class of signals considered here we find from Eqs. (3.2.30) ff. that the joint and conditional d.d.s of $\boldsymbol{X}$ and a_0 are specifically

$$\begin{aligned} w_n(\boldsymbol{x}, a_0) &= \sigma(a_0)W_n(\boldsymbol{x}|a_0) \\ &= \sigma(a_0)f_0(\boldsymbol{x})I_0\left(a_0\Psi^{*1/2}_{x-\mathrm{inc}}\right)e^{-a_0^2\Psi^*_{s\text{-inc}}/2}, \end{aligned} \tag{5.3.1}$$

with $f_0(\boldsymbol{x}) = (2\pi\psi)^{-J/2}(\det \boldsymbol{k}_N)^{-1/2}e^{-1/2\tilde{\boldsymbol{x}}\boldsymbol{k}_N\boldsymbol{x}}$. Here $\sigma(a_0)$ is given specifically by

$$\left.\begin{aligned} &\text{Peak power:} && \sigma(a_0) = \frac{1}{P_{\mathrm{M}}^{1/2}}, \quad O < a_0 < P_{\mathrm{M}}^{1/2}; \quad \left(a_0^2\right)_m = P_{\mathrm{M}} \\ &\text{Average power:} && \sigma(a_0) = \frac{a_0e^{-a_0^2/2P_O}}{P_O}, \quad O < a_0 < \infty; \quad \overline{a_0^2} = 2P_O \end{aligned}\right\}. \tag{5.3.2}$$

From Eqs. (3.2.31) ff., we have, respectively,

$$\begin{aligned} \Psi^*_{s-\mathrm{inc}}|_{\mathrm{n.b.}} &\equiv \left(\tilde{\boldsymbol{a}}\boldsymbol{k}_N^{-1}\boldsymbol{a} + \tilde{\boldsymbol{b}}\boldsymbol{k}_N^{-1}\boldsymbol{b}\right)/2 \doteq \left(\tilde{\boldsymbol{a}}+\tilde{\boldsymbol{b}}\right)\boldsymbol{k}_N^{-1}(\boldsymbol{a}+\boldsymbol{b})/2 = \left(\tilde{\boldsymbol{s}}\boldsymbol{k}_N^{-1}\boldsymbol{s}\right)_{\mathrm{n.b.}}/2; \\ \Psi^*_{x-\mathrm{inc}}|_{\mathrm{n.b.}} &\equiv \tilde{\boldsymbol{x}}\boldsymbol{k}_N^{-1}\left(a\tilde{a}+b\tilde{b}\right)\boldsymbol{k}_N^{-1}\boldsymbol{x} \doteq \boldsymbol{k}_N^{-1}(a+b)(a+b)\boldsymbol{k}_N^{-1}\boldsymbol{x} = \tilde{\boldsymbol{x}}\boldsymbol{k}_N^{-1}s\tilde{s}\boldsymbol{k}_N^{-1}x = \left(\tilde{\boldsymbol{x}}\boldsymbol{k}^{-1}\boldsymbol{s}\right)^2_{\mathrm{n.b.}} \end{aligned} \tag{5.3.3}$$

and $\boldsymbol{S}$ here in $s_{\mathrm{n.b.}} \equiv (a+b)$. Inserting Eq. (5.3.1) into Eq. (5.2.2) then gives the desired Bayes estimator of (normalized) amplitude, *with quadratic cost function*, when reception is incoherent with epoch P uniformly distributed over an RF cycle. We write accordingly

$$T_{\mathrm{R}}^{(N)}(\boldsymbol{x}) = a_0^*(\boldsymbol{x})_{P_{\mathrm{M}}} = \frac{\int_{-\infty}^{\infty} a_0\sigma(a_0)I_0\left(a_0\Psi^{*1/2}_{x-\mathrm{inc}}\right)e^{-a_0^2\Psi^*_{s\text{-inc}}}da_0}{\int_{-\infty}^{\infty}\sigma(a_0)I_0\left(a_0\Psi^{*1/2}_{x-\mathrm{inc}}\right)e^{-a_0^2\Psi^*_{s\text{-inc}}}da_0}. \tag{5.3.4}$$

Let us consider as our first case where $\sigma(a_0)$ is uniform $0 < a_0 < (a_0)_{\max}$, [cf. Eq. (5.3.2)]. Then Eq. (5.3.3) becomes

$$a_0^*(\boldsymbol{x})_{P_{\mathrm{M}}} = \frac{\int_0^{P_{\mathrm{M}}^{1/2}} a_0I_0\left(a_0\Psi^{*1/2}_{x-\mathrm{inc}}\right)e^{-a_0^2\Psi^*_{s\text{-inc}}}da_0}{\int_0^{P_{\mathrm{M}}^{1/2}} I_0\left(a_0\Psi^{*1/2}_{x-\mathrm{inc}}\right)e^{-a_0^2\Psi^*_{s\text{-inc}}}da_0}. \tag{5.3.4a}$$

Although a closed form analytic evaluation of (5.3.4) is not generally possible, we can obtain useful results in the weak and strong signal cases. Thus, for *weak signals*, we may use a threshold development, obtained by expanding $W_n(\mathbf{X}|a_0)$ in powers of a_0, averaging over a_0 in both numerator and denominator, and then developing the resulting fraction. The result for the uniform d.d. of Eq. (5.3.2) is

$$a_0^*(x)_{P_M} = \frac{1}{2}\left(3\overline{a_{0u}^2}\right)^{1/2}\left(1 - \frac{\overline{a_{0u}^2}}{4}\Psi^*_{s-\text{inc}} + \cdots\right) + \frac{\sqrt{3}}{16}\left(\overline{a_{0u}^2}\right)^{1/2}\Psi^*_{s-\text{inc}} + O\left(\overline{a_0^5}, x^4\right), \tag{5.3.5}$$

where $\overline{a_{0u}^2} = P_M/3$.

With *strong signals* as an upper limit, on the other hand, we may set $P_M \to \infty$ in Eq. (5.3.4) without seriously altering the result. Evaluation of the integrals then follows directly giving us

$$a_0^*(\mathbf{x})_{P_M\to\infty} = \left(\frac{\pi\Psi^*_{s-\text{inc}}}{2}\right)^{-1/2}\begin{cases} {}_1F_1\left(\dfrac{1}{2};\ -\dfrac{\Psi^*_{x-\text{inc}}}{2\Psi^*_{s-\text{inc}}}\right)^{-1}, & (5.3.6) \\ \text{or} & \\ e^{\Psi^*_{x-\text{inc}}/4\Psi^*_{s-\text{inc}}}\,\boldsymbol{I}_0\left(\dfrac{\Psi^*_{x-\text{inc}}}{4\Psi^*_{s-\text{inc}}}\right)^{-1}, & (5.3.6a) \end{cases}$$

which is essentially

$$\frac{\Psi^{*1/2}_{x-\text{inc}}}{\Psi^*_{s-\text{inc}}} \tag{5.3.6b}$$

as long as $\Psi^*_{x-\text{inc}} \gg \Psi^*_{s-\text{inc}}$. Figure 5.3 shows Eq. (5.3.6) as a function of x, where $x = \Psi^*_{x-\text{inc}}/4\Psi^*_{s-\text{inc}}$. Note that in the weak-signal case a_0^* depends linearly on $\Psi^*_{x-\text{inc}}$ [cf. Eq. (5.3.5)].Similar results are obtained for the Rayleigh d.d. of amplitudes of Eq. (5.3.2),

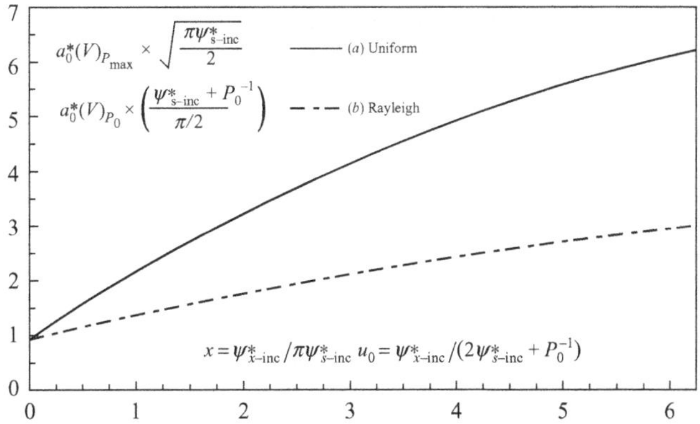

FIGURE 5.3 Bayes estimators of amplitude with incoherent reception, a quadratic cost function, for (a) peak-value constants and $P_M \to \infty$, and (b) mean-square value constraint, all signals, Eq. (5.3.2).

corresponding to the maximum power constraint $\overline{a_{0R}^2} = 2P_0$. The Bayes estimator [Eq. (5.3.3)] now becomes

$$a_0^*(\boldsymbol{x})_{P_0} = \frac{\int_0^\infty a_0^* I_0\left(a_0 \Psi_{x-\text{inc}}^{1/2}\right) e^{-a_0^2\left(\Psi_{s-\text{inc}}^* + P_0^{-1}\right)/2} da_0}{\int_0^\infty I_0\left(a_0 \Psi_{x-\text{inc}}^{1/2}\right) e^{-a_0^2\left(\Psi_{s-\text{inc}}^* + P_0^{-1}\right)/2} da_0}. \tag{5.3.7}$$

With *weak signals*, we can use Eq. (5.3.5) directly, where for the Rayleigh d.d. $\overline{a_{0R}^m} = (2P_0)^{m/2}\Gamma(m/2+1)$, or we can evaluate Eq. (5.3.7) and apply the condition $P_0^2 \ll 1$. The threshold estimator is found to be

$$a_0^*(\boldsymbol{x})_{P_0} \doteq \frac{\sqrt{\pi}}{2}\left(\overline{a_{0R}^2}\right)^{1/2}\left(1 - \frac{\overline{a_{0R}^2}}{4}\Psi_{s-\text{inc}}^* + \cdots\right) + \frac{\sqrt{\pi}}{16}\left(\overline{a_{0R}^2}\right)^{3/2}\Psi_{s-\text{inc}}^* + O\left(a_0^5, x^4\right), \tag{5.3.8}$$

which is seen to be nearly the same as $a_0^*(\boldsymbol{x})_{P_M}$ for weak signals where $\left(\overline{a_0^2}\right)_u = \left(\overline{a_0^2}\right)_R$, that is, when the mean power in the signal is the same under both constraints.

The general expression for $a_0^*(\boldsymbol{x})_{P_0}$ is found once more to be specifically for all signal strengths

$$a_0^*(\boldsymbol{x})_{P_0} = \frac{\sqrt{\pi}}{\sqrt{2}}\left(P_0^{-1} + \Psi_{s-\text{inc}}^*\right)^{-1/2}\begin{cases} {}_1F_1\left(-\frac{1}{2};\, 1;\, -u_0\right) \\ \text{or} \\ e^{-u_0/2}\left[(1+u_0)I_0(u_0/2) + u_0 I_1(u_0/2)\right] \end{cases}, \tag{5.3.9}$$

where $u_0 \equiv \Psi_{x-\text{inc}}^* / \left(2\Psi_{s-\text{inc}}^* + P_0^{-1}\right)$ (cf. Fig. 5.3). For *strong signals P*, this becomes

$$a_0^*(\boldsymbol{x})_{P_0} \cong \frac{\Psi_{x-\text{inc}}^*}{\Psi_{s-\text{inc}}^*} \tag{5.3.9a}$$

as in the case of the uniform d.d. above when $\Psi_{x-\text{inc}}^* \gg \Psi_{s-\text{inc}}^*$ [cf. Eq. (5.3.6b)].

The structure of the Bayes receivers for estimation of amplitude $\left(a_0^*\right)$, when $p(H_1) = 1$, that is, signal surely present in the data $\boldsymbol{X}$, is readily deduced from the results (5.3.5), (5.3.6), and (5.3.6), (5.3.9). These may be summarized as follows:

$$\textit{Weak signals:}\quad a_0^*(\boldsymbol{x})\,{}_{P_M}\Big|_{P_0\ \text{weak}} \doteq B_0\Big|_{\substack{P_M \\ P_0}} + B_2\Big|_{\substack{P_M \\ P_0}} \Psi_{x-\text{inc}}^*, \tag{5.3.10a}$$

while for the strong signal cases one has

$$\textit{Strong signals:}\quad \left.\begin{aligned} a_0^*(\boldsymbol{x})\Big|_{P_M}\Big|_{\text{strong}} &\cong C_{P_M}\,{}_1F_1\left(\frac{1}{2};\, 1;\, -\Psi_{x-\text{inc}}^*/2\Psi_{s-\text{inc}}^*\right)^{-1}; \\ &\cong C_{P_0}\,{}_1F_1\left(-\frac{1}{2};\, 1;\, -u0\right); \\ \text{i.e.,}\ \alpha^*(\boldsymbol{x})|_{P_M,\,P_0}|_{\text{strong}} &\cong \sqrt{\Psi_{x-\text{inc}}^*/\Psi_{s-\text{inc}}^*}, \end{aligned}\right\} \tag{5.3.10b}$$

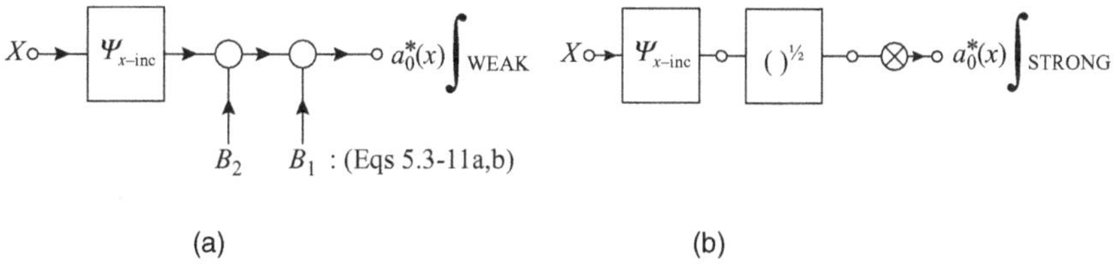

FIGURE 5.4 The weak (a) and strong (b) signal cases of amplitude estimation for incoherent reception, Eqs. (5.3.10a)–(5.3.11b).

where B_0, B_2, C_{P_M}, and C_{P_0} are respectively

$$B_0|_{P_M} \doteq \frac{1}{2}\left(3\overline{a_{0u}^2}\right)^{1/2}\left(1 - \frac{\overline{a_{0u}^2}}{4}\Psi^*_{s-\text{inc}} + \cdots\right);\ B_2|_{P_M} \doteq \frac{\sqrt{3}}{16}\left(\overline{a_{0u}^2}\right)^{3/2}; \quad C_{P_M} \cong \left(\pi\Psi^*_{s-\text{inc}}/2\right)^{-1/2}, \tag{5.3.11a}$$

$$B_1|_{P_0} \doteq \frac{\sqrt{2}}{2}\left(\overline{a_{0R}^2}\right)^{1/2}\left(1 - \frac{\overline{a_{0R}^2}}{4}\Psi^*_{s-\text{inc}} + \cdots\right);\ B_2|_{P_0} \doteq \frac{\sqrt{\pi}}{16}\left(\overline{a_{0R}^2}\right)^{3/2}. \quad C_{P_0} \cong (\pi/2)^{-1/2}\left(P_0^{-1} + \Psi^*_{s-\text{inc}}\right)^{-1/2} \tag{5.3.11b}$$

The diagrams of Fig. 5.4 illustrate the data processing operations required for the example of incoherent amplitude estimation. These may alternatively be expressed in terms of W-K filters by the results of Section 3.3. We have

$$\left.\begin{array}{ll}\text{(i)} & \text{Signal: } \boldsymbol{a}(\text{or } \boldsymbol{b}),\ \text{i.e., } \boldsymbol{s} = \boldsymbol{a} = Re\,\hat{s} = Re\left[A_j \exp\left(w_0 t - \Phi_j\right)\right], \\ \text{(ii)} & \Psi^*_{s-\text{inc}}:\ \therefore\ \Psi^*_{s-\text{inc}} = \tilde{\boldsymbol{a}}\boldsymbol{k}_N^{-1}\boldsymbol{a}\left(= \tilde{\boldsymbol{b}}\boldsymbol{k}_N^{-1}\boldsymbol{b}\right) - \tilde{\boldsymbol{s}}\boldsymbol{k}_N^{-1}\boldsymbol{s};\ (\boldsymbol{s} = \boldsymbol{s}_{\text{inc}} \text{ here}).\end{array}\right\}. \tag{5.3.12a}$$

From Eq. (5.2.15b) the associated W-K filter is seen to be

$$w^{(a)} = \boldsymbol{k}_N^{-1}\boldsymbol{a}C_0 = \boldsymbol{k}_N^{-1}\boldsymbol{s}/\tilde{\boldsymbol{s}}\boldsymbol{k}_N^{-1}\boldsymbol{s} = \boldsymbol{k}_N^{-1}\boldsymbol{s}/\Psi^*_{s-\text{inc}}. \tag{5.3.13}$$

For this we must use both **a** and **b**, since $\Psi^*_{s-\text{inc}} = \left(\tilde{\boldsymbol{x}}\boldsymbol{k}_N^{-1}\boldsymbol{a}\right)^2 + \left(\tilde{\boldsymbol{x}}\boldsymbol{k}_N^{-1}\boldsymbol{b}\right)^2$. The result is

$$\Psi^*_{x-\text{inc}} = \left[\left(\tilde{\boldsymbol{w}}^{(a)}\boldsymbol{x}\right)^2 + \left(\tilde{\boldsymbol{w}}^{(b)}\boldsymbol{x}\right)^2\right]\Psi^{*2}_{s-\text{inc}}, \quad \text{with } w^{(a,b)} = \boldsymbol{k}_N^{-1}(\boldsymbol{a} \text{ or } \boldsymbol{b})/\Psi^*_{s-\text{inc}}, \tag{5.3.14}$$

which expresses $\Psi^*_{x-\text{inc}}$ in terms of the W-K filters, which are required for both $\boldsymbol{a} = \lfloor A_j \cos\Phi_j \rfloor$ and $\boldsymbol{b} = \left[A_j \sin\Phi_j\right]$. Fig. 5.5 shows Fig. 5.4 a and b in terms of these filters.

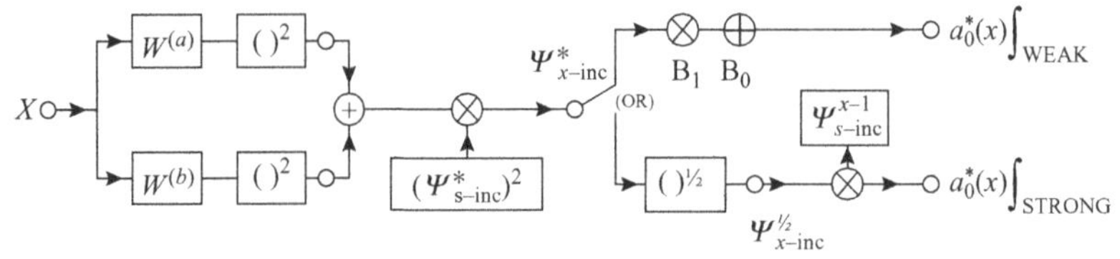

FIGURE 5.5 Estimation of amplitude: equivalent diagrams to Fig. 5.4 a and b, in terms of the W-K filters $w^{(a)}$, $w^{(b)}$; QCF.

5.3.2 "Simple" Cost Functions SCF$_1$ (Incoherent Estimation)

We repeat the analysis of the preceding section, Section 5.3.1, now with the nonstrict simple cost function SCF$_1$ of Eqs. (5.1.12) and (5.1.12a), in place of the quadratic cost function of Eq. (5.1.18). From Eq. (5.1.14), we know that the Bayes estimator of amplitude here is the unconditional maximum likelihood estimator $\hat{a}_0^*(\boldsymbol{X})$, which is obtained by applying Eq. (5.1.40) to the particular likelihood function $L(\boldsymbol{X}, a_0) = \sigma(a_0)\left\langle W_J(\boldsymbol{X} - a_0\sqrt{\psi}\boldsymbol{s})_N\right\rangle_\varepsilon = w_{Jx2}(\boldsymbol{X}, a_0)$ [Eq. (5.3.1)]. Carrying out the indicated operations, we have

$$\frac{\partial}{\partial a_0}\log L(X, a_0)\big|_{a_0=\hat{a}_0^*} = \left[\frac{\sigma'(a_0)}{\sigma(a_0)} + \Psi^{*1/2}_{x-\text{inc}}\frac{I_1\left(a_0\Psi^{*1/2}_{x-\text{inc}}\right)}{I_0\left(a_0\Psi^{*1/2}_{x-\text{inc}}\right)} - \Psi^*_{s-\text{inc}}a_0\right]_{a_0=\hat{a}_0^*} = 0 \quad (5.3.15)$$

as the relation from which the desired UML estimator is obtained. With the uniform and Rayleigh d.d. [Eq. (5.3.2a)] of amplitude, we can write Eq. (5.3.15) specifically as

$$(1)\ \textit{Uniform:}\quad \lambda\frac{I_1(\lambda\hat{z}^*)}{I_0(\lambda\hat{z}^*)} = \hat{z}^* \quad \lambda \equiv \left(\frac{\Psi^*_{x-\text{inc}}}{\Psi^*_{s-\text{inc}}}\right)^{1/2}, \quad \hat{z}^* \equiv \hat{a}_0^*\left(\Psi^*_{s-\text{inc}}\right)^{1/2} \quad (5.3.16a)$$

$$(2)\ \textit{Rayleigh:}\quad \frac{1}{\hat{z}^*} - (1+\eta)\hat{z}^* + \lambda\frac{I_1(\lambda\hat{z}^*)}{I_0(\lambda\hat{z}^*)} = 0, \quad \eta \equiv \left(P_0\Psi^*_{s-\text{inc}}\right)^{-1} \quad (5.3.16b)$$

The solutions of Eqs. (5.3.16a) and (5.3.16b) are the desired Bayes estimators $\hat{z}^*$ (or $\hat{a}_0^*$).

In the *threshold cases*, these optimum estimators can be put in the form $\hat{a}_0^* = \hat{B}_0 + \hat{B}_2\tilde{\boldsymbol{v}}\boldsymbol{G}_0\boldsymbol{v}$, as above [cf. Eqs. (5.3.5) and (5.2.8)]. For the uniform amplitude d.d., the coefficients $\hat{B}_0, \hat{B}_2$ are conveniently obtained by curve fitting to the general relation (5.3.16a), shown in Fig. 5.6. With the Rayleigh d.d., one can easily find the threshold development of this kind directly from the weak-signal expansion of Eq. (5.3.16b). The results in each instance are

$$(1)\ \textit{Uniform:}\quad \hat{z}^* \doteq \hat{b}_{0u} + \hat{b}_{2u}\lambda^2 \quad \text{or} \quad \hat{a}_{0u}^*(\boldsymbol{X}) \doteq \left(-\frac{0.80\cdot 2}{\Psi^*_{s-\text{inc}}}\right) + \frac{0.80\Psi^*_{x-\text{inc}}}{\Psi^*_{s-\text{inc}}} \quad (5.3.17a)$$

$$(2)\ \textit{Rayleigh:}\quad \hat{a}_{0R}^*(X) \doteq \frac{\lambda}{\sqrt{1+\eta}}\left[\Psi^{*-1/2}_{s-\text{inc}} + \Psi^{*-3/2}_{s-\text{inc}}\Psi^*_{x-\text{inc}}/4(1+\eta)\right], \ \text{all}\ \eta > 0 \quad (5.3.17b)$$

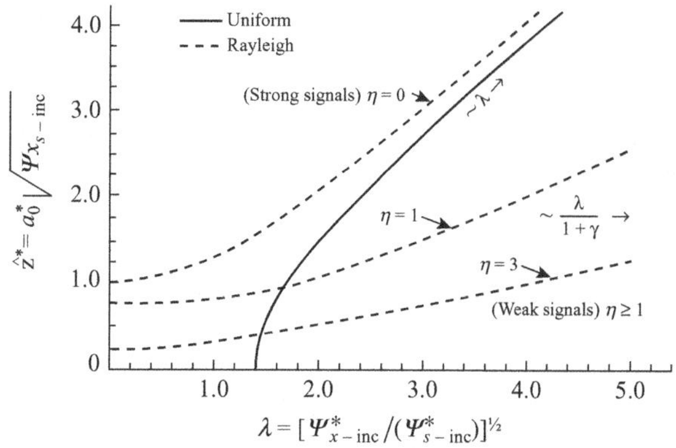

FIGURE 5.6 Bayes estimator of amplitude; incoherent reception, simple cost function, for (a) uniform d.d., and (b) Rayleigh d.d..

For *strong signals*, on the other hand, where $\lambda \gg 1$ on the average[23], one has

$$\text{(1) } \textit{Uniform:} \quad \hat{z}^* \cong \lambda - \frac{1}{2\lambda} + O\left(\frac{-1}{\lambda^3}\right) \quad \text{or} \quad \hat{a}^*_{0u}(\boldsymbol{X}) \cong \frac{\Psi^*_{x-\mathrm{inc}}}{\Psi^*_{s-\mathrm{inc}}} \tag{5.3.18a}$$

$$\text{(2) } \textit{Rayleigh:} \quad \hat{z}^* \cong \lambda - \frac{\lambda}{1+\eta} + O\left(\frac{1+\eta}{\lambda^3}\right) \quad \text{or} \quad \hat{a}^*_{0R}(X) \cong \frac{\Psi^*_{x-\mathrm{inc}}}{\Psi^*_{s-\mathrm{inc}}}, \quad \Psi^*_{s-\mathrm{inc}} \to \infty \tag{5.3.18b}$$

Figure 5.6 shows $\hat{z}^*$ as a function of the received data ($\sim \lambda$) for both (1), (2) above. Note that for the uniform pdf of amplitudes, $\hat{z}^*$ and λ are uniquely related, while for the Rayleigh pdf various relationships are possible, depending on the mean signal power P_0. On the average, the curve for $\eta = 1$ corresponds to a comparatively weak-signal state, while that for $\eta = 0$ exhibits $\hat{z}$ in the limit of very strong signals. Observe that, as we would expect, the strong-signal estimators are essentially independent of whether or not the amplitudes are uniformly or Rayleigh distributed. For weak signals, on the other hand, the character of the amplitude distribution becomes significant. In fact, with a uniform pdf we must reject all λ's less than $\sqrt{2}$, so that, when $\sigma(a_0)$ is unknown to the observer and when λ is actually found to be less than $\sqrt{2}$, it is clear that the subcriterion leading to the pdf (1), employing a peak-value constraint, is not then appropriate. The criterion (2), leading to the Rayleigh pdf, however, may be acceptable. The structure of these Bayes receivers is sketched once more in Figs. 5.4 and 5.5. The only change is in the computer, which now performs its operation on $\Psi^*_{x-\mathrm{inc}}$ according to Eqs. (5.3.16a) and (5.3.16b), rather than Eqs, (5.3.6) or (5.3.7).

Since these Bayes estimators are UML estimators also, we can apply Eq. (21.14) of Ref. [2] at once for *threshold reception* to obtain the expected distributions of $\hat{a}^*_{0u}, \hat{a}^*_{0R}$. These are accordingly asymptotically normal, with means $\overline{\hat{a}^*_{0u}}, \overline{\hat{a}^*_{0R}}$ and variances σ^*_u, σ^*_R, respectively, in the large-sample threshold theory, where the variances may be found as in Eqs. (21.118) of Ref. [2].

[23] Note that $\lambda \gg 1$ on the average implies strong signal. Since $\boldsymbol{x} = a_0\boldsymbol{s} + \boldsymbol{n}$, we have $\Psi^*_{x-\mathrm{inc}} = E_{x|H_1}\{\tilde{\boldsymbol{x}}\boldsymbol{k}_N^{-1}(\boldsymbol{a}\tilde{\boldsymbol{a}} + \boldsymbol{b}\tilde{\boldsymbol{b}})\boldsymbol{k}_N^{-1}\boldsymbol{x}\} = \overline{a_0^2}k_s k_N^{-1} + 2\Psi^*_{s-\mathrm{inc}} \gg \Psi^*_{s-\mathrm{inc}}$ with $\overline{a_0^2} \gg 1$.

The Bayes risk, or cost, associated with the incoherent estimators of Section 5.3 are more complex than those for the case of coherent reception (Section 5.2). However, for the situation of *strong signals* the optimum estimators for both the Rayleigh and uniform pdfs of amplitude in the incoherent cases, *and* for the quadratic and simple cost functions, have the same limiting form, namely,

$$a_0^*(\boldsymbol{X}) \cong \left(\Psi_{x-\text{inc}}^*\right)^{1/2} / \Psi_{s-\text{inc}}^*, \tag{5.3.19}$$

from (5.3.9a), (5.3.10b), (5.3.18a,b). The Bayes risk in the case of the QCF is found to have the form

$$QCF: \quad R_{(u \text{ or } R)}^* \cong C_0 \overline{\left(a_0^* - a_0\right)^2_{(u \text{ or } R)}} \cong C_0 / \Psi_{s-\text{inc}}^*, \tag{5.3.20}$$

with $1 \leq \Psi_{x-\text{inc}}^* / \Psi_{s-\text{inc}}^{*2} < 2$ (cf. p. 989, Eq. (21.120e) et seq., of Ref. [2]). For the simple cost function (SCF_1) (5.1.12a), the corresponding Bayes risk is given by (5.1.15) and is found from

$$SCF: \quad R_{(u \text{ or } R)}^* \cong C_0 \left[A_1 - \int_{\boldsymbol{\Gamma}} \sigma\left(a_0^*\right) W_J\left(\boldsymbol{X} | a_0^*\right) d\boldsymbol{X} \right] \quad (>0), \tag{5.3.21a}$$

with $\Psi_{s-\text{inc}}^* = \tilde{\boldsymbol{a}} \boldsymbol{k}_N^{-1} \boldsymbol{a} = \tilde{\boldsymbol{b}} \boldsymbol{k}_N^{-1} \boldsymbol{b}$ and $\Psi_{x-\text{inc}}^* = \left(\tilde{\boldsymbol{x}} \boldsymbol{k}_N^{-1} \boldsymbol{a}\right)^2 + \left(\tilde{\boldsymbol{x}} \boldsymbol{k}_N^{-1} \boldsymbol{b}\right)^2$ from (5.3.12a) et seq. for these narrowband inputs. Note that while $R_{(u \text{ or } R)}^*$ is independent of the *a priori* distribution of the amplitudes $\boldsymbol{a}_0$ being estimated, it is not independent of the choice of cost function. See Sections 21.3.2 and 21.3.3 of Ref. [2] for a more detailed discussion of the incoherent (n.b.) cases, especially for the weak-signal forms of the optimum estimator a_0^*.

5.4 WAVEFORM ESTIMATION (RANDOM FIELDS)

When the signal, as well as the accompanying noise, is represented by an entirely random field, the approach outline in Sections 5.2 and 5.3 for amplitude estimation can be readily extended to the estimation of signal waveforms. In fact, in the case of Gaussian signals and additive background noise fields, the Bayes estimators are often simpler than their deterministic counterparts. Here we illustrate those cases with the following prototypical example of *simple estimation*, now in space and time, where $\boldsymbol{\gamma}_\mu = \gamma_{mn}$, $t_\lambda = t_n$, $0 < |\boldsymbol{\gamma}_m| \leq |\boldsymbol{R}|$, $0 \leq t_n \leq T$, and when quadratic and simple cost functions [Eqs. (5.1.18) and (5.1.12a)] are employed.

5.4.1 Normal Noise Signals in Normal Noise Fields (Quadratic Cost Function)

Let us begin by choosing a quadratic cost function $C(\boldsymbol{S}, \gamma) = \|\boldsymbol{S} - \gamma_\sigma\|^2$ [cf. Eq. (5.1.24)] and applying Eq. 5.1.26a) to obtain first the Bayes estimator $T_R^{(N)}(\boldsymbol{X}) = \boldsymbol{\gamma}_0^*$ of waveform $\boldsymbol{S}$ in $0, |\boldsymbol{R}|; 0, T$, when simple estimation procedures are employed, that is when $\boldsymbol{\gamma}_\mu = \gamma_m$, and t_n, $(m, n = 11, \ldots, MN = J)$. Here specifically, we have

$$\sigma(\boldsymbol{S}) = \left[(2\pi)^J \det \boldsymbol{K}_S\right]^{-1/2} e^{-1/2\, \tilde{\boldsymbol{S}} \boldsymbol{K}_S^{-1} \boldsymbol{S}} \tag{5.4.1a}$$

$$\text{and} \quad F_J\left(\boldsymbol{X} | \boldsymbol{S}\right) = W_J(\boldsymbol{X} - \boldsymbol{S})_N = \left[(2\pi)^J \det \boldsymbol{K}_N\right]^{-1/2} e^{-1/2 (\tilde{\boldsymbol{X}} - \boldsymbol{S}) \boldsymbol{K}_N^{-1} (\boldsymbol{X} - \boldsymbol{S})} \tag{5.4.1b}$$

Applying these to Eq. (5.1.26a), we can write for the desired optimum estimator (of each component $S_{j=mn}$ of $\boldsymbol{S}$)

$$\gamma_0^*(\boldsymbol{X})_S = \frac{\int_{\boldsymbol{\Omega}_S} \boldsymbol{S} \exp\left[-1/2\tilde{\boldsymbol{S}}(\boldsymbol{K}_S^{-1} + \boldsymbol{K}_N^{-1})\boldsymbol{S} + \tilde{\boldsymbol{X}}\boldsymbol{K}_N^{-1}\boldsymbol{S}\right] d\boldsymbol{S}}{\int_{\boldsymbol{\Omega}_S} \exp\left[-1/2\tilde{\boldsymbol{S}}(\boldsymbol{K}_S^{-1} + \boldsymbol{K}_N^{-1})\boldsymbol{S} + \tilde{\boldsymbol{X}}\boldsymbol{K}_N^{-1}\boldsymbol{S}\right] d\boldsymbol{S}}, \tag{5.4.2}$$

where $\boldsymbol{\Omega}_S$ is the region P for each $(-\infty, \infty)$, S_j, $j = mn = 11, \ldots, MN = J$. To evaluate Eq. (5.4.2), let us consider the conditional characteristic function

$$F_J(i\xi|\boldsymbol{X})_S \equiv \int_{\boldsymbol{\Omega}_S} e^{i\tilde{\xi}\boldsymbol{S}} \sigma(\boldsymbol{S}) F_J(\boldsymbol{X}|\boldsymbol{S}) d\boldsymbol{S}. \tag{5.4.3}$$

Then Eq. (5.4.2) becomes alternatively

$$\boldsymbol{\gamma}_0^*(\boldsymbol{X})_S = -i\frac{d}{s\boldsymbol{\xi}} \log F_J(i\boldsymbol{\xi}|\boldsymbol{X}))S|_{\xi=0}. \tag{5.4.4}$$

Writing $\boldsymbol{M} \neq \boldsymbol{K}_S^{-1} + \boldsymbol{K}_N^{-1}$, we obtain from Eqs. (5.4.1a) and (5.4.1b), in Eq. (5.4.3)

$$\begin{aligned} F_J(i\boldsymbol{\xi}|\boldsymbol{X})_S &= \left[(2\pi)^J \det \boldsymbol{K}_S\boldsymbol{K}_N \det \boldsymbol{M}\right]^{-1/2} \exp\left(1/2\, \tilde{\boldsymbol{X}}\boldsymbol{K}_N^{-1}\boldsymbol{M}^{-1}\boldsymbol{K}_N^{-1}\boldsymbol{X}\right) \\ &\quad \cdot \exp\left(1/2\, \tilde{\boldsymbol{\xi}}\boldsymbol{M}^{-1}\boldsymbol{\xi} + i\boldsymbol{\xi}\boldsymbol{M}^{-1}\boldsymbol{K}_N^{-1}\boldsymbol{X}\right). \end{aligned} \tag{5.4.5}$$

Applying this to Eq. (5.4.4), we find directly that

$$T_{\mathrm{R}}^{(N)}\{\boldsymbol{X}\} = \gamma_0^*(\boldsymbol{X})_S = \boldsymbol{S}^* = \boldsymbol{M}^{-1}\boldsymbol{K}_N^{-1}\boldsymbol{X} = \boldsymbol{Q}_{\mathrm{E}}\boldsymbol{X}, \tag{5.4.6}$$

where $\boldsymbol{Q}_{\mathrm{E}}$ is $\boldsymbol{M}^{-1}\boldsymbol{K}_N^{-1}$, or the equivalent expressions

$$\boldsymbol{Q}_{\mathrm{E}} = \boldsymbol{K}_S(\boldsymbol{K}_N + \boldsymbol{K}_S)^{-1} = (\boldsymbol{I} + \boldsymbol{K}_N\boldsymbol{K}_S)^{-1} = \left[(\boldsymbol{I} + \boldsymbol{K}_S\boldsymbol{K}_N^{-1}) \cdot (\boldsymbol{K}_S\boldsymbol{K}_N^{-1})\right]^{-1} \neq \tilde{\boldsymbol{Q}}_{\mathrm{E}}. \tag{5.4.7}$$

Thus, the Bayes estimator of $\boldsymbol{S}$ with respect to the *quadratic cost function* (5.1.24) is given by Eq. (5.4.6).

5.4.2 Normal Noise Signals in Normal Noise Fields ("Simple" Cost Functions)

As we see directly on comparison with Eq. 21.29 of Ref. [2], the Bayes estimator for the QCF, Eq. (5.4.6), is also the unconditional maximum likelihood estimator of $\boldsymbol{S}$, when $\sigma(\boldsymbol{S})$ is given by Eq. (5.4.1a), for example,

$$\boldsymbol{S}^* = \hat{\boldsymbol{S}} = \boldsymbol{Q}_{\mathrm{E}}\boldsymbol{X}, \text{ where } \boldsymbol{Q}_{\mathrm{E}} = \boldsymbol{M}^{-1}\boldsymbol{K}_N^{-1} \text{ or } \boldsymbol{M}^{-1} = \boldsymbol{Q}_{\mathrm{E}}\boldsymbol{K}_N. \tag{5.4.8}$$

Moreover, from Eqs. (5.1.41) and (5.1.4) we know that $\hat{\boldsymbol{S}}$ can be interpreted as a Bayes estimator with respect to the *"strict" simple cost function* (SCF_2) (5.1.42), so that, in effect, we can write $\hat{\boldsymbol{S}} \equiv \hat{\boldsymbol{S}}^* = \boldsymbol{S}^*$ of Eq. (5.4.8). Finally, one can also show that, in terms of the *less strict simple cost function* (SCF_1), S_k^* is equal to $(\boldsymbol{Q}_E\boldsymbol{X})_k$ (all $k = 1, \ldots, n$), and consequently $\hat{\boldsymbol{S}} \equiv \hat{\boldsymbol{S}}^* = \boldsymbol{Q}_E\boldsymbol{X}$ here as well. This is another example of a system that is optimum with respect to two or more different criteria, involving in this case three different cost functions. That such an invariance can occur is clearly strongly dependent on the statistical structure of the noise and signal processes (fields), as well as on the choice of cost function.

5.4.2.1 ***Bayes Risk*** Before determining the Bayes risk, let us first find the characteristic functions of $\boldsymbol{S}$, of the estimator $\boldsymbol{S}^*$, and of $\boldsymbol{S}^* - \boldsymbol{S}$. Since $\boldsymbol{S}$ and $\boldsymbol{S}^*(\boldsymbol{X})$ are both normal with zero means, the former with variance $\boldsymbol{K}_S$ and the latter with $\boldsymbol{K}_{S^*} = \boldsymbol{E}_{X|H_1}\left\{\boldsymbol{S}^*\tilde{\boldsymbol{S}}^*\right\} = \boldsymbol{Q}_E\boldsymbol{E}_{X|H_1}\{\boldsymbol{X}\tilde{\boldsymbol{X}}\}\tilde{\boldsymbol{Q}}_E = \boldsymbol{K}_S\tilde{\boldsymbol{Q}}_E$ [Eq. (5.4.7)], their characteristic functions are

$$F_J(i\xi)_S = \boldsymbol{E}_S\left(e^{i\tilde{\xi}\boldsymbol{S}}\right) = e^{-1/2\,\tilde{\xi}\boldsymbol{K}_S\xi} \tag{5.4.9}$$

$$\text{and}\quad F_J(i\xi)_{S^*} = \boldsymbol{E}_{X|H_1}\left(e^{i\tilde{\xi}\boldsymbol{S}^*}\right) = e^{-1/2\,\tilde{\xi}\boldsymbol{K}_S\boldsymbol{Q}_E\xi}. \tag{5.4.10}$$

For the c.f. of $\boldsymbol{S}^* - \boldsymbol{S}$, we must remember that $\boldsymbol{S}$ and $\boldsymbol{S}^*$ are *not* independent: $\boldsymbol{S}^* = \boldsymbol{Q}_E\boldsymbol{X} = \boldsymbol{Q}_E(\boldsymbol{S} + \boldsymbol{N})$, but since both $\boldsymbol{S}$ and $\boldsymbol{N}$ or $\boldsymbol{S}$ and $\boldsymbol{X}$ are normally distributed, we again expect a normal d.d. for $\boldsymbol{S}^* - \boldsymbol{S}$. Specifically the means of $\boldsymbol{S}^* - \boldsymbol{S}$ vanish, and the covariance matrix is

$$\boldsymbol{K}_{S^*-S} \equiv \boldsymbol{E}_{S,N}\left\{\left(\boldsymbol{S}^* - \boldsymbol{S}\right)\left(\tilde{\boldsymbol{S}}^* - \tilde{\boldsymbol{S}}\right)\right\} = \boldsymbol{K}_S(\boldsymbol{I} - \boldsymbol{Q}_E), \tag{5.4.11}$$

where we have used $\boldsymbol{E}_{X|H_1}\left\{\boldsymbol{S}^*\tilde{\boldsymbol{S}}^*\right\} = \boldsymbol{K}_S\tilde{\boldsymbol{Q}}_E$ and Eq. (5.4.7). Again the Gaussian character of these distributions follows from the linear nature of the Bayes estimator $\boldsymbol{S}^*\left(= \hat{\boldsymbol{S}}^*, \text{ etc.}\right)$ and the normal properties of both the original signal and noise processes. Note also that the pdf of a single estimate S_j^* is normal, with zero mean and variance $(\boldsymbol{M})_{jj}^{-1}$, since $F_1\left(i\xi_j\right)_S = F_J(0, 0, \ldots, i\xi_j, \ldots, 0)_{S^*}$, and so on.

The *Bayes risk* for $\hat{\boldsymbol{S}}^*$ in the case of the quadratic cost function is now easily found from Eq. (5.4.11). Let us consider the Bayes risk associated with the single estimation of waveform S_j at time $t_\lambda = t_n$ and positions $\boldsymbol{\gamma}_\mu = \boldsymbol{\gamma}_m$ where $t_n (n = 1, \ldots, N)$ is any one of the instants and $\boldsymbol{\gamma}_m$, $m = 1, \ldots, M$ is a sample point in space, at which the data $\boldsymbol{X}$ are acquired. From Eqs. (5.1.27a) and (5.4.8), we have for the Bayes risk here

$$R_j^* \equiv C_0\boldsymbol{E}_{X,S_j}\left\{\left(S_j - S_j^*\right)^2\right\} = C_0\left(\boldsymbol{K}_S - \boldsymbol{K}_S\tilde{\boldsymbol{Q}}_E\right)_{jj}. \tag{5.4.12}$$

The Bayes risk associated with the simple estimation of, say, L-values of the waveform, for example $S_{j_1}, \ldots, S_{j_L}$, $11 \leq j_1, \ldots, j_L < J = MN$, where j_l, and so on, is any one data-sampling instant in the interval $(0, T)$, becomes

$$R_{[j]}^* = \sum_{l=1}^{j} R_{j_l}^* = C_0\sum_{l=1}^{j}\left[(\boldsymbol{K}_S)_{j_lj_l} - \sum_{p=1}^{p}(\boldsymbol{K}_S)_{j_lp}\left(\tilde{\boldsymbol{Q}}_E\right)_{pj_l}\right], \tag{5.4.13a}$$

and if $j = p$, that is, if all points $(1 \leq j \leq p)$ are considered, we have

$$R^*_{[p]} \equiv C_0 \operatorname{tr}\left(\boldsymbol{K}_S - \boldsymbol{K}_S\tilde{\boldsymbol{Q}}_{\mathrm{E}}\right). \tag{5.4.13b}$$

The Bayes risk for $j = mn$ points $(\gamma_m, t_n)\,(l = 1, \ldots, j)$ follows similarly from Eq. (5.4.13a). We have

$$R^*_{\lambda[j]} = C_0 \sum_{l=1}^{j} J_T[(\gamma_m, t_n)_l, (\gamma_n, t_n)_l] \quad 11 \leq j \leq J, \tag{5.4.14}$$

with a corresponding result for $R^*_{\lambda[p]}$ [Eq. (5.4.13b)].

Note that, as $p \to \infty$, the Bayes risk $R^*_{\lambda[p]}$ for the values of S at the j sampling points P also becomes infinite, since the Bayes risk for each sampling instant t_λ in the interval $(0, |\boldsymbol{R}|)$, $(0, T)$ is itself finite and there are now an infinite number of such points in this interval. However, by modifying the definition of $R^*_{\lambda[p]}$ so as to define instead the Bayes risk $\mathsf{R}^*_{\lambda[p]}$ *per sampling point*, by

$$\mathsf{R}^*_{\lambda[p]} \equiv R^*_{\lambda[p]}/n, \tag{5.4.15a}$$

we obtain in the limit

$$\mathsf{R}^*_{\lambda[p]}, \tag{5.4.15b}$$

a finite quantity which may be used as a measure of the expected performance of the Bayes estimator S^*.

5.4.2.2 Minimax Estimators Finally, we observe here that since the Minimax estimator is Bayes for a uniform *a priori* signal distribution (Sec. 5.1.4) we can obtain the specific Minimax extractor on setting $\boldsymbol{K}_S^{-1} = 0$, the null matrix. Then, $\boldsymbol{Q}_{\mathrm{E}} \to \boldsymbol{Q}_{\mathrm{M}} = \boldsymbol{I}$ [cf. Eq. (5.4.7)], and consequently

$$\boldsymbol{S}^* \to \boldsymbol{S}^*_{\mathrm{M}} = \hat{\boldsymbol{S}}^*_{\mathrm{M}} = \boldsymbol{X}, \tag{5.4.16}$$

that is, the Minimax estimates are just the sampled values themselves. The Minimax average risks $R^*_{\mathrm{Mk}}, R^*_{\mathrm{M}[j]}$ [Eqs. (5.4.13a) and (5.4.13b)] are readily computed from Eq. (5.4.12) with the help of Eq. (5.4.16). They are, respectively, $\psi\left(= \overline{N^2}\right)$ and $J\psi$. Since $\boldsymbol{S}^*$ for quadratic cost functions is also equal to the UMLE $\hat{\boldsymbol{S}}^*$, these remarks apply equally well for the corresponding simple cost functions and the associated Minimax estimators.

5.4.2.3 Extensions The temporal extension of the theory to *smoothing* and *prediction*, for which the sample points in time are t_λ, where t_λ represents *interpolation*, that is, $t_n < t_\lambda < t_{n+1}$, or *extrapolation* ($t_1 > t_\lambda$, or $t_N < t_\lambda$, i.e., *prediction*), is discussed in Sections 21.4.2 and 21.4.3, of Ref. [2]. The further extension of this theory to space–time sampling is formally made by introducing the space–time points $\left(\boldsymbol{\gamma}_\mu, t_\lambda\right)$. The (μ, λ) now designates points in the space–time manifold which are not coincident with the sample points (γ_m, t_n) which select the data upon which the estimate at $\left(\boldsymbol{\gamma}_\mu, t_\lambda\right)$ is made. Again, depending on the choice of cost function and then *a priori* distributions, these estimators (predictors and extrapolators) may be linear or nonlinear, even if the noise (and signal) fields are Gaussian.

5.5 SUMMARY REMARKS

Estimation theory is necessarily less "compact" than the corresponding theory for detection, since the system designer has more degrees of freedom at his disposal, in particular the choice of cost function, and frequently the quantity to be estimated. However, it is possible to make certain general observations about the results obtained in Sections 5.1–5.4:

(1) Weak-signal or threshold estimators usually provide asymptotically sufficient estimates, and the pdfs of these estimators are themselves usually normal. This is true for maximum likelihood estimators of amplitude and waveform, which are Bayes relative to simple cost functions (Sections 5.1.1 and 5.1.2), and also for corresponding systems based on quadratic cost functions (Section 5.1.3).
(2) Strong-signal estimators of these cost functions also exhibit a similar invariance: regardless of whether or not a simple or a quadratic cost function is chosen, the Bayes system has the same structure in this limiting situation [although the Bayes *risks* may (of course) be different].
(3) Finally, the maximum likelihood approach is found to have a broader interpretation from the viewpoint of decision theory, as Section 5.4.1 has indicated.

A brief review of some of the implications of (3) gives further insight into the significance of maximum likelihood estimation. This may be seen in the case of waveform estimation if we set $\boldsymbol{\gamma} = \boldsymbol{S}$ therein, according to Eq. (5.1.4). Then we have

$$p(\boldsymbol{\gamma}|\boldsymbol{S}) = p(\boldsymbol{\gamma} = \boldsymbol{S}|\boldsymbol{S}) = \int_{\boldsymbol{\Gamma}} F_J(\boldsymbol{X}|\boldsymbol{S})\delta[\boldsymbol{S} - \boldsymbol{\gamma}_S(\boldsymbol{X})]d\boldsymbol{X}, \tag{5.5.1}$$

which is the probability of a correct decision when the signal is $\boldsymbol{S}$. It is clearly greatest when for each $\boldsymbol{X}$ we choose $\boldsymbol{\gamma}_S(\boldsymbol{X})$ equal to the maximum conditional likelihood estimate $\hat{\boldsymbol{\gamma}}_S$. In other words, the *maximum (conditional) likelihood estimator maximizes the probability of a correct decision*, without regard to incorrect decisions or their cost. By taking the average of both sides of Eq. (5.5.1) with respect to the *a priori* signal distribution $\sigma(\boldsymbol{S})$, we obtain the average probability of a correct estimate,

$$\int_{\boldsymbol{\Omega}_S} p\Big(\boldsymbol{S}|\boldsymbol{S})\sigma(\boldsymbol{S})\, d\boldsymbol{S} = \int_{\boldsymbol{\Gamma}}\int_{\boldsymbol{\Omega}_S} \sigma(\boldsymbol{S})F_J(\boldsymbol{X}|\boldsymbol{S})\delta[\boldsymbol{S} - \boldsymbol{\gamma}_\sigma(\boldsymbol{X})]\, d\boldsymbol{S}\, d\boldsymbol{X}, \tag{5.5.2}$$

where $p(\boldsymbol{S}|\boldsymbol{S})$ represents $p(\boldsymbol{\gamma} = \boldsymbol{S}|\boldsymbol{S})$. By the same reasoning as for Eq. (5.5.2), we see that Eq. (5.5.2) is largest when, for each $\boldsymbol{X}$, $\boldsymbol{\gamma}_\sigma(\boldsymbol{X})$ is chosen as the particular value of $\boldsymbol{S}$ that makes the unconditional likelihood function (or *a posteriori* probability of $\boldsymbol{S}$, given $\boldsymbol{X}$) a maximum. Thus, *the maximum* unconditional *likelihood estimator* $\hat{\boldsymbol{S}}$ *maximizes the* average *probability of a correct decision* (here an estimate, under $p(H_1 = 1)$), when all possible signals are taken into account, and again without particular regard for incorrect decisions and their costs.

Because the maximum likelihood estimator effectively assigns the greatest probabilities to the least costs, this cost of a correct decision is always less than that of any other decisions. Thus, the closest estimates have the greatest probabilities, and may be expected to minimize the average risk for cost functions other than the simple one provided that certain symmetries are present in the joint density $p(\mathbf{S}, \mathbf{X}) = \sigma(\mathbf{S})F_J(\mathbf{X}|\mathbf{S})$ and in the cost function itself. Specifically, in the case of the squared-error cost function (5.1.18), it is found that, if *the joint distribution of* $\mathbf{X}$ *and* $\mathbf{S}$ *is symmetrical about the unconditional maximum likelihood estimate* $\hat{\mathbf{S}}$ *for every* $\mathbf{X}$ (remember that $\hat{\mathbf{S}}$ depends on $\mathbf{X}$), *then* $\hat{\mathbf{S}}$ *is a Bayes solution with respect to this cost function.*[24]

A receiver which presents the *a posteriori* probability $p(\mathbf{S}|\mathbf{X})$ as a function of $\mathbf{S}$ at its output may also be operated as a decision system by taking the maximum value of the output as the decision or estimate. From the above, we see that such a receiver maximizes the probability of a correct decision for each $\mathbf{S}$, and also maximizes the average probability of a correct decision [cf. Eq. (5.5.2)] when the various $\mathbf{S}$ appear at random. The receiver's limitation is that it ignores the possibly different relative importance of the various system errors, that is, discrepancies between $\boldsymbol{\gamma}$ and $\mathbf{S}$. This is equivalent to saying that it is a Bayes extractor with the simple cost function of Section 5.1. On the other hand, as we have just seen, under certain symmetry conditions the Bayes extractor for the squared-error cost function is also the (unconditional) maximum likelihood estimator $\hat{\mathbf{S}}$, so that from this point of view the maximum likelihood receiver may be considered as one which penalizes incorrect decisions according to $C(\mathbf{S}, \boldsymbol{\gamma}_0) = C_0\|\mathbf{S} - \boldsymbol{\gamma}_0\|^2$ and simultaneously maximizes the average probability of a correct decision. It is evident, therefore, that under certain conditions *an extraction system may minimize average risk for more than one cost function*, just as in the case of detection (cf. Section 5.4), where a given system may be optimum for many cost assignments provided that the threshold K is held fixed. In fact, as we have demonstrated in Section 5.4, it is possible for the same system $\mathbf{S}^* = \hat{\mathbf{S}} = \hat{\mathbf{S}}^*$, to be optimum with respect to three different cost functions, when, for example, the signal as well as the noise is a normal process[25]. Note, again, cf. Section 5.4.2.2, that the conditional and unconditional maximum likelihood estimators are also identical for uniform distributions of the quantities being estimated.

REFERENCES

1. D. Middleton and D. Van Meter, Detection and Extraction of Signals in Noise from the Point of View of Statistical Decision Theory, *J. Soc. Ind. and Appl. Math.*, **3** (Part I), 192–253 (1955); **4** (Part II), 86–119 (1956).
2. D. Middleton, *An Introduction to Statistical Communication Theory*, McGraw-Hill, New York, 1960–1972; IEEE Press, *Classic Reissue*, Piscataway, NJ, 1996.
3. J. L. Hodges and E. L. Lehmann, Some Applications of the Cramér-Rao Inequality, *Proceedings of the 2d Berkeley Symposium on Mathematical Statistics and Probability*, July–August, 1950.
4. N. Wiener, *The Extrapolation, Interpolations, and Smoothing of Stationary Time Series*, John Wiley & Sons. Inc., New York, 1949.

[24] Wald [12] uses a similar approach to show that the maximum conditional likelihood estimate $\boldsymbol{\gamma}_S$ (or notation) is Minimax when the cost function depends only on the error, signal and noise are additive (in our terminology), the signal is one-dimensional ($\mathbf{S} = S$), and the *a priori* signal distribution is uniform (over a bounded interval).

[25] Theorem II (Section 5.1.5) states a similar result, now, however, without specifying the particular nature of the signal and noise processes, but with respect to the Bayes estimator derived from a quadratic cost function.

5. S. Sherman, Non-Mean-Square Error Criteria, *IRE Trans. Inform. Theory*, **IT-4**,125 (1958).
6. L. A. Zadeh and J. R. Ragaziuni, An Extension of Wiener's Theory of Prediction, *J. Appl. Phys.*, **26**, 645 (1950).
7. J. L. Hodges and E. L. Lehmann, Some Problems in Minimax Point Estimation, *Am. Math. Stat.*, **21**, 182 (1950).
8. E. J. G. Pitman, The Estimation of Location and Scale Parameters of a Continuous Population in Any Given Form, *Biometicka*, **30**, 391 (1939).
9. M. A. Gershick and L. J. Savage, Bayes and Minimax Estimates for Quadratic Loss Functions, *Proceedings of the 2d Berkeley Symposium on Mathematical Statistics and Probability*, p. 53 (1950).
10. L. A. Zadeh, General Filters for Separation of Signal and Noise, *Symposium on Information Networks*, April 1954, p. 3.
11. N. Ashby, On the Extractions of Noise-Like Signals from a Noisy Background from the Risk Point of View, *Air Force Cambridge Research Center Technical Report*, AFCRC-TR, 26–123, December 1956.
12. A. Wald, Contributions to the Theory of Statistical Estimation and Testing Hypotheses, Theorems 5 and 6, *Am. Math. Stat.*, **10**, 299, (1939).

6

JOINT DETECTION AND ESTIMATION, $p(H_1) \leq 1$[1]: I. FOUNDATIONS

In most applications of Statistical Decision Theory (SDT) to communication problems (and to other applications involving hypothesis testing and statistical estimation), the basic reception processes of signal detection and parameter estimation are treated as separate and distinct operations, albeit within the common framework provided by SDT (Chapters 1–5 preceding).[2] Accordingly, detection is considered as necessarily carried out under uncertainty as to the presence or absence of the *received* signal in the accompanying noise, with *a priori* probability $p(= p[H_1]) < 1$, but independently of any associated signal waveform or parameter estimation (cf. remarks, beginning of Chapter 2). On the other hand, for the estimation process it is usually postulated that the signal is surely present, that is, $p(H_1) = 1$, along with the noise, so that there is no requirement for detection in such cases. It has been within this conceptual framework that an extensive and effective body of fundamental theory and practical application has been developed, along parallel but essentially separate tracks, as exemplified in detail by the concepts, methods, and results of Chapters 1–5 above.

However, in many actual situations, namely those limited by accompanying noise and interference, we cannot assume *a priori* that the received signal, whose features we wish to estimate, for example by the general methods of Chapter 5, is surely present in the received data. Rather, we have to acknowledge the fact that the received signal itself is not *a priori* surely known to be present, that is, in reality $p(H_1) < 1$. Then joint estimation *and* detection

[1] We include the previously treated cases of estimation where $p(H_1) = 1$, Chapters 1 and 5, now regarded as limiting forms of the more general situations $p(H_1) \leq 1$ of Chapter 6.

[2] An exception is the practice of obtaining signal parameters of unknown prior distributions by *conditional* maximum likelihood estimation (CMLE), cf. Sections 5.1.5. There are, however, conceptual and quantity difficulties with this technique.

become essential: *estimation*, which now must suitably account for the bias introduced by the presence of noise alone in a certain fraction $[q(H_0) = 1 - p(H_1) > 0]$ of the received data samples, and *detection*, which is needed to provide a probabilistic measure of the signal's presence, since $p(H_1) < 1$ here. With the latter we are able to decide whether to retain or reject the estimate,[3] depending on our desired, preselected probability of the signal presence. The explicit structure of the detection and estimation algorithms will depend both on the nature and extent of the coupling between the two processes, ranging from no coupling to strong coupling, as we shall see in the following sections.[4]

Accordingly, the central theme of this chapter is the extension of our earlier SDT treatment of minimum average risk or Bayesian (i.e., optimum) and suboptimum detection and estimation discussed in Chapters 1–5. This includes specifically now the important class of problems where detection (D) and estimation (E) are carried out jointly under the common condition of uncertainty as to the presence of the received signal, for example, $p(H_1) < 1$. A significant feature of the analysis, treated subsequently in Chapter 7, is the generalization to include *adaptive procedures*, that is, "learning with and without teacher" via sequences of decisions, and extensions to include multiple hypothesis states in the detection portions of the joint D and E process. The analysis here is based largely on the early work of Middleton and Esposito [1], who initially developed the coupled (Bayes) theory for this binary "on–off" situation explicitly from the viewpoint of SDT. This has permitted the quantitative joining of the D and E components of the decision process in a general way. This joining in turn is accomplished through the introduction of suitable cost functions, along with the subsequent derivation of the desired optimal algorithms and associated performance measures. Extensions to include adaptive procedures and sequences of decision when $p(H_1) < 1$ [2], and multiple alternative detection regimes (Chapter 4), the latter carried out originally by Fredriksen [3], are described in Chapter 7 following.

Following a concise general formulation in Section 6.1 below, we develop the theory, beginning in Section 6.2 with the simplest cases: joint but uncoupled detection and estimation under $p(H_1) \leq 1$, based on a data sample of fixed size, that is, "one-shot" decision making. Two common and important criteria for optimal estimation are employed here: (1) *minimum mean square error* (MMSE) (Section 5.2.3; Section 21.2.2, [4]), and (2) *unconditional maximum likelihood estimation* (UMLE) (Section 5.1.2; Section 21.1.2, [4]). The former employs a quadratic cost function (QCF) and the latter is based on "simple" cost functions (SCFs). Next, in Section 6.3 we broaden the analysis to include the general case of coupled detection and estimation. Section 6.4 provides some explicit examples, for which fully developed analytic results are possible (in the uncoupled cases). These give quantitative insights regarding the sensitivity of the estimators to departures of $p(H_1)$ from

[3] In the case of sequences of decisions, that is, estimates and detections, we may wish to retain individually rejected estimates for composite evaluation, for example, for "tracking," and other sequential decision making, cf. Chapter 7 ff.

[4] Earlier (i.e., largely pre-1968) analyses, such as those presented in classical statistics, for example in Kendall and Stewart [5], have not fully acknowledged the interdependence of the estimation and detection phases of the joint D and E processes, while suggesting that detection be performed first, before estimation. Such "decision-directed" methods [6] are necessarily suboptimum because of the unavoidable possibility of errors in the first stage of this two-stage procedure, and their effects on the second decision. In fact, as we shall see [cf. Section 6.3.1 ff.] optimal procedures can require that estimations be considered *before*, or in parallel with, detection. Recognition of the complexities of the joint D and E problem is, of course, not new: for specifics see [4], Sections 18.2.3, 21.2.5, respectively, and Section 6.3 here, as well as Refs [6–8,12,13] in Ref. [1] of this chapter, along with Refs [3,11–14] in Ref. [2], also of chapter 6. Details of the earlier history of the problem are more fully discussed in the Introductions (Section 1) of Refs. [1,2].

unity, the classical limiting case in common practice. The above analyses are then extended in Chapter 7 to include for $p(H_1) < 1$ a number of estimation problems treated earlier in Ref. [4] under $p(H_1) = 1$. Further extensions to the adaptive or sequential "learning" D and E processes complete Chapter 7. Several final sections in each chapter summarize various results for both the nonadaptive ("one-shot") systems and the adaptive learning cases where multiple alternative hypothesis states for detection occur. A brief discussion of the implications of the results, along with problems and selected references [5]-[14], particularly to relevant earlier work, complete our presentation of this particular subject in this chapter.

6.1 JOINT DETECTION AND ESTIMATION UNDER PRIOR UNCERTAINTY $[p(H_1) \leq 1]$: FORMULATION

We accordingly begin first with a general model of the reception situation, as sketched in Fig. 6.1, specializing it subsequently to various simpler, special cases. The (real, vector) coupling operator $\boldsymbol{T}_{\mathrm{AR}}$, cf. Fig. 1.1, includes space–time sampling of the input field α and is represented by the scalar (aperture) or vector (array) operators, $\hat{R}, \hat{\boldsymbol{R}}$:

$$\boldsymbol{T}_{\mathrm{AR}} = Re\left[\hat{R}\alpha_n\right] = [X(t_n)]\text{: time-sampled input to the detector-estimator system from a receiving aperture;} \tag{6.1.1a}$$

$$= Re\left[\hat{\boldsymbol{R}}a_n\right] = [X(\mathbf{r}_m, t_n)] = \left[X_j\right]\text{: same as (6.1.1a), now with spatial sampling by an array of sensors.} \tag{6.1.1b}$$

As noted earlier, cf. Section 1.3.1, for time samples on the data acquisition interval (T) we have $n = 1, \ldots, N$, and for space–time sampling, $j = mn = 1, \ldots, J = MN$, in the usual way, cf. Eq. (1.3.1), with $\mathbf{r}_m$ the location of the mth sensor element of the array. (See Chapter 1 for details.)

We postulate next an on–off detection situation. Then the sampled input to the system is either in *state* H_1: an arbitrary received signal $\boldsymbol{S}(\boldsymbol{\theta})$, corrupted in some fashion by a noise process ($\boldsymbol{N}$), or in *state* H_0: the noise field alone, for example,

$$H_1\text{: } \boldsymbol{X} = \boldsymbol{S} \otimes \boldsymbol{N};\ H_0\text{: } \boldsymbol{X} = \boldsymbol{N}, \tag{6.1.2}$$

where $\otimes$ indicates "combination," which may be additive, or multiplicative in the case of active systems, for example. In general, the purpose of the system depicted in Fig. 6.1 is to

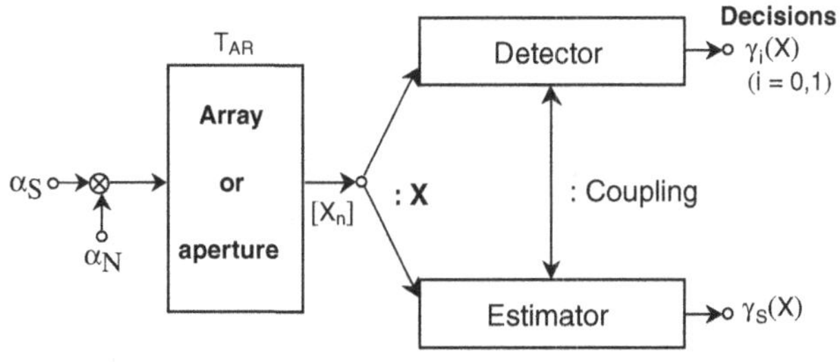

FIGURE 6.1 A general model of a coupled joint detection–estimation processor; fixed observation interval; $\left(\boldsymbol{X} = \hat{\boldsymbol{R}}a\right)$.

provide a double decision at its output: a detection decision as to the presence or absence of the signal and, possibly, an estimate of the signal waveform $\boldsymbol{S}$, or of the signal parameters $\boldsymbol{\theta}$. As indicated in Fig. 6.1, the detector and the estimator are in general functionally related. For the moment we note that Fig. 6.1 includes, among other cases: (1) the usual detection problem when the estimator is not present; (2) the usual estimation problem when the signal is present with probability unity in the observation interval (and therefore no detection is necessary); and (3) a new type of estimation problem when there is no detection operation involved,[5] but an estimate has to be made without certainty as to the presence of the signal, that is, $p(H_1) < 1$.

As we have mentioned at the beginning of the chapter, we shall use the familiar Bayesian approach outlined in Chapter 1, which includes the concepts and methods of SDT to determine system structure and performance. Accordingly, we observe again that structure is embodied in the decision rule δ and performance is evaluated as an *average risk* or *cost* $R(\sigma, \delta)$. Specifically, we express the average risk associated with the decisions $\boldsymbol{\gamma}$, for both detection and estimation, in the compact form

$$\begin{aligned} R(\sigma, \boldsymbol{\delta})_{\mathrm{D}\otimes\mathrm{E}} &= \langle C(\boldsymbol{S}, \boldsymbol{\gamma})\rangle_{\Delta,\Gamma,\Omega_S} \\ &= \int_{\Omega_S} \sigma_J(\boldsymbol{S})\mathbf{d}\boldsymbol{S} \int_{\Gamma} F_J(\boldsymbol{X}|\boldsymbol{S})\mathbf{d}\boldsymbol{X} \int_{\Delta} \delta(\boldsymbol{\gamma}|\boldsymbol{X}) C(\boldsymbol{S}, \boldsymbol{\gamma})\mathbf{d}\boldsymbol{\gamma}, \end{aligned} \tag{6.1.3a}$$

where $\sigma(\boldsymbol{S})$ once more is the *a priori* probability density (pdf)[6] of signal waveform $\boldsymbol{S} = (S_1, \ldots, S_J)$; $\sigma(\boldsymbol{\theta})$ is the corresponding pdf of the L signal parameters $\boldsymbol{\theta} = (\theta_1, \ldots, \theta_L)$, for $\boldsymbol{S} = \boldsymbol{S}(\boldsymbol{\theta})$, namely,

$$\begin{aligned} R(\sigma, \delta)_{\mathrm{D}\otimes\mathrm{E}} &= \langle C(\boldsymbol{\theta}, \boldsymbol{\gamma})\rangle_{\Delta,\Gamma,\Omega_\theta} \\ &= \int_{\Omega_\theta} \sigma_L(\boldsymbol{\theta})\mathbf{d}\boldsymbol{\theta} \int_{\Gamma} F_J(\boldsymbol{X}|\boldsymbol{S}(\boldsymbol{\theta}))\mathbf{d}\boldsymbol{X} \int_{\Delta} \delta(\boldsymbol{\gamma}|\boldsymbol{X}) C(\boldsymbol{\theta}, \boldsymbol{\gamma})\mathbf{d}\boldsymbol{\gamma}. \end{aligned} \tag{6.1.3b}$$

The pdfs $F_J(\boldsymbol{X}|\boldsymbol{S})$, $F_J(\boldsymbol{X}|\boldsymbol{S}[\boldsymbol{\theta}])$ are respectively the conditional pdfs of the received data $\boldsymbol{X}$ given $\boldsymbol{S}$ or $\boldsymbol{S}(\boldsymbol{\theta})$, while $C(\boldsymbol{S},\boldsymbol{\gamma})$, $C(\boldsymbol{\theta},\boldsymbol{\gamma})$ are the cost functions relating the decisions $\boldsymbol{\gamma} = (\gamma_1, \ldots, \gamma_{J \text{ or } L})$ about $\boldsymbol{S}$ or $\boldsymbol{\theta}$ in some quantitative way, cf. Section 1.4. In any case $\boldsymbol{X}$ and $\boldsymbol{S}$ are N or $J(= MN)$ dimensional column vectors, given by (6.1.1a) and (6.1.1b), where

$$\boldsymbol{S} = \boldsymbol{S}(\mathbf{r}_m, t_n; \boldsymbol{\theta}), \text{ or } \boldsymbol{S}(t_n; \boldsymbol{\theta}). \tag{6.1.3c}$$

The symbol $\mathrm{D} \otimes \mathrm{E}$ indicates that the decisions regarding detection (D) and estimation (E) may be coupled and thus interactive, depending on the choice of cost functions C, as noted in Fig. 6.1. We shall see a general example of such coupling in Section 6.3 following. In any case, the decision rule $\delta(\boldsymbol{\gamma}|\boldsymbol{X})$ relating the decisions $\boldsymbol{\gamma}$ to the data $\boldsymbol{X}$ represents the system structure and may be specified *a priori* or obtained under various optimality conditions,

[5] As we shall see presently, however, for the results of (3) to be used effectively, a separate detection process *is* required, even though this process itself does not influence the actual quantitative value of the estimate.

[6] Unless otherwise indicated we again use the convention that pdfs and functions of different arguments are different functions or have different values, for example, $\sigma(\boldsymbol{S}) \neq \sigma(\boldsymbol{\theta})$, and so on.

usually by minimizing the average risk, for example, symbolically ($\lim_{\delta \to \delta^*} R$). The function spaces for received signal ($\boldsymbol{S}$ and/or parameter $\boldsymbol{\theta}$), received data $\boldsymbol{X}$, and decisions $\boldsymbol{\gamma}$ are again denoted by Ω, Γ, Δ respectively (a full exposition is given in Chapter 1 at the beginning of the book).

Equations (6.1.3) are valid both for detection and estimation considered separately or jointly, once appropriate cost functions have been specified. It is also useful to observe that the decision vector $\boldsymbol{\gamma}$ is allowed to assume only a discrete set of values in detection theory, for instance, $\boldsymbol{\gamma} = (\gamma_0 = 0, \gamma_1 = 1)$ for an on–off situation [cf. Section 1.6]. In estimation problems the decision rule $\boldsymbol{\gamma}$ may generally have a continuum of values $\delta(\boldsymbol{\gamma}|\boldsymbol{X}) = \delta(\boldsymbol{\gamma} - \boldsymbol{\gamma}_\theta(\boldsymbol{X}))$, and the estimators $\boldsymbol{\gamma}_\theta$ are L-dimensional vectors $\boldsymbol{\gamma}_\theta = [\gamma_\theta(\boldsymbol{X})_\ell], \ell = 1, \ldots, L$, corresponding to the L parameters $\boldsymbol{\theta} = [\theta_\ell]$ being estimated; cf. Chapter 5.

Usually we seek an optimum, that is, here a Bayes system, by suitable choice of decision rule δ, that is, we seek the *optimum structure*, δ^*, such that $(\delta \to \delta^*)R \to R^*(\sigma, \delta^*)$, when an appropriate pdf σ is assigned for $\boldsymbol{S}$ or $\boldsymbol{\theta}$. Accordingly, we write[7]

$$\sigma(\boldsymbol{S}) = q\delta(\boldsymbol{S} - \boldsymbol{0}) + pw_L(\boldsymbol{S}), \tag{6.1.4}$$

where once more $p = p(H_1), q = q(H_0) = 1 - p$ are the *a priori* probabilities that the data sample does or does not contain a signal; $\delta(\bullet)$ is the Dirac delta function; and $w_L(\boldsymbol{S})$ is the pdf of $\boldsymbol{S}$ under hypothesis H_1. By analogy with (6.1.4) we can also write[7]

$$\sigma(\boldsymbol{\theta}) = q\delta(\boldsymbol{\theta} - \boldsymbol{0}) + pw_L(\boldsymbol{\theta}) \tag{6.1.4a}$$

for the pdf of those parameters (like the signal amplitude), which are directly related to the received signal energy, so that $\boldsymbol{S} = \boldsymbol{0}$ implies $\boldsymbol{\theta} = \boldsymbol{0}$.[7]

A similar relation cannot be written, however, for other parameters (like phase or frequency) whose values are not defined when $\boldsymbol{S} = \boldsymbol{0}$: in this case it is meaningless to use the transformation $\sigma(\boldsymbol{S})d\boldsymbol{S} = \sigma(\boldsymbol{\theta})d\boldsymbol{\theta}$ under the hypothesis H_0, since *the pdf of the* $\boldsymbol{\theta}$, $w_L(\boldsymbol{\theta})$, *is only defined under hypothesis* H_1. This, of course, does not prevent our using (6.1.4b) below in our estimation of $\boldsymbol{\theta}$ when $p(H_1) < 1$, with the understanding that $q(H_0)$ always refers to $\boldsymbol{S} = \boldsymbol{0}$ regardless of $\boldsymbol{\theta}$, namely,

$$\sigma(\boldsymbol{\theta}) = q\delta(\boldsymbol{S} - \boldsymbol{0}) + pw_L(\boldsymbol{\theta}). \tag{6.1.4b}$$

As noted above we are concerned here with the problem of Bayes systems *for joint signal detection and estimation*, where in all cases the prior probability $p(H_1)$ is less than (or equal to) unity, that is, $0 \le p(H_1) \le 1$. We begin with a number of reasonable assumptions and restrictions, most of which can either be removed, modified, or mitigated in subsequent extensions of the theory. These assumptions are

(1) a fixed amount of *a priori* data: no adaptivity or "learning";

(2) a single set of terminal decisions;

(3) binary detection of a signal in noise (H_1) versus noise alone (H_0) in the detection portion of the decision process;

[7] The number of random variables in $\sigma(S)$ is J and in $\sigma(\boldsymbol{\theta})$ is L.

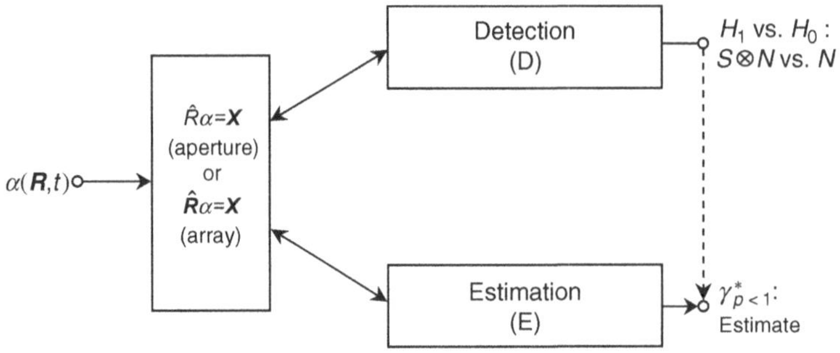

FIGURE 6.2 Joint detection and estimation *without coupling*. The dotted line represents acceptance or rejection of the *biased* (under H_1) estimator $\gamma^*_{p<1}$ upon the detection decision H_1 or H_0. (D) and (E) here are separate operations based on the same data input and signal, for fixed observation intervals.

(4) at least one signal parameter unknown at the receiver (so that signal estimation is a meaningful operation);

(5) $0 \leq p(H_1) \leq 1$: This is the critical new feature of the theory. [The limiting situation $p(H_1) = 1$ is the classical estimation case, cf. Chapter 5.]

It is worth noting that for estimation under prior uncertainty as to the presence of the signal, for example, $p < 1$, *two* data ensembles are now potentially involved, as in the usual case of binary detection: one set, $\{X\}_{H_1}$, under hypothesis state H_1: $S \otimes N$, and the other, $\{X\}_{H_0}$, under H_0: N. In applications, however, only *one representation*, $\boldsymbol{X}$, from one or the other ensemble, is actually employed. Which one, of course, is not known *a priori*. This is the new feature for estimation when $p < 1$.

As mentioned previously, we seek an optimum, or Bayes system for the combined operations of detection and estimation, with appropriate choices of cost functions and coupling between the two operations. Because of the many ways in which detection and extraction can be coupled together, we next briefly note a series of situations of increasing complexity. There are several types of coupling that we can assume. In the present section we discuss only the following cases because of their comparative analytical and conceptual simplicity:[8]

6.1.1 Case 1: No Coupling

The two operations of detection and estimation are done in parallel and independently; (see Fig. 6.2). Since the two operations are independent, the total average risk (6.1.3) is clearly

$$R = R_D + R_E, \tag{6.1.5}$$

and the two components of the total risk can be minimized separately. The Bayes optimum or minimization of the average detection risk $p < 1$ (with the usual constant cost functions, cf. Section 2.2.1) yields the well-known *generalized likelihood ratio* (GLR) *test*,[9] Eqs. (2.17a)

[8] See Section 6.3 ff. for the general formulation.

[9] This is also sometimes called the "average likelihood ratio (ALR) test." When the unknown parameters ($\boldsymbol{\theta}$) are replaced by maximum likelihood estimators ($\hat{\boldsymbol{\theta}}$) in $\log \Lambda(\boldsymbol{X})$, this latter quantity in (6.1.6) is then sometimes called a "generalized likelihood ratio test," not to be confused with (6.1.6)–(6.1.6b) here. However, conceptual problems arise because of the assumption that estimates under $p(H_1) = 1$ can be safely used *in detection*, where $p(H_1) < 1$ necessarily. See Section 1.8 for a discussion.

and (2.17b):

$$\begin{aligned}&\text{Decide } H_1,\ \text{signal and noise:}\ S\otimes N,\ \text{if}\ \log\Lambda(X)\geq\log K,\ \text{or}\\&\text{Decide } H_0,\ \text{noise alone:}\qquad N,\ \text{if}\quad \Lambda(X)<\log K,\end{aligned}\tag{6.1.6}$$

where the "threshold" log K is a simple function of the assigned costs of correct and incorrect decisions:

Eq. (1.7.3):

$$\log K\equiv\log\left[\frac{C_1^{(0)}-C_0^{(0)}}{C_0^{(1)}-C_1^{(1)}}\right]=\log\left[\frac{C_\alpha-C_{1-\alpha}}{C_\beta-C_{1-\beta}}\right],\quad C_1^{(0)}\equiv C_\alpha>C_{1-\alpha};C_0^{(1)}\equiv C_\beta>C_{1-\beta},\text{etc.,}\tag{6.1.6a}$$

and the GLR is

$$\left.\begin{aligned}&\text{Eq. (1.7.2):}\quad \Lambda(X)\equiv\mu\langle F_J(X|S(\theta))\rangle_\theta/F_J(X|\mathbf{0}),\mu\equiv p/q,\\&\textit{with}\quad \langle F_J(X|S(\theta))\rangle_\theta=\int_\Omega F_J(X|S(\theta))w_L(\theta)d\theta=\int_\Omega W_{J\times L}(X,\theta)d\theta\end{aligned}\right\},\tag{6.1.6b}$$

which defines $w_{J\times L}(X,\theta)$, where Λ is the familiar *generalized likelihood ratio*,[10] Eq. (1.7.2). Here $\alpha^*=\alpha(S\otimes N|H_0)^*$, $\beta^*=\beta(N|H_1)^*$ are respectively the (optimum) conditional probabilities of Type I and II errors, namely, of deciding $H_1=S\otimes N$ or H_0: N when the reverse (H_0, or H_1) is true.

The (constant) costs in (6.1.6a) are

$$\left\{\begin{aligned}&C_\alpha\equiv C_1^{(0)}:\ \text{cost of deciding incorrectly that a signal is present when only noise occurs}\\&C_{1-\alpha}\equiv C_0^{(0)}:\ \text{cost of deciding correctly that only noise occurs;}\end{aligned}\right\}\tag{6.1.6c}$$

and

$$\left\{\begin{aligned}&C_\beta\equiv C_0^{(1)}:\ \text{cost of deciding incorrectly that only noise is present when actually a signal and noise occurs}\\&C_{1-\beta}\equiv C_1^{(1)}:\ \text{cost of correctly deciding a signal is present in the noise.}\end{aligned}\right\}\tag{6.1.6d}$$

Of course, for meaningful results the costs of decision errors must exceed those of correct decisions, cf. (6.1.6a) and Section 1.6.

For optimal detection, $\Lambda(X)$ is the processing algorithm,[11] that is, the detector, and it embodies the operations to be performed on the received data X to achieve either of the decisions H_1 or H_0, according to (6.1.6). Even when there is *weak coupling*, by proper choice of the cost functions coupling detection and estimation it is still possible separately to minimize the average risks R_D, R_E (6.1.5), and use (6.1.6) for detection, with appropriate

[10] See footnote 7.

[11] It is convenient for most analytical and practical purposes to use log Λ rather than Λ, the former being monotonic in Λ and as shown in Section 1.9.1.1, the error probabilities α^*, β^*, and so on, remain unchanged.

adjustments of the threshold K; see Section 6.3 ff. and Section 4.1 of Ref. [1]. Specifically, note that the Bayes risk itself, $R_D^* = (\lim_{\delta \to \delta^*} R_D)$ for detection here, may be written explicitly [Eq. (1.7.1)] as

$$
\begin{aligned}
R_D^* &= \Re_0 + p\big(C_\beta - C_{1-\beta}\big)\int_\Gamma \delta^*(\gamma_0|\mathbf{X})(\Lambda - K)\mathbf{dX} \\
&= \Re_0 + p\big(C_\beta - C_{1-\beta}\big)(K\alpha^* + \beta^*), \text{ with } \Re_0 \equiv qC_\alpha + pC_{1-\beta}.
\end{aligned}
\tag{6.1.7}
$$

Now the decision is to make $\delta^*_{1,0}$ obey $\delta_0^* + \delta_1^* = 1$, with $\delta_0^* = 1, \delta_1^* = 0$, for H_0, and $\delta_1^* = 1$, $\delta_0^* = 0$, for H_1, from which the decision process (6.1.6) above follows. The details are developed in Section 1.7.

The minimization of the estimation risk R_E, however, involves an essentially new problem, since the estimator's performance must now be optimized in the face of uncertainty as to the signal's presence, for example, $p(H_1 \leq 1)$. Results for this problem are developed in Section 6.2 ff.

6.1.2 Case 2: Coupling

In this case there is coupling between the detection and estimation, and it consists in the estimation being directed by the result of the detection operation. By this we mean that either the estimator does not act on the data unless the detector has decided that the signal is present, or that the two operations are still done in parallel, but that the estimate is rejected if the detector's decision is that only noise is present at the input. These two cases are illustrated in Fig. 6.3a and b. Different orders of complexity are possible here, according to the amount of knowledge that the detector has of the estimator's structure and according to the selection of the cost functions. In general, the total risk cannot be written as the sum of two independent components, and its minimization leads to a joint optimization. This, in turn, involves tests that have more complicated statistics than the usual generalized likelihood ratio.

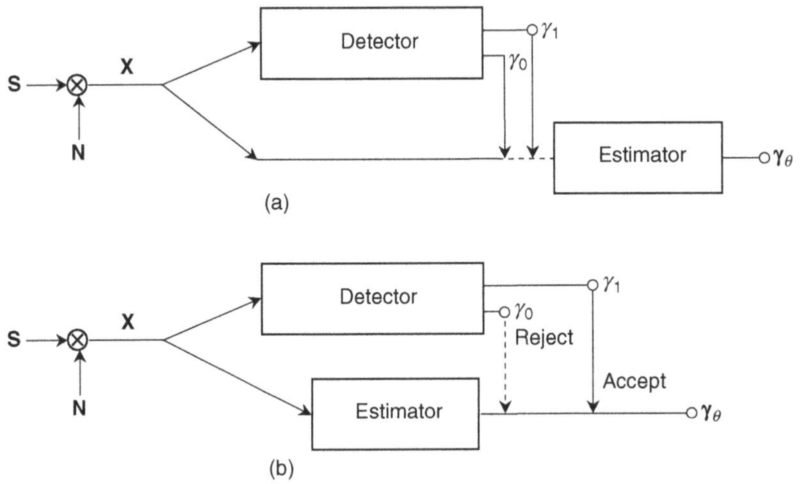

FIGURE 6.3 "No coupling": (a) detection-directed estimate, $p < 1$; (b) detection-directed estimation with decision rejection.

The general problem is examined in Section 6.3 ff. Its specific applications to joint D and E problems in telecommunications largely remain to be explored.

Accordingly, we begin our analysis with a treatment of the "no-coupling" cases (Section 6.2 ff.), and then extend it to the general situations of "strong" and "weak" coupling (Section 6.3 ff.).

6.2 OPTIMAL ESTIMATION [$P(H_1) \leq 1$]: NO COUPLING

"No coupling" is defined as the situation when $p(H_1 \leq 1)$ where the costs of detection do not depend on the signal or its parameters *and* where the detection (D) and estimation (E) portion of the average cost decision (6.1.5) can be independently treated, that is, minimized in most instances. ("Weak coupling," on the other hand, permits cost functions in detection that do depend on the signal, while also maintaining a detector structure independent of the estimator, vide Section 6.3.2 ff.). Figure 6.3 illustrates the uncoupled cases of *detection-directed estimation*. This means that estimation is either carried out after the detector has decided that the signal is present, or that the estimation (done in parallel with detection) is then accepted or rejected, depending on the detector's decisions. In either instance, the D cost functions do not depend on the signal, so that the D and E processes are uncoupled.

Thus, for "no coupling" the average risk R_{E} associated with signal or parameter estimation follows at once from [(6.1.3a) and (6.1.3b)] and (6.1.5). The average risk R_{D} attributable to the detection process, with the usual constant cost assignments of [(6.1.6a), (6.1.6c), (6.1.6d)] is determined in the customary way from (6.1.6), (6.1.6a–d), as described more fully in Section 1.7 earlier. For this reason we direct our attention here solely to the estimation portion of the overall decision process in this new situation where $p \leq 1$.

Accordingly, let $C'(\boldsymbol{\gamma})$ and $C(\boldsymbol{S}, \boldsymbol{\gamma})$ be the cost functions under hypothesis H_0 and H_1, respectively, and apply the decision rules $\delta(\boldsymbol{\gamma}|\boldsymbol{X}) = \delta(\boldsymbol{\gamma} - \boldsymbol{\gamma}_S(\boldsymbol{X}))$ for waveforms, integrated over decision space Δ, for example, $\int_\Delta(\ )\delta(\boldsymbol{\gamma}|\boldsymbol{X})\boldsymbol{dX}$ in Eq. (6.1.3). Then, the average risk R_{E} becomes for waveform estimation, $p < 1$:

$$R_{\mathrm{E}} = R_S(\sigma, \delta) = \int_\Gamma \left\{ [qF_J(\boldsymbol{X}|\boldsymbol{S})C'(\boldsymbol{\gamma})]_{\boldsymbol{S}=\boldsymbol{0}} + p\int_{\Omega_S} F_J(\boldsymbol{X}|\boldsymbol{S})C(\boldsymbol{S}, \boldsymbol{\gamma})w_J(\boldsymbol{S})\mathbf{d}\boldsymbol{S} \right\} \mathbf{d}\boldsymbol{X}. \tag{6.2.1a}$$

The corresponding expression for the average risk associated with the estimation of the L signal parameters $\boldsymbol{\theta}(= \theta_1, \ldots, \theta_L)$, where now the decision rule is analogously $\delta(\boldsymbol{\gamma}|\boldsymbol{X}) = \delta(\boldsymbol{\gamma} - \boldsymbol{\gamma}_{\boldsymbol{\theta}}(\boldsymbol{X}))$, is explicitly

$$R_{\mathrm{E}} = R_{\boldsymbol{\theta}}(\sigma, \delta) = \int_\Gamma \left\{ \begin{array}{l} [qF_J(\boldsymbol{X}|\boldsymbol{S}(\boldsymbol{\theta}))C'(\boldsymbol{\gamma}_{\boldsymbol{\theta}})]_{\boldsymbol{S}=\boldsymbol{0}} \\ +p\displaystyle\int_{\Omega_\theta} F_J(\boldsymbol{X}|\boldsymbol{S}(\boldsymbol{\theta}))C(\boldsymbol{\theta}, \boldsymbol{\gamma}_{\boldsymbol{\theta}})w_L(\boldsymbol{\theta})\mathbf{d}\boldsymbol{\theta} \end{array} \right\} \mathbf{d}\boldsymbol{X}. \tag{6.2.1b}$$

To obtain the Bayes estimators of waveform $\boldsymbol{S}$, namely $\boldsymbol{\gamma}_S^* = \boldsymbol{\gamma}^*(\boldsymbol{S}|\boldsymbol{X})$, or of the signal parameters $\boldsymbol{\theta}$, for example, $\boldsymbol{\gamma}_\theta^* = \boldsymbol{\gamma}^*(\boldsymbol{\theta}|\boldsymbol{X})$, we first evaluate the variation $\boldsymbol{\delta R}_{\mathrm{E}}$ of the average risk, (6.2.1a) or (6.2.1b), setting it equal to zero for an extremum in the usual way, namely,

$$\boldsymbol{\delta R}_S = \int_\Gamma \boldsymbol{\delta\gamma} \left\{ q\left[F_J \frac{\partial C'}{\partial \boldsymbol{\gamma}}\right]_{\boldsymbol{S}=\boldsymbol{0}} + p\left\langle F_J \frac{\partial C}{\partial \boldsymbol{\gamma}} \right\rangle_{\boldsymbol{S}\neq\boldsymbol{0}} \right\} \mathbf{d}\boldsymbol{X} = \boldsymbol{0}, \tag{6.2.2a}$$

which becomes the (J or L) set of equations[12] in $\boldsymbol{\gamma}$

$$\left[q\left(F_J \frac{\partial C'}{\partial \boldsymbol{\gamma}} \right)_{\mathbf{S}=\mathbf{0}} + p\left\langle F_J \frac{\partial C}{\partial \boldsymbol{\gamma}} \right\rangle_{\mathbf{S} \text{ or } \boldsymbol{\theta}} \right]_{\boldsymbol{\gamma} \to \boldsymbol{\gamma}^*} = \mathbf{0}. \tag{6.2.2b}$$ [13]

The solutions (when they exist) of the Eqs. (6.2.2a) and (6.2.2b) are not only extremal solutions but are minimizing solutions for the average risk R_E, that is, $R_\mathrm{E} \to R_\mathrm{E}^*$, the minimum average or Bayes risk, provided the Hessians of the integrands $G(\boldsymbol{\gamma})$ of [(6.2.1a) and (6.2.1b)] here are positive, that is, $\boldsymbol{H}(G) = \left[\partial G^{(j)}/\partial \gamma_j\right] > 0$, with $G^{(j)} = \left[\partial G/\partial \gamma_j\right]$, the Hessian being the Jacobian of the derivative of $G = G(\boldsymbol{\gamma}_1, \ldots, \boldsymbol{\gamma}_{J \text{ or } L})$, namely, $H(G) = \left[(\partial^2/\partial\gamma_i\partial\gamma_j)G\right]$.

Here we note as expected that when $p = 1$, $(q = 0)$, we have the "classical" cases of estimation when the signal is *a priori* known surely to be present, so that [(6.2.2a) and (6.2.2b)] reduce at once to the familiar relations for determining $\boldsymbol{\gamma}^*_{p=1}$, cf. Section 5.2 preceding. Alternatively, when $p = 0$ and $q = 1$, only the first terms of (6.2.2) determine $\boldsymbol{\gamma}^*$. Then $\boldsymbol{\gamma}^*_{p=0}$ must vanish, since it is surely known that there is no signal present. This puts a condition on the cost function C', namely, $\partial C'/\partial \boldsymbol{\gamma}|_{\boldsymbol{\gamma}=0} = 0$, a condition that is satisfied by most cost functions. We note that in the case of waveform or amplitude estimation it may also be reasonable to assume that $C'(\boldsymbol{\gamma}_\theta) = C(\mathbf{0}, \boldsymbol{\gamma}_\theta)$. This assumption introduces some simplification in the solutions of (6.2.2). (The examples of Section 6.4 ff. illustrate this point.) Finally, because there is no coupling between estimator and detector we can estimate as well as detect on the whole data space Γ, rejecting or putting aside an estimate if the detector decides H_0. (See also the discussion in Section 6.3.2.)

6.2.1 Quadratic Cost Function: MMSE and Bayes Risk

We next specialize the extremal relation (6.2.2b) to the cases of waveform and parameter estimation with a QCF, that is, from (6.1.3a) and (6.1.3b) on averaging over decision space Δ, C_S becomes

$$C_S = C_0|\mathbf{S} - \boldsymbol{\gamma}_S|^2, \text{ or } C_{\hat{\theta}} = C_0|\hat{\boldsymbol{\theta}} - \boldsymbol{\gamma}_{\hat{\theta}}|^2. \tag{6.2.3}$$ [14]

Here $\hat{\boldsymbol{\theta}} = \left[\hat{\theta}_\ell\right], \ell = 1, \ldots, L$, are the L parameters to be estimated out of $L_\mathrm{M} (\geq L)$ parameters $\hat{\boldsymbol{\theta}} = (\hat{\boldsymbol{\theta}}, \boldsymbol{\theta}')$, $(L_\mathrm{M} = L + L')$, where $\boldsymbol{\theta}'$ are other random parameters not subject to estimation here ([4], p. 966), with $C'(\boldsymbol{\gamma}) = C(\mathbf{0}, \boldsymbol{\gamma})$. These quadratic cost functions yield MMSE estimators.

6.2.1.1 *The Bayes Estimator* $\gamma^*_{p\leq 1}|_\mathrm{QCF}$

Putting the second relation of (6.2.3) into (6.2.2b) and solving for $\boldsymbol{\gamma}^*_{\hat{\theta}}$, we readily obtain the Bayes estimator $\boldsymbol{\gamma}^*_{p\leq 1}$ of $\hat{\boldsymbol{\theta}}$. Our initial result is

$$\gamma^*_{p\leq 1}(\hat{\boldsymbol{\theta}}|\boldsymbol{X})_\mathrm{QCF} = p\int_{\hat{\Omega}} \hat{\boldsymbol{\theta}} W_{J\times L}(\boldsymbol{X}, \hat{\boldsymbol{\theta}})\, d\hat{\boldsymbol{\theta}} \Big/ \left\{ qW_J(\boldsymbol{X})_\mathbf{0} + p\int_{\hat{\Omega}} W_{J\times L}(\boldsymbol{X}, \hat{\boldsymbol{\theta}})\, d\hat{\boldsymbol{\theta}} \right\}, \tag{6.2.4}$$

[12] It is assumed that $\partial C/\partial \boldsymbol{\gamma}$ exists and C is a convex cost function, so that the decision rule δ is not randomized [Section 5.1 and Section 21.2, p. 961 of Ref. [4]; also 6.3.1.1].

[13] Depending on the structure of the cost functions, the solutions of (6.2.2b) are either simultaneous joint solutions (cf. Section 6.2.2.2) or L separate solutions (Section 6.2.2.3).

[14] For convenience $\hat{\boldsymbol{\theta}}, \boldsymbol{\theta}', \boldsymbol{\gamma}^*_{\hat{\theta}}$, and so on, are all normalized to be dimensionless and C_0 is simply a "cost"; $\mathbf{S}$ and $\boldsymbol{\gamma}_S$ may not be normalized, so that C_0 has the dimensions of ("cost/waveform").

where the various pdfs $W_{J\times L}$, W_J are defined by the equivalent forms

$$\left.\begin{aligned} &W_{J\times L}(\boldsymbol{X},\hat{\boldsymbol{\theta}}) \equiv F_J\left(\boldsymbol{X}|\boldsymbol{S}(\hat{\boldsymbol{\theta}})\right)w_L(\hat{\boldsymbol{\theta}}) \equiv W_J(\boldsymbol{X}|\hat{\boldsymbol{\theta}})w_L(\hat{\boldsymbol{\theta}}),\\ &\therefore\ W_J(\boldsymbol{X})_{\hat{\boldsymbol{\theta}}} \equiv \int_{\hat{\Omega}} W_{J\times L}(\boldsymbol{X},\hat{\boldsymbol{\theta}})\mathbf{d}\hat{\boldsymbol{\theta}} = \left\langle F_J\left(\boldsymbol{X}|\boldsymbol{S}(\hat{\boldsymbol{\theta}})\right)\right\rangle_{\hat{\boldsymbol{\theta}}};\\ &W_J(\boldsymbol{X})_{\mathbf{0}} \equiv F_J(\boldsymbol{X}|\boldsymbol{S}=\mathbf{0}) \equiv F_J(\boldsymbol{X}|\mathbf{0})_{\boldsymbol{S}=\mathbf{0}} \neq F_J(\boldsymbol{X}|\boldsymbol{S}(\mathbf{0})) \end{aligned}\right\}. \tag{6.2.4a}$$

We note also that the GLR (6.1.6b) may be expressed here alternatively as

$$\Lambda(\boldsymbol{X}) = \mu W_J(\boldsymbol{X})_{\hat{\boldsymbol{\theta}}}/W_J(\boldsymbol{X})_{\mathbf{0}} = \mu\left\langle F_J\left(\boldsymbol{X}|\boldsymbol{S}(\hat{\boldsymbol{\theta}})\right)\right\rangle_{\hat{\boldsymbol{\theta}}}\Big/F_J(\boldsymbol{X}|\mathbf{0}), \tag{6.2.4b}$$

with $\mu = p/q$ as before. The frequently occurring denominator of (6.2.4), (6.1.6b) et. seq.,

$$W_J\left(\boldsymbol{X}|_{p,q}\right) \equiv qW_J(\boldsymbol{X})_{\mathbf{0}} + p\int_{\hat{\Omega}} W_{J\times L}(\boldsymbol{X},\hat{\boldsymbol{\theta}})d\hat{\boldsymbol{\theta}}, \tag{6.2.4c}$$

may be interpreted physically as the pdf of the *a priori* uncertainty-weighted data.

Equation (6.2.4) can be put into a much more compact and revealing form with the help of the classical result of (5.1.22) or (21.62) [4]):

$$\boldsymbol{\gamma}^*_{p=1}(\boldsymbol{X}|\hat{\boldsymbol{\theta}})_{\text{QCF}} = \int_{\hat{\Omega}} \hat{\boldsymbol{\theta}} W_{J\times L}(\boldsymbol{X},\hat{\boldsymbol{\theta}})d\hat{\boldsymbol{\theta}}\Big/\int_{\hat{\Omega}} W_{J\times L}(\boldsymbol{X},\hat{\boldsymbol{\theta}})d\hat{\boldsymbol{\theta}}. \tag{6.2.5}$$

We see that (6.2.4) can now be written

$$\boldsymbol{\gamma}^*_{p\leq 1}\Big|_{\text{QCF}} = \int_{\hat{\Omega}} \frac{\hat{\boldsymbol{\theta}} W_{J\times L}(\boldsymbol{X},\hat{\boldsymbol{\theta}})\mathbf{d}\hat{\boldsymbol{\theta}}}{W_J(\boldsymbol{X})_{\hat{\boldsymbol{\theta}}}} \cdot \frac{pW_J(\boldsymbol{X})_{\hat{\boldsymbol{\theta}}}}{qW_J(\boldsymbol{X})_{\mathbf{0}}\left[1+\mu W_J(\boldsymbol{X})_{\hat{\boldsymbol{\theta}}}/W_J(\boldsymbol{X})_{\mathbf{0}}\right]}, \tag{6.2.6}$$

or,

$$\boxed{\boldsymbol{\gamma}^*(\hat{\boldsymbol{\theta}})_{p\leq 1|\text{QCF}} = \frac{\Lambda}{1+\Lambda}\boldsymbol{\gamma}^*(\hat{\boldsymbol{\theta}})_{p=1|\text{QCF}},} \tag{6.2.7}$$

where once more Λ is the GLR (6.1.6b). (Equation (6.2.7) was originally obtained by Middleton and Esposito [1] in 1968.)

An alternative representation of $\boldsymbol{\gamma}^*_{p=1}|_{\text{QCF}}$, (6.2.5), for estimating the $\left[\hat{\theta}_\ell\right], \ell = 1,\ldots,L$, of parameters of $\boldsymbol{S}(\boldsymbol{\theta})$ from the complete set $\boldsymbol{\theta} = \left(\hat{\boldsymbol{\theta}},\boldsymbol{\theta}'\right)$, is given by

$$\boldsymbol{\gamma}^*_{p=1}(\hat{\boldsymbol{\theta}}|\boldsymbol{X})_{\text{QCF}} = \int_{\hat{\Omega}} \hat{\boldsymbol{\theta}} w_L(\hat{\boldsymbol{\theta}})e^{\ell^{(21)}(\boldsymbol{X}|\hat{\boldsymbol{\theta}})_{\text{QCF}}}d\hat{\boldsymbol{\theta}}, \tag{6.2.8}$$

with[15]

$$\ell^{(21)}(\boldsymbol{X}|\hat{\boldsymbol{\theta}})_{\text{QCF}} \equiv \log\left\langle F_J(\boldsymbol{X}|\boldsymbol{S}(\hat{\boldsymbol{\theta}},\boldsymbol{\theta}'))\right\rangle_{\boldsymbol{\theta}'} - \log\left\langle F_J\left(\boldsymbol{X}|\boldsymbol{S}(\hat{\boldsymbol{\theta}})\right)\right\rangle_{\boldsymbol{\theta}'}, \tag{6.2.8a}$$

[15] Of course, if the $\boldsymbol{\theta}'$ are deterministic, that is, are *a priori* known, the pdf $w_{L'}(\boldsymbol{\theta}') = \delta(\boldsymbol{\theta}'-\boldsymbol{\theta}) = \Pi_k\delta(\boldsymbol{\theta}'_\kappa - \boldsymbol{\theta}_{o\kappa})$, with a consequent simplification of [(6.2.8a) and (6.2.8b)].

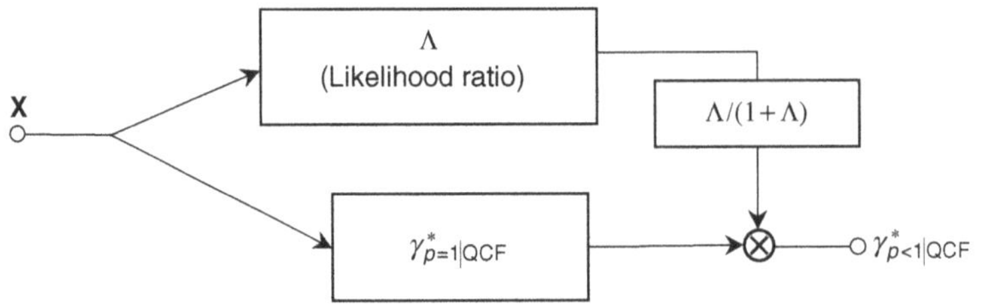

FIGURE 6.4 No-coupling case: Bayes estimation with QCF of parameters $\hat{\boldsymbol{\theta}}$, or waveform $\boldsymbol{S}$, when $p \leq 1$.

where specifically

$$\langle F_J \rangle_{\boldsymbol{\theta}'} = \int_{\Omega'} F_J\big(\boldsymbol{X}|\boldsymbol{S}(\hat{\boldsymbol{\theta}},\boldsymbol{\theta}')\big) w_{L'}(\boldsymbol{\theta}')\boldsymbol{d\theta}' \equiv W_J\big(\boldsymbol{X}|\hat{\boldsymbol{\theta}}\big), \text{cf.}(6.2.4\text{a}). \tag{6.2.8b}$$

Our result (6.2.7) applies as well for estimators of waveform $\boldsymbol{S}$ on replacing $\hat{\boldsymbol{\theta}}$ by $\boldsymbol{S}$ in (6.2.5a) for the components of $\boldsymbol{\gamma}^*_{p\leq 1}|_{\text{QCF}}$, namely,

$$\left.\begin{array}{c} \boldsymbol{\gamma}^*_{p=1}(\boldsymbol{S}|\boldsymbol{X})_{\text{QCF}} = \displaystyle\int_{\Omega} \boldsymbol{S} W_{J\times J}(\boldsymbol{X},\boldsymbol{S})\boldsymbol{dS} \Big/ \int_{\Omega} W_{J\times J}(\boldsymbol{X},\boldsymbol{S})\mathbf{d}\boldsymbol{S}, \\ \text{with } W_{J\times J}(\boldsymbol{X},\boldsymbol{S}) = F_J(\boldsymbol{X}|\boldsymbol{S})w_J(\boldsymbol{S}) \end{array}\right\}, \tag{6.2.9}$$

where specifically the GLR (6.2.4b) becomes directly

$$\Lambda(\boldsymbol{X}) = \mu W_J(\boldsymbol{X})_{\boldsymbol{S}} / W_J(\boldsymbol{X})_{\mathbf{0}} = \mu \langle F_J(\boldsymbol{X}|\boldsymbol{S})\rangle_{\boldsymbol{S}} / F_J(\boldsymbol{X}|\mathbf{0}). \tag{6.2.9a}$$

Equation (6.2.7) shows that to obtain the Bayes or optimum estimate under a QCF, of the general parameters $\hat{\boldsymbol{\theta}}$ when the signal, $\boldsymbol{S}(\hat{\boldsymbol{\theta}},\boldsymbol{\theta}')$, is not known to be surely present, for example, $p < 1$, we must first determine the "classical" $(p = 1)$ Bayes estimate, (6.2.5a) or (6.2.8), and weight it with a scalar factor that is a simple function of what a binary on–off Bayes detector produces for the same input data $\boldsymbol{X}$, just before the detector's decision δ_1, "yes" or δ_0, "no." The schema of this estimation strategy, along with the accompanying detection process, is shown in Fig. 6.4. Observe that when $\mu \to \infty$ (i.e., $p \to 1$) we obtain the expected classical result, $\boldsymbol{\gamma}^*_{p=1}$, and that when $\mu \to 0$ (i.e., $p \to 0$) we have the expected null estimator $\boldsymbol{\gamma}^*_{p=0} = \mathbf{0}$, since only noise is *a priori* known to be present at all times.

6.2.1.2 The Bayes Risk, with QCF The estimator $\boldsymbol{\gamma}^*_{p\leq 1}|_{\text{QCF}}$, (6.2.7), is a minimum mean square estimator, which minimizes the average risk $R_{\text{E}}(\sigma,\delta)$, [(6.2.1a) and (6.2.1b)]. The resultant Bayes risk is accordingly given by

$$R^*_{\text{E}}(\sigma,\delta^*)_{p\leq 1|\text{QCF}} = C_0\left\langle \left|\boldsymbol{\gamma}^*_{p\leq 1|\text{QCF}} - \hat{\boldsymbol{\theta}}\right|^2\right\rangle_H, \tag{6.2.10}$$

where $C_0(>0)$ is an appropriate cost factor,[16] and where now R^*_{E} has the expanded form

$$R^*_{\text{E}}(\sigma,\delta^*)_{p\leq 1|\text{QCF}} = C_0\hat{E}_H\left|\boldsymbol{\gamma}^*_{p\leq 1|\text{QCF}} - \hat{\boldsymbol{\theta}}\right|^2 = C_0\hat{E}_H\left\{\boldsymbol{\gamma}^*\cdot\boldsymbol{\gamma}^* - 2\hat{\boldsymbol{\theta}}\cdot\boldsymbol{\gamma} + \hat{\boldsymbol{\theta}}\cdot\hat{\boldsymbol{\theta}}\right\}_{p\leq 1|\text{QCF}}, \tag{6.2.10a}$$

[16] See footnote, Eq. (6.2.3).

in which the averaging (or "expectation") operator $\hat{E}_H$ is now explicitly

$$\hat{E}_H = \langle\,\rangle_H = q\langle\,\rangle_{H_0:\, \boldsymbol{X},\boldsymbol{S}=\boldsymbol{0}} + p\langle\,\rangle_{H_0:\, \boldsymbol{X},\boldsymbol{S}}. \tag{6.2.11}$$

In more detail, we can write (6.2.11), with the help of (6.1.4), (6.1.4a), as

$$\left.\begin{aligned}\hat{E}_H &= \int_\Gamma d\boldsymbol{X}\int_{\hat{\Omega}} F_J(\boldsymbol{X}|\boldsymbol{S}(\hat{\boldsymbol{\theta}}))\sigma_L(\hat{\boldsymbol{\theta}})(\,)_{\boldsymbol{X},\hat{\boldsymbol{\theta}}}d\hat{\boldsymbol{\theta}}\\ &= \int_\Gamma d\boldsymbol{X}\int_{\hat{\Omega}} F_J(\boldsymbol{X}|\boldsymbol{S}(\hat{\boldsymbol{\theta}}))\left[q\delta(\hat{\boldsymbol{\theta}}-\boldsymbol{0})|_{\boldsymbol{S}=\boldsymbol{0}} + pw_L(\hat{\boldsymbol{\theta}})\right](\,)_{\boldsymbol{X},\hat{\boldsymbol{\theta}}}d\hat{\boldsymbol{\theta}}\end{aligned}\right\}, \tag{6.2.11a}$$

where the component averages are

$$\left.\begin{aligned}\langle\,\rangle_{H_0} &\equiv \int_\Gamma F_J(\boldsymbol{X}|\boldsymbol{0})_{\boldsymbol{S}=\boldsymbol{0}}(\,)_{\boldsymbol{X}}d\boldsymbol{X};\\ \langle\,\rangle_{H_1} &\equiv \int_\Gamma d\boldsymbol{X}\int_\Omega F_J(\boldsymbol{X}|\boldsymbol{S}(\hat{\boldsymbol{\theta}}))w_L(\hat{\boldsymbol{\theta}})(\,)_{\boldsymbol{X},\hat{\boldsymbol{\theta}}}d\hat{\boldsymbol{\theta}}.\end{aligned}\right. \tag{6.2.11b}$$

In the case of waveform estimation (6.2.10a) becomes

$$R_{\mathrm{E}}^{*}(\sigma,\delta^{*})_{p\leq 1}\Big|_{\mathrm{QCF}} = C_0\hat{E}_H\{\boldsymbol{S}^{*}\cdot\boldsymbol{S}^{*} - 2\boldsymbol{S}\cdot\boldsymbol{S}^{*} + \boldsymbol{S}\cdot\boldsymbol{S}\}_{p\leq 1:\,\mathrm{QCF}}, \tag{6.2.12}$$

and the expectation operator (6.2.11a) is correspondingly

$$\hat{E}_H|_{\boldsymbol{S}} = \int_\Gamma d\boldsymbol{X}\int_{\hat{\Omega}} F_J(\boldsymbol{X}|\boldsymbol{S})\left[q\delta(\boldsymbol{S}-\boldsymbol{0})|_{\boldsymbol{S}=0} + pw_J(\boldsymbol{S})\right](\,)_{\boldsymbol{X},\boldsymbol{S}}d\boldsymbol{S}. \tag{6.2.12a}$$

A major task here, of course, is the evaluation of the estimators (6.2.7) and the associated Bayes risk (6.2.10), (6.2.12). This is usually not achievable in explicit analytic form, although some special cases of interest can be carried out, as discussed in Section 6.4. Practicable analytic results are generally obtainable only in the threshold or weak-signal regimes. However, numerical results in specific cases are always achievable in principle, by computational methods, once the governing probability distributions (pdfs) have been specified.

6.2.2 Simple Cost Functions: UMLE and Bayes Risk

As we have already seen in Section 5.2.2 earlier, "simple" cost functions are particularly important in the classical theory of $R_{\mathrm{E}}(\sigma,\delta)$ described in Chapter 5, since they lead to several forms of unconditional maximum (i.e., Bayes) likelihood estimators. There are two main types of SCF: the "nonstrict" simple cost function for the parameters $R_{\mathrm{E}\theta}(\sigma,\delta)$, namely,

$$SCF_1 \qquad C(\hat{\boldsymbol{\theta}},\boldsymbol{\gamma})_1 \equiv C_0\sum_{\ell=1}^{L}\left[A_\ell - \delta(\gamma_\ell - \hat{\theta}_\ell)\right], \quad \boldsymbol{\gamma}\rightarrow\boldsymbol{\gamma}_{SCF_1}, \tag{6.2.13a}$$

and the "strict" form of simple cost function

$$\text{SCF}_2 \qquad \begin{aligned} C(\hat{\boldsymbol{\theta}}, \boldsymbol{\gamma})_2 &\equiv C_0\left\{A_\ell - \delta(\boldsymbol{\gamma} - \hat{\boldsymbol{\theta}})\right\} \\ &= C_0\left\{A_L - \prod_{\ell=1}^{L} \delta(\gamma_\ell - \hat{\theta}_\ell)\right\}, \end{aligned} \quad \boldsymbol{\gamma} \to \boldsymbol{\gamma}_{\text{SCF}_2}. \tag{6.2.13b}$$

Although the former SCF_1 (6.2.13a) is less strict and does not impose so severe a penalty as can the strict form SCF_2, (6.2.13b), we shall begin our extension of the classical theory here using the latter. We do this since the analysis is somewhat more involved, and can in turn be more easily specialized to the simpler cost function (6.2.13a). [Again, unless otherwise indicated, we require $\boldsymbol{\gamma}$ and $\hat{\boldsymbol{\theta}}$ to be suitably normalized, so that the δ-functions and A_L are dimensionless.]

6.2.2.1 Bayes Risk (SCF) Equation (6.2.13b) is now applied to (6.2.1), where we have averaged over decision space Δ, in (6.1.3b) noting that now that this average becomes $\left\langle \delta(\boldsymbol{\gamma} - \boldsymbol{\gamma}_\theta^*)\delta(\boldsymbol{\gamma} - \hat{\boldsymbol{\theta}}) \right\rangle \Delta$: $\boldsymbol{\gamma} = \delta(\boldsymbol{\gamma}_\theta^*(\mathbf{X}) - \hat{\boldsymbol{\theta}})$, cf. (6.2.1b). This yields a Bayes risk of the form

$$\left\langle C(\hat{\boldsymbol{\theta}}, \boldsymbol{\gamma})_2 \right\rangle_{\hat{\Omega},\Gamma,\Delta} = R_{\text{E}}(\sigma, \delta^*)^*_{\text{SCF}_2} = C_0 \cdot \left\{A_L - \left\langle \delta(\boldsymbol{\gamma}_{\boldsymbol{\theta}}^*(\mathbf{X}) - \hat{\boldsymbol{\theta}}) \right\rangle_H \right\}, \tag{6.2.14}$$

where now $\langle \ \rangle_H (\equiv \hat{E}_H)$ is given by (6.2.11) and [(6.2.11a) and (6.2.11b)] or (6.2.12a). The Bayes risk for $p \leq 1$ then becomes specifically

$$R_{\text{E}}^*(\sigma, \delta^*)\Big|_{\substack{\text{SCF}_2 \\ p\leq 1}} = C_0 \cdot \left\{A_L - \left[q\left\langle \delta(\boldsymbol{\gamma}^*_{\text{SCF}_2} - \mathbf{0}) \right\rangle_{H_0:\, \mathbf{X}} + p\left\langle \delta(\boldsymbol{\gamma}^*_{\text{SCF}_2} - \hat{\boldsymbol{\theta}}) \right\rangle_{H_0:\, \mathbf{X},\hat{\boldsymbol{\theta}}}\right]\right\}, \tag{6.2.14a}$$

which may be expressed in still more detail as

$$R_{\text{E}}^*(\sigma, \delta^*)\Big|_{\substack{\text{SCF}_2 \\ p\leq 1}} = C_0 \cdot \left\{A_L - \int_\Gamma \left[\begin{array}{c} qF_J(\mathbf{X}|\mathbf{0})_{\mathbf{S}=\mathbf{0}}\delta(\boldsymbol{\gamma}^*_{p\leq 1}(\mathbf{X}) - \mathbf{0}) \\ + p\int_{\hat{\Omega}} F_J(\mathbf{X}|\hat{\boldsymbol{\theta}}) w_L(\hat{\boldsymbol{\theta}})\delta(\boldsymbol{\gamma}^*_{p\leq 1}(\mathbf{X}) - \hat{\boldsymbol{\theta}}) d\hat{\boldsymbol{\theta}} \end{array}\right] d\mathbf{X}\right\} \tag{6.2.15a}$$

$$= C_0 \cdot \left\{A_L - \int_\Gamma \left[\begin{array}{c} qF_J(\mathbf{X}|\mathbf{0})_{\mathbf{S}=\mathbf{0}}\delta(\boldsymbol{\gamma}^*_{p\leq 1}(\mathbf{X}) - \mathbf{0}) \\ + pF_J(\mathbf{X}|\boldsymbol{\gamma}^*_{p\leq 1}(\mathbf{X}))w_L(\boldsymbol{\gamma}^*_{p\leq 1}(\mathbf{X})) \end{array}\right] d\mathbf{X}\right\}, \tag{6.2.15b}$$

which reduces, as required (see Section 5.2 or Eq. (21.85) of Ref. [4]) when $q = 0(p = 1)$.

Equations (6.2.14), [(6.2.15a) and (6.2.15b)] can be formally evaluated by using [(6.2.4a) and (6.2.4b)] as needed, with [(6.2.13a) and (6.2.13b), by employing the integral representation of the δ-functions:

$$\delta(\boldsymbol{\gamma}^*_{p\leq 1} - \hat{\boldsymbol{\theta}}) = \int_{-\infty}^{\infty} \!\!\cdots\! \int \exp\left[i\sum \xi_\ell \boldsymbol{\gamma}^*_{p\leq 1}(\mathbf{X}) - i\sum \eta_\ell \hat{\boldsymbol{\theta}}_\ell\right] d\boldsymbol{\xi}\, d\boldsymbol{\eta}/(2\pi)^{2L}. \tag{6.2.16}$$

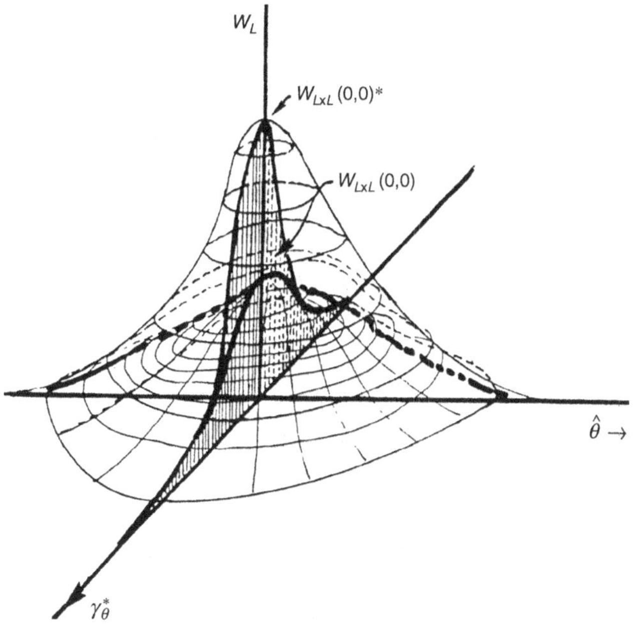

FIGURE 6.5 Schematic illustration of $\left(w_{1\times 1}\middle|\gamma^*,\hat{\theta}\right)$, showing $\left[w_{1\times 1}^* > w_{1\times 1}\right]_{\mathbf{0},\mathbf{0}}$.

Averaging over $\boldsymbol{X} \in \Gamma$ and $\hat{\boldsymbol{\theta}} \in \hat{\Omega}$, using [(6.2.4a) and (6.2.4b)] in [(6.2.15a) and (6.2.15b)], we see at once that the $(\boldsymbol{\xi}, \boldsymbol{\eta})$-integral of δ is simply the characteristic function (c.f.)

$$F_{L\times L}\left(i\xi, -i\eta\right)_{\gamma_\theta^*,\hat{\theta}} = \int_{\hat{\Omega}} d\hat{\theta} \int_{\Gamma} d\boldsymbol{X}\, W_{J\times L}\left(\boldsymbol{X},\hat{\theta}\right) e^{i\Sigma_\ell\left(\xi_\ell \gamma_{p\leq 1}^*(\boldsymbol{X}) - \eta_\ell \hat{\theta}_\ell\right)}, \tag{6.2.17}$$

so that

$$\begin{aligned} \left\langle \delta\left(\boldsymbol{\gamma}_{p\leq 1}^* - \hat{\boldsymbol{\theta}}\right)\right\rangle_{H_1} &= \int_{-\infty}^{\infty}\!\!\cdots\!\!\int F_{L\times L}(i\boldsymbol{\xi}, -i\boldsymbol{\eta})\, d\boldsymbol{\xi}\, d\boldsymbol{\eta}/(2\pi)^{2L} \\ &= w_{L\times L}\left(\boldsymbol{\gamma}_{p\leq 1}^*, \hat{\boldsymbol{\theta}}\right)_{\left(\boldsymbol{\gamma}_{p\leq 1}^*, \hat{\boldsymbol{\theta}}\right)=\mathbf{0}} = w_{L\times L}(\mathbf{0},\mathbf{0}). \end{aligned} \tag{6.2.18}$$

This is just the joint pdf of $\boldsymbol{\gamma}_{p\leq 1}^*, \hat{\boldsymbol{\theta}}$ at $(\mathbf{0},\mathbf{0})$, sketched schematically in Fig. 6.5 for $L = 1$. Since $\left(\boldsymbol{\gamma}^*, \hat{\boldsymbol{\theta}}\right)$ are dimensionless, cf. remarks following (6.2.13b) above, $w_{L\times L}$ is dimensionless, as required. Similarly, we obtain for the first term of [(6.2.15a) and (6.2.15b)]

$$\left\langle \delta\left(\boldsymbol{\gamma}_{p\leq 1}^* - \mathbf{0}\right)\right\rangle_{H_0} = \int_{-\infty}^{\infty}\!\!\cdots\!\!\int F_L(i\boldsymbol{\xi})\, d\boldsymbol{\xi}/(2\pi)^L = w_L\left(\boldsymbol{\gamma}_{p\leq 1}^*\right)_{\left(\boldsymbol{\gamma}_{p\leq 1}^*=0\right)} = w_L(\mathbf{0})_{\boldsymbol{\gamma}_{p\leq 1}^*}. \tag{6.2.19}$$

Accordingly, we can now write for the desired Bayes risk for the "strict" cost function (SCF_2) (6.2.13b),

$$R_{\mathrm{E}}^*(\sigma,\delta^*)\Big|_{\substack{SCF_2 \\ p\leq 1}} = C_0 \cdot \left\{A_L - \left[q w_L(\mathbf{0})_{\boldsymbol{\gamma}_{p\leq 1}^*} + p w_{L\times L}(\mathbf{0},\mathbf{0})_{\boldsymbol{\gamma}_{p\leq 1}^*,\hat{\boldsymbol{\theta}}}\right]\right\}. \tag{6.2.20}$$

Evaluation of (6.2.20) depends on our ability to specify the estimators $\boldsymbol{\gamma}^*_{p\leq 1}$ and then to obtain w_L and $w_{L\times L}$ explicitly. This is generally a difficult task. A less difficult and usually tractable problem is offered in threshold operation.

6.2.2.2 *Bayes Estimators* $\boldsymbol{\gamma}^*_{p\leq 1}|_{\mathrm{SCF}}$

To obtain $\boldsymbol{\gamma}^*_{p\leq 1}|_{\mathrm{SCF}}$, which minimizes the average risk, we must *maximize* the integrand of $\int_\Gamma(\ )d\boldsymbol{X}$ in (6.2.15b). This is equivalent to the condition[17]

$$\max_{\boldsymbol{\gamma}\to\boldsymbol{\gamma}^*}\left\{qW_{J\times L}\left(\boldsymbol{X},\boldsymbol{\gamma}_{p\leq 1}|\boldsymbol{S}=\boldsymbol{0}\right)_{\hat{\boldsymbol{\theta}}=\boldsymbol{0}}\right\}+pF_J\left(\boldsymbol{X}|\boldsymbol{S}\left(\boldsymbol{\gamma}_{p\leq 1}\right)\right)w_L\left(\boldsymbol{\gamma}_{p\leq 1}\right)_{\hat{\boldsymbol{\theta}}}, \tag{6.2.21}$$

where explicitly from (6.2.15b)

$$W_{J\times L}\left(\boldsymbol{X},\boldsymbol{\gamma}_{p\leq 1}|\boldsymbol{S}=\boldsymbol{0}\right)_{\hat{\boldsymbol{\theta}}=\boldsymbol{0}}\equiv F_J(\boldsymbol{X}|\boldsymbol{0})\prod_{\ell=1}^{L}\delta\left(\gamma_{p\leq 1}(\boldsymbol{X})_\ell-0\right);\, w_L(\hat{\boldsymbol{\theta}})\to w_L\left(\boldsymbol{\gamma}_{p\leq 1}\right)_{\hat{\boldsymbol{\theta}}}, \tag{6.2.21a}$$

$$W_{J\times L}\left(\boldsymbol{X},\boldsymbol{\gamma}_{p\leq 1}\right)=F_J\left(\boldsymbol{X}|\boldsymbol{S}\left(\boldsymbol{\gamma}_{p\leq 1}\right)\right)w_L\left(\boldsymbol{\gamma}_{p\leq 1}\right), \tag{6.2.21b}$$

in which $W_{J\times L}$ (6.2.21a) here is the probability that the system makes the (estimation) decisions $\boldsymbol{\gamma}_{p\leq 1}$ when $\boldsymbol{S}=\boldsymbol{0}$ and therefore $\hat{\boldsymbol{\theta}}=\boldsymbol{0}$ and where $w_L\left(\boldsymbol{\gamma}_{p\leq 1}\right)$ is the pdf of $\hat{\boldsymbol{\theta}}$ with $\hat{\boldsymbol{\theta}}$ replaced by the estimators $\boldsymbol{\gamma}_{p\leq 1}$. Both $W_{J\times L}$ and w_L are, of course, functions of the received data $\boldsymbol{X}$. From (6.2.21) it is at once evident that unless the first term vanishes, $\boldsymbol{\gamma}^*_{p\leq 1}$ is not the classical UMLE of Sections 5.1.2 and 5.2.1. However, (6.2.21) is a generalized form of maximum likelihood, wherein both terms of (6.2.21) taken together can be considered as an extension of the classical form for $\boldsymbol{\gamma}^*_{p=1}$. Here the operation $\max\left(\boldsymbol{\gamma}\to\boldsymbol{\gamma}^*\right)$ requires that this likelihood function be maximized for all possible choices of $\boldsymbol{\gamma}$. Any monotonic function G of the expression in brackets [ ] in [(6.2.15a) and (6.2.15b)] may be chosen, for example,

$$\left[\frac{\partial}{\partial\boldsymbol{\gamma}}G\{\ \}\right]_{\boldsymbol{\gamma}=\boldsymbol{\gamma}^*}=\boldsymbol{0} \tag{6.2.22}$$

in the usual way, where the choice of G is a matter of analytic convenience. In the present case $(p<1)$, $G(y)=y$ appears to be the simplest choice, whereas for the classical situation $(p=1)$, $G(y)=\log y$ is usually selected, cf. Section 5.1.2. Accordingly, even when $p<1$ and amplitude or waveform estimation is carried out, the SCF once again leads to a maximum likelihood interpretation of the resulting Bayes estimators (with the Hessian of G, $H(G)<0$ now, cf. remarks after (6.2.2b). Thus, the maximizing conditions determining the L-component vector $\boldsymbol{\gamma}^*_{p\leq I}|_{\mathrm{SCF}_2}$ is now from (6.2.21) the set of L equations:

$$\mathrm{SCF}_2\text{:}\ \left[q\frac{\partial W_{J\times L}\left(\boldsymbol{X},\boldsymbol{\gamma}_P|\boldsymbol{S}=\boldsymbol{0}\right)}{\partial\boldsymbol{\gamma}_P}+p\frac{\partial W_{J\times L}\left(\boldsymbol{X},\boldsymbol{\gamma}_P\right)}{\partial\boldsymbol{\gamma}_P}\right]_{\boldsymbol{\gamma}_P=\boldsymbol{\gamma}_{p\leq 1}\to\boldsymbol{\gamma}^*_{p\leq 1}}=\boldsymbol{0};\ \boldsymbol{\gamma}_{p\leq 1}=\boldsymbol{\gamma}_{p\leq 1}(\boldsymbol{X}) \quad P\equiv\boldsymbol{S}\text{ or }\hat{\boldsymbol{\theta}}. \tag{6.2.23a}$$

[17] See the discussion in the paragraph following Eq. (6.1.4a). Note that the integrations over $\boldsymbol{X}$ should be deleted from the original derivation in Eqs. (5.15)–(5.17) of Ref. [1].

These L-extremal equations of (6.2.23a) can be alternatively written (for $\boldsymbol{\gamma}_S$ or $\boldsymbol{\gamma}_{\hat{\theta}}$):

$$\left[\frac{\partial}{\partial \boldsymbol{\gamma}}\delta(\boldsymbol{\gamma}-\mathbf{0})+\frac{\partial \hat{\Lambda}_L}{\partial \boldsymbol{\gamma}}\right]_{\boldsymbol{\gamma}\to\boldsymbol{\gamma}^*_{p\leq 1}}=\mathbf{0}. \tag{6.2.23b}$$

Note that in Eqs. [(6.2.23a) and (6.2.23b)] a *joint* optimization of the $\gamma_1,\ldots,\gamma_L$ components of $\boldsymbol{\gamma}$ is indicated, requiring the simultaneous solution for the optimum estimators $\gamma_1^*,\ldots,\gamma_L^*$. This is, of course, the consequence of the choice of the "strict" form of the simple cost function SCF_2, (6.2.13b). Here the likelihood ratio *form* $\hat{\Lambda}_L$ is specifically

$$\hat{\Lambda}_L(\boldsymbol{X}|\boldsymbol{\gamma}_{p=1})\equiv \mu F_J(\boldsymbol{X}|\boldsymbol{\gamma}_{p\leq 1})w_L(\boldsymbol{\gamma}_{p\leq 1})_{\hat{\boldsymbol{\theta}}}/F_J(\boldsymbol{X}|\boldsymbol{S}=\mathbf{0})\equiv\Lambda. \tag{6.2.23c}$$

(Note that $\hat{\Lambda}_L\neq\Lambda$, (6.2.4b), where Λ is the GLR for binary (on–off) *detection.* $\hat{\Lambda}_L$ is a "*likelihood function*" for estimation here, which only under special circumstances quantitatively equals Λ (6.1.6b), vide Chapter 5).

To obtain the structure of the L-component vector estimator $\boldsymbol{\gamma}^*_{p\leq 1}$, let us begin by setting[18]

$$\boldsymbol{\gamma}^*_{p\leq 1}|_{H_0}=a\boldsymbol{Q}^*(\boldsymbol{X});\boldsymbol{\gamma}^*_{p\leq 1}|_{H_1}=b\boldsymbol{\gamma}^*_{p=1}(\boldsymbol{X}),\ (a,b\geq 0), \tag{6.2.24}$$

and apply the expectation operator $\hat{E}_H$, [(6.2.11a) and (6.2.11b)] to (6.2.24) to get directly

$$\langle\boldsymbol{\gamma}^*_{p\leq 1}\rangle_H=q\langle a\boldsymbol{Q}^*\rangle_{H_0}+p\langle b\gamma^*_{p\leq 1}\rangle, p+q=1,\ (1\geq p,q\geq 0). \tag{6.2.25}$$

Next, incorporating the condition that $\boldsymbol{\gamma}^*_{p\leq 1}$ be *unbiased* under $\langle\ \rangle_H$, or *unconditionally unbiased*, for example,

$$\langle\boldsymbol{\gamma}^*_{p\leq 1}\rangle_H=p\langle\hat{\boldsymbol{\theta}}\rangle, \tag{6.2.26}$$

cf. the discussion in Section 6.3.4 ff, we have (on absorbing a into $\boldsymbol{Q}^*$):

$$q\langle\boldsymbol{Q}^*\rangle_{H_0}+(b-1)p\langle\boldsymbol{\gamma}^*_{p=1}\rangle=\mathbf{0}, \tag{6.2.27}$$

since $\langle\boldsymbol{\gamma}^*_{p=1}\rangle_{H_1}=\langle\hat{\boldsymbol{\theta}}\rangle$, cf. Eq. (6.2.26)). When $(p=0,q=1)$, it follows that

$$q\langle\boldsymbol{Q}^*\rangle_{H_0}=\langle\boldsymbol{Q}^*\rangle_{H_0}=\mathbf{0},\ q=1, \tag{6.2.28a}$$

and

$$\therefore\ \ b=1,\ (0\leq p\leq 1),\ \text{with}\ \therefore\ \langle\boldsymbol{\gamma}^*_{p\leq 1}\rangle_H=p\langle\boldsymbol{\gamma}^*_{p=1}\rangle_{H_1}=p\langle\hat{\boldsymbol{\theta}}\rangle. \tag{6.2.28b}$$

Consequently, (6.2.25), (6.2.28b) give

$$\left.\begin{aligned}\boldsymbol{\gamma}^*_{p\leq 1}=q\boldsymbol{Q}^*+p\boldsymbol{\gamma}^*_{p=1},\ \text{or}\ \boldsymbol{\gamma}^*_{p\leq 1}&=p\boldsymbol{\gamma}^*_{p=1}|_{H_1}\\ &=q\boldsymbol{Q}^*|_{H0}\end{aligned}\right\}, \tag{6.2.29}$$

[18] Estimation is carried out on the whole data space Γ, for H_0 and H_1 each, it being understood that the estimate is rejected, or not used, if the detector decides $H_0, \boldsymbol{X}\in\Gamma$. See the discussion following Eqs. (6.1.6a) and (6.1.6b).

for the structure of the optimum estimator $p \leq 1$. We now note from (6.2.28a) that

$$\langle \boldsymbol{Q}^* \rangle_{H_0} = \int_\Gamma \boldsymbol{Q}(X)^* F_J(X|\mathbf{0})_{S=\mathbf{0}} dX = \mathbf{0}, \tag{6.2.30}$$

and since $F_J > 0$ (for at least some $X \in \Gamma$), it follows that $\boldsymbol{Q}^* = \mathbf{0}$, all $X \in \Gamma$, or $\boldsymbol{Q}^*, X \in \Gamma$, is both positive and negative in such a way that $\langle \boldsymbol{Q}^* \rangle_{H_0}$ is zero. In either case $\boldsymbol{Q}^*$ is unbiased $\langle \boldsymbol{Q}^* \rangle_{H_0} = q\langle \hat{\boldsymbol{\theta}} \rangle = \mathbf{0}$, under H_0 from (6.2.30) or equally from the fact that $\langle \boldsymbol{\gamma}^*_{p\leq 1} \rangle_{H}, \langle \boldsymbol{\gamma}^*_{p=1} \rangle_{H_1}$ are both unbiased, cf. (6.2.28b) and therefore $\langle \boldsymbol{Q}^* \rangle_{H_0}$ must be also. In fact, it is at once evident that $\boldsymbol{Q}^*$ vanishes, all $X \in \Gamma$: H_0, by observing from [(6.2.23a) and (6.2.23b)] that here $\delta'(\boldsymbol{\gamma} - \mathbf{0}) \to \delta'(\boldsymbol{Q}^* - \mathbf{0}) = \mathbf{0}$.[19]

Accordingly, Eq. (6.2.29) allows us to write

$$\boxed{\boldsymbol{\gamma}^*_{p\leq 1}|_{\text{SCF}_2} = p\boldsymbol{\gamma}^*_{p=1}|_{\text{SCF}_2} = \left[\boldsymbol{\gamma}^*_{p=1} - q\boldsymbol{\gamma}^*_{p=1}\right]_{\text{SCF}_2}, q = 1 - p}\,, \tag{6.2.31}$$

where $-q\boldsymbol{\gamma}^*_{p=1}(X)_{\text{SCF}_2}$ *is the sample bias vis-à-vis* H_1, for example, and the (average) *bias* $= -q\langle \boldsymbol{\gamma}^*_{p=1} \rangle = -q\langle \hat{\boldsymbol{\theta}} \rangle$. A nonzero bias (vis-à-vis H_1) is the result of assuming that $X \in \Gamma$: H_1, that is, the received data contain a signal, under estimation, whereas actually (with *a priori* probability $q(>0)$) the data do not contain a signal, cf. remarks in Section 6.3.4 ff. The presence of the bias, of course, affects the Bayes risk, through $qw_L(\mathbf{0})_{\boldsymbol{\gamma}^*_{p\leq 1}}$ in (6.2.20), with $\boldsymbol{\gamma}^*_{p\leq 1}|_{\text{SCF}}$ now given by (6.2.31). The actual structure of $\boldsymbol{\gamma}^*_{p\leq 1}|_{\text{SCF}_2}$ is obtained from the "classical" form $\boldsymbol{\gamma}^*_{p=1}|_{\text{SCF}_2}$ [cf. Section 6.4.1 ff.], which is repeated here, namely,

$$p = 1: \qquad \left\{ \frac{\partial}{\partial \boldsymbol{\gamma}} W_{J\times L}(X, \boldsymbol{\gamma}) \right\}_{\boldsymbol{\gamma} \to \boldsymbol{\gamma}^*_{p=1}|_{\text{SCF}_2}} = \mathbf{0}, \text{ cf. (6.2.23b)}. \tag{6.2.32}$$

An illustrative analytic example for $\boldsymbol{\gamma}^*_{p\leq 1}|_{\text{SCF}_2}$, along with the associated Bayes risk, is included in Section 6.4.2 ff.

6.2.2.3 Nonstrict "Simple" Cost Function SCF$_1$: Bayes Risk and Estimators We return now to the cost function (6.2.13a), which is the less strict form of simple cost function. This is more commonly employed in practice because it does not excessively penalize the user by yielding too high an average risk in the minimization process (see the comments following Eq. (5.1.37)). The Bayes risk (6.2.15b) is now modified to

$$R^*_E(\sigma, \delta^*)\Big|_{\substack{\text{SCF}_1 \\ p\leq 1}} = C_0 \sum_{\ell=1}^{L} \left\{ A_L - \int_\Gamma \left[\begin{aligned} &qF_J(X|\mathbf{0})_{S=\mathbf{0}} \delta\left(\gamma_{p\leq 1}(X)^*_\ell - 0\right) \\ &+ p\int_\Omega W_J(X|\hat{\theta}_\ell) w_1(\hat{\theta}_\ell) \delta\left(\boldsymbol{\gamma}_{p\leq 1}(X)_\ell - \hat{\theta}_\ell\right) d\hat{\theta}_\ell \end{aligned} \right] dX \right\}. \tag{6.2.33}$$

[19] We assume throughout that $\delta(y - a)$ is symmetric positive about a, that is, is not one-sided. Thus $\delta'(y - a)$ is antisymmetric and passes through zero at $y = a$; similar remarks apply to $\delta(\boldsymbol{y} - \boldsymbol{a})$ in (6.2.13) et seq.

The extreme (maximizing) condition determining $\gamma^*_{p\leq 1}|_\ell$ is found from (6.2.23b), which is similarly modified to

$$\left[\frac{\partial}{\partial \gamma_\ell}\delta(\gamma_\ell - 0) + \frac{\partial \hat{\Lambda}_1(\boldsymbol{X}, \gamma_\ell)}{\partial \gamma_\ell}\right]_{\gamma_\ell \to \gamma^*_{p\prec 1}|_\ell} = 0; \ell = 1, \ldots, L, \tag{6.2.34}$$

where

$$\hat{\Lambda}_1\left(\boldsymbol{X}|\gamma_{p\leq 1,\ell}\right) \equiv \mu W_J\left(\boldsymbol{X}|\gamma_{p\leq 1,\ell}\right) w_1\left(\gamma_{p\leq 1,\ell}\right) / F_J(\boldsymbol{X}|\boldsymbol{S} = \boldsymbol{0}) \tag{6.2.34a}$$

[cf. (6.2.23b) and (6.2.23c)] with $W_J w_1 = W_{J\times 1}(\boldsymbol{X}, p \leq 1)$. Finally, $\gamma^*_{p\leq 1,\ell}|_{\mathrm{SCF}_1}$ is obtained by the procedure of (6.2.24)–(6.2.31), now for each component $\ell = 1, \ldots, L$ separately, as a consequence of our choice of the "nonstrict" form of the simple cost function, SCF_1, (6.2.13a). We now have the scalar form of (6.2.31), which is the solution of (6.2.34), namely,[20]

$$\boxed{\left(\gamma^*_{p\leq 1}\right)_\ell|_{\mathrm{SCF}_1} = p \cdot \left(\gamma^*_{p=1}\right)_\ell|_{\mathrm{SCF}_1}; \ell = 1, \ldots, L.} \tag{6.2.35}$$

The evaluation of (6.2.33) parallels Eqs. (6.2.16)–(6.2.19) above: the resulting forms are

$$\begin{aligned}
\left\langle \delta\left(\gamma^*_{p\leq 1,\ell} - \hat{\theta}_\ell\right)\right\rangle_{H_1} &= w^*_{1\times 1}\left(\gamma^*_\ell, \hat{\theta}_\ell\right)\Big|_{\gamma^*_\ell = \hat{\theta}_\ell = 0} \equiv w^*_{1\times 1}(0,0)_{(\gamma^*_\ell, \hat{\theta}_\ell)}, \\
\left\langle \delta\left(\gamma^*_{p\leq 1,\ell} - 0\right)\right\rangle_{H_0} &= w^*_1(\gamma^*)_{\gamma^*_\ell = 0} = w^*_1(0)_{\gamma^*_\ell},
\end{aligned} \tag{6.2.36}$$

with the Bayes risk (6.2.33) now represented compactly here by

$$R^*_{\mathrm{E}}(\sigma, \delta^*)\Big|_{\substack{\mathrm{SCF}_1 \\ p \leq 1}} = C_0 \sum_{\ell=1}^{L} \left\{ A_L - \left[q w^*_1(0)_{\gamma^*_\ell} + p w^*_{1\times 1}(0,0)_{\gamma^*_k, \hat{\theta}_\ell} \right]_{p\leq 1} \right\}. \tag{6.2.37}$$

Of course, the quantitative evaluation of R^*_{E}, (6.2.37), still requires that we obtain the joint pdfs of γ^*_ℓ and $\hat{\theta}_\ell$ from the corresponding versions of (6.2.16)–(6.2.20) above. The task is easier with the SCF_1, (6.2.13a), than for the stricter cost function SCF_2, (6.2.13b), but can still be formidable, particularly for non-Gaussian environments.

PROBLEMS

P6.1 Carry out the evaluations of $\left\langle \delta\left(\boldsymbol{\gamma}^*_{p\leq 1} - \hat{\boldsymbol{\theta}}\right)\right\rangle_{H_0, H_1}$, outlined in Eqs. (6.2.16)–(6.2.20).

P6.2 Show that the Hessian $H(G)|_{\boldsymbol{\gamma}^*} < 0$, $G = qW_{J\times L}(\boldsymbol{X}, \boldsymbol{\gamma}_P|\boldsymbol{S} = 0) + pW_{J\times L}(\boldsymbol{X}, \boldsymbol{\gamma}_P)$, ($P = \boldsymbol{S}$ or $\boldsymbol{\theta}$) and discuss the conditions on G that ensure that $H(G)|_{\boldsymbol{\gamma}^*}$ is negative.

P6.3 Carry out the evaluation in the case of the nonstrict cost function of Bayes risk $R^*_{E|\mathrm{SCF}_1}$, obtaining Eq. (6.2.36) in the text.

[20] Compare (6.2.34), (6.2.34a) with Eq. (6.2.23b).

6.3 SIMULTANEOUS JOINT DETECTION AND ESTIMATION: GENERAL THEORY[21]

We are now ready to extend the simpler, no-coupling analysis of Section 6.2 above to the general formulation of *coupled detection* (D) *and estimation* (E), still on the fixed observation interval (0,*T*). In this situation, we do *not* assume that the average risks for D and E necessarily can be separately minimized.

We begin with the general model of the joint detection and estimation situation to be considered here, shown in Fig. 6.6. As before, we assume two possible, mutually exclusive hypotheses $H_{0,1}$ for the received data in the interval (0,*T*): either (1) H_0: the signal is absent, and therefore only noise is present or (2) H_1: the signal is present together with the noise process. Again, let N be the noise process and let the signal be described by the functional form $\mathbf{S} = [S(\mathbf{r}_m, t_n; \boldsymbol{\theta})]$ or $\mathbf{S} = [S(t_n, \boldsymbol{\theta})]$, cf. after (6.1.3a), where $\boldsymbol{\theta}$ is a vector representing the set of parameters which are to be estimated. At the end of the observation interval (0,*T*) the receiver is required to make *two functionally related decisions*: (1) a detection decision, $\gamma_i (i = 0 \text{ or } 1)$, as to the presence or absence of the signal and (2) an estimation decision, $\boldsymbol{\gamma_\theta}$, as to the value of the (parameter) vector $\boldsymbol{\theta}$. As explained in Section 6.1 above, the relation between the two decisions can be of different types. For example, the two decision processes may be entirely independent, cf. Fig. 6.2 (but even in this case estimation must be performed without certainty that the signal is present, that is, $p(H_1) < 1$). Or the decision rules may couple estimation and detection in such a way that the two operations are mutually dependent.

6.3.1 The General Case: Strong Coupling

Our first step is to select a reasonable set of cost functions, $\{C_{k\ell}\}$, $(k, \ell = 0, 1)$, defined as follows [2].

$$
\left.
\begin{array}{ll}
C_{00}\left[=C_{1-\alpha}=C_0^{(0)}\right]: & \text{cost of correctly deciding, when } H_0 \text{ is true; } f_{00}=0, \\
C_{10}(\boldsymbol{\gamma}_\theta)=C_\alpha\left(=C_1^{(0)}\right)+f_{10}[\boldsymbol{\gamma}_\theta(\mathbf{X})]: & \text{cost of incorrectly deciding } \delta_1=H_1 \textit{ and } \text{making an estimate, } \boldsymbol{\gamma_\theta}=\boldsymbol{\gamma_\theta}(\mathbf{X}), \text{ when } H_0 \text{ is true;} \\
C_{01}(\boldsymbol{\theta})=C_\beta\left(=C_0^{(1)}\right)+f_{01}[\mathbf{S}(\boldsymbol{\theta})]: & \text{cost of incorrectly deciding } \delta_0=H_0, \text{ when in fact } H_1 \text{ is true and } \theta \text{ is the true value of the signal parameters: } C_\beta \equiv C_1, \text{(6.1.6a);} \\
C_{11}(\boldsymbol{\theta},\boldsymbol{\gamma}_\theta)=C_{1-\beta}\left(=C_1^{(1)}\right)+f_{11}[\boldsymbol{\theta},\boldsymbol{\gamma}_\theta(\mathbf{X})]: & \text{cost of correctly deciding } \delta_1=H_1 \textit{ and } \text{making an estimate, } \boldsymbol{\gamma_\theta}=\boldsymbol{\gamma_\theta}(\mathbf{X}), \text{ when } H_1 \text{ is true and } \boldsymbol{\theta} \text{ is the true value of the signal parameters.}
\end{array}
\right\}
\quad (6.3.1)^{22}
$$

[21] In this Section and subsequently, unless otherwise indicated, $\boldsymbol{\theta}$ represents only those parameters that are to be estimated, for example, the $\boldsymbol{\theta}$ here are equivalent to the $\hat{\boldsymbol{\theta}}$ of Section 6.2.1 above. Any other parameters ($\boldsymbol{\theta}'$) are not explicitly represented, unless they are specifically needed in the analysis.

[22] Equation (6.3.1) is equivalent to items (1) and (2), Section 4.2 [1].

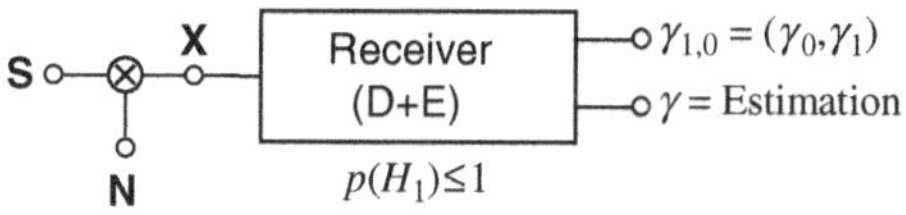

FIGURE 6.6 The general coupled joint detection and estimation situation, for fixed observation intervals.

With these general cost functions in hand, we proceed concisely as follows. Again, we let $p = p(H_1)$ be the *a priori* probability of the state H_1 and let $q = q(H_0)$ be the *a priori* probability of the state H_0, with $p + q = 1$. Next, define also the following probabilities and probability density functions (pdfs): $\delta_0(H_0|\boldsymbol{X})$, probability of choosing H_0 if $\boldsymbol{X}$ is the vector of the observed data; $\delta_\gamma(\boldsymbol{\gamma}_\theta|\boldsymbol{X})$, the pdf for the estimate $\boldsymbol{\gamma}_\theta$ when $\boldsymbol{X}$ is the vector of the observed data. Then δ_0, δ_γ are, in effect, decision rules. Clearly, since at least a detection decision is made, $\delta_0(H_0|\boldsymbol{X}) + \delta_1(H_1|\boldsymbol{X}) = 1$, where $\delta_1(H_1|\boldsymbol{X})$ is the probability of making the decision that H_1 is true. Furthermore, we make the reasonable postulate that

$$\delta_\gamma(\boldsymbol{\gamma}_\theta|\boldsymbol{X})d\boldsymbol{\gamma}_\theta = \delta_1(H_1|\boldsymbol{X}) = 1 - \delta_0(H_0|\boldsymbol{X}), \tag{6.3.2}$$

that is, an estimate is accepted as an output if and only if, at the same time, the detector declares that the signal is present. (Note that for the time being, $\delta_0, \delta_1, \delta_\gamma$ may be randomized decision rules.)

Next, for notational convenience in developing the general case, let us condense our earlier definitions [(6.2.4a) and (6.2.4b)], and rewrite them as

$$F_0(\boldsymbol{X}) \equiv F_J(\boldsymbol{X}|\boldsymbol{0}) = \text{the unconditional pdf of the data } \boldsymbol{X} \text{ under } H_0\text{: } N, \qquad (6.3.3\text{a})^{23}$$

and

$$F_1(\boldsymbol{X}|\boldsymbol{\theta}) \equiv F_J(\boldsymbol{X}|\boldsymbol{S}(\boldsymbol{\theta})) = \text{the pdf of the data} \boldsymbol{X} \text{ under } H_1\text{: } S \otimes N, \qquad (6.3.3\text{b})^{23}$$

and $w_L(\boldsymbol{\theta})$ as before, cf. (6.2.4a), is the *a priori* pdf of the L parameters $\boldsymbol{\theta}$ when H_1 is true. With these definitions, the average cost per decision [(6.1.3a), (6.1.3b)] for this general coupled detection and estimation procedure is

$$\begin{aligned} R_{D\otimes E} = {} & q\int_\Gamma \delta_0(H_0|\boldsymbol{X})C_{00}F_0(\boldsymbol{X})d\boldsymbol{X} + q\int_\Delta\int_\Gamma C_{10}(\boldsymbol{\gamma}_\theta)\delta_\gamma(\boldsymbol{\gamma}_\theta|\boldsymbol{X})F_0(\boldsymbol{X})d\boldsymbol{X}d\boldsymbol{\gamma}_\theta \\ & + p\int_\Omega\int_\Gamma w_L(\boldsymbol{\theta})\delta_0(H_0|\boldsymbol{X})C_{01}(\boldsymbol{\theta})F_1(\boldsymbol{X}|\boldsymbol{\theta})d\boldsymbol{X}d\boldsymbol{\theta} \\ & + p\int_\Delta\int_\Omega\int_\Gamma w_L(\boldsymbol{\theta})C_{11}(\gamma_0,\boldsymbol{\theta})\delta_\gamma(\boldsymbol{\gamma}_\theta|\boldsymbol{X})F_1(\boldsymbol{X}|\boldsymbol{\theta})d\boldsymbol{X}d\boldsymbol{\theta}d\boldsymbol{\gamma}_\theta \end{aligned} \tag{6.3.4}$$

[23] Thus, we have equivalently, with $\hat{\boldsymbol{\theta}} \to \boldsymbol{\theta}$ here

$$F_0(\boldsymbol{X}) = F_J(\boldsymbol{X}|\boldsymbol{S} = \boldsymbol{0}) = F_J(\boldsymbol{X}|\boldsymbol{0})_{\boldsymbol{S}=\boldsymbol{0}} = w_J(\boldsymbol{X})_{\boldsymbol{0}}, \text{ and} \qquad (6.3.\text{i}),$$

$$F_1(\boldsymbol{X}|\boldsymbol{\theta}) = F_J(\boldsymbol{X}|\boldsymbol{S}(\boldsymbol{\theta})) = W_{J\times L}(\boldsymbol{X},\boldsymbol{\theta})/w_L(\boldsymbol{\theta}), \qquad (6.3.\text{ii}).$$

where Γ, Ω, Δ are respectively the (usual) data, signal (parameter), and decision spaces ([4], Chapter 18, and Section 6.1 here). Note that from Bayes' theorem, we can write for the

$$F_{\boldsymbol{\theta}}(\boldsymbol{\theta}|\boldsymbol{X}) = F_1(\boldsymbol{X}|\boldsymbol{\theta})w_L(\boldsymbol{\theta})/F_1(\boldsymbol{X}) \equiv w_L(\boldsymbol{\theta}|\boldsymbol{X}), \tag{6.3.5}$$

with

$$F_1(\boldsymbol{X}) = \int_{\Omega} F_1(\boldsymbol{X}|\boldsymbol{\theta})w_L(\boldsymbol{\theta})d\boldsymbol{\theta}, \tag{6.3.6}$$

where the unconditional pdf of $\boldsymbol{\theta}$ under H_1 is $w_L(\boldsymbol{\theta})$, with $w_1(\boldsymbol{\theta}|\boldsymbol{X})$ the conditional or *a posteriori* pdf of θ, given $\boldsymbol{X}$.

With the help of (6.3.4) we can alternatively express the average risk in a more convenient form as

$$R_{D\otimes E} = q\int_{\Gamma} \delta_0(H_0|\boldsymbol{X})C_{00}F_0(\boldsymbol{X})d\boldsymbol{X} + p\int_{\Gamma}\int_{\Omega} \delta_0(H_0|\boldsymbol{X})C_{01}(\boldsymbol{\theta})F_1(\boldsymbol{X}|\boldsymbol{\theta})w_L(\boldsymbol{\theta})d\boldsymbol{X}d\boldsymbol{\theta}$$
$$+ q\int_{\Delta}\int_{\Gamma} \delta_\gamma(\gamma_{\boldsymbol{\theta}}|\boldsymbol{X})\left[qC_{10}(\boldsymbol{\gamma_\theta})F_0(\boldsymbol{X}) + pF_1(\boldsymbol{X})\int_{\Omega} C_{11}(\gamma_0,\boldsymbol{\theta})F_0(\boldsymbol{\theta}|\boldsymbol{X})d\boldsymbol{\theta}\right]d\boldsymbol{\gamma_\theta}d\boldsymbol{X}. \tag{6.3.7}$$

Minimization of this average risk yields the two simultaneous equations determining the jointly optimum decision rules δ_0^{**} (or δ_1^{**}, by (6.3.2)), and δ_γ^{**}, respectively, for detection and estimation, namely,

$$\delta_{\boldsymbol{\gamma}} \to \delta_{\boldsymbol{\gamma}}^{**}[\delta R_{D\oplus E}] = 0, \quad \text{so that} \quad \underset{\delta_{\boldsymbol{\gamma}} \to \delta_{\boldsymbol{\gamma}}^{**}}{\delta}\left[R_{D\otimes E} + \lambda_0\int \delta_\gamma(\boldsymbol{\gamma_\theta}|\boldsymbol{X})d\boldsymbol{\gamma_\theta}\right] = 0, \tag{6.3.8}$$

where we must include the logical constraint (6.3.2); here λ_0 is a Lagrange multiplier.

Although the simultaneous solution of (6.3.8) yields the desired $\delta_0^*, \delta_\gamma^{**}$, an equivalent and more revealing approach is chosen in what follows, which employs a "series" or step-by-step optimization. For this we need first to establish the result that the optimum estimation rule δ_γ^* for an arbitrarily selected detection decision rule $\delta_0 (= 1 - \delta_1)$ is not randomized, and can therefore be written

$$\delta_\gamma^*(\boldsymbol{\gamma_\theta}|\boldsymbol{X}) = \delta_1(H_1|\boldsymbol{X})\delta(\boldsymbol{\gamma_\theta} - \boldsymbol{\gamma_\theta^*}), \tag{6.3.9}$$

where $\delta(a - b)$ is the usual Dirac delta function and $\gamma_{\boldsymbol{\theta}}^*$ is the optimum estimator (a relative optimum, because δ_1 is a specified decision once δ_0 is).

6.3.1.1 ***Nonrandomized Decision Rules: $\boldsymbol{\delta_\gamma^*}, \boldsymbol{\delta_\gamma^{**}}$*** Let us now show that the optimum decision rules (δ_γ^*,δ_γ^{**}) for estimation are nonrandomized here, that is, that δ_γ^*, δ_γ^{**} obey Eq. (6.3.9) above. We start with (6.3.8) and use that portion of (6.3.7) containing δ_γ, namely the third term. For notational simplicity, let us write $\delta_\gamma(\boldsymbol{\gamma}|\boldsymbol{X}) = g(\boldsymbol{\gamma}|\boldsymbol{X}) = g$. We next fix δ_0 (and therefore δ_1) for all $\boldsymbol{X}$. Then, (6.3.8) becomes, compactly

$$\delta_g\left[\int_{\Delta} d\boldsymbol{\gamma}\int_{\Gamma} g(\boldsymbol{\gamma}|\boldsymbol{X})H(\boldsymbol{\gamma},\boldsymbol{X})d\boldsymbol{X} + \lambda_0\int_{\Delta} g d\boldsymbol{\gamma}\right] = 0. \tag{6.3.10}$$

Carrying out the variation and noting that $\boldsymbol{\delta} g = (\partial g/\partial \boldsymbol{\gamma})\boldsymbol{\delta\gamma}$ since g is a function of $\boldsymbol{\gamma}$, we obtain

$$\int_{\Delta} \boldsymbol{\delta\gamma}\, \boldsymbol{d\gamma} \left[\int_{\Gamma} \frac{\partial}{\partial \boldsymbol{\gamma}} (gH) \boldsymbol{dX} + \lambda_0 \frac{\partial g}{\partial \boldsymbol{\gamma}} \right] = 0, \tag{6.3.10a}$$

and for arbitrary $\boldsymbol{\delta\gamma}(\neq 0)$ the extremal relation, now on $\boldsymbol{\gamma}$, becomes

$$\left[\int_{\Gamma} \frac{\partial}{\partial \boldsymbol{\gamma}} (gH) \boldsymbol{\delta X} + \lambda_0 \frac{\partial g}{\partial \boldsymbol{\gamma}} \right]_{\boldsymbol{\gamma}=\boldsymbol{\gamma}^*} = 0. \tag{6.3.10b}$$

This determines a $\boldsymbol{\gamma}^*$, which is taken to be that value (there may be others) which absolutely minimizes the third term of (6.3.7).

Next, we note from the postulate (6.3.8), for example, the constraint in (6.3.8), that

$$\int_{\Delta} g \boldsymbol{d\gamma} = \delta_1 (H_1 | \boldsymbol{X}) = \text{constant; here } (0 \leq \delta_1 \leq 1). \tag{6.3.11}$$

We have

$$\therefore \quad \delta\gamma \int_{\Delta} g \boldsymbol{d\gamma} = 0 \text{ or } \int \frac{\partial g}{\partial \boldsymbol{\gamma}} \delta_{\gamma} \boldsymbol{d\gamma} = 0, \tag{6.3.12}$$

and since $\delta\boldsymbol{\gamma} \neq 0$, we have for any $\boldsymbol{\gamma}$, and particularly for $\boldsymbol{\gamma}^*$, as determined by (6.3.10), the relation

$$\left(\frac{dg}{d\boldsymbol{\gamma}} \right)_{\boldsymbol{\gamma}=\boldsymbol{\gamma}^*} = 0. \tag{6.3.13}$$

But since there is no constraint on the magnitude of the estimators ($\boldsymbol{\gamma}^*$), that is, the domain of each component γ_ℓ^* is $-\infty < \gamma_\ell^* < +\infty$, Eq. (6.3.13) in conjunction with (6.3.11) implies that

$$g = \delta_1 (H_1 | \boldsymbol{X}) \delta(\boldsymbol{\gamma} - \boldsymbol{\gamma}^*) = \delta \boldsymbol{\gamma}^*, \tag{6.3.14}$$

which shows that the estimator rule is nonrandomized. Incidentally, applying (6.3.13) to (6.3.10b) gives at once the determining relation for $\boldsymbol{\gamma}^*$ [see Eq. (6.3.8)]:

$$\left(\frac{dH}{d\boldsymbol{\gamma}} \right)_{\boldsymbol{\gamma}=\boldsymbol{\gamma}^*=\boldsymbol{\gamma}^{**}} = 0. \tag{6.3.15}$$

Here $\boldsymbol{\gamma}^* = \boldsymbol{\gamma}^{**}$, since the decision rule $\boldsymbol{\gamma}_\gamma$ for estimation is always proportional to a constant (since δ_1 is specified) $\times\, \delta(\boldsymbol{\gamma} - \boldsymbol{\gamma}^*)$, namely, $\boldsymbol{\gamma}_\gamma$ is functionally independent of the detection decision rules $\delta_{0,1}$.

6.3.1.2 An Alternative Approach to Eq. (6.3.8) Returning now to our alternative approach to the direct evaluation of Eq. (6.3.8), having established that for estimation, decision rules δ_γ^*, δ_γ^{**} are nonrandomized, we see that if (6.3.9) is inserted in (6.3.7), the average risk (now Bayes risk with respect to estimation) can be rewritten as

$$\begin{aligned} R_{\mathrm{D\otimes E}} = {} & q\int_\Gamma \delta_0(H_0|\boldsymbol{X})C_{00}F_0(\boldsymbol{X})d\boldsymbol{X} + q\int_\Gamma \delta_1(H_1|\boldsymbol{X})C_{10}(\boldsymbol{\gamma}_\theta^*)F_0(\boldsymbol{X})d\boldsymbol{X} \\ & + p\int_\Omega\int_\Gamma \delta_0(H_0|\boldsymbol{X})w_L(\boldsymbol{\theta})C_{01}(\boldsymbol{\theta})F_1(\boldsymbol{X}|\boldsymbol{\theta})d\boldsymbol{X}\,d\boldsymbol{\theta} \\ & + p\int_\Omega\int_\Gamma \delta_1(H_1|\boldsymbol{X})w_L(\boldsymbol{\theta})C_{11}(\boldsymbol{\theta},\boldsymbol{\gamma}_\theta^*)F_1(\boldsymbol{X}|\boldsymbol{\theta})d\boldsymbol{X}\,d\boldsymbol{\theta}\ . \end{aligned} \tag{6.3.16}$$

We want further to minimize $R^*_{\mathrm{D\otimes E}}$, by choosing the optimum detection rules for a fixed $\boldsymbol{\gamma}_\theta^{**}$. As in the standard treatment [cf. [4], Section 19.1.2], we observe from (6.3.16) that the optimum detection rules are also not randomized and are

$$\begin{aligned} &\text{Choose } \delta_1^{**}(H_1|\boldsymbol{X}) = 1, \text{ if } \Lambda_g \geq 1, \\ &\text{Choose } \delta_0^{**}(H_0|\boldsymbol{X}) = 1, \text{ if } \Lambda_g < 1, \end{aligned} \tag{6.3.17}$$

with $\delta_1^{**} + \delta_0^{**} = 1$ (6.2.13). Here Λ_g is now a modified likelihood ratio (functional) having the form, from (6.3.2):

$$\Lambda_g = \mu\frac{\int w_L(\boldsymbol{\theta})\left[C_{01}(\boldsymbol{\theta}) - C_{11}(\boldsymbol{\theta},\boldsymbol{\gamma}_\theta^{**})\right]F_1(\boldsymbol{X}|\boldsymbol{\theta})d\boldsymbol{\theta}}{\left[C_{10}(\boldsymbol{\gamma}_\theta^{**}) - C_{00}\right]F_1(\boldsymbol{X})} \tag{6.3.18}$$

with $\mu = p/q$ as before, (6.1.6b). It is clear from (6.3.9) here that now the optimum estimator $\boldsymbol{\gamma}_\theta^{**}$ corresponds to the optimum decision rule δ_1^{**} for detection, that is,

$$\delta_\gamma^{**}(\boldsymbol{\gamma}_\theta|\boldsymbol{X}) = \delta_1^{**}(H_1|\boldsymbol{X})\delta\left(\boldsymbol{\gamma}_\theta - \boldsymbol{\gamma}_\theta^{**}\right). \tag{6.3.19}$$

The form of $\boldsymbol{\gamma}_\theta^{**}$ can be derived from solving (6.3.8). This corresponds to the minimization equation

$$(\boldsymbol{\gamma}_\theta \to \boldsymbol{\gamma}_\theta^{**})\delta\left\{qC_{10}(\boldsymbol{\gamma}_\theta)F_0(\boldsymbol{X}) + pF_1(\boldsymbol{X})\int_\Omega C_{11}(\boldsymbol{\theta},\boldsymbol{\gamma}_\theta)F_\theta(\boldsymbol{\theta}|\boldsymbol{X})d\boldsymbol{\theta}\right\} = \mathbf{0}, \tag{6.3.20}$$

cf. Eq. (6.3.15) above, which is the generalization of (6.2.21) and (6.2.23) to the strong coupling cases. It is important to note again that the structure of the estimator does not depend on the detection rule (except for the relation (6.3.9)), whereas the value of the optimum estimate $\boldsymbol{\gamma}_\theta^{**}$ must be employed in the evaluation of the GLR, Λ_g, (6.3.18). Thus, *in these strongly coupled cases it is the operation of estimation (E) which must logically be performed first, before detection (D).*[24] Of course, if the detector's decision is that H_0 is the true hypothesis state, then the estimate is not accepted (but may be stored for "tracking" in the sequential decision situations, cf. Chapter 7).

[24] For an earlier, alternative development of the strong coupling situation, see Section 4.2 of Ref. [1], also Problem 6.11 ff.

Finally, a number of interesting observations can be made regarding the processing structures (6.3.18) and (6.3.20):

(1) Together these structures comprise the receiver of Fig. 6.6. The output of this receiver is either the decisions that a signal is present, along with an estimate of the parameters $\boldsymbol{\theta}$, or the decision that only noise is present. From (6.3.18), we can expect structures $(\sim\Lambda_g)$ much more complex than those required for detection alone, without accompanying estimation.

(2) We see also when C_{01}, C_{10}, and C_{11}, (6.3.1) are constants, that the extended GLR (6.3.18), as expected, reduces to the familiar GLR (6.1.6b) of independent Bayes detectors.

Whatever the choice of cost functions, the structure of the detector and estimator are embodied in their respective decision rules.

6.3.2 Special Cases I: Bayes Detection and Estimation With Weak Coupling

If we set $f_{10} = C'_{10}$, a constant, and chose $f_{11} = 0$ in (6.3.1), so that these cost functions reduce to

$$C_{00} = C_{1-\alpha}; C_{11} = C_{1-\beta}; C_{10} = C_\alpha + C'_{10}; C_{01} = C_\beta + f_{01}[\boldsymbol{S}(\boldsymbol{\theta})], \tag{6.3.21a}$$

we have at once the *weak-coupling cases*, so defined by the fact that the *coupling* of the detector and estimator now is independent of the estimator's structure. Accordingly, the average risks R_{D} and R_{E} associated with detection and estimation can be minimized separately, as is done in Section 6.2 above, cf. Eqs. (6.1.5) et seq. It is thus easily shown (Problem 6.5) that the decision rule for detection is nonrandomized and is specifically a modified form of a GLR test, where a likelihood ratio functional Λ'_g is compared with a threshold of unity:

$$\Lambda'_g(\boldsymbol{X}) = K'\Lambda + \int_\Omega f_{01}[\boldsymbol{S}(\boldsymbol{\theta})]\hat{\Lambda}'(\boldsymbol{X}|\boldsymbol{\theta})w_L(\boldsymbol{\theta})d\boldsymbol{\theta}/(C_\alpha - C_{1-\alpha} + C'_{10})\begin{Bmatrix} \geq 1\text{: decide } H_1 \\ < 1\text{: decide } H_0 \end{Bmatrix}. \tag{6.3.22}$$

Here again, like $\hat{\Lambda}_L$ (6.2.23c), $\hat{\Lambda}'(\boldsymbol{X}|\boldsymbol{\theta})$ is a likelihood function ($\neq \Lambda$, (6.1.6b)), defined now by

$$\hat{\Lambda}'(\boldsymbol{X}|\boldsymbol{\theta}) \equiv \mu F_J(\boldsymbol{X}|\boldsymbol{S}(\boldsymbol{\theta}))/F_J(\boldsymbol{X}|\boldsymbol{0}), \mu \equiv p/q. \tag{6.3.22a}$$

The scale factor K' is the (positive) cost ratio

$$K' \equiv (C_\beta - C_{1-\beta})/(C_\alpha - C_{1-\alpha} + C'_{10}). \tag{6.3.22b}$$

As the results of Problem 6.6 show, the Bayes risk for detection using Λ'_g as test statistic in these weakly coupled cases can be expressed in several forms:

$$R^*_{\text{D-weak}} = qC_{1-\alpha} + p\Re_{01} + q(C_\alpha - C_{1-\alpha} + C'_{10})\int_\Gamma F_0(\boldsymbol{X})[\Lambda'_g(\boldsymbol{X}) - 1]\delta^*_1(H_1|\boldsymbol{X})d\boldsymbol{X}, \tag{6.3.23a}$$

or as

$$R^*_{\text{D-weak}} = qC_{1-\alpha} + p\mathfrak{R}_{10} + q\left(C_\alpha - C_{1-\alpha} + C'_{10}\right) + \int_\Gamma F_0(\boldsymbol{X})\left[1 - \Lambda'_g(\boldsymbol{X})\right]\delta_0^*(H_0|\boldsymbol{X})d\boldsymbol{X}, \tag{6.3.23b}$$

where

$$\mathfrak{R}_{01} \equiv \int_\Gamma\int_\Omega \{C_\beta + f_{01}(\boldsymbol{S}(\boldsymbol{\theta}))\}w_L(\boldsymbol{\theta})F_1(\boldsymbol{X}|\boldsymbol{\theta})d\boldsymbol{X}d\boldsymbol{\theta} \; ; \; \mathfrak{R}_{10} \equiv \int_\Gamma F_0(\boldsymbol{X})\left[\Lambda'_g(\boldsymbol{X}) - 1\right]d\boldsymbol{X}. \tag{6.3.23c}$$

Clearly, if $f_{01} = C'_{01}$, a constant, (6.3.22) reduces to the usual form of a generalized likelihood ratio, with the modified threshold

$$K'' = \left(C_\alpha - C_{1-\alpha} + C'_{10}\right)/\left(C_\beta - C_{1-\beta} + C'_{01}\right), \tag{6.3.24}$$

in which C'_{10} and C'_{01} must be chosen such that "failure" $\left(\equiv C_\alpha + C'_{10}, C_\beta + C'_{01}\right) >$ "success" $\left(\equiv C_{1-\alpha}, C_{1-\beta}\right)$, respectively. This is the simplest joint detection and estimation situation and the one considered in Section 6.2 explicitly. Note that now with the appropriately weak or no coupling here [C'_{01}, C'_{10} constant or zero, cf. (6.3.8)], *then both operations (D and E) can be performed simultaneously, or in any order*, cf. *Figs. 6.2 and 6.4*. This is not true in the general case described in Section 6.3.1, cf. remarks after Eq. (6.3.20).

Proceeding next to the estimation process, we see that because the detection process is now independent of estimator structure, we can use estimator-dependent cost functions separately in the estimation process. For the latter, we accordingly employ $C'_{10}(\boldsymbol{\gamma}_\theta)$ and $C'_{11}(\boldsymbol{\theta}, \boldsymbol{\gamma}_\theta) = C_{1-\beta} + f_{11}[\boldsymbol{\theta}, \boldsymbol{\gamma}_\theta(\boldsymbol{X})]$, the respective costs of estimating the signal when it is not or is present, cf. (6.3.1). The structure of the optimum estimator $\boldsymbol{\gamma}^*_{p\leq 1}$ follows in the usual way from minimizing the average risk R_{10}, cf. (6.3.5), where $\sigma(\boldsymbol{\theta})$ [(6.1.4a) and (6.1.4b)] is again applied in (6.1.3). We write

$$R_{\text{E}} = q\int_{\Gamma'} F_J(\boldsymbol{X}|0)[C_{10}(\boldsymbol{\gamma}_\theta)]d\boldsymbol{X} + p\int_{\Gamma'}\int_\Omega F_J(\boldsymbol{X}|\boldsymbol{S})C_{11}[\boldsymbol{\theta}, \boldsymbol{\gamma}_\theta(\boldsymbol{X})]w_L(\boldsymbol{\theta})d\boldsymbol{\theta}\, d\boldsymbol{X}, \tag{6.3.25}$$

where $\Gamma' \varepsilon \Gamma$ is the data region specified by the decision rule of the detector's operation, when the detector's decision is H_1. The first integral represents the average risk due to a (Bayes) error of the first kind (probability $q\alpha$) in the detector and resulting in the estimate of a signal which is not present. The second integral is the average cost of estimating the signal when indeed it is present in the observation interval (with probability $p(1-\beta)$), vide. Section 6.4.3 ff, for example.

Remembering that the cost assignments $C_\alpha, C_{1-\beta}$, as well as $C_{1-\alpha}, C_\beta$, are *constant* costs [(6.1.6a) et seq.], we observe that minimization of (6.3.7) over $\boldsymbol{\gamma}_\theta$ gives us directly

$$\boldsymbol{\delta} R_{\text{E}} = q\int_{\Gamma'} F_J(\boldsymbol{X}|0)\frac{\partial f_{10}}{\partial \boldsymbol{\gamma}}\Big|_{\boldsymbol{\gamma}=\boldsymbol{\gamma}^*} d\boldsymbol{X}\boldsymbol{\delta\gamma}_\theta + p\int_{\Gamma'}\int_\Omega F_J(\boldsymbol{X}|\boldsymbol{\theta})w_L\frac{\partial f_{11}}{\partial \boldsymbol{\gamma}}\Big|_{\boldsymbol{\gamma}=\boldsymbol{\gamma}^*} d\boldsymbol{\theta}\, d\boldsymbol{X}\boldsymbol{\delta\gamma}_\theta = \boldsymbol{0}. \tag{6.3.26}$$

The optimum (vector) estimator $\boldsymbol{\gamma}_{\boldsymbol{\theta}}$, when it exists, is then the one which satisfies the set of equations

$$\left[qF_J(\mathbf{X}|0)\frac{\partial f_{10}}{\partial \boldsymbol{\gamma}} + p\int_{\Omega} d\boldsymbol{\theta}\, w_L(\boldsymbol{\theta})F_J(\mathbf{X}|\boldsymbol{\theta})\frac{\partial f_{11}}{\partial \boldsymbol{\gamma}}\right]_{\boldsymbol{\gamma}_{\boldsymbol{\theta}}=\boldsymbol{\gamma}^*_{p\le 1}} = \mathbf{0}. \qquad (6.3.27)$$

We observe that (6.3.27) is similar to the previous result [(6.2.2a) and (6.2.2b)] derived for the optimum estimator with uncertainty as to the presence of the signal. In fact, (6.3.27) and (6.2.2b) are identical if we set

$$\left.\begin{aligned} f_{10}(\boldsymbol{\gamma}_{\boldsymbol{\theta}}) &= C'(\boldsymbol{\gamma}) = C(0, \boldsymbol{\gamma})\,; \\ f_{11}(\boldsymbol{\theta}, \boldsymbol{\gamma}_{\boldsymbol{\theta}}) &= C(\boldsymbol{\theta}, \boldsymbol{\gamma}) \end{aligned}\right\}, \qquad (6.3.28)$$

vide (6.2.2b) et seq. Consequently, the results of Section 6.2 above also apply specifically for Bayes estimation with the QCF and SCF choices, yielding the same estimators and corresponding Bayes risks R^*_{E}. However, the Bayes risk for detection is modified to $R^*_{\text{D-weak}}$ (6.3.23), because of $f_{01}(\boldsymbol{\theta})$ in $C_{01}(\boldsymbol{\theta})$ (6.2.22).

We also note that since the detector's decision rule does not depend on the estimate, the structure of the optimum estimator $\boldsymbol{\gamma}^*_{p\le 1}$ is not a function of the data domain of Γ, as is evident from (6.3.27). Consequently, the structure of the estimator and its associated average risk do not depend on the form of the "weak" coupling used here in the detection-directed estimation. *One can choose, for instance, to estimate only when the detector has decided that the signal is present or, instead, to estimate on the whole data space* Γ, *rejecting the estimate if the detector's decision is* H_0. In both cases the optimum estimator is given by (6.3.27) and the associated minimum average risk by (6.3.25) when $\boldsymbol{\gamma}_{\boldsymbol{\theta}}$ is replaced by $\boldsymbol{\gamma}^*_{p\le 1}$.

The same results apply for waveform ($\mathbf{S}$-) estimation, with obvious changes. If quadratic cost functions are used, for instance in amplitude estimation, and we set $f_{10} = f_{11}(0, \mathbf{S})$, a very reasonable choice, we can derive again the result of (6.2.7), that is, also in this case the optimum estimator has the form (6.2.7) and the interpretation shown in Fig. 6.4.

6.3.3 Special Cases II: Further Discussion of $\boldsymbol{\gamma}^*_{p<1|\mathrm{QCF}}$ for Weak or No Coupling

The general structure of the estimator (6.3.27) is of great interest, namely a structure that is independent of whether or not a detection is actually performed, cf. Eq. (6.3.28) et seq. However, we shall confine our detailed treatment here specifically to the two interesting special cases involving *weak* or *no coupling*. These occur specifically when (1) $C_{11}(\boldsymbol{\theta}, \boldsymbol{\gamma}_{\boldsymbol{\theta}})$ is a quadratic cost function, in which it is meaningful to assume that $[C'(\boldsymbol{\gamma}) \equiv]$ $C_{10}(\boldsymbol{\gamma}_{\boldsymbol{\theta}}) = C_{11}(\boldsymbol{\theta}, \boldsymbol{\gamma}_{\boldsymbol{\theta}})$ and (2) where the UMLE, $p < 1$, in the weak (or no coupling) case, is obtained (Section 6.3.2).

The condition $C_{10} = C_{11}$ on these QCFs estimators deserves some discussion. We recall (6.3.1) that $C_{10}(\boldsymbol{\gamma}_{\boldsymbol{\theta}})$ is the cost of declaring the signal present (and therefore making the estimate $\boldsymbol{\gamma}_{\boldsymbol{\theta}}$) when only noise is present, while $C_{11}(\boldsymbol{\theta}, \boldsymbol{\gamma}_{\boldsymbol{\theta}})$ is the cost of a correct detection decision and of the estimate $\boldsymbol{\gamma}_{\boldsymbol{\theta}}$ when $\boldsymbol{\theta}$ is the true value of the parameter. We observe that the above condition is only reasonable for the so-called "energy" parameters, i.e., for those parameters $\boldsymbol{\theta}$ for which $\mathbf{S} = 0$ implies $\boldsymbol{\theta} = \mathbf{0}$ and vice versa. These parameters, such as scale factors, waveform values, or signal duration, are all of paramount importance. On the other hand, the above condition is not reasonable for those other parameters (the so-called "nuisance" parameters in statistical terminology), such as phase or frequency, where $\boldsymbol{\theta} = \mathbf{0}$ does not necessarily imply $\boldsymbol{S} = \mathbf{0}$.

The solution of (6.3.27) in the uncoupled cases, or equivalently, from (6.2.2) earlier, as expected is seen to be

$$\textit{Eq. (6.2.7)}: \qquad \boldsymbol{\gamma}^*_{p\leq 1|\mathrm{QCF}} = \frac{\Lambda}{\Lambda + 1}\boldsymbol{\gamma}^*_{p=1|\mathrm{QCF}}, \tag{6.3.29}$$

cf. Section 5.1 of Ref. [1], discussed already in Section 6.2.1 and illustrated by the examples of Section 6.4 ff.

Equation (6.3.29) can also be written in a different but equally suggestive form if we note that the correction factor $\Lambda/(\Lambda + 1)$ is the *a posteriori probability that the signal is present*. This can easily be established by observing that

$$\frac{\Lambda}{\Lambda + 1} = \frac{p\int_\Omega F_1(\boldsymbol{X}|\boldsymbol{\theta}) w_L(\boldsymbol{\theta}) \boldsymbol{d\theta}}{p\int_\Omega F_1(\boldsymbol{X}|\boldsymbol{\theta}) w_L(\boldsymbol{\theta}) \boldsymbol{d\theta} + qF_0(\boldsymbol{X})} = \frac{pF_1(X)}{pF_1(X) + qF_0(X)}. \tag{6.3.30}$$

Next, according to Bayes' theorem we can write

$$\frac{pF_1(X)}{pF_1(X) + qF_0(X)} = P(H_1|X), \tag{6.3.31}$$

where $P(H_1|\boldsymbol{X})$ is *the a posteriori probability that the signal is present*, given the data $\boldsymbol{X}$. Thus, (6.3.29) can be rewritten in the form

$$\boldsymbol{\gamma}^*_{p\leq 1}\Big|_{\mathrm{QCF}} = P(H_1|\boldsymbol{X})\boldsymbol{\gamma}^*_{p=1}\Big|_{\mathrm{QCF}}. \tag{6.3.32}$$

6.3.3.1 Relation to Sherman's Theorem [7] For the "classical" cases, where it is *a priori* certain that a signal is present $(p(H_1) = 1)$, $\gamma^*_{p=1}|_{\mathrm{QCF}}$ is also known to be optimal for other useful cases of cost function (with, of course, different resulting Bayes risks), as a consequence of Sherman's theorem ([7], and Chapter 21 of Ref. [4]). Accordingly, it is useful to see whether and under what conditions $\gamma^*_{p\leq 1}|_{\mathrm{QCF}}$ (6.3.29) and (6.3.32), $p \leq 1$ (obtained under the conditions $C_{10}(\boldsymbol{\gamma_\theta}) = C_{11}(\boldsymbol{\theta}, \boldsymbol{\gamma_\theta}) = C_0|\boldsymbol{\gamma_\theta} - \boldsymbol{\theta}|^2$, (6.2.3)) may still be valid for other classes of cost functions, for instance the class of cost functions which are even in $\boldsymbol{\gamma_\theta} - \boldsymbol{\theta}$ and monotonic in $|\boldsymbol{\gamma_\theta} - \boldsymbol{\theta}|$, when $p < 1$.

As we shall see below, this extension of (6.3.29) and (6.3.32) unfortunately does not hold for other than the quadratic cost function. The consequence is that *Sherman's theorem* [7] and *Theorem II*, p. 974 of Ref. [4], which is so useful in conventional estimation, $(p = 1)$, does not generalize to this new situation of estimation under uncertainty.

To show this, let $\boldsymbol{X}$ in the usual fashion be the vector of the received data, $\boldsymbol{\theta}$ an unknown parameter to be estimated, $\boldsymbol{\gamma_\theta}$ the corresponding estimator, and $C(\boldsymbol{\theta}, \boldsymbol{\gamma_\theta})$ the chosen cost function. It is well known that if $C(\boldsymbol{\theta}, \boldsymbol{\gamma_\theta})$ is a QCF, that is, $C(\boldsymbol{\theta}, \boldsymbol{\gamma}) = C_0|\boldsymbol{\theta} - \boldsymbol{\gamma_\theta}|^2$, the optimum estimator $\boldsymbol{\gamma}^*_{\boldsymbol{\theta}}|_{\mathrm{QCF}}$ is the *a posteriori* mean

$$\boldsymbol{\gamma}^*_{\boldsymbol{\theta}}(\boldsymbol{X})_{\mathrm{QCF}} = \int_\Omega \boldsymbol{\theta}\, W_1(\boldsymbol{\theta}|\boldsymbol{X}) \boldsymbol{d\theta}, \tag{6.3.33}$$

where $W_1(\boldsymbol{\theta}|\boldsymbol{X})$ is the posterior pdf of $\boldsymbol{\theta}$ given the data $\boldsymbol{X}$. Sherman's theorem states that if the pdf $W_L(\boldsymbol{\theta}|\boldsymbol{X})$ is unimodal and symmetric about the mode, the estimator (6.3.33) is still optimum for the class of cost functions $C(\boldsymbol{\theta}, \boldsymbol{\gamma}_\theta)$ satisfying the following conditions: (i) $C(\boldsymbol{\theta}, \boldsymbol{\gamma}_\theta)$ is an even function of $\boldsymbol{\theta} - \boldsymbol{\gamma}$ and (ii) $C(\boldsymbol{\theta}, \boldsymbol{\gamma}_\theta)$ is a nondecreasing function of $\boldsymbol{\theta}, \boldsymbol{\gamma}_\theta$. Here, the aim is to prove that if $W_1(\boldsymbol{\theta}|\boldsymbol{X})_\theta$ is not symmetric, then (6.3.30) and (6.3.32) do not hold for the general class of cost functions mentioned above. This follows immediately if we note that in the situation of estimation under uncertainty ($p < 1$) the *a posteriori* pdf $W_1(\boldsymbol{\theta}|\boldsymbol{X})$ must contain a delta function at zero (and therefore cannot be symmetric) since a delta function $\boldsymbol{\theta} = \mathbf{0}$ is present in the corresponding *a priori* distribution.

Now we chose $C(\boldsymbol{\theta}, \boldsymbol{\gamma}_\theta)$ to be a cost function satisfying the two conditions mentioned above, and assume that it posses a Taylor expansion[25] around $\boldsymbol{\gamma}_\theta$. Then we can write

$$C(\boldsymbol{\theta}, \boldsymbol{\gamma}_\theta) = \sum_{n=1}^{\infty} a_{2n} (\boldsymbol{\theta} - \boldsymbol{\gamma}_\theta)^{2n}. \tag{6.3.34}$$

For each $\boldsymbol{X}$ the conditional risk $R_{\mathrm{E}}(\boldsymbol{X})$ corresponding to the estimation process is

$$R(\boldsymbol{X})_{\mathrm{E}} = \sum_{n=1}^{\infty} a_{2n} \int_{\Omega} (\boldsymbol{\theta} - \boldsymbol{\gamma}_\theta)^{2n} W_1(\boldsymbol{\theta}|\boldsymbol{X}) d\boldsymbol{\theta}. \tag{6.3.35}$$

The form of the optimum estimator follows from the extremal condition $\partial R(\boldsymbol{X})/\delta \boldsymbol{\gamma}_\theta = \mathbf{0}$, or

$$\frac{\partial R_{\mathrm{E}}}{\partial \boldsymbol{\gamma}_\theta} = \sum_{n=1}^{\infty} 2n\, a_{2n} \int_{\Omega} (\boldsymbol{\theta} - \boldsymbol{\gamma}_\theta)^{2n-1} W_1(\boldsymbol{\theta}|\boldsymbol{X}) d\boldsymbol{\theta} = \mathbf{0}. \tag{6.3.36}$$

Since this relation must be satisfied for any arbitrary set of coefficients $\{a_{2n}\}$, it is clear that (6.3.33) cannot hold in this case, unless the summations in (6.3.34)–(6.3.36) collapse to the single term $n = 1$, namely for the QCF.

6.3.3.2 *A Limited Generalization* There is, however, a less useful generalization of (6.3.29) [or (6.3.32)] which can be rather easily derived: it is possible to show that *if* $\boldsymbol{\gamma}_{p=1}$ *is any* (not necessarily optimum, e.g., Bayes) estimator yielding an unbiased estimate of $\boldsymbol{\theta}$, an "energy" parameter under $p = 1$, then this property of unbiasedness is preserved in a situation of uncertainty as to the signal's presence ($p < 1$) if the vector $\boldsymbol{\theta}$ is estimated by an estimator of the form (6.3.29), that is,

$$\boldsymbol{\gamma}_{p \le 1} = \frac{\Lambda}{\Lambda + 1} \boldsymbol{\gamma}_{p=1}, \tag{6.3.37}$$

where $\boldsymbol{\gamma}_{p=1}$ is an arbitrary estimator, unbiased under H_1.

Proof: We express this unknown, arbitrary estimator, which is designed to keep the estimate unbiased under certainty ($p = 1$) as $\gamma_{\theta,p\le 1} = H(\boldsymbol{X}) \gamma_\theta|_{p=1}$, where $H(\boldsymbol{X})$ is an

[25] Other expressions can be employed if a Taylor's series is not possible, for example, a Fourier series. The proof remains essentially unchanged.

unknown function of the data X, to be determined. Since it is required that this estimator be unbiased, under $H, p < 1$, as well, we have

$$\int_\Gamma H(X)\boldsymbol{\gamma}_{\boldsymbol{\theta},p=1}(X)\left[qF_0(X|\boldsymbol{\theta}) + p\int_\Omega F_1(X|\boldsymbol{\theta})w_L(\boldsymbol{\theta})d\boldsymbol{\theta}\right]dX \equiv \left\langle \boldsymbol{\gamma}_{\boldsymbol{\theta},p\leq 1}(X)\right\rangle_{H_0+H_1}, \quad (6.3.38)$$

where the brackets indicate an averaging operation under the joint hypotheses $H = (H_0 + H_1)$ (6.2.11). However, since the $\boldsymbol{\theta}$ are energy parameters, their expected value under the hypothesis H_0 is zero. Therefore

$$\left\langle \boldsymbol{\gamma}_{\boldsymbol{\theta},p\leq 1}(X)\right\rangle_{H_0+H_1} = p\bar{\boldsymbol{\theta}} = p\langle \boldsymbol{\gamma}_{\boldsymbol{\theta},p=1}\rangle_{H_1}. \quad (6.3.39)$$

From (6.3.38), (6.3.39), and the expression for Λ (6.2.4b), we have

$$\int_\Gamma qF_0(X|0)\left[(\Lambda + 1)H(X)\boldsymbol{\gamma}_{\boldsymbol{\theta},p=1}(X) - \Lambda\boldsymbol{\gamma}_{\boldsymbol{\theta},p=1}(X)\right]dX = \mathbf{0}. \quad (6.3.40)$$

Since $F_0(X|0)$ is a pdf and therefore a non-negative quantity, it follows that

$$(\Lambda + 1)H(X) - \Lambda = 0, \text{ or } H(X) = \frac{\Lambda}{\Lambda + 1}, \quad (6.3.41)$$

which is our stated result (6.3.37).

Finally, the proof has been presented in Section 6.3.1.1 that the optimum decision rules δ^*, δ^{**} (6.3.8) generally and in the uncoupled cases in particular, are nonrandomized.

6.3.4 Estimator Bias ($p \leq 1$)

There remains the important question of *estimator bias*. By definition, if the average of the estimator $\boldsymbol{\gamma}_{\boldsymbol{\theta}}$ under the appropriate hypothesis, for example, H_1 if $p(H_1) = 1$ or $H(= H_0 + H_1)$ if $p(H_1) \leq 1$, is equal to the expected value of the quantity(ies) being estimated, namely, that is, $\langle \boldsymbol{\gamma}_\theta\rangle_{H_1} = \langle \boldsymbol{\theta}\rangle$ when $p = 1$, or $\langle \boldsymbol{\gamma}_\theta\rangle_H = p\langle \boldsymbol{\theta}\rangle$ when $p \leq 1$, then we say that the estimator is *unbiased*. Otherwise, it is *biased*, that is, $\langle \boldsymbol{\gamma}_\theta\rangle \neq p\langle \boldsymbol{\theta}\rangle_H$, and the resulting estimate can be noticeably distant from the "true" (i.e., ensemble) value $(\sim\langle \boldsymbol{\theta}\rangle)$. However, if the bias is a known function of the parameters, $\boldsymbol{b}(\boldsymbol{\theta})$, the bias can be removed from the data so that the new estimator, $\boldsymbol{\gamma}_{\boldsymbol{\theta}} - \boldsymbol{b}(\boldsymbol{\theta})$, is then unbiased.[26]

In any case, to determine estimator bias, it is necessary to establish the relations

$$\langle \boldsymbol{\gamma}_{\boldsymbol{\theta}}\rangle_{H_1} = \langle \boldsymbol{\theta}\rangle, p = 1; \langle \boldsymbol{\gamma}_{\boldsymbol{\theta}}\rangle_H = p\langle \boldsymbol{\theta}\rangle, p \leq 1; H = H_0 + H_1, \text{ as before.} \quad (6.3.42)$$

Thus, we need to evaluate $\langle \boldsymbol{\gamma}_{\boldsymbol{\theta}}\rangle_{H_{()}}$ derived for the cost functions, $C_{()}(\boldsymbol{\theta}, \boldsymbol{\gamma})$, (6.3.1), which are appropriate to the type of coupling employed between detector and estimator, whether "strong" (Section 6.3.1), "weak" (Section 6.3.2), or "none" (Section 6.2). For the "simple" cost functions (SCF), (6.2.13), $p \leq 1$, Eq. (6.2.26) is a condition imposed upon $\boldsymbol{\gamma}_{p\leq 1}|_{\text{SCF}}$, but for the QCF of Section 6.2.1 we need to determine whether or not $\boldsymbol{\gamma}_{p\leq 1}|_{\text{QCF}}$ is indeed unbiased.

[26] Bias and its lack play an important role in establishing *maximum* lower bounds on the estimator's variance. See Sections 5.1.1.1 and 5.1.1.2.

6.3.4.1 No Coupling Accordingly, let us consider first the case of "no coupling" (Section 6.2) and thus apply the averaging operator $\hat{E}_H = \hat{G}_H = \langle\ \rangle_{H:\,X,\boldsymbol{\theta}}$ (6.2.11) to the classical relation (6.2.9) for $\boldsymbol{\gamma}^*_{p=1|\mathrm{QCF}}$, as (6.3.42) requires. This gives us

$$\left\langle \boldsymbol{\gamma}^*(\boldsymbol{\theta})_{p\le 1|\mathrm{QCF}}\right\rangle_H = \int_\Gamma d\boldsymbol{X}\frac{\Lambda}{\Lambda+1}\boldsymbol{\gamma}^*_{p=1|\mathrm{QCF}}\left[qF_J(\boldsymbol{X}|\boldsymbol{0})_{\boldsymbol{S}=\boldsymbol{0}} + p\int_\Omega F_J(\boldsymbol{X}|\boldsymbol{S}(\boldsymbol{\theta}))w_L(\boldsymbol{\theta})d\boldsymbol{\theta}\right]. \tag{6.3.43a}$$

Multiplying and dividing by $F_J(\boldsymbol{X}|\boldsymbol{0})_{\boldsymbol{S}=\boldsymbol{0}}$, and using (6.2.4) with (6.2.4a–c) for $\boldsymbol{\gamma}^*_{p=1|\mathrm{QCF}}$, we obtain

$$\begin{aligned}\left\langle \boldsymbol{\gamma}^*(\boldsymbol{\theta})_{p\le 1|\mathrm{QCF}}\right\rangle_H &= \int_\Gamma d\boldsymbol{X}\frac{\Lambda}{\Lambda+1}\int_\Omega \boldsymbol{\theta}\, W_{J\times L}(\boldsymbol{X}|\boldsymbol{\theta})d\boldsymbol{\theta}\, p\left\{\frac{qF_J(\boldsymbol{X}|\boldsymbol{0})_{\boldsymbol{S}=\boldsymbol{0}}}{p\langle F_J(\boldsymbol{X}|\boldsymbol{S}(\boldsymbol{\theta}))\rangle_{\boldsymbol{\theta}}}\right\}[\Lambda+1]\\ &= p\int_\Gamma\int_\Omega \boldsymbol{\theta}\, W_{J\times L}(\boldsymbol{X}|\boldsymbol{\theta})d\boldsymbol{X}\, d\boldsymbol{\theta} = p\langle\boldsymbol{\theta}\rangle_H,\end{aligned} \tag{6.3.43b}$$

which establishes the desired result. A similar procedure for the (vector) estimator $\boldsymbol{\gamma}^*_S$ of the received waveform $\boldsymbol{S}$, (6.2.2b), produces the analogous relation here for these no-coupling cases, namely,

$$\left\langle \boldsymbol{\gamma}^*(\boldsymbol{S})_{p\le 1|\mathrm{QCF}}\right\rangle_H = p\langle\boldsymbol{S}\rangle_H, \tag{6.3.43c}$$

(cf. Problem 6.8).

The estimators (6.3.32) are thus *unconditionally unbiased.* From the viewpoint of the H_1: $S\otimes N$ state alone, however, it is clear that the estimators are biased, since we are assuming that the signal, and the parameters to be estimated, are present in all ensemble members, that is, $p=1$, when actually they are not. The bias stems from the fact that a fraction ($q>0$) of the data ensemble contains noise only. The bias under H_1 is directly found by writing $\boldsymbol{\gamma}^*_{p\le 1|\mathrm{QCF}}$ alternatively as

$$\boldsymbol{\gamma}^*_{p\le 1|\mathrm{QCF}} = \left[1-\frac{1}{\Lambda(\boldsymbol{X})+1}\right]\boldsymbol{\gamma}^*_{p=1|\mathrm{QCF}} = \boldsymbol{\gamma}^*_{p=1|\mathrm{QCF}} - B(\boldsymbol{X})\boldsymbol{\gamma}^*_{p=1|\mathrm{QCF}}, \tag{6.3.44}$$

$$B(\boldsymbol{X}) \equiv (\Lambda+1)^{-1}, \quad 0\le B(\boldsymbol{X})\le 1. \tag{6.3.44a}$$

Clearly, $B(\boldsymbol{X})$ vanishes when $p=1$ (or $\mu=\infty$) and is unity when $p=1$ ($\mu=0$), as required for these limiting cases. Here $B(\boldsymbol{X})\boldsymbol{\gamma}^*_{p=1|\mathrm{QCF}}$ is the *sample bias* of $\boldsymbol{\gamma}^*_{p=1|\mathrm{QCF}}$ due to the fact that $p\le 1$. The sample bias (6.3.44a) depends on the particular sample $\boldsymbol{X}$, and, of course, structurally on the cost functions chosen,[27] cf. (6.2.2b). Similarly, we may write (6.3.43b) as

$$\left\langle \boldsymbol{\gamma}^*_{p\le 1|\mathrm{QCF,SCF}}\right\rangle_{H_0+H_1} = p\bar{\boldsymbol{\theta}} = \bar{\boldsymbol{\theta}} - q\bar{\boldsymbol{\theta}}, \tag{6.3.44b}$$

[27] For the SCF, the explicit structure of $\boldsymbol{\gamma}^*_{p\le 1} = p\boldsymbol{\gamma}^*_{p=1}$ depends in turn on the explicit statistical form of $\hat{\Lambda}$, (6.2.23c), unlike our result (6.2.7) for $\boldsymbol{\gamma}^*_{p\le 1|\mathrm{QCF}}$, which depends on Λ (6.1.6b).

where now $q\bar{\boldsymbol{\theta}}$ is the *average bias*, that is, the average under H_1 of the sample bias, or more simply "the bias," from the usual definition [cf., Sections 5.1.1.1 and 5.1.1.2].

Thus, the effect of neglecting the average bias $q\bar{\boldsymbol{\theta}}$, whatever the cost function employed in estimation here, is always erroneously to *increase* the true *average* estimate of the parameter(s) $\boldsymbol{\theta}$ under $H = H_0 + H_1$, by the amount $q\bar{\boldsymbol{\theta}}$, which can be significantly large when q is large (≤ 1). Neglect of the sample bias $B\boldsymbol{\gamma}^*_{p=1}$ can likewise erroneously increase the true sample estimate, namely $\boldsymbol{\gamma}^*_{p<1}(\boldsymbol{X})$, by the amount $B\boldsymbol{\gamma}^*_{p=1}$, which can lie in the range $(0-1)\cdot B\boldsymbol{\gamma}^*_{p=1}$ for the QCF, cf. (6.3.44). Analogous increases (errors) occur in the sample bias for $\text{SCF}_{1,2}$, [cf. (6.2.31) and (6.2.35)]. The reductions of $\boldsymbol{\gamma}^*_{p=1}$ when $p < 1$ represent compensation for the *a priori* absence ($q > 0$) of the signal in the received data $\boldsymbol{X}$. Section 6.4 following presents some numerical examples with the QCF and $\text{SCF}_{1,2}$ under the no-coupling condition.

6.3.4.2 Weak Coupling Our results above apply also in the situation of *weak coupling* when the cost functions of (6.3.28) are employed, specifically for the QCF and SCF of Section 6.2, cf. (6.3.27), (6.3.28). However, with other cost functions (f_{10}, f_{11}) it is not evident that the resulting estimators are unbiased (under H), although we suspect that the Bayesian optimality requirement may so ensure. Each situation has to be explored individually, for the given statistical distributions Λ and $\hat{\Lambda}'$, (6.3.22a), (6.2.4b).

6.3.4.3 Strong Coupling The generally unbiased nature of the (coupled) estimators here is similarly an open question. As in the cases above we suspect that unbiasedness may also be conditioned on Bayesian optimality, but no proof or disproof appears as yet available.

6.3.5 Remarks on Interval Estimation , $p(H_1) \leq 1$

A frequently useful concept in evaluating the accuracy of our estimates $\boldsymbol{\gamma}^{(*)}_{p\leq 1}$ of model parameters $\boldsymbol{\theta}$ is the *interval estimate*, which we have already discussed briefly in Section 5.1 for the classical situation where the signal containing the set of parameters in question is *a priori* known surely to be present, that is, $\boldsymbol{\gamma}^*_{p=1}$. This estimate set is defined now as the (joint) probability P_I, that the particular set of estimates under $p(H_1) = 1$ falls within $(\mathbf{1} \pm \boldsymbol{\lambda})$ 100% of the true value of the parameter(s) $\hat{\boldsymbol{\theta}}\left(\text{of } \hat{\boldsymbol{\theta}}, \boldsymbol{\theta}'\right)$ being estimated, [8]. Here we extend the concept of the interval estimate P_I to the uncertain case, $p < 1$, namely,

$$\left.\begin{aligned} P_{I(p\leq 1)} &= P_I\left[(\mathbf{1}-\boldsymbol{\lambda})\cdot\hat{\boldsymbol{\theta}} \leq \boldsymbol{\gamma}_\theta(\boldsymbol{X}|\boldsymbol{\theta})_{p\leq 1} \leq (\mathbf{1}+\boldsymbol{\lambda})\cdot\hat{\boldsymbol{\theta}}\right] \\ &= \int_{(1-\lambda_1)\hat{\theta}_1}^{(1+\lambda_1)\hat{\theta}_1} (d\lambda_\theta)_1 \cdots \int_{(1-\lambda_L)\hat{\theta}_L}^{(1+\lambda_L)\hat{\theta}_L} (d\lambda_\theta)_L P_L\left(\boldsymbol{\gamma}_\theta|\hat{\boldsymbol{\theta}}\right)_{p\leq 1} \end{aligned}\right\}, \tag{6.3.45}$$

where $1 \pm \lambda_\ell$, $\ell = 1, \ldots, L$ denote the bounds of the various intervals chosen for the different parameters $\boldsymbol{\gamma}_{\theta|p\leq 1} = (\gamma_{\theta 1}, \ldots, \gamma_{\theta L})_{p\leq 1}$. Frequently, the same interval factors are used, that is, $\lambda_\ell = \lambda(> 0)$. For weak or no coupling, where detection and estimation are separate operations, cf. Sections 6.2 and 6.3.2, the conditional pdf P_L of the estimators, given $\hat{\boldsymbol{\theta}}$, may be obtained formally from the inversion of the following characteristic function $e^{i\boldsymbol{\xi}\cdot\boldsymbol{\gamma}_{\theta|p\leq 1}}$, namely,

$$P_L\left(\boldsymbol{\gamma}_\theta|\hat{\boldsymbol{\theta}}\right)_{p\leq 1} = \int_{-\infty}^{\infty} e^{-i\boldsymbol{\xi}\cdot\boldsymbol{\gamma}_\theta}\frac{d\boldsymbol{\xi}}{(2\pi)^L}\int_{-\infty}^{\infty} e^{i\boldsymbol{\xi}\cdot\boldsymbol{\gamma}_\theta(\boldsymbol{X}|\boldsymbol{\theta})_{p\leq 1}}\left[qF_J(\boldsymbol{X}|\boldsymbol{\theta})_{\hat{\boldsymbol{\theta}}=0} + pF_J\left(\boldsymbol{X}|\boldsymbol{S}(\hat{\boldsymbol{\theta}})\right)\right]d\boldsymbol{X}. \tag{6.3.46}$$

In our applications one possible choice of "true values" of the parameters $\hat{\boldsymbol{\theta}}$ may be their sample means, $p\langle\hat{\boldsymbol{\theta}}\rangle_{\text{sample}}$, cf. (6.3.42), established from the data for which positive detection has been registered. For example, if our particular value of $\left(\boldsymbol{\gamma}_{\theta|p\leq 1}\right)_\ell$ lies inside $\left(1 \pm \theta_\ell\right)\hat{\theta}_\ell$, and we choose a particular value of λ_ℓ, say 0.95, namely 95% confidence limits, *and* the detector decides "yes, $S \oplus N$: signal present," we accept the estimate $\left(\boldsymbol{\gamma}_{\theta|p\leq 1}\right)_\ell$ as a satisfactory value. Otherwise, we can reject the estimate, even though the detector indicates the presence of a signal, cf. Section 6.3.6 ff. Alternatively, if we can show that our estimates are unconditionally unbiased (Section 6.3.4), then $\langle\boldsymbol{\gamma}_\theta\rangle_{H_0+H_1} = p\bar{\boldsymbol{\theta}}$, $0 \leq p \leq 1$, and $\lim_{J\to\infty} \text{var.}_{H_0+H_1}\boldsymbol{\gamma}_{\theta|p\leq 1} \to \text{var.}\,\boldsymbol{\theta} = \mathbf{0}$. This supports the notion that it is reasonable to replace the usually unknown "true" values $\hat{\boldsymbol{\theta}}$ with their ensemble mean equivalents. (See Problem 6.16 for the resulting interval estimate for the example of Section 6.4.4, where one selects for $\boldsymbol{\gamma}_\theta(\boldsymbol{X}|H) = a^*_{p\leq 1}(\boldsymbol{X})_{\text{SCF}}$, Eq. (6.4.20).)

6.3.6 Detection Probabilities

As we have pointed out in preceding sections, in all cases where estimation is carried out with uncertainty as to the presence of the signal ($p(H_1) < 1$), we must also conduct a parallel detection. This concomitant detection operation, performed simultaneously for weak or no coupling or subsequently with strong coupling to estimation, is embodied in the various likelihood ratios $\Lambda_{(\,)} = \left(\Lambda, \Lambda_g, \Lambda'_g\right)$ in our present cases of binary on–off detection.[28] Whatever the degree of coupling in our Bayesian formulation, the detection process still involves the comparison of some form of likelihood ratio with an appropriate threshold $K(>0)$.

Accordingly, we can write for detection here the following set of binary decisions:

1. *No Coupling*: Decide H_1: $S \otimes N$, signal and noise; Decide H_0: N, noise alone

$$\left.\begin{array}{l}\text{if } \log\Lambda \geq \log K \\ \text{if } \log\Lambda < \log K\end{array}\right\}, \quad \text{Eq.(6.1.6)}; \tag{6.3.47a}$$

2. *Weak Coupling*: Decide H_1: $S \otimes N$; Decide H_0: N

$$\left.\begin{array}{l}\text{if } \log\Lambda'_g \geq \log K = 0 \\ \text{if } \log\Lambda'_g < \log K = 0\end{array}\right\}, \quad \text{Eq.(6.3.22)}; \tag{6.3.47b}$$

[28] $\hat{\Lambda}_L$, Eq. (6.2.23c), and $\hat{\Lambda}_1$, Eqs. (6.2.34) and (6.2.34a), are likelihood ratios which, however, do not arise in the detection process, unlike $\Lambda, \Lambda_g, \Lambda'_g$, cf. (6.3.47).

5. *Strong Coupling*: Decide H_1: $S \otimes N$, Decide H_0: N

$$\left.\begin{array}{l}\text{if } \log \Lambda_g \geq \log K = 0 \\ \text{if } \log \Lambda_g < \log K = 0\end{array}\right\}, \quad \text{Eqs.(6.3.17), (6.3.18)} \qquad (6.3.47c)$$

The decision processes for detection in (1) and (2) are independent of the accompanying estimation procedures, cf. Sections 6.2 and 6.3.2, but depend on the (optimum) estimator $\boldsymbol{\gamma}_\theta^{**}$, (6.3.5), in Case 5 above as noted in Section 6.3.1. In all cases $\log\Lambda$ embodies the detector structure or data ($\boldsymbol{X}$) processing algorithm (see Section 3.1). We also observe here that the accompanying estimate is accepted provided the probability of correct detection P_D^* is equal to or greater than some preassigned lower limit $P^*_{\text{D-accept.}}$, namely,

$$\begin{aligned} P_\text{D}^* &= p\left(1 - \beta^{(*)}\right) \geq P^*_{\text{D-accept.}} \\ &= p p^*_{\text{D-accept.}} \end{aligned} \qquad (6.3.48)$$

The probability P_D^* then depends on the (conditional) false rejection probability, or Type II error probability $\beta(S \otimes N|N)^*$, as well as directly on $p(H_1)$.

In the *no-coupling cases* (6.3.47a), $\log K$ is nonzero, generally, which implies a preset value of the "false alarm" probability α_F in what is then a *Neyman–Pearson* (NP) class of detector (Section 3.3.1). As we have seen in Section 3.3.1, such a detector minimizes the adjustable part of the Type II error probability contained in the average risk R_D (e.g., that part which contains $\delta_1(H_1|\boldsymbol{X})$), according to

$$\delta_1 \overset{\min}{\longrightarrow} \delta_1^*(p\beta + \lambda_0 \alpha_\text{F}) = p\beta^*_\text{NP} + Kq\alpha_\text{F}; \quad \lambda_0 \to K, \beta \to \beta^*_\text{NP}. \qquad (6.3.49)$$

On the other hand, for the *weak*- and *strong-coupling* cases [(6.3.47b) and (6.3.47c)], the threshold K is unity, which in turn indicates that these detectors belong to the *Ideal Observer* (IO) class, cf. Section 3.3.2. This requires *joint* minimization of the Type I (α) and Type II (β) error probabilities in the associated Bayes risk R_D^*, (6.3.7), (6.1.7), now according to

$$\delta_1 \overset{\min}{\longrightarrow} \delta_1^*(p\beta + q\alpha) = p\beta_I^* + q\alpha_I^*; \quad \lambda_0 \to K = 1. \qquad (6.3.50)$$

[All the decision rules $\delta_1, \delta_1^*, \delta_1^{**}$, and so on, are nonrandom, as we have already shown in preceding sections.]

From Section 3.2.1 we have the following generic forms for the various (Bayes) error probabilities $\alpha = \alpha_\text{F}^*, \beta^*$, and so on, above:

$$\alpha_\text{F}^* = \int_{\log K}^{\infty} Q_1(x)dx; \quad 1 - \beta^* = \int_{\log K}^{\infty} P_1(x)dx; \text{ etc.,} \qquad (6.3.51)$$

where

$$\left.\begin{aligned} Q_1(x) = Q_1(x|H_0) &= \int_\Gamma F_J(\boldsymbol{X}|\boldsymbol{\theta})_{S=0}\, \delta\{x - \log \Lambda(\boldsymbol{X})\} d\boldsymbol{X}, \\ P_1(x) = P_1(x|H_1) &= \int_\Gamma \left\langle F_J(\boldsymbol{X}|\boldsymbol{\theta})_S \right\rangle_{\boldsymbol{\theta}} \delta\{x - \log \Lambda(\boldsymbol{X})\} d\boldsymbol{X}, \\ \text{with } P_1(x|H_1) = \mu^{-1} e^x Q_1(x|H_0) &\text{ or } P_1(x) = \mu^{-1} e^x Q_1(x), \end{aligned}\right\} \qquad (6.3.51a)$$

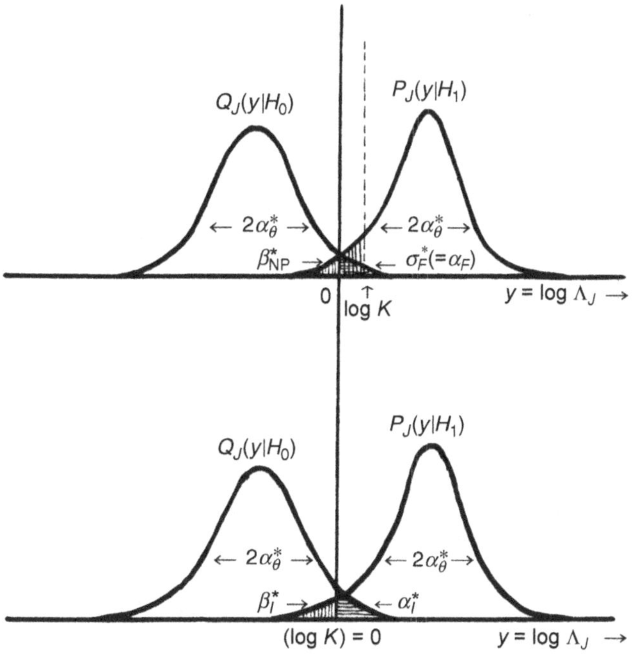

FIGURE 6.7 Conditional error probabilities $\alpha^*_{()}, \beta^{[*]}_{()}$, Eq. (6.3.51), for the on–off signal detection cases (6.3.47) of *no coupling*, $K(> 1)$; *weak and strong coupling* $K = 1$.

these last from Section 3.4.1. The NP detectors (6.3.49) are commonly used in radar and sonar applications. On the other hand, the IO detectors ($K = 1$), for which $p = q(= 1/2)$ is associated with the so-called "symmetric channel," are typically required in telecommunications, where a common aim is to minimize bit-error probabilities P_e. Figure 6.7 shows the pdfs Q_1 and P_1 for both types of detector $\Lambda = \Lambda_{NP}$, or $\Lambda_{IO} = \Lambda' g, \Lambda_g$ (6.3.47).

6.3.7 Waveform Estimation ($p \leq 1$): Coupled and Uncoupled D and E

In Sections 6.3.1–6.3.6, we have focused primarily on parameter estimation when $p \leq 1$. An important special case arises when waveform $\boldsymbol{S} \equiv [S, \ldots, S_J]$ itself is the desired "parameter." The preceding analysis of the coupled D and E cases is easily carried over to handle this situation.

Formally, the following transformations are indicated:

$$\text{In (6.3.7) et seq.:} \quad \begin{aligned} &\boldsymbol{\gamma}_\theta^* \to \boldsymbol{\gamma}_S^*;\ F_0(\boldsymbol{\theta}|\boldsymbol{X})d\boldsymbol{\theta} \to F_0(\boldsymbol{S}|\boldsymbol{X})d\boldsymbol{S}, \text{ with} \\ &F_1(\boldsymbol{X}|\boldsymbol{\theta}) \to F_1(\boldsymbol{X}|\boldsymbol{S}) \text{ and} \\ &F_1(\boldsymbol{X}) = \int_\Omega F_1(\boldsymbol{X}|\boldsymbol{S}) w_J(\boldsymbol{S}) d\boldsymbol{S} \quad (6.3.6). \end{aligned} \tag{6.3.52}$$

The cost functions (6.3.1) are modified now to

$$\left.\begin{array}{lll} C_{00}\left[= C_{1-\beta} = C_0^{(0)}\right] & : & \text{cost of correctly deciding } \delta_0 = H_0, \text{ when } H_0 \text{ is true;} \\ C_{10}(\boldsymbol{\gamma}_S) = C_\alpha\left(= C_1^{(0)}\right) + f_{10}[\boldsymbol{\gamma}_S(\boldsymbol{X})] & : & \text{cost of incorrectly deciding } \delta_1 = H_1 \text{ and making an estimate } \boldsymbol{\gamma}_S = \boldsymbol{\gamma}_S(\boldsymbol{X}) \text{ when } H_0 \text{ is true;} \\ C_{01}(\boldsymbol{S}) = C_\beta\left(= C_0^{(1)}\right) + f_{01}[\boldsymbol{S}] & : & \text{cost of incorrectly deciding } \delta_0 = H_0, \text{ when in fact } H_1 \text{ is true and } \boldsymbol{S} \text{ is the true value of the signal waveform;} \\ C_{11}(\boldsymbol{S}, \boldsymbol{\gamma}_S) = C_{1-\beta}\left(= C_1^{(1)}\right) = f_{11}[\boldsymbol{S}, \boldsymbol{\gamma}_S(\boldsymbol{X})] & : & \text{cost of correctly deciding } \delta_1 = H_1 \textit{ and } \text{making an estimate } \boldsymbol{\gamma}_S = \boldsymbol{\gamma}_S(\boldsymbol{X}), \text{ when } H_1 \text{ is true and } \boldsymbol{S} \text{ is the true value of the signal waveform.} \end{array}\right\} \tag{6.3.53}$$

Applying the above modifications [(6.3.52) and (6.3.53)] to (6.3.7) et seq., we see that, finally, the optimum detection process is still represented by (6.3.17), where Λ_g, Eq. (6.3.18), becomes explicitly here for the detection process in waveform estimation when $(p < 1)$:

$$\Lambda_g = \Lambda_g(\boldsymbol{X}) = \mu \frac{\int_\Omega w_J(\boldsymbol{S})\left[C_{01}(\boldsymbol{S}) - C_{11}\left(\boldsymbol{S}, \boldsymbol{\gamma}_S^{**}\right)\right] F_1(\boldsymbol{X}|\boldsymbol{S})\, d\boldsymbol{S}}{\left[C_{10}\left(\boldsymbol{\gamma}_S^{**}\right) - C_{00}\right] F_1(\boldsymbol{X})}. \tag{6.3.54}$$

The corresponding minimization equation for $\boldsymbol{\gamma}_S^{**}(\boldsymbol{X})$ under $(p \leq 1)$ is given by Eq. (6.3.20), with $\boldsymbol{\gamma}_{\boldsymbol{\theta}} \to \boldsymbol{\gamma}_S, \boldsymbol{\theta} \to \boldsymbol{S}, d\boldsymbol{\theta} \to d\boldsymbol{S}$ therein [cf. (6.3.15) and the steps (6.3.10) et seq.]. Again, the estimation operation must be carried out *before* detection in these strongly coupled cases. For *weak coupling* we have $\Lambda_g \to \Lambda'_g$, Eq. (6.3.22), with $\boldsymbol{S}(\boldsymbol{\theta}) \to \boldsymbol{S}, \boldsymbol{\theta} \to \boldsymbol{S}$, and so on, along with (6.3.22a) and (6.3.22b). Here the D and E operations, with $(p \leq 1)$, can be performed in any order as before, since now the detection process is independent of estimator structure. Moreover, estimator-dependent cost functions, vide (6.3.53), can still be employed. Equations (6.3.25)–(6.3.28) likewise hold on replacing $\boldsymbol{\theta}$ by $\boldsymbol{S}$ formally.

The same formal substitutions involving $\boldsymbol{\theta} \to \boldsymbol{S}$, applied to the material of Sections 6.3–6.6, similarly provide the explicit general structures and results for waveform estimation. Table 6.1 remains applicable, with the appropriate formal substitution of $\boldsymbol{S}$ for $\boldsymbol{\theta}$, as do the various degradation criteria (vis-à-vis $p = 1$) of Section 6.3.9.1. For the uncoupled D and E cases of Section 6.2, the indicated replacement of $\boldsymbol{\theta}$ by $\boldsymbol{S}$ are made to provide analogous results for estimation as well as detection (Section 6.1). An analytic example is presented in Section 7.1 presently, but for the most part numerical evaluation is required for quantitative results, as in the more common situations involving parameter estimation.

6.3.8 Extensions and Modifications

The present general theory is now extended to include a threshold on the estimators, which in turn are used to (1) improve detection and (2) provide estimates of the parameters themselves. Estimation is still required preceding detection, in which the estimates are

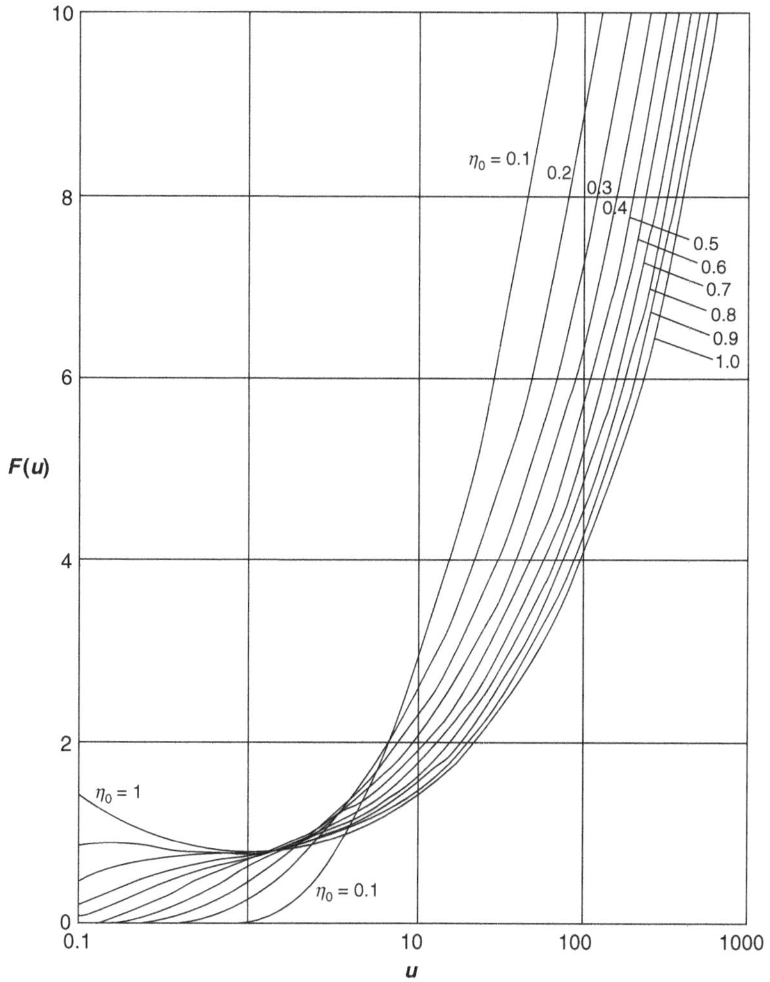

FIGURE 6.8 The scaling function $F(u)$, Eq. (6.4.26b), for the SCF examples, and for weak $(\eta_o \sim 0.1)$ to strong fading $(\eta_o = 1)$; $u \equiv \sigma^2 \Phi_s$, (6.4.15).

found with the help of a detector, and are accepted (or rejected) according to whether or not they are passed by the M-ary detector used now as an estimator. They are also used to improve detection and are accepted (or rejected) according to the binary outcome of the main detection. The extension, and modification (the M-ary detector), is shown in Fig. 6.8.

Let us expatiate on $E \rightarrow D_M$. As in the general theory the estimation process must be done first, consistent with the strong coupling condition of the general theory, that is, Eq. (6.3.18) applies: $E \rightarrow D_{\text{Binary}}$. However, for the $E \rightarrow D_M$ we have K_E, and for this *we chose a maximum likelihood estimator, which is for* $p < 1$, $p\gamma_p^{**}$ (for the M-ary estimator (= detector) we refer the reader to Chapter 4). Accordingly, for K_E we select m estimators out of M for which $p\left(\boldsymbol{\gamma}_{p=1}^{**}\right)_m$ is greater than K_E, or equivalently, m estimators $\left(\gamma_{p=1}^{**}\right)_m \geq K_E/p$. In this way, we dodge the issue of having to know p, putting it into the new threshold $K_E/p = K'_E$. (Of course, it is still in K'_E, but we can adjust K'_E to whatever we like, subject to whatever is the false alarm probability, in conjunction with what is an acceptable probability of p_{D}, the probability of a signal (see the remarks in the last paragraph of this section).)

Thus, having selected $0 \leq m \leq M$ estimators, we form the *binary* detector

$$D_{\text{Binary}} = \Lambda\left(\boldsymbol{X}, \left(\boldsymbol{\gamma}^{**}_{p<1}\right)_m\right)$$

$$= \mu \int_{(\boldsymbol{\theta}_m)} \frac{w_m([\boldsymbol{\theta}]_m)\{C_{01}\}([\theta]_m) - C_{11}\left([\theta]_m, \boldsymbol{\gamma}^{**}_{\theta\, m}\right) F_1\left(\boldsymbol{X}|(\theta)_m\right) d\boldsymbol{\theta}_m}{C_{10}\left(\left(\left[\boldsymbol{\gamma}^{**}_{p<1}\right]_m\right) - C_{00}\right) F_0(\boldsymbol{X}|0)} \gtrless K_{\mathrm{D}} \left| \begin{array}{l} \text{decide} \\ H_1: S \otimes N \\ \text{or} \\ H: N \end{array} \right. \qquad (6.3.55)$$

in which $\mu = p/q = p/(1-p)$. Equation (6.3.55) is now modified by inserting the estimates $\left(\boldsymbol{\gamma}^{**}_{p<1}\right)_m$ into $w_m\left[(\boldsymbol{\theta})_m\right]$ and $\boldsymbol{S}([\boldsymbol{\theta}]_m)$, by means of the relation

$$w_m\left[(\boldsymbol{\theta})_m\right] = \delta\left(\boldsymbol{\theta}_m - \left(\boldsymbol{\gamma}^{**}_{p<1}\right)_m\right), \left(\boldsymbol{\gamma}^{**}_{p=1}\right)_m \geq K'_{\mathrm{E}}, \qquad (6.3.56)$$

to obtain the required binary test for acceptance or rejection of the detection process, and including the presence of those estimates which exceed the threshold K'_{E}. The estimators are found from

$$\max_{\gamma_\theta} \to \boldsymbol{\gamma}^{**}_{p<1}\left\{\frac{w_M(\boldsymbol{\theta}) F_J(\boldsymbol{X}|\boldsymbol{\theta})_m}{F_J(\boldsymbol{X}|\boldsymbol{0})}\right\}_m = \boldsymbol{0}, \text{(2.1.54a) of Ref. [4];} \quad \left(\gamma^{**}_{p=1}\right)_m \geq K'_{\mathrm{E}}. \quad (6.3.57)$$

[We can also proceed as follows: we first apply to *all* the estimates

$$\max_{\gamma_\theta} \to \boldsymbol{\gamma}^{**}_{p<1}\left\{\frac{w_M(\boldsymbol{\theta}) F_J(\boldsymbol{X}|\boldsymbol{\theta})}{F_J(\boldsymbol{X}|\boldsymbol{0})}\right\}_M = \boldsymbol{0}, \quad \text{(2.1.54c) of Ref. [4]}, \qquad (6.3.57a)$$

and then select the m estimates which have the threshold K'_{E}.]

Finally, we need to discuss the thresholds K'_{E}, $K'_{\mathrm{D}} = K_{\mathrm{D}}/\mu$. If $K'_{\mathrm{E}} > K'_{\mathrm{D}}$ we have a situation where too few estimates are included and the binary detector is deprived of the needed number, while the reverse is the case for $K'_{\mathrm{E}} < K'_{\mathrm{D}}$. The solution is to equate the two:

$$K'_{\mathrm{E}} = K'_{\mathrm{D}} \text{ or } K_{\mathrm{E}} = pK_{\mathrm{D}}/\mu = (1-p)K_{\mathrm{D}}. \qquad (6.3.58)$$

This also exhibits the required limiting behavior:

$p = 1$: $K_{\mathrm{E}} = 0 = K_{\mathrm{D}}$: *no detection is needed, and* $\therefore$ *no threshold on the estimates.*
$p = 0$: $K_{\mathrm{E}} = K_{\mathrm{D}}$ and $K'_{\mathrm{E}} = K_{\mathrm{E}}/p \to \infty$, and $K'_{\mathrm{D}} = K_{\mathrm{D}}/\mu \to \infty$ $\therefore$ *no signal present and thus no estimator at all and only noise is passed. Therefore,*
H_0 is decided, consistent with $K'_{\mathrm{E}} \to \infty$.

(6.3.58a)

Equation (6.3.58) establishes the required relationship between the M-ary threshold K_{E} and the estimators, and the accept/reject threshold region in the binary detector. We note that only for the case of maximum likelihood estimators, where the dependence on p is linear, does this approach work. For example, it does not work for estimators based on the quadratic cost function, as then the threshold is not a linear function of p.

A simpler variant of the general case, where the cost functions are all constant in the binary detector, can be constructed. Equation (6.3.55) is now

$$D_{\text{Binary}} = \Lambda\left(\boldsymbol{X}, \left(\boldsymbol{\gamma}^{**}_{p<1}\right)_{m=1}\right) = \mu \int_{(\theta_m)} \frac{w_m\left([\boldsymbol{\theta}]_m\right) F_1\left(\boldsymbol{X}|(\theta)_m\right) d\boldsymbol{\theta}_m}{F_0(\boldsymbol{X}|\boldsymbol{0})} \begin{array}{|l} \geq K_{\text{D}}\colon H_1\colon S \otimes N \\ < K_{\text{D}}\colon H_0\colon N \end{array} \tag{6.3.58b}$$

but (6.3.56) remains, with (6.3.57) and (6.3.57a) as do the relations (6.3.58).

An example procedure for implementing this estimator-detector is the following. Select the kth Doppler scenario if it has one energy that exceeds all the rest ($\ell = 1, \ldots, k, \neq k$). The threshold for the estimator is K'_{E} and the maximum estimate above this threshold is regarded as a $p = 1$ estimate. Now, $K'_{\text{E}} = K'_{\text{D}}$ so there is *one* threshold for the two operations (D, E). This estimate is fed into the detector, with all other Dopplers in the kth scenario, since once one is known, the others are determined. The detector then employs $\log \Lambda(\boldsymbol{X}|\boldsymbol{S}); \gamma^{(1)}_{p=1}, \ldots, \gamma^{(k)}_{p=1}$ in making its decision. The environment against which it works is *not* the same for the estimator, hence for the same threshold $K'_{\text{D}}(=K'_{\text{E}})$ we obtain a certain false alarm probability $(\alpha_{\text{F}})_{\text{D}}$ which in turn leads to a certain detection probability p_{D}. For one acceptable value of $(\alpha_{\text{F}})_{\text{D}}, p_{\text{D}}$ may be too low, or it may be acceptable. If too low, lower the threshold $K'_{\text{D}}(=K'_{\text{E}})$ and get a larger value, at some larger false alarm rate. The reverse procedure is used if p_{D} is larger than just acceptable, and K'_{D} can then be made smaller. So, adjusting $K'_{\text{D}}(=K'_{\text{E}})$ between an acceptable value of $(\alpha_{\text{F}})_{\text{D}}$ may yield an acceptable p_{D}. Of course, it may not, and in such a case the detector rejects the estimate.

6.3.9 Summary Remarks

In the preceding three sections, we have presented in some detail a theory of the joint binary detection and estimation of signals and their parameters, when there is prior uncertainty as to the presence of the signal [1,2]. The basic approach is Bayesian, that is, fully implemented by *a priori* probabilities, and based on the methods of SDT. Here optimization is determined by averaging appropriately selected cost functions to obtain a minimum average cost or "risk" (i.e., a Bayes risk). Detection is now required, to permit the observer the choice of accepting or rejecting (or not using) the accompanying estimate, based on some preselected probability of a correct decision that the signal is truly present in the accompanying noise.

So far, our attention here has been directed to the generic fixed sample, or "single-shot" decision situation. This extension of the classical fixed sample theory (Chapters 1–5) generally leads to more complex detectors and estimators than the classical Bayes approaches provide, since now the detection and estimation processes can mutually interact and influence the two types of process. Even when there is no explicit coupling between the two, the fact that there is prior uncertainty regarding signal presence in the received data can strongly influence the resulting estimates, i.e., can create a significant and unknown bias in the estimate, unless the effect of $p(H_1) < 1$ is properly accounted for. Table 6.1 summarizes our results in the case of single, fixed sample sizes, showing the type of decision, choices of cost function, reference to detector and estimator structures, and the joint D and E strategies involved. As to the acceptance or rejection (or filing) of the

TABLE 6.1 **Joint Bayes Detection and Estimation, $p \leq 1$: H_1 Versus H_0**

No Coupling: (Section 6.2) Cost Functions (C.F.s)	Weak Coupling: (Section 6.3.2) Cost Functions	Strong Coupling: (Section 6.3.1) Cost Functions
I. Detection		
δ_1^*: $C_1^{(0)} = C_\alpha (> C_{1-\alpha})$ vs. (H_0, q)	δ_1^*: $C_{10} = C_\alpha + C'_{10}$ vs. (H_0, q)	δ_1^{**}: $C_{10}(\boldsymbol{\gamma_\theta}) = C_\alpha + f_{10}[\boldsymbol{\gamma}_\theta]$ vs. (H_0, q)
δ_0^*: $C_0^{(1)} = C_\beta (> C_{1-\beta})$ vs. (H_1, p)	δ_0^*: $C_{01}(\boldsymbol{\theta}) = C_\beta + f_{01}[\boldsymbol{S}(\boldsymbol{\theta})]$ vs. (H_1, p)	δ_0^{**}: $C_{01}(\boldsymbol{\theta}) = C_\beta + f_{01}[\boldsymbol{S}(\boldsymbol{\theta})]$ vs. (H_1, p)
δ_0^*: $C_0^{(0)} = C_{1-\alpha}; (H_0, q)$	δ_0^*: $C_{00} = C_{1-\alpha}; (H_0, q)$	δ_0^{**}: $C_{00} = C_{1-\alpha}; (H_0, q)$
δ_1^*: $C_1^{(1)} = C_{1-\beta}; (H_1, p)$	δ_1^*: $C_{11}(\boldsymbol{\theta}, \boldsymbol{\gamma_\theta}) = C_{1-\beta}; (H_1, p)$	δ_1^{**}: $C_{11}(\boldsymbol{\theta}, \boldsymbol{\gamma_\theta}) = C_{1-\beta} + f_{11}[\boldsymbol{\theta}, \boldsymbol{\gamma_\theta}]; (H_1, p)$
Structure + test $\log\Lambda \begin{cases} \geq \log K: H_1: \delta_1^* \\ < \log K: H_0: \delta_0^* \end{cases}$ $K = (C_\alpha - C_{1-\alpha})/(C_\beta - C_{1-\beta}) > 0$ Eqs. (6.1.6a) and (6.1.6b)	$\log \Lambda'_g \begin{cases} \geq 0: H_1: \delta_1^* \\ < 0: H_0: \delta_0^* \end{cases}$ Eq. (6.3.22)	$\log \Lambda_g \begin{cases} \geq 0: H_1: \delta_1^{**} \\ < 0: H_0: \delta_0^{**} \end{cases}$ Eq. (6.3.18)
II. Estimation		
$\underset{(H1)}{p}, q$: $C(\boldsymbol{\theta}, \boldsymbol{\gamma_\theta}) = C_0 \lvert\boldsymbol{\theta} - \boldsymbol{\gamma_\theta}\rvert^2$, QCF $= \begin{bmatrix} C_0\left\{A_L - \prod_{\ell=1}^{L} \cdot \delta(\gamma_\ell - \hat{\theta}_\ell)\right\} \\ C_0\left\{\sum_{\ell=1}^{L} \cdot [A_\ell - \delta(\gamma_\ell - \hat{\theta}_\ell)]\right\} \end{bmatrix}$	$(H_1), C'_{10}(> 0) = f_{10}$; vs. (H_0, q) $(H_0), C_{01}(> 0) = f_{01}(S(\theta))$; vs. (H_1, p)	$\boldsymbol{\gamma}_{\boldsymbol{\theta}}^{**}$: $f_{10}[\boldsymbol{\gamma_\theta}]$ vs. (H_0, q) $\boldsymbol{\gamma}_{\boldsymbol{\theta}}^{**}$: $f_{01}[\boldsymbol{S}(\boldsymbol{\theta})]$ vs. (H_1, p) $\boldsymbol{\gamma}_{\boldsymbol{\theta}}^{**}$: $f_{00} = 0; (H_0, q)$ $\boldsymbol{\gamma}_{\boldsymbol{\theta}}^{**}$: $f_{11} = f_{11}(\boldsymbol{\theta}, \boldsymbol{\gamma_\theta}); (H_1, p)$
$\underset{(H_0)}{q}$: $C'(\boldsymbol{\gamma_\theta}) = C(\boldsymbol{0}, \boldsymbol{\gamma_\theta})$ Eqs. (6.2.3), [(6.2.13a) and (6.2.13b)]		$\boldsymbol{\gamma}_{\boldsymbol{\theta}}^{**}$ into Λ_g, Eq. (6.3.18) Eq. (6.3.20)

$\boldsymbol{\gamma}^*_{p\le 1\|QCF} = \frac{\Lambda}{\Lambda+1}\boldsymbol{\gamma}^*_{p=1\|QCF}$: (6.2.7) $\boldsymbol{\gamma}^*_{p\le 1\|SCF} = p\boldsymbol{\gamma}^*_{p=1\|SCF}$: (6.2.31), (6.2.35)	q: $C_{10}(\boldsymbol{\gamma_\theta}) = C_\alpha + f_{10}(\boldsymbol{\gamma_\theta})$ vs. (H_0) p: $C_{11}(\boldsymbol{\theta},\boldsymbol{\gamma_\theta}) = C_{1-\beta} + f_{11}(\boldsymbol{\theta},\boldsymbol{\gamma_\theta}), (H_1)$ reduces to no-coupling cases if $\left.\begin{array}{l} q\text{: } C_{10}(\boldsymbol{\gamma_\theta}) \to C(0,\boldsymbol{\gamma_\theta}) = C'(\boldsymbol{\gamma_\theta}) \\ p,q\text{: } C_{11}(\boldsymbol{\theta},\boldsymbol{\gamma_\theta}) \to C(\boldsymbol{\theta},\boldsymbol{\gamma_\theta}); \end{array}\right\}$ QCF, $SCF_{1,2}$	
III. Bayes Risk $R^*_{D\otimes E}$ $\boldsymbol{R}^*_{\mathbf{D}\otimes\mathbf{E}} = \boldsymbol{R}^*_{\mathbf{D}} + \boldsymbol{R}^*_{\mathbf{E}}$	$\boldsymbol{R}^*_{\mathbf{D}\otimes\mathbf{E}} = \boldsymbol{R}^*_{\mathbf{D}} + \boldsymbol{R}^*_{\mathbf{E}}$ $\boldsymbol{R}^*_{\mathbf{D}}$ involves $\Lambda'_g = F(\langle f_{01}(\boldsymbol{S}(\boldsymbol{\theta}))\rangle_{\boldsymbol{\theta}})$	$\boldsymbol{R}^*_{\mathbf{D}\otimes\mathbf{E}}$ = D and E not separable
IV. Remarks D, E in any order E = detection-directed	$\left\{\begin{array}{l}\text{D and E uncoupled; in any order} \\ \text{E-C.F.s independent of D-C.F.s}\end{array}\right.$ E = detection-directed	E first, then D: D is estimation-directed (Problem 6.11)
V. 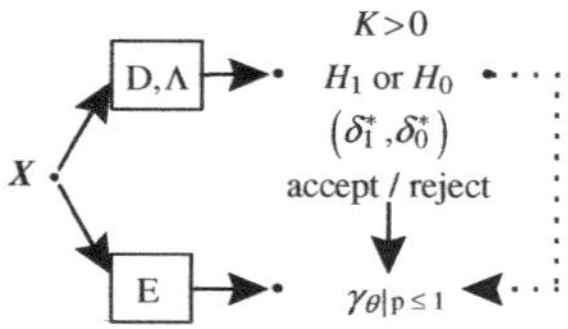 	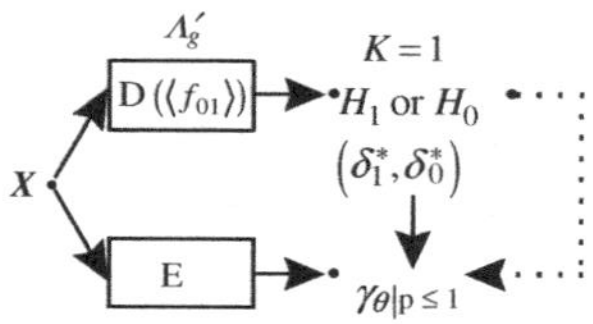 	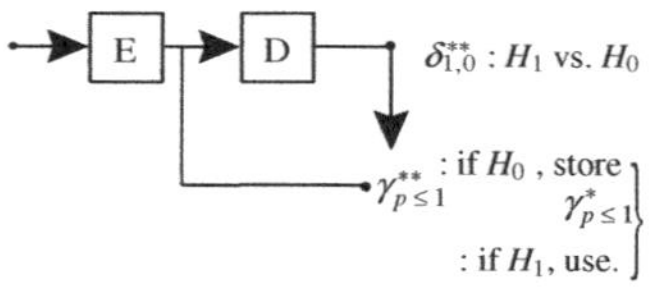

estimate following the decision H_0 of the detector, this is governed by what our choice may be of an acceptable probability (P_D or p_D) for the D and E process in question. This is discussed above in Section 6.3.6. We shall illustrate these methods with some specific analytic examples in Section 6.4 following.

6.3.9.1 Performance Degradation Finally, there is the question of the extent to which performance, that is, the estimation process here, is degraded by the familiar assumption that the quantities to be estimated are actually present, that is, that the signal containing them is present in the noise, $p(H_1) = 1, q(H_0) = 0$. Clearly, optimal estimators under $p = 1$ are not optimal under $H = H_0 + H_1 (p < 1)$. Thus, a natural measure of degradation compares the average errors ($\sim R_E$) incurred by using the optimum estimators for $p = 1$, now suboptimum under $H, p < 1$, with the Bayes errors $(\sim R_E^*)$ produced by the optimal estimators under the true state of operation, H, $(p \leq 1, q \geq 0)$. For this purpose we introduce the *relative Bayes error*, $\Delta R^*_{E|p\leq 1}$, defined by

$$\Delta R^*_{E|p\leq 1} \equiv \left(R_{E|p\leq 1} - R^*_{E|p\leq 1}\right)/R^*_{E|p\leq 1} \qquad (\geq 0), \tag{6.3.59}$$

now with the suboptimum estimators specified by $\boldsymbol{\gamma}_{\theta|p\leq 1} = \boldsymbol{\gamma}^*_{\theta|p=1}$ here.

In general, $R^*_{E|p\leq 1}$ is usually more difficult to evaluate, cf. Section 6.4.2.3, than $R_{E|p\leq 1}$. This suggests that we use a simpler criterion of performance degradation. One such is the *fractional mean difference* or *mean relative error* between the estimators $\boldsymbol{\gamma}^*_{\boldsymbol{\theta}|p\leq 1}$ and $\boldsymbol{\gamma}_{\theta|p\leq 1}\left(= \boldsymbol{\gamma}^*_{\boldsymbol{\theta}|p=1}\right)$, defined by

$$D\boldsymbol{\gamma}^*_{\boldsymbol{\theta}} \equiv \left[\left\langle\left(\boldsymbol{\gamma}^*_{\boldsymbol{\theta}|p=1}\right)_\ell\right\rangle_H - \left\langle\left(\boldsymbol{\gamma}^*_{\boldsymbol{\theta}|p\leq 1}\right)_\ell\right\rangle_H \Big/ \left\langle\left(\boldsymbol{\gamma}^*_{\boldsymbol{\theta}|p\leq 1}\right)_\ell\right\rangle_H\right], \qquad \ell = 1, \ldots, L. \tag{6.3.60}$$

When $\boldsymbol{\gamma}^*_{\boldsymbol{\theta}|p\leq 1}$ is unbiased under H, i.e., $\left\langle\boldsymbol{\gamma}^*_{\boldsymbol{\theta}|p=1}\right\rangle_H = p\langle\boldsymbol{\theta}\rangle$, (6.3.43b), as is the case for the QCF and $SCF_{1,2}$ with no coupling, the mean relative error (6.3.60) becomes

$$D\boldsymbol{\gamma}^*_{\boldsymbol{\theta}} = \left[\frac{\left\langle\left(\gamma^*_{\theta|p=1}\right)_\ell\right\rangle_H - p\langle\theta_\ell\rangle}{p\langle\theta_\ell\rangle}\right] = \left[\frac{\left\langle\left(\gamma^*_{\theta|p=1}\right)_\ell\right\rangle}{p\langle\theta_\ell\rangle} - 1_\ell\right], \quad \ell = 1, \ldots, L. \tag{6.3.60a}$$

Similar relations can be employed to compare the performance of a pair of suboptimum systems when detection and estimation are separable, namely,

$$\Delta R^{(21)}_{E|p\leq 1} = \left|R^{(2)}_{E|p\leq 1} - R^{(1)}_{E|p\leq 1}\right| / R^{(1)}_{E|p\leq 1} \qquad (\geq 0), \tag{6.3.61}$$

with the corresponding form for (6.3.60a).

Finally, we observe that when detection and estimation are coupled, cf. Section 6.3.1 above, measures of degradation like (6.3.59)–(6.3.61) must be suitably extended, to include the effects of the joint detection process. This means that $R_{E|p\leq 1}$, $R^*_{E|p\leq 1}$ (6.3.59), and so on, are to be replaced by $R_{D\otimes E|p\leq 1}$, $R^*_{D\otimes E|p\leq 1}$, vide (6.3.14) et. seq. Again, the choice of the coupling cost functions C_{10}, C_{11} (6.3.1) is critical. Theory (and application) here remain to be developed. In the above, whether D and E are coupled or not, we must not forget the requirement of an accompanying detection process, discussed in Section 6.3.6, which establishes our acceptance or rejection of the estimate in question. We shall encounter some illustrative examples in Section 6.4 following.

PROBLEMS

P6.4 Obtain the Hessian associated with Eq. (6.3.10b) and with (6.3.13), (6.3.15), (6.3.20), and verify that the resulting estimators $\boldsymbol{\gamma}^*, \boldsymbol{\gamma}^{**}$ do indeed minimize the Bayes risk (6.3.7). Discuss the conditions necessary and sufficient for this purpose.

P6.5 Show that the optimum detection rules $\delta_0^{**}, \delta_1^*$, (6.3.16), (6.3.17), are nonrandomized.

P6.6 Obtain the results Eqs. (6.3.23) and (6.3.24) of the text, and give the condition that the results of (6.3.27) are optimal, that is, yield the minimum (average or Bayes) risk R_{E}^* here.

P6.7 If the cost function $C(\boldsymbol{\theta}, \boldsymbol{\gamma}_{\boldsymbol{\theta}})$ in Eq. (6.3.34) does not posses a Taylor's expansion, show that Sherman's theorem ([7]; [4], p. 970) cannot be attained.

P6.8 Show that for the MMSE vector estimate $\boldsymbol{\gamma}_S^*$, (6.3.43c), of the received waveform $\boldsymbol{S}$, (6.2.2b), $p \leq 1$, is

$$\left\langle \boldsymbol{\gamma}^*(\boldsymbol{S})_{p\leq 1|\mathrm{QCF}} \right\rangle_{H=H_0+H_1} = p\langle \boldsymbol{S}\rangle_H. \tag{1}$$

P6.9 Obtain the class of cost functions (f_{10}, f_{11}) in the weak coupling cases, $p \leq 1$, such that the resulting Bayes estimators are unbiased. (See Section 6.3.4.2)

P6.10 Explore the condition on the cost function in the strong coupling cases $(p \leq 1)$, such that the resulting Bayes estimators remain unbiased.

P6.11 **(a)** Starting with the average risk (6.3.4) for the strong coupling cases (Section 6.3.1), show that

$$R_{\mathrm{D}\otimes\mathrm{E}} = \int_\Gamma A_\sigma(\boldsymbol{X})d\boldsymbol{X} + \int_\Gamma d\boldsymbol{X}\delta(\boldsymbol{\gamma}_1|\boldsymbol{X})\{B_\sigma(\boldsymbol{\gamma}, \boldsymbol{X}) - A_\sigma(\boldsymbol{X})\}, \tag{1}$$

where

$$\left.\begin{aligned} B_\sigma(\boldsymbol{\gamma}, \boldsymbol{X}) &= qC_{10}F_0(\boldsymbol{X}) + pC_{1-\beta}\int_\Omega w_L(\boldsymbol{\theta})F_1(\boldsymbol{X}|\boldsymbol{\theta})d\boldsymbol{\theta} \\ &\quad + p\int_\Omega f_{11}[\boldsymbol{S}(\boldsymbol{\theta}), \boldsymbol{\gamma}_\theta(\boldsymbol{X})]F_1(\boldsymbol{X}|\boldsymbol{\theta})w_L(\boldsymbol{\theta})d\boldsymbol{\theta} + qf_{10}(\boldsymbol{\gamma}(\boldsymbol{X})) \end{aligned}\right\}, \tag{2a}$$

and

$$A_\sigma(\boldsymbol{X}) = \delta_0(H_0|\boldsymbol{X})\left\{qC_{1-\alpha}F_0(\boldsymbol{X}) + pC_\beta\int_\Omega \{F_1(\boldsymbol{X}|\boldsymbol{\theta}) + f_{01}(\boldsymbol{S}(\boldsymbol{\theta}))F_1(\boldsymbol{X}|\boldsymbol{\theta})\}w_L(\boldsymbol{\theta})d\boldsymbol{\theta}\right\}. \tag{2b}$$

The choice of cost functions ensures that A_σ, B_σ are always positive.

(b) For the Bayes risk from the estimation process, show that $\boldsymbol{\gamma} = \boldsymbol{\gamma}^*$ is determined by

$$\frac{\partial B_\sigma(\boldsymbol{\gamma}, \boldsymbol{X})}{d\boldsymbol{\gamma}}\Big|_{\boldsymbol{\gamma}=\boldsymbol{\gamma}^*} = \mathbf{0}, \tag{3}$$

so that even with strong coupling the structure of the estimator is identical to that derived for the case of weak coupling [cf. Eq. (6.3.27) and remarks following it].

(c) Show, also, that the optimum decision rules $\delta_{0,1}^{**}$ for the detection portion of these coupled joint D and E operations are

$$\left.\begin{array}{l} \text{Decide } \delta^{**}(H_1|\boldsymbol{X}) = 0\text{: no signal, if } B_\sigma - A_\sigma > 0 \\ \text{Decide } \delta^{**}(H_1|\boldsymbol{X}) = 1\text{: signal present, if } B_\sigma - A_\sigma \leq 0 \end{array}\right\}. \tag{4}$$

The detector's structure is in general very complex. This is evident even when one employs the simple cost assignments $C_{1-\alpha} = C_{1-\beta} = C_{00} = f_{10} = 0$. Then, show that the resulting likelihood ratio becomes

$$\Lambda''_g = \frac{C_\beta}{C_\alpha}\Lambda + \frac{1}{C_\alpha}\int_\Omega f_{01}[\boldsymbol{S}(\boldsymbol{\theta})] - f_{11}\left[\boldsymbol{S}(\boldsymbol{\theta}), \boldsymbol{\gamma}_\theta^*(\boldsymbol{X})\right]\hat{\Lambda}(\boldsymbol{X}|\boldsymbol{\theta})w_L(\boldsymbol{\theta})d\boldsymbol{\theta}, \tag{6}$$

which is a generalization of Λ'_g, Eq. (6.3.22). Show also that the strong coupling case of Section 6.3.1, cf. Eq. (6.3.18), reduces to the weak coupling case Λ'_g, Eq. (6.3.22), on setting $C_{00} = 0$; $f_{01} = C_{01} + (\boldsymbol{S}(\boldsymbol{\theta}))$; $f_{10} = C_{10}$, independent of $\boldsymbol{\gamma}$, with $C_\alpha = C_{1-\beta}$, in Eq. (1) above.

P6.12 Carry out the various steps indicated in the explicit development of the joint, strongly coupled D and E formalism outlined in Section 6.3.7 for estimation of waveform, $\boldsymbol{S}$.

6.4 JOINT D AND E: EXAMPLES–ESTIMATION OF SIGNAL AMPLITUDES [$P(H_1) \leq 1$]

The purpose here is to illustrate the general theory of joint detection and estimation, outlined in preceding Sections, with a selection of specific examples which are both analytically tractable and useful. Accordingly, theses examples are limited to the no-coupling cases treated generally in Section 6.2 above. In addition, these examples provide some quantitative as well as qualitative insight into the degradation of the estimation process when there is uncertainty regarding the presence of the signal in the received data sample, namely, when the *a priori* probability $p(H_1)$ is less than unity. Accordingly, one of the important desired results here is the dependence of the accuracy of estimation on $p(H_1)$ when $p(H_1) < 1$, as well as when $p(H_1) = 1$. Again, when $p < 1$, acceptance of the estimates depends on an acceptable probability P_D of correct detection of signal presence in the noise, cf. Section 6.3.6 above.

The simplest illustrative and useful examples involve estimation of (normalized) signal amplitude a_o and signal intensity $I_0 = a_0^2$, under the following specific conditions:

(i) Additive signal and homogeneous and stationary normal noise $N(r,t)$, with zero mean $\langle N(r,t)\rangle = 0$, intensity $\psi = \langle N^2\rangle$, and covariance $K_N(r_2 - r_1, t_2 - t_1)$.
(ii) Coherent observation, that is, signal epoch ε_0 is known a priori.
(iii) All other features of the signal are also *a priori* specified.
(iv) Estimation (E) and detection (D) are uncoupled, *cf*. Table 6.1.
(v) Signal amplitude a_0 is normally distributed about $\overline{a_0}$, with variance $\sigma^2 = \overline{a_0^2} - \overline{a_0}^2$ and $-\infty \leq a_0 \leq \infty$.
(vi) Signal and noise fields are uniform over the receiving array or aperture.

(6.4.0)

These conditions are not so academic as they may appear at first glance. Normal noise is quite prevalent; so is amplitude fading, from small amounts to large; $\overline{a_0} = 0$ is a common phenomenon, and a Gaussian model here is often reasonable. Coherent reception is achievable in many applications. Although one is dealing generally with space–time fields in practical cases (vide Chapter 1 et seq.), one important subclass of operation occurs when noise and signal fields can be treated as being essentially uniform over the receiving array or aperture, that is, $\psi^{(m)} = \psi$ and $a_0^{(m)} = a_0$. Conditions (iii) and (vi) are invoked simply to keep the treatment less complex, without in any way diluting the essential concepts involved. Likewise, condition (iv) allows a fully analytic implementation of the basic analysis and serves as the starting point for more advanced, and largely numerical investigations.

Accordingly, under (i)–(v) above, we can write for the pdf of signal amplitude a_0 explicitly

$$\textit{Condition (v)}: \qquad w_1(a_0) = \left(2\pi\sigma^2\right)^{-1/2} e^{-(a_0-\overline{a_0})^2/2\sigma^2}, \sigma^2 = \overline{a_0^2} - \overline{a_0}^2, \tag{6.4.1}$$

$$\text{with } \lim_{\sigma^2 \to 0} w_1(a_0) = \delta(a_0 - \overline{a_0}). \tag{6.4.1a}$$

The pdf of the $J(= MN)$ samples of additive signal ($a_0\boldsymbol{s}$) and normal noise ($\boldsymbol{N}$) gives the well-known relation[29]

$$\textit{Condition (i)–(iii), (vi)}: \quad \begin{aligned} &F_J(\boldsymbol{x}|a_0\boldsymbol{s}) = w_J(\boldsymbol{x} - a_0\boldsymbol{s}) = \frac{e^{-(\tilde{\boldsymbol{x}}-a_0\tilde{\boldsymbol{s}})\boldsymbol{k}_N^{-1}(\boldsymbol{x}-a_0\boldsymbol{s})/2}}{\left[(2\pi)^J \det \boldsymbol{k}_N\right]^{1/2}}; \\ &\left.\begin{aligned} \theta = a_0 s = S/\sqrt{\psi} \\ x = X/\sqrt{\psi} \end{aligned}\right\}, \end{aligned} \tag{6.4.2}$$

where now $\theta = a_0$ is the (only) quantity to be estimated here, for example, $L = 1$, and $\boldsymbol{k}_N = \boldsymbol{K}_N/\psi = [K_N(\boldsymbol{r}_m - \boldsymbol{r}_{m'}, t_n - t_{n'})/\psi]$, $[j,j'] = (mn, m'n') \leq (J, J')$, is the normalized noise covariance. With no signal (6.4.2) reduces at once to

$$w_J(\boldsymbol{x}) = \frac{e^{-\left(\tilde{\boldsymbol{x}}\boldsymbol{k}_N^{-1}\boldsymbol{x}\right)/2}}{\left[(2\pi)^J \det \boldsymbol{k}_N\right]^{1/2}} = w_J(\boldsymbol{x}|\boldsymbol{0}) = [F_J(\boldsymbol{x}|\boldsymbol{S} = \boldsymbol{0})], \quad \text{cf. (6.2.34a)}, \tag{6.4.3}$$

which is needed in determining the specific detection statistic (6.1.6b) in the detection phase of the joint D and E operation here.

Combining (6.4.1) and (6.4.3) then gives

$$w_{J\times 1}(\boldsymbol{x}, a_0) = \frac{\exp\left\{-(1/2)(\tilde{\boldsymbol{x}}\text{-}a_0\tilde{\boldsymbol{s}})\boldsymbol{k}_N^{-1}(\boldsymbol{x} - a_0\boldsymbol{s}) - (1/2)(a_0 - \bar{a}_0)^2/2\sigma^2\right\}}{\left[2\pi\sigma^2 \cdot (2\pi)^J \det \boldsymbol{k}_N\right]^{1/2}}. \tag{6.4.4}$$

This relation is also needed in determining the estimators of the scale parameter a_0 under the familiar condition $p(H_1) = 1$, both for the minimum mean square error estimator from

[29] Note that here, and throughout Section 6.4, we use the normalized data form $\boldsymbol{x} = \boldsymbol{X}/\sqrt{\psi}$.

the QCF and for the unconditional maximum likelihood estimator from the "simple" cost functions ($\text{SCF}_{1,2}$). [These latter, as seen above in (6.2.13a) and (6.2.13b) are equivalent here.] The desired Bayes estimators $a^*_{p\leq 1}$ then follow directly from (6.2.7) and (6.2.31), respectively. A second, more complex evaluation is that of the minimum expected (Bayes) error in the estimator $a^*_{p\leq 1}$, obtained from the relations (6.2.10), (6.2.11) for the QCF and from (6.2.20) or (6.2.37), all for the SCF, with $L = 1$.

6.4.1 Amplitude Estimation,[30] $p(H_1) = 1$

We begin with a summary treatment of the classical cases where $p(H_1) = 1$. The results are then to be applied to the general situation of signal uncertainty, $p < 1$.

6.4.1.1 Bayes Estimators From (6.4.4) in (6.2.5) with $q = 0, \hat{\theta} = \theta = a_0$, on completing the square (in a_0) we readily obtain the earlier result for the MMSE (Bayes) estimator of a_0:

$$p = 1: \quad \gamma^*(a_0|\mathbf{x})_{\text{QCF}} \equiv a^*_{p=1}(\mathbf{x})_{\text{QCF}} = \frac{\sigma^2\Phi_x + \bar{a}_0}{\sigma^2\Phi_s + 1}, \text{ with } \Phi_x \equiv \tilde{\mathbf{x}}\mathbf{k}_N^{-1}\mathbf{s}, \Phi_s \equiv \tilde{\mathbf{s}}\mathbf{k}_N^{-1}\mathbf{s}, \tag{6.4.5}$$

cf. (21.97) of Ref. [4] and Problem 6.13. Here $\mathbf{x} = \mathbf{n} + \boldsymbol{\theta} = \mathbf{n} + a_0\mathbf{s}$, with $\mathbf{n} = \mathbf{N}/\sqrt{\psi}$ for the normalized noise.

In a similar fashion we may use the following modified forms of (6.2.32), (6.2.34a), with (6.4.3), to obtain the UMLE of a_0, namely,

$$p = 1: \quad \left[\frac{\partial}{\partial\gamma_1}\left(\log \mu^{-1}\hat{\Lambda}_1\right)\right]_{\gamma_1=\gamma_1^*} = 0 \quad \text{or} \quad \left[\frac{(\partial/\partial a_0)\log W_{J\times 1}(\mathbf{x}, a_0)}{W_J(\mathbf{x}|\mathbf{0})}\right]_{a_0=a_0^*} = 0, L = 1, \tag{6.4.6a}$$

which becomes

$$\left[\frac{\partial}{\partial a_0}\left\{-\frac{1}{2}\left[-2a_0\Phi_x + a_0^*\Phi_s\right] - \frac{1}{2}(a_0 - \bar{a}_0)^2/\sigma^2\right\}\right]_{a_0=a_0^*} = 0, \tag{6.4.6b}$$

or

$$p = 1: \ \gamma * (a_0|\mathbf{x})_{\text{SCF}_{1,2}} \equiv a_0(\mathbf{x})^*_{\text{SCF}} = \frac{\sigma^2\Phi_x + \bar{a}_0}{\sigma^2\Phi_s + 1} = a_0(\mathbf{x})^*_{\text{QCF}}, \text{cf. (6.4.5)}. \tag{6.4.6c}$$

Thus, for $p = 1$ the UMLE and MMSE of amplitude a_0 are equal. Moreover, when $\sigma^2 \to \infty$, then $a_0^*|_{\text{SCF,QCF}} = \Phi_x/\Phi_s = a_0^*|_{\text{minimax}}$: the *a priori* pdf of a_0 is uniform, cf. (21.72) of Ref. [4]. On the other hand, when $\sigma^2 \to 0$, $a_0^*|_{\text{SCF,QCF}} = \bar{a}_0$, as expected, since this is the only value of a_0^* and therefore no estimation is needed, only detection.

6.4.1.2 Pdfs of $a^*_{p=1}(\mathbf{x})$ To evaluate performance, as represented by the Bayes risk or error R^*_{E}, for $p = 1$ and subsequently for $p < 1$ (Section 6.4.2 ff.), we shall require various

[30] For further analytic details and discussion here, see Section 21.3.1 of Ref. [4].

first-order pdfs of $a^*_{p=1}(=y)$, among them $w_1(y|H_0)$, $w_1(y|H_1)$, and $w_1(y|a_0)$. These are readily obtained for the present example, since from (6.4.5), and (6.4.6c), $a^*_{p=1}$ is a *linear* function of the data set $\boldsymbol{x}$ and is consequently itself normally distributed, in as much as both $\boldsymbol{x}$ and a_0 are postulated to obey normal statistics, cf. (6.4.1), (6.4.3), (6.4.4). We note, *en passant* and also as expected, that $a^*_{p=1}$, (6.4.6c) is *unbiased* ($p=1$):

$$\langle a^*_{p=1}\rangle_{H_1} = \frac{\left[\sigma^2\left\langle(\tilde{\boldsymbol{n}}+a_0\tilde{\boldsymbol{s}})\boldsymbol{k}_N\boldsymbol{s}\right\rangle_{\boldsymbol{n},a_0} + \bar{a}_0\right]}{(\sigma^2\Phi_s+1)} = \bar{a}_0. \tag{6.4.7}$$

Moreover, in the strong-signal cases, that is, when $\sigma^2\Phi_s >> 1(\text{and}\,\sigma^2\Phi_s >> \bar{a}_0)$ it is readily shown that $a^*_{p=1}$ is $a_0\cdot\left(1+\mathrm{O}\left[(\sigma^2\Phi_s)^{-1}\right]\right)$, cf. Problem 6.13b.

We begin with the first-order pdfs of $y=a_0(\boldsymbol{x})^*_{p=1}$ and $y-a_0$, under H_1, which are found to be (Problem 6.13),

$$\left.\begin{aligned} w_1(y|H_1) &\equiv w_1\left(a^*_{p=1}|H_1\right) = \frac{e^{-\left(a^*_0-\bar{a}_0\right)^2/2\sigma^2_0}}{\left(2\pi\sigma^2_0\right)^{1/2}}, \text{ with } \langle a^*_0\rangle = \bar{a}_0, \sigma^2_0 \equiv \overline{a^2_0} - \bar{a}^2_0 = \frac{\sigma^4\Phi_s}{(1+\sigma^2\Phi_s)} \\ &= \left(\frac{\lambda}{2\pi\sigma^4\Phi_s}\right)^{1/2} e^{-\lambda(y-\bar{a}_0)^2/2\sigma^4\Phi_s}, \lambda \equiv 1+\sigma^2\Phi_s. \end{aligned}\right\} \tag{6.4.8}$$

The first and second moments of y under H_1 are respectively

$$\langle y\rangle_{H_1} = \bar{a}_0, \left\langle y^2\right\rangle_{H_1} = \bar{a}^2_0 + \frac{\sigma^4\Phi_s}{(1+\sigma^2\Phi_s)}. \tag{6.4.8a}$$

Similarly, the first-order pdf of $\Delta\, a^*_0\left(\equiv a^*_{p=1}-a_0\right)$ becomes (Problem 6.13)

$$w_1(y-a_0|H_1) \equiv w_1\left(\Delta\, a^*_0|\boldsymbol{x}\right)_{p=1} = \frac{e^{-\left(\Delta a^*_0\right)^2/2\Delta\sigma^{*2}_0}}{\left(2\pi\Delta\sigma^{*2}_0\right)^{1/2}}, \text{ with } \Delta\sigma^{*2}_0 = \frac{\sigma^2}{(\sigma^2\Phi_s+1)}, \tag{6.4.9a}$$

$$= \left(\frac{2\pi\sigma^2}{\lambda}\right)^{-1/2} e^{-\lambda(y-a_0)^2/2\sigma^2}, \tag{6.4.9b}$$

which reduces as $\sigma^2\to\infty$ to the minimax pdf

$$\lim_{\sigma^2\to\infty} w_1\left(\Delta\, a^*_0|\boldsymbol{x}\right)_{p=1} = \left(\frac{\Phi_s}{2\pi}\right)^{1/2} e^{-\left(a^*_0-\Phi_x/\Phi_s\right)^2\Phi_s/2} = w_1\left(y-a_0|H_1\right)\big|_{\sigma^2\to\infty}, \tag{6.4.10}$$

from (21.103c) of Ref. [4]. Other needed pdfs of $y\left[=a^*_0(\boldsymbol{x})_{p=1}\right]$ may be obtained in the same way [cf. Problem 6.13]. We have

$$w_1(y|H_0) = \left(2\pi\sigma^4\Phi_s\right)^{-1/2}\lambda e^{-(\lambda y-\bar{a}_0)^2/2\sigma^4\Phi_s}, y\equiv a^*_{p=1}, \tag{6.4.11}$$

and with the help of (6.4.4), we obtain

$$w_1(y|a_0)_{H_1} = \frac{w_1(y, a_0|H_1)}{w_1(a_0)} = \left(2\pi\sigma^4\Phi_s\right)^{-1/2} \lambda e^{-\left(\lambda y - \bar{a}_0 - a_0\sigma^2\Phi_s\right)^2/2\sigma^4\Phi_s} \tag{6.4.12}$$

and

$$w_1(y, a_0|H_1) = \lambda\left(2\pi\sigma^3\Phi_s\right)^{-1} e^{-\left[\left(\lambda y - \bar{a}_0 - a_0\sigma^2\Phi_s\right)^2 + (\bar{a}_0 - a_0)^2\sigma^2\Phi_s\right]/2\sigma^4\Phi_s}. \tag{6.4.13}$$

(The associated characteristic functions are specified in Problem 6.14, Table 6.2.)

The quantity $\sigma^2\Phi_s = \left(\overline{a_0^2} - \bar{a}_0^2\right)\tilde{\mathbf{s}}\mathbf{k}_N\mathbf{s}$, cf. (6.4.5), has an important physical interpretation for the present (coherent) D and E example: it is the "*output signal-to-noise ratio for estimation*," namely $(S/N)^2_{\text{out}|\text{E}}$ for this example. Thus, using the *fading factor* η_0, defined by

$$\eta_0 \equiv 1 - \frac{\bar{a}_0^2}{\overline{a_0^2}}, \text{ so that } \sigma^2 = \eta_0\overline{a_0^2}, \quad \therefore \quad \bar{a}_0^2 = (1 - \eta_0)\overline{a_0^2}, \tag{6.4.14}$$

we can write

$$\sigma^2\Phi_s = \eta_0\overline{a_0^2}\Phi_s \equiv \left(\frac{S}{N}\right)^2_{\text{out}|\text{E}} = \eta_0\Phi_s\left(\frac{S}{N}\right)^2_{\text{in}} \equiv \Pi^*_{\text{E-coh}}\left(\frac{S}{N}\right)^2_{\text{in}}; \left(\frac{S}{N}\right)^2_{\text{in}} \equiv \overline{a_0^2}. \tag{6.4.15}$$

The quantity $\Pi^*_{E\text{-}coh}(= \eta_0\Phi_s)$ is a *processing gain* associated with coherent estimation here. The limits $(0, 1)$ within which η_0 is confined need clarification. For "deep fading," $\overline{a_0} = 0$ and $\therefore \eta_0 = 1, \sigma^2 = \overline{a_0^2}$. As we shall see presently (Section 6.4.3), detection can be carried out at all levels of $(S/N)^2_{\text{in}} = \overline{a_0^2}(> 0)$, so that $\eta_0 = 1$ is a valid value of the fading factor for the present example of this section, including the accompanying estimation process. On the other hand, for $\eta_0 = 0$, that is, $\sigma^2 = 0$, we have the case where $a_0 = \bar{a}_0$, and so on, from the limiting pdf $w_1(a_0) = \delta(a_0 - \bar{a}_0)$, (6.4.1a), which states that the parameter a_0 to be estimated is in fact *a priori* known, cf. remarks after Eq. (6.4.6c). There is thus no longer an estimation problem, as reflected in the fact that now $(S/N)^2_{\text{out}|\text{E}} = 0$, but only one of detection, cf. Section 6.4.3 ff. Accordingly, for joint detection and estimation here η_0 must satisfy the condition

$$0 < \eta_0 \leq 1, \tag{6.4.16}$$

with $(S/N)^2_{\text{out}|\text{E}}$ defined in (6.4.15) above.

6.4.1.3 Bayes Risk: Minimum Average Error

In order to compare our results when $p < 1$ with the classical cases $(p = 1)$ we need the latter to begin with. Most of these are available in Section 21.3.1 of Ref. [4]. We have (Problem 6.15):

Eq. (21.102*c*):
$$R^*_{\text{E}|\text{QCF}}\big|_{p=1} = C_0\left[\left\langle a_0^{*2}\right\rangle_{H_1} - 2\left\langle a_0^* a_0\right\rangle_{H_1} + \overline{a_0^2}\right] = \frac{C_0\sigma^2}{(1 + \sigma^2\Phi_s)}, \tag{6.4.17a}$$

Eq. (21.102*d*): $$R^*_{\mathrm{E|QCF:\ minimax}} = C_0 \Phi_s^{-1}, p = 1; \tag{6.4.17b}$$

Eqs. (6.2.20), (6.2.37):
$$\begin{aligned} R^*_{\mathrm{E|SCF}}\big|_{p=1} &= C_0\left[A_1 - w_1(y, a_0|H_1)\big|_{a_0^*=a_0=0}\right] \\ &= C_0\left[A_1 - \frac{(1+\sigma^2\Phi_s)}{2\pi\sigma^3\Phi_s^{1/2}} e^{-\bar{a}_0^2(1+\sigma^2\Phi_s)/2\sigma^4\Phi_s}\right], \end{aligned} \tag{6.4.17c}$$

this last with the help of (6.4.13). We note that although $a^*_{0\text{-QCF}} = a^*_{0\text{-SCF}}$ here, vide (6.4.6a), the corresponding Bayes risks, or minimum average errors, are generally quite different, as a consequence of the strongly different cost functions chosen. As expected, $R^*_{\mathrm{E}} \to 0$ when $\sigma \to 0$, since $a_0^* \to \bar{a}_0$, a known quantity. Also, $\left(R^*_{\mathrm{E|SCF}}\right)_{\mathrm{minimax}} \doteq \left(2\pi\sigma\Phi_S^{1/2}\right)^{-1}, \sigma^2 \gg 1$, here, with $\exp\left(-\bar{a}_0^2/2\sigma^2\Phi_s\right) \doteq 1$, cf. (6.4.17b) and Section 6.4.2.2 ff.

6.4.2 Bayes Estimators and Bayes Error, $p(H_1) \leq 1$

We are now ready to extend the analyses of our illustrative examples of Section 6.4.1 above to the critical cases when $p(H_1) \leq 1$, namely, estimation of signal amplitude when the signal is not surely known to be present in the noise. In addition, we confine our attention to the uncoupled cases of Section 6.2, where estimation is detection-directed, but where both operations can be independently carried out. Accordingly, for this we need first the optimum detector structure, which is obtained by applying (6.4.4) to the likelihood ratio (6.1.6b). This in turn gives us the Bayes "on–off" detection processor expressed in terms of the Bayes estimators $a_0^*(\boldsymbol{x})_{\mathrm{QCF,\ SCF}}$ (6.4.6c). Thus we write

$$\left.\begin{aligned} \Lambda(\boldsymbol{x}) &= \mu \int_{[a_o]} \frac{\text{Eq.}(6.4.4)\, da_o}{\text{Eq.}(6.4.3)} \\ &= \frac{\mu}{\sqrt{1+\sigma^2\Phi_s}} e^{-\bar{a}_0^2/2\sigma^2 + (1+\sigma^2\Phi_s)a^*_{p=1}(x)^2/2\sigma^2}, \ (\mu = p/q); \sigma > 0 \\ &= \frac{\mu}{\sqrt{1+\sigma^2\Phi_s}} e^{-\bar{a}_o^2/2\sigma^2 + (\sigma^2\Phi_x + \bar{a}_o)^2/2\sigma^2(1+\sigma^2\Phi_s)} \ . \end{aligned}\right\} \tag{6.4.18}$$

Equation (6.1.6) provides the desired optimal binary test for signal presence or absence in the usual way. Equation (6.4.18) applies in the uncoupled situation: no coupling between detection and estimation. For the cases of weak coupling (Section 6.3.2), the more complicated likelihood ratio $\Lambda'_g(\boldsymbol{x})$ (6.3.22), must be employed, namely,

$$\Lambda'_g(\boldsymbol{x}) = K'\Lambda(\boldsymbol{x}) + \mu \int_{[a_o]} f_{01}[\boldsymbol{S}(a_o)] \frac{\dfrac{\text{Eq.}(6.4.4)}{\text{Eq.}(6.4.3)}}{(C_\alpha - C_{1-\alpha} + C'_{10})} da_o, \tag{6.4.19}$$

which depends of course on our choice of f_{01}. We shall limit our present analysis here to the uncoupled D and E modes. In any case, the procedures of Section 6.3.6 above govern detection and the acceptance or rejection of the associated estimates.

Finally, when $\sigma \to 0$ (no fading), (6.4.1a) applies, with $\eta_o = 0$, cf. (6.4.14) and $\therefore a^*_{p=1} = a_o = \bar{a}_o$ are known *a priori*. Consequently there is no need for the estimation process, and the Bayes detector (6.4.18) here then reduces to the exact and well-know classical result[31]

$$Eq.\,(20.117a), [4]: \quad \Lambda(\boldsymbol{x})|_{\sigma=0} = \mu e^{-a_o^2\Phi_s/2 + a_o\Phi_x}, \quad \text{with } a_o = \bar{a}_o = \sqrt{a_o^2}, \text{etc.} \tag{6.4.19a}$$

In fact, we obtain from (6.4.18) through $O(\sigma^4)$ the result

$$\begin{aligned}\log \Lambda = &\left(\log \mu - \bar{a}_o^2\Phi_s/2\right) + \bar{a}_o\Phi_x + \frac{\sigma^2}{2}\left[(\Phi_x - \bar{a}_o\Phi_s)^2 - \Phi_s\right] \\ &- \frac{\sigma^4\Phi_s}{2}\left[(\Phi_x - \bar{a}_o\Phi_s)^2 - \frac{\Phi_s}{2}\right] + O(\sigma^6),\end{aligned} \tag{6.4.19b}$$

which shows the complex structure (in Φ_x) of the threshold expansion (in σ^2). [See the remarks following Section 6.4.4 ff.]

6.4.2.1 *Bayes Estimators, p* ≤ 1 For the Bayes estimators of amplitude $a^*_{p\leq 1}$ here we apply the results (6.4.6c) under the condition $p = 1$ for $a^*_{p\leq 1|\text{SCF}}$ to (6.2.31), $L = 1$, or (6.2.35), and (6.4.5) for $a^*_{p\leq 1|\text{QCF}}$ to (6.2.7), to write at once

$$\text{SCF}_{1,2}: \qquad a^*_{p\leq 1}(\boldsymbol{x})_{\text{SCF}} = p\left(\frac{\sigma^2\Phi_s + \bar{a}_o}{\sigma^2\Phi_s + 1}\right) = pa^*_{p=1},\ 0 \leq p \leq 1; \tag{6.4.20}$$

$$\begin{aligned}\text{QCF:} \qquad a^*_{p\leq 1}(\boldsymbol{x})_{\text{QCF}} &= \frac{\Lambda(\boldsymbol{x})}{1 + \Lambda(\boldsymbol{x})} a^*_{p=1}(\boldsymbol{x})_{\text{QCF}} \\ &= a^*_{p=1}(\boldsymbol{x})_{\text{QCF}} \cdot \left\{1 + \frac{\sqrt{\lambda}}{\mu} e^{\bar{a}_o^2/2\sigma^2 - \left(\sigma^2\Phi_x + \bar{a}_o\right)^2/2\sigma^2\lambda}\right\}^{-1}, \\ &\qquad \lambda \equiv \sigma^2\Phi_s + 1;\ \mu = p/q,\end{aligned} \tag{6.4.21}$$

$$= \frac{a^*_{p=1}(\boldsymbol{x})^*_{\text{QCF}}}{1 + C(\boldsymbol{x})} \leq a^*_{p=1|\text{QCF}}, \tag{6.4.21a}$$

so that alternatively

$$C(\boldsymbol{x}) = \frac{\sqrt{\lambda}}{\mu} e^{\bar{a}_o^2/2\sigma^2 - a^*_{p\leq 1}(\boldsymbol{x})_{\text{QCF}}/2\sigma^2} = \Lambda(\boldsymbol{x})^{-1}. \tag{6.4.21b}$$

[31] We can also obtain (6.4.19a) by expanding $(1 + \sigma^2\Phi_s)^{-1} \doteq 1 - \sigma^2\Phi_s + \cdots$ in (6.4.18).

The data dependent quantity $C(\boldsymbol{x})$ may also be expressed from (6.3.44) in terms of the coefficient $B(\boldsymbol{x})$, (6.3.44a), representing the QCF sample bias here, as

$$C(\boldsymbol{x}) = \frac{B(\boldsymbol{x})}{(1 - B(\boldsymbol{x}))}, 0 \leq B(\boldsymbol{x}) \leq 1, \quad \text{and} \therefore \ 0 \leq C(\boldsymbol{x}) \leq \infty, \tag{6.4.21c}$$

where from (6.3.44a) and (6.4.18)

$$B(\boldsymbol{x}) = [1 + \Lambda(\boldsymbol{x})]^{-1} = \left[1 + \frac{\mu}{\sqrt{\lambda}} e^{-\bar{a}_o^2/2\sigma^2 + \left(\sigma^2\Phi_x + \bar{a}_o\right)^2/2\sigma^2\lambda}\right]^{-1} \tag{6.4.22}$$

explicitly in this example. Note that although $a^*_{p=1}(\boldsymbol{x})_{\text{QCF}}$ is linear in the data $\boldsymbol{x}$, $a^*_{p<1}(\boldsymbol{x})_{\text{QCF}}$ is not. On the other hand, $a^*_{p\leq 1}(\boldsymbol{x})_{\text{SCF}}$ is linear in $\boldsymbol{x}$, all $0 \leq p \leq 1$.

From (6.3.44b), or using (6.2.11a) directly for the averages, *we see that the estimators of a_o are unconditionally unbiased*, that is,

$$\left\langle a^*_{p\leq 1|\text{QCF}} \right\rangle_H = \left\langle a^*_{p\leq 1|\text{SCF}_{1,2}} \right\rangle_H = p\bar{a}_0; \quad H = H_0 + H_1. \tag{6.4.23}$$

The corresponding *average bias* is now $q\bar{a}_0$, from (6.3.44b). These results hold not only for the no-coupling cases but also for the situation where the coupling is weak, in the sense of Section 6.3.2 above, provided the cost functions used in (6.3.26) are employed. [For other cost functions and for the cases of strong coupling (Section 6.3.1, where the D and E processes are not separable), it is not generally known whether the estimators remain unconditionally unbiased, that is, under $\langle\,\rangle_H$ each example must be examined separately, Sections 6.3.4.2 and 6.3.4.3.]

6.4.2.2 Bayes Error, SCF$_{1,2}$; $p \leq 1$ We begin with the analytically simpler case of the SCF Bayes error associated with $a^*_{p\leq 1|\text{SCF}}$. From (6.2.37), $L = 1$, and $w_1(y, a_o|H_1)$, (6.4.13), $w_1(y|H_0)$, (6.4.11), we note that if we set $z \equiv a^*_{p\leq 1|\text{SCF}}$, we obtain from (6.4.20) the relation $z = py$ and therefore the new pdfs of z, which are needed now in the determination of $R^*_{\text{E}:\text{SCF}}$, (6.2.37), where $p \leq 1$. Specifically, from the relation $y = z/p$ these are

$$W_1(z, a_o|H_1) = p^{-1}w_1\left(y = \frac{z}{p}, a_o|\, H_1\right); W_1(z|H_0) = p^{-1}w_1\left(y = \frac{z}{p}|\, H_0\right), \tag{6.4.24}$$

so that the Bayes error for the SCF becomes here, since $\mu \equiv p/q$ (6.1.6b):

$$\begin{aligned} R^*_{\text{E}|\text{SCF}}|_{p\leq 1} &= C_0\left[A_1 - \left\{\mu^{-1}W_1(z|H_0)_{z=0} + W_1(z, a_o|H_1)_{z=a_o=0}\right\}_{p\leq 1}\right] \\ &= C_0\left[A_1 - B(p)^*\left(\frac{1+\sigma^2\Phi_s}{\sigma^2\Phi_s^{1/2}}\right)\frac{e^{-\bar{a}_o^2/2\sigma^4\Phi_s}}{\sqrt{2\pi}}\right] \end{aligned} \tag{6.4.25}$$

in which with (6.4.14),

$$B(p)^* \equiv \mu^{-1} + \left(2\pi\sigma^2\right)^{-1/2} e^{-\bar{a}_o^2/2\sigma^2} = \mu^{-1} + \left(2\pi\eta_o\bar{a}_o^2\right)^{-1/2} e^{-(1-\eta_o)/2\eta_o}, \mu = \text{p}/q. \tag{6.4.25a}$$

[Observe that when $p = 1$, that is, $\mu \to \infty$, $R^*_{\text{E}|\text{SCF}}|_{p\leq 1}$ here reduces to (6.4.17c), as required.]

The actual evaluation of the Bayes error (6.4.17c) here, as defined in (6.2.14) for the SCF_2, and in (6.2.33) for SCF_1, is somewhat arbitrary, depending as it does on the choice of the constant $A_{L=1}$. However, the choice of A_1 is not quantitatively very critical, since the concern is primarily with the *relative* minimum error, which in turn is based on comparisons with the classical Bayes estimator $a^*_{p=1}$ under H_1. This, of course, is suboptimum under $H = H_0 + H_1, p < 1$. A useful reasonable choice of A_1 is made by selecting some (finite) maximum value of the output signal-to-noise ratio $(S/N)^*_{\text{out}} = \sigma^2\Phi_s$, (6.4.15), such that $R^*_{\text{E}}(\sigma^2\Phi_{s-\max}) = 0$, with $R^*_{\text{E}}(\sigma^2\Phi_s = 0) = C_0 A^*_1 (> 0)$, so that, necessarily, smaller values of $(S/N)^*_{\text{out}}$ produce larger errors, on average. It is not difficult to show (Problem 6.15) that this is achieved by using (6.4.14) in (6.4.25) and setting

$$A^*_1 \equiv B(p)^* F(u_o), u_o = \sigma^2\Phi_{\max}, \tag{6.4.26a}$$

so that

$$F(u) \equiv \left(\frac{1+u}{\sqrt{u}}\right) \frac{e^{-(1-\eta_o)/2\eta_o u}}{\left(2\pi\eta_o \overline{a_o^2}\right)^{1/2}}, F(u_o) \geq F(u) \geq 0, \tag{6.4.26b}$$

since $0 \leq u \leq u_o$, with $u = \sigma^2\Phi_s = (S/N)^2_{\text{out}|\text{E}}$ (6.4.15). The result is that the Bayes error here for $SCF_{1,2}$, (6.4.25), can be written compactly as

$$R^*_{\text{E}|\text{SCF}}\big|_{p \leq 1} = C_0 B(p)^* [F(u_0) - F(u)]_{u_o > u} > 0; \quad = 0, \quad u = u_o. \tag{6.4.27}$$

[Note incidentally that $B(p)^* = \mu^{-1} + (1/2)F(1)$.] Figure 6.8 illustrates the function $F(u)$ (6.4.26b).

6.4.3 Performance Degradation, $p < 1$

We are interested in most instances in the error incurred when we assume that the signal containing the parameters being estimated is present, that is, under the assumption that $p = 1$. When this is actually not the case, that is, $p < 1$, we naturally choose the $H_{1-\text{optimum}}$ estimator $(p = 1)\gamma^*_\theta|_{p=1}$ for what is now the suboptimum estimator under $H = H_0 + H_1$, $p < 1$. Accordingly, for this purpose we use the *relative Bayes error* $\Delta R^*_{\text{E}|p \leq 1|\text{SCF}_1}$, which from (6.3.59) is defined by the relation

$$\Delta R^*_{\text{E}|p \leq 1} \equiv \left(\frac{R_{\text{E}|p \leq 1} - R^*_{\text{E}|p \leq 1}}{R^*_{\text{E}|p \leq 1}} (\geq 1)\right), \tag{6.4.28}$$

where $R^*_{\text{E}|p \leq 1}$ is the average error associated with the suboptimum estimator γ_θ (or estimators, $\boldsymbol{\gamma}_\theta$) in question. The definition (6.4.28) is quite general and applies for all estimators, optimum and suboptimum for which it is possible to optimize the D and E processes separately, as discussed in Sections 6.3.2–6.3.6.

6.4.3.1 Relative Bayes Error; $p \leq 1$ Accordingly, for the present example involving the SCF estimators (6.4.23), where now $y = a^*_{p=1|H_1}$ is chosen to be the suboptimum estimator under $H = H_0 + H_1$, we must use the pdfs w_1 of (6.4.11), (6.4.13) in (6.2.37), $L = 1$, in place of $SCF_{1,2} = W_1|_{H_0,H_1}$, (6.4.24). Then,

$$B(p) = pB(p)^*, \tag{6.4.29}$$

and with $A_1 = A_1^*$, (6.4.26a), the suboptimum average error is here

$$R_{E|p\leq 1|SCF} = C_0 B(p)^* [F(u_0) - pF(u)\ ] \tag{6.4.30}$$

for the now suboptimum estimator $a_{p=1}^*$ *under H*.

Applying (6.4.27) and (6.4.30) to (6.4.28) gives us the relative Bayes error in simple form:

$$SCF_{1,2}: \qquad \Delta R_{E|p\leq 1|SCF}^* = (1-p)\left[\frac{F(u_0)}{F(u)} - 1\right]^{-1}, \tag{6.4.31}$$

cf. (6.4.26b) for $F(u)$, $0 \leq u \leq u_0$. As expected, when $p = 1$, $\Delta R_{E|p\leq 1|SCF} = 0$, $(u < u_0)$: $a_{p=1}^*$ is now optimum, while for $p = 0$, ΔR_E^* is maximal and $a_{p=1}^*$ can be noticeably suboptimum. Figure 6.9 illustrates the behavior of ΔR_E^* as a function of the *a priori* signal probability p.

A number of general and expected results[32] can be noted at once for estimation of amplitude and intensity $\left(= a_{p\leq 1}^2\right)$ here: (1) increasing $u = (S/N)_{out}^2 (\leq u_0)$ leads to lower Bayes error and lower suboptimum average error (6.4.30), but a slower rate for the latter, as $u \to u_o$, since $p\,F(u) \leq F(u)$. (The *relative* Bayes error here (6.4.31), however, *increases* with increasing $(S/N)_{out}^2$ as $u \to u_o$ (Fig. 6.9)[33], because of the way the average risks are defined); (2) larger values of $p(\leq 1)$ also yield less (relative) Bayes error, for given $u = (S/N)_{out}^2$; (3) $p = O(0.9 - 1.0)$ can give values of the estimators often acceptably close to the usually postulated situation of $p = 1$, provided $(S/N)_{out}^2$ is large enough. This in turn implies a concomitant probability of detection (P_D) which is equal to or above the threshold of acceptability $(P_{D\text{-accept.}})$, (6.3.47) et seq.

In addition, the suboptimum estimator $a_{p<1}\left(= a_{p=1}^*\right)$ here for the $SCF_{1,2}$ is biased with respect to $H = H_0 + H_1$, since

$$\langle a_{p\leq 1|SCF}\rangle_H = p\bar{a}_0 + q\frac{\bar{a}_0}{(\sigma^2\Phi_s + 1)} = \langle a_{p\leq 1|SCF}^*\rangle + q\frac{\bar{a}_0}{\lambda}, \tag{6.4.32}$$

where the bias is $q\bar{a}_0/\lambda$, unlike the optimum estimator (6.4.23).

6.4.3.2 Bayes Error and Relative Bayes Error, p ≤ 1, QCF This quantity is

$$QCF: \qquad R_{E|p\leq 1|QCF}^* = C_0\left\{\overline{a_0^2} - 2\langle a_0 a_{p\leq 1}^*\rangle_H + \langle a_{p\leq 1}^{*2}\rangle_H\right\}_{QCF}, \tag{6.4.33a}$$

[32] Ref. [8] contains a wider spectrum of numerical results.

[33] Note from (6.4.15) that $u = 4\Phi_s$ for the two fading cases illustrated in Fig. 6.9.

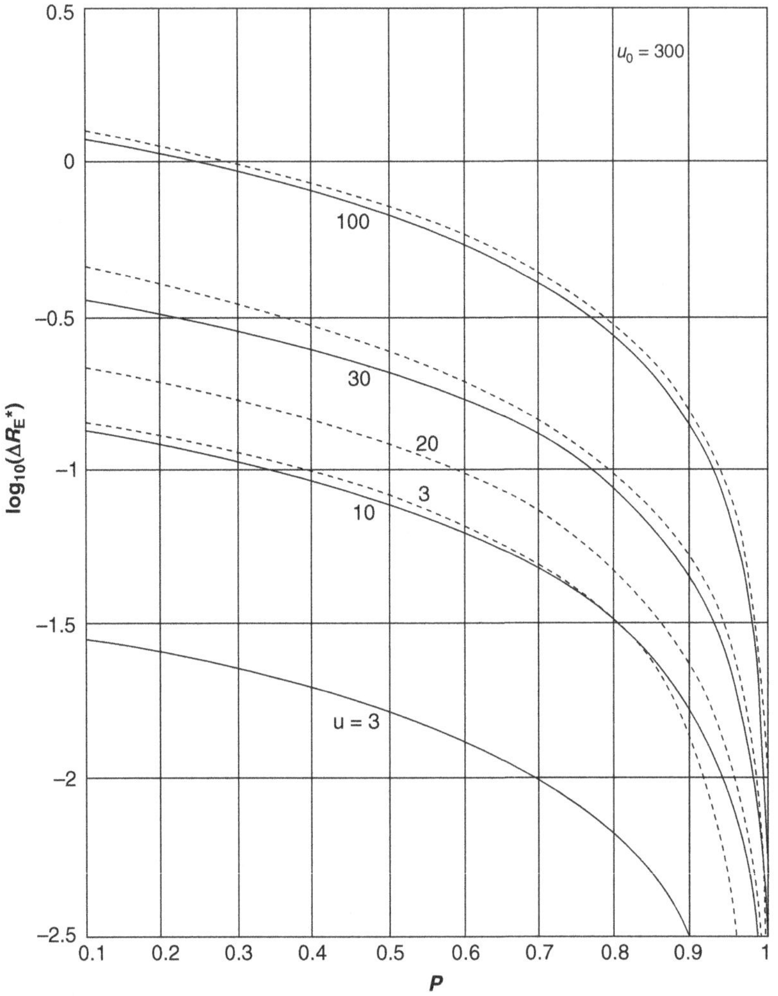

FIGURE 6.9 The relative (SCF) Bayes error (6.4.31) versus p, for the cases $\overline{a_o^2} = 40$, $\eta_o = 0.1$ (weak fading ——), $\therefore$ $a_o = 6$, and $\overline{a_o^2} = 4$, $\eta_o = 1$ (heavy fading ------), $\therefore \bar{a}_o = 0$, cf. (6.4.14); $u \equiv \sigma^2 \Phi_s$, (6.4.15).

where $a^*_{p \le 1|\mathrm{QCF}}$ is given by (6.4.21) and (6.4.21a–c). This in turn can be expressed compactly by

$$R^*_{\mathrm{E}|p \le 1|\mathrm{QCF}} = C_0 \Big[q I_2(H_0) + p \Big\{ \overline{a_o^2} - I_{11}(H_1) + I_2(H_1) \Big\} \Big], \tag{6.4.33b}$$

following [(3.27a) and (3.27b)] of Ref. [1], or by direct substitution using [(6.4.21) and (6.4.21a–c)]. The components of (6.4.33b) are explicitly

$$I_2(H_0) \equiv \overline{\left[\frac{a^*_{p=1|\mathrm{QCF}}}{1 + \Lambda^{-1}} \right]^2}^{H_0} = \int_{-\infty}^{\infty} y^2 \left(1 + \frac{A}{\mu} e^{-y^2/2B^2} \right)^{-2} w_1(y|H_0)\, dy, \tag{6.4.34a}$$

$$I_{11}(H_1) \equiv \overline{\left[\frac{2a_0 a^*_{p=1|QCF}}{1+\Lambda^{-1}}\right]}^{H_1} = \int_{-\infty}^{\infty} 2y^2\left(1+\frac{A}{\mu}e^{-y^2/2B^2}\right)^{-1} w_1(y|H_1)\,dy, \quad (6.4.34b)$$

$$I_2(H_1) \equiv \overline{\left[\frac{a^*_{p=1|\mathrm{QCF}}}{1+\Lambda^{-1}}\right]^2}^{H_1} = \int_{-\infty}^{\infty} y^2\left(1+\frac{A}{\mu}e^{-y^2/2B^2}\right)^{-2} w_1(y|H_1)\,dy, \quad (6.4.34c)$$

in which, cf. (6.4.21), (6.4.21b):

$$A \equiv \sqrt{\lambda}e^{\bar{a}_o^2/2\sigma^2};\, B^2 = \frac{\sigma^2}{\lambda};\, \lambda \equiv 1+\sigma^2\Phi_s, \quad (6.4.34d)$$

and where we have used $w_1(y|H_1) = \int w_1(y|a_o)w(a_o)da_o$ to integrate over a_o in (6.4.34b), remembering that $\boldsymbol{x}|H_1 = \boldsymbol{n} + a_o\boldsymbol{s}$ in $a^*_{p=1|\mathrm{QCF}}$. The pdfs $w_1(y|H_0)$ and $w_1(y|H_1)$ are given respectively by (6.4.11), (6.4.8). When $A/\mu < 1$, and in particular for $p >> q$, that is, large μ in relation to A, termwise integration of $I_2(H_0)$, and so on is practical. However, generally (for all finite A/μ) numerical integration is recommended and is readily carried out [9],[34] some results of which are shown in Fig. 6.10. When $\mu \to \infty$, that is, $p = 1$, (6.4.33a) and (6.4.33b) reduce at once to $a^*_{p\leq 1} \to a^*_{p=1}$, as required, and $\therefore\ R^*_{\mathrm{E}|p\leq 1|\mathrm{QCF}} \to C_0\sigma^2/(1+\sigma^2\Phi_s)$, (6.4.17a), also as expected. (This may be verified by observing that $R^*_{\mathrm{E}|p=1|\mathrm{QCF}} = C_0\left[\overline{a_0^2} - \langle y^2\rangle_{H_1}\right]$ and by evaluating $\langle y^2\rangle_{H_1}$ directly, noting that $\overline{a_o^2} = \bar{a}_o^2 + \sigma^2$, cf. Problem 6.13.)

Next, we determine the *suboptimum average risk* $R_{\mathrm{E}|p\leq 1|\mathrm{QCF}}$ for the now suboptimum estimator $a_{p\leq 1|\mathrm{QCF}} = a^*_{p=1|\mathrm{QCF}}$. Applying [(6.2.11)–(6.2.13)] to $a_{p\leq 1|\mathrm{QCF}}$ we easily show (Problem 6.15) that the QCF analogue here of (6.4.30) is

$$\left.\begin{aligned} R_{\mathrm{E}|p\leq 1|\mathrm{QCF}} &= C_0\left\langle \left(a_o - a^*_{p=1}\right)^2_{\mathrm{QCF}}\right\rangle_H = C_0\left\{q\left\langle\left(a_o - a^*_{p=1}\right)^2\right\rangle_{H_0} + p\left\langle\left(a_o - a^*_{p=1}\right)^2\right\rangle_{H_1}\right\}_{\mathrm{QCF}} \\ &= C_0\left\{q\frac{\sigma^4\Phi_s + \bar{a}_o^2}{(\sigma^2\Phi_s+1)^2} + p\frac{\sigma^2}{(\sigma^2\Phi_s+1)}\right\}_{\mathrm{QCF}}. \end{aligned}\right\} \quad (6.4.35)$$

Figure 6.11 illustrates the suboptimum case $R_{\mathrm{E}|p\leq 1|\mathrm{QCF}}$ (6.4.35), for the same parameters used in Fig. 6.10. The suboptimum estimator $a_{p=1|\mathrm{QCF}}$ under H is also biased, since

$$\langle a_{p\leq 1|\mathrm{QCF}}\rangle_H = p\bar{a}_o + q\frac{\bar{a}_o}{(\sigma^2\Phi_s+1)} = \langle a_{p\leq 1|\mathrm{SCF}}\rangle_H, \quad (6.4.36)$$

which is the same bias as $a_{p\leq 1|\mathrm{SCF}}$, cf. (6.4.32). This follows, of course, from (6.4.6c) above, namely from the fact that $a^*_{p=1|\mathrm{SCF}} = a^*_{p=1|\mathrm{QCF}}$ here. (We remark that $R_{\mathrm{E}|p=0} \gtrless R_{\mathrm{E}|p=1}$ according to $\eta_o\overline{a_o^2} \gtrless \bar{a}_o^2$.)

[34] The author is indebted to Dr. A. H. Nuttall, NUWC, for the calculations of Figs. 6.8–6.12.

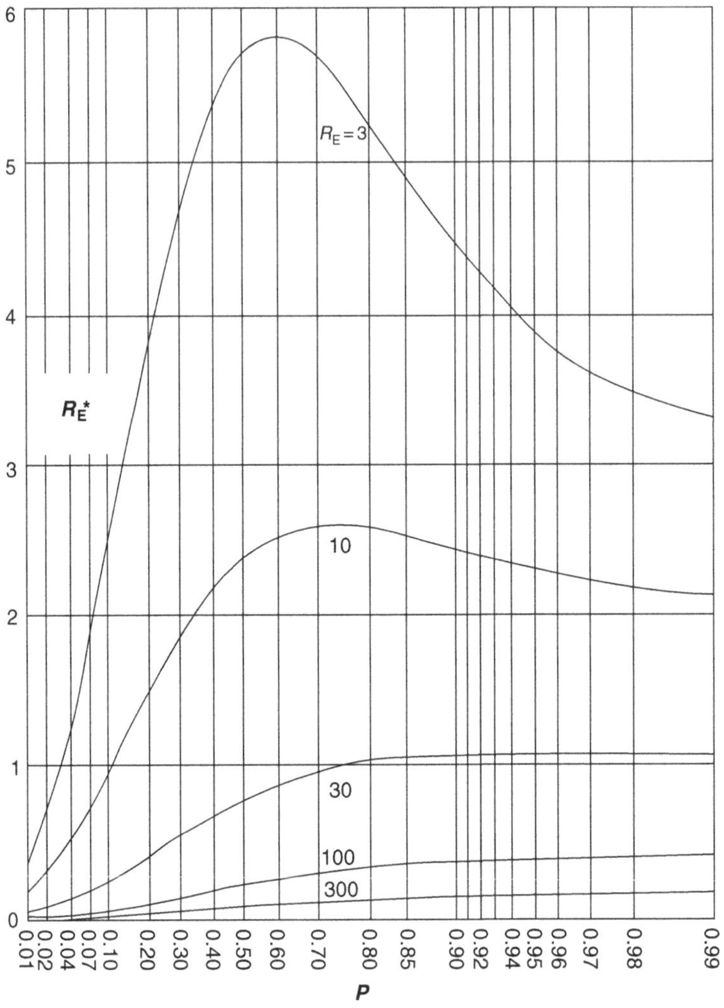

FIGURE 6.10 Bayes risk, $R^*_{E|p\leq 1|\mathrm{QCF}}$, Eq. (6.4.33b), for $a^*_{p\leq 1|\mathrm{QCF}}$, (6.4.21), as a function of the *a priori* probability p, with $\overline{I_0} = a_o^2\Phi_s = (S/N)^2_{\mathrm{out}|E}/\eta_o$, (6.4.15). Here $(S/N)^2_{\mathrm{out}|E} = 10I_0$; $\sigma^2 = 4, \overline{a_o} = 6$; $\therefore \overline{a_o^2} = 40$; $\eta_o = 0.1$.

Applying (6.4.33b) and (6.4.35) to (6.4.28) gives us *the relative Bayes error* for the amplitude estimator under the QCF:

$$\Delta R^*_{\mathrm{E}|p\leq 1|\mathrm{QCF}} = \frac{p\sigma^2/\lambda + q\left(\sigma^4\Phi_s + \bar{a}_o^2\right)/\lambda^2}{p\left\{\overline{a_o^2} - I_{11}(H_1) + I_2(H_1)\right\} + qI_2(H_0)} - 1 \quad (\geq 0). \tag{6.4.37}$$

[It is not difficult to show that $\Delta R^*_{\mathrm{E}} = 0$ when $p = 1$ $(q = 0)$, as required from the definition (6.4.28). At the other extreme, when $p = 0$ $(q = 1)$, then $I_2(H_0)|_{q=1} \to 0$ and $\Delta R^*_{\mathrm{E}|p=0|\mathrm{QCF}} \to \infty$, as indicated in Figs. 6.12a–c ff.]

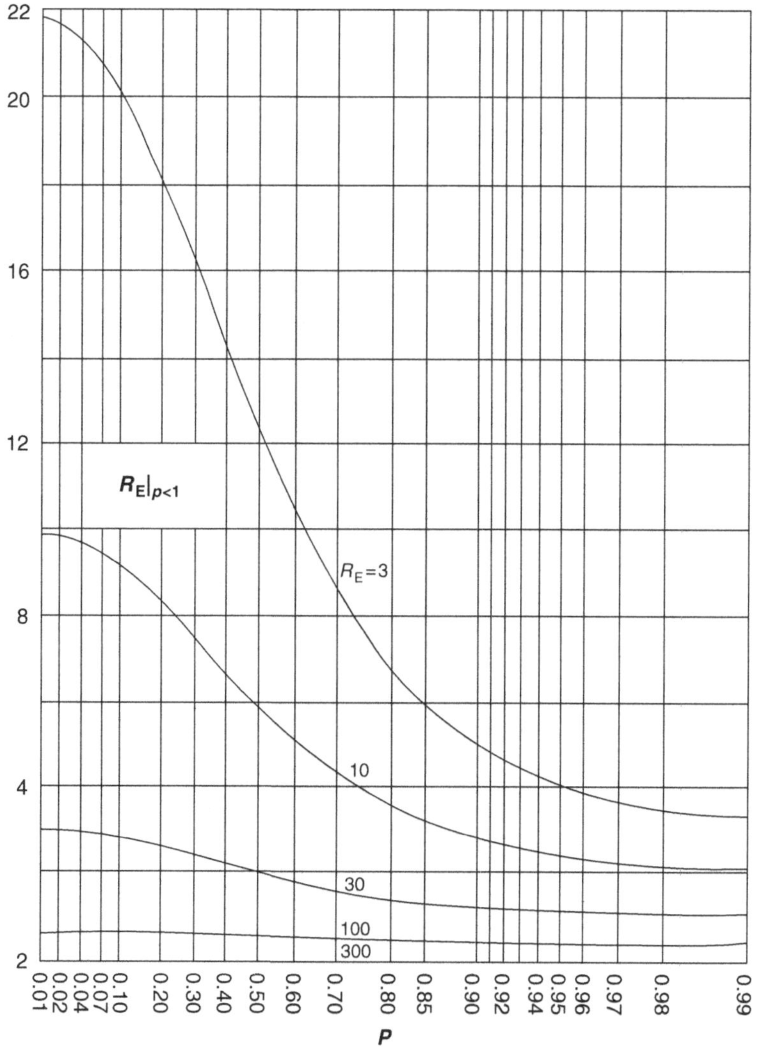

FIGURE 6.11 Suboptimum average risk $R_{E|p\le 1|QCF}$, (6.4.35), for $a_{p\le 1|QCF} = a^*_{p=1|QCF}$ under H. Same parameters as Figure 6.10.

For the Figs. 6.10–6.12a–c ff., it is easily seen from (6.4.15) that

$$I_0 \equiv a_o^2 \Phi_s = \frac{(S/N)^2_{out|E}}{\eta_o}; \quad \Pi^*_{E\text{-coh}} \equiv \eta_o \Phi_s, \text{with } (S/N)^2_{out|E} = \eta_o \overline{a_o^2} \Phi_s. \tag{6.4.38}$$

(The specific numerical values employed are $\sigma^2 = 4, \overline{a_o} = 6; \therefore \ \eta_o = 0.1, \overline{a_o^2} = 40 = \left(S/N_{in}^2\right)$.) Π may be interpreted as a processing gain; see Section 3.4.9.

Several observations can be made: as processing gain, $\left(\sim \Phi_s \text{ or } (S/N)^2_{out}\right)$, is increased, the average quadratic error decreases for the Bayes and suboptimum cases, for all values of the *a priori* probability $p \le 1$. This is to be expected, inasmuch as the interfering effects

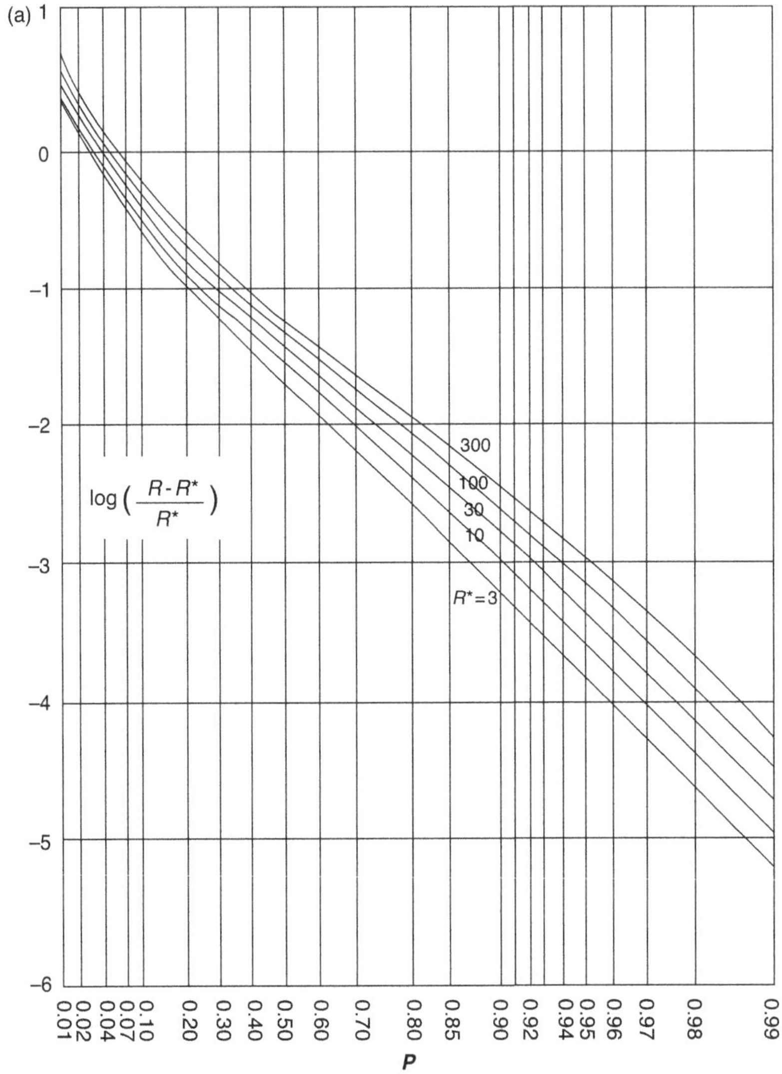

FIGURE 6.12 (a) The relative Bayes error $R^*_{\mathrm{E}|\underline{p} \leq 1|\mathrm{SCF}}$ (6.4.37) versus the *a priori* probability p, for heavy fading: $\overline{a_o} = 0$ and $\therefore \eta_o = 1.0$; $I_o = \overline{a_o^2}\Phi_s = 4\Phi_s = (S/N)^2_{\mathrm{out|E}}/\eta_o$ (6.4.15). (b) Relative Bayes risk (QCF), (6.4.37), vs. *a priori* probability p, for $\overline{a_o} = 6$, $\overline{a_o^2} = 40 \therefore \sigma^2 = 4$ and $\eta_o = 0.1$; $I_0 = a_o^2\Phi_s = (S/N)^2_{\mathrm{out|E}}/\eta_o, =3, 10, 30$; $(S/N)^2_{\mathrm{out|E}} = 10I_0$, cf. Figs. 6.10 and 6.11. (c) Same as *Figure 6.12b*; $I_0 = 30$, 100, 300.

of the noise on estimation decrease relative to the signal. However, there is always some (average) error as long as $(S/N)^2_{\mathrm{out}}$ is finite, even as $p \to 1$, again due to the presence of the noise. For the suboptimum situation (Fig. 6.11), the average risk increases *nonmonotonically* with decreasing values of p, also, for the same reasons, in spite of the pdf, $w_1(a_0)$, of the amplitude, which does specifically exhibit its influence in the optimum cases (Fig. 6.10), through the nonmonotonic behavior of R^*_{E} with p. The relative Bayes error $\Delta R^*_{\mathrm{E|QCF}}$ (6.4.37)

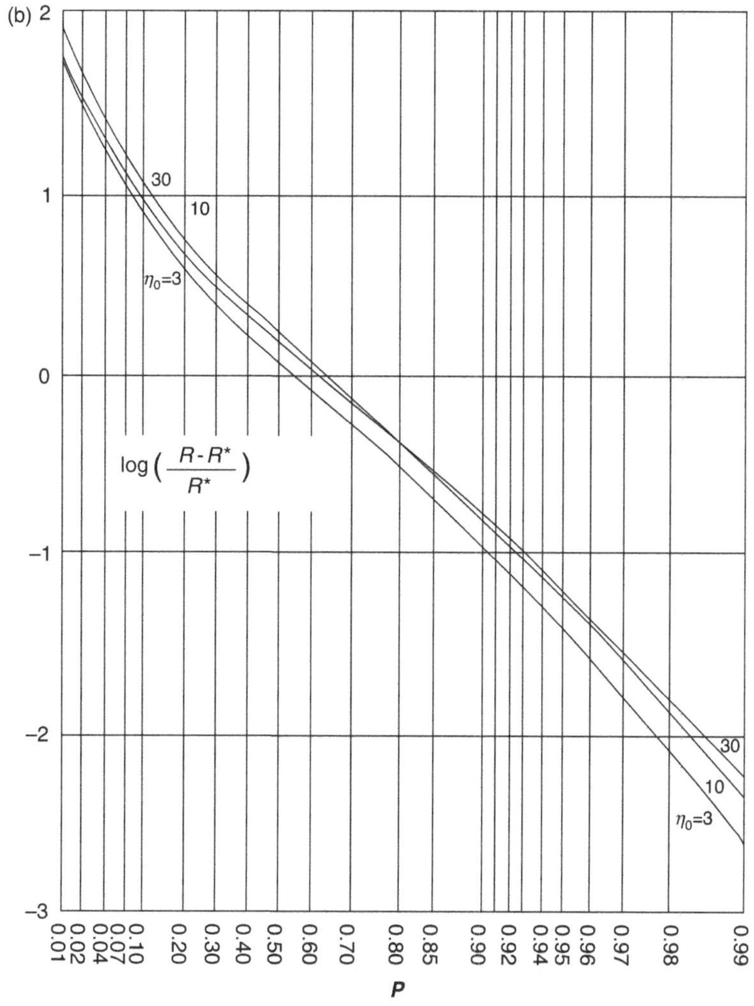

FIGURE 6.12 (*Continued*)

here behaves generally as expected: as $p \rightarrow 1$, $\Delta R^*_{E|QCF} \rightarrow 0$, and as $p \rightarrow 0$, $\Delta R^*_{E|QCF} \rightarrow \infty$. However, there is relatively small variation with $(S/N)^2_{out|E}$ throughout as p changes, indicating a certain measure of robustness of the relative Bayes error to changes in $(S/N)^2_{out|E}$ for $0 < p \leq 1$. (Similar behavior is noted [9] for other parameter values of the pdf $w_1(a_o)$, for example, $\bar{a}_o = 0, \overline{a_o^2} = \sigma^2 = 4 = (S/N)^2_{in}; \eta_o = 1$ (heavy fading)).

6.4.3.3 ***Performance Degradation: Mean Relative Error*** A usually much simpler and often useful measure of performance degradation is *the mean relative error* between a suboptimum and optimum estimator, defined above by Eqs. (6.3.60) and (6.3.60a). Here the suboptimum estimator, under $p < 1$, is chosen to be the optimum one under $p = 1$, reflecting the fact that frequently estimation is made under the assumption that the signal is truly present in the data sample when actually it is not.

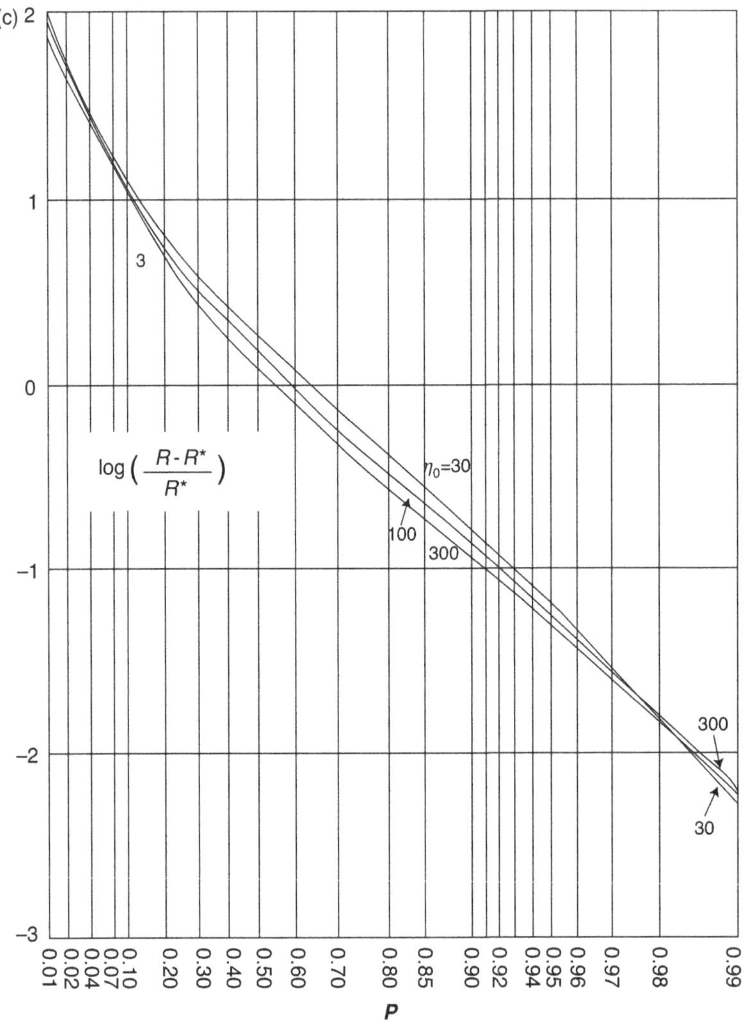

FIGURE 6.12 (*Continued*)

For our present example of optimum amplitude estimation, cf. (6.4.6c), for $p = 1$ and the SCF and QCFs, we readily show that (Problem 6.15d), vide (6.4.36):

$$\langle a^*_{p=1}\rangle_H = q\left(\frac{\bar{a}_o}{1+\sigma^2\Phi_s}\right) + p\bar{a}_o = \frac{\bar{a}_o(1+p\sigma^2\Phi_s)}{1+\sigma^2\Phi_s}, \text{ with } \langle a^*_{p\leq 1}\rangle_H = p\bar{a}_o. \quad (6.4.39)$$

Consequently, applying (6.4.39) to the simpler measure of degradation (6.3.56a) gives the simple result

$$Da^*_o = \left[\mu\left(1+\sigma^2\Phi_s\right)\right]^{-1}, \mu \equiv p/q = \frac{p}{(1-p)}, \quad (6.4.40)$$

such that $(\lim p \to 1)Da^*_o = 0$ and $(\lim p \to 1)Da^*_o = \infty$, both of which limits are entirely reasonable. We note also that Da^*_o vanishes when $(S/N)^2_{\text{out}|\text{E}}(= \sigma^2\Phi_s)$, (6.4.15), becomes indefinitely large, that is, when $\sigma^2\Phi_s \to \infty$. Under this condition it is shown in Section 6.4.4 following that the associated (Bayes) detection probability (6.4.47b), $P_D^* = pp_D^* \to 1$, with false alarm probability $\alpha_F^* \to 0, \sigma^2 > 0$. Thus, when $\sigma^2\Phi_s \to \infty$, *on the average* $\langle a_{p=1}^* \rangle_H = \langle a_{p\leq 1}^* \rangle_H = p\bar{a}_o^*$, the unconditionally unbiased estimate (6.4.23) for both the SCF and QCF. In practical cases, however, at best $1 \ll \sigma^2\Phi_s < \infty$, so that $\langle a_{p=1}^* \rangle_H \cong \langle a_{p\leq 1}^* \rangle_H = p\bar{a}_o^*$ and $P_D^* \leq 1, \alpha_F^* \geq 0$, with the latter quantitatively described by (6.4.46) ff.

The role of the cost function is implicit in the degradation result (6.4.40) and in the structure of the estimators themselves, for example, as used to obtain $a_{p=1}^*$, (6.4.5), (6.4.6c), and $a_{p\leq 1}^*$, (6.4.20), (6.4.21), respectively for the SCF and QCF employed here. The more complete analysis, of course, is given by the calculation of the average risks or errors R_E^*, R_E, respectively, namely, (6.4.27), (6.4.30) for the SCF and (6.4.33b), (6.4.35) for the QCF. The latter averages are usually computationally much more intensive, as we would expect.

6.4.4 Acceptance or Rejection of the Estimator: Detection Probabilities

It remains to apply the results of Section 6.3.6 to determine whether or not the estimate in question is to be accepted or rejected for our present example, described at the beginning of Section 6.4. Again, the explicit analysis is limited to the cases of "no coupling," cf. (6.3.47a), and to the optimum and suboptimum estimation of the (normalized) amplitude a_o, when the signal (6.4.2) is coherently observed in an additive, correlated Gauss noise process, and a_o is normally distributed according to (6.4.1). The appropriate likelihood ratio $\Lambda(x)$ here is given by (6.4.18). Its associated pdfs $Q_1(u), P_1(u)$, under H_0 and H_1, respectively, are to be determined from (6.3.51). The desired (conditional) false alarm probability α_F^* and the (conditional) probability of correct detection p_D^* are then to be obtained from (6.3.50). It turns out, as we shall see below, that exact results can be obtained here, essentially due to the original Gaussian nature of Φ_x and $a_{p=1}^*$ in $\Lambda(x)$, cf. (6.4.18), as well as to the normal pdf (6.4.1) of the amplitude a_o.

We begin by observing that $\Lambda(x)$, (6.4.18) here, can be expressed in the more convenient logarithmic form

$$\log \Lambda(\boldsymbol{x}) = \left(\log \mu - \log \sqrt{\lambda} - \frac{\bar{a}_o^2}{2\sigma^2}\right) + \frac{\lambda y(\boldsymbol{x})^2}{2\sigma^2} \equiv u(\boldsymbol{x}), \text{ with } y(\boldsymbol{x}) \equiv a_{p=1}^* = \frac{\sigma^2\Phi_x + \bar{a}_o}{1+\sigma^2\Phi_s}, \tag{6.4.41}$$

where $y(\boldsymbol{x})$ is governed by the pdfs $w_1(y|H_0)$, (6.4.11), and $w_1(y|H_1)$, (6.4.8), again with $\lambda \equiv 1 + \sigma^2\Phi_s$, cf. (6.4.34d). The decision process itself is described by (6.3.47a) for these uncoupled cases.

Applying these pdfs [(6.4.8) and (6.4.11)] according to (6.3.51), to (6.4.41), completing the square in the exponential, and so on, give us for the pdfs of $u = \log \Lambda(x)$

$$Q_1(u) = \int_{-\infty}^{\infty} \frac{e^{-i\xi(u-B_0)-F} \cdot e^{1/\lambda_0 p(1-i\xi/p)}}{\sqrt{1-i\xi/p}} \frac{d\xi}{2\pi}; P_1(u) = \int_{-\infty}^{\infty} \frac{e^{-i\xi(u-B_0)-F} \cdot e^{1/\lambda_0 \hat{p}(1-i\xi/\hat{p})}}{\sqrt{1-i\xi/\hat{p}}} \frac{d\xi}{2\pi}, \tag{6.4.42a}$$

where

$$\left.\begin{aligned} \rho \equiv \frac{\lambda}{\sigma^2 \Phi_s} = \lambda \hat{\rho};\, \mathrm{F} \equiv \frac{\bar{a}_0^2}{2\sigma^4 \Phi_s} = \frac{\hat{F}}{\lambda};\, \lambda_0 \equiv \frac{\sigma^6 \Phi_s^2}{\lambda\, \bar{a}_o^2};\, \lambda \equiv 1 + \sigma^2 \Phi_s, \text{ and} \\ \mathrm{B}_0 \equiv \log \mu - \log \sqrt{\lambda} - \frac{\bar{a}_o^2}{2\sigma^2} \end{aligned}\right\}. \tag{6.4.42b}$$

[Note at once from the integrands of (6.4.42a) that the characteristic functions of $u|H_0, H_1$ are explicitly

$$F_1(i\xi|H_0)_u = \frac{e^{-i\xi B_0 - F} \cdot e^{1/\lambda_0 \rho(1 - i\xi/\rho)}}{(1 - i\xi/\rho)^{1/2}};\, F_1(i\xi|H_1)_u = \frac{e^{-i\xi B_0 - \hat{F}} \cdot e^{1/\lambda_0 \hat{\rho}(1 - i\xi/\hat{\rho})}}{(1 - i\xi/\hat{\rho})^{1/2}}, \tag{6.4.42c}$$

and that $1/\lambda_0 \rho - F = 0$; $1/\lambda_0 \hat{\rho} - \hat{F} = 0$, from (6.4.42b), thus ensuring that $F_1(0|H_{0,1})_u = 1$. Clearly, $(\lim(\xi) \to \infty) F_1 \to 0$: there are no steady or oscillating components in the c.f.s and consequently no delta functions in the pdfs. With the transformation $p = -i\xi$, the relations (6.4.42a) take the more familiar equivalent forms [10]

$$Q_1(u) = \int_{-\infty}^{\infty} e^{p(u - B_0) - F} \frac{e^{1/\lambda_0(\rho + p)}}{\sqrt{\rho + p}} \frac{dp}{2\pi i};\quad P_1(u) = \int_{-\infty}^{\infty} e^{p(u - B_0) - \hat{F}} \frac{e^{1/\lambda_0(\hat{\rho} + p)}}{\sqrt{\hat{\rho} + p}} \frac{dp}{2\pi i}, \tag{6.4.43}$$

which are given at once by No. 651, p.78, of Campbell and Foster's tables [10], namely,[35]

$$\left.\begin{aligned} Q_1(u) &= \sqrt{\frac{\rho}{\pi}} e^{-F} \left\{ \frac{1}{\sqrt{u - B_0}} e^{-\rho(u - B_0)} \cosh 2\sqrt{\frac{(u - B_0)}{\lambda_0}} \right\},\; u - B_0 \geq 0+;\; = 0 \text{ elsewhere,} \quad (6.4.44a) \\ P_1(u) &= \sqrt{\frac{\hat{\rho}}{\pi}} e^{-\hat{F}} \left\{ \frac{1}{\sqrt{u - B_0}} e^{-\hat{\rho}(u - B_0)} \cosh 2\sqrt{\frac{(u - B_0)}{\lambda_0}} \right\},\; u - B_0 \geq 0+;\; = 0 \text{ elsewhere} \quad (6.4.44b) \end{aligned}\right\}$$

A more convenient form for evaluating the false alarm probability α_F^* and the conditional detection probability $p_D^* = 1 - \beta^*$ (6.3.51) is obtained by making the

[35] The relations (6.4.44a) and (6.4.44b) can readily be obtained by observing that

$$F_1(i\xi)_x = \frac{e^{i\xi B_0 - F} \cdot e^{1/a(1 - bi\xi)}}{(1 - bi\xi)^{1/2}} = \sum_{m=0}^{\infty} \left(\frac{1}{a}\right)^m \frac{e^{i\xi B_0 - F}}{(1 - bi\xi)^{m + 1/2}} \text{ and that } \int_{-\infty}^{\infty} e^{-i\xi x} F_1(i\xi)_x \frac{d\xi}{d\pi}$$

$$= e^{-F} \sum_{m=0}^{\infty} \frac{(x - B_0)^{m + 1/2}}{\Gamma(m + 1/2)} e^{-(x - B_0)} \bigg|_{|x - B > 0+} \equiv e^{-F} \sum_{m=0}^{\infty} w_1(x - B_0|m)_\Gamma$$

where the pdfs $w_1(x - B_0|m)_\Gamma$ are recognized as Γ-pdfs. Summing the series with the help of $\Gamma(m + 1/2) = (2m)!\sqrt{m}/2^{2m} m!$, the (next to the last) relation above gives the desired results [(6.4.44a) and (6.4.4b).

transformations $u - B_o = z = v^2$ so that Q_1, P_1 [(6.4.44a) and (6.4.44b)] become alternatively

$$\left.\begin{aligned} Q_1(v) &= \sqrt{\frac{\rho}{\pi}}\left\{e^{-\left(\sqrt{\rho}v-1/\sqrt{\lambda_0\rho}\right)^2} + e^{-\left(\sqrt{\rho}v+1/\sqrt{\lambda_0\rho}\right)^2}\right\}, v \ge 0+; \quad =0 \text{ elsewhere} \quad &(6.4.45a)\\ P_1(v) &= \sqrt{\frac{\hat{\rho}}{\pi}}\left\{e^{-\left(\sqrt{\hat{\rho}}v-1/\sqrt{\lambda_0\hat{\rho}}\right)^2} + e^{-\left(\sqrt{\hat{\rho}}v+1/\sqrt{\lambda_0\hat{\rho}}\right)^2}\right\}, v \ge 0+; \quad =0 \text{ elsewhere} \quad &(6.4.45b)\end{aligned}\right\}$$

since $1/\lambda_0\rho - F = 1/\lambda_0\hat{\rho} - \hat{F} = 0$, from (6.4.42b). [Note that Q_1, P_1 are non-Gaussian pdfs in v.] Next, applying (6.4.45a) and (6.4.45b) to (6.3.50) yields by obvious transformations to the desired detection probabilities:

$$\textit{False alarm prob}:\ \alpha_F^* = \int_{(\log K - B_0)^{1/2}} Q_1(v)dv = 1 - \frac{1}{2}\left\{\Theta\left[C_{H_0}^{(+)}\right] + \Theta\left[C_{H_0}^{(-)}\right]\right\}, \quad (6.4.46)$$

with

$$C_{H_0}^{(\pm)} \equiv \left\{\frac{\log(K/\mu) + (1/2)\log\lambda + \bar{a}_o^2/2\sigma^2}{\sigma^2\Phi_s/\lambda}\right\}^{1/2} \pm \left\{\frac{\bar{a}_o^*}{2\sigma^4\Phi_s}\right\}^{1/2}, \quad (6.4.46a)$$

and

$$\textit{Detection prob}:\ p_D^* = 1 - \beta^* = \int_{(\log K - B_0)^{1/2}} P_1(v)dv = 1 - \frac{1}{2}\left\{\Theta\left[C_{H_1}^{(+)}\right] + \Theta\left[C_{H_1}^{(-)}\right]\right\}, \quad (6.4.47)$$

with

$$C_{H_1}^{(\pm)} \equiv \left\{\frac{\log(K/\mu) + (1/2)\log\lambda + \bar{a}_o^2/2\sigma^2}{\sigma^2\Phi_s}\right\}^{1/2} \pm \left\{\frac{\lambda\bar{a}_o^*}{2\sigma^4\Phi_s}\right\}^{1/2}, \quad (6.4.47a)$$

and

$$P_D^* = pp_D^* = p(1-\beta^*), \quad (6.4.47b)$$

in which, as usual, $\Theta(x) \equiv (2/\sqrt{\pi})\int_0^x e^{-t^2}dt = erf(x)$ is the familiar error function. Equations (6.4.46) and (6.4.47) apply for $\sigma^2 > 0$, that is, $\eta_o > 0$, cf. (6.4.14), and $\bar{a}_o^2 \ge 0$.

Given α_F^* we must determine the corresponding threshold (K) numerically, since (6.4.46) does not invert analytically. However, when $\bar{a}_o = 0$, namely in the cases of deep fading $\eta_o = 1$, then $C_{H_0,H_1}^{(+)} = C_{H_0,H_1}^{(-)}$, respectively, and we easily find that

$$\bar{a}_o = 0: \qquad \log K = \log\mu - \frac{1}{2}\log\lambda + \frac{\sigma^2\Phi_s}{\lambda}\left[\Theta^{-1}(1-\alpha^*)\right]^2; \ \sigma^2 > 0, \quad (6.4.48a)$$

with

$$\bar{a}_o = 0: \qquad P_D^*\big|_{\bar{a}_o=0} = 1 - \Theta[C_{H_1}] = 1 - \Theta\left[\left\{\log(K/\mu) + \frac{1}{2}\log\lambda\right\}^{1/2} \Big/ \sqrt{\sigma^2\Phi_s}\right]. \quad (6.4.48b)$$

On using (6.4.48a) in (6.4.48b), we can show directly that the latter reduces to the simple result

$$P^*_{\mathrm{D}}|_{\bar{a}_o=0} = p\left\{1 - \Theta\left[\Theta^{-1}(1-\alpha^*_{\mathrm{F}})/\sqrt{1+\sigma^2\Phi_s}\right]\right\}. \tag{6.4.48c}$$

We also observe when $(S/N)^2_{\mathrm{out|E}} (= \sigma^2\Phi_s, (6.4.15))$ is very large, that is, $\sigma^2\Phi_s >> 1$, for a given $\log(K/\mu)$, then $\alpha^*_{\mathrm{F}} \to 0$ and $p^*_{\mathrm{D}} \to 1$ as required: the detection process is *consistent* (p. 59) and of course it is Bayes optimal in view of (6.4.41). Finally, when $\sigma^2 \to 0, (\eta_o = 0), \bar{a}^2_o > 0$, Eq. (6.4.41) reduce to Eq. (6.4.19a) here: just detection, no estimation, since $\bar{a}_o = a_o$, and so on, are now *a priori* known. Similarly, Eqs. (6.4.46) and (6.4.47), as well as (6.4.19a), reduce to well-known results. We recall from (6.4.6c) and (6.4.18) that these detection results are also appropriate to the uncoupled estimation process associated with the estimators $a^*_{p=1}$ for the SCF and QCF.

Figure 6.13, based on (6.4.48c), illustrates the simple special case of deep fading: $\bar{a}_o = 0, \therefore \eta_o = 1$, so that I_0 is the effective output signal-to-noise ratio for estimation (E) *and* detection (D), cf. (6.4.38). In addition, results for a number of selected values of the false alarm probability α^*_{F} are included, for each choice of I_0. As expected, increasing $(S/N)^2_{\mathrm{out}}$ increases P^*_{D}, and for a specified $(S/N)^2_{\mathrm{in}} \equiv \overline{a^2_0}$ this means increasing the processing gain $(\sim \Phi_s)$. Similarly, smaller false alarm probabilities reduce P^*_{D}. Again, we accept the estimate

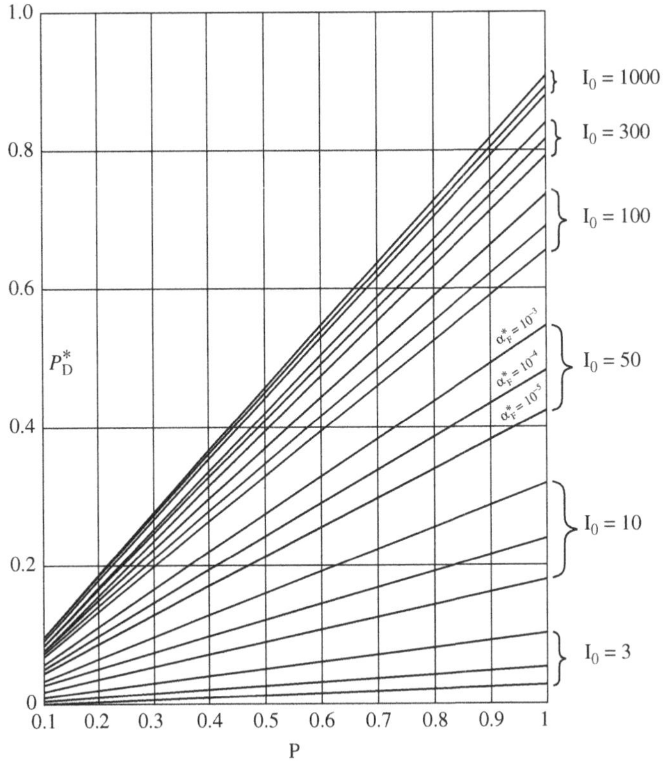

FIGURE 6.13 Probability of detection P^*_D vs. *a priori* probability p under deep fading (6.4.48c), $\overline{a_o} = 0 \therefore \eta_o = 1$ and $I_0 \equiv (S/N)^2_{\mathrm{out}|E}$, for the false alarm probabilities $\alpha^*_F = 10^{-3}, 10^{-4}, 10^{-5}$.

when P_D^* is at or above some reasonable prechosen value $P_{\text{D-accept}}$, cf. (6.3.48) and the discussion in Section 6.3.6 earlier.

Extension to the coupled cases becomes more complex as we extend the analysis to the "weak coupling" case, cf. Sections 6.3.2 and 6.3.3. The procedure and results are further modified by the selection of the additional cost functions (6.3.1). For reasonable choices of P_D^*, namely those for which P_D^*, (6.3.48), can be evaluated in suitably manageable analytic form, governed in turn by the ability to obtain the needed pdfs of $\alpha^{(*)}, \beta^{(*)}$ analytically, we may expect results analogous to the examples above. Otherwise, the weakly coupled, and certainly the strong-coupled cases, present major technical problems and consequently require numerical methods. These topics are reserved for future investigation.

6.4.5 Remarks on the Estimation of Signal Intensity $I_o \equiv a_o^2$

Instead of estimating the (normalized) signal amplitude the estimate of the corresponding signal intensity $I_o (\equiv a_o^2)$ may be desired. Here we may take advantage of the general observation that any monotonic function of an optimum algorithm, that is, a Bayes likelihood ratio $\Lambda(\boldsymbol{x})$ for detection, or a Bayes estimator γ_θ^* for optimally estimating a parameter θ, is also Bayes optimal. Well-known examples are the familiar log Λ representation in detection (e.g., (6.1.6b)) and $\log \hat{\Lambda}_1$, (6.4.6a) for estimation. Since I_o is simply monotonic in a_o, many of the specific results of Sections 6.4.1–6.4.4 above are immediately transformed to corresponding relations for I_o through the relation $a_o = I_o^{1/2}$, with optimality preserved. Thus, we see that

$$p = 1\colon \gamma^*(I_o|\boldsymbol{x})_{\text{SCF,QCF}} = I_o^*|_{\text{SCF,QCF}} = a_{p=1}^{*2} = \left(\frac{\sigma^2\Phi_x + \overline{a_o}}{\sigma^2\Phi_s + 1}\right)^2 = y(x)^2, \lambda = 1 + \sigma^2\Phi_s, \tag{6.4.49}$$

from (6.4.5), (6.4.6c).

However, the estimator I_o^* is biased under H_1, since from (6.4.8a)

$$p = 1\colon \quad \left\langle I_o^*|_{\text{SCF,QCF}}\right\rangle_{H_1} = \langle y^2\rangle_{H_1} = \bar{a}_o^2 + \sigma^4\Phi_s/(\sigma^2\Phi_s + 1) = \langle a_{p=1}^{*2}\rangle + \langle B(\boldsymbol{x})\rangle_{H_1}, \tag{6.4.50}$$

cf. (6.4.7), with an *average bias*

$$\langle B(\boldsymbol{x})\rangle_{H_1} = \sigma^4\Phi_s/(\sigma^2\Phi_s + 1), \tag{6.4.50a}$$

usually called "the bias," vide the discussion following Eqs. (6.3.43a–c). In addition, the Bayes risks for $I_o^*|_{p=1}$ are analogous to (6.4.17a–c) for $a_{p=1}^*$, and are also not monotonic in the Bayes risks for $a_{p=1}^* \left(= \sqrt{I_0^*}\right)_{p=1}$, as can be seen from calculations using the transformed pdfs of $I_o^*(= y^2)$, $I_o^*(= a_o^2)$, namely,

$$\left.\begin{aligned}
\hat{w}_1\left(I_o^*|_{H_{0,1}}\right) &= \frac{1}{2\sqrt{I_o^*}} w_1\left(\sqrt{I_o^*}|_{H_{0,1}}\right)_y, \text{ Eqs. (6.4.11) and (6.4.8);} \\
\hat{w}_1\left(I_o^*, I_o|_{H_1}\right) &= \frac{1}{4\sqrt{I_o^* I_o}} w_1\left(\sqrt{I_o^*}, \sqrt{I_o}|_{H_1}\right)_{y,a_o}, \text{ Eq. (6.4.13);} \\
\hat{w}_1(I_o) &= \frac{1}{2\sqrt{I_o}} w_1\left(\sqrt{I_o}\right)_{a_o}, \text{ Eq. (6.4.1)}
\end{aligned}\right\}, I_o^* \geq 0, \tag{6.4.51}$$

or by direct calculations based on $a^*_{p=1}$ itself. For instance, in the case of the QCF we have now for the Bayes or minimum average error

$$p = 1: \quad \begin{aligned} R^*_{E|QCF|p=1} &= C_0\left\{\left\langle I_o^{*^2}\right\rangle_{H_1} - 2\left\langle I_o^* I_o\right\rangle_{H_1} + \overline{I_o^2}\right\} \\ &= C_0\left\{\left\langle a_o^{*^4}\right\rangle_{H_1} - \left\langle 2a_o^{*^2} a_o^2\right\rangle_{H_1} + \overline{a_o^4}\right\}_{p=1}, \end{aligned} \tag{6.4.52}$$

which on the application of (6.4.8) and (6.4.1), can easily be shown to differ significantly from the square of (6.4.17a). Similar remarks hold for comparisons of optimum and suboptimum estimators of signal intensity, $p = 1$.

By the same reasoning, in the more general situation for $p \leq 1$ we see that Eqs. (6.4.20), (6.4.21) for $a^*_{p\leq 1|QCF,SCF}$ may be extended directly, like (6.4.4a) for $p = 1$, to the case $p \leq 1$:

$$\left\{\begin{aligned} & I_o^*\Big|_{SCF|p\leq 1} = pI_o^*\Big|_{SCF|p=1} = p\left(\frac{\sigma^2\Phi_x + a_o}{\sigma^2\Phi_s + 1}\right)^2 = py(\boldsymbol{x})^2 && \text{(6.4.53a)} \\ & I_o^*\Big|_{QCF|p\leq 1} = \frac{\Lambda(\boldsymbol{x})}{1+\Lambda(\boldsymbol{x})} I_o^*\Big|_{QCF|p=1} = I_o^*\Big|_{QCF|p=1} \cdot \left\{1 + \frac{\sqrt{\lambda}}{\mu} e^{\bar{a}_o^2/2\sigma^2 - \left(\sigma^2\Phi_x + \bar{a}_o\right)^2/2\sigma^2\lambda}\right\}^{-1} && \text{(6.4.53b)} \end{aligned}\right.$$

since $\Lambda(\boldsymbol{x})$, (6.4.18), here is invariant of the estimator(s) $\left(I_o^* \text{ or } I_o\right)$, depending only on the original pdfs (6.4.1) of a_o and the received data $\boldsymbol{x}$ (6.4.2), (6.4.3). This also means that the associated detection process [specified in Section 6.4.4 and depending on the (log) likelihood ratio of (6.4.18) specifically], which is required for acceptance or rejection of an estimate (Section 6.3.6) is similarly invariant of the estimators employed in the generic example of this section. But performance and comparisons, measured here by Bayes error, when $p \leq 1$ also do not simply scale according to the relation between I_o^* and a_o^*. We may expect results for Bayes error to be much more complex for the former than for the latter, by extrapolation from the results for $p = 1$, cf. (6.4.52). In summary, then, we may state generally that: (1) quantities to be estimated that are monotonic in other estimatible quantities, for example, $I_o = a_o^2$ here, have Bayes estimators monotonic in the corresponding other Bayes estimators. However, (2) the associated measures of performance (Bayes error or risk) are not similarly monotonic. Moreover, (3) the new Bayes estimators are generally biased, even when the original ones are unbiased, as has been shown by our analysis throughout the chapter up to this point. Performance and performance comparisons, along the lines of Section 6.3.6, must be separately calculated: they are not simply scaled by the monotonicity relationship exemplified here by (6.4.49).

PROBLEMS

P6.13 **(a)** Obtain the results Eq. (6.4.5), $p = 1$, for γ^*_{QCF}, and for γ^*_{SCF}, Eq. (6.4.6c), of the normalized amplitude a_0, obeying the pdfs of Eqs. (6.4.1) and (6.4.2).

(b) Obtain in the strong signal cases the optimum amplitude estimator $a^*_{p=1} \cong a_0$, $\sigma^2\Phi_s >> 1$, cf. (6.4.7).

(c) Obtain the pdf $w_1(y|H_1)$ under H_1 of $y \equiv a_o^*(\mathbf{x})_{p=1}$, as given by Eq. (6.4.8), and show that accordingly (6.4.8a) represents $\langle y \rangle_{H_1}$ and $\langle y^2 \rangle_{H_1}$ here. *Hint*: use the Gaussian nature of a_o and a_o^* along with their various characteristic functions (Problem 6.14).

(d) Obtain the pdf under H_1 of $\Delta a_o^* = y - a_o$, Eq. (6.4.9a).

(e) Show that $w_1(y|H_0)$, $w_1(y|a_o)$, $w_1(y, a_o|H_1)$, are respectively given by Eqs. (6.4.11), (6.4.12), (6.4.13).

P6.14 For the Gaussian example of Section 6.4, the principal aim is to obtain Bayes estimators and optimal D and E performance of the (normalized) Gaussian amplitude a_o. The signal waveform $\boldsymbol{s}$ is known *a priori*, namely, $\boldsymbol{S} = a_0 \boldsymbol{s}\sqrt{\psi}$, as is the covariance function $K_N(r_1, t_1; r_2, t_2)(= \psi_N \boldsymbol{k}_N)$ of the additive normal noise N, with $\bar{N} = 0$, which accompanies the signal. Here the pdf $w_1(a_o)$ of a_o is specified by Eq. (6.4.1), the noise pdf by (6.4.3), with the normalized data vector $\boldsymbol{x}(= [\boldsymbol{X}/\sqrt{\psi}])$ in the usual way.

(a) Show that if the likelihood ratio is $\Lambda(\boldsymbol{x}) = \mu\, D(\boldsymbol{x}) >> 1$ or $<< 1$, then here the Bayes estimator (6.4.21) of a_o for the QCF becomes

$$\left\{\begin{array}{r} 0 \leq \Lambda << 1\colon a_{p\leq 1}(\boldsymbol{x})^*_{\mathrm{QCF}} \doteq (\Lambda - \Lambda^2 + \cdots) a^*_{p=1}(\boldsymbol{x})^*_{\mathrm{QCF}} \doteq \Lambda\, a^*_{p=1}(\boldsymbol{X})^*_{\mathrm{QCF}}, \\ \text{Eqs.}(6.4.5) \text{ and } (6.4.6c) \\ 1 << \Lambda \leq \infty\colon \cong (1 - \Lambda^{-1} + \Lambda^{-2} + \cdots) a^*_{p=1}(x)^*_{\mathrm{QCF}} \cong a^*_{p=1}(\boldsymbol{X})^*_{\mathrm{QCF}} \end{array}\right\} \tag{1}$$

where $\Lambda(\boldsymbol{x})$ now is given explicitly by (6.4.18).

(b) Show for H_0: $\boldsymbol{x} = \boldsymbol{n}$; $H_1 = \boldsymbol{x} = \boldsymbol{n} + a_o\boldsymbol{s}$, where $\Phi_x \equiv \tilde{\boldsymbol{x}}\boldsymbol{k}_N^{-1}\boldsymbol{s}$; $\Phi_s \equiv \tilde{\boldsymbol{s}}\boldsymbol{k}_N^{-1}\boldsymbol{s}$, that

$$\langle \Phi_x \rangle_{H_0} = 0;\ \langle \Phi_x^2 \rangle_{H_0} = \Phi_s;\ \langle \Phi_x \rangle_{H_1} = \bar{a}_0 \Phi_s;\ \langle \Phi_x^2 \rangle_{H_1} = \Phi_s + \overline{a_o^2}\Phi_s^2, \tag{2}$$

in which $\langle\,\rangle_{H_0} = \langle\,\rangle_{\boldsymbol{n}}$; $\langle\,\rangle_{H_1} = \langle\,\rangle_{\boldsymbol{n},\,a_o}$.

(c) For the pdfs of column one of Table 6.2 below, obtain the associated characteristic function and moments:

P6.15 **(a)** Using the results of Eqs. (6.4.8)–(6.4.13) and Problem 6.14(b), obtain the Bayes risks R_E^*, (6.4.17), $p = 1$.

(b) Verify the results (6.4.26) and (6.4.27).

(c) Verify the results (6.4.35) for the average risk $R^*_{E|p\leq 1|\mathrm{QCF}}$.

(d) Obtain $\langle a_{p\leq 1|\mathrm{QCF}} \rangle$, Eq. (6.4.36).

P6.16 Apply the relations [(6.3.45) and (6.3.46)] for *interval estimation* of $a^*_{p|\mathrm{SCF}}$ when $p \leq 1$, (6.4.20), to the example discussed in Section 6.4:

(a) Show that the pdf (6.3.45) of the estimator $\gamma_\theta = a^*_{p\leq 1|\mathrm{SCF}}$, Eq. (6.4.20), of the true value a_o here is given by

TABLE 6.2 $[\lambda \equiv 1 + \sigma^2\Phi_s, (6.4.8)]$

pdf : $y \equiv a_{p=1}(x)^*$	log (c.f.) $= \log F_1(i\xi\|H_{0,1})_y$	$\langle y \rangle$	$\langle y^2 \rangle - \langle y \rangle^2 = \sigma^2$
$w_1(y\|H_0)$, Eq. (6.4.11)	$i\bar{a}_o\xi/\lambda - \sigma^4\Phi_s\xi^2/2\lambda^2$	$\dfrac{\bar{a}_o}{\lambda}$	$\dfrac{\sigma^4\Phi_s}{\lambda^2}$
$w_1(y\|H_1)$, Eqs. (6.4.8) and (6.4.8a)	$i\bar{a}_o\xi - \sigma^4\Phi_s\xi^2/2\lambda^2$	$\bar{a}_o$ (5.4.7)	$\dfrac{\sigma^4\Phi_s}{\lambda}$
$w_1(y - a_o\|H_1)$, Eqs. (6.4.9a) and (6.4.9b)	$-\sigma^2\xi^2/2\lambda$	$\langle \Delta a_o^* \rangle_{H_1} = \langle y - a_o \rangle_{H_1} = 0$	$\dfrac{\sigma^2}{\lambda}$
$w_1(y\|a_o)_{H_1}$, Eq. (6.4.12)	$i\left(\dfrac{\bar{a}_o + a_o\sigma^2\Phi_s}{\lambda}\right) \times \xi - \dfrac{\sigma^4\Phi_s\xi^2}{2\lambda^2}$	$\bar{a}_o$	$\dfrac{\sigma^4\Phi_s}{\lambda^2}$
$w_1(y, a_o\|H_1)$, Eq. (6.4.13)	$i\left(\dfrac{\bar{a}_o + a_o\sigma^2\Phi_s}{\lambda} + \bar{a}_o\right) \times \xi - \dfrac{\xi^2}{2}\left(\dfrac{\sigma^4\Phi_s}{\lambda^2} + \sigma^2\right)$	$2\bar{a}_o$	$\dfrac{\sigma^4\Phi_s}{\lambda^2} + \sigma^2$
$w_1(a_o)$, Eq. (6.4.1)	$i\bar{a}_o\xi - \sigma^2\xi^2/2$	$\bar{a}_o$	σ^2

$$P_L\left(a^*_{p\leq 1|\text{SCF}}|a_0\right) = q \cdot \left\{ \exp \frac{\left[-(a_o^* - pa_o/\lambda)^2/2\sigma^4\Phi_s p^2/\lambda^2\right]}{\left(2\sigma^4\Phi_s p^2/\lambda^2\right)^{1/2}} \right\} + p\left\{ \exp \frac{\left[-(a_o^* - pa_o)^2/2\sigma^4\Phi_s/p^2/\lambda\right]}{\left(2\sigma^4\Phi_s p^2/\lambda\right)^{1/2}} \right\}, \tag{1}$$

with c.f.

$$F_1\left(i\xi|H\right)_{py} = q \exp\left[ip\bar{a}_o\xi/\lambda - \sigma^4\Phi_s p^2\xi^2/2\lambda^2\right] + p \exp\left[ip\bar{a}_o\xi - \sigma^4\Phi_s p^2\xi^2/2\lambda\right]. \tag{1a}$$

(b) Show that the "interval estimate" (6.3.45), or probability that $a^*_{p\leq 1|\text{SCF}}$ lies in the interval $\left[\left(1 - \hat{\lambda}\right), \left(1 + \hat{\lambda}\right)\right]$, is here

$$P_{I|p\leq 1} = \frac{q}{2}\left\{\Theta\left[\frac{\left(1+\hat{\lambda}\right)a_o - p\bar{a}_o/\lambda}{\left(2\sigma^4\Phi_s p^2/\lambda^2\right)^{1/2}}\right] - \Theta\left[\frac{\left(1-\hat{\lambda}\right)a_o - p\bar{a}_o/\lambda}{\left(2\sigma^4\Phi_s p^2/\lambda^2\right)^{1/2}}\right]\right\} + \frac{p}{2}\left\{\Theta\left[\frac{\left(1+\hat{\lambda}\right)a_o - p\bar{a}_o}{\left(2\sigma^4\Phi_s p^2/\lambda\right)^{1/2}}\right] - \Theta\left[\frac{\left(1-\hat{\lambda}\right)a_o - p\bar{a}_o}{\left(2\sigma^4\Phi_s p^2/\lambda\right)^{1/2}}\right]\right\}. \tag{2}$$

Hence, for the unconditionally unbiased estimator $\langle a^*_{p\le 1|\mathrm{SCF}}\rangle_H = p\bar{a}_o$, cf. (6.4.23), we may reasonably replace the unknown value of a_0 by $p\bar{a}_0$, so that the interval estimate (2) now becomes

$$P_{I(p\le 1)} = \frac{q}{2}\left\{\Theta\left[\frac{\left(1+\hat{\lambda}-\lambda^{-1}\right)p\bar{a}_o}{\left(2\sigma^4\Phi_s p^2/\lambda^2\right)^{1/2}}\right] - \Theta\left[\frac{\left(1-\hat{\lambda}-\lambda^{-1}\right)p\bar{a}_o}{\left(2\sigma^4\Phi_s p^2/\lambda^2\right)^{1/2}}\right]\right\}$$

$$+\, p\Theta\left[\frac{\hat{\lambda}p\bar{a}_o}{\left(2\sigma^4\Phi_s p^2/\lambda\right)^{1/2}}\right], \tag{3}$$

which reduces further to the simple result

$$P_{I|p=1} = \Theta\left[\frac{\hat{\lambda}p\bar{a}_o}{\left(2\sigma^4\Phi_s p^2/\lambda\right)^{1/2}}\right],\ q=0. \tag{3a}$$

P6.17 The Bayes risk for the intensity estimator $I_0^*(\boldsymbol{x})_{p=1}\left(=a^{*2}_{p=1}\right)$, (6.4.49), is given by Eq. (6.4.52) for the QCF, namely,

$$R^*_{E|\mathrm{QCF}|p=1} = C_0\left[\left\langle I_0^{*2}\right\rangle_{H_1} - 2\langle I_0^* I_0\rangle_{H_1} + \overline{I_0^{*2}}\right] \tag{1}$$

Show that

$$\begin{cases} \overline{I_0^{*2}} = \bar{a}_o^4 + 6\sigma^2\bar{a}_o^2 + 3\sigma^4/4;\ \left\langle I_0^{*2}\right\rangle_{H_1} = \left\{3 + 6\bar{a}_o\lambda/2\sigma^2\Phi_s + 4\bar{a}_o^4\lambda^2/\left(2\sigma^2\Phi_s\right)^2\right\}\dfrac{\sigma^3\Phi_s^2}{\lambda^{3/2}}, & \text{(2a)} \\ -2\langle I_0^* I_0\rangle_{H_1} = -\dfrac{2}{\lambda^2\sqrt{\pi}}\{2\sigma^2 A_2 + 6\bar{a}_o A_3 + 12\sigma^2\bar{a}_o^2 A_4/4 + 3\sigma^4 A_4/4 & \end{cases}$$

$$+\left(A_2\bar{a}_o^2 + A_3\bar{a}_o^3 + A4_2\bar{a}_o^4\right)\}, \tag{2b}$$

with

$$\left.\begin{aligned} A_2 &= \sigma^4\Phi_s/2 + \bar{a}_o^2; \\ A_3 &= 2\bar{a}_o\sigma^2\Phi_s; \\ A_4 &= \sigma^4\Phi_s^2. \end{aligned}\right\} \tag{2c}$$

P6.18 Show for $u(\boldsymbol{x}) = B_0 + \frac{\lambda}{2\sigma^2}\ell_0^*(\boldsymbol{x})$ in Eq. (6.4.41) that the pdfs $Q_1\left(\ell_0^*\right)$ and $P_1\left(\ell_0^*\right)$ are given by Eqs. (6.4.44a) and (6.4.44b) on setting $(u-B_0)\frac{(2\sigma^2/\lambda)}{\lambda} = \ell_0^*$ therein. Hence show that p_D^* (and P_D^*), as well as α^*, remain unchanged from the general results (6.4.46) and (6.4.47).

P6.19 (a) Evaluate the likelihood ratio for detection Λ_g' for the example of Section 6.4, in the weakly coupled D and E cases, when the cost function $f_{01}(\boldsymbol{S}(a_0))$ in Eq. (6.3.21a) is given by

$$\text{A. } f_{01} = a_o^2 A_{01}; \quad \text{B. } f_{01} = a_o B_{01}. \tag{1}$$

Thus, show that Λ_g' is explicitly

$$\begin{aligned}\text{A.} \quad \Lambda_g'(\boldsymbol{x}) &= K'\Lambda \cdot \left\{ 1/\sqrt{\lambda} + \frac{A_{01}\mu}{[C_\beta - C_{1-\beta}]} \frac{\sigma^2}{\lambda}\left[1 + \frac{\lambda}{\sqrt{2}\sigma^2} a_{p=1}^{*2}(\boldsymbol{x})\right]\right\} \\ &= \Lambda \cdot \left\{ a_2 + b_2 y(\boldsymbol{x})^2 \right\}\end{aligned} \tag{2}$$

$$\text{B.} \quad \Lambda_g'(\boldsymbol{x}) = \Lambda \cdot \left\{ K' + B_{01} a_{p=1}^*(\boldsymbol{x}) \right\} = \Lambda \cdot [a_1 + b_1 y(\boldsymbol{x})]. \tag{3}$$

(b) Obtain the pdfs $Q_1(u)$, $P_1(u)$ for $u = u(\boldsymbol{x}) =$ a *monotonic equivalent to* $\log \Lambda_g'(\boldsymbol{x})$ under H_0 and H_1, for example,

$$\text{A. } \boldsymbol{u}(\boldsymbol{x}) = \log \Lambda + a_2 + b_2 y^2(\boldsymbol{x}); \text{ B. } \boldsymbol{u}_1(\boldsymbol{x}) = \log \Lambda + a_1 + b_1 y(\boldsymbol{x}) \tag{4}$$

(c) Obtain the probabilities p_D^* and of performance by obvious modifications of the analysis of Section 6.4.4. Thus, (6.4.41) becomes

$$\begin{aligned}&\text{A. } \log \Lambda_g'' = B_0 + a_2 + (\lambda/2\sigma^2 + b_2)y^2; \\ &\text{B. } \log \Lambda_g'' = B_0 + a_1 + b_1 y + \frac{\lambda}{2\sigma^2} y^2.\end{aligned} \tag{6}$$

(For B. complete the square and modify the pdfs of y (Eqs. (6.4.8) and (6.4.11) correspondingly.) Note that the decision process is now: decide H_1: $S \otimes N$ if $u_{1,2}(\boldsymbol{x}) \geq A(> 0)$, or H_0: N if $u_{1,2}(\boldsymbol{x}) < A$, where the threshold $A(= A(K'))$ is determined by presetting α_F^* at the desired level.

P6.20 (a) Prove that

Theorem: If the function $F_J(\boldsymbol{X}|\boldsymbol{S})$ of the independent variable S is bounded and continuous for every α_F^* and for every value of the parameter $S \in \Omega$, then for each $\boldsymbol{X} \in \Gamma$ there exists a value $\boldsymbol{X}$ such that

$$\int_\Omega F_J(\boldsymbol{X}|\boldsymbol{S}) w_L(\boldsymbol{S}) d\boldsymbol{S} = F_J\left(\boldsymbol{X}|\hat{\boldsymbol{S}}(\boldsymbol{X})\right). \tag{1}$$

or equivalently, in terms of likelihood ratio $\langle \Lambda_J(\boldsymbol{X}|\boldsymbol{S}) \rangle_S = \Lambda_J\left(\boldsymbol{X}|\hat{\boldsymbol{S}}\right)$. Thus the $\hat{\boldsymbol{S}}(\boldsymbol{X})$ is a (fixed) signal (for some $\boldsymbol{X}$). The $\hat{\boldsymbol{S}}(\boldsymbol{X})$ is a "pseudo" estimate, with little general value in itself except with respect to the actual detector's performance.

(b) Let $\boldsymbol{N}$ and $\boldsymbol{S}$ be additive Gaussian noises, with zero means, and let $\boldsymbol{N}$ have the covariance $\boldsymbol{K}_N$. Then (1), written in terms of likelihood ratios, is given by

$$\int_\Omega \exp\left(-\frac{1}{2} (\tilde{\boldsymbol{S}} \boldsymbol{K}_N^{-1} \boldsymbol{S}) + \tilde{\boldsymbol{X}} \boldsymbol{K}_N \boldsymbol{S} \right) w_J(\boldsymbol{S}) d\boldsymbol{S} = \exp\left[-\frac{1}{2} \left(\tilde{\hat{\boldsymbol{S}}} \boldsymbol{K}_N^{-1} \hat{\boldsymbol{S}} \right) + \tilde{\boldsymbol{X}} \boldsymbol{K}_N^{-1} \hat{\boldsymbol{S}} \right]. \tag{2}$$

We next give $\boldsymbol{S}$ the Gaussian distribution $w_J(\boldsymbol{S}) = (2\pi)^{-J/2} (\det \boldsymbol{K}_S)^{-1/2}$ $\exp\left(-\frac{1}{2} \left(\hat{\boldsymbol{S}} \boldsymbol{K}_S^{-1} \hat{\boldsymbol{S}} \right) \right)$. Then show that the likelihood ratio becomes

$$\Lambda = \mu \det\left(\boldsymbol{I} + \boldsymbol{K}_S\boldsymbol{K}_N^{-1}\right)^{1/2} \exp\left\{\frac{1}{2}\tilde{\boldsymbol{X}}\boldsymbol{K}_N^{-1}\left(\boldsymbol{K}_S^{-1} + \boldsymbol{K}_N^{-1}\right)\boldsymbol{K}_N^{-1}X\right\}. \tag{3}$$

The structure of the estimator is

$$\hat{\boldsymbol{S}}(\boldsymbol{X}) = \left(\boldsymbol{K}_S^{-1} + \boldsymbol{K}_N^{-1}\right)\boldsymbol{K}_N^{-1}\boldsymbol{X}, \tag{4}$$

which is recognized as a minimum variance estimator of the stochastic signal $\boldsymbol{S}$, under the hypothesis (H_1) that the signal is indeed present in the observation interval $(p = 1)$. Thus, the interpretation is given that the average likelihood ratio (the right most part of (2)) consists of a fixed bias term depending only on *a priori* information and an operation of cross-correlation of the data $\boldsymbol{X}$ with a minimum variance estimate $\hat{\boldsymbol{S}}(\boldsymbol{X})$ of the stochastic signal $\boldsymbol{S}$. (This is essentially the Price–Kailath result, specialized to the binary problem above [11]. See Esposito [12] for some further explanation of these results.)

P6.21 In the Bayes theory of optimum detection, it is illuminating in the context of joint detection and estimation as discussed in this chapter to point out an *interpretation* of detection alone that also incorporates the concept of an estimator. This is a form of minimum variance estimator $\boldsymbol{S}$ of the signal, which is obtained from the average (i.e., the generalized) likelihood ratio, Λ, where now it is assumed that the signal estimator $\hat{\boldsymbol{S}}$ is perfectly known, that is, obtained under $H_1\,(p = 1)$. For additive Gauss noise, this estimator is specifically a quantity linearly related to the logarithmic gradient of the (average) likelihood ratio Λ and conversely, this (average) likelihood ratio is expressible uniquely in terms of this minimum variance estimator, $\hat{\boldsymbol{S}}$, *obtained under* $H_1\,(p = 1)$. Expressed analytically, this result is

$$\hat{\boldsymbol{S}} = \boldsymbol{K}_N \nabla \log \Lambda \quad \text{and} \quad \Lambda = \exp\left(\tilde{\boldsymbol{X}}\boldsymbol{K}_N\hat{\boldsymbol{S}} - \int \tilde{\boldsymbol{X}}\boldsymbol{K}_N d\hat{\boldsymbol{S}} + C\right), \tag{1}$$

where C is a constant independent of $\boldsymbol{X}$, and where specifically

$$\Lambda = \frac{p\langle W_J(\boldsymbol{X} - \boldsymbol{S})_N\rangle_S}{q w_J(\boldsymbol{X})} = \mu \int_\Omega \exp\left[-\frac{1}{2}(\tilde{\boldsymbol{X}} - \tilde{\boldsymbol{S}})\boldsymbol{K}_N^{-1}(\boldsymbol{X} - \boldsymbol{S})\right] w_J(\boldsymbol{S}) d\boldsymbol{S}$$

$$\Big/ \exp\left(-\frac{1}{2}\tilde{\boldsymbol{X}}\boldsymbol{K}_N^{-1}\boldsymbol{X}\right), \quad \mu = p/q. \tag{2}$$

It is emphasized that this estimator $\hat{\boldsymbol{S}}$ is useful only with respect to the detection process here. Hence, we may alternatively call it a "*pseudoestimator*," to distinguish it from the estimators under $H_1\,(p < 1)$, which occur in the joint D + E problems considered in the present chapter. (If the Bayes estimator $\gamma^*(\boldsymbol{S})_{p<1}$ is desired when $p < 1$, see, e.g., Sections 6.2.1, 6.2.2, 6.3, and 6.3.7.)

Accordingly, show that $\hat{\boldsymbol{S}}$ is given by (1) and that the estimator–correlator is explicitly represented in the Gaussian cases by the first term in the exponent of Λ, (1).

Hints: note that the minimum variance (MMSE) of $\boldsymbol{S}$ under $H_1(p = 1)$ is

$$\hat{\boldsymbol{S}} = \hat{\boldsymbol{S}}(\boldsymbol{X}) = \int_{\Omega} \boldsymbol{S} \exp\left\{-\frac{1}{2}(\tilde{\boldsymbol{X}} - \tilde{\boldsymbol{S}})\boldsymbol{K}_N^{-1}(\boldsymbol{X} - \boldsymbol{S})\right\} w_J(\boldsymbol{S}) \Big/ \int_{\Omega} \exp\left\{-\frac{1}{2}(\tilde{\boldsymbol{X}} - \tilde{\boldsymbol{S}})\boldsymbol{K}_N^{-1}(\boldsymbol{X} - \boldsymbol{S})\right\} w_J(\boldsymbol{S}) d\boldsymbol{S}. \tag{3}$$

Define a function $G(\boldsymbol{X})$ equal to the numerator of (2). Using (3), obtain $G^{-1}\nabla G = -\boldsymbol{K}_N^{-1}(\boldsymbol{X} - \boldsymbol{S})$ where from (2) get $\nabla G = \left(-\boldsymbol{K}_N^{-1}\boldsymbol{X}\Lambda + \nabla\Lambda\right) \exp\left(-\frac{1}{2}\tilde{\boldsymbol{X}}\boldsymbol{K}_N^{-1}\boldsymbol{X}\right)$, and divided by $G^{-1}\nabla G$, using (2) again, obtain

$$G^{-1}\nabla G = \boldsymbol{K}_N^{-1}\boldsymbol{X} + \Lambda^{-1}\nabla\Lambda. \tag{4}$$

From (4) and $G^{-1}\nabla G$ obtain the desired result

$$\hat{\boldsymbol{S}} = \boldsymbol{K}_N \nabla \log \Lambda; \; H(p = 1). \tag{5}$$

Thus, the transformation mentioned above is linear and is embodied by $\boldsymbol{K}_N$, the covariance matrix of the afore-mentioned Gaussian noise. To establish the converse, Λ, Eq. (1), one can always express this generalized (i.e., averaged over $\boldsymbol{S}$) likelihood ratio Λ as

$$\Lambda = \exp\left\{\boldsymbol{K}_N^{-1} \int \hat{\boldsymbol{S}} d\boldsymbol{X} + C\right\}, \tag{6}$$

which on integration by parts is just (1). Thus, Λ can always be represented by an "estimator–correlator," namely, the first term in (1), and an additional bias term that depends on the estimator and on the data. Finally, we remark that the estimator $\hat{\boldsymbol{S}}$ in (3) is *noncasual*, since $\boldsymbol{S}$ is estimated as a whole given all data $\boldsymbol{X}$ in the sample. (These results are due to Esposito [13], also [12]; see also Refs [11,14].)

6.5 SUMMARY REMARKS, $p(H)_1 \leq 1$: I — FOUNDATIONS

The fundamental problem addressed here, with extensions in this chapter, is parameter estimation[36] when it is not surely known *a priori* that a signal containing the parameter (or parameters) to be extracted is present in a noise background. As mentioned earlier, the dominating assumption in the application of estimation procedures is almost always that the signal, with its parameters, is present in the received data $X(t)$, namely that $p(H_1) = 1$. Often there is little justification for this, particularly when detection for one or another reason is required, which is an admission that the signal detection probability P_D is not strictly unity. The principal results of assuming $P_D = 1$, that is, that $p(H_1)$ is correspondingly unity, is that the estimators, including optimal or Bayes estimators, are degraded, namely in reality are *suboptimum* and are moreover *biased*, with an unknown bias.

If $p(H_1)$ is close to unity, or if P_D is likewise nearly unity, it may be acceptable to assume $p(H_1) = 1$, and so on, and regard the resultant estimator as effectively optimum or as an

[36] Including the signal itself.

effective processor if it is suboptimum to begin with. The key question here, clearly, is how close to unity must $p(H_1)$ be for such a decision to be made with acceptably small error. This is ultimately a subjective choice, of course. However, the quantitative results of the examples of Section 6.4 strongly suggest that p should be $O(\geq 0.90)$ at least. Such values indicate small false alarm probabilities and require large values of $(S/N)_{\text{out}}$, which in turn may demand large processing gains [$\Pi^{(*)}$, cf. ___] in the situations involving weak received signals.

The structure of the Bayes estimators can also be noticeably influenced by the degree of coupling between estimator and detector, as noted in Table 6.1. With "no coupling," detection (D) and estimation (E) are carried out independently. The processing required for each does not affect the other. With "weak coupling," D and E are again independent, but the detection algorithm, Λ'_g, depends on a cost function f_{01} containing the signal, cf. Eq. (6.3.22). The order in which D and E are carried out is arbitrary once more. In the case of strong coupling, however, estimation must precede detection (Section 6.3.1), with the detection process influenced by the estimate and vice versa (Table 6.1).

In several cases of practical interest it is possible to carry through the formal operation indicated by the general theory and thus obtain explicit analytic results for the estimators under $p < 1$, and for performance (expected risk, etc.). This is evident for amplitude estimation (Section 6.4), under rather restrictive conditions (6.4.0). The latter, however, can be considerably mitigated in the threshold regimes, allowing extensions to non-Gaussian and nonadditive signal and noise. Chapter 7 provides a number of extensions of the joint theory discussed in the this chapter. These include the role of incoherent reception in amplitude estimation under $p \leq 1$; an example of waveform estimation; multiple alternative D and E [3]; and extension of the present generic "one-shot" (i.e., single-observation interval) theory to include "multishot" or joint sequential detection and estimation under the condition $p(H_1) \leq 1$ [2].

Finally, it is emphasized that in specific applications quantitative results here in most cases must be obtained by computational methods, usually sooner rather than later in the analysis. Because of the present memory, speed, and economy of available computational assets, this is not the inhibiting factor that it was at the initial development of the theory in 1968 [1], 1970 [2].

REFERENCES

1. D. Middleton and R. Esposito, Simultaneous Optimum Detection and Estimation of Signals in Noise, *IEEE Trans. Inform. Theory*, **IT-14**(3), 434–440 (1968).
2. D. Middleton and R. Esposito, New Results in the Theory of Simultaneous Optimum Detection and Estimation of Signals in Noise, *Prob. Peredachi Inf. (Prob. Inform. Transm.)*, **6**(2), 3–20, (1970), Ak. Nauk, Moscow, USSR; English translation published Jan. 1973, pp. 93–106, Consultants Bureau, Plenum Publishing, New York. (Originally published as a Raytheon Technical Memorandum (T-838), Oct. 31, 1969; a concise version of the above paper was presented by R. E. at the Symposium on Information Theory, Dubna, USSR, June 1969.)
3. A. Fredriksen, D. Middleton, and D. Vandelinde, Simultaneous Signal Detection and Estimation Under Multiple Hypotheses, *IEEE Trans. Inform. Theory*, **IT-18**(5), 607–614 (1972). This material represents a portion of Fredriksen's Ph.D. dissertation, May 1970.
4. D. Middleton, *An Introduction to Statistical Communication Theory*, McGraw-Hill, New York, 1960; IEEE Press, *Classic Reissue*, Piscataway, NJ, 1996.

5. M. G. Kendall and A. Stewart, *The Advanced Theory of Statistics*, 2nd ed., Vol. 2, Hafner Publishing Co., New York, 1967, p. 161.
6. J. A. Proakis, P. R. Drouilhet, Jr., and R. Price, Performance of Coherent Detection Systems Using Decision-Directed Channel Measurement, *IEEE Trans. Commun. Syst.*, **12**, 54–63 (1964).
7. S. Sherman, Non-Mean-Square Error Criteria, *IRE Trans. on Inform. Theory*, **IT-4**, 125 (1958).
8. D. Middleton, *Topics in Communication Theory*, McGraw-Hill, New York, 1965–1972.
9. D. Middleton and A. H. Nuttall, Estimation of Signal Amplitude and Intensity Under Prior Uncertainty, Technical Report, Naval Undersea Warfare Center (NUWC), Newport, RI, 1999.
10. G. A. Campbell and R. M. Foster, *Fourier Intregrals for Practical Applications*, 2nd Printing, D. Van Nostrand, New York, NY, 1948.
11. R. Price, Optimum Detection of Random Signals in Noise, With Applications to Scatter-Multipath Communications, Part I, *IRE Trans. Inform. Theory* **IT-2**, 125–135 (1956), and see Ref. [14].
12. R. Esposito, A Class of Estimators for Optimum Adaptive Detection, *Inform. Control*, **10**, 137–148 (1967).
13. R. Esposito, On a Relation between Detection and Estimation in Decision Theory, *Inform. Control*, **12**, 116–120 (1968).
14. T. Kailath, Adaptive Matched Filters, in R. Bellman (ed.), *Mathematical Optimization Techniques*, Dowden, Hutchinson, and Ross, Inc., Stroudsburg, PA, 1963.

7

JOINT DETECTION AND ESTIMATION UNDER UNCERTAINTY, $p_k(H_1) < 1$. II. MULTIPLE HYPOTHESES AND SEQUENTIAL OBSERVATIONS

In Chapter 6, we have described from a Bayesian viewpoint a theory of simultaneous binary detection (H_1: $S \oplus N$ vs. H_0: N) and estimation of a single signal (or one or more of its parameters). In all cases, only one signal out of a set of one or more possible signals can be present, subject to simultaneous detection and estimation.[1] The coupling strategies between detection and estimation in these cases, and their results, depend on three different measures of coupling, namely, no coupling, weak coupling, and strong coupling, (cf. Sections 6.1.1, 6.1.2 and 6.3.1, and Table 6.1). The treatment in Chapter 6 deals with the nonsequential single binary on–off decision situation, based on discrete sample data received in the finite space–time interval $(0, \boldsymbol{R})$; $(0, T)$.

Here, we extend the analysis to topic (1), the single decision cases involving simultaneous optimum detection and estimation ($p(H_1) < 1$) under multiple hypotheses, and topic (2), to joint detection and estimation for both sequential observations and multiple decisions. These extensions of the original study of Middleton and Esposito [2, 3], are based mainly on the work of Fredriksen [4, 5], particularly in Ref. [4], extended more recently here to include discrete sampling in space as well as time. Accordingly, this chapter is organized as follows: Section 7.1 treats topic (1) above. Section 7.2 presents a solution of examples relevant to topic (1). Topic (2) on sequential operations is discussed in Section 7.3. Section 7.4

[1] The important situation of D and E when multiple signals can be present *at the same time*, although requiring no fundamentally new theory, is clearly much more complex and is properly the subject of special analysis, cf. Bar-Shalom [1], for example.

Non-Gaussian Statistical Communication Theory, David Middleton.

TABLE 7.1   Topical Summary

		Joint Optimum Detection and Estimation with Multiple Signal States, Overlapping Parameter Spaces, and Various Degrees of Coupling	
One sample interval: $r = 1$	Unsupervised learning	**Section 7.1** *Multiple Hypotheses*: $H_i : S_i(\boldsymbol{\theta}) \oplus N$; $i, k = 1, \dots, K$; $i = 0$: H_0: N	
		7.1.1 Formulation; 7.1.1.1 Optimum Decision Rules	
		7.1.2 Specific Cost Functions; 7.1.2.1 QCFs; 7.1.2.2 QCFs + Gaussian Statistics	
		7.1.2.3 Specific Cost Functions; **II:** (1) SCF$_2$; (2) SCF$_1$	
		7.1.3 Special Cases: Binary Detection and Estimation	
		7.1.3.1 Binary On-Off Detection	
		7.1.3.2 Binary On-Off Detection and Estimation – Strong Coupling	
		7.1.3.3 Binary D and E: $H_1 : S_1(\boldsymbol{\theta}) \oplus N$ vs. $H_2 : S_2(\boldsymbol{\theta}) \oplus N$	
Multiple sample intervals: $r \geq 2$		**Section 7.2** Uncoupled D + E: Multiple Hypotheses: QCF and SCF$_{1,2}$	
		Section 7.3 Sequential D + E: Recursive Distributions, Multiple Hypotheses, Overlapping Parameter Domains	
		7.3.1 Sequential Observations; r - Data Intervals; **I:** Binary Signals	
		$H_1 : S_1(\boldsymbol{\theta}) \oplus N$ vs. H_0: N	
		7.3.1.1 and 7.3.1.2 Uncoupled Detection and Estimation, QCF	
		7.3.1.3 Uncoupled Detection and Estimation, SCF$_{1,2}$	
		7.3.2 Unsupervised Learning: Coupled D + E, r - Data Intervals, Binary On-Off states	
		7.3.2.1 Strong Coupling	
		7.3.2.2 Weak Coupling	
		7.3.3 Unsupervised Learning and Overlapping Hypothesis Classes: r - Intervals Data	
		7.3.3.1 Strongly Coupled D + E: QCF	
		7.3.3.2 Strongly Coupled D + E: SCF$_{1,2}$	
		7.3.3.3 Weak Coupling: D + E: SCF$_{1,2}$	
		7.3.3.4 Multiple Signals / Data Interval: Multiple Tracking	
		7.3.4 Unsupervised Learning: D+E with Multiple Hypotheses and No Coupling	
	Supervised learning	**7.3.5** Supervised Learning (An Introduction)	
		Section. 7.4 Concluding Remarks	
		(1) QCF versus SCF$_{1,2}$: LMS Estimates versus UMLE Estimates: Coupling versus No Coupling	
		(2) Waveform Prediction in Unsupervised Learning	
		(3) Simplification of Recursive Formulations by Simple Markoff (i.e., first Order) Processes	
		(4) Sequential Recursive Formulation versus Wald Theory	
		References	

concludes with some brief discussions of existing and related topics, as well as problems yet to be solved. Table 7.1 provides a guide to the contents of the chapter.

7.1   JOINTLY OPTIMUM DETECTION AND ESTIMATION UNDER MULTIPLE HYPOTHESES, $p(H_1) \leq 1$

Throughout this section we assume that the received data are discretely sampled, according to (1.6.2a) and (1.6.2b), only during one space–time observation interval of duration $|\boldsymbol{R}|T$. We also assume that to each member of a finite transmitted message, set $\{\mathsf{M}_k\}$, $k = 0, 1, \dots, K$, consisting of $K + 1$ elements, there is a distinct received signal selected from the sampled set $\{S_k(\boldsymbol{r}_m, t_n), k = 0, 1, \dots, M\}$. The relations between a message symbol M_k and a (discrete) received signal $\{S_k\}$ can also require the additional step involving the estimation of the descriptive parameters, represented by a vector $\boldsymbol{\theta}_k$, which is associated with the set of signal samples $\{S_k(\boldsymbol{r}_m, t_n|\boldsymbol{\theta}_k)\}$. Thus, when a message symbol M_k is supplied by the message source, a corresponding (continuous) signal is injected into the channel, with the result that the *received signal* (after discrete sampling, refer to the beginning of Section 1.6), becomes in general $\boldsymbol{X}_k(\boldsymbol{r}_m, t_n|\boldsymbol{\theta}_k) = \boldsymbol{N}_k(\boldsymbol{r}_m, t_n|\boldsymbol{\theta}_k) \oplus S_k(\boldsymbol{r}_m, t_n|\boldsymbol{\theta}_k)$ or $\boldsymbol{X}_k = \boldsymbol{N}_k \oplus S_k, j = m, n, m = 0, 1, \ldots, M, n = 0, 1, \ldots, N$ (cf. Section 1.6). As before, $\oplus$ represents "combination," which may be additive, multiplicative, or other physically possible association of signal and noise. The received discrete sampled waveform $\boldsymbol{X}_k$ accordingly depends on the unknown parameter vector $\boldsymbol{\theta}_k$. Here, we have assumed that both the transmitted and received signals are deterministic. This, however, is not a restriction: random signal waveforms may be included as well. The formalism here essentially follows that of the treatment of multiple alternative detection summarized in Chapter 4, but with the added complication of joint parameter or waveform estimates.

7.1.1 Formulation

To allow the possibility that no signal is present during the observation interval $|\boldsymbol{R}|T$ and thus that the received data consist of noise alone, we let one of the message symbols (M_0) correspond to the null signal $\{S_0\} \equiv 0$. The parameter set corresponding to the null signal is $\boldsymbol{\theta} = \boldsymbol{0}$, namely, we require that $\boldsymbol{\theta} = \boldsymbol{0}$ implies $S(\gamma_m, t_n|\boldsymbol{\theta}) \equiv 0$. This restriction limits the parameters considered to be nonnuisance or energy-dependent parameters (amplitude and duration of a signal are examples of such parameters).

The space of possible observed data vectors $\boldsymbol{X}$ is Γ and the space of all possible received signal parameters $\boldsymbol{\theta}$ is Ω, refer to Chapter 1. The spaces Γ and Ω are in general quite arbitrary and the probabilities are to be interpreted as appropriately defined measures on these spaces. The subsets $\Omega_0, \Omega_1, \ldots, \Omega_K$ of the signal space Ω that are assigned to the message symbols $\mathsf{M}_0, \mathsf{M}_1, \mathsf{M}_i, \ldots, \mathsf{M}_K$ are called *signal classes* or *hypothesis classes*.

The decision space is the discrete set $\boldsymbol{\gamma} = \{\gamma_0, \gamma_1, \gamma_i, \ldots, \gamma_K\}$:

γ_0 The decision that hypotheses H_0 is true: noise alone is present, and thus no estimate is required;

γ_i The decision that hypotheses H_i is true, namely, a signal of class Ω_i is present, and an estimate $\hat{\theta}(\boldsymbol{X})$ of the signal parameters θ is required.

Next, we let p_i denote the *a priori* probability of occurrence of signals of class Ω_i, where each signal in class i has the *same* $L(\neq K)$ set of parameters $\boldsymbol{\theta} = [\theta_1, \theta_2, \ldots, \theta_\ell, \ldots, \theta_L]$. (These parameters, however, have *different* pdf values $w_i(\boldsymbol{\theta}) \neq w_k(\boldsymbol{\theta})$, which enables us to deal with the cases where θ_ℓ may be zero, while θ_i and so on, may be nonvanishing). Thus, $w_i(\boldsymbol{\theta})$ implies both $w_i(\boldsymbol{\theta})$ and the set $\boldsymbol{\theta}$ for class i, that is, $\boldsymbol{\theta} \equiv \boldsymbol{\theta}^{(i)}$, which can be different for class $k(\neq i)$. Moreover, $\boldsymbol{\theta}$ may include the original parameter set of the transmitted signals and any parameters introduced by the medium. If we assume that elements of the signal class Ω_k are distributed according to the probability density $w_k(\boldsymbol{\theta})$, the total *a priori* probability density function $w(\boldsymbol{\theta})$ can be expressed as

$$w(\boldsymbol{\theta}) = \sum_{i=0}^{K} p_i w_i(\boldsymbol{\theta}). \tag{7.1.1}$$

Minimizing an average risk or cost associated with the combined operation of detection and estimation, as described in Chapters 4 and 6, provides the basis for analyzing multiple alternative reception here. Accordingly, our analysis begins with the expression of the average risk for the combined operations of detection and estimation and is given by

$$R_{D\oplus E} = \sum_{k=0}^{K} \int_{\Gamma} dX \int_{\Omega} d\boldsymbol{\theta} C(\boldsymbol{\theta}, \gamma_k) P(\gamma_k|\boldsymbol{X}) W(\boldsymbol{X}|\boldsymbol{\theta}) w(\boldsymbol{\theta}). \tag{7.1.2}$$

For *disjoint signal classes*, it is always possible to determine whether a correct or an incorrect decision has been made, given the actual parameter vector $\boldsymbol{\theta}$ associated with a signal and the decision γ made by the receiver. If a cost $C(\boldsymbol{\theta}, \gamma)$ is now associated with each parameter–decision pair $(\boldsymbol{\theta}, \gamma)$, it can be interpreted as the cost of either a correct or an incorrect decision and thus the presence, whether true or not, of a signal of class $\Omega_{(i \text{ or } k)}$.

As we have seen in Chapter 6, the cost functions here play a particularly significant role in joint detection (D) and estimation (E) under uncertainty, that is, when the *a priori* probabilities as to the presence of signals is less than unity, since they determine the extent to which the two operations are coupled together. Accordingly, we find it useful to establish the following category of cost functions in some detail, where typically,

$$\left. \begin{array}{l} C_{ik}^{\beta} = \text{cost of deciding a signal of class } \Omega_k\text{: } S_k(\boldsymbol{\theta}) \otimes N \text{ is present,} \\ \qquad \text{when actually a signal of class } \Omega_i\text{: } S_i(\boldsymbol{\theta}) \otimes N \text{ occurs, including} \\ \qquad \text{the null signal } \Omega_0\text{: } N,\ i = 0, 1, \ldots, K, i \neq k, i = k. \end{array} \right\} \tag{7.1.2a}$$

where the symbol $\otimes$, denotes *no coupling*, *weak coupling*, *strong coupling*, respectively, that is, $D + E, D \otimes E, D \otimes E$, between detection D and estimation E. Explicitly, we have in more detail the following:

I. No coupling:

$$\left. \begin{array}{l} C_{ik}^{(D+E)} = C_{ik}^{(D)} + C_{ik}^{(E)}, \text{ where } C_{ik}^{(D)}, C_{ik}^{(E)} \text{ are constant (positive or zero)} \\ \qquad \text{costs for detection and estimation. (Since detection} \\ \qquad \text{and estimation are uncoupled, we can, at a later stage} \\ \qquad \text{of separate optimizations, replace } C_{ik}^{(E)} \text{ by } C_{ik}^{(E)}(\boldsymbol{\theta}, \hat{\boldsymbol{\theta}}), \\ \qquad \text{where } \boldsymbol{\theta} = \hat{\boldsymbol{\theta}}(\boldsymbol{X}) \text{ is the estimator of } \boldsymbol{\theta}. \end{array} \right\} \tag{7.1.2b}$$

Here $C_{ik}^{(D \text{ or } E)} = a_{ik}^{(D)}, a_{ik}^{(E)}$, the costs of classifying $\boldsymbol{\theta}$ in Ω_k, when $\boldsymbol{\theta} \in \Omega_k$. (Since detection and estimation are uncoupled, we can at a later stage of separate optimizations, replace $C_{ik}^{(E)}$ by $C_{ik}^{(E)}(\boldsymbol{\theta}, \hat{\boldsymbol{\theta}})$, where $\hat{\boldsymbol{\theta}} = \hat{\boldsymbol{\theta}}(\boldsymbol{X})$ is the estimator of $\boldsymbol{\theta}$.) Similarly, for weak coupling, we can write (see Section 7.2 following):

II. Weak coupling:

$$\left. \begin{array}{ll} C_{ik}^{(D\oplus E)} = a_{ik}^{(D\oplus E)} + C_{ik}^{(D\oplus E)} & = \text{cost of classifying } \boldsymbol{\theta} \in \Omega_k, \text{ when } \boldsymbol{\theta} \in \Omega_k, \\ \qquad = \left(a_{ik}|_{\text{weak}} + f_{ik}(\boldsymbol{\theta})_{\text{weak}}\right) & \quad \text{in detection and estimation. We observe that} \\ & \quad \text{coupling is weak, in that detection does not} \\ & \quad \text{depend on the estimator structure.} \end{array} \right\} \tag{7.1.2c}$$

III. Strong coupling:

$$C_{ik}^{(D \otimes E)} = \left[a_{ik} + f_{ik}(\boldsymbol{\theta}, \hat{\boldsymbol{\theta}})\right]^{(D \otimes E)} = \left.\begin{array}{l}\text{is the strong coupling cost } a_{ik} \text{ of classifying } \boldsymbol{\theta} \\ \text{in } \Omega_k, \text{ when again, } \boldsymbol{\theta} \in \Omega_i, \text{ and consequently} \\ \text{choosing } H_k\text{: } S_k(\boldsymbol{\theta}) \otimes N \text{ versus the true state} \\ H_i\text{: } S_i(\boldsymbol{\theta}) \otimes N. \text{ Here, } f_{ik}^{(D \otimes E)} \text{ represents the the} \\ \text{added cost of presenting } \boldsymbol{\theta} \text{ as an estimator} \\ \text{of the parameters } \boldsymbol{\theta} \in \Omega_k \text{ when } \boldsymbol{\theta} \in \Omega_i.\end{array}\right\} \tag{7.1.2d}$$

However, *the signal classes often are not disjoint*, that is, they may overlap (cf. Section 1.1). Then, it is impossible to decide when an error or a correct decision has been made, because the same signal may belong to two different hypothesis classes (cf. Section 1.10), and therefore the interpretation of $C(\boldsymbol{\theta}, \gamma)$ as the cost of a correct or incorrect decision is no longer valid. Reinterpreting $C(\boldsymbol{\theta}, \gamma_i)$ as the cost of assigning $\boldsymbol{\theta}$ to the hypothesis class Ω_i allows us again to define an average risk and thus to determine decision rules that minimize it. Accordingly, the cost function to be used here is given by

$$C(\boldsymbol{\theta}, \gamma_R) = \sum_{i=0}^{K} C_{ik}(\boldsymbol{\theta}, \hat{\boldsymbol{\theta}}) p_i w_i(\boldsymbol{\theta}) / w(\boldsymbol{\theta}), \tag{7.1.3}$$

where $w(\boldsymbol{\theta})$ is defined by (7.1.1) and $C_{ik}(\boldsymbol{\theta}, \hat{\boldsymbol{\theta}})$ is the cost of making decision γ_k when $\boldsymbol{\theta}$ is contained in Ω_i. (For nonoverlapping hypothesis classes, (7.1.3) reduces to $C(\boldsymbol{\theta}, \gamma_k) = C(\boldsymbol{\theta}, \hat{\boldsymbol{\theta}})$, since $w_{i=k}(\boldsymbol{\theta})/w(\boldsymbol{\theta}) = 1$, in as much as $w(\boldsymbol{\theta}) = w_k(\boldsymbol{\theta})$ (cf. 7.1.1).) We will consider a cost assignment for which the cost of misclassification and incorrect estimation are separate; that is,

$$C_{ik}(\boldsymbol{\theta}, \hat{\boldsymbol{\theta}}) \equiv a_{ik} + f_{ik}(\boldsymbol{\theta}, \hat{\boldsymbol{\theta}}) \tag{7.1.4}$$

where a_{ik} is the cost of classifying $\boldsymbol{\theta}$ in Ω_k when $\boldsymbol{\theta} \in \Omega_i$, and $f_{ik}(\boldsymbol{\theta}, \hat{\boldsymbol{\theta}}(\boldsymbol{X}))$ is the additional cost of presenting $\hat{\boldsymbol{\theta}} = \hat{\boldsymbol{\theta}}(\boldsymbol{X})$ as an estimate of a parameter of class Ω_k when actually $\boldsymbol{\theta} \in \Omega_i$. (This cost function is a generalization of a cost assignment originally introduced by Ogg [6].) *The dependence of $C_{ik}(\boldsymbol{\theta}, \hat{\boldsymbol{\theta}})$ on $\boldsymbol{\theta}$, and implicitly on $\boldsymbol{X}$ through $\hat{\boldsymbol{\theta}}(\boldsymbol{X})$, has the effect of coupling the estimation and detection operations.* This dependence, exemplified by (7.1.4), is the generalization of *strong coupling* (Section 6.3.1) to the multisignal cases considered here in Chapter 7.

The denominator of $C(\boldsymbol{\theta}, \gamma_k)$ in (7.1.3) cancels $w(\boldsymbol{\theta})$ appearing in (7.1.2), the total average risk. Since we require here that a definite (detection) decision be made at the end of the observation interval, we have then

$$P(\gamma_0|\boldsymbol{X}) = 1 - \sum_{k=1}^{K} P(\gamma_k|\boldsymbol{X}). \tag{7.1.5}$$

Furthermore, if (7.1.3) and (7.1.5) are then substituted into (7.1.2), the total average risk becomes

$$R_{D \oplus E} = \int_\Gamma dX \left[\sum_{i=1}^{K} p_i \int_{\Omega_i} d\boldsymbol{\theta} C_{i0}(\boldsymbol{\theta}, \hat{\boldsymbol{\theta}}) W(\boldsymbol{X}|\boldsymbol{\theta}) w_i(\boldsymbol{\theta}) \right]$$
$$+ \int_\Gamma dX \left\{ \sum_{k=1}^{K} P(\gamma_i|\boldsymbol{X}) \left[\begin{array}{l} \sum_{i=1}^{K} p_i \int_{\Omega_i} d\boldsymbol{\theta} \left[C_{ik}(\boldsymbol{\theta}, \hat{\boldsymbol{\theta}}) - C_{i0}(\boldsymbol{\theta}, \hat{\boldsymbol{\theta}}) \right] \\ \cdot W(\boldsymbol{X}|\boldsymbol{\theta}) w_i(\boldsymbol{\theta}) \end{array} \right] \right\}. \tag{7.1.6}$$

We assume in addition that $C_{ik}(\boldsymbol{\theta}, \hat{\boldsymbol{\theta}}) \geq 0$ in (7.1.4) for all i and k, with $C_{00} = 0$. Also, when γ_0 is decided, no estimate is required. Therefore, we can also assume that $f_{i0}(\boldsymbol{\theta}, \hat{\boldsymbol{\theta}}(\boldsymbol{X})) = 0$ and consequently $C_{i0}(\boldsymbol{\theta}, \hat{\boldsymbol{\theta}}) = a_{i0}$. From these comments, it is clear that the first term on the right-hand side of (7.1.6) is a positive constant independent of any decision rule. Thus, we need only consider the second term in the determination of the optimum (i.e., Bayes) decision rule.

7.1.1.1 Optimum Decision Rule In order to determine this optimum decision rule for detection and require the analysis to be similar to the multiple alternative situation of "pure" detection, we define the following modified average-likelihood ratio:

$$L_{ik}(\boldsymbol{X}) \equiv \frac{p_i \int_{\Omega_i} d\boldsymbol{\theta} \left[C_{i0}(\boldsymbol{\theta}, \hat{\boldsymbol{\theta}}) - C_{ik}(\boldsymbol{\theta}, \hat{\boldsymbol{\theta}}) \right] W(\boldsymbol{X}|\boldsymbol{\theta}) w_i(\boldsymbol{\theta})}{C_{0k}(\boldsymbol{\theta}, \hat{\boldsymbol{\theta}}) p_0 W(\boldsymbol{X}|\boldsymbol{0})}. \tag{7.1.7}$$

The average likelihood ratio $L_{ik}(\boldsymbol{X})$ corresponds formally to the modified likelihood ratio Λ_g(6.3.18). For notational convenience, we now make the following definitions:

$$A_0(\boldsymbol{X}) \equiv \sum_{i=0}^{K} p_i \int_{\Omega_i} d\boldsymbol{\theta}\, C_{i0}\left(\boldsymbol{\theta}, \hat{\boldsymbol{\theta}}\right) W(\boldsymbol{X}|\boldsymbol{\theta}) w_i(\boldsymbol{\theta}) \tag{7.1.8a}$$

$$C_k(\boldsymbol{X}) \equiv C_{0k}\left(\boldsymbol{\theta}, \hat{\boldsymbol{\theta}}\right). \tag{7.1.8b}$$

These definitions, plus the observation that at least one of the parameters $\boldsymbol{\theta}$ is an energy-dependent parameter and therefore we can set $w_0(\boldsymbol{\theta}) = \delta(\boldsymbol{\theta} - 0)$, allow us to express the average risk as

$$R_{D+E} = \int_\Gamma dX A_0(\boldsymbol{X}) + \int_\Gamma dX p_0 W(\boldsymbol{X}|0) \cdot \sum_{k=1}^{K} P(\gamma_k|\boldsymbol{X}) C_k(\boldsymbol{X}) \left[1 - \sum_{i=1}^{K} L_{ik}(\boldsymbol{X}) \right]. \tag{7.1.9}$$

Since $C_{0k}(\boldsymbol{\theta}, \hat{\boldsymbol{\theta}}) \geq 0$ and $C_{00}(\boldsymbol{\theta}, \hat{\boldsymbol{\theta}}) = 0$, we see that $C_k(\boldsymbol{X}) \geq 0$ for all $\boldsymbol{X}$ and k. Since p_0 and $W_1(\boldsymbol{X}|0)$ are probabilities (and probability densities) and therefore positive, the following nonrandom decision rule for detection minimizes the average risk:

Decide hypothesis H_k if signal S_k and estimator $\gamma_0 = \boldsymbol{\theta}, \hat{\boldsymbol{\theta}}(\boldsymbol{X})$ are correctly decided when

$$\text{(i)} \quad \sum_{i=1}^{K} L_{ik}(\boldsymbol{X}) > 1$$

and

$$\text{(ii)} \quad C_k(X)\left[1-\sum_{i=1}^{K} L_{ik}(X)\right] \le C_\ell(X)\left[1-\sum_{i=1}^{K} L_{i\ell}(X)\right]. \tag{7.1.10}$$

At this point, we have completed only the first stage of a two-stage minimization process. Our strategy in the two-stage minimization procedure is to determine first a decision rule that depends on an estimator $\hat{\boldsymbol{\theta}}(X)$ as if it were optimum. The second stage of the minimization process consists of finding the (presumably unique) function $\hat{\boldsymbol{\theta}}^*(X)$ that further minimizes the average risk (7.1.9). This minimization process becomes, when we take into account the decision rule (7.1.10),

$$R_{D+E} = \int_\Gamma dX A_0(X) + \sum_{k=1}^{K} \int_{\Gamma_k} dX p_0 W(X|0) C_k(X) \cdot \left[1-\sum_{i=1}^{K} L_{ik}(X)\right], \tag{7.1.11}$$

where Γ_k denotes the region of the observation space Γ for which γ_k is decided. The *optimum estimator* that results from the second minimization procedure, for the case where γ_k is decided after the first stage, is denoted by $\hat{\boldsymbol{\theta}}^*(X)$. This in turn is determined from the condition

$$\min_{\hat{\boldsymbol{\theta}} \to \hat{\boldsymbol{\theta}}_k^*(X)} = \int_{\Gamma_k} dX p_0 W(X|0) C_k(X)\left[1-\sum_{i=1}^{K} L_{ik}(X)\right], \; k \neq 0. \tag{7.1.12}$$

We have included a subscript k in the optimum estimator to indicate that γ_k was decided, which is equivalent to deciding that $\boldsymbol{\theta} \in \Omega_\kappa, k \neq 0$; and therefore the resulting estimate is the best obtainable estimate of the signal of class Ω_k.

In order to obtain the final optimum detection rule, we substitute the optimum estimator back into the detection rule (7.1.10) originally obtained. The result is

$$L_{ik}(X) = \mu_i \int_{\Omega_i} d\boldsymbol{\theta} \left[\frac{W(X|\boldsymbol{\theta})}{W(X|0)}\right] w_i(\boldsymbol{\theta}) \mathsf{C}(\boldsymbol{\theta}, \hat{\boldsymbol{\theta}}^*), \tag{7.1.12a}$$

where

$$\mathsf{C}(\boldsymbol{\theta}, \hat{\boldsymbol{\theta}}^*) = \frac{C_{i0}(\boldsymbol{\theta}, \hat{\boldsymbol{\theta}}^*) - C_{ik}(\boldsymbol{\theta}, \hat{\boldsymbol{\theta}}^*)}{C_{0k}(\boldsymbol{\theta}, \hat{\boldsymbol{\theta}}^*)}, \quad \mu_i = p_i/p, \quad p = \sum_{i=1}^{K} p_i. \tag{7.1.12b}$$

This optimum decision rule for the detector as given in (7.1.10) depends on the structure of the optimum estimator. However, we should not conclude from this that estimation must precede detection. *It is emphasized that the detection and estimation operations are performed simultaneously*, although the two operations have certain processing features in common. This redundancy is eliminated from Figs. 7.1 and 7.2, which show the sequence of operations for "one shot," multiple alternative, joint detection and estimation. Figure 7.1 illustrates the sequence of operations involved in the decision rule for detection given in (7.1.10). The output of block $B_k(X)$, $k = 1, 2, \ldots, \cdots K$, is

$$B_k(X) \equiv 1 - \sum_{i=1}^{K} L_{ik}(X) \tag{7.1.12c}$$

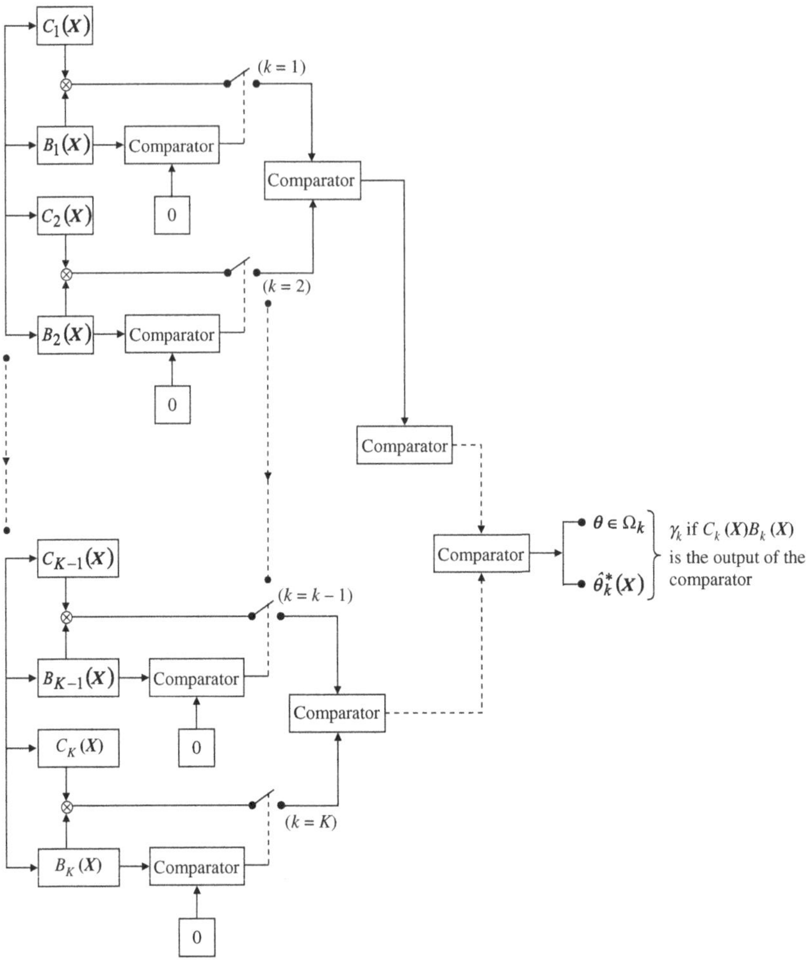

FIGURE 7.1 Schematic diagram shows the sequence of operations resulting in the decision γ_k, and thus that a signal of class k is correctly and optimally detected with optimum estimates of its parameters $\hat{\boldsymbol{\theta}}_k^*$, as described by (7.1.10). The block labeled "comparator" has an output that is the smaller of its two inputs. A more detailed description of the blocks $B_k(\boldsymbol{X}), k = 1, 2, \ldots, K$, is given in Figure 7.2. A nonzero output of the comparator immediately following any of the B–blocks closes the switch to which it is connected.

as shown in Figure 7.2. The operations indicated by (i) of the decision rule (7.1.10) are performed by the comparators whose inputs are from the B-blocks and blocks labeled 0. A nonzero output from any of these comparators closes the switch to which it is connected. Closing the switch in the kth branch allows the product $C_k(\boldsymbol{X})B_k(\boldsymbol{X})$, which results from operations involved in part (ii) of (7.1.10), to be put into another comparator that compares the output of a branch with that of the previous branch. The final output is a detection decision that $\boldsymbol{\theta} \in \Omega_k$, say, and an estimation decision $\hat{\boldsymbol{\theta}}_k(\boldsymbol{X})$.

In Ref. [4], it can be shown that the decision rules determined above reduce for special binary cases to known results. Thus, in case of pure detection for the binary on–off case (Chapter 1), we obtain the well-known generalized likelihood ratio test. Also, for strongly

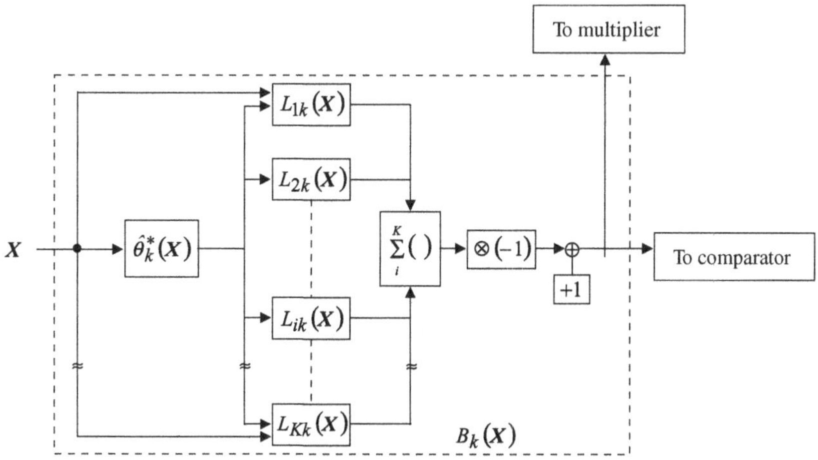

FIGURE 7.2 Schematic diagram of the operations performed by the blocks $B_k(X), k = 1, 2, \ldots, K$ of Figure 7.1. The block labeled $\hat{\boldsymbol{\theta}}_k^*(X)$ is necessary because the likelihood ratios $L_{ik}(X)$ depend on $\hat{\boldsymbol{\theta}}_k^*(X)$ through the cost function.

coupled joint detection and estimation (Section 6.3.1), the results obtained in the binary on–off case are identical with those obtained in Ref. [2] and in Chapter 6.

7.1.2 Specific Cost Functions

In order to make further progress with the analysis, specific cost assignments must be made and estimates $\hat{\boldsymbol{\theta}}^*$ determined in accordance with the minimization operation indicated in (7.12.12a), (7.12.12b), and (7.12.12c). An example illustrates the general procedure and provides the functional form or structure of a specific class of estimator and detector. Although the cost functions are specified in what follows, the noise statistics here are general.

As mentioned previously, the usual costs of detection decisions are modified [cf. (7.1.4)] to take into account the influence such decisions have on whether or not an estimate is presented at the output of the joint detection and estimation procedure.

7.1.2.1 Quadratic Cost Functions (QCFs) (Strong Coupling: Nonoverlapping Hypothesis Classes) The specific cost function chosen for this example will be

$$C_{ik}(\boldsymbol{\theta}, \hat{\boldsymbol{\theta}}) = a_{ik} + b_{ik}\left[\boldsymbol{\theta} - \hat{\boldsymbol{\theta}}(X)\right]^T \boldsymbol{E}_{ik}\left[\boldsymbol{\theta} - \hat{\boldsymbol{\theta}}(X)\right], \quad \boldsymbol{\theta} \in \Omega_i \text{ and } i, k = 0, 1, \ldots, K, \tag{7.1.13}$$

where $\boldsymbol{E}_{ik}$ are positive-definite *matrices* and $(\;)^T$ indicates "transpose." The cost factors a_{ik} are the usual constant cost assignments associated with the detection operation (Chapters 19 and 23 of Ref. [7], and Chapter 4), namely, the cost of deciding that hypothesis H_k occurs when actually H_i is true. The cost factor a_{ik} is associated with a correct detection decision, where we shall make the standard assumption that a correct detection decision costs nothing, that is, $a_{ii} = 0$ for $i = 0, 1, \ldots, K$.

We next discuss these b_{ik} factors in more detail. Assuming that $\boldsymbol{\theta}$ is not a nuisance parameter, that is, $\boldsymbol{\theta} = 0$ implies $S = 0$, we say that $\boldsymbol{\theta} \in \boldsymbol{\Omega}_0$ implies H_0 is true. If, in addition, we decide γ_0, then a correct decision has been made, and no estimate is presented at the output. It is therefore reasonable to choose $b_{00} = 0$. This combined with $a_{00} = 0$ results in $C_{00}(\boldsymbol{\theta}, \hat{\boldsymbol{\theta}}) = 0$, if $\boldsymbol{\theta} \in \boldsymbol{\Omega}_0$. Now if γ_0 is decided and $\boldsymbol{\theta} \in \boldsymbol{\Omega}_i (i \neq 0)$, an incorrect decision has been made, but nevertheless no estimate is presented at the output. Considering this joint detection and estimation procedures as one stage in an adaptive process, we observe that no updating takes place as a result of this decision. We next arbitrarily assume that such a mistake is not so serious as the use of erroneous estimates for updating the device, which would be the case if γ_k is decided and $\boldsymbol{\theta} \in \boldsymbol{\Omega}_i (i \neq 0)$. Accordingly, the following relations between the relative magnitudes of the cost factors b_{ik} are reasonable:

$$\begin{aligned} &b_{0k} \geq b_{kk} > b_{00} = 0, \quad && k \neq 0 \\ &b_{ik} \geq b_{kk} && i, k = 1, 2, \ldots, K. \end{aligned} \tag{7.1.14}$$

Taking (7.1.14) into account, we see that the cost functions $C_{ik}(\boldsymbol{\theta}, \hat{\boldsymbol{\theta}})$ for various combinations of signal classes Ω_i and decisions γ_k become

$$\begin{aligned} C_{00}(\boldsymbol{\theta}, \hat{\boldsymbol{\theta}}) &= 0, \\ C_{0k}(\boldsymbol{\theta}, \hat{\boldsymbol{\theta}}) &= a_{0k} + b_{0k} \hat{\boldsymbol{\theta}}^T \boldsymbol{E}_{ik} \hat{\boldsymbol{\theta}} \qquad && k \neq 0, \\ C_{ik}(\boldsymbol{\theta}, \hat{\boldsymbol{\theta}}) &= a_{ik} + b_{ik} (\boldsymbol{\theta} - \hat{\boldsymbol{\theta}})^T \boldsymbol{E}_{ik} (\boldsymbol{\theta} - \hat{\boldsymbol{\theta}}) && i \neq 0. \end{aligned} \tag{7.1.15}$$

Our goal is to determine that $\hat{\boldsymbol{\theta}}^*$ that minimizes the integral in (7.2.12) and that we designate it by I_k. This integral can be expressed as

$$I_k = p_0 a_{0k} - \sum_{i=1}^{K} (a_{i0} - a_{ik}) p_i \int_{\Omega_i} d\boldsymbol{\theta}\, W(\boldsymbol{X}|\boldsymbol{\theta}) w_i(\boldsymbol{\theta}) + \int_{\Gamma_k} Z(\boldsymbol{X}, \hat{\boldsymbol{\theta}})\, d\boldsymbol{X}, \tag{7.1.16a}$$

where

$$Z_k(\boldsymbol{X}, \hat{\boldsymbol{\theta}}) \equiv \sum_{i=1}^{K} (b_{ik} - b_{i0}) p_i \int_{\Omega_i} d\boldsymbol{\theta} (\boldsymbol{\theta} - \hat{\boldsymbol{\theta}})^T \boldsymbol{E}_{ik} (\boldsymbol{\theta} - \hat{\boldsymbol{\theta}}) W(\boldsymbol{X}|\boldsymbol{\theta}) w_i(\boldsymbol{\theta}). \tag{7.1.16b}$$

Relations (7.1.16) are obtained by substituting the cost factors (7.1.15) into (7.1.12) along with $w_0(\boldsymbol{\theta}) = \delta(\boldsymbol{\theta} - 0)$ and $C_{00}(\boldsymbol{\theta}, \hat{\boldsymbol{\theta}}) = 0$. The first two terms on the right-hand side of (7.1.16a) are independent of $\hat{\boldsymbol{\theta}}$ and thus do not affect the optimum estimator. From relations (7.1.14), it follows that $b_{ik} - b_{i0} > 0$ for $k \neq 0$: therefore, the function $Z_k(\boldsymbol{X}, \hat{\boldsymbol{\theta}})$ is positive for all $\boldsymbol{X}$. Hence, any function $\hat{\boldsymbol{\theta}}$ that minimizes $Z_k(\boldsymbol{X}, \hat{\boldsymbol{\theta}})$ will also minimize $\int_{\Gamma_k} d\boldsymbol{X} Z_k(\boldsymbol{X}, \hat{\boldsymbol{\theta}})$. A necessary condition that a minimizing function $\hat{\boldsymbol{\theta}}^*$ must satisfy is

$$\left[\frac{\partial Z_k}{\partial \hat{\boldsymbol{\theta}}_1}, \frac{\partial Z_k}{\partial \hat{\boldsymbol{\theta}}_2}, \ldots, \frac{\partial Z_k}{\partial \hat{\boldsymbol{\theta}}_k},\right] \Big|_{\hat{\boldsymbol{\theta}} = \hat{\boldsymbol{\theta}}^*} = 0, \tag{7.1.17}$$

where $\hat{\boldsymbol{\theta}}_\iota$ is the ith component of $\hat{\boldsymbol{\theta}}$. If the Hessian matrix of second partial derivatives of $Z_k(\boldsymbol{X}, \hat{\boldsymbol{\theta}})$ with respect to the components of $\hat{\boldsymbol{\theta}}$, evaluated at $\hat{\boldsymbol{\theta}}^*$, is positive definite, then the function $\hat{\boldsymbol{\theta}}^*$, determined from condition (7.1.17), is the function that minimizes $Z_k(\boldsymbol{X}, \hat{\boldsymbol{\theta}})$.

Accordingly, from the definition of $Z(\boldsymbol{X},\hat{\boldsymbol{\theta}})$ given in (7.1.16b), it follows that the Hessian matrix $\boldsymbol{E}_{ik}$ is

$$2\sum_{i=0}^{K}(b_{ik}-b_{i0})p_i\int_{\Omega_i} d\boldsymbol{\theta}\, W(\boldsymbol{X}|\boldsymbol{\theta})w_i(\boldsymbol{\theta})\,\boldsymbol{E}_{ik}. \tag{7.1.17a}$$

Since $\boldsymbol{E}_{ik}$ are by definition positive definite and the scalar factors multiplying them are positive, the Hessian is positive definite and therefore the solution of (7.1.17) minimizes $\int_{\Gamma_k} d\boldsymbol{X}\, Z_k(\boldsymbol{X},\hat{\boldsymbol{\theta}}(\boldsymbol{X}))$.

Next, substituting (7.1.16b) into (7.1.17) and observing that the gradient of $(\boldsymbol{\theta}-\hat{\boldsymbol{\theta}})^T \boldsymbol{E}_{ik}(\boldsymbol{\theta}-\hat{\boldsymbol{\theta}})$ is $-2\boldsymbol{E}_{ik}(\boldsymbol{\theta}-\hat{\boldsymbol{\theta}})$, we find that the function $\hat{\boldsymbol{\theta}}_k^*$ is a solution of

$$\sum_{i=0}^{K}[b_{ik}-b_{i0}]p_i\int_{\Omega_i} d\boldsymbol{\theta}\,\boldsymbol{E}_{ik}\cdot\left[\boldsymbol{\theta}-\hat{\boldsymbol{\theta}}_k^*\right]W(\boldsymbol{X}|\boldsymbol{\theta})w_i(\boldsymbol{\theta})=0. \tag{7.1.17b}$$

Solving for $\hat{\boldsymbol{\theta}}_k^*(\boldsymbol{X})$, we obtain

$$\hat{\boldsymbol{\theta}}_k^*(\boldsymbol{X}) = \left[\sum_{i=0}^{K}[b_{ik}-b_{i0}]p_i\int_{\Omega_i}\boldsymbol{\theta}\,W(\boldsymbol{X}|\boldsymbol{\theta})w_i(\boldsymbol{\theta})d\boldsymbol{\theta}\,\boldsymbol{E}_{ik}\right]^{-1} \cdot\sum_{i=0}^{K}[b_{ik}-b_{i0}]p_i\int_{\Omega_i}\boldsymbol{E}_{ik}\,d\boldsymbol{\theta}\,W(\boldsymbol{X}|\boldsymbol{\theta})w_i(\boldsymbol{\theta}). \tag{7.1.18a}$$

Recalling that $w_0(\boldsymbol{\theta})=\delta(\boldsymbol{\theta}-0)$ and $b_{00}=0$, and that if $\boldsymbol{E}_{ik}$ is not a function of i, Eq. (7.1.18a) can also be written as

$$\hat{\boldsymbol{\theta}}_k^*(\boldsymbol{X}) = \sum_{i=1}^{K}\frac{(b_{ik}-b_{i0})\Lambda_i(\boldsymbol{X})/b_{0k}}{1+\sum_{i=1}^{K}[(b_{ik}-b_{i0})\Lambda_i(\boldsymbol{X})/b_{0k}]}\hat{\Theta}_i^*(\boldsymbol{X}). \tag{7.1.18b}$$

Here, $\Lambda_i(\boldsymbol{X})$ is the familiar *average-likelihood ratio* (i.e., GLR) and $\hat{\Theta}_i^*(\boldsymbol{X})$ is the *least-squares* or *minimum-variance estimator* for a signal of class $\boldsymbol{\Omega}_i$ in the presence of *certainty* ($p_i=1$), namely,

$$\Lambda_i(\boldsymbol{X}) \equiv p_i\int_{\Omega_i} W(\boldsymbol{X}|\boldsymbol{\theta})w_i(\boldsymbol{\theta})\,d\boldsymbol{\theta}/p_0 W(\boldsymbol{X}|\boldsymbol{0}), \tag{7.1.19a}$$

$p_i < 1$, and

$$\hat{\Theta}_i^*(\boldsymbol{X}) \equiv \frac{\int_{\Omega_i}\boldsymbol{\theta}\,W(\boldsymbol{X}|\boldsymbol{\theta})w_i(\boldsymbol{\theta})\,d\boldsymbol{\theta}}{\int_{\Omega_i}W(\boldsymbol{X}|\boldsymbol{\theta})w_i(\boldsymbol{\theta})\,d\boldsymbol{\theta}}. \tag{7.1.19b}$$

If we now assume that $b_{i0}=0$ and $b_{ik}=b_{0k}>0$ for $i=1,\ldots,K$, (7.1.18a) reduces to the simpler expression

$$\hat{\boldsymbol{\theta}}_k^*(\boldsymbol{X}) = \sum_{i=1}^{K}\Lambda_i(\boldsymbol{X})\left[1+\sum_{i=1}^{K}\Lambda_i(\boldsymbol{X})\right]^{-1}\hat{\Theta}_i^*(\boldsymbol{X}), \tag{7.1.20}$$

which can also be put in the more revealing form:

$$\hat{\boldsymbol{\theta}}_k^*(\boldsymbol{X}) = \sum_{i=1}^{K} P(H_i|\boldsymbol{X})\hat{\Theta}_i^*(\boldsymbol{X}), \tag{7.1.21}$$

where $P(H_i|\boldsymbol{X})$ is the *a posteriori* probability that hypothesis H_i is true. It can easily be shown [4] by several applications of Bayes' rule and the definition of a marginal probability density that

$$P(H_i|\boldsymbol{X}) = \Lambda_i(\boldsymbol{X})\left[1 + \sum_{i=1}^{K} \Lambda_i(\boldsymbol{X})\right]^{-1}. \tag{7.1.21a}$$

The estimator given in (7.1.20) is less complicated than that given in (7.1.18b). It also has the reasonable interpretation (7.1.21) of being a weighted sum of least-squares estimators when there is no uncertainty about the various hypotheses, with the weights being the *a posteriori* probabilities of the respective hypotheses. From its symmetry, the estimator given by (7.1.20) is seen to be independent of the index k. Hence, whatever decision $\gamma_k(k \neq 0)$ is made, the same estimate is presented at the output and, therefore, for this case we could have dropped the subscript k on $\hat{\boldsymbol{\theta}}_k^*(\boldsymbol{X})$. This is not the case with (7.1.18b), which is derived under less restrictive assumptions on the cost assignments. Therefore, if we are fortunate and the costs lead to an estimator like (7.1.20), the joint detection and estimation device need only be constructed with one estimator structure instead of K structures.

In this example, the optimum decision rule results when we substitute the estimator (7.1.18b) or (7.1.20) into the decision rule obtained originally. It is seen from (7.1.10) that the important factor necessary for the explicit determination of the optimum detector structure is

$$B_k = C_k(\boldsymbol{X})\left[1 - \sum_{k=1}^{K} L_{ik}(\boldsymbol{X})\right]. \tag{7.1.21b}$$

This factor is given in (7.1.22a) for the case when the cost assignments are such that they lead to the estimator given by (7.1.20), including the assumption that $b_{ik} \equiv 1$. It can be derived in a straightforward manner with some tedious algebra. We obtain finally

$$\begin{aligned} C_k(\boldsymbol{X})\left[1 - \sum_{i=1}^{K} L_{ik}(\boldsymbol{X})\right] &= a_{0k} + \sum_{i=1}^{K}[a_{ik} - a_{i0}]\Lambda_i(\boldsymbol{X})H_i(\boldsymbol{X}) - \left(1 - \sum_{i=1}^{K}\Lambda_i(\boldsymbol{X})\right) \\ &\quad \cdot \sum_{i=1}^{K}\sum_{\ell=1}^{K} P(H_i|\boldsymbol{X})P(H_\ell|\boldsymbol{X})\left(\widetilde{\hat{\Theta}_\ell^*}\right)\hat{\Theta}_i^*, \end{aligned} \tag{7.1.22a}$$

where

$$H_i(\boldsymbol{X}) \equiv \frac{\int_{\Omega_i} \tilde{\boldsymbol{\theta}}\boldsymbol{\theta}\, W(\boldsymbol{X}|\boldsymbol{\theta})w_i(\boldsymbol{\theta})d\boldsymbol{\theta}}{\int_{\Omega_i} W(\boldsymbol{X}|\boldsymbol{\theta})w_i(\boldsymbol{\theta})d\boldsymbol{\theta}}, \tag{7.1.22b}$$

and (7.1.19b) define $\hat{\Theta}_i^*(\boldsymbol{X})$.

7.1.2.2 Quadratic Cost Functions and Gaussian Statistics (also Section 3.3 of Ref. [4]) This example specializes further the results of Section 7.1.2.1. We assume that the probability densities $w_i(\hat{\boldsymbol{\theta}})$ for $i = 1, 2, \ldots, K$ are Gaussian with means m_i and covariance matrices $\boldsymbol{K}_i$, as may be the case if the components of $\boldsymbol{\theta}$ are a set of *discrete–time samples of a signal waveform* $\boldsymbol{X}$. Furthermore, we assume that the noise is Gaussian with zero mean and covariance matrix $\boldsymbol{K}_{\mathrm{N}_i}$ in which $\boldsymbol{K}_{\mathrm{N}_i}$ is a $\mathrm{J} \times \mathrm{J}$ positive definite (symmetrical matrix), as is $\boldsymbol{K}_i$ for the number of signals (here $\boldsymbol{\theta}_i = S$). Under these assumptions, we have

$$H_i(\boldsymbol{X}) = \mathrm{tr}(\boldsymbol{Q}_i) + \left(\tilde{\hat{\boldsymbol{\theta}}}{}_\iota^*\right)\left(\hat{\boldsymbol{\theta}}_\iota^*\right), \tag{7.1.23a}$$

where the matrix

$$\boldsymbol{Q}_i = \left(\boldsymbol{K}_{\mathrm{N}_i}^{-1} + \boldsymbol{K}_i^{-1}\right)^{-1}, \tag{7.1.23b}$$

with $\mathrm{tr}(\boldsymbol{Q}_i)$ denoting the trace of $\boldsymbol{Q}_i$. The optimum detector structure for $b_{ik} = 1$ is given by

$$\begin{aligned} C_k(\boldsymbol{X})\left[1 - \sum_{i=1}^{K} L_{ik}(\boldsymbol{X})\right] &= a_{0k} + \sum_{i=1}^{K}\left[(a_{ik} - a_{i0}) + tr(\boldsymbol{Q}_i) + \left(\tilde{\hat{\Theta}}{}_i^*\right)\hat{\Theta}_i^*\right]\Lambda_i \\ &\quad + \left(1 - \sum_{i=1}^{K}\Lambda_i\right)\sum_{i=1}^{K}\sum_{\ell=1}^{K} P(H_i|\boldsymbol{X})P(H_\ell|\boldsymbol{X})\left(\tilde{\hat{\Theta}}{}_\ell^*\right)\hat{\Theta}_i^*. \end{aligned} \tag{7.1.23c}$$

Since

$$P(H_i|\boldsymbol{X}) = \Lambda_i(\boldsymbol{X}) \cdot \left[1 + \sum_{i=1}^{K}\Lambda_i(\boldsymbol{X})\right]^{-1}, \tag{7.1.23d}$$

the detector structure can be expressed as a function of the generalized likelihood ratios and the least-squares estimators of the various signals in the absence of uncertainty. In addition, the left-hand side of (7.1.23c) can be expressed solely as a function of the generalized likelihood ratios, $\Lambda_i(\boldsymbol{X})$, for $i = 1, \ldots, K$, or as a function of the least-squares estimates $\hat{\Theta}_i^*$ for $i = 1, \ldots, K$. This last observation follows from some relations discussed in Refs. [8,9].

7.1.2.3 Simple Cost Functions (SCFs)$_{1,2}$ In Sections 7.1.2.1 and 7.1.2.2, we have used a quadratic cost function (QCF) to illustrate the general theory of joint detection and estimation under multiple hypotheses. Another important cost assignment often used in estimation theory is the simple cost function, of which there are two principal forms, namely, the "strict" SCF (SCF_2) and the "nonstrict" SCF (SCF_1), refer to Section 6.2.2. The reason for their importance is that they lead to unconditional maximum-likelihood estimators (UCMLEs). This section is devoted to a discussion of joint detection and estimation under multiple hypotheses when a simple cost function is assumed.

SCF$_2$ We start with the strict SCF (SCF_2), which imposes a more severe penalty than SCF_1 on derivations from the exact estimates of $\boldsymbol{\theta}$. This simple cost function is given by

$$C(\boldsymbol{\theta}, \gamma_k) = a_{ik} + b_1\left(c_1 - \delta(\hat{\boldsymbol{\theta}} - \boldsymbol{\theta})\right), \quad \theta \in \boldsymbol{\Omega}_i, \tag{7.1.24a}$$

where

$$\delta\left(\hat{\boldsymbol{\theta}}(X) - \boldsymbol{\theta}\right) = \prod_{\ell=1}^{L} \delta\left(\hat{\boldsymbol{\theta}}_\ell(X) - \boldsymbol{\theta}_\ell\right). \tag{7.1.24b}$$

In this last relation, $\delta(\cdot)$ is the δ-function, $\hat{\boldsymbol{\theta}}_\ell$ is the ℓth component of $\hat{\boldsymbol{\theta}}$, and $\hat{\boldsymbol{\theta}}$ and $\boldsymbol{\theta}$ are L-component vectors. The factors a_{ik} are the usual constant cost assignments, that is, the cost of deciding that a signal of class $\boldsymbol{\Omega}_i$ is present. The factors b_1 and c_1 are related to the cost of correct and incorrect *estimation* decisions and a discussion of them is found in Chapter 23 of Ref. [7]; see also Sections 6.2.2.2 and 6.2.2.3.

Substitution of (7.1.24a) into relation (7.1.6) for the average risk R_{D+E} results in

$$\begin{aligned} R_{D+E} = & \int_\Gamma dX\left[\sum_{i=0}^{K}(a_{i0} + b_i c_i)p_i \int_{\Omega_i} W(X|\boldsymbol{\theta})w_i(\boldsymbol{\theta})d\boldsymbol{\theta}\right] \\ & + \int dX\left[\sum_{k=0}^{K} P(\boldsymbol{\gamma}_k|X)\left\{\sum_{i=0}^{K}(a_{ik} + a_{i0})p_i \int_{\Omega_i} W(X|\boldsymbol{\theta}) \cdot w_i(\boldsymbol{\theta})d\boldsymbol{\theta}\right\}\right] \\ & - \int_\Gamma dX\left[b_0 p_0 W(X|\boldsymbol{\theta})\delta(\hat{\boldsymbol{\theta}} - \mathbf{0}) + \sum_{i=1}^{K} b_i p_i W(X|\boldsymbol{\theta}) w_i(\hat{\boldsymbol{\theta}})\right]. \end{aligned} \tag{7.1.25}$$

The first term on the right-hand side of (7.1.25) is a positive constant independent of the decision rule or the estimator. The second term depends only on the decision rule and not on the estimator, whereas the third term depends on the estimator $\hat{\boldsymbol{\theta}}(X)$ and not on the decision rule. The integrand of the third term is always positive, and therefore the integral will also be positive. The integral is preceded by a minus sign, and therefore that part of the average risk will be minimized if we determine $\hat{\boldsymbol{\theta}}^*(X)$ by

$$\max_{\hat{\boldsymbol{\theta}}(X)} \int_\Gamma dX\left[b_0 p_0 W(X|\mathbf{0})\delta(\hat{\boldsymbol{\theta}} - \mathbf{0}) + \sum_{i=1}^{K} b_i p_i W(X|\boldsymbol{\theta}) w_i(\hat{\boldsymbol{\theta}})\right]. \tag{7.1.26}$$

If the factors b_i are all equal to C_0, condition (7.1.26) becomes

$$\max_{\hat{\boldsymbol{\theta}}(X)} \int_\Gamma dX\left[C_0 p_0 W(X|\mathbf{0})\delta(\hat{\boldsymbol{\theta}} - \mathbf{0}) + C_0 \sum_{i=1}^{K} p_i W(X|\boldsymbol{\theta}) w_i(\hat{\boldsymbol{\theta}})\right]. \tag{7.1.27}$$

In the binary on–off case where both signal and noise or the noise alone is present during the observation integral and $K = 1$, (7.1.27) reduces to

$$\max_{\hat{\boldsymbol{\theta}}(X)} \int_\Gamma dX\left[p_0 W(X|\mathbf{0})\delta(\hat{\boldsymbol{\theta}} - \mathbf{0}) + p_i W(X|\boldsymbol{\theta}) w_i(\hat{\boldsymbol{\theta}})\right], \tag{7.1.28}$$

which is equivalent to (6.2.21) et seq. The optimum decision rule is to decide γ_k if

$$\left.\begin{array}{ll}\text{(i)} & E_k(\boldsymbol{X}) < 0 \\ \text{and} & \\ \text{(ii)} & E_k(\boldsymbol{X}) \le E_\ell(\boldsymbol{X}), \quad \ell = 1, \ldots, K; (\neq k)\end{array}\right\}, \tag{7.1.29}$$

otherwise decide γ_0. Here,

$$E_k(\boldsymbol{X}) \equiv \sum_{i=0}^{K} (a_{ik} - a_{i0}) p_i \int_{\Omega_i} W(\boldsymbol{X}|\boldsymbol{\theta}) w_i(\boldsymbol{\theta}) d\boldsymbol{\theta}. \tag{7.1.29a}$$

***SCF*$_1$** When the other simple cost function SCF$_1$ is used instead of SCF$_2$, we replace the vector delta function in (7.1.24a) now by the alternative (6.2.13a), extended to the multiple signal case. The resulting cost function becomes

$$C(\boldsymbol{\theta}, \gamma_k)_1 = a_{ik} + b_i \sum_{\ell=1}^{L} \left\{ A_{i\ell} - \delta\left(\gamma_\ell - \hat{\boldsymbol{\theta}}_\ell[\boldsymbol{X}]\right) \right\} = a_{ik} + b_i \left\{ c_i - \sum_{i=1}^{L} \delta\left(\gamma_\ell - \hat{\boldsymbol{\theta}}_\ell[\boldsymbol{X}]\right) \right\}, \tag{7.1.30}$$

so that $c_i = \sum_\ell^L A_{i\ell}$. The desired optimizing equations are readily obtained by replacing the vector delta function in (7.1.24b) by the sum of scalar delta functions $\delta\left(\gamma_\ell - \hat{\boldsymbol{\theta}}_\ell[\boldsymbol{X}]\right)$, $\ell = 1, \ldots, L$. We see that the average risk R_{D+E}, Eq. (7.1.25), is accordingly

$$\begin{aligned} R_{D+E} = & \int_\Gamma d\boldsymbol{X} \left[\sum_{i=0}^{K} (a_{i0} + b_i c_i) p_i \int_{\Omega_i} W(\boldsymbol{X}|\boldsymbol{\theta}) w_i(\boldsymbol{\theta}) d\boldsymbol{\theta} \right] \\ & + \int d\boldsymbol{X} \left[\sum_{k=0}^{K} P(\boldsymbol{\gamma}_k|\boldsymbol{X}) \left\{ \sum_{i=0}^{K} (a_{ik} + a_{i0}) p_i \int_{\Omega_i} W(\boldsymbol{X}|\boldsymbol{\theta}) \cdot w_i(\boldsymbol{\theta}) d\boldsymbol{\theta} \right\} \right] \\ & - \int_\Gamma d\boldsymbol{X} \left[b_0 p_0 W(\boldsymbol{X}|\boldsymbol{0}) \sum_{\ell=1}^{K} \delta(\hat{\boldsymbol{\theta}} - \boldsymbol{0}) + \sum_{i=1}^{K} b_i p_i \sum_{\ell=1}^{K} W(\boldsymbol{X}|\hat{\boldsymbol{\theta}}_\ell) w_i(\hat{\boldsymbol{\theta}}_\ell) \right]. \end{aligned} \tag{7.1.31}$$

As before, refer to Eq. (7.1.25), the first term of (7.1.31) is a positive constant that is independent of the decision process and the estimator, with the second term depending only on the decision rule and not on the estimator. The third term, however, is a function of the estimators θ_ℓ, $\ell = 1, \ldots, L$, and does not depend on the decision rule. Since the integrand of this third term is always positive, the resulting integral is positive. Because this term is preceded by a minus sign, this part of the average risk R_{D+E} is to be minimized. When $\hat{\boldsymbol{\theta}}^*(\boldsymbol{X}) \left(= \sum_\ell^L \hat{\boldsymbol{\theta}}_\ell(\boldsymbol{X}) \right)$ is maximized, for each $\hat{\boldsymbol{\theta}}_\ell$, we have the relation

$$\max_{\left\{\hat{\boldsymbol{\theta}}_\ell \to \hat{\boldsymbol{\theta}}_\ell^*\right\}} \int_\Gamma d\boldsymbol{X} \left[b_0 p_0 W(\boldsymbol{X}|\boldsymbol{\theta}) \sum_{\ell=1}^{L} \left[\delta\left(\hat{\boldsymbol{\theta}}_\ell(\boldsymbol{X}) - 0\right) \right] + \sum_{i=1}^{K} b_i p_i \sum_{\ell=1}^{L} W\left(\boldsymbol{X}|\hat{\boldsymbol{\theta}}_\ell(\boldsymbol{X})\right) w_i(\hat{\boldsymbol{\theta}}_\ell) \right], \tag{7.1.32}$$

(each $\hat{\boldsymbol{\theta}}_\ell \to \hat{\boldsymbol{\theta}}_\ell^*$ separately from the others).

A useful and simpler case arises when the cost factors $\{b_i\}$ are all equal to $C_0(>0)$, so that the optimizing condition (7.1.32) becomes explicitly

$$\max_{\{\hat{\boldsymbol{\theta}}_\ell \to \hat{\boldsymbol{\theta}}_\ell^*\}} \int_\Gamma dX \left[p_0 W(X|\boldsymbol{\theta}) \sum_{\ell=1}^{L} \left[\delta\left(\hat{\boldsymbol{\theta}}_\ell(X) - 0\right)\right] + \sum_{k=1}^{K} P_k \sum_{\ell=1}^{L} W\left(X|\hat{\boldsymbol{\theta}}_\ell(X)\right) w_i(\hat{\boldsymbol{\theta}}_k) \right], \tag{7.1.33a}$$

which is independent of the scaling cost C_0. In the binary on–off case $(K = 1)$, in which both signal and noise (H_1) or the noise alone (H_0) is present during the observation period, Eq. (7.1.33a) reduces to

$$\max_{\{\hat{\boldsymbol{\theta}}_\ell \to \hat{\boldsymbol{\theta}}_\ell^*\}} \int_\Gamma dX \left[\left\{ p_0 W(X|\boldsymbol{\theta}) \sum_{\ell}^{L} \left[\delta\left(\hat{\boldsymbol{\theta}}_\ell(X) - 0\right)\right] + p_1 \sum_{\ell}^{L} W\left(X|\hat{\boldsymbol{\theta}}_\ell(X)\right) w_i(\hat{\boldsymbol{\theta}}_\ell) \right\} \right], \tag{7.1.33b}$$

which is likewise seen to be equivalent to Eq. (6.2.33) et seq.

Consequently, in (7.1.32) the optimum decision rule here is to decide γ_k when

$$\text{(i)}\quad G_k(X) < 0 \quad \text{and} \quad \text{(ii)}\quad G_k(X) \leq G_{k'}(X), \quad k' = 1, \ldots, K, \ (k' \neq k); \tag{7.1.34}$$

otherwise decide γ_0. Here, $G_k(X)$ is given by

$$G_k(X) \equiv \sum_{k=1}^{K} (a_{ik} - a_{i0}) p_i \int_{\Omega_i} W(X|\boldsymbol{\theta}) w_i(\boldsymbol{\theta}) \boldsymbol{d\theta}. \tag{7.1.34a}$$

7.1.3 Special Cases: Binary Detection and Estimation

We comment now on several special cases of the general results obtained above. First on our list is the familiar binary signal examples treated at length in Chapters 1, 3 and 6. For example, the case of pure binary detection H_1: $S \oplus N$ versus H_0: N, where there is no estimation process involved, reduces to the well-known and expected result in terms of the generalized likelihood ratio $\Lambda(X)$.

7.1.3.1 Binary On–Off Detection Only

$$\Lambda(X) = \mu \int_{\Omega_1} W(X|\boldsymbol{\theta}) w_i(\boldsymbol{\theta}) \boldsymbol{d\theta} / W(X|\boldsymbol{\theta}) \quad \begin{matrix} \geq \\ \\ < \end{matrix} \Bigg\} K \quad \begin{matrix} \text{: decide } H_1 \\ \text{or} \\ \text{: decide } H_0 \end{matrix} \tag{7.1.35}$$

where $K = (C_{01} - C_{00})/(C_{10} - C_{11})(> 0)$ is the decision threshold (expressed in the notation of Chapter 7), refer to Section 1.6.3.

7.1.3.2 Strongly-Coupled, Joint Detection and Estimation: Binary On–Off Signals Here, estimates are a part of the decision process where it is decided that a signal and noise, other than noise alone, are present, and thus estimation of the possible signal's parameters is required. Accordingly, we have to include in our cost assignments a factor that accounts for the additional cost of estimation. The cost functions chosen have the following form:

$$\left.\begin{aligned}
C(\boldsymbol{\theta}\in\Omega_0,\gamma_0) &= C(\mathbf{0},\gamma_0) = C_{00}^{(1)} + C_{00}^{(2)}(\boldsymbol{\theta},\hat{\boldsymbol{\theta}}) = C_{1-a} + C_{00}\\
C(\boldsymbol{\theta}\in\Omega_0,\gamma_1) &= C(\mathbf{0},\gamma_1) = C_{01}^{(1)} + C_{01}^{(2)}(\boldsymbol{\theta},\hat{\boldsymbol{\theta}}) = C_a + f_{10}\left(\hat{\boldsymbol{\theta}}\right)\\
C(\boldsymbol{\theta}\in\Omega_1,\gamma_0) &= C(\boldsymbol{\theta},\gamma_0) = C_{10}^{(1)} + C_{10}^{(2)}(\boldsymbol{\theta},\hat{\boldsymbol{\theta}}) = C_\beta + f_{01}(\boldsymbol{\theta})(=f_{01}[S(\boldsymbol{\theta})])\\
C(\boldsymbol{\theta}\in\Omega_1,\gamma_1) &= C(\boldsymbol{\theta},\gamma_1) = C_{11}^{(1)} + C_{11}^{(2)}(\boldsymbol{\theta},\hat{\boldsymbol{\theta}}) = C_{1-\beta} + f_{11}(\boldsymbol{\theta},\hat{\boldsymbol{\theta}})
\end{aligned}\right\}. \tag{7.1.36}$$

The cost assignments on the extreme right are those employed in Eq. (6.3.1). Again, since we are dealing with the binary on–off case and since $(i,k)=0,1$, there is only one likelihood ratio to deal with, namely, $L_{11}(X)$, which is given by Eq. (7.1.7). Substitution of the cost assignments (7.1.36) with $\hat{\boldsymbol{\theta}}(X)$ replaced by $\hat{\boldsymbol{\theta}}^*(X)$ yields finally

$$L_{11}(X) = \frac{(C_\beta - C_{1-\beta})\Lambda(X) + \mu\int_{\Omega_1}\left[f_{01}(\boldsymbol{\theta}) - f_{11}\left(\boldsymbol{\theta},\hat{\boldsymbol{\theta}}^*\right)\right]\Lambda'(X,\boldsymbol{\theta})w_1(\boldsymbol{\theta})d\boldsymbol{\theta}}{(C_\alpha - C_{1-\alpha}) + f_{10}(\hat{\boldsymbol{\theta}}^*) - C_{00}} \tag{7.1.37a}$$

where $\Lambda(X)$ is the generalized likelihood ratio given by (7.1.35) and $\Lambda'(X,\boldsymbol{\theta})$ is the "ordinary" or unaveraged likelihood ratio, given by $\mu(W(X,\boldsymbol{\theta}))/W(X,\mathbf{0})$. If we let $C_{1-\alpha} = C_{1-\beta} = C_{00} = f_{10}(\hat{\boldsymbol{\theta}}^*) = 0$, then

$$L_{11}(X) = \frac{C_\beta}{C_\alpha}\Lambda(X) + \frac{1}{C_\alpha}\int_{\Omega_1}\left[f_{01}(\boldsymbol{\theta}) - f_{11}(\boldsymbol{\theta},\hat{\boldsymbol{\theta}}^*)\right]\Lambda'(X,\boldsymbol{\theta})W_1(\boldsymbol{\theta})d\boldsymbol{\theta}. \tag{7.1.37b}$$

This last expression is identical to H_1: $S\oplus N$ in Chapter 6 (also Eq. (1) of Ref. [5]). When $L_{11}(X) > 1$, we decide that a signal was present during the observation period *and* that the optimum estimate of $\boldsymbol{\theta}$ is $\hat{\boldsymbol{\theta}}^*(X)$. On the other hand, when $L_{11}(X) \le 1$, we decide H_0: N, no signal is present. In this instance no estimate is required.

7.1.3.3 Joint Detection and Estimation: The Strongly Coupled, Binary Case H_1: $S_1 \oplus N$ versus H_2: $S_2 \oplus N$, where $\boldsymbol{\theta}_1 \neq 0$ versus $\boldsymbol{\theta}_2 \neq 0$. Here, we use the earlier result (Eq. 7.1.2), excluding the null signal case, to illustrate the general situation of K distinct, nonzero signals when detection and estimation are strongly coupled. From this, we may easily specialize the result to the binary, two-signal situation. Thus, when the null hypothesis is excluded, we begin by observing that $w_0(X)p_0 = 0$ and $P(\gamma_0|X) = 0$. The average risk R_{D+E} (7.1.2) then becomes

$$R_{D+E} = \int_\Gamma dX\left\{\sum_{k=1}^{K}P(\gamma_k|X)\left[\sum_{i=1}^{K}p_i\int_{\Omega_i}d\boldsymbol{\theta}\,C(\boldsymbol{\theta},\gamma_k)W(X|\boldsymbol{\theta})w_i(\boldsymbol{\theta})\right]\right\}. \tag{7.1.38}$$

The condition that a definite decision is required is expressed by

$$P(\gamma_{k'}|X) = 1 - \sum_{\substack{k=1\\k\neq k'}}^{K}P(\gamma_k|X). \tag{7.1.39}$$

When this relation is substituted into (7.1.38) for R_{D+E}, we see that

$$R_{D+E} = \int_{\Gamma} dX \left\{ \sum_{i=1}^{K} p_i \int_{\Omega_i} d\boldsymbol{\theta}\, C(\boldsymbol{\theta}, \gamma_k) W(\boldsymbol{X}|\boldsymbol{\theta}) w_i(\boldsymbol{\theta}) + \sum_{\substack{k=1 \\ k \neq k'}}^{K} P(\gamma_k|\boldsymbol{X}) B_{kk'}(\boldsymbol{X}) \right\} \quad (7.1.40a)$$

where

$$B_{kk'}(\boldsymbol{X}) \equiv \sum_{i=1}^{K} p_i \int_{\Omega_i} \left(C(\boldsymbol{\theta}, \gamma_k) - C(\boldsymbol{\theta}, \gamma_{k'}) \right) W(\boldsymbol{X}|\boldsymbol{\theta}) w_i(\boldsymbol{\theta})\, d\boldsymbol{\theta}. \quad (7.1.40b)$$

The first term is a positive constant independent of the probabilities $P(\gamma_{k'}|\boldsymbol{X})$ that embody the decision rule. Clearly, the following nonrandom decision rule for detection may be chosen:

I. Detection

$$\left.\begin{array}{l} \text{(i) Decide: } \gamma_k (k \neq k') \quad \text{or} \quad P(\gamma_k|\boldsymbol{X}) = 1, \quad \text{if} \begin{cases} \text{(a) } B_{kk'}(\boldsymbol{X}) < 0 \\ \text{(b) } B_{kk'}(\boldsymbol{X}) \le B_{ik'}(\boldsymbol{X}) \quad \text{for all } i (i \neq k') \end{cases} \\ \text{(ii) Otherwise decide: } \gamma_{k'(k' \neq k)} \text{ or } P(\gamma_{k'}|\boldsymbol{X}) = 1 \end{array}\right\}. \quad (7.1.41)$$

Again let Γ_k be that region of the observation space for which γ_k is decided. The average risk (7.1.40a), with the decision rule for detection (7.1.41), becomes

$$R_{D+E} = \int_{\Gamma} dX \left[\sum_{i=1}^{K} p_i \int_{\Omega_i} d\boldsymbol{\theta}\, C(\theta, \gamma_{k'})\, W(\boldsymbol{X}|\boldsymbol{\theta}) w_i(\boldsymbol{\theta}) \right] + \sum_{\substack{k=1 \\ k \neq k'}}^{K} \int_{\Gamma_k} d\boldsymbol{X}\, B_{kk'}(\boldsymbol{X}). \quad (7.1.42)$$

The optimum estimator for the case when γ_k is decided, $\hat{\theta}_k^*(\boldsymbol{X})_{p_k<1}$, is that function defined over Γ_k that minimizes the expression above for the average risk, namely, (7.1.42). Since the first term of (7.1.42) involves only the decision $\gamma_{k'}$ and is accordingly independent of $\hat{\theta}_k^*(\boldsymbol{X})_{p_k<1}$, the condition that determines this estimator becomes

$$\min_{\hat{\boldsymbol{\theta}}_k(\boldsymbol{X})} \int_{\Gamma_k} dX \left[\sum_{i=1}^{K} p_i \int_{\Omega_i} \left(C(\boldsymbol{\theta} - \gamma_k) - C(\boldsymbol{\theta} - \gamma_{k'}) \right) W(\boldsymbol{X}|\boldsymbol{\theta}) w_i(\boldsymbol{\theta})\, d\boldsymbol{\theta} \right]. \quad (7.1.43a)$$

Since $C(\theta, \gamma_{k'})$ is independent of $\hat{\theta}_k(\boldsymbol{X})$, this last relation becomes

$$\min_{\hat{\boldsymbol{\theta}}_k(\boldsymbol{X})} \int_{\Gamma_k} dX \left[\sum_{i=1}^{K} p_i \int_{\Omega_i} C(\boldsymbol{\theta}, \gamma_k) W(\boldsymbol{X}|\boldsymbol{\theta}) w_i(\boldsymbol{\theta}) d\boldsymbol{\theta} \right]. \quad (7.1.43b)$$

If $P(\gamma_k|X) = 0$ for all $k \neq k'$, that is, $\gamma_k \neq \gamma_{k'}$ is decided, then the average risk is given by only the first term in (7.1.42). The optimum estimator, $\hat{\theta}_k^*(X)_{p_k<1}$, for this case is determined from the following condition:

II. Estimation

$$\min_{\hat{\theta}_{k'}(X)} \int_{\Gamma_{k'}} dX \left[\sum_{i=1}^{K} p_i \int_{\Omega_i} C(\boldsymbol{\theta}, \gamma_{k'}) W(X|\boldsymbol{\theta}) w_i(\boldsymbol{\theta}) d\boldsymbol{\theta} \right]. \tag{7.1.44}$$

This last relation is identical to (7.1.43b) if we allow $k = k'$. The optimum detection rule is now given by (7.1.41) with $\hat{\theta}_k^*(X)_{p_k<1}(k = 1, \ldots, K)$, substituted for the various unknown estimators involved in (7.1.41).

If we now specialize to the binary two-signal case, we may take $k = 1$ and $k' = 2$. Then, we have

$$B_{12}(X) = \sum_{i=1}^{2} p_i \int_{\Omega_i} (C(\boldsymbol{\theta}, \gamma_1) - C(\boldsymbol{\theta}, \gamma_2)) W(X|\boldsymbol{\theta}) d\boldsymbol{\theta}, \tag{7.1.45}$$

and the estimators are determined from the following conditions:

$$\text{Estimation:} \quad \min_{\hat{\theta}_k(X)} \int_{\Gamma} dX \left[\sum_{i=1}^{2} p_i \int_{\Omega_i} C(\boldsymbol{\theta}, \gamma_k) W(X|\boldsymbol{\theta}) w_i(\boldsymbol{\theta}) d\boldsymbol{\theta} \right], \quad (k = 1, 2). \tag{7.1.46}$$

Indicating by $B_{12}^*(X)$, the factor $B_{12}(X)$, (7.1.45), with $\hat{\theta}_1^*(X)_{p_1<1}$ and $\hat{\theta}_2^*(X)_{p_2<1}$ substituted in place of $\hat{\theta}_1(X)$ and $\hat{\theta}_2(X)$, respectively, we see that the optimized decision rule for the aforementioned binary detection case is given by the following for the optimum estimators:

$$\left.\begin{array}{l} \text{(i) Choose } \hat{\boldsymbol{\theta}}_1^*, \quad \text{when } P^*(\gamma_1|X) = 1 \quad \text{if } B_{12}^*(X) < 0 \\ \quad \text{or} \\ \text{(ii) Choose } \hat{\boldsymbol{\theta}}_2^*, \quad \text{when } P^*(\gamma_2 X) = 1 \quad \text{if } B_{12}^*(X) \geq 0 \end{array}\right\}, \tag{7.1.47}$$

which is the required solution. (We could have derived the result (7.1.47) from (7.1.2) by modifying the results of Section 7.1.1 directly, but the approach above is simpler.) Note that here the hypothesis classes H_1 and H_2 overlap, as postulated from Eq. (7.2.3) and other equations.

7.1.3.4 Remarks on Extrapolation and Filtering The previous results are easily extended to include extrapolation and filtering. This section briefly discusses how this extension is to be achieved. In extrapolation, estimates of some function of the waveform are desired at instants in time outside the space–time observation interval $(R - T)$, whereas in filtering such estimates are desired for space–time instants within the observation interval.

Consider least-squares prediction of a signal waveform. Here we are interested in obtaining an estimate of $S(\boldsymbol{r}, \tau)$, for $\boldsymbol{r} > R, \tau > T$, based on the observed data sample $\boldsymbol{X}$. Let $\tilde{\theta} = [\boldsymbol{S}(\boldsymbol{r}, \tau)]$ where the components of $\boldsymbol{S}$ are discrete space–time samples of $S(T_{r,\tau})$ for $(\boldsymbol{r}, t) \in [R - T]$. Under the same assumptions that resulted in (7.1.21), we find that

$$\hat{S}^*(\boldsymbol{X}; \boldsymbol{r}, \tau) = \sum_{i=1}^{K} P(H_1|\boldsymbol{X}) \hat{S}_i(\boldsymbol{X}; \boldsymbol{r}, \tau), \tag{7.1.48}$$

where

$$\hat{\boldsymbol{S}}_i(\boldsymbol{X};\boldsymbol{r},\tau) \triangleq \frac{\int_{\Omega_1}\boldsymbol{S}(\boldsymbol{r},\tau)W(\boldsymbol{X}|\boldsymbol{S})W_i(\boldsymbol{S})d\boldsymbol{S}}{\int_{\Omega_1}W(\boldsymbol{X}|\boldsymbol{S})W_i(\boldsymbol{S})d\boldsymbol{S}}. \tag{7.1.48a}$$

Optimum joint detection and estimation under multiple hypotheses have been studied. Specific estimator and detector structures are determined when that part of the total cost assignment associated with estimation is a quadratic function of the estimation error. It has been shown that in the presence of uncertainty ($P(H_i) < 1$), one can choose an optimum estimator that is a weighted sum of least-squares estimators in the absence of uncertainty. The weighting coefficients are functions of all the generalized likelihood ratios and the cost-of-estimation coefficients b_{ik}.

In the case when $b_{ik} = b_{0k}(i = 1,\ldots,K)$, the weighting coefficients reduce to the posterior probabilities of the various hypotheses. The jointly optimum detector structure in this instance has also been determined to have the nonlinear form given in (7.1.22a), which becomes (7.1.23) when the noise and prior densities of the signal parameters are assumed to be Gaussian. The detector structure given by (7.1.23) is a form of correlation detector, consisting of cross-correlation between the various minimum-variance estimators in the absence of uncertainty and cross-correlation of the received data and these same estimators. Also, for an assignment of simple cost function for the cost of estimation, we see that the optimum estimator can be interpreted as a generalized maximum-likelihood estimator. Finally, we have indicated how the results obtained can be applied to optimum prediction and filtering. (The details are given in Ref. [4].)

7.2 UNCOUPLED OPTIMUM DETECTION AND ESTIMATION, MULTIPLE HYPOTHESES, AND OVERLAPPING PARAMETER SPACES

Even though uncoupled Bayesian detection and estimation has a larger minimum average risk, that is, is "less efficiently" optimum than the coupled cases with essentially similar cost assignments, it nevertheless deserves attention because of its greater structural simplicity, as well as its lesser computational burden. In addition, one can take direct advantage of the results of Sections 7.1 and 7.1.3 to obtain the desired results, here for the "single-shot" or single space–time data acquisition interval $|\boldsymbol{R}|T$.

Accordingly, we start with Eq. (7.1.6) and use (7.1.2a) for the separate D and E cost functions, which allow to write for the total uncoupled average risk (7.1.6), now in two parts:

$$R_D+R_E=\begin{cases} R_D = \int_\Gamma d\boldsymbol{X}\left[\sum_{i=0}^{K}p_i\int_{\Omega_i}C_{i0}^{(D)}W(\boldsymbol{X}|\boldsymbol{\theta})w_i(\boldsymbol{\theta})d\boldsymbol{\theta}\right] \\ \quad +\int_\Gamma d\boldsymbol{X}\left[\sum_{k=1}^{K}P(\gamma_k|\boldsymbol{X})\cdot\left\{\sum_{i=0}^{K}p_i\int_{\Omega_i}\left(C_{ik}^{(D)}-C_{i0}^{(D)}\right)W(\boldsymbol{X}|\boldsymbol{\theta})w_i(\boldsymbol{\theta})d\boldsymbol{\theta}\right\}\right] & (7.2.1a) \\ +R_E = \int_\Gamma d\boldsymbol{X}\left[\sum_{i=0}^{K}p_i\int_{\Omega_i}C_{i0}^{(E)}W(\boldsymbol{X}|\boldsymbol{\theta})w_i(\boldsymbol{\theta})d\boldsymbol{\theta}\right] \\ \quad +\int_\Gamma d\boldsymbol{X}\left[\sum_{k=1}^{K}P(\gamma_k|\boldsymbol{X})\cdot\left\{\sum_{i=0}^{K}p_i\int_{\Omega_i}\left(C_{ik}^{(E)}-C_{i0}^{(E)}\right)W(\boldsymbol{X}|\boldsymbol{\theta})w_i(\boldsymbol{\theta})d\boldsymbol{\theta}\right\}\right], & (7.2.1b) \end{cases}$$

since $C_{()}^{(D)}$ and $C_{()}^{(E)}$ do not depend on $\boldsymbol{\theta}$ and the estimator $\hat{\boldsymbol{\theta}} = \hat{\boldsymbol{\theta}}(\boldsymbol{X})$ are themselves independent constants. Thus, R_D and R_E are uncoupled and capable of separate optimization (i.e., here minimization). Possibly overlapping hypothesis classes Ω_k and Ω_i in signal parameter spaces are accounted for by the cost functions $C^{(D)}, C^{(E)}$, which are given by (7.1.3), with (7.1.1), now with $C_{ik}^{(D)}, C_{ik}^{(E)}$ as constants, that is, $C_{ik}^{(D,E)} = a_{ik}^{(D,E)}$ from (7.1.4). Paralleling the steps (7.1.7)–(7.1.11), we obtain for the Bayes detector here, from (7.2.1a):

I. Detection Decide H_k if

$$\sum_{i=1}^{K} L_{ik}^{(D)}(\boldsymbol{X}) > 1, \quad \text{with } C_k^{(D)} \cdot \left[1 - \sum_{i=1}^{K} L_{ik}^{(D)}(\boldsymbol{X})\right] \leq C_{i\ell}^{(D)} \cdot \left[1 - \sum_{i=1}^{K} L_{i\ell}^{(D)}(\boldsymbol{X})\right], \tag{7.2.2}$$

where (7.1.7) is modified to

$$L_{ik}^{(D)}(\boldsymbol{X}) = \mu_i C_{ik}^{(D)} \int_{\Omega_i} \frac{W(\boldsymbol{X}|\boldsymbol{\theta}) w_i(\boldsymbol{\theta}) d\boldsymbol{\theta}}{W(\boldsymbol{X}|0)} = C_{ik}^{(D)} \Lambda_1(\boldsymbol{X}) \tag{7.2.2a}$$

with b_{ik}

$$C_{ik}^{(D)} \equiv \left[C_{i0}^{(D)} - C_{ik}^{(D)}\right] / C_{0k}^{(D)}; \quad \Lambda_1(\boldsymbol{X}) = \mu_i \int_{\Omega_i} \frac{W(\boldsymbol{X}|\boldsymbol{\theta})}{W(\boldsymbol{X}|0)} w_i(\boldsymbol{\theta}) d\boldsymbol{\theta}. \tag{7.2.2b}$$

(This is an alternative form of the results of $K(\equiv M)$-ary detection in Section 4.1 obtained previously.)

For the Bayes estimation portion R_E of $R_D + R_E$, (7.2.1a) and (7.2.1b), since the estimation and detection operations are independent (refer to Section 7.1.1, as well as Section 6.2 in the binary on–off cases), we can replace $C_{ik}^{(E)}$ by $C_{ik}^{(E)}(\boldsymbol{\theta}, \hat{\boldsymbol{\theta}})$, where $\boldsymbol{\theta} = \hat{\boldsymbol{\theta}}(\boldsymbol{X})$ is the estimator of $\boldsymbol{\theta}$. Then, we can obtain the optimum estimator $\hat{\boldsymbol{\theta}}^*(\boldsymbol{X})$ by replicating (7.1.1)–(7.1.12) now with $C_k^{(E)} = C_{0k}^{(E)}$ replaced by $C_{0k}^{(E)}(\boldsymbol{\theta}, \hat{\boldsymbol{\theta}})$ in (7.1.12), but *not* using the old L_{ik}, (7.1.7) in (7.1.10) or substituting the result of the general condition (7.1.12) in (7.1.10) to further optimize the detector's decision rule (7.2.2).

The results for the specific cost functions $C_{0k}^{(E)}(\boldsymbol{\theta}, \hat{\boldsymbol{\theta}})$, that is, here for the QCF (7.1.14) and (7.1.15), become in this uncoupled case.

II. Estimation QCF: Eq. (7.1.20):

$$\hat{\boldsymbol{\theta}}_k^*(\boldsymbol{X}) = \sum_{i=1}^{K} \Lambda_i(\boldsymbol{X}) \hat{\Theta}_i^*(\boldsymbol{X}) / [1 + \Lambda_i(\boldsymbol{X})] \tag{7.2.3}$$

under the modified cost functions, with $b_{i0} = 0, b_{ik} > b_{0k} > 0, i = 1, \ldots, K$, with $b_{00} = 0$. Here, $\boldsymbol{E}_{ik}$ is not a function of i. Again, $\hat{\Theta}_i^*(\boldsymbol{X})$, Eq. (7.1.19b), is the classic LMS or minimum-variance estimator under the *a priori* condition of certainty $p_i = 1$, each $i \geq 1$, and $\Lambda_i(\boldsymbol{X})$ is the corresponding classic binary on–off detector for the signals of class $i, p_i < 1$, refer to Eq. (7.1.19a). Equation (7.1.18b) represents a slightly more general case, when $(b_{ik} - b_{i0}) > 0, b_{0k} > 0, \boldsymbol{E}_{ik} = \boldsymbol{E}_k$ (see Section 7.1.2.1). The binary on–off and binary two signal cases are discussed in Section 7.1.3.

When $SCF_{1,2}$ are the chosen cost functions, the detection and estimation operations are naturally uncoupled by this choice, as the results of Sections 7.1.3.1 and 7.1.3.2, have demonstrated. Here, "no coupling" is the rule for $SCF_{1,2}$, whereas for the QCF the strongly coupled cases give a smaller Bayes risk than either weak or no coupling. Thus, we see that the choice of cost function plays a crucial role in optimum performance: (1) This choice can either yield a smaller Bayes risk if it permits strong coupling, although further optimization of the detector, or (2) Bayes performance is already independent of coupling, that is, no coupling is already the state of operation. The results for the QCF are an example of (1), while the results for the $SCF_{1,2}$ are an example of (2).

7.2.1 A Generalized Cost Function for *K*-Signals with Overlapping Parameter Values

A reasonable cost function suggested by Ogg [6], and discussed by Middleton Section 2.2, pp. 22 and 23 of Ref. [10]) for the case of overlapping signal classes in Bayes detection theory may be defined by

$$C(\boldsymbol{\theta}, \gamma_k) = \frac{\sum_{i=0}^{K} C_{ik}^{(1)} p_i w_i(\boldsymbol{\theta})}{\sum_{i=0}^{K} p_i w_i(\boldsymbol{\theta})}. \tag{7.2.4}$$

When the hypothesis classes are nonoverlapping, this cost function clearly reduces to the usual constant cost of misclassification. Moreover, $C(\boldsymbol{\theta}, \gamma_k)$ has the property of being continuous in the prior densities. When the hypothesis classes overlap, the cost function (7.2.4) also has the plausible and desirable property that a less probable decision costs more than a more probable one, that is, if $i \neq k, \boldsymbol{\theta}$, the parameter set is contained in both Ω_i and Ω_k and $p_i w_i(\boldsymbol{\theta}) < p_k w_k(\boldsymbol{\theta})$, then $C(\boldsymbol{\theta}, \gamma_k) < C(\boldsymbol{\theta}, \gamma_i)$. Accordingly, the cost function employed here for the case of overlapping hypothesis classes is a direct analogue of (7.2.4) and is given by

$$C(\boldsymbol{\theta}, \gamma_k) = \sum_{i=0}^{K} \left[C_{ik}^{(1)} + C_{ik}^{(2)}\left(\boldsymbol{\theta}, \hat{\boldsymbol{\theta}}(\boldsymbol{X})\right)\right] p_i w_i(\boldsymbol{\theta}) \Big/ \sum_{i=0}^{K} p_i w_i(\boldsymbol{\theta}). \tag{7.2.4a}$$

Here, $C_{ik}^{(1)}$ is the cost of classifying $\hat{\boldsymbol{\theta}}$ in Ω_k, when $\hat{\boldsymbol{\theta}} \in \Omega_i$, and $C_{ik}^{(2)}(\boldsymbol{\theta}, \hat{\boldsymbol{\theta}}(\boldsymbol{X}))$ is the additional cost of presenting $\hat{\boldsymbol{\theta}}(\boldsymbol{X})$ as an estimate of a parameter of class Ω_k when $\boldsymbol{\theta} \in \Omega_i$.

Again, an average risk may be defined: It is the same expression as that given previously in (7.1.2a)–(7.1.2d). If the structure of the decision space Δ is the same as in Section 7.1 and the cost assignment is assumed to be of the form (7.2.4a), then (7.1.2) becomes

$$R_{D+E} = \int_{\Gamma} d\boldsymbol{X} \left\{ \sum_{k=0}^{K} P(\gamma_\ell|\boldsymbol{X}) \left[\sum_{i=0}^{K} p_i \int_{\Omega} d\boldsymbol{\theta} \left[C_{ik}^{(1)} + C_{ik}^{(2)}\left(\boldsymbol{\theta}, \hat{\boldsymbol{\theta}}(\boldsymbol{X})\right)\right] W(\boldsymbol{X}|\boldsymbol{\theta}) w_i(\boldsymbol{\theta}) \right] \right\}. \tag{7.2.5}$$

This expression for the average risk is seen to be identical to (7.1.6), however, with $C(\boldsymbol{\theta}, \gamma_k)$ expressed in more detail and with *the integral extended over the entire signal parameter space* Ω *, rather than just over the subspaces* Ω_i. Extending the integral to Ω takes into account the possibility that the subspaces, Ω_i, may overlap and perhaps all may be identical with Ω. Therefore, this special choice of cost function given by (7.2.4a) leads in the case of the overlapping hypothesis class to an average risk identical with that for the nonoverlapping case, provided $C(\boldsymbol{\theta}, \gamma_k)$ in (7.2.4) is now identified with $C_{ik}^{(1)} + C_{ik}^{(2)}\left(\boldsymbol{\theta}, \hat{\boldsymbol{\theta}}(\boldsymbol{X})\right)$ when $\boldsymbol{\theta} \in \Omega_i$, for example, with (7.2.4a). Now let

$$C'(\boldsymbol{\theta}, \gamma_k) \equiv C_{ik}^{(1)} + C_{ik}^{(2)}\left(\boldsymbol{\theta}, \hat{\boldsymbol{\theta}}(\boldsymbol{X})\right), \quad \boldsymbol{\theta} \in \Omega_i. \tag{7.2.6}$$

Recalling that the objective is to find the decision rule that minimizes the average risk, we see that the optimum decision rule obtained in the overlapping case is identical with that obtained in the nonoverlapping case, provided we replace $C(\boldsymbol{\theta}, \gamma_k)$ by $C'(\boldsymbol{\theta}, \gamma_k)$ (Eq. (7.2.6)). Postulating that $C'(\boldsymbol{\theta}, \gamma_0) = 0$ when $\boldsymbol{\theta} \in \Omega_0$, we find that

$$C_k\left(\boldsymbol{X}, \hat{\boldsymbol{\theta}}(\boldsymbol{X})\right) = \left(C_{0k}^{(1)} - C_{00}^{(1)}\right) + C_{0k}^{(2)}\left(\boldsymbol{\theta}, \hat{\boldsymbol{\theta}}(\boldsymbol{X})\right); \quad \mu_i = p_i/p_0; \tag{7.2.7a}$$

$$B_k\left(\boldsymbol{X}, \hat{\boldsymbol{\theta}}(\boldsymbol{X})\right) = 1 - \sum_{i=1}^{K} \frac{\mu_i \int_\Omega d\boldsymbol{\theta}[C'(\boldsymbol{\theta}, \gamma_0) - C'(\boldsymbol{\theta}, \gamma_k)] W(\boldsymbol{X}|\boldsymbol{\theta}) w_i(\boldsymbol{\theta})}{C_k\left(\boldsymbol{X}, \hat{\boldsymbol{\theta}}(\boldsymbol{X})\right) W(\boldsymbol{X}|\boldsymbol{0})} \tag{7.2.7b}$$

where $C_{ik}^{(1)}$ and $C_{ik}^{(2)}(\boldsymbol{\theta}, \hat{\boldsymbol{\theta}}(\boldsymbol{X}))$ are determined as in (7.2.4a). The resulting optimum estimator is determined by the following condition:

$$\min_{\hat{\boldsymbol{\theta}}(\boldsymbol{X})} \int_{\Gamma_k} d\boldsymbol{X} p_0 W(\boldsymbol{X}|\boldsymbol{0}) C_k\left(\boldsymbol{X}, \hat{\boldsymbol{\theta}}(\boldsymbol{X})\right) B_k\left(\boldsymbol{X}, \hat{\boldsymbol{\theta}}(\boldsymbol{X})\right) \tag{7.2.8}$$

and the optimum detector structure is given by

Decide γ_k, that is, $P^*(\gamma_k|\boldsymbol{X}) = 1$ if:

$$\begin{aligned} &\text{(i) } C_k\left(\boldsymbol{X}, \hat{\boldsymbol{\theta}}_k^*(\boldsymbol{X})\right) B_k\left(\boldsymbol{X}, \hat{\boldsymbol{\theta}}_k^*(\boldsymbol{X})\right) < 0 \\ &\text{(ii) } C_k\left(\boldsymbol{X}, \hat{\boldsymbol{\theta}}_k^*(\boldsymbol{X})\right) B_k\left(\boldsymbol{X}, \hat{\boldsymbol{\theta}}_k^*(\boldsymbol{X})\right) \le C_i\left(\boldsymbol{X}, \hat{\boldsymbol{\theta}}_i^*(\boldsymbol{X})\right) B_i\left(\boldsymbol{X}, \hat{\boldsymbol{\theta}}_i^*(\boldsymbol{X})\right), \text{ for all } i. \end{aligned} \tag{7.2.8a}$$

7.2.2 QCF: Overlapping Hypothesis Classes

As in the binary signal cases of Chapters 1, 3 and 6, the results of Section 7.2.1 show that when the hypothesis classes overlap, it is possible to choose a reasonable cost function so that the resulting detector and estimator structures are identical with those obtained in the nonoverlapping case. As indicated in Section 7.1.2.2, this permits the use of convenient prior probability functions such as the Gaussian, refer to Eq. (7.1.13), in the analysis of these Bayesian detection and estimation systems. Thus, it is evident that choosing the factor $C_{ik}^{(2)}(\boldsymbol{\theta}, \hat{\boldsymbol{\theta}})$ in (7.2.4a) to have the following quadratic dependence on $\boldsymbol{\theta}$ and $\hat{\boldsymbol{\theta}}(\boldsymbol{X})$, that is,

$$C_{ik}^{(2)}\left(\boldsymbol{\theta}, \hat{\boldsymbol{\theta}}(\boldsymbol{X})\right) \equiv C_{ik}^{(2)}\left[\boldsymbol{\theta} - \hat{\boldsymbol{\theta}}(\boldsymbol{X})\right]^T G_{ik}\left[\boldsymbol{\theta} - \hat{\boldsymbol{\theta}}(\boldsymbol{X})\right], \tag{7.2.9}$$

results in the same estimator and detector structures as with the nonoverlapping ones (7.1.13), with the specific forms obtained in the quadratic cost function example discussed in Section 7.1.2. Consequently, the following estimator structures are obtained in the *case of overlapping signal classes when the cost assignment is given by* (7.2.4a) *and* $C_{ik}^{(2)}(\boldsymbol{\theta}, \hat{\boldsymbol{\theta}}(\boldsymbol{X}))$ *is given by* (7.2.9):

$$\hat{\boldsymbol{\theta}}_k^*(\boldsymbol{X})_{p_k<1} = \sum_{i=1}^{K} \frac{\left[\left(C_{ik}^{(2)} - C_{i0}^{(2)}\right)/C_{0k}^{(2)}\right]\Lambda_i(\boldsymbol{X})}{1 + \sum_{i=1}^{K}\left[\left(C_{ik}^{(2)} - C_{i0}^{(2)}\right)/C_{0k}^{(2)}\right]\Lambda_i(\boldsymbol{X})} \hat{\Theta}_i^*(\boldsymbol{X})_{p_i=1}, \tag{7.2.10}$$

where $C_{0k}^{(2)} \geq C_{kk}^{(2)} > C_{k0}^{(2)} \geq C_{00}^{(2)} = 0\ (k \neq 0)$ and $C_{ik}^{(2)} \geq C_{kk}^{(2)}\ (i, k = 1, 2, \ldots, K)$. If in addition we set $C_{i0}^{(2)} = 0$ and $C_{ik}^{(2)} = C_{0k}^{(2)}$ for $i, k \neq 0$, then

$$\hat{\boldsymbol{\theta}}_k^*(\boldsymbol{X})_{p_k<1} = \sum_{i=1}^{K} \frac{\Lambda_i(\boldsymbol{X})}{1 + \sum_{i=1}^{K}\Lambda_i(\boldsymbol{X})} \hat{\Theta}_i^*(\boldsymbol{X}). \tag{7.2.10a}$$

The generalized likelihood ratios, $\Lambda_i(\boldsymbol{X})$, and the estimators in the absence of uncertainty, $\hat{\boldsymbol{\theta}}_i^*(\boldsymbol{X})_{p_i=1}$, are now given by

$$\Lambda_i(\boldsymbol{X}) = \frac{p_i \int_\Omega d\boldsymbol{\theta}\, W(\boldsymbol{X}|\boldsymbol{\theta}) w_i(\boldsymbol{\theta})}{p_0 W(\boldsymbol{X}|\boldsymbol{0})} \tag{7.2.11a}$$

and

$$\hat{\Theta}_i^*(\boldsymbol{X}) = \frac{\int_\Omega \boldsymbol{\theta}\, W(\boldsymbol{X}|\boldsymbol{\theta}) w_i(\boldsymbol{\theta})\, d\boldsymbol{\theta}}{\int_\Omega W(\boldsymbol{X}|\boldsymbol{\theta}) w_i(\boldsymbol{\theta})\, d\boldsymbol{\theta}}, \tag{7.2.11b}$$

refer to Eq. (7.1.19b), where the range of integration has been extended to the whole signal parameter space Ω.

If we require next that $C_{ik}^{(2)} = 1$, then the estimator structure is still given by (7.2.10a) and the detector structure is now represented by (7.2.8a). This is expressed below for the case in which the *parameter vector is the waveform vector*, that is, $\boldsymbol{\theta} \rightarrow \boldsymbol{S}$, where $\boldsymbol{S} = [S_k] = [S_k(\boldsymbol{r}_m, \boldsymbol{t}_m)], k = 1, \ldots, K$, and k represents, as before, the kth signal sample:

$$\begin{aligned} C_k\left(\boldsymbol{X}, \hat{\boldsymbol{S}}_k^*(\boldsymbol{X})\right) B_k\left(\boldsymbol{X}, \hat{\boldsymbol{S}}_k^*(\boldsymbol{X})\right) = \\ C_{0k}^{(1)} + \sum_{i=1}^{K}\left[C_{ik}^{(1)} - C_{i0}^{(1)}\right]\Lambda_i(\boldsymbol{X}) + \sum_{i=1}^{K}\Lambda_i(\boldsymbol{X})\left(\hat{\boldsymbol{S}}_i^{*T}\hat{\boldsymbol{S}}_i^*\right)_{p_i=1} \\ + \left[1 + \sum_{i=1}^{K}\Lambda_i(\boldsymbol{X})\right]\sum_{i=1}^{K}\sum_{k=1}^{K} P(H_i|\boldsymbol{X})P(H_{k'}|\boldsymbol{X})\left(\hat{\boldsymbol{S}}_k^*(\boldsymbol{X})_{p_k=1}\right)^T \hat{\boldsymbol{S}}_i^*(\boldsymbol{X})_{p_i=1}, \end{aligned} \tag{7.2.12}$$

where

$$\left(\hat{\boldsymbol{S}}_i^{*T}(\boldsymbol{X})\hat{\boldsymbol{S}}_i^*(\boldsymbol{X})\right)_{p_i=1} \equiv \frac{\int_\Omega \boldsymbol{S}^T\boldsymbol{S}\, W(\boldsymbol{X}|\boldsymbol{S}) w_i(\boldsymbol{S})\, d\boldsymbol{S}}{\int_\Omega W(\boldsymbol{X}|\boldsymbol{S}) w_i(\boldsymbol{S})\, d\boldsymbol{S}}. \tag{7.2.12a}$$

7.2.2.1 D + E in Normal Noise In Section 7.1.1, we halted the general analysis of the detector structure with Eq. (7.1.10) because at that point specific assumptions concerning the prior probability density functions $w_i(\boldsymbol{\theta}), w_i(\boldsymbol{S})$ were required. Now that these prior probability densities are no longer restricted to a subspace Ω_i of the signal parameter space Ω, we may carry the analysis further. For example, if we assume that the noise is Gaussian with zero mean and that $w_i(\boldsymbol{S})$ are also Gaussian, that is,

$$w_i(\boldsymbol{S}) = G(\boldsymbol{S}; \boldsymbol{m}_i, \boldsymbol{K}_i) \equiv (2\pi)^{-J/2}(\det \boldsymbol{K}_i)^{-1/2} \exp\left[-\frac{1}{2}(\boldsymbol{S}-\boldsymbol{m}_i)^T \boldsymbol{K}_i^{-1}(\boldsymbol{S}-\boldsymbol{m}_i)\right] \tag{7.2.13a}$$

and

$$W(\boldsymbol{X}|\boldsymbol{S}) = G(\boldsymbol{X}; \boldsymbol{S}, \boldsymbol{K}_i), \text{ that is, } \boldsymbol{K}_S \equiv \boldsymbol{K}_N, \tag{7.2.13b}$$

then we may evaluate $\left(\hat{\boldsymbol{S}}_i^{*T} \hat{\boldsymbol{S}}_i^*\right)_{p_i=1}$ and thus completely determine the explicit dependence of the detector on the received data $\boldsymbol{X}$. Both numerator and the denominator of (7.2.12a) contain the factor $W(\boldsymbol{X}|\boldsymbol{S})w_i(\boldsymbol{S})$, which under our Gaussian assumptions (7.2.13a) and (7.2.13b) becomes $G(\boldsymbol{X}; \boldsymbol{S}, \boldsymbol{K}_\mathrm{N})G(\boldsymbol{S}; \boldsymbol{m}_i, \boldsymbol{K}_i)$. This product is represented by

$$G(\boldsymbol{X}; \boldsymbol{S}, \boldsymbol{Q}_i)G(\boldsymbol{S}; \boldsymbol{m}_i, \boldsymbol{K}_i) = \frac{G(\boldsymbol{X}; \boldsymbol{0}, \boldsymbol{Q}_i)G(\boldsymbol{m}_i; \boldsymbol{0}, \boldsymbol{K}_i)}{G\left(\hat{\boldsymbol{S}}_i^*; \boldsymbol{0}, \boldsymbol{K}_i\right)}, \tag{7.2.14}$$

where now (cf. (7.1.23a))

$$\boldsymbol{Q}_i \equiv \left(\boldsymbol{K}_N^{-1} + \boldsymbol{K}_i\right)^{-1} \tag{7.2.14a}$$

$$\hat{\boldsymbol{S}}_i^* \equiv \boldsymbol{Q}_i\left(\boldsymbol{K}_N^{-1} + \boldsymbol{X} + \boldsymbol{K}_i^{-1}\boldsymbol{m}_i\right). \tag{7.2.14b}$$

The quantity $\hat{\boldsymbol{S}}_i^*$ in (7.2.14b) indicates that this is an optimum least-squares estimate of the waveform vector *in the absence of uncertainty* and is equivalent to $\hat{\boldsymbol{S}}_i^*(\boldsymbol{X})_{p_i=1}$ in the quadratic cost function case. Since $G\left(\boldsymbol{S}; \hat{\boldsymbol{S}}_i^*, \boldsymbol{Q}_N\right)$ is a probability density function, its integral over the signal space Ω is unity. Substituting (7.2.14b) in (7.2.14a) gives

$$\left(\hat{\boldsymbol{S}}(\boldsymbol{X})^{*T}\hat{\boldsymbol{S}}(\boldsymbol{X})^*\right)_{p_i=1} = \int_{-\infty}^{\infty} \cdots \int \boldsymbol{S}^T \boldsymbol{S}\, G\left(\boldsymbol{S}; \hat{\boldsymbol{S}}_i^*, \boldsymbol{Q}_i\right) d\boldsymbol{S}. \tag{7.2.15a}$$

From the way in which the elements of the covariance matrix of a Gaussian probability density are defined, one sees that the integral on the right-hand side of this last relation is given by

$$\int_{-\infty}^{\infty} \cdots \int\int \boldsymbol{S}^T \boldsymbol{S}\, G\left(\boldsymbol{S}; \hat{\boldsymbol{S}}_i^*, \boldsymbol{Q}_i\right) d\boldsymbol{S} = \mathrm{Tr}(\boldsymbol{Q}_i) + \left(\hat{\boldsymbol{S}}_i^*\right)^T \hat{\boldsymbol{S}}_i^*, \tag{7.2.15b}$$

where again $\mathrm{Tr}(\boldsymbol{Q}_i)$ indicates the trace of the matrix $\boldsymbol{Q}_i$. Accordingly, we have (from (7.2.15a)

$$\left(\hat{\boldsymbol{S}}(\boldsymbol{X})^{*T}\hat{\boldsymbol{S}}(\boldsymbol{X})^*\right)_{p_i=1} = \mathrm{Tr}(\boldsymbol{Q}_i) + \left(\hat{\boldsymbol{S}}_i^*\right)^T \hat{\boldsymbol{S}}_i^*. \tag{7.2.16}$$

Substituting this expression into (7.2.12), we also obtain the following expression for the detector structure, with $\boldsymbol{\theta} \to \boldsymbol{S}$. Thus, we have

$$
\begin{aligned}
C_k\left(\boldsymbol{X}, \hat{\boldsymbol{S}}_k^*(\boldsymbol{X})\right) B_k\left(\boldsymbol{X}, \hat{\boldsymbol{S}}_k^*(\boldsymbol{X})\right)
&= C_{0k}^{(1)} + \sum_{i=1}^{K}\left[\left(C_{ik}^{(1)} - C_{i0}^{(1)}\right) + \mathrm{Tr}(\boldsymbol{Q}_i) + \left(\hat{\boldsymbol{S}}_i^*(\boldsymbol{X})_{p_i=1}\right)^T \hat{\boldsymbol{S}}_i^*(\boldsymbol{X})_{p_i=1}\right]\Lambda_i(\boldsymbol{X}) \\
&\quad - \left(1 + \sum_{i=1}^{K}\Lambda_i(\boldsymbol{X})\right)\sum_{i=1}^{K}\sum_{k=1}^{K} P(H_i|\boldsymbol{X})P(H_k|\boldsymbol{X})\left(\hat{\boldsymbol{S}}_k^*(\boldsymbol{X})_{p_k=1}\right)^T \hat{\boldsymbol{S}}_i^*(\boldsymbol{X})_{p_i=1}
\end{aligned}
\tag{7.2.17}
$$

where we have equated $\hat{\boldsymbol{S}}_i^*$ with $\hat{\boldsymbol{S}}_i^*(\boldsymbol{X})_{p_i=1}$.

Since $P(H_i|\boldsymbol{X}) = \Lambda_i(\boldsymbol{X}) \cdot \left[1 + \sum_{i=1}^{K}\Lambda_i(\boldsymbol{X})\right]^{-1}$, we observe that the detector structure can be expressed as a function of the generalized likelihood ratios and the least-squares estimators of the various signals in the absence of uncertainty. Furthermore, the right-hand side of (7.2.17) can be expressed solely as a function of the generalized likelihood ratios, $\Lambda_i(\boldsymbol{X})$ for $i = 1, \ldots, K$, or as a function of the least-squares estimators, $\hat{\boldsymbol{S}}_i^*(\boldsymbol{X})_{p_i=1}$ for $i = 1, \ldots, K$. This last follows from some relations discussed in Refs. [8, 9]. In particular, it is shown in Ref. [9] that the least-squares or minimum-variance estimator $\hat{\boldsymbol{S}}_i^*(\boldsymbol{X})_{p_i=1}$ can be represented by the following vector function of $\Lambda_i(\boldsymbol{X})$:

$$
\hat{\boldsymbol{S}}_i^*(\boldsymbol{X})_{p_i=1} = \boldsymbol{K}_{Ni}\nabla \log \Lambda_i(\boldsymbol{X}), \tag{7.2.18}
$$

where in particular $\nabla\log\Lambda_i(\boldsymbol{X})$ is a column vector whose components are $\left[(1/\Lambda_i)(\partial\Lambda_i/\partial \boldsymbol{X}_j)\right], j = 1, \ldots, K$. When Eq. (7.2.18) is substituted into (7.2.17), the detector structure is now expressed solely in terms of the generalized likelihood ratios Λ_i. Conversely, the generalized likelihood ratios can be expressed as a function of the various least-squares estimators, $\hat{\boldsymbol{S}}_i^*(\boldsymbol{X})_{p_i=1}$, namely,

$$
\Lambda_i(\boldsymbol{X}) = e^{\left(\boldsymbol{X}^T \boldsymbol{K}_{Ni}^{-1}\hat{\boldsymbol{S}}_i^*(\boldsymbol{X})_{p_i=1} + b_i\right)} \tag{7.2.19}
$$

where $b_i \equiv C - \int \boldsymbol{X}^T \boldsymbol{K}_{N_i}^{-1}\, d\boldsymbol{S}_i$ is a *bias term* (see Problem 6.21 and Eq. 8 in Ref. [9]) to ensure the vanishing of the error probabilities as sample size increases without limit. Substitution into (7.2.17) of the various generalized likelihood ratios, expressed as functions of the received data and the least-squares estimators, as in (7.2.19), provides a nonlinear detector structure that is a function of only the least-squares estimators and the received data. In addition, the detector is seen to be a type of estimator–correlator, the basic operations being correlation of the received data with the various least-squares estimators in the absence of uncertainty ($p_i = 1$), and cross-correlation of these same estimators with themselves.

7.2.3 Simple Cost Functions ($\mathbf{SCF}_{1,2}$): Joint $D + E$ with Overlapping Hypotheses Classes

Because of the way in which these cost functions are defined (refer to Eq. (7.1.3), and extended by Eq. (7.2.4a), now with $C_{ik}^{(1)} + C_{ik}^{(21)}\left(\boldsymbol{\theta}, \hat{\boldsymbol{\theta}}(\boldsymbol{X})\right)$ replaced specifically by (7.1.24a) and (7.1.24b), we see that the analysis and results of Section 7.1.2.3 still apply. Here, however, the domain of a particular parameter space Ω_i is extended to include the entire space Ω, that is, $\Omega_i \to \Omega$ when Ω_i *may* extend over all Ω or may vanish over part of Ω. (See the discussions above following Eq. (7.2.5).)

7.3 SIMULTANEOUS DETECTION AND ESTIMATION: SEQUENCES OF OBSERVATIONS AND DECISIONS [2–5]

In Sections 7.1 and 7.2, we have discussed joint detection (D) and estimation (E) based on data from a single space–time observation interval. Here, we introduce the important extension to the "multishot" or sequence of more than one observation period, where now we are concerned with optimal or Bayes processing and decision making based on such a succession of intervals and decisions. As before, we have to deal with the problems of no- and strong-coupling between the detection and estimation portions of the data processing, where the results of estimation depend strongly on the choices of cost functions as well as on the results of the concomitant detection. Among the many obvious applications of the analysis here is target-tracking, improving the estimation of unknown signal parameters generally, and signal discrimination, as well as various other aspects of classification. As expected, this is usually a much more intensive set of data and processing tasks than those involving the "one-shot" cases discussed mainly in this book. Accordingly, this section should be regarded as a partial analytic (algorithmic) introduction to the subject, emphasizing the relatively new Bayesian features of joint detection and estimation under uncertainty [2–5]. Moreover, the general subject of target tracking and estimation has been extensively treated elsewhere (see, for example, Ref. [1]).

Accordingly, in Section 7.3, we consider first the common case of separate (i.e., uncoupled) detection and estimation (Section 7.3.1), in which, however, the estimates are now modified by the fact that they are made under uncertainty as to whether or not the particular signal is actually present (refer to the binary cases discussed in Section 6.2). Next (Section 7.3.2), we examine the situation where there is strong coupling between detection and estimation, which influences *both* the detection and estimation structures and decisions. Both unsupervised and supervised leaning modes, that is, "learning without" and "learning with" teacher, are considered here.

7.3.1 Sequential Observations and Unsupervised Learning: I. Binary Systems with Joint Uncoupled $D + E$

In our earlier work [2, 3], and in Chapter 6, as well as in previous sections of Chapter 7, we have constructed a Bayes, that is, minimum-average risk, theory of simultaneous detection and estimation of a single signal in noise when a single space–time interval $|\boldsymbol{R}|T$ is available at the receiver for the joint reception process. This is a situation of considerable importance and a necessary starting point for further developments. Accordingly, our next step is to consider the more general situation when N_0 intervals (each of duration $|\boldsymbol{R}|T$) are available to the receiver, where at the end of each interval we require an appropriate set of decisions as to the detection and estimation of the single signal at that stage. In each interval ($r = 1, 2, \ldots$), the signal, if it is present, has the form $S_r(\boldsymbol{r}_m, t_n|\boldsymbol{\theta})$, where $\boldsymbol{\theta} = (\boldsymbol{\theta}_\ell, \ldots, \boldsymbol{\theta}_L)$ is a set of parameters to be estimated. In each interval (r) of duration $|\boldsymbol{R}|T$, there are the same probabilities p and $1 - p$ that the signal is present or

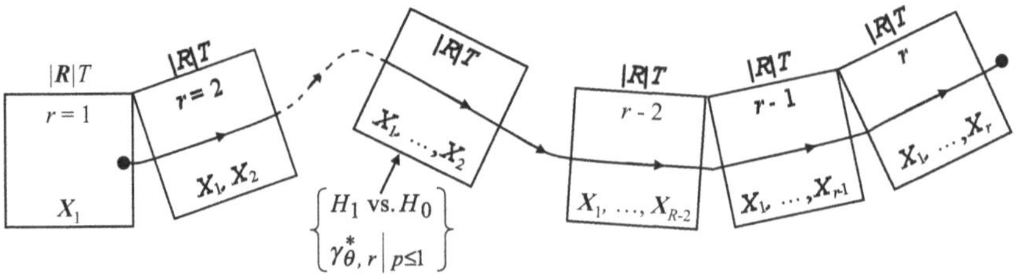

$$\text{decide}: \begin{Bmatrix} H_1: \quad \text{vs.} \quad H_0: \\ \gamma^*_{\theta,r}\big|_{p\le 1} \neq 1; = 0 \end{Bmatrix} \text{each interval; k decisions in } k|\boldsymbol{R}|T.$$

FIGURE 7.3 Schematic diagram of binary (on–off) uncoupled simultaneous detection and estimation sequence. **A.** D and E at end of r intervals ___ (Section 7.3.1.2). **B.** "Tracking" D and E for each interval (Section 7.3.1.3). This is for Estimation **I**.

absent, that is, that only noise is present. The parameters $\boldsymbol{\theta}$ are unknown, but their values are required to be constant throughout the set of observation intervals $N_o|\boldsymbol{R}|T$. This situation may be considered as a model for adaptive or "learning" processes in the sense that the receiver gathers information regarding the values of $\boldsymbol{\theta}$ during each observation interval and uses this information, along with the preceding data, to improve the detection and the estimation performance. The underlying assumption here, which makes the model useful for practical situations, is that the space–time constant of what may be in reality space–time varying parameters $\boldsymbol{\theta}(\boldsymbol{r}, t)$ is longer than the total observation interval $N_o|\boldsymbol{R}|T$. Figure 7.3 illustrates a typical data sequence.

Although the theory of multiple alternative joint detection and estimation discussed earlier in this chapter assumes a "one-shot" reception situation, it could have been considered as one stage in a sequential decision system. It therefore should be noted at this point that estimator–correlator structures, such as the one under discussion, have a built-in adaptive feature. A discussion of this point is to be found in Ref. [11].

We present now an initial step toward the extension of the theory to this more general situation, including optimum, that is, Bayes, procedures. Specifically, we consider the case *where binary detection and estimation are uncoupled*, and where estimation must be performed in the face of uncertainty as to the signal's presence ($p < 1$). We also assume that the unknown parameters $\boldsymbol{\theta}$ to be estimated are such that it is appropriate to assume $C_{10}(\gamma_{\boldsymbol{\theta}}) = C_{11}(\boldsymbol{\gamma}_{\theta}, 0)$, that is, $\boldsymbol{\theta}$ are "energy" parameters (as in Eq. (6.1.4a)). The relatively simple situation of *uncoupled detection and estimation* is needed to provide the necessary insight into the more involved cases where coupling occurs.

7.3.1.1 Detection Here we consider first the classical binary on–off detector, tended not to include ($r > 1$) received data intervals ($\boldsymbol{X}_1, \ldots, \boldsymbol{X}_{r-1}$), where detection decisions are made recursively at the end of $r = 1, 2, \ldots, r$ space–time intervals, of duration $r^2|\boldsymbol{R}|T$, by comparison with a threshold K_r (> 0) in the manner of Section 6.2. We begin first with the optimum (i.e., Bayes) adaptive receiver for detection alone in the binary on–off case, that is, H_1: $S \otimes N$ versus H_0: N. This situation has been analyzed initially by Fralick [12]. His solution is especially interesting, since it has a recursive form. Let $\boldsymbol{X}_1, \boldsymbol{X}_2, \ldots, \boldsymbol{X}_r$ be the vectors of the observed data, respectively in the first, second, and rth interval. Then, it is shown in Ref. [12] that one of the operations of the optimum adaptive detector is the recursive calculation of $w(\boldsymbol{\theta}|\boldsymbol{X}_1, \boldsymbol{X}_2, \ldots, \boldsymbol{X}_k)$, which is pdf of $\boldsymbol{\theta}$

given the observations $X_1, X_2, \ldots, X_k$. Specifically, the recursive algorithm has the form

$$w_r(\boldsymbol{\theta}) \equiv w(\boldsymbol{\theta}|\mathbf{X}_1, \mathbf{X}_2, \ldots, \mathbf{X}_r) = \frac{W(\mathbf{X}_r|\boldsymbol{\theta},\, \mathbf{X}_1, \mathbf{X}_2, \ldots, \mathbf{X}_{r-1})}{W(\mathbf{X}_r|\mathbf{X}_1, \ldots, \mathbf{X}_{r-1})} w(\boldsymbol{\theta}|\mathbf{X}_1, \ldots, \mathbf{X}_{r-1}). \quad (7.3.1)$$

Equation (7.3.1) is readily derived by successive applications of Bayes' theorem, as shown by the following set of relations:

$$\begin{aligned}
r = 1:&\quad w(\boldsymbol{\theta}|\mathbf{X}_1)W(\mathbf{X}_1) = w(\boldsymbol{\theta})W(\mathbf{X}_1|\boldsymbol{\theta})\\
r = 2:&\quad w(\boldsymbol{\theta}|\mathbf{X}_1)W(\mathbf{X}_2|\boldsymbol{\theta}, \mathbf{X}_1) = w(\boldsymbol{\theta}|\mathbf{X}_1, \mathbf{X}_2)W(\mathbf{X}_2|\mathbf{X}_1)\\
r = 3:&\quad w(\boldsymbol{\theta}|\mathbf{X}_1, \mathbf{X}_2)W(\mathbf{X}_3|\boldsymbol{\theta}, \mathbf{X}_1, \mathbf{X}_2) = w(\boldsymbol{\theta}|\mathbf{X}_1, \mathbf{X}_2, \mathbf{X}_3)W(\mathbf{X}_3|\mathbf{X}_1, \mathbf{X}_2)\\
\vdots\quad&\qquad\qquad\vdots\qquad\qquad\qquad\qquad\qquad\qquad\vdots\\
r = r:&\quad w(\boldsymbol{\theta}|\mathbf{X}_1, \ldots, \mathbf{X}_{r-1})W(\mathbf{X}_r|\boldsymbol{\theta}, \mathbf{X}_1, \mathbf{X}_2, \ldots, \mathbf{X}_{r-1}) = w(\boldsymbol{\theta}|\mathbf{X}_1, \ldots, \mathbf{X}_{r-1})\\
&\qquad\qquad\qquad\qquad\qquad\qquad\qquad\cdot W(\mathbf{X}_r|\mathbf{X}_1, \mathbf{X}_2, \ldots, \mathbf{X}_{r-1}),
\end{aligned} \quad (7.3.1a)$$

from which (7.3.1) follows directly. The pdf $w_r(\boldsymbol{\theta})$ is then employed in the rth interval to form the required (generalized) likelihood ratio Λ_{r+1} for the $r+1$ stage, for the optimum binary on–off test against the threshold K_{r+1} in the usual way. With the help of (7.3.1), this then explicitly becomes

$$\left.\begin{aligned}
&\textbf{Decide } H_1\text{: S} \otimes \text{N if: } \Lambda_{r+1}(\mathbf{X}_1,\mathbf{X}_2, \ldots,\mathbf{X}_{r+1}) = \int_\Omega \Lambda_{r+1}(\mathbf{X}_{r+1}|\boldsymbol{\theta})w_r(\boldsymbol{\theta})d\boldsymbol{\theta}\\
&\qquad\qquad\qquad\qquad\qquad\qquad\qquad\qquad = \mu\int_\Omega \frac{W(\mathbf{X}_{r+1}|\boldsymbol{\theta})w_r(\boldsymbol{\theta})d\boldsymbol{\theta}}{W(\mathbf{X}_{r+1}|\mathbf{0})}\\
&\textbf{Decide } H_0\text{: N if: } \Lambda_{r+1} < \mathcal{K}_{r+1}, \quad \mu = p/q.
\end{aligned}\right\} \geq \mathcal{K}_{r+1} \quad (7.3.2)$$

(The threshold $\mathcal{K}_{r+1}$ is a positive constant ratio of (positive or zero) cost functions in the manner of Chapters 1 and 3.) Here, for the initial interval ($r = 1$), Eq. (7.3.2) reduces to

$$\Lambda_1(\mathbf{X}_1) = \int_\Omega \Lambda_r(\mathbf{X}_1|\boldsymbol{\theta})w_0(\boldsymbol{\theta})d\boldsymbol{\theta} = \mu\int_\Omega W(\mathbf{X}_1|\boldsymbol{\theta})w_0(\boldsymbol{\theta})d\boldsymbol{\theta}/W(\mathbf{X}_1|\mathbf{0}) = \Lambda(\mathbf{X}), \quad (7.3.2a)$$

which is GLR for detection for the single sample data $\mathbf{X}$ for binary detection, as discussed in Chapters 1–6, with $w_0(\boldsymbol{\theta})$ the initial *a priori* probability of $\boldsymbol{\theta}$.

It can also be demonstrated [13] that the detector structure (7.3.2) is *stable*, that is, the sequence of likelihood ratios $\Lambda_{r+1}, r \geq 0$ converges to a finite limit with probability one. In addition, it can be shown that the sequence of pdf values $w_R(\boldsymbol{\theta})$, appearing above (7.3.1) and subsequently (cf. (7.3.3), etc.) converges to the true value $\boldsymbol{\theta}_0$ of the unknown parameters $\boldsymbol{\theta}$, namely, $\lim_{R\to\infty} w_R(\boldsymbol{\theta}) = \delta(\boldsymbol{\theta} - \boldsymbol{\theta}_0)$ (see Section 7.3.1.4.). It is important to observe here that each of the pdf values $w_0(\boldsymbol{\theta}), w_1(\boldsymbol{\theta}), \ldots, w_r(\boldsymbol{\theta})$ in the sequence of distributions represents in any of the r intervals the pdf of $\boldsymbol{\theta}$ *under the hypothesis* H_1, conditional on past observations, refer to Eq. (7.3.1).

7.3.1.2 Uncoupled Estimators for r Intervals, with QCF Let us now turn our attention to the estimation problem, and use the *quadratic cost function (QCF)*. Here, we have the situation of H_1 alone: a signal is known to be present and no detection is required. It is clear that we must consider the class of adaptive estimators whose structure changes at each interval. There are at least two different estimators that can be considered here.[2] First, (7.3.3) suggests that it is possible to design an estimator that is *consistent* by using, in the rth interval for example, the pdf $w_{r-1}(\boldsymbol{\theta})$, (7.3.1), recursively evaluated for the detection procedure. This is the vector of Bayes adaptive estimators of $\boldsymbol{\theta}$. At the end of the rth interval, it is given by the set of relations:

$$\textbf{QCF:} \quad \gamma^*_{\boldsymbol{\theta},r}(X,\ldots,X_r)_{\text{QCF}} = \int \boldsymbol{\theta}\, w(\boldsymbol{\theta}|X_1,\ldots,X_r)d\boldsymbol{\theta} = \int \boldsymbol{\theta}\, w_{r-1}(\boldsymbol{\theta})\, d\boldsymbol{\theta}, \tag{7.3.3}$$

where it is readily seen from (7.3.1), that this vector becomes

$$\left.\gamma^*_{\boldsymbol{\theta},r}\right|_{\substack{\text{QCF}\\ p=1}} = \frac{\int_\Omega \boldsymbol{\theta}\, W(\boldsymbol{\theta}|X_1,\ldots,X_r)d\boldsymbol{\theta}}{\int_\Omega W(\boldsymbol{\theta}|X_1,\ldots,X_r)d\boldsymbol{\theta}} = \int_\Omega \frac{\boldsymbol{\theta}\, w(\boldsymbol{\theta}|X_1,\ldots,X_r)W(X_1,\ldots,X_r)d\boldsymbol{\theta}}{W(X_1,\ldots,X_r)} \tag{7.3.3a}$$

$$= \int_\Omega \boldsymbol{\theta}\, w_r(\boldsymbol{\theta})d\boldsymbol{\theta}. \tag{7.3.3b}$$

Since $w(\boldsymbol{\theta};X,\ldots,X_r) = w(\boldsymbol{\theta}|X_1,\ldots,X_{r-1})\, w(X,\ldots,X_r)$, it follows that

$$\left.\boldsymbol{\gamma}^*_{\boldsymbol{\theta},r}\right|_{\substack{\text{QCF}\\ p=1}} = \int_\Omega \boldsymbol{\theta}\, w_{r-1}(\boldsymbol{\theta})d\boldsymbol{\theta} = \boldsymbol{\gamma}^*_{\boldsymbol{\theta},r}(X_1,\ldots,X_r)_{\substack{\text{QCF}\\ p=1}}. \tag{7.3.4}$$

As we shall show presently, for this estimator, we have

$$\lim_{r\to\infty} \left.\boldsymbol{\gamma}^*_{\boldsymbol{\theta},r}\right|_{\text{QCF}} = \boldsymbol{\theta}_0, \quad \text{probability 1}, \tag{7.3.5}$$

that is, the vector *estimator* $\boldsymbol{\gamma}^*_{\boldsymbol{\theta},r}$ *is consistent*. The proof of this is outlined in Section 7.3.1.4. Note, moreover, that at the end of each interval r, the estimator $\boldsymbol{\gamma}^*_{\boldsymbol{\theta}}|_{p=1}$ yields the optimum estimates of $\boldsymbol{\theta}$ (with respect to a quadratic cost function) for the whole sequence $X_1, X_2, \ldots, X_r$ and not for the specific interval r considered.

If, on the other hand, an (vector) estimate of $\boldsymbol{\theta}$ is required for the *specific interval* r, during which the signal *may be absent* (with probability $1-p$), then we need a different approach to the extraction problem (see **II** below). However, that situation is similar to the one examined for the single-interval case.

We now look at both these estimators in more detail.

I. Uncoupled Estimator for r Intervals ($p = 1$): Let us start with the estimator mentioned above (7.3.4). The point of view here is that we desire an optimum

[2] See Ref. [14] for an example of another possible estimator, designed with the constraint that it be linear.

(vector) estimate of $\boldsymbol{\theta}$ at the end of rth-stage of the process. In this stage, the *a priori* pdf of $\boldsymbol{\theta}$, conditional on the past observations $\boldsymbol{X}_1, \boldsymbol{X}_2, \ldots, \boldsymbol{X}_{r-1}$, is $w_{r-1}(\boldsymbol{\theta})$ and is computed recursively from (7.3.1) as in the detection operation, according to (7.3.3) et seq.). Then as we have seen above, at the end of rth interval, the Bayes vector estimator of $\boldsymbol{\theta}$ has the form (7.3.4), namely,

$$\boldsymbol{\gamma}^*_{\theta,r}(\boldsymbol{X}_1, \ldots, \boldsymbol{X}_r)\Big|_{\substack{\text{QCF}\\ p=1}} = \int \boldsymbol{\theta}\, w_{r-1}(\boldsymbol{\theta}) d\boldsymbol{\theta} = \boldsymbol{\gamma}^*_{\theta,r}\Big|_{p=1}. \tag{7.3.5a}$$

We shall show presently (Section 7.3.1.4) that this estimator is also *stable*, that is, converges to a finite limit, as well as consistent. Figure 7.3 illustrates this case.

II. Uncoupled Estimator for rth Interval (QCF): As mentioned at the beginning of Section 7.3.1.2 following (7.3.5), there is another estimator that can be considered where $p \leq 1$, that is, the signal is not surely known to be present. Now the point of view is that we require an optimum estimate of the parameter $\boldsymbol{\theta}$ during a specific part of the process, say in rth interval. This estimate only refers to the interval r considered, although, of course, the information gathered from previous intervals is still put to use.

The key point here is that for a specific interval, there is never certainty as to the presence of the signal. Consider, for example, rth interval. Let $w(\boldsymbol{\theta}|\boldsymbol{X}_1, \ldots, \boldsymbol{X}_{r-1}) \equiv w_{r-1}(\boldsymbol{\theta})$ be, namely, the *a priori* pdf of $\boldsymbol{\theta}$, conditional on the past observations $\boldsymbol{X}_1, \ldots, \boldsymbol{X}_{r-1}$. The situation is similar to the case of a single observation $\boldsymbol{X}$ analyzed already in Section 6.3, the only difference being that the *a priori* pdf $w(\boldsymbol{\theta})$ is now replaced by the conditional *a priori* pdf $w_{r-1}(\boldsymbol{\theta})$. Thus, we can immediately extend the general results of Section 6.3 to this situation. In particular, for QCF, we get specifically the desired estimator (in these uncoupled cases), refer to Eqs. (6.2.7) and (6.3.29):

$$\boldsymbol{\gamma}^*_{\theta,r}\Big|_{p\leq 1} = \frac{\Lambda_r}{\Lambda_{r+1}}\boldsymbol{\gamma}^*_{\theta,r}\Big|_{p\equiv 1,\, p=1}, \tag{7.3.6}$$

where now the accompanying detector that authorizes the acceptance or rejection of the estimate is given by Λ_r, the average likelihood ratio for rth interval. Its form is also given by an expression similar to (7.3.2), for example,

$$\Lambda_r \equiv \Lambda_r(\boldsymbol{X}_1, \ldots, \boldsymbol{X}_r) = \int \Lambda_r(\boldsymbol{X}_r|\boldsymbol{\theta}) w_{r-1}(\boldsymbol{\theta}) d\boldsymbol{\theta} \tag{7.3.7}$$

where $\boldsymbol{\gamma}^*_{\theta,r}|_{p=1}$ is now the estimator $\boldsymbol{\gamma}^*_{\theta,r}$ (refer to Eq. (7.3.4).

The asymptotic properties of (7.3.6) as $r \to \infty$ are of interest. Reasoning similar to the one used above to discuss the stability and consistency of (7.3.4), Fralick [12] has shown that as $r \to \infty$, Λ_r tends to the likelihood ratio $\Lambda_\infty = \Lambda(\boldsymbol{X}_\infty|\boldsymbol{\theta}_0)$, which is appropriate for the detection of the perfectly known signal $S(\boldsymbol{r}_m, t_n|\boldsymbol{\theta}_0)$. Thus, from (7.3.5), we obtain

$$\lim_{r\to\infty} \boldsymbol{\gamma}_{\theta,r|p<1} = \frac{\Lambda(\boldsymbol{X}_\infty|\boldsymbol{\theta}_0)}{\Lambda(\boldsymbol{X}_\infty|\boldsymbol{\theta}_0)+1}\boldsymbol{\theta}_0, \quad \text{with} \quad \boldsymbol{X}_\infty = \lim_{r\to\infty}(\boldsymbol{X}_1, \boldsymbol{X}_2, \ldots, \boldsymbol{X}_r). \tag{7.3.8}$$

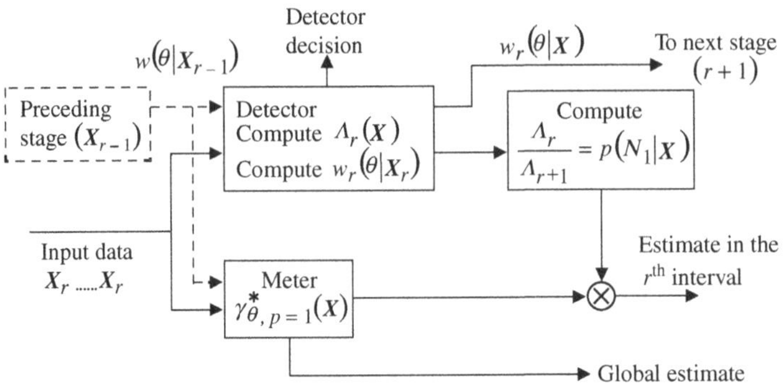

FIGURE 7.4 Multistage uncoupled detection and estimation with QCF. Block diagram of the rth stage.

The Bayes risk corresponding to this estimator is at rth stage, where now we take the average $E^{(r)}$ over *all the interval data* $(X_1, \ldots, X_r)$, as represented by $E^{(r)} \equiv E^{(r)}(X_1, X_2, \ldots, X_r)$:

$$R^*_{r,p<1} = E^{(r)}\left\{\left(\theta - \gamma^*_{\theta,r}|_{p<1}\right)^2\right\} = E^{(r)}\left\{\theta_0 \cdot \theta_0 - 2\theta_0 \cdot \gamma^*_{\theta,r}|_{p<1} + \left(\gamma^*_\theta \cdot \gamma^*_\theta\right)_{r|p-1}\right\}. \tag{7.3.9}$$

Asymptotically, from (7.3.6), (7.3.8), and (7.3.9), we have finally

$$\lim_{r\to\infty} R^*_{r,p<1} = E^{[\infty]}\left[\frac{\theta_0}{\Lambda(X_\infty|\theta_0)+1}\right]^2 = \frac{\theta_0 \cdot \theta_0}{E_{X_\infty}\left\{\left[\Lambda(X_\infty|\theta_0)+1\right]^{-2}\right\}^{-1}}, \tag{7.3.10}$$

where $E^{(\infty)} \equiv \lim_{r\to\infty} E^{(r)}(X_1, X_2, \ldots, X_r)$.

On the basis of the above **(II)**, we can obtain Fig. 7.4 for the combined uncoupled multistage detection and estimation process (with QCF) considered in Section 7.3.1.

7.3.1.3 Uncoupled Estimation (I) and (II): (SCF)$_{1,2}$ When the "simple" cost functions $(\text{SCF})_{1,2}$ of Section 6.2 are employed instead of QCF above, the corresponding Bayes estimators of the parameters $\boldsymbol{\theta}$, namely, unconditional maximum-likelihood estimators under the condition of uncertainty ($p < 1$), have much less complex structures than QCF counterparts.

For the first class of estimator (**I**) determined above under a QCF, the corresponding estimators are as follows:

$$\textbf{I:}\ \gamma^*_{\theta,r}|_{p=1}|_{\text{SCF}_1} = \left[\gamma^*_\theta\left(\theta_\ell|X_1, \ldots, X_r\right)_{p=1}\right];\ \gamma^*_{\theta,r}|_{p=1}|_{\text{SCF}_2} = \gamma^*_\theta\left(\theta|X_1, \ldots, X_r\right)_{p=1} \tag{7.3.11}$$

respectively). In the first case, the individual components of the resulting UMLE estimator $\gamma^*_{\theta,r}|_{p=1}|_{\text{SCF}_1}$ are optimized separately of each other, whereas for $(\text{SCF})_2$, the components of $\gamma^*_{\theta,r}|_{p=1}|_{\text{SCF}_2}$ are optimized simultaneously.

This applies not only to the single interval data but also to the multiple interval cases of the recursive estimators $r > 1$ of type **I** (Section 7.3.1.2).

For the estimator of rth individual interval **(II)**, where there is uncertainty as to the signal's presence, we have in place of (7.3.4,5), $\boldsymbol{\gamma}^*_{\boldsymbol{\theta},r}|_{p<1}|_{\mathrm{SCF}_{1,2}}$, the optimum estimators:

$$\textbf{II:}\quad \left.\begin{aligned} \boldsymbol{\gamma}^*_{\boldsymbol{\theta},r}|_{p<1}|_{\mathrm{SCF}_1} &= p\left[\boldsymbol{\gamma}^*_{\boldsymbol{\theta},r}(\boldsymbol{\theta}_\ell|\boldsymbol{X}_1,\ldots,\boldsymbol{X}_r)_{p=1}\right] \\ \boldsymbol{\gamma}^*_{\boldsymbol{\theta},r}|_{p<1}|_{\mathrm{SCF}_2} &= p\boldsymbol{\gamma}^*_{\boldsymbol{\theta}}(\boldsymbol{\theta}|\boldsymbol{X}_1,\ldots,\boldsymbol{X}_r)_{p=1} \end{aligned}\right\}, \tag{7.3.12}$$

which are the counterparts of $\boldsymbol{\gamma}^*_{\boldsymbol{\theta},r}|_{\mathrm{QCF}}$ of **(II)** above.

7.3.1.4 Consistency and Convergence of the Estimators Here, we outline a demonstration that the estimators in Sections 7.3.1.2 and 7.3.1.3 are both consistent and stable, namely,

$$\left.\begin{aligned} &\text{Consistency: } \lim_{r\to\infty}\int G(\boldsymbol{\theta})dP_r(\boldsymbol{\theta}) = G(\boldsymbol{\theta}_0); \\ &\text{Stability: } \lim_{r\to\infty}\boldsymbol{\gamma}^*_{\boldsymbol{\theta},r} = \boldsymbol{\gamma}^*_{\boldsymbol{\theta},\infty}, \quad \text{with } \boldsymbol{E}\left\{|\boldsymbol{\gamma}^*_{\boldsymbol{\theta},\infty}|\right\} \le C_0 \end{aligned}\right\}, \tag{7.3.13}$$

where $C_0\ (>0)$ is a positive constant and $\boldsymbol{E}$ is the expectation operator.

We begin with a sketch of the proof first for stability. For this we need the following theorem [14]:

Theorem: Any sequence $(a_1, a_2, \ldots, a_{r_0+1})$ *such that* $a_r \le \int G(\boldsymbol{\theta})dP_r(\boldsymbol{\theta})$, *where* $P_r(\boldsymbol{\theta}) = P(\boldsymbol{\theta}|\boldsymbol{X}_1,\ldots,\boldsymbol{X}_r)$ *is a probability measure*, that is, a cumulative distribution function corresponding to the pdf $w_r(\boldsymbol{\theta}) = w(\boldsymbol{\theta}|\boldsymbol{X}_1,\ldots,\boldsymbol{X}_r)$, *is a bounded martingale* [13] *if*

$$\begin{aligned} &(1)\ G(\boldsymbol{\theta}) \textit{ is any nonnegative Lebesque measurable function}; \\ &(2)\ \max G(\boldsymbol{\theta}) \le L_0 < \infty. \end{aligned} \tag{7.3.13a}$$

Then, the sequence $\boldsymbol{\gamma}^*_{\boldsymbol{\theta},1}, \boldsymbol{\gamma}^*_{\boldsymbol{\theta},2}, \ldots, \boldsymbol{\gamma}^*_{\boldsymbol{\theta},r}$ in which each term is defined by (7.3.4) is a bounded martingale. According to another theorem of Doob [13], the limit of a bounded martingale exists with probability 1, that is,

$$\lim_{r\to\infty}\boldsymbol{\gamma}^*_{\boldsymbol{\theta},r} = \boldsymbol{\gamma}^*_{\boldsymbol{\theta},\infty}, \quad \text{and} \quad \boldsymbol{E}\left\{\left|\boldsymbol{\gamma}^*_{\boldsymbol{\theta},\infty}\right|\right\} \le L_0. \tag{7.3.14}$$

Accordingly, the adaptive estimation procedure (7.3.4) is *stable*.

In order to establish *consistency*, let us define the indicator function I_θ such that

$$I_\theta = \begin{cases} 1 & \text{if } \boldsymbol{\theta} \in \boldsymbol{\Theta}, \\ 0 & \text{if } \boldsymbol{\theta} \notin \boldsymbol{\Theta}. \end{cases} \tag{7.3.15}$$

Then the sequence

$$P_r(\boldsymbol{\Theta}) = \int I_\theta dP_r(\boldsymbol{\Theta}) \tag{7.3.16}$$

is a bounded martingale. Hence, with probability 1,

$$\lim_{r\to\infty} P_r(\boldsymbol{\Theta}) = P_\infty(\boldsymbol{\Theta}). \tag{7.3.17}$$

It is well known [15] that if the sequence $\{a_1, a_2, \ldots, a_{r_0}, a_{r_0+1}\}$ is a bounded martingale, then $E\{a_{r_0+1}|a_1, \ldots, a_{r_0}\}$ converges to a_{r_0+1} with probability 1. Therefore, $P_r(\boldsymbol{\Theta})$ converges to 1 when $\boldsymbol{\theta}_0 \in \boldsymbol{\Theta}$ and to zero when $\boldsymbol{\theta}_0 \notin \boldsymbol{\Theta}$. Consequently, $P_\infty(\boldsymbol{\Theta})$ must be a step function at $\boldsymbol{\theta} = \boldsymbol{\theta}_0$. The consistency of the estimator follows from the above considerations and from the fact that if $G(\boldsymbol{\theta})$ is any arbitrary continuous function, then

$$\lim_{r \to \infty} \int G(\boldsymbol{\theta}) dP_r(\boldsymbol{\theta}) = G(\boldsymbol{\theta}_0), \tag{7.3.18}$$

Thus, the Bayes estimator (7.3.4) is *stable* (by (7.3.15) and *consistent* (by (7.3.18). The Bayes risk at rth stage has the expression

$$R_r^* = \boldsymbol{E}^{(r)}\left\{\left(\boldsymbol{\theta}_0 - \overset{*}{\boldsymbol{\gamma}}_{\boldsymbol{\theta},r}\Big|_{P=1}\right)\right\}^2 = \boldsymbol{E}\left(\overset{*}{\boldsymbol{\gamma}}_{\boldsymbol{\theta},r} \cdot \overset{*}{\boldsymbol{\gamma}}_{\boldsymbol{\theta},r} - 2\overset{*}{\boldsymbol{\gamma}}_{\boldsymbol{\theta},r} \cdot \boldsymbol{\theta}_0 + \boldsymbol{\theta}_0 \cdot \boldsymbol{\theta}_0\right) \tag{7.3.19}$$

where $\boldsymbol{E}^{(r)} = \boldsymbol{E}(\boldsymbol{X}_1, \ldots, \boldsymbol{X}_r)$ is the expectation operator over the random data $\boldsymbol{X}_1, \ldots, \boldsymbol{X}_r$. Quite obviously by the above

$$\lim_{r \to \infty} R_r^* = 0. \tag{7.3.20}$$

Accordingly, on the basis of the above, we have constructed the schematic diagram of Fig. 7.4 for the combined binary multistage detection and estimation process (with QCF) considered in Section 7.3.1.2. The $D + E$ process may be described as follows:

At each stage r of the process, the detector uses the pdf $w_{r-1}(\boldsymbol{\theta})$ evaluated at the previous stage, together with the new data $\boldsymbol{X}_r$ to compute the likelihood ratio Λ_r, (Eq. (7.3.2)). This is then compared with a threshold for the detection decision at that stage. At the same time, the pdf $w_{r-1}(\boldsymbol{\theta})$ is also used in the estimator $\overset{*}{\boldsymbol{\gamma}}_{\boldsymbol{\theta},r|p=1}$, together with the new data $\boldsymbol{X}_r$ to compute an estimate (7.3.3), which in Fig. 7.4 we may call "global." According to our discussion of Section 7.3.1.2 for **II**, this is the Bayes estimate of $\boldsymbol{\theta}$ given the data $\boldsymbol{X}_1, \ldots, \boldsymbol{X}_r$. At the output of the likelihood computer, the receiver also evaluates the quantity $\Lambda_r/(\Lambda_r + 1)$: this number is then multiplied by the estimate $\boldsymbol{\gamma}_{\boldsymbol{\theta},r|p=1}$ to yield $\overset{*}{\boldsymbol{\gamma}}_{\boldsymbol{\theta},r|p<1}$. According to our discussion of Section 7.3.1.2, **II**, *this is the binary Bayes signal estimate* of $\boldsymbol{\theta}$ for rth interval. Furthermore, the detector also uses the pdf $w_{r-1}(\boldsymbol{\theta})$ to evaluate a new, updated version of the pdf (see Eq. (7.3.1)). This result is then used in the next stage, together with the next data vector $\boldsymbol{X}_{r+1}$ for the next set of detection and estimation processes.

We conclude this discussion noting that the simple relations, discovered by Esposito [9] and Kailath [11], between $\log\Lambda_r$ and $\overset{*}{\boldsymbol{\gamma}}_{\boldsymbol{\theta},r|p=1}$ may greatly simplify the practical implementation of the receiver depicted in Fig. 7.4. In this regard, see also Problems 6.20 and 6.21. The same discussion also applies in this case for the $\mathrm{SCF}_{1,2}$, except that now the estimation portion of the $D + E$ process employs the optimum estimators Eqs. (7.3.11) and (7.3.12)), with Fig. 7.4 modified accordingly. Finally, we note that estimator I is also stable and consistent, by (7.3.13)–(7.3.20), as well. Here, however, $p = 1$ and no detection is required.

7.3.2 Sequential Observations and Unsupervised Learning: II. Joint $D + E$ for Binary Systems with Strong and Weak Coupling

In this section, we use the results of Sections 6.3.1 and 6.3.2 to extend the uncoupled analysis of Section 7.3.1 to include the important cases where the coupling is now between detection and estimation. Here the coupling is strong that is, the cost functions of Eq. (6.3.1) apply, and in the cases where the coupling is weak (Section 6.3.2), in which only C_{01} depends on the parameters $\boldsymbol{\theta}$, through $S(\boldsymbol{\theta})$, Eq. (6.3.21a). The latter implies that (the weak) coupling between detector and estimator is independent of the *structure* of the estimator, so that the resulting total average risk $R_D + R_E$ can be separately minimized for detection and estimation, that is, $R_D \to R_D^*$, $R_E \to R_E^*$, separately, as shown above in Section 7.3.1.

7.3.2.1 Strong Coupling We begin by applying (6.3.18) to (7.3.2) for the resulting detector's modified likelihood ratio functional for rth interval in the sequence of decisions, now on the basis of all the preceding data $\boldsymbol{X}_1, \ldots, \boldsymbol{X}_r$, at the end of which a detection decision H_1: $S \otimes N$ or H_0: N for this interval is made. If the former (H_1), the corresponding optimum estimator $\boldsymbol{\gamma}^{**}_{\boldsymbol{\theta},r|p<1}$ of the parameters $\boldsymbol{\theta}$ is accepted. The binary on–off decision process in question now becomes, by extension of the results of Section 7.3.1.1 for the detector (**I**) and for the estimator (**II**) of Section 7.3.1.2:

$$\left.\begin{array}{ll}\text{Decide } H_1\text{: } S \otimes N \text{ if: } \Lambda_g^{(r)}(\boldsymbol{X}_1, \ldots, \boldsymbol{X}_r) & = \displaystyle\int_\Omega \Lambda_g^{(r)}(\boldsymbol{X}_r|\boldsymbol{\theta}) w_{r-1}(\boldsymbol{\theta}) d\boldsymbol{\theta} \\ & = \mu \displaystyle\int_\Omega \frac{W(\boldsymbol{X}_r|\boldsymbol{\theta})}{W(\boldsymbol{X}_r|\boldsymbol{0})} w_{r-1}(\boldsymbol{\theta}) d\boldsymbol{\theta}\, C^{(r)}\left(\boldsymbol{\theta}, \boldsymbol{\gamma}^{**}_{\boldsymbol{\theta},r|p<1}\right) \geq 1 \\ \text{Decide } H_1\text{: } S \otimes N \text{ if:} & \Lambda_g^{(r)} < 1 \end{array}\right\}, \tag{7.3.21}$$

where $\mu = p/q$ as before and $w_{r-1}(\boldsymbol{\theta}) = w(\boldsymbol{\theta}|\boldsymbol{X}_1, \ldots, \boldsymbol{X}_{r-1})$. The composite cost function C is now

$$C^{(r)}\left(\boldsymbol{\theta}, \boldsymbol{\gamma}^{**}_{\boldsymbol{\theta},r|p<1}\right) \equiv \frac{C_{01}(\boldsymbol{\theta}) - C_{11}^{(r)}\left(\boldsymbol{\theta}, \boldsymbol{\gamma}^{**}_{\boldsymbol{\theta},r|p<1}\right)}{C_{10}^{(r)}\left(\boldsymbol{\theta}, \boldsymbol{\gamma}^{**}_{\boldsymbol{\theta},r|p<1}\right) - C_{00}} (>0). \tag{7.3.21a}$$

The component cost functions (Eq. (6.3.1)) are now specifically modified to include the recursive character of $w_{r-1}(\boldsymbol{\theta})$, refer to Eq. (7.3.21), through $\boldsymbol{\gamma}^{**}_{\boldsymbol{\theta},r|p<1}$. We have

$$\left.\begin{array}{ll} C_{00} = \text{constant}(\geq 0)\text{:} & \text{the cost of correctly deciding } H_0, \text{noise only, with} f_{11} = 0 \\ C_{10}^{(r)} = C_1^{(0)} + f_{10}^{(r)}\left(\boldsymbol{\gamma}^{**}_{\boldsymbol{\theta},r}\right)\text{:} & \text{the cost of incorrectly deciding } H_1 \text{ and making the} \\ & \text{estimate } \boldsymbol{\gamma}^{**}_{\boldsymbol{\theta},r} \text{ when } H_0 \text{ is true} \\ C_{01} = C_0^{(1)} + f_{01}[S(\boldsymbol{\theta})]\text{:} & \text{the costs of incorrectly deciding } H_0 \text{ when } H_1 \\ & \text{is true and } \boldsymbol{\theta} \text{ are the true parameters values} \\ C_{11}^{(r)} = C_1^{(1)} + f_{11}^{(r)}\left[\boldsymbol{\theta}, \boldsymbol{\gamma}^{**}_{\boldsymbol{\theta},r}\right]\text{:} & \text{cost of correctly deciding } H_1 \text{ and making the estimate} \\ & \boldsymbol{\gamma}^{**}_{\boldsymbol{\theta},r} \text{ when } H_1 \text{ is indeed the case and } \boldsymbol{\theta} \text{ are indeed} \\ & \text{the true parameter values.} \end{array}\right\}. \tag{7.3.22}$$

(Note that $f_{10}^{(r)}, f_{11}^{(r)}$ are functions of r, because of $\boldsymbol{\gamma}^{**}_{\boldsymbol{\theta},r}$, whereas C_{00}, C_{01} are not.) Here, $\boldsymbol{\gamma}^{**}_{\boldsymbol{\theta},r} = \boldsymbol{\gamma}^{**}_{\boldsymbol{\theta},r}(\boldsymbol{X}_1, \ldots, \boldsymbol{X}_r)$ and is determined from the appropriate extremal equation (min. or max.) depending on the cost functions $C_{10}^{(r)}, C_{11}^{(r)}$, modified here from (6.3.8) to include all the data preceding $(r+1)$-*st* decision:

II: Estimation:

$$\boldsymbol{\gamma}_{\theta,r} \xrightarrow{\delta} \boldsymbol{\gamma}^{**}_{\theta,r}\left\{ qC_{10}^{(r)}(\boldsymbol{\gamma}_{\theta,r})W(\boldsymbol{X}_1,\ldots,\boldsymbol{X}_r|0) + pW(\boldsymbol{X}_1,\ldots,\boldsymbol{X}_r)\int_\Omega C_{11}^{(r_1)}(\boldsymbol{\theta},\boldsymbol{\gamma}_{\theta,r})w_r(\boldsymbol{\theta})d\boldsymbol{\theta}\right\}, \tag{7.3.23}$$

where

$$w_r(\boldsymbol{\theta}) = w(\boldsymbol{\theta}|\boldsymbol{X}_1,\ldots,\boldsymbol{X}_r), \quad \text{refer to Eq. (7.3.1); } W(\boldsymbol{X}_1,\ldots,\boldsymbol{X}_r) \equiv \int_\Omega W(\boldsymbol{X}_1,\ldots,\boldsymbol{X}_r|\boldsymbol{\theta})w_r(\boldsymbol{\theta})\,d\boldsymbol{\theta}. \tag{7.3.23a}$$

Note that these cost assignments remain structurally the same as r changes: for a specified r, they depend on all the data preceding and including rth interval, because of the recursive "updating" of the detector and estimator. The cost portions of (7.3.22), of course, do not change with r.

As in the single interval cases ($r = 1$), refer to Section 6.3.1 preceding, estimation, when $r \geq 1$, must logically be performed before detection. If the decision of the detector is that H_0 is the true hypothesis state, that is, no signal is present, then the estimate is not accepted. However, in the common instance of *tracking*, the estimate is stored in the course of the tracking sequence of decisions and may be used to "fill-in" or enhance the track of the signal source. Note also that when the costs C_{01}, C_{10}, and C_{11} are constant (with the usual requirement that the costs of "failure" exceed the costs of "success"), the extended GLR (7.3.21) reduces for the simple interval ($r = 1$), as expected, to the familiar GLR of Section 6.1 and Chapters 1 and 3.

7.3.2.2 Weak Coupling We next extend the results of Section 6.3.2, as in Section 7.3.2.1, to include decision making for rth interval, based as well on the data of the preceding $r - 1$ intervals. However, the "weak-coupling" used here requires that now the four cost functions (7.3.22) of the composite cost function $\mathcal{C}$, (7.3.21a) above, be replaced by their reduced forms, for example:

$$\text{(6.3.21a): } C_{00} \equiv C_{1-\alpha}; \quad C_{11} = C_{1-\beta}; \quad C_{10} = C_\alpha + C'_{10}; \quad \text{and} \quad C_{01} = C_\beta = f_{01}[\boldsymbol{S}(\boldsymbol{\theta})], \tag{7.3.24}$$

where $C'_{10} = f_{10}$ (a constant). The extended version of this weakly coupled detector, with the associated decision process, refer to Eqs. (6.3.22a) and (6.3.22b), becomes explicitly

$$\Lambda_g^{(r')}(\boldsymbol{X}_1,\ldots,\boldsymbol{X}_r) = K'_r\Lambda_r + \int_\Omega f_{01}[S(\boldsymbol{\theta})]\frac{\hat{\Lambda}'_r(\boldsymbol{X}_r|\boldsymbol{\theta})w_{r-1}(\boldsymbol{\theta})d\boldsymbol{\theta}}{[\boldsymbol{C}_\alpha - C_{1-\alpha} + C'_{10}]}\begin{cases} \geq 1: \text{decide } H_1 \\ \text{or} \\ < 1: \text{decide } H_0 \end{cases}, \tag{7.3.25}$$

where

$$\Lambda_r \equiv \Lambda_r(\boldsymbol{X}_1,\ldots,\boldsymbol{X}_r) = \int \Lambda_r(\boldsymbol{X}_r|\boldsymbol{\theta})w_{r-1}(\boldsymbol{\theta})d\boldsymbol{\theta} = \mu\int_\Omega \frac{W(\boldsymbol{X}_r|\boldsymbol{\theta})}{w(\boldsymbol{X}_r|\boldsymbol{0})}w_{r-1}(\boldsymbol{\theta})d\boldsymbol{\theta}, \tag{7.3.25a}$$

and (from (6.3.22a)),

$$\Lambda'_r \equiv \hat{\Lambda}'_r(\boldsymbol{X}_r|\boldsymbol{\theta})w_{r-1}(\boldsymbol{\theta}) = \hat{\Lambda}'_r(\boldsymbol{\theta};\boldsymbol{X}_1,\ldots,\boldsymbol{X}_r) = \mu\frac{W(\boldsymbol{X}_r|\boldsymbol{\theta})}{w(\boldsymbol{X}_r|\boldsymbol{0})}w_{r-1}(\boldsymbol{\theta}). \quad (7.3.25b)$$

The scale factor K_r is the (positive) constant cost ratio (cf. (6.3.22b)), independent of

$$K'_r \equiv (C_\beta - C_{1-\beta})/(C_\alpha - C_{1-\alpha} + C'_{10}) = K'. \quad (7.3.25c)$$

Finally, we note that with weak or no coupling, C'_{01}, C'_{10} are constants, that is, $C'_{01} = C_\beta + f_{01}$ where f_{01} is also now a constant. Then, *both* $D+E$ *operations can be performed simultaneously or in any order*, as before in the case of the single sample $(r \equiv 1)$ (Section 6.3.2).

For the optimum estimator in these weak coupling cases, we have directly the desired extension of the results (6.3.26) and (6.3.27) et seq. to the sequence of γ ($\geq$1) data intervals here. The results are for the optimum estimator $\overset{*}{\boldsymbol{\gamma}}_{\boldsymbol{\theta},r|p<1}$, which now satisfies the (vector) set of equations (when the estimator exists):

$$\left[qW(\boldsymbol{X}_1,\ldots,\boldsymbol{X}_r|\boldsymbol{0})\frac{\partial f_{10}^{(r)}}{\partial\boldsymbol{\gamma}_{\boldsymbol{\theta}}} + p\int_\Omega d\boldsymbol{\theta}\, W(\boldsymbol{X}_1,\ldots,\boldsymbol{X}_r|\boldsymbol{\theta})w_r(\boldsymbol{\theta})\frac{\partial f_{11}^{(r)}}{\partial\boldsymbol{\gamma}_{\boldsymbol{\theta}}}\right]_{\boldsymbol{\gamma}_{\boldsymbol{\theta}}\to\overset{*}{\boldsymbol{\gamma}}_{\boldsymbol{\theta},r|p<1}} = \boldsymbol{0}. \quad (7.3.26)$$

Here, $f_{10}^{(r)} = C(0,\boldsymbol{\gamma}_{\boldsymbol{\theta}}) = f_{10}$ and $f_{11}^{(r)} = C(\boldsymbol{\theta},\boldsymbol{\gamma}_{\boldsymbol{\theta}})$ (cf. (6.3.28) and remarks following). We observe, moreover, that the solutions of (7.3.23) and (7.3.26) are generally different, that is, $\overset{*}{\boldsymbol{\gamma}}_{\boldsymbol{\theta},r|p\leq 1} \neq \overset{**}{\boldsymbol{\gamma}}_{\boldsymbol{\theta},r|p<1}$. However, the results of (7.3.26) are identical with the results (6.2.2a) and (6.2.2b), extended to r-intervals, specifically for the Bayes estimation with QCF and $\text{SCF}_{1,2}$, although the resulting Bayes risks are different. (We refer the reader to the last two paragraphs of Section 6.3.2, extended to the r-interval case.)

7.3.2.3 Multiple Signals In Each Interval: Multiple Tracking The results of Section 7.3.2, in particular Sections 7.3.2.1 and 7.3.2.2 may be applied directly to the multiple tracking problem, where we have one or more signals $S_i^{(q)}, q = 1,\ldots,Q$, of $i = 1,\ldots,k,\ldots,K$ types present in each data interval $(r \geq 1)$. We only need to use the results of Section 7.3.2 for *each* track $(q = 1,\ldots,Q)$ produced by these separate signals $S_k^{(q)}$, mainly Q tracks in all. Thus, we have Q separate parallel applications of the algorithms in Sections 7.3.2.1 and 7.3.2.2. This is illustrated schematically in Fig. 7.5. For each track, it is assumed that the speed of the target is essentially constant and its direction is comparatively slowly changing from interval to interval, which latter can be accounted for by adjusting the beam illuminating the target, to keep it in the beam. The amount of the adjustment then measures the change in direction. Other more refined methods (prediction, multiple beams, etc.), are also available and are embodied in most modern methods of tracking targets [1].

7.3.3 Sequential Observations and Unsupervised Learning: III. Joint $D+E$ Under Multiple Hypotheses with Strong and Weak Coupling and Overlapping Hypotheses Classes

With the help of the results of Sections 7.3.1 and 7.3.1.2, we now extend the analysis of Sections 7.2.1 and 7.2.2 to the situation of r-data intervals and single signals out of K possible

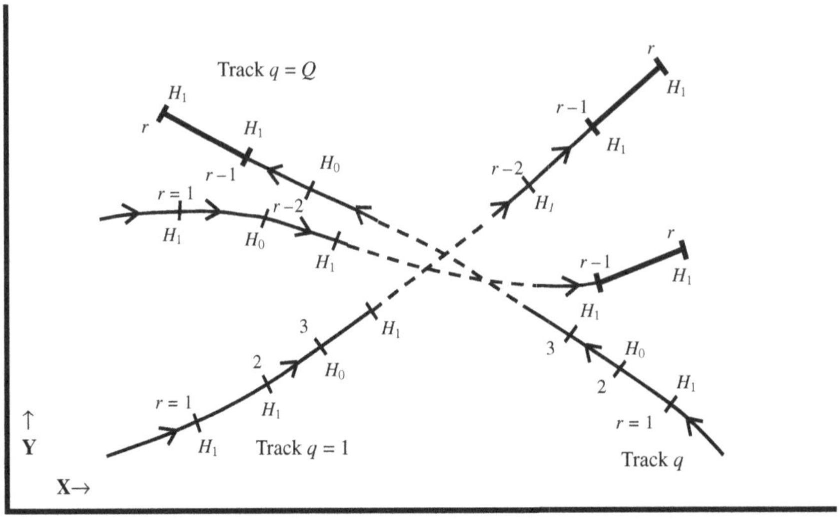

FIGURE 7.5 Two-dimensional diagram of multiple tracks ($q = 1, 2, \ldots, Q$), with r-data intervals (⊢⊣, ⊢⊣), with decisions at the end of each interval, including the latest interval (r); $\Delta_r = [r - (r-1)]\Delta = \Delta$, $r = 1, 2, \ldots, r$ and showing estimator rejections $H_0 : N$ and acceptances $H_1 : S \otimes N (r \geq 1)$on each track (Unsupervised Learning), Section 7.3.2.

signals. Our starting point is the Bayes risk (7.1.11) with the optimum estimator (7.1.12) substituted therein, namely:

$$R_{D+E}^{(r)*} = \int_{\Gamma^{(r)}} A_0(\boldsymbol{X}_1, \ldots, \boldsymbol{X}_r) \boldsymbol{dX}_1 \cdots \boldsymbol{dX}_r + \sum_{k=1}^{K} \int_{\Gamma^{(r)}} \boldsymbol{dX}_1 \cdots \boldsymbol{dX}_r \\ \cdot p_0 \, W(\boldsymbol{X}_r|\boldsymbol{0}) C_{0k}^{(r)} \left(\boldsymbol{\theta}, \hat{\boldsymbol{\theta}}_{k,r}^{*}\right) \cdot \left[1 - \sum_{k=1}^{K} L_{ik}(\boldsymbol{X}_1, \ldots, \boldsymbol{X}_r)\right]. \tag{7.3.27}$$

Here, $A_0(\boldsymbol{X}_1, \ldots, \boldsymbol{X}_r)$ is the extended version of (7.1.8a) and f_{01}, (Eq. (7.1.8b)). The generalized likelihood ratio L_{ik} is given by the extended version of (7.1.7), (7.1.12a), and (7.1.12b), namely,

$$L_{ik}(\boldsymbol{X}_1, \ldots, \boldsymbol{X}_r) = \mu \int_{\Omega_i} \frac{W(\boldsymbol{X}_r|\boldsymbol{\theta}) w_i^{(r-1)}(\boldsymbol{\theta}) \boldsymbol{d\theta}}{W(\boldsymbol{X}_r|\boldsymbol{0})} C_{ik}^{(r)}\left(\boldsymbol{\theta}, \hat{\boldsymbol{\theta}}_r^{*}\right); \quad \mu_i \equiv p_i/p_0, \tag{7.3.28a}$$

with

$$C_{ik}^{(r)}\left(\boldsymbol{\theta}, \hat{\boldsymbol{\theta}}_r^{*}\right) = \frac{C_{i0}^{(r)}\left(\boldsymbol{\theta}, \hat{\boldsymbol{\theta}}_{k,r}^{*}\right) - C_{ik}^{(r)}\left(\boldsymbol{\theta}, \hat{\boldsymbol{\theta}}_{k,r}^{*}\right)}{C_{0k}^{(r)}\left(\boldsymbol{\theta}, \hat{\boldsymbol{\theta}}_{k,r}^{*}\right)}; \quad \hat{\boldsymbol{\theta}}_{k,r}^{*} \equiv \hat{\boldsymbol{\theta}}_k^{*}(\boldsymbol{X}_1, \ldots, \boldsymbol{X}_r) \tag{7.3.28b}$$

and

$$w_i^{(r-1)}(\boldsymbol{\theta}) = w_i(\boldsymbol{\theta}|\boldsymbol{X}_1, \ldots, \boldsymbol{X}_r); \text{cf. (7.3.21) et seq.} \tag{7.3.28c}$$

For detection we have from (7.1.10) (after $\hat{\boldsymbol{\theta}}_r^*$, the optimum estimator has been applied to the extended version of (7.1.10)), for the kth hypothesis class H_k: $S_k \otimes N$:

I. Detection:

$$\text{Decide } H_k\text{: } S_k \otimes N \text{ if: } \begin{cases} (1)\ \sum_{i=1}^{K} L_{ik}^{(r)}(\boldsymbol{X}_1, \ldots, \boldsymbol{X}_r) > 1 \\ (2)\ C_{0k}^{(r)}\left(\boldsymbol{\theta}, \hat{\boldsymbol{\theta}}_{k,r}^*\right)\left[1 - \sum_{i=1}^{K} L_{ik}\right] \le C_{0\ell}^{(r)}\left(\boldsymbol{\theta}, \hat{\boldsymbol{\theta}}_{k,r}^*\right)\left[1 - \sum_{i=1}^{K} L_{i\ell}\right]. \end{cases} \tag{7.3.29}$$

Optimum estimation in these strongly coupled cases is found from the extension of condition (7.1.12), which becomes

II. Estimation:

$$\begin{cases} \min\limits_{\boldsymbol{\theta} \to \hat{\boldsymbol{\theta}}_{R,r}^*} \int_{\Gamma_R^{(r)}(R \text{ times})} p_0 W(\boldsymbol{X}_r|\boldsymbol{0}) C_{0k}^{(r)}\left(\boldsymbol{\theta}, \hat{\boldsymbol{\theta}}_{k,r}(\boldsymbol{X}_1, \ldots, \boldsymbol{X}_r)\right) \\ \qquad \cdot \left[1 - \sum_{i=1}^{K} L_{ik}(\boldsymbol{X}_1, \ldots, \boldsymbol{X}_r)\right], \quad k \ne 0 \end{cases} \tag{7.3.30}$$

Figures 7.1 and 7.2 also show the sequence of operations, applying here to the $\hat{\boldsymbol{\theta}}_{k,r}^*$ stage, where now (7.1.12c) becomes $B_k(\boldsymbol{X}) \to B_k^{(r)}(\boldsymbol{X}_1, \ldots, \boldsymbol{X}_r)$. (See the discussion after Eq. (7.1.12), to the end of Section 7.1.1.)

7.3.3.1 The Quadratic Cost Function: Strong Coupling At this point, to proceed further, we must specify the cost functions in detail, as is done in Sections 7.1.2.1 and 7.1.2.3. For QCF, we have here (7.1.15) now extended to

$$\text{QCF:} \quad \left.\begin{aligned} C_{00}^{(r)}\left(\hat{\boldsymbol{\theta}}, \hat{\boldsymbol{\theta}}_r\right) = 0;\ C_{0k}^{(r)}\left(\hat{\boldsymbol{\theta}}, \hat{\boldsymbol{\theta}}_r\right) &= a_{0k} + b_{0k}\tilde{\boldsymbol{\theta}}\, \boldsymbol{E}_{ik}\boldsymbol{\theta}, \quad k \ge 1 \\ C_{ik}^{(r)}\left(\hat{\boldsymbol{\theta}}, \hat{\boldsymbol{\theta}}_r\right) &= a_{ik} + b_{ik}\left(\tilde{\boldsymbol{\theta}} - \tilde{\hat{\boldsymbol{\theta}}}_r\right)\boldsymbol{E}_{ik}\left(\boldsymbol{\theta} - \hat{\boldsymbol{\theta}}_r\right), \quad i \ge 1 \end{aligned}\right\}, \tag{7.3.31}$$

before the second optimization, that is, before $\hat{\boldsymbol{\theta}}_r^* \to \hat{\boldsymbol{\theta}}_{k,r}^*$, which minimizes the integral in (7.1.12). This second optimization is described in (7.1.16a)–(7.1.18b), specifically here for rth interval, recursively including all the data from the preceding $r - 1$ intervals, as well as $\boldsymbol{X}_r$. The result is easily seen to be for these costs functions, with the constants (a_{ik}, b_{ik}) again obeying the conditions (7.1.14),

$$\hat{\boldsymbol{\theta}}_{k,r}^*(\boldsymbol{X}_1, \ldots, \boldsymbol{X}_r)_{\text{QCF}} = \frac{\sum_{k=1}^{K} p_i (b_{ik} - b_{i0}) \boldsymbol{E}_{ik} W(\boldsymbol{X}_r|\boldsymbol{\theta}) w_i^{(r-1)}(\boldsymbol{\theta}) d\boldsymbol{\theta}}{\sum_{i=1}^{K} (b_{ik} - b_{i0}) p_i \int_{\Omega_i} \boldsymbol{\theta}\, W(\boldsymbol{X}_r|\boldsymbol{\theta}) w_i^{(r-1)}(\boldsymbol{\theta}) \boldsymbol{E}_{ik} d\boldsymbol{\theta}}. \tag{7.3.32}$$

When $\boldsymbol{E}_{ik}$ is not a function of i and recalling that $w_0^{(r-1)}(\boldsymbol{\theta}) = \delta(\boldsymbol{\theta} - \boldsymbol{0})$, with $b_{00} = 0$ now, we find that (7.3.32) reduces to the simpler result:

$$\hat{\boldsymbol{\theta}}^*_{k,r}(\boldsymbol{X}_1, \ldots, \boldsymbol{X}_r)_{\text{QCF}} = \left\{ \sum_{i=1}^{K} \frac{(b_{ik} - b_{i0})\Lambda_i^{(r)}(\boldsymbol{X}_1, \ldots, \boldsymbol{X}_r)/b_{0k}}{1 + \sum_{i=1}^{K}(b_{ik} - b_{i0})\Lambda_i^{(r)}(\boldsymbol{X}_1, \ldots, \boldsymbol{X}_r)/b_{0k}} \right\} \Theta_i^{(r)}(\boldsymbol{X}_1, \ldots, \boldsymbol{X}_r)^*_{\text{QCF}}. \tag{7.3.33}$$

Here, $\Lambda_i^{(r)}(\boldsymbol{X}_1, \ldots, \boldsymbol{X}_r)$ is the recursive GLR, that is,

$$\Lambda_i^{(r)} \equiv \Lambda_i^{(r)}(\boldsymbol{X}_1, \ldots, \boldsymbol{X}_r) \equiv \int_{\Omega_i} \frac{W(\boldsymbol{X}_r|\boldsymbol{\theta})w_i^{(r-1)}(\boldsymbol{\theta})d\boldsymbol{\theta}}{W(\boldsymbol{X}_r|\boldsymbol{0})}, \tag{7.3.33a}$$

and $\boldsymbol{\Theta}_i^{(r)*}|_{\text{QCF}}$ is the recursive LMS or minimum variance estimator for $\boldsymbol{\theta}_k$, of a signal S_i of class H_i: Ω_i, when there is no uncertainty as to the presence of S_i, $1 \le i \le k$, refer to the discussion following Eq. (7.1.2). Equation (7.3.33b) gives $\boldsymbol{\Theta}_i^{(r)*}|_{\text{QCF}}$:

$$\begin{aligned} \boldsymbol{\Theta}_i^{(r)*}|_{\text{QCF}} &\equiv \boldsymbol{\Theta}_i^{(r)}(\boldsymbol{X}_1, \ldots, \boldsymbol{X}_r)^*_{\text{QCF}} \\ &\equiv \int_{\Omega_i} \boldsymbol{\theta}\, W(\boldsymbol{X}_r|\boldsymbol{\theta})w_i^{(r-1)}(\boldsymbol{\theta})d\boldsymbol{\theta} \Big/ \int_{\Omega_i} W(\boldsymbol{X}_r|\boldsymbol{\theta})w_i^{(r-1)}(\boldsymbol{\theta})d\boldsymbol{\theta}. \end{aligned} \tag{7.3.33b}$$

The quantity $C_{0k}^{(r)} \cdot \left[1 - \sum_i L_{ik}^{(r)}\right]$, $(k; k \to \ell)$, is the key quantity needed for determining the optimum detector structure. Setting $b_{ik} = 1$, $b_{i0} = 0$, $b_{ik} = b_{0k} = 1$, $i = 1, \ldots, K$ in the general cost assignments for QCF (7.3.31), we find the optimum detection conditions (7.3.30) become here explicitly

$$\begin{aligned} C_{0k}^{(r)}(\boldsymbol{X}_1, \ldots, \boldsymbol{X}_r)\left[1 - \sum_{i=1}^{K} L_{ik}^{(r)}(\boldsymbol{X}_1, \ldots, \boldsymbol{X}_r)\right] &= a_{0k} + \sum_{i=1}^{K}(a_{ik} - a_{i0})\Lambda_i^{(r)}H_i^{(r)} - \left[1 + \sum_{i=1}^{K}\Lambda_i^{(r)}\right] \\ &\quad \cdot \sum_{i=1}^{K}\sum_{\ell=1}^{K} P(H_i|\boldsymbol{X}_1, \ldots, \boldsymbol{X}_r)P\big(H_k|\boldsymbol{X}_1, \ldots, \boldsymbol{X}_p\big) \\ &\quad \cdot \tilde{\boldsymbol{\Theta}}_k^{(r)*}\boldsymbol{\Theta}_i^{(r)**}, \end{aligned} \tag{7.3.34}$$

where

$$H_i(\boldsymbol{X}_1, \ldots, \boldsymbol{X}_r) \equiv \int_{\Omega_i} \tilde{\boldsymbol{\theta}}\boldsymbol{\theta}\, W(\boldsymbol{X}_r|\boldsymbol{\theta})w_i^{(r-1)}(\boldsymbol{\theta})d\boldsymbol{\theta} \Big/ \int_{\Omega_i} W\Big(\boldsymbol{X}_r|\boldsymbol{\theta})w_i^{(r-1)}(\boldsymbol{\theta})d\boldsymbol{\theta} \tag{7.3.34a}$$

and

$$P(H_i|\boldsymbol{X}_1, \ldots, \boldsymbol{X}_r) = \Lambda_i^{(r)} \Big/ \left(1 + \sum_{i=1}^{K}\Lambda_i^{(r)}\right), \quad \Lambda_i^{(r)} = \Lambda_i^{(r)}(\boldsymbol{X}_1, \ldots, \boldsymbol{X}_r). \tag{7.3.34b}$$

Here, $P(H_i|\boldsymbol{X}_1, \ldots, \boldsymbol{X}_r)$ is the posterior probability that the hypothesis H_i is true. Again, refer to Eqs. (7.3.21) and (7.3.21a), it is easily shown by a number of applications of Bayes theorem with the definition of marginal probability density that (7.3.34b) is valid. Also, once more (cf. (7.3.21a) et seq.) we have with the assumption that $b_{i0} = 0$, $b_{ik} = b_{0k} > 0$, $i = 1, 2, \ldots, K$ that (7.3.33) reduces to

$$\hat{\boldsymbol{\theta}}^*_{k,r}\Big|_{\text{QCF}} = \sum_{i=1}^{K} \left[\Lambda_i^{(r)} \Big/ \left(1 + \sum_{i=1}^{K} \Lambda_i^{(r)}\right) \right] \Theta_i^{(r)*}\Big|_{\text{QCF}} = \sum_{i=1}^{K} P(H_i|\boldsymbol{X}_1, \ldots, \boldsymbol{X}_r)\Theta_i^{(r)*}\Big|_{\text{QCF}}. \tag{7.3.35}$$

Thus, in this somewhat simpler case, the optimum estimator (7.3.35) is seen to be the weighted sum of least-squares estimators under no uncertainty regarding the various hypotheses, where the weights are the *a posteriori* probabilities of the respective hypotheses. (See the comments following Eq. (7.1.21).)

Thus, Section 7.3.3 is a direct extension of Fredriksen's work [4, 5]. The series of general results (7.3.29)–(7.3.32), and slightly more specialized relations (7.3.33)–(7.3.35), describe the joint detection and estimation of a signal's parameters $\boldsymbol{\theta}$ when the two operations are strongly coupled together, explicitly through the cost functions (7.3.31), which represent a generalization of cost functions assigned to the original binary on–off problem considered by Middleton and Esposito [2, 3].

7.3.3.2 "Simple" Cost Functions ($SCF_{1,2}$) As noted previously, in Sections 5.1.5 and 7.1.2.3, the simple cost function plays an important role in estimation theory. There are two variations of principal interest—the strict SCF_2 and the nonstrict SCF_1, both of which lead to unconditional maximum-likelihood estimators. Sections 7.1.3.2 and 7.3.1.2 have considered these extensions for multiple hypothesis classes $(k = 1, \ldots, K)$ and for sequences $(r \geq 1)$ of decisions.

Here we treat the combined situation, where in place of the generalized QCF, (7.1.15), we employ the following extended "simple" cost functions:

$$\textbf{SCF}_1\text{:}\quad C_{ik}^{(r)}\left(\boldsymbol{\theta}, \hat{\boldsymbol{\theta}}_k\right)_1 = a_{ik} + b_i \left[c_i - \sum_{\ell=1}^{L} \delta\left(\boldsymbol{\theta}_\ell - \hat{\boldsymbol{\theta}}_{\ell,r}\right) \right] \tag{7.3.36a}$$

$$\textbf{SCF}_2\text{:}\quad C_{ik}^{(r)}\left(\boldsymbol{\theta}, \hat{\boldsymbol{\theta}}_k\right)_2 = a_{ik} + b_i \left[c_i - \delta\left(\boldsymbol{\theta} - \hat{\boldsymbol{\theta}}_r\right) \right]. \tag{7.3.36b}$$

The average risk to be minimized is given by the extension of (7.1.25) to the case of $r \geq 1$ intervals as before (cf. (7.1.6)–(7.1.11) and Section 7.3.3.1), with the result that (7.1.26) is now the condition that $\hat{\boldsymbol{\theta}}_r \to \hat{\boldsymbol{\theta}}^*_r(\boldsymbol{X}_1, \ldots, \boldsymbol{X}_r)$ is determined by the following:

II. Estimation:

$$\max_{\boldsymbol{\theta} \to \hat{\boldsymbol{\theta}}^*} \int_{\Gamma^{(r)} = (\Gamma x \Gamma x \cdots \Gamma)_r} d\boldsymbol{X}_1 \cdots d\boldsymbol{X}_r \left\{ \begin{array}{l} b_0 p_0 W(\boldsymbol{X}_r|\boldsymbol{0}) \left[\displaystyle\sum_{\ell=1}^{L} \delta\left(\hat{\boldsymbol{\theta}}_{\ell,r} - \boldsymbol{0}\right) \right] \\ + \displaystyle\sum_{k=1}^{K} \sum_{\ell=1}^{L} b_i p_i W\left(\boldsymbol{X}_r|\hat{\boldsymbol{\theta}}_{\ell,r}\right) w_i^{(r-1)}\left(\hat{\boldsymbol{\theta}}_{\ell,r}\right) \end{array} \right\} \textbf{SCF}_1 \tag{7.3.37a}$$

and

$$\max_{\boldsymbol{\theta}\to\hat{\boldsymbol{\theta}}^*} \int_{\Gamma^{(r)}=(\Gamma x\Gamma x\cdots\Gamma)_r} d\boldsymbol{X}_1\cdots d\boldsymbol{X}_r \left\{ \begin{array}{l} b_0p_0W(\boldsymbol{X}_r|\boldsymbol{0})\left[\delta(\hat{\boldsymbol{\theta}}_r-\boldsymbol{0})\right] \\ +\sum_{k=1}^{K} b_ip_iW(\boldsymbol{X}_r|\hat{\boldsymbol{\theta}}_r)w_i^{(r-1)}(\hat{\boldsymbol{\theta}}_r) \end{array} \right\} \mathbf{SCF}_2 \quad (7.3.37b)$$

where $w_i^{(r-1)}(\hat{\boldsymbol{\theta}}_r) \equiv w_i(\hat{\boldsymbol{\theta}}_r|\boldsymbol{X}_1,\ldots,\boldsymbol{X}_r)$. Equation (7.3.37) is the "simple" cost function counterpart of (7.3.30) for QCF cost function (7.1.15), extended to r-intervals for $C_{ik}^{(r)}(\boldsymbol{\theta},\hat{\boldsymbol{\theta}}_r^*)$, now given by (7.3.36a) and (7.3.36b).The optimal detection counterpart to (7.3.3.1) is (cf. (7.3.29) or (7.1.28)) given by the following:

I. Detection:

$$\left\{ \begin{array}{lll} (1)\ \text{Decide } H_k \text{ if:} & D_k^{(r)}(\boldsymbol{X}_1,\ldots,\boldsymbol{X}_r) < 0 & \\ & D_k^{(r)}(\boldsymbol{X}_1,\ldots,\boldsymbol{X}_r) \le D_\ell^{(r)}(\boldsymbol{X}_1,\ldots,\boldsymbol{X}_r), & \text{all } \ell = 1,\ldots,K \\ (2)\ \text{Otherwise, decide } H_0: & & \end{array} \right\} \quad (7.3.38)$$

In this instance we have specifically

$$D_k^{(r)}(\boldsymbol{X}_1,\ldots,\boldsymbol{X}_r) = \sum_{i=0}^{K}(a_{ik}-a_{i0})p_i\int_{\Omega_i} W\left(\boldsymbol{X}_r|\boldsymbol{\theta}\right)w_i^{(r-1)}(\boldsymbol{\theta})d\theta, \quad (7.3.38a)$$

with $w_i^{(r-1)}(\boldsymbol{\theta}) = w_i(\boldsymbol{\theta}|\boldsymbol{X}_1,\ldots,\boldsymbol{X}_r)$. Here we observe again *that optimum detection and estimation are uncoupled by these choices of cost function* (7.3.36a) and (7.3.36b).

7.3.3.3 Weak Coupling In case of multiple hypotheses and sequences of decision intervals ($r \ge 1$) discussed here (Section 7.3.3), let us examine what happens when the coupling between detection and estimation is weak, in the sense of Sections 6.2.3 and 7.3.2.2. This is not possible with the cost functions (7.3.28) and (7.3.21), unless they are all suitable (positive or zero) constants that obey the generic condition that "failure" must be more expensive than "success." However, with the choices below, which are modeled on (6.3.21a), we can obtain representative results, analogous to those in Section 7.3.2.2. These cost functions are now

$$\left. \begin{array}{l} C_{00} = a_{00}(\ge 0); \quad C_{0k} = a_{0k} + C'_{0k} \\ C_{i0} = a_{i0} + f_{i0}(S_i(\boldsymbol{\theta})), \quad i \ge 1; \quad C_{ik} = a_{ik} + C'_{ik}, \quad k \ge 1 \end{array} \right\}, \quad (7.3.39)$$

where as before the first subscript denotes the true state and the second the chosen state. Applying these cost functions to (7.3.28) gives us the desired weak coupling results for the detection phase of the joint D and E procedure here. We find that the specific result in a simple extension of the binary on–off case is as follows:

I. Detection: Decide H_k: $S_k \otimes N$ when

$$\begin{array}{l} \text{(i)}\ \sum_{i=1}^{K}\hat{L}_{ik}^{(r)}(\boldsymbol{X}_1,\ldots,\boldsymbol{X}_r) > 1 \\ \text{(ii)}\ C_{0k}\cdot\left[1-\sum_{i=1}^{K}\hat{L}_{ik}^{(r)}\right] \le C_{0\ell}\cdot\left[1-\sum_{i=1}^{K}\hat{L}_{i\ell}^{(r)}\right] \end{array} \quad (7.3.40)$$

(and not $H_{k'}$ otherwise), where now

$$\hat{L}_{ik}^{(r)}(\boldsymbol{X}_1, \ldots, \boldsymbol{X}_r)_{\text{weak}} \equiv K_{ik}^{(r)}\Lambda_i^{(r)} + \mu_i \int_{\Omega_i} \frac{f_{i0}(\boldsymbol{S}(\boldsymbol{\theta}))\hat{\Lambda}_i^{(r)}(\boldsymbol{X}_r|\boldsymbol{\theta})w_i^{(r-1)}(\boldsymbol{\theta})d\boldsymbol{\theta}}{(C_{0k} - C_{00})}, \quad (7.3.41)$$

where

$$\Lambda_i^{(r)} \equiv \mu_i \int_{\Omega_i} \frac{W(\boldsymbol{X}_r|\boldsymbol{\theta})}{W(\boldsymbol{X}_r|\boldsymbol{0})} w_i^{(r-1)}(\boldsymbol{\theta})d\boldsymbol{\theta} = \int_{\Omega_i} \hat{\Lambda}_i^{(r)}(\boldsymbol{X}_r|\boldsymbol{\theta})w_i^{(r-1)}(\boldsymbol{\theta})d\boldsymbol{\theta}, \quad (7.3.41a)$$

where

$$\hat{\Lambda}_i^{(r)} \equiv \mu_i \frac{W(\boldsymbol{X}_r|\boldsymbol{\theta})}{W(\boldsymbol{X}_r|\boldsymbol{0})}, \quad (7.3.41b)$$

$$K_{ik}^{(r)} \equiv (a_{i0} - C_{ii})/(C_{0k} - C_{00}). \quad (7.3.41c)$$

Again (cf. (7.2.25c), in (7.3.39) with weak or no coupling, C'_{0k}, C'_{ik} are constants when $f_{ik} (= C'_{ik})$ is also a constant. Also, both the D and E operations can be performed in any order or simultaneously, as apart from the constant cost functions in (7.3.41). Here, $\hat{L}_{ik}^{(r)}|_{\text{weak}}$ is *independent of the estimator structure*. In fact, the optimum estimator is seen to obey the following condition:

II. Estimation:

$$\left(\min_{\boldsymbol{\theta} \to \boldsymbol{\theta}^*_{k,r}}^{\delta}\right) \int_{\Gamma^{(r)}_{k=(\Gamma_k x \Gamma_k \cdots x \Gamma_k)}} C_{0k} p_0 W(\boldsymbol{X}_r|0) \left[1 - \sum_{i=1}^{K} \hat{L}_{ik}^{(r)}\right], \quad k \geq 1, \quad (7.3.42)$$

from which it is also evident that optimization of the estimator is now independent of the detectors optimization (see the discussion in Section 6.3.3, in particular Eqs. (6.3.27) and (6.3.28), for evaluating (7.3.42)).

7.3.4 Sequential Observations and Overlapping Multiple Hypothesis Classes: Joint $D + E$ with No Coupling

Here we may extend the results of Section 7.2 to the case of sequential *sets* of $r \geq 1$ observations, with the help of the results of Section 7.3.3. From (7.2.2)–(7.2.2b), we can write the extended results for QCF. These are for the uncoupled Bayes detector (D) here:

I. Detection (D): Decide H_k: $S_k \otimes N$ if:

$$\left.\begin{aligned} &(1)\ \sum_{i=1}^{K} L_{ik}^{(D)}(\boldsymbol{X}_1, \ldots, \boldsymbol{X}_r) > 1 \\ &(2)\ C_k^{(D)} \cdot \left[1 - L_{ik}^{(D)}(\boldsymbol{X}_1, \ldots, \boldsymbol{X}_r)\right] \leq C_\ell^{(D)} \cdot \left[1 - \sum_{i=1}^{K} L_{i\ell}(\boldsymbol{X}_1, \ldots, \boldsymbol{X}_r)\right] \end{aligned}\right\}, \quad (7.3.43)$$

where

$$L_{ik}^{(D)}(\boldsymbol{X}_1,\ldots,\boldsymbol{X}_r) \equiv \mu_i C_{ik}^{(D)} \int_{\Omega_i} \frac{W(\boldsymbol{X}_r|\boldsymbol{\theta}) w_i^{(r-1)}(\boldsymbol{\theta}) d\boldsymbol{\theta}}{W(\boldsymbol{X}_r|0)} \quad (>0), \qquad (7.3.43a)$$

$$\mu_i \equiv p_i/p_0; \quad \text{same for } k \to \ell$$

with

$$C_{ik}^{(D)} = \frac{C_{i0}^{(D,r)} - C_{ik}^{(D,r)}}{C_{0k}^{(D,r)}} (>0); \quad w_i^{(r-1)}(\boldsymbol{\theta}) = w_i(\boldsymbol{\theta}|\boldsymbol{X}_1,\ldots,\boldsymbol{X}_r), \text{ refer to (7.3.1)};$$

$$\Lambda_i^{(r)} = \text{Eq.(7.3.46)}. \qquad (7.3.43b)$$

II. Estimation (*E*):

$$\min_{\boldsymbol{\theta}_k \to \hat{\boldsymbol{\theta}}_{k,r}^*} \int_{\Gamma^{(r)}} p_0 W(\boldsymbol{X}_r|\boldsymbol{0}) C_{0k}^{(E;r)} \left(\boldsymbol{\theta}, \hat{\boldsymbol{\theta}}(\boldsymbol{X}_1,\ldots,\boldsymbol{X}_r)\right) \left[1 - \sum_{i=1}^{K} L_{ik}(\boldsymbol{X}_1,\ldots,\boldsymbol{X}_r)\right], \quad k \neq 0, \qquad (7.3.44)$$

which by the argument given in the third paragraph of Section 7.2 (and essentially following the analysis of (7.1.13)–(7.1.20) translates (7.3.44) into the optimum estimator for the QCF here, namely,

$$\boldsymbol{\theta}_{k,r}^*(\boldsymbol{X}_1,\ldots,\boldsymbol{X}_r)_{\text{QCF}} = \sum_{i=1}^{K} \Lambda_i^{(r)} \left(1 + \Lambda_i^{(r)}\right)^{-1} \Theta_i^{(r)}(\boldsymbol{X}_1,\ldots,\boldsymbol{X}_r)_{\text{QCF}}^*. \qquad (7.3.45)$$

Here, $\Lambda_i^{(r)}(\boldsymbol{X}_1,\ldots,\boldsymbol{X}_r)$ is once again, refer to Eq. (7.3.33a), the recursive generalized likelihood ratio (GLR), namely,

$$\Lambda_i^{(r)} = \Lambda_i^{(r)}(\boldsymbol{X}_1,\ldots,\boldsymbol{X}_r) \equiv \mu_i \int_{\Omega_i} \frac{W(\boldsymbol{X}_r|\boldsymbol{\theta}) w_i^{(r-1)}(\boldsymbol{\theta}) d\boldsymbol{\theta}}{W(\boldsymbol{X}_r|0)}, \qquad (7.3.46)$$

and $\Theta_i^{(r)}\big|_{\text{QCF}}^*$ is given by (7.3.33b), which is explicitly for these recursive relations $(r > 1)$:

$$\Theta_i^{(r)}(\boldsymbol{X}_1,\ldots,\boldsymbol{X}_r)^* \equiv \int_{\Omega_i} \boldsymbol{\theta}\, W(\boldsymbol{X}_r|\boldsymbol{\theta}) w_i^{(r-1)}(\boldsymbol{\theta})\, d\boldsymbol{\theta} \Big/ \int_{\Omega_i} W(\boldsymbol{X}_r|\boldsymbol{\theta}) w_i^{(r-1)}(\boldsymbol{\theta})\, d\boldsymbol{\theta}. \qquad (7.3.47)$$

Because of the nature of the optimization of the estimator here (refer to Eqs. (7.1.16a)–(7.1.18b), the result depends only on the coefficients $(b_{ik} - b_{i0})$ of the quadratic form, $(\tilde{\boldsymbol{\theta}} - \hat{\boldsymbol{\theta}}) \tilde{\boldsymbol{E}}_{ik} (\boldsymbol{\theta} - \hat{\boldsymbol{\theta}})$, which applies only to the estimator $\hat{\boldsymbol{\theta}}_k$, since estimation and detection are here uncoupled. The result is, of course, that the estimator is optimized for QCF and the detector is optimized only for the constant cost assignments, representing now the independence of the average risks R_D, R_E in the total $R_D + R_E$. Note that $\Theta_i^{(r)*}$ is again an LMS estimator and that the $\Lambda_i^{(r)}$ are the recursive forms of the familiar GLR for optimum binary on-off detection.

7.3.5 Supervised Learning (Self-Taught Mode): An Introduction ([4], Section 4.5)

The learning problems discussed at length in Section 7.3 are all examples of unsupervised learning or "learning without a teacher". Another approach to the learning problem is supervised learning or "learning with" a teacher [16]. This method postulates the existence of a set of learning observations, in that we are given a sequence of observations, $\boldsymbol{X}_1, \boldsymbol{X}_2, \ldots, \boldsymbol{X}_r$, along with the proper classification of each of these observations. An observation is thus considered to be classified if we can identify the signal class from which it originated. Given a sequence of classified observations involving the same signal class, we are thus able in principle to compute a posterior probability density for any initially unknown or partially known signal parameters, which in turn may be used as a prior density for subsequent observation intervals. This approach, like the previous recursive method of unsupervised learning in Section 7.3, has the advantage that the number of component densities in any total posterior density is fixed. This is a result of the fact that all possible classifications of the received data need not be considered if correct classifications are known.

Here we consider the following sequential problem with the processes of *detection and estimation being performed jointly and strongly coupled.* The updated prior probability densities will be determined by the supervised learning approach, the supervision in this case being supplied by the detector's output. Thus, if the detection operation decides a signal of class Ω_i is present, then only those data sets $(\boldsymbol{X}_{\hat{r}})$ in the total prior density for $(X_1, \ldots, X_r)$ will be updated that pertain to the parameters of class Ω_i. These data sets we indicate by $\hat{r} \geq 1$, that is, $\left(\hat{\boldsymbol{X}}_1, \hat{\boldsymbol{X}}_2, \ldots, \hat{\boldsymbol{X}}_{\hat{r}}\right)$. Thus, for example, out of $(\boldsymbol{X}_1, \ldots, \boldsymbol{X}_r)$ data sets, we see that for $(\boldsymbol{X}_2, \ldots, \boldsymbol{X}_6, \boldsymbol{X}_7, \ldots, \boldsymbol{X}_{r-1})$, the detector decides Ω_i: $S_i(\boldsymbol{\theta})$, and so we use the latter set, renumbering it $X_2 \equiv \hat{X}_1, X_6 \equiv \hat{X}_2, \ldots X_{r-1} \equiv \hat{X}_{\hat{r}}$ as a new consecutive sequence.

For the learning problem now under consideration, we consider the $\hat{r} + 1$st data set $\boldsymbol{X}_{r+1}$ and make an estimate of $\boldsymbol{\theta}$, based on the detector's decisions that $\hat{X}_{\hat{r}+1}$ contains the signal S_i. The decision space Δ for this problem consists of three decision elements, γ_0, γ_1, and γ_2, where

$$\left.\begin{aligned}
\gamma_0 &= \text{decision that } \boldsymbol{\theta} = \mathbf{0}\text{:} && \text{we accept } H_0\text{: } S_i \text{ is not present} \\
\gamma_1 &= \text{decision that } \boldsymbol{\theta} \neq \mathbf{0}\text{:} && \text{but the receiver is unable to identify or observe} \\
& && \text{the signal } S_i\text{, that is, we accept the hypothesis } H_1 \\
\gamma_2 &= \text{decision that } \boldsymbol{\theta} \neq \mathbf{0}\text{:} && \text{and } S_i \text{ is identified, namely, we accept hypothesis } H_2
\end{aligned}\right\}. \tag{7.3.48}$$

Here H_0: N, H_1: $S \otimes N$, but S_i is effectively zero for the observer, and H_2: S_i and S_i is distinguished in the accompanying noise. Thus, the observation space Γ is partitioned into three mutually disjoint subspaces Γ_0, Γ_1, and Γ_2, where

$$\Gamma_i = \left\{\boldsymbol{X} | P^*(\gamma_i|\boldsymbol{X}) = 1\right\}; \quad i = 0, 1, 2; \quad \Gamma = \sum_{r=0}^{2} \Gamma_i, \tag{7.3.49}$$

where $P^*\left(\gamma_1|\hat{X}_{\hat{r}+1}\right)$ are determined so as to minimize the conditional risk $R_c\left(\hat{X}_{\hat{r}+1}\right)$:

$$R_c\left(\hat{X}_{\hat{r}+1}\right) = \int_{\Delta} \boldsymbol{d\gamma} P\left(\boldsymbol{\gamma}|\hat{X}_{\hat{r}+1}\right) \int_{\Omega} \boldsymbol{d\theta}\, W\left(\boldsymbol{\theta}|\hat{X}_{\hat{r}+1}\right), \tag{7.3.50}$$

where $C(\boldsymbol{\theta}, \boldsymbol{\gamma})$ are cost functions. Here, in more detail, $W(\boldsymbol{\theta}|\hat{X}_{\hat{r}+1})$ consists of the combination

$$\left.\begin{aligned} W(\boldsymbol{\theta}|\hat{X}_{\hat{r}+1}) &= P(H_0|\hat{\boldsymbol{X}}_{\hat{r}+1})\delta(\boldsymbol{\theta}-\mathbf{0}) + P(H_1|\hat{\boldsymbol{X}}_{\hat{r}+1})\, W_1(\boldsymbol{\theta}|\hat{\boldsymbol{X}}_{\hat{r}+1})_1 \\ &\quad + P(H_2|\hat{\boldsymbol{X}}_{\hat{r}+1})\, W_2(\boldsymbol{\theta}|\hat{\boldsymbol{X}}_{\hat{r}+1})_2 \end{aligned}\right\}. \tag{7.3.51}$$

Since the decision space is the discrete set, $\{\gamma_0, \gamma_1, \gamma_2\}$, the integral over Δ reduces to a sum, and hence (7.3.50) may be rewritten as

$$R_c(\boldsymbol{X}) = \sum_{i=0}^{2} P(\gamma_i|\hat{\boldsymbol{X}}_{r+1}) \int_{\Omega_i} d\boldsymbol{\theta}\, C(\boldsymbol{\theta}, \boldsymbol{\gamma}_i) W(\boldsymbol{\theta}|\hat{\boldsymbol{X}}_{r+1}). \tag{7.3.52}$$

Since we also require that a definite decision be made for each observation interval $\hat{\boldsymbol{X}}_{r+1}$, then the conditional risk $R_c(\boldsymbol{X})$ can be expressed as follows:

$$\left.\begin{aligned} R_c(\boldsymbol{X}) &= \int_{\Omega_i} d\boldsymbol{\theta}\, C(\boldsymbol{\theta}, \gamma_0) W(\boldsymbol{\theta}|\hat{\boldsymbol{X}}_{r+1})_0 \\ &\quad + \sum_{i=1}^{2} W(\gamma_i|\hat{\boldsymbol{X}}_{r+1}) \int_{\Omega_i} d\boldsymbol{\theta} [C(\boldsymbol{\theta}, \boldsymbol{\gamma}_i) - C(\boldsymbol{\theta}, \gamma_0)] W(\boldsymbol{\theta}|\hat{\boldsymbol{X}}_{r+1})_i \end{aligned}\right\}. \tag{7.3.53}$$

Example: QCFs:

To obtain explicit results, we must use specific cost functions appearing in (7.3.53). We shall accordingly select the following quadratic cost functions:

$$\left\{\begin{aligned} C(\boldsymbol{\theta}, \gamma_0) &= C_{00}^{(2)} && \text{when } H_0 \text{ is true} \\ C(\boldsymbol{\theta}, \gamma_0) &= C_{10}^{(1)} + C_{10}^{(2)} \cdot \left(\tilde{\boldsymbol{\theta}} - \tilde{\hat{\boldsymbol{\theta}}}\right)(\boldsymbol{\theta} - \hat{\theta}) && \text{when } H_1 \text{ is true} \\ C(\boldsymbol{\theta}, \gamma_0) &= C_{20}^{(1)} + C_{20}^{(1)} \cdot \left(\tilde{\boldsymbol{\theta}} - \tilde{\hat{\boldsymbol{\theta}}}\right)(\boldsymbol{\theta} - \hat{\theta}) && \text{when } H_2 \text{ is true} \end{aligned}\right\} \tag{7.3.54a}$$

$$\left\{\begin{aligned} C(\boldsymbol{\theta}, \gamma_1) &= C_{01}^{(1)} + C_{01}^{(2)} \cdot \tilde{\hat{\boldsymbol{\theta}}}\hat{\boldsymbol{\theta}} && \text{when } H_0 \text{ is true} \\ C(\boldsymbol{\theta}, \gamma_1) &= C_{11}^{(2)} \cdot \left(\tilde{\boldsymbol{\theta}} - \tilde{\hat{\boldsymbol{\theta}}}\right)(\boldsymbol{\theta} - \hat{\boldsymbol{\theta}}) && \text{when } H_1 \text{ is true} \\ C(\boldsymbol{\theta}, \gamma_1) &= C_{21}^{(1)} + C_{02}^{(2)} \cdot \tilde{\hat{\boldsymbol{\theta}}}\hat{\boldsymbol{\theta}} && \text{when } H_2 \text{ is true} \end{aligned}\right\} \tag{7.3.54b}$$

$$\left\{\begin{aligned} C(\boldsymbol{\theta}, \gamma_2) &= C_{02}^{(1)} + C_{02}^{(2)} \cdot \tilde{\hat{\boldsymbol{\theta}}}\hat{\boldsymbol{\theta}} && \text{when } H_0 \text{ is true} \\ C(\boldsymbol{\theta}, \gamma_2) &= C_{12}^{(1)} + C_{02}^{(2)} \cdot \left(\tilde{\boldsymbol{\theta}} - \tilde{\hat{\boldsymbol{\theta}}}\right)\left(\boldsymbol{\theta} - \hat{\boldsymbol{\theta}}\right) && \text{when } H_1 \text{ is true} \\ C(\boldsymbol{\theta}, \gamma_2) &= C_{22}^{(2)} \cdot \left(\tilde{\boldsymbol{\theta}} - \tilde{\hat{\boldsymbol{\theta}}}\right)\left(\boldsymbol{\theta} - \hat{\boldsymbol{\theta}}\right) && \text{when } H_2 \text{ is true} \end{aligned}\right\} \tag{7.3.54c}$$

For convenience, let us next define a combination of costs:

$$C_i \equiv \int_{\Omega} d\boldsymbol{\theta}[C(\boldsymbol{\theta}, \gamma_i) - C(\boldsymbol{\theta}, \gamma_0)] W(\boldsymbol{\theta}|\boldsymbol{X}), \quad i = 1, 2. \tag{7.3.55}$$

After $W(\boldsymbol{\theta}|\boldsymbol{X})$, as given by (7.3.51), and the above cost factors are substituted into the definition (7.3.55) of C_1 and C_2, it is easily seen that

$$\left.\begin{aligned} C_1 &= P(H_0|\boldsymbol{X})\left[C_{01}^{(1)} + \left(C_{01}^{(2)} - C_{00}^{(2)}\right)\tilde{\hat{\boldsymbol{\theta}}}\hat{\boldsymbol{\theta}}\right] \\ &= P(H_1|\boldsymbol{X})\left[-C_{10}^{(1)} + \left(C_{11}^{(2)} - C_{10}^{(2)}\right)\int_{-\infty}^{\infty}\left(\tilde{\boldsymbol{\theta}} - \tilde{\hat{\boldsymbol{\theta}}}\right)\left(\boldsymbol{\theta} - \hat{\boldsymbol{\theta}}\right)W(\boldsymbol{\theta}|\boldsymbol{X})_1 d\boldsymbol{\theta}\right] \\ &= P(H_2|\boldsymbol{X})\left[C_{21}^{(1)} - C_{20}^{(1)} + \left(C_{21}^{(2)} - C_{20}^{(2)}\right)\int_{-\infty}^{\infty}\left(\tilde{\boldsymbol{\theta}} - \tilde{\hat{\boldsymbol{\theta}}}\right)\left(\boldsymbol{\theta} - \hat{\boldsymbol{\theta}}\right)W(\boldsymbol{\theta}|\boldsymbol{X})_2 d\boldsymbol{\theta}\right] \end{aligned}\right\} \tag{7.3.56a}$$

$$\left.\begin{aligned} C_2 &= P(H_0|\boldsymbol{X})\left[C_{02}^{(1)} + \left(C_{02}^{(2)} - C_{00}^{(2)}\right)\tilde{\hat{\boldsymbol{\theta}}}\hat{\boldsymbol{\theta}}\right] \\ &= P(H_1|\boldsymbol{X})\left[C_{12}^{(1)} + -C_{10}^{(1)} + \left(C_{12}^{(2)} - C_{10}^{(2)}\right)\int_{-\infty}^{\infty}\left(\tilde{\boldsymbol{\theta}} - \tilde{\hat{\boldsymbol{\theta}}}\right)\left(\boldsymbol{\theta} - \hat{\boldsymbol{\theta}}\right)W(\boldsymbol{\theta}|\boldsymbol{X})_1 d\boldsymbol{\theta}\right] \\ &= P(H_2|\boldsymbol{X})\left[-C_{20}^{(1)} + \left(C_{22}^{(2)} - C_{20}^{(2)}\right)\int_{-\infty}^{\infty}\left(\tilde{\boldsymbol{\theta}} - \tilde{\hat{\boldsymbol{\theta}}}\right)\left(\boldsymbol{\theta} - \hat{\boldsymbol{\theta}}\right)W(\boldsymbol{\theta}|\boldsymbol{X})_2 d\boldsymbol{\theta}\right] \end{aligned}\right\}. \tag{7.3.56b}$$

To carry the analysis further, let us assume that $W(\boldsymbol{\theta}|\boldsymbol{X})$ is Gaussian, for example, in order to obtain closed-form specific results. We have

$$W(\theta|\boldsymbol{X}) = G\left(\theta; \mu, \sigma^2\right), \quad \mu = \mu_0 \text{ or } \mu_1; \quad \sigma^2 = \sigma_0^2 \text{ or } \sigma_1^2, \tag{7.3.57}$$

respectively, for $W\left(r_0 \text{ or } r_1|\hat{\boldsymbol{X}}_{\hat{r}+1}\right)$ and let q_0, q_1 be the *a priori* probabilities associated with states represented by G_0 and G_1, remembering that G represents the data set $\boldsymbol{X} = \left[X_j\right]$. We observe next that the factor C_2 contains an unknown estimator $\hat{\boldsymbol{\theta}}(\boldsymbol{X})$. The optimum Bayes estimator, denoted by $\boldsymbol{\theta}^*(\boldsymbol{X})_{q_0>0}$, is determined in such a way as to minimize C_2. This estimator is easily seen to be equal to the conditional mean, namely,

$$\boldsymbol{\theta}^*(\boldsymbol{X}) = \int_{\Omega} \boldsymbol{\theta}\, W(\boldsymbol{\theta}|\boldsymbol{X})d\boldsymbol{\theta} = P(H_1|\boldsymbol{X})\mu_0 + P(H_2|\boldsymbol{X})\mu_1. \tag{7.3.57a}$$

This estimator is substituted in place of $\hat{\boldsymbol{\theta}}(\boldsymbol{X})_{\mu_0>0}$ in C_2 to obtain

$$\left.\begin{aligned} C_1^* &= P(H_0|\boldsymbol{X})\left[C_{01}^{(1)} + \left(C_{01}^{(2)} - C_{00}^{(2)}\right)\tilde{\hat{\boldsymbol{\theta}}}\hat{\boldsymbol{\theta}}\right] \\ &= P(H_1|\boldsymbol{X})\left[-C_{10}^{(1)} + \left(C_{11}^{(2)} - C_{10}^{(2)}\right)\left(T_r(\sigma_0^2) + \tilde{\boldsymbol{\mu}}_0\boldsymbol{\mu}_0 - 2\tilde{\boldsymbol{\mu}}_0\boldsymbol{\theta}_{q_0>0} + \left(\tilde{\hat{\boldsymbol{\theta}}}^*_{q_0>0}\right)\left(\hat{\boldsymbol{\theta}}^*_{q_0>0}\right)\right)\right] \\ &= P(H_2|\boldsymbol{X})\left[C_{21}^{(1)} - C_{20}^{(1)} + \left(C_{21}^{(2)} - C_{20}^{(2)}\right)\left(T_r(\sigma_1^2) + \tilde{\boldsymbol{\mu}}_1\boldsymbol{\mu}_1 - 2\tilde{\boldsymbol{\mu}}_1\hat{\boldsymbol{\theta}}^*_{q_0>0} + \left(\tilde{\hat{\boldsymbol{\theta}}}^*_{q_0>0}\right)\left(\hat{\boldsymbol{\theta}}^*_{q_0>0}\right)\right)\right] \end{aligned}\right\}, \tag{7.3.58a}$$

$$\left.\begin{aligned} C_2^* &= P(H_0|\boldsymbol{X})\left[C_{02}^{(1)} + \left(C_{02}^{(2)} - C_{00}^{(2)}\right)\tilde{\hat{\boldsymbol{\theta}}}\hat{\boldsymbol{\theta}}\right] \\ &= P(H_1|\boldsymbol{X})\left[C_{12}^{(1)} - C_{10}^{(1)} + \left(C_{12}^{(2)} - C_{10}^{(2)}\right)\left(T_r(\sigma_0^2) + \tilde{\boldsymbol{\mu}}_0\boldsymbol{\mu}_0 - 2\tilde{\boldsymbol{\mu}}_0\boldsymbol{\theta}_{q_0>0} + \left(\tilde{\hat{\boldsymbol{\theta}}}^*_{q_0>0}\right)\left(\hat{\boldsymbol{\theta}}^*_{q_0>0}\right)\right)\right] \\ &= P(H_2|\boldsymbol{X})\left[-C_{20}^{(1)} + \left(C_{22}^{(2)} - C_{20}^{(2)}\right)\left(T_r(\sigma_1^2) + \tilde{\boldsymbol{\mu}}_1\boldsymbol{\mu}_1 - 2\tilde{\boldsymbol{\mu}}_1\hat{\boldsymbol{\theta}}^*_{q_0>0} + \left(\tilde{\hat{\boldsymbol{\theta}}}^*_{q_0>0}\right)\left(\hat{\boldsymbol{\theta}}^*_{q_0>0}\right)\right)\right] \end{aligned}\right\}. \tag{7.3.58b}$$

The optimum decision rule is now readily seen to be

$$W\left(r_0|\hat{\boldsymbol{X}}_{\hat{r}+1}\right)^* = \begin{cases} 1, & \text{if } W\left(r_1|\hat{\boldsymbol{X}}_{\hat{r}+1}\right) = W\left(r_0|\hat{\boldsymbol{X}}_{\hat{r}+1}\right) = 0 \\ 0, & \text{otherwise} \end{cases} \tag{7.3.59a}$$

$$W\left(r_1|\hat{\boldsymbol{X}}_{\hat{r}+1}\right)^* = 1, \ \text{if } C_1^* < 0 \ \text{ and } \ C_1^* < C_2^*; \quad = 0 \ \text{ otherwise}; \tag{7.3.59b}$$

$$W\left(r_2|\hat{\boldsymbol{X}}_{\hat{r}+1}\right)^* = 1, \ \text{if } C_2^* > 0 \ \text{ and } \ C_1^* < C_2^*; \quad = 0 \ \text{ otherwise}. \tag{7.3.59c}$$

Furthermore, the sequence of updated means $\boldsymbol{\mu}_1 (= [\boldsymbol{\mu}_{1i}]); i = 1, \ldots, \hat{r}+1$, is stable, since

(1) We would reject all the data in the most conservative case, with the conclusion that this case contained no signal. In such a case, each $\mu_i = \mu_0$, the original value with which we started.

(2) At the other extreme, we would accept all the data. Then, μ_i would converge to zero when $\boldsymbol{\theta} = 0$ to $\boldsymbol{\theta}/p$ if $\boldsymbol{\theta} \neq \mathbf{0}$.[3]

7.4 CONCLUDING REMARKS

The preceding sections provide a foundation for "one-shot" $(r = 1)$ and "multishot" $(r > 1)$ sequences of optimum decisions: for both detection and estimation, jointly and independently, where both operations are performed under conditions of uncertainty. Once more, we desire optimality, in the sense of minimizing the average risk $R_D + R_E$ (independently) and $R_D \oplus R_E$ (coupled together). The former represents weak or no coupling (refer to Sections and 7.3.3.3), and the latter strong coupling (refer to Sections 7.3.3.1 and 7.3.3.2), for both binary and multiple hypothesis states. All multishot cases here are examples of unsupervised learning or "learning without teacher." However, examples of supervised learning are given in Section 7.3.5. As we have seen in Chapter 6, in Section 7.1, Eq. (7.1.13), and subsequently, it is the choice of cost function that determines the degree of coupling, for example, Eqs. (7.1.4), (7.3.21a) and (7.3.29). The cost functions are also generalized in the manner of Eqs. (7.3.3) and (7.3.4) to account

[3] See Ref. [4], Appendix B, for the results of computer simulations.

explicitly for overlapping hypothesis classes (Section 7.2). Specific results for the quadratic cost function and the "simple" cost function are illustrated, the former results characteristic of least mean square estimators and the latter of unconditional maximum-likelihood estimators, the most common instrument of Bayes risk (or cost) in estimation.

Since Bayes estimations under uncertainty ($0 \leq p < 1$) for binary on–off detection always produces positively biased estimators (Section 6.3.4) and similarly under $H_i, i = 1, \ldots, K$ in the multiple hypothesis cases $0 \leq p_0 < 1$, $0 < p_i < 1$ of Sections and 7.3.3, estimation on the assumption of certainty as to the presence of a signal is always suboptimum and produces a larger average estimate and hence a larger average risk than the optimum (Bayes) cases. For example, if $\boldsymbol{\gamma}_{\theta}$ here is the estimator under the (false) assumption of certainty, we have, letting $\boldsymbol{\gamma}_{\theta}^{*}|_{\text{false}}$ equal this false optimum estimator,

$$\boldsymbol{\gamma}_{\theta}^{*}|_{\text{false}} = \boldsymbol{\gamma}_{\boldsymbol{\theta},p<1}^{*}, \quad \text{or} \quad \left.\begin{aligned} \boldsymbol{\gamma}_{\boldsymbol{\theta}}^{*}|_{\text{false}} &= \left(\frac{1+\Lambda}{\Lambda}\right)\boldsymbol{\gamma}_{\boldsymbol{\theta},p=1}^{*} \\ \boldsymbol{\gamma}_{\boldsymbol{\theta}}^{*}|_{\text{false}} &= \boldsymbol{\gamma}_{\boldsymbol{\theta},p=1}^{*}/p \end{aligned}\right\} > \boldsymbol{\gamma}_{\boldsymbol{\theta},p=1}^{*} \begin{cases} \text{QCF} & (7.4.1\text{a}) \\ \text{SCF}_{1,2} & (7.4.1\text{b}) \end{cases}$$

from (7.3.6), (7.3.11) and (7.3.12), for instance, in the binary cases, with corresponding examples by obvious extension to the multiple sample, multiple hypothesis situations. Accordingly, making the erroneous assumption of certainty as to the presence of the signal always leads on the average to too large an estimate. This is one justification of joint and coupled detection and estimation, which applies for strong as well as weak signals. A second one is that this procedure can further help detection *and* estimation when one encounters weak or threshold signals, particularly when the coupling is strong. As always, the accompanying increase in data processing must be considered, but with the current high speed and minimal costs of modern computing, achieving "on-line" results should be easily accomplished.

Calculating the Bayes risks in the above situations provides another measure of the superiority of the Bayesian approach vis-à-vis the suboptimum results for the falsely chosen estimators in Eqs. (7.4.1a) and (7.4.1b). In fact, coupling D and E minimizes the Bayes risk further. Strong coupling results are more effective than weak coupling, which in turn is better in this respect than uncoupled D and E (see section following Eq. (6.3.20)). Accordingly, we have for the average risks and their minimization the inequalities

$$(R_D + D_E) > (R_D^* + R_E), (R_D + R_E^*) > (R_D^* + R_E^*) > (R_D^* \otimes R_E^*) > (R_D^* \otimes R_E^*) \tag{7.4.1c}$$

where the first four terms involve no coupling, the fifth weak coupling, and the last, strong coupling.

Our results in the preceding sections are also easily extended to the important cases of extrapolation, interpolation, and filtering ([7]; Sections 21.4.2 and 21.4.3 for $p = 1$). For applications to tracking, we are particularly concerned with prediction (i.e., forward extrapolation), where the estimates of the same function of the signal waveform or the waveform itself at space–time instants outside the data interval $(|\boldsymbol{R}|, \Delta t)$ are desired, and where joint D and E are employed. This is to be done not only for a single data sample ($r = 1$) but also where D and E for ($r \geq 1$) samples depend recursively on the preceding data samples ($r = 1, \ldots, r-1$) (Sections 7.3.2 and 7.3.3).

As an example, let us consider the situation of Section 7.3.3.1 for strong coupling and multihypotheses where we wish to determine the minimum least-square prediction of signal waveform at $\boldsymbol{r}_\lambda, t_\lambda$ beyond the last data interval received (and processed). If we let $\boldsymbol{\theta}^*_{k,r} = [\boldsymbol{S}, S(\boldsymbol{r}_\lambda, t_\lambda)]$, where $\boldsymbol{S} = [S(\boldsymbol{r}_m, t_n)]$ and $S_\lambda = [S(\boldsymbol{r}_\lambda, t_\lambda)]$, then (7.3.35) can be directly modified to yield on the basis of quadratic cost functions

$$\hat{S}^*(\boldsymbol{X}_1, \ldots, \boldsymbol{X}_r | \boldsymbol{r}_\lambda, t_\lambda)\Big|_{\text{QCF}} = \sum_{i=1}^{K} \left[\Lambda_i^{(r)} \Big/ \left(1 + \sum_{i=1}^{K} \Lambda_i^{(r)} \right) \right] \hat{S}_i^*(\boldsymbol{X}_1, \ldots, \boldsymbol{X}_r | \boldsymbol{r}_\lambda, t_\lambda). \quad (7.4.2)$$

Here,

$$\left. \begin{aligned} \hat{S}_i^*(\boldsymbol{X}_1, \ldots, \boldsymbol{X}_r | \boldsymbol{r}_\lambda, t_\lambda) \equiv \int_{\Omega_i} & S(\boldsymbol{r}_\lambda, t_\lambda) W(\boldsymbol{X}_r | \boldsymbol{S}_1, \ldots, \boldsymbol{S}_r) \\ & \cdot w_i^{(r-1)}(\boldsymbol{S}_1, \ldots, \boldsymbol{S}_r | \boldsymbol{X}_1, \ldots, \boldsymbol{X}_{r-1}) d\boldsymbol{S}_1 \cdots d\boldsymbol{S}_r \end{aligned} \right\}, \quad (7.4.2a)$$

and Λ_i is given by (7.3.33a), modified to

$$\left. \begin{aligned} \Lambda_i^{(r)} = \Lambda_i^{(r)}(\boldsymbol{X}_1, \ldots, \boldsymbol{X}_r) = \mu_i \int_{\Omega_i} & \frac{W(\boldsymbol{X}_r | \boldsymbol{S}_1, \ldots, \boldsymbol{S}_r)}{W(\boldsymbol{X}_r | 0)} \\ & \cdot w_i^{(r-1)}(\boldsymbol{S}_1, \ldots, \boldsymbol{S}_r | \boldsymbol{X}_1, \ldots, \boldsymbol{X}_{r-1}) d\boldsymbol{S}_1 \cdots d\boldsymbol{S}_r \end{aligned} \right\} \quad (7.4.2b)$$

refer to Eq. (7.3.28c), where (7.4.2) is subject to the assumptions leading to Eqs. (7.3.34) et seq. This, of course, is the recursive extension for the case (7.3.20)–(7.3.21a), considered originally for a single data sample ($r = 1$) by Fredriksen et al. [5].

All these cases, and particularly those with recursive distributions (refer to Eqs. (7.3.1) and (7.3.2)), are increasingly complex with respect to the distributions involved. They can present formidable computational problems, especially if the number (r) of data sets become at all large, that is, $r >> 1$. However, this problem can be mitigated if we recall that most physical random fields are *simple Markoff*, that is, their most recent values depend only on a preceding observation, that is,

$$W_J(\alpha_J, u_J | \alpha_{J-1}, u_{J-1}; \ldots; \alpha_1, u_1) = W_2(\alpha_J, u_J | \alpha_{J-1}, u_{J-1}), (j = 11, \ldots, MN), \quad (7.4.3)$$

where[4] $\mu_1 \leq \mu_2 \leq \cdots \mu_j \leq \cdots \leq \mu_J, J = MN$. A still simpler field statistically is the *purely random field*, where

$$W_1(\alpha_J, u_J | \ldots; \alpha_1, u_1) = W_1(\alpha_J, u_J), \quad (7.4.3a)$$

which is approximated experimentally by sampling at space–time intervals large enough to avoid significant correlations between neighboring points in the space–time manifold. For the former, we can easily show ([7], Problem 1.16, and Section 1.5.2, ibid., Eq. (1.95b)) the following:

Simple Markoff:

$$W_J(\alpha_J, u_J; \ldots; \alpha_1, u_1) = \prod_{j=2}^{J} W_2(\alpha_j, u_j | \alpha_{j-1}, u_{j-1}) \Big/ \prod_{j=2}^{J-1} W_1(\alpha_j, u_j), \quad J = MN; \quad (7.4.4a)$$

[4] Here $\alpha_{j\,(=mn)}$, for example, is a space–time field $\alpha(\boldsymbol{r}_m, t_n)$ and the μ_j refer to the independent variables $\boldsymbol{r}_m$ and t_n.

Purely Random Markoff:

$$W_J = \prod_{j=2}^{J-1} W_1\left(\alpha_j, u_j\right). \tag{7.4.4b}$$

The above assumes that the random field is fully described by a single random field $\alpha(\mathbf{r}_m, t_n)$ and is not simply a *projection* of a more complex field. In physical problems, this is usually not the case, $\alpha(\mathbf{r}_m, t_n)$ is usually a projection of a more complex random field, involving space and time derivatives of the field as well for a full statistical description, that is, $W_2\left(\alpha_j|u_j; \dot{\alpha}_{j-1}|u_{j-1}; \ldots, \text{etc.}\right)$ of a simple Markoff field. The question of how many other components are needed for a full description depends on the underlying physics of the problem, expressed, for example, by the appropriate Langevin equation (cf. Chapter 10 of Ref. [7]) or set of random dynamical equations, which can completely describe the field. (These questions are touched upon briefly in Ref. [7], Chapter 10, and references in the following chapters.)

The point to which we wish to call attention here is that the physical problem may lead us to replace the full recursive distributions, such as $w_1^{(r)}, w_1^{(r-1)}$ in Section 7.3, by a suitable simple Markoff field (or process), in the appropriate dimensions (i.e., $\alpha, \dot{\alpha}$, etc.) and thus simplify the calculations in the practical realizations of the above results. For example, the Markoff conditions (7.4.4a) and (7.4.4b) are written for $w_r(\boldsymbol{\theta})$, (7.3.1):

$$w_r(\boldsymbol{\theta}) = W(\boldsymbol{\theta}|\mathbf{X}_r) = W(\mathbf{X}_r|\boldsymbol{\theta})w(\boldsymbol{\theta})/W(\mathbf{X}_r)\text{: purely random field} \tag{7.4.5a}$$

$$= W(\boldsymbol{\theta}|\mathbf{X}_{r-1}, \mathbf{X}_r) = W(\mathbf{X}_r|\boldsymbol{\theta}, \mathbf{X}_{r-1})w(\boldsymbol{\theta}|\mathbf{X}_{r-1})/W(\mathbf{X}_r|\mathbf{X}_{r-1})\text{: purely Markoff field.} \tag{7.4.5b}$$

It should be noted that the sequential operations of Section 7.3 are not the same as those employed in the Wald theory [17] of sequential (binary) detection. The principal differences are that in the Wald theory, the conditional probabilities of error (α, β) are fixed and the sample size $(\sim J)$ is allowed to vary, and that there are two thresholds for terminating the test, as distinct from the single threshold employed in the usual binary problems. In our sequential scheme here, the data sequences are of fixed size (although they may vary from interval to interval), and β or (α, β) are minimized for each interval.

Again we emphasize the introductory nature of our treatment of joint $D + E$ with respect to the broad problem of target tracking and similar applications. At the same time, we call attention to possible improvements in performance produced by strongly coupled detection and estimation, even for strong signals, as discussed briefly in the second paragraph above.[5] Thus, strong coupling is generally more effective than weak or no coupling in minimizing the Bayes risk and the choice of cost function is critical in achieving optimality. When the "simple cost functions" are employed, that is, those

[5] Finally, we refer the reader to Ref. [4] for additional material, particularly to Appendices A–C therein. Here, results are presented for computer simulations of an appropriate form of unsupervised learning, for some simulations of supervised learning, and a discussion is given of the convergence of the approximate form of unsupervised learning, treated in Section 4.4 of Ref. [4]. This approximate form requires all updated probability derivatives to have the same formed structure, unlike the exact recursive forms used here, which are characterized by exponential growth as the number (r) of data intervals increase, resulting in great computation burdens at the time. The latter was a formidable problem computationally about 1970, but is not now in 2000s.

representing unconditional maximum likelihood estimates, detection and estimations are automatically uncoupled, the estimators are formally simpler than those required for the LMS estimators (i.e., using QCF). However, the results in each case are not generally comparable: only explicit calculation of the Bayes risks can provide a quantitative comparison. In any case, many problems remain for further study, extending into the learning processes for neural networks [18], pattern recognition and related fields of applications, including space–time systems for communications [19], some of which are treated here in the remaining chapters of the present book.

REFERENCES

1. Y. Bar-Shalom and X. R. Li, *Multitarget–Multisensor Tracking: Principles and Techniques*, OPAMP Technical Books, Los Angeles, CA, 1995.
2. D. Middleton and R. Esposito, Simultaneous Optimum Detection and Estimation of Signals in Noise, *IEEE Trans. Inform. Theory*, **IT-14**(3), 434–444 (1968).
3. D. Middleton and R. Esposito, New Results in the Theory of Simultaneous Optimum Detection and Estimation of Signals in Noise, *Problemy Peredachi Informatsii* (*Problems of Information Transmission*), **6**(2), Ak. Nauk, Moscow, (1970), pp. 3–20. English translation published Jan. 1973, pp. 93–106, Consultants Bureau, Plenum Publishing, New York. (Originally published as a Raytheon Technical Memorandum (T–838), Oct. 31, 1969; a concise version of the above paper was presented (by R.E.) at the Symposium on Information Theory, Dubna, USSR, June 1969.)
4. A. Fredriksen, Simultaneous Signal Detection and Estimation Under Multiple Hypotheses, PhD dissertation, Department of E.E., Johns Hopkins University, Baltimore, MD, May 1970.
5. A. Fredriksen, D. Middleton, and D. Vandelinde, Simultaneous Signal Detection and Estimation Under Multiple Hypotheses, *IEEE Trans. Inform. Theory*, **IT-18**(5), 607–614 (1972).
6. F. C. Ogg, Jr., A Note on Bayes Detection of Signals, *IEEE Trans. Inform. Theory*, **IT-10**, 57–60, (January 1964).
7. D. Middleton, *An Introduction to Statistical Communication Theory*, IEEE Press, Classic Reissue, Piscataway, NJ, 1996.
8. R. Esposito, A Class of Estimators for Optimum Adaptive Detection, *Information and Control*, **10**, 137–148, (February 1967).
9. R. Esposito, On a Relation Between Detection and Estimation in Decision Theory, *Inform. Control*, **12**, 116–120 (February 1968).
10. D. Middleton, *Topics in Communication Theory*, McGraw-Hill, New York, 1965.
11. T. Kailath, Adaptive Matched Filters, in R. Bellman, Ed., *Mathematical Optimization Techniques*, University of California Press, Berkeley, CA, 1963, Chapter 6.
12. S. C. Fralick, Learning to Recognize Patterns Without a Teacher, *IEEE Trans. Inform. Theory*, **13**(1), 57–64 (1967).
13. J. L. Doob, *Stochastic Processes*, John Wiley & Sons, Inc., New York, 1953, p. 319.
14. R. F. Daly, The Adaptive Binary Detection Problem on the Real Line, Stanford Electronics Laboratory, Report TR-2003-3 (February 1962).
15. H. J. Scudder, Adaptive Communication Receivers, *IEEE Trans. Inform. Theory*, **IT-11**, 167–174 (April 1965).
16. (a) N. Abramson and D. J. Braverman, Learning to Recognize Patterns in a Random Environment, *IEEE Trans. Inform. Theory*, **IT-8,** 58–63 (September 1962); (b) D. J. Braverman,

Machine Learning and Automatic Pattern Recognition, Technical Report No. 2003-1, Stanford Electronics Labs, Stanford, CA, February 1961.

17. A. Wald, *Sequential Analysis*, John Wiley & Sons, Inc., New York, 1947; reprinted by Dover Publications, New York, 1973.
18. S. Haykin, *Neural Networks: A Comprehensive Foundation*, MacMillan, New York, 1994.
19. D. Middleton, Improvement Under Unsupervised Learning for Joint Strongly Coupled Bayes Detection and Estimation Vis-à-Vis the Uncoupled Cases, *Proceedings of CISS*, paper 285, Princeton University, Princeton, NJ, March 2004.

8

THE CANONICAL CHANNEL I: SCALAR FIELD PROPAGATION[1] IN A DETERMINISTIC MEDIUM

The ultimate purpose of Chapters 8–9, is to establish the required statistics of the physical channels that connect transmitter and receiver. These include not only the omnipresent background noise and interfering signals but also the self-generated scatter noise, as well as the distortions of the signal produced by the inhomogeneous character of the medium itself. As noted frequently earlier, the canonical channel considered here consists of the medium of scalar propagation and the coupling to it provided by the transmitting and receiving apertures or arrays. The latter convert the input signal process into space–time fields and the received signal and noise fields back into a temporal process. Signal processing, such as detection and estimation, as embodied in the sequence of operations $\hat{\mathbf{T}}_{\mathrm{AT}}\hat{\mathbf{T}}_{M}^{(N)}\hat{\mathbf{T}}_{\mathrm{AR}}$ is shown symbolically in Fig. 8.1. We emphasize the scalar nature here of the channel and the resulting propagation of signals and noise.

Our program in Chapters 8–9 is to present analytic structures of the scalar field and canonical scalar channel, to illustrate their physical properties, and to quantify the resulting fields $\alpha(\mathbf{r}, t)$ and the channel.

Accordingly, Chapter 8 provides results for a variety of deterministic media, both bounded and unbounded, to which deterministic signals are applied [1–8]. These results are needed in the "classical" treatment of *random (scalar) channels* (the so-called statistical–physical (S-P) approach [9]) discussed subsequently in Chapter 9 following [10–14]. In future work we will present the elements of a new approach, for example the physical-statistics (P-S) equivalent [9,15], also based on the underlying physics but interpreted almost

[1] Some vector field propagation is briefly considered in Section 8.6.3.

Non-Gaussian Statistical Communication Theory, David Middleton.

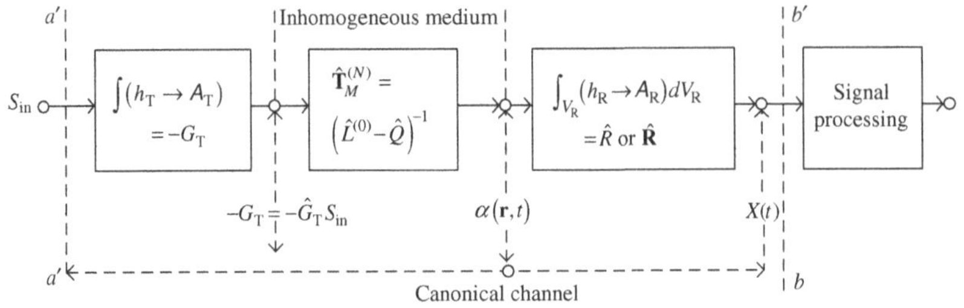

FIGURE 8.1 The canonical scalar channel: operational schematic of an inhomogeneous (linear) medium, with source and receiver coupling $\left(h_{(\,)}, A_{(\,)}\right)$, and $\alpha(\mathbf{r}, t)$ the resulting scalar field. The quantity $\hat{Q}$ is the *inhomogeneity operator* associated with this inhomogeneous (linear) medium. [The medium here is deterministic and scalar, but is formally generalized to random media in the general situation where, as a consequence of randomness, $\hat{Q}$ is a stochastic operator, cf. Chapter 9.

entirely by statistical models. These enable us to obtain *probability distributions* of such complex phenomena as multiple scatter, unlike the "classical" approaches, which are in practice limited to lower order moments of the scattered field. Moreover, these scattered fields, to all orders of single- or multiple scatter, are generally *non-Gaussian*, with correspondingly much more complex analytic representations [9,15].

From a more general viewpoint, scattering can be regarded as an important example of "feedback", where interaction generates modifying reaction, a feature of most natural phenomena. An analytical description of this is given here by appropriate integral equations. We shall encounter examples, both in the deterministic cases of this chapter and in following work devoted to their stochastic extensions.[2] Thus, this chapter provides the structural detail needed to represent the random canonical channels encountered later in Chapter 9. For subsequent applications, this chapter provides foundations, that is, "macroalgorithms," for computational evaluation in specific situations.[3]

We remark that the deterministic cases of propagation and scattering occur when everything pertinent is known about the physical situation encountered and the desired "solution" can then be uniquely determined. These "solutions" in fact are a property of the initial information and are implied by that information. There is no randomness involved. From the broader viewpoint of probability theory the deterministic sample or *representation*, constitutes an ensemble of one member, with probability unity for all components of the representation. Classical models of this type we call *a priori models*: they are completely determined in their initial formulation and are particularly useful when this representation is specified. Thus, the treatment of these deterministic problems is *nonpredictive*: all elements

[2] We remark that one difference between conventional engineering usage and the space–time physical formulations is in the dimensionality of the "feedback" operation. The former is usually one dimensional (i.e., involving time only), whereas the latter is four dimensional. The former is, *au fond*, described by ordinary, differential (and integral) equations; the latter require partial differential and corresponding integral equations.

[3] For the analytic background on which much of the present deterministic exposition in this chapter (and to a lesser extent in Chapter 9) depends, we refer the reader to the classic treatise of Morse and Feshbach [1], and to the book by Lindsay [2], as well as to numbers of other well-known works [3–5]. The "classical" treatment of propagation in random media is concisely surveyed from an operator viewpoint in Chapter 9. It is partially based on the review chapter of Frisch [10] and the books by Tatarskii [11], Ishimaru [12], and Bass and Fuks [13]. More recently, there is the four volume set (1978), first published in English in 1989, of Rytov et al. [14]. Many additional references are noted in the pertinent places (e.g., [16,17]) to these topics in optic and acoustics, applied to special media—such as the ocean, the atmosphere, and space, and so on, cf., Chapter 9.

for the solution are provided initially. However, for a great many applications in many fields—certainly in Communication Theory—we must deal explicitly with randomness. Here the ensemble of representations has many members, each with nonunity probability measures. Such models we call *a posteriori models*, since we do not *a priori* know with which ensemble member we have to deal. Such models are *predictive*, or nondeterministic. Their "solutions" are expressed in terms of moments and distributions. These models will be discussed and applied throughout the rest of this book (except for this chapter).

This chapter now presents some of the results of the standard nonrandom, that is deterministic, classical mathematical physics of propagation in linear homogeneous media. This is then extended to the inhomogeneous situation, including bounded media as well as the infinite unbounded cases. In brief, the topics discussed here are the canonical scalar channel for the following:

Section 8.1	Propagation in ideal, unbounded media (except for distributed sources), including the associated Green's functions (GFs), and special solutions, as well as conditions on physical causality, extensions to boundedness, and the Generalized Huygens Principal.
Section 8.2	Equivalent time-variable (linear) filters, that is "engineering representations" of the classical space–time field and channel; and the conditions on the equivalence of these filters to the underlying physical quantities.
Section 8.3	Deterministic scatter from inhomogeneous media, with operational solutions, as a prelude to the treatment of the random channels of Chapter 9.
Section 8.4	The deterministic scattered field of Section 8.2, in wave number–frequency space.
Section 8.5	Extensions and innovations: global operators and integral equations, including multimedia and feedback formulations.
Section 8.6	Energy considerations.
Section 8.7	Summary remarks and next steps.

Accordingly, let us begin our journey in this chapter through the hierarchy of physical situations outlined above as prelude to those described in the subsequent Chapter 9.

8.1 THE GENERIC DETERMINISTIC CHANNEL: HOMOGENEOUS[4] UNBOUNDED MEDIA

We begin with a short formal overview of the generic canonical channel. The channel and its constitutive elements are here treated as nonrandom, or deterministic. This includes not only

[4] In customary mathematical usage "homogeneous" refers to a field with no sources present, that is, the field in a source-free region, whereas inhomogeneous denotes a field in the presence of sources, that is, a nonzero source (G_T). The partial differential equation describing the field in the former case is a *homogeneous* partial differential equation, whereas the latter is an *inhomogeneous* partial differential equation (PDE) (See also pp. 792, 793 of Ref. [1] for these terms applied to boundary conditions.)

However, here we shall use these terms to describe the physical properties of the medium in question: a medium whose properties are independent of position (and time), is "homogeneous", and one where the properties change with position and time is "inhomogeneous". The former is represented by a propagation equation with constant parameters while the latter has position-dependent parameters.

the medium itself but also the coupling to it by the aperture or arrays, in the manner of Fig. 8.1. Our initial aim is to develop canonical expressions for the channel, which by definition have a similar basic structure and are therefore not restricted to any particular physical application. This includes beam forming (in the manner of Section 2.5) for transmission and reception, homogeneous and inhomogeneous media, deterministic and ultimately random signal and noise fields in Chapter 9. For this we make liberal use of operator formulations. These are compact and emphasize the large-scale structure and interactions of the channel as well as providing a vehicle for numerical evaluation. Having done this we can go on in many cases (cf. [13,14]) to the more detailed analysis implied by these operational forms.

8.1.1 Components of the Generic Channel: Coupling

We consider first the coupling to the medium and the input signal source. This is represented by the aperture weighting (or Green's) functions. These are, in effect, (linear) space–time filters, h_T, h_R, which embody their associated "beam patterns", A_T, A_R, in the manner of Sections 2.5, namely, the double Fourier Transforms of h_T, h_R. The input signal source $S_{in}(t)$ and $\hat{G}_T$, the local aperture operator, is such that the *source density function* G_T per unit element $\mathbf{d\xi}$ of its associated aperture, at a point $\boldsymbol{\xi}$ is,

$$G_T(\boldsymbol{\xi}, t) = \hat{G}_T S_{in}(t) = \int_{-\infty}^{\infty} h_T(\boldsymbol{\xi}, t-\tau) S_{in}(\boldsymbol{\xi}, \tau) d\tau. \tag{8.1.1}$$

In some situations the aperture weighting is time-variable, namely, the *structure* of h_T changes with time (t), in addition to the "memory" (τ) of the filter, so that h_T (and similarly h_R) are now time-varying filters, with weighting functions $h_{T,R}(\xi; t, t-\tau)$. In a similar fashion the output of the receiving aperture h_R when $\alpha(\mathbf{R}, t)$ is the input field to the receiver, is given by

$$X(t) = \hat{R}\alpha(\mathbf{R}, t) = \int_{V_R(\eta)} dV_R(\boldsymbol{\eta}) \int_{-\infty}^{\infty} h_R(\boldsymbol{\eta}(\mathbf{R}); t-\tau, t) \alpha(\mathbf{R}, t) d\tau. \tag{8.1.2}$$

Here $\eta(\mathbf{R})$ embodies the spatial coördinates of the field with respect to a reference system located at the receiver. The quantity $\hat{R}$ is the aperture operator for the received signal:

$$\hat{R} \equiv \int_{V_R} dV_R(\boldsymbol{\eta}) \int_{-\infty}^{\infty} h_R(\boldsymbol{\eta}[\mathbf{R}]; t-\tau, t)(\,)_{\mathbf{R},\tau} d\tau, \tag{8.1.3}$$

and $dV(\boldsymbol{\eta})$ is a volume element, which can be replaced by a surface or line element, depending on the physical distribution of the aperture in space. A similar interpretation of $S_{in}(\boldsymbol{\xi}, \tau)$ in (8.1.1) applies, as well, depending on the distribution of the applied signal to the transmitting aperture represented by h_T. For apertures, as distinct from arrays, it is assumed that the applied signal is distributed to a continuum of spatial element $\mathbf{d\xi}$, $\mathbf{d\eta}$.

For arrays the situation is somewhat different: here they are approximated by point-elements, as described in Section 2.5.3. Thus, for transmission we have for the source producing the field $\alpha(\mathbf{R},t)$ external to the source,

$$\int_{V_T} \hat{G}_T S_{in}(\boldsymbol{\xi}, \tau) d\boldsymbol{\xi} \rightarrow \sum_{m=1}^{M} \int_{-\infty}^{\infty} h_T^{(m)}(\boldsymbol{\xi}_m; t-\tau, t) S_{in}(\boldsymbol{\xi}_m, \tau) d\tau, \tag{8.1.4}$$

with the $(m = 1, \ldots, M)$ point-filters, $h_T^{(m)}$, all within a domain $V_T^{(M)}$ in space. For reception, the continuous aperture operator $\hat{R}$ becomes the discrete element operator

$$\hat{R} \rightarrow \tilde{\mathbf{1}}\hat{\mathbf{R}}_M, \quad \text{with} \quad X_M(t) = \tilde{\mathbf{1}}\hat{\mathbf{R}}_M \alpha(\mathbf{R}, t) = \sum_{m=1}^{M} \int_{-\infty}^{\infty} h_T^{(m)}(\boldsymbol{\eta}_m(\mathbf{R}); t-\tau, t) \alpha(\mathbf{R}, \tau) d\tau, \tag{8.1.5}$$

where $\mathbf{1}$ is a column vector of $\mathbf{1}$'s and $\hat{\mathbf{R}}_M$ is the column vector $[\hat{R}_m = \int_{-\infty}^{\infty} h_R^{(m)} \alpha \, d\tau]$, as indicated above in Eq. (8.1.5). The beam patterns that these arrays produce depend of course on the placement of the M sensors in space and on their combination at some point in reception to produce $X(t)$, and in transmission to produce the field $\alpha(\mathbf{R}, t)$ in the medium. We shall illustrate this phenomenon of field generation for a variety of simple, common homogeneous media presently in Section 8.2.

We can represent the channel compactly by Fig. 8.1, in terms of the following (mostly) linear operators:

$$X(t) = \hat{\mathbf{T}}_{AR} \hat{\mathbf{T}}_M^{(N)} \hat{\mathbf{T}}_{AT}\{S_{in}\} = \hat{\mathbf{T}}_{AR}\{\alpha\} : \left\{ \begin{array}{l} \hat{\mathbf{T}}_{AT} = -\hat{G}_T \,;\; G_T = \hat{G}_T\{S_{in}\} \\ \hat{\mathbf{T}}_{AR} = \hat{R} \text{ (or } \hat{\mathbf{R}}) \\ \alpha = \hat{\mathbf{T}}_M^{(N)} \hat{\mathbf{T}}_{AT}\{S_{in}\} = \mathbf{T}_M^{(N)}\{-G_T\} \end{array} \right\}. \tag{8.1.6}$$

Accordingly, $\mathbf{X}'(t) = \hat{\mathbf{T}}_S\{X(t)\}$ is the discrete, that is (vector) input, to subsequent temporal signal processing when $X(t)$ is the (continuous) input to the samples.

8.1.2 Propagation in An Ideal Medium[5]

As a necessary introduction to the more realistic situations of propagation in nonideal media, where the nonideality is manifest by inhomogeneity (including boundaries), scattering, and randomness, we consider first the comparatively simple cases of the *Ideal Medium*. This is defined by the following properties: unbounded, deterministic, and homogeneity. (The medium is unbounded in the sense that an (outward) propagating wavefront never encounters a bounding surface.) When a source is included, such media remain "ideal", exclusive of the source, even if the source itself is stochastic and the resulting field is therefore random.

Accordingly, we begin by describing the propagation equations canonically, in terms of differential operators, these are specifically identified by the particular medium they represent, now for the time being restricted to the ideal channel and nonrandom signals. The channel in turn, includes the couplings to the medium by apertures or arrays, as

[5] The "classical" analytical foundations of this chapter are provided here principally by appropriate Chapters of Refs. [1–5]: specifically, by Chapter 7, [1], as well as Chapters 11 and 12 of Ref. [1]; also Section 8.11 of Ref. [5] and Chapters 8–10 of Ref. [2].

illustrated schematically in Fig. 8.1. The coupling operators are introduced in Sections 2.5 and are given specifically above by Eqs. (8.1.1)–(8.1.5). For the Ideal Medium we write canonically,[6]

$$\left.\begin{aligned} \hat{L}^{(0)}\alpha_{\mathrm{H}} &= -\hat{G}_{\mathrm{T}}S_{\mathrm{in}} = -G_{\mathrm{T}}(\neq 0 \text{ in } V_{\mathrm{T}}); \\ &= 0, \text{ elsewhere.} \end{aligned}\right]_{+\mathbf{E}} . \tag{8.1.7}$$

Here **E** represents the initial and boundary conditions which must be incorporated in the solution of (8.1.7) and which are determined by the specific physics of the application at hand.

The quantity α_{H} is the homogeneous field generated by the source density G_{T} and $\hat{G}_{\mathrm{T}}$ in the associated *source density operator*. We observe specifically from (8.1.1) that

$$G_{\mathrm{T}}(\mathbf{R},t) = \int_{-\infty}^{\infty} h_{\mathrm{T}}(t, t-\tau|\mathbf{R})S_{\mathrm{in}}(t,\mathbf{R})d\tau = h_{\mathrm{T}} \otimes S_{\mathrm{in}} \tag{8.1.8}$$

where $\otimes$ here denotes the convolution of the input applied at point **R** of the aperture weighting function h_{T}, and $\hat{G}_{\mathrm{T}} \equiv \int h \cdot (\,)_\tau d\tau$ is an aperture operator. The Ideal Medium operator $\hat{L}^{(0)}$ usually has the form (cf. the example in [(8.1.7) and 8.1.8a)][7]

$$\hat{L}^{(0)} = \hat{L}^{(0)}\left(\nabla^2, \frac{\partial^2}{\partial t^2}; \nabla, \frac{\partial}{\partial t}; a_1, \ldots, a_n\right), \tag{8.1.8a}$$

in rectangular coordinates, where the parameters $\mathbf{a} = a_1, \ldots, a_n$ are (real) constants independent of time and position, $\left(\mathbf{R} = \hat{\mathbf{i}}_1 x + \hat{\mathbf{i}}_2 y + \hat{\mathbf{i}}_3 z\right)$. We have written $\mathbf{E} = \text{I.C.} + \text{B.C.}$ to remind us here that specific solutions always require physically meaningful *initial conditions* (I.C.s) and *boundary conditions* (B.C.s). Of course, in the present situations of the unbounded Ideal Medium as we have defined it here, the boundaries are at S_0 and at S (or **R**) $\to \infty$ for no boundaries on the external field. The geometry of this source is sketched

[6] For a rigorous treatment of linear operators like these used in Chapters 8–9 and elsewhere in this book, see for example Kolmogoroff and Fomin [18], also Mukherjea and Pothoven, [19].

[7] From a purely mathematical viewpoint the physical propagation equations represented by (8.1.7), embodied operationally in $\hat{L}^{(0)}$, (8.1.7) and (8.1.8a), and given explicitly in Table 8.1 below, are general second-order linear partial differential equations. They take the form

$$F\left(\alpha; x_1, \ldots, x_K; \frac{\partial \alpha}{\partial x_1}, \ldots, \frac{\partial \alpha}{\partial x_K}\right) + \sum_{ij}^{K} a_{ij}(x_1, \ldots, x_K)\frac{\partial^2}{\partial x_i \cdot \partial x_j}\alpha(x_1, \ldots, x_K) = 0 \tag{I}$$

where the $x_i, i = 1, 2, \ldots, K$, are the independent variables. Here the $a_{ij}, i, j = 1, 2, \ldots, K$ are real, continuously differentiable functions of the $\boldsymbol{x}_i$, F is real, continuously differentiable in each of its arguments, and $\alpha = \alpha(x_1, \ldots, x_K)$ is the solution to (I). The a_{ij} are defined to be symmetric, that is, $a_{ij} = a_{ji}$. The classification of such equations (I) is made from selected properties of the *principal part* $\left(\equiv \sum_{ij}^{K} a_{ij}\partial^2\alpha/\partial x_i \partial x_j\right)$. It can be shown ([5], Section 8.11, pp. 498–501) that, according to the eigenvalues of the matrix $A = \left[a_{ij}\right]$ at any K-dimensional point $x_1', x_2', \ldots, x_K'$, the equation (I) is called

(a) *elliptic*, if all eigenvalues are positive, or negative;

(b) *hyperbolic*, if eigenvalues all have one sign, except one, which has the opposite sign;

(c) *parabolic*, when some eigenvalues are zero.

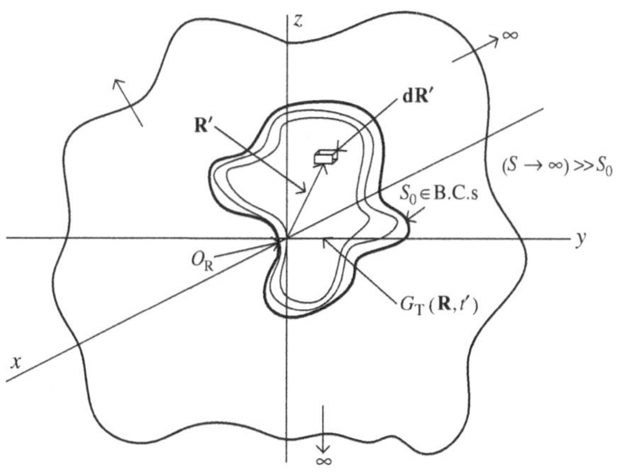

FIGURE 8.2 Unbounded ideal medium ($S \to \infty$) with a bounded source, G_T, for which boundary conditions and initial conditions apply, [$\mathbf{E} \equiv$ B.C. + I.C.], cf. Eq. (8.1.7) and Eq. (8.1.39) et. seq.

in Fig. 8.2. Equation (8.1.7) is technically an *inhomogeneous* propagation equation because of the presence of a source, but satisfies the homogeneous condition[8] outside the source regime ([1], p. 834). In Table 8.1 we list the most common types of propagation encountered in acoustic and electromagnetic[9] applications with their linear operators $\hat{L}^{(0)}$, which we use throughout the book. Here and elsewhere throughout this book, c_0 denotes a constant speed of propagation of a wave front in an ideal medium.

The relations in Table 8.1 all apply for rectangular coördinates in three dimensional spaces. The solution for the one- and two-dimensional cases are presented as problems at the end of the chapter. (We add Type 6 because of its close relation to 3, the wave equation with radiation absorption.)

8.1.3 Green's Function for the Ideal Medium

The Green's function (GF) for the ideal (i.e., unbounded) medium is defined as the *impulse response of this medium* to a four-dimensional "impulse" or delta function, in direct extension of the electrical engineering definitions for the weighting function of a linear time-invariant filter. Thus, the defining relation here is given by

$$\hat{L}^{(0)} g_\infty^{(0)} = -\boldsymbol{\delta}(\mathbf{R}' - \mathbf{R})\delta(t' - t) \equiv -\boldsymbol{\delta}_{\mathbf{R}'\mathbf{R}}\delta_{t't}, \tag{8.1.9a}$$

$$\therefore\ g_\infty^{(0)} = \hat{L}^{(0)-1}(-\boldsymbol{\delta}_{\mathbf{R}'\mathbf{R}}\delta_{t't}) \equiv \hat{M}_\infty \boldsymbol{\delta}_{\mathbf{RR}'}\delta_{tt'}, \tag{8.1.9b}$$

[8] In our terminology the presence of a boundary constitutes an inhomogeneity, since it can produce a scattered, albeit here a deterministic field.

[9] Types 1–3 apply for acoustical propagation, Types 1, 2 also for the scalar representations of electromagnetic waves, (like the dominant E-field component) as well as the propagation of sound in solids, fluids, and gases (Type 4), and related Types (5, 6) for diffusion problems. In all cases the parameters $\mathbf{a} = (a_1, a_{2,\ldots})$ have values appropriate to the media involved.

TABLE 8.1 Linear Differential Operators $\hat{L}^{(0)}$ for Common Types of Propagation Equations[a,b]

Type[c]	PDE's: $\hat{L}^{(0)}$	Classification	Parameters
1. *The time-dependent Helmholtz Eq.*	$\nabla^2 - \frac{1}{c_0^2}\frac{\partial^2}{\partial t^2}$	Hyperbolic	$a_1 = c_0^{-2}$
1a (Wave equation without absorption); *Harmonic or classical Helmholtz* Eq. ([1]; pp. 834–840) pp. 893–894; [5], pp. 511–514.	$\nabla^2 + k^2$	Elliptic	$a_1 = k^2$
2. *Wave equation with radiation absorption.* [1], pp. 865–	$\nabla^2 - a_R^2\frac{\partial}{\partial t} - \frac{1}{c_0^2}\frac{\partial^2}{\partial t^2}$	Hyperbolic	$a_1 = a_R^2$; $a_2 = c_0^{-2}$
3. *Wave equation with relaxation absorption*	$\left(1 + \hat{\tau}_0\frac{\partial}{\partial t}\right)\nabla^2 - \frac{1}{c_0^2}\frac{\partial^2}{\partial t^2}$	Hyperbolic	$a_1 = \hat{\tau}_0$; $a_2 = c_0^{-2}$
4. *Wave equation with relaxation and absorption*	$\left(1 + \hat{\tau}_0\frac{\partial}{\partial t}\right)\nabla^2 - a_R^2\frac{\partial}{\partial t} - \frac{1}{c_0^2}\frac{\partial^2}{\partial t^2}$	Hyperbolic	$a_1 = \hat{\tau}_0$; $a_2 = a_R^2$; $a_3 = c_0^{-2}$
5. *Diffusion equations* (Heat Flow, etc.) [1], Section 7.4; pp. 860–861, [5], pp. 516, 517.	$\nabla^2 - b_0^2\frac{\partial}{\partial t}$	Parabolic	$a_1 = b_0^2$
6. *Maximum velocity of heat absorption* ([1], pp. 865–869). (extension of 2.).	$\nabla^2 - a_H^2\frac{\partial}{\partial t} - \frac{1}{c_0^2}\frac{\partial^2}{\partial t^2}$	Hyperbolic	$a_1 = a_H$; $a_2 = c_0^{-2}$

[a] Rectangular coördinates, in 3 space dimensions. For example, $\nabla^2 = \left(\frac{\partial^2}{\partial x^2} + \frac{\partial^2}{\partial y^2} + \frac{\partial^2}{\partial z^2}\right)$, $\nabla = \left(\hat{\mathbf{i}}_1\frac{\partial}{\partial x} + \hat{\mathbf{i}}_2\frac{\partial}{\partial y} + \hat{\mathbf{i}}_3\frac{\partial}{\partial z}\right)$.

[b] These propagation equations are usually applicable *approximations*. By their derivative from conservation laws, the equations of state and other considerations, specific approximations are obtained. These are briefly discussed, for example, in Section 1.2.1 of Chew [6] for the acoustic cases. A much fuller treatment is given in Chapters 5 and 10 of Flaté [20]. The ocean is a well-known and important example of a highly inhomogeneous (and anisotropic) acoustic medium, particularly over long ranges and significant depths. These more complex cases require computational solutions. These, in turn, may be obtained in a first step from the initial operational formulations (or "macro algorithms") here in Chapter 8 (Sections 8.3 and 8.4) and particularly in Chapter 9 following. We shall also use throughout the notation $g_\infty^{(0)}, \hat{M}_\infty^{(0)}$, interchangeably with $g_\infty, \hat{M}_\infty$, when there is no ambiguity in meaning; see, for example, Section 8.3.3 ff. where $g_\infty^{(0)} \neq g_\infty^{(1)}, g_\infty^{(k)}$, and so on.

[c] See footnote 9.

where $\hat{M}_\infty$ is the inverse operator to the differential operator $\hat{L}^{(0)}$, that is $\hat{M}_\infty^{-1} = \hat{L}^{(0)}$ formally, where $g_\infty^{(0)}$ is the desired Green's function for the ideal medium.[10] Accordingly, $\hat{M}_\infty$ is an *integral operator, which now includes initial and boundary conditions* (**E**), to be described explicitly below, (cf. Section 8.1.6.3 and Table 8.3, ff.). Integrating (8.1.9b) over

[10] The subscript (∞) denotes the ideal, unbounded medium.

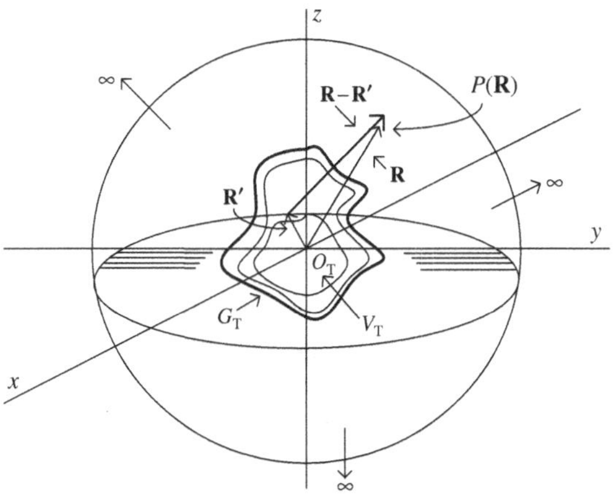

FIGURE 8.3 The radiative domain of the Green's function $g^{(0)}(\mathbf{R}, t|\mathbf{R}', t')$, is everywhere outside the space–time point $P(\mathbf{0}, 0)$ at O_T. The radiative domain of a distributed source G_T about O_T is the volume outside V_T, that $|\mathbf{R}| \geq \lim_{V\to\infty}(V - V_\mathrm{T})$ and $t > t_0-$.

$(\mathbf{R}', t')$ shows at once that $\hat{M}_\infty$ and $g_\infty^{(0)}$ are related by

$$\hat{M}_\infty^{(0)}(\mathbf{R}, t|\mathbf{R}', t') = \int_{-\infty}^{\infty} dt' \int_{-\infty}^{\infty} g_\infty^{(0)}(\mathbf{R}, t|\mathbf{R}', t')(\,)_{\mathbf{R}',t'}\mathbf{dR}' = \hat{L}^{(0)^{-1}};\ \text{with } \mathbf{dR}' \equiv dx'dy'dz'; t' \geq t_0^{(-)};$$
$$= 0, \quad t' < t_0^{(-)}; \tag{8.1.10}$$

where now the initial condition for g_∞ (and ∴ for $\hat{M}_\infty^{(0)}$) is 0, $\left(t < t_0^{(-)}\right)$, for the point source $-\delta_{\mathbf{R}'\mathbf{R}}\delta_{t't}$ producing this impulse response g_∞, Eq. (8.1.9a) of the medium. The initial condition represents *temporal causality*, whereby the medium can not respond to the point excitation before the excitation is initiated, clearly a reasonable (macroscopic) condition. From the above it is also clear that $\hat{M}_\infty$ is an *integral operator*, whose kernel is $g_\infty^{(0)}$, cf. Eqs. (8.1.9a and 8.1.10). Figure 8.3 shows the geometry of $g_\infty^{(0)}(\mathbf{R}, t|\mathbf{R}', t')$ and the domain of a source G_T.[11]

The importance of the Green's function lies in the fact that, knowing it and the initial and boundary conditions (I.C.s + B.C.s) *on the field one can directly obtain from* (8.1.7) *particular solutions for any arbitrary source distribution* G_T. Thus, from [(8.1.7), (8.1.8)] and (8.1.9b) one has the field α_H, subject to boundary conditions (B.C.s) and initial conditions (I.C.s) on the source function $G_\mathrm{T}(\mathbf{R}', t')$. We shall give the general result in Section 8.1.6. Note that the Green's Function[11] g_∞, being a δ-function *point* source, has no preset boundary conditions and an initial condition which depends only on a range of specific times (t'). Also, we observe finally, that the integral Green's function[11] (8.1.10) *can be viewed as a projection operator* $(\hat{M}_\infty)$, whereby all points $\mathbf{R}$ in the medium can receive the field generated by a distributed source (G_T). Thus, $\hat{M}_\infty$, in conjunction with the G_T, represent the field at *all* points $\left(|R| > V_\mathrm{T}, t' > t_0^{1-1}\right)$, produced by this source in V_T at time t'.

[11] Henceforth we shall drop the superscript (0) on $g_\infty^{(0)}$ and $\hat{M}_\infty^{(0)}$ until we have to distinguish between homogeneous and inhomogeneous media, beginning with Section 8.3 ff. and where appropriate in subsequent chapters.

The calculation of Green's functions from (8.1.9a) is readily carried out by Fourier and Laplace transforms. These can represent a spatial transform pair (cf. Section 2.5.1 et seq.):

$$B(\mathbf{k}) = \mathsf{F}_{\mathbf{R}}\{b(\mathbf{R})\} = \int_{-\infty(\mathbf{R})}^{\infty} e^{i\mathbf{k}\cdot\mathbf{R}} b(\mathbf{R})\mathbf{dR} \text{ and } b(\mathbf{R}) = \mathsf{F}_{\mathbf{k}}\{B(\mathbf{k})\} = \int_{-\infty(\mathbf{k})}^{\infty} e^{-i\mathbf{k}\cdot\mathbf{R}} B(\mathbf{k})\mathbf{dk}/(2\pi)^3, \tag{8.1.11}$$

with $\mathbf{dR} = dx\,dy\,dz$, $\mathbf{R} = \hat{\mathbf{i}}_1 x + \hat{\mathbf{i}}_2 y + \hat{\mathbf{i}}_3 z$, and $\mathbf{k} = \hat{\mathbf{i}}_1 k_x + \hat{\mathbf{i}}_2 k_y + \hat{\mathbf{i}}_3 k_z$ in rectangular coördinates. Temporal transforms are given as usual by

$$A(s) = F_t\{a(t)\} = \int_{-\infty}^{\infty} e^{-st} a(t)dt,\ \mathrm{Re}(s) \geq 0, t > 0-; \quad = 0, t \leq 0-, \tag{8.1.12a}$$

$$a(t) = F_s\{A(s)\} = \int_{-i\infty+d}^{i\infty+d} A(s)e^{s(t-t')}ds/2\pi i = \int_{Br_1(s)} A(s)e^{s(t-t')}ds/2\pi i, \tag{8.1.12b}$$

where $s = i\omega + c(\geq 0)$, $c < d$ and the straight line $(d - i\infty,\ d + i\infty)$ $d \geq 0$, defines the *Bromwich contour* (cf. Chapter IV of Ref. [21]).

By suitable choice of d this allows us to include transients, namely $a(t) > 0$ when $\left(t \geq t_0^-\right)$ and $a(t) = 0$, $t < t_0^-$. In steady-state regimes, in which $t' \to -\infty$, we choose $d = 0$ and use $s = i\omega = 2\pi i f$, with ω, f real frequencies, and with appropriate indentations of the Bromwich contour Br_1 to include any singularities to the left of and on $\mathrm{Re}(s) = 0$. In this case [(8.1.12a) and (8.1.12b)] is a Fourier transform pair. Otherwise, for $(t' > -\infty)$ it is a Laplace transform pair. Thus, the Bromwich contour (Br_1) combines steady state or transient behavior, by an appropriate selection of d, cf. Fig. 8.4.

Following the examples of [(8.1.11) and (8.1.12)] we can also obtain a general space–time transform pair by combining these equations and replacing $\mathbf{R}$ by $\mathbf{R} - \mathbf{R}' \equiv \boldsymbol{\rho}$ and t by $t - t' \equiv \tau$. This gives relations like

$$g_\infty = \mathsf{F}_{\mathbf{R}}\mathsf{F}_t\{g_\infty(\mathbf{R}, t|\mathbf{R}', t')\} \equiv g_\infty(\mathbf{k}, s|\mathbf{R}', t'). \tag{8.1.13a}$$

The inverse of the (double) transform of Eq. (8.1.13a) is thus explicitly

$$g_\infty(\mathbf{R}, t|\mathbf{R}', t') = \mathsf{F}_s\mathsf{F}_{\mathbf{k}}\{g_\infty(\mathbf{k}, s\ \mathbf{R}', t')\} = \int_{Br_1(s)}\int_{\mathbf{k}} g_\infty e^{-i\mathbf{k}\cdot(\mathbf{R}-\mathbf{R}')+s(t-t')}\mathbf{dk}|ds\Big/(2\pi)^4 i. \tag{8.1.13b}$$

Equation (8.1.9a) becomes now (see, e.g., pp. 513, 514 of Ref. [5]):

$$\begin{aligned}\hat{L}^{(0)} g_\infty &= \int_{\mathbf{k}}\int_{Br_1(s)} L_0(\mathbf{k}, s) g_\infty e^{-i\mathbf{k}\cdot(\mathbf{R}-\mathbf{R}')+s(t-t')}\mathbf{dk}ds/(2\pi)^4 i \\ &= -\iint e^{-i\mathbf{k}\cdot(\mathbf{R}-\mathbf{R}')+s(t-t')} \cdot \frac{\mathbf{dk}ds}{(2\pi)^4 i}\end{aligned} \tag{8.1.14}$$

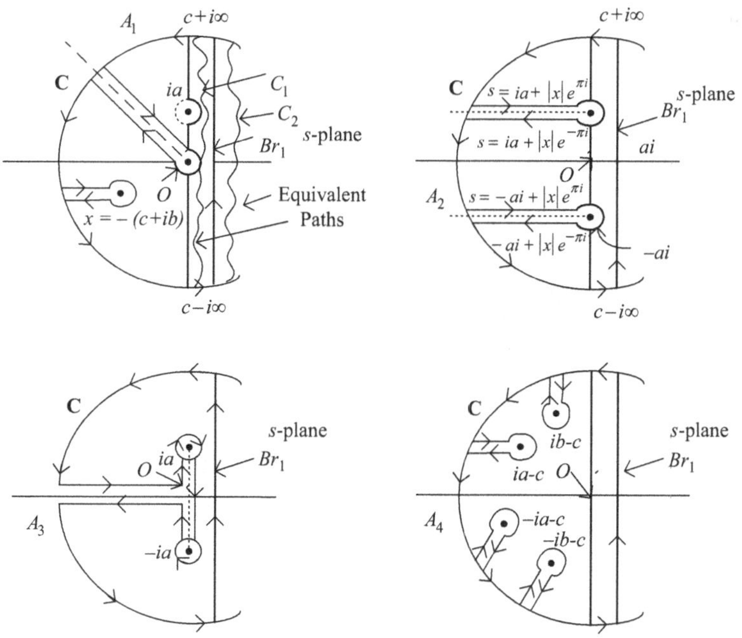

FIGURE 8.4 Bromwich contours Br_1 for a variety of equivalent contours (**C**) (cf. [21]); $\mathbf{A}_1$, simple pole on imaginary s-axis, with branch point at $s=0$ and simple pole at $-(c+ib)$; $\mathbf{A}_2$, double branch points at $s=\pm\, ia$; $\mathbf{A}_3$, double branch points at $s=\pm\, ia$; $\mathbf{A}_4$, simple poles at $\pm\, ia-c$ and $\pm\, ib-d$. (The equivalent contours $\mathbf{C}+Br_1=0$ because $Br_1=-\mathbf{C}$, with branch barriers given by the dotted lines to the branch points at $s=0,\ \pm\, ia$.)

Here $\hat{L}_0=\hat{L}^{(0)}(\nabla,\nabla^2,\ldots,\partial/\partial t,\partial^2/\partial t^2,\ldots)\Rightarrow L_0(\mathbf{k},s)$, which is obtained by differentiating $e^{-i\mathbf{k}\cdot\mathbf{R}+st}$ in the integrand of (8.1.14) and is the *coefficient* of $\exp(-i\mathbf{k}\cdot\mathbf{R}+st)$, that is,[12]

$$\hat{L}_0\left(\nabla,\nabla^2,\ldots;\frac{\partial}{\partial t},\frac{\partial^2}{\partial t^2},\mathbf{k}\right)e^{-i\mathbf{k}\cdot\mathbf{R}+st}\Rightarrow L_0\big(-i\mathbf{k},-k\cdot k,\ldots;s,s^2,\ldots\big)\equiv L_0(\mathbf{k},s). \tag{8.1.15}$$

Accordingly, we obtain from (8.1.14)

$$\begin{aligned} g_\infty &= -\int_{\mathbf{k}}\int_{s} e^{-i\mathbf{k}\cdot(\mathbf{R}-\mathbf{R}')+s(t-t')}\frac{\mathbf{dk}ds}{(2\pi)^4 i} \\ &= \int_{\mathbf{k}}\int_{Br_1(s)} L_0(\mathbf{k},s)g_\infty(\mathbf{k},s|\mathbf{R}',t')_\infty e^{-i\mathbf{k}\cdot(\mathbf{R}-\mathbf{R}')+s(t-t')}\frac{\mathbf{dk}ds}{(2\pi)^4 i}. \end{aligned} \tag{8.1.16}$$

On comparing integrands in (8.1.16) we see from (8.1.14), (8.1.13a) and (8.1.13b) that, on setting

$$\mathsf{Y}_0(\mathbf{k},s)\equiv -L_0(\mathbf{k},s)^{-1},\ \text{and}\ \therefore\ g_\infty(\mathbf{R},t|\mathbf{R}',t')=\mathsf{F}_{\mathbf{k}}\mathsf{F}_s\{Y_0(\mathbf{k},s)\}, \tag{8.1.17a}$$

[12] We note that $L_0(\mathbf{k},s)$ is the result of applying the operator $\hat{L}_0$ to a given function (here exp $(\sim i\mathbf{k}\cdot\mathbf{R}+st)$ and is thus itself a function, *not* an operator.)

or in detail

$$g_\infty(\mathbf{R}, t|\mathbf{R}', t') = \int_{Br_1(s)} e^{s(t-t')} \frac{ds}{2\pi i} \int_{\mathbf{k}} e^{-i\mathbf{k}\cdot(\mathbf{R}-\mathbf{R}')} [\mathrm{Y}_0(\mathbf{k}, s)] \frac{d\mathbf{k}}{(2\pi)^3} = g_\infty(\mathbf{R}-\mathbf{R}', t-t'), \tag{8.1.17b}$$

where now specifically the (double) Fourier transform $F_{\mathbf{k},s}$ here for these *Hom-Stat* media are with respect to $\boldsymbol{\rho} \equiv \mathbf{R} - \mathbf{R}'$ and $\Delta t \equiv t - \tau'$.

Equation (8.1.17b) is the general expression for determining the Green's function for these ideal media, and the one we shall use in Section 8.1.5 to obtain the specific results of Table 8.2. The corresponding *integral Green's function* operator $\hat{M}_\infty$, Eqs. (8.1.9b) and (8.1.10) is explicitly here

$$\hat{M}_\infty(\mathbf{R}, t|\mathbf{R}', t') = \int_{t_0^{(-)}}^{\infty} dt' \int_{Br_1(s)} \frac{e^{s(t-t')} ds}{2\pi i} \int_{-\infty}^{\infty} d\mathbf{R}'(\ )_{\mathbf{R}', t'} \int_{\mathbf{k}} e^{-i\mathbf{k}\cdot(\mathbf{R}-\mathbf{R}')} [Y_0(\mathbf{k}, s)] \frac{d\mathbf{k}}{(2\pi)^3}, \tag{8.1.18}$$

$$= \hat{M}_\infty(\mathbf{R}-\mathbf{R}', t-t'). \tag{8.1.18a}$$

For a source density $G_T(\mathbf{R}, t)$, Eq. (8.1.8), we have at once the resulting homogeneous field $\alpha(\mathbf{R}, t)_H = \hat{M}_\infty(-G_T)$ in this medium. In fact, we see *from the defining relations* [(8.1.9a) and (8.1.9b)], *for a (here deterministic) homogeneous and stationary* (i.e. Hom-Stat) *medium*[13] *the Green's function (and its integral operated form* $\hat{M}_\infty$*) are always a function of the differences* $\boldsymbol{\rho} = \mathbf{R} - \mathbf{R}'$, $\Delta t = t - t'$. Specific examples are given presently in Section 8.1.5 ff. (This is also true for random media which are *Hom-Stat*, cf. Section 9.1.2 ff.)

The possible presence of the gradient operator ∇ (and odd powers of ∇) in these scalar equations of propagation in some such form as $\mathbf{a}_0 \cdot \nabla$, and so on, which produces $\mathbf{a}_0 \cdot \mathbf{k}$ in $\mathrm{Y}_0(\mathbf{k}, s)$, represents an *inhomogeneous element*. (Since the media here are postulated to be homogeneous, we exclude the presence of ∇, etc.)

Equations (8.1.17b) and (8.1.18) can be put into more convenient equivalent forms for evaluation if we now employ a form of spherical coordinates $\mathbf{k} = (k, w(= \cos\theta), \phi)$. The result is

$$g_\infty(\mathbf{R}, t|\mathbf{R}', t') = \int_{Br_1(s)} e^{s(t-t')} \frac{ds}{2\pi i} \cdot \frac{1}{(2\pi)^3} \int_0^\infty k^2 dk \int_{-1}^{+1} dw \int_0^{2\pi} e^{-ik|\mathbf{R}-\mathbf{R}'|w} [Y_0(k, s)] d\phi. \tag{8.1.19}$$

With the extension of the limits on k to $(-\infty \le k \le \infty)$, we have directly

$$g_\infty(\mathbf{R}, t|\mathbf{R}', t') = \int_{Br_1(s)} e^{s\Delta t} \frac{ds}{2\pi i} \cdot \int_{-\infty}^{\infty} \frac{k^2 dk}{(2\pi)^2} \left(\frac{e^{ik\rho} - e^{-ik\rho}}{ik\rho} \right) \mathrm{Y}_0(\mathbf{k}, s), \tag{8.1.20}$$

[13] From a probabilistic point of view a deterministic medium may be considered as constituting an ensemble of *one* representation.

where

$$\rho \equiv |\mathbf{R} - \mathbf{R}'|;\ \Delta t \equiv t' - t. \tag{8.1.20a}$$

8.1.3.1 Example: The Helmholtz Equation in an Ideal Unbounded Medium As a useful example, let us evaluate g_∞ for the *time-dependent Helmholtz equation* (No. (1) in Table 8.1). Its Green's function is found as follows: from (8.1.15) we have

$$\nabla^2 - \frac{1}{c_0^2}\frac{\partial^2}{\partial t^2} \to -\Upsilon_0\big|_{\text{Helm}} = -\mathbf{k}\cdot\mathbf{k} - s^2/c_0^2 = -\left(k^2 + s^2/c_0^2\right). \tag{8.1.21}$$

Equation (8.1.20) becomes specifically on integration over w, with the additional observation that the k-integrand is even in k:

$$g_\infty(\mathbf{R}, t|\mathbf{R}', t')_{\text{Helm}} = \frac{1}{(2\pi)^2 i\rho}\int_{Br_1(s)} e^{s\Delta t}\frac{ds}{2\pi i}\int \frac{1}{\mathbf{C}_k k^2 + s^2/c_0^2} k e^{-ik\rho}\,dk, \tag{8.1.22}$$

cf. (8.1.17b). At this point we can evaluate either the k- or the s-integral first. For the former or the latter, we have

$$\frac{1}{k^2 + s^2/c_0^2} = \frac{1}{(k - is/c_0)(k + is/c_0)} \text{ or } \frac{c_0^2}{(s - ikc_0)(s + ikc_0)}. \tag{8.1.23}$$

In each case the respective poles at $k = -is/c_0$ and $s = ik_0 c_0$ around $\mathbf{C}$ are the ones that yield "outgoing" waves and with the condition $\Delta t < 0$ also represent space–time causality.[14] Applying (8.1.23) to (8.1.22) gives us after a little manipulation the well-known result

$$g_\infty(\mathbf{R}, t|\mathbf{R}', t')_{\text{Helm}} = \frac{1}{4\pi\rho}\left\{\begin{matrix} \displaystyle\int_{-\infty}^{\infty} e^{s(\Delta t - \rho/c_0)}\frac{ds}{2\pi i}, & ① \\ \displaystyle\int_{-\infty}^{\infty} e^{ikc_0(\Delta t - \rho/c_0)}\frac{c_0 dk}{(2\pi)}, & ② \end{matrix}\right\} = \frac{\delta(\Delta t - \rho/c_0)}{4\pi\rho}. \tag{8.1.24}$$

Here Fig. 8.5a and b illustrate the contours $\mathbf{C}_k$ of integration.

We observe that the first resolution of the integrand in (8.1.23) into the poles of k gives $k = \omega/c_0$ for the required outgoing propagation. This is the linear relation between k and ω characteristic of *nondispersive* media. In fact, when k is nonlinear in ω, our outgoing wave solutions for k indicate the *dispersive* but still homogeneous nature of the media. See examples (2)–(6), in Table 8.2.

[14] In Fig. 8.5a, for the contribution of the simple poles on $\mathbf{C}(-\infty i, \infty i)$, the first at $k = -is/c_0$ integrates in the counter clockwise direction (+) to $i\int_\pi^0 d\theta = -\pi i$ and the second, at $k = -is/c_0$, integrates in the clockwise direction $(-) + 0 - \int_0^\pi i d\theta = -\pi i$, for a total of $-2\pi i$. The total contribution is $(\mathbf{C}_k - 2\pi i) + \mathbf{C}' = 0$ or $\mathbf{C}_k = 2\pi i$, since $\mathbf{C}'$ vanishes at infinity. Moreover, we see that the Bromwich contour Br_1, cf. Fig. 8.4, requires that all singularities in $\mathbf{s}$ lie to its left. This ensures that transients $\left(t \geq t_0^-\right)$ are included, as well as steady-state results $(-\infty < t < \infty)$.

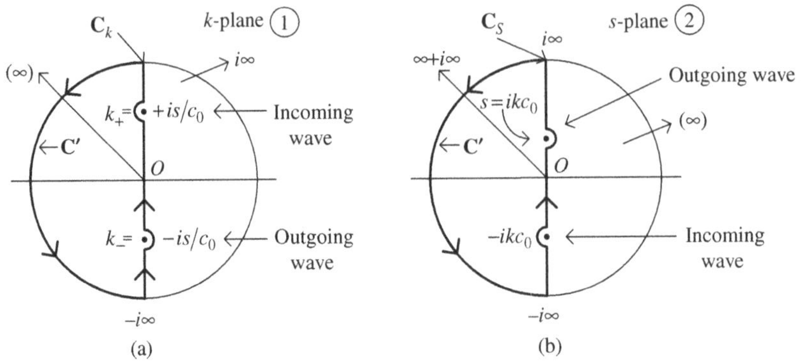

FIGURE 8.5 (a and b) Contours of integration $\mathbf{C}_k, \mathbf{C}_s$ for causality: $(\Delta t - \rho/c_0) \geq 0$, for the outgoing wave in (8.1.23) for the Helmholtz equation, cf. 1, Table 8.1.

8.1.4 Causality, Regularity, and Reciprocity of the Green's Function

The causality condition on g_∞ exhibited by the choice of poles (8.1.23) et. seq. in our Helmholtz example above can be alternatively expressed by the relations:

(i) *Space–Time Causality or Radiation Condition:*[15]

$$\lim_{R \to \infty} R\left(\hat{\mathbf{n}} \cdot \nabla - \frac{\partial}{c_0 \partial t}\right) g_\infty = 0, \; R = |\mathbf{R}|, \tag{8.1.25}$$

which ensures that only outgoing waves from the source $(\sim G_\mathrm{T})$ are propagated in this unbounded medium. Generally, this causality condition is obtained by appropriately modifying the contours in the complex k- or s-planes—the Helmholtz case above ((8.1.22)b) et seq.) is an example, cf. Fig. 8.5 to ensure only the singularities that yield the required outgoing waves. Other conditions and properties are

(ii) *Regularity at Infinity:*[15]

This ensures that g_∞ vanishes as $R \to \infty$ in such a way that

$$\lim_{R \to \infty} R g_\infty \to 0, \tag{8.1.26}$$

(which is not necessarily guaranteed by (8.1.25).) This condition is a reflection of the fact that (here) the point source has finite energy and consequently its field, generated by g_∞, must also, for all $R \equiv |\mathbf{R}|$.

There are two other conditions on g_∞:

(iii) *Uniqueness:*

At all points exterior to a closed surface S, g_∞ satisfies

$$\hat{L}^{(0)} g_\infty = -\delta_{\mathbf{R}'\mathbf{R}} \delta_{t't}, \tag{8.1.27}$$

[15] Ref. [3], the extension of Eq. (17) from Eq. (20).

and g_∞ obeys the homogeneous boundary conditions

$$ag_\infty + b\,\hat{\mathbf{n}} \cdot \nabla g_\infty = 0. \tag{8.1.28}$$

(iv) This is also seen from the fact that it is reasonable to postulate that the *initial conditions*[16] are $g_\infty = 0$, $\partial g_\infty/\partial t = 0$, t (and t') ≤ 0. The boundary conditions become $g_\infty = \hat{\mathbf{n}} \cdot \nabla g_\infty = 0$, as well here.

(Therefore, conditions (i) – (iv) also determine the uniqueness of the field α_{H} consequent upon the Green's function g_∞, and B.C.s and I.C.s.)

Finally, there is another important consequence of the causality condition on the Green's function:

(v) *Reciprocity:*[17]

$$g_\infty(\mathbf{R}, t|\mathbf{R}', t') = g_\infty(\mathbf{R}', -t'|\mathbf{R}, -t), \tag{8.1.29}$$

which in addition requires g_∞ to satisfy the initial conditions (iv), namely,

$$g_\infty(\mathbf{R}, t|\mathbf{R}', t') = 0; \ \frac{\partial g_\infty}{\partial t} = 0; \ t \leq t'(-) \tag{8.1.29a}$$

Reciprocity (8.1.29) is determined by (1) the fact that the *same* boundary conditions apply to the Green's functions in (8.1.29) and (2) by causality for the initial conduits (8.1.29a) cf. [5].[17] Simply stated, (8.1.29) allows one to interchange the location of the impulsive source with that of the field at point **R**, that is, ($\mathbf{R}' \to \mathbf{R}$, $\mathbf{R} \to \mathbf{R}'$), with the respective times reversed, that is, $t' \to -t$ and $t \to -t'$.

We remark that some of the propagation equations listed in Table 8.1 (and for the associated media themselves) are not invariant under a time reversal ($t = -t'$) but do distinguish between past and future, that is, $G(\mathbf{R}, t|\mathbf{R}', t') = 0, t \leq t'-$. This is characteristic of diffusion-dominated propagation, for example, Items 4 and 5 in Table 8.1. They still obey reciprocity, Eq. (8.1.29), however, since it embodies causality.

8.1.5 Selected Green's Functions

For the media types cited in Table 8.1 above we may still apply the general relation $L_0(\mathbf{k}, s)$ to the more general results (8.1.19) and (8.1.20). Similar contour selections to those used in (8.1.21)–(8.1.24), although more complex in the cases involving branch points as well as simple poles, yield the specific results given below. These include not only the desired Green's functions but also their space- and time-Laplace transforms.[18]

[16] For the $(1/c_0^2)\partial^2/\partial t^2$ and $\partial/\partial t$ terms we use the reciprocity relation (8.1.29), representing causality, for example, $g_\infty(\mathbf{R}, t|\mathbf{R}', t') = g_\infty(\mathbf{R}', -t'|\mathbf{R}, t)$, where now $-t'$ is earlier than $-t$. Accordingly, $g_\infty^{(0)}(\mathbf{R}', -t'|\mathbf{R}, t)$ and $\partial g^{(0)}/\partial t = 0$, and only the lower limit $t = t_0$ is possibly nonzero, representing nonzero initial conditions. For the details see [2], pp. 835–837.

[17] For [5], see pp. 511, 512, Eqs. (8.176)–(8.179) and for proofs and discussion; also [1], pp. 835, 836.

[18] Most of the details are available in Sections 7.3 and 7.4 of Ref. [1], along with the table, pp. 890–894 [1]. See also [5], Section 8.11, p. 498, with examples **a** – **e**, pp. 501–517, and Table 8.3, p. 529, references, p. 530. An appropriate use of the well-known Table 1 of Campbell and Foster [22] is also recommended, as are the tables of the Bateman Manuscript Project [23].

Our summary of results here, for the media listed in Table 8.1, accordingly becomes for these unbounded cases.[19]

(1) *Time-Dependent Helmholtz Equation*:[20]

$$\left(\nabla^2 - \frac{1}{c_0^2}\frac{\partial^2}{\partial t^2}\right) g_\infty = -\delta(\mathbf{R}' - \mathbf{R})\delta(t' - t) \equiv -\delta_{\mathbf{RR}'}\delta_{tt'}. \tag{8.1.30a}$$

$$-L_0^{-1} = \mathrm{Y}_0(\mathbf{k}, s) = \left(k^2 + s^2/c_0^2\right)^{-1}; \quad Y_0(\mathbf{R}, s|\mathbf{R}')$$
$$= Y_0(\rho, s) = \frac{e^{-s\rho/c_0}}{4\pi\rho}; \quad \rho \equiv |\mathbf{R} - \mathbf{R}'|, \Delta t = t - t'. \tag{8.1.30b}$$

$$g\left(\mathbf{R}, t|\mathbf{R}', t'\right)_\infty \equiv G(\rho, \Delta t)_\infty = \delta(\Delta t - \rho/c_0)/4\pi\rho, \Delta t \geq 0. \tag{8.1.30c}$$

(2) *Wave Equation with Radiation Absorption*:

$$\left(\nabla^2 - a_R^2\frac{\partial}{\partial t} - \frac{1}{c_0^2}\frac{\partial^2}{\partial t^2}\right) g_\infty = -\delta_{\mathbf{RR}'}\delta_{tt'}. \tag{8.1.31a}$$

$$-L_0^{-1} = \mathrm{Y}_0(\mathbf{k}, s) = \left(k^2 + s^2/c_0^2 + a_R^2 s\right)^{-1}; Y_0(\rho, s) = e^{-(\rho s/c_0)\left(1 + c_0^2 a_R^2/s\right)^{1/2}} \Big/ 4\pi\rho \tag{8.1.31b}$$

$$g(\mathbf{R}, t|\mathbf{R}', t')_\infty \equiv G(\rho, \Delta t)_\infty = \frac{e^{-\left(\frac{1}{2}\right)a_R^2 c_0^2 \Delta t}}{4\pi\rho}\left\{\delta(\Delta t - \rho/c_0) + \right.$$
$$\left. + \frac{a_R^2\rho}{2\sqrt{(\rho/c_0)^2 - \Delta t^2}} J_1\left(\frac{a_R^2 c_0^2}{2}\sqrt{(\rho/c)^2 - \Delta t^2}\right) 1_{\Delta t - \rho/c_0}\right\} 1_{\Delta t},$$

or equivalently (8.1.31c)

$$G(\rho, \Delta t)_\infty = \frac{e^{-(1/2)a_R^2 c_0^2 \Delta t}}{4\pi\rho}$$
$$\times \left\{\delta(\Delta t - \rho/c_0) + \frac{a_R^2 c\rho}{2\sqrt{\Delta t^2 - (\rho/c_0)^2}} I_1\left(\frac{a_R^2 c_0^2}{2}\sqrt{\Delta t^2 - (\rho/c_0)^2}\right) \cdot 1_{\Delta t - \rho/c_0}\right\} 1_{\Delta t}, \tag{8.1.31d}$$

([1], pp. 866–868 or pair no. 863.1, $\sigma = 0$, of Ref. [23])

[19] We remark that it is generally simpler to integrate over k first, and then s, as we can see from the results of 1–6 below. Note, however, that because of the causality condition one (or more) of the poles, branch points, and so on, which lie on the k-axis $(-i\infty, i\infty)$, must be excluded. The notation $1_{\Delta t}$ denotes a unit step function, with $\Delta t \geq 0-$. Similarly, 1_{a-b} = unity for $(a - b) \geq 0-$, and so on.

[20] Note again that $L^{(0)}$ here is a specific *function*, not the operator $\hat{L}^{(1)}$, cf. footnote for Eq. (8.1.15).

(3) *Wave Equation with Relaxation Absorption*:

$$\left(\left\{1+\hat{\tau}_0\frac{\partial}{\partial t}\right\}\nabla^2-\frac{1}{c_0^2}\frac{\partial^2}{\partial t^2}\right)g_\infty=-\delta_{\mathbf{RR}'}\delta_{tt'}. \tag{8.1.32a}$$

$$-L_0^{-1}=\mathrm{Y}_0(\mathbf{k},s)=\left[(1+\hat{\tau}_0 s)k^2+s^2\big/c_0^2\right]^{-1};\, Y_0(\rho,s)=\frac{e^{-(\rho s/c_0)(1+\hat{\tau}_0 s)^{-1/2}}}{4\pi\rho(1+\hat{\tau}_0 s)} \tag{8.1.32b}$$

$$g_\infty\equiv G(\rho,\Delta t)_\infty=\frac{1}{4\pi\rho}\left\{1-\frac{1}{\sqrt{\pi}}\sum_{n=0}^{\infty}\frac{(-1)^n(1/2)_n}{(2n+1)!}\left(\frac{\rho}{c_0\sqrt{\hat{\tau}_0\Delta t}}\right)^{2n+1}\right.$$
$$\left.\cdot\, e^{-\Delta t/\hat{\tau}_0}\,{}_1F_1(-1-2n;1/2-n;\Delta t/\hat{\tau}_0)\right\}1_{\Delta t} \tag{8.1.32c}$$

(use pair no. 581.4 of Ref. [23].)

(4) *Wave Equation with Relaxation and Radiation Absorption*:

$$\left[\left(1+\hat{\tau}_0\frac{\partial}{\partial t}\right)\nabla^2-a_{\mathrm{R}}^2\frac{\partial}{\partial t}-\frac{1}{c_0^2}\frac{\partial^2}{\partial t^2}\right]g_\infty=-\delta_{\mathbf{RR}'}\delta_{tt'}. \tag{8.1.33a}$$

$$-L_0^{-1}=\mathrm{Y}_0(\mathbf{k},s)=\left[(1+\hat{\tau}_0 s)k^2+a_{\mathrm{R}}^2 s+s^2/c_0^2\right]^{-1};\quad Y_0(\rho,s)$$
$$=\left\{\frac{\exp\left[\left(-(\rho s/c_0)(1+(a_{\mathrm{R}}^2c^2/s)/(1+\hat{\tau}_0 s)\right)^{1/2}\right]}{4\pi\rho(1+\hat{\tau}_0 s)}\right\} \tag{8.1.33b}$$

$$g_\infty\equiv G(\rho,\Delta t)_\infty \tag{8.1.33c}$$

(5) *The Diffusion Equation*: $$\left(\nabla^2-b_0^2\frac{\partial}{\partial t}\right)g_\infty=-\delta_{\mathbf{RR}'}\delta_{tt'}. \tag{8.1.34a}$$

$$-L_0^{-1}=\mathrm{Y}_0(\mathbf{k},s)=\left(k^2+b_0^2 s\right)^{-1};\; Y_0(\rho,s)=\frac{e^{-b_0\rho\sqrt{s}}}{4\pi\rho}; \tag{8.1.34b}$$

$$\therefore\; g_\infty=G(\rho,\Delta t)_\infty=\frac{b_0}{8(\pi\Delta t)^{3/2}}e^{-b_0^2\rho^2/4\Delta t}1_{\Delta t}. \tag{8.1.34c}$$

((Eq. 8.188, [5], p. 517; [1], p. 860, 861))

(6) *Diffusion: Maximum Velocity of Heat Absorption*: See (**2**) above.

The quantities Y_0 and Y_0 are the kernels of the following Fourier and Laplace transforms.[21] Thus, we have from (8.1.13a)–(8.1.22), when $\mathbf{k}=k$ and we use (8.1.19) and (8.1.20a):

$$G(\rho,\Delta t)_\infty = \mathsf{F}_s\mathsf{F}_{\mathbf{k}\to k}\left\{\mathsf{Y}_0(\mathbf{k},s)e^{-st'+i\mathbf{k}\cdot\mathbf{R}'}\right\} = \int_{Br_1(s)} e^{s\Delta t}\frac{ds}{2\pi i}\int_{-\infty}^{\infty}\mathsf{Y}_0(k,s)e^{-ik\rho}\frac{kdk}{(2\pi)^2 i\rho}; \tag{8.1.36a}$$

$$G(\rho,\Delta t)_\infty = \mathsf{F}_s\left\{Y_0(\rho,s)e^{-st'}\right\} = \int_{Br_1(s)}^{0} e^{s\Delta t}Y_0(\rho,s)ds/2\pi i;\ \rho\equiv|\mathbf{R}-\mathbf{R}'|;\ \Delta t = t-t'; \tag{8.1.36b}$$

$$Y_0(\rho,s) = \mathsf{F}_k\left\{\mathsf{Y}_0(k,s)k^2e^{ik\mathbf{R}'}\right\} = \int_{-\infty}^{\infty}\mathsf{Y}_0(k,s)e^{-ik\rho}\frac{kdk}{(2\pi)^2 i\rho} \tag{8.1.36c}$$

with the inverses

$$\mathsf{Y}_0(k,s)k^2e^{-st'} = \int_{-\infty}^{\infty} e^{ik\rho}\frac{d\rho}{2\pi}\int_{-\infty}^{\infty} e^{-st}G(\rho,\Delta t)_\infty dt = \mathsf{F}_\rho\mathsf{F}_t\{G(\rho,\Delta t)_\infty\}; \tag{8.1.37a}$$

$$Y_0(\rho,s)e^{-st'} = \int_{-\infty}^{\infty} e^{-st}G(\rho,\Delta t)_\infty dt = \int_{t'\le t-}^{\infty} e^{-st}G(\rho,\Delta t)_\infty dt = \mathsf{F}_t\{G(\rho,\Delta t)_\infty\}; \tag{8.1.37b}$$

$$Y_0(\rho,s) = \mathsf{F}_{\mathbf{R}}\{G(\rho,\Delta t)_\infty\} = \int_{-\infty}^{\infty} e^{ik\rho}G(\rho,\Delta t)\frac{d\rho}{2\pi}. \tag{8.1.37c}$$

Note that the Green's function here is always a function of $\rho = |\mathbf{R}-\mathbf{R}'|$. This is a direct consequence of the postulated *homogeneity of the medium*, and in fact represents an even stronger property, namely *isotropy*. These are established analytically by the symmetrical nature of the point-source $(\delta_{\mathbf{RR}'}\delta_{tt'})$, generally the basic response of the medium. When $k\to\mathbf{k}$, that is, when $L_0 = L_0(\mathbf{k},s)$, (cf. remarks following Eq. (8.1.20)), we have the general relations (8.1.13a), which are given by (8.1.19) in a form of rectangular coördinates. The specific results above (8.1.30)–(8.1.37) are much more complex. Since the Green's function $G(\rho,\Delta t)_\infty$ and the fields resulting from them are all real, we have the additional relations for

[21] Depending on d in the Bromwich contour $Br_1(s)$, (8.1.12b) and discussion following.

their transforms (8.1.36) and (8.1.37)

$$Y_0(\rho, s) = Y_0(\rho, -s)^* \quad \text{and} \quad \mathsf{Y}_0(\mathbf{k}, s) = \mathsf{Y}_0(-\mathbf{k}, -s)^*. \tag{8.1.37d}$$

8.1.5.1 Dispersion I Next, we observe for those unbounded homogeneous (and isotropic) media that solving $\mathsf{Y}_0(\mathbf{k}, s)$ for k_{singular} associated with the *outgoing wave*, that is, the causal condition in these physical problems, gives us one measure of the medium's *dispersion*. The dispersion here, of course, is the result of energy dissipation in the medium, represented analytically by the presence of the operator $(\partial/\partial t)$ in $\hat{L}^{(0)}$, cf. Eq. (8.19) and Table 8.1 and correspondingly, in the Green's functions (8.1.31b)–(8.1.34b), and Table 8.2. The result specifically is to spread the wave number–frequency spectrum $\mathsf{Y}_0(\mathbf{k}, s)$ vis-à-vis that of the nondissipative medium. Mathematically, this type of dispersion is represented by a *nonlinear* relation between (angular) wave number k and (angular) frequency ω, namely,

$$k = F(\omega)/c_0; F(\omega) \neq \omega \tag{8.1.38}$$

for otherwise homogeneous isotropic media, which without dissipation support a linear relation, that is, $k = \omega/c_0$. Media for which (8.1.38) applies are termed *frequency dispersive*. (For further discussion see Section 8.3.4.3.)

Table 8.2 summarizes the previous properties and (some of) the areas of application of these propagation equations of the resulting field $\alpha_H = \hat{M}_\infty(-G_T)$, Eq. (8.1.11), obtained from the appropriate Green's function above and described in Section 8.1.6 immediately following.

8.1.6 A Generalized Huygens Principle: Solution for the Homogeneous Field $\boldsymbol{\alpha}_H$

In the simplest terms, the Huygens Principle assumes that each point on a given wavefront behaves like a point source that in turn radiates a spherical wave, traveling at a speed c_0. The resulting field at a specified point at a later time is then the sum of the fields of these individual point sources. The envelope of the waves from all such synchronous points constitutes the next wave front. Nothing, however, is said specifically about the original source itself and its *associated boundary and initial conditions* C(B.C.s and I.C.s). Here we take these into account. The result we call a *Generalized Huygens Principle* (GHP), which now gives us the complete solution for the field $\alpha_H(\mathbf{R}, t)$ in this infinite homogeneous medium, produced by the source (density) G_T distributed about the origin O_T in the manner of and contained by, the surface S_0, cf. Fig. 8.6 below.

8.1.6.1 GHP for the Field $\boldsymbol{\alpha}_H$ We begin by incorporating the boundary and initial conditions (C) explicitly in the solution of its defining relation (8.1.9a). Its reciprocal property is a consequence, as we have seen (Eq. (8.1.29), of *causality* and the fact that both $g_\infty(\mathbf{R}, t|\mathbf{R}', t')$ and its reciprocal $g_\infty(\mathbf{R}', -t'|\mathbf{R}, -t)$ obey the same boundary conditions.[22] Now by a similar method used to establish reciprocity of the Green's function, we begin with

[22] For details, see pp. 835, 836 of [1].

TABLE 8.2 Types of Homogeneous Isotropic Dispersive Unbounded Media

Med. Types: Eq. Class.	Wave Number: $k_+ = -i(s/c_0)[1+\alpha_0(s)] = -i\gamma(s)$	Feature[a]	$\alpha_0(s)$	Physical Applications[b]
1. Time-dep. *Helmholtz* (hyperbolic) (generalized Helmholtz)	$-is/c_0 = \omega/c_0$	Non-D	$= 0$ (no dispersion)	Sonar; radar; telcom; space
2. Radiation absorption (hyperbolic)	$-(is/c_0)\left(1+a_R^2c_0^2/s\right)^{1/2} = \omega/c_0\left(1-ia_R^2c_0^2/\omega\right)^{1/2}$	D	$= \left(1+a_R^2c_0^2/s\right)^{1/2} - 1 \begin{cases} \cong a_R^2c_0^2/2s \\ \doteq 1 \end{cases}$	Radar; telcom.
3. Relaxation absorption (hyperbolic)	$-(is/c_o)(1+\hat{\tau}_0 s)^{-1/2} = \omega/c_0(1+i\hat{\tau}_0\omega)^{-1/2}$	D	$= (1+\hat{\tau}_0 s)^{1/2} - 1 \begin{cases} \simeq 0 \\ = -\hat{\tau}s/2 \end{cases}$	Sonar, telcom; water, air, space
4. Radiation and relaxation absorption (hyperbolic)	$-(is/c_o)\left(\dfrac{1+a_R^2c_0^2/s}{1+\hat{\tau}_0 s}\right)^{1/2} = (\omega/c_0)\left(\dfrac{1-ia_R^2c_0^2/\omega}{1+i\hat{\tau}_0\omega}\right)^{1/2}$	D	=	Water/air; sonar
5. Diffusion (parabolic)	$-(is/c_0)b_0c_0/\sqrt{s} = (\omega/c_0)\left(\dfrac{b_0c_0e^{i\pi/4}}{\sqrt{\omega}}\right)$	D	=	Heat flow; solids liquids
6. Diffusion and radiation absorption (hyperbolic)	$-(is/c_o)\left(1+a_0^2c_0^2/s\right)^{1/2} = (\omega/c_0)\left(1-ia_0^2c_0^2/\omega\right)^{1/2}$	D	=	Heat flow; solids liquids (max. velocity)

[a] D = dispersive ≡ dissipation.
[b] In electromagnetic applications, only the dominant scalar E-field.

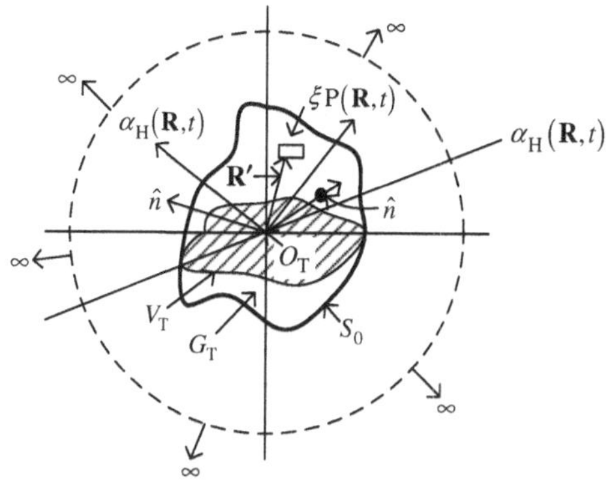

FIGURE 8.6 Geometry of a finite distributed source (G_T in V_T) bounded by the surface S_0, and in an infinite homogeneous medium ($V - V_T \to \infty$).

the relations (8.1.7) and (8.1.9a), namely,

$$\hat{L}^{(0)}\alpha_H = -G_T\}_C \quad \text{and} \quad \hat{L}^{(0)}g_\infty = -\delta(\mathbf{R}' - \mathbf{R})\delta(t' - t) \tag{8.1.39}$$

and multiply the former by g_∞ and the latter by α_H and subtract the result, to obtain

$$g_\infty\hat{L}^{(0)}\alpha_H - \alpha_H\hat{L}^{(0)}g_\infty = -g_\infty G_T + \alpha_H\delta_{\mathbf{R}'\mathbf{R}}\delta_{t't}. \tag{8.1.39a}$$

Next, we integrate over the volume V_T containing the source and over time $t' = t_0$ to t, to get

$$\int_{t_0}^{t} dt' \int_{V_T} \mathbf{dR}'\left\{g_\infty\hat{L}^{(0)'}\alpha'_H - \alpha'_H\hat{L}^{(0)'}g_\infty\right) = -\int_{t_0}^{t} dt' \int_{V_T} \mathbf{dR}' g_\infty G'_T + \alpha_H(\mathbf{R}, t). \tag{8.1.40}$$

To proceed further we must now introduce the explicit structure of $\hat{L}^{(0)}$, which from the examples of Section 8.1.5 has two or three additive terms involving $[\nabla^2, (\partial/\partial t), (1/c_0^2)(\partial^2/\partial t^2)]$. Specific results are then obtained from the application of Green's theorem (see, e.g., [24]) in expressions which contain ∇^2 and thus exhibits the effects of the boundary conditions *on any sources on* $(S_0^1 \to)S'_0$. We have[23]

$$\int_{V_T} \left[A_2\nabla^{2'}A'_1 - A_1\nabla^{2'}A'_2\right]\mathbf{dR}'$$

$$= \oint_{S_0} \left\{\left[A_2\nabla' A'_1 - A_1\nabla' A'_2\right]\cdot\hat{\mathbf{n}}'\right\}_{S'_0} dS'_0 \equiv \oint_{S_0} \left[\mathbf{L}_1(g_\infty, \alpha'_H)\cdot\hat{\mathbf{n}}'\right]_{S'_0} dS'_0, \tag{8.1.41a}$$

with $\nabla' A_1 \cdot \hat{\mathbf{n}}' \equiv \hat{\mathbf{n}}' \cdot \nabla' A_1 = \frac{\partial A_1}{\partial n'}$ and so on. The integrand is taken *on the bounding surface* $(\hat{\mathbf{n}}' dS' = \mathbf{dS}'_0)$, which we indicate explicitly by writing $S'_0 = S_0(\mathbf{R}', t')$. *Here* $\hat{\mathbf{n}}'$ *is an*

[23] In the following, the primes refer to the variables of integration, in $\mathbf{R}'$ and to $(\mathbf{R}', \mathbf{t}')$—dependent quantities, that is, $S'_0 = S_0(\mathbf{R}', t')$, and so on.

outwardly drawn normal. Thus, we have

$$\left.\begin{aligned}&\oint_{S_0}\mathbf{L}_1(g_\infty,\alpha'_{\mathrm{H}})dS'_0=\oint_{S'_0}(g_\infty\hat{\mathbf{n}}'\cdot\nabla'\alpha'_{\mathrm{H}}-\alpha'_{\mathrm{H}}\hat{\mathbf{n}}'\cdot\nabla'g)dS'_0\\&g_\infty\hat{\mathbf{n}}'\cdot\nabla'\phi'dS'_0=g_\infty\cdot(\nabla\phi)'\cdot\hat{\mathbf{n}}'dS'_0=g_\infty\cdot(\nabla\phi)'\cdot\mathbf{dS}'_0,\end{aligned}\right\}\tag{8.1.41b}$$

where alternatively $\hat{\mathbf{n}}'\cdot\nabla\equiv\partial/\partial n'$.

For the terms containing $\partial/\partial t,\partial^2/c_0^2\partial t^2$, and thus embodying the *initial conditions*, we have for these components of $\hat{L}^{(0)}$

$$\int\limits_{V_\mathrm{T}}\mathbf{L}_2\left(g_\infty,\alpha'_{\mathrm{H}}\right)_{V_\mathrm{T}\times T}\mathbf{dR}'$$
$$=\left\{\begin{aligned}&-\frac{1}{c_0^2}\int\limits_{V_\mathrm{T}}\cdot\mathbf{dR}'\left[\frac{\partial g_\infty}{\partial t'}\alpha'_{\mathrm{H}}-\frac{\partial\alpha'_{\mathrm{H}}}{\partial t'}g_\infty\right]_{(\mathbf{R}_0,t_0)}:\text{ for }\frac{\partial^2}{\partial t'^2}\\&+a\int_{V_\mathrm{T}}\mathbf{dR}'\left[g_\infty\alpha'_{\mathrm{H}}\right]_{(\mathbf{R}_0,t_0)}:\text{ for }\frac{\partial}{\partial t'}+(\text{space}-\text{time terms }\mathbf{L}_{S_0\times T},\text{ if any}),\end{aligned}\right.\tag{8.1.41c}$$

this last for the initial conditions on α'_{H} in $\mathbf{L}_{S_0\times T}$, where $[\,]_{(\mathbf{R}_0,t_0)}$ represents the initial condition at $(\mathbf{R}_0,t_0)$. Here $\oint_{S'_0}$ represents the integral over the closed surface $\left(S'_0\right)$ and $\mathbf{dR}'=dx'dy'dz'$ is a volume element in V_T. Combining (8.1.40)–(8.1.41b) we have finally the three components for the complete solution for the field α_H:

$$\alpha_H(\mathbf{R},t)=\int\limits_{t_0}^{t}dt\int\limits_{V_\mathrm{T}}\mathbf{dR}'g_\infty(\mathbf{R},t|\mathbf{R}',t')G_\mathrm{T}(\mathbf{R}',t')$$
$$+\left\{\begin{matrix}\displaystyle\int\limits_{t_0}^{t}dt\oint\limits_{S_0}\left[\mathbf{L}_1\left(g_\infty,\alpha'_{\mathrm{H}}\right)\cdot\hat{\mathbf{n}}'\right]_{S'_0}dS'_0 & \displaystyle+\frac{1}{c_0^2}\int\limits_{V_\mathrm{T}\times T+S'_0\times T}\mathbf{L}_2\left(g_\infty,\alpha'_{\mathrm{H}}\right)|_{t_0-}\mathbf{dR}'\\ \leftarrow\quad(\text{B.C.s})\quad\rightarrow & \leftarrow\quad(\text{I.C.s})\quad\rightarrow\end{matrix}\right\},\ (\text{B.C.}+\text{I.C.}\equiv\mathsf{C})\tag{8.1.41d}$$

with the governing conditions (8.1.41b) and (8.1.41c) embodied in C. Thus, the first term of (8.1.41d) gives that part of the field in the infinite space beyond S'_0 due to the source within V_T, the second term is the contribution to the field attributable to any sources *on the bounding surface* S'_0, and the third term is the component representing the initial impulses applied to V_T at time t_0. The source density G_T contains the coupling by the aperture or array, of the signal input to this homogeneous infinite medium in which the field $\alpha_\mathrm{H}(\mathbf{R},t)$ is propagating.

From (8.1.10) in (8.1.41b), we see that the generic form of the integral Green's function operator $\hat{M}_\infty$ is specifically for volumes and surfaces, respectively, in these cases of the infinite homogeneous medium,

$$\hat{M}_\infty|_V=\int dt'\int_{V_T}g_\infty\Big(\mathbf{R},t|\mathbf{R}',t')(\,)_{R',t'}\mathbf{dR}'\tag{8.1.41e}$$

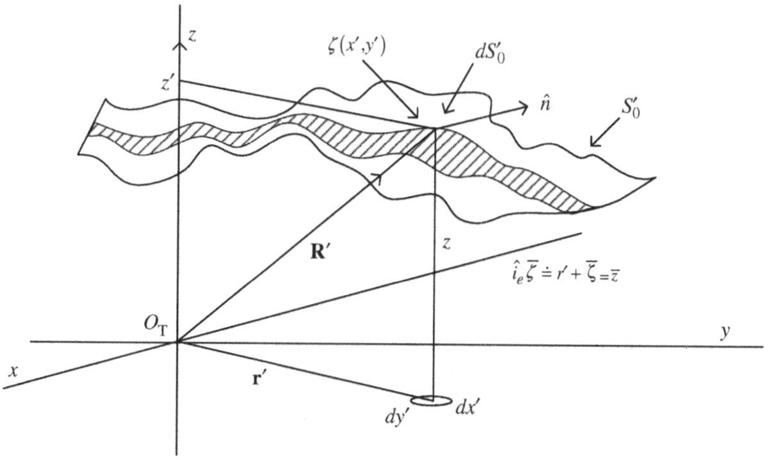

FIGURE 8.7 An irregular surface S_0 with surface element $dS_0'(\mathbf{R}')$, Eq. (8.1.42b), (on an irregular strip).

and
$$\hat{M}_\infty|_{S_0'} = \int dt' \oint_{S_0'} \left(g_\infty^{(0)}\hat{\mathbf{n}}' \cdot \nabla' - \left[\hat{\mathbf{n}}' \cdot \nabla' g_\infty\right]\right)(\,)_{R',t'} dS_0'(\mathbf{R}) \tag{8.1.41f}$$

vis-à-vis the reference plane $z' = 0$. The last term in (8.1.41d), (in more detail in 8.1.41b), represents the effects of the initial conditions, which can often be omitted because $\left[\alpha_H, \frac{\partial \alpha_H}{\partial t}\right]_{t_0} = 0$.

8.1.6.2 Evaluation of the Surface Integrals in Eqs. (8.1.41a–d); (8.1.42) To carry out evaluations of the surface integral $\oint(\,)dS_0'$, we need to determine $dS_0' = dS_0'(\mathbf{R}', t')$ in more detail. For these deterministic examples we begin first with the assumption that the surface S_0' is fixed relative to the reference coordinates (x', y', z'), i.e. $dS_0'(\mathbf{R}', t') = dS_0'(\mathbf{R}')$. The surface S_0' may be irregular, so that the (outward drawn) normal $\hat{\mathbf{n}}'$ to it can have a variety of directions depending on the locations of the element dS_0'. Then, we can write $\zeta(\mathbf{R}')$ for the surface elevation with respect to the xy-plane, at a point $\mathbf{R}'$ from O_T, as shown in Fig. 8.7 with a typical portion of the (fixed) surface S_0'. The normal $\hat{\mathbf{n}}'$ at the element $dS(\mathbf{R}')$ varies with location of the irregularity and is given by[24]

$$\hat{\mathbf{n}}' = \hat{\mathbf{i}}_x n_x' + \hat{\mathbf{i}}_y n_y' + \hat{\mathbf{i}}_z n_z', \text{ with } n_x' = \zeta_x n_z';\; n_y' = \zeta_y n_z';\; n_z' = \left(1 + \zeta_{x'}^2 + \zeta_{y'}^2\right)^{-1/2} \tag{8.1.42a}$$

and for the surface element[25]:

$$dS_0' = dx'dy'/n_z' = dx'dy' \cdot \left(1 + \zeta_{x'}^2 + \zeta_{y'}^2\right)^{1/2}, \text{ with } \zeta_{x'} = \frac{\partial \zeta}{\partial x'};\; \zeta_{y'} = \frac{\partial \zeta}{\partial y'};\; \zeta_{z'} = \frac{\partial \zeta}{\partial z'} = 1 \tag{8.1.42b}$$

[24] It is assumed that the surface is continuous, with at least continuous first and second derivatives.

[25] Eqs. (8.1.42a–c) follow directly from $n_x' = \zeta_{x'}/F; n_y' = \zeta_{x'}/F; n_z' = (\zeta_{z'} = 1)/F$, where F is a normalization factor, obtained from $|\mathbf{n}'| = \left(n_x'^2 + n_y'^2 + n_z'^2\right)^{1/2} = \left(\zeta_{x'}^2 + \zeta_{y'}^2 + 1/2\right)/\left(F^2\right)^{1/2} = 1$, so that $F = \left(1 + \zeta_{x'}^2 + \zeta_{y'}^2\right)^{1/2}$ and $n_z' = F^{-1}$.

and $\zeta = \zeta(\mathbf{R}') = \mathbf{z}', \zeta_{x'} = \zeta_{x'}(\mathbf{R}')$, etc. Here $\hat{\mathbf{n}}' = \hat{\mathbf{n}}'(\mathbf{R}')$, and[25]

$$\therefore n'_x = \frac{\zeta_{x'}}{\sqrt{1+\zeta_{x'}^2+\zeta_{y'}^2}};\; n'_y = \frac{\zeta_{y'}}{\sqrt{1+\zeta_{x'}^2+\zeta_{x'}^2}};\; n'_z = \frac{1}{\sqrt{1+\zeta_{x'}^2+\zeta_{y'}^2}}. \tag{8.1.42c}$$

From Fig. 8.7 we see that for (a fixed position) $\mathbf{r}' = \hat{\mathbf{i}}_x x' + \hat{\mathbf{i}}_y y'$,

$$\mathbf{R}' = \mathbf{r}' + \boldsymbol{\zeta} \quad \therefore \boldsymbol{\zeta} = \hat{\mathbf{i}}_z \zeta(\mathbf{R}') = \zeta(\mathbf{r} + \zeta) \doteq \zeta(r + F);\; \hat{\mathbf{n}}' = \hat{\mathbf{n}}'(\mathbf{R}'). \tag{8.1.42d}$$

We now have in detail the needed results for evaluating the various surfaces integrals encountered in the GHP, namely,

$$\left.\begin{aligned}
\frac{\partial}{\partial n'} &= \hat{\mathbf{n}}' \cdot \nabla' = \frac{(\zeta_x(\partial/\partial x') + \zeta_y(\partial/\partial y') + (\partial/\partial z'))}{\sqrt{1+\zeta_{x'}^2+\zeta_{y'}^2}} \\
&= \hat{J}'(\mathbf{R}'), \quad \text{and} \quad \therefore \quad dS'_0(\mathbf{R}') = \mathbf{dr}'\left(1+\zeta_{x'}^2+\zeta_{y'}^2\right)^{1/2}, \\
&\text{so that explicitly} \\
\hat{\mathbf{n}}' \cdot \nabla'(\,)dS'_0 &\equiv \hat{J}(\mathbf{R}' = \mathbf{r}' + \boldsymbol{\zeta})(\,)_{\mathbf{R}'} dx'dy' \\
&= \left(\zeta_{x'}\frac{\partial}{\partial x'} + \zeta_{y'}\frac{\partial}{\partial y'} + \frac{\partial}{\partial z'}\right)(\,)_{\mathbf{R}'}\mathbf{dr}', d\mathbf{r}' = dx'dy',
\end{aligned}\right\} \tag{8.1.42e}$$

which accordingly defines the local surface operator $\hat{J}' = \hat{J}'(\mathbf{R}')$.

This becomes approximately

$$\hat{J}' \equiv \left(\zeta_{x'}\frac{\partial}{\partial x'} + \zeta_{y'}\frac{\partial}{\partial y'} + \frac{\partial}{\partial z'}\right) = (\nabla_{\mathbf{R}'}\zeta) \cdot \nabla_{\mathbf{R}'}, \tag{8.1.42f}$$

since ζ is normally small enough for $\hat{J}$ to depend only on $\mathbf{r}' + \bar{\zeta}$. This may be expressed by requiring the variance of ζ to be small compared to the magnitude of $|\bar{\zeta}|$, that is, var $\zeta \ll |\bar{\zeta}|^2$, *and* to $\mathbf{r}'$, namely, var $\zeta \ll \bar{\zeta}^2, r'$ or r'. Note that if $z' = 0$, the condition becomes var $\zeta \ll r'$. The former condition applies for surface roughness small compared to the radius of curvature (ρ) of the surface, while the latter applies for flat mean surfaces ($\bar{\rho} = 0$) characteristic of ocean surfaces, for example, and sometimes ocean bottoms, with analogous cases for land.

If the surface is moving, for example, the ocean with respect to a stationary or moving source, *then* $\zeta(\mathbf{R}') \to \zeta(\mathbf{R}', t')$ and $dS'_0(\mathbf{R}') \to dS'_0(\mathbf{R}', t')$, so that $\hat{\mathbf{n}}'$ and ∇' *are also functions of time* (t') *as well as location* ($\mathbf{R}'$). The result (8.1.42e) is directly extended to (8.1.42g) below:

$$\left.\begin{aligned}
\frac{\partial}{\partial n'}(\,)' dS'_0 = \hat{\mathbf{n}}' \cdot \nabla'(\,)_{\mathbf{R}',t'} dS'_0(\mathbf{R}', t') &\doteq \left(\zeta_{x'}\frac{\partial}{\partial x'} + \zeta_{y'}\frac{\partial}{\partial y'} + \frac{\partial}{\partial z'}\right)_{\mathbf{R}',t'}(\,)_{\mathbf{R}',t'}\mathbf{dr}' \\
&= (\nabla_{\mathbf{R}'}\zeta) \cdot \nabla_{\mathbf{R}'}(\,)_{\mathbf{R}',t'}\mathbf{dr}'.
\end{aligned}\right\} = \hat{J}' \cdot (\,)_{\mathbf{R}',t'}\mathbf{dr}', \tag{8.1.42g}$$

If the motion of the surface is large at $\mathbf{R}'$, the position of the point $\boldsymbol{\zeta}$ on the surface changes significantly. However again, in most physical situations ζ changes only a small amount with respect to the mean surface, so that $\mathbf{R}' \doteq \mathbf{r}' + \bar{\zeta}$, where appropriate to the operator $\hat{J}'$, and the quantities it operates upon. Then we find from (8.1.42f) and (8.1.42g) that the operators [(8.1.42a) and (8.1.42b)] can be written with acceptable approximation as

$$\hat{M}_\infty|_V = \int dt' \int_{V_T} g_\infty \cdot (\,)_{\mathbf{R}',t'} \mathbf{dR}'; \quad \hat{M}_\infty|_{S_0'} \doteq \int dt' \oint_{S_0} \left[\left\{ g_\infty \hat{J}' - \left[\hat{J}' g_\infty \right] \right\}_{\mathbf{R}',t'} (\,)_{\mathbf{R}',t'} \right]_{\mathbf{R}',t'} \mathbf{dr}', \tag{8.1.42h}$$

with $\hat{J}'$ now given explicitly by (8.1.42f) and (8.1.42g). We also note once more that these systems are deterministic: there are no random elements. In Chapter 9, we shall discuss what happens when randomness is introduced, to represent the many important *a priori* situations which now require a probabilistic description for their quantification.

8.1.6.3 GHP Examples Inhomogeneous Media If there are no sources on the surface S_0, then $\alpha_H|_{S_0'} = 0$ and $n' \cdot \nabla' \alpha_H|_{S_0'} = 0$, Eqs. (8.1.42a) and (8.1.42b), and the surface integral accordingly vanishes. (As we shall see in Section 8.3.3 following, however, if there are inhomogeneities in the medium outside the source contained in S_0', the resulting (deterministic) *backscatter* produces an additional, reflected scatter contribution from S_0',. Eq. (8.3.20). This in turn may be regarded as equivalent to scatter from virtual sources on S_0'. Here we shall include the case of sources on S_0', whether actual or virtual.) Explicit results for the propagating field $\alpha(\mathbf{R}, t)$are now obtained from the Green's functions of Section 8.1.5. We have accordingly for the integral of the surface integrals, cf. (8.1.42e) and (8.1.42f):

$$\hat{g}_S^{(0)} \equiv \left(g_\infty \hat{J}' - \left[\hat{J}' g_\infty \right] \right) (\,)_{\mathbf{R}',t'}; \quad \text{with} \quad dS_0' = dS_0(\mathbf{R}', t'). \tag{8.1.42i}$$

Note that in the case of the surface integrals the operator kernel g_∞ for volumes becomes a local *surface operator* $\hat{g}_{S'}^{(0)}$. The corresponding (global) integral operator for the surface component is $\int dt \oint \hat{g}_S^{(0)} (\,)_{\mathbf{R}',t'} dS_0'$.

(1) *Time-Dependent Helmholtz Equation:*

$$\begin{aligned} \alpha_{\mathbf{H}}(\mathbf{R}, t) = & \int_{t_0}^{t} dt' \int_{V_T} g_\infty(\mathbf{R}, t|\mathbf{R}', t) G_T(\mathbf{R}', t') \mathbf{dR}' \\ & + \int_{t_0}^{t} dt' \oint_{S_0'} \hat{g}_S^{(0)} \alpha_H' dS_0'(\mathbf{R}', t') \\ & + \frac{1}{c_0^2} \int_{V_T} \left[\left(\frac{\partial g_\infty}{\partial t'} \alpha_H' \right)_{\mathbf{R}_0, t_0} - \left(g_\infty \frac{\partial \alpha_H'}{\partial t'} \right) \right]_{\mathbf{R}_0, t_0} \mathbf{dR}', \end{aligned} \tag{8.1.43}$$

where specifically

$$\left. \begin{aligned} \left(\frac{\partial g_\infty}{\partial t'}\alpha_{\mathrm{H}}\right)_{\mathbf{R}_0,t_0} &= \left[\frac{\partial'}{\partial t_0} g_\infty(\mathbf{R},t|\mathbf{R}_0,t_0)\right]\alpha_{\mathrm{H}}(\mathbf{R}_0,t_0); \\ \left(g_\infty \frac{\partial \alpha_{\mathrm{H}}}{\partial t'}\right)_{\mathbf{R}_0,t_0} &= g_\infty(\mathbf{R},t|\mathbf{R}_0,t_0)\frac{\partial \alpha_{\mathrm{H}}(\mathbf{R},t|\mathbf{R}_0,t_0)}{\partial t_0} \end{aligned} \right\}; \tag{8.1.43a}$$

(2) *Wave Equation with Radiation Absorption:*

$$\alpha_{\mathrm{H}}(\mathbf{R},t) = \int_{t_0}^{t} dt' \int_{V_{\mathrm{T}}} g_\infty G'_{\mathrm{T}} \mathbf{dR}' + \int_{t_0}^{t} dt' \oint_{S'_0} \hat{g}_S^{(0)} \alpha'_{\mathrm{H}} dS'_0$$
$$+ a_R \int_{V_{\mathrm{T}}} (g_\infty \alpha')_{t_0,R_0} \mathbf{dR}'_0 - \frac{1}{c_0^2}$$
$$\times \int_{V_{\mathrm{T}}} \left[\left(\frac{\partial g_\infty}{\partial t_0}\right)_{t_0,R_0} \alpha_{\mathrm{H}}(\mathbf{R}_0,t_0) - g_\infty(\mathbf{R},t|\mathbf{R}'_0,t_0)\left(\frac{\partial \alpha_{\mathrm{H}}}{\partial t_0}\right)_{t_0,R_0}\right] \mathbf{dR}'_0; \tag{8.1.44}$$

(3) *Wave Equation with Relaxation Absorption:*

$$\alpha_{\mathrm{H}}(R,t) = \int_{t_0}^{t} dt' \int_{V_{\mathrm{T}}} g_\infty G'_{\mathrm{T}} dR'$$
$$+ \int_{t_0}^{t} dt' \oint_{S'_0} \left[g_\infty \left(1 + \hat{\tau}_0 \frac{\partial}{\partial t'}\right) \hat{J}' \alpha'_{\mathrm{H}} - \alpha'_{\mathrm{H}}\left(1 + \hat{\tau}_0 \frac{\partial}{\partial t'}\right)(\hat{J}' g_\infty)\right] dS'_0$$
$$- \frac{1}{c_0^2} \int_{V_{\mathrm{T}}} \left[g'_\infty \frac{\partial \alpha'_{\mathrm{H}}}{\partial t'} - \alpha'_{\mathrm{H}} \frac{\partial g'}{\partial t'}\right]_{t_0,\mathbf{R}_0}^{t} \mathbf{dR}'_0; \tag{8.1.45}$$

(4) *Wave Equation with Relaxation and Radiation Absorption:*

$$\alpha_{\mathrm{H}}(\mathbf{R},t) = \int_{t_0}^{t} dt' \int_{V_{\mathrm{T}}} g_\infty G'_{\mathrm{T}} \mathbf{dR}' + \int_{t_0}^{t} dt' \oint_{S_0} \left[g'_\infty \cdot \left(1 + \hat{\tau}_0 \frac{\partial}{\partial t'}\right) \hat{J}' \alpha'_{\mathrm{H}}\right.$$
$$\left. - \alpha'_{\mathrm{H}} \cdot \left(1 + \hat{\tau}_0 \frac{\partial}{\partial t'}\right)(\hat{J}' g'_\infty) dS'_0\right]_{S_0} + a_R^2 \int_{V_{\mathrm{T}}} (g\alpha'_{\mathrm{H}})_{t_0,\mathbf{R}_0} \mathbf{dR}'_0$$
$$- \frac{1}{c_0^2} \int_{V_{\mathrm{T}}}^{t'} \left[g'_\infty \frac{\partial \alpha'_{\mathrm{H}}}{\partial t'} - \alpha'_{\mathrm{H}} \frac{\partial g'_\infty}{\partial t'}\right]_{t_0,\mathbf{R}_0}^{t} \mathbf{dR}'_0; \tag{8.1.46}$$

(5) *The Diffusion Equation:*

$$\alpha_{\mathrm{H}}(\mathbf{R},t) = \int_{t_0}^{t} dt' \int_{V_{\mathrm{T}}} g_\infty G'_{\mathrm{T}} \mathbf{dR}' + \int_{t_0}^{t} dt' \oint_{S_0} \cdot \hat{g}'_S \alpha'_{\mathrm{H}} dS'_0 + b_0^2 \int_{V_{\mathrm{T}}} \left[g_\infty^{(0)} \alpha'_{\mathrm{H}} \right]_{t_0,\mathbf{R}_0} \mathbf{dR}'_0; \tag{8.1.47}$$

(6) *Diffusion: Maximum Velocity of Heat Absorption:* (See (2), Eq. (8.1.44) above).

Here the primes (′) refer to the variables of integration, that is, as noted earlier $\left(\nabla'\alpha'_{\mathrm{H}} = \hat{i}_1\left({}^{\partial}/_{\partial x'}\right) + \hat{i}_2\left({}^{\partial}/_{\partial y'}\right)\hat{i}_3\left({}^{\partial}/_{\partial z'}\right)\right)\alpha_{\mathrm{H}}(\mathbf{R}',t')$, and so on, $g_\infty = g(\mathbf{R},t|\mathbf{R}',t')$, of course. In all of these relations (1)–(6) the Green's function g_∞ can obey a variety of boundary conditions, according to Table 8.3 below.

8.1.6.4 The Simplest Case—Homogeneous Media and Volume Sources Only: Zero Field Initially For most of the cases of interest to us here there are no surface sources, and the domain $(V - V_{\mathrm{T}})$ of the field is infinite so that $\int dt' \oint_{S_0} (\ ) dS'_0 = 0$. Furthermore, for the initial conditions on field generation it is reasonable to set $\alpha_{\mathrm{H}} = 0, \partial\alpha_{\mathrm{H}}/\partial t = 0$ at $t = t_0$, with $t = t_0$ as the initial time. The result for the field is a very considerable simplification, that is, only the first term generally in (8.1.42) remains nonvanishing (with similar effect in the specific cases 1–6 above). Accordingly, we now have the "simple" result

$$\alpha_{\mathrm{H}}(\mathbf{R},t) = \int_{t_0}^{\infty} dt' \int_{V_{\mathrm{T}}} g_\infty(\mathbf{R},t|\mathbf{R}',t')_\infty G_{\mathrm{T}}(\mathbf{R}',t')\mathbf{dR}', \tag{8.1.49}$$

generally. In particular, for the homogeneous and isotropic media discussed here in Section 8.1 [cf. 1–6, Eqs. (8.1.30)–(8.1.34) above], Eq. (8.1.38) with (8.1.36a) becomes finally for any (real) source density G_{T},

$$\alpha_{\mathrm{H}}(\mathbf{R},t) = \int_{t_0}^{\infty} dt' \int_{V_{\mathrm{T}}} G(\rho,\Delta t)_\infty G_{\mathrm{T}}(\mathbf{R}',t')\mathbf{dR}', \tag{8.1.49a}$$

$$= \int_{t_0} dt' \int_{V_{\mathrm{T}}} \mathbf{dR}' G_{\mathrm{T}}(\mathbf{R}',t') \int_{Br_1(s)} e^{s\Delta t} Y_0(\rho,s) ds/2\pi i, \tag{8.1.49b}$$

$$= \int_{t_0}^{\infty} dt' \int_{V_{\mathrm{T}}} dR' G_{\mathrm{T}}(\mathbf{R}',t') \int_{Br_1(s)} e^{s\Delta t} \frac{ds}{2\pi i} \int_{-\infty}^{\infty} \mathrm{Y}_0(k,s) \frac{e^{-i\rho k} k dk}{(2\pi)^2 i\rho}; \left.\begin{matrix} \rho = |\mathbf{R} - \mathbf{R}'| \\ \Delta t = t - t' \end{matrix}\right\}, \tag{8.1.49c}$$ [26]

[26] Note that $\mathbf{k} \to k$ here. (Generally, we have $\mathbf{k} \neq k$ in the analysis.)

TABLE 8.3 Types of Scalar Boundary Conditions ($g = g_\infty$)[a]

	Dirichlet	Neumann	Mixed (Cauchy)
Homogeneous[b] ≡ "no sources on boundary"	$\left.\begin{array}{l} g\|_{S_0} = 0 \\ \alpha_H\|_{S_0} = 0 \end{array}\right\}\text{on } S_0'$	$\left.\begin{array}{l} \left(\frac{\partial g}{\partial n}\right)_{S_0} = \hat{\mathbf{n}}\cdot\nabla' g\|_{S_0} = 0 \\ \left(\frac{\partial \alpha_H}{\partial n}\right)_{S_0} = \hat{\mathbf{n}}\cdot\nabla' \alpha'_H = 0 \end{array}\right\}\text{on } S_0'$	$a\begin{bmatrix} g \\ \alpha_H \end{bmatrix} + b\begin{bmatrix} \left(\frac{\partial g}{\partial n}\right) \\ \left(\frac{\partial \alpha_H}{\partial n}\right) \end{bmatrix}_{S_0} = 0 \text{ on } S_0'$
Inhomogeneous[b] ≡ "surface layer of sources"	$\left.\begin{array}{l} g\|_{S_0} \neq 0 \\ \alpha_H\|_{S_0} \neq 0 \end{array}\right\}\text{on } S_0'$	$\left.\begin{array}{l} \left(\frac{\partial g}{\partial n}\right)_{S_0} = f(\neq 0) \\ \left(\frac{\partial \alpha_H}{\partial n}\right)_{S_0} = F(\neq 0) \end{array}\right\}\text{on } S_0'$	$a\begin{bmatrix} g \\ \alpha_H \end{bmatrix} + b\begin{bmatrix} \frac{\partial g}{\partial n} \\ \frac{\partial \alpha_H}{\partial n} \end{bmatrix}_{S_0} = \alpha_H \text{ on } S_0', \neq 0$

[a] [1], pp. 495; 678, 679; 792, 793.

[b] In the general usage in this book these terms refer to the nature of the medium: "homogeneous" means "no boundaries" (other than those bounding sources), with "inhomogeneous" signifying discontinuities in the medium. The physics of the situation will dictate the appropriate boundary conditions (cf., e.g., the discussion on pp. 792, 793 and pp. 806, 807 of Ref. [1]).

and on replacing the Green's function and source density by their (double) Fourier transforms $\mathsf{F}_{\mathbf{R}'}\mathsf{F}_{t'}(\,)$ and integrating over $\mathbf{R}'$ and t' to obtain

$$= \int_{Br_1} e^{st}\frac{ds}{2\pi i}\int_{-\infty}^{\infty} \mathsf{Y}_0(\mathbf{k},s)G_{00}(\mathbf{k},s)_{\mathrm{T}}e^{-i\mathbf{k}\cdot\mathbf{R}}\frac{\mathbf{dk}}{(2\pi)^3}. \tag{8.1.49d}$$

where Y_0 and Y_0 are given explicitly in (1)–(6) of Section 8.1.5. We remind the reader *that the (distributed) source density* G_{T}, *given by* (8.1.1), *or* (8.1.4), *contains the coupling aperture or array for the input signal.* The isotropy, and with it homogeneity, of these media is at once evident from $\rho \equiv |\mathbf{R}-\mathbf{R}'|$. For finite sources of duration T (in time), measured from $t'=0$, we simply insert the factor $(1-e^{-sT})$ in (8.1.49b) and (8.1.49c). For (8.1.49), consequently, the integral over t' is for the interval (0, T). The source density $G_{\mathrm{T}}(\rho,\Delta t)$ embodies the finite interval in space through $\mathbf{R}'$, for example $0\le \mathbf{R}'\le(\mathbf{R}_{\mathrm{T}})$.

As expected, when the propagation equations are supported by different physical media, for example atmospheres versus oceans, and so on, their parameters, of course, assume different values (and dimensions). Thus, for heat flow (5, 6 above and Table 8.2) the speed c_0 of propagation is different in air from that in a solid or fluid—namely the speed of sound in that particular medium. In like fashion, the speed of propagation of electromagnetic waves is different from that of heat, or sound, in similar media. The same is true for the other parameters $(a_{\mathrm{R}}, a_0, \hat{\tau}_0, \ldots)$ which appear in these equations.

8.1.7 The Explicit Role of the Aperture or Array

In addition, for a more complete description of the propagated field, we must include the coupling aperture or array explicitly, as required by our definitions of the *channel. The aperture, of course, determines the directional characteristics of the radiated energy from the source.* We observe, for example from Eqs. (2.5.9a), (2.5.9b), (2.5.12a), and (2.5.12b), that we can write for the source density

$$G_{\mathrm{T}}(\mathbf{R},t) = S_{\mathrm{T}}(\mathbf{R},t)_{\Delta} = \int_{-\infty}^{\infty} Y_{\mathrm{T}}(\mathbf{R},f)S_{\mathrm{in}}(\mathbf{R},f)_{\Delta}e^{i\omega t}df \tag{8.1.50}$$

with $S_{\mathrm{in}-\Delta}\neq 0, (t\in\Delta, \mathbf{R}\in V_{\mathrm{T}})$ and zero elsewhere. Here Y_{T} is the *aperture transfer* or *system function*, cf. (Eq. (2.8.1a) and $S_{\mathrm{in}-\Delta}$ represents the input signal to it, of finite duration Δ, cf. Section 2.5.1. We can accordingly rewrite (8.1.49c) for the radiated field, again where the only source is in V_{T} and with zero initial conditions, to obtain

$$\begin{aligned}\alpha_{\mathrm{H}}(\mathbf{R},t) = &\int_{Br_1(s)} e^{st}\frac{ds}{2\pi i}\int_{-\infty}^{\infty}\mathsf{Y}_0(k,s)\frac{e^{-i\rho k}kdk}{(2\pi)^2 i\rho}\\ &\cdot\int_{V_{\mathrm{T}}}\mathbf{dR}'\int_{-\infty}^{\infty}dt'\int_{-\infty}^{\infty}Y_{\mathrm{T}}(\mathbf{R}',f')S_{\mathrm{in}}(\mathbf{R}',f')_{\Delta}e^{(s'-s)t'}\frac{ds'}{2\pi i}\\ s = 2\pi i f;\; s' = 2\pi i f'.&\end{aligned} \tag{8.1.51}$$

Since

$$\int_{-\infty}^{\infty} e^{(s'-s)t'} \frac{dt'}{2\pi i} = \delta(s'-s), \quad \text{and} \quad \rho = |\mathbf{R} - \mathbf{R}'|, \tag{8.1.51a}$$

we readily obtain for (8.1.51), on writing $\mathbf{R}' \equiv \boldsymbol{\xi}$ for the spatial variable of the source elements in the source density of the transmitter (Section 2.5.2), the desired result

$$\alpha_H(\mathbf{R}, t)|_{\text{aperture}} = \int_{Br_1(s)} e^{st}\left(1 - e^{-sT}\right)\frac{ds}{2\pi i}\int_{V_T} Y_T(\boldsymbol{\xi}, s/2\pi i) S_{\text{in}}(\boldsymbol{\xi}, s/2\pi i)_\Delta \mathbf{d}\boldsymbol{\xi}$$

$$\times \int_{-\infty}^{\infty} \mathrm{Y}_0(k, s)\frac{e^{-i|\mathbf{R}-\boldsymbol{\xi}|k} k dk}{(2\pi)^2 i|\mathbf{R} - \boldsymbol{\xi}|}. \tag{8.1.52}$$

[In (8.1.52), we have replaced the finite duration ($0 \leq t \leq T$) of the input signal $S_{\text{in}-\Delta}$ by the equivalent time truncation factor $(1 - e^{-sT})$.] Here, of course, $\mathrm{Y}_0(k, s) = -L_0^{-1}$, given explicitly in Section 8.1.5. The corresponding result for (8.1.49b) [and (8.1.52)] is easily found to be

$$\alpha_H(\mathbf{R}, t)|_{\text{aperture}} = \int_{Br_1(s)} e^{st}\frac{(1 - e^{-sT})ds}{2\pi i}\int_{V_T} Y_T(\boldsymbol{\xi}, s/2\pi i) S_{\text{in}}(\boldsymbol{\xi}, s/2\pi i) Y_0(|\mathbf{R} - \boldsymbol{\xi}|, s)\mathbf{d}\boldsymbol{\xi}.$$

$$\tag{8.1.53}$$

In the case of arrays of discrete (point) sensors in space, we use Eqs. (8.1.4) and (8.1.5) for G_T in place of Eq. (8.1.1), (8.1.50) or Eqs. (2.5.9a), (2.5.9b), (2.5.12a), and (2.5.12b), namely,

$$G_T(\mathbf{R}, t)|_{\text{array}} = \sum_{m=1}^{M}\int_{-\infty}^{\infty} h_T^{(m)}(\boldsymbol{\xi}_m; t - \tau) S_{\text{in}}^{(m)}(\boldsymbol{\xi}_m, \tau)_\Delta d\tau, \tag{8.1.54a}$$

$$= \sum_{m=1}^{M}\int_{-\infty}^{\infty} Y_T^{(m)}(\boldsymbol{\xi}_m; s/2\pi i) S_{\text{in}}^{(m)}(\boldsymbol{\xi}_m, s/2\pi i)_\Delta e^{st} ds/2\pi i. \tag{8.1.54b}$$

Here the sensor's filter weighting (i.e., Green's) function or equivalently, its system function (cf. Section 2.5.1 et. seq.), is not time-variable, that is, $h_T^{(m)}(\boldsymbol{\xi}_m; \tau) \neq h_T^{(m)}(\boldsymbol{\xi}_m; t, \tau)$, in keeping with the usual condition that the input signal and its receiving sensor are not in relative motion.[27] The expressions for the field $\alpha_H(\mathbf{R}, t)$ correspond to Eqs. (8.1.52) and (8.1.53), but employ (discrete) arrays instead of (continuous) apertures. This follows directly from G_T [(8.1.54a) and (8.1.54b)] in place of G_T, (8.1.50). We have now, again for

[27] Relative motions, however, might occur here in towed or otherwise transmitting arrays, for example, for active radar and sonar applications. In such cases time-varying filters, with $h_T = h_T(\boldsymbol{\xi}; t, \tau)$, would be appropriate for the array structure.

input signals of duration (0, *T*):

$$\alpha_H(\mathbf{R},t)\big|_{\text{array}} = \int_{Br_1(s)} e^{st}\left(1-e^{-sT}\right)\frac{ds}{2\pi i}\sum_{m=1}^{M}\int_{-\infty}^{\infty} Y_T^{(m)}(\boldsymbol{\xi}_m;s/2\pi i)S_{\text{in}}^{(m)}(\boldsymbol{\xi}_m,s/2\pi i)_\Delta \cdot \int_{-\infty}^{\infty} Y_0(k,s)\frac{e^{-i|\mathbf{R}-\boldsymbol{\xi}_m|k}kdk}{(2\pi)^2 i|\mathbf{R}-\boldsymbol{\xi}_m|}. \tag{8.1.55}$$

Similarly, we obtain for (8.1.53), the array counterpart

$$\alpha_H(\mathbf{R},t)\big|_{\text{array}} = \int_{Br_1(s)} e^{st}\left(1-e^{-sT}\right)\frac{ds}{2\pi i}\sum_{m=1}^{M} Y_T^{(m)}(\boldsymbol{\xi}_m;s/2\pi i)S_{\text{in}}^{(m)}(\boldsymbol{\xi}_m,s/2\pi i)_\Delta Y_0(|\mathbf{R}-\boldsymbol{\xi}_m|,s). \tag{8.1.56}$$

and of course (8.1.55) equals (8.1.56), just as (8.1.52) equals (8.1.53). However, this is not necessarily the case for $\alpha_{H-\text{aperture}} = \alpha_{H-\text{array}}$.

We observe again [cf. remarks following Eq. (8.1.37c), as well as those preceding Eq. (8.1.40)] that whereas the *media* considered so far are homogeneous and isotropic, as exemplified by their Green's functions, the *fields* produced in them are directional, that is, are anisotropic. This occurs because of the nonuniform beam patterns produced by the aperture or array employed in their generation, cf. [(8.1.52) and (8.1.53)], [(8.1.55) and (8.1.56)]. Accordingly, the relations (8.1.52) and (8.1.53) for the directional fields here now represent *inhomogeneous*, unbounded (scalar) fields (produced by the directional source at V_T, independent of propagation in the homogeneous but possibly dispersive, (i.e., absorptive) media, cf. Table 8.2 above. In many physical applications, such as radar in water vapor or rain, or sonar in the ocean at moderate to long ranges, dispersion can be a problem, as well as for the signal frequencies employed. Furthermore, the medium itself can produce scattering and distortions in directions of propagation, which can also be significant, so that these comparatively simple results no longer apply. We shall address some of these problems presently.

8.2 THE ENGINEERING APPROACH: I—THE MEDIUM AND CHANNEL AS TIME-VARYING LINEAR FILTERS (DETERMINISTIC MEDIA)

In the previous section, we have presented a general physical description of a propagating field in an unbounded homogeneous (linear) medium, which may or may not be dispersive. The central element here is the appropriate Green's function of this medium, or equivalently, the space-time weighting function, $g_\infty = G(\rho,\Delta t)_\infty$, cf. (8.1.30a) et. seq. In addition, for the *channel* we include a general signal source and the coupling of the source to the medium, the latter in the form of a suitable aperture or discrete array.[28] The resulting *field* $\alpha_H(\mathbf{R},t)$ is thus

[28] Strictly speaking, we usually define the channel as having both a transmitting and a receiving coupling, cf. Fig. 8.1.

represented generally by Eqs. (8.1.42a), or equally by Eqs. (8.1.43)–(8.1.48) above, which simplify to [(8.1.52) and (8.1.53)] and [(8.1.55) and (8.1.56)].

A frequent alternative or "engineering" approach is to represent the medium and the channel by appropriate linear filters. These filters can be time-invariant or time-variable. The former represents the medium and the coupling to it when there is no selective motion of the source and its projection by aperture or array, as well as motion of the medium itself. The latter case arises when there is relative motion of one or more of these elements and usually appears as a Doppler effect. The weighting function of these elements, $h(\tau)$ or $h(t, \tau)$, are also the Green's functions discussed above in Section 8.1 which fully describe propagation for a given source when initial and boundary conditions are specified.[29] Here, for $h(\tau, t)$ in the aforementioned case of relative motion, t describes the filters time variability due to this motion, which τ represents the "memory" of the filter (see the comments in Section 3.5.2 1, Eqs. (3.5.15a), (3.5.15b), (3.5.15c). In this instance, however, there is no relative motion between the components of the channel, so that time-*invariant* filters appear to be reasonable candidates for possible equivalents to the physical, space–time Green's function and the resulting canonical channel.

8.2.1 Equivalent Temporal Filters

The task now is to establish the possible validity and the conditions under which these commonly used filter equivalents are employed. We begin with the general operator forms for the field, and the channel (which includes the receiving coupling, $\hat{h}_R$):

$$\alpha_H(\mathbf{R}, t) = \hat{h}_M\left(\hat{h}_T S_{in}\right) \equiv \hat{h}_{MT} S_{in} \equiv \hat{h}_F S_{in}; \quad X(t) = \hat{h}_R \alpha_H = \hat{h}_R \hat{h}_M \hat{h}_T S_{in} \equiv \hat{h}_C S_{in} \tag{8.2.1a}$$

with the assumptions to be tested here, namely,[30]

$$\alpha_H(\mathbf{R}, t)\left(\overset{?}{=}\right) \int_{-\infty}^{\infty} h_F(\tau|R) S_{in}(t-\tau)d\tau; \quad X(t)\left(\overset{?}{=}\right) \int_{-\infty}^{\infty} h_C(\tau|R) S_{in}(t-\tau)d\tau. \tag{8.2.1b}$$

Note that we at once require the necessary condition *that* $S_{in}(= S_{in}(t))$ *be independent of its spatial location* ($\boldsymbol{\xi}$) *in the aperture or array* (denoted by $\hat{h}_T$ in $\hat{h}_F$), that is, that the same signal be applied to each sensor element in the source function. (See, e.g., Section 2.5.1, 2.5.2, Eq. (2.5.17c) and Section 2.5.3.1). Figure 8.1 illustrates the canonical channel, which is also discussed in Section 8.1.1. This is established explicitly from Eq. (8.2.2b) immediately following. For the unbounded homogeneous media considered here we consider first

[29] We remark that these temporal filters are the solutions of an *ordinary* differential equation (ODE), whose coefficient may be time-variable , cf. Middleton [25]. Sections 2.2.1 and 2.2.2, Eqs. (2.26), (2.27) and the rest of Section 2.2 therein. On the other hand, the physical medium here can be interpreted as a linear space–time filter, whose weighting function is the Green's function $g_\infty(\mathbf{R}, t|\mathbf{R}', t')$, Eq. (8.1.9a)–(8.1.20b) et seq., where the coefficients are functions of the independent variables and partial differential operators, in the manner of (8.1.9a), cf. footnote 2.

[30] For the general time-varying case, (8.2.1b) is modified by replacing $h_F(\tau)$ by $h_F(t, \tau)$ and $h_C(\tau|R)$ by $h_C(t, \tau|R)$.

the results Eqs. (8.1.52) and (8.1.53) for a finite sample (0, *T*) of the field:

Eq. (8.2.1b):
$$\alpha_H(\mathbf{R},t)_\Delta = \hat{h}_F S_{in} \overset{(?)}{=} \int_{Br_1} e^{st}(1-e^{-sT})Y_F(s/2\pi i|R)S_{in}(s/2\pi i)_\Delta \frac{ds}{2\pi i} \tag{8.2.2a}$$

Eqs. (8.1.52) and (8.1.53):
$$= \int_{Br_1} e^{st}(1-e^{-sT}) \int_{V_T} \mathrm{Y}_T(\boldsymbol{\xi}, s/2\pi i|R)S_{in}(\boldsymbol{\xi}, s/2\pi i)_\Delta \mathbf{d}\boldsymbol{\xi} \cdot \int_{-\infty}^{\infty} \mathrm{Y}_0(k,s)\frac{e^{-ik|\mathbf{R}-\boldsymbol{\xi}|}kdk}{|\mathbf{R}-\boldsymbol{\xi}|(2\pi)^2 i} \tag{8.2.2b}$$

where $Y_F(f|R) = \mathsf{F}_t\{h(\tau|R)\}$. From (8.2.2a) it is immediately evident that we must have S_{in} independent of position $\boldsymbol{\xi}$ in the aperture or array, that is, $S_{in}(\boldsymbol{\xi}, s/2\pi i)_\Delta = S_{in}(s/2\pi i)_\Delta$, namely, that the *same* signal is applied to each element ($\mathbf{d}\boldsymbol{\xi}$) of the aperture or array.

Moreover, we must impose the far-field (FF) *or* Fraunhofer constraint on the radiated field, so that the size of the source, $V_T(\boldsymbol{\xi})$, is ignorable in determining this field: the field appears to be coming from a single-point in space, namely. O_T in Fig. 2.15, that is from $|\mathbf{R}-\boldsymbol{\xi}| \doteq R$ in the denominator of (8.2.2b). However, a directional beam is still generated, effectively now at O_T. However, for this we must retain $\boldsymbol{\xi}$ in the phase factor of exp $(-ik|\mathbf{R}-\boldsymbol{\xi}|)$, which becomes in the far-field approximation $k|\mathbf{R}-\boldsymbol{\xi}| \doteq k\left(R - \hat{\mathbf{i}}_T \cdot \boldsymbol{\xi}\right)$, cf. Section 2.5.2.1. This factor still depends on $\boldsymbol{\xi}$ and R and provides the directionality of the radiated signal, as determined by the resulting beam pattern in (8.2.2b), namely

$$\mathsf{A}_T(\boldsymbol{\nu}_T - \boldsymbol{\nu}_{OT}; s/2\pi i)_{FF} = \int_{-\infty}^{\infty} Y_T(\boldsymbol{\xi}, s/2\pi i)e^{ik\boldsymbol{\xi}\cdot(\hat{\mathbf{i}}_T - \hat{\mathbf{i}}_{OT})}\mathbf{d}\boldsymbol{\xi}, \quad \text{with} \quad \boldsymbol{\nu}_T \equiv k\hat{\mathbf{i}}_T/2\pi; \boldsymbol{\nu}_{OT} = k\hat{\mathbf{i}}_{OT}/2\pi. \tag{8.2.3}$$

Here we have inserted a *steering-vector* ν_{OT} [cf. (2.5.16a) et seq.] into the phase factor of the aperture. This modifies $Y_0(\rho, s)$ to $Y_0(R, s)$, the Fourier transform of Eq. (8.1.36c), which is now the far-field form

$$Y_0(\rho,s) \doteq Y_0(R,s)_{FF} = \int_{-\infty}^{\infty} \mathrm{Y}_0(k,s)_{FF} e^{-ikR}\frac{kdk}{(2\pi)^2 iR} = \frac{1}{4\pi R}\int_{-\infty}^{\infty} \mathrm{Y}_0(k,s)_{FF} e^{-ikR}\frac{kdk}{\pi i}. \tag{8.2.4}$$

The quantity $k \to k_+(= -i\gamma(s))$ noted above in Table 8.2, represents the complex wave number associated with dissipation, that is, for dispersive media, obtained by solving the denominator of (8.2.4) for the singularity representing outgoing radiation from the source. Equation (8.2.4) thus determines $Y_0(R,s)_{FF}$. Specific values of k_+ correspond to different types of media. Combining now (8.2.3) and (8.2.4) in (8.2.2a) with the position-free input S_{in} yields directly the desired results for the far-field $\alpha_H|_{FF}$:

$$\alpha_H(\mathbf{R},t)\Big|_{\substack{FF \\ S_{in}\neq S_{in}(\boldsymbol{\xi},-)}} \doteq \int_{Br_1} e^{st}(1-e^{-sT})S_{in}(s/2\pi i)_\Delta\{\mathsf{A}_T(\boldsymbol{\nu}_T[h] - \boldsymbol{\nu}_{OT}[h]; s/2\pi i)_{FF}F(s)_{FF} \cdot \left[\frac{e^{-\gamma(s)R}}{2\pi R}\right]\Big\}\frac{ds}{2\pi i}. \tag{8.2.5}$$

Here F_{FF} and k_+ and $\gamma(s)$ are specifically, with $s = i\omega$:

$$F(s)_{FF} = \int_{-\infty}^{\infty} Y_0(k,s)kdk/\pi i, \quad \text{with} \quad k \to k_+ = -i\gamma(s) = -i\left(\frac{s}{c_0}\right)[1+\alpha_0(s)],$$
$$\text{and } \boldsymbol{\nu}_T = \boldsymbol{\nu}_T(s) = \boldsymbol{\nu}_T(\gamma(s)), \text{ and so on .} \qquad (8.2.5a)$$

Note that for the Helmholtz medium that $-i\gamma(s) = \omega/c_0$, characteristic of non-dispersive media, and $\alpha_0(s) = 0$, so that $F(s)_{FF} = 1$, cf. Example, 8.1.3.1. In general, however, if the medium is dissipative, k_+ for the required outgoing wave from the source, is no longer linear in ω (or s), and the values of $F(s)_{FF}$ and A_T are accordingly modified.

Next, comparing (8.2.5) and (8.2.2a) we see that $Y_F(s/2\pi i)_{FF}$ is represented in detail by (8.2.5), namely

$$\left.\begin{aligned} Y_F(R,s)_{FF} &\equiv A_T(\nu_T(\gamma(s)) - \boldsymbol{\nu}_{OT}(\gamma(s); s/2\pi i))_{FF}\left(F(s)_{FF}\frac{e^{-\gamma(s)R}}{4\pi R}\right), \\ &= A_T(s/2\pi i)_{FF}F(s)_{FF}e^{-\gamma(s)R}/4\pi R, \end{aligned}\right\} \qquad (8.2.6)$$

or by its Fourier transform

$$h_F(\tau|R)_{FF} \equiv F_f\{h_F\} = \int_{-\infty}^{\infty} Y_F(R,s)e^{2\pi i f\tau}df = A_T(s/2\pi i)_{FF}\left(F(s)_{FF}\frac{e^{-\gamma(s)R}}{4\pi R}\right). \qquad (8.2.6a)$$

The field $\alpha_H(\mathbf{R},t)$, (8.2.5), becomes finally for the continuous sample on (0, T), from $S_{in} \neq (0,T)$ and zero elsewhere, namely,

$$\alpha_H(\mathbf{R},t)\Big|_{\substack{FF \\ S_{in}(t)_\Delta \\ \Delta}} \doteq \int_{Br_1} e^{st}S_{in}(s/2\pi i)_\Delta Y_F(s/2\pi i|R)_{FF}\frac{ds}{2\pi i} = \int_{-\infty}^{\infty} h_F(\tau|R)_{FF}S_{in}(t-\tau)_\Delta d\tau.$$
$$(8.2.7)$$

This result (8.2.7) shows that it is indeed possible to represent the physical (Green's function assisted) solution for these homogeneous fields in unbounded media by a suitable (here) time-invariant linear filter, [(8.2.6) and (8.2.6a)], which in turn establishes the equivalence of (8.2.2a) and (8.2.2b).

The necessary and sufficient $(n+s)$ conditions together are that:

Equivalency Conditions for the Field, Eq. (8.2.7):

(8.2.8)

I. $S_{in}(\xi,t)_\Delta = S_{in}(t)_\Delta$ *at each element* $\mathbf{d\xi}$ *of the source function* $G_T = \hat{G}_T\{S_{in}\}$ in $V_T(\boldsymbol{\xi})$, *the region occupied by the aperture, that is, the driving signal is the same at each element.*

II. *The region for which an equivalent time-invariant filter represents the corresponding physical solution is the far-field or Fraunhofer region.*

III. *This filter depends parametrically on range* ($|\mathbf{R}|$) *in a specified way, cf.* (8.2.6) and (8.2.6*a*).

Whether or not the medium is dispersive, as long as it is homogeneous, that is, is unbounded and without scattering elements, this equivalence is not altered, provided the conditions (8.2.8) are obeyed.

The above procedure, based on [(8.2.2a) and (8.2.2b)] for the channel as a whole, namely $\hat{h}_R\alpha_H$, cf. (8.2.1), is readily carried out if we require that the $n+s$ conditions (8.2.8) apply for the reception process as well. From Section 2.5.4 with $\delta(\hat{\boldsymbol{\nu}}-0)$, we have at once the received far-field signal in $(0, T)$

$$X(t|R)_\Delta = \hat{h}_R\alpha_H\Big|_{\substack{\Delta \\ \mathrm{FF} \\ \alpha_H \mathrm{indep\cdot of}\ \boldsymbol{\eta}}}$$

$$= \int_{Br_1} e^{st}\left(1-e^{-sT}\right)\mathsf{A}_{\mathrm{Rec}}\left(\boldsymbol{\Delta}\hat{\mathbf{i}}_R s/2\pi i c_0, s/2\pi i\right)S_\alpha(s/2\pi i|R)_{\substack{\Delta \\ \mathrm{FF}}}\frac{ds}{2\pi i}. \tag{8.2.9}$$

Here α in S_α indicates the received signal, that is, $\alpha_H|$ at the receiving aperture, and $\boldsymbol{\Delta}\hat{\mathbf{i}}_R = \hat{\mathbf{i}}_R - \hat{\mathbf{i}}_{OR}$, with R regarded as essentially constant in a sufficiently small interval ΔR about any fixed range in the far-field. Since $h_R = \mathsf{F}_f\{\mathsf{A}_{\mathrm{Rec}}\}$ we can write (8.2.9) alternatively

$$X(t|R)\Big|_{\substack{\Delta \\ \mathrm{FF}}} = \int_{-\infty}^{\infty} h_R(\tau)S_\alpha(t-\tau|R)\Big|_{\substack{\Delta \\ \mathrm{FF}}} e^{-i\omega\tau}, \ \omega = 2\pi f. \tag{8.2.9a}$$

in which (I) of (8.2.8) applies, so that S_α is also independent of the location of the receiving aperture element $\mathbf{d}\boldsymbol{\eta}$. The (time-invariant) filter h_R is explicitly

$$h_R(\tau) = \int_{-\infty}^{\infty} \mathsf{A}_{\mathrm{Rec}}\left(\boldsymbol{\Delta}\hat{\mathbf{i}}_R f/c_0, f\right)_{\mathrm{FF}} e^{i\omega\tau}df. \tag{8.2.10}$$

Accordingly, combining this with (8.2.7) and (8.2.9a) gives directly

$$X(t|R)_\Delta = \iint_{-\infty}^{\infty} h_R(\tau)h_F(\tau'|R)_{\mathrm{FF}}S_{\mathrm{in}}(t-\tau-\tau')_\Delta d\tau\, d\tau' \tag{8.2.11}$$

with (8.2.6a) for $h_F\left(\tau'|R\right)_{\mathrm{FF}}$. We see that the result (8.2.11) cannot be expressed in the form of a time-varying filter operating on $S_{\mathrm{in}-\Delta}$, but is rather as a double filtering operation, here involving *two* time-invariant filters h_R, h_F. Consequently, the expression (8.2.1a) cannot apply for the *channel*, but only for the source field α_H, that is, $h_R h_F \neq h_C$, even when the conditions (8.2.8) are obeyed.

An exception occurs, however, if in the further and often acceptable approximation, h_R represents an *all-pass filter* with respect to the input signal, in the manner of Fig. 8.8 (for narrowband inputs):

$$h_R = \mathsf{A}_{\mathrm{Rec}}\left(\boldsymbol{\Delta}\hat{\mathbf{i}}_R f_0/c_0, f_0\right)\delta(\tau-0), \tag{8.2.12a}$$

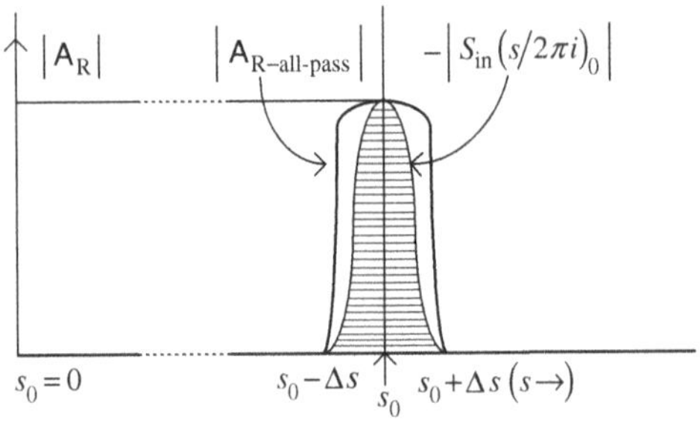

FIGURE 8.8 (Modulus of a) narrowband signal $S_{in}(s'/2\pi i)_0$ with a narrowband (i.e., all-pass beam pattern in $(-\Delta s', \Delta s')$), or aperture response of the receiver.

so that (8.2.11) becomes

$$X(t|R)_\Delta \doteq \mathsf{A}_{Rec}\Big|_{f_0} \cdot \int_{-\infty}^{\infty} h_F(\tau|R)\Big|_{\substack{FF\\ f_0}} S_{in}(t-\tau)d\tau \tag{8.2.12b}$$

with h_F given explicitly by (8.2.6a). Then we can indeed write (8.2.1b)

$$X(t|R)_\Delta \doteq \int_{-\infty}^{\infty} h_C(\tau|R)\Big|_{\substack{FF\\ \text{all-pass}}} S_{in}(t-\tau)d\tau \tag{8.2.13a}$$

where now explicitly

$$h_C(\tau|R)\Big|_{\substack{FF\\ \text{all-pass}}} \doteq \mathsf{A}(f_0)\Big|_{\substack{Rec\\ \text{all-pass}}} \cdot h_F(\tau|R)_{FF}. \tag{8.2.13b}$$

[The all-pass approximation (8.2.12a) and (8.2.12b) may also be acceptable in transmission, where $s\rightarrow s_0$ (or $f\rightarrow f_0$) in $h_F(\tau|R) \doteq h_F|_{\text{all-pass}}$, Eq. (8.2.6a).]

Note that when an *array* of (point) sensors is employed, the results of Section 2.5.3.1 are at once applicable for the field α_H in the above [cf. Eq. (8.2.5) et. seq.]. For the channel, the results of Sections 2.5.4 and 2.5.5 are similarly applicable.

Finally, when there is relative motion of the transmitting and receiving platforms, and/or of the medium itself, the equivalence of the "engineering" model is again established for the field α_H. The conditions of (8.2.8) apply provided these are extended to require *time-variable filters*[31] in place of time-invariant ones. Similarly, a pair of time-variable filters are needed in representing the channel, extending (8.2.10) and (8.2.11) appropriately. Again, whenever the

[31] As noted earlier (cf. Chapter 3) in the context here of an equivalent description of the propagating field and the associated channel, when there is relative motion, the time-variable filter represents a mapping of spatial variations into time (t), with the "memory" of the filter expressed by τ, that is, h (variability: t; memory: τ) $= h(t, \tau)$, the notation being consistent with our convention of expressing dependency on space first and on time second, that is, "space–time". The mapping involved occurs here because $R = R(t)$ and $R(t) = c_0 t$, or some similar relation: for a specific distance R there is a specified time in the field. The geometry of the observer in relation to a fixed point in space then easily determines $\mathbf{R} = c_0 t$. If the point R is in (small) motion relative to source or receiver, then $R = ct = (c_0 + v_1)t$, $|v_d|/c << 1$, where $\pm v_d$ is the Doppler speed of relative motion, depending on the frame of reference of the observer.

approximation of an all-pass receiving aperture is acceptable, this can be further reduced to a single, now time-variable filter, in the manner of Eq. (8.2.13) above for the channel as well. In summary, the conditions for the time-variable or time-invariant linear filter h_C equivalently to represent the *channel* are now as follows:

Equivalency Conditions for the Channel (h_C) *:*

$$(8.2.14)\quad \begin{cases} \textbf{I.}\ \ S_{in}(\boldsymbol{\xi}, t)_\Delta = S_{in}(t)_\Delta\text{: the same input signal must be applied to each, element } d\boldsymbol{\xi} \text{ of the aperture (or array).} \\ \textbf{II.}\ \ \text{The far-field or Fraunhofer region of the field and for the channel must apply.} \\ \textbf{III.}\ \ \text{This filter } (h_C) \text{ is parametrically a function of range (R).} \\ \textbf{IV.}\ \ \text{The receiver portion of the coupling must be essentially an all-pass network; that is, } h_R \doteq \delta(t-0) \end{cases}$$

8.2.2 Causality: Extensions of the Paley–Wiener Criterion [26]

Just as there is a variety of conditions which the space–time Green's functions of these deterministic linear media must obey mathematically [cf. Section 8.1.4], and to be physically applied in practice, so also is there an analogous set of conditions for their engineering equivalents. In addition to the requirement that the time-invariant and time-variable filters (h_C), (8.2.7) represent the field in the Fraunhofer or far-field region, the input signal must be similarly applied to each sensor or sensor element. To represent the channel as well now with a single filter, $h_C(\tau)$, the receiver coupling (h_R) must be an all-pass filter in the spectral domain of the input signal, cf. Fig. 8.8, in the manner of (8.2.12a)–(8.2.13a) above. Again, for moving platforms and media the time-invariant channel filter is replaced by a time-variable filter.

Finally, for realizability, or equivalently *causality*, these filters must operate only on the "past" of their inputs, as well as being *stable*, i.e., the weighting function must obey the condition[32]

$$\int_{-\infty}^{\infty} |h(\tau)|d\tau < \infty\text{: } \textit{bounded inputs} \Rightarrow \textit{bounded outputs}. \tag{8.2.15}$$

The *causality condition* for these linear *time-invariant filters* is given by the well-known Paley–Wiener condition ([26]; also pp. 96, 97; p. 702 of Ref. [25]):

$$|J_{P-W}| = \int_0^{\infty} \frac{\log|Y_{oF}(iu)_{FF}|^2}{1+u^2} du < \infty;\ u = (\text{normalized angular frequency} \sim \omega = 2\pi f), \tag{8.2.16a}$$

where Y_{oF}, Eq. (8.2.6) is the system function, that is, $\mathsf{F}_t\{h_F\}$, of the equivalent filter defined here by (8.2.6a). When a time-variable filter is required, the extended version of (8.2.16a)

[32] See Sections. 2.2, 2.2.1, and 2.2.2 of Ref. [25], for further discussion in the case of linear discrete lumped constant or variable parameter networks obeying linear ordinary differential equations, cf. Eq. (2.28) of Ref. [25]. Here, however, although analogous, we have the discrete networks replaced by continuous ones, except for the couplings to the field when arrays are used. The ODEs then become linear partial differential equations.

becomes

$$|J_{P-W}(\hat{u})| \equiv \int_0^\infty \frac{\log\left|Y_{oF}(i\hat{u}, iu)_{FF}\right|^2}{1+u^2} du < \infty, \tag{8.2.16b}$$

with $\hat{u}$ now a normalized frequency, $\hat{u} = 2\pi\hat{\nu}/\hat{\nu}_0$. Since $h_F(t,\tau) = h_F(t, t-\tau)$, with $h_F = 0$ we have $t - \tau \leq 0$ and $h_F(t,\tau) = 0, \tau \leq 0$, in $t \leq 0$ when $\tau \leq 0$. Thus, causality represented by $h_F = 0, \tau \leq 0$ implies $t < 0$ also: the time-constraint on $h(t,\tau)$ implies a similar constraint for space expressed in terms of time, that is, $t(\mathbf{R}) < 0$, when $\tau \leq 0$. Here the spatial variations caused by Doppler effects expressed in time changes are experienced as frequency variations. In Eqs. (8.2.16a) and (8.2.16b), Y_{oF} and Y_{oF} are the Fourier transforms

$$\textit{No Doppler:}\; Y_{oF}(i\omega|R) = \mathsf{F}_\tau\{h_F\} = \int_{-\infty}^{\infty} h_F(\tau|R)e^{-i\omega\tau}d\tau, \quad \text{with} \quad h_F(\tau|R) \equiv \int_{-\infty}^{\infty} h_F(t,\tau|R)dt. \tag{8.2.17a}$$

$$\textit{Doppler:}\; Y_0(i\hat{\omega}, i\omega|R)_F = \mathsf{F}_t\mathsf{F}_\tau\{h_F\} = \int_{-\infty}^{\infty} dt \int_{-\infty}^{\infty} d\tau\, h_F(t,\tau|R)e^{2\pi i\hat{\nu}t - i\omega\tau}, (\omega = 2\pi f; \hat{\omega} = 2\pi\hat{\nu}). \tag{8.2.17b}$$

Thus, the causality conditions [(8.2.16a) and (8.2.16b)] for the linear temporal filter equivalents here embody the physical requirement of outgoing radiation from the sources, mathematically exhibited by the appropriate singularity on the contour $(-\infty, \infty)$ of integration, in the manner of (8.1.22) and (8.1.23), cf. Fig. 8.5a and b.

For future use we shall need the complete set of Fourier transforms

$$Y(t,f) = \mathsf{F}_\tau\{h_F\} = \int_{-\infty}^{\infty} h(t,\tau)e^{-2\pi i f\tau}d\tau \equiv \textit{time variable system or response function},^{33} \tag{8.2.18a}$$

$$H(\hat{\nu},\tau) = \mathsf{F}_t\{h_F\} = \int_{-\infty}^{\infty} h(t,\tau)e^{2\pi i\hat{\nu}t}dt \equiv \textit{spreading function}, \tag{8.2.18b}$$

$$\mathsf{Y}(\hat{\nu},f) = \mathsf{F}_t\mathsf{F}_\tau\{h_F\} = \iint_{-\infty} h_\tau(t,\tau)e^{2\pi i\hat{\nu}t - 2\pi i f\tau}dtd\tau \equiv \textit{bifrequency response function}, \tag{8.218c}$$

[33] For general usage we employ the notation of (8.2.18a)–(8.2.20), now without the range parameter (R), it being implicitly assumed when referring to a physical field or channel unless otherwise indicated. For specialized descriptions, we may also include various sub- and superscripts, as well as changes of variables, like ω for f, $\hat{\omega}$ for $\hat{\nu}$, and others. It will be clear from the context what the notation signifies.

The inverse relations (i.e., transform pairs) for the time-variable *weighting function* for the Green's function, here of the medium are

$$\left.\begin{aligned} h_{\mathrm{F}}(t,\tau) = \mathsf{F}_{\mathrm{f}}\{Y\} &= \int_{-\infty}^{\infty} Y(t,f)e^{2\pi i\tau f}df, && (8.2.19a)\\ = \mathsf{F}\hat{\nu}\{H\} &= \int_{-\infty}^{\infty} H(\hat{\nu},\tau)e^{-2\pi it\hat{\nu}}d\hat{\nu}, && (8.2.19b)\\ = \mathsf{F}_{\hat{\nu},f}\{Y\} &= \iint_{-\infty}^{\infty} Y(\hat{\nu},f)e^{-2\pi it\hat{\nu}+2\pi i\tau f}d\hat{\nu}df, && (8.2.19c) \end{aligned}\right\}$$

and with the cross-relations such as $\mathsf{F}_{\hat{\nu}}\{H\} = \mathsf{F}_f\{Y\} = h(t,\tau)$, and so on. *Causality* in these cases is determined by the extension of the Paley-Wiener condition, (8.2.16), and *stability* by

$$\iint_{-\infty}^{\infty} |h(t,\tau)|dt\,d\tau < \infty, \quad \text{or} \quad \iint_{-\infty}^{\infty} |\mathrm{Y}(\hat{\nu},f)|d\hat{\nu}\,df < \infty, \quad \text{etc}\,. \tag{8.2.20}$$

(an application of Parseval's Theorem). Physically (8.2.20) is an expression of the fact that finite inputs produce finite outputs. These relations have their counterparts for the Green's function, respectively, in (8.1.25) and (8.1.26) above [Section (2.2) of Ref. [25] treats at considerable length the corresponding properties of *time-invariant* linear filters.] The extension to the time-variable cases has been given in detail by Bello ([27], also in Goldberg [28]), as well as Blahut et.al. [29], Grünbaum et.al. [30], Kailath [31], and Kennedy [32].

8.3 INHOMOGENEOUS MEDIA AND CHANNELS—DETERMINISTIC SCATTER AND OPERATIONAL SOLUTIONS [6]

We note again that the salient feature of the deterministic state is one in which everything is known about our model of reality—there are no random elements. Here, the channel is completely specified and the desired calculation of the field $\alpha(\mathbf{R},t)$ is thus uniquely determined. From a probabilistic viewpoint our model constitutes an ensemble of one member with probability measure unity. For deterministic inhomogeneous media we may expect much greater complexity in determining the Green's function than for the homogeneous cases treated previously, even for unbounded media like those discussed in Sections 8.1 and 8.2. Furthermore, in the bounded cases we have the additional feature of the scattered radiation from the boundaries, which are also postulated to contain inhomogeneities in addition to the medium itself. These inhomogeneities may form a continuum or may be discrete, but their treatment is handled in essentially similar fashion.

For the Green's function in these cases it turns out, rather surprisingly at first glance, that reciprocity—the equivalent interchange of source and observer—is preserved. This may be explained succinctly by the observation that the Green's function has an omni-directional beam pattern, so that the scattering produced by these inhomogeneities are always contained in such a beam pattern: there can be no scattering "out-of-the-beam." This is not the case, however, for typical fields from distributed sources G_T. These have *directional* beams, which vanish effectively inside the 4π ster-radians of the omni-directional beam. Some or all the scattered energy with respect to these beams, in both transmission and reception, will be lost outside the beam and consequently some or all of the equivalence of reciprocity will be destroyed. Figures 8.9 and 8.10 ff. illustrates these effects. (These remarks apply equally to dispersive and nondispersive media, and are discussed at greater length in Sections 8.3.1, 8.3.2, 8.3.3 below.)

Other difficulties in the treatment of inhomogeneous media are the problems of explicitly accounting for the different classes of multiple scatter, that is, single scatter, pair scattering, and so on, and in addition including boundaries for reflection and transmission of the propagating waves. If, for example, we designate $Q(\mathbf{R}, t)$ as the local density of the inhomogeneous effect, in the volume or at an interface, then conventional techniques can fail in handling them, since such media do not generally support space- and/or time-harmonic solutions (the latter involving steady-state solutions), unless $Q = Q(\mathbf{R})$ or a space-harmonic representation when $Q = Q(t)$. Even the "classical" perturbation and variational techniques (cf. Chapter 9 of Ref. [1]) are only applicable when Q is deterministic. This is not the usual situation which we encounter in the oceans and atmosphere[34], whose local and distributed properties are essentially random in both space and time.

However, the deterministic cases, examples of which are discussed here, do supply *typical or representative members of an appropriate ensemble,* when suitable probability measures are assigned to the set. These, in turn, enable us to obtain the desired moments and pdfs that quantitatively describe the random fields of the physical phenomena under consideration. Thus, the deterministic cases are preludes[35] to the general stochastic situation of propagation in the ocean, the atmosphere, and in space. An extensive *deterministic* or nonprobabilistic treatment of these problems of obtaining (single) representative ensemble members has been provided by Chew [6], for both linear homogeneous and inhomogeneous media, who considers both acoustic and electromagnetic propagation and deterministic scattering with methods of evaluating both direct and inverse problems.

We note in addition here that as long as these deterministic propagation equations are linear in the field, the role of the (deterministic) scattering elements does not alter the field's linear character. However, as Chew ([6], Chapter 2), for example, has pointed out, the scattered field here is *nonlinear with respect to the scatterers themselves*. Situations of this type belong to the class of *inverse problems*. In this case, the field solutions are still linear but are no longer unique (Chew [6], p. 511, etc). They must be selected with the help of additional information, particular to the problem at hand. We shall see examples of this in Section 8.3.1 following, in the use of series methods of solution. [Chew's work, [6], is entirely "classical," i.e., deterministic.] Finally, inhomogeneous (scattered) and boundaries (themselves inhomogeneous) produce nonlinear fields, but only for the random media, represented by a stochastic (i.e., Langevin) equation, or propagation, as discussed presently in Chapter 9.

[34] The ocean is highly inhomogeneous and thus requires varieties of approaches of different complexities discussed in the above reference. The atmosphere can be similarly complex acoustically and electromagnetically, also depending on the frequencies used. Space is comparatively free for propagation, except, of course, for large radiating bodies (the sun) and the electromagnetic (EM) fields it generates, as well as other types of ambient radiation, including magnetic fields, charged particles and so on, on a much larger scale then oceans and atmospheres. All these phenomena can and do interfere to varying degrees with reception.

[35] From the more general viewpoint of probability theory, a deterministic example represents an ensemble of one known member. See also the discussion in Chapter 9.

Although exact analytical results in the more general extension to random field, as well as in most of the general deterministic cases have been elusive and their numerical solutions constrained by limited computer power in the recent past and earlier, various approximate methods have been, and are, available for treating these media. Among them is *ray-tracing*, a comparatively high frequency technique (for example, see Tolstoy and Clay [33], and Brekhovskhikh [34]), and *modal analysis*, a comparatively low-frequency method [33]. An extension of these ideas, combined with path-integral methods and super-eikond procedures, has also been developed and applied, for instance, to the study of sound speed fluctuations in nonisotropic oceans [20] and in subsequent studies, by Jensen et al. [35]. All these methods are necessarily approximate, where a skillful approximation is one of the major keys to success.

In our present account of the general scattering problem, here in Chapter 9 immediately following, our philosophy is to preserve the exact formal solutions, which are presented in a canonical operator development, reserving approximations to the last possible moment. This has the great conceptional advantage of allowing us to identify for given situations the disposable or nonrelevant parts of the general formulation. In any case approximations are almost inevitable: we seek to postpone them as long as possible. Accordingly, from the general formulation we may find manageable analytic solutions, with some measure of the limitations on their applicability. In turn, the operator formulation can provide us with typical representations of a random ensemble, or *Langevin equation* (cf. Chapter 10 of Ref. [25]), whose solutions in turn are statistical moments and probability distributions (cf. Chapter 9). It is here, ultimately, that online computational techniques based on the operational solutions or "macro-algorithms" can give us the desired numerical results (even in the deterministic cases, cf. Chew [6], Chapters 8 and 9).

8.3.1 Deterministic Volume and Surface Scatter: The Green's Function and Associated Field $\boldsymbol{\alpha}^{(Q)}$

Let us begin with *volume scatter*, by determining the GF of the *unbounded* medium containing a continuum of scatterers that constitute the inhomogeneities in the infinite medium ($V \to \infty$). This is shown in Fig. 8.9 with the bounding surfaces S_0, and S' now removed to infinity, that is, $S' > S = \infty$: all scatterers are included in the volume defined by the surface $S - S_0 = \infty$; $S_0 > 0$, where S_0 is the finite closed surface bounding the source function in V_T, and the point sources at $(\mathbf{R}', t')$ within V_T. The GF $g_\infty^{(Q)}(\mathbf{R}, t|\mathbf{R}', t')$ is determined by the operator equation, cf. [(8.1.9a) and (8.1.9b)], extended here to[36]

$$\hat{L}^{(1)} g_\infty^{(Q)} = -\delta_{\mathbf{R}'\mathbf{R}}\delta_{t't}, \quad \text{where} \quad \hat{L}^{(1)} \equiv \hat{L}^{(0)} - \hat{Q} \text{ for } V - V_{\mathrm{T}}. \tag{8.3.1}$$

The operator $\hat{Q} \equiv \hat{Q}(\mathbf{R}, t)$ represents the local density operator for these volume scatterers. They in turn act as secondary sources whose radiative contribution depends on the primary source in V_T. Similarly, the operator equation for the *field* $\alpha_\infty^{(0)}$ is given by

$$\hat{L}^{(1)} \alpha_\infty^{(Q)} = \left(\hat{L}^{(0)} - \hat{Q}\right)\alpha_\infty^{(Q)} = -G_{\mathrm{T}}(\mathbf{R}, t)]_{+\mathrm{C}}; \neq 0, G_{\mathrm{T}} \in V_{\mathrm{T}}; = 0 \text{ elsewhere}, \tag{8.3.2}$$

[36] The subscripts on $g_\infty^{(Q)}, \alpha_\infty^{(Q)}$ indicate the infinite medium (outside V_T), and the superscript (Q) denotes the presence of inhomogeneities. We shall modify or omit these indicators in the following wherever there is no confusion in doing so.

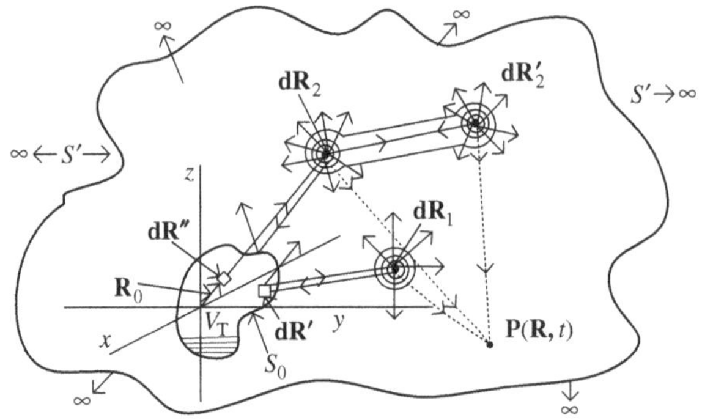

FIGURE 8.9 Schematic of elementary emitters $\mathbf{dR}', \mathbf{dR}''$ as part of distributed source in V_T, exciting inhomogeneous elements, $\mathbf{dR}_1$, and coupled pairs $\mathbf{dR}_2, \mathbf{dR'}_2$, each of which scatter in all directions: single scatter ($k = 1$) is omni-directional, and pair scatters ($k = 2$) obey a dipole pattern of energy redistribution, with these and higher order (multiple) scatter from k-tuples ($k \geq 3$). Here the deterministic scatterers ($k \geq 1$) represent elements in a continuum of the medium in the semi-infinite volume V (where $S \to \infty$) excluding the primary source within its boundary surface S_0. (Backscatter from S_0 is also indicated.)

cf. (8.1.7), respectively for (8.3.1) and (8.3.2). This latter pair is the extension of the former for the now inhomogeneous medium, as contrasted with the homogeneous cases considered previously in Section 8.1. Both Eqs. (8.3.1) and, (8.3.2) are subject to appropriate initial and boundary conditions, the latter associated here with the surface of the source function[37] G_T; (note remarks following Eq. (8.1.42), etc.).

The formal solutions are obtained as before [(cf. 8.1.9a), (8.1.9b), (8.1.10)]. We begin with (8.3.1) and multiply both sides of Eq. (8.3.1) by $\hat{M}_\infty^{(0)}$. This is the *integral* Green's function operator for the corresponding infinite *inhomogeneous medium,* which is over all space and time, including the impulsive point sources at $\mathbf{R}' = \mathbf{R}$ and time $t' = t$. We obtain

$$\hat{M}_\infty^{(0)}\left(\hat{L}^{(0)} - \hat{Q}\right)g_\infty^{(Q)} = -\hat{M}_\infty^{(0)}\delta_{\mathbf{R}'\mathbf{R}}\delta_{t't}, \text{ which becomes } \left(\hat{1} - \hat{M}_\infty^{(0)}\hat{Q}\right)g_\infty^{(Q)} = g_\infty^{(0)}. \tag{8.3.3}$$

This gives us directly the desired result

$$g_\infty^{(Q)} = \left(\hat{1} - \hat{M}_\infty^{(0)}\hat{Q}\right)^{-1} g_\infty^{(0)} = \left(\hat{1} - \hat{\eta}_\infty\right)^{-1} g_\infty^{(0)}, \tag{8.3.4a}$$

with

$$\hat{\eta}_\infty \equiv \hat{M}_\infty^{(0)}\hat{Q}. \tag{8.3.4b}$$

The integration implied in $\hat{M}_\infty^{(0)}$ is over all space contained in S'. The quantity $\hat{\eta}$ is similar to the "*mass operator*" (or the "*field renormalization operator*" in quantum electrodynamics

[37] For the moment we shall neglect the backscatter from the surface S_0 bounding the source. We consider here only the scatter, locally embodied in $\hat{Q}$, which is attributable to the inhomogeneities in the medium outside the source in V_T. Backscatter is considered in Section 8.3.4.

or quantum field theory).[38] In control theory it is analogous to the "loop-cycle" or "loop-iteration operator". Where a $\hat{Q}$ is a local operator, $\hat{M}_{\infty}^{(0)}$ and $\hat{\eta}$ are nonlocal or global, that is, they function over the whole (here infinite) domain of $S' \to \infty$.

Note that Eq. (8.3.4) is also equivalent to the integral equation

$$g_{\infty}^{(Q)} = g_{\infty}^{(0)} + \eta_{\infty} g_{\infty}^{(Q)}, \tag{8.3.4c}$$

which is particularly suited to those cases where strong as well as weak scattering occurs.

We can also express (8.3.4) in a more revealing representation by writing the operator $\left(\hat{1} - \hat{\eta}\right)^{-1}$ in series form:

$$g_{\infty}^{(Q)} = \sum_{k=0}^{Q \to \infty} \hat{\eta}_{\infty}^{(k)} g_{\infty}^{(0)} = g_{\infty}^{(0)} + \sum_{k=1}^{\infty} \hat{\eta}_{\infty}^{(k)} g_{\infty}^{(0)} = g_{\infty}^{(0)} + \sum_{k=1}^{\infty} g_{\infty}^{(k)};\ \left\| \hat{\eta}_{\infty}^{(k)} \right\| = \left\| \hat{\eta}_{\infty} \hat{1} \right\| < \infty, \tag{8.3.5}$$

$$g_{\infty}^{(k)} \equiv \hat{\eta}_{\infty}^{(k)} = g_{\infty}$$

This series solution for the inhomogeneous Green's function $g^{(Q)}$ is called the *perturbation series solution* (PSS). Here $\left\| \hat{\eta}_{\infty}^{(k)} \right\|$ represents[39] the *norm* of $\hat{\eta}_{\infty}^{(k)}$. In this series form we must have $\left\| \hat{\eta}_{\infty} \right\| < 1$ for convergence. For weak scatter ($\left\| \hat{\eta}_{\infty} \right\| << 1$) this series expression is particularly useful (cf. the Born approximation, discussed in Section 9.3.)

The physical interpretation of (8.3.4), is that the point source at $(\mathbf{R}', t')$ of the Green's function propagates omni directionally and a portion of this field in turn may be scattered by an inhomogeneity at $(\mathbf{R}_1, t_1)$, represented by $g_{\infty}^{(1)}$. The process can repeat itself with pairs $(k=2)$ and triples $(k=3)$ and more $(k>3)$, of interacting scatterers. Thus, $k(\geq 1)$ expresses the *order* of the scattering, namely, a "k-tuple". All orders can potentially occur: for weak scatter only $k=1$ (and possibly $k=2$) need to be considered. For strong scatter, k-tuples ranging from 1 to large or essentially infinite values may need to be taken into account. Note also that the *form* $\left(\hat{1} - \hat{\eta}_{\infty}\right)^{-1}$ in these results suggests some kind of "feedback" operation, by analogy with classical control theory. We shall use this feature to outline briefly one potential method of (numerical) solution in Section 8.5.3.

Finally, we readily obtain similar results to [(8.3.4a) and (8.3.5)] for *surfaces*, if we start with the well-known scattering relation

$$g_{S}^{(Q)} = g_{S}^{(0)} + \hat{\eta}_{S} g_{S}^{(Q)}, \ \hat{\eta}_{S} \equiv \hat{M}_{S}^{(0)} \hat{Q}_{S}. \tag{8.3.5a}$$

From (8.3.5a) we observe that a formalism corresponding to (8.3.3) et. seq. can be defined, namely,

$$g_{S}^{(Q)} = \left(\hat{1} - \hat{\eta}_{S}\right)^{-1} g_{S}^{(0)}, \text{ implying } L_{S}^{(1)} = \hat{L}_{S}^{(0)} - \hat{Q}_{S}; \hat{M}_{S}^{(0)} = \hat{L}_{S}^{(0)-1}. \tag{8.3.5b}$$

[38] Here we modify the terminology somewhat and call $\hat{\eta}_{\infty}$ the global mass operator or global field renormalization operator, and $\hat{Q}$ the (local) mass operator, with Q the kernel of this local mass operator.

[39] We shall not attempt a rigorous treatment of linear operator theory here, referring the reader instead to several well-known texts on the subject [18,19].

from which follow relations analogous to (8.3.2)—(8.3.5), now for (deterministic), surface scatter here.

8.3.2 The Associated Field and Equivalent Solutions for Volumes and Surfaces

A similar interpretation can be made, for example, for the field from a distributed source (in V_T), with appropriate initial and boundary conditions, and produced in the infinite scattering medium ($S \to S' \to \infty$) of Fig. 8.9.

We begin with the relation (8.3.2) and carry out a set of operations similar to those used for obtaining the GF relations (8.3.4) and (8.3.5). As before, cf. 8.3.1, we shall see that there are two equivalent forms of solutions: an integral equation and the perturbation series solution. The formal result is easily found from the relations

$$\hat{L}^{(1)}\alpha_\infty^{(Q)} = -G_T \quad \text{or} \quad \alpha_\infty^{(Q)} = \left(\hat{1} - \hat{\eta}_\infty\right)^{-1}\alpha_H;\ \alpha_H = \hat{M}_\infty^{(0)}(-G_T). \tag{8.3.6}$$

From (8.3.6), we have respectively

$$\alpha_\infty^{(Q)} = \alpha_H + \hat{\eta}_\infty \alpha_\infty^{(Q)} \quad \text{and} \quad \alpha_\infty^{(Q)} = \alpha_H + \sum_{k=1}^{\infty} \hat{\eta}_\infty^{(k)} \alpha_H. \tag{8.3.7a}$$

From (8.3.4b) we exhibit the integral Green's function and local inhomogeneous operators that comprise the mass operator $\hat{\eta}_\infty$:

$$\alpha_\infty^{(Q)} = \hat{M}_\infty^{(0)}(-G_T) + \hat{M}_\infty^{(0)}\hat{Q}\alpha_\infty^{(Q)};\ \alpha_\infty^{(Q)} = \alpha_H + \sum_{k=1}^{\infty}\left(\hat{M}_\infty^{(0)}\hat{Q}\right)^{(k)};\alpha_H,\ \|\hat{\eta}_\infty\| < 1 \tag{8.3.7b}$$

for the two forms of solution. For the first, this integral equation is not restricted by the convergence condition $\|\eta\| < 1$, only that $\|\eta\| < \infty$, whereas the perturbation form in (8.3.7b) usually requires $\|\hat{\eta}_\infty\|$ to be comparatively small (< 1) for useful results.

It is clear from Eq. (8.3.6)–(8.3.7b) that the scattered field $\alpha_\infty^{(Q)}$ is *nonlinearly* dependent on the local scattering element, represented by the operator $\hat{Q}$. However, the linearity of the medium (and hence for the field α propagation in it) is still evident from the following: consider the additive field $\alpha_{12} = \alpha_1 + \alpha_2$. From (8.3.7b) it follows at once that

$$\left.\begin{aligned} \alpha_\infty^{(Q)}\big|_{12} &= \alpha_{H_1} + \sum_{k=1}^{\infty}\eta_\infty^{(k)}\alpha_{H_1} + \alpha_{H_2} + \sum_{k=2}^{\infty}\eta_\infty^{(k)}\alpha_{H_2} = (\alpha_{H_1} + \alpha_{H_2}) + \sum_{k=1}^{\infty}\eta_\infty^{(k)}(\alpha_{H_1} + \alpha_{H_2}), \\ &= \alpha_{\infty-1}^{(Q)} + \alpha_{\infty-2}^{(Q)} \end{aligned}\right\} \tag{8.3.7c}$$

Since $\eta_{\infty;1,2}^{(k)} = \hat{M}_\infty^{(0)}(\hat{Q}_1 + \hat{Q}_2)$ and $\hat{Q}_1 = \hat{Q}_2 = \hat{Q}$. This last relation holds because the *same* scatterers are now producing the total scattered field $\alpha_\infty^{(Q)}|_{12}$, that is, in this deterministic case the scatterers are unchanged, or "frozen" in this nonrandom medium. On the other hand, for random media, this argument no longer holds, vide Chapter 9. We also observe, by the same argument, that only in the important cases of *random* fields, do boundaries matter. Therefore, there is nonlinearity. This occurs because $\hat{\eta}$ boundary is different for each for each representation in the ensemble. When the randomness is removed by averaging, that is, the result is now deterministic, and the propagation of the average result is then linear, superposition then applies.

In addition, from (8.1.42a) for the global (or integral) Green's function operator $\hat{M}_\infty^{(0)}$ we have specifically the following generic forms for volume and for general (closed) surfaces in V:

$$\hat{M}_\infty^{(0)}|_V = \int dt' \int_{V_T} g_\infty^{(0)}(\mathbf{R}, t|\mathbf{R}', t') V\,(\,)_{\mathbf{R}',t'} dR';$$

$$\hat{M}_\infty^{(0)}|_S = \int dt' \oint_{S'} \left(g_\infty^{(0)} \hat{\mathbf{n}}' \cdot \nabla' - \left[\hat{\mathbf{n}}' \cdot \nabla' g_\infty^{(0)} \right] \right) (\,)_{(\mathbf{R}',t') \in S'} dS', \tag{8.3.8}$$

The (local) inhomogeneity operator $\hat{Q}(B, t')_{V \text{ or } S}$ depends generally on space–time but is, by definition, not a projection operator like $\hat{M}_\infty|_{V,S}$. However, $\hat{M}_\infty^{(0)}|_S$ and its kernel $g_\infty^{(0)}|_S$ are also functions *of the surface geometry*, specifically embodied in J'(8.1.42i).

The local inhomogeneity operators $\hat{Q}_{V,S}(\mathbf{R}', t')$ depends on the medium in question. It is a local volume or surface quantity, which in the case of surfaces is usually a reflection or transmission kernel $Q_S = R_0(\mathbf{R}', t')$ or $T_0(\mathbf{R}', t')$. The mass operator $\hat{\eta}$, cf. (8.3.4a) and (8.3.4b), is for surfaces or volumes

$$\hat{\eta}_\infty|_S = \hat{M}_S^{(0)} \hat{Q}_S \quad \text{or} \quad \hat{\eta}_\infty|_V = \hat{M}_V^{(0)} \hat{Q}_V \tag{8.3.8a}$$

In particular, we now find that these mass operators become, in detail from [(8.1.42a)–(8.1.42f)] and [(8.1.42h) and (8.1.42i)] with the added local inhomogeneity kernels Q_V, Q_S,

$$\hat{\eta}_\infty|_V = \hat{M}_V^{(0)} \hat{Q}_V = \int dt' \int_{V_T} g^{(0)} \left(\mathbf{R}, t|\mathbf{R}', t') Q_V(\mathbf{R}', t')(\,)_{\mathbf{R}',t'} \mathbf{dR}' \right. \tag{8.3.9a}$$

$$\hat{\eta}_\infty|_S = \hat{M}_S^{(0)} \hat{Q}_S = \int dt' \oint_{S_0} \left\{ g_\infty^{(0)} J' - \left[J' g_\infty^{(0)} \right] \right\} \left[Q_S(\mathbf{R}', t')(\,)_{\mathbf{R}',t'} \right] \mathbf{dR}' \tag{8.3.9b}$$

where now $\alpha^{(Q)} \to Q_S \alpha^{(Q)}$, cf. also Eq. (8.3.25b). Here the surface kernel J' is explicitly

$$\hat{J}'(\mathbf{R}', t') = \left(\zeta_{x'} \frac{\partial}{\partial x'} + \zeta_{y'} \frac{\partial}{\partial y'} + \frac{\partial}{\partial z'} \right)_{\mathbf{R}',t'} \doteq \hat{J}'(\mathbf{r}', t'), \tag{8.3.9c}$$

In particular, we have used the results of Section 8.1.5.2 to evaluate $\hat{\mathbf{n}}' \cdot \nabla'(\,) dS'$. The quantity to be used in the parentheses $(\,)_{\mathbf{R}',t'}$ of (8.3.9a) and (8.3.9b) is for the PSS form of solution the *unperturbed field* α_H, cf. Sections 8.1.6.1, and 8.1.6.3, where $\alpha_\infty^{(k)} = \hat{\eta}_\infty^{(k)} \alpha_\text{H}$. On the other hand, for the integral equation equivalent in (8.3.7a) and (8.3.7b) one uses the *scattered field* $\alpha^{(Q)}$. In either case the total scattered field is $\alpha^{(Q)}$, Eqs. (8.3.6) and (8.3.7). Equivalent results for $\alpha^{(Q)}$ are obtained where we employ the GHP and those of Sections 8.3.4.1 and 8.3.4.2 following. Here one more in the PSS form α_H is the homogeneous or unscattered part of the total field $\alpha^{(Q)}$, and $\hat{\eta}(1 - \hat{\eta})^{-1} \alpha_\text{H} \left(= \sum_{k=1}^{\infty} \eta^{(k)} \alpha_\text{H} \right)$ is the scattered portion.

8.3.2.1 Example As a simple volume example, using the (time-dependent) Helmholtz equation in this unbounded *inhomogeneous* medium and neglecting backscatter, we have explicitly (with the boundary and initial conditions that accompany (8.3.2)):

$$\left(\nabla^2 - \frac{1}{c_0^2}[1+\varepsilon(\mathbf{R},t)]\frac{\partial^2}{\partial t^2}\right)\alpha^{(Q)}(\mathbf{R},t) = -G_{\mathrm{T}}(\mathbf{R},t), \neq 0, G_{\mathrm{T}} \in V_{\mathrm{T}}; = 0, \text{ elsewhere}, + \mathrm{C} \tag{8.3.10}$$

Now

$$\hat{L}_V^{(0)} \equiv \nabla^2 - \frac{1}{c^2}\frac{\partial^2}{\partial t^2} \quad \text{and} \quad \hat{Q}_V(\mathbf{R},t) \equiv \frac{\varepsilon(\mathbf{R},t)}{c_0^2}\frac{\partial^2}{\partial t^2}, \quad \text{with} \quad \hat{L}_V^{(1)} \equiv \hat{L}_V^{(0)} - Q_V, \text{ cf. (8.3.2)}, \tag{8.3.11}$$

subject, say, to the initial conditions $t < t_0^{(-)} = 0, \frac{\partial^2 \alpha^{(Q)}}{\partial t^2} = \frac{\partial \alpha^{(Q)}}{\partial t} = \alpha^{(Q)} = 0$ and with no surface sources on the boundaries[40] of V_{T}. Thus, $\alpha^{(k)} = \hat{\eta}^{(k)}\alpha_{\mathrm{H}}, k \geq 1$, represent the various orders of k-tuples as well as their contributions to the resulting scatter components. Equation (8.3.6) can be written alternatively

$$\alpha_\infty^{(Q)} \equiv \hat{M}_\infty^{(Q)}\alpha_{\mathrm{H}} = \hat{M}_\infty^{(Q)}\left(\hat{M}_\infty(-G_{\mathrm{T}})\right) \therefore \hat{M}_\infty^{(Q)} = \left(\hat{1} - \hat{\eta}_\infty\right)^{-1} = 1 + \sum_{k=1}^{\infty}\hat{\eta}_\infty^{(k)}, \tag{8.3.12}$$

The integral Green's function for the inhomogeneous medium here is $\hat{M}_\infty^{(Q)} = \hat{M}_V^{(Q)}$ for the infinite volume $(V_S - V_T) \to \infty$, which is the inhomogeneous counterpart of $\hat{M}_\infty^{(0)}$ in the homogeneous cases discussed in Section 8.1 above. Since there is negligible backscatter assumed in the present case, $\hat{Q}_S = 0$ and $\therefore\ \hat{\eta}_S = 0$. There is no surface-generated component of scatter. Only volume scatter in the medium occurs, with $\hat{\eta}_V$, Eq. (8.3.9a) the only mass operator activated. Thus, the scattered field is due solely to activated scatterers in the volume.

Table 8.4 provides specific examples of Eqs. (8.3.20) and (8.3.22), extensions of the homogeneous cases of Section 8.1.6.3 to include inhomogeneous media.

Although this table applies here for the deterministic cases considered in Chapter 8, it also applies for random media, where now the parameters $\epsilon, \hat{a}_0$, and so on are *random* and the resulting ensemble of propagation equations (1)–(6) constitute the corresponding Langevin equations, cf. Chapter 9.

At this point, we need to consider the various operators used in Sections 8.3.1 and 8.3.2 in more detail. Their formal properties are treated vigorously in Ref. [18,19]. We proceed therefore with some further specific results below.

8.3.3 Inhomogeneous Reciprocity

It is possible now to show that the Green's function for the inhomogeneous volume and surface environments are reciprocal, that is, the field remains unchanged if source location and the observation point are interchanged. Analytically, as we have seen earlier (cf. 8.1.29) for the homogeneous, unbounded medium, this means that now $g_\infty^{(k)}(\mathbf{R},t|\mathbf{R}',t') = g_\infty^{(k)}(\mathbf{R}',-t'|\mathbf{R},-t)$, and $\lim_{R\to\infty} g_\infty^{(b)} = g_\infty^{(Q)}$, cf. Sections 8.1 and 8.2.

[40] An in the case of the GHP discussed in Section 8.1.5 above for the homogeneous media.

TABLE 8.4 Propagation in Deterministic Inhomogeneous Media: The Inhomogeneity Operator $\hat{Q}$

		Reflection and Transmission	
Type $\hat{L}^{(1)} \equiv L^{(0)} - \hat{Q}$	$\hat{Q} = \hat{Q}_V$	$\hat{Q}_S : \mathsf{R}_0, \mathsf{T}_0$	Remarks
(1) Time-dependent Helmholtz: $\nabla^2 - \frac{(1+\varepsilon)}{c_0^2}\frac{\partial^2}{\partial t^2}$	$\frac{\varepsilon(\mathbf{R},t)}{c_0^2}\frac{\partial^2}{\partial t^2}$	$\mathsf{R}_0(\mathbf{R},t), \mathsf{T}_0(\mathbf{R},t)$ *(Example)*	cf. Eq. (8.1.30a); Section 8.3.4.2
(2) Wave equation with radiation absorption $\nabla^2 - (a_{0R} + \hat{a}_R)\frac{\partial}{\partial t} - \frac{1}{c_0^2}\frac{\partial^2}{\partial t^2}$	$\hat{a}_R(\mathbf{R},t)\frac{\partial}{\partial t}$	$\mathsf{R}_0(\mathbf{R},t), \mathsf{T}_0(\mathbf{R},t)$	Eq. (8.1.31a); Section 8.3.4.2
(3) Wave equation with relaxation absorption $\left[1 + (\hat{\tau}_0 + \hat{\tau})\frac{\partial}{\partial t}\right]\nabla^2 - \frac{1}{c_0^2}\frac{\partial^2}{\partial t^2}$	$-\hat{\tau}(\mathbf{R},t)\frac{\partial}{\partial t}\nabla^2$	$R_0(\mathbf{R},t), T_0(\mathbf{R},t)$	Eq. (8.1.32a); Section 8.3.4.2
(4) Wave equation with relaxation and radiation absorption $\left[1 + (\hat{\tau}_0 + \hat{\tau})\frac{\partial}{\partial t}\right]\nabla^2 - (a_{0R} + \hat{a}_R)\frac{\partial}{\partial t} - \frac{1}{c_0^2}\frac{\partial^2}{\partial t^2}$	$-\hat{\tau}(\mathbf{R},t)\frac{\partial}{\partial t}\nabla^2 + \hat{a}_R(\mathbf{R},t)\frac{\partial}{\partial t}$	$R_0(\mathbf{R},t), T_0(\mathbf{R},t)$	Eq. (8.1.33a); Section 8.3.4.2
(5) Propagation by diffusion $\nabla^2 - \left(b_0 + \hat{b}_0\right)^2\frac{\partial}{\partial t}$	$\left(2b_0\hat{b}_0 + \hat{b}_2^2\right)\frac{\partial}{\partial t}$	$\mathsf{R}_0(\mathbf{R},t), \mathsf{T}_0(\mathbf{R},t)$	Eq. (8.1.34a); Section 8.3.4.2
(6) Diffusion: maximum velocity of heat absorption $\nabla^2 - \left(b_{0R} + \hat{b}_R\right)\frac{\partial}{\partial t} - \frac{1}{c_0^2}\frac{\partial^2}{\partial t^2}$	$\hat{b}_{0R}\frac{\partial}{\partial t}$	$\mathsf{R}_0(\mathbf{R},t), \mathsf{T}_0(\mathbf{R},t)$	Eq. (8.1.31a); Section 8.3.4.2

8.3.3.1 Volume[41] To verify this important feature for *inhomogeneous* media some preliminaries are required. First, we observe that for a representative k-tuple we obtain (all $k \geq 1$)

$$\left.\begin{aligned} k \geq 1: g_{\infty}^{(k)}(\mathbf{R}_k, t_k|\mathbf{R}', \mathbf{t}') &= \iint g_{\infty}^{(Q)}(\mathbf{R}_k, t_k|\mathbf{R}'_{k-1}, t'_{k-1})Q(\mathbf{R}'_{k-1}, t'_{k-1}) \\ &\quad \times g^{(k-1)}(\mathbf{R}'_{k-1}, t'_{k-1}|\mathbf{R}', t')\mathbf{dR}'_{k-1}dt'_{k-1} \\ &= \hat{M}_{\infty}^{(0)}\hat{Q}g_{\infty}^{(k-1)} = \hat{\eta}g_{\infty}^{(k-1)} = \eta^{(k)}g_{\infty}^{(0)};\ (V_{\infty} - V_T) \rightarrow \infty \text{ here.} \end{aligned}\right\} \tag{8.3.13}$$

We have already shown that $g_{\infty}^{(0)}$ is reciprocal, including symmetry in $\mathbf{R}, \mathbf{R}'$, that is, $g^{(0)}(\mathbf{R}, t|\mathbf{R}', t') = g_{\infty}^{(0)}(\mathbf{R}', -t'|\mathbf{R}, -t)$, [(v), Section 8.1.4, by the method outlined in (8.1.39)–(8.1.42)]. Accordingly, for $k = 1$, we get from (8.3.13)

$$g_{\infty}^{(1)}(\mathbf{R}_1, t_1|\mathbf{R}', t') = \int_{t_0}^{t_1} dt'_0 \int_{V_T(R_0)} g_{\infty}^{(0)}(\mathbf{R}_1, t_1|\mathbf{R}'_0, t'_0)Q(\mathbf{R}'_0, t'_0)g_{\infty}^{(0)}(\mathbf{R}'_0, t'_0|\mathbf{R}', t')\,\mathbf{dR}'_0 dt'_0, \tag{8.3.14}$$

which on applying reciprocity to the Green's functions $g_{\infty}^{(0)}$ in the right-hand members of (8.3.14) allows us to write

$$\iint g_{\infty}^{(0)}(\mathbf{R}'_0, -t'_0|\mathbf{R}_1, -t_1)Q(\mathbf{R}'_0, -t'_0)g_{\infty}^{(0)}(\mathbf{R}', -t'|\mathbf{R}'_0, -t'_0)\mathbf{dR}'_0 dt'_0 = g_{\infty}^{(1)}(\mathbf{R}', -t'|\mathbf{R}_1, -t_1), \tag{8.3.15a}$$

with $-t' > -t'_0 > -t_1$, consistent with the interchange of sources and the sequence of times of emission from the (single) scatterers here. Since $g_{\infty}^{(0)}|_{\text{recip}} = g_{\infty}^{(0)}$, the left member of (8.3.15a) becomes again the right member of (8.3.14) and we have the desired reciprocal relation for $k = 1$:

$$g_{\infty}^{(1)}(\mathbf{R}_1, t_1|\mathbf{R}', t') = g^{(1)}(\mathbf{R}', -t'|\mathbf{R}, -t_1) = \hat{\eta}g_{\infty}^{(0)} = \hat{M}_{\infty}^{(0)}\hat{Q}g_{\infty}^{(0)}(\mathbf{R}, t|\mathbf{R}', t'). \tag{8.3.15b}$$

Proceeding in the same way, we obtain for $k = 2$

$$\begin{aligned} g_{\infty}^{(k)} = g_{\infty}^{(2)}(\mathbf{R}_2, t_2|\mathbf{R}', t') &= \int_V \cdots \int g_{\infty}^{(0)}(\mathbf{R}_2, t_2|\mathbf{R}'_1, t'_1)Q(\mathbf{R}'_1, t'_1)g_{\infty}^{(0)}(\mathbf{R}'_1, t'_1|\mathbf{R}'_2, t'_2) \\ &\quad Q(\mathbf{R}'_2, t'_2)g_{\infty}^{(0)}(\mathbf{R}'_2, t'_2|\mathbf{R}', t') \cdot \mathbf{dR}'_1\mathbf{dR}'_2 dt'_1 dt'_2, \end{aligned} \tag{8.3.15c}$$

or

$$\begin{aligned} g_{\infty}^{(2)}(-\mathbf{R}', t'|\mathbf{R}_2, -t_2) &= \int \cdots \int g_{\infty}^{(0)}(\mathbf{R}'_1, -t'_1|\mathbf{R}_2, -t_2)Q(\mathbf{R}'_2, t'_2)g^{(0)}(\mathbf{R}'_2, -t'_2|\mathbf{R}'_1, -t'_1) \\ &\quad Q(\mathbf{R}'_1, t'_1)g_{\infty}^{(0)}(\mathbf{R}', -t'|\mathbf{R}'_2, -t'_2),\ \cdot \mathbf{dR}'_1\mathbf{dR}'_2 dt'_1 dt'_2 \end{aligned} \tag{8.3.15d}$$

[41] For simplification we omit the subscript (V) on the Green's functions an don the inhomogeneity kernel $Q(R', t')$, and so on.

on applying reciprocity to the $g_\infty^{(0)}$. Changing the variable of integration $R'_1 \rightleftarrows R'_2$, $t'_1 \rightleftarrows t'_2$, noting that $t' > -t'_2 > -t'_1$ as required, and involving reciprocity once more, we see that the right-hand members of (8.3.15d) are equal to (8.3.15c). Thus, the left-hand members of (8.3.15c) and (8.3.15d) are equal and hence $g_\infty^{(2)}$ is reciprocal.

Repeating for $k \geq 3$, using (8.3.13) in each case, we see that $g_\infty^{(k)}|_{\text{Recip}} = g_\infty^{(k)}$, that is, $g^{(k)}(R_k, t_k|R', t') = g_{\text{Recip}}^{(k)} = g^k(R', -t'|R_k, -t_k)$, (8.3.15b) and by extension $g_\infty^{(Q)}|_{\text{Recip}} = g_\infty^{(Q)}$ as $k \to \infty$. Thus, we have in abbreviated form

$$\begin{aligned} g_\infty^{(Q)} &= g_\infty^{(Q)}|_{\text{recip}}, \text{ with} \\ g_\infty^{(Q)} &= \lim_{k\to\infty} g_\infty^{(k)} = \iint g_\infty^{(k)} \hat{Q} g_\infty^{(k-1)} \mathbf{dR}_{k-1} dt_{k-1}; \\ g_\infty^{(Q)}|_{\text{recip}} &= \lim_{k\to\infty} \left(\hat{\eta} g_\infty^{(k-1)}\right) = \sum_{k=0}^{\infty} \eta^{(k)} g_\infty^{(0)} \end{aligned} \tag{8.3.16}$$

cf. (8.3.13). This leads directly to the extension of the Generalized Huygens Principal discussed in Section 8.1.6, to the situation here, also with an unbounded but inhomogeneous medium. (This medium, however, is so far still deterministic.)

8.3.3.2 Surfaces Surfaces are handled in similar fashion. Instead of the (local) Green's function kernel for volumes $g_\infty^{(0)}(\mathbf{R}, t|\mathbf{R}', t')_V$, we have now the local Green's function operator for surfaces:

$$\hat{g}_\infty^{(0)}|_S \equiv \left(g_\infty^{(0)} \hat{J}' - \left[\hat{J}' g_\infty^{(0)}\right]\right)_J \hat{g}_\infty^{(0)}(\mathbf{R}, t|\mathbf{R}', t')_S, \tag{8.3.17}$$

where the local surface operation component $\hat{J}$ is given by (8.1.42i), namely,

$$\hat{J}' = \hat{J}'(\mathbf{R}', t') = \left(\zeta_{x'} \frac{\partial}{\partial x'} + \zeta_{y'} \frac{\partial}{\partial y'} + \frac{\partial}{\partial z'}\right)_{\mathbf{R}', t'}; \ \zeta' = \zeta(\mathbf{r}', t'). \tag{8.3.17a}$$

[See Section 8.1.6.2 for a detailed treatment.] We may then proceed with $\hat{g}_\infty^{(0)}|_S$ in place of $g_\infty^{(0)}\left(= g_\infty^{(0)}|_V\right)$ in (8.3.13)–(8.3.16), above, to show reciprocity for the surface Green's functions.

Although reciprocity in (linear) *inhomogeneous* media holds for the Green's functions for such media, cf. Eq. (8.3.13) et. seq., this property is no longer strictly true when a distributed source $G_{\text{OT}}(\mathbf{R}', t')$ is employed, which as a result has a directional "beam" (or "beams"), (cf. Section 2.5, Section 2.5.1.1). Thus, interchanging source location and observation point, using the now directional source G_{OT} and directing it at the original source location, in the manner of Figure 8.10, shows that some radiated energy at least escapes the main beam, apart from additional energy scattered outside the overall beam pattern. On the other hand, the beam pattern of a Green's function is uniform (in the appropriate number of dimensions) and thus completely encloses the point of observation and all point scatterers.

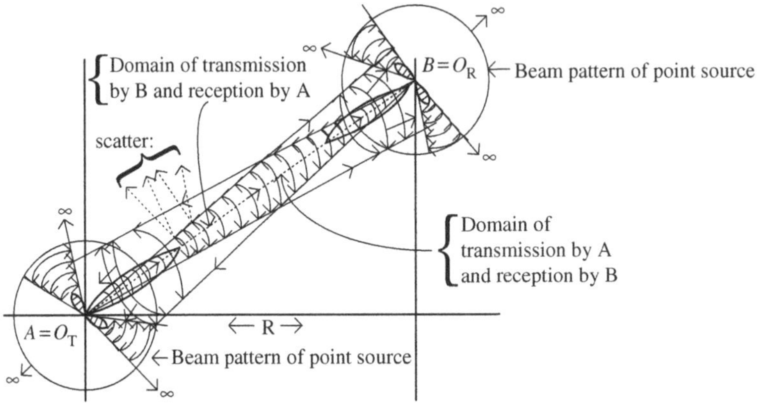

FIGURE 8.10 Partial reciprocity of identical directional sources O_{T} versus O_{R}. T and R are reciprocal points of observation, as are transmitter and receiver location. [Two-dimensional section.]

8.3.4 The GHP for Inhomogeneous Deterministic Media including Backscatter

We proceed next to consider in detail the structure of the field in these now inhomogeneous cases. We begin with the GHP for this case and observe first that the linear operators $\hat{L}^{(0)}$ and $\hat{L}^{(1)}$, from Eqs. (8.3.1) and (8.3.2) above, consist of a *linear combination* of two or more of the following components[42]

$$\hat{L}^{(0)} = \hat{L}^{(0)}\left(\nabla^2, \frac{\partial\nabla^2}{\partial t}; \frac{\partial}{\partial t}, \frac{\partial^2}{\partial t^2}\right);\ \hat{L}^{(1)} = \hat{L}^{(1)}\left(\nabla^2, \frac{\partial\nabla^2}{\partial t}; \frac{\partial}{\partial t}, \frac{\partial}{\partial t^2}; \hat{Q}\right) = \hat{L}^{(0)} - \hat{Q}, \tag{8.3.18}$$

constituting the equations of propagation here (namely, Table 8.4 above):

$$\hat{L}^{(1)} g_{\infty}^{(Q)} = -\delta_{\mathbf{R}'\mathbf{R}}\delta_{t't};\ \hat{L}^{(1)}\alpha_{\infty}^{(Q)} = -G_{\mathrm{T}}; \neq 0, \mathbf{R}\varepsilon V_{\mathrm{T}}; \quad = 0 \text{ elsewhere} \tag{8.3.18a}$$

Again [cf. (8.1.39)–(8.1.42)], forward multiplying the first relation in (8.3.18a) by $\alpha_{\infty}^{(Q)}$ and the second by $g_{\infty}^{(Q)}$ subtracting the first from the second, and integrating over the source volume (V_{T}) and time (t) from initiation of the field $\alpha_{\infty}^{(Q)}$ to the present (t), gives[43]

$$\int_{t_0^-}^{t^+} dt' \int_{V_T} \mathbf{dR}'\left(g_{\infty}^{(Q)}\hat{L}^{(1)'}\alpha_{\infty}^{(Q)'} - \alpha_{\infty}^{(Q)}\hat{L}^{(1)'}g_{\infty}^{(Q)'}\right) = -\int_{t_0^-}^{t^+} dt' \int_{V_T} \mathbf{dR}' g_{\infty}^{(Q)} G_{\mathrm{T}} + \alpha_{\infty}^{(Q)}, \tag{8.3.19a}$$

[42] Since $\frac{\partial}{\partial t}$, $\frac{\partial^2}{\partial t^2}$, ∇^2, and so on, are bounded linear operators, and $\hat{L}^{(0)}$, $\hat{L}^{(1)}$ are linear combinations of these operators in a linear sum, they are also bounded. Rectangular coordinates also are used throughout, unless otherwise indicated.

[43] The t^+ and t_0^- indicate that entire delta functions about t (>0) and t_0 (>0) are included.

with $\mathbf{dR}' = dx'dy'dz' \equiv dV_{\mathrm{T}}$ as before. From (8.3.18) this becomes explicitly for the field $\alpha_{\infty}^{(Q)}$

$$\alpha_{\infty}^{(Q)}(\mathbf{R},t) = \int_{t_0}^{t'} dt' \int_{V_{\mathrm{T}}} \mathbf{dR}' g_{\infty}^{(Q)} G_{\mathrm{T}}(\mathbf{R}',t') + \int_{t_0}^{t^+} dt' \int_{V_{\mathrm{T}}} \mathbf{dR}' \left(g_{\infty}^{(Q)} \hat{L}^{(0)'} \alpha_{\infty}^{(Q)'} - \alpha_{\infty}^{(Q)} \hat{L}^{(0)'} g_{\infty}^{(Q)} \right)$$

$$- \int_{t_0}^{t^+} dt' \int_{V_T} \mathbf{dR}' \left(g_{\infty}^{(Q)} \hat{Q}' \alpha_{\infty}^{(Q)'} - \alpha_{\infty}^{(Q)} \hat{Q}' g_{\infty}^{(Q)'} \right). \tag{8.3.19b}$$

From the treatment in Section 8.1.6 we note that the second and third terms of (8.3.19b) can be converted respectively with the help of Green's theorem [24] and time integrations. These become in turn surface integrals, which embody the boundary condition, and initial conditions for the latter. We see this directly from the specific character of $\hat{L}^{(0)}$, which allows us to write (8.3.19b), for the moment in abbreviated form:

$$\alpha^{(Q)} = \int_{t_0}^{t^+} dt' \int_{V_{\mathrm{T}}} \mathbf{dR}' \left[g_{\infty}^{(Q)} G_{\mathrm{T}} + \mathbf{L}_1 \left(g^{(Q)}, \alpha_{\infty}^{(Q)'}; \nabla'^2, \hat{a}(\mathbf{R}',t') \frac{\partial}{\partial t'} \nabla'^2 \right) \right.$$

$$\left. + \mathbf{L}_2 \left(g_{\infty}^{(Q)}, \alpha_{\infty}^{(Q)'}; \frac{\partial}{\partial t'}, \frac{\partial^2}{\partial t'^2} \right) - \mathbf{L}_3 \left(g_{\infty}^{(Q)}, \alpha_{\infty}^{(Q)'}; \hat{Q}' \right) \right]. \tag{8.3.19c}$$

The second term in (8.3.19c) is transformed into a combined surface–time integral (because of terms like $\hat{\tau}(\partial/\partial t)\nabla^2$) and thus includes the boundary conditions for any (primary) source distribution on the surface of V_{T}. The third and fourth terms (in $(\partial/\partial t)(\partial^2/\partial t^2)$) represent the initial conditions. Note that $\hat{Q}' = 0$ in the last term of (8.3.19c), for the initial condition at $(\mathbf{R}' = \mathbf{R}_0, t' = t_0)$ so that $\mathbf{L}_3 = 0$. This is because the field $\left(\alpha_{\infty}^{(Q)} = \alpha_{\infty}^{(0)} \right)$ has not at this instant reached the inhomogeneities, which are in the volume V, external to V_T. Similarly, for the Green's function here, we have since $t > t' \geq t_0^-$, $g^{(Q)}(\mathbf{R},t|\mathbf{R}',t') \to g_{\infty}^{(0)}(\mathbf{R},t|\mathbf{R}_0,t_0)$, that is, $k = 0, (\mathbf{R}' = \mathbf{R}_0, t' = t_0)$ for the initial conditions explicit in the third term of (8.3.19c). From the results of Section 8.1.6, (8.1.41b)–(8.1.41d), (8.1.42) we see that (8.3.19c) becomes finally

$$\alpha_{\infty}^{(Q)}(\mathbf{R},t) = \int_{t_0^-}^{t^+} dt' \Bigg(\int_{V_T} \mathbf{dR}' g_{\infty}^{(Q)}(\mathbf{R},t|\mathbf{R}',t') G_{\mathrm{T}}(\mathbf{R}',t') \overbrace{\oint_{\mathbf{S}_0} \left[g_{\infty}^{(Q)}(1+a_0)\hat{\mathbf{n}}' \cdot \nabla' \alpha_{\infty}^{(Q)'} - \alpha_{\infty}^{(Q)}(1+a_0)\hat{\mathbf{n}}' \cdot \nabla' g_{\infty}^{(Q)'} \right] dS_0'}^{\leftarrow \mathbf{L}_1 \rightarrow} \Bigg)$$

$$\Bigg\{ \overbrace{\left(\int_{V_{\mathrm{T}}} \mathbf{dR}' \left[g^{(Q)} a(\mathbf{R}',t') \nabla^{2'} \alpha_{\infty}^{(Q)'} - \alpha_{\infty}^{(Q)} \nabla^{2'} g_{\infty}^{(\infty)} a(\mathbf{R}',t') \right] \right)_{R_0,t_0}}^{\leftarrow \mathbf{L}_2 \rightarrow} - \overbrace{\int_{V_{\mathrm{T}}} \mathbf{dR}' \left[\left(\frac{\partial g_{\infty}^{(Q)}}{\partial t} \alpha_{\infty}^{(Q)'} \right)_{R_0,t_0} - \left(g_{\infty}^{(Q)} \frac{\partial \alpha_{\infty}^{(Q)'}}{\partial t} \right)_{R_0,t_0} \right]}^{\leftarrow \mathbf{L}_3 \rightarrow} \Bigg\} \tag{8.3.20}$$

The factor $(1 + a_0)$ in the $\mathbf{L}_1$ term represents the constant part of $\hat{a} = a_0 + a(\mathbf{R}, t)$, if any, cf. (3) in (8.1.45), for example, $a_0 = \hat{\tau}_0$. We note, in addition, that the third term in (8.3.20) is the portion of the second term in (8.3.19b) for which Green's theorem does not convert to a surface integral, for example, in those cases where $\hat{L}^{(0)}$ contains the term $a(\mathbf{R}', t')(\partial/\partial t)\nabla^2$. It is instead a component of the initial conditions as designated by the subscripts $(\mathbf{R}_0, t_0)$.

Physically, the various terms of (8.3.20) represent the following behavior:

(8.3.21)

(1) The first term includes an unscattered or homogeneous outward field $(k = 0)$, plus most of the energy of all the scattered components $(k \geq 1)$, because of the directionality of the aperture (or array) included in G_T. Most of this energy is directed away from the source in V_T and constitutes *forward scatter*.

(2) Some of the total radiation is scattered "out of the beam," that is, out of the main beam.

(3) A small amount of the total radiation constitutes *backscatter* $\mathbf{L}_1$, from all the multiple scatter components $(k \geq 1)$, which is added to any original source on the surface (S_0) of G_T. Backscatter is from a secondary source, wholly dependent on the original source in V_T and possibly sources on S_0. It is, of course, reflected from S_0 and eventually dissipated in the unbounded medium surrounding V_T. The second term in (8.3-20) represents the (possibly) original $(k = 0)$ and secondary scatter sources $(k \geq 1)$.

(4) The two $\mathbf{L}_2$ terms embody the initial conditions and are usually; zero (see below).

Observe that when the medium contains no inhomogeneities, $\hat{Q}_V = 0$, and $g_\infty^{(Q)} \to g_\infty^{(0)}$, $\alpha_\infty^{(Q)} = \alpha_\infty^{(0)} = \alpha_H$ in (8.3.20): the result, as expected, reduces to the homogeneous case discussed in Section 8.1.6 above. Figure 8.11 illustrates the remarks in (8.3.21).

The explicit GHPs listed in (1)–(6) of Section 8.1.6 apply specifically for these deterministic inhomogeneous media on replacing g_∞ by $g_\infty^{(Q)}$ and α_H by $\alpha_\infty^{(Q)}$ in the four terms of (8.3.20). The last two terms embody the initial conditions (I.C.s) and accordingly

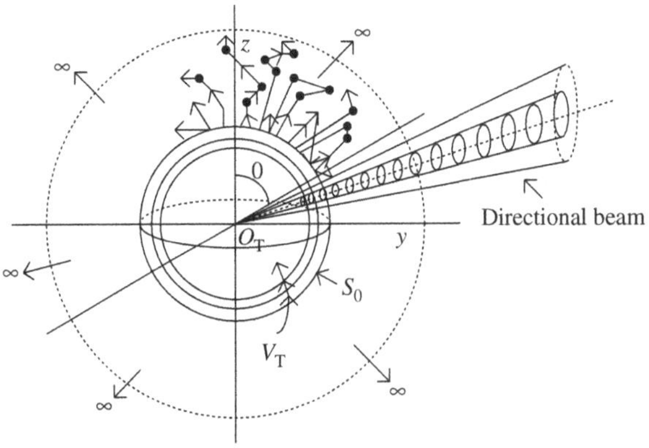

FIGURE 8.11 Radiating volume V_T containing source T, in an unbounded *inhomogeneous* medium, which produces scattering, some of which is backscattered from the bounding surface or outward S_0 of V_T, as well as primarily forward scattered by a directional aperture or array (in G_T).

requires only $g_\infty^{(0)}$ and $\alpha_\infty^{(0)}$ at $(R', t') = (R_0, t_0)$. Usually, we have $\alpha_\infty^{(Q)} = 0, \partial\alpha_\infty^{(Q)}/\partial t = 0$, with no primary sources on the surface S_0 of V_T. Then just the first two terms contribute in these examples, with the surface integral now producing only secondary signals, that is, backscatter from S_0. Without restricting the generality of (8.3.20) noticeably, we may employ these initial conditions and write finally for the deterministic field here

$$\alpha_\infty^{(Q)}(\mathbf{R}, t) = \int_{t_0^-}^{t^+} dt' \int_{V_T} d\mathbf{R}' g^{(Q)}(\mathbf{R}, t|\mathbf{R}', t') G_\mathrm{T}(\mathbf{R}', t') + \int_{t_0^-}^{t^+} dt' \oint_{S_0} \Big[g_\infty^{(Q)} \cdot (1 + a_0)\hat{\mathbf{n}}' \cdot \nabla' \alpha_\infty^{(Q)'} - \alpha_\infty^{(Q)'} \cdot (1 + a_0)\hat{\mathbf{n}}' \cdot \nabla' g_\infty^{(Q)'} \Big]_{S_0'} dS_0' \tag{8.3.22}$$

with $\alpha_\infty^{(0)} = \partial\alpha_\infty^{(0)}/\partial t = 0$ at (R_0, t_0) and no primary source on S_0, just the secondary sources of backscatter. (In many applications, cf. (1), (2), (5), (6), Eqs. (8.1.43), (8.1.44), (8.1.47), (8.1.48), a_0 above can also be set equal to zero.)

8.3.4.1 Integral Equations[44] Formally more compact and offering an alternative solution to the series solutions discussed in Section 8.2.1, Equation (8.3.20) is an inhomogeneous *Fredholm integral equation of the second kind*,[44] here over the semi-infinite regime V, cf. Fig. 8.11. (If we assume nonvanishing initial conditions, we have in place of (8.3.22), (8.3.20).) In any case from (8.3.7a) and (8.3.7b), we have at once the resulting integral equation

$$(\hat{1} - \hat{\eta}_\infty)\alpha_\infty^{(Q)} = \alpha_\mathrm{H} \Rightarrow \alpha_\infty^{(Q)} = \alpha_\mathrm{H} + \hat{\eta}_\infty \alpha_\infty^{(Q)}. \tag{8.3.23}$$

The integral equation (8.3.23) is equivalent to the integral equation (8.3.22) represented by the GHP described above in Section 8.3.4. It is, however, formally much simpler than Eq. (8.3.22), with only the first power of the mass operator $\hat{\eta}_\infty$. Furthermore $\hat{\eta}_\infty$ for volumes and surfaces is given explicitly by Eqs. (8.3.9a) and (8.3.9b), where $\hat{\eta}_\infty = \hat{M}_\infty^{(Q)}\hat{Q}$, with $\hat{M}_\infty^{(Q)}, \hat{Q}$ represented by (8.3.8) *with the total resulting field $\alpha^{(Q)}$ appearing now in the brackets* $(\ )_{\mathbf{R}', t'}$ instead of the homogeneous field α_H of the source (G_T). Note the coupling to the medium is embodied in the source density G_T and is given explicitly by the results of Section 8.3.1 in Eqs. (8.3.2), (8.3.6), (8.3.7) and in Section 8.3.4. See also Section 8.1.7 Eqs. (8.1.52) and (8.1.53) for continuous source distribution (i.e., apertures) and Eqs. (8.1.54)–(8.1.56) for arrays. For the surface integrals one has $\hat{M}_S$, Eq. (8.3.9), and the results of Section 8.3.2 to facilitate the calculations.

The relation (8.3.23), however, does not explicitly reveal the *backscatter* from S_0, which appears here as the second term of (8.3.22). This feature is exhibited if we analyze the anatomy of the operator $\hat{\eta}_V$ in more detail: the scattering process in this case has two components $\alpha_\infty^{(Q_V)}$ and $\alpha_\infty^{(Q_{V+S_0})}$, represented by

$$(\hat{1} - \hat{\eta}_V)\alpha_\infty^{(Q_V)} = \alpha_H \quad \text{and} \quad (1 - \hat{\eta}_{S_0})\alpha_\infty^{(Q_{V+S_0})} = \alpha_\infty^{(Q_V)} \tag{8.3.23a}$$

[44] For a treatment of integral equations see Ref. [1], Chapter 8, also pp. 904, 959, 991–999, Lovitt [36], and in even more detail, Chew [6]. Useful and extensive tables of solutions and methods are also given in Polyanin and Manzhirov [7]. Actually, this integral equation is usually an integral-differential equation, depending on the structure of the propagation equation, cf. $\hat{Q}$ in Table 8.4.

and

$$\therefore \left(\hat{1}-\hat{\eta}_{S_0}\right)(1-\hat{\eta}_V)\alpha^{(Q_{V+S_0})}=\alpha_H;\ (Q_V, Q_{V+S_0}\equiv Q\to\infty). \tag{8.3.23b}$$

Note that $\alpha^{\left(Q_{V+S_0}\right)}=\alpha^{(Q)}$ *represents the total scattered field in V and from* S_0, because the total number of scattering elements remains constant since this is an inherent property of the medium. In addition, reflection is a continuation of the same field, modified by the interface. Only the apportionment of the total (constant) energy among them is altered by the mixing process between surface and volume scatter. (This applies even if the reflectivity of the surface S_0 is less than unity ($S_{0R}<1$), and some energy is transmitted into the second medium, namely that bounding the primary source of α_H.) This gives us directly

$$\alpha^{(Q)}\equiv\alpha^{(Q_{V+S_0})}=\left(1-\hat{\eta}_{S_0}\right)^{-1}\left(1-\hat{\eta}_V\right)^{-1}\alpha_H=\left[1-\left(\hat{\eta}_{S_0}+\hat{\eta}_V\right)+\hat{\eta}_{S_0}\hat{\eta}_V\right]^{-1}\alpha_H, \tag{8.3.24}$$

where $\hat{\eta}_S, \hat{\eta}_V$ do not ordinarily commute, that is, $\hat{\eta}_S\hat{\eta}_V\neq\hat{\eta}_V\hat{\eta}_S$, in as much as the operator factors of $\hat{\eta}_V=\hat{M}_V^{(0)}\hat{Q}_V\neq\hat{Q}_V\hat{M}_V^{(0)}$, and so on, do not themselves commute.[45] Thus, the operator $\hat{\eta}_\infty$ in (8.3.23) is explicitly for the mixed scatter and backscatter in the volume V,

$$\hat{\eta}_\infty=\hat{\eta}_{S_0}+\hat{\eta}_V-\hat{\eta}_{S_0}\hat{\eta}_V\doteq\hat{\eta}_{S_0}+\hat{\eta}_V+O\left(\|\hat{\eta}_V\hat{\eta}_{S_0}\|\ll\|\hat{\eta}_{V_0}\|\,\|\hat{\eta}_{S_0}\|\right). \tag{8.3.24a}$$

Usually, the interactive component $\hat{\eta}_{S_0}\hat{\eta}_V$ is negligible compared to $\hat{\eta}_{S_0}$ and $\hat{\eta}_V$. The relative strength of the surface and volume scatter depends on the inhomogeneous natures of these two media, that is, on their "roughness" or departure from homogeneity.

The integral equation (8.3.23) can now be expressed in detail, on using the approximation $\hat{\eta}_\infty\doteq\hat{\eta}_V+\hat{\eta}_S$. We have for the volume component

$$\hat{\eta}_V=\hat{M}_V^{(0)}\hat{Q}_V=\int_{t_0^+}^{t}dt'\int_{V_T}\mathbf{dR}'g_\infty^{(0)}(\mathbf{R},t|\mathbf{R}',t')Q_V(\mathbf{R}',t')(\,)_{\mathbf{R}',t'} \tag{8.3.25a}$$

with Q_V obtained from $\hat{L}^{(1)}$(8.3.2). Also, the surface contribution to the backscatter is

$$\hat{\eta}_{S_0}=\hat{M}_{S_0}\hat{Q}_{S_0}=\int_{t_0^+}^{t}dt'\oint_{S_0'}\left\{g_\infty^{(0)}\frac{\partial}{\partial n'}-\left[\frac{\partial}{\partial n'}g^{(0)}\right]\right\}\left[\hat{Q}'_{S'}(\,)'_{\mathbf{R}',t'}\right]dS',\quad\text{with}\quad \hat{Q}_{S_0'}=\mathcal{R}_0\hat{1} \tag{8.3.25b}$$

where $\mathcal{R}_0(=\mathcal{R}_0(\mathbf{R}',t'))$ is the plane wave reflection coefficient for each point $\mathbf{R}'$ (at time t') on the surface S_0, cf. Fig. 8.11. Accordingly, the integral equation (8.3.23) becomes compactly:

$$\alpha_\infty^{(Q)}(\mathbf{R},t)\doteq\alpha_H+(\hat{\eta}_V+\hat{\eta}_S)\alpha^{(Q)}=\alpha_H(\mathbf{R},t)+\left(\hat{M}_V^{(0)}\hat{Q}_V+\hat{M}_{S_0}\hat{Q}_{S_0}\right)\alpha_\infty^{(Q)}(\mathbf{R}',t'). \tag{8.3.26}$$

[45] They and $\hat{\eta}_{00,S}, \hat{\eta}_{00,V}$, however, do commute in their convolutional form. See also Sections 8.3.1, 8.3.2, and 8.4.2.

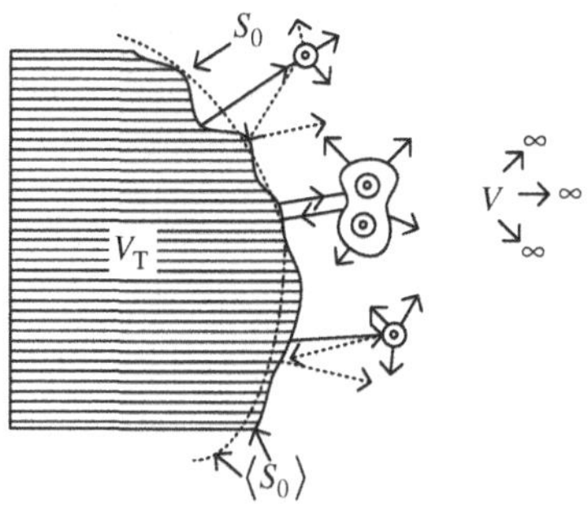

FIGURE 8.12 Section of V_T and its boundary surface S_0, with backscatter into V from S_0 due to scatter from inhomogeneities in V.

The homogeneous component α_H, if any, is given by the first term in (8.3.20), (8.3.22). It is assumed that there are no primary sources on S_0, only those produced by the backscatter from the (deterministic) inhomogeneities in the medium V itself, which in turn reradiate on reflection from S_0. Figure 8.12 provides a schematic illustration of the phenomenon.

Equation (8.3.26) as stated above, with (8.3.25a) and (8.3.25b), is an approximate solution where the surface $\otimes$ volume interactions are omitted on the usual observation that they are physically ignorable. If this is not the case, Eq. (8.3.26) is, from (8.3.24a), modified to the "exact" form[46]

$$\alpha_{\infty}^{(Q)}(\mathbf{R},t) \doteq \alpha_H(\mathbf{R},t) + \left(\hat{M}_V\hat{Q}_V + \hat{M}_{S_0}\hat{Q}_{S_0} - \hat{\eta}_{S_0}\hat{\eta}_V\right)\alpha_{\infty}^{(Q)}(\mathbf{R}',t'), \tag{8.3.26a}$$

in which $\hat{\eta}_{S_0}\hat{\eta}_V = \hat{M}_{S_0}^{(0)}\hat{Q}_{S_0}\hat{M}_V^{(0)}\hat{Q}_V$ requires now a double set of integrations based on (8.3.25a) and (8.3.25b) to quantify the contributions of the surface–volume interactions. We emphasize again the fact that here $Q = Q_{V+S_0}$, that is, $\alpha_{\infty}^{(Q)} = \alpha_{\infty}^{Q_{V+S_0}}$ cf. (8.3.23b) represents the total scattered field in V, due to scatterers in V *and* to the backscatter from the interface S_0, in addition to any contribution from the unscattered field α_H, in V. Equation (8.3.26a), in addition, includes the "interaction field" $\hat{\eta}_{S_0}\hat{\eta}_{V_0}\alpha_{\infty}^{(Q)}$ when it becomes significant. We determine the general level of complexity of the scattered field $\alpha^{(Q)}$ directly by noting the number of different surface and volume global operations $\eta_{S'}, \eta_{V'}$ *and their combinations* in the integral equation for $\alpha^{(Q)}$. Thus, in the example (8.3.26) we have a "2-0" complexity, while for (8.3.26a) the complexity number is "2-1", including the interactive terms. For additional examples, see Sections 8.5.1.1and 8.5.1.2.

8.3.4.2 Example: The Time-Dependent Helmholtz Equation A useful illustration of the general results above is provided by the time-dependent Helmholtz equation, whose GF $g_\infty = g_\infty^{(0)}$ is given specifically by (8.1.30c). Let as assume again the common situation where the initial conditions vanish, that is, $\alpha^{(Q)} = \partial\alpha^{(Q)}/\partial t|_{R_0,t_0} = 0$, and the inhomogeneous (densities) associated with V and S_0 are respectively from (8.3.25b)

$$\hat{Q}_V = \frac{\varepsilon(\mathbf{R}',t')}{c_0^2}\frac{\partial^2}{\partial t^2} \quad \text{and} \quad \hat{Q}_S = \mathsf{R}_0(\mathbf{R}',t')\hat{1}. \tag{8.3.27}$$

[46] The approximation sign ($\doteq$) occurs because of the inherent approximation in replacing the truly exact relation (8.3.24) by the approximate "exact" Eq. (8.3.24a).

The quantity R_0 is the plane-wave reflection coefficient. A common set of boundary conditions (between gas and liquid, or liquid and solid in acoustical and certain electromagnetic applications) ([33]; also [11] and [17], is here (for complete reflections at the boundary)

$$\alpha_\infty^{(Q_{V+S})} = \mathcal{R}_0 \alpha^{(Q_V)}; \quad \frac{\partial}{\partial n}\alpha_\infty^{(Q_{V+S})} = \frac{-\partial}{\partial n}\mathcal{R}_0 \alpha^{(Q_V)} \tag{8.3.28}$$

or equivalently in detail

$$\alpha_\infty^{(Q_{V+S})}\Big|_{\substack{\text{reflect.}\\ \text{backscat}}} = \mathcal{R}_0 \alpha_\infty^{(Q_V)}\Big|_{\substack{\text{incident}\\ \text{scat.}}}; \frac{\partial}{\partial n}\alpha_\infty^{(Q_{V+S})}\Big|_{\substack{\text{reflect.}\\ \text{backscat}}} = -\frac{\partial}{\partial n}\left(\mathcal{R}_0 \alpha_\infty^{(Q_V)}\right)\Big|_{\substack{\text{incident.}\\ \text{scat.}}} \tag{8.3.28a}$$

Inserting (8.3.27) into the general conditions (8.3.28) and (8.3.28a) gives the reflected, scattered field[47] *in terms of the incident field* $\alpha^{(Q_V)}$, which may or may not be itself a scattered field. Then we have here $\alpha_\infty^{(Q_{V+S})} = \mathcal{R}_0 \alpha_\infty^{(Q_V)} = \mathcal{R}_0 \alpha_\infty^{(Q)}$. In this case the Green's function is explicitly from (8.1.30c)

$$g_\infty^{(0)}(\mathbf{R}, t|\mathbf{R}', t') = \frac{\delta(t' - [t - \rho/c_0])}{4\pi\rho}, \quad \rho \equiv |\mathbf{R} - \mathbf{R}'|. \tag{8.3.29}$$

The integrand of the surface integral in (8.3.25b) now yields the following results when we reverse the order of integration. For applications when volume ⊗ surface scatter can be neglected, we can now employ (8.3.26). Here S_0 refers to the interface between two different media in general. Using the boundary conditions [(8.3.28) and (8.3.28a)] we obtain

$$\begin{aligned} \hat{\eta}_{S_0}\alpha_\infty^{(Q)}\big|_{\text{Helm.}} &= -\int_{t_0^-}^{t^+} dt' \oint_{S_0'} \left[\frac{\partial}{\partial n'}\left(g_\infty^{(0)}\mathcal{R}'_0\alpha_\infty^{(Q)'}\right)\right]_{S_0'} dS_0' \\ &= -\oint_{S_0'}\left[\frac{\partial}{\partial n'}\frac{\mathcal{R}_0(\mathbf{R}', t - \rho/c_0)\alpha_\infty^{(Q)}(\mathbf{R}', t - \rho/c_0)}{4\pi\rho}\right]_{S_0'} dS_0'. \end{aligned} \tag{8.3.30}$$

Similarly, for (8.3.25a) we also obtain from (8.3.27)

$$\begin{aligned} \hat{\eta}_V\alpha_\infty^{(Q)}\big|_{\text{Helm.}} &= \frac{1}{c_0^2}\int_{t_0^-}^{t^+} dt' \int_V g_\infty^{(0)}\varepsilon(\mathbf{R}', t')\frac{\partial^2\alpha_\infty^{(Q)}}{\partial t'^2}\,\mathbf{dR}' \\ &= \frac{1}{c_0^2}\int_V \frac{\varepsilon(\mathbf{R}', t - \rho/c_0)}{4\pi\rho}\left[\frac{\partial^2}{\partial t'^2}\alpha_\infty^{(Q)}(\mathbf{R}', t')\right]_{t'=t-\rho/c_0} \mathbf{dR}'. \end{aligned} \tag{8.3.31}$$

[47] We shall usually designate the medium containing inhomogeneities specifically, namely, $\alpha^{(Q_V)}$ or $\alpha^{(Q_{V+S})}$, and so on, or $\alpha^{(V)}$, $\alpha^{(V_0+S_0)}$, and so on, cf. Section (8.5.1).

with $\mathbf{dR}' = dx'dy'dz'$, the volume element here for the semi-infinite space V. We combine (8.3.30) and (8.3.31), now specialized to the deterministic inhomogeneous Helmholtz medium, given in (8.3.26) explicitly. Here the Green's function $g_{\infty}^{(0)}$ is still given by [(8.1.30b) and (8.1.30c)]. We see directly that the total field in V is given by the "exact" relations.

$$\alpha_{\infty}^{(Q)}(\mathbf{R},t)\big|_{\text{Helmholtz}} \doteq \overbrace{\int_{V_T} \frac{G_T(\mathbf{R}',t-\rho/c_0)\,\mathbf{dR}'}{4\pi\rho}}^{\leftarrow \alpha_H(\mathbf{R},t) \rightarrow} + \frac{1}{c_0^2}\int_V \frac{\varepsilon(\mathbf{R}',t-\rho/c_0)}{8\pi\rho}\left[\frac{\partial}{\partial t'}\alpha_{\infty}^{(Q)}(\mathbf{R}',t')\right]_{t'=t-\rho/c_0}\mathbf{dR}'$$
$$-\oint_{S_0'}\left[\frac{\partial}{\partial n'}\left\{\mathcal{R}_0(\mathbf{R}',t-\rho/c_0)\alpha_{\infty}^{(Q)}(\mathbf{R}',t-\rho/c_0)\right\}\right]_{S_0'} dS_0', \qquad (8.3.32)$$

where, as before, we have postulated that the initial conditions vanish, that is, $\alpha_{\infty}^{(Q)} = \partial\alpha_{\infty}^{(Q)}/\partial t = 0, t = t_0^-$, (cf. (iv), footnote[48]). Here the surface integral in (8.3.32) can be pressed more fully if we use the results (8.1.42 f) and (8.1.42 g). These give us explicitly

$$\oint_{S_0'}\left\{\left[\zeta_{x'}(\mathbf{R}',t')\frac{\partial}{\partial x'} + \zeta_{y'}(\mathbf{R}',t')\frac{\partial}{\partial y'} + \frac{\partial}{\partial z'}\right]\mathrm{R}_0(\mathbf{R}',t')\alpha^{(Q_{V+S})}(\mathbf{R}',t')\right\}\mathbf{dr}'$$
$$\mathbf{R}' = \mathbf{r}' + \boldsymbol{\zeta};\quad t' = t - \rho'/c_0;\quad \rho' = \mathbf{R} - (\mathbf{r}' + \boldsymbol{\zeta}'),$$
$$(\doteq \mathbf{R} - \mathbf{r}' = \mathbf{R} - \mathbf{R}') \qquad (8.3.32a)$$

where $\mathbf{r}' = \hat{i}_x x' + \hat{i}_y y'$, $\mathbf{dr}' = dx'dy'$, $\boldsymbol{\zeta} = \boldsymbol{\zeta}(\mathbf{R},t) = \hat{\mathbf{i}}_2\zeta$; see Fig. 8.7. In addition, the couplings to the (here) deterministic inhomogeneous Helmholtz medium (namely, (1) in Table 8.4) are embodied in the source density function G_T. For continuous apertures and

[48] The second term of (8.3.32), which represents scattering in the infinite volume (V), can be reduced to a simpler result under the zero initial conditions assumed here. Writing $u \equiv g_{\infty}^{(0)}\varepsilon, \alpha \equiv \alpha_{\infty}^{(Q)}$ in general, we begin with the identity

$$\frac{\partial(u\alpha)}{\partial t} \equiv \alpha\frac{\partial u}{\partial t} + u\frac{\partial \alpha}{\partial t} \quad \therefore \quad \frac{\partial}{\partial t}\left(\alpha\frac{\partial u}{\partial t}\right) + \frac{\partial}{\partial t}\left(u\frac{\partial \alpha}{\partial t}\right) \equiv \frac{\partial}{\partial t}\left(\alpha\frac{\partial u}{\partial t}\right) + \frac{\partial u}{\partial t}\frac{\partial \alpha}{\partial t} + u\frac{\partial^2\alpha}{\partial t^2} \qquad \text{(i)}$$

$$\therefore \quad u\frac{\partial^2\alpha}{\partial t^2} = \frac{\partial^2(u\alpha)}{\partial t^2} - \frac{\partial}{\partial t}\left(\alpha\frac{\partial u}{\partial t}\right) - \frac{\partial u}{\partial t}\frac{\partial \alpha}{\partial t}, \text{ with } \int_{t_0}^{t}\frac{\partial u}{\partial t}\frac{\partial \alpha}{\partial t}dt = \frac{1}{2}\left[\frac{u\partial\alpha}{\partial t} + \frac{\alpha\partial u}{\partial t}\right] = \frac{1}{2}\frac{\partial}{\partial t}(u\alpha)\Big]_{t_0}^{t}, \qquad \text{(ii)}$$

since $\int_{t_0}^{t}\frac{\partial u}{\partial t}\frac{\partial \alpha}{\partial t}dt = \left[u\frac{\partial\alpha}{\partial t}\right.$ or $\left.\alpha\frac{\partial u}{\partial t}\right]_{t_0}^{t} = \frac{1}{2}\frac{\partial}{\partial t}(u\alpha)\Big|_{t_0}^{t}$. Consequently, we obtain

$$\int_{t_0}^{t} u\frac{\partial^2\alpha}{\partial t^2}dt = \left[\frac{\partial}{\partial t}(u\alpha) - \alpha\frac{\partial u}{\partial t} - \frac{1}{2}u\frac{\partial\alpha}{\partial t} - \frac{1}{2}\alpha\frac{\partial u}{\partial t}\right]_{t_0}^{t} = \frac{1}{2}\left[u\frac{\partial\alpha}{\partial t} - \alpha\frac{\partial u}{\partial t}\right]_{t_0}^{t}. \qquad \text{(iii)}$$

This reduces (because $\alpha = \partial\alpha/\partial t = 0$), in the case of the Helmholtz medium, cf. (8.3.31), (8.3.32) with $u = 1$ here, to

$$\frac{1}{4\pi}\varepsilon\frac{\partial^2}{\partial t'^2}\alpha_{\infty}^{(Q)} = \frac{\varepsilon(\mathbf{R}',t-\rho/c_0)}{8\pi\rho}\left[\frac{\partial}{\partial t'}\alpha^{(Q)}(\mathbf{R}',t')\right]_{t'=t-\rho/c_0}. \qquad \text{(iv)}$$

discrete arrays these are

$$\left.\begin{aligned}&[\text{Eq.8.1.50})]: \qquad G_{\mathrm{T}}(\mathbf{R}',t')_{\text{aperture}} = \int_{-\infty}^{\infty} Y_{\mathrm{T}}(\mathbf{R}',f')S_{\text{in}}(\mathbf{R}',f')_{\Delta}e^{i\omega't'}df' \\ &[\text{Eq.8.1.54}a)]: \qquad G_{\mathrm{T}}(R',L')_{\text{array}} = \sum_{m=1}^{M}\int_{-\infty}^{\infty} h_{\mathrm{T}}^{(m)}(\boldsymbol{\xi}_m,t-\tau)S_{\text{in}}^{(m)}(\boldsymbol{\xi}_m,\tau)_{\Delta}d\tau\end{aligned}\right\}; \qquad (8.3.33)$$

Our result (8.3.32), (8.3.32a) includes backscatter from the surface S_0' (third term), as well as scatter from the inhomogeneities in V_{T} (second term). In the far-field of V_{T} (and hence of S_0') this backscatter will be negligible compared to that produced in V_{T} by the primary source in V_{T}. The unperturbed field α_{H} from the distributed source is given by the first term of (8.3.32).

In addition, we have assumed that the mutual scattering between volume and surface scatter is negligible vis-à-vis the volume and surface scatter considered separately, that is, $\left|\hat{\eta}_V\hat{\eta}_S\alpha_\infty^{(Q)}\right| \ll \left|\hat{\eta}_V\alpha_\infty^{(Q)}\right|\left|\hat{\eta}_S\alpha_\infty^{(Q)}\right|$, cf. (8.3.25). It is also assumed that there are no primary sources on S_0', just the effect of the backscatter from elements in the medium V.[49] Of course, if it is expected that there is significant interaction between volume and surface scatter, we must use 8.3.26a. Note that when there are no inhomogeneities in V, that is, $\hat{Q}_V = 0$ and $\therefore\ \hat{Q}_S = 0$—there is no backscatter—only the first term (α_{H}) of (8.3.32) remains. The "exact" analytic evaluation of (8.3.32) depends on (1), the far-field condition (2) the specific form of ε (3) the simplicity of the surface geometry (S_0') and (4) on the complexity of $\alpha_\infty^{(Q)}$, (8.3.26), which in most cases with irregular geometries as it stands is usually intractable. Other means such as numerical methods (see Section 8.5.3, following) must be employed.

8.3.4.3 Dispersion II In a preceding Section 8.1.5.1, we observed that for a homogeneous medium which is absorbent, that is, dissipates energy, the resulting wave numbers k (or $\nu = k/2\pi$) is a nonlinear function of frequency, cf. Eq. (8.1.38), namely,

$$(8.1\text{-}38): \quad k = F(\omega)/c_0 \neq \omega/c_0;\ \nu = k/2\pi. \qquad (8.3.34)$$

The accompanying partial differential equation of space–time propagation has a dissipative term containing the operator $\partial/\partial t$ responsible for the dissipative contribution. This effect is called *frequency dispersion* and is the result of intrinsic mechanisms in the molecular structure of the medium. For this reason it is also called *intrinsic dispersion.* Here the constant c_0 represents the phase velocity of propagation of each frequency component of the propagated wave. The governing PDE in these cases obey the operational equation for homogeneous media Eq. (8.1.39)

$$L^{(0)}\alpha_{\mathrm{H}} = -G_{\mathrm{T}},\ (\mathbf{R} \in V_{\mathrm{T}}; = 0, R \notin V_{\mathrm{T}}),\ t > t_0^- \qquad (8.3.35)$$

and in more detail, the GHP, cf. Section 8.1.6, which include boundary and initial conditions.

[49] Observe again that V_{T} refers to the volume occupied by the primary or original source. This includes the transmitting aperture or array, and $V(\notin V_{\mathrm{T}})$ is here (the infinite) volume containing the deterministic (i.e., fixed) inhomogeneities.

There are also other types of dispersion. For example, when the phase velocity c_0 is inhomogeneous, that is, becomes

$$c_0 \rightarrow c_m = \omega/k_m(\omega) \quad \text{or} \quad c(\mathbf{R}) = \omega/k(\mathbf{R}) = c_0 + c_1(\mathbf{R}) \tag{8.3.36}$$

(with ω ($=2\pi f$) a specified angular frequency component of the signal).

The first relation represents the phase velocity c_m of discrete modes in a wave guide, ($m=0,1,2,\ldots$). The second represents a phase velocity that depends on position in space. Both of these exhibit *spatial* or *geometric dispersion.* For instance, models of acoustic propagation in the ocean often use the approximation $c_0 = c_0 + c_1(z)$ for a depth (z)-dependent variation in phase velocity. More extreme cases of spatial dispersion can include refraction or reflection at interfaces, which can result from range-dependent propagation speeds. In fact, when inhomogeneity of the medium is described by the spatial dependence of the parameters $\boldsymbol{\theta}$ of the propagation equation, cf. Eq. (8.3.2),

$$\left[\hat{L}^{(0)} - \hat{Q}(\boldsymbol{\theta})\right]\alpha^{(Q)}(\mathbf{R}, t) = 0, \quad \mathbf{R} \notin V_{\mathrm{T}}; \boldsymbol{\theta} = (a_1, a_2, \ldots, a_n) = \boldsymbol{\theta}(\mathbf{R}), \tag{8.3.37}$$

we may expect spatial dispersion, along with frequency dispersion if there is a dissipative term in the PDE of propagation. Similar remarks apply for propagation in the atmosphere. Media that support spatial dispersion are obviously inhomogeneous. In fact, this includes (deterministic) scatter, as well, as shown in Section 8.3.4 and generally in Section 8.3 itself.

8.3.5 Generalizations and Remarks

There is a variety of extensions for the above analyses. We list a number of them here, which the reader can easily obtain. We note specifically:

(1) Preset sources on the closed surface S_0 bounding the volume V_{T}, for example, the second term of (8.3.20).
(2) Initial conditions other than the usual $\alpha = \partial\alpha/\partial t = 0$.
(3) Other media, which are described by Green's functions of the type (2)–(6), cf. (8.1.31) and (8.1.33). These are considerably more complex than the GF for the Helmholtz medium.
(4) The explicit role of the (continuous) aperture or (discrete) array, contained in the source density $G_{\mathrm{T}}(\mathbf{R}', t')$, cf. (8.1.49) et. seq. and the unscattered component α_{H}, (8.1.49a) et. seq., and Section 8.1.6 in detail. Section 8.4 also gives these desired forms in the treatment of the GHP results in the Fourier transform domain.
(5) For the *channel*, as defined in the present volume, we must include the receiver, in the manner of Section 8.2.

We have so far discussed the space–time field generated by the primary sources and the scattering by and on the deterministic inhomogeneities in the infinite medium (contained in V, cf. Fig. 8.11, and by the bounding surface S_0' of V_{T}, cf. Fig. 8.12). The backscatter by the surface S_0' is primarily noticeable in the relative proximity of this surface and then falls off to a negligible intensity a few mean-free path lengths of the scatter from the surface, as shown in

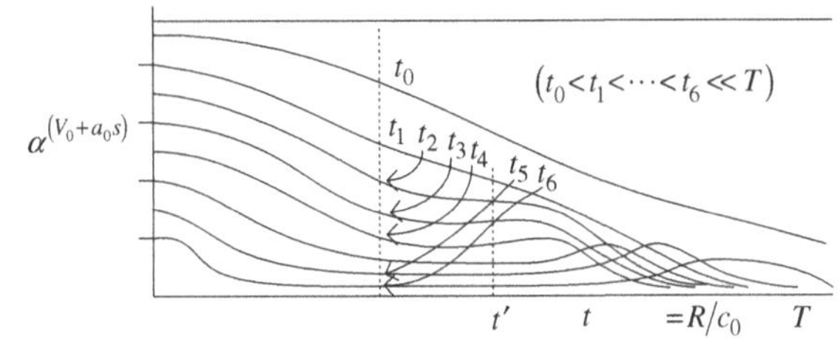

FIGURE 8.13 Sketch of transient response of the Green's function of a typical medium, with changes in time $t_1, t_2, \ldots, t_5 \ll T$, the effective duration of the response, showing the evolving scatter interactions as time progressed during the response time (T).

Fig. 8.13 (cf. remarks following Eq. (8.3.32)). The unscattered, that is, coherent component α_H, however, is specularly reflected in several directions, depending on the roughness of the surface and the directionality of the aperture or array. These unscattered specular components represent "resolvable multipath," a familiar phenomenon in telecommunications, sonar, and radar. In the next section (8.4), we examine the wave number–frequency equivalents of the space–time field, and include reception, whose output is the data sample $X(\mathbf{R}, t)$, which is then subject to the optimal and near-optimal process of signal extraction, the elements of which have been treated in Chapters 1–7.

8.4 THE DETERMINISTIC SCATTERED FIELD IN WAVE NUMBER–FREQUENCY SPACE: INNOVATIONS

It is instructive and sometimes analytically simpler to represent the space–time fields $\alpha_\infty^{(0)}, \alpha_\infty^{(Q)}$ in terms of their Fourier and Laplace transforms. In part, this can occur because the components $\hat{M}_\infty^{(0)}, \hat{Q}$ of the *mass operator* $\hat{\eta}_\infty \left(= \hat{M}_\infty^{(0)} \hat{Q}\right)$ *commute* in wave number–frequency space, although they clearly do not commute in the space–time domain cf. Section 8.3.3.1. This is particularly true when the general aperture/array coupling of the input signal to the medium is included cf. (Eqs. (8.1.49), (8.1.54a), and (8.1.54b)).

For this purpose, we begin with a variety of transform equivalents, some represented in Sections 8.1–8.3. Thus, for the integral Green's functions operator $\hat{M}_\infty^{(0)}$, (8.1.9b) and, (8.1.10), we may write

$$\hat{M}_\infty^{(0)}(\mathbf{R},t|\mathbf{R}',t') = \mathsf{F}_{\mathbf{k}}\mathsf{F}_s\left\{\hat{\mathsf{M}}^{(0)}\right\} = \mathsf{F}_s\left\{\left[\int \mathbf{dZ}'\right] Y_0(\mathbf{R},s|\mathbf{R}')e^{-st'}(\,)_{\mathbf{R}',t'}\right\}; \mathsf{F}_s\{\,\} \equiv \int_{Br_1} e^{st}\frac{ds}{2\pi i}(\,)_s \tag{8.4.1a}$$

$$= \mathsf{F}_{\mathbf{k}}\mathsf{F}_s\left\{\left[\int \mathbf{dZ}'\right] \mathsf{Y}_0(\mathbf{k},s)e^{i\mathbf{k}\cdot\mathbf{R}'-st'}(\,)_{\mathbf{R},t}\right\}; \mathsf{F}_{\mathbf{k}}\mathsf{F}_s\{\,\} \equiv \int_{Br_1} e^{st}\frac{ds}{2\pi i}\int_{-\infty}^{\infty}\frac{\mathbf{dk}}{(2\pi)^3}(\,)_{\mathbf{k},s}. \tag{8.4.1b}$$

Here, we have

$$Y_0(\mathbf{R},s|\mathbf{R}') = Y_0(\rho,s), \rho \equiv |\mathbf{R}-\mathbf{R}'|, \text{ (with different explicit values, vide, Section 8.1.5, (1)–(6))} \tag{8.4.2a}$$

and

$$\mathsf{Y}_0(\mathbf{k},s) = -L_0(k,s)^{-1}, \text{ Eqs. (8.1.15) and (8.1.17), and } \left[\left[\int \mathbf{dZ}'\right]\right] \equiv \int_{t_0^-}^{t^+} dt' \int_{V_T} \mathbf{dR}'(\,)_{R',t'}. \tag{8.4.2b}$$

Similarly, we find that the space–time transforms of the integral Green's function operator $\hat{M}_\infty^{(0)}$ are

$$\hat{\mathsf{M}}_\infty^{(0)} = \mathsf{F_R F}_t\left\{\hat{M}^{(0)}\right\} = -\mathsf{F_R F}_t\left\{\left[\left[\int \mathbf{dZ}'\right]\right] g^{(0)}(\mathbf{R},t|\mathbf{R}',t')(\,)_{\mathbf{R}',t'}\right\} = \left\{\begin{array}{l}\int \mathbf{dZ}' e^{-i\mathbf{k}\cdot\mathbf{R}'+st'}\mathsf{Y}_0(\mathbf{k},s)\odot_{\mathbf{k},s} \\ \equiv \hat{\mathsf{Y}}_0 \odot (\,)_{\mathbf{k},s}\end{array}\right\} \tag{8.4.3a}$$

with the single time transform for $\hat{M}_\infty^{(0)}$ given by

$$\hat{Y}_0 = \mathsf{F}_t\left\{\hat{M}_\infty^{(0)}\right\} = \mathsf{F}_t\left\{\left[\left[\int \mathbf{dZ}'\right]\right] g^{(0)}(\mathbf{R},t|\mathbf{R}',t')(\,)_{\mathbf{R}',t'}\right\} = \left\{\begin{array}{l}\left[\left[\int \mathbf{dZ}'\right]\right] Y_0(\mathbf{R},s|\mathbf{R}')e^{+st'}(\,)_{\mathbf{R},s} \\ \equiv \hat{Y}_0(\mathbf{R},s)\otimes\end{array}\right\}. \tag{8.4.3b}$$

For the inhomogeneity operator $\hat{Q}(\mathbf{R},t)$ (cf. (8.3.1) et seq.), we have

$$\hat{Q}(\mathbf{R},t) = \mathsf{F}_s\{\hat{Q}_0(\mathbf{R},\mathbf{s})\} = \mathsf{F_k F}_s\{\hat{Q}_{00}(\mathbf{k},s)\}, \tag{8.4.4a}$$

with the inverses

$$\hat{Q}_0(\mathbf{R},s) = \mathsf{F}_t\{Q(\mathbf{R},t)\};\ Q_{00}(\mathbf{k},s) = \mathsf{F_R F}_t\{Q(\mathbf{R},t)\}. \tag{8.4.4b}$$

The *mass or field renormalization operator* $\hat{\eta}_\infty$ (cf. Eq. (8.3.4) et seq.) is the *separable* product of the global operator $\hat{M}_\infty^{(0)}$, namely, the integral Green's function operator, and the local inhomogeneity operator $\hat{Q}$, that is,

$$\hat{\eta}_\infty = \hat{M}_\infty^{(0)}\hat{Q} = \hat{M}_\infty \cdot \hat{Q}\left(\neq \hat{Q}\hat{M}_\infty^{(0)}\right):\quad \hat{M}_\infty^{(0)} \text{ and } \hat{Q} \text{ do } \textit{not} \text{ commute.} \tag{8.4.5}$$

As before and throughout, $\hat{M}_\infty^{(0)}$ is represented for volumes by

$$\hat{M}_\infty^{(0)} \Rightarrow \hat{M}_V^{(0)}(\mathbf{R},t|\mathbf{R}',t') = \int_{t_0^-}^{t^+} dt' \int_{-\infty}^{\infty} g_\infty^{(0)}(\mathbf{R},t|\mathbf{R}',t')(\,)_{\mathbf{R}',t'}\mathbf{dR}',\ \text{(Eq. 8.1.10)}, \tag{8.4.5a}$$

and for surfaces by

$$\begin{aligned}\hat{M}_\infty^{(0)} \Rightarrow \hat{M}_S^{(0)} = \int_{t_0^-}^{t^+} dt' \oint_{S_0'} \Bigg[& g_\infty^{(0)}\left\{1;1+a_1\frac{\partial}{\partial t'},\text{ and so on}\right\}\frac{\partial}{\partial n'}(\,)_{R',t'} \\ & -(\,)_{R',t'}\left\{1;1+a_1\frac{\partial}{\partial t'},\text{ and so on}\right\}\frac{\partial g_\infty^{(0)}}{\partial n'}\Bigg]_{on\ S_0'} dS_0',\end{aligned} \tag{8.4.5b}$$

refer to Section 8.1.6.3, (1)–(6). The local inhomogeneity operator generally is $\hat{Q}(\mathbf{R},t)$, which with the field $\alpha_\infty^{(Q)}$ appears in $(\;)_{R',t'}$ along with $\hat{M}_V^{(0)}$ and $\hat{M}_S^{(0)}$ for $\left(\hat{\eta}_\infty = \hat{M}_\infty \hat{Q}\right)_{V \text{ or } S}$, as shown typically by (8.3.25a) and (8.3.25b).

8.4.1 Transform Operator Solutions

Let us consider the integral equation (8.3.23) first and determine its Fourier-Laplace transforms. We have, on representing each quantity by its various transforms,

$$\mathsf{F}_t\left\{\alpha_\infty^{(Q)} = \alpha_{\mathrm{H}} + \hat{\eta}_\infty \alpha_\infty^{(Q)}\right\}: \quad \alpha_0^{(Q)} = \alpha_{0\mathrm{H}} + \mathsf{F}_t\left\{\hat{M}_\infty^{(0)} \hat{Q} \alpha_\infty^{(Q)}\right\}$$
$$= \alpha_{0\mathrm{H}} + \mathsf{F}_t\left\{\mathsf{F}_s(\hat{\eta}_0) \otimes \mathsf{F}_s\left(\alpha_\infty^{(Q)}\right)\right\} \tag{8.4.6a}$$

$$\therefore \quad \alpha_0^{(Q)} = \alpha_{0\mathrm{H}} + \hat{\eta}_0 \otimes \alpha_0^{(Q)}, \tag{8.4.6b}$$

where $\otimes$ *denotes convolution of one variable* (here in frequency) and $\mathsf{F}_t\mathsf{F}_s = 1$, with $\hat{\boldsymbol{\eta}}_0 \otimes = \hat{M}_0 \otimes \hat{Q}_0 \otimes$. Note that here $\hat{\mathsf{M}}^{(0)} = \mathsf{F}_t\left\{\hat{M}_\infty^{(0)}\right\} = \hat{Y}_0'$, where

$$\hat{\mathsf{M}}_0^{(0)} = \hat{Y}_0' = \hat{Y}_0(\mathbf{R}, s|\mathbf{R}') e^{+st'}(\;)_{\mathbf{R},t} \quad \text{with}$$
$$\hat{\eta}_0 = \hat{\eta}_0(\mathbf{R}, s|\mathbf{R}', s') \quad : F_t\left\{\hat{\eta}_\infty \otimes \alpha_\infty^{(Q)}\right\} = \hat{Y}_0' \otimes \hat{Q} \otimes \alpha_0^{(Q)}, \tag{8.4.6c}$$

and the subscripts (0) indicate the transform variable s here, that is, $\alpha_0^{(Q)} = \alpha^{(Q)}(\mathbf{R}, s)$, $\alpha_{0_{\mathrm{H}}} = \alpha_{0_{\mathrm{H}}}(\mathbf{R}, s)$ and so on. Consequently, the integral equation in (8.3.23) becomes in its *space–frequency form.*

(I) Space–Frequency (S–F):

$$\alpha_{0_{\mathrm{H}}} + \hat{Y}_0' \otimes \hat{Q}_0 \otimes \alpha_0^{(Q)} = \alpha_0^{(Q)}; \quad \alpha_{0_{\mathrm{H}}}\left(= \hat{Y}_0' \otimes G_{0\mathrm{T}}\right)$$
$$= \int_{V_{\mathrm{T}}} d\mathbf{R}' \int_{\mathrm{Br}_1} e^{s'(t-t')} \frac{ds'}{2\pi i} Y_0(\mathbf{R}, s'|\mathbf{R}') G_0(\mathbf{R}', s - s')_{\mathrm{T}}. \tag{8.4.7}$$

Exhibiting the role of backscatter, this result becomes in more detail (Eq. (8.3.26), Figs. 8.10 and 8.11):

(II) Space-Frequency with Backscatter:

$$\alpha_{0_{\mathrm{H}}}(\mathbf{R}, s) + \left[\hat{Y}_{0V}' \otimes \hat{Q}_{0V} \otimes \alpha_\infty^{(Q)}(\mathbf{R}, s) + \hat{Y}_{0S}' \otimes \hat{Q}_{0S} \otimes \alpha_\infty^{(Q)}(\mathbf{R}, s)\right]$$
$$\doteq \alpha_0^{(Q)}(\mathbf{R}, s); \; \alpha_{0_{\mathrm{H}}}(\mathbf{R}, s)\left(= \hat{Y}_0' \otimes G_{0\mathrm{T}}\right), \tag{8.4.7a}$$

where α_{0_H} is explicitly given by (8.4.7), above. For example, Eqs. (8.3.6) and (8.3.7b) are now represented by

$$\alpha_0^{(Q)} = \left(\hat{1} - \hat{\eta}_0\right)^{-1} \otimes \alpha_{0_H} = \alpha_{0_H} + \sum_{R=1}^{\infty} \hat{\eta}_0^{(k)} \otimes \alpha_{0_H} = \alpha_{0_H} + \sum_{R=1}^{\infty} \left[\hat{Y}'_0 \otimes \hat{Q}_0\right]^{(k)} \otimes \alpha_{0_H}, \tag{8.4.8}$$

subject, to the appropriate convergence conditions (Eq. (8.3.7)).

In a similar fashion to the above for the space–frequency forms of (8.3.23), we have for the wave number–frequency representation:

(III) Wave Number–Frequency

$$\mathsf{F}_{\mathbf{R}}\mathsf{F}_t\left\{\alpha_\infty^{(Q)} = \alpha_{0_H} + \hat{\eta}\alpha_\infty^{(Q)}\right\} : \alpha_{00}^{(Q)} = \alpha_{00\mathrm{H}} + \mathsf{F}_{\mathbf{R}}\mathsf{F}_t\left\{\hat{M}^{(0)}\hat{Q}\alpha_\infty^{(Q)}\right\}$$
$$= \alpha_{00\mathrm{H}} + \mathsf{F}_{\mathbf{R}}\mathsf{F}_t\left\{\mathsf{F}_{\mathbf{k}}\mathsf{F}_s(\hat{\eta}_{00}) \odot \mathsf{F}_{\mathbf{k}}\mathsf{F}_s\left(\alpha_{00}^{(Q)}\right)\right\}, \tag{8.4.9a}$$

where $\odot$ is the double convolution involving the functions of $(\mathbf{k}, s)$. Here, $\alpha_{00}^{(Q)} \equiv \alpha_\infty^{(Q)}(\mathbf{k}, s)$, $\alpha_{00H} \equiv \alpha_\infty(\mathbf{k}, s)_H$ and so on are the double Fourier transforms indicated by $\mathsf{F}_{\mathbf{R}}\mathsf{F}_t\left\{\alpha_\infty^{(Q)}\right\}$, and so on. Equation (8.4.9a) reduces finally to

$$\alpha_{00}^{(Q)} = \alpha_{00,\mathrm{H}} + \hat{\eta}_{00} \odot \alpha_{00}^{(Q)}, \tag{8.4.9b}$$

now with

$$\left.\begin{aligned} \mathsf{F}_{\mathbf{R}}\mathsf{F}_t\left\{\hat{M}^{(0)}\hat{Q}\alpha_\infty^{(Q)}\right\} &= \hat{\eta}_{00} \odot \alpha_{00}^{(Q)} = \hat{\mathsf{Y}}'_{00} \odot \hat{Q}_{00} \odot \alpha_{00}^{(Q)}, \text{ where} \\ \hat{\mathsf{Y}}'_{00}\odot &= \int \mathbf{dZ}'\mathsf{Y}_0(\mathbf{k}, s)_\infty e^{-i\mathbf{k}\cdot\mathbf{R}'+st'} \odot (\,)_{\mathbf{k},s};\ \hat{Q}_{00} \equiv \hat{Q}_{00}(\mathbf{k}, s), \text{ etc.} \end{aligned}\right\}, \tag{8.4.9c}$$

with $\mathbf{dZ}' \equiv dt'\mathbf{dR}'$, refer to Eq. (8.4.2b) above. Furthermore, the double Fourier transform of the undisturbed field (discussed as before, in Section 8.1.6) explicitly becomes

$$\alpha_{00,\mathrm{H}}(\mathbf{k}, s) = \hat{\mathsf{Y}}'_{00} \odot G_{00,\mathrm{T}} = \int_{-\infty}^{\infty} \frac{\mathbf{dk}'}{(2\pi)^3} \int_{\mathrm{Br}_1} \frac{ds'}{2\pi} \mathsf{Y}_0(\mathbf{k}', s) G_{00}(\mathbf{k} - \mathbf{k}', s - s')_{\mathrm{T}} \tag{8.4.10}$$

from (8.4.9c), with

$$\hat{Y}_{00} \equiv \mathsf{Y}_0, \tag{8.4.10a}$$

The full spectral, that is, wave number–frequency, results in the case of *backscatter*, (cf. (8.4.7a)), and is now obtained directly from (8.3.26) with the help of (8.4.9b) and (8.4.10).

(IV) Full Spectrum with Backscatter:

$$\alpha_{00,\mathrm{H}}(\mathbf{k},f) + \left[\hat{Y}'_{00} \odot \hat{Q}_{00} \odot \alpha^{(Q)}(\mathbf{k},s)^{+}\right]_V + \left[\hat{Y}'_{00} \odot \hat{Q}_{00} \odot \alpha^{(Q)}_{00}(\mathbf{k},s)\right]_S$$

$$= \alpha^{(Q)}_{00}(\mathbf{k},s); \quad \alpha_{00,\mathrm{H}} = \left[\hat{Y}'_{00} \odot G_{00\mathrm{T}}\right]_{V_T}. \tag{8.4.11a}$$

Here again, the subscript V and S refer to the infinite domain of the deterministic inhomogeneous media and the (here) purely reflective surface bounding the source volume V_{T}. The series solution analogous to (8.4.8)

$$\alpha^{(Q)}_{00}(\mathbf{k},f) = \left[\hat{1} - \hat{\eta}_{00}\odot\right]^{-1} \alpha_{00|\mathrm{H}}(\mathbf{k},f) = \alpha_{00,\mathrm{H}}(\mathbf{k},f)$$

$$+ \sum_{k=1}^{\infty} \left(\hat{Y}'_{00} \odot \hat{Q}_{00}\odot\right)^{(k)} \alpha_{00,\mathrm{H}}(\mathbf{k},f) \quad \text{with} \quad \|\hat{\eta}_{00}\| < 1. \tag{8.4.11b}$$

from (8.4.9c)–(8.4.11a).

8.4.2 Commutation and Convolution

Although the space–time operators $\hat{M}^{(0)}_{\infty}$ and $\hat{Q}$ do not commute (Section 8.3.4.1), their Fourier transforms do, since $\mathsf{F}_{\mathbf{R}}\mathsf{F}_t\left\{\hat{M}^{(0)}\right\} = -\hat{Y}_{00}\odot$ and $\mathsf{F}_{\mathbf{R}}\mathsf{F}_t\{\hat{Q}\} = \hat{Q}_{00}\odot$, so that $\hat{\eta}_{00}\odot$, (8.4.9c) and (8.4.11b), represent a pair of convolutions, for example,

$$\hat{\eta}_{00}\odot = \hat{Y}_{00} \odot \hat{Q}_{00} \cdot \odot = \hat{Q}_{00} \odot \hat{Y}_{00}\odot = \hat{M}^{(0)} \odot \hat{Q}_{00}, \text{ and so on.} \tag{8.4.12}$$

These do commute, since generally

$$\hat{A}(\mathbf{k},s) \odot \hat{B}(\mathbf{k},s)\odot = \int_{-\infty}^{\infty} \frac{\mathbf{dk}'}{(2\pi)^3} \int_{Br_1} \frac{ds'}{2\pi i} A(\mathbf{k}',s')B(\mathbf{k}-\mathbf{k}',s-s')(\,)_{\mathbf{k}',s'}$$

$$= \int_{-\infty}^{\infty} \frac{\mathbf{dk}'}{(2\pi)^3} \int_{Br_1} \frac{ds'}{2\pi i} B(\mathbf{k}',s')A(\mathbf{k}-\mathbf{k}',s-s')(\,)_{\mathbf{k}',s'} = \hat{B}(\mathbf{k},s) \odot \hat{A}(\mathbf{k},s)\otimes, \tag{8.4.12a}$$

on change of variables: $(s - s' = s'', \mathbf{k} - \mathbf{k}' = \mathbf{k}'')$, then with $(s'' \to s', k'' \to k')$. Accordingly, we see from (8.4.12a) that $\hat{\eta}^{(k)}_{00}\odot, k = 2$, is

$$\hat{\eta}^{(2)}_{00}\odot = \hat{\eta}_{00,1} \odot \hat{\eta}_{00,2} = \hat{\eta}_{00,2} \odot \hat{\eta}_{00,1} = \hat{Y}'^{(1)}_{00} \odot \hat{Y}'^{(2)}_{00} \odot \hat{Q}^{(1)}_{00} \odot \hat{Q}^{(2)}_{00} \odot \text{ and so on.} \tag{8.4.13}$$

The general result for $k \geq 2$ is

$$\left.\begin{aligned}\left(\hat{\boldsymbol{\eta}}^{(k)}\odot\right) = \prod_{l=1}^{k}\left[\hat{M}_{00,l}^{(0)}\odot\hat{Q}_{00}^{(l)}\odot\right] &= \hat{\eta}_{00,1}\odot\hat{\eta}_{00,2}\odot\cdots\hat{\eta}_{00,k}\odot = \prod_{l=1}^{k}\hat{\mathrm{Y}}_{00}^{(l)}\odot\hat{Q}_{00}^{(l)}\odot \\ &= \left(\hat{\mathrm{Y}}_{00}\odot\hat{\mathrm{Y}}_{00}\odot\cdots\hat{\mathrm{Y}}_{00}\right)_k\odot\left(\hat{Q}_{00}\odot\hat{Q}_{00}\odot\cdots\hat{Q}_{00}\odot\right)_k, \\ &= \hat{\eta}_{00}^{(k)}\odot\hat{Y}_{00}^{(k)}\odot\hat{Q}_{00}^{(k)}\odot,\end{aligned}\right\} \tag{8.4.14}$$

in various combinations of factors, which all commute with one another. More compactly, from the alternative results (8.3.20) and (8.3.22), we can replace the series representation (8.4.11b) by

$$\alpha_{00}^{(Q)}(\mathbf{k},f) = \alpha_{00,\mathrm{H}}(\mathbf{k},f) + \hat{\eta}_{00}^{(Q)}\odot\alpha_{00}^{(Q)}(\mathbf{k},f), \text{ where } \hat{\eta}_{00}^{(Q)}\odot \equiv \sum_{k=1}^{\infty}\hat{\eta}^{(k)}\odot, \tag{8.4.15}$$

with $\alpha_{00,\mathrm{H}}$ given by (8.4.10), where the kernel of $\hat{\eta}_{00}^{(Q)}$ is $g_{00}^{(Q)} = \mathsf{F}_{\mathbf{R}}\mathsf{F}_t\left\{g_{\infty}^{(Q)}\right\}$, from (8.3.16). In this way, a variety of equivalent expressions for $\alpha_{00}^{(Q)}$ can be derived.

Thus, we have the frequently encountered case of commutation in transform space (**k**, *s*), which when further transformed from this transform space, results once more in non-commuting factors in space–time (**R**, *t*). Operating in the transform space (**k**, *s*) may often lead to simpler results or approximations than direct efforts in the (**R**, *t*) domain.

8.5 EXTENSIONS AND INNOVATIONS, MULTIMEDIA INTERACTIONS

This section contains a number of extensions of the analysis Sections 8.3 and 8.4. The emphasis, as before, is on operational forms and methods. These in turn can be used as basic structures for computational results, as well as for the limited detailed analytic results directly obtainable in these complex situations. The presence of inhomogeneities in the physical medium, in addition to absorption (dispersion) along with complex boundaries, creates major technical problems, which though functionally solvable present considerable analytic difficulties (Chapters 8 and 9 of Ref. [6]). Here the channel and its components are deterministic—that is, these are no random elements and everything is known *a priori* except the field itself, which is then uniquely determined by the propagation model and the associated coupling to the medium in question. Equivalently, the deterministic case is an ensemble of a single representation of probability unity.

Topics examined here are also prelude to the random treatment required in Chapter 9 following.

(I) *Boundaries as Distributed Inhomogeneities*
(II) *Multimedia Interactions: The Deterministic Mass Operator or* $\hat{\eta}$*-form*
(III) *The Feedback Formulation*
(IV) *The Engineering Approach, II.*

For I, we have already quantitatively described the role of the surface effects for homogeneous media in Section 8.1.6, Eqs. (8.1.38)–(8.1.46), for inhomogeneous media in

Section 8.3.1 and 8.3.3, cf. Eqs. (8.3.17)–(8.3.22), and the Helmholtz example in Section 8.3.3. In general, the role of the boundary conditions is revealed in the surface component of the resulting field, as illustrated here by Figs. 8.11 and 8.12. The initial conditions are embodied in the driving elements of the signal source in the volume V_T, cf. Fig. 8.11.

8.5.1 The $\hat{\eta}$-Form: Multimedia Interactions

For II above, we have already presented a preview in Section 8.3.3.1 of how to handle the case where two contiguous media are involved and one is perfectly reflecting. We give here additional examples of the multimedia treatment, or $\hat{\eta}$-form, where two or three media are considered. Typical physical situations are as follows:

(1) *Atmosphere/ocean/bottom*, in the case of underwater acoustics (sonar, communications, etc.)

(2) *Atmosphere/ground* (radar, communications, etc.) and *water/bottom* and *water/atmosphere* (acoustics)

Variations of these with different source positions may also occur. One purpose of this material is to demonstrate the various interactions between the (here deterministic) media that can occur in the process of signal propagation. The $\hat{\eta}$-form allows us to account qualitatively (and ultimately quantitatively) for possible couplings between different scattering mechanisms, in realistically complex situations. It also provides a convenient way to identify and discuss those interactions that are not significant, thus allowing the usually much needed simplifications in the solutions. Figures 8.12 and 8.14a illustrate the two media cases discussed in Section 8.3.4.1, while Figs. 8.14b and 8.15 illustrate the more general three media configurations. The general situation when both reflection and transmission occur at and through media boundaries can be quite complex, as the following examples show.

8.5.1.1 Example: Radar and Telecommunications—One- and Two-Media Models

We begin with several two-medium cases, mostly appropriate to radar and telecommunications. The former operates usually in a "monostatic" regime, that is, where transmitters (T) and receivers (R) are located on the same platform ($R = T$). The latter requires a "bistatic" regime by definition, where receiver and transmitter are located at different places ($R \neq T$). The radar configuration is really a combination of the two if there is a target (Tg) present ($T \to \mathrm{Tg} \to R$). Figure 8.14a and b illustrate schematically a typical radar situation, as well as T and R in the telecommunication case.

The scattering analysis for Fig. 8.14a is given in operator form by the following[50]:

$$\left.\begin{array}{ll} \left(1-\hat{\eta}_{V_0}\right)\alpha^{(V_0)} = \alpha_H: & \text{scattered field in volume } V_0, \text{ due to scatterers in } V_0 \text{ only;} \\ \left(1-\hat{\eta}_{S_{0R}}\right)\alpha^{\left(V_0+R_{0/0}\right)} = \alpha^{(V_0)}: & \text{scattered field in } V_0, \text{ due to scatterers on } S_0 \text{ and in } V_0, \text{ with (perfect) reflection at } S_0. \end{array}\right\} \tag{8.5.1}$$

[50] Here, $R_{0/T}$ represents the portion of the scattered field in V_0 impinging on the interface S_0 between V_0 and V_1 and then being perfectly reflected (R) back into V_0 by S_0. Similarly, $R_{0/T}$ represents that portion of the scattering from V_0 that is transmitted (T) through the interface into V_1, and so on.

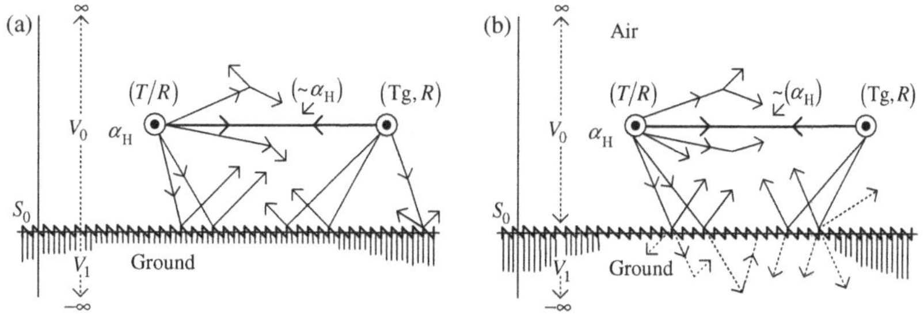

FIGURE 8.14 (a) Schematic of a radar monostatic configuration $T \to \mathrm{Tg} \to R(=T)$ and communication link $T \to R$, with scattering and backscatter = total scatter in V_0, when S_0 is perfectly reflecting $\left(R_{0/S} = 1\right)$. (b) Same as (a), but with S_0 partially reflecting and transmitting into V_1. Total scatter in V_0 unchanged, but at lower energy than in Fig. 8.14a. Some energy lost in V_1. Here $R_{0/S} < 1$, $\mathsf{T}_{0/S} > 0$.

Note that $\alpha^{(V_0+R_{0/R})}$ includes all scatterers on S_0 and in V_0, and the multiple interactions between the multiple reflected scatter from S_0. (The multiple backscatter from the first backscatter are not included in the figures.) However, as discussed in Section 8.3.4.1, after Eq. (8.3.23b), the number of scattering elements remaining on the surface S_0 and in the volume V_0 increases the number of coupled, that is, interacting, scatterers, creating a more complex scattering environment, mainly in the neighborhood of the interface. In this first case, only reflection takes place, that is, $R_{0/S} = 1$. In the second case $\left(R_{0/S} < 1\right)$, with some transmission through the interface S_0, the same sort of scenario in V_0 and on S_0, now with a few additional scatterers in V_1, contributes to those in V_0, but with a loss of energy in V_0, due to scattering out of $V_0\left(R_{0/T} > 0\right)$ and the scattering from V_0 mostly remaining in V_1. The number of scatterers in V_0, V_1 and on S_0 remains constant.

Returning to (8.5.2), we easily see that[51]

$$\alpha^{(V_0)} = \alpha^{(V_0+R_{0/0})} = \left(1 - \hat{\eta}_{S_{0/0}}\right)^{-1}\left(1 - \hat{\eta}_{V_0}\right)\alpha_H \doteq \left[1 - \left(\hat{\eta}_{S_{0_R}} + \hat{\eta}_{V_0}\right)\right]^{-1}\alpha_{\mathrm{H}}, (0 \leq s_{OR} \leq 1), \tag{8.5.2}$$

where $\left\|\hat{\eta}_{S_0}\hat{\eta}_{V_0}\right\| << \left\|\hat{\eta}_{S_0} + \hat{\eta}_{V_0}\right\|$, by our usual assumption that interactions between types of scatter are generally negligible vis-à-vis the primary effects themselves. The series solution for (8.5.2) becomes (cf. 8.3.8) et seq.)

$$\begin{aligned} \alpha^{(V_0)} &= \alpha^{\left(V_0+R_{0/R}\right)} \doteq \alpha_{\mathrm{H}} + \sum_{k=1}^{Q\to\infty}\left(\hat{\eta}_{R_{0/R}} + \hat{\eta}_{V_0}\right)^{(k)}\alpha_{\mathrm{H}}^{(k)} \\ &\doteq \alpha_{\mathrm{H}} + \left\{\sum_{k=1}^{\infty}\hat{\eta}_{R_{0/R}}^{(k)} + \sum_{k=1}^{\infty}\hat{\eta}_{V_0}^{(k)}\right\}\alpha_{\mathrm{H}}, \quad \left(0 \leq R_{0/S} \leq 1\right), \end{aligned} \tag{8.5.3}$$

which is the approximate result, neglecting all interactions between surface and volume scatter of the second and higher orders. Equation (8.5.3) is one form of the solution to the integral equation (8.3.23).

If the boundary S_0 allows transmission $\left(R_{0/T} > 0\right)$ as well as reflection $0 < R_{0/R} < 1$, cf. Fig. 8.14b, a third relation for (8.5.1) giving the transmitted field through S_0 and

[51] See the remarks regarding commutability in footnote 45.

into V_1 becomes[51]

$$\alpha^{(V_1)} \equiv \alpha^{\left(V_0+R_{0/T}+V_1\right)} = \left(1-\hat{\eta}_{V_1}\right)^{-1}\left(1-\hat{\eta}_{R_{0/T}}\right)^{-1}\left(1-\hat{\eta}_{V_0}\right)^{-1}\alpha_{\mathrm{H}}. \tag{8.5.4}$$

This reduces to

$$\begin{aligned}\alpha^{(V_1)} &= \left(1-\hat{\eta}_M\right)^{-1}\alpha_{\mathrm{H}} \\ &= \left[1-\left(\hat{\eta}_{V_1}+\hat{\eta}_{R_{0/T}}+\hat{\eta}_{V_0}\right)+\left(\hat{\eta}_{V_1}+\hat{\eta}_{R_{0/T}}+\hat{\eta}_{V_1}\hat{\eta}_{V_0}+\hat{\eta}_{R_{0/T}}\hat{\eta}_{V_0}\right)-\hat{\eta}_{V_1}\hat{\eta}_{R_{0/T}}\hat{\eta}_{V_1}\right]^{-1}\alpha_{\mathrm{H}},\end{aligned} \tag{8.5.5}$$

where $\hat{\eta}_M$ is the composite operator described in full in the second equation of (8.5.4). Making the usual approximations, namely, neglecting all higher orders of scattering interactions $\left(\hat{\eta}_{V_1}\hat{\eta}_{R_{0/T}}, \text{ etc.}\right)$, we obtain the extension of (8.5.3):

$$\alpha^{(V_1)} = \alpha^{\left(V_0+R_{0/T}+V_1\right)} \doteq \alpha_{\mathrm{H}} + \left\{\sum_{k=1}^{\infty} \hat{\eta}_{V_1}^{(k)} + \hat{\eta}_{R_{0/T}}^{(k)} + \hat{\eta}_{V_0}^{(k)}\right\}\alpha_{\mathrm{H}}\left(=\alpha^{(Q)}\right), \quad \left(0 < R_{0/T} < 1\right), \tag{8.5.6}$$

with (8.5.3) representing the field, including reflections, in V_0. Of course, when scattering interactions cannot be neglected—not a usual situation in the types of communication treated here—we must use part or all of $\hat{\eta}_M$. In many cases, the mass operators for the different media and their interfaces are additive to a satisfactory approximation.

8.5.1.2 Example: Ocean Environments—Three Media Here, we extend our discussion to the important case where three media are involved and where the primary source is located in the middle one (V_1), in the manner of Fig. 8.15a and b.

The scattering analysis for these examples may be established in the same way. Now we have besides medium V_1, possible reflections from and transmission through the interfaces S_0 and S_1, Fig. 8.15a. The various fields associated with the different media are given by

$$V_0\!: \alpha^{(V_0)} = \alpha^{(V_1+R_{0/T}+V_0)} = \left(1-\hat{\eta}_{V_0}\right)^{-1}\left(1-\hat{\eta}_{R_{0/T}}\right)^{-1}\left(1-\hat{\eta}_{V_1}\right)^{-1}\alpha_H; \tag{8.5.7a}$$

$$\begin{aligned}V_1\!: \alpha^{(V_1)} = \alpha^{\left(V_1+S_{1R}+S_{0R}\right)} &= \left(1-\hat{\eta}_{R_{1/R}}\right)^{-1}\left(1-\hat{\eta}_{V_1}\right)^{-1}\alpha_{\mathrm{H}} + \left(1-\hat{\eta}_{R_{0/R}}\right)^{-1}\left(1-\hat{\eta}_{V_1}\right)^{-1}\alpha_{\mathrm{H}}; \\ &= \left[\left(1-\hat{\eta}_{R_{1/R}}\right)^{-1} + \left(1-\hat{\eta}_{R_{0/R}}\right)^{-1}\right]\left(1-\hat{\eta}_{V_1}\right)^{-1}\alpha_{\mathrm{H}};\end{aligned} \tag{8.5.7b}$$

$$V_2\!: \alpha^{(V_2)} = \alpha^{\left(V_1+R_{1/T}+V_2\right)} = \left(1-\hat{\eta}_{V_2}\right)^{-1}\left(1-\hat{\eta}_{R_{2/T}}\right)^{-1}\left(1-\hat{\eta}_{V_1}\right)^{-1}\alpha_{\mathrm{H}}. \tag{8.5.7c}$$

These relations formally apply also for the inhomogeneous case of Fig. 8.15b, where the particular inhomogeneity occurs in the speed of propagation $c(z)$ (or more generally, $c(\mathbf{R}, t')$), which can cause a very nonhomogeneous spatial concentration of the source

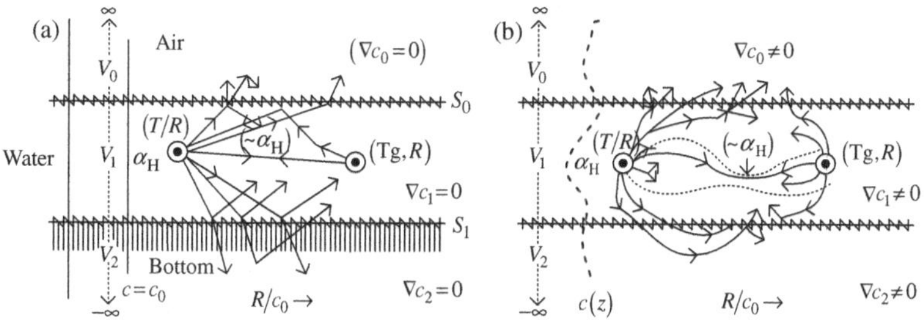

FIGURE 8.15 (a) Schematic of a monostatic sonar in a gradient-free (i.e., $\nabla c(z) = 0$) environment ($c = c_0$) and a communication link ($T \to R$), with scattering, backscatter, and penetration of other media, and scattering therein, ($0 \leq S_{OR_1}, S_{OR_2} < 1$). (b) Same as (a), except for the presence of velocity gradients in all three media.

energy and the scattering produced in the medium by the source. Note from Eqs. (8.3.9a) and (8.3.9b) that $\hat{\eta}_V, \hat{\eta}_S$ throughout contains $\hat{Q}_V$, Table 8.4, and $\hat{Q}_S = \mathsf{R}_0\hat{1}$ or $\mathsf{T}_0\hat{1}$, the (plane wave) reflection or transmission operators at the various interfaces in Figs. 8.14a, 8.14b, 8.15a and 8.15b.

The unscattered component of the original field α_H from the primary source, if any, is reduced in energy by the amount created in the scattered fields. Thus, α_H represents the unscattered component at any time after its initiation and subject to the scattering elements in the various media in Figs. 8.14 and 8.15. For example, a deterministic pulse of energy injected into the medium will have its energy progressively reduced by scattering, until after enough time it has become essentially all scattered energy. Any of the original structural waves, at various ranges (i.e., times R/c_0), will obey the geometry of the medium, that is, reflection and transmission at interfaces and in the medium itself. The reflections constitute "multipath," that is, resolvable and organized "scatter," and transmission is regular outward propagation from the sources.

Finally, from the above multiple boundary effects and the argument given earlier (Eq. (8.3.7c)), it is evident that because of the boundaries the fields in the various regions of propagation are no longer linear, in the sense that superposition holds. This can be seen directly by application of (8.3.7c) to the fields in each region of the examples in (8.5.1)–(8.5.6). We leave the proof to the reader.

8.5.2 The Feedback Operational Representation and Solution

The form of Eqs. (8.3.7a), (8.3.7b), and (8.3.23a), and other similar expressions (Section 8.5.1) suggest that these representations and their series solutions for the present linear *deterministic* media may be interpreted as a generalization of the familiar one-dimensional feedback loop of engineering practice. Its output is the feedback (operational) solutions, (FOS) for the scattered field. Thus, from (8.3.23) we have

$$\begin{aligned} \alpha^{(Q)} &= \left(\hat{1} - \hat{M}_{\infty}^{(0)}\hat{Q}\right)^{-1}\alpha_H, \quad \text{with} \quad \hat{\eta} = \hat{M}_{\infty}^{(0)}\hat{Q} \\ \alpha_H &= \hat{M}_{\infty}^{(0)}(-G_T),\ G_T = G_T(\mathbf{R}', t') \in V_T; = O, G_T \notin V_T, \end{aligned} \tag{8.5.8}$$

where α_H is the unscattered or unperturbed source field. Again, $\hat{\eta}$ is analogous to the "mass operator" or "field renormalization" operator of quantum field theory. The composite operator $\hat{\eta} = \hat{M}_\infty^{(0)}\hat{Q}$ is global and embodies the rescaling of the original field required as a result of the scattering produced in the medium and its boundaries. Here again, $\hat{M}_\infty^{(0)}$ is the integral Green's function operator, refer to Eq. (8.1.10), and $\hat{Q}$ represents any (local) inhomogeneity in the medium.[52] Also note again that $\hat{M}_\infty^{(0)}$ is a global operator, whereas $\hat{Q}$ is local: the former is also a *projection operator*, propagating the nonlocal source density G_T, $(\mathbf{R}' \in V_T)$, to all permitted points $(\mathbf{R}, t) \notin V_T$ outside the source V_T, where $\hat{Q} = \hat{Q}(\mathbf{R}', t')$ is defined at a point $(\mathbf{R}', t')$. Other equivalent forms of (8.3.23) are the solutions:

$$\alpha^{(Q)} = \alpha_{\mathrm{H}} + \frac{\hat{\eta}_\infty}{\hat{1} - \hat{\eta}_\infty}\alpha_{\mathrm{H}} = \sum_{k=0}^{\infty}\left(\hat{M}_\infty^{(0)}\hat{Q}\right)^{(k)}\alpha_{\mathrm{H}} = \alpha_{\mathrm{H}} + \alpha_1^{(Q)} \tag{8.5.9a}$$

$$\therefore \quad \alpha_1^{(Q)} = \sum_{k=1}^{\infty}\left(\hat{M}_\infty^{(0)}\hat{Q}\right)^{(k)}\alpha_H = \frac{\hat{\eta}_\infty}{\hat{1} - \hat{\eta}_\infty}\alpha_{\mathrm{H}}\text{: inhomogeneous field component of } \alpha^{(Q)} \tag{8.5.9b}$$

These series are often called the *perturbation series solutions* (PSS) to the dynamical equation (8.3.23). As before (Section 8.3), the superscript (k) denotes the kth *iteration* of the operator(s) in question. *Note that the* kth *order mass operator* $\hat{\eta}^{(k)}\alpha_{\mathrm{H}}$ *contains all the* kth *order, and no other interactions of the set of k scattering elements, illuminated by the incident field* α_{H}. *Clearly, all the* kth *order interactions or "k-tuples" are physically independent of all other orders of coupled scatters.* (This critically important observation assits in construction of the probability distributions of the scattered field in the random environments encountered in most practical applications.)

The *feedback operational representation (FOR)*[53] of (8.5.9a) and (8.5.9b) is illustrated in Fig. 8.16.

If in Fig. 8.16 we regard α_1 as the input field to $M_\infty^{(0)}$, with α_2 the field input to $\hat{Q}$ and α_{F} the "feedback" field to G_{T}, then Fig. 8.16 represents a "field circuit" diagram, from which the functional equations relating the various fields $(\alpha, \alpha_1, \alpha_2, \alpha_{\mathrm{F}})$ can be immediately written:

$$\alpha_1 = -G_T + \alpha_{\mathrm{F}}; \quad \alpha_2 = \alpha; \quad \alpha = \hat{M}_\infty\alpha_1; \quad \alpha_{\mathrm{F}} = \hat{Q}\alpha_2, \tag{8.5.10}$$

These are then directly solved for α to give us the resulting integral equation:

$$\alpha = \hat{M}_\infty\left(-G_{\mathrm{T}} + \hat{Q}\alpha\right), \text{ that is } \alpha - \hat{M}_\infty^{(0)}\hat{Q}\alpha = -\hat{M}_\infty^{(0)}G_{\mathrm{T}} = \alpha_{\mathrm{H}}, \tag{8.5.10a}$$

[52] Exclusive of boundaries, which are considered separately as defining a limitation on the medium in question. As noted in **I.** (beginning of Section 8.4.1), boundaries are a continuously limiting form of distributed inhomogeneities.

[53] This "loop formulation" (or FOR here) is usually called the "classical approach" in modern control theory. The more modern approach is based on "state variables," which in turn are differential equations derived from the optimization of the Lagrange, Hamilton, and Euler equations of classical mechanisms, based on the energy of the system in question. The resulting Euler or dynamical equations, here space–time (partial) differential equations, are the equations of propagation, which we have already seen in this chapter, for both Hom-Stat and non-Hom-Stat media. For random fields, the ensemble, with its associated probability measures of such dynamical equations, constitutes the Langevin equation (cf. Chapter 9). For an extension treatment of dynamical equations of the resulting fields, see in particular Chapters 2 and 3 of Ref. [1] in addition to Ref. [37–40]. Examples of both the direct approach based on calculation of the energies from the equation of propagation and a concise outline of the more general Lagrange–Hamiltonian–Euler approach are given in Section 8.6.

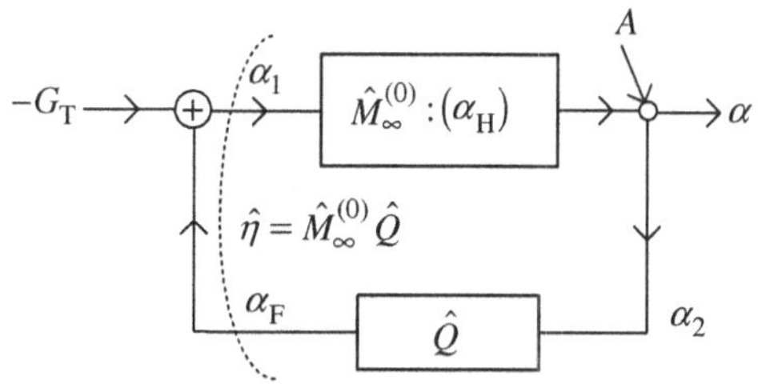

FIGURE 8.16 Feedback operational representation, Eq. (8.5.9a) and (8.5.9b) for the deterministic propagation equation (8.5.8), whose solution (FOS) is expressed in the various equivalent forms (8.5.9a) and (8.5.9b).

or

$$\alpha = \alpha_H + \hat{M}_\infty^{(0)} \hat{Q}\alpha = \alpha_H + \hat{\eta}_\infty \alpha, \quad \alpha = \alpha^{(Q)}, \tag{8.5.10b}$$

the latter being just the original propagation equation (8.5.9) where we have written $\alpha = \alpha^{(Q)}$, to emphasize the inhomogeneous character of the resultant field. Accordingly, the closed loop, feedback diagram Fig. 8.16 also embodies the equivalent (8.5.9).

The interpretation of the FOS, Eq. (8.5.9), is straightforward in terms of a simulated iteration process. Conceptually, we start the loop operating by injecting the signal $-G_T$, which gives $\alpha = \alpha^{(0)}$ at A in Fig. 8.16. This, in turn, is fed back through $\hat{Q}$ and then forward through $\hat{M}_\infty$ to give $\alpha^{(1)}$ at A (hence the terms "feedback" operator for $\hat{Q}$ and "feedforward" operator for $\hat{M}_\infty$). The sequence is clearly as follows:

0th iteration: $$\alpha^{(0)} = -\hat{M}_\infty^{(0)} G_T = \alpha_H, \left(\hat{Q} = 0\right). \tag{8.5.11a}$$

1st iteration: $$\alpha^{(1)} = \alpha^{(0)} + \hat{M}_\infty^{(0)} \hat{Q}\alpha^{(0)} = \left(1 + \hat{M}_\infty^{(0)} \hat{Q}\right)\alpha_H. \tag{8.5.11b}$$

2nd iteration: $$\alpha^{(2)} = \alpha^{(0)} + \hat{M}_\infty^{(0)} \hat{Q}\alpha^{(1)} = \left(1 + \hat{M}_\infty^{(0)} \hat{Q} + \left(\hat{M}_\infty^{(0)} \hat{Q}\right)^{(2)}\right)\alpha_H. \tag{8.5.11c}$$

$$\vdots \qquad \vdots \qquad \vdots \qquad \vdots$$

kth iteration: $$\alpha^{(k)} = \alpha^{(0)} + \hat{M}_\infty^{(0)} \hat{Q}\alpha^{(k-1)} = \sum_{m=0}^{k} \left(\hat{M}_\infty^{(0)} \hat{Q}\right)^{(m)} \alpha_H, \quad m \geq 0; \left(\alpha^{(-1)} \equiv 0\right) \tag{8.5.11d}$$

Thus, as $k \to \infty$, we obtain the same series, (8.5.9a), *which is moreover termwise identical to the* PSS or, equivalently, to the FOS. The series in (8.5.9a) as $k \to \infty$ is precisely the summed series of partial solutions (8.5.11) obtained "classically" by iteration of succeeding approximations in the dynamical (integral) equation $\alpha_\infty^{(k)} - \hat{M}_\infty^{(0)} \hat{Q}\alpha_\infty^{(k-1)} = \alpha_\infty^{(0)}$, namely the PSS (8.5.9). This establishes the equivalence of FOS, FOR, and PSS (also we shall see in Chapter 9 ff., the associated Feynman diagram series.

The integral Green's function operator $\hat{M}_{\infty}^{(0)}$(8.1.42h) for volume and surface are explicitly (cf. (8.3.8))

$$\hat{M}_{\infty}^{(0)}|_V = \int dt' \int_{V_T} g_{\infty}^{(0)}(\mathbf{R}, t|\mathbf{R}', t')(\,)_{\mathbf{R}',t'}\mathbf{dR}'; \hat{M}_{\infty}^{(0)}|_S$$
$$= \int dt' \oint_{S'_0} \left\{ g_{\infty}^{(0)}\hat{\mathbf{n}}' \cdot \nabla'(\,) - \left[\hat{\mathbf{n}}' \cdot \nabla' g_{\infty}^{(0)}\right] \right\} (\,)_{(\mathbf{R}',t')\in S_0} dS'_0. \tag{8.5.12}$$

As noted in Section 8.3.1.1, the local operator $\hat{Q}(\mathbf{R}', t')_{V \,\mathrm{or}\, S}$ is a property of the scattering medium, for example, the kernels

$$Q_V = Q(\mathbf{R}, t)_V, \quad Q_S = \mathsf{R}_0(\mathbf{R}', t'), \mathsf{T}_0(\mathbf{R}', t'), \tag{8.5.12a}$$

the reflection and transmission coefficient of the reflecting or transmitting interface. From (8.3.9a) and (8.3.9b) $Q_{V,S}$ is to be inserted into the last brackets $(\,)_{\mathbf{R}',t}$ of (8.5.12); see remarks after Eq. (8.3.9).)

In all cases, the two factors of the mass operator $\hat{\boldsymbol{\eta}}_{\infty}$ are the "feedforward" operators, which are always the global or integral operator $\hat{M}_{\infty}^{(0)}$ and the local inhomogeneity operator $\hat{Q}$. These two operators are always distinct, that is, separate factors of $\hat{\boldsymbol{\eta}}_{\infty}$, not only for volumes but also for surfaces as well, as indicated by (8.5.11a) and (8.5.11b) in these deterministic cases.[54] (In the terminology of control theory, the field renormalization or mass operator $\hat{\boldsymbol{\eta}}_{\infty}\left(\equiv \hat{M}_{\infty}^{(0)}\hat{Q}\right)$, refer to Eq. (8.5.3), is called the *loop cycle-* or *loop-iteration, operator*.[54]) As examples of *FOS*/*FOR* representations (Fig. 8.17), we may apply them to the cases illustrated in Section 8.5.1, as shown in Fig. 8.14a and b (and by extension to 8.15a and b).

The loop iterations of (8.5.11) indicate one potentially exact method of actually evaluating the FOS and hence the PSS. The purely computational effort is expected to be very large: at each iteration the entire $(k-1)$st iterated field is required in order to obtain the kth-order iteration at any one field point $P(\mathbf{R}, t)$. However, as k is made larger, the output $\alpha^{(k)}$ approaches the true or "equilibrium" value[55] (for all $(\mathbf{R}, t)$, even though we may be interested only in $\alpha^{(k)}$ at a single, preselected point $P(\mathbf{R}, t)$). This simulation is nontrivial from another viewpoint: physical boundary, initial, and radiation conditions (cf. Sections 8.1 and 8.3.) must also be suitably imposed to ensure physically meaningful solutions. In addition, we must be alert to possible local and distributed "pathologies".

Rather than attempting to work with a continuum of field values, we employ multidimensional sampling theory and related techniques, some of them already discussed in previous chapters (see also Refs. [41–43]). Discrete sampling is needed for the required

[54] For example, see Refs. [37–40]. During the late 1950s through the 1970s control theory was being developed to include statistical phenomena.

[55] It should be emphasized that the numerical steps in the iteration process (8.5.11) involved in this simulation are not generally isomorphic with the physical propagation events that are being modeled by the mathematical iteration (8.5.11). They are instead an artificial decomposition into sequence of interactions, which do not necessarily coincide with the temporal and spatial progress of the actual propagation and scattering processes that occur in the medium and at the interfaces. It is only the final "equilibrium" values (as $k \to \infty$) that represent the field at specified points $P(\mathbf{R}_m, t_n)$.

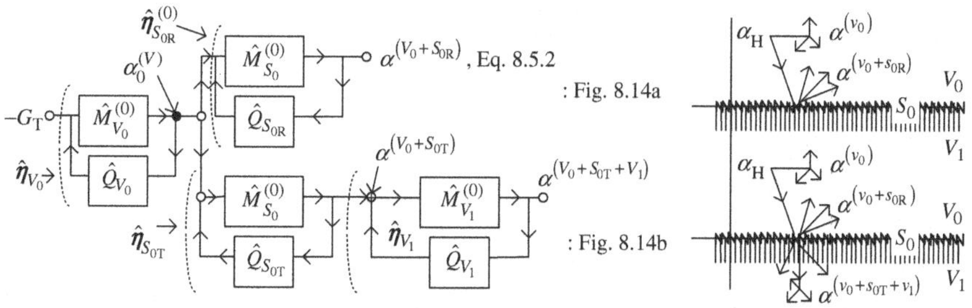

FIGURE 8.17 *FOR* (and *FOS*) for the examples of Section 8.5.1.1, Fig. 8.14a and b, Eqs. (8.5.1), (8.5.2), (8.5.4), and (8.5.5).

digital data and for data reduction to manageable proportions. The numerical problem is the order of that demanded for modern weather analysis and prediction. Modern methods here should yield FOS results on acceptable timescales once the "macroalgorithms" of the physical formulating, for instance, described in this chapter, have been translated into suitable software. One such approach is suggested here by again using control theory methods, conceptually outlined in Section 8.5.3.

8.5.3 An Estimation Procedure for the Deterministic Mass Operators $\hat{Q}$ and $\hat{\eta}$

When the elements $\left(-G_T, \hat{M}^{(0)}\right)$ of the "input–output" structure of Fig. 8.16 are available, estimates of the mass operator $\hat{Q}$ may be made, in principle to any degree of accuracy from the deterministic empirical or *a priori* field $\alpha^{(Q)}$, in a controlled way with determinable error. This is a form of "system identification" problem, typical of control theory.[56] Here, in essence, one introduces a "black box" or system with a controllable input and observes the resulting output(s). One then attempts to infer the operations of the system that relate the two. One key feature here is that the system in question is known to be linear, so that unique relations are established.

For Eq. (8.5.8) to be a *physically* useful device, as opposed to a mathematical relation, we must be able to indicate a physical procedure for determining $\hat{Q}$. One way is according to the scheme shown in Fig. 8.18. This approach produces a converging series of estimates $\hat{Q}_{\text{est}}$ of $\hat{Q}$ by iteration, when α and G_T are known empirically and when $\hat{M}_\infty^{(0)}$ is either

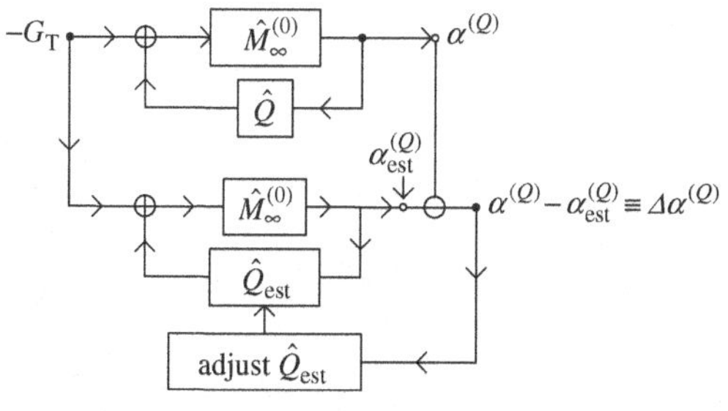

FIGURE 8.18 Feedback operational method for obtaining an estimate of the (deterministic) medium's inhomogeneity operator $\hat{Q}$, refer to Fig. 8.16.

[56] See, for example, Refs [37,40], which are discussed briefly in footnote 52.

experimentally or analytically given. The basic approach is conceptually quite simple: with a starting estimate $\hat{Q}_{\text{est}}$ of $\hat{Q}$, one runs through the second loop, using the (experimentally) observed α_{est}, the given source $-G_{\text{T}}$, and the known global descriptions of the homogeneous component of the medium ($\hat{M}_{\infty}^{(0)}$ here). The result is then $\alpha_{\text{est}}^{(Q)}$, which is then subtracted from the observed $\alpha^{(Q)}$ from the first loop, which in turn leads to an adjustment of $\hat{Q}_{\text{est}}$, following a second comparison, with a further reduction in the error $\Delta\alpha^{(Q)} = \alpha^{(Q)} - \alpha_{\text{est}}^{(Q)}$, resulting in $\Delta\hat{Q}^{(Q)} \equiv \hat{Q} - \hat{Q}_{\text{est}}$. Various optimum and suboptimum schemes from control theory (e.g., Wiener filtering and gradient climbing, now extended to *four* dimensions) may be used to drive the errors $\Delta\alpha^{(Q)}, \Delta\hat{Q}^{(Q)}$ essentially to zero, so that $\hat{Q}_{\text{est}} \to \hat{Q}$. This technique may also suggest acceptable approximations in experimental situations. For example, in the deterministic case of scattering volume that has a complex surface boundary (in the manner of Fig. 8.12), the easiest way to estimate its scattering effect may be to obtain $\hat{Q}_{S_0}$ by experiment, using the technique suggested by Fig. 8.18. This approach may also be extended to the more general situations of Fig. 8.17 by successive determinations of $\alpha^{(V_0)}$ and $\hat{Q}_{V_0}$, $\alpha^{(V_0,+R_{0/S})}$ and $\hat{Q}_{V_0+R_{0/S}}$, and so on, where we also use the information embodied in $\hat{\boldsymbol{\eta}}_{V_0+R_{0/R}} = \hat{\boldsymbol{\eta}}_{R_{0/R}} + \hat{\boldsymbol{\eta}}_{V_0} - \hat{\boldsymbol{\eta}}_{S_0}\hat{\boldsymbol{\eta}}_{V_0} \doteq \hat{\boldsymbol{\eta}}_{S_0} + \hat{\boldsymbol{\eta}}_{V_0}$, and so on. Here, estimating $\hat{Q}$ is of course equivalent to estimating $\hat{\eta}(\equiv \hat{M}\hat{Q})$ since $\hat{M}$ is known. The exploitation of these methods, however, is outside the scope of this book. In any case, because of the dimensionality (i.e., iterated space–time values), these methods will be computationally very large. See Chapters 8 and 9 of Ref. [6]. For other specific applications, see Refs [8,37–40].

8.5.4 The Engineering Approach II: Inhomogeneous Deterministic Media[57]

The "engineering approach" to the field representation in case of homogeneous media, outside the finite region V_{T} of the source, has been shown to be equivalent to the physically derived result from basic principles (i.e., propagation equations, boundary and initial conditions, etc.), as discussed in Section 8.2.1. This gives a rather general result, *provided the conditions (8.2.8) and (8.2.10) hold* for the field and its reception in the full situation of the channel. The latter requires an additional (linear) filter $h_{\text{R}}(\tau)$, or more generally, a time-variable filter $h_{\text{R}}(\tau, t)$, when relative motion of source, medium, and receiver occurs. When we have to deal with an *inhomogeneous* medium of propagation, as we have observed from Section 8.3, although the medium in question is still linear, it is characterized by an additive, deterministic, source-dependent scatter or "noise." This "noise" has largely lost its coherent structure, but the original signal and field are still assumed to be deterministic. The received field is, for example, given by (8.3.8), (8.5.9a) and (8.5.9b):

$$\alpha^{(Q)}(\mathbf{R}, t) = \alpha_{\text{H}}(\mathbf{R}, t) + (1 - \hat{\eta})^{-1}\hat{\eta}\alpha_{\text{H}}(\mathbf{R}', t') = \sum_{k=0}^{\infty} \hat{\eta}^{(k)}\alpha_{\text{H}}(\mathbf{R}', t'), \tag{8.5.13}$$

where the conditions (8.2.8) and (8.2.10) apply for $k = 0$. For $k \geq 1$, the result is noise, that is,

$$\alpha^{(Q)} = \sum_{k=1}^{\infty} \left(\hat{M}_{\infty}^{(0)}\hat{Q}\right)^{(k)} \alpha_{\text{H}}, \quad \text{with} \quad \hat{M}_V^{(0)}\hat{Q}_V = \hat{M}_S^{(0)}Q_S. \tag{8.5.13a}$$

[57] See Section 8.2 for a treatment of the homogeneous cases.

respectively, for volume or surface, refer to Eqs. (8.3.25a) and (8.3.25b). In more detail, expressed in (**k**, s) space, we have for the unscattered, that is signal, portion of the field

$$\alpha_{00}(\mathbf{k}, s)_{\mathrm{H}} = \int_{-\infty}^{\infty} \frac{d\mathbf{k}'}{(2\pi)^3} \int_{Br_1} \frac{ds'}{2\pi i} \mathrm{Y}_0(k', s') G_{00}(\mathbf{k} - \mathbf{k}', s - s')_T, \quad \text{refer to Eq. (8.4.10)}, \tag{8.5.13b}$$

where $\mathrm{Y}_0(\mathbf{k}, s) \equiv -L_0(\mathbf{k}, s)$, from (8.1.17b). Here $\hat{M}_{00}^{(0)}, \hat{Q}_{00}$ commute, whereas in (**R**, t) space they do not (Section 8.4.2).

For the ambient and scatter noise component α_N of the received field, we can write an ad hoc linear filter equivalent

$$\alpha_{\mathrm{N}}(\mathbf{R}, t) = \int_{-\infty}^{\infty} h_{\mathrm{N}}(\tau|R)_{\mathrm{FF}} S_{\mathrm{in}}(t - \tau)_{\Delta} d\tau \ \text{ or } \int_{-\infty}^{\infty} h_{\mathrm{N}}(\tau, t|R)_{\mathrm{FF}} S_{\mathrm{in}}(t - \tau)_{\Delta} d\tau \tag{8.5.14}$$

generally, to take account of any Doppler. The presence of the range (R) is to indicate a change in magnitude due to the spreading effect of the noise field as it becomes more distant from its primary source in V_{T}. Accordingly, the presence of the inhomogeneities when a field α_H is generated produces the sum field $\alpha^{(Q)} = \alpha_{\mathrm{H}} + \alpha_{\mathrm{N}}$, both of which can be described by linear (generally time-variable) filters representing the scattering medium, ambient sources, and the coherent (signal) components, when they are present and not entirely reduced to scatter. In effect, conditions I–IV of (8.2.14) apply automatically for the ambient and scatter noise fields (α_{H}), with the far-field requirement (II) usually dominant. The rôle of $\alpha_{\mathrm{amb.}}$ versus $\alpha_{\mathrm{scat.}}$ is decided by the geometry of the signal source and the particular distribution of the ambient sources. The causality (realizability) conditions (Section 8.2.2) for the equivalent channel filters h_{N}, and so on, remain basically unchanged.

Finally, we note that the *energy* in the unscattered and scattered components is not explicitly indicated or quantified in these relations and, in fact, in any of our analysis above. We shall discuss energy relations in Section 8.6. We note, however, that a calculation of the energy expended in the scattering component at the expense of the energy originally in the unscattering field depends on the following:

(1) The nature of the medium itself, that is, whether or not it is dissipative, elastic, viscous, and so on, or air, water, earth, sand or rock, and so on.
(2) The type of propagation, that is, longitudinal (pressure) waves, transverse waves, for example, electromagnetic as well as mechanical, elastic, and so on.
(3) Quantum mechanical, and so on.

We shall present below a brief treatment of energy calculations for the various types of wave equations treated in the preceding sections of this chapter.

8.6 ENERGY CONSIDERATIONS

Here we are concerned with the energy in the propagated wave $\alpha(\mathbf{R}, t)$, or more specifically, with the *energy density*. The nature of the propagation will, of course, depend on the physical characteristic of the medium in question, whether gas, liquid, or solid, or even vacuum. Gases and liquids do not support a shear, while solids do. These last are *deformable*: two points in such media undergo a change in their separation on the application of a force such as a change in pressure, and remain deformed when the force is removed. This is in contrast to an *elastic medium* that regains its original shape when the force is removed. Thus, it is deformable *and* elastic. A continuous elastic medium is one in which the (temporary) deformation (i.e., *strain*) is not localized and supports propagation throughout the medium. The result is "elastic or mechanical radiation." The force acting to produce the deformation, usually a *pressure* (force per unit volume or area), is called a *stress* and the resulting deformation a *strain*. Accordingly, we may say in these instances that *stress produces strain* or deformation, temporary or permanent, depending on the nature of the body to which the stress is applied. In case of an applied pulse, the strain is propagated in the gas, liquid, or solid. An important exception is the nonmechanical, electrodynamical situation where the medium can be a vacuum. The propagating phenomenon is solely an (EM) field, which, unlike the example above, needs no physical medium for its initiation and maintenance, although such fields can still propagate—however, weakly in some instances—in material media.

The above allows us to summarize the different media supporting propagation in a limited hierarchy of increasing density:

(8.6.1)

- (1) Vacuum: Empty space, no matter present: only EM propagation possible
- (2) Gas: Low-density continuous media (for our range of frequency most employed in this book) Such environments do not support a shear, that is, $\nabla \times d = 0$, $\mathbf{d} =$ *a displacement* field. Here $\mathbf{d}$ is said to be irrotational.
- (3) Liquid: Usually of much greater density than gas, for example, water. Such media also does not support a shear
- (4) Solids: These are sufficiently denser to maintain shape or at most to undergo minor distortion (shear) from the resting state, which is restored when the applied force (stress) is removed. Such bodies do support a shear, that is, $\nabla \times \mathbf{d} \neq 0$. We distinguish elastic and nonelastic media as they lead to different classes of propagation.

8.6.1 Outline of the Variation Method

We now distinguish the two main classes of radiation, which depend on how they are produced. One class is the so-called "mechanical radiation," which is generated by the medium itself when subjected to a deformation. The second does not depend on a physical medium for resulting field, although such media can modify its properties. Examples of the latter class are the electromagnetic field and the gravitational field associated with matter. Both classes have vector and scalar manifestation. We shall present examples of both in the following section.

There are two principal methods of evaluating the energy flux of these two classes of fields. One approach is the direct calculation from the appropriate field itself. The second is the more general *variational method:* The latter procedure begins with canonical expressions for the energy, and (usually) the energy density, in several forms. These start with the basic relations

$$\left.\begin{aligned} &\textit{Lagrangian function}:\quad L = K + \mathsf{V} \\ &\textit{Hamiltonian function}:\quad H = 2K - L, \text{ when the total energy is} \\ &\qquad\qquad\qquad\qquad E = K + \mathsf{V} = 2K - L, \end{aligned}\right\} \tag{8.6.2}$$

where $\mathsf{E} = \mathsf{K} + \mathsf{V} = \mathsf{H}$ is the corresponding total energy density and $E = H$. Here K and V represent the *kinetic* and *potential energy*, respectively, in the relations (8.6.2) and K, V, and so on are densities. Applications of the variational procedure

$$\delta\int_t (2K - H)dt = \delta\int_t Ldt \quad \left(\text{or } S^2\int_t Ldt = 0\right). \tag{8.6.3}$$

thus yields the *Euler equation*, which is just the equation of propagation for α. (See Chapter 3 of Ref. [1], which provides an extensive discussion of this variational principal. A brief outline of the variation method is provided in Section 8.6.4.)

Thus, the *Lagrange density* $\mathcal{L}$ here is a function of the field variables $\alpha(g)$, $(i = 1, \ldots, n)$ and the associated *gradients*, that is, $\nabla\alpha_{ij} = \partial\alpha_i/\partial g_j$. In our present treatment, we limit j to the canonical coordinate, in particular to the (z_1, z_2, z_3) spatial coordinates, and to $z_4 = t$ time. The *Lagrange integral* L is invariant and is in this case

$$\mathsf{L} = \int_{a_1}^{k_1} \cdots \int_{a_4}^{k_4} \mathcal{L}[\boldsymbol{\alpha}(\mathbf{z}), \{\nabla\boldsymbol{\alpha}\}]d\mathbf{z}. \tag{8.6.3}$$

We next seek an extremism that vanishes, namely, $\delta\mathsf{L} = 0$, or

$$\delta\mathsf{L} = \sum_{j=1}^{4} \frac{\partial}{\partial z_j}\left(\frac{\partial\mathcal{L}}{\partial\alpha_{ij}}\right) - \frac{\partial\mathcal{L}}{\partial\alpha_i} = 0, \quad i = (1, \ldots, 4) \tag{8.6.4}$$

for the field variable $\{\alpha_i\}$, all i. (If $\delta\mathsf{L} = 0$ and this results in a (multidimensional) point of *inflexion*, we proceed formally by determining whether $\delta^2\mathsf{L} > 0$—a minimum, or $\delta^2\mathsf{L} < 0$—a maximum in the usual way.)

The dependence of L on $\alpha_{i4} (i = 1, 2, 3)$ can be shown to lead to two principal cases: (*I*) L is *linear* in α_{i4} (*II*) L is quadratic in α_{i4}. We have for the linear case

I. $\mathcal{L}$ linear in α_{i4}:

$$p_i = \delta\mathcal{L}/\partial a_{i4}\text{: } p_i \text{ and the Hamiltonian } \mathcal{H}, \text{ Eq. (8.6.2), are independent of } \alpha_{i4}\}. \tag{8.6.5}$$

In addition, the *stress energy tensor* (or *dyadic*) $\mathcal{B}$ has the components

$$\mathcal{B} = [\mathcal{B}_{mj}] = \left[\sum_i^4 \alpha_{im}\frac{\partial\mathcal{L}}{\partial\alpha_{ij}} - \delta_{mj}\right], \tag{8.6.6}$$

where the Hamiltonian (density) is $H = 2\mathcal{K} - \mathcal{L}$, Eq. (8.6.2). The (4, 4) component here is the *energy density*

$$\mathcal{B}_{44} = \mathcal{H} = \sum_{i=1}^{n} p_i \alpha_{i4} - \mathcal{L}. \tag{8.6.7}$$

Furthermore, if the momenta $\{p_i\}$ depend on α_{i4}, it is then possible to remove the term α_{14} from $\mathcal{B}_{44}$, thus obtaining the *Hamiltonian energy density* $\mathcal{H}$, refer to Eq. (8.6.2). This in turn depends on the *canonical momentum density* α_{14}, refer to Eq. (8.6.5), and their spatial dimensions ($\mathbf{z}$). Accordingly, we see that the *equations* of *motion* can be alternatively written in the canonical form

$$\alpha = \alpha_{14} = \frac{\partial \mathcal{H}}{\partial p_i}; \quad p_i = \frac{\partial p}{2t} = \sum_{j=1}^{3} \frac{\partial}{\partial z_j}\left(\frac{\partial \mathcal{H}_i}{\partial \alpha_{ij}}\right) - \frac{\partial \mathcal{H}}{\partial \alpha_i}. \tag{8.6.7a}$$

The following equation applies only when $\mathcal{L}$ contains a quadratic function of α_{i4}'s. Thus, we note that

$$II.\ \mathcal{L}\ \textit{quadratic in}\ \alpha_{i4}: \qquad \therefore \quad p_i = \frac{\partial \mathcal{L}_i}{\partial \alpha_{14}}, \tag{8.6.8}$$

refer to Eq. (8.6.5), is also a linear function of α_{14}. For a full treatment, see chapter 3, with examples in Sections 3.1 and 3.3, with scalar and vector fields, respectively, in Sections 3.3 and 3.4.

8.6.2 Preliminary Remarks

In Section 8.6, we consider the role of energy and energy flux in two fundamentally different classes of media. Class I supports only "mechanical" radiation (the central theme of Lindsay's book [2]). Class II, on the other hand, can itself exist as a radiation field, in a vacuum and in material media. This can in turn modify the radiation according to their specific characteristics as a gas, a liquid, or a solid. Class I media directly produces and alters the propagation that they themselves generate in response to a deformation. Class II is represented by the electromagnetic field, although alternated by material media, and is essentially undistorted, as long as the medium is not ferromagnetic or in the more extreme state of plasma. (For an extended treatment of electromagnetic theory, see also Ref. [3].)

We begin with a (comparatively) simple ideal non-Gaussian isotropic class I material medium, which supports a shear. For *class I cases*, the generic field equations for propagations are shown to be in Section 6.3 of Ref. [2], for nondissipative media,

$$\rho\ddot{\boldsymbol{\alpha}}_d = (B_{\mathrm{M}} + 4\mu/3)\nabla\nabla\cdot\boldsymbol{\alpha}_d - \mu\nabla\times\nabla\times\boldsymbol{\alpha}_d = (B_{\mathrm{M}} + \mu/3)\nabla\nabla\cdot\boldsymbol{\alpha}_d + \mu\nabla\cdot\nabla\boldsymbol{\alpha}_d \tag{8.6.9}$$

(cf. (8.6.37a) ff.) Here the displacement field $\boldsymbol{\alpha}_d = \boldsymbol{\alpha}_d(\mathbf{R}, t)$ is the elastic element $\mathbf{d}\,(=\alpha)$ and $\mathbf{d} = \boldsymbol{\alpha}_d = \mathbf{v}(\mathbf{R}, t)$ is the element (or "positive") velocity. Similarly, $\ddot{\mathbf{d}} = \dot{\mathbf{v}} = \ddot{\boldsymbol{\alpha}}_d$ represents the element's acceleration (where as before $(\cdot) \equiv \partial/\partial t$). The parameters B_M, μ (and $s = -\nabla\cdot d = -D_0$) are respectively the *bulk modulus* and the *shear modulus*. The condensation s and *the dilation strain* D_{0_i} will be included presently in our treatment of more complex but still nondissipative media. This example considers a deformable elastic medium that is fully restored to its undistorted equilibrium state when the stresses are removed.

Two special cases of interest can occur here:

(i) $\nabla \times \mathbf{d}(= \nabla \times \boldsymbol{\alpha}_d) = 0$: $\mathbf{d}$ is an *irrotational vector displacement*, resulting in propagation of a longitudinal wave with a vertical deformation of the medium and its release to equilibrium, in the direction of propagation. There are no shear stresses and therefore no shear strains. Equation (8.6.3) reduces here to

$$\rho\ddot{\boldsymbol{\alpha}}_d = (B_{\rm M} + 4\mu/3)\nabla^2\boldsymbol{\alpha}_d, \quad \text{or equivalently,} \quad \nabla^2\boldsymbol{\alpha}_d = \frac{1}{c_{\rm p}^2}\ddot{\boldsymbol{\alpha}}_d, \quad \text{with}$$
$$c_{\rm p} = \sqrt{(B_M + \mu/3)/p}. \tag{8.6.10}$$

We may call the resulting waves from (8.6.4) P- or longitudinal waves.

(ii) $\nabla \cdot \mathbf{d} = \nabla \cdot \boldsymbol{\alpha} = 0$: $\mathbf{d} = \mathbf{d}_y$ is a *solenoidal vector displacement*, representing a vertical deformation and its release, now transversal to the wave's direction of propagation. Equation (8.6.3) reduces to

$$\rho\ddot{\boldsymbol{\alpha}}_d = \mu\nabla^2\boldsymbol{\alpha}_d \quad \text{or} \quad \nabla^2\boldsymbol{\alpha}_d = \frac{1}{c_s^2}\ddot{\boldsymbol{\alpha}}_d, \quad \text{with} \quad c_s \equiv \sqrt{\mu/p}. \tag{8.6.11}$$

Clearly, from (8.6.10) and (8.6.11), we have

$$c_p/c_s = \sqrt{(B_{\rm M} + \mu/3)/p} > 1, \quad \mu > 0. \tag{8.6.12}$$

This type of medium is accordingly seen to support two types of wave: a pressure or P-wave and a S-wave or shear wave. These travel with different speeds $\left(c_p \neq c_s\right)$. In fact, we can separate the two Helmholtz equations that govern the respective propagations at the two different speeds c_p and c_s, refer to Eqs. (8.6.10) and (8.6.11), by observing that the equation of motion (8.6.10) is an additive combination of the two. Thus, we have

Thus, P-waves travel faster than S-waves, as is well known, for example, in case of earthquakes.

Case II *Electromagnetic Radiation:* Waves of this type are always transverse, both E- and H-field propagations, and both E and H waves travel with the same velocity (see Section 8.6.3.2, and in particular Chapter 1 of Ref. [3].[58] With these two classes, we shall obtain the desired energy flow relations, which is the principal aim of Section 8.6.2.

Here, however, we shall use the direct method of calculating the energy in the field $\alpha(\mathbf{R}, t)$. This begins with the propagation equation itself for $\alpha(\mathbf{R}, t)$. This is the approach used for a variety of examples of the field $\alpha(\mathbf{R}, t)$, refer to (8.1.45) and Sections 8.3 and 8.4, associated with the last 3 media in (8.6.1). The important case of the vector electromagnetic field (1) is provided here also.

[58] Case I media and propagation therein are treated in much more detail in [2], Sections 6.3, pp. 149–153; 9.11, pp. 261–266; 10.2 (Moving, Compressive, Viscous, Thermally Conducting Fluid), pp. 303–310; see Eqs. (5), (10), et seq. And finally, refer to Section 12.5, pp. 354–360, for the Navier–Stokes equation and its perturbation solutions.

8.6.3 Energy Density and Density Flux: Direct Models—A Brief Introduction

For a simple case of the direct method here, let us determine the energy density associated with the field $\alpha(\mathbf{R}, t)$ in a given volume external to the region V_T of a source (or sources), in the manner of Fig. 8.6. In particular, we also need to determine the energy flux through a surface element in a section of the surfaces S bounding V_T, as shown in Fig. 8.9.

We begin with the assumption of a scalar field in an ideal medium, which is homogeneous and stationary (Hom-Stat), typical of those discussed in Section 8.1. After this, we shall discuss a number of examples of increasing complexity, including anisotropic and dissipative media.

8.6.3.1 The Ideal Incompressible Fluid Medium Here the fluid in question (including gas obeying the conditions here) is incompressible and irrotational, that is, has no vortices. Using a velocity potential $\boldsymbol{\Psi}_v$ (of dimensions $[L^2/T]$), we observe that the velocity of the fluid is represented by $\mathbf{v}$, where now

$$\mathbf{v} = \nabla\Psi_v, \text{ so that the kinetic energy density is } \mathsf{K} = \frac{1}{2}\rho_0\mathbf{v} - \mathbf{v}, \text{ or equivalently,} \tag{8.6.13a}$$

$$\mathsf{K} = \frac{1}{2}\rho_0\{|\nabla\Psi_v|\}^2 = \frac{1}{2}\rho_0\left\{\left(\frac{\partial\Psi_v}{\partial x}\right)^2 + \left(\frac{\partial\Psi_v}{\partial y}\right)^2 + \left(\frac{\partial\Psi_v}{\partial z}\right)^2\right\},$$
$$\text{with} \quad \nabla\times\mathbf{v} = \nabla\times\nabla\Psi_v = 0 \tag{8.6.13b}$$

Here, ρ_0 is the density of the fluid of gas that is in motion. The potential energy (density) P is an (adjustable) constant, because of the incompressibility of the medium, that is, $P = C_0$ (≥ 0), so that the total energy density is therefore

$$E = K + P = \frac{1}{2}\rho_0(\nabla\Psi_\mathbf{v})^2 + C_0 = \frac{1}{2}\rho_0\nabla\Psi_\mathbf{v}\cdot\nabla\Psi_\mathbf{v} + C_0. \tag{8.6.14}$$

Because of incompressibility, no energy is stored or released in the fluid, so that we regard the constant potential energy C_0 as zero here.

The energy flux, or more precisely, the *energy density fluid*, in E, through a unit surface element ΔS boundary the volume $V(= V - V_T)$ (Fig. 8.19), namely,

$$\dot{E} = \dot{K} = \rho_0\mathbf{v}\cdot\dot{\mathbf{v}} = \rho_0\nabla\Psi_{\dot{\mathbf{v}}}\cdot\nabla\dot{\Phi}_\mathbf{v}, \quad \text{where} \quad \mathbf{v} = \nabla\Psi_v. \tag{8.6.15}$$

The Euler or propagation operation is found to be the Laplacian from the following argument. Start with the *equation of continuity* here, which is in this case

$$\frac{\partial\rho}{\partial t} = \nabla\cdot\rho\mathbf{v}, \quad \text{with} \quad \mathbf{v} = \nabla\Psi_v \tag{8.6.15a}$$

Since here the fluid (or gas) density is constant, that is, $\rho = \rho_0$, no fluid is added or subtracted from an element dV as a consequence of the "distortion" in the medium.[59] Consequently,

[59] In Section 8.6, we make a number of simplifying assumptions (unless otherwise stated), for example, $\rho \doteq \rho_0$, a constant. For a much fuller account of the actual combinations (in the ocean, for example), see Chapter 1 of Ref. [18].

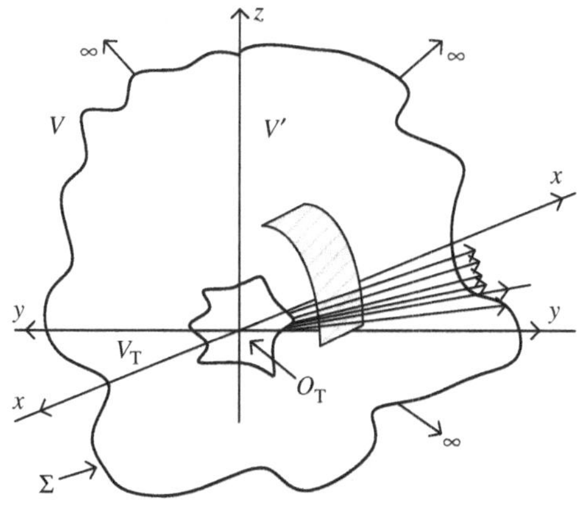

FIGURE 8.19 A volume V containing a source V_T, where $V' \equiv V - V_T$ is source-free region, bounded by an infinitely distant closed surface $\Sigma(\rightarrow \infty)$. An energy density is shown flux from a surface element dS of Σ (i.e., dS $\rightarrow \infty$), of the volume $V'(\rightarrow \infty)$.

$$\left.\begin{array}{c} \dfrac{\partial \rho}{\partial t}\left(=\dfrac{\partial \rho_0}{\partial t}\right)=0 \;\therefore\; 0=\nabla\cdot\rho\mathbf{v}=\rho_0\nabla\cdot\mathbf{v}=\rho_0\nabla\cdot\nabla\Phi_v \\ \text{or} \\ \therefore\; \nabla^2\Phi_v \equiv \nabla^2\alpha_v = 0,\, \Psi_\mathbf{v}=\alpha_\mathbf{v} \end{array}\right\}, \tag{8.6.16}$$

where Φ_v is the propagating velocity *potential* of the velocity field itself, for this example of an ideal *incompressible* fluid or gas. In general, Φ_v is a nonvanishing scalar. Its velocity potential represents, for example, the propagation of an impulse of energy from a series of contiguous elements, without disturbing each element. This is analogous to a light ball impacting a series of heavy balls hanging side by side, through which the impact energy now travels without moving the heavy balls themselves. The medium (the heavy balls here) do not move, while the energy impulse travels through them.

8.6.3.2 The Ideal Compressible Nonviscous Fluid: Helmholtz Field We begin our analysis of the energy and its flux by first establishing the associated fields. Here, the fluid or gas is shear but nonviscous and compressible, to which comparatively weak pressure source is applied, producing a small change in an otherwise constant pressure that then occurs

$$p = -\rho|\dot{\mathbf{v}}| = -\rho_0\frac{\partial \Phi_\mathbf{v}}{\partial t} = -\rho_0\frac{\partial^2 \Phi_d}{\partial t^2} \tag{8.6.17}$$

(The minus sign $(-)$ represents the (slight) expansion of the associated volume element dV from equilibrium that results.) Here, as above, $\mathbf{v} = \dot{\mathbf{d}}$ is a velocity, where $\mathbf{d}$ is the consequent small displacement of the fluid element dV. Applying Hooke's law next, we can write

$$p = -B_M\boldsymbol{\nabla}\cdot\mathbf{d} = -B_M\boldsymbol{\nabla}\cdot\boldsymbol{\nabla}\Phi_d \tag{8.6.18}$$

Combining (8.6.7) and (8.6.8), we obtain the desired relation for the propagation of the perturbed displacement field in terms of its displacement potential Φ_d:

$$B_{\mathrm{M}}\nabla^2\Phi_d = \rho_0\frac{\partial^2\Phi_d}{\partial t^2} \quad \text{or} \quad \nabla^2\Phi_d = \frac{1}{c_\Phi^2}\frac{\partial^2\Phi_d}{\partial t^2}, \qquad \text{with } B_M > 0, \tag{8.6.19}$$

where, in V, the speed of propagation of a wavefront is given by

$$e_\Phi = \sqrt{B_{\mathrm{M}}/\rho_0}. \tag{8.6.19a}$$

Here, B_{M} is the compressibility or bulk modulus of the fluid (or gas), which depends on the particular medium.

It is easily seen that the pressure p and the displacement $\mathbf{d}$ also obey a Helmholtz equation. We begin with ∂^2/dt^2 of Φ_d and write from (8.6.8) the approximate relations (e.g., $\rho \doteq \rho_0$):

$$\ddot{p} \doteq -\rho_0\frac{\partial^2\Phi_d}{\partial t^2} \quad \text{and} \quad \nabla^2 p \rightarrow \nabla\cdot\nabla\frac{\partial^2\Phi_d}{\partial t^2}, \tag{8.6.20a}$$

so that

$$\left(\nabla^2 p - \frac{1}{c_p^2}\ddot{p}\right) = B_{\mathrm{M}}\frac{\partial^2}{\partial t^2}\left(\nabla^2\Phi_d - \ddot{\Phi}_d/c_\Phi^2\right) = 0, \qquad \text{from (8.6.9)} \tag{8.6.20b}$$

and

$$\therefore\ \nabla^2 p - \frac{1}{c_p^2}\ddot{p} = 0 \quad \text{or} \quad \hat{L}^{(0)}p = 0, \quad \text{where} \quad \hat{L}^{(0)} \equiv \nabla^2 p - \frac{1}{c_p^2}\frac{\partial^2}{\partial t^2} \tag{8.6.20c}$$

and the pressure p is here a *scalar*. In a similar fashion, we also observe that for the vector displacement $\mathbf{d}$,

$$\nabla^2\boldsymbol{\alpha}_d - \frac{1}{c_d^2}\frac{\partial^2\boldsymbol{\alpha}_d}{\partial t^2} = 0, \qquad \text{with} \quad \boldsymbol{\alpha}_d \equiv \mathbf{d}, \boldsymbol{\alpha}_d = \boldsymbol{\alpha}_d(\mathbf{R}, t); \mathbf{d} = \mathbf{d}(\mathbf{R}, t), \tag{8.6.21}$$

which now represents a vector field $\alpha(\mathbf{R}, t)$, also obeying a Helmholtz equation of propagation. We remark that the respective speeds of these propagating fields are all equal, namely, c_0:

$$c_0 \equiv c_d = c_p = c_\Phi = \sqrt{B_{\mathrm{M}}/\rho_0} \quad \text{or} \quad B_{\mathrm{M}} = \rho_0 c_{(\Phi,d,p)}^2 = \rho_0 c_0^2. \tag{8.6.22}$$

Their equality stems from the fact that each represents different aspects of the same phenomenon.

8.6.4 Equal Nonviscous Elastic Media

Our aim, as stated at the beginning of this section, is now to obtain relations that represent the energy, or more precisely, the energy density and energy flux in the radiation field represented by the above propagation equations. To achieve this, we recall that these relations are essentially dynamical or force equations, expressing a dynamic equilibrium between impressed forces (*stresses*) and the resulting local *strains* (or deformations) produced in the media. It is the energy in the field that completes the physical description of the propagation process. We accordingly begin with the following.

8.6.4.1 Ideal Shearless Media ($\mu = 0$) We start by first postulating an ideal, that is, lossless, homogeneous and isotropic *elastic* medium, which does not support a shear ($\mu = 0$), where μ is the *shear modulus*. Such media are gases or liquids, which are accordingly *irrotational*, so that "particle displacement" obeys $\nabla \times \boldsymbol{\alpha}_d = \mathbf{0}$ and the familiar Helmholtz relation (8.6.21), with (8.6.22), outrides the region of sources.

We wish now to obtain a relation for the energy and its transfer, that is, flux. Accordingly, if we determine the scalar product of both sides of Eq. (8.6.11) multiplied by the particle velocity $\dot{\alpha}_d$, the result is recognized as the time rate of change of quantities that have the dimensions of energy per unit volume or energy density. Since $\nabla\nabla \cdot \boldsymbol{\alpha}_d = \nabla^2 \boldsymbol{\alpha}_d$ here (in view of $\nabla \times \boldsymbol{\alpha}_d = 0$), we can write

$$\rho_0 \dot{\boldsymbol{\alpha}}_d \cdot \ddot{\boldsymbol{\alpha}}_d = B_{\mathrm{M}} \dot{\alpha}_d \cdot \nabla(\nabla \cdot \boldsymbol{\alpha}_d). \tag{8.6.23}$$

Using the relation between the *dilation D* and the *condensation* that is specifically

$$D = \nabla \cdot \boldsymbol{\alpha} \quad \text{and} \quad s \equiv -D_0 = -\nabla \cdot \boldsymbol{\alpha}_d, \tag{8.6.23a}$$

namely, the fractional change in density associated with the resulting volume deformation (*strain*), we get

$$\boldsymbol{\alpha}_d \cdot \nabla(\nabla \cdot \boldsymbol{\alpha}_d) = -\boldsymbol{\alpha}_d \cdot \nabla s, \quad \text{and} \quad \therefore \rho_0 \dot{\boldsymbol{\alpha}}_d \cdot \ddot{\boldsymbol{\alpha}}_d = -B_{\mathrm{M}} \dot{\boldsymbol{\alpha}}_d \cdot \nabla s. \tag{8.6.24}$$

From the identity $\mathbf{A} \cdot \nabla B = \nabla \cdot (\mathbf{A}B) - B\nabla \cdot \mathbf{A}$ applied to the second relation in (8.6.24), we next obtain

$$\rho_0 \dot{\boldsymbol{\alpha}}_d \cdot \ddot{\boldsymbol{\alpha}}_d = -B_{\mathrm{M}} \nabla \cdot (s\dot{\boldsymbol{\alpha}}_d) + s\nabla \cdot \boldsymbol{\alpha}_d \tag{8.6.25a}$$

and equivalently the derivation of flux density

$$\frac{\partial}{\partial t}\left[\frac{1}{2}\rho_0(\dot{\boldsymbol{\alpha}}_d \cdot \dot{\boldsymbol{\alpha}}_d)\right] + \frac{\partial}{\partial t}\left(\frac{1}{2}B_{\mathrm{M}} s^2\right) = -B_{\mathrm{M}} \nabla \cdot (s\dot{\boldsymbol{\alpha}}_d). \tag{8.6.25b}$$

We observe from the above relations involving s, D, B_{M}, and $\boldsymbol{\alpha}_d$ that

$$\nabla \cdot \dot{\boldsymbol{\alpha}}_d = \frac{\partial}{\partial t}(\nabla \cdot \boldsymbol{\alpha}_d) = -\frac{\partial s}{\partial t} = -\dot{s}. \tag{8.6.25c}$$

It is at this point that we integrate (8.6.25b) over a volume $V - V_{\mathrm{T}} \equiv V_0$ enclosed by a surface Σ_0, refer to Fig. 8.19, excluding any sources (or sinks, that is, "receivers"). For this we shall also need Gauss's theorem, and for later applications, Stokes' theorem. We have altogether the well-known relations:

$$\textbf{(I)}\ \textit{Gauss's Theorem}: \quad \int_{V_0} \nabla \cdot \mathbf{A} dV_0 = \int_{\Sigma} \mathbf{A} \cdot \hat{\mathbf{n}} dS, \tag{8.6.26}$$

and the extension, which is known as Green's theorem, refer to Eq. (8.1.41a):

$$\textbf{(II)}\ \textit{Green's Theorem}: \quad \int_{V_0} (A\nabla \cdot \nabla B - B\nabla \cdot \nabla A) dV_0 = \int_{\Sigma} (A\nabla B - B\nabla A) \cdot \hat{\mathbf{n}} dS. \tag{8.6.27}$$

$$\textbf{(III)}\ \textit{Stokes' Theorem}: \quad \int_{\Sigma} (\nabla \times A) \cdot \hat{\mathbf{n}} dS = \oint_C \mathbf{A} \cdot \hat{\mathbf{l}} dl \quad (= 0), \tag{8.6.28}$$

where C is the closed boundary line.

In the above, $\hat{n}$ is an outward drawn, that is positive, unit normal vector and $\oint$ is the border of a segment of a closed surface, in the counter-clockwise direction, with a unit fragment $\hat{\mathbf{l}}$ along the line segment in the positive direction. From (8.6.26) with $\mathbf{A} = \rho\mathbf{v}$, we see the following:

$$\textbf{(IV)}\ \textit{Equation of Continuity}: \quad \int_{V_0} \nabla \cdot \rho\mathbf{v} dV_0 = -\int_t \frac{\partial \rho}{\partial t} dt \quad \text{or}$$

$$\nabla \cdot \rho\mathbf{v} = -\frac{\partial \rho}{\partial t} = \nabla \cdot \rho\dot{\mathbf{d}} \quad (\doteq \rho_0 \nabla \cdot \dot{\mathbf{d}}). \tag{8.6.29}$$

Now applying Gauss's theorem to (8.6.25b), we can write for the rate of change of the total energy E in the volume V:

$$\dot{E} \equiv \int_V \frac{\partial}{\partial t}\left[\frac{1}{2}\rho_0 \dot{\boldsymbol{\alpha}}_d \cdot \dot{\boldsymbol{\alpha}}_d + \frac{1}{2} B_M s^2\right] dV = -\int_V B_M \nabla \cdot (\dot{\boldsymbol{\alpha}}_d s)\, dV. \tag{8.6.30}$$

Using Gauss's theorem (8.6.26), we can directly convert the volume integral for the outword energy flux to the more explicit form:

$$\dot{E} = \int_V \frac{\partial}{\partial t}\left[\frac{\rho_0}{2} \dot{\boldsymbol{\alpha}}_d \cdot \dot{\boldsymbol{\alpha}}_d + \frac{1}{2} B_M s^2\right] dV = -\int_{\Sigma} (B_M (\dot{\boldsymbol{\alpha}} s) \cdot \hat{n})\, dS \tag{8.6.30a}$$

The left-hand members of (8.6.30) are identified respectively as the time rate of changes of the kinetic and potential energies stored in the field $\boldsymbol{\alpha}_d$, which is in the volume V. The right-hand member of (8.6.30) is the total energy flux *out of the volume* V, through the bounding surface Σ. The surface integral in (8.6.30) represents the rate at which work is done by the fluid medium simile V enclosed by Σ. This is the rate at which energy is instantaneously leaving V through the surface Σ.

Accordingly, we may express the integrands of (8.6.30) similarly as the *kinetic energy density* U_k, the potential energy density U_p, and the energy flux density $\mathbf{S}_p$ per unit volume of the field $\boldsymbol{\alpha}_d$. These relations are specifically

$$U_k \equiv \frac{1}{2}\rho_0 (\dot{\boldsymbol{\alpha}}_d \cdot \dot{\boldsymbol{\alpha}}_d);\ U_p \equiv \frac{1}{2} B_M s^2 = \frac{1}{2} B_M (\nabla \cdot \dot{\boldsymbol{\alpha}}_d)^2 = \frac{1}{2}\rho_0 c_0^2 D_0, \tag{8.6.31}$$

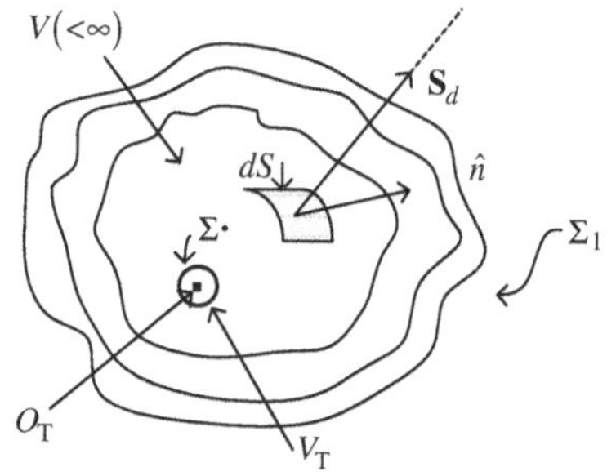

FIGURE 8.20 A (virtual) surface $\Sigma(=\Sigma_0+\Sigma_1)$ enclosing both a source, centered at O_T and designated in V_T, and the (finite) volume $V = V' - V_T$ containing the (displacement) field $\boldsymbol{\alpha}_d$ and radiating energy, with the flux S_d, (8.6.30). (There are no source and sinks in V.)

where the energy density flux is

$$\mathbf{S}_{\mathrm{p}} \equiv B_{\mathrm{M}} s \dot{\boldsymbol{\alpha}}_d = -B_{\mathrm{M}} \dot{\boldsymbol{\alpha}}_d \nabla \cdot \dot{\boldsymbol{\alpha}}_d. \tag{8.6.31a}$$

Thus, (8.6.30) can be written compactly

$$\dot{E} = \int_{\mathrm{V}} \dot{\mathsf{E}}\, dV = \int_{\mathrm{V}} \left(\dot{U}_{\mathrm{K}} + \dot{U}_{\mathrm{P}}\right) dV = -\int_{\Sigma} \mathbf{S}_{\mathrm{P}_y} \cdot \hat{n} dS, \quad \text{with} \quad \dot{\mathsf{E}} \equiv \dot{U}_{\mathrm{K}} + \dot{U}_{\mathrm{P}_y} \tag{8.6.32}$$

as the temporal rate of change of the total energy density E. (We use the notation $\mathbf{S}_{\mathrm{P}_y}$ to remind us that we are dealing with a *vector* field $\boldsymbol{\alpha}_d$ here. Moreover, as we shall observe in more detail in an example below, $\mathbf{S}_{\mathrm{P}_y}$ is an *acoustic poynting vector* representing the energy flux density out of the (virtual) surface Σ.

8.6.4.2 Acoustic Waves We can use the results of Section 8.6.3 above specifically for propagation of acoustic waves in media similar to those discussed there [2]. In particular, these media are specifically the atmosphere and the fluid portions of the ocean. We summarize these results below, refer to Section 8.6.3.2, and begin with the ambient pressure $p(\mathbf{R}, t)$, which is specifically

$$p(\mathbf{R}, t) = \rho_0 c_0^2 s(\mathbf{R}, t) = -\rho_0 c_0^2 \nabla \cdot \boldsymbol{\alpha}_d, \quad \text{with} \quad B_M = \rho_0 c_0^2, \tag{8.6.33}$$

as an expression of Hooke's law for small amplitudes. Using the *displacement potential* Φ_d (8.6.18)–(8.6.20b), $\nabla\Phi_d = \mathbf{d} = \boldsymbol{\alpha}_d$, we can alternatively express (8.6.17) as

$$p(\mathbf{R}, t) = -\rho_0 \ddot{\Phi}_d = -\rho_0 c_0^2 \nabla^2 \Phi_d, \tag{8.6.34}$$

and note that, as expected, both p and Φ_d obey scalar wave equations.

In terms of displacement potential Φ_d, we can alternatively and equivalently write the associated energy and flux densities in this case as

$$U_K \equiv \frac{1}{2}\rho_0 \left(\nabla\dot{\Phi}_d \cdot \nabla\dot{\Phi}_d\right); \quad U_{\mathrm{P}} = \frac{1}{2}\rho_0 c_0^2 \left(\nabla^2 \dot{\Phi}_d\right)^2; \quad \text{and} \quad \mathbf{S}_{\mathrm{P}_y} = -\rho_0 c_0^2 \dot{\Phi}_d \nabla \dot{\Phi}_d. \tag{8.6.35a}$$

This yields the energy flux density

$$\dot{E} = \frac{\partial}{\partial t}\left[\frac{1}{2}\rho_0 \dot{\boldsymbol{\alpha}}_d \cdot \dot{\boldsymbol{\alpha}}_d + \frac{1}{2}\rho_0 c_0^2 (\nabla \cdot \dot{\boldsymbol{\alpha}}_d)^2\right] = \rho_0\left[\ddot{\boldsymbol{\alpha}}_d \cdot \dot{\boldsymbol{\alpha}}_d + c_0^2 (\nabla \cdot \dot{\boldsymbol{\alpha}}_d)(\nabla \cdot \dot{\boldsymbol{\alpha}}_d)\right], \quad (8.6.36a)$$

or

$$\dot{E} = \rho_0\left[\nabla \ddot{\Phi}_d \cdot \nabla \dot{\Phi}_d + (\nabla \dot{\Phi}_d)\ddot{\Phi}_d\right] = \rho_0 \nabla \cdot \left(\ddot{\Phi}_d \nabla \dot{\Phi}_d\right). \quad (8.6.36b)$$

These equivalent forms are also useful, as we shall see later [ ] in determining field intensities when the medium is not homogeneous. The present results are useful for the comparatively simple models commonly used [33].

8.6.4.3 Elastic Wave Propagation in an Ideal Solid or Elastic Fluid Medium with Shear **($\boldsymbol{\mu} \geq \mathbf{0}$)** This is the case of propagation in a medium, like that already considered above, which is nondissipative but can now support a shear, that is, has a positive shear modulus $\mu > 0$. This situation applies to radiation in *solids and elastic fluids*, which here are still homogeneous and isotropic, and still "ideal," in the sense that there are no physical mechanisms for *energy dissipation*. The fundamental dynamical equation for the field propagated in such media is (from Section 6.4, of Ref. [2]), apart from sources and sinks, refer to Fig. 8.19:

$$\rho_0 \ddot{\boldsymbol{\alpha}}_d = (B_M + 4\mu/3)\nabla\nabla \cdot \boldsymbol{\alpha}_d - \mu\nabla \times \nabla \times \boldsymbol{\alpha}_d = (B_M + \mu/3)\nabla\nabla \cdot \boldsymbol{\alpha}_d - \mu\nabla \cdot \nabla\boldsymbol{\alpha}_d, \quad (8.6.37)$$

where we have employed the vector identity and other vector relations, namely:

$$\nabla \times (\nabla \times \mathbf{A}) = \nabla\nabla \cdot \mathbf{A} - \nabla \cdot \nabla\mathbf{A}, \quad \text{and}$$
$$\nabla \cdot \nabla\mathbf{A} = \hat{\mathbf{i}}_x \nabla^2 A_x + \hat{\mathbf{i}}_y \nabla^2 A_y + \hat{\mathbf{i}}_z \nabla^2 A_z, \quad \text{with} \quad \nabla \cdot \nabla(\,) \equiv \nabla^2 \quad (8.6.37a)$$

(in rectangular coördinates). These elastic solids are deformable and return to their equilibrium shape where the impressed stress is removed. For example, some ocean bottoms as well as land surfaces to a good approximation begin in this fashion.

To obtain an energy equation from the propagation equation (8.6.31) above, we proceed as in (8.6.13) et seq. in Section 8.6.3. Using the relation for the *condensation* $s = -\nabla \cdot \boldsymbol{\alpha}_d$ (8.6.23) and the following vector and tensor (dyadic) identities,

$$\nabla \cdot (\mathbf{A}B) = \mathbf{A} \cdot \nabla B + B\nabla \cdot \mathbf{A}; \quad \nabla \cdot (\mathbf{A} \cdot \nabla\mathbf{C}) = \mathbf{A}(\nabla \cdot \nabla C) + \nabla\mathbf{A}\colon \nabla\mathbf{C}, \quad (8.6.37b)$$

refer to Eq. 4 in Section 6.4, of Ref. [2], and from Eq. (5) [2], we have the (2nd rank covariant) tensor, that is, dyadic

$$\nabla\mathbf{A}\colon \nabla\mathbf{C} \equiv \sum_{k,l} \frac{\partial A_k}{\partial x_l}\frac{\partial C_l}{\partial x_k}, \quad \text{with} \quad \{x_l\} = x, y, z \quad \text{as} \quad l = 1, 2, 3;$$
$$A_k = \hat{i}_k \cdot A \text{ and so on.} \quad (8.6.37c)$$

This allows us to rewrite the scalar product of $\dot{\boldsymbol{\alpha}}_d$ and Eq. (8.6.31) explicitly as

$$\rho_0\dot{\boldsymbol{\alpha}}_d\ddot{\boldsymbol{\alpha}}_d = -G[\nabla\cdot(s\dot{\boldsymbol{\alpha}}_d) - s\nabla\cdot\dot{\boldsymbol{\alpha}}_d] + \mu[\nabla\cdot(\dot{\boldsymbol{\alpha}}_d\cdot\nabla\boldsymbol{\alpha}_d) - \nabla\dot{\boldsymbol{\alpha}}_d\colon\nabla\boldsymbol{\alpha}_d], \quad \text{with } G \equiv B_M + \mu/3. \tag{8.6.38}$$

Integrating over a finite volume of this elastic (solid) medium, combining terms, and using Gauss's theorem (8.6.26) once more, we now obtain the more general version of (8.6.30a), namely:

$$\left.\begin{aligned} \dot{E} = \int_V \dot{E}dV &= \int_V \frac{\partial}{\partial t}\left[\frac{\rho_0}{2}(\dot{\boldsymbol{\alpha}}_d\cdot\dot{\boldsymbol{\alpha}}_d) + \left\{\frac{1}{2}(B_M+\mu/3)s^2 + \frac{\mu}{2}\nabla\boldsymbol{\alpha}_d\colon\nabla\boldsymbol{\alpha}_d\right\}\right]dV \\ &= -\int_\Sigma [(B_M+\mu/3)s\dot{\boldsymbol{\alpha}}_d - \mu\dot{\boldsymbol{\alpha}}_d\cdot\nabla\boldsymbol{\alpha}_d]\cdot\hat{\mathbf{n}}\,dS \end{aligned}\right\}. \tag{8.6.39}$$

The component $(\mu/2)\nabla\boldsymbol{\alpha}_d\colon\nabla\boldsymbol{\alpha}_d$ in the potential energy is the contribution to elastic strain potential energy from the shear strains. Similarly, the component $\int_\Sigma(\mu\dot{\boldsymbol{\alpha}}_d\cdot\nabla\boldsymbol{\alpha}_d)\cdot\hat{\mathbf{n}}\,dS$ contributes to the flow of elastic energy across Σ for this medium, which is subject to shear as well as compressibility.

The kinetic and potential energy densities of the field in this type of solid are from (8.6.34) directly, since $\mu\nabla\dot{\boldsymbol{\alpha}}_d\colon\nabla\boldsymbol{\alpha}_d = (\partial/\partial t)((\mu/2)\nabla\boldsymbol{\alpha}_d\colon\nabla\boldsymbol{\alpha}_d)$:

$$U_k = \rho_0\ddot{\boldsymbol{\alpha}}_d\cdot\dot{\boldsymbol{\alpha}}_d = \rho_0(\dot{\boldsymbol{\alpha}}_d\cdot\dot{\boldsymbol{\alpha}}_d)_2; \quad U_p = \frac{1}{2}\left[(B_M+\mu/3)s^2 + \mu\nabla\boldsymbol{\alpha}_d\colon\nabla\boldsymbol{\alpha}_d\right], \tag{8.6.40a}$$

with the total density flux $\mathbf{S}_d$ is given by

$$\mathbf{S}_d = -\int_\Sigma\left[\left(B_M+\frac{\mu}{3}\right)s\dot{\boldsymbol{\alpha}}_d - \mu(\dot{\boldsymbol{\alpha}}_d\cdot\nabla)\boldsymbol{\alpha}_d\right]\cdot\hat{\mathbf{n}}\,dS = -\int\mathbf{S}_{\mathbf{p}_y}\cdot\hat{\mathbf{n}}\,dS(>0); \; s = -\nabla\cdot\boldsymbol{\alpha}_d. \tag{8.6.40b}$$

More compactly, we have again for the total energy flux out of the volume

$$\dot{E} = \int_V(\dot{U}_k + \dot{U}_\mathbf{p})\,dV = -\int_\mathbf{n}\mathbf{S}_{\mathbf{p}_y}\cdot\hat{\mathbf{n}}\,dS. \tag{8.6.40c}$$

At this point it is instructive to interpret the various components here in some detail:

(1) U_k = The kinetic energy density stored in the volume V at time t; $U_\mathbf{p}$ = the potential density, similarly stored at time t.

(2) The integral $\int_V(\,)dV$ = total energy per unit volume, at any instant t; $\partial/\partial t(\,)$ = rate of change of (1) in V.

(3) $\int_\Sigma(\,)dV$ = rate at which work is done by the fluid (or solid) in V, bounded by the surface Σ, namely, the rate at which energy is instantaneously leaving the volume, through the surface Σ.

(4) The elastic dyadic stress $(\mu/2)\nabla\boldsymbol{\alpha}_d\colon\nabla\boldsymbol{\alpha}_d$ = contribution to the elastic stress potential energy, via the shear strains. Thus, $\int_\Sigma(\mu\dot{\boldsymbol{\alpha}}_d\cdot\nabla\boldsymbol{\alpha}_d)\cdot\hat{\mathbf{n}}$ is the contribution

to the rate of flow of elastic energy across Σ, of the medium here that is shearable and compressible.

(5) $\mathbf{S}_d = -\int_\Sigma \mathbf{S_p}\cdot\hat{\mathbf{n}}\,dS$ (Eq. 8.6.40b) = Total flow of instantaneous energy per unit area in the elastic radiation field; $\mathbf{S_p}\cdot\hat{\mathbf{n}}$ represents the total instantaneous rate of flow per unit area in the elastic radiation field.

Furthermore, we observe that $\mathbf{S}_d$, is here the expression of the acoustic poynting's theorem for this type of liquid (or solid). The integrand of $\mathbf{S}_d$ is the corresponding acoustic poynting vector $\mathbf{S_p}$, namely,

$$\mathbf{S_p} = (B_M + \mu/3)\dot{\boldsymbol{\alpha}}_d s - \mu\dot{\boldsymbol{\alpha}}_d \cdot \nabla\boldsymbol{\alpha}_d (> 0): \quad (8.6.41a)$$

[Elastic medium (fluid or solid), with shear ($\mu > 0$)].

$$\text{Acoustic poynting vector reduces to } \mathbf{S_p} = B_M \dot{\boldsymbol{\alpha}}_d s, \quad (8.6.41b)$$

where $s = -\nabla\cdot\boldsymbol{\alpha}_d$ again, and $\mu = 0$ [Elastic medium, no shear ($\mu = 0$)].

Thus, we see that the overall potential energy density in the solid medium is augmented by the contribution to the elastic strain (potential) energy derived from the shear strains produced in the solid medium. Similarly, the second term in the flux vector $\mathbf{S}_d$ is the modification to the energy density flux vector in the elastic radiation field attributable to the shear strains. The relations yield E explicitly, by inspection (See Section 6.4 of Ref. [2], which extends this analysis to anisotropic media).

8.6.5 Energy Densities and Flux Densities in the Dissipative Media

In Section 8.6.1, we have provided examples for ideal media, which are characteristically Non-Lossy, that is, nondissipative. Here, we extend the discussion to the more complex physical cases where energy is lost in heat in the process of propagation in dissipative media. We begin our discussion by extension to gases and liquids that are *viscous*, through which compressional waves are transmitted. This generalization (particularly of the results in (5) above) to include fluid frictional (e.g., viscosity) effects requires a number of essential modifications of the relation (8.6.41).

8.6.5.1 Elastic Wave Propagation in Viscous Media: An Introduction to the Navier–Stokes Equation[60] Using a generalization of the propagation equation for an ideal elastic *fluid (in gas)*, we can show that it is possible to extend the connection between stress and strain in such media (refer to (5) above) in the following manner[60], (Section 12.5, of Ref. [2]. The generalization in question requires that the fluid displacement $\mathbf{d}$ in the statement of Hooke's law, Eq. (8.6.18), be replaced by the *flow velocity* $\mathbf{v}$ ($=\dot{\mathbf{d}}$) and that the elastic constant (μ and B_M) be replaced by coefficient of *viscosity* (β_1, β_2), that is, ($\mu \to \beta_2$ and $B_M \to \beta_2$).

Thus, the elastic constant as for shear becomes the *shear viscosity* β_2, and the *bulk modulus* B_M is replaced by the *bulk viscosity* β_1. In addition, since the fluid motion obeys a

[60] See Ref. [2], Section 9.11, pp. 261–269; Section 12.4, pp. 351–354, and in particular, Section 12.5, pp. 354–360. (i) of Section 8.6.2 is back for the most part in Section 2.5 of Ref. [2]. See also Chapter 2 of Ref. [1].

pressure gradient, a term $-\nabla p$ is included. Accordingly, (8.6.41) is now replaced by the general elastic equation of propagation:

$$\rho\{(\ddot{\boldsymbol{\alpha}}_d + \dot{\boldsymbol{\alpha}}_d \cdot \nabla)\dot{\boldsymbol{\alpha}}_d\} = (\beta_1 + 4\beta_2/3)\nabla\nabla \cdot \dot{\boldsymbol{\alpha}}_d - \beta_1 \nabla \times \nabla \times \dot{\boldsymbol{\alpha}} - \nabla_p; \quad \boldsymbol{\alpha}_d = \mathbf{d}. \tag{8.6.42}$$

Note that we are dealing only with a fluid (i.e., gas or liquid), not with solids. The physical interpretation of the bulk viscosity β_1 is seen to be the ratio of the stress $\beta_1 \nabla \cdot \dot{\mathbf{d}}$ to the time rate of change of the total dilation, $\nabla \cdot \dot{\mathbf{d}}$.

Setting $\dot{\mathbf{d}} = \boldsymbol{v}$ in the above relation (8.6.42) and using the *equation of continuity* (8.6.29) $\nabla \cdot \rho\boldsymbol{v} = \dot{\rho} = -\partial\rho/\partial t$, we see that Eq. (8.6.36a) is a form of Navier–Stokes equation of fluid flow in a viscous, that is, frictional, fluid flow. The result is specifically

$$\frac{\partial(\rho\boldsymbol{v})}{\partial t} + \rho[\boldsymbol{v} \cdot \nabla\boldsymbol{v}] + \boldsymbol{v}\nabla \cdot (\rho\boldsymbol{v}) = (\beta_1 + 4\beta_2/3)\nabla\nabla \cdot \boldsymbol{v} - \beta_1 \nabla \times \nabla \times \boldsymbol{v} - \nabla p. \tag{8.6.43a}$$

Note that unlike the energy relations of nondissipative media in (1)–(5) of Section 8.6.4 above, this equation is nonlinear, which is characteristic of the various forms of the Navier–Stokes equation. The left member of (8.6.43a) can be shown to represent a normalized form U'_{K}/ρ of the kinetic energy. The right member is the complete *acceleration*, with dissipative (i.e., friction) associated with this viscous fluid flow. Remembering that the displacement field $\boldsymbol{\alpha}_d = \mathbf{d}$ and $\therefore\ \dot{\alpha}_d = \boldsymbol{v}$, we can write (8.6.43a) alternatively:

$$\frac{\partial(\rho\dot{\boldsymbol{\alpha}}_d)}{\partial t} + \rho(\dot{\boldsymbol{\alpha}}_d \cdot \nabla)\dot{\boldsymbol{\alpha}}_d + \dot{\boldsymbol{\alpha}}_d\nabla \cdot (\rho\dot{\boldsymbol{\alpha}}_d) = (\beta_1 + 4\beta_2/3)\nabla\nabla \cdot \dot{\boldsymbol{\alpha}}_d - \beta_1 \nabla \times \nabla \times \dot{\boldsymbol{\alpha}}_d - \nabla p. \tag{8.6.43b}$$

Next, to determine the corresponding energy relations in the usual way, we multiply by $(\cdot\,\dot{\alpha}_d)$ from the right-hand side of both left and right members of (8.6.43b) and observe that

$$\left.\begin{aligned} &\frac{\partial(\rho\dot{\boldsymbol{\alpha}}_d)}{\partial t} \cdot \dot{\boldsymbol{\alpha}}_d = 2\ddot{\rho}\left(\frac{U'_{\mathrm{K}}}{\rho}\right) + 2\rho\frac{\partial}{\partial t}\left(\frac{U'_{\mathrm{K}}}{\rho}\right); \quad \rho(\dot{\boldsymbol{\alpha}}_d \cdot \nabla\dot{\boldsymbol{\alpha}}_d) \cdot \dot{\boldsymbol{\alpha}}_d = 2\rho(\dot{\alpha} \cdot \nabla)\left(\frac{U'_{\mathrm{K}}}{\rho}\right): \\ &\qquad \dot{\boldsymbol{\alpha}}_d\nabla \cdot (\rho\dot{\boldsymbol{\alpha}}_d) \cdot \dot{\boldsymbol{\alpha}}_d = 2(\nabla \cdot \rho\dot{\boldsymbol{\alpha}}_d)\left(\frac{U'_{\mathrm{K}}}{\rho}\right), \quad \text{with} \\ &\qquad U'_{\mathrm{K}} \equiv \rho\left[\frac{\dot{\boldsymbol{\alpha}}_d \cdot \dot{\boldsymbol{\alpha}}_d}{2}\right] (\neq U_{\mathrm{K}}) \end{aligned}\right\} \tag{8.6.44}$$

For the right member of (8.6.43a), we obtain for the accelerated energy

$$-[\beta_1 + \beta_2/3]\nabla\dot{s} \cdot \dot{\boldsymbol{\alpha}}_d + \beta_2\dot{\boldsymbol{\alpha}}_d\nabla^2 \mathbf{2}d - \dot{\alpha} \cdot \nabla p; \quad \dot{s} = -\nabla \cdot \dot{\boldsymbol{\alpha}}_d, \text{refer to Eq. (8.6.25c).} \tag{8.6.45}$$

Combining (8.6.44) and (8.6.45) in (8.6.43b) gives us finally the desired result in terms of kinetic energy and acceleration:

$$\left[2\ddot{\rho} + 2\rho\frac{\partial}{\partial t} + 2\rho(\dot{\boldsymbol{\alpha}}_d \cdot \nabla) + 2(\nabla \cdot \rho\dot{\boldsymbol{\alpha}}_d)\right]\left(\frac{U'_{\mathrm{K}}}{\rho}\right)$$
$$= (\beta_1 + \beta_2/3)\nabla\dot{s} \cdot \dot{\boldsymbol{\alpha}}_d + \beta_2\dot{\boldsymbol{\alpha}}_d \cdot \nabla^2\dot{\boldsymbol{\alpha}}_d - \dot{\boldsymbol{\alpha}}_d \cdot \nabla p. \qquad (8.6.46)$$

This general result can be expressed alternatively in terms of *kinetic energy density*, *energy flux* from the bounding surface Σ, and the fluid energy acceleration (i.e., the right-hand member of (8.6.46). Again, we note the nonlinear terms $O(\dot{\alpha}^3)$, $[2\rho\dot{\alpha}_d \cdot \nabla + 2\nabla \cdot (\rho\dot{\alpha}_d)]\{\rho((\dot{\alpha} \cdot \dot{\alpha})/2\rho)\}$. This nonlinearity greatly complicates the solution for $\dot{\boldsymbol{\alpha}}_d$, from the propagation equation (8.6.43b) of the displacement field $\dot{\boldsymbol{\alpha}}_d$ (or specifically here $\dot{\boldsymbol{\alpha}}_d = \dot{\mathbf{d}} = \boldsymbol{v}$, for the fluid velocity). Thus, an alternative form is

$$2\rho\frac{\partial}{\partial t}\overline{U}_K + \{2\ddot{\rho} = \dot{\boldsymbol{\alpha}}_d \cdot \nabla\overline{U}_{\mathrm{K}} + 2\nabla \cdot \rho\dot{\boldsymbol{\alpha}}_d\}\overline{U}_{\mathrm{K}} = (\beta_1 + \beta_2/3)\dot{\boldsymbol{\alpha}}_d \cdot \nabla\dot{s} + \beta_2\dot{\boldsymbol{\alpha}}_d \cdot \nabla^2\dot{\boldsymbol{\alpha}}_d - \dot{\boldsymbol{\alpha}}_d \cdot \nabla p, \qquad (8.6.47a)$$

where $U_K \equiv U'_{\mathrm{K}}/\rho$, or more compactly,

$$H_1(p)\dot{\overline{U}}_{\mathrm{K}} + H_0(\rho, \dot{\boldsymbol{\alpha}}_d)\overline{U}_{\mathrm{K}} = J_{12}(\beta_1, \beta_2; \dot{\boldsymbol{\alpha}}_d, \nabla\dot{s}) \qquad (8.6.47b)$$

$$H_1(\rho) = 2\rho; \quad H_0(\rho, \dot{\boldsymbol{\alpha}}_d) = 2[\ddot{\rho} + \rho\dot{\boldsymbol{\alpha}} \cdot \nabla + \nabla \cdot \rho\dot{\boldsymbol{\alpha}}_d]. \qquad (8.6.47c)$$

Finally, we point out a perturbational approach to solutions for $\dot{\boldsymbol{\alpha}}(\equiv \nu)$, Eqs. (8.6.42) and (8.6.43). This is usually effective in most instances because of the comparatively small deviations from linearity (and ideality) vis-à-vis the idealized media considered first in (1)–(5), of Section 8.6.4. The procedure begins by using a series of first- and higher-order terms for various quantities of interest, namely,

$$p_e = p_1 + p_2 + \cdots; \rho_e = \rho_1 + \rho_2 + \cdots; \nu = \dot{\boldsymbol{\alpha}} = \dot{\boldsymbol{\alpha}}_1 + \dot{\boldsymbol{\alpha}}_2 + \cdots. \qquad (8.6.48)$$

The method is described in detail in Ref. [2], pp. 356–360, and leads to such features as *acoustic streaming* and the appropriately modified equations of continuity, the dynamic equation of state, the relaxation mechanism, and so on. We shall not pursue these topics here, however, leaving them to the reader's choice. Rather, we shall consider the other energies involved in other types of propagation, namely, electromagnetic regimes, which can propagate in vacuum as well as in physically defined "mechanical" media, for example, gases, liquids, and solids.

8.6.5.2 Electromagnetic Energy Considerations Another important example is that provided by the calculation of the energy, and in particular, the energy density in an electromagnetic field, where there is also dissipation. "Energy" equations can be constructed similar to those for the particle or element displacement field $\boldsymbol{\alpha}_d$ discussed in Section 8.6.1 for material media and the induced "mechanical" radiation therein. As we shall see, the familiar Poynting relation (Sec. 2.10 of Ref. [3]) is obtained for the flux intensity density of the radiation out of the bounding surface of closed volume V, refer to Fig. 8.20. Here, we consider a homogeneous and isotropic medium (which need not be a vacuum) that is also stationary and nonferromagnetic. The medium is, therefore, free of hysteresis effects and is also free of deformable bodies.[61]

Accordingly, we begin with the EM field equations for this (stationary) medium (Section A.9 of Ref. [3]). Physical examples of such media, as mentioned earlier, are provided by the atmosphere, ocean and solids, as well as the comparative vacuum of space, under appropriate conditions of the type mentioned above. The field equations are specifically

$$\left.\begin{array}{ll}(1)\ \nabla\times\mathbf{E}+\dfrac{\partial\mathbf{B}}{\partial t}=\mathbf{O};\ (2)\ \nabla\times\mathbf{H}-\dfrac{\partial\mathbf{D}}{\partial t}=\mathbf{J}\\ (3)\ \nabla\cdot\mathbf{B}=\mathbf{O}; \qquad (4)\ \nabla\cdot\mathbf{D}=\rho\end{array}\right\}. \tag{8.6.49}$$

Here $\mathbf{E}$ and $\mathbf{D}$ are respectively the *electric field strength* and the *electric displacement vector*. Analogously, $\mathbf{H}$ and $\mathbf{B}$ are the magnetic field strength and magnetic *indication vector*. The quantities $\mathbf{J}$ and $\boldsymbol{\rho}$ are likewise the (vector) current density and the (scalar) charge density (Chapter 1 of Ref. [3]). (Note that the dimensions of $\mathbf{E}\cdot\mathbf{J}$ and $\mathbf{H}\cdot\mathbf{B}$ are [power/volume] = [energy E/time volume] = [W/m^3] = [E/s m^3] = $\dot{\varepsilon}$), refer to Eqs. (8.6.31) and (8.6.31a). We next multiply (1) by $\mathbf{H}$ and (2) by $\mathbf{E}$ and subtract the latter from the former to obtain

$$\mathbf{H}\cdot\nabla\times\mathbf{E}+\mathbf{H}\cdot\dot{\mathbf{B}}-\mathbf{E}\cdot\nabla\times\mathbf{H}+\mathbf{E}\cdot\dot{\mathbf{D}}+\mathbf{E}\cdot\dot{\mathbf{J}}=0. \tag{8.6.50a}$$

Using the vector identity $\nabla\cdot(\mathbf{E}\times\mathbf{H})=\mathbf{H}\cdot\nabla\times\mathbf{E}+\mathbf{E}\cdot\nabla\times\mathbf{H}$, we can write (8.6.50a) (8.6.43a) in the form

$$\mathbf{E}\cdot\dot{\mathbf{D}}+\mathbf{H}\cdot\dot{\mathbf{B}}+\mathbf{E}\cdot\mathbf{J}=-\nabla\cdot(\mathbf{E}\times\mathbf{H}). \tag{8.6.50b}$$

When this is integrated over a finite volume, as before, where the boundary is abstract (not a reflecting or scattering surface), it becomes with the help of Gauss theorem (8.6.26) the desired total energy flux $\dot{E}$:

$$\dot{E}=\int_V\left[\mathbf{E}\cdot\dot{\mathbf{D}}+\mathbf{H}\cdot\dot{\mathbf{B}}+\mathbf{E}\cdot\mathbf{J}\right]dV=-\int_\Sigma S_{\mathrm{P_y}}\cdot\hat{n}dS, \quad \text{with} \quad S_{\mathrm{P_y}}\equiv\mathbf{E}\times\mathbf{H},\ \left[\mathrm{W/m^3}\right], \tag{8.6.51}$$

which reduces to

$$\dot{E}=\int_V\frac{\partial}{\partial t}\left[\frac{\varepsilon E\cdot E}{2}+\frac{\mu H\cdot H}{2}\cdot\int^t\mathbf{J}(t')dt'\right]dV=-\int_\Sigma S_{\mathrm{P_y}}\cdot\hat{n}\,dS, \tag{8.6.51a}$$

for a Hom-Stat and isotropic EM field. (Here, of course, $(\,\dot{}\,)=\partial/\partial t$.) The first two terms of the volume integral represent the time sets of change of the electric and magnetic energy stored in the electromagnetic field in the volume V. The third term, $\mathbf{E}\cdot\mathbf{J}$, is the *power* dissipated in heat (J/s) in the volume, which is an irreversible transformation. The surface integral $\left(\int_\Sigma\right)$ is the total net EM energy flux out of the (virtual) boundary surface Σ. (We have rewritten the dissipated power in the volume (V) as the time derivative of the dissipated energy density, at time t.)

[61] Deformable bodies would permit the transformation of electromagnetic energy into elastic energy of the resulting stressed medium ($\mu > 0$).

From Poynting's theory, the quantity $\mathbf{S}_E(=\mathbf{E}\times\mathbf{H})$ is called the *electromagnetic poynting's vector*, as derived by him in 1884.[62] This is represented by the integral over the surface Σ, as noted above, and represents the total effective energy flux out of this (virtual) surface Σ.

For isotropic media and in the absence of ferromagnetic materials in V, we can also write $D=\varepsilon E$ and $B=\mu H$, where (ε, μ)[63] are the respective electric and magnetic inductive capacities of the medium. Moreover, it is usually the case that the vector density[63] is given by $\mathbf{J}=\sigma E$, where σ is the medium's conductivity. The electromagnetic energy flux (Eq. (8.6.51) takes a more familiar form (8.6.51a) where $\mathbf{S}_E\left(\equiv S_{p_y}\right)$ is the electromagnetic poynting vector, which measures the energy flux density leaving the volume V through the virtual surface Σ. (Harmonic fields are assumed for the moment and the factor (1/2) accounts for one complete cycle of **E** and **H**). At any instant $\dot{E}$, the factors 1/2 are removed.). In any case, the first two terms in the integrand represent $\dot{\varepsilon}$ for the electric and magnetic energy flux. The third integral is once more the *Julian heat loss*, per unit time in the volume V.

Finally, we note that the field equations (8.6.49) can also be expressed in terms of the *Hertz vector potentials* $\boldsymbol{\Pi}^{(1)}$ and $\boldsymbol{\Pi}^{(2)}$, here for a general homogeneous isotropic and conducting medium V, where the free change ρ is always zero in V because of the extreme density of the relaxation time. If this medium is free of fixed polarization $(\mathbf{P}_0, \mathbf{M}_0)$, then (8.6.49) becomes generally

$$\left.\begin{array}{ll}\mathbf{E}=\nabla\times\nabla\times\boldsymbol{\Pi}^{(1)}-\mu\nabla\times\dfrac{\partial}{\partial t}\boldsymbol{\Pi}^{(2)}; & \mathbf{H}=\nabla\times\left[\varepsilon\dfrac{\partial}{\partial t}\boldsymbol{\Pi}^{(1)}+\sigma\boldsymbol{\Pi}^{(1)}\right]+\nabla\times\nabla\times\boldsymbol{\Pi}^{(2)} \\ \mathbf{D}=\varepsilon\mathbf{E}; & \mathbf{B}=\mu\mathbf{H}\end{array}\right\} \tag{8.6.52a}$$

where $\boldsymbol{\Pi}^{(1)}$ and $\boldsymbol{\Pi}^{(2)}$ obey

$$\left.\begin{array}{l}\nabla\times\nabla\times\boldsymbol{\Pi}^{(1)}-\nabla\nabla\cdot\boldsymbol{\Pi}^{(1)}+\mu\varepsilon\dfrac{\partial^2}{\partial t^2}\boldsymbol{\Pi}^{(1)}+\mu\varepsilon\dfrac{\partial}{\partial t}\boldsymbol{\Pi}^{(1)}=0 \\ \text{and} \\ \nabla\times\nabla\times\boldsymbol{\Pi}^{(2)}-\nabla\nabla\cdot\boldsymbol{\Pi}^{(2)}+\mu\varepsilon\dfrac{\partial^2}{\partial t^2}\boldsymbol{\Pi}^{(2)}+\mu\varepsilon\dfrac{\partial}{\partial t}\boldsymbol{\Pi}^{(1)}=0\end{array}\right\} \tag{8.6.52b}$$

[62] This also dates from Oliver Heavyside, who derived it later the same year ([3], p. 132).

[63] We have $\mu=[MLQ^{-2}]$, magnetic inductive capacity, in henry/m; $c_0=1/\sqrt{\mu_0\varepsilon_0}=2.998\times10^8(\doteq3\times10^8)$m/s; ($\mu_0, \varepsilon_0$ = in free space); $\sigma_0=0$, Q = Coulomb; $\varepsilon=[M^{-1}L^{-3}T^2Q^2]$ electric conductive capacity, in farads/m; $\sigma=M^{-1}L^{-3}TQ^{-2}$, mho/m in vacuum, and $\sigma\sim O(10^7)$ mho/m for most metals, with $\sigma=O(10^{-7}-10^{-15})$ for most dielectrics. The values of ε are expressed by $K_e=\varepsilon/\varepsilon_0=O(10^{-20})$, where $\sigma_O=8.85\times10^{-12}$ farad/m and $\mu_0=1.257\times10^{-6}$ henry/m, and so on (see pp. 19–23; Appendix I, pp. 601–603; Appendix III, pp. 605–607, of Ref. [3] for details. These dimensions are given in terms of mass $[M]$, length $[L]$, time $[T]$, and charge $[Q]$.

In addition, when the medium has no conduction currents and no free charges, energy solution of the vector Hertzian propagation $\boldsymbol{\Pi}^{(1)}$, establishes an EM field according to

$$\left.\begin{aligned} &\mathbf{B} = \mu\varepsilon\nabla \times \frac{\partial}{\partial t}\boldsymbol{\Pi}^{(1)}, \quad \text{and} \quad \therefore \mathbf{H} = \varepsilon\frac{\partial}{\partial t}\boldsymbol{\Pi}^{(1)} = \mathbf{B}/\mu \\ &\mathbf{E} = \nabla\nabla\cdot\boldsymbol{\Pi}^{(1)} - \mu\varepsilon\frac{\partial^2}{\partial t^2}\boldsymbol{\Pi}^{(1)}, \quad \text{with} \quad \mathbf{D} = \varepsilon\mathbf{E} \end{aligned}\right\}. \tag{8.6.53}$$

The rectangular coordinates reduce here to

$$\nabla^2\boldsymbol{\Pi}^{(1),(2)} - \mu\varepsilon\frac{\partial^2}{\partial t^2}\boldsymbol{\Pi}^{(1),(2)} = 0. \tag{8.6.53a}$$

For details, see Ref. [3]. Section 1.1.

8.6.5.3 Other Equations of Propagation: Energy, Energy Flux, in V and Through Σ

As in Sections 8.6.2 and 8.6.3, we calculate the energy and energy flux for other basic equations of propagation. These equations are listed in Table 8.1 and the corresponding energies and fluxes are summarized in Table 8.5. One scalar and one vector identity will prove useful here. These are as follows:

(I) *Scalar* α_d: $\nabla\cdot(\dot{\boldsymbol{\alpha}}_d\cdot\nabla\boldsymbol{\alpha}_d) = \dot{\boldsymbol{\alpha}}_d\cdot\nabla\cdot\nabla\boldsymbol{\alpha}_d + \nabla\dot{\boldsymbol{\alpha}}_d\colon\nabla\boldsymbol{\alpha}_d.$ (8.6.54a)

Here, $\alpha_d = [\mathbf{d}]$ = scalar displacement of the moving element from equilibrium (Eq. 8.6.11 et seq.).

(II) *Vector* $\boldsymbol{\alpha}_d$: $\nabla\cdot(\dot{\boldsymbol{\alpha}}_d\cdot\nabla\boldsymbol{\alpha}_d) = \dot{\boldsymbol{\alpha}}_d\cdot\nabla\cdot\nabla\boldsymbol{\alpha}_d + \nabla\dot{\boldsymbol{\alpha}}_d\colon\nabla\boldsymbol{\alpha}_d.$ (8.6.54b)

We have as examples the six field equations of ____ and their associated energies shown in Table 8.5.

It should be emphasized that the above equations of propagation are strictly approximates only, but nevertheless still acceptably accurate in many applications. Chapter 2 of Ref. [1] treats fluids (and gases), as well as solids in much more detail than Section 8.6 of this book. Section 8.6 is intended simply to be introductory.

8.6.6 Extensions: Arrays and Finite Duration Sources and Summary Remarks

In the preceding development of examples of energy and energy flux densities, we have assumed that the vector and scalar fields $[\boldsymbol{\alpha}(\mathbf{R}, t), \alpha(\mathbf{R}, t)]$ are already "on" in the medium in question. Here, we extend our results (Sections 8.6.1–8.6.5) to include explicitly the role of the source of a general signal of finite duration for the two basic classes of radiation noted in Section 8.6.2, namely, "mechanical radiation" and radiation that does not depend on the medium for its generation, that is, EM in particular. We limit our discussion of scalar fields to the ideal, unbounded medium of Section 8.1 and to the (deterministic) scattering cases of Section 8.3. We then extend these results to vector fields.

TABLE 8.5 Selected Models of Energy Propagation and Energy Flux in Gases, Liquids, and Solids (Based on element displacement d(R, t))

Generic Models: Propagation Equations and Type	Medium G,L,S	U_K	U_P	U_D (Dissipative)	$\mathbf{S}_{Py}$: (Acoustic/EM) poynting Vector
Section 8.6.2					
(1) Ideal, incompressible, nonelastic, fluid $\nabla^2\Psi_V = \nabla^2\boldsymbol{\alpha}_V = 0$	G,L	$\rho_0\lvert\Psi_V\rvert^2/2$ $\rho_0\nabla\Psi_V\cdot\nabla\Psi_V$	C = constant (=0)	–	–
(2) Ideal, compressible (nonviscous shearless) (Helmholtz field) $\nabla^2\boldsymbol{\alpha}_d = \frac{1}{c_0^2}\ddot{\boldsymbol{\alpha}}_d$ $c_0 = \sqrt{B_M/\rho_0};\ B_M = \rho_0 c_0^2;$ Eqs. (8.6.21) and (8.6.22)(8.6.22) (Also for p, Φ_d, refer to Eqs. (8.6.20b)	G, L	$\rho_0(\dot{\boldsymbol{\alpha}}\cdot\dot{\boldsymbol{\alpha}})/2$	–	–	–
(3) and (4) Ideal elastic, compressible, (Nonviscous, shearless) $\nabla^2\boldsymbol{\alpha}_d = \frac{1}{c_0^2}\ddot{\boldsymbol{\alpha}}_d$; Eq. (8.6.21) and (8.6.23)	G, L	$\rho_0(\dot{\boldsymbol{\alpha}}\cdot\dot{\boldsymbol{\alpha}})/2$ $2U_k/\partial t$	$\frac{1}{2}B_M\sigma^2 = \frac{1}{2}B_M(\nabla\cdot\boldsymbol{\alpha}_d)^2$ $\frac{\partial}{\partial t}\left(\frac{B_M}{2}s^2\right) = \dot{U}_p$	–	$B_M\dot{\boldsymbol{\alpha}}_d s$; (Eq. 8.6.26) $= -B_M\dot{\boldsymbol{\alpha}}_d\nabla\cdot\boldsymbol{\alpha}_d$
(5) Elastic, compressible, nonviscous, with shear					
$\left\{\begin{array}{l}\rho_0\ddot{\boldsymbol{\alpha}}_d = (B_M+4\mu/3)\nabla\cdot\nabla\boldsymbol{\alpha}_d - \mu\nabla\times\nabla\times\boldsymbol{\alpha}_d \\ = (B_M+\mu/3)\nabla\nabla\cdot\boldsymbol{\alpha}_d + \mu\nabla\cdot\nabla\boldsymbol{\alpha}_d \\ = \text{Eq. (8.6.37)}\end{array}\right.$	S	$\rho_0(\dot{\boldsymbol{\alpha}}\cdot\dot{\boldsymbol{\alpha}})/2$	$[(B_M+\mu/3)s^3/2$ $+\mu\nabla\boldsymbol{\alpha}_d : \nabla\boldsymbol{\alpha}_d]$	–	$(B_M+\mu/3)s\dot{\boldsymbol{\alpha}}_d - \mu(\dot{\boldsymbol{\alpha}}_d\cdot\nabla)\boldsymbol{\alpha}_d$

(a) *Irrotational medium*: $\nabla \times \boldsymbol{\alpha}_d = 0$ $\rho_0 \ddot{\boldsymbol{\alpha}}_d = (B_M + 4\mu/3)\nabla \cdot \nabla \boldsymbol{\alpha}_d$; $c_p \quad = (B_M + 4\mu/3)^{1/2}$	S	$\rho_0 \left(\frac{\dot{\boldsymbol{\alpha}} \cdot \dot{\boldsymbol{\alpha}}}{2} \right)$	$c_P^2 \nabla \dot{\boldsymbol{\alpha}}_d : \nabla \boldsymbol{\alpha}_d$	–	$-c_P^2 (\dot{\boldsymbol{\alpha}}_d \cdot \nabla) \boldsymbol{\alpha}_d$
(b) *Solenoidal medium*: $\nabla \cdot \boldsymbol{\alpha}_d = 0$; $\frac{\ddot{\boldsymbol{\alpha}}}{c_S^2} = \nabla \cdot \nabla \boldsymbol{\alpha}_d; c_S = \sqrt{\mu/\rho_0} < c_P$	S	$\rho_0 \left(\frac{\dot{\boldsymbol{\alpha}} \cdot \dot{\boldsymbol{\alpha}}}{2} \right)$	$c_S^2 \nabla \dot{\boldsymbol{\alpha}}_d : \nabla \boldsymbol{\alpha}_d$	–	$-c_S^2 (\dot{\boldsymbol{\alpha}}_d \cdot) \nabla \boldsymbol{\alpha}_d$
Section 8.6.3: compressible elastic media with Dissipation (1) Navier–Stokes:	L	$\left(H_1(\rho) \frac{1}{U_k} + H_0(\rho_0 \dot{\boldsymbol{\alpha}}_d) \quad \overline{U}_k = J_{12} \right)$			
([2], pp. 354–360); $\overline{U}_K \equiv U_K/\rho$ See also Section 10.2 of Ref. [2] for *moving thermally conducting, viscous medium* $\rho\{\ddot{\boldsymbol{\alpha}}_d + (\dot{\boldsymbol{\alpha}}_d \cdot \nabla)\boldsymbol{\alpha}_d\}$ $= (\beta_1 + \beta_2/3)\nabla \cdot \nabla \dot{\boldsymbol{\alpha}}_d - \beta_1 \nabla \times \nabla \times \dot{\boldsymbol{\alpha}}_d$		$\overline{U}_K \equiv U_K/\rho$	$\rho =$ (variable density)		

(*continued*)

TABLE 8.5 *(Continued)*

Generic Models: Propagation Equations and Type	Medium G,L,S	U_K	U_P	U_D (Dissipative)	$\mathbf{S}_{Py}$: (Acoustic/EM) poynting Vector
(a) Electromagnetic energy and flux density (isotropic, nonferrous conducting linear media)	G,L,S + Vacuum				
$\left\{\begin{array}{l}\nabla\times\mathbf{E}+\dfrac{\partial\mathbf{B}}{\partial t}=\mathbf{0};\nabla\times\mathbf{H}+\dfrac{\partial\mathbf{D}}{\partial t}=\mathbf{J}\\ \nabla\cdot\mathbf{B}=\mathbf{0};\nabla\cdot\mathbf{D}=\rho\end{array}\right\}$		$\frac{\varepsilon}{2}\mathbf{E}\cdot\mathbf{E}$	$\frac{\mu}{2}\mathbf{H}\cdot\mathbf{H}$	$\mathbf{E}\cdot\int{}^{t}\mathbf{J}(t')dt'$	$\mathbf{S}_{P_y}=\mathbf{E}\times dt$ $=E=-\int_{\Sigma}\mathbf{S}_{P_y}\cdot\hat{\mathbf{n}}\,ds$
Eq. 8.6.49					
$\dot{E}=\int_V\frac{\partial}{\partial t}\left[\varepsilon\cdot\dot{\mathbf{D}}+\mu\cdot\dot{\mathbf{B}}+\mathbf{E}-\int^{t}\mathbf{J}(t')dt\right]$					
(b) Electromagnetic field (Hom-Stat isotropic)	G,L,S, Vol.				
$\left[\nabla\cdot\nabla-\mu\sigma\frac{\partial}{\partial t}-\mu\varepsilon\frac{\partial^2}{\partial t^2}\right]\mathbf{A}=\mathbf{0}$ $\mathbf{A}=\mathbf{E},\mathbf{H},\mathbf{\Pi}^{(1),(2)}$, (8.6.42a),		$\epsilon\mathbf{E}\cdot\mathbf{E}/2$	$\mu\mathbf{H}\cdot\mathbf{H}/2$	$\sigma\mathbf{E}\cdot\mathbf{H}$	$\mathbf{S}_{P_y}=\mathbf{E}\times\mathbf{H}$

(3)[a] Radiation dissipation: scalar + EM vector fields						
$\left.\left(\nabla^2 - \frac{1}{\alpha_R}\frac{\partial}{\partial t} - \frac{1}{c_0^2}\frac{\partial^2}{\partial t^2}\right)\right\} \begin{matrix} \alpha = 0 \\ \boldsymbol{\alpha} = \mathbf{0} \end{matrix}$	G,L,S, Vacuum	$\begin{matrix} \rightarrow \\ \rightarrow \end{matrix}\left.\right\} \begin{matrix} \dot{\alpha}^2/2 \\ \dot{\alpha}\cdot\dot{\alpha}/2 \end{matrix}$	$\left.\begin{matrix} c_0^2 \frac{\dot{\nabla}\alpha = \nabla\alpha}{2} \\ c_0^2 \frac{\dot{\alpha}^2 = \nabla\alpha}{2} \end{matrix}\right\}$	$(c_0^2/a_R)\left\{\begin{matrix} \lvert\dot{\boldsymbol{\alpha}}\rvert^2 \\ \boldsymbol{\alpha}\cdot\boldsymbol{\alpha} \end{matrix}\right.$	$-c_0^2\left\{\begin{matrix} \dot{\alpha}\,\nabla\dot{\boldsymbol{\alpha}} \\ (\dot{\boldsymbol{\alpha}}\cdot\nabla)\dot{\boldsymbol{\alpha}} \end{matrix}\right.$	
(4)[a] Relaxation dissipation: scalar + EM vector fields						
$\left.\left(\left(1+\hat{\tau}_0\frac{\partial}{\partial t}\right)\nabla^2 - \frac{1}{c_0^2}\frac{\partial^2}{\partial t^2}\right)\right\} \begin{matrix} \alpha = 0 \\ \boldsymbol{\alpha} = \mathbf{0} \end{matrix}$	G,L,S, Vacuum	$\begin{matrix} \rightarrow \\ \rightarrow \end{matrix}\left.\right\} \begin{matrix} \dot{\alpha}^2/2 \\ \dot{\alpha}\cdot\dot{\alpha}/2 \end{matrix}$	$c_0^2\left\{\begin{matrix} \nabla\dot{\alpha} : \nabla\alpha/2 \\ \nabla\dot{\boldsymbol{\alpha}} : \nabla\dot{\boldsymbol{\alpha}}/2 \end{matrix}\right.$	–	$-c_0^2(1+\hat{2})\left\{\begin{matrix} \dot{\alpha}\,\nabla\alpha \\ \dot{\boldsymbol{\alpha}}\cdot\nabla\dot{\boldsymbol{\alpha}} \end{matrix}\right.$	
$(\nabla \times \boldsymbol{\alpha} = 0; \nabla\cdot\boldsymbol{\alpha} = 0)$						
(5)[a] Relaxation and radiation dissipation						
$\left.\left(\left(1+\hat{\tau}_0\frac{\partial}{\partial t}\right)\nabla^2 - \frac{1}{\alpha_R}\frac{\partial}{\partial t} - \frac{1}{c_0^2}\frac{\partial^2}{\partial t^2}\right)\right\} \begin{matrix} \alpha = 0 \\ \boldsymbol{\alpha} = \mathbf{0} \end{matrix}$	G,L,S Vacuum	$\begin{matrix} \rightarrow \\ \rightarrow \end{matrix}\left.\right\} \begin{matrix} \dot{\alpha}^2/2 \\ \dot{\alpha}\cdot\dot{\alpha}/2 \end{matrix}$	$c_0^2\left\{\begin{matrix} \nabla\dot{\alpha} : \nabla\alpha/2 \\ \nabla\dot{\boldsymbol{\alpha}} : \nabla\dot{\boldsymbol{\alpha}}/2 \end{matrix}\right.$	$\frac{c_0^2}{a_R}\left.\begin{matrix} \dot{\boldsymbol{\alpha}}^2 \\ \dot{\boldsymbol{\alpha}}\cdot\dot{\boldsymbol{\alpha}} \end{matrix}\right\}$	$-c_0^2(2+\hat{\varepsilon}_0)\left\{\begin{matrix} \dot{\alpha}\,\nabla\alpha \\ \dot{\boldsymbol{\alpha}}\cdot\nabla\boldsymbol{\alpha} \end{matrix}\right.$	
(6)[a] Diffusion:						
$\frac{1}{b^2}\nabla^2\left\{\begin{matrix} \alpha \\ \boldsymbol{\alpha} \end{matrix}\right\} = \begin{matrix} 0 \\ \mathbf{0} \end{matrix}$	G,L,S, Vacuum	$\begin{matrix} \rightarrow \\ \rightarrow \end{matrix}\left.\right\} \begin{matrix} \dot{\alpha}^2/2b^2 \\ \dot{\alpha}\cdot\dot{\alpha}/2b^2 \end{matrix}$	–	–	$-\left\{\begin{matrix} \dot{\alpha}\nabla\alpha/2b^2 \\ \dot{\boldsymbol{\alpha}}\nabla\boldsymbol{\alpha}/2b^2 \end{matrix}\right.$	

[a] When the dimensions of $[\alpha]$ here are $[M/L^2]$, that is $[\alpha] = [\rho_0 L]$; $U_{K,P}$ = energy = $[\mathsf{E}dV]$; E = energy density and $[E]$ = energy $[ML/J^2]$ $\dot{\mathsf{E}}, \dot{E}$ = energy density flux, in V; Eqs. (3) – (6) are *Energy analogies*. Also, $[B_M] = [\mu] = [ML/T^2]$ here.

8.6.6.1 Scalar Fields For scalar fields, continuous apertures, and signals of finite duration Δ, $(0 \le t \le T)$, we have from Eqs. (8.1.53) and (8.1.55) the field

$$\begin{array}{c}\text{Signal}\\ \leftarrow\text{"window"}\rightarrow\leftarrow\text{singnal source}\rightarrow\leftarrow\text{medium}\rightarrow\end{array}$$

$$\alpha_{\mathrm{H}}(\mathbf{R},t)=\int_{Br_1(s)} e^{si}\left\{\left(\frac{1-e^{-sT}}{2\pi i}\right)ds\int_{V_{\mathrm{T}}} Y_{\mathrm{T}}(\boldsymbol{\xi},s/2\pi i)_{\Delta}Y_0(\rho,s'/2\pi i)\mathbf{d}\boldsymbol{\xi}\right\}, \tag{8.6.55a}$$

where Y_0 is the double Fourier transform of the Green's function $G_0(\rho,\Delta t)$, (8.1.36a)–(8.1.36c); (8.1.37a)–(8.1.37c) and Y_{T} represents the Fourier transform of the (continuous) source $G_{\mathrm{T}}(\mathbf{R},t)$ in V_{T}, that is, $Y_{\mathrm{T}}=\int G(\mathbf{R},t)e^{-i\omega t}dt$. In operator language, α_{H} is the solution of

$$\hat{L}^{(0)}\alpha_{\mathrm{H}}=-G_{\mathrm{T}}|_{\mathbf{C}},\,\text{Eq. (8.1.5)}, \tag{8.6.55b}$$

where $\mathbf{C}$ represents the boundary (here the surface $S\rightarrow\infty$) and initial conditions, for example, $\alpha_{\mathrm{H}}=\partial\alpha_H/\partial t=0$). In the case of discrete ("point") emitting sensors of the source, $G_{\mathrm{T}}, \boldsymbol{\xi}\rightarrow\boldsymbol{\xi}_m, m=1,\ldots,M$. The volume integral $\int_{V_{\mathrm{T}}}(\ )\mathbf{d}\boldsymbol{\xi}$ in (8.6.55a) is replaced by the sum:

$$\int_{V_{\mathrm{T}}}(\ )\mathbf{d}\boldsymbol{\xi}\rightarrow\sum_{m=1}^{M}Y_{\mathrm{T}}(\boldsymbol{\xi}_m,s/2\pi i)S_{\mathrm{in}}(\boldsymbol{\xi},s/2\pi i)_{\Delta}Y_0(\rho_m,s/2\pi i),\ \rho_m=|\mathbf{R}-\boldsymbol{\xi}_m|. \tag{8.6.56}$$

For both (8.6.55a) and (8.6.55b)Y_0 is specifically

$$Y_0=\int_{-\infty}^{\infty}\frac{\mathsf{Y}_0(k,s)e^{ik|\mathbf{R}-\boldsymbol{\xi}_{(m)}|}}{|\mathbf{R}-\boldsymbol{\xi}_m|}\frac{kdk}{(2\pi)^2}. \tag{8.6.57}$$

In the important case of inhomogeneous (and unbounded) media, with scattering (cf. Section 8.3, Eqs. (8.3.6), (8.3.7a), (8.3.7b), (8.3.8), etc.) the extension of (8.6.55) and (8.6.56) become

$$\left.\begin{array}{l} L^{(1a)}\alpha_{(\infty)}^{(Q)}--G_T;\hat{L}^{(1)}\equiv\hat{L}^{(0)}=Q\ \text{ and }\ \therefore\alpha_{(\infty)}^{(Q)}=\alpha_{\mathrm{H}}+\hat{\eta}_{(\infty)}\alpha_{(\infty)}^{(Q)};\ \text{ with }\ \hat{\eta}_{(\infty)}\hat{M}_{(\infty)}^{(Q)}Q\\ \alpha_{(\infty)}^{(Q)}=\alpha_{\mathrm{H}}+\sum_{n=1}^{\infty}\hat{\eta}_{(\infty)}^{(n)}=\alpha_{\mathrm{H}};\ \left\|\hat{\eta}_{(\infty)}\right\|<1\end{array}\right\}. \tag{8.6.58}$$

Here, as before, $\hat{M}_{(\infty)}=\hat{M}_{(\infty)}^{(0)}(\mathbf{R},t|\mathbf{R}',t')$ represents the *integral* Green's function operator, (8.3.8), for surface or volumes. The operator $\hat{Q}=\hat{Q}(\mathbf{R},t)$ embodies the local inhomogeneity, here the scattering element, refer to Eq. (8.3.8) for volumes and surfaces. $\hat{\boldsymbol{\eta}}_{(\infty)}$ is the "mass operator" or field renormalization operator. See the discussion in Section 8.3.2 and examples. Similar, more complex expressions result when there are boundaries (Section 8.5).

8.6.6.2 Vector Fields The extension to vector fields $\alpha(\mathbf{R}, t)$ is readily made. In place of scalar operators, for example $\hat{\boldsymbol{\eta}}_{(\infty)}, \hat{M}_{(\infty)}, \hat{Q}$, we have instead the (free space) *integral dyadic, Green's function* $\hat{\mathbf{M}}^{(D)}_{(\infty)}$, namely:

$$\hat{\mathbf{M}}^{(D)}_{(\infty)} \equiv \hat{\mathbf{i}}_x\hat{\mathbf{M}}^{(x)}_{(\infty)} + \hat{\mathbf{i}}_y\hat{\mathbf{M}}^{(y)}_{(\infty)} + \hat{\mathbf{i}}_z\hat{\mathbf{M}}^{(z)}_{(\infty)}, \quad \text{or} \quad \hat{\mathbf{i}}_1\hat{\mathbf{M}}^{(1)}_{(\infty)} + \hat{\mathbf{i}}_2\hat{\mathbf{M}}^{(2)}_{(\infty)} + \hat{\mathbf{i}}_3\hat{\mathbf{M}}^{(3)}_{(\infty)} \tag{8.6.59}$$

with its components

$$\hat{\mathbf{M}}^{(k)}(\mathbf{R}, t|\mathbf{R}', t') = \hat{i}_k\hat{M}^{(0)}_{(\infty)}(\mathbf{R}, t|\mathbf{R}', t'), \quad k = (1,2,3) \quad \text{or} \quad (x, y, z). \tag{8.6.59a}$$

The $\hat{\mathbf{M}}^{(k)} = \hat{\mathbf{M}}^{(x,y,z)}(\mathbf{R}, t|\mathbf{R}', t')$ represent a *source* pointed in x, y, or z direction, that is,

$$\hat{L}^{(k)}g^{(k)}_{\infty} = -\hat{\mathbf{i}}_k\boldsymbol{\delta}(\mathbf{R}-\mathbf{R}')\delta(t-t') \quad \text{or}$$

$$\hat{\mathbf{M}}^{(k)}_{(\infty)} = \hat{\mathbf{i}}_k\hat{M}^{(k)}_{(\infty)} = \hat{i}_k\int_{\Delta} dt'\int_{V_T} d\mathbf{R}g^{(k)}_{\infty}\left(\mathbf{R}, t|\mathbf{R}', t'\right)(\,)_{\mathbf{R}',t'}. \tag{8.6.59b}$$

Since $\hat{M}^{(k)}_{(\infty)} = \hat{L}^{(k)-1}_{(\infty)}$, we have, accordingly for the source-free integral Green's function dyadic:

$$\hat{\mathbf{M}}^{(D)}_{(\infty)} = \mathbf{I}^{(D)}\int_{\Delta} dt'\int_{V_T} d\mathbf{R}'g^{(0)}_{\infty}(\mathbf{R}, t|\mathbf{R}', t')(\,)_{\mathbf{R}',t'}, \quad \text{with} \quad \mathbf{I}^{(D)} = \left[\delta_{kj}\right] = \begin{bmatrix} 1 & 0 & 0 \\ 0 & 1 & 0 \\ 0 & 0 & 1 \end{bmatrix}, \tag{8.6.60}$$

the familiarity *idem factor*.

Since α_{H} now is given by $\boldsymbol{\alpha}_{\mathrm{H}} = \hat{\mathbf{M}}^{(D)}G_{\mathrm{T}}$, we have from the extension of (8.6.58) to the vector field case:

$$\boldsymbol{\alpha}^{(Q)}_{\infty} = \boldsymbol{\alpha}_{\mathrm{H}} + \sum_{n=1}^{\infty}\left(\hat{\eta}^{(D)}_{\infty}-\right)^{(n)}\boldsymbol{\alpha}_{\mathrm{H}}, \quad \text{with}$$

$$\left[\hat{\eta}^{(D)}_{\infty}-\right]^{n} = \left[\hat{\mathbf{M}}^{(D)}_{(\infty)}Q-\right]^{n} = \left[\hat{\eta}^{(D)}_{\infty}\cdot\hat{\eta}^{(D)}_{\infty}\cdot\cdots\cdot\hat{\eta}^{(D)}_{\infty}\cdot\right], \tag{8.6.61}$$

which is the direct result.

We note, finally, that the radiation condition here becomes

$$\lim_{R\to\infty} R\left(\nabla\times\mathbf{g}^{(D)} - \frac{\hat{\mathbf{i}}_{\mathrm{R}}}{c_0}\frac{\partial}{\partial t}\mathbf{g}^{(D)}\right) = 0, \tag{8.6.62}$$

which is the vector extension of the scalar case. The energy and energy flux densities for both the scalar and vector fields follow directly from the results for $\alpha_{\mathrm{H}} = \alpha_d$ and $\boldsymbol{\alpha}_{\mathrm{H}}, \alpha^{(Q)}_{(\infty)}, \boldsymbol{\alpha}^{(Q)}_{(\infty)}$ in Section 8.6.1–8.6.5.

Summarizing the principal results of this Section 8.6, we have provided here a rather condensed set of relations, for determining the energetic measure for scalar and vector field $\alpha_d\left(= \alpha_{\mathrm{H}}, \alpha^{(Q)}_{(\infty)}\right), \boldsymbol{\alpha}_d = \boldsymbol{\alpha}_{\mathrm{H}}, \boldsymbol{\alpha}^{(Q)}_{(\infty)}$. This has been done for a variety of physical media, including a vacuum, as well as gases, liquids, and solids, which in turn support mechanical as well as electrodynamic radiation (Table 8.6).

TABLE 8.6 Source, Field, and Aperture Dimensions

Physical Quantity	Dimensions	Dimensions: $a' = 2$	Acoustic: $a = d - 3$; $b = 1$; $c = -2$	E: $a = d - 1$; $b = 1$; $c = 2$ H: $a = d - 3$; $b = 0$; $c = -1$
1. Green's functions: $g_\infty, g_\infty^{(Q)}$	$L^{a'-3}T^{-1}$	$L^{-1}T^{-1}$	$L^{-1}T^{-1}$	$L^{-1}T^{-1}$
2. Integral GF's: $\hat{M}_\infty, \hat{M}_\infty^{(Q)}$	$L^{a'-d}$	L^{2-d}	L^{2-d}	L^{2-d}
3. FRO's: $\hat{\eta}_\infty, \hat{\eta}_\infty^{(k)}$	[1]	[1]	[1]	[1]
4. Field forms: $\alpha_{\mathrm{H}}, \alpha^{(k)}, \alpha^{(Q)}$	$\boxed{L^{a+a'-a}M^bT^c}$	$\boxed{L^{a+2-d}M^bT^c}$	$\boxed{M/LT^2}$	$\boxed{\mathbf{E} = LMT^2Q^{-1} \vdots \mathbf{H} = L^{-1}T^{-1}Q}$
5. Source function: $G_T = \hat{G}_T S_{in}$	$L^aM^bT^c$	$L^aM^bT^c$	$L^{d-3}MT^{-2}$	$L^{d-1}MT^2Q^{-1} \vdots L^{d-3}T^{-1}Q$
6. Aperture filters (trans): h_T, A_T	$L^aM^bT^{c-1}S^{-1}$	$L^aM^bT^{c-1}S^{-1}$	$L^{d-3}MT^{-3}S^{-1}$	$L^{d-1}MTQ^{-1} \vdots L^{d-3}T^{-2}QS^{-1}$
7. Beam pattern (trans): A_T	$L^{a+3-d}M^bT^cS^{-1}$	$L^{a+3-d}M^bT^cS^{-1}$	$MT^{-2}S^{-1}$	$L^2MT^2Q^{-1}S^{-1} \vdots L^0T^{-1}QS^{-1}$
8. Aperture Filters (Rec): h_R, A_R	$L^{2d-a-a'-3}M^{-b}T^{-c-1}S$	$L^{3-d-a-5}M^{-b}T^{c-1}S$	$L^{d-2}M^{-1}TS$	$L^{d-4}M^{-1}T^{-3}Q^{-1}S \vdots L^{d-2}T^0QS$
9. Beam pattern (Rec): A_R	$L^{d-a-a'}M^{-b}T^{-c}S$	$L^{d-a-2}M^{-b}T^{-c}S$	$LM^{-1}T^2S$	$L^{-1}M^{-1}T^{-2}QS \vdots LTQS$
10. $\mathsf{A}_T\mathsf{A}_R$	$L^{3-a'}$	L	L	L
11. Gen. Huygens Principle (GHP)	$\boxed{a' = 2;}$	$(a' = 2)$	$(a' = 2)$	$(a' = 2)$

8.7 SUMMARY: RESULTS AND CONCLUSIONS

Our general aim in Chapter 8 has been to provide an overall framework in which to describe the propagation field and in turn the channel, which includes the reception process. This has been done here for the deterministic or nonrandom situation of a given single, typical member of a set or ensemble of such members. As such, it is the precursor to a probabilistic treatment of the ensemble itself, on whose random properties a statistical communication theory is based (see, for example, Ref. [25]). The vehicle of our analysis is the linear operator, which enable us to consider multimedia models and deterministic scattering phenomena easily in a formal fashion. This operator formulation gives us a qualitative picture of the various interactions within and between media and provides the analytic "macroalgorithms" to guide their ultimate quantification by computer. Although the computational tasks may be great, often on the scale of weather prediction, they are now in practice well within the speed and capacity of modern machines. An additional advantage of the operational approach is its comparative compactness and formal simplicity, allowing us a ready overview of the often complex interactions encountered in the physically more realistic communication environments, that is, telecommunications, radar, and sonar.

A brief review of the table of contents provides a concise and self-explanatory account of the material in Chapter 8. A few observations are noted here for added emphasis. In the case of homogeneous (unbounded) media we have determined the Green's function (i.e., the four-dimensional filter response or weighting function of the medium) for a variety of different media, including dissipative, that is, dispersive cases. These media are also isotropic as well as homogeneous. When the media are inhomogeneous, as we discussed in Sections 8.3 and 8.4, isotropy is destroyed. However, the reciprocity (of the Green's function) still holds (Section 8.3.3) in such cases. The presence of directional beams and the presence of inhomogeneously distributed sources destroys the overall reciprocity in these cases, except for a partial effect most evident in the direction of the main beam (Fig. 8.10). Inhomogeneity of the medium also causes scattering of the signal source, both in the medium and at and through boundaries (Sections 8.3, 8.3.4 and 8.5.1). The scattering, of course, is deterministic, since the inhomogeneities of the medium and so on are postulated to be fixed here in Chapter 8.

The often used "engineering approach" of replacing the channel (in the general case) by a linear time variable temporal filter is shown here to be a valid equivalent to the actual physical channel only if a number of important conditions are obeyed:

(1) The received field must be in the transmitter's far-field or Fraunhofer region.
(2) The transmitted signal must be the same, applied to each element of the transmitting aperture or array.
(3) The time-variable filter's scale or amplitude parametrically depends on range (R).
(4) The receiver portion of the coupling of field to aperture (or array) must be essentially an all-pass network (for details, see Section 8.2.1).

One innovation here is the use of the *feedback concept* (Section 8.5.2) in these four-dimensional "circuits," suggesting a possible method for evaluating the effects of the distributed inhomogeneities, embodied in the mass operator $\hat{Q}$. Here, $\hat{M}_{\infty}^{(0)}$ is the "feed forward" and $\hat{Q}$ is the "feedback" operation. Another innovation is the $\hat{\eta}_{\infty}$-operator form, which allows us to describe operator interactions between boundaries, volumes, and multiple media systematically. We note that $\hat{M}^{(0)}$ and $\hat{Q}^{(0)}$ are separable factors for both volumes and surfaces in these deterministic cases. In addition, they can be shown to commute in frequency–wave number space, with the possibility of simplified evaluation (cf. Section 8.4.2), although they do not commute ordinarily in the usual space–time formulation.

In Chapter 9, we shall examine the extension of the deterministic propagation results of this chapter to the random cases. Here our task is to predict, on the average, from suitable statistical models the future behavior of random propagation phenomena. Perhaps the most important of these, from the viewpoint of communication through the channel, is scatter noise, when it occurs. In addition is the omnipresent ambient noise field that accompanies and that sets ultimate limits to all communication. This kind of noise is also an important limiting factor on our ability to extract weak signals, when scatter noise is sufficiently weak or absent (i.e. *a posteriori* models, vide the introduction to Section 8.1).

REFERENCES

1. P. M. Morse and H. Feshbach, *Methods of Theoretical Physics*, 1st ed., Vols. I and II, McGraw-Hill, New York, 1953.
2. R. B. Lindsey, *Mechanical Radiation*, McGraw-Hill, New York, 1960.
3. J. A. Stratton, *Electromagnetic Theory*, McGraw-Hill, New York, 1941.
4. J. D. Jackson, *Classical Electrodynamics*, John Wiley & Sons, Inc., New York, 1962, 1975.
5. J. T. Cushing, *Applied Mathematics for Physical Scientists*, John Wiley & Sons, Inc., New York, 1975.
6. W.C. Chew, *Waves and Fields in Inhomogeneous Media*, IEEE Press, Piscataway, NJ, 1995.
7. A. D. Polyanin and A. V. Manzhirov, *Handbook of Integral Equations*, CRC Press, Boca Raton, FL, 1998.
8. S. Chandrasekhar, *Radiative Transfer*, Oxford University Press, 1950; Dover Pub., New York, 1960.
9. D. Middleton, New Physical-Statistical Methods and Models for Clutter and Reverberation: The KA-Distribution and Related Probability Structures, *IEEE J. Oceanic Eng.* **24**, (1), (1999), and Errata, **26**, (2), 300–301, 2001. See also, New Results in Applied Scattering Theory: The Physical-Statistical Approach, including Strong Multiple Scatter vs. Classical Statistic-Physical Methods, and the Born vs. Rytov Approximations vs. Exact Strong Scatter Probability Distributions, *Waves Random Media* (Great Britain), **12**, 99–144, (2002)
10. U. Frisch, Wave Propagation in Random Media, in A. T. Bharucha-Reid (Ed.), *Probabilistic Methods in Applied Mathematics*, Vol. 1, Academic Press, New York, 1968, pp. 76–198.
11. V. I. Tatarskii, *The Effects of the Turbulent Atmosphere on Wave Propagation*, Vol. TT-68-50464, U.S. Dept. of Commerce, NTIS, Springfield, VA, 1971.
12. A. Ishimaru, *Wave Propagation and Scattering in Random Media*, Vols. I and II, Academic Press, New York, 1978.
13. F. G. Bass and I. M. Fuks, *Wave Scattering from Statistically Rough Surfaces* (translated by C. B. and J. F. Vesecky), Vol. 93, International Series in Natural Philosophy, Pergamon Press, New York, 1979.
14. S. M. Rytov, Yu. A. Kravtsov, and V. I. Tatarskii, *Principles of Statistical Radio Physics*, 4 Vols., (translated from the Russian Edition in 1989), Springer, New York, 1978. (See especially Vols. 3, 4.).
15. D. Middleton, Remarks on the Born and Rytov Scatter Approximations and Related Exact Models and Probability Distributions, *J. Opt. Soc. Am.* **A**,**16**, 2213–2218, (1999).

16. I. Tolstoy, *Wave Propagation*, McGraw-Hill, New York, 1973.
17. L. Brekhovskikh and V. Goncharov, *Mechanics of Continua and Wave Dynamics*, Springer Series in Wave Phenomena, Springer, New York, 1985 (Russian Edition, Nauka Publishing, Moscow, 1982).
18. A. N. Kolmogroff and S. V. Fomin, *Introductory Real Analysis* (Transl. by R. A. Silverman), Dover, New York, 1970. Chapter 6.
19. A. Mukherjea and K. Pothoven, *Real and Functional Analysis*, Plenum Press, New York, 1978.
20. S. M. Flatté, R. Dashen, W. H. Munk, K. M. Watson, and F. Zachariasen, *Sound Transmission Through a Fluctuating Ocean*, Cambridge University Press, New York, 1979. (See also the bibliography therein.).
21. N. W. McLachlin, *Complex Variables and Operational Calculus*, Cambridge University Press, London, 1939.
22. G. A. Campbell and R. M. Foster, *Fourier Integrals for Practical Applications*, D. Van Nostrand, New York, 1948.
23. H. Bateman, in A. Erdelyi, F. Oberhettinger, and F. G. Tricomi (Eds.), *Tables of Integral Transforms*, Vols. 1 and 2, McGraw-Hill, New York, 1954.
24. H. Margenau and G. M. Murphy, *The Mathematics of Physics and Chemistry*, D. Van Nostrand, New York, 1943, Chapter 4, Sections 4.14–4.19.
25. D. Middleton, *An Introduction to Statistical Communication Theory*, Classic Reissue, IEEE Press, Piscataway, NJ, 1996.
26. R. E. A. C. Paley and N. Wiener, *Fourier Transforms in the Complex Domain*, Vol. XIX, Ann. Math. Soc., College Park, MD, American Mathematical Society, New York, 1934; vide Theorem XII also.
27. P. Bello, Characterization of Randomly Time-Variant Linear Channels, *IEEE Trans. Comm. Sys.*, **S-11**,(4) (1963). See also the first paper in Ref. [28] ff.
28. B., Goldberg,(Ed.), *Communication Channels: Characterization and Behavior*, IEEE Selected Reprint Series, IEEE Press, Piscataway, NJ, 1975.
29. R. E. Blahut, W. Miller, Jr., and C. H. Wilcox, *Radar and Sonar, Part I, Institute for Mathematics and Its Applications (IMA)*, Vol. 32, Springer-Verlag, New York, 1991.
30. F. A., Grünbaum, M., Bernfield, and R. E., Blahut,(Eds.), *Radar and Sonar, Part II, IMA*, Vol. 39, Springer-Verlag, New York, 1992.
31. T., Kailath, (Ed.), *Linear Least-Square Estimation*, Vol. 17 of Benchmark Papers in Electrical Engineering, Dowden, Hutchinson, and Ross Inc., 1977.
32. R. J. Kennedy, *Fading, Dispersive Communication Channels*, John Wiley & Sons, Inc., New York, 1969.
33. I. Tolstoy and C. S. Clay, *Ocean Acoustics* (Theory and Experiment in Underwater Sound), McGraw-Hill, New York, 1966.
34. L. M. Brekhovskikh, *Waves in Layered Media*, 2nd ed., Academic Press, New York, 1980.
35. F. B. Jensen, W. A. Kuperman, M. B. Porter, and H. Schmidt, *Computational Ocean Acoustics*, American Institute Press, New York, 1994.
36. W. A. Lovitt, *Linear Integral Equations*, McGraw-Hill, New York, 1924, Dover Reprint Ed., New York, 1950.
37. V. A. Belakrishnan and V. Peterka, Identification in Automatic Control Systems, *Automatica*, **5**,817, (1969).
38. V. A. Balakrishnan, V. Peterka, and M. Athans, *Optimal Control*, McGraw-Hill, New York, 1966.
39. J. C. Williams, *The Analysis of Feedback Systems*, MIT Press, Cambridge, MA, 1971.

40. M. G. Safonov, *Stability and Robustness of Multivariable Feedback Systems*, MIT Press, Cambridge, MA, 1980. (See also for additional references.).
41. D. P. Petersen and D. Middleton, Sampling and Reconstruction of Wave-Number Limited Functions in *N*-Dimensional Euclidian Spaces, *Inform. Control*, **5**, 279–323, (1962); also, On Representative Observations, *Tellus*, 15, 387–405, (1963) (Sweden).
42. D. P. Petersen, Linear Sequential Coding of Random Space–Time Fields, *Inform. Sci.*, **10**, 217–241, (1976), and references therein and in Ref. [41] above.
43. V. I. Belyayev, *Processing and Theoretical Analysis of Oceanographic Observations (in Russian)*, Nauka Dumka, Kiev, 1973.

9

THE CANONICAL CHANNEL II: SCATTERING IN RANDOM MEDIA;[1] "CLASSICAL" OPERATOR SOLUTIONS

In Chapter 8 we have presented elements of the classical theory of propagation in a deterministic medium, with and without boundaries and with and without inhomogeneities. For deterministic inhomogeneous media, this includes deterministic scatter (Section 8.3). Our treatment is based on linear operators (Section 8.3.2.1), which provide formally exact results in terms of a "mass operator" $\hat{Q}$ and its generalizations, a *global mass operator*, $\hat{\eta} = \hat{M}\hat{Q}$ (Section 8.3.1). We now extend this approach directly to the more complex and important case of random media. The key concept here, of course, is the *ensemble* of representative members, subject to an appropriate probability measure, which are the stochastic solutions of this set of dynamical propagation equations. These equations are governed by appropriated boundary and initial conditions, which describe the possible propagation trajectories and scattering events associated with the ensemble. The set itself thus forms a stochastic equation, called a Langevin equation.

[1] The literature on this subject is vast and growing, not only because of scientific interest but also because of the many practical areas where scattering is a significant problem, especially in communications. A selection of references, mostly books, can provide a rich background on the subject. We cite again the work of Chew [1], Felsen and Marcuvitz [2] for nonrandom inhomogeneous media, cf. Chapter 8 of this book. For the random inhomogeneous media considered here note the papers and books of Frisch [3], Tatarskii [4], Ishimaru [5], Bass and Fuks [6], Rytov et al. [7], Twersky [8], Dence and Spense, [9], Klayatskin, [10], Flatté, [11], Furutsu [12], Oglivy [13], Uscinski [14], and more recently, Tatarskii et al. [15], and Wheelon [16]. This list is by means exhaustive, but it is suggestive of the intense interest over the last 50 years and up to and including the present in general scattering problems that occur in variety of physical situations, especially in acoustic and electromagnetic applications.

Non-Gaussian Statistical Communication Theory, David Middleton.

This ensemble of (partial) differential equations is the extension of the Langevin equations for the ensemble of ordinary differential equations describing simple dynamical systems, cf. Chapter 10 of Ref. [1]. Thus, the deterministic inhomogeneity operator $\hat{Q}$ of Chapter 8, cf. Eq. 8.3.2 et seq. now becomes the stochastic inhomogeneity operator $\hat{Q}$. The field "solution" of the Langevin equation is likewise a set of statistical results based on the ensembles. Consequently, it is the statistics of this random field that constitutes the desired solutions in these cases. These solutions are specifically the various moments and the probability distributions (or probability densities) of the random field. (See, for instance, the examples of the Fokker–Planck equation and the more general equations of Smoluchoroski and Boltzmann, of Chapter 10 of Ref. [1].) Because of this formulation, we call the standard treatment the "classical" *statistical-physics (S-P) approach* to the problem of scattering in random media.

Thus, the main purpose in this chapter is not to obtain explicit quantitative solutions in "classical" terms. It is rather to provide an operational structure that can be used (1) to present an initial framework for computational methods for numerical results, (2) to offer physical insights, and (3) to present quantitative results in probabilistic terms, that is, moments and pdfs. Classical solutions and approximations are amply discussed in many of the references, cf. [2–18]. For example, RKT [7], Parts 3 and 4, is particularly recommended.

Unfortunately, from the viewpoint of statistical communication theory (SCT), classical approaches, even including those based here on an operator formulation, do not in most cases yield analytical results for the *probability densities* required by SCT. This is particularly true for the non-Gaussian statistics that physically represent most scatter phenomena. Even so, an operator formulation is still useful in a number of ways. Besides simplifying the formal analysis, making it more compact and thereby helping to classify the interactive mechanisms involved, the operator formulation in the classical regime does provide us with some useful quantitative results. These include here the first- and second-order moments of the scatter process (i.e., Dyson's equation and a form of Bethe–Salpeter's equation), as well as guidance to a variety of useful approximations. Moreover, we introduce these, classical methods, (i) not only because of their historical importance but also (ii) because they are currently being pursued, and finally (iii) we employ them because their description is needed here to understand the place of new approaches and in what important ways they differs from classical methods. It provides a framework in the form of "macroalgorithms" for direct computation solutions in real-time (i.e., "on-line"). This "mini-weather prediction" task appears to be well within the capability of modern computing.

Accordingly, the organization of this chapter consists of the following principal topics. (1) After a brief description of the principal channel components in Section 9.1.1, including the random inhomogeneities and ambient noise, Section 9.1 introduces the first-order and feedback representation moments (Dyson's equation). Section 9.2 the second-order moments (Bethe–Salpeter equation), as well as higher order moments and the Transport equation. Section 9.3 follows with diagram equivalents (i.e., Feynman diagrams). The principal results are then summarized in Section 9.4, leading in turn to the new methods. Finally, our treatment is presented in terms of general signals and aperture or arrays, so that propagation and reception are represented by general waveforms in the medium and in the receiver.

9.1 RANDOM MEDIA: OPERATIONAL SOLUTIONS—FIRST- AND SECOND-ORDER MOMENTS

Before we begin the evaluation of the first- and second-order moments of general scatter, namely, scatter from rough or random boundaries and from local inhomogeneities in the volume, let us again consider an overview of the channel, as we have defined it throughout. It is the medium with its array or aperture couplings. The anatomy of the canonical channel is indicated by the schematics of Figs. 9.1 and 9.2 below.

From Eq. (8.5.13), we can write at once the generic relations for the inhomogeneous field in its various operator forms. This Langevin equation in differential form that seems in integral form, is accordingly represented by

$$\left(\hat{L}^{(0)} - \hat{Q}\right)\alpha^{(Q)} = -G_{\mathrm{T}} + \boldsymbol{C}, \tag{9.1.1}$$

$$\alpha^{(Q)}(\mathbf{R},t) = \left(\hat{1}-\hat{\eta}_{\infty}\right)^{-1}\alpha_{\mathrm{H}}(\mathbf{R},t) = \sum_{k=0}^{\infty}\hat{\eta}_{\infty}^{(k)}\alpha_{\mathrm{H}} = \alpha_{\mathrm{H}}(\mathbf{R},t) + \left(\hat{1}-\hat{\eta}_{\infty}\right)^{-1}\hat{\eta}_{\infty}\alpha_{\mathrm{H}}(\mathbf{R},t), \tag{9.1.2}$$

The mass operator $\hat{\eta}_{\infty} = \hat{M}_{\infty}\hat{Q}$ is now a random operator, because of the random local inhomogeneity operator $\hat{Q}$. We can describe this mass operator in more detail for volumes and surfaces by the general random forms

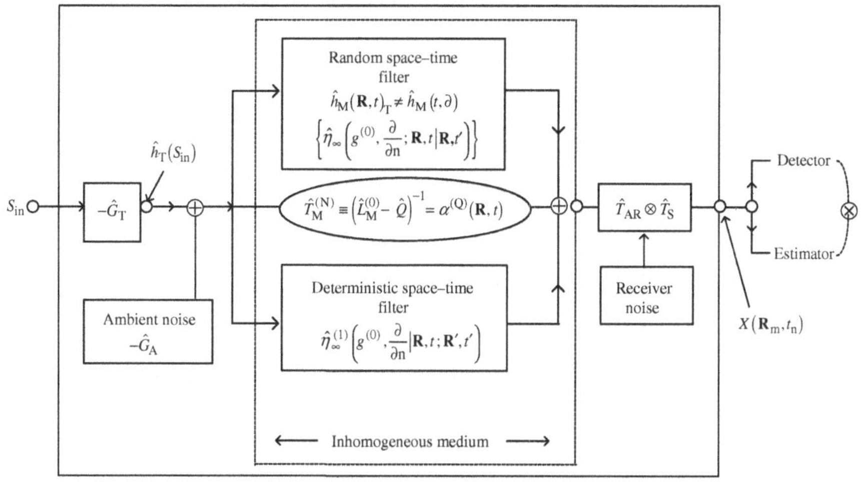

FIGURE 9.1 Schematic anatomy of the canonical channel, with ambient and receiver noise mechanism. The deterministic and random space–time filters $\left(\hat{\eta}_{\infty}, \hat{\eta}_{\infty}^{(1)}\right)$ are the Green's functions for the volume and surface of the random medium, cf. Sections 8.3.2 and 9.1.1. Here $\alpha^{(Q)}$ is the (ensemble of the) received field, with possible detection (D) and estimation (E) coupling $\otimes$.

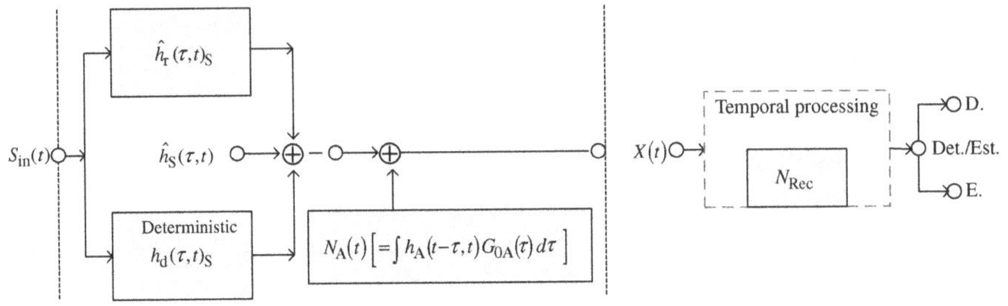

FIGURE 9.2 Temporal channel equivalent of the space–time generalized channel of *Fig. 9.1* (valid under the conditions (8.2.8), (8.2.14), extended to random fields).

$$\hat{\eta}_\infty = \hat{M}_V^{(0)}\hat{Q}_V = \hat{H}_\infty\left(g_\infty^{(0)}, \frac{\partial}{\partial n} : \mathbf{R}', \mathbf{t}'\right)\hat{Q}(\mathbf{R}', t')|_{V \text{ or } S}$$
$$= \hat{\eta}_\infty|_V = -\int dt' \int_V g_\infty^{(0)}(\mathbf{R}, t|\mathbf{R}', t')Q_V(\mathbf{R}', t')(\,)_{\mathbf{R}', t'}\mathbf{dR}', \text{ or} \tag{9.1.3a}$$

$$= \hat{\eta}_\infty|_S = \hat{M}_S^{(0)}\hat{Q}_S = \int dt' \oint_S \left[g_\infty^{(0)}\frac{\partial}{\partial n}(\,) - (\,)\frac{\partial}{\partial n}g_\infty^{(0)}\right]\hat{Q}_S(\mathbf{R}', t')(\,)_{\mathbf{R}', t'}dS(\mathbf{R}' \text{ or } S') \tag{9.1.3b}$$

Here the now random surface dS is specified by the results of Section 8.1.6.2, where the surface variations are regarded as statistical. Here also $g_\infty^{(0)}$ (and its derivatives) are still deterministic, whereas the integral Green's function operator $\hat{M}_\infty$, is not for random surfaces, since in (9.1.4b) $\partial/\partial n = \hat{n} \cdot \nabla$, $\hat{Q}_{V \text{ or } S}$, and dS are all now random quantities. The scattered field $\alpha^{(Q)}$, of course, is also random, as is $\hat{\eta}_\infty$.

Figure 9.3 illustrates the various scattering surfaces and the volume domains involved in the basic scenarios where V is finite and infinite, that is S is the outer bounding surface. In the case of $V - V_T \to \infty$, $S_1 \to \infty$. The inhomogeneities lie in $V - V_T$ and possibly on the surfaces S_0 and S_1, which can produce backscatter. Note again that $g_\infty^{(0)}$ is the kernel of the nonlocal propagator $\hat{M}_\infty^{(0)}$, which is applied to each inhomogeneous element in the volume $V(\neq V_T)$ and on the surfaces S_0, S.

Finally, we must note also here that the so-called Engineering Approach above, (9.1.1), requires that certain conditions on such filters must be observed [cf. (8.2.8) and (8.2.14)], the principal ones now being "same signal" applied to all elements of the transmitting array (aperture) and far-field operation, along with all-pass reception. In addition, the filters must obey the realizability (i.e., causality) conditions[2] of Section 8.2.2.

[2] Of course, for noncasual filters, we must wait for all the received data in the observation interval to be accumulated before processing.

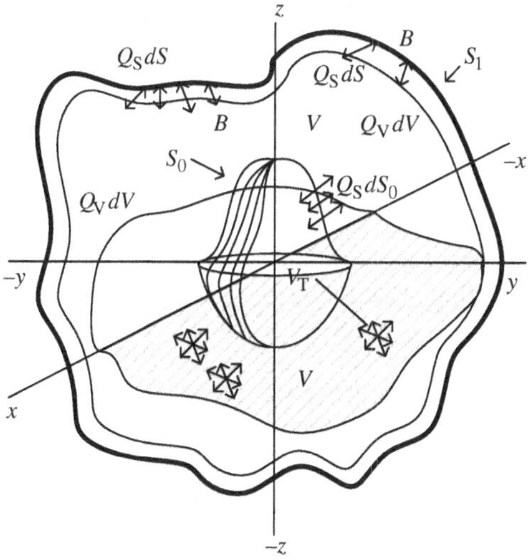

FIGURE 9.3 Binary sources in V_T and sources (scatterers) on the surfaces S_0, S_1, with sources and/ or scatterers) in the volume V, between S_0 and S_1. The $Q_S ds_0, Q_S ds_1, Q_V dv$ represent scattering elements on S_0, S_1, and in V, with B indicating backscatter from irradiated inhomogeneities on S_0, S_1. See also Fig. 8.9.

9.1.1 Operator Forms: Moment Solutions and Dyson's Equation

We start with the generic ensemble represented formally here by the feedback operational solution (FOS), namely, the random, that is, Langevin equivalent of Eq. (8.5.8) et. seq.

$$(FOS){:}\alpha^{(Q)} = \left(\hat{1}-\hat{\eta}_\infty\right)^{-1}\alpha_H = \sum_{k=0}^{\infty}\hat{\eta}_\infty^{(k)}\alpha_H, \text{with} \alpha_H = \hat{M}_\infty(-G_T) \text{and} \hat{\eta}_\infty \equiv \hat{M}_\infty\hat{Q}; \|\hat{\eta}_\infty\| < 1, \tag{9.1.4}$$

where $\hat{\eta}_\infty$ and $\alpha^{(Q)}$ are random quantities. This can be equivalently interpreted as an ensemble of feedback loops whose output is $\alpha^{(Q)}$ and where the global scattering operator is equivalently the feedback operator $\hat{\eta}_\infty$. This is shown in Fig. 8.16, now extended to each member of the random field $\alpha^{(Q)}$ in the defining relations (9.1.3). The relation (9.1.4) between the random $\alpha^{(Q)}$ and the usually not originally random input field α_H is also called the feedback operation relation (FOR) between the $\text{FOS}\left(\alpha^{(Q)}\right)$ and α_H, which may be deterministic or random. We remark that (9.1.3a) may apply for both surfaces and volume scattering, however with different structures: $\eta_V \neq \eta_S$, and in detail $M_V \neq M_S, Q_V \neq Q_S$ as Eqs. (8.3.8) and (8.3.9a–c) now extended to the random situation, shows. The feedback diagram of Fig. 8.16 is typical here, as well, but as we shall see presently the precise form of the FOS depends on whether or not the integral Green's function operator $\hat{M}_\infty$ and the local inhomogeneity or scattering operator $\hat{Q}$ are statistically independent. The relations (9.1.4) also lead naturally to equivalent FOR representations, which can act as a first step in the former evaluation of the scattered field. Finally, because of the convergence condition $\|\hat{\eta}_0\| < 1$ on the expansion it is also called the perturbation series solution (PSS), where some form of stochastic convergence is required for the random series, for example convergence in the mean-square (CMS) or the more restrictive almost certain or strong convergence (ACC); see Ref. [1], Section 2.1.1.

The FOR (and FOS) are also another way of expressing the governing Langevin equation (9.1.2) in the form of the (ensemble) of integral equations[3], from (9.1.5a):

$$\alpha^{(Q)}(\mathbf{R}, t) = \alpha_{\mathrm{H}}(\mathbf{R}, t) + \hat{\eta}_{\infty}\alpha^{(Q)}(\mathbf{R}, t);\ \hat{\eta}_{\infty}|_{V}, \hat{\eta}_{\infty}|_{S}, \tag{9.1.5}$$

where the only restrictions on $\hat{\eta}_{\infty}$ are that its kernel η_{∞} be positive definite, bounded, and not "intrinsically singular". Equations (9.1.5a) and (9.1.5b) apply for both scattering in volumes and from surfaces, cf. Sections 8.3.1, 8.3.4, and so on. There are two principal problems to be treated, which are associated with these stochastic Langevin equations. One is the direct problem, given $\hat{\eta}_{\infty}$, to obtain the scattered field $\alpha^{(Q)}$. The other is the *inverse problem*[4], of determining the nature of the scattering inhomogeneities from the scattered field. This is more difficult because the scattered field is *nonlinearly* related to the scattering elements in question, and because solutions of inverse problems are notoriously *nonunique*. These two features considerably complicate the task of obtaining closed form results, which lead us to iteration methods exemplified by the alternative perturbation techniques (PSS) of the FOS (9.1.5a). In the random or "predictive" cases, one therefore has to confront both the direct problem of determining the statistics of the scattered field $\alpha^{(Q)}$ and the indirect task of inferring the statistical properties of the scatterers, represented by $\hat{Q}$ (in the mass operator $\hat{\eta}_{\infty}$) (9.1.3a). For the latter, one possibility is the extension of the measurement procedures outlined earlier in Section 8.5.3 for the deterministic cases. This approach now includes an empirical ensemble, from which an estimate of an average scattering kernel $\left(Q^{(d)}\right)$ can be obtained.

Before we go on to consider specific solutions to the Langevin equation namely, *moments* of $\alpha^{(Q)}$, let us consider the FOS, (9.1.4), again. The critical feature to note here is that $\hat{\eta}_{\infty}^{(k)}\alpha_{\mathrm{H}}$ represents the ensemble or set of all the kth-order "quasiparticles" in scatter theory (see Ref. [18]). It is at once evident that all interactions of different orders $(k \neq k')$ are statistically independent at any given instant, since a kth-order quasiparticle is always different from a k'-order quasiparticle, and any non- kth-order ones. (We shall can use this property to construct probability distributions of the scattered field, as distinct from the moment calculations of classical theory here.) More immediate to the present treatment is the observation that the global massoperators $\hat{\eta}_V$ and $\hat{\eta}_S$ for volumes and surfaces are not only different, as we would expect, but the different local mass operators $\hat{Q}_V, \hat{Q}_S$, and projection operators $\hat{M}_V$, $\hat{M}_S$ may be statistically related, with different consequences for the moment solutions of the Langevin equations.

[3] Equation (9.1.3b) is usually a Voltera integral equation of the second kind. See Chapter 8 of Ref. [19]; also Lovitt [20].

[4] See Chew [1] for a treatment of the inverse scattering problem, for a representative member of the ensemble, that is, the deterministic case wherein the ensemble contains only one member function. For the random situation considered here, the ensemble is extended to an infinite number of similarly but randomly generated numbers, with an appropriate probability measure or measures, as required.

9.1.1.1 Dyson's Equation and the Equivalent Deterministic Mass Operator Let us develop a (formal) solution directly from the FOS by averaging (9.1.3a) termwise: we have at once for the first moment

$$\left\langle \alpha^{(Q)} \right\rangle = \left\langle \left(\hat{1} - \hat{\eta}_\infty\right)^{-1} \right\rangle_{\alpha_\mathrm{H}} = \sum_{k=0}^{\infty} \left\langle \hat{\eta}_\infty^{(k)} \right\rangle \alpha_\mathrm{H} \sum_{k=0}^{\infty} \langle \hat{M}_\infty \hat{Q} \rangle^{(k)} \alpha_\mathrm{H} \tag{9.1.6a}$$

where $\hat{\eta}_\infty^{(k)} = \left(\hat{M}_\infty^{(0)} \hat{Q}\right)^{(k)}$ can take on a number of different average forms. These depend on how the integral Green's function $\left(\hat{M}_\infty^{(0)}\right)$ and the inhomogeneities $(\hat{Q})$ in the medium (V) and on the boundaries (S) are activated. Specifically, we note that, for $k \geq 1$:

Volumes: $\left\langle \hat{\eta}_\infty^{(k)} |_V \right\rangle = \hat{M}_V^{(0)^{(k)}} \left\langle Q_V^{(k)} \right\rangle$: the integral Green's function $\hat{M}_V^{(0)}$ is deterministic as is its kernel, $g_\infty^{(0)}$ while the local inhomogeneities (Q_V) are random in the volume: cf. (8.1.42a), (8.3.9a, 25a). We may usually expect the scatterers to reradiate independently of their orientation in the medium;

Surfaces: $\left\langle \hat{\eta}_\infty^{(k)} |_S \right\rangle = \left\langle \left(\hat{M}_S^{(0)^{(k)}} Q_S\right)^{(k)} \right\rangle$: $q_{00|s}$ remains nonrandom, but the integral surface Green's function $\hat{M}_S^{(0)}$ is stochastic, because of the random nature of the surface here (cf. (8.1.42) and Section 8.1.4, [(8.3.9a) and (8.3.25b)]. In general, we expect that $\hat{M}_S$ and $\hat{Q}_S$ are statistically dependent because the scattering of radiation from the random surface will be affected by the aspect of the surface vis-à-vis the illuminating sources.

(9.1.6b)

Therefore, we must adjust the averages, specifically in Section 9.1.1.4 following.

$\left\langle \alpha_1^{(Q)} \alpha_2^{(Q)} \right\rangle = \langle \alpha(\mathbf{R}_1, t_1) \alpha(\mathbf{R}_2, t_2) \rangle$, and so on. Now, however, we encounter multiple series and multiple order moments of the mass operator $\hat{\eta}_\infty$, with an increasing difficulty of obtaining convergence conditions.

For these reasons, including the comparative complexity of the results of the direct expansion (9.1.6a), we use the second equivalent relation (9.1.3b). This replaces the FOS (and PSS) and the ensuing averages (9.1.6a) by equivalent *deterministic mass operators* $\hat{\eta}_1^{(d)}$ and $\left\langle \hat{\eta}^{(d)} \right\rangle$, defined by the relations for volumes and surfaces: for volumes $\hat{M}_{1|V}$ is nonrandom, for surfaces $\hat{M}_{1|S}$ is random:

$$\left.\begin{array}{l} \therefore\ \hat{\eta}_{1/V}^{(d)} \langle \alpha^{(Q)} \rangle \equiv \langle \hat{\eta}_V \alpha^{(Q)} \rangle = \langle \hat{M}_{1/V} \hat{Q}_{1/V} \alpha^{(Q)} \rangle \\ \therefore\ \hat{\eta}_{1/S}^{(d)} \langle \alpha^{(Q)} \rangle \equiv \langle \hat{\eta}_S \alpha^{(Q)} \rangle = \langle \hat{M}_{1/S} \rangle \hat{Q}_{1/S} \langle \alpha^{(Q)} \rangle \end{array}\right\} \begin{array}{l} ;\ \hat{M}_{1/V} \text{ nonrandom.} \\ ;\ \therefore \hat{\eta}_S = \hat{M}_{1/S} \hat{Q}_{1/S};\ \text{both factors random.} \end{array} \tag{9.1.7a}$$

with $\hat{\eta}_\infty$ subject to the various statistical operations on $\hat{M}_\infty$ and $\hat{Q}$ implied by (9.1.6b) above. We now obtain the mean field $\langle \alpha^{(Q)} \rangle$, using (9.1.7a)

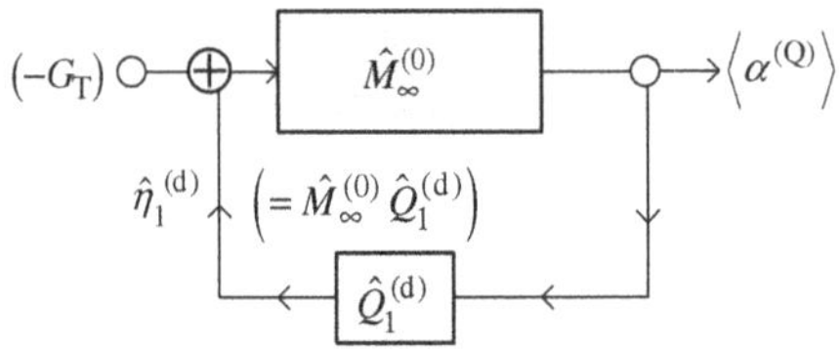

FIGURE 9.4 The FOR$\langle\alpha^{(Q)}\rangle$ for the deterministic integral for volumes Eq. (9.1.7b), where the FOR is given in the Figure. See Eqs. (8.5.10a) and (8.5.11a–d) for the derivations of the FOS.

$$\left\langle\alpha^{(Q)}\right\rangle=\alpha_{\rm H}+\left\langle\hat{\eta}_\infty\alpha^{(Q)}\right\rangle=\alpha_{\rm H}+\left(\hat{\eta}_{1/V}^{(d)}\,\text{or}\left\langle\hat{\eta}^{(d)}\right\rangle_{1/S}\right)\langle\alpha\rangle^{(Q)}=\alpha_{\rm H}+\sum_{n=1}^{\infty}\left\langle\left(\hat{M}^{(0)}\hat{Q}\right)^{(n)}\right\rangle_{V\,\text{or}\,S}\alpha_{\rm H}, \tag{9.1.7b}$$

cf. 8.5.11d, $m\rightarrow\infty$, with the appropriate averages, Eq. (9.1.7a) and convergence conditions for these deterministic results.[5] The second equality in (9.1.7b) is a form of the (first-order) *Dyson equation*, and $\hat{\eta}_1^{(d)}$ is the equivalent deterministic (global) mass operator (DMO). In terms of the PSS, (9.1.7b) becomes alternatively the deterministic PSS

$$\left\langle\alpha^{(Q)}\right\rangle=\alpha_{\rm H}+\sum_{n=1}^{\infty}\left[\hat{\eta}_1^{(d)}\right]^{(n)}\alpha_{\rm H},\ \left\|\hat{\eta}_1^{(d)}\right\|<1 \tag{9.1.7c}$$

equivalent to (9.1.7b), but subject to stricter convergence conditions.[5] In line with the DMO we call the medium supporting the average field $\langle\alpha^{(Q)}\rangle$ the (first-order) equivalent deterministic medium (EDM$_1$). Figure 9.4 shows the FOR (and at the same time the FOS), on replacing the $\hat{Q}$ in Fig. 8.16 by $\hat{Q}_1^{(d)}$. We note that the integral equation in $\hat{\eta}_1^{(d)}$, (9.1.7b), is a *canonical expression*: it applies for all media and surfaces (interfaces) for which the *local* mass operator $\hat{Q}\left(\neq\hat{Q}_1^{(d)}\right)$ is a random function, statistically related or not to the global Green's function $\hat{M}_\infty$, cf. (9.1.6b).

9.1.1.2 The Equivalent Mass Operator for Volumes $\hat{Q}_{1/V}^{(d)}$ When $\hat{Q}_1$ is random (and $\hat{M}_\infty$ is not), it is necessary to obtain $\hat{Q}_1^{(d)}$ to put (9.1.7b) into a form more convenient for calculation and in some cases to obtain a closed analytic form for the solution. We begin with the identity Eq. (9.1.7a) and since we are dealing with volume scatter here, $\hat{M}_\infty^{(0)}$ is not a random function, (9.1.6b), as it is for surfaces. With the definitions

$$\hat{\eta}_\infty\equiv\hat{M}_\infty^{(0)}\hat{Q};\quad \hat{\eta}_1^{(d)}\equiv\hat{M}_\infty^{(0)}\hat{Q}_1^{(d)}, \tag{9.1.8}$$

we obtain directly from (9.1.6a) the following identity

$$\hat{Q}_1^{(d)}\left\langle\left(\hat{1}-\hat{\eta}_\infty\right)^{-1}\right\rangle\equiv\left\langle\hat{Q}\left(\hat{1}-\hat{\eta}_\infty\right)^{-1}\right\rangle, \tag{9.1.8a}$$

$$\hat{Q}_1^{(d)}=\hat{M}_\infty^{-1}\langle x\rangle/(1+\langle x\rangle)=\hat{M}_\infty^{-1}\sum_{m=0}^{\infty}(-1)^m\langle x\rangle^{m+1},\ \text{with}\ \langle x\rangle\equiv\sum_{k=1}^{\infty}\left\langle\hat{\eta}^{(k)}\right\rangle,\ \|\langle x\rangle\|<1. \tag{9.1.8b}$$

[5] See Chapter 8 of Ref. [19]. Note that on a term-by-term basis, Eq. (9.1.7b) and Eq. (9.1.6a) are not equal.

It is clear that since $\hat{\eta}_\infty = \hat{M}_\infty^{(0)}\hat{Q}$ incorporates the projection operators $\hat{M}_\infty^{(0)}, \hat{Q}_1^{(d)}$ (8.1.10), (9.1.10), we can write $\hat{\eta}_\infty = \hat{\eta}_\infty(\mathbf{R}, t|\mathbf{R}', t')$, for both volumes or surfaces, in the form

$$\hat{\eta}_1^{(d)} = \sum_{m=0}^{\infty} \hat{A}_m^{(1)}(\mathbf{R}, t|\mathbf{R}', t') = \int_V \hat{\eta}_1^{(d)}(\mathbf{R}, t|\mathbf{R}', t')_V (\,)_{\mathbf{R}',t'} \mathbf{dR}'dt' \tag{9.1.9a}$$

$$= \int_V \hat{M}_\infty^{(0)}(\mathbf{R}, t|\mathbf{R}_0, t_0)\mathbf{dR}_0 dt_0 \int_V \hat{Q}_1^{(d)}(\mathbf{R}_0, t_0|\mathbf{R}', t')_V (\,)_{\mathbf{R}',t'} \mathbf{dR}'dt'. \tag{9.1.9b}$$

Requiring each term in the PSS (which takes the form (9.1.8a) to be of the same order in the local inhomogeneity operator $\hat{Q}$ results in the series, also expressed as the integral:

$$\hat{Q}_1^{(d)} \equiv \sum_{m=0}^{\infty} \hat{B}_m^{(1)}(\mathbf{R}, t|\mathbf{R}', t') = \int_V \hat{Q}_1^{(d)}(\mathbf{R}, t|\mathbf{R}', t')(\,)_{\mathbf{R}',t'} \mathbf{dR}'dt'. \tag{9.1.10}$$

Equation (9.1.10) shows as we expect, that $\hat{Q}_1^{(d)}$ *is* (except for the initial term $\hat{B}_0^{(1)}$) *a global operator* (since it contains $\hat{M}_\infty^{(0)}$), and equivalently since $\hat{Q}_1^{(d)}$ depends on $\langle x\rangle$(9.1.8b). The components of the sum in (9.1.10) are then found—somewhat laboriously—from (9.1.8b) to be,[6]:

$$\left.\begin{aligned}
\hat{B}_0^{(1)} = \langle Q\rangle;\quad \hat{B}_1^{(1)} &= \langle \hat{Q}\hat{M}\hat{Q}\rangle - \langle\hat{Q}\rangle\hat{M}\langle\hat{Q}\rangle;\\
\hat{B}_2^{(1)} &= \left\langle \hat{Q}(\hat{M}\hat{Q})^{(2)}\right\rangle - 2\langle\hat{Q}\rangle\left\langle (MQ)^{(2)}\right\rangle + \langle\hat{Q}\rangle\langle M\hat{Q}\rangle^2;\\
\hat{B}_3^{(1)} &= \left\langle \hat{Q}(\hat{M}\hat{Q})^{(3)}\right\rangle - \langle\hat{Q}\rangle\left(\hat{M}\langle\hat{Q}\rangle\right)^3 + 4\langle\hat{Q}\rangle\hat{M}\langle\hat{Q}\rangle\langle\hat{Q}\hat{M}\hat{Q}\rangle;\\
&\quad -\langle\hat{Q}\hat{M}\hat{Q}\rangle^2 - 2\langle\hat{Q}\rangle\left\langle(\hat{M}\hat{Q})^3\right\rangle;\ \text{etc.}
\end{aligned}\right\} \tag{9.1.11}$$

Similarly, the expression determining $\hat{\eta}_1^{(d)}$ in terms of the global mass operator $\hat{\eta}_\infty$ and its averages is obtained here by multiplying [(9.1.8a) and (9.1.8b)] by $\hat{M}_\infty$ and using the series (9.1.8b) with (9.1.9) to derive the $\hat{A}_m^{(1)}$ operators. Thus, now one has

$$\hat{\eta}_1^{(d)} = \langle x\rangle/(1+\langle x\rangle) = \sum_{m=0}^{\infty}(-1)^m\langle x\rangle^{m+1},\ 0 \le \|\langle x\rangle\| < 1, \tag{9.1.12}$$

and

$$\begin{aligned}
\therefore\ A_0^{(1)} = \langle\hat{\eta}_\infty\rangle;\ A_1^{(1)} &= \left\langle\hat{\eta}_\infty^{(2)}\right\rangle - \langle\hat{\eta}_\infty\rangle^2;\\
A_2^{(1)} &= \left\langle\hat{\eta}_\infty^{(3)}\right\rangle - 2\langle\hat{\eta}_\infty\rangle\left\langle\hat{\eta}_\infty^{(3)}\right\rangle + \langle\hat{\eta}_\infty\rangle^3;\\
A_3^{(1)} &= \left\langle\hat{\eta}_\infty^{(4)}\right\rangle - \langle\hat{\eta}_\infty\rangle^4 + 4\langle\hat{\eta}_\infty\rangle^2\left\langle\hat{\eta}_\infty^{(2)}\right\rangle - \left\langle\hat{\eta}_\infty^{(2)}\right\rangle^2 - 2\langle\hat{\eta}_\infty\rangle\langle\eta^{(3)}\rangle;\ \text{etc.}
\end{aligned} \tag{9.1.13}$$

[6] We may drop from time to time the subscripts (∞) on $\hat{M}_\infty^{(0)}$ and the superscripts (^{0}), as well, to reduce the complexity of these symbols.

As we have noted above (Eq. (9.1.8b) et. seq.) the relations (9.1.8)–(9.1.13) apply for scattering from both volumes or surfaces.

9.1.1.3 Vanishing Means $\langle Q\rangle = 0$ and Symmetrical pdfs of $\hat{Q}$ In the important special cases when the mean scattering operator vanishes, that is, $\langle Q\rangle = 0$, the results for $Q_1^{(d)}$ simplify considerably. We see directly from Eqs. (9.1.11) and (9.1.13) that

$$\left.\begin{aligned}
&\hat{B}_0^{(1)} = 0; \hat{B}_1^{(1)} = \langle \hat{Q}\hat{M}\hat{Q}\rangle; \quad \hat{B}_2^{(1)} = \left\langle \hat{Q}(\hat{M}\hat{Q})^{(2)}\right\rangle; \hat{B}_3^{(1)} = \left\langle Q(MQ)^{(3)}\right\rangle - \langle \hat{Q}\hat{M}\hat{Q}\rangle^2;\\
&\hat{B}_4^{(1)} = \left\langle Q(MQ)^{(4)}\right\rangle - 2\langle QMQ\rangle\left\langle Q(MQ)^{(2)}\right\rangle;\\
&\hat{B}_5^{(1)} = \left\langle Q(MQ)^{(5)}\right\rangle + 2\langle QMQ\rangle\left\langle (MQ)^{(4)}\right\rangle - \left\langle Q(MQ)^{(2)}\right\rangle^2 + \langle Q\hat{M}\hat{Q}\rangle^3; \text{ etc.}
\end{aligned}\right\} \tag{9.1.14}$$

In terms of the global operator $\hat{\eta}_\infty$, where $\langle \hat{\eta}_\infty\rangle = 0$, we obtain

$$\left.\begin{aligned}
&\hat{A}_0^{(1)} = 0; A_1^{(1)} = \hat{\eta}_\infty^{(2)}; \quad \hat{A}_2^{(1)} = \left\langle \hat{\eta}_\infty^{(3)}\right\rangle; \hat{A}_3^{(1)} = \left\langle \hat{\eta}_\infty^{(4)}\right\rangle - \left\langle \eta_\infty^{(2)}\right\rangle^2;\\
&\hat{A}_4^{(1)} = \left\langle \hat{\eta}_\infty^{(5)}\right\rangle - 2\left\langle \eta_\infty^{(2)}\right\rangle\left\langle \hat{\eta}_\infty^{(3)}\right\rangle;\\
&\hat{A}_5^{(4)} = \left\langle \eta_\infty^{(6)}\right\rangle + 2\left\langle \eta_\infty^{(2)}\right\rangle\left\langle \eta_\infty^{(4)}\right\rangle - \left\langle \hat{\eta}_\infty^{(3)}\right\rangle^2 + \langle \eta^{(2)}\rangle^3; \text{ etc.}
\end{aligned}\right\} \tag{9.1.14a}$$

A further simplification occurs when $\hat{Q}$ and $\therefore \hat{\eta}_\infty$ have symmetrical pdfs about $\langle \hat{Q}\rangle = 0$, that is, $\langle Q^{2m+1}\rangle = 0$. Then [(9.1.14) and (9.1.14a)] reduce further to

$$\left.\begin{aligned}
&\hat{B}_0^{(1)} = 0; \hat{B}_1^{(1)} = \langle \hat{Q}\hat{M}\hat{Q}\rangle; \quad \hat{B}_2^{(1)} = 0; \hat{B}_3^{(1)} = \left\langle \hat{Q}(\hat{M}\hat{Q})^{(3)}\right\rangle - \langle \hat{Q}\hat{M}\hat{Q}\rangle^2;\\
&\hat{B}_4^{(1)} = 0; \hat{B}_5^{(1)} = \left\langle \hat{Q}(\hat{M}\hat{Q})^{(5)}\right\rangle + 2\langle \hat{Q}\hat{M}\hat{Q}\rangle\left\langle (\hat{M}\hat{Q})^{(4)}\right\rangle;\\
&\quad\vdots \qquad \vdots \qquad -\left\langle \hat{Q}(\hat{M}\hat{Q})^{(2)}\right\rangle^2 + \langle \hat{Q}\hat{M}\hat{Q}\rangle^3; \text{ etc.,}\\
&\text{which suggests:} \quad \hat{B}_{2m}^{(1)} = 0; \hat{B}_{2m+1}^{(1)} \neq 0; = \left\langle \hat{Q}(\hat{M}\hat{Q})^{2m+1}\right\rangle + \left\langle 0\left(\hat{Q}^{2m+2}\right)\right\rangle, m \geq 3.
\end{aligned}\right\} \tag{9.1.15a}$$

The corresponding development for $\hat{\eta}_d^{(1)}$ is

$$\left.\begin{aligned}
&\hat{A}_0^{(1)} = 0; \hat{A}_1^{(1)} = \left\langle \hat{\eta}^{(2)}\right\rangle \quad \hat{A}_2^{(1)} = 0; \hat{A}_3^{(1)} = \left\langle \hat{\eta}^{(4)}\right\rangle - \left\langle \hat{\eta}^{(2)}\right\rangle^2;\\
&\hat{A}_4^{(1)} = 0; \hat{A}_5^{(1)} = \left\langle \hat{\eta}^{(6)}\right\rangle + 2\left\langle \hat{\eta}^{(2)}\right\rangle\left\langle \hat{\eta}^{(1)}\right\rangle + \left\langle \hat{\eta}^{(2)}\right\rangle^3;\\
&\quad\vdots \qquad \vdots\\
&\hat{A}_{2m}^{(1)} = 0; \hat{A}_{2m+1}^{(1)} \neq 0; = \left\langle \hat{\eta}^{(2m+2)}\right\rangle + 0\left\langle \hat{\eta}^{(2m)}\right\rangle, \left\langle \hat{\eta}^{(2m-2)}\right\rangle, \ldots, \left\langle \hat{\eta}^{(2)}\right\rangle;\\
&\qquad \text{in products } \left\langle 0\left(\hat{\eta}^{2m+2}\right)\right\rangle.
\end{aligned}\right\} \tag{9.1.15b}$$

In most instances where the PSS is used, $\|\langle\hat{\eta}_\infty\rangle\|$ is sufficiently small vis-à-vis unity on physical grounds, that is, the series converge reasonably quickly.[7]

For volumes we have for the average global, from (8.1.10)

$$\hat{M}_\infty^{(0)}(\mathbf{R},t|\mathbf{R}',t')_V = \int_V g^{(0)}(\mathbf{R},t|\mathbf{R}_0,t_0)_V(\,)_{\mathbf{R}_0,t_0}\mathbf{dR}_0 dt_0, \tag{9.1.16}$$

where $\hat{M}_\infty^{(0)}$ is of course nonrandom.

9.1.1.4 Structure of $\hat{\eta}_1^{(d)}$ for Surface Scatter From the general expressions [(9.1.9) and (9.1.10)] we need now to specialize the results above (9.1.11)–(9.1.15b) to account for the different possible statistical situations represented by (9.1.6b) and the different physical situations encountered by scatterers in the volume and on a surface.

However, when we are dealing with a *random* Green's function, where the integral operator $\hat{M}_\infty^{(0)}|_S$ that occurs in the case of scatter from random surfaces (cf. Section 8.3.4 extended to the ensemble of random surfaces), we have a more complex picture. The random global Green's function operator here is defined now by

$$\hat{M}_\infty^{(0)}\left(\mathbf{R},t|\mathbf{R}_0,t_0\right)_S \equiv \int dt_0 \oint_S \left\{g_\infty^{(0)} J(\mathbf{r}_0,t_0)_\zeta - \left[J(\mathbf{r}_0,t_0)_\zeta g_\infty^{(0)}\right]\right\}(\,)_{\mathbf{r}_0,t_0}\mathbf{dr}_0. \tag{9.1.17}$$

In more detail, the elements of (9.1.17) are (Section 8.1.5.2)

$$\left.\begin{aligned} & g_\infty^{(0)} = g_\infty^{(0)}\left(\mathbf{R},t\middle|\mathbf{r}_0,t_0\right);\ \left[\frac{\partial}{\partial n}(\,)\right] ds = [\hat{\mathbf{n}}\bullet\nabla(\,)]\,dS(\mathbf{r}_0,t_0) = J(\mathbf{r}_0,t_0)\mathbf{dr}_0;\mathbf{dr}_0 \equiv dx_0 dy_0 \\ & \text{with} \\ & \left\{\begin{aligned} & \hat{\mathbf{n}}\cdot\nabla \equiv J_0(\mathbf{r}_0,t_0)_\zeta = \left(\zeta_x\frac{\partial}{\partial x_0}+\zeta_y\frac{\partial}{\partial y_0}+\frac{\partial}{\partial z_0}\right)\Big/\left(1+\zeta_x^2+\zeta_y^2\right)^{1/2}; dS(\mathbf{r}_0,t_0) = \mathbf{dr}_0\sqrt{1+\zeta_x^2+\zeta_y^2}; \\ & \zeta = \zeta(\mathbf{r}_0,t_0) = \hat{\mathbf{i}}_z\zeta(\mathbf{r}_0,t_0);\zeta_x\frac{\partial\zeta}{\partial x_0};\zeta_y\frac{\partial\zeta}{\partial y_0};\mathbf{r}_0 = \hat{\mathbf{i}}_x x_0 + \hat{\mathbf{i}}_y y_0 \end{aligned}\right. \end{aligned}\right\}, \tag{9.1.17a}$$

as shown in Fig. 9.5. Equations (9.1.17) and (9.1.17a) represent the integral random Green's function or projection operator, for any (as yet unspecified) sources on the random surface $S(\mathbf{r}_0,t_0)$, radiating into the space outside this surface.

For inhomogeneities distributed on S that can scatter the incident radiation, or sources therein on, we obtain the desired mass operator $\hat{\eta}_\infty$ by combining $\hat{M}_S^{(0)}$ and $\hat{Q}_S$ in the usual way, to obtain from (9.1.17)

$$\hat{\eta}_{1/S} = \int dt_0 \oint_S \left\{g_\infty^{(0)} J - \left[J g^{(0)}\right] Q_S\right\}(\,)_{\mathbf{r}_0,t_0}\mathbf{dr}_0, \tag{9.1.18a}$$

[7] See Chapter 8.

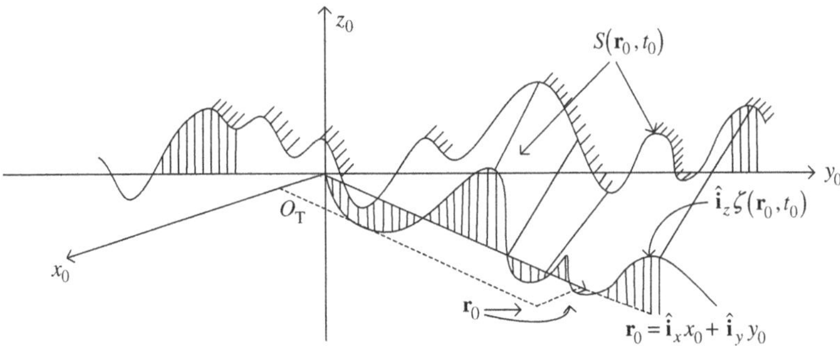

FIGURE 9.5 Surface elevation $\hat{\mathbf{i}}_z\zeta(\mathbf{r}_0, t_0)$ along $\mathbf{r}_0$ in the x_0y_0-plane.

which in more detail is

$$\hat{\eta}_{1/S} = \int dt_0 \oint_S \left\{ g_\infty^{(0)}(\mathbf{R}, t|\mathbf{r}_0, t_0) J(\mathbf{r}_0, t_0)_\zeta - \left[J(\mathbf{r}_0, t_0)_\zeta g_\infty^{(0)}(\mathbf{R}, t|\mathbf{r}_0, t_0) \right] \right\} Q(\mathbf{r}_0, t_0)_\zeta (\,)_{\mathbf{r}_0, t_0} \mathbf{dr}_0. \tag{9.1.18b}$$

Equations (9.1.18a) and (9.1.18b) apply for each member of the ensemble of random surfaces S_0 (in $\hat{M}_S$) and the local random inhomogeneities $Q(\mathbf{r}_0, t)$ on these surfaces. Thus, both J and Q are random while the free-space Green's function $g_\infty^{(0)}$ is not.

We are now ready to consider the average field $\langle \alpha^{(Q)} \rangle$, appropriate to the Dyson equation (9.1.7b), for surface scatter. We distinguish two cases, (I) where the random surface S and the scattering inhomogeneities on it are statistically independent and (II) in which they are statistically related. First, we note that here from (9.1.14), (9.1.18b) both $\hat{M}_{\infty|S}^{(0)}$ and $\hat{Q}_S$ are random, and therefore $\hat{\eta}_S$ is also. We begin with the surface counterpart to (9.1.16), using (9.1.9b), to obtain, with the help of (9.1.18b)

$$\hat{\eta}_{1/S}^{(d)} = \int dt_0 \oint_S \mathbf{dr}_0 \left\{ g_\infty^{(0)} J - \left[J g_\infty^{(0)} \right] \int dt' \oint_S Q_{1/S}^{(d)} \right\} (\,)_{\mathbf{r}', t'} \mathbf{dr}', \tag{9.1.19a}$$

or in more detail,

$$\begin{aligned} \hat{\eta}_{1/S}^{(d)} &= \int dt_0 \oint_S \mathbf{dr}_0 \left\{ g_\infty^{(0)}(\mathbf{R}, t|\mathbf{r}_0, t_0) J(\mathbf{r}_0, t_0) \int dt' \oint_S Q_{1/S}^{(d)}(\mathbf{r}_0, t_0|\mathbf{r}', t') (\,)_{\mathbf{r}', t'} \right. \\ &\quad \left. - \left[J_0(\mathbf{r}_0, t_0) g_\infty^{(0)}(\mathbf{R}, t|\mathbf{r}_0, t_0) \right] \int dt' \oint_S Q_{1/S}^{(d)}(\mathbf{r}_0, t_0|\mathbf{r}', t') \right\} (\,)_{\mathbf{r}', t'} \mathbf{dr}', \end{aligned} \tag{9.1.19b}$$

with $\mathbf{dr}' = dx'dy'$ and $r' = \hat{\mathbf{i}}_x x' + \hat{\mathbf{i}}_y y'$(9.1.17a). Here the average, that is the deterministic operation $\eta_{1/S}^{(d)}$ is obtained from [(9.1.9a) and (9.1.9b)], which is explicitly represented by the various series (9.1.11), (9.1.14), (9.1.15) and implicitly by (9.1.13a), with $\hat{\eta}_1^{(d)} \to \hat{\eta}_{1/S}^{(d)}$. For Cases I and II we see finally that

$$\textit{Case I: } \hat{\eta}_{1/S}^{(d)} \to \left\langle \hat{\eta}_{1/S}^{(d)} \right\rangle_{\zeta/Q} \quad \text{or} \quad \textit{Case II: } \hat{\eta}_{1/S}^{(d)} \to \left\langle \hat{\eta}_{1/S}^{(d)} \right\rangle_{(\zeta_1, Q_S)}, \tag{9.1.20}$$

where ζ/Q indicate the average over the elevation ζ, independent of the average over the $\hat{Q}_S$ comprising $\hat{Q}_{1/S}^{(d)}$. In a similar fashion, (ζ, Q) denotes the joint average over ζ and Q_S. In Case I, one replaces J by $\langle J\rangle_\zeta$. For Case II, the average is taken over J and $\hat{Q}_{1,S}^{(d)}$ jointly, in (9.1.19b). Here Q_S usually represents the plane-wave reflection and/or transmission coefficient at a point, $\mathsf{R}_0(\mathbf{r}_0, t_0), \mathsf{T}_0(\mathbf{r}_0, t_0)$, Section 8.3.4. Accordingly, with (9.1.11) and (9.1.19), averaged as indicated in (9.1.20), we are formally equipped to evaluate *Dyson's equation*, namely the (deterministic) integral equation[8] (9.1.7b)

$$\textit{Dyson's equation}: \left\langle \alpha^{(Q)}(\mathbf{R}, t)\right\rangle_{V \text{ or } S} = \alpha_{\mathrm{H}}(\mathbf{R}, t) + \left\langle \hat{\eta}_1^{(d)}\right\rangle_{V \text{ or } S}\left\langle \alpha^{(Q)}(\mathbf{R}, t)\right\rangle_{V \text{ or } S} \tag{9.1.21}$$

for random scattering in volumes (V), from surface (S), or combinations of both, as outlined in Section 8.4 in the deterministic cases of a given member of the ensemble postulated there. As before, (9.1.21) represents a *feedback* or *FOR system*, which is now expressed in a modified version of Fig. 9.4,[9] namely, Fig. 9.6, representing the more general situation where $\hat{M}_S^{(0)}$ and $\hat{Q}_S$ are statistically related.

Although the deterministic nature of the *equivalent deterministic medium* is implicit in the mass operators $\hat{\eta}_{1/V}^{(d)}$ (9.1.16), $\left\langle \hat{\eta}_{1/S}^{(d)}\right\rangle$, cf. (9.1.20), its suggested exploitation by the FOS formulation above (9.1.21), as indicated by the FOR of Figs. 9.2 and 9.4 is potentially new, at least in the context of the computer-aided evaluations recommended here. The reduction of the ensemble to a deterministic form, cf. (9.1.21) above, allows us to employ all the technical aperture outlined in Chapter 89 preceding, for dealing with the classical nonrandom inhomogeneous medium [2]. However, there is a price to pay: in going from the ensemble $\{\alpha^{(Q)}\}$ to the deterministic result $\langle\alpha^{(Q)}\rangle$ we have destroyed most of the statistical information inherent in the ensemble representing the scattered field. Here (in the first-order case) all we have remaining is the average field $\langle\alpha^{(Q)}\rangle$ (9.1.21). Expressed alternatively, given $\langle\alpha^{(Q)}\rangle$ we cannot reconstruct the ensemble $\{\alpha^{(Q)}\}$. Equivalently, we cannot infer $\hat{Q}$ from the measure $\hat{Q}^{(d)}$, or more generally, the mass operator $\hat{\eta}_\infty$ from $\left\langle \hat{\eta}_{1|V \text{ or } S}^{(d)}\right\rangle$. In particular, when $\alpha_{\mathrm{H}}, \hat{M}_{V,S}^{(0)}$ (Eq. 9.1.17, are known, we can estimate $\hat{Q}_{1|V \text{ or } S}^{(d)}$, in principle, to any degree of accuracy in a controlled way, with a determinable error, by the method (or its extensions to measure $\left\langle \hat{\eta}_{1|V \text{ or } S}^{(d)}\right\rangle$ directly, knowing α_{H} and $\langle\alpha^{(Q)}\rangle$ of Section 8.5.3.

9.1.2 Dyson's Equation in Statistically Homogeneous and Stationary Media

Next, we illustrate how the integral equations (9.1.5b), (9.1.7b) may be solved analytically, without recourse to the perturbation method (i.e., the PSS of (9.1.7c), for example). We begin with the generic case of an infinite volume V of continuous inhomogeneities (external to the distribution (source $-G_T$), and we neglect the backscatter (Section 8.3.4.2). Accordingly, let

[8] Depending on the boundary and initial conditions this may be an integro-differential equation. See Chapter 9 of Ref. [20].

[9] The loop relations for this FOR are $\alpha_1 = \alpha_{\mathrm{H}} + \alpha_{\mathrm{F}}$; $\alpha_2 = \langle\alpha\rangle$; $\alpha_{\mathrm{F}} = \langle\hat{\eta}\rangle\alpha_2 = \langle\hat{\eta}\rangle\alpha$, which becomes $\langle\alpha\rangle = \alpha_{\mathrm{H}} + \langle\hat{\eta}\rangle\langle\alpha\rangle$, first iteration, which upon repetition leads to $\alpha^{(Q)} = \alpha_{\mathrm{H}} + \langle\hat{\eta}\rangle\alpha^{(Q)}, Q \to \infty$, cf. (8.5.11) for details.

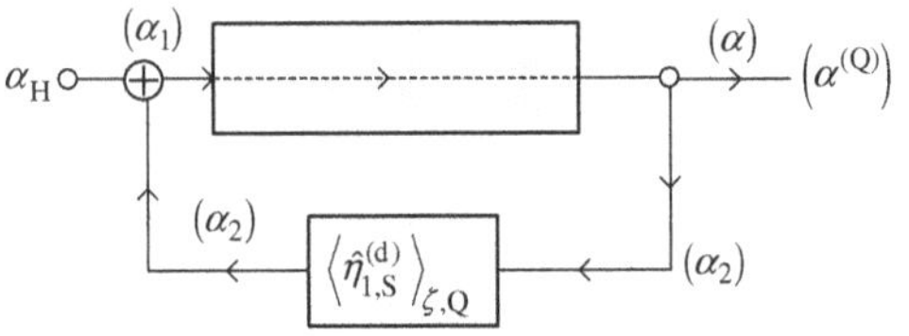

FIGURE 9.6 The FOR of $\langle \alpha^{(Q)} \rangle$ for Dyson's equation (9.1.21), with the averaged integral mass operator of Eq. (9.1.20), for *surface scatter*, cf. Fig. 9.4.

us write Dyson's equation (9.1.7a) for this situation, with the help of (9.1.9a), in the more explicit form

$$\left\langle \alpha^{(Q)}(\mathbf{R},t)_V \right\rangle = \left[\alpha_H(\mathbf{R},t) + \vdots \iint_{V_T} dt' \mathbf{dR}' g_\infty^{(0)}(\mathbf{R}',t' | \mathbf{R}_0, t_0)_V \iint_V Q_1^{(d)}(\mathbf{R}',t' | \mathbf{R}_0,t_0) \vdots \left\langle \alpha^{(Q)}(\mathbf{R}_0,t_0) \right\rangle \mathbf{dR}_0 dt_0 \right]_V$$

$$\vdots \leftarrow \quad \hat{M}_\infty|_V \quad \rightarrow \ \vdots \leftarrow \quad Q_1 \quad \rightarrow \ \vdots \leftarrow \ \left\langle \alpha^{(Q)} \right\rangle \ \rightarrow \qquad (9.1.22)$$

Here $Q_1^{(d)}$ (or equivalently $g_\infty^{(0)} Q_1^{(d)} \equiv \eta_1^{(d)}$, cf. (9.1.8)) is the *kernel* associated with the mass operator for the average field $\langle \alpha^{(Q)} \rangle$. It is, of course, a deterministic quantity, a statistic of the random field[10] $\alpha^{(Q)}$ in V.

9.1.2.1 Dyson's Equation: Hom-Stat Media—Volume Scatter In this case, we can obtain closed-form solutions for Dyson's equations [(9.1.7b) and (9.1.21)]. Let us consider the particular example of (9.1.21) above where V is the infinite volume containing scattering elements and let the ensemble of individual representations, that is, the Langevin equation (9.1.3), be Hom-Stat. This in turn means that the kernels $Q_1^{(d)}$ and $g_\infty^{(0)}$ in (9.1.9) and (9.1.10) individually exhibit Hom-Stat properties. Accordingly, $Q_1^{(d)}$ *and* $g_\infty^{(0)}$ *depend only on the differences* $(\mathbf{R}'-\mathbf{R}_0,\ t'-t_0)$ so that

$$\left. \begin{aligned} Q_{1,V}^{(d)}|_{Hom\text{-}Stat} &= Q_1^{(d)}(\mathbf{R}'-\mathbf{R}_0, t'-t_0)_V = \int_V \mathbf{dR}' \int_{t_0}^{t'} q_1^{(d)}(\mathbf{R}'-\mathbf{R}_0, t'-t_0)_V dt' \\ &= \mathsf{F}_k \mathsf{F}_s \left\{ g_{00,1}^{(d)}|_V \right\} = \int_{Br_1} \frac{ds}{2\pi i} \int_{-\infty}^{\infty} \frac{\mathbf{dk}}{(2\pi)^3} g_{00,1}^{(d)}(\mathbf{k},s)_V e^{-i\mathbf{k}\cdot(\mathbf{R}'-\mathbf{R}_0)+s(t'-t_0)} \end{aligned} \right\}. \qquad (9.1.23)$$

Here the double Fourier transform[11] is

$$\left. \begin{aligned} g_{00,1}^{(d)}|_V \equiv \mathsf{F}_R \mathsf{F}_t \left\{ Q_{1,V}^{(d)} \right\}, \text{ with } Q_1^{(d)}(\mathbf{R}'-\mathbf{R}_0, t'-t_0)_V = Q_{1|V}^{(d)}, \text{ with } \mathbf{R}\rightarrow\mathbf{R}', \mathbf{R}'\rightarrow\mathbf{R}_0 \\ t\rightarrow t', t'\rightarrow t_0 \end{aligned} \right\} \qquad (9.1.23a)$$

[10] We remark that (9.1.22) is a generalized version of the result (4.15), (4.28), and so on, in RKT [7]. It is a generalization to an arbitrary field of a distributed source and the (possibly) non-Gaussian statistics, vis-à-vis a point source in a medium with a Gaussian index of refraction.

[11] For the Hom-Stat cases the Fourier transforms F_k, F_s are with respect to the *differences* $\rho = \mathbf{R}_0 - \mathbf{R}'$, $\Delta t = t_0 - t'$.

in the above. For the Dyson equations (9.1.7b) and (9.1.22), the Hom-Stat assumption represents the necessary and sufficient condition for the factorability of $Q_{1|V}^{(d)}$ and $g_{\infty}^{(0)}|_V$ and consequently the reduction of (9.1.22) to a closed analytic solution. Similar remarks apply for the (deterministic) global mass operator $\hat{\eta}_{1|V}^{(d)}$, (9.1.6b) here. Accordingly, we can write (9.1.9)[12]

$$\left.\begin{aligned}\hat{\eta}_1^{(d)}\left(\mathbf{R}',t\middle|\mathbf{R}_0,t_0\right)_V &= \int_V dR_0 \int_{t_0^{(-)}}^{t} \eta_1^{(d)}\left(\mathbf{R}'-\mathbf{R}_0,t'-t_0\right)_V dt_0 \\ &= \int_{Br_1} e^{s\left(t'-t_0\right)}\frac{ds}{2\pi i}\int_{-\infty}^{\infty}\frac{\mathbf{dk}}{(2\pi)^3}\eta_{00,1}^{(d)}(\mathbf{k},s)_V e^{-i\mathbf{k}\cdot(\mathbf{R}'-\mathbf{R}_0)}\end{aligned}\right\} \tag{9.1.24}$$

Here the kernel $\eta_{1|V}^{(d)} = \left[g_{\infty}^{(0)} Q_1^{(d)}\right]_V$, where $g_{\infty}^{(0)}$ is deterministic. However, in the more general situation where $g_{\infty}^{(0)}$ is a statistical quantity, as in the case of sources on a random surface (Section 9.1.1.4), either primary or generated by the source α_{H} and/or scattering sources in the volume outside of V_T, the kernel $\eta_{1/S}^{(d)}$ is determined by the additional average in (9.1.20), cf. Section 9.1.3 ff.

We next use the space–time transforms of α_{H}, $\left\langle \alpha_V^{(Q)}\right\rangle$, and $\left[\hat{M}_{\infty}^{(0)}\hat{Q}_1^{(d)}\right]_V\left(=\hat{H}_{1/V}^{(d)}\right)$:

$$\alpha_H(\mathbf{R},t) = \int_{Br_1} e^{st}\frac{ds}{2\pi i}\int_{-\infty}^{\infty} \mathrm{Y}_0(\mathbf{k},s)G_{00}(\mathbf{k},s)_T e^{-i\mathbf{k}\cdot\mathbf{R}}\frac{\mathbf{dk}}{(2\pi)^3}, \tag{9.1.25a}$$

$$\left\langle \rho\alpha^{(Q)}(\mathbf{R},t)_V\right\rangle = \int_{Br_1} e^{st}\frac{ds}{2\pi i}\int_{-\infty}^{\infty}\left\langle \alpha_{00}^{(Q)}(\mathbf{k},s)_V\right\rangle e^{-i\mathbf{k}\cdot\mathbf{R}}\frac{\mathbf{dk}}{(2\pi)^3}, \tag{9.1.25b}$$

and

$$\left[\hat{M}_{\infty}^{(0)}\hat{Q}_1^{(d)}\left\langle \alpha^{(Q)}\right\rangle\right]_V = \int_{Br_1} e^{st}\frac{ds}{2\pi i}\int_{-\infty}^{\infty} \mathrm{Y}_0(\mathbf{k},s)g_{00,1}^{(d)}(\mathbf{k},s)_V\left\langle \alpha_{00}^{(Q)}(\mathbf{k},s)_V\right\rangle e^{-i\mathbf{k}\cdot\mathbf{R}}\frac{\mathbf{dk}}{(2\pi)^3}, \tag{9.1.25c}$$

From these we obtain the desired closed-form result for Dyson's equation (9.1.22)[13] here which is a result of the postulated Hom-Stat character of the medium, namely,

$$\left\langle \alpha^{(Q)}(\mathbf{R},t)_V\right\rangle = \int_{Br_1} e^{st}\frac{ds}{2\pi i}\int_{-\infty}^{\infty}\left(\frac{\mathrm{Y}_0 G_{00,T}}{1-\mathrm{Y}_0 g_{00,1}^{(Q)}}\right)_{(\mathbf{k},s),V} e^{-i\mathbf{k}\cdot\mathbf{R}}\frac{\mathbf{dk}}{(2\pi)^3}, \tag{9.1.26}$$

[12] For these Hom-Stat cases the Fourier transforms F_k, F_s are with respect to the *differences* $\rho = \mathbf{R}-\mathbf{R}_0$, $\Delta t = t-t_0$.

[13] This is the extension of Eq. (4.36), p. 132, Vol. 4, [7] to arbitrary signal inputs, including time, as well as space, general Green's functions, as well as the array or aperture.

We can also express this last more compactly, since the kernel $\eta_{00}^{(d)}, (\mathbf{k}, s)$(9.1.24) is the (double) Fourier transform of $\left(M^{(0)}Q_1^{(d)}\right)_V$, namely,

$$\left\langle \alpha^{(Q)}(\mathbf{R}, t)_V \right\rangle = \int_{Br_1} e^{st} \frac{ds}{2\pi i} \int_{-\infty}^{\infty} \left(\frac{\mathsf{Y}_0 G_{00,\mathrm{T}}}{1-\eta_{00,1}^{(d)}} \right)_{(\mathbf{k},s)|V} e^{-i\mathbf{k}\cdot\mathbf{R}} \frac{d\mathbf{k}}{(2\pi)^3}, \qquad (9.1.26a)$$

where specifically we have

$$\left.\begin{aligned} \eta_{00,1}^{(d)}(\mathbf{k}, s)_V &\equiv \mathsf{Y}_0(\mathbf{k}, s)_\infty q_\infty^{(d)}, (\mathbf{k}, s)_V, \text{ in which } \mathsf{Y}_0(\mathbf{k}, s)_\infty \\ &= -L_0^{-1} = \mathsf{F}_\mathbf{R}\mathsf{F}_\mathrm{T}\{\mathbf{G}\}, \text{ Eqs. (8.1.17a) and (8.1.17b)} \\ q_\infty^{(d)} &= \text{Eq. (9.1.23) and (9.1.23a).} \end{aligned}\right\} \qquad (9.1.27)$$

and by definition, cf. Eq. (8.1.50):

$$\begin{aligned} G_{00,\mathrm{T}}(\mathbf{k}, s)_V &= \mathsf{F}_\mathrm{R}\mathsf{F}_t\{G_\mathrm{T}(\mathbf{R}, t)_V\} \\ &= \int_{-\infty}^{\infty} d\mathbf{R} \int_{-\infty}^{\infty} dt \int_{-\infty}^{\infty} Y_\mathrm{T}(R_1, s/2\pi i) S_\mathrm{in}(R, s/2\pi i)_\Delta e^{i\mathbf{k}\bullet\mathbf{R}-st} \frac{ds}{2\pi i}. \end{aligned} \qquad (9.1.27a)$$

Here $S_\mathrm{in}(\mathbf{R}, t)_\Delta$ is the truncated (i.e., finite duration) input signal [cf. Eq. (8.1.50)], which vanishes outside the space–time interval Δ. The quantity Y_0 (cf. 8.1.17a) is the (double) Fourier transform of the Green's function, with $G_{00,\mathrm{T}}$ a similar transform of the source density $G_\mathrm{T}(\mathbf{R}, t)$, which contains the aperture or array structure of the distributed sources in V_T.

The integral cf. [(9.1.26) and (9.1.26a)] is the *wave number–frequency(WNF) amplitude spectrum of the mean scattered field.* The denominator, $1-\eta_{00,1}^{(d)}$, (9.1.27), is the field renormalization factor, proportional to the local mass operator kernel $Q_1^{(d)}|_V$ for the volume of scatterers, illustrated in Fig. 9.7. The quantity $Q_1^{(d)}$ depends on the density of scatterers and the other characteristics of the scattering medium. In most applications $Q_1^{(d)}$ is known only approximately, but can in principle be found by suitable application of the methods outlined in Section 8.5.3, where now the deterministic local mass operator Q is replaced by $Q_1^{(d)}|_V$. Again, this involves a four-dimensional computational model. We observe (from Section (8.1.7)) that since the aperture or array is embodied in the original source density $G(\mathbf{R}, t)$, we can further exhibit its role explicitly in specifying the injected signal in space, as well as time, by using the relations (8.1.53) or (8.1.56). For example, in the case of arrays, which are here characterized by discrete "point"-elements, we see that the result (9.1.27a) becomes for input signals of finite duration Δ:

$$G_{00,\mathrm{T}}(\mathbf{k}, s)|_\mathrm{array} = \sum_{m=1}^{M} e^{i\mathbf{k}\cdot\xi_m} Y_\mathrm{T}^{(m)}(\xi_m, s/2\pi i) S_\mathrm{in}(\xi_m, s/2\pi i)_\Delta (1-e^{-sT}), \qquad (9.1.28a)$$

and for apertures

$$G_{00,\mathrm{T}}(\mathbf{k}, s)|_\mathrm{aperture} = \int_{-\infty}^{\infty} e^{i\mathbf{k}\cdot\mathbf{R}} Y_\mathrm{T}^{(m)}(\mathbf{R}, s/2\pi i) S_\mathrm{in}(\mathbf{R}, s/2\pi i)_\Delta (1-e^{-sT}) d\mathbf{R}. \qquad (9.1.28b)$$

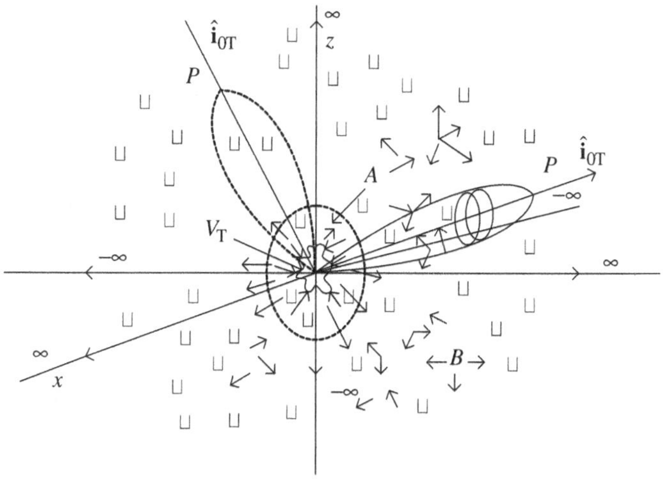

FIGURE 9.7 Scattering by inhomogeneities in an infinite medium excited by a source in V_T, with a moveable beam pattern $P\left[\sim \sum_m(\,)_m\right]$. B is the region of significant scatter, with negligible backscatter (A).

From Section 2.5.2, and more particularly Section 8.1.7 we now obtain for the Dyson equation [(9.1.26) and (9.1.26a)] a result that contains the explicit array structure as well as the finite duration of the input signal:

$$\begin{aligned}\left\langle \alpha^{(Q)}(\mathbf{R},t)_V\right\rangle &= \int_{Br_1} e^{st}\left(1-e^{-sT}\right)\frac{ds}{2\pi i}\int_{-\infty}^{\infty}\left(\frac{\mathsf{Y}_0(\mathbf{k},s)_\infty\,\mathbf{dk}/(2\pi)^3}{1-\eta_{00,1}^{(d)}(\mathbf{k},s)}\right)_V \\ &\quad\cdot\, e^{-i\mathbf{k}\cdot\mathbf{R}}\sum_{m=1}^{M} e^{-i\mathbf{k}\cdot\xi_m} Y_{\mathrm{T}}^{(m)}(\xi_m, s/2\pi i)S_{\mathrm{in}}(\xi_m, s/2\pi i)_\Delta\end{aligned} \qquad (9.1.29)$$

Figure 9.7 shows the average field in the scattering situation of an infinite volume containing discrete scatterers with a source (in V_T) producing an adjustable beam pattern. To evaluate (9.1.29) we must choose representative values of $\eta_{00,1}^{(d)}$, or equivalently, $q_{00,1}^{(d)}\left(\text{or } Q_1^{(d)}\right)$. In general this can be done only approximately (cf. Section 9.3 following). The sum in Eq. (9.1.29) is seen from Section 2.5.3 to be the generalized beam pattern $\mathsf{A}_{M|S_{in}}$, that is, one which depends on the input signal $S_{in-\Delta}$ as well as on the filters associated with each (point) sensor. Inserting a steering vector $\mathbf{k}_{OT} = 2\pi\boldsymbol{\nu}_{OT,T}$ we have

$$\sum_{m=1}^{M} Y_{\mathrm{T}} S_{\mathrm{in}-\Delta} e^{i\mathbf{k}\cdot\xi} \rightarrow \sum_{m=1}^{M} Y_{\mathrm{T}}^{(m)} S_{\mathrm{in}-\Delta} e^{i(\mathbf{k}-\mathbf{k}_{OT})\cdot\xi_m} \equiv \mathsf{A}_M(\Delta\boldsymbol{\nu}_{\mathrm{T}}(s), s/2\pi i|s_{\mathrm{in}-\Delta}), \qquad (9.1.30a)$$

where

$$\Delta\boldsymbol{\nu}_{\mathrm{T}}(s) \equiv \boldsymbol{\nu}_{\mathrm{T}}-\nu_{\mathrm{T}}, \boldsymbol{\nu}_{\mathrm{T}}, \nu_{\mathrm{T}} = f(s), \text{ and } \mathsf{A}_M|_{S_{\mathrm{in}}} = \sum_{m=1}^{M}\mathsf{A}_m(\Delta\boldsymbol{\nu}_{\mathrm{T}}, s/2\pi i)|_{S_\Delta}, \text{ with } \mathsf{A}_m \equiv Y_{\mathrm{T}}^{(m)} S_{\mathrm{in}-\Delta}. \qquad (9.1.30b)$$

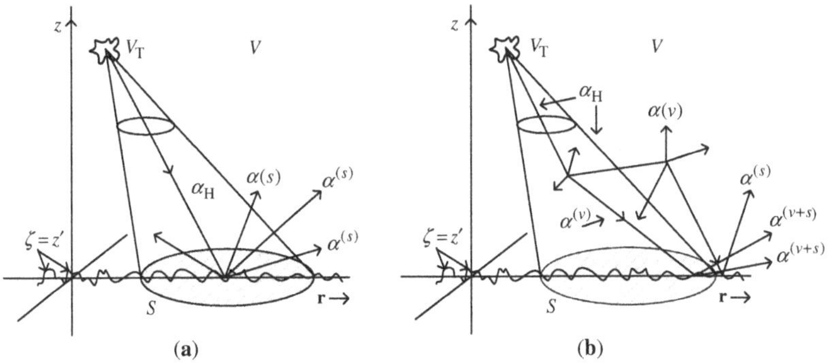

FIGURE 9.8 (*a*) Case I. Scattering from completely reflecting surface (*S*) only, into the homogeneous medium *V*. (*b*) Case II. Same as (a), except that now *V* is a scattering medium. ($S \times V$ scattering is neglected.)

If the same signal is applied to each sensor in the transmitting array, the $\mathsf{A}_m|_{S_{\text{in}}} Y_{\text{T}}^{(m)}$ and $A_M|_{S'_{\text{in}}}$ factors into $S_{\text{in}}(s/2\pi i)_{\Delta}\mathsf{A}_M$, cf. Section 2.5.3.1. (For a further discussion, including reception, see Sections 2.5.4 and 2.5.5.) Note finally, that the double Fourier transform of $\langle \alpha(\mathbf{R}, t) \rangle_V$, Eq. (9.1.29), namely the wave number–frequency amplitude spectrum of the average field, is the integrand

$$\left\langle \alpha_{00}^{(Q)}(\mathbf{k}, s)_V \right\rangle = \left(\frac{Y_0(\mathbf{k}, s)_{\infty} \mathsf{A}_M(\Delta\mathbf{k}_{\text{T}}/2\pi s/2\pi i)}{1 - \eta_{00,1}^{(d)}(\mathbf{k}, s)} \Big|_{S_{\text{in}-\Delta}} \right)_V, \tag{9.1.31}$$

in which $\Delta\mathbf{k}_{\text{T}} \equiv \mathbf{k} - \mathbf{k}_{O\text{T}} = 2\pi\Delta\boldsymbol{\nu}_{\text{T}}(s)$, [(Eqs. (9.1.30a) and (9.1.30b)], contains the steering vector. An important additional feature is the *frequency dependence of wave number*, that is, $\Delta\boldsymbol{\nu}_{O\text{T}} = \mathbf{f}_{O\text{T}}(s)$, which exhibits the "entanglement" of space and time in propagation (see also Sections 2.5.2 and 2.5.6, item 2, and examples in Chapter 8).

9.1.2.2 Dyson's Equation, Cont'd.: Hom-Stat Media—Surface Scatter In the common case where we have scattering from a surface or interface between two media capable of reflection and transmission (air–water, water-bottom, etc.), the canonical Dyson equations (9.1.7b) and (9.1.21) are handled essentially in the same way above for volume, except that the primary signal source remains in the volume. The main scattering mechanism is provided by the interface or reflecting surface, from above and below, as shown in Fig. 9.8a and b. Equation (9.1.21) is modified to

$$\left\langle \alpha^{(Q)}(\mathbf{R}, t)_S \right\rangle = \alpha_{\text{H}}(\mathbf{R}, \mathbf{t}) + \hat{\eta}_{1/S}^{(d)} \left\langle \alpha^{(Q)}(\mathbf{R}, t)_S \right\rangle; \hat{\eta}^{(d)} \tag{9.1.32}$$

where $\hat{\eta}_1^{(d)}|_{V+S}$ is established presently (9.1.36). The exciting field $\alpha_{\text{H}}(R, t)$ is represented as before by Eq. (9.1.25a).

We distinguish now two types of scenario: *Type I*, where a source in a homogeneous medium (*V*) is scattered by inhomogeneities on a random, perfectly reflecting surface (*S*) and *Type II*, in which the medium (*V*) contains inhomogeneities in addition to those on the scattering surface (S). The situations are illustrated in Fig. 9.8a and b. Here we employ the results of Section 8.3.4.1 and specifically Eq. (8.3.26), with (8.3.25a) and (8.3.25b), since we are neglecting the higher order of scattering interactions embodied in (8.3.26a). Now we have *the added requirement that the random surface* $\zeta(\mathbf{R}', t')$ *be Hom-Stat*, as well as $g_\infty^{(0)}$ and the inhomogeneity component $Q_{1/S}^{(d)}$, for example

$$\left.\begin{aligned} Q_{1/S}^{(d)}\big|_{\text{Hom-Stat}} &= Q^{(d)}(\mathbf{R}'-\mathbf{R}_0, t'-t_0)_S = \int dt' \oint_{S'} q_1^{(d)}(\mathbf{R}'-\mathbf{R}_0, t'-t_0)_S dS'(\mathbf{R}', t')_\zeta \\ &= \mathsf{F}_\mathbf{k}\mathsf{F}_s\left\{q_{00,1/S}^{(d)}\right\} = \int_{Br_1} \frac{ds}{2\pi i} \int \frac{\mathbf{dk}}{(2\pi)^3} q_{00,1}^{(d)}(\mathbf{k}, s)_{S'} e^{-i\mathbf{k}\cdot(\mathbf{R}'-\mathbf{R}_0)+s(t'-t_0)} \end{aligned}\right\}, \tag{9.1.33}$$

with the inverse, like (9.1.22),

$$q_{00,1}^{(d)}|_S = \mathsf{F}_\rho \mathsf{F}_{\Delta t}\left\{Q_{1/S}^{(d)}\right\}, \text{ with } Q_{1/S}^{(d)} \text{ above.} \tag{9.1.34}$$

[See Eq. (9.1.23) and (9.1.23a) and footnote 2. The *average* global mass operator (9.1.7a), analogous to (9.1.24), is given by

$$\left.\begin{aligned} \hat{\eta}_1^{(d)}\left(\mathbf{R}', t' \middle| \mathbf{R}_0, t_0\right)_S &= \int_{t_0^{(-)}}^{t'} dt_0 \oint_{S'} \hat{\eta}_1^{(d)}(\mathbf{R}'-\mathbf{R}_0|t'-t_0)\mathbf{dS}'(\mathbf{R}_0)_S \\ &= \int_{Br_1} \frac{ds}{2\pi i} e^{s(t'-t_0)} \oint_S \hat{\eta}_{00,1}^{(d)}(\mathbf{k}, s)_S e^{-i\mathbf{k}\cdot(\mathbf{R}'-\mathbf{R}_0)} \frac{\mathbf{dk}}{(2\pi)^3} \end{aligned}\right\}, \tag{9.1.35}$$

with $\hat{\eta}_1^{(d)} = \left\langle \hat{M}_{1/s}^{(0)} \right\rangle \hat{Q}_{1/s}^{(0)}$, from the appropriate averages, cf. (9.1.32), (9.1.7a), and remarks following Eq. (9.1.24).

We begin next with the relations [(8.3.23a) and (8.3.23b)], which on taking the appropriate averages to convert to the Dyson equations, becomes

$$I: \begin{cases} \left(1-\hat{\eta}_S^{(d)}\right)\left\langle \alpha^{(S)} \right\rangle = \alpha_H \\ \left\langle \alpha^{(S)} \right\rangle = \alpha_H + \hat{\eta}_S^{(d)} \left\langle \alpha^{(S)} \right\rangle \end{cases};$$

$$II: \begin{cases} \left(1-\hat{\eta}_S^{(d)}\right)\left\langle \alpha^{(V+S)} \right\rangle = \left\langle \alpha^{(V)} \right\rangle = \left(1-\hat{\eta}_S^{(d)}\right)\left(1-\hat{\eta}_V^{(d)}\right)\alpha_H \doteq \left(1-\hat{\eta}_S^{(d)}-\hat{\eta}_V^{(d)}\right)\alpha_H \\ \left\langle \alpha^{(V+S)} \right\rangle \doteq \alpha_H + \left(\hat{\eta}_S^{(d)}+\hat{\eta}_V^{(d)}\right)\left\langle \alpha^{(V+S)} \right\rangle \doteq \alpha_H + \hat{\eta}_{V+S}^{(d)}\left\langle \alpha^{(V+S)} \right\rangle \end{cases}, \tag{9.1.36}$$

which last is in more detail similar to Eq. (8.3.26), where we neglect the comparatively smaller scatter product terms $\hat{\eta}_S^{(d)}\hat{\eta}_S^{(d)}\langle\alpha_H\rangle$. Following now with the same procedure as we used to obtain (9.1.26), and so on, we find for (I and II) above, shown schematically in Fig. 9.8a and b, that the Dyson equations are respectively:

$$\text{I. } \left\langle \alpha^{(Q)}(\mathbf{R},t)_S \right\rangle = \int_{\mathrm{Br}_1} e^{st}\left(1-e^{sT}\right)\frac{ds}{2\pi i}\int_{-\infty}^{\infty}\left(\frac{\left[\hat{Z}_0(\mathbf{k},s)G_{00,\mathrm{T}}(\mathbf{k},s)\right]}{1-\hat{Z}_0(\mathbf{k},s)g_{00,1}^{(d)}(\mathbf{k},s)}\right)_{S'}\frac{e^{-i\mathbf{k}\cdot\mathbf{R}}\mathbf{dk}}{(2\pi)^3}. \tag{9.1.37}$$

Here $\hat{Z}_0$ is a deterministic (i.e., average) local operator, which is the surface counterpart to Y_0 for volumes in (9.1.26), and so on. It is given explicitly by

$$\hat{Z}_0(\mathbf{k},s) = \int dt'\int \mathbf{dr}'\hat{h}(\boldsymbol{\rho},\Delta t)_S(\ )_{\mathbf{R}',t'}e^{i\mathbf{k}\cdot\boldsymbol{\rho}-s\Delta t},\ \text{with } \boldsymbol{\rho} = R-R';\Delta t = t-t', \tag{9.1.38a}$$

where $\hat{h}$, and thus $\hat{Z}_0$ by (9.1.38a), are represented by the (deterministic) relations[14]

$$\left.\begin{aligned}\hat{h}(\boldsymbol{\rho},\Delta t)_{S'} = \hat{g}^{(0)}(R-R',t-t') &= \left\langle g_\infty^{(0)}J' - \left[J'g_\infty^{(0)}\right]\right\rangle_{\mathbf{R},\zeta}\\ &= \left\langle g_\infty^{(0)}\nabla'\zeta\cdot\nabla' - \nabla'\zeta\cdot\nabla g_\infty^{(0)}\right\rangle_{\mathbf{R},\zeta}\end{aligned}\right\} \tag{9.1.38b}$$

with $\zeta(=\zeta'(\mathbf{R}',t')\doteq\zeta'(\mathbf{r}',t'))$ the random surface elevation, cf. Eq. (8.1.42g). The global mass operator is specifically (in transform space) and in its space–time form

$$\hat{\eta}_{00,1}^{(d)}(\mathbf{k},s) \equiv \hat{Z}_0(\mathbf{k},s)_S q_{00,1}^{(d)}(\mathbf{k},s)_S, \tag{9.1.38c}$$

with the components [(9.1.38a) and (9.1.38b)]. In the relation (9.1.37) $\hat{Z}_{00,1}$, Eqs. (9.1.38a) and (9.1.38b), operates on $G_{00,\mathrm{T}}$, the (transform of the) source density $G_\mathrm{T}(\mathbf{R}',t')$ of the transmitter, producing the original field $\alpha_\mathrm{H}(\mathbf{R},t)$. Similarly, $\hat{Z}_{0,S}$ also modifies the (transform of the) average of the effect of the local inhomogeneities, embodied in $q_{00,1}^{(d)}$ (or equivalently, $Q_{1/S}^{(d)}$). The role of the transmitter's array A_M, with applied signal, is obtained by substituting A_M in Eq. (9.1.30a), and so on, for $G_{00,\mathrm{T}}$, that is

$$\text{I. } \left\langle \alpha^{(S)}(\mathbf{R},t)_S \right\rangle = \int_{Br_1} e^{st}\left(1-e^{-sT}\right)\frac{ds}{2\pi i}\int_{-\infty}^{\infty}\left(\frac{\left[\hat{Z}_0(\mathbf{k},s)_S\mathsf{A}_M(\Delta\mathbf{k}/2\pi,s/2\pi i)\right]}{1-\eta_{00,1}^{(d)}(\mathbf{k},s)}\right)_{S_{\mathrm{in}-\Delta}} e^{-i\mathbf{k}\cdot\mathbf{R}}\frac{\mathbf{dk}}{(2\pi)^3}, \tag{9.1.39}$$

[14] Because of the assumed Hom-Stat properties of the random surface $\zeta(\mathbf{R}',t')$

$$\hat{J}' \equiv \frac{\partial\zeta}{\partial x'}\frac{\partial}{\partial x'} + \frac{\partial\zeta}{\partial y'}\frac{\partial}{\partial y'} + \frac{\partial}{\partial z'} = \zeta_x\frac{\partial}{\partial x'} + \zeta_y\frac{\partial}{\partial y'} + \frac{\partial}{\partial z'} = \hat{f}(\zeta);\ \hat{g}_S^{(0)} = \left[g_\infty^{(0)}\langle\hat{f}\rangle - \langle\hat{f}\rangle g_\infty^{(0)}\right],$$

$g_S^{(0)}$ (and $\hat{M}_{1/S}^{(0)}$) are also Hom-Stat, provided ζ, and $Q_V\langle\alpha_V\rangle$ have at least continuous first derivatives.

cf. [(9.1.29) and (9.1.31)] for volumes. Equations (9.1.37) et. seq. and (9.1.39) is the formal solution to Dyson's equation for the scattered field in the homogeneous volume (V), attributable to the scattering from the random rough surface (S), here on perfect reflection, when the higher order scatter $\hat{\eta}_S \hat{\eta}_V \alpha^{(V+S)}$ is negligible. Here $\hat{Z}_0$, unlike its counterpart for Y_0 for volume, is seen to be a (global) operator (9.1.38a).

For the more complex Dyson situation II, (9.1.36), we need now the contribution to the volume scatter $\hat{\eta}_V^{(d)} \langle \alpha^{(V+S)} \rangle$, cf. Fig. 9.8b. This is readily obtained from (9.1.26) et. seq. by noting that $\eta_{V+S}^{(d)} \doteq \eta_V^{(d)} + \eta_S^{(d)}$. Thus, the contribution of volume scatter can be considered separately from that due to the surface, when the "cross-scatter" component $\left(\sim \eta_S^{(d)} \eta_V^{(d)}\right)$ is neglected here, cf. (9.1.36). Accordingly, we may use our previous result (9.1.26) et. seq. and write for the added volume component directly, here $\eta_V^{(d)} \langle \alpha^{(V+S)} \rangle$, and for the surface contribution $\eta_S^{(d)} \langle \alpha^{(V+S)} \rangle$, (9.1.32), to obtain directly

$$
\begin{aligned}
\textbf{II.}\ \left\langle \alpha^{(V+S)}(\mathbf{R},t) \right\rangle &\doteq \alpha_{\mathrm{H}}(\mathbf{R},t) + \left\langle \alpha^{(S)}(\mathbf{R},t)_S \right\rangle + \left\langle \alpha^{(V)}(\mathbf{R},t)_V \right\rangle \\
&= \alpha_{\mathrm{H}}(\mathbf{R},t) + \Bigg[\mathbf{I}.\left\langle \alpha^{(Q)}(\mathbf{R},t)_S \right\rangle \\
&= \int_{\mathrm{Br}_1} e^{st}\left(1-e^{sT}\right)\frac{ds}{2\pi i}\int_{-\infty}^{\infty}\left(\frac{\left[\hat{Z}_0(\mathbf{k},s)G_{00,\mathrm{T}}(\mathbf{k},s)\right]}{1-\hat{Z}_0(\mathbf{k},s)g_{00,1}^{(d)}(\mathbf{k},s)}\right)_{S'} \\
&\quad \times \frac{e^{-i\mathbf{k}\cdot\mathbf{R}}d\mathbf{k}}{(2\pi)^3}\Bigg] + \Bigg[\left\langle \alpha^{(Q)}(\mathbf{R},t)_V \right\rangle \\
&= \int_{Br_1} e^{st}\frac{ds}{2\pi i}\int_{-\infty}^{\infty}\left(\frac{\mathsf{Y}_0 G_{00,T}}{1-\eta_{000,1}^{(d)}}\right)_{(\mathbf{k},s)|V} e^{-i\mathbf{k}\cdot\mathbf{R}}\frac{d\mathbf{k}}{(2\pi)^3}\Bigg],
\end{aligned}
\tag{9.1.40}
$$

As before, $\alpha_{\mathrm{H}}(\mathbf{R},t)$, the homogeneous component, if any, is given by (9.1.25a) above. In short, the Dyson equation for Case **II**, Fig. 9.8b, is the sum of the two right hand members of Eq. (9.1.36), that is, $\alpha_{\mathrm{H}} + \hat{\eta}_{V+S}^{(d)} \langle \alpha^{(V+S)} \rangle$, as required. The explicit incorporation of the generalized beam patters A_M can be included, as indicated in (9.1.31) and (9.1.39), on applying the results above to $\hat{\eta}_{V+S}^{(d)} \equiv \hat{\eta}_S^{(d)} \hat{\eta}_V^{(d)}$ explicitly. Similar to the case of volume scattering (9.1.2.1), in the Hom-Stat situations of surface scatter (Section 9.1.2.2) we have obtained formal, closed-form results, cf. (9.1.37) and (9.1.39). As we have seen, the global operator $\hat{Q}_{1/S}^{(d)}$ for the surface inhomogeneities is also found from the results (9.1.9a)–(9.1.15b), applied specifically to surface cases. However, $\hat{Q}_{1/S}^{(d)}$ itself does not posses an explicit closed form but must be approximated. Various types of such approximations are considered in Section 9.3.

9.1.2.3 Dyson's Equation—The Purely Deterministic Hom-Stat Case Our results for $\langle \alpha^{(Q)}(\mathbf{R},t) \rangle$ when the medium is considered to be deterministic and Hom-Stat (Section 8.3 et. seq.) may be obtained at once from Section 9.1.2.2, Eq. (9.1.26) for scatter in volumes and (9.1.37) and from surfaces. Since the medium now constitutes a one-member ensemble, the associated probability measure is unity. Thus, $\langle \alpha^{(Q)} \rangle = \alpha^{(Q)}$: no averages are required. The solution for the scattered field (9.1.26) becomes, for volumes:

$$\alpha^{(Q)}(\mathbf{R},t)_V = \mathsf{F}_k\mathsf{F}_s\{\mathrm{Y}_0 G_{00,\mathrm{T}}/(1-\mathrm{Y}_0 q_{00,\mathrm{T}})\}_V; q_{00,1}^{(d)}|_V = q_{00,1}|_V = \mathsf{F}_{\mathbf{R},t}\{Q_{1/V}\}, \quad (9.1.41)$$

this last from (9.1.8a) and (9.1.23a). Similarly, $\eta_{00,1}^{(d)} = \eta_{00,1} = Y_0 q_{00,1}$, for volumes [(9.1.27) and (9.1.27a)]. Equation (9.1.41) represents the solution of the integral equation for these deterministic cases

$$\alpha^{(Q)}(\mathbf{R},t) = \alpha_{\mathrm{H}}(\mathbf{R},t) + \hat{\eta}_\infty \alpha^{(Q)}(\mathbf{R},t), (S \text{ or } V), \quad (9.1.42)$$

which in the Hom-Stat regime postulated here can be effectuated by simple Fourier transforms. Here "Hom-Stat" requires a steady state in time and homogeneity in space, while the signal causing the scatter from the medium's inhomogeneities is "on". This, in turn, is determined by the transmitted signal's duration $\Delta(\leq \infty)$, which is expressed in $G_{00,\mathrm{T}}$, (the double transform) of the signal density $G_{\mathrm{T}}(\mathbf{R}',t')$, vide Section 8.1.7 for example. [Equivalent forms of solution to (9.1.41) are (9.1.29) and (9.1.31).

Similar modifications of Dyson's equation are made for deterministic scatter from surfaces. The result, from (9.1.37) et. seq. above, is

$$\alpha^{(Q)}(\mathbf{R},t)_S = \mathsf{F}_{\mathbf{k}}\mathsf{F}_s\left\{\left([\hat{Z}_0 G_{00,\mathrm{T}}]/(1-\hat{Z}_0 q_{00,1})\right)_S\right\}, \quad (9.1.43)$$

which is the solution to (9.1.42) for surfaces, with the operator kernel $\hat{h}$, cf. (9.1.38b), *without the average* here.

9.1.3 Example: The Statistical Structure of the Mass Operator $\hat{Q}_1^{(d)}$, with $\langle \hat{Q} \rangle = 0$

In this instance the mean (Langevin) mass operator $\left\langle \eta_1^{(d)} \left(= \hat{M}_\infty^{(0)} \left\langle \hat{Q}_1^{(d)} \right\rangle \right) \right\rangle$, Eqs. (9.1.14a), (9.1.15a), (9.1.15b) simplified considerably [cf. Section 9.1.1.3], when $\langle \hat{Q} \rangle$ vanished and is symmetrically distributed about $\langle Q \rangle = 0$. We outline here as an example of the calculations involved a more detailed development of (9.1.15a). In this case, generally, we start with the appropriate development of $\hat{Q}_1^{(d)}$ embodied in the relation (9.1.10) and Section 9.1.1.4 for surface scatter, namely,

$$\hat{Q}_1^{(d)} = \sum_{m=0}^{\infty} \hat{B}_m(\mathbf{R},t|\mathbf{R}',t')_{V \text{ or } S} = \int_{V \text{ or } S} \hat{Q}_1^{(d)}(\mathbf{R},t|\mathbf{R}',t')(\,)_{\mathbf{R}',t'} d\mathbf{R}' dt', \quad (9.1.41)$$

where the operator series $\hat{B}_m$ is explicitly obtained from (9.1.15a), with the corresponding relation for $\eta_1^{(d)}$ from (9.1.15b) and the results of Section 9.1.1.4.

To proceed further we must use a specific form of Langevin equation, appropriate to the class of propagation we are considering (Table 8.4). In addition, we must treat $\hat{Q}$ as a stochastic operator in this Langevin equation:

9.1.3.1 Example: The Stochastic Time-Dependent Helmholtz Equation. We begin first with propagation in volumes.

$$\left(\nabla^2 - \frac{1}{c_0^2}[1+\varepsilon_R(\mathbf{R},t)]\frac{\partial^2}{\partial t^2}\right)\alpha^{(Q)}(\mathbf{R},t) = -G_T, \quad \mathbf{R}\varepsilon V_T; = 0, \text{ elsewhere}, \tag{9.1.42}$$

subject to the familiar boundary and initial condition. (Dirichlet, i.e., no primary sources on the boundaries (Table 8.3), and $\partial\alpha^{(Q)}/\partial t = \partial^2\alpha^{(Q)}/\partial t^2 = 0$, with $\alpha^{(Q)} \in V - V_T$, $V \to \infty$ (infinite medium)). Accordingly, we see at once that

$$\hat{Q} \equiv \frac{\varepsilon(\mathbf{R},t)}{c_0^2}\frac{\partial^2}{\partial t^2}. \tag{9.1.43}$$

Applying this to the first term of (9.1.22) yields directly, for volumes[15]:

$$\left\langle \hat{Q}_1\left(\hat{F}_{12}|\hat{Q}_2\right)_V\right\rangle = \left\langle \frac{\varepsilon_R(\mathbf{R}_1,t_1)}{c_0^2}\int_{t_0^-}^{t^+}\int_{V_T}\left\{\frac{\partial^2}{\partial t^2}g_\infty^{(0)}(\mathbf{R}_1,t_1|\mathbf{R}_2,t_2)\right\}\frac{\varepsilon_R(\mathbf{R}_2,t_2)}{c_0^2}\right\rangle$$
$$\times \frac{\partial^2}{\partial t_2^2}(\,)_{\mathbf{R}_2,t_2} dR_2, dt_2\Bigg|_{R_2=R';\, t_2=t'}, \tag{9.1.44a}$$

notationally in view of (9.1.22). We easily see that[16]

$$\langle \hat{Q}_1\hat{F}_{12}|_V\hat{Q}_2\rangle = \frac{1}{c_0^2}\int_{t_0^-}^{t^+}\int_{V_T} K_{\varepsilon_R}(\mathbf{R}_1,t_1;\mathbf{R}_2,t_2)\frac{\partial^2}{\partial t_1^2}g^{(0)}\left(\mathbf{R}_1,t_1\middle|\mathbf{R}_2,t_2\right)\frac{\partial^2}{\partial t_2^2}(\,)_{\mathbf{R}_2,t_2}\mathbf{dR}_2, dt_2, \tag{9.1.44b}$$

where $K_{\varepsilon_R} = \langle \varepsilon_R(\mathbf{R}_1,t_1)\varepsilon_R(\mathbf{R}_2,t_2)\rangle$ is the covariance of ε_R.

Similarly, for scattering from surfaces in this case for backscatter from the *surface* $S = S_0$ enclosing V_T we obtain[16]

$$\langle \hat{Q}_1\hat{F}_{12}|_S\hat{Q}_2\rangle = \int\int_{S_0} K_{R_0}(\mathbf{R}_1,t_1;\mathbf{R}_2,t_2)\left\{\frac{\partial^2}{\partial t_1^2}\left[g_\infty^{(0)}\frac{\partial}{\partial n}(\,) - (\,)\frac{\partial g_\infty^{(0)}}{\partial n}\right]\right\}\frac{\partial^2}{\partial t_2^2}(\,)_{\mathbf{R}_2,t_2}dS_0(\mathbf{R}_2,t_2), \tag{9.1.45}$$

[15] We use the well-known relation for differentiating under the integral sign

$$\frac{\partial}{\partial c}\int_a^b f(x,c)dx = \int_a^b \frac{\partial}{\partial c}f(x,c)dx + f(b,c)\frac{\partial b}{\partial c} - f(a,c)\frac{\partial a}{\partial c},$$

where the last two terms here are zero.

[16] By inserting a factor $1_{T\times V} = 1$ where $t_0 \le t \le t$ and $V \in V_T$, and 0 elsewhere in the integrand, we can replace the finite limits by $(-\infty,\infty)$.

in which $K_{R_0} = \langle R_0(\mathbf{R}_1, t_1) R_0(\mathbf{R}_2, t_2)\rangle$ is the covariance of the (plane-wave) reflection coefficient.

We follow the same procedure for the forth-order moment of $\hat{Q}_V$, cf. (9.1.22), and find that

$$\hat{A}_3^{(1)} = \left\langle \hat{Q}(\hat{F}_V\hat{Q})^{(3)} \right\rangle = \left(\iint_{(t_R)}\right)^{(3)} K_{\varepsilon_R}(\mathbf{R}_1, t_1; \ldots; \mathbf{R}_4, t_4) \frac{\partial^2}{\partial t_1^2} \hat{F}_{12}\left(R_1, t_1 \middle| R_2, t_2\right)_V$$
$$\cdots \frac{\partial^2}{\partial t_3^2} \hat{F}_{24}\left(R_3, t_3 \middle| R_4, t_4\right)_V \bullet \frac{\partial^2}{\partial t_4^2}(\,)_{R_4, t_4} \mathbf{dR}_2 \ldots \mathbf{dR}_4 dt_2 \ldots dt_4. \tag{9.1.46}$$

In fact, the general formula for the $2m$th moment of $\hat{Q}_V$, from $\hat{B}_{2m-1}^{(1)}, m \geq 1$, is now readily seen to be

$$\left\langle \hat{Q}(\hat{F}_V\hat{Q})^{(2m-1)} \right\rangle = \left(\iint_{(t)}\right)^{(2m-1)} K_{\varepsilon_R}(\mathbf{R}_1, t_1; \ldots; \mathbf{R}_{2m}, t_{2m}) \left(\prod_{i=1}^{2m-1} \frac{\partial^2}{\partial t_i^2} \hat{F}_{i,iH}|_V\right) \frac{\partial^2}{\partial t_{2m}^2}(\,)_{R_{2m}, t_{2m}} \prod_{i=2}^{2m} \mathbf{dR}_i dt_i, \tag{9.1.47a}$$

with $m \geq 1$. For surface scatter, we have alternatively

$$\left\langle \hat{Q}(\hat{F}_{S_0}\hat{Q})^{(2m-1)} \right\rangle = \left(\iint_{(t)S_0}\right)^{(2m-1)} K_{R_0}(\mathbf{R}_1, t_1; \ldots; \mathbf{R}_{2m}, t_{2m}) \prod_{i=1}^{2m-1}$$
$$\bullet \left(\frac{\partial^2}{\partial t_i^2}\left[g_\infty^{(0)} \frac{\partial}{\partial n}(\,) - (\,)\frac{\partial}{\partial n} g_\infty^{(0)}\right]\right)_{i,i+1} \frac{\partial^2}{\partial t_{2m}^2}(\,)_{\mathbf{R}_{2m}, t_{2m}} \prod_{i=2}^{2m} dS_0(\mathbf{R}_i, t_i), \tag{9.1.47b}$$

where

$$\left.\begin{aligned} &\left[g_\infty^{(0)} \frac{\partial}{\partial n}(\,) - (\,)\frac{\partial}{\partial n} g_\infty^{(0)}\right]_{i,i+1} = g_\infty^{(0)}\left(\mathbf{R}_i, t_i \middle| \mathbf{R}_{i+1}, t_{i+1}\right) \frac{\partial}{\partial n_i}(\,)_i - (\,)_i \frac{\partial}{\partial n_i} g_\infty^{(0)}\left(\mathbf{R}_i, t_i \middle| \mathbf{R}_{i+1}, t_{i+1}\right) \\ &\text{with} \quad \frac{\partial}{\partial n_i} \equiv \hat{\mathbf{n}}_i \bullet \nabla_i; \quad \left(\int\oint_{(t)\,S_0}\right)^{(2m-1)} = \int_{(t_1)}\oint_{S_0(\mathbf{R}_2, t_2)} \cdots\cdots\cdots \int\oint_{S_0(\mathbf{R}_{2m}, t_{2m})} (\,). \end{aligned}\right\} \tag{9.1.47c}$$

These higher order even moments of $\hat{Q}$, represented here by $K_{\varepsilon_R;2,\ldots,2m}$ and $K_{R_0;2,\ldots,2m}$, are the *higher order covariance* which constitute the *mass operator* $\hat{Q}_1^{(d)}$, along with the first-order covariances $K_{\varepsilon_0}(\mathbf{R}_1, t_1; \mathbf{R}_2, t_2), K_{R_0}(\mathbf{R}_1, t_1; \mathbf{R}_2, t_2)$. As we can see from [(9.1.28a) and (9.1.28b)] their complexity increases with the order (m). For the odd moments (in $\hat{Q}$) $m \geq 0$, we have a similar behavior:

$$\left\langle \hat{Q}(\hat{F}_V\hat{Q})^{(2m)}\right\rangle = \left(\iint\right)^{(2m)} K_{\varepsilon_R}(\mathbf{R}_1,t_1;\ldots;\mathbf{R}_{2m+1},t_{2m+1})\prod_{i=1}^{2m}\hat{F}_{i,i+1}|_V\frac{\partial^2}{\partial t_{2m+1}^2}()_{\mathbf{R}_{2m+1},t_{2m+1}}\prod_{i=2}^{2m+1}\mathbf{dR}_i dt_i \tag{9.1.48a}$$

for volume scatter, $m \geq 1$. (For $m = 0$, we have $\langle \hat{Q}\rangle = 0$, cf. (9.1.14).) For surface scatter the odd moments ($m \geq 1$) become explicitly

$$\left\langle \hat{Q}(\hat{F}_{S_0}\hat{Q})^{(2m)}\right\rangle = \left(\iint\right)^{(2m)} K_{R_0}(\mathbf{R}_1,t_1;\ldots;\mathbf{R}_{2m+1},t_{2m+1})\prod_{i=1}^{2m}\frac{\partial^2}{\partial t_i^2}\left[g_\infty^{(0)}\frac{\partial}{\partial n}()-()\frac{\partial}{\partial n}g_\infty^{(0)}\right]_{i,i+1} \cdot \frac{\partial^2}{\partial t_{2m+i}^2}()_{\mathbf{R}_{2m+1},t_{2m+1}}\prod_{i=2}^{2m+1} dS_0(\mathbf{R}_i,t_i), m \geq 1, \tag{9.1.48b}$$

again with $\langle \hat{Q}\rangle = 0$ here.

9.1.3.2 The Gaussian Case We note first from the above that although $\langle \hat{Q}\rangle$ vanishes, the higher order ($m \geq 1$) odd moments generally do not, so that $\langle \hat{Q}\hat{F}_{V,S}\hat{Q}\rangle$ is the lowest order of nonvanishing moment comprising the statistic $Q_1^{(d)}$. The mass operator $Q_1^{(d)}$, (9.1.22), in this time-dependent Helmholtz is accordingly given by the combination of odd and even terms above, corresponding to the development in (9.1.13).

Consequently, $\hat{Q}_1^{(d)}$ for volumes (V_T) and surfaces (S_T) is seen to require for a complete description *all* the moments of $\hat{Q}$, of all orders. This is to be expected, since $\hat{Q}_1^{(d)}$ is nonlinear in $\hat{Q}$, because of the general scattering, which involves all orders of interaction. This results in a different, if not impossible, analytic task, particularly for strong scattering. Approximations are accordingly required and are inherent in any practical approach. (Some of these are noted in Section 9.3.) Even when Gaussian statistics for the local inhomogeneities ($\hat{Q}$) are employed, where the higher order covariances reduce to sums of products of second-order statistics, (9.1.30a), this remains the case. Effective approximate methods for obtaining estimates of the mass operator $\hat{\eta}_1^{(d)}\left(= M\hat{Q}_1^{(d)}\right)$ require experimental and numerical methods, most readily applied to the governing integral equation (9.1.7a). This is described briefly in Section 8.5.3 and is discussed in Chew [1].

We also observe Table 8.4 of the preceding chapter that the inhomogeneity operator $\hat{Q}$ for other common types of media, with their characteristic propagation equations (cf. 2-6 in the table), have the same simple linear structural product relations between the random parameters $\alpha_{12}^2, \hat{\tau}_{01}$, and so on, and the various differential operators $((\partial/\partial t)$, $(\partial/\partial t)\nabla^2, \ldots$ etc). This allows us to replace in (9.1.23) – (9.1.28b) the local operators $\left(\partial^2/\partial t_i\right)$ for the time-dependent Helmholtz equation by the various others, for example, $(\partial/\partial t), (\partial/\partial t_i)\nabla^2$, and so on, and the statistical moments K_ε, K_{R_0}, by $K_{\alpha_{1\infty}^2}, K\hat{\tau}\ldots$, and so on, of these other random parameters.

Finally, we obtain our basic desired result: a generalized Huygens Principle (GHP) for random scattering in an infinite volume of scattering,[17] cf. Section 8.3.4. This is now

[17] Here we neglect the backscatter effect (cf. Sections 8.3.3.1. and 8.3.3.2 and Figs. 8.1 and 8.2), and for surface scatter, the array volume contributions.

provided by (9.1.28a) and (9.1.22) to give $\eta_\mu^{(k)}\left(=\hat{M}_\infty Q_{1,1}^{(d)}\right)$ for the resulting mean field $\left\langle \alpha_V^{(Q)} \right\rangle$ [Dyson's equation (9.1.7a) or equivalently (9.1.7a) in expanded form]. A similar GHP is obtained for surface scatter $\left\langle \alpha_S^{(Q)} \right\rangle$ when (9.1.22) is used to yield $\eta_S^{(d)} = \hat{M}_{\infty,S} Q_{1,S}^{(d)}$.

9.1.4 Remarks

Section 9.1 has introduced the extension of the deterministic results of Section 8.3 et. seq. to random fields. These earlier results produce representative members of the ensemble which constitute the resulting Langevin equation, namely a set of dynamical equations whose solution is described by statistical moments, and more fully, by the probability distribution of the random field.

The discussion here has largely been devoted to the first moment of the scattered field, obtained from the Dyson equation. This includes the physically approximate situation where the scatterers can be considered to be homogeneous and stationary, mathematically represented by (9.1.23) and (9.1.24) and for volume by (9.1.34) and (9.1.35). The extension to a four-dimensional feedback system by the FOR of 9.1.1.1 by analogy with the familiar one-dimensional one of electrical engineering practice, is now made for these average Langevin equations, which are schematically illustrated in Figs. 9.4 and 9.6. (See also Fig. 8.18 for a typical representation of the ensemble.) These results are themselves extensions of the more familiar relations of the "classical" scattering theory [1–7, 17], for example, in the following respects, which include

i. broadband signals and transients, as well as steady state examples;
ii. distributed sources (V_T);
iii. general aperture and arrays;
iv. general (linear) random media, both non-Hom-Stat and Hom-Stat, in which the Green's functions (both deterministic and random) represent a variety of possible media impulse responses;
v. scattering from surfaces as well as volumes, including from both;
vi. the statistics of the inhomogeneities, represented by $\hat{Q}$ are general, that is, non-Gaussian;
vii. the mass operator $\hat{Q}^{(d)}$ (and hence $\hat{\eta}^{(d)}$) are nonlocal, as in $\hat{\eta}$;
viii. the generalized Dyson equations (9.1.26) and (9.1.37) are *nonlinear* in $\hat{Q}^{(d)}$ (and $\therefore\ \hat{\eta}^{(d)}$), although they are linear in the average field $\langle \alpha^{(Q)} \rangle$.

The (integral) Dyson equation discussed in Section 9.1.1 is the first-order Dyson equation, for the mean scattered field $\langle \alpha^{(Q)} \rangle$. It is formally exact, but depends on knowledge of $\hat{Q}_1^{(d)}$ that is not usually available. This is true even for Hom-Stat fields, for which exact closed form solutions of the basic integral equation (9.1.7a) are available (9.1.26, 37) et. seq. Approximate forms for $Q^{(d)}$ offer one approach to quantitative results. The numerical methods of the feedback approach outlined in Chapter 8 offer another, albeit a computational intensive one, but one within the capabilities of modern methods.

9.2 HIGHER ORDER MOMENTS OPERATIONAL SOLUTIONS FOR THE LANGEVIN EQUATION

The Dyson equation can be formally generalized to higher order moments, for example, $\left\langle \alpha_1^{(Q)} \ldots \alpha_\mu^{(Q)} \right\rangle$. Their evaluation can be obtained in fashion similar to that discussed in Section 9.1.2. As expected, they become progressively more coupled as their order ($\mu \geq 2$) increases. We consider first the second-order medium ($\mu = 2$), that is, one constructed from the ensemble

$$\alpha^{(2)} = \alpha_1\alpha_2 = \alpha(\mathbf{R}_1, t_1)\alpha(\mathbf{R}_2, t_2) \equiv \alpha^{(2)}(\mathbf{R}_1, t_1; \mathbf{R}_2, t_2), \tag{9.2.1}$$

which is then generalized further in Section 9.2.3 for the case $\mu > 2$.

9.2.1 The Second-Order Moments: Analysis of the Bethe–Salpeter Equation (BSE)

Accordingly, let us examine in particular the second-order medium ($\mu = 2$), with the help of the FOS (Section 9.1.1). We use now the basic ensemble relation (9.1.5b) and consider the ensemble at two field points $P(\mathbf{R}_1, t_1)$ and $P(\mathbf{R}_2, t_2)$, namely,

$$\alpha_1 = \alpha_{\mathrm{H1}} + \hat{M}_1\hat{Q}_1\alpha_1; \alpha_2 = \alpha_{\mathrm{H2}} + \hat{M}_2\alpha_2, \quad \text{with} \quad \alpha_1 = \alpha(\mathbf{R}_1, t_1); \alpha_2 = \alpha(\mathbf{R}_2, t_2), \tag{9.2.2}$$

where $\hat{M}_1 = \hat{M}_\infty(\mathbf{R}, t|\mathbf{R}_1, t_1)$ and $\hat{Q}_1 = \hat{Q}(\mathbf{R}_1, t_1)$, and so on. Their average product is the second-order, second moment

$$(m = 2) : \mathrm{M}_{12}^{(\alpha)} \equiv \langle \alpha_1\alpha_2 \rangle = \hat{M}_1\langle \hat{Q}_1\alpha_1\hat{M}_2\hat{Q}_2\alpha_2 \rangle + \alpha_{\mathrm{H2}}\langle \hat{M}_1\hat{Q}_1\alpha_1 \rangle$$

$$+ \alpha_{\mathrm{H1}}\langle \hat{M}_2\hat{Q}_2\alpha_2 \rangle + \alpha_{\mathrm{H1}}\alpha_{\mathrm{H2}}, \text{ or} \tag{9.2.3a}$$

$$= \langle \hat{\eta}_1\alpha_1\hat{\eta}_2\alpha_2 \rangle + \alpha_{\mathrm{H1}}\langle \hat{\eta}_1\alpha_1 \rangle + \alpha_{\mathrm{H2}}\langle \hat{\eta}_2\alpha_2 \rangle + \alpha_{\mathrm{H1}}\alpha_{\mathrm{H2}}. \tag{9.2.3b}$$

Because the similar operators ($\sim\hat{\eta}$) commute and clearly do the different fields $(\alpha_1, \alpha_2, \alpha_{\mathrm{H1}}, \alpha_{\mathrm{H2}})$, we can write $\langle \hat{\eta}_1\alpha_1\hat{\eta}_2\alpha_2 \rangle = \langle \hat{\eta}_1\hat{\eta}_2\alpha_1\alpha_2 \rangle = \hat{M}_1\hat{M}_2\langle \hat{Q}_1\hat{Q}_2\alpha_1\alpha_2 \rangle$, since $\hat{M}_2\hat{Q}_1 = \hat{Q}_1\hat{M}_2$, since $\hat{M}_2$ and $\hat{Q}_1$ are also entirely different functions. By extension of (9.1.6) and so on in the (first-order) Dyson formulation this enables us to define a (deterministic) *second-order mass-operator* or "*intensity operator*" $\hat{Q}_{12}^{(d)}$ ([7], p. 128, Eq. 4.22):

$$\hat{Q}_{12}^{(d)}\langle \alpha_1\alpha_2 \rangle \equiv \langle \hat{Q}_1\hat{Q}_2\alpha_1\alpha_2 \rangle, \quad \text{where} \quad \hat{Q}_{12}^{(d)} = \hat{Q}_{12}^{(d)}\left(\mathbf{R}_1, t_1; \mathbf{R}_2, t_2 \middle| \mathbf{R}_1', t_1'; \mathbf{R}_2', t_2'\right), \tag{9.2.4}$$

$\alpha_1 = \alpha_1\left(\mathbf{R}_1', t_1'\right)$ and so on, in (9.2.4). Equivalently, the second-order global intensity operator (GIO) is given by

$$\hat{\eta}_{12}^{(d)}\langle \alpha_1\alpha_2 \rangle \equiv \langle \hat{\eta}_1\hat{\eta}_2\alpha_1\alpha_2 \rangle \doteq \hat{\eta}_{12}^{(d)}\left(\mathbf{R}_1, t_1; \mathbf{R}_2, t_2 \middle| \mathbf{R}_1', t_1'; \mathbf{R}_2', t_2'\right) = \hat{M}_1\hat{M}_2\langle \hat{Q}_1\hat{Q}_2\alpha_1\alpha_2 \rangle, \tag{9.2.5}$$

$$\equiv \hat{M}_1\hat{M}_2\hat{Q}_{12}^{(d)}\langle \alpha_1\alpha_2 \rangle \tag{9.2.5a}$$

These operators $\left(\hat{Q}_{12}^{(d)}, \hat{\eta}_{12}^{(d)}\right)$ embody the inhomogeneity effects of the now equivalent second-order deterministic medium. The relations (9.2.4) and (9.2.5), as we shall see presently, allow us to calculate $\hat{Q}_{12}^{(d)}$ and $\hat{\eta}_{12}^{(d)}$.

Using the defining relations (9.1.6), (9.2.5), with (9.1.9),

$$\left.\begin{aligned}\hat{\eta}_1^{(d)}\langle\alpha_1\rangle \equiv \langle\hat{\eta}_1\alpha_1\rangle; \hat{\eta}_2^{(d)}\langle\alpha_2\rangle \equiv \langle\hat{\eta}_2\alpha_2\rangle; \text{ and } \hat{\eta}_{12}^{(d)}\langle\alpha_1\alpha_2\rangle \equiv \langle\hat{\eta}_1\hat{\eta}_2\alpha_1\alpha_2\rangle, \text{ and}\\ \hat{\eta}_{12}^{(d)} \equiv \hat{\eta}_{12}^{(d)}\left(\mathbf{R}_1,t_1;\mathbf{R}_2,t_2\middle|\mathbf{R}_1',t_1';\mathbf{R}_2',t_2'\right)\end{aligned}\right\} \tag{9.2.6}$$

in (9.2.3a), or (9.2.3b) allows us to express the second moment of the field at different positions and times, namely (9.2.6) $\hat{M}_{12}^{(\alpha)}(=\langle\alpha_1\alpha_2\rangle)$, entirely in terms of the deterministic global operators (9.2.6), with the help of the following equivalent of (9.2.3b), from (9.1.6), (9.1.8b), et. seq., to obtain finally

$$\mathrm{M}_{12}^{(d)} \equiv \langle\alpha_1\alpha_2\rangle = \left(1-\hat{\eta}_{12}^{(d)}\right)^{-1}\left(1-\hat{\eta}_1^{(d)}\hat{\eta}_2^{(d)}\right)\langle\alpha_1\rangle\langle\alpha_2\rangle. \tag{9.2.7}$$

Equation (9.2.7) is the FOS form of an analogue to the BSE, arising originally in quantum field theory [see, for example, Chapter 4 of Ref. [7], and pp. 117, 129, 130, also Section 4.3]. As in the first-order cases discussed in Section 9.1, the averaging process destroys information: from $\langle\alpha_1\alpha_2\rangle$ we cannot reconstruct the ensemble $\{\alpha_1\alpha_2\}$, with all its statistical properties. We can, of course, obtain the (product of the) mean values in the limit,

$$\mathrm{M}_1^{(d)}\mathrm{M}_2^{(d)}z \lim_{t_2-t_1\to\infty}\langle\alpha_1\alpha_2\rangle \Rightarrow \left\langle(1-\eta_1)^{-1}\right\rangle\alpha_{H1}\left\langle(1-\eta_2)^{-1}\right\rangle\alpha_{H2}\Big|_{t_2-t_1\to\infty} = \langle\alpha_1\rangle\langle\alpha_2\rangle\Big|_{t_2-t_1\to\infty}. \tag{9.2.8}$$

Equally important to note, moreover, is that Eq. (9.2.7) is in the form of a Dyson's equation (9.1.7), albeit with a different "mass operator" $\hat{Q}_{12}^{(d)}$ as feedback source and different feedforward operator, analogous to $\hat{M}_1(=\hat{M}_\infty)$. In fact, if we set

$$\left.\begin{aligned}B_{12}^{(1)} &\equiv \left(1-\hat{\eta}_1^{(d)}\hat{\eta}_2^{(d)}\right)\langle\alpha_1\rangle\langle\alpha_2\rangle \equiv \hat{H}_{12}^{(1)}\langle\alpha_1\rangle\langle\alpha_2\rangle, \text{ i.e.}\\ \hat{H}_{12}^{(1)} &\equiv 1-\hat{\eta}_1^{(d)}\hat{\eta}_2^{(d)}\end{aligned}\right\}, \tag{9.2.9a}$$

then Equation (9.2.7) becomes

$$\mathrm{M}_{12}^{(\alpha)} \equiv \langle\alpha_1\alpha_2\rangle = \left(1-\hat{\eta}_2^{(d)}\right)^{-1}\hat{H}_{12}^{(1)}\langle\alpha_1\rangle\langle\alpha_2\rangle, \tag{9.2.9b}$$

which is the second-(order) second moment of the scattered field $\alpha\left(=\alpha^{(Q)}\right)$. Accordingly, the covariance of this field is given by the integral equation

$$K_{12}^{(\alpha)} \equiv \langle\alpha_1\alpha_2\rangle-\langle\alpha_1\rangle\langle\alpha_2\rangle = \left\{\left(1-\hat{\eta}_{12}^{(d)}\right)^{-1}\hat{H}_{12}^{(1)}-1\right\}\langle\alpha_1\rangle\langle\alpha_2\rangle = K_{12}^{(\alpha)}(\mathbf{R}_1,t_1;\mathbf{R}_2,t_2). \tag{9.2.10}$$

This covariance is also readily seen to be from (9.2.9a)–(9.2.10)

$$\left.\begin{aligned}\left(1-\hat{\eta}_{12}^{(d)}\right)K_{12}^{(\alpha)} &= \left(\hat{\eta}_{12}^{(d)}-\hat{\eta}_{1}^{(d)}\hat{\eta}_{2}^{(d)}\right)\langle\alpha_1\rangle\langle\alpha_2\rangle, \text{ or}\\ K_{12}^{(\alpha)} &= \left(1-\hat{\eta}_{12}^{(d)}\right)^{-1}\left(\hat{\eta}_{12}^{(d)}-\hat{\eta}_{1}^{(d)}\hat{\eta}_{2}^{(d)}\right)\langle\alpha_1\rangle\langle\alpha_2\rangle.\end{aligned}\right\} \tag{9.2.10a}$$

These higher order integral equations, analogous to (9.1.7a), formally represent particular moment solutions to the Langevin equation (9.2.3a) (without the average). For the intensity $\langle\alpha_1^2\rangle$ and variance of the scattered field we have at once from (9.2.9b) and (9.2.10):

$$I_{11}^{(\alpha)} = \mathrm{M}_{11}^{(\alpha)} = \langle\alpha_1^2\rangle = \left(1-\hat{\eta}_{11}^{(d)}\right)^{-1}\left(1-\hat{\eta}_{1}^{(d)(2)}\right)\langle\alpha_1\rangle^2, \tag{9.2.11a}$$

$$K_{11}^{(\alpha)} = \langle\alpha_1^2\rangle-\langle\alpha_1\rangle^2 = \left\{\left(1-\hat{\eta}_{11}^{(d)}\right)^{-1}\hat{H}_{11}^{(1)}-1\right\}\langle\alpha_1\rangle^2 \tag{9.2.11b}$$

(Note that $I^{(\alpha)} = \alpha^2$ is different from $I_{11}^{(\alpha)} = \langle\alpha_1^2\rangle$)

On comparison with (9.1.2) it is evident that (9.2.9b) is in the indicated Dyson form, cf. (9.1.7a). Here $\hat{H}_{12}^{(1)}\langle\alpha_1\rangle\langle\alpha_2\rangle = B_{12}^{(1)}$ takes the place respectively of the original, unperturbed field α_{H}, and $\hat{\eta}_{12}^{(d)}$ is now the new field renormalization operator, where $\hat{H}_{12}^{(1)}\langle\alpha_1\rangle\langle\alpha_2\rangle$ is effectively the new field. The desired output is $\langle\alpha_1\alpha_2\rangle$, now the space–time second-order second moment of the total field. The FOR diagram for this second-order case, namely the analogue of Fig. 9.2, is shown to be from the equivalent circuit diagram approach of Section 8.5 to be Fig. 9.9.

Correspondingly, these perturbation series solutions of (9.2.7) and (9.2.9b) are canonically represented by the integral equation

$$PSS: \qquad \mathrm{M}_{12}^{(\alpha)} = \langle\alpha_1\alpha_2\rangle = \alpha_{\mathrm{H}1}\alpha_{\mathrm{H}2} + \sum_{k=1}^{\infty}\left(\hat{\eta}_{12}^{(d)}\right)^{(k)}\hat{H}_{12}^{(1)}\langle\alpha_1\rangle\langle\alpha_2\rangle. \tag{9.2.12}$$

When the input signals are random, or has random components (ϕ), the complete ensemble average is

$$\langle\langle\alpha_1\alpha_2\rangle\rangle_{\phi} = \langle\alpha_{\mathrm{H}1}\alpha_{\mathrm{H}2}\rangle_{\phi} + \sum_{k=1}^{\infty}\left(\hat{\eta}_{12}^{(d)}\right)^{(k)}\hat{H}_{12}^{(1)}\bullet\langle\alpha_1\rangle\langle\alpha_2\rangle. \tag{9.2.12a}$$

It is also seen from (9.2.7)–(9.2.11a) that calculation of the higher order moments, for example $\langle\alpha_1\alpha_2\rangle$ here, depends explicitly on the determination of the lower order moment.

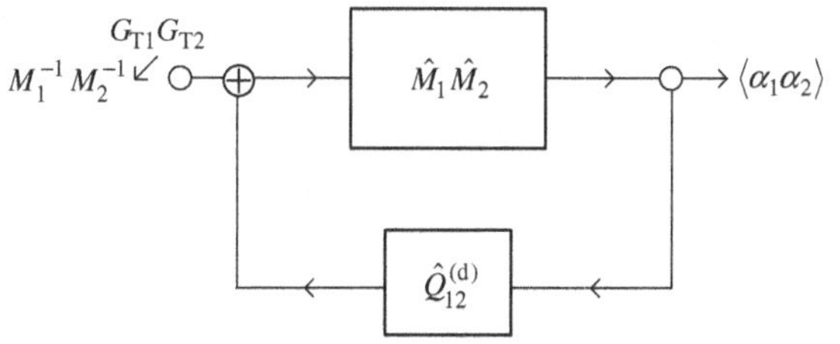

FIGURE 9.9 Feedback operational representation for the second order, that is BSE (9.2.7), (9.2.46), or equivalently here, the second-order Dyson equation (9.2.7).

Finally, all the above results apply, of course, for the Fourier transform (FT) expressions for α_1, α_2, that is, $\alpha_{01}, \alpha_{001}$, and so on, with $\hat{\eta}_1^{(d)} \to \hat{\mathsf{Y}}_{001}\hat{Q}_{001}^{(d)}$, double FT, and so on, with $\hat{Q}_{00,12}^{(d)} = \mathsf{F}_{\mathbf{R}_2,t_2}\mathsf{F}_{\mathbf{R}_1,t_1}$, cf. (9.2.4) and so on. A possible advantage in employing these transformed relations is that the operator $\hat{M}_1\hat{Q}_1, \hat{M}_2\hat{Q}_2$, commutes in (**k**,*s*)-space, Section 8.4.2. For the *intensity spectrum of the scattered field* α, we use the results of Chapter 2 for the space–time Wiener–Khintchine (W-K) relations, in particular for the discretely sampled field data $\{\alpha(\mathbf{R}_m, t_n)\}$. (See also Section 2.5, when apertures and arrays provide the coupling of the input signal to the medium, here a random scattering environment.) Accordingly, we have for the resulting wave number–frequency intensity spectrum in the general case for non-Hom-Stat scattered fields and ensembles of infinite duration, that is, $\Delta = \infty$:[18]

$$\mathsf{W}_\alpha(\boldsymbol{\nu},f)_\infty = 2\mathsf{F}_{\mathbf{R},t}\left\{\left\langle K_{11}^{(\alpha)}\right\rangle\right\} = 2\int_{-\infty}^{\infty}\left\langle K_{11}^{(\alpha)}(\mathbf{p}_1,\mathbf{p}_1-\mathbf{p})\right\rangle_\alpha e^{2\pi i\mathbf{q}\cdot\mathbf{p}}d\mathbf{p},\ d\mathbf{p} = d\mathbf{R}dt \tag{9.2.13a}$$

$\mathbf{p} = (\mathbf{R}, t), \mathbf{p}_1 \equiv (\mathbf{R}_1, t_1), \mathbf{p}_2 \equiv (\mathbf{R}_2, t_2)$, with $\mathbf{q} = (\boldsymbol{\nu}, f)$. The inverse transform is in turn

$$\left\langle K_{11}^{(\alpha)}(\mathbf{p}_1,\mathbf{p}_1-\mathbf{p})\right\rangle_\alpha = \frac{1}{2}\mathsf{F}_{\boldsymbol{\nu},f}^{-1}\{\mathsf{W}_\alpha\}\left\{\left\langle K_{11}^{(\alpha)}\right\rangle\right\} = \frac{1}{2}\int_{-\infty}^{\infty}\mathsf{W}_\alpha(\boldsymbol{\nu},f)_\infty e^{-2\pi i\mathbf{q}\cdot\mathbf{p}}d\mathbf{q},\ d\mathbf{q} = d\boldsymbol{\nu}\,d\mathbf{f} \tag{9.2.13b}$$

The conditions for which (9.2.13a) and (9.2.13b) hold are given in Sections 2.3–2.4. For truncated non-Hom-Stat continuous ensembles (where $|\Delta| < \infty$), relations similar to (9.2.13a) and (9.2.13b) are also given there. For these scattered fields K_α is specified by (9.2.10). When the ensemble is Hom-Stat we can drop the averages $(\langle\,\rangle_\alpha)$ in (9.2.13a) and (9.2.13b). For discretely sampled scattered fields $\alpha(\mathbf{R}_m, t_n)$ the structure of the covariance K_α is given by (9.2.10) and the second-order second moment by (9.2.9b).

9.2.2 The Structure of $\hat{Q}_{12}^{(d)}$

This is similar to the first-order cases ($m = 1$) where the inhomogeneity operator $\hat{Q}_1^{(d)}$ appears in the evaluation of the mean field $\langle\alpha\rangle$. Section 9.1.1.2, for the case of the second-order moments ($m = 2$) of the scattered field $\langle\alpha_1\alpha_2\rangle$, [(9.2.3a) and (9.2.3b)], the inhomogeneity operator $\hat{Q}_{12}^{(d)}$ is also always an integral operator. Its structure may also be developed explicitly in terms of feedforward and feedback operators, as additional solutions of the original Langevin equation here.

The procedure for obtaining $\hat{Q}_{12}^{(d)}$ is an obvious, although more involved, extension of the method used to obtain $\hat{Q}_1^{(d)}$ (and $\hat{\eta}_1^{(d)}$), Section 9.1.1.2. Here one starts with the defining relations (9.2.5) for $\hat{\eta}_{12}^{(d)}$ and then uses (9.2.5a) to get $\hat{Q}_{12}^{(d)}$. Equation (9.2.5) is explicitly

$$\begin{aligned}&\hat{\eta}_{12}^{(d)}\langle\alpha_1\alpha_2\rangle \equiv \langle\hat{\eta}_1\hat{\eta}_2\alpha_1\alpha_2\rangle\\&\hat{\eta}_{12}^{(d)}(1-\hat{\eta}_1)^{-1}(1-\hat{\eta}_2^{-1})\alpha_{H1}\alpha_{H2} = \langle\hat{\eta}_1\hat{\eta}_2(1-\hat{\eta}_1^{-1})(1-\hat{\eta}_2^{-1})\alpha_{H1}\alpha_{H2}\rangle\end{aligned} \tag{9.2.14a}$$

[18] The redundant statistical average $\langle\,\rangle_\alpha$ is simply to remind the reader of the averages associated with the particular random mechanism associated with the non-Hom-Stat phenomenon involved here. Usually it is implicit in the non-Hom-Stat covariance K_α itself and is consequently omitted, cf. Sections 2.3 and 2.4.

$$\therefore\ \hat{\eta}_{12}^{(d)} = \left\langle \hat{\eta}_1\hat{\eta}_2\left(1+\sum_1^\infty \hat{\eta}_1^{(n)}\right)\left(1+\sum_1^\infty \hat{\eta}_2^{(n)}\right)\Big/\left(1+\sum_1^n \hat{\eta}_1^{(n)}\right)\left(1+\sum_1^n \hat{\eta}_2^{(n)}\right)\right\rangle,$$
$$\|\hat{\eta}_1\|, \|\hat{\eta}_2\| < 1, \qquad (9.2.14b)$$

where the commutation property of η_1, η_2, and η_1 versus η_2 have been invoked [cf. Chapter 8]. Equation (9.2.14b) can be put in a more convenient form if we let

$$\hat{y}_1 \equiv \sum_1^\infty \hat{\eta}_1^{(n)} \quad \text{and} \quad \hat{y}_2 \equiv \sum_1^\infty \hat{\eta}_2^{(n)}$$

and use $(1+y)^{-1} = 1-y+y^2-\ldots$, to obtain

$$\hat{\eta}_{12}^{(d)} = \langle \hat{\eta}_1\hat{\eta}_2(1+\hat{y}_1+\hat{y}_2+\hat{y}_1\hat{y}_2)\rangle\left\langle 1-(\hat{y}_1+\hat{y}_2+\hat{y}_1\hat{y}_2)+(\hat{y}_1+\hat{y}_2+\hat{y}_1\hat{y}_2)^2-(\,)^3+\ldots\right\rangle \qquad (9.2.14c)$$

After some tedious but basically straightforward algebra, we obtain the desired result below through $\left\langle O\left(\hat{\eta}^{(s)}\right)\right\rangle$:

$$\left.\begin{aligned}\hat{\eta}_{12}^{(d)} = {} & [\langle\hat{\eta}_1\hat{\eta}_2\rangle]+\left[\left\langle\hat{\eta}_1^{(2)}\hat{\eta}_2\right\rangle+\left\langle\hat{\eta}_1\hat{\eta}_2^{(2)}\right\rangle\right]+\left[\left\langle\hat{\eta}_1^{(3)}\hat{\eta}_2\right\rangle+\left\langle\hat{\eta}_1\hat{\eta}_2^{(3)}\right\rangle+\left\langle\hat{\eta}_1^{(2)}\hat{\eta}_2^{(2)}\right\rangle-\langle\hat{\eta}_1\hat{\eta}_2\rangle^2\right]\\ & +\left[\left\langle\hat{\eta}_1^{(4)}\hat{\eta}_2\right\rangle+\left\langle\hat{\eta}_1^{(3)}\hat{\eta}_2^{(2)}\right\rangle+\left\langle\hat{\eta}_1^{(2)}\hat{\eta}_2^{(3)}\right\rangle+\left\langle\hat{\eta}_1\hat{\eta}_2^{(4)}\right\rangle\right.\\ & \left.-(\langle\hat{\eta}_1+\hat{\eta}_2\rangle)\left(\left\langle\hat{\eta}_1^{(3)}\hat{\eta}_2\right\rangle+\left\langle\hat{\eta}_1^{(2)}\hat{\eta}_2^{(2)}\right\rangle+\left\langle\hat{\eta}_1\hat{\eta}_2^{(3)}\right\rangle\right)\right]+\left\langle 0\left(\hat{\eta}^{(6)}\right)\right\rangle\end{aligned}\right\} \qquad (9.2.15)$$

$$= \hat{A}_0^{(12)}+\hat{A}_1^{(12)}+\hat{A}_2^{(12)}+\hat{A}_3^{(12)}+\hat{A}_4^{(12)}+O\left(\hat{A}_5^{(12)}\right),\ \|\hat{\eta}_{1,2}\|<1, \qquad (9.2.15a)$$

where $A_m^{(12)}, (m \geq 0)$, represent the respective quantities in the various brackets, [ ]. Using (9.2.5a) in (9.2.15) we see directly that the intensity operator $\hat{Q}_{12}^{(d)}$ is represented by

$$\left.\begin{aligned}\hat{Q}_{12}^{(d)} = {} & \hat{M}_2^{-1}\hat{M}_1^{-1}\hat{\eta}_{12}^{(d)} = [\langle\hat{Q}_1\hat{Q}_2\rangle]+[\langle(\hat{Q}_1\hat{M}_1\hat{Q}_1)\hat{Q}_2\rangle+\langle\hat{Q}_1(\hat{Q}_2\hat{M}_2\hat{Q}_2)\rangle]\\ & +[\langle(\hat{Q}_1\hat{M}_1\hat{Q}_1\hat{M}_1\hat{Q}_1)\hat{Q}_2\rangle+\langle\hat{Q}_1(\hat{Q}_2\hat{M}_2\hat{Q}_2\hat{M}_2\hat{Q}_2)\rangle+\langle(\hat{Q}_1\hat{M}_1\hat{Q}_1)(\hat{Q}_2\hat{M}_2\hat{Q}_2)\rangle]\\ & -\langle\hat{Q}_1\hat{Q}_2\rangle^2+\hat{M}_2^{-1}\hat{M}_1^{-1}([\,]_4+[\,]_5)+\ldots\end{aligned}\right\}$$

$$= \hat{B}_0^{(12)}+\hat{B}_1^{(12)}+\hat{B}_2^{(12)}+\text{etc.} \qquad (9.2.16)$$

When $\langle\hat{\eta}_1\rangle = \langle\hat{\eta}_2\rangle = 0$ or $\langle\hat{Q}_1\rangle = \langle\hat{Q}_2\rangle = 0$ there is no noticeable reduction in the complexity of the operators $\hat{\eta}_{12}^{(d)}, \hat{Q}_{12}^{(d)}$, unlike the first-order cases, (9.1.13a) and (9.1.14).

[If, however, the distribution (density) or pdf of $\hat{\eta}_1$ (or $\hat{Q}_1$) and/or $\hat{\eta}_2$ (or $\hat{Q}_2$) are symmetric, we see that these results (9.2.15), (9.2.16) reduce to

$$\hat{\eta}_{12}^{(d)} = [\langle\hat{\eta}_1\hat{\eta}_2\rangle] + \left[\left\langle\hat{\eta}_1^{(2)}\hat{\eta}_2^{(2)}\right\rangle - \langle\hat{\eta}_1\hat{\eta}_2\rangle^2\right] + \left\langle O\left(\hat{\eta}_1^{(3)}\hat{\eta}_2^{(3)}\right)\right\rangle = \hat{A}_0^{(12)} + \hat{A}_2^{(12)} - O\left(\hat{A}_4^{(12)}\right). \tag{9.2.16a}$$

Only the even in terms ($m = 2m' \geq 0$) are nonvanishing, greatly simplifying the expressions for these second-order operators $\hat{\eta}_{12}^{(d)}, \hat{Q}_{1,2}^{(d)}$. In any case, these operators, which play a central role in the evaluation of the various moments of the scattered field, namely moment solutions to the generalized Langevin equation representing the ensemble of propagation equations $\{L^{(1)}\alpha^{(Q)} = -G_T\}$, and so on, are vastly more complex (i.e. "difficult") than those representing homogeneous media. It is this situation that motivates our efforts to obtain the purely statistical solution.

9.2.3 Higher-Order Moment Solutions ($m \geq 3$) and Related Topics

Solutions for the higher-order moments $\langle\alpha_1, \ldots, \alpha_m\rangle$ may be developed formally in similar fashion to that described in Section 9.2. As expected, the resulting structures become progressively more complex. The principal formal result here is that *all these moments may be expressed in the FOS form of a Dyson-type equation, cf.* (9.1.7), (9.2.7), *or "first-order" form*, of the type

$$\left.\begin{aligned} \mathrm{M}_{12\ldots m,\alpha} &\equiv \langle\alpha_1\ldots\alpha_m\rangle = \left(\hat{1} - \hat{\eta}_{12\ldots m}^{(d)}\right)^{-1} B_{12\ldots m}^{(m-1)} \\ \text{with} \quad \hat{H}_{12\ldots m}^{(m-1)} &= \left(\hat{1} - \hat{\eta}_{12\ldots m}^{(d)}\right)^{-1} \end{aligned}\right\} m \geq 2. \tag{9.2.17}$$

Similarly, one has

$$B_{12\ldots m}^{(m-1)} \equiv \hat{H}_{12\ldots m}^{(m-1)}\langle\alpha_1\rangle\ldots\langle\alpha_m\rangle;\ \hat{\eta}_{12\ldots m}^{(d)} \equiv \hat{\eta}_{12\ldots m}^{(d)}(R_1, t_1; \ldots; R_m, t_m | R_1', t_1'; \ldots; R_m', t_m')$$
$$\equiv \hat{M}_1 \ldots \hat{M}_m \hat{Q}_{12\ldots m}^{(d)} \tag{9.2.17a}$$

and $\hat{H}_{12\ldots m}^{(m-1)} = \hat{f}\left(\hat{\eta}_1^{(d)}, \ldots, \hat{\eta}_{12\ldots m-1}^{(d)}\right)$. Again, $\hat{\eta}_{12\ldots m,\infty}^{(d)}$ is the equivalent (mth-order) field renormalization operator, cf. (9.2.5), wherein $\hat{Q}_{12\ldots m}^{(d)}$ is the corresponding mth-order equivalent deterministic mass operator, defined by[19]

$$\hat{Q}_{12\ldots m}^{(d)}\langle\alpha_1\ldots\alpha_m\rangle \equiv \langle\hat{Q}_1\ldots\hat{Q}_m\alpha_1\ldots\alpha_m\rangle, m \geq 2 \tag{9.2.18}$$

Here in detail, $\hat{Q}_{12\ldots m}^{(d)}$ is the deterministic function

$$\hat{Q}_{12\ldots m}^{(d)} = \hat{Q}(\mathbf{R}_1, t_1; \mathbf{R}_2, t_2; \ldots; \mathbf{R}_m, t_m | \mathbf{R}_1', t_1'; \ldots; \mathbf{R}'_m, t_m') \tag{9.2.18a}$$

[19] Note that for $m = 1$, $\langle\alpha_1\rangle \to \alpha_H(= -\hat{G}_T(S_{in}))$, obtain (9.1.7a). For $m = 2$ we also obtain (9.2.8b), as expected.

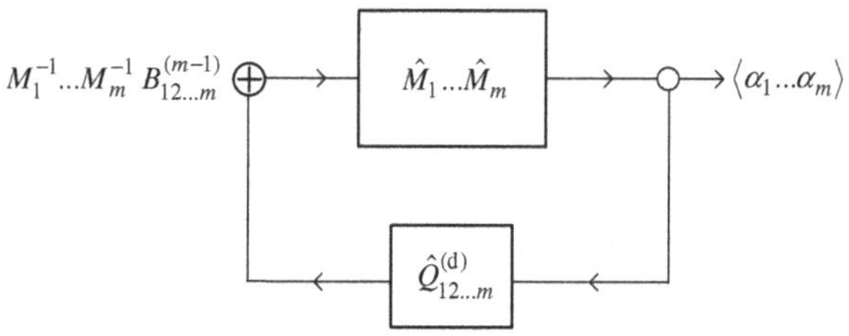

FIGURE 9.10 *m*th-order Dyson equation: FOR for the *m*th-order ($m \geq 2$) FOS, where (9.2.17) provides the mth-order (total) field moment $\langle\alpha_1...\alpha_m\rangle$.

cf. (9.2.4). This type of definition of the "mass operator" $\hat{Q}^{(d)}$ as noted above[20], cf. Eq. (9.2.7) ff., in fact as we have already seen from Section 9.1.2.2, is the basis for the ensuing "first-order", or Dyson form, (9.2.17), in general case. Figure 9.10 shows the generalized FOR for the FOS (9.2.14a).

The expressions for $B_{12...m}^{(m-1)}$(9.2.17a) form a hierarchy of increasing complexity as m increases, each new relation (m) depending on the various lower order results ($m-1$, $m-2, \ldots, 1$). Following the procedures leading to Eq. (9.2.2)–(9.2.7), for each m (≥ 3), we obtain after considerable (operational) algebra the explicit results

$$B_1^{(0)} = 1\langle\alpha_1\rangle; \hat{H}_1^{(0)} = \hat{1};\ \hat{\eta}_{1,\infty}^{(d)}, \tag{9.2.19a}$$

$$B_{12}^{(1)} = \hat{H}_{12}^{(1)}\langle\alpha_1\rangle\langle\alpha_2\rangle;\ \hat{H}_{12}^{(1)} = \hat{1}-\hat{\eta}_1^{(d)}\hat{\eta}_2^{(d)};\ \text{Eq. (9.2.9a)}, \tag{9.2.19b}$$

$$\begin{aligned} B_{123}^{(2)} &= \hat{H}_{123}^{(2)}\langle\alpha_1\rangle\langle\alpha_2\rangle\langle\alpha_3\rangle, \\ B_{1234}^{(3)} &= \hat{H}_{1234}^{(3)}\langle\alpha_1\rangle\langle\alpha_2\rangle\langle\alpha_3\rangle\langle\alpha_4\rangle, \end{aligned} \tag{9.2.19c}$$

cf. Eq. (9.2.17), et. seq. Here specifically we have written for conciseness $\hat{A}_1 \equiv \hat{\eta}_1^{(d)}$, $\hat{A}_{12} \equiv \hat{\eta}_{12}^{(d)}$, and so on so that $\hat{H}_{123}^{(2)}$, $\hat{H}_{1234}^{(3)}$ are explicitly

$$\begin{aligned} \hat{H}_{123}^{(2)} = {} & \left(\hat{1}-\hat{A}_{12}\right)^{-1}\left(\hat{1}-\hat{A}_3\right)\hat{A}_{12}\hat{H}_{12}^{(1)} + \left(\hat{1}-\hat{A}_{13}\right)^{-1}\left(\hat{1}-\hat{A}_2\right)\hat{A}_{13}\hat{H}_{13}^{(1)} \\ & + \left(\hat{1}-\hat{A}_{23}\right)^{-1}\left(\hat{1}-\hat{A}_1\right)\hat{A}_{23}\hat{H}_{23}^{(1)} + \left[\hat{1}-\hat{A}_1\hat{A}_2-\hat{A}_1\hat{A}_3-\hat{A}_2\hat{A}_3+2\hat{A}_1\hat{A}_2\hat{A}_3\right]; \end{aligned} \tag{9.2.20}$$

$$\begin{aligned} \hat{H}_{1234}^{(3)} = {} & \Big\{\Big[\left(\hat{1}-\hat{A}_{123}\right)^{-1}\left(\hat{1}-\hat{A}_4\right)\hat{A}_{123}\hat{H}_{123}^{(2)} + \left(\hat{1}-\hat{A}_{124}\right)^{-1}\left(\hat{1}-\hat{A}_3\right)\hat{A}_{124}\hat{H}_{124}^{(2)} \\ & + \left(\hat{1}-\hat{A}_{134}\right)^{-1}\left(\hat{1}-\hat{A}_2\right)\hat{A}_{134}\hat{H}_{134} + \left(\hat{1}-\hat{A}_{234}\right)^{-1}\left(\hat{1}-\hat{A}_1\right)\hat{A}_{234}\hat{H}_{234}\Big] \\ & + \Big[\left(\hat{1}-\hat{A}_{34}\right)^{-1}\left(\hat{1}-\hat{A}_1\right)\left(\hat{1}-\hat{A}_2\right)\hat{A}_{34}\hat{H}_{34}^{(1)} + \left(\hat{1}-\hat{A}_{24}\right)^{-1}\left(\hat{1}-\hat{A}_1\right)\left(\hat{1}-\hat{A}_3\right)\hat{A}_{24}\hat{H}_{24}^{(1)} \end{aligned}$$

[20] The result (9.2.7), which stems directly from (9.2.4), is a simpler definition of the mass operator than one finds, for example, in the original literature ([4], ¶ 60, pp. 343–347 and Eq. (36a). The $\hat{Q}_{12}^{(d)}$ derived in Section 9.2.2 generalizes much more easily in these higher moment cases, cf. (9.2.17), et. seq.

$$+ (\hat{1}-\hat{A}_{23})^{-1}(\hat{1}-\hat{A}_1)(\hat{1}-\hat{A}_4)\hat{A}_{23}\hat{H}_{23}^{(1)} + (\hat{1}-\hat{A}_{14})^{-1}(\hat{1}-\hat{A}_2)(\hat{1}-\hat{A}_3)\hat{A}_{14}\hat{H}_{14}^{(1)}$$
$$+ (\hat{1}-\hat{A}_{13})^{-1}(\hat{1}-\hat{A}_2)(\hat{1}-\hat{A}_4)\hat{A}_{13}\hat{H}_{13}^{(1)} + (\hat{1}-\hat{A}_{12})^{-1}(\hat{1}-\hat{A}_3)(\hat{1}-\hat{A}_4)\hat{A}_{12}\hat{H}_{12}^{(1)}\Big]$$
$$+ \Big[\hat{1}-(\hat{A}_1\hat{A}_2+\hat{A}_1\hat{A}_3+\hat{A}_1\hat{A}_4+\hat{A}_2\hat{A}_3+\hat{A}_2\hat{A}_4+\hat{A}_3\hat{A}_4)$$
$$+2(\hat{A}_1\hat{A}_2\hat{A}_3+\hat{A}_1\hat{A}_2\hat{A}_4+\hat{A}_2\hat{A}_3\hat{A}_4+\hat{A}_1\hat{A}_3\hat{A}_4)-3\hat{A}_1\hat{A}_2\hat{A}_3\hat{A}_4\Big]\Big\}. \qquad (9.2.21)$$

These higher order moments are important if we with to examine such quantities as the statistics of the field intensity $I^{(\alpha)} = \alpha^2$, namely its mean square, $\langle I_\alpha^2\rangle = \langle\alpha^4\rangle$, its fluctuation $\langle I_\alpha^2\rangle - \langle I_\alpha^2\rangle = \langle\alpha^4\rangle - \langle\alpha^2\rangle^2$, and its covariances, $\langle(\alpha_1^2-\langle\alpha_1^2\rangle)(\alpha_2^2-\langle\alpha_2^2\rangle)\rangle$, and so on. The higher order moments are also important in assessing the departure from Gauss statistics. Other moments of interest are the central moments $\Delta I_\alpha^{(12\ldots m)} \equiv \langle(\alpha_1-\langle\alpha_1\rangle)(\alpha_2-\langle\alpha_2\rangle)\ldots(\alpha_m-\langle\alpha_m\rangle)\rangle, m \geq 1$, which are readily obtained as various combinations of $M_{12\ldots m,\alpha}$ (9.3.1), on expansion of $\Delta I_\alpha^{(12\ldots m)}$, explicitly with the help of (9.3.3), (9.3.4), (9.3.5). Similarly, (cf. Sections 9.1.1.1 and 9.2.2), one can obtain the associated deterministic medium operators $\hat{Q}_{12\ldots m}^{(d)}$, (9.3.2), with, of course, a very considerable increase in the complexity of the resulting structure. This once more reinforces the argument for computational solutions. Our operational formulation here presents an initial step, from the mathematical formulation to operational procedures for calculation. At the same time it provides a physical background, motivating use of a generalized feedback process by which ultimately numerical results may be obtained.

9.2.4 Transport Equations

The concept of the equivalent deterministic medium (EDM), as developed above in Section 9.1.1.1, allows us to obtain an appropriate form of Dyson equation, cf. (9.1.7a), (9.2.9), (9.2.10), and (9.2.17). This is now the basic integrodifferential equation (of the FOR) governing *the propagation of the field statistic in question*, for example, $\langle\alpha\rangle$, $\langle\alpha_1\alpha_2\rangle$, and so on. The corresponding FOS and PSS accordingly represent the desired formal solutions for these field statistics. Conversely, the actual propagation equations for the field statistics are obtained by operator inversion of the FOR. This is a form of "transport" equation, where the quantity transported or "propagated" is some field statistic, and the medium supporting it is now the EDM for that statistic.

As a first example, let us consider the propagation of the mean field $\langle\alpha(\mathbf{R},t)\rangle$. Since $\hat{M}_\infty^{-1}\hat{\eta}_{1,\infty}^{(d)} = \hat{Q}_1^{(d)}$, the EDM here, we may apply $\hat{M}_\infty^{-1}$ ($\equiv \hat{L}^{(0)}$, cf. Section 8.1.5) to (9.1.7), to get directly

$$\left(\hat{L}^{(0)}-\hat{Q}_1^{(d)}\right)\langle\alpha\rangle = \hat{M}_\infty^{-1}\alpha_{\mathrm{H}} = -G_{\mathrm{T}}\ [+\text{b.c.s}+\text{i.c.s}] \qquad (9.2.22)$$

for this *first-order transport equation*. However, this is not an ordinary dynamical equation for the propagation of the field α, which is usually a partial differential equation, but is rather *an integro-differential equation for the propagation of* $\langle\alpha\rangle$, since $\hat{Q}_1^{(d)}$ is always an integrodifferential operator [cf. (9.1.12)]. The initial conditions are those of the source (G_{T}), and the boundary conditions are those incorporated in $\hat{M}_\infty$, namely, an *unbounded medium*. Boundary effects, if any, are included in the inhomogeneity operator, $\hat{Q}$, contained in $\hat{Q}^{(d)}$, cf. (9.1.10). Sometimes it may be easier to work with the wave number–frequency-form. This gives the WNF transport equation (with the help of Section 8.4.1):

$$\left(\hat{L}_o^{(0)} - \hat{Q}_{oo}^{(d)}\right) \oplus \langle \alpha_0 \rangle = -G_{oT}, \tag{9.2.23}$$

where $\hat{L}_o^{(0)} \equiv \mathsf{F}_R^{-1}\mathsf{F}_t\hat{L}^{(0)}$. With kernels of the type (Footnote 3), closed form solutions like the (integrand of) [(9.1.18a), (9.1.21) are possible here.

An a second example of a transport equation, let us now consider the general second-order cases examined in Section 9.2.1, where the field statistic is now

$$\mathrm{M}_{12}^{(\alpha)}(\mathbf{r}_1, t_1; \mathbf{r}_2, t_2) \equiv \langle \alpha_1 \alpha_2 \rangle. \tag{9.2.24}$$

The formal solutions for $\mathrm{M}_{12}^{(\alpha)}$ is the FOS (9.2.7), in the indicated Dyson form (9.2.7b). The associated *transport equation* for $\mathrm{M}_{12}^{(\alpha)}$ is found after forward multiplication by $\hat{M}_{2,\infty}^{-1}\hat{M}_{1,\infty}^{-1}$, with (9.2.7a), to be

$$\begin{aligned}
\left(\hat{L}_2^{(0)}\hat{L}_1^{(0)} - \hat{Q}_{12}^{(d)}\right)\hat{M}_{12}^{(\alpha)} &= \hat{M}_{2,\infty}^{-1}\hat{M}_{1,\infty}^{-1}\left(\hat{1} - \hat{\eta}_{1,\infty}^{(d)}\hat{\eta}_{2,\infty}^{(d)}\right)\langle\alpha_1\rangle\langle\alpha_2\rangle + [\text{b.c.s} + \text{i.c.s}] \\
&= \hat{L}_2^{(0)}\langle\alpha_2\rangle\hat{L}_1^{(0)}\langle\alpha_1\rangle - \hat{Q}_1^{(d)}\hat{Q}_2^{(d)}\langle\alpha_1\rangle\langle\alpha_2\rangle + [\text{b.c.s} + \text{i.c.s}] \quad , \\
&= G_{\mathrm{T}1}G_{t2} - \left\{G_{\mathrm{T}2}\hat{L}_1^{(0)}\langle\alpha_1\rangle + G_{\mathrm{T}1}\hat{L}_2^{(0)}\langle\alpha_2\rangle\right\} + [\text{b.c.s} + \text{i.c.s}]
\end{aligned} \tag{9.2.25}$$

since $\hat{Q}_1^{(d)}\langle\alpha_1\rangle = \hat{L}_1^{(0)}\langle\alpha_1\rangle + G_{\mathrm{T}1}$ from Eq. (9.2.22). Again, $\hat{Q}_{12}^{(d)}$ is always an integral operator (cf. Section 9.2.2); the initial conditions are those of the source (G_{T}), for an effectively unbounded medium $(\hat{M}_\infty)$ here, with boundary effects once more included in $\hat{Q}$, as a component of $\hat{Q}_{12}^{(d)}$, cf. (9.2.15). The transport equation (9.2.25) for $\langle\alpha_1\alpha_2\rangle$ is clearly symmetrical in its indexes (1, 2). It is an integrodifferential equation, of a more complex character than Eq. (9.2.25) for $\langle\alpha\rangle$, and, in fact, requires the solution of (9.2.22) in order to obtain $\mathrm{M}_{12}^{(\alpha)} = \langle\alpha_1\alpha_2\rangle$. As a special case of (9.2.25), we find that the transport equation for the *field intensity* $I_\alpha\left(= \langle\alpha_1^2\rangle\right)$ becomes

$$\left(\hat{L}_1^{(0)^2} - \hat{Q}_{11}^{(d)}\right)I_\alpha = G_{\mathrm{T}1}^2 - \left\{2G_{\mathrm{T}1}\hat{L}_1^{(0)}\langle\alpha_1\rangle\right\} + [\text{b.c.s} + \text{i.c.s}], \tag{9.2.26}$$

with the integral

$$I_{11}^{(\alpha)} \equiv \langle\alpha_1^2\rangle = \left(1 - \hat{\eta}_{11}^{(d)}\right)^{-1}\left(1 - \hat{\eta}_1^{(d)(2)}\right)\langle\alpha_1\rangle^2. \tag{9.2.26a}$$

Transport equations for higher order moments of the field may be established in similar fashion from the results of Section 9.2.3. These, in turn, require a hierarchy of lower order solutions, with their associated lower order transport equations. Because of their integral character transport equations for field statistics are always global relations, like their equivalents the FOR, as distinct from the (usually) differential equations directly governing propagation of the field, which are essentially local. The integral of a transport equation is the corresponding statistic of the scattered field itself, that is Eqs. (9.2.22, 9.2.25) for $\langle\alpha_1\rangle$ and $\langle\alpha_1\alpha_2\rangle$ here, as is easily demonstrated by direct integration of [(9.2.22)–(9.2.25)]. A possible practical utility of the transport equation vis-à-vis the corresponding FOS is that it may be easier to approximate and to solve numerically. In general, this is a computational task on the order of that for large-scale weather prediction; indeed, what we are seeking is a kind of "weather-prediction" in the medium, as represented by the various field statistics themselves.

9.2.5 The Gaussian Case

When the statistics of the random parameters in the Langevin equations are normal, that is, Gaussian, the kernels $Q_1^{(d)}|_{V,S}$ simplify considerably, provided $\langle Q_{V,S}\rangle = 0$, cf. (9.1.14). This is usually a reasonable assumption for most applications, from the view point of the Central Limit Theorem (CLT); (see Section 9.1.3.1, for instance). The corresponding reduction in complexity of the mass operator kernels $Q_1^{(d)}$ then follows. The higher order pdfs of the mass operator kernels now obey the well-known relations ([17], Section 7.3.3 ibid)

$$\left.\begin{aligned} E\{z_1z_2,\ldots,z_{2m}\} &= \sum_{\text{all pairs } k\neq l}\prod^{m}\overline{z_kz_l} = \sum_{\text{all pairs}}\left(\overline{z_kz_l}-\overline{z_pz_r},\ldots,\overline{z_qz_s}\right)_{k\neq l,p\neq r,\ldots,q\neq s,\text{ etc}} \\ E\{z_1z_2,\ldots,z_{2m+1}\} &= 0;\; z = Q_{V,S}, \end{aligned}\right\} \tag{9.2.27}$$

The number of averages over pairs in the above is equal to the number of different ways $2m$ different variables $z_1,\ldots,z_{2m}$ can be saluted in pairs. This number is[21] $(2m)!/2^m m!$. For example, when $m=2$, we get three ways into which $\overline{z_1z_2z_3z_4}$ can be factored into products of (first-order) covariances, that is,

$$m=2:\quad \mathbf{E}\{z_1z_2z_3z_4\} = \overline{z_1z_2}\bullet\overline{z_3z_4}+\overline{z_1z_3}\bullet\overline{z_2z_4}+\overline{z_1z_4}\bullet\overline{z_2z_3} \tag{9.2.28}$$

And for $m=3$, we see that the number of pairs is now 15, with an (approximately) exponential increase as the order (m) increases. We note in passing that $\overline{z_1z_2,\ldots,z_m}$ may also be used to derive other moments in the Gaussian case. For example, consider

$$\langle z_1z_2z_3z_4\rangle_{z_2\to z_1} = \langle z_1^2z_3z_4\rangle = \langle z_1^2\rangle\bullet\langle z_3z_4\rangle+2\langle z_1z_2\rangle\bullet\langle z_1z_4\rangle,\text{ or} \tag{9.2.28a}$$

$$\langle z_1z_2z_3z_4\rangle_{z_1\to z_2\to z_3}\langle z_3^3z_4\rangle,\text{ etc.,} \tag{9.2.28b}$$

again with $z = Q_V$ or Q_S and $\langle z\rangle = \langle Q\rangle = 0$.

[21] There are $(2m)!$ permutations, but 2^m interchanges of argument and $m!$ permutations of the factors give no additional separations into new pairs.

With Gaussian statistics for the kernel Q, and the condition $\langle Q\rangle = 0$, only the even order terms involving Q in (9.1.14) are nonvanishing, that is $m+11 = 2n$, $m \geq 1$, obeying (9.2.27). Accordingly, we have

$$m=1(n=2):\left\langle \hat{Q}(\hat{M}\hat{Q})^{(m)}\right\rangle = \langle \hat{Q}\hat{M}\hat{Q}\rangle = \langle \hat{Q}(\mathbf{R}_1,t_1)\hat{M}(\mathbf{R}_1,t_1|\mathbf{R}',t')\hat{Q}(\mathbf{R}',t')\rangle = \hat{M}_1\langle \hat{Q}_1\hat{Q}_2\rangle$$
$$= \int g^{(0)}\left(\mathbf{R}_1,t_1|\mathbf{R}_2,t_2)K_Q(\mathbf{R}_1,t_1;\mathbf{R}',t')()_{\mathbf{R}',t'}\mathbf{dR}'dt' \qquad (9.2.29a)$$

where $K_Q = \langle Q_1Q_2\rangle, (\mathbf{R}_2,t_2) = (\mathbf{R}',t')$ is the covariance of Q. Similarly, we see that by (9.2.27)

$$m=2(\neq 2n):\quad \left\langle \hat{Q}(\hat{M}\hat{Q})^{(2)}\right\rangle = 0, \qquad (9.2.29b)$$

and

$$m=3(n=2):\left\langle \hat{Q}(\hat{M}\hat{Q})^{(3)}\right\rangle = \langle \hat{Q}(\hat{M}\hat{Q}\hat{M}\hat{Q}\hat{M}\hat{Q})\rangle$$
$$= \langle \hat{Q}(\mathbf{R}_1,t_1)\hat{M}(\mathbf{R}_1,t_1|\mathbf{R}_2,t_2)\hat{Q}(\mathbf{R}_2,t_2)\ldots\hat{Q}(\mathbf{R}',t')\rangle$$
$$= \int\ldots\int g_1^{(0)}g_2^{(0)}g_3^{(0)}\langle \hat{Q}(\mathbf{R}_1,t_1)\ldots\hat{Q}(\mathbf{R}',t')\rangle()_{\mathbf{R}',t'}\mathbf{dR}_2\ldots\mathbf{dR}'dt_2\ldots dt'$$

where

$$\left\langle \hat{Q}(\mathbf{R}_1,t_1)\ldots\hat{Q}(\mathbf{R}',t')\right\rangle = K_{Q_{12}}K_{Q_{34}} + K_{Q_{13}}K_{Q_{24}} + K_{Q_{14}}K_{Q_{23}} \text{ by } (9.2.28), \qquad (9.2.29c)$$

in which $K_{Q_{12}} = \left\langle \hat{Q}(\mathbf{R}_1,t_1)\hat{Q}(\mathbf{R}_2,t_2)\right\rangle = K_{Q_{12}}(\mathbf{R}_1,t_1;\mathbf{R}_2,t_2)$, and so on. The higher order moments are computed in the same fashion, with an ever increasing complexity of sums of covariance products. The result is a simplification of moment structure: only first-order covariances are involved, but the overall moment structure (i.e., the left member of (9.2.29c)) is more complicated when its Gaussian character is explicitly exhibited. In the non-Gaussian cases these higher moments do not factor, however, and are thus more compact, that is "simpler." The drawback here is that they are often unknown. When they are calculated the result can provide a simplification.

Finally, observe that the results above for volume scatter in the Gaussian case carry over formally for surface scatter if we simply replace $\int g^0 Q_V(\,)_{\mathbf{R}}\mathbf{dR}'$ in the above by

$$g^{(0)}|_V \to g_\infty^{(0)}\frac{\partial}{\partial n}(\,) - (\,)\frac{\partial g_\infty}{\partial n};\ \int_{V_T}(\,)\mathbf{dR}' \to \oint_S(\,)dS(\mathbf{R}'\in S), \quad \text{and} \quad Q_V \to Q_S, \qquad (9.2.30)$$

cf. (9.1.46). See also Sections 9.1.2.2 and 9.1.3. However, even when $\hat{Q}$ is Gaussian, the random scatter is usually not.

9.2.6 Very Strong Scatter: Saturation $\|\hat{\eta}\| \simeq 1$

In all our treatment above we are dealing with a random perturbation system, equivalently represented by an ensemble of integrodifferential equations, namely, a Langevin propagation equation, where the governing condition on the global scattering operator is $\|\hat{\eta}_\infty\| < 1$, [cf. Section 8.2.2, Eqs. (8.3.12) et. seq.]. What happens when very strong scatter is produced, that is, $\|\eta\| \to 1$? More particularly, what happens under this condition to the Langevin equation of propagation, namely the randomized version of Eqs. (8.3.20) and (8.3.22)?

9.2.6.1 An Example As a simple but often useful case, let us assume that the medium with no surface sources on S_0 (cf. 8.1.47) is highly dissipative and still homogeneous. We begin with the governing equation, which is here the modified, time-variable Helmholtz relation with absorption, that is, the Green's function (G.F.) (8.1.31), extended to an ensemble of such equations by the random nature of the constant absorption parameter α_R. We write for the G.F. $\left(g_\infty^{(0)}\right)$:

$$\left\{\left(\nabla^2 - a_R^2 - \frac{1}{c_0^2}\frac{\partial^2}{\partial t^2}\right) g_\infty^{(0)} = -\delta(\mathbf{R}-\mathbf{R}')\delta(t-t')\right\}, \tag{9.2.31}$$

where the braces { } explicitly remind us (for the moment[22]) of the ensemble nature of this Langevin equation. For a general distributed source (density) $G_T(\mathbf{R}, t)$ we have the corresponding field, in the case of the infinite (here ideal) medium cf. (8.1.44) with vanishing initial conditions. Here the governing Langevin equation from which $\alpha^{(Q)}$ derived is the first relation in (8.1.38), namely,

$$\left.\begin{aligned}\left(\nabla^2 - a_R^2\frac{\partial}{\partial t} - \frac{1}{c_0^2}\frac{\partial^2}{\partial t^2}\right)\alpha^{(Q)} &= -G_T(\mathbf{R}, t), \mathbf{R} \in V_T, \\ &= 0, \qquad \mathbf{R} \notin V - V_T,\end{aligned}\right\} t_0^- \le t \le t_R \tag{9.2.32}$$

for the initial period $(t_R - t_0)$ of field buildup. (For a discussion of G_T, see Sections 8.1.1 and 8.1.5.)

Next, for saturation or very strong scattering, we have the condition that at some time t_R later than the initiation time t_0^- of the field $\alpha^{(Q)}(\mathbf{R}, t)$

$$\overline{\left| a_0^2 \frac{\partial \alpha^{(Q)}}{\partial t}\right|} >> \overline{\left|\frac{1}{c_0^2} - \frac{\partial^2}{\partial t^2}\right|}, \text{ or more simply, that } c_0 \sim \infty, \text{ for } t_R \le t. \tag{9.2.33}$$

Now for these times

$$\left.\begin{aligned}\left(\nabla^2 - a_R^2\frac{\partial}{\partial t}\right)\alpha^{(Q)} &= -G_T(\mathbf{R}, t), \mathbf{R} \in V_T \\ &= 0, \qquad \mathbf{R} \notin V_T\end{aligned}\right\}; \ t_R \le t, \tag{9.2.34a}$$

is the governing equation. Applying $c_0 \sim \infty$, (with proper care for the limit in the exponential of the G.F. $G(\rho, \Delta t)$ of (8.1.31c)), we obtain:

$$\left.\begin{aligned} G(\rho, \Delta t) \doteq \frac{a_R\left(e^{-a_R^2\rho^2/4\Delta t}\right)}{8(\pi\Delta t)^{3/2}} 1_{\Delta t}; \quad t_R \le t; \ \Delta t_R = t - t_R \\ \rho = \left|\mathbf{R} - \mathbf{R}'\right| \end{aligned}\right\}; \tag{9.2.34b}$$

[22] For the most part we drop the braces in Chapters 8–9, it being understood that we are always dealing here with ensembles. As stated above, a_R^2 is a real, random constant with respect to the ensemble (9.2.31), namely a particular constant value for each ensemble representation.

$b_0 = a_{\mathrm{R}}$ is a random constant with respect to the random G.F. (9.2.34b). Correspondingly, the resulting random diffusions field $\alpha^{(Q)}(\mathbf{R},t)$ here is (from (8.1.47) with no surface sources on S_0 and $\alpha_H = 0$ initially) the stochastic solution

$$\alpha^{(Q)}(\mathbf{R},t) = \int_{t_{\mathrm{R}} > t_0^-}^{t} \frac{a_{\mathrm{R}} dt'}{8(\pi\Delta t)^{3/2}} \int_{V_{\mathrm{T}}} e^{-a_{\mathrm{R}}^2(\mathbf{R}-\mathbf{R}')^2/4\Delta t} G_{\mathrm{T}}(\mathbf{R}',t')\mathbf{dR}'. \tag{9.2.35}$$

The initial random field $a^{(Q)}(t_0 \le t \le t_R)$ is the solution of (9.2.32), which is obtained from the first term of (8.1.44), now with the randomized G.F. (8.1.31c). (Particular solutions of (9.2.34a) and (9.2.32) are given directly by (9.2.35), and (9.2.32), with the corresponding particular values of $a_R = a$ constant.)

The diffusion mechanism, of course, depends on the nature of the medium and the field applied to it. In all cases it takes a finite time (t_R) for the applied field to reach a saturated state, where diffusion is achieved. For gases and liquids (the atmosphere, water, and similar media) the time to equilibrium (t_R) as a diffusion process is a function of the speed of the propagation heat, and this accordingly depends on the mean free path λ_g of the gas or liquid molecules. Here this speed is that of sound in the medium, provided the excitation is acoustic. When the excitation is electromagnetic, propagation involves different mechanisms, impeding the free-space speed—that of light—due to scattering and in many cases the intrinsic material characteristics (molecular and atomic), particularly for solids, dielectrics and even conductors (resistance), with a conversion into heat. In any case, diffusion can become a significant component of the propagation process, if not the only one. (For details see Section 2.5 of Ref. [19]; Sections 9.12, 10.4, and 12.1 of Ref. [21]; and Sections 7.5 and 7.7 of Ref. [22].)

In a more general model the random parameter a_R is a function of position and time, $a_R = a_{\mathrm{R}}(\mathbf{R},t)$, with a corresponding adjustment of the saturation condition (9.2.33) above, condensed into a simple requirement that $c_0 \sim \infty$, for $t \ge t_R$. If there are other random parameters, that is $c = c(\mathbf{R},t)$ with

$$1/c^2(\mathbf{R},t) = n(\mathbf{R},t)/c_0^2 = [1+\varepsilon(\mathbf{R},t)]/c_0^2, \tag{9.2.36}$$

(where $n(\mathbf{R},t)$ is often called the *index of refraction*, with $n(\mathbf{R},t)$, or equivalently $\varepsilon(\mathbf{R},t)$ a random quantity), the Langevin equation (9.2.32) is now inhomogeneous and becomes specifically

$$\left.\begin{aligned} \left(\nabla^2 - a_R^2(\mathbf{R},t)\frac{\partial}{\partial t} - \left[\frac{1+\varepsilon(\mathbf{R},t)}{c_0^2}\frac{\partial^2}{\partial t^2}\right]\right)\alpha^{(Q)} &= -G_T(\mathbf{R},t), \mathbf{R} \in V_{\mathrm{T}} \\ &= 0, \qquad \mathbf{R} \in V - V_T \end{aligned}\right\}, \; t_0^- < t < t_{\mathrm{R}} \tag{9.2.37}$$

with the left member alternatively expressible as

$$\left[\left(\nabla^2 - a_{0\mathrm{R}}^2\frac{\partial}{\partial t} - \frac{1}{c_0^2}\frac{\partial^2}{\partial t^2}\right) - \left(\hat{a}_{\mathrm{R}}^2/(\mathbf{R},t)\frac{\partial}{\partial t} + \frac{\varepsilon(\mathbf{R},t)}{c_0^2}\frac{\partial^2}{\partial t^2}\right)\right]\alpha^{(Q)} \equiv \left(\hat{L}^{(0)} - \hat{Q}\right),\alpha^{(Q)}, t_0 < t_{\mathrm{R}}. \tag{9.2.38}$$

Here $\hat{Q}$ is the random inhomogeneity operator in this case, because of $\hat{a}_{\mathrm{R}}^2(> 0)$ and ε, (with $\bar{\varepsilon} = 0$ usually). When the saturation condition (i.e., the extended version of (9.2.33), or $c_0 \sim \infty$ for simplicity is achieved), and when the fluctuations in $\hat{a}_{\mathrm{R}}^2$ are small vis-à-vis

a_{0R}^2, with $t \geq t_R$, then it is clear that [(9.2.37) and (9.2.38)] reduce to the diffusion equation (9.2.34a) where now $a_R\left(=\sqrt{a_{0R}^2+\hat{a}_R^2}\right)=a_R(\mathbf{R},t)(>0)$ is a random variable and furthermore is a function of position and time, obeying (9.2.33). In fact, *whenever the absorption term*$(\sim(\partial/\partial t)\alpha)$*greatly exceeds the other time dependent terms in the propagation equation, one arrives at a diffusion equation after a certain term* $t \geq t_R$. In other words, *a state of saturation exists*. Alternatively, an intermediate state during $\left(t_0^- \leq t < t_R\right)$ occurs, where propagation obeys a wave equation of the type (9.2.37), and where individual solutions of the form [(8.3.20) and (8.3.22)] represent the resulting stochastic field $\alpha^{(Q)}(\mathbf{R},t)$. In terms of the mass operator $\hat{\eta}_\infty$, for the latter one has $0 \leq \|\hat{\eta}\| < 1$ and for the former, $\|\hat{\eta}\| \to 1-\varepsilon$, where ε is a small positive number.

9.2.6.2 Solutions to the Homogeneous Langevin Equation in the Diffusion Limit $\|\eta\| \sim 1$ Because the diffusion field (9.2.35) is nonrecursive, that is, is not subject to feedback (cf. Fig. 8.18), the mass operator $\hat{Q}$ vanishes. Accordingly, the nth-order moment of (9.2.35) can be written

$$\left\langle\alpha_1^{(Q)},\ldots,\alpha_n^{(Q)}\right\rangle = \left\langle\alpha^{(Q)}(\mathbf{R}_1,t_1)\ldots\alpha^{(Q)}(\mathbf{R}_n,t_n)\right\rangle$$
$$= \left\langle\prod_{m=1}^{n}\int_{t_0^{(1)-}}^{t_1}\frac{dt^{(1)}}{8(\pi\Delta t_m)^{3/2}}\int_{V_T^{(m)}}\mathbf{dR}^{(m)}G_T\left(\mathbf{R}^{(m)},t^{(m)}\right)a_R e^{-a_R^2\rho_m^2/4\Delta t_m}\right\rangle_{w_1(a_R)}, \tag{9.2.39}$$

with

$$\rho_m=|\mathbf{R}_m-\mathbf{R}^{(m)}|;\ \Delta t_m=t^{(m)}-t_R(>0), m=1,2,\ldots,n. \tag{9.2.39a}$$

In the case of a *Rayleigh distribution* of the average diffusion parameter a_R, that is,

$$w_1(a_R)=w_1(x)=\frac{x}{\psi}e^{-x^2/2\psi}, x\geq 0;\ \psi=\overline{x^2}. \tag{9.2.39b}$$

we can write

$$M_{12\ldots n}^{(\alpha)}\equiv\left\langle\alpha_1^{(Q)}\ldots\alpha_n^{(Q)}\right\rangle=\left[\frac{1}{8(\pi)^{3/2}}\right]^n\int_{t_R}^{t_1}\frac{dt^{(1)}}{(\Delta t)^{3/2}}\cdots\int_{t_R}^{t_n}(\Delta t)^{3/2}dt^{(n)}\int_{V_T^{(1)}}\mathbf{dR}^{(1)}G_T\left(R^{(1)},t^{(1)}\right)$$
$$\cdots\int_{V_T^{(1)}}\mathbf{dR}^{(1)}G_T\left(R^{(n)},t^{(n)}\right)\cdot\left\langle\frac{x^{n+1}e^{-A_n x^2}}{\psi}\right\rangle_{w_1(x)} \tag{9.2.40a}$$

in which

$$A_n=1+2\psi\sum_{m=1}^{n}\rho_m^2/4\Delta t_m;\quad \rho_m^2=|\mathbf{R}_m-\mathbf{R}^{(m)}|^2. \tag{9.2.40b}$$

Since

$$\frac{1}{\psi}\int_0^\infty x^{n+1}e^{-x^2A_n}dx=\frac{(2\psi)^{n/2}\Gamma(n/2+1)}{\left[1+2\psi\sum_{m=1}^{n}\rho_m^2/4\Delta t_m\right]^{n/2+1}}, \tag{9.2.41}$$

we have finally

$$M_{1,2\ldots n}^{(\alpha)}=\left\langle\alpha_1^{(Q)}\ldots\alpha_n^{(Q)}\right\rangle=\frac{\psi^{n/2}\Gamma(n/2+1)}{2^{5n/2}\pi^{3n/2}}\prod_{m=1}^{n}\int_{t_0^-}^{t_m}\frac{dt^{(m)}}{[\Delta t^{(m)}]^{3/2}}\int_{V_T^{(m)}}\mathbf{dR}^{(m)}\frac{G_T(\mathbf{R}^{(m)},t^{(m)})}{\left[1+2\psi\sum_{m=1}^{n}\rho_m^2/4\Delta t_m\right]^{n/2+1}}. \tag{9.2.42}$$

From Eq. (9.2.42) we obtain at once specific analytic results for the mean ($n=1$) and second-moment function ($n=2$) for the above example:

$$(n=1):\quad M_1^{(d)}=\left\langle\alpha^{(Q)}\right\rangle=\frac{\psi^{1/2}}{2^{7/2}\pi}\int_{t_0^{-1}}^{t}\frac{dt^{(1)}}{(\Delta t_1)^{3/2}}\int_{V_T}\frac{\mathbf{dR}^{(1)}G_T(\mathbf{R}^{(1)},t^{(1)})}{\left(1+2\psi\rho_1^2/4\Delta t_1\right)^{3/2}}, \tag{9.2.43}$$

$$(n=2):\quad M_{12}^{(d)}=\left\langle\alpha_1^{(Q)}\alpha_2^{(Q)}\right\rangle=\frac{\psi}{2^5\pi^3}\int_{t_0^{-1}}^{t_1}\frac{dt^{(1)}}{[\Delta t_1]^{3/2}}\int_{t_0^{-1}}^{t_2}\frac{dt^{(2)}}{[\Delta t_2]^{3/2}}\int_{V_T^{(1)}}\int_{V_T^{(2)}}\frac{\mathbf{dR}^{(1)}\mathbf{dR}^{(2)}G_T(\mathbf{R}^{(1)},t^{(1)})G_T(\mathbf{R}^{(2)},t^{(2)})}{\left[1+2\psi(\rho_1^2/4\Delta t_1+\rho_2^2/4\Delta t_2)\right]^2}. \tag{9.2.44}$$

For $n\geq 3$ the analytical burden although explicit enough, cf. (9.2.42), increases greatly, but it is still less than that encountered in evaluating the operator solutions analytically. Those higher moment numerical solutions require computer solutions, in both cases. Other examples of saturation can occur in types 3 and 4 [Eqs. (8.1.32) and (8.1.33)] and, in fact, when the term $a(\partial\alpha/\partial t)$ is dominant in the propagation equation. We also note that the covariance of this example is given by

$$K_{12}^{(\alpha)}\equiv\langle\alpha_1\alpha_2\rangle-\langle\alpha_1\rangle\langle\alpha_2\rangle=M_{12}^{(\alpha)}-M_1^{(\alpha)}M_2^{(\alpha)},\ \text{Eqs. (9.2.43) and (9.2.44)}, \tag{9.2.45}$$

with the associated intensity spectrum (9.2.13a), provided the medium is homogeneous and stationary, at least in the equilibrium or diffusion phase considered here.

9.2.7 Remarks

In Section 9.2, we have presented a formal extension of our operator approach to the calculation of the second and higher order moments of the scattered field, under rather general conditions. These require the evaluation of the appropriate order of mass operator $\hat{Q}_{12}^{(d)}$ or its extension $\hat{\eta}_{12}^{(d)}$, and so on. The procedure is outlined in some detail for the covariance, and second-order second moment, in Section 9.2.1. (These moments are a form of Bethe–Salpeter equation employed in quantum mechanics. Unlike the first-order moments (Dyson's equation), it is not susceptible to a closed form solution [Eqs. (9.1.18) and (9.1.21)] obtained by appropriate Laplace transforms, ever in the case of homogeneous, stationary medias (RKT, [7], in Chapter 4 of Volume 4, 3rd paragraph; also, Section 4.3 ibid.). Higher order moments (Section 9.2.3) are treated in similar fashion, with a further increase of complexity as well as a further extension of the FOR representation [cf. Fig. 9.10].

In Section 9.2.4 the concept of the transport equation is introduced, as an integrodifferential equation which describes the *propagation of a statistic* of the field, for example, the propagation of the mean field $\langle \alpha^{(Q)}(\mathbf{R}, t) \rangle$, of the second-order average field, and so on. Just as the mean mass operator $\hat{Q}_1^{(d)}$ and the second-order average mass operator $\hat{Q}_{12}^{(d)}$ are global operators (e.g., function of $\hat{M}_\infty$), so also are the extended forms $\hat{\eta}_1^{(d)} \left(= \hat{M}_\infty \hat{Q}_1^{(d)} \right)$ and so on. The integrals of the transport equations are the corresponding statistics of the scattered field.

In both Sections 9.1 and 9.2, the statistics are generally *non-Gaussian*, even when the statistics of the parameters of the Langevin equation are themselves Gaussian. Finally, we have seen that when the Langevin equation has a dominant absorption term, it becomes after a certain time a diffusion field. This is a result of *saturation* and it is no longer described by a feedback mechanism (FOR), Section 9.2.6. This in turn is only possible if a source is continuously applied to the scattering medium and the dissipation term becomes large relative to the propagation term $(c_0 \to \infty)$. The details are discussed in Sections 9.1 and 9.2. Equivalent diagram methods illustrate our results above, in Section 9.3.

9.3 EQUIVALENT REPRESENTATIONS: ELEMENTARY FEYNMAN DIAGRAMS

As is well known [4, 7, 12, 18], diagram representations and methods are often useful and compact forms for describing field interactions. Moreover, they are also useful and insightful in the development of approximations, which we must almost always seek if we are to obtain practical solutions. This is particularly true in the case of strong scattering, and in defining the extent to which weak-scattering models may be appropriate. Here we extend the usual deterministic Feynman diagrams to include ensemble diagrams as well, defining the symbol "vocabulary" in a consistent fashion[23] as we proceed to describe the various ensemble and moment solutions discussed in Sections 9.1 and 9.2. In Section 9.3.1, we define the vocabulary of our diagram elements and use them to represent the operational results developed in Sections 9.1 and 9.2. Our treatment here generalized to some extent earlier representations [4, 7, 9], in that arbitrary (linear) media, broadband signals, and various boundary discontinuities, involving surface scatter, and feedback forms [cf. Figs. 9.4 and 9.6] are considered. Our treatment is an elementary survey.

[23] There are, of course, as many diagram "vocabularies" as there are users [18], equally valid as long as they are logical, consistent, and simple. Our symbol choices overlap somewhat with others ([4], p. 60), [9], and Mattuck [18], for example.

9.3.1 Diagram Vocabulary

Let us begin our discussion of diagram methods first with a description of some of the principal elements and their interpretation in terms of the operator formulations of the three preceding sections. Table 9.1 provides these data. The elements in the table are largely self explanatory. We shall use them to obtain the diagram equivalents of the principal results of Sections 9.1–9.3 in what follows.

9.3.1.1 The FOR and FOS in the Deterministic Case We begin with the feedback operational representation and solutions (i.e., FOR and FOS) for the deterministic situation in Chapter 8, in particular for deterministic scatter discussed in Section 8.3. Here we have only one member function, where everything needed *a priori* to determine the scatter produced in the medium. From Table 9.1 and Figs. 9.3 and 9.5, respectively for volume and surface effects, are that

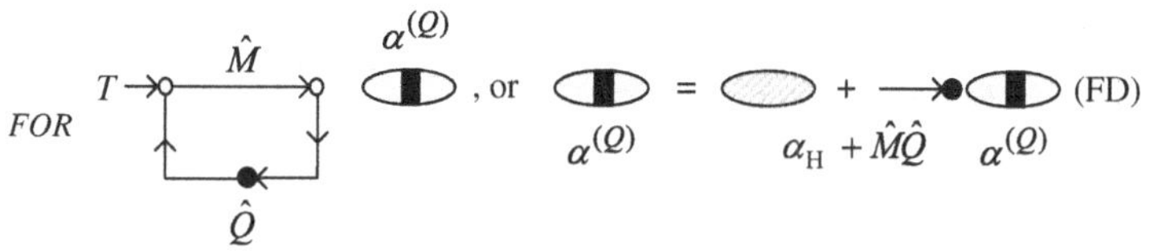

(9.3.1)

FOS; (*FD*)

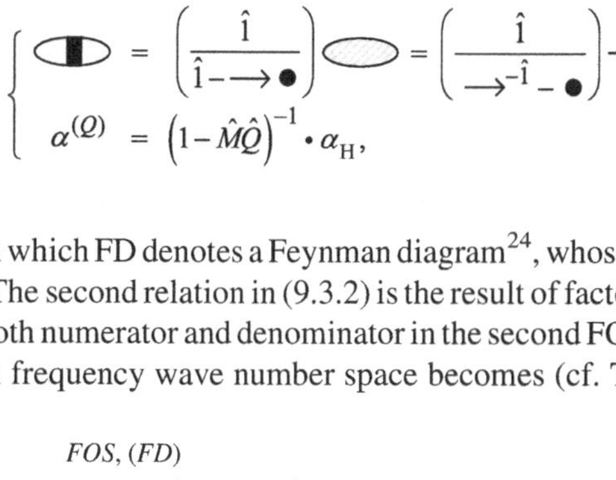

$$\rightarrow T: \alpha^{(Q)} = \alpha_H + \sum_{k=1}^{\infty} \left(\hat{M}\hat{Q}\right)^k \alpha_H \quad (9.3.2)$$

$$\alpha^{(Q)} = \left(1-\hat{M}\hat{Q}\right)^{-1} \bullet \alpha_H,$$

in which FD denotes a Feynman diagram[24], whose PSS is given by the last relation of (9.3.2). (The second relation in (9.3.2) is the result of factoring out the integral operator $\hat{M}(= \rightarrow)$ in both numerator and denominator in the second FOS diagram.) The FOS equivalent of (9.3.2) in frequency wave number space becomes (cf. Table 9.1 and (8.4.11b)).

FOS, (*FD*)

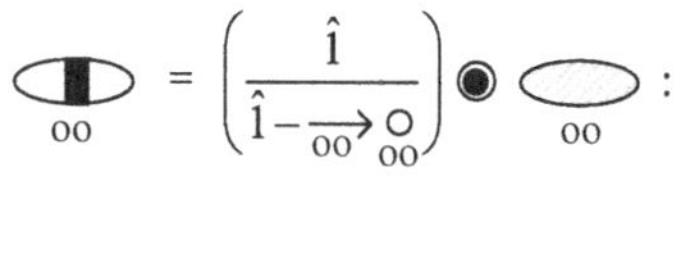

$$: \alpha_{oo}^{(Q)} = \alpha_H + \sum \left[\hat{Y}_o' \odot \hat{Q}_{oo} \odot\right]^k \alpha_{oo,H}, \text{ with} \quad (9.3.3)$$

$$= \underset{oo}{\rightarrow} \circledcirc \; T_{oo}$$

Equation (9.3.3) is the deterministic version of the *integrands* of Eqs. (9.1.26) and (9.1.37), et. seq. without any averaging provided $\hat{Q}^{(d)} \rightarrow \hat{Q}$ is Hom-Stat, that is, has the form $\hat{Q}(\mathbf{R}-\mathbf{R}', t-t')$, cf. (9.1.23). We also note that $G_{00,T} =$ has the appropriately equivalent Feynman diagrams. Note that, in general, the feed forward ($\rightarrow$) and feedback ($\bullet$) operators do not commute, except as convolutions in transform ($\mathbf{k}$, s) space, cf. Section 8.4. The multiple interaction terms ($k \geq 2$) are explicitly revealed by the expansion (9.3.2), namely,

[24] For our purposes here we shall use this term for all diagrams.

TABLE 9.1 A Dictionary of Space–Time Diagram Elements

Deterministic Elements	Random Elements
(1) $\rightarrow\ \equiv \hat{M}$ "feedforward" (integral operator); $\therefore \rightarrow^{-1} = \hat{L}^{(0)}$ $\hat{M}_V, \hat{M}_S$ differential operator	(1) $\Rightarrow\ = \hat{M}$ random surface integral operator
(2) $\leftarrow\!\bullet$ or $\bullet$ $\hat{Q}$: "feedback" (inhomogeneity) Op.	(2) $\leftarrow\!\circ$ or $\circ \equiv \hat{Q}\left(=\left\{\hat{Q}\right\}\right)$; ensemble, or random *feedback* operator: led inhom. operator
	(2a) $\langle \overset{(d)}{\circ} \rangle = \overset{(d)}{\bullet}$ = global sum of averages of inhomogeneity operator $\hat{Q}$; Eq. (9.1.10). et seq
(3) $T = -G_{\mathrm{T}}$: deterministic source function (density).	(3) $\Pi \equiv \{T\} \equiv$ random source functions
(4) ⬭ $\equiv \alpha_{\mathrm{H}}$: homogeneous deterministic field $\left(=\rightarrow\ = -\hat{M}G_{\mathrm{T}}\right)$	(4) ⬭ $\equiv\ \rightarrow T = -\hat{M}_{\infty}G_{\mathrm{T}}$; homogeneous $=$ deterministic field
(5) ⬭ (with filled centre) $\equiv \alpha^{(Q)}$: inhomogeneous (det.) field	(5) ⬭ $= \left(\alpha^{(Q)}\right)\left(=\alpha^{(Q)}\right)$ random scattered field (Hom-Stat or not)
(6) $\equiv$ average or $\mathbf{E}\{\ \}$	(6) ⬭ (with filled arrow) $= \hat{\eta}_{\Sigma}$ sum of all terms, $k \geq 2$ multiple scatter
(7) $\circledbullet_1 = \hat{Q}_1^{(d)}$; ("feedback") operator	(7) ⬭ (overlined) = average random field
(8) $\Rightarrow = \hat{M}_{\mathrm{S}}$; $\rightarrow = \hat{M}_{\mathrm{V}}$: "feedforward" operators	(8) $\Rightarrow (= \rightarrow)$ for surfaces; $\overset{(d)}{-\!\bullet\!-} = \hat{\eta}_{\infty}^{(d)}$
(9) $\underset{1\ 2}{\cap}, \cup$ = stat. av. over two elements (arguments); $\underset{1\ 2\ 3\ 4\ \ldots\ n}{\cap\cap\cap\ldots\cap}$ = stat. av. over n elements (or arguments); $\underset{o\ o}{\cap} = \left\langle \hat{Q}_1\hat{Q}_2 \right\rangle$, cf. 8)	(9) $\frown \equiv \langle\ \rangle\ (\ \ )$; average;
$\rightarrow\langle o\rangle \rightarrow \langle o\rangle = \begin{cases} \hat{M}_1\left\langle \hat{Q}_1\right\rangle \hat{M}_2\left\langle \hat{Q}_2\right\rangle \\ \left\langle \hat{N}_1\right\rangle\left\langle \hat{N}_2\right\rangle \end{cases}$	$\overset{1}{o} \rightarrow \overset{2}{o} \rightarrow \overset{3}{o} \rightarrow \overset{4}{o} = \left\langle \hat{Q}_1\hat{M}_2\left\langle \hat{Q}_2\hat{M}_3\hat{M}_3\right\rangle_{23}\hat{M}_4\hat{Q}_4\right\rangle_{14}$ $\left\langle \hat{Q}_1\left\langle \hat{N}_2\hat{N}_3\right\rangle \hat{N}_4\right\rangle$
$\rightarrow \overset{\frown}{o \rightarrow o} = \hat{M}\left\langle \hat{Q}_1\hat{M}_2\hat{Q}_2\right\rangle = \left\langle \hat{M}_1\hat{Q}_1\hat{M}_2\hat{Q}_2\right\rangle = \left\langle \hat{N}_1\hat{N}_2\right\rangle$ (M deterministic or random if random)	$\overset{1}{o} \rightarrow \overset{2}{o} \rightarrow \overset{3}{o} \rightarrow \overset{4}{o} = \left\langle \hat{Q}_1\hat{M}_2\left\langle \hat{Q}_2\hat{M}_3\hat{Q}_3\hat{M}_4\right\rangle \hat{Q}_4\right\rangle_{(2\text{–}4)}$ $\overset{1}{o} \rightarrow \overset{2}{o} \rightarrow \overset{3}{o} \rightarrow \overset{4}{o} = \left\langle \hat{Q}_1\hat{M}_2\hat{Q}_2\right\rangle \hat{M}_3\left\langle \hat{Q}_3\hat{M}_3\hat{Q}_4\right\rangle$, etc. etc. etc.
(10) $\hat{o}$: subscript $\equiv \mathbf{F}$, $\left(\mathbf{F}_R^{-1}\text{or } \mathbf{F}_t\right)$; $\alpha_0 = \mathbf{F}\ \{\alpha\}$	(10) $\hat{o}$: subscript $\equiv \mathbf{F}$; e,g. $\{\alpha_0\} = \mathbf{F}_t\{\{\alpha\}\}$
$\hat{o}\hat{o}$: double subscript $\equiv \mathbf{F}_{\mathbf{R}}^{-1}\mathbf{F}_t$ $\left(\text{or } \mathbf{F}_t\,\mathbf{F}_{\mathbf{R}}^{-1}\right)$; $\odot$ $\odot$: double convolution	$\hat{o}\hat{o}$: double subscript $\equiv \mathbf{F}_{\mathbf{R}}^{-1}\mathbf{F}_t$, etc.
$\vec{o} = \hat{M}_0 = \mathbf{F}_t$ or $\mathbf{F}_{\mathbf{R}}^{-1}\{\hat{M}\} = \hat{Y}_0$, etc.	⬭(o) $= \mathbf{F}_t$ or $\mathbf{F}_{\mathbf{R}}^{-1}\{\{\alpha\}\}$
$\underset{o}{⬭} = \mathbf{F}_t$ or $\mathbf{F}_{\mathbf{R}}^{-1}\{\alpha_H\}$,	⬭(oo) $= \mathbf{F}_{\mathbf{R}}^{-1}\mathbf{F}_t\{\{\alpha\}\}$
$\underset{oo}{⬭} = \mathbf{F}_t\,\mathbf{F}_{\mathbf{R}}^{-1}\left(\text{or } \mathbf{F}_{\mathbf{R}}^{-1}\mathbf{F}_t\right)\{\alpha_H\}$,	
(11) $\langle \underset{12}{⬭} \rangle_{12} \equiv \langle \alpha_1\alpha_2 \rangle$ ↑α random	(11) ⬭(12) $\equiv \{\alpha_1\alpha_2\}$
$\langle ⬭ \rangle_{12\text{-}m} \equiv \langle \alpha_1\alpha_2 \ldots \alpha_m\rangle, m \geq 1.$	⬭(0,12) $\equiv \{\alpha_{01}\alpha_{02}\}$, etc.
$\langle ⬭ \rangle_{0,12} \equiv \langle \alpha_{01}\alpha_{02}\rangle$	⬭(00,12) $\equiv \{\alpha_{01}\alpha_{02}\}\left(=\{\alpha_{00,1}\alpha_{00,2}\}\right)$ (double (t, $\mathbf{R}$) transformer)
$\langle ⬭ \rangle_{0,12\text{-}m} \equiv \langle \alpha_{01}\alpha_{02}\ldots\alpha_{0m}\rangle, m \geq 1$, etc.	
$\langle ⬭ \rangle_{00,12\text{-}m} \equiv \langle \alpha_{00,1}\alpha_{00,2}\ldots\alpha_{00,m}\rangle$	

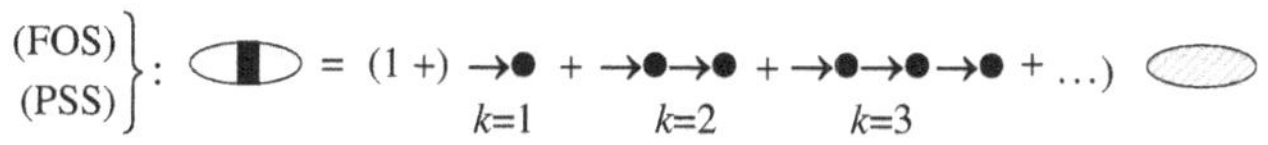

$$\left.\begin{array}{l}\text{(FOS)}\\\text{(PSS)}\end{array}\right\} : \quad = (1+) \underset{k=1}{\to\bullet} + \underset{k=2}{\to\bullet\to\bullet} + \underset{k=3}{\to\bullet\to\bullet\to\bullet} + \ldots) \qquad (9.3.4)$$

We can use *diagram iteration* alternatively to solve the FOR of (9.3.1) and obtain

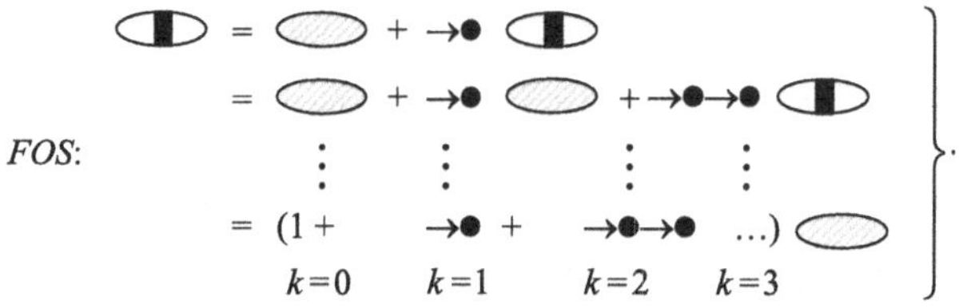

$$FOS: \qquad\qquad \left.\begin{array}{l} = + \to\bullet \\ = + \to\bullet + \to\bullet\to\bullet \\ \vdots \quad \vdots \quad \vdots \quad \vdots \\ = (1+ \underset{k=0}{\quad} \underset{k=1}{\to\bullet} + \underset{k=2}{\to\bullet\to\bullet} \underset{k=3}{\ldots}) \end{array}\right\}. \qquad (9.3.5)$$

This corresponds to "iterating the loop" indefinitely in the feedback representation and is obviously identical to the PSS $\Sigma_k(\hat{M}Q)^k \alpha_H$ in (9.3.4). These diagrams show that their elements can be added, subtracted, multiplied, and inverted (i.e., "divided"), cf. (9.3.2) and (9.3.3), as well as Fourier transformed (and by other linear operators), as well.

9.3.1.2 Equivalents for Random Fields: Basic Diagrams

We next consider the more general cases of random fields. We begin with the following basic *ensemble diagrams*. These are the ensemble analogies of the deterministic cases of Section 9.3.2, based on the generic integral equation for scatter in random inhomogeneous media:

$$\alpha^{(Q)}(\mathbf{R}, t) = \alpha_H(\mathbf{R}, t) + \hat{M}_\infty \hat{Q} \alpha^{(Q)}(\mathbf{R}, t), \qquad (9.3.6)$$

In various forms, these diagram represent the operational expressions for the scattered field. The "solutions" here of (9.3.6) are, of course, the various moments and distributions which describe the probabilistic status of the ensemble. Accordingly, the basic diagram may be summarized as follows:

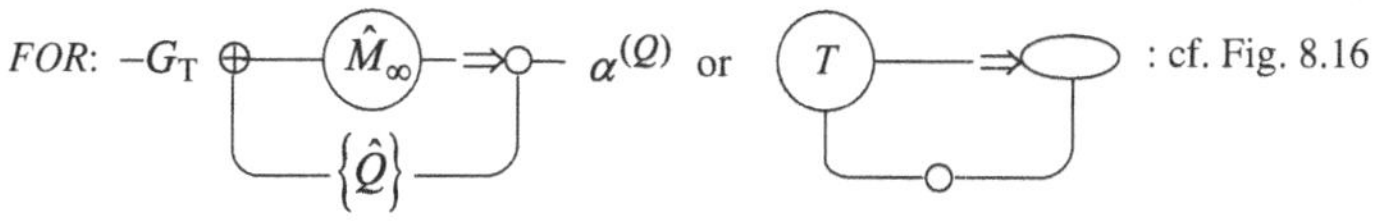

$$FOR: \quad -G_T \oplus -(\hat{M}_\infty) - \Rightarrow \circ - \quad \alpha^{(Q)} \text{ or } (T) - \Rightarrow \quad : \text{ cf. Fig. 8.16} \qquad (9.3.7a)$$

$$FD: \quad = + \Rightarrow \quad \left.\begin{array}{l} \\ \{\alpha\} = \alpha_H + \hat{M}_\infty (\hat{Q}\alpha) \end{array}\right\} \begin{array}{l}\text{Eq. (9.3.6)}\end{array} \qquad (9.3.7b)$$

$$FOS: \quad = \left(\frac{\hat{1}}{\hat{1}-\Rightarrow-\circ}\right) , \text{ with } = \Rightarrow (T) = \hat{M}_\infty(-G_T) \left.\begin{array}{l}\\ (T) = -G_T\end{array}\right\} \qquad (9.3.7c)$$

$$PSS: \quad = [\hat{1}+\Rightarrow\circ + \Rightarrow\circ\Rightarrow\circ + \ldots] , \text{ cf. (9.3.4)} \qquad (9.3.7d)$$

Here we have $\Rightarrow (\equiv \hat{M}_\infty)$ representing the feedforward operator, that is, the integral Green's function for random surface (S), to be replaced by its nonrandom counterpart for volume (V), namely $\Rightarrow$. Here O denoted the local ensemble feedback or random scattering operator, $\hat{Q}$. The ensemble Feynman diagram equivalent (FD) of the FOR, (9.3.7a), is precisely (9.3.7b),

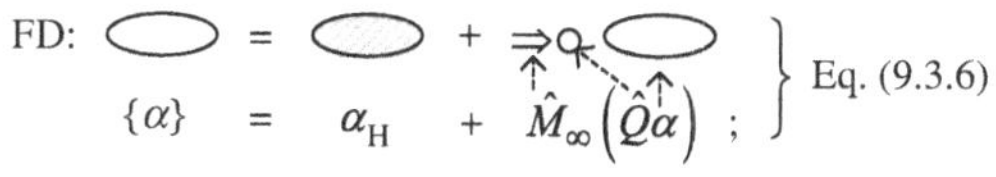

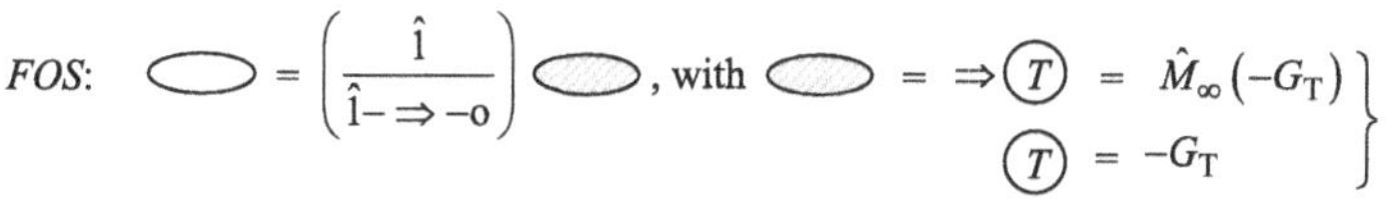

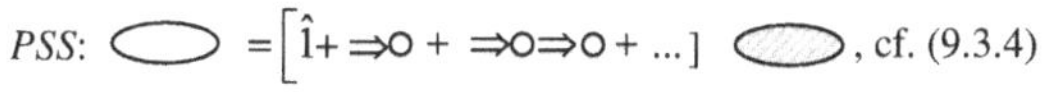

while (9.3.7c) gives the corresponding ensemble FOS, and (9.3.7d) the PSS, which is readily obtained by iterating the FD, (9.3.7b), or as the unaveraged form (9.1.6a) of (9.1.21).

9.3.1.3 First-Order Dyson Equation and the Associated Deterministic Medium The (deterministic) first-order Dyson equation, Eq. (9.1.21), is now easily written:

Dyson's equations :

$$\left\langle \alpha^{(Q)} \right\rangle = \alpha_H + \hat{M}_\infty \hat{Q}_1^{(d)} \left\langle \alpha^{(Q)} \right\rangle \text{, Eq. (9.1.21)} \tag{9.3.8}$$

where now we denote the average by the overhead bar, applied here to the corresponding ensemble symbol. From Table 9.1, our general diagram conventions are to represent ensemble quantities as "open" forms, e.g. ⬭, o, etc., and average and otherwise deterministic quantities by solid forms, e.g., •, $\overset{(d)}{\bullet}$, with field averages denoted by loop ∩ brackets, (——), for example, $\overline{⬭} = \langle \alpha \rangle$, and so on. The aim is consistency of notation and simplicity with a minimal evolution of complexity.

We note once more that diagrams can be manipulated in the same fashion as are operator equations [cf. (9.1.21), etc]. One also observes the same positioning rules and conditions of inversion, for example, for multiplication and "division." Addition and subtraction are also directly equivalent to the addition and subtraction of operators and operator-derived (algebraic) quantities [cf. (9.1.2) et. seq. for volume, and (9.1.37) and (9.1.39) for surfaces], as indicated in the examples below. Finally, *transforms of diagrams are the corresponding diagrams of the transforms* [cf. Section 8.4, now applied to the various operator elements of the direct diagrams]. This also permits an explicit diagrammatic representation, with the help of Table 9.1, of the various wave number–frequency spectra of the scattered field, particularly the amplitude and intensity spectra in operator form (cf. Section 8.4).

Accordingly, the corresponding diagrams for the *equivalent medium* [cf. Section 9.1.1.1] become:

Equiv. Med. FOR :
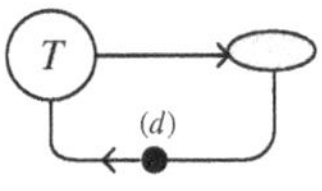

, cf. (9.1.21); (9.3.8a)

Equiv. Med. FOS :
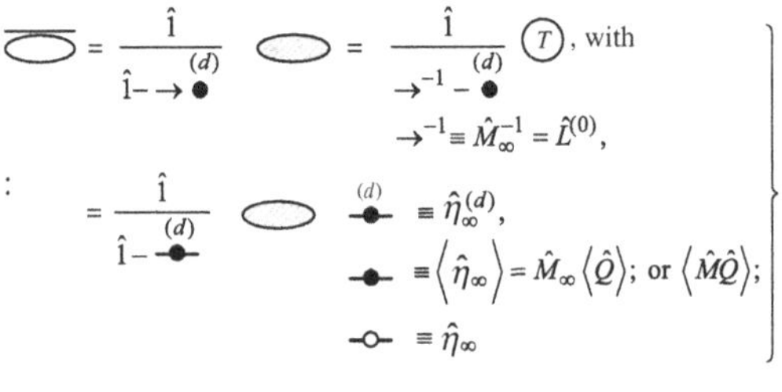

(9.3.8b)

Equiv.Med.FD:

First-Order Dyson Equations :

(9.3.8c)

PSS:

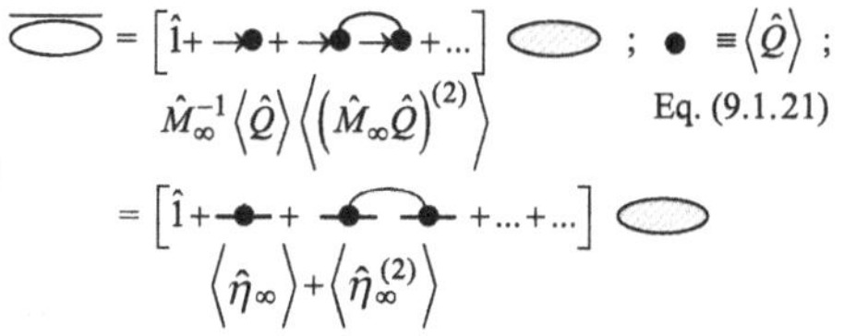

$$\overline{\bigcirc} = \left[\hat{1} + \rightarrow\!\bullet + \rightarrow\!\bullet\!\rightarrow\!\bullet + ...\right]\bigcirc \quad ; \quad \bullet \equiv \langle\hat{\varrho}\rangle ;$$

$$\hat{M}_\infty^{-1}\langle\hat{\varrho}\rangle\left\langle\left(\hat{M}_\infty\hat{\varrho}\right)^{(2)}\right\rangle \qquad \text{Eq. (9.1.21)}$$

$$= \left[\hat{1} + \!\bullet\!\bullet\!- + \!-\!\bullet\!\frown\!\bullet\!- + ...+\right]\bigcirc$$

$$\langle\hat{\eta}_\infty\rangle + \langle\hat{\eta}_\infty^{(2)}\rangle$$

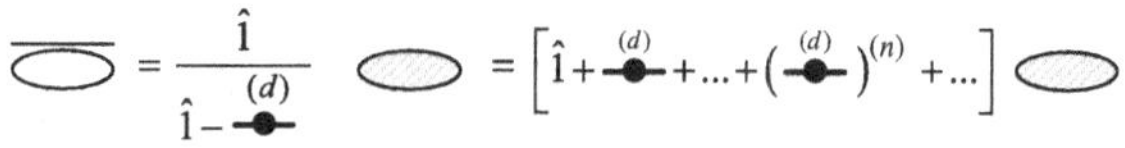

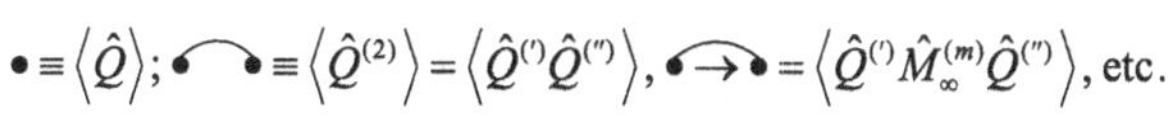

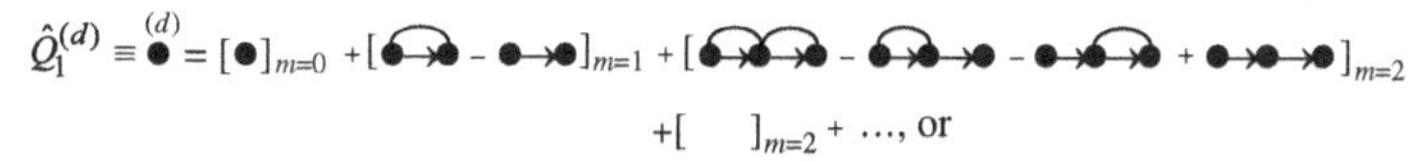

(9.3.8d)

FOS + *PSS*:

$$\overline{\bigcirc} = \frac{\hat{1}}{\hat{1} - \overset{(d)}{\bullet\!\bullet}} \bigcirc = \left[\hat{1} + \overset{(d)}{\bullet\!\bullet} + ... + \left(\overset{(d)}{\bullet\!\bullet}\right)^{(n)} + ...\right]\bigcirc$$

(9.3.8e)

Here $\bigcap\!\!\overset{...}{}\!\!\bigcap_{(n\geq2)}$ denotes the k-fold average:

$$\bullet \equiv \langle\hat{\varrho}\rangle ; \quad \frown\!\bullet \equiv \langle\hat{Q}^{(2)}\rangle = \langle\hat{Q}^{('')}\hat{Q}^{('')}\rangle, \quad \bullet\!\rightarrow\!\bullet = \langle\hat{Q}^{('')}\hat{M}_\infty^{(m)}\hat{Q}^{('')}\rangle, \text{ etc.}$$

(9.3.8f)

Equations (9.3.8e) follow from the iteration of the FD (9.4.8c).

In the same way we find that the ED$_1$M, cf. [(9.1.7b) and (9.1.7c)] as developed in Sections 9.1.2 and 9.1.3, can be expressed diagrammatically as

$$\hat{Q}_1^{(d)} \equiv \overset{(d)}{\bullet} = [\bullet]_{m=0} + [\frown\!\!\rightarrow\!\!\bullet - \bullet\!\rightarrow\!\bullet]_{m=1} + [\frown\!\!\rightarrow\!\!\bullet - \frown\!\!\rightarrow\!\!\bullet - \bullet\!\frown\!\!\rightarrow\!\!\bullet + \bullet\!\rightarrow\!\!\bullet\!\rightarrow\!\bullet]_{m=2}$$

$$+ [\quad]_{m=2} + ..., \text{ or}$$

(9.3.9a)

$$= [\bullet]_{m=0} + [\frown\!\bullet\!- - \bullet\!\bullet\!-]_{m=1} + [\bullet\!\frown\!\bullet\!- - (\frown\!\bullet\!\bullet) - (\bullet\!\frown\!\bullet) + (\bullet\!\bullet\!\bullet)]_{m=2}$$

(9.3.9b)

this last in terms of $\hat{\eta}_\infty$ ($\equiv -\!\circ\!-$), where for example, $\langle\hat{\eta}_\infty^{(2)}\rangle = \langle -\!\circ\!- -\!\circ\!- \rangle = -\!\bullet\!\frown\!\bullet\!-$, and so on, cf. (9.3.8b) above.

9.3.1.4 Second Order Diagrams

Second-order diagrams are handled in the same way. Thus, for the second-order Dyson (Bethe-Salpeter Equations) form for $\langle\alpha_1\alpha_2\rangle$, ([(9.2.3a) and (9.2.3b)], etc. cf., Section 9.2.1), we readily construct its FD:

Second-order Dyson equation: (Bethe–Salpeter analogue)

$$\overline{\bigcirc}_{12} = \bigcirc + \overset{d}{\rightarrow\!\bullet}\,\overline{\bigcirc} \quad ; \quad \rightarrow \equiv \rightarrow \rightarrow$$

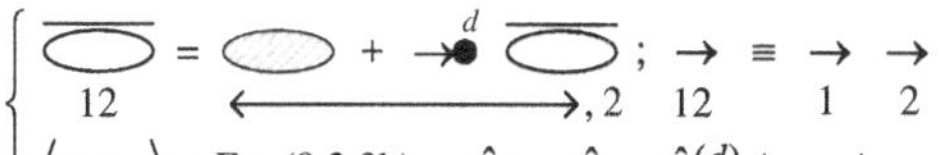

$$\langle\alpha_1\alpha_2\rangle = \text{Eq. (9.2.3b)} + \hat{M}_{1,\infty}\,\hat{M}_{2,\infty}\,\hat{Q}_{12}^{(d)}\langle\alpha_1\alpha_2\rangle,$$

(9.3.10a)

with (9.2.5a), where $\rightarrow_{1,2} \equiv \hat{M}(R_{12}, t_{1_2}|R''', t''')_\infty$, and so on, and where, from (9.2.7) we obtain

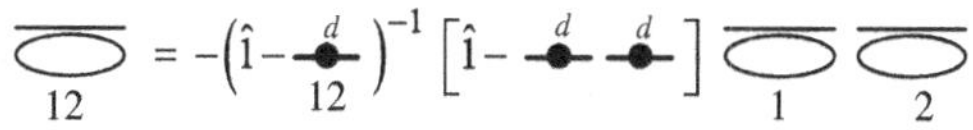

or

$$\overline{\bigcirc}_{12} = -\left(\hat{1} - \overset{d}{\underset{12}{\bullet}}\right)^{-1}\left[\hat{1} - \overset{d}{\bullet}\overset{d}{\bullet}\right]\overline{\bigcirc}_1\,\overline{\bigcirc}_2$$

$$\hat{M}_{12}^{(d)} \equiv \langle\alpha_1\alpha_2\rangle = \left(\hat{1} - \hat{\eta}_{12}^{(d)}\right)^{-1}\left(\hat{1} - \hat{\eta}_{1,\infty}^{(d)}\hat{\eta}_{2,\infty}^{(d)}\right)\langle\alpha_1\rangle\langle\alpha_2\rangle$$

(9.3.10b)

We have also the equivalent (second-order) medium FOR and FOS, cf. Fig. 9.9:

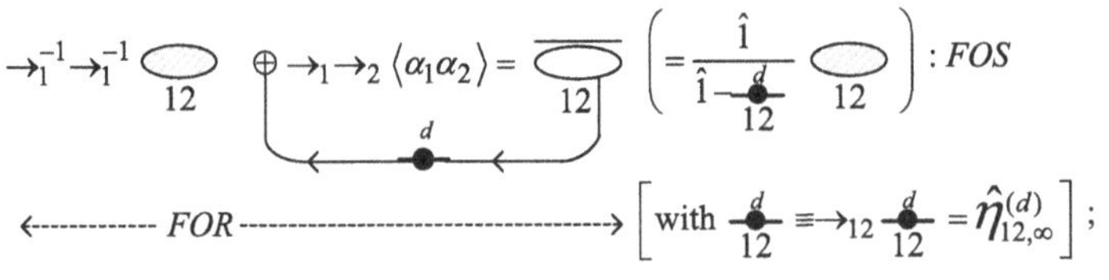

$$\rightarrow_1^{-1} \rightarrow_1^{-1} \underset{12}{\bigcirc} \quad \oplus \rightarrow_1 \rightarrow_2 \langle \alpha_1 \alpha_2 \rangle = \overline{\underset{12}{\bigcirc}} \left(= \frac{\hat{1}}{\hat{1} - \underset{12}{\overset{d}{\bullet}}} \underset{12}{\bigcirc} \right) : FOS \tag{9.3.11}$$

$$\longleftarrow ------ \; FOR \; ------------------\longrightarrow \left[\text{with } \underset{12}{\overset{d}{\bullet}} \equiv \rightarrow_{12} \underset{12}{\overset{d}{\bullet}} = \hat{\eta}_{12,\infty}^{(d)} \right];$$

PSS$_2$: Eqs. (9.2.12) and (9.2.12a):

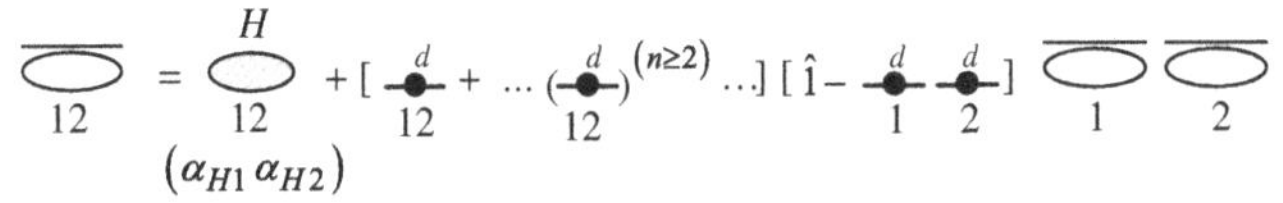

$$\overline{\underset{12}{\bigcirc}} = \overset{H}{\underset{12}{\bigcirc}} + [\underset{12}{\overset{d}{\bullet}} + \cdots (\underset{12}{\overset{d}{\bullet}})^{(n \geq 2)} \cdots] [\hat{1} - \underset{1}{\overset{d}{\bullet}} \underset{2}{\overset{d}{\bullet}}] \overline{\underset{1}{\bigcirc}} \; \overline{\underset{2}{\bigcirc}} \tag{9.3.12}$$

$$(\alpha_{H1} \alpha_{H2})$$

The explicit structure of the second-order equivalent deterministic medium or operator $\hat{Q}_{12}^{(d)}$ as given by (9.2.16), and so on, is diagrammatically,

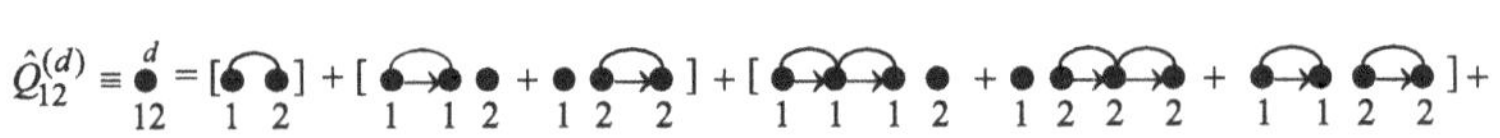

$$\hat{Q}_{12}^{(d)} \equiv \underset{12}{\overset{d}{\bullet}} = [\underset{1 \; 2}{\bullet \bullet}] + [\underset{1 \; 1 \; 2}{\bullet \bullet \bullet} + \underset{1 \; 2 \; 2}{\bullet \bullet \bullet}] + [\underset{1 \; 1 \; 1 \; 2}{\bullet \bullet \bullet} + \underset{1 \; 2 \; 2 \; 2}{\bullet \bullet \bullet} + \underset{1 \; 1 \; 2 \; 2}{\bullet \bullet \bullet}] + \tag{9.3.13}$$

Accordingly, $\eta_{12,\infty}^{(d)}$, (9.2.15), is represented by $\rightarrow_{12} \underset{12}{\overset{d}{\bullet}} (\equiv \underset{12}{\overset{d}{\bullet}})$ that is $\hat{M}_2^{-1}, M_1^{-1}$, (i.e., $\rightarrow_{21}^{-1}$) times the diagram (9.3.13).

Higher-order moments (Section 9.2.3) may be diagrammed in the same way. Of course, each is based on the hierarchy of diagrams below its order, but the generalized Dyson form (9.2.17) and FOR (Fig. 9.10) permits an immediate formal generalization of the results (9.3.8a)–(9.3.12); $\hat{Q}_{12\ldots m}^{(d)}$, [(9.2.18) and (9.2.18a)], however, is progressively more complex. One diagrammatic advantage of the generalized Dyson form (9.2.17) is that it explicitly eliminates the intricate ladder, vertex, and lattice diagrams originally constructed for $\langle \alpha_1 \alpha_2 \rangle$, $(m = 2)$ [cf. [4], ¶s60, 61, for the classical Helmholtz equation (9.1.42), $\varepsilon = \varepsilon$ ($\mathbf{R}, t$). These latter, of course, are subsumed in the hierarchy embodied in (9.2.17). As we see it is possible to obtain explicit statistics for its various moments, for example, $\langle \hat{Q}_1 \hat{Q}_2 \rangle$, and so on, and accordingly analyze the resulting diagram structure, although the results can still be highly complex, requiring knowledge of the higher order covariance of the random quantities involved.

9.3.2 Diagram Approximations

A major problem, as always for strong-scatter situations, is the approximate evaluation of the FOS or its PSS equivalent. Even for the "exact" analytic solutions in the Hom-Stat cases, cf. Section 9.1.2, it is necessary to use approximations of the mass operator $\hat{Q}_1^{(d)}$. Another important question arises in truncating the PSS in the comparatively weak-signal regimes. Here, for example, we can ask what is the effect of stopping at a given term in the PSS and what is the overall contribution of the remaining terms, and what criteria of truncation are reasonable. In any event, various approximations are needed for quantitative results.

We consider briefly here two useful classes of approximation: (I) series modification and (II) truncation. In the former, the character of the equivalent deterministic medium, or its "mass" operator, is modified. In the latter, the PSS is truncated: a finite number of orders of interaction is retained, while the higher orders are discarded. We begin with Case I.

9.3.2.1 Approximation: Series Modification—The First Order Dyson Equation   As applied to the various forms of the Dyson equation [cf. (9.1.2), (9.1.7a) and (9.1.7b), (9.3.8) (9.3.11)], the form of the equivalent deterministic scattering operators (EDSO), $\hat{Q}_1^{(d)}, \hat{Q}_{12}^{(d)}, \ldots, \hat{Q}_{12\ldots m}^{(d)}$, cf. (9.1.7b), (9.1.7c), (9.1.12), is modified in some fashion, suggested by the physics of the problem. This EDSO is altered, usually by truncation, while the infinite operator series implied by $\left(1 - \hat{\eta}_{1,\infty}^{(d)}\right)^{-1}$, and so on, cf., [(9.1.7b) and (9.1.7c)] and so on, remains.

Let us examine, for example, the first order Dyson equation [(9.1.7b) and (9.1.7c)] for the mean field $\langle \alpha \rangle$ and postulate a first-order independent structure for $\hat{Q}$:

Ia. *First-Order Independence (Volumes and Surfaces):*

$$\hat{Q}^{(d)} = \prod_{n=1}^{k} \langle \hat{Q} \rangle^n = \langle \hat{Q}_1 \rangle \langle \hat{Q}_2 \rangle \ldots \langle \hat{Q}_n \rangle = (\bullet\bullet \ldots \bullet)_{12} \qquad (9.3.14)$$

(9.1.27) in (9.1.26) for volumes and from (9.3.14) in (9.1.38a,b,c) and then in (9.1.38c) for surfaces we see at once that the average mass operator (V and S) $\hat{Q}_1^{(d)}$ is directly [(from 9.1.11)], $\langle \hat{Q} \rangle \neq 0$ (V or S)

$$\hat{Q}_1^{(d)}\Big|_{\text{indep.}} = \hat{A}_0^{(1)} = \langle \hat{Q} \rangle = \bullet; \; \hat{A}_m^{(1)} = 0, m \geq 1; \; \therefore \; \hat{\eta}_{1,\infty}^{(d)} \equiv \hat{M}_\infty \langle \hat{Q} \rangle = \rightarrow \bullet \qquad (9.3.15)$$

where $\hat{Q}_1^{(d)}$ reduces to a single term, so that [(9.1.7b) and (9.1.7c)] still becomes an integral equation,

$$\langle \alpha \rangle = \alpha_{\text{H}} + \hat{M}_\infty \langle \hat{Q} \rangle \langle \alpha \rangle = \left(\hat{1} - \hat{M} \langle \hat{Q} \rangle\right)^{-1} \alpha_{\text{H}}, \qquad (9.3.16)$$

with the following equivalent (deterministic) diagrams [cf. (9.3.8), (9.3.8a)]:

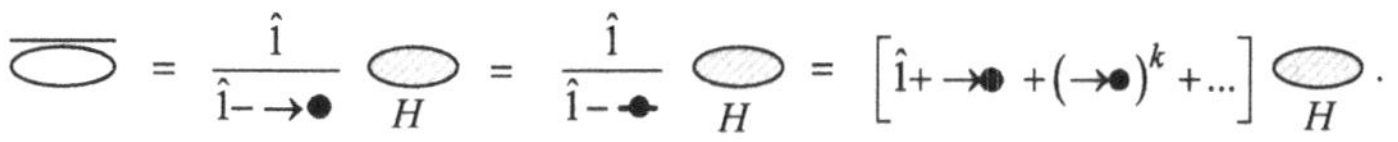

$$\overline{\bigcirc} = \frac{\hat{1}}{\hat{1} - \rightarrow\bullet} \underset{H}{\bigcirc} = \frac{\hat{1}}{\hat{1} - \bullet\!\!\leftarrow} \underset{H}{\bigcirc} = \left[\hat{1} + \rightarrow\bullet + (\rightarrow\bullet)^k + \ldots\right] \underset{H}{\bigcirc}. \qquad (9.3.16a)$$

However, equation (9.3.16) can still in principle be solved *exactly* analytically [cf. (9.1.26), and so on, for volume and (9.1.36) et seq. for surfaces], since the mass operator $\hat{Q}_1^{(d)}$ has only a finite number of terms. In the case of no volume or surface scatter with $\langle Q \rangle = 0$ and $\therefore \hat{Q}_1^{(d)} = 0$, as expected, $\langle \alpha \rangle$ then reduce to α_H. This clearly shows, as for example $\bar{\varepsilon} = a$ in the Helmholtz equation (9.1.42), with all higher order terms vanishing cf. (9.3.15), that the postulate $\langle Q \rangle = 0$ (9.3.14) here is too restrictive to provide meaningful results in such cases.

In a similar way, we may employ a more sophisticated, second-order assumption on the statistical character of $\hat{Q}$ which is often used. This is the second-order "independent" or (generalized) *Bourret (B)* or *bilocal approximation*, involving second-order covariances, or "nearest (B) neighbor" correlations or equivalently.

Ib. *Second-Order Independence (Volumes and Surfaces):*

$$\prod_{n=1}^{k/2, k+1/2} \langle \hat{Q}^{(2)} \rangle^n \begin{cases} = \langle \hat{Q}_1 \hat{Q}_2 \rangle \langle Q_3 Q_4 \rangle \cdots \langle Q_{k-1} Q_k \rangle = [\;\frown\;\frown\;\cdots\;\frown\;]_{k=\text{even}} & (9.3.17a) \\ = \langle \hat{Q}_1 \hat{Q}_2 \rangle \langle Q_3 Q_4 \rangle \cdots \langle Q_{k-2} Q_{k-1} \rangle \langle Q_k \rangle = [\;\frown\;\frown\;\cdots\;\frown\!\bullet\;]_{k=\text{odd}} & (9.3.17b) \end{cases}$$

Alternatively, for the odd-order case we can also set, to[25]

$$\Pi_k \left\langle \hat{Q}^{(2)} \right\rangle^n = \langle \hat{Q}_1 \rangle \prod_{n=1}^{(k-1)/2} \left\langle Q^{(2)} \right\rangle^n \langle \hat{Q}_1 \rangle \left[\langle \hat{Q}_2 \hat{Q}_3 \rangle \langle \hat{Q}_4 \hat{Q}_5 \rangle \cdots \langle Q_{k-1} Q_k \rangle \right] = [\bullet \frown \frown \cdots \frown]_{(n)}.$$

(9.3.17c)

Consequently, the mass operator, $Q_1^{(d)}$, (9.1.27) becomes here from (9.1.11) for $\langle \hat{Q} \rangle \neq 0, = 0$:

$$\xleftarrow{\quad} \hat{B}_0^{(1)} \xrightarrow{\quad} \quad \xleftarrow{\quad} \hat{B}_1^{(1)} \xrightarrow{\quad} \quad \xleftarrow{\quad\quad} \hat{B}_2^{(1)} \xrightarrow{\quad\quad} \quad \xleftarrow{\quad} \hat{B}_3^{(1)} \xrightarrow{\quad}$$

$$\hat{Q}_1^{(d)} = \langle \hat{Q} \rangle + \left[\langle \hat{Q} \hat{M} \hat{Q} \rangle - \langle \hat{Q} \rangle \hat{M} \langle \hat{Q} \rangle \right] + \left[\left\langle \hat{Q} (\hat{M} \hat{Q})^2 \right\rangle - 2 \langle Q \rangle \left\langle (\hat{M} \hat{Q})^2 \right\rangle + \langle \hat{Q} \rangle \langle M \hat{Q} \rangle^2 \right]$$

$$+ \left[\left\langle \hat{Q} (\hat{M} \hat{Q})^3 \right\rangle - \langle Q \rangle \left(\hat{M} \langle \hat{Q} \rangle \right)^3 + 4 \langle \hat{Q} \rangle \hat{M} \langle Q \rangle \langle \hat{Q} \hat{M} \hat{Q} \rangle - \left(\hat{Q} \hat{M} \hat{Q} \right)^2 - 2 \langle \hat{Q} \rangle \left\langle (\hat{M} \hat{Q})^3 \right\rangle \right]$$

(9.3.18a)

When $\langle \hat{Q} \rangle = 0$ and has symmetrical pdfs about $\langle \hat{Q} \rangle = 0$, we obtain the much simpler result, cf. (9.1.15a):

$$\xleftarrow{\quad} \hat{B}_1^{(1)} \xrightarrow{\quad} \quad \xleftarrow{\quad} \hat{B}_3^{(1)} \xrightarrow{\quad} \quad \xleftarrow{\quad\quad} \hat{B}_5^{(1)} \xrightarrow{\quad\quad}$$

$$\left. \hat{Q}_1^{(d)} = \langle \hat{Q} \hat{M} \hat{Q} \rangle + \left[\left\langle \hat{Q} (\hat{M} \hat{Q})^{(3)} \right\rangle - \langle \hat{Q} \hat{M} \hat{Q} \rangle^2 \right] + \left[\left\langle \hat{Q} (\hat{M} \hat{Q})^5 \right\rangle + 2 \langle \hat{Q} \hat{M} \hat{Q} \rangle^2 \left\langle (\hat{M} \hat{Q})^4 \right\rangle - \left\langle \hat{Q} (\hat{M} \hat{Q})^{(2)} \right\rangle^2 + \langle \hat{Q} \hat{M} \hat{Q} \rangle^3 \right] + \cdots \right. $$
$$\text{with}$$
$$\hat{B}_{2m}^{(1)} = 0, \hat{B}_{2m+1}^{(1)} \neq 0, = \left\langle \hat{Q} (\hat{M} \hat{Q})^{2m+1} \right\rangle + \left\langle 0 \left(\hat{Q}^{2m+2} \right) \right\rangle, m \geq 3. $$

(9.3.18b)

In diagram form the first few terms are

$$\hat{Q}_1^{(d)} \equiv \overset{(d)}{\underset{\bullet}{}} \Big|_B = [\bullet]_{m=0} + [\,\text{⬭}\!-\!\bullet\!\rightarrow\!\bullet\,]_{m=1} + [\bullet\!\rightarrow\!\text{⬭}\!+\!\bullet\!\rightarrow\!\bullet\!\rightarrow\!\bullet\,]_{m=2}$$

$$[\,\text{⬭}\!\rightarrow\!\bullet\!+\!\bullet\!\rightarrow\!\bullet\!\rightarrow\!\bullet\,]_{m=2}$$

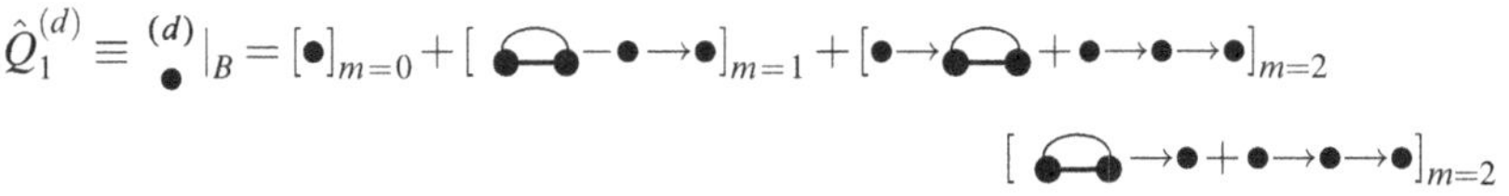

(9.3.18c)

Now, unlike the purely independent case (9.3.14), (9.3.15), the series for $\hat{Q}_1^{(d)}$ does not terminate, so that in this second-order situation we are generally forced to truncate the series to obtain manageable approximations. In the Hom-Stat cases for the first-order Dyson equation of Section 9.1.2, for the Fourier transforms of the kernels $Q_1^{(d)}|_{V \text{ or } S}$, we have from (9.1.23a) and (9.1.34) respectively,

$$q_{00,1}^{(d)}(\mathbf{k}, s)|_{V \text{ or } S} = \mathsf{F}_\rho \mathsf{F}_{\Delta t} \left\{ Q^{(d)}(\rho, \Delta t)_{V \text{ or } S} \right\},$$

(9.3.19)

which appear explicitly in $\langle \alpha^{(Q)} \rangle$, (9.1.26) et seq. for volumes and (9.1.37) et seq. for surfaces.

Note *that when* $\langle \hat{Q} \rangle = 0$, which represents purely volume scatter effects without boundaries or interfaces, the bilocal or *Bourret approximation* (9.3.17a–c) *for* $\hat{Q}_1^{(d)}$ *reduces to a*

---

[25] These are not the only arrangements: we can have $\bullet$ in any position without changing the nth-order ($n = $ odd) contribution.

single term:

$$\langle \hat{Q} \rangle = \bullet = 0 : \hat{Q}_1^{(d)}|_{\text{Bourret:B}} = \langle \hat{Q} \hat{M}_\infty \hat{Q} \rangle \left(= \hat{A}_1^{(1)} \right); \; \hat{A}_m^{(1)} = 0, m \geq 2 := \; \text{⊶⊷}, \quad (9.3.20)$$

for this first-order equivalent deterministic medium. [As we have just seen (cf. (9.3.18a) and (9.3.18b)), with interfaces the Bourret approximation [(9.3.17a) and (9.3.17b)] yields a non terminating series.] This compact result suggests the utility of the bilocal approximation in applications, wherever it appears reasonable to describe the principal scattering effects of the medium through second-order spatial (and temporal) correlations.

The various mean fields $\langle \alpha^{(Q)} \rangle$ are diagrammed here in (9.3.8) when $\hat{Q}_1^{(d)}|_B$ is given either by (9.3.18), or by (9.3.20) in the unbounded (unlayered) cases. For the latter, we have explicitly

$$\langle \hat{\varrho} \rangle = 0: \; \overline{\bigcirc}\Big|_{\text{Bourret}} = \frac{\hat{1}}{\hat{1} - \text{→⊶⊷}} \; \underset{H}{\bigcirc} = \left\{ 1 + (\text{→⊶⊷})^{\Sigma(n) \geq 1} \left(\hat{\eta}_1^{(d)}|_B = \text{→⊶⊷} \right) = \text{⊶⊷} \right\} \overline{\bigcirc}$$

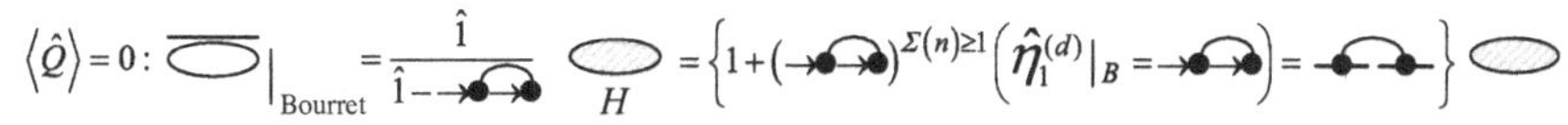

$$(9.3.21)$$

for the associated FOS and PSS. Unlike the strictly independent case Ia above, here $\langle \alpha^{(Q)} \rangle \neq \alpha_H$ and $\langle \alpha^{(Q)} \rangle$ now represents a meaningful expression for the mean scattered field when there are no interfaces, cf. remarks following Eq. (9.3.16). Equation (9.3.21) is a generalization of Tatarskii's Eq. (26), ¶61, [7], here based on the "classical" Helmholtz equation, Section 8.3.1.1, Eq. (8.3.10) et seq., now with time-dependent index refractions $\varepsilon = \varepsilon(\mathbf{R}, t')$, to arbitrary (linear) media and input signals. Similarly, the corresponding analytic solutions are given here by (9.1.26) et seq. and (9.1.37) et seq., for volumes and surfaces, respectively.

Ic. *Series Modification: Second-Order "Dyson Equation"*: We may extend the above next to the space-time correlation function of the (total) field, $\langle \alpha_1 \alpha_2 \rangle$, cf. [(9.3.10a) and (9.3.10b)]. If we use the postulated independent structure of $\hat{Q}$, cf. (9.3.14), we find from (9.3.15) and (9.3.16), (9.3.16a) that

(1) *First-Order Independence*:

$$\hat{Q}_{12}^{(d)}|_{\text{ind}} = \langle \hat{Q}_1 \rangle \langle \hat{Q}_2 \rangle = \underset{1 \; 2}{\bullet \; \bullet} = \hat{A}_o^{(12)}; \hat{A}_m^{(12)} = 0, \; m \geq 1;$$

$$(9.3.22)$$

$$\therefore \hat{\eta}_{12}^{(d)}|_{\text{ind}} \hat{M}_1 \hat{M}_2 \langle \hat{Q}_1 \rangle \langle \hat{Q}_2 \rangle = \underset{1}{\to \bullet} \underset{2}{\to \bullet} = \underset{1}{\overset{(d)}{\to \bullet}} \underset{2}{\overset{(d)}{\to \bullet}} = \hat{\eta}_1^{(d)} \hat{\eta}_2^{(d)}|_{\text{ind}}.$$

Accordingly, with (9.2.3a) and (9.2.3b) et seq., (9.3.10b) and (9.3.16), Eq. (9.3.22) reduces to the not unexpected result

$$\langle \alpha_1 \alpha_2 \rangle_{\text{ind}} = \langle \alpha_1 \rangle_{\text{ind}} \cdot \langle \alpha_2 \rangle_{\text{ind}}; \; \text{diagram } 12|_{\text{ind}} = \text{diagram } 1|_{\text{ind}} \times \text{diagram } 2|_{\text{ind}}$$

$$(9.3.16)_1, \text{ etc.} \qquad\qquad (9.3.16a)_1, \text{ etc.} \qquad\qquad (9.3.23)$$

Again, when $\langle \hat{Q} \rangle = 0$, the postulate I is still too restrictive for both surfaces and volumes [cf. (9.3.16) and following remarks].

(2) *Second-Order Dependence*:
   On the other hand, if we use the bilocal assumption (9.3.17a–c) here, we see that the equivalent deterministic scattering operator $\hat{Q}_{12}^{(d)}$, (9.1.16) is given by (9.3.13) diagrammatically, with third- and higher-order correlations, for example, ⌒⌒⌒, and so on, reducing to ⊶•⊷. In the important case of volume or surface

scatter ($\bullet = \langle Q \rangle = 0$) (with all odd moments of Q), we find that the infinite series (9.2.16) for $\hat{Q}_{12}^{(d)}$ now contains only terms of pairs $\langle Q_1 Q_2 \rangle$ and so on,

$$\hat{Q}_{12}^{(d)} = \langle \hat{Q}_1 \hat{Q}_2 \rangle + \left\langle \hat{Q}_1 (\hat{M}_1 \hat{Q}_1)^2 \hat{Q}_2 \right\rangle + \ldots = \hat{B}_0^{(12)} + \hat{B}_2^{(12)} + \ldots, \qquad (9.3.24)$$

$$\langle \hat{Q} \rangle^{2m+1} = \bullet^{2m+1} = 0: \quad \hat{Q}_{12}^{(d)}\big|_B = \underset{\underset{\hat{B}_0^{(12)}\,+}{1\ \ 2}}{\bullet\ \ \bullet} + [\ \underset{\underset{\hat{B}_2^{(12)}}{1\ 1\ 1\ 2}}{\text{⌢⌢}} + \underset{1\ 2\ 2\ 2}{\text{⌢⌢}} + \underset{1\ 1\ 2\ 2}{\text{⌢⌢}}\ ] + \ldots + \hat{B}_{2m}^{(12)}\big|_{m \geq 2} \qquad (9.3.25a)$$

$$\therefore \hat{\eta}_{12}^{(d)}\big|_B = \left[\langle \hat{\eta}_1^{(2)} \hat{\eta}_2^{(2)} \rangle - \langle \hat{\eta}_1 \hat{\eta}_2 \rangle^2 \right] + \underset{\hat{A}_0^{(12)}\,+}{} \quad \underset{\hat{A}_2^{(12)}}{} \quad + \hat{A}_{2m}^{(12)}\big|_{m \geq 2}$$

$$= \underset{1\ \ 2}{\text{⌢}} + [\ (\underset{1}{(-\bullet-)^{(2)}}\,(\underset{2}{-\bullet-)^{(2)}}) - (\underset{1\ \ 2}{-\bullet-\bullet-})^{(2)} + \ldots\ ] \qquad (9.3.25b)$$

The general second-moment function of the field (= the covariance + $\langle \alpha_1^{(Q)} \rangle \langle \alpha_2^{(Q)} \rangle$), without approximations, becomes diagrammatically now from (9.2.7) as well as (9.2.9b), cf. Section 9.2.1 above and (9.3.10b):

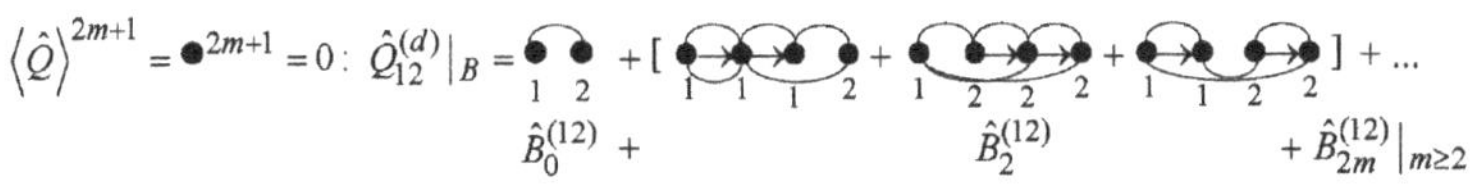

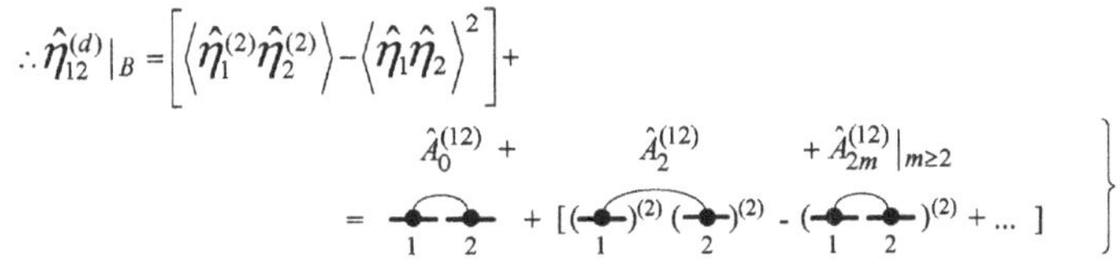

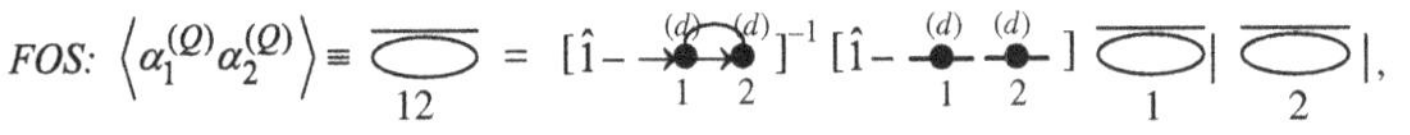

$$FOS: \quad \left\langle \alpha_1^{(Q)} \alpha_2^{(Q)} \right\rangle \equiv \underset{12}{\text{⬭}} = [\hat{1} - \underset{1\ \ 2}{\overset{(d)\ \ (d)}{\rightarrow\bullet\bullet\leftarrow}}]^{-1} [\hat{1} - \underset{1\ \ 2}{\overset{(d)\ \ (d)}{-\bullet-\ \ -\bullet-}}]\, \underset{1}{\text{⬭}}\big|\, \underset{2}{\text{⬭}}\big|, \qquad (9.3.26)$$

where $\underset{1 \text{ or } 2}{\text{⬭}}$ is given explicitly by Eq. (9.3.8). [Other diagram equivalents are (9.3.11) and (9.3.12). The same *form* applies for the Bourret approximation, except that now this approximation is substituted for $\eta_{1,2}^{(d)} \cdot \eta_{12}^{(d)}$ (and correspondingly for $\hat{Q}_1^{(d)}, \hat{Q}_{1,2}^{(d)}$), as well as for $\text{⬭} \rightarrow \text{⬭}\big|_B$, and so on. Even with these simplifying conditions, the development for $\left\langle \alpha_1^{(Q)} \alpha_2^{(Q)} \right\rangle_B$, (9.3.26) is considerably more complex than that for $\langle \alpha_1^{(Q)} \rangle_B$, (9.3.21) here. Our results (9.3.10), (9.3.12), and (9.3.26) with $\langle \bullet^{2m+1} \rangle = 0$, are generalizations of Tatarskii's Eq. (27–49), ¶61, [7], to general (linear) media and arbitrary signals, expressed now in the Dysonian hierarchy of Section 9.2.3.

(3) *Some Higher Order Relations*

As expected when $\hat{Q}$ obeys *first-order independence* [Id, (1)], (9.3.22), we can show that for (9.2.18)

$$\hat{Q}_{12\ldots m}^{(d)}\big|_{ind} = \prod_{n=1}^{m} \langle \hat{Q}_n \rangle = A_0^{(m)} \ ; A_{n \geq 1}^{(m)} = 0, \quad \text{and} \quad \hat{\eta}_{12\ldots m}^{(d)}\big|_{ind} = \prod^{m} \hat{M}_{n,\infty} \langle \hat{Q}_n \rangle = \prod^{m} \hat{\eta}_{n,\infty}^{(d)}$$

$$= \underset{1\ 2\ 3\ \ \ m}{\bullet\bullet\bullet\ldots\bullet} \qquad\qquad\qquad\qquad = \underset{1\ \ \ 2\ \ \ \ m}{\rightarrow\bullet\rightarrow\bullet\ldots\rightarrow\bullet} \qquad (9.3.27)$$

The corresponding mth order field moment can likewise be demonstrated to be [from (9.2.19)–(9.2.21)]

$$\langle \alpha_1 \ldots \alpha_m \rangle_{ind} = \langle \alpha_1 \rangle \langle \alpha_2 \rangle \ldots \langle \alpha_m \rangle, \quad m \geq 1, \qquad (9.3.28)$$

cf. (9.3.23). This again is entirely to be expected, since the scattering operator is completely space–time independent, namely, (9.3.27). [With $\langle \hat{Q} \rangle^{2m+1} = 0, m \geq 1$, we have again too strict a postulate (9.3.14) to give meaningful results physically.]

In the *bilocal cases* (II), (9.3.14), however, when $\langle \hat{Q} \rangle \neq 0$, that is, there are interfaces producing specular reflections; noticeable space–time correlation in the resulting multiple scattering can occur, as we have already seen (for Bourret). No simplifications occur generally in the formal structure of $Q_{12\ldots m}^{(d)}$ and in the FOS and associated diagrams, except that all third- and higher order scattering correlations reduce to products of pairs and means (of $\hat{Q}$). However, when $\langle \hat{Q}^m \rangle, m \geq 1 = 0$, it can be shown that

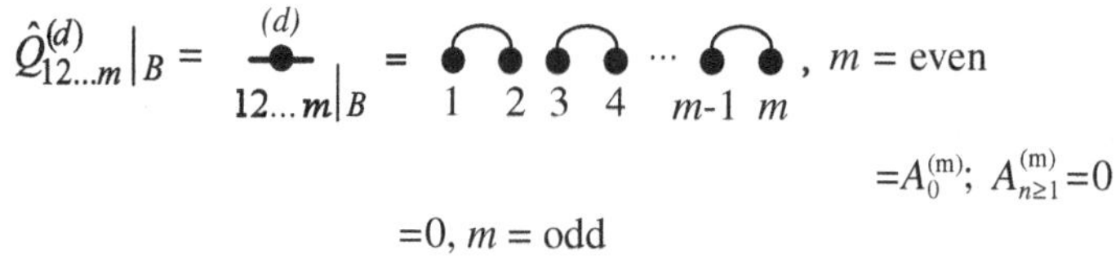

$$\hat{Q}_{12\ldots m}^{(d)}\big|_B = \overset{(d)}{\underset{12\ldots m|B}{\bullet\!-\!\bullet}} = \overset{\frown}{\underset{1}{\bullet}}\,\overset{\frown}{\underset{2}{\bullet}}\,\overset{\frown}{\underset{3}{\bullet}}\,\overset{\frown}{\underset{4}{\bullet}}\,\cdots\,\overset{\frown}{\underset{m\text{-}1}{\bullet}}\,\overset{\frown}{\underset{m}{\bullet}}\,, \quad m = \text{even}$$

$$= A_0^{(\mathrm{m})};\; A_{n\geq 1}^{(\mathrm{m})} = 0 \tag{9.3.29a}$$

$$= 0,\, m = \text{odd}$$

so that

$$\hat{\eta}_{12\ldots m}^{(d)}\big|_B = \hat{\eta}_{12}^{(d)}\big|_B\,\hat{\eta}_{34}^{(d)}\big|_B\cdots\hat{\eta}_{m-1,m}^{(d)}\bigg|_{\substack{B\\ m=\text{even}}} = \overset{\rightarrow}{\underset{1}{\bullet}}\overset{\rightarrow}{\underset{2}{\bullet}}\,\overset{\rightarrow}{\underset{3}{\bullet}}\overset{\rightarrow}{\underset{4}{\bullet}}\cdots\overset{\rightarrow}{\underset{m\text{-}1}{\bullet}}\overset{\rightarrow}{\underset{m}{\bullet}}$$

$$= 0,\, m = \text{odd}. \tag{9.3.29b}$$

The *m*th-order diagram for $\langle \alpha_1 \ldots \alpha_m \rangle$, (9.2.17a), is compactly.

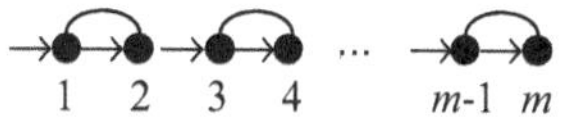

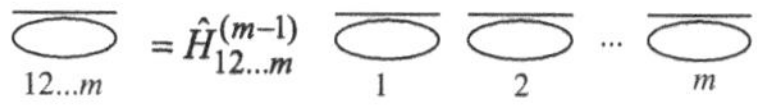

$$\overset{\frown}{\underset{12\ldots m}{\smile}} = \hat{H}_{12\ldots m}^{(m-1)}\,\overset{\frown}{\underset{1}{\smile}}\,\overset{\frown}{\underset{2}{\smile}}\cdots\overset{\frown}{\underset{m}{\smile}} \tag{9.3.30}$$

where $\hat{H}_{12\ldots m}^{(m-1)}$ is structured specifically according to Section 9.2.3. Again, we remark that these results apply formally for both volumes and surfaces.

II. *Series Truncation:*

In the case cited above the "mass operator" $Q_1^{(d)}$ and so on, is approximated either as an infinite or terminated series [cf. (9.3.18), (9.3.15), (9.3.20), (9.3.27), (9.3.29a)], thus altering the form but not the fact that the resulting approximate solution is still an infinite though subseries of the original PSS. On the other hand, *truncation* replaces the exact (or previously approximated) infinite operator series with a finite operator series. Of these latter there are two principal types: *Taylor series approximations*, which yield analytic forms directly and *Born approximations*, which constitute a hierarchy of truncated operator series.

Let us consider Taylor first:

A. Taylor Series Approximations:

Although the exact solution $\langle \alpha^{(Q)} \rangle$ of, say, the first-order Dyson equation [(9.1.7b) and (9.1.7c)], (9.3.8), requires an infinite PSS, cf. [(9.1.7b) and (9.1.7c)], we can reduce the evaluation of $\langle \alpha^{(Q)} \rangle$ to a purely algebraic solution by means of a Taylor series expansion with truncation. Thus, if we expand $\langle \alpha^{(Q)} \rangle$, for example

$$\left\langle \alpha^{(Q)}(\mathbf{R}',t') \right\rangle = \left\langle \alpha^{(Q)}(\mathbf{R},t) \right\rangle + (\mathbf{R}'-\mathbf{R}) \cdot \nabla_{\mathbf{R}} \left\langle \alpha^{(Q)}(\mathbf{R},t) \right\rangle + (t'-t)\frac{\partial}{\partial t}\left\langle \alpha^{(Q)}(\mathbf{R},t) \right\rangle + \ldots, \tag{9.3.31}$$

and keep only the leading term for the generic integral equation

$$\left\langle \alpha^{(Q)} \right\rangle = \alpha_{\mathrm{H}} + \eta^{(d)} \alpha^{(Q)}, \tag{9.3.31a}$$

we have directly the first-order approximation form

$$\hat{\eta}_1^{(d)} \alpha^{(Q)} = \left[\hat{M}_\infty \hat{Q}_1^{(d)} \left\langle \alpha^{(Q)} \right\rangle \right]_1 \doteq \langle \alpha(\mathbf{R}, t) \rangle \hat{M}_\infty Q_1^{(d)} 1. \tag{9.3.32}$$

Here $\hat{M}_\infty \hat{Q}_1^{(d)} 1$ is simply an (analytic) function of $(\mathbf{R}, t)$, namely,

$$\begin{aligned} \hat{\eta}_1^{(d)}|_1 1 = \hat{M}_\infty \hat{Q}_1^{(d)} \hat{1} &= \int M_\infty(\mathbf{R}, t | \mathbf{R}', t') d\mathbf{R}' dt' \int \hat{Q}_1^{(d)}(\mathbf{R}', t' | \mathbf{R}'', t'') 1 dR'' dt'' \\ &\equiv N_1^{(d)}(\mathbf{R}, t). \end{aligned} \tag{9.3.33}$$

Applying [(9.3.32) and (9.3.33)] to (9.3.31a) gives the first-order approximate *analytic* result

$$\left\langle \alpha^{(Q)}(\mathbf{R}, t) \right\rangle_1 \doteq \frac{\alpha_{\mathrm{H}}(\mathbf{R}, t)}{1 - N_1^{(d)}(\mathbf{R}, t)}, \tag{9.3.34}$$

where the exact or various approximate expressions for the equivalent deterministic medium operator $\hat{Q}_1^{(d)}$ may be employed, cf. (9.1.27), (9.3.15), (9.3.20). Thus, for the fully independent (Ia, 1) and bilocal approximations Ib, Ic-(2), $\left\langle \hat{Q}^{(2m+1)} \right\rangle|_{m \geq 0} = 0$, we have specifically (and exactly) here

$$N_{1-\mathrm{ind}}^{(d)} = \hat{M}_\infty \langle \hat{Q} \rangle 1;\ N_{1-\mathrm{ind}}^{(d)} = \langle \hat{Q} \hat{M}_\infty \hat{Q} \rangle 1, \langle Q^{2m+1} \rangle_{m \geq 0} = 0. \tag{9.3.35}$$

Clearly, there may be some points (foci) or lines (caustics) in space where $\langle \alpha^{(Q)} \rangle_1 \to \infty$, when the denominator of (9.3.34) vanishes. The reality of these singularities depends on how well (9.3.34) approximates $\langle \alpha^{(Q)} \rangle$ itself. The true foci, and so on may not be infinite, but may be located close to the singularities of the approximation. Thus, (9.3.35) may serve as a guide as to potentially real foci, and so on.

Higher order yield a combination of analytic forms and simple differential operators, as expected from the Taylor series (9.3.31). Thus, the second-order approximation for $\langle \alpha^{(Q)} \rangle$ becomes directly, in the same fashion

$$\left\langle \alpha^{(2)}(\mathbf{R}, t) \right\rangle \doteq \left(\hat{1} - N_1^{(d)} \hat{1} - N_2^{(d)} \right)^{-1} \alpha_{\mathrm{H}} \doteq \alpha_{\mathrm{H}} + \sum_{k=1}^{m} \left(N_1^{(d)} + N_2^{(d)} \right)^{(k)} \alpha_{\mathrm{H}}, \tag{9.3.36}$$

where now

$$\hat{N}_2^{(d)} \equiv \left[\hat{\eta}_1^{(d)} \times (\mathbf{R}' - \mathbf{R}) \right] \cdot \nabla_{\mathbf{R}} + \left[\hat{\eta}_1^{(d)} \times (t' - t) \right] \frac{\partial}{\partial t}, \tag{9.3.36a}$$

and truncation is chosen after $m(\geq 1)$ terms. The presence of these and subsequent high-order terms stems, of course, directly from the higher order terms in the original Taylor series (9.3.31) for $\langle \alpha^{(Q)} \rangle$.

The error in using (9.3.34) for example, is generally difficult to assess, because the true value of $\langle \alpha^{(Q)} \rangle$ is not usually available. However, in the cases where the Hom-Stat conditions of Section 9.1.2 and so on, are obeyed we can use the exact result (9.1.26) and (9.1.37) to determine the error vis-à-vis using (9.3.34) for $\langle \alpha \rangle_{V \text{ or } S}$.

In a similar way, we may develop a Taylor series approximation for the second-moment function $\left\langle \alpha_1^{(Q)} \alpha_2^{(Q)} \right\rangle$, (9.2.7), (9.2.9b), using a two-variable expansion of $\left\langle \alpha_1^{(Q)} \alpha_2^{(Q)} \right\rangle$ about $(\mathbf{R}_1, t_1; \mathbf{R}_2, t_2)$, cf. (9.3.31). The first-order result here is the analytic expression

$$\left.\begin{aligned} \left\langle \alpha_1^{(Q)} \alpha_2^{(Q)} \right\rangle &\doteq \left\{ \frac{1 - N_1^{(d)} N_2^{(d)}}{\left(1 - N_{12}^{(d)}\right)\left(1 - N_1^{(d)}\right)\left(1 - N_2^{(d)}\right)} \right\} \alpha_{\mathrm{H}1} \alpha_{\mathrm{H}2}; \\ N_{1,2}^{(d)} &= N_{12}^{(d)} 1 = \hat{M}_{1,\infty} \hat{M}_{2,\infty} \hat{Q}_{12}^{(d)} 1, \end{aligned}\right\} \tag{9.3.37}$$

where $N_1^{(d)} = N_1^{(d)}(\mathbf{R}_1, t_1)$, and so on, cf. (9.3.33). Again, various approximate forms for $\hat{Q}_{12}^{(d)}$, $\hat{Q}_{1,2}^{(d)}$ may be used, cf. (9.3.22), [9.3.23a), (9.3.23b)] and (9.3.35) above. Higher order terms (cf. (9.3.31)) in the Taylor expansion may also be sought, as in [(9.3.36) and (9.3.36a)] for $\langle \alpha^{(Q)} \rangle$. The error, again, is difficult to determine, because of the complexity of the exact solution.

IIb. *Born Approximations:*

Unlike the above procedures I, II in Section 9.3.2.1, which essentially approximate the "mass operator", $Q^{(d)}$ with an infinite or truncated series, cf. (9.3.18a,b,c or (9.3.20), (9.3.25), we may truncate the PSS directly. The resulting finite series is called a *Born approximation* series, whose order depends on the stage at which truncation is applied. Thus, for the first-order Dyson equation (9.3.8d) for the mean field $\langle \alpha^{(Q)} \rangle$ we can construct the following hierarchy of approximations and associated diagrams:

$(PSS)_k$:

$k = 0$: 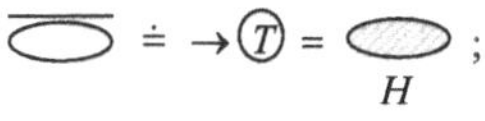 ; $\left\langle \alpha^{(Q)} \right\rangle \doteq \alpha_H$; (9.3.37a)

$k = 1$:

First Born approximation 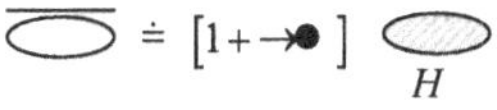 ; $\alpha^{(Q)} = \alpha_{\mathrm{H}} + \hat{M} \left\langle \hat{Q} \right\rangle \bullet \alpha_{\mathrm{H}} \, (FOS)_1$

$k = 2$:

(9.3.37b)

Second Born approximation 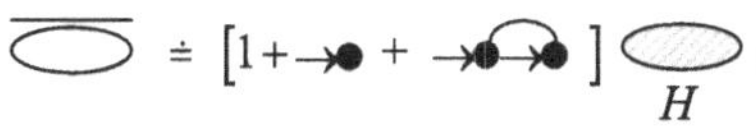 ;

$$\left\langle \alpha^{(Q)} \right\rangle \doteq \sum_{k=0}^{2} \left\langle \left(\hat{M}_\infty \hat{Q} \right)^{(k)} \right\rangle \alpha_{\mathrm{H}}; (FOS)_2 \tag{9.3.37c}$$

$\vdots$ etc. $\vdots$ etc. etc. $(k \geq 3)$

The kth-order

Born approximation is 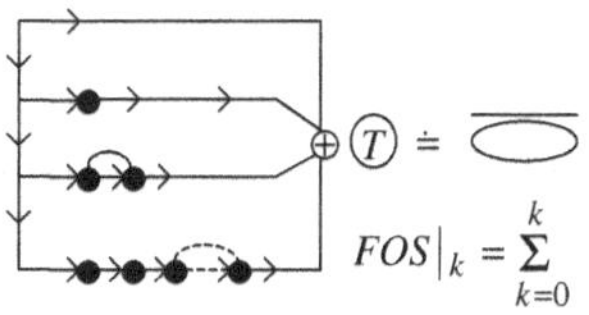 $\left\langle \alpha^{(Q)} \right\rangle$: (9.3.37d)

$$FOS|_k = \sum_{k=0}^{k} \left\langle \left(\hat{M}_\infty \hat{Q} \right)^{(k)} \right\rangle \alpha_{\mathrm{H}}.$$

Unlike the approaches above which use modifications of the "mass operator," $Q^{(d)}$, etc., but where the resulting PSS remain infinite or where equivalently, the corresponding FOR and FOS [(9.3.21), etc.] contain closed feedback loops, all Born approximations contain only feedforward loops, as indicated in (9.3.37d). This is the immediate consequence, of course, of the truncation of the direct PSS (9.3.8d).

More involved Born approximations are readily constructed by truncation of the Dyson equation, cf. (9.3.8). Here we replace $\langle\hat{Q}\rangle, \langle\hat{Q}\ldots\hat{Q}\rangle$, and so on in the direct PSS by $\hat{Q}^{(d)},\ldots,\left(\hat{Q}^{(d)}\ldots\hat{Q}^{(d)}\right)$, and so on, where $\hat{Q}^{(d)}_{(\,)}$ is given by (9.3.9a), and (9.3.9b) which is an infinite (sub-) series of operators, k each feedforward loop of $(FOR)_k$, (9.3.37d). In the important cases of no interfaces (i.e. no coherent terms) $\left\langle\hat{Q}^{2m+1}\right\rangle_{n\geq 0} = 0$ where, for example, a bilocal approximation (truncation) of $\hat{Q}^{(d)}$ is chosen to represent the correlation structure of the scattering operator $\hat{Q}$, we see at once from (9.3.20) that each $\hat{Q}$ in the $(FOR)_k$ of (9.3.37d) is replaced by $\hat{Q}^{(d)} \doteq$ , (e.g., $\hat{\eta}^{(d)} = \bullet \rightarrow$ ), $k = 1$;  becomes , $k = 2$; and for k,  becomes , with $2k\,\hat{Q}$ and $\hat{M}$, operators in each feedforward link. Because of the bilocal assumption, after $k = 2$, we lose the higher order correlations. However, for $0 < k \leq 2$, we have generated a more sophisticated Born approximation. This is because of our use of the Dyson equation, rather than the direct average of the PSS for $\langle\alpha^{(Q)}\rangle$, cf. (9.3.8d), and our implementation of the bilocal approximation of the mass operator, $\hat{Q}^{(d)}$, cf. (9.3.20).

As in all these approximate situations, the evaluation of error remains a major and difficult problem, because the exact results are generally unknown. We must rely, instead, on our physical intuition, aided by the (essentially) local physics governing the inhomogeneity operator $\hat{Q}$ in question. This is one reason why the physical "anatomization" of the so-far canonical $\hat{Q}$ is critically important in specific cases and is often the principal modeling subject. In the special situations where an exact solution is available [for example, in Section 9.1.2 above], the error in the various Born approximations $(k \geq 1)$ can be evaluated, relatively conveniently, at least for small k.

9.3.3 A Characterization of the Random Channel: First- and Second-Order-Moments I

Our interest in these first- and second-order moments stems from their usefulness in describing the physical events of scattering and ambient noise in acoustic and electromagnetic channels and our ability to quantify their structure adequately in many practical situations. We mention in particular such quantities in the covariance case as the intensity spectrum, as well as other measures of noise (and signal) energizes, and their expansive coherent component. Accordingly, we may use the results of Sections 9.1 and 9.2 to characterize the general channel $(T_M^{(N)}$ of Fig. 9.1), specifically in terms of the homogeneous and inhomogeneous (random) operators $\hat{M}$ and $\hat{\eta}_\infty (\equiv \hat{M}_\infty\hat{Q})$, cf. [(9.1.6a), (9.1.6b), (9.1.7a), (9.1.7b)]. In addition to scattering and deterministic inhomogeneities, we include now *ambient fields*.

9.3.3.1 Operator Structure of the Random Channel: Ambient Fields The channel operator $\mathbf{T}_M^{(N)}\left(= \left[\hat{1}-\hat{\eta}_\infty\right]^{-1}\hat{M}_\infty\right)$, cf. Fig. 9.1, since from $\alpha = \mathbf{T}_M^{(N)}\{-G_T\}$ is conveniently written now

$$T_M^{(N)} = \left(\hat{1} + \frac{\hat{\eta}_\infty}{\hat{1} - \hat{\eta}_\infty}\right)\hat{M}_\infty \equiv \hat{M}_\infty + \hat{\mathsf{I}}_\infty \equiv \hat{\mathsf{M}}_\infty. \tag{9.3.38a}$$

$$\therefore \quad \hat{\mathsf{I}}_\infty = \left(\hat{1} - \hat{\eta}_\infty\right)^{-1}\hat{\eta}_\infty \hat{M}_\infty. \tag{9.3.38b}$$

Here we can immediately separate the homogeneous $(\hat{M}_\infty)$ from the inhomogeneous $\left(\hat{\mathsf{I}}_\infty\right)$ operations involved because of the assumed linearity of the medium. [Of course, for an ideal medium, $\hat{\mathsf{I}}_\infty = 0$ and $\mathbf{T}_M^{(N)} = \hat{M}_\infty$.] Moreover, when a variety of scattering mechanisms is excited we use the $\hat{M}$-form to account for their possible interactions.

When there are noticeable *ambient sources*, of source density $-G_A$, these may be added to the desired signal source $(-G_\mathrm{T})$, so that the basic Langevin equation becomes

$$\{(\hat{L}^{(0)} - \hat{Q})\alpha^{(Q)} = -G_\mathrm{T} - G_\mathrm{A} + [\text{b.c.s} + \text{i.c.s}]\}, \text{ or} \tag{9.3.39a}$$

$$\left(\hat{1} - \hat{\eta}_\infty\right)\alpha^{(Q)} = \alpha_\mathrm{H} + \alpha_\mathrm{A}, \quad \text{with} \quad \alpha_\mathrm{A} = \hat{M}_\infty(-G_\mathrm{A}), \tag{9.3.39b}$$

$$\text{and} \quad \therefore \quad \alpha = \mathbf{T}_M^{(N)}\{-G_\mathrm{T} - G_\mathrm{A}\}, \tag{9.3.39c}$$

where $-G_\mathrm{A}$ (like $-G_\mathrm{T}$) is the source function describing the ambient field, and where $-G_\mathrm{A}$ may be localized or distributed. As sources extraneous to that of the desired signal, the ambient field mechanism has a different physical (and hence statistical) structure from that of the signal *and* the secondary scattering sources produced by the interaction of the original signal with inhomogeneities. The key differences lie in the fact that the ambient sources are usually independently emitting as well as independently spatially distributed, unlike the scatter, which is initiated by the desired signal source and is radiatively coupled (e.g., multiple scatter). We can, of course, have scatter produced by the ambient sources, but these effects are usually small, except for deterministic multipaths. In any case, it is clear from [(9.3.38a) and (9.3.38b)] and [(9.3.39a) and (9.3.39b)] that the *medium operator* $\mathbf{T}_M^{(N)}$ $(= \hat{\mathsf{M}}_\infty$ and hence $\hat{\mathsf{I}}_\infty$, also) remain unchanged by the presence of sources, ambient and desired, as we would expect as long as these sources themselves do not alter the medium.

9.3.3.2 Elementary Statistics of the Channel Operator $\mathbf{T}_M^{(N)}$ Here the principal channel operator statistics, like those for the field (cf. Sections. 9.1 and 9.2), are the lower order moments and covariance functions, for example, $\langle\alpha\rangle$, $\langle\alpha_1\alpha_2\rangle$, $\langle\alpha_1\alpha_2\rangle - \langle\alpha_1\rangle\langle\alpha_2\rangle$, and so on. By direct averaging of (9.3.38a) we find directly that

$$\left\langle \mathbf{T}_M^{(N)} \right\rangle = \left\langle \hat{\mathsf{M}}_\infty \right\rangle = \hat{M}_\infty + \left\langle \hat{\mathsf{I}}_\infty \right\rangle; \hat{K}_{\infty,I} = \left\langle \hat{\mathsf{I}}_1\hat{\mathsf{I}}_2 \right\rangle - \left\langle \hat{\mathsf{I}}_{\infty 1} \right\rangle\left\langle \hat{\mathsf{I}}_{\infty 2} \right\rangle, \tag{9.3.40a}$$

$$\therefore \left\langle \mathbf{T}_{M1}^{(N)}\mathbf{T}_{M2}^{(N)} \right\rangle = \left\langle \hat{\mathsf{M}}_{\infty 1}\hat{\mathsf{M}}_{\infty 2} \right\rangle = \hat{M}_{\infty 1}\hat{M}_{\infty 2} + \hat{M}_{\infty 1}\left\langle \hat{\mathsf{I}}_{\infty 2} \right\rangle + \hat{M}_{\infty 2} = \left\langle \hat{\mathsf{I}}_{\infty 1} \right\rangle + \left\langle \hat{\mathsf{I}}_{\infty 1}\hat{\mathsf{I}}_{\infty 2} \right\rangle, \tag{9.3.40b}$$

with higher order moments determined in the same way. In more detail diagrammatically with the help of (9.3.38b) and (9.3.7a–d), (9.3.8a–d), we can readily write

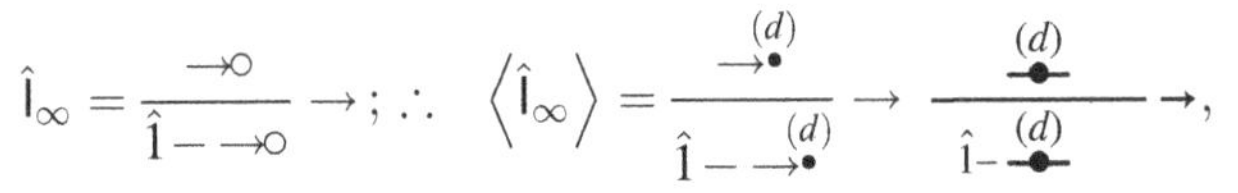

$$
\hat{I}_\infty = \frac{\rightarrow\circ}{\hat{1}-\rightarrow\circ}\rightarrow \;;\; \therefore\; \left\langle \hat{I}_\infty \right\rangle = \frac{\overset{(d)}{\rightarrow\bullet}}{\hat{1}-\overset{(d)}{\rightarrow\bullet}}\rightarrow\frac{\overset{(d)}{\bullet}}{\hat{1}-\overset{(d)}{\bullet}}\rightarrow, \tag{9.3.41a}
$$

so that

$$
\left\langle \mathbf{T}_M^{(N)} \right\rangle = \left\langle \hat{M}_\infty \right\rangle = \rightarrow + \frac{\overset{(d)}{\bullet}}{\hat{1}-\overset{(d)}{\bullet}}\rightarrow, \text{ etc., cf. (9.3.8a–d).} \tag{9.3.41b}
$$

The covariance function $\hat{K}_{\infty,\hat{\imath}}$, (9.3.40a), is likewise represented diagrammatically by the PTS expansion

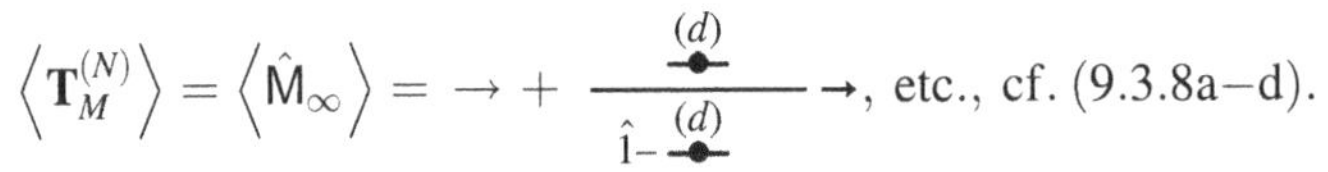

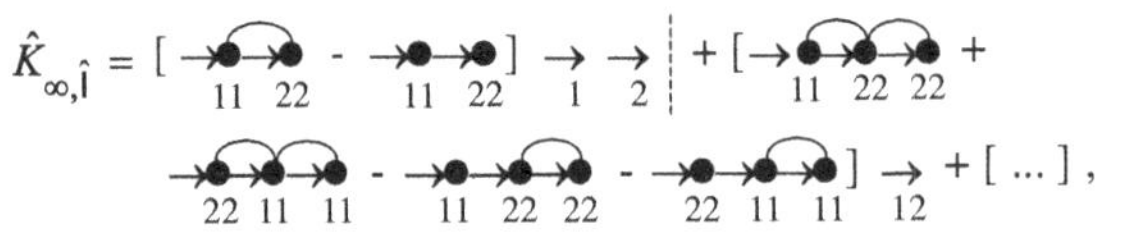

$$
\hat{K}_{\infty,\hat{\imath}} = [\underset{11\ 22}{\rightarrow\bullet\bullet} - \underset{11\ 22}{\rightarrow\bullet\rightarrow\bullet}]\underset{1\ \ 2}{\rightarrow\ \rightarrow} \;\bigg|\; + [\underset{11\ 22\ 22}{\rightarrow\bullet\bullet\bullet} +
$$
$$
\underset{22\ 11\ 11}{\rightarrow\bullet\bullet\bullet} - \underset{11\ 22\ 22}{\rightarrow\bullet\rightarrow\bullet\bullet} - \underset{22\ 11\ 11}{\rightarrow\bullet\rightarrow\bullet\bullet}]\underset{12}{\rightarrow} + [\,...\,], \tag{9.3.42a}
$$

or more compactly,

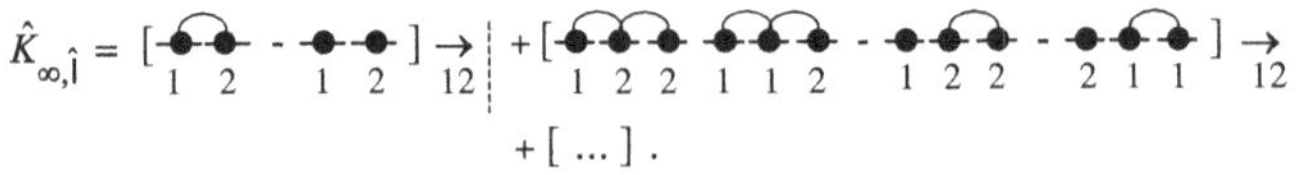

$$
\hat{K}_{\infty,\hat{\imath}} = [\underset{1\ 2}{\bullet\bullet} - \underset{1\ 2}{\bullet\rightarrow\bullet}]\underset{12}{\rightarrow} \;\bigg|\; + [\underset{1\ 2\ 2}{\bullet\bullet\bullet} + \underset{1\ 1\ 2}{\bullet\bullet\bullet} - \underset{1\ 2\ 2}{\bullet\rightarrow\bullet\bullet} - \underset{2\ 1\ 1}{\bullet\rightarrow\bullet\bullet}]\underset{12}{\rightarrow}
$$
$$
+ [\,...\,]. \tag{9.3.42b}
$$

cf. [(9.3.9a) and (9.3.9b)]. Similarly, for the *mean intensity operator* $\left\langle \hat{I}_{\infty,1} \right\rangle \left(\equiv \left\langle \hat{I}_{\infty,1}^{(2)} \right\rangle\right)$ of the random channel, we get at once from (9.3.41a)

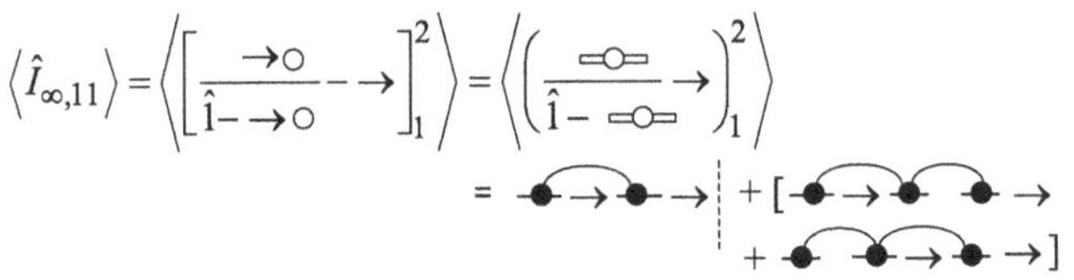

$$
\left\langle \hat{I}_{\infty,11} \right\rangle = \left\langle \left[\frac{\rightarrow\circ}{\hat{1}-\rightarrow\circ} - \rightarrow\right]_1^2 \right\rangle = \left\langle \left(\frac{\Longleftrightarrow}{\hat{1}-\Longleftrightarrow}\rightarrow\right)_1^2 \right\rangle \tag{9.3.43a}
$$
$$
= \bullet\rightarrow\bullet\rightarrow \;\bigg|\; + [\bullet\rightarrow\bullet\;\;\bullet\rightarrow
$$
$$
+ \bullet\;\;\bullet\rightarrow\bullet\rightarrow]
$$

$$
+ [\,...\,], \text{ etc.} \tag{9.3.43b}
$$

on direct multiplication, cf. [(9.3.9a) and (9.3.9b)]. Higher order operator moments like $\left\langle \hat{I}_{\infty,12} \right\rangle \left(\equiv \left\langle \hat{I}_{\infty 1} \right\rangle\left\langle \hat{I}_{\infty 2}\right\rangle\right)$, $\left\langle \hat{I}_{\infty,1234} \right\rangle \left(\equiv \left\langle \hat{I}_{\infty 1|}, \hat{I}_{\infty 2|}, \hat{I}_{\infty 3|}, \hat{I}_{\infty 4} \right\rangle\right)$ are obtained in similar fashion. Of course, the principal practical use of these expansions occurs when the higher-order terms can be neglected (e.g., Born approximations, cf. Eqs. (9.3.37a) et seq. above, where, for example, all components to the right of the vertical dashed line in [(9.3.42a) and (9.3.42b)], [(9.3.43a) and (9.3.43b)] are dropped).

9.3.4 Elementary Statistics of the Received Field

The received waveform $X(t)$ following the receiver aperture $\left(\hat{R}\right)$ but before any subsequent signal processing, is now readily found from (9.3.38a) applied to the sources in [(9.3.39a) and (9.3.39b)]. We have

$$X = \hat{R}\alpha^{(Q)} - \hat{R}\left(\hat{M}_\infty + \hat{\mathsf{I}}_\infty\right)(-G_T - G_A) = \hat{R}\left[\hat{1} + \left(\hat{1} - \hat{\eta}_\infty\right)^{-1}\hat{\eta}_\infty\right](\alpha_H + \alpha_A), \quad (9.3.44)$$

cf. (9.3.39b). Consequently, the mean received waveform is

$$\langle X\rangle = \hat{R}\left\langle\alpha^{(Q)}\right\rangle = \hat{R}\left(\hat{M}_\infty + \left\langle\hat{\mathsf{I}}_\infty\right\rangle\right)(-G_T - G_A) = \hat{R}\hat{\mathsf{M}}_\infty(-G_T - G_A). \quad (9.3.45)$$

The *incoherent component* of $\langle X\rangle$ is

$$\langle X\rangle_{\text{inc}} = \langle X\rangle - \hat{R}(\alpha_H + \alpha_A) = \hat{R}\left\langle\hat{\mathsf{I}}_\infty\right\rangle(-G_T - G_A) \neq 0. \quad (9.3.45a)$$

The nonvanishing character of $\langle X\rangle_{\text{inc}}$ is generally evident: although this component cannot be (time-) correlated with the input signal, for example, this cross-correlation function vanishes, the ensemble average does not, since $\left\langle\hat{\mathsf{I}}_\infty\right\rangle \neq 0$, cf. [(9.3.41a) and (9.3.41b)], which represents a mean renormalization of the original field $\alpha_H(+\alpha_A)$.

The covariance and second-order moment of the received wave are, similarly,

$$\begin{aligned} K_X(t_1, t_2) &= \langle X_1X_2\rangle - \langle X_1\rangle\langle X_2\rangle = \hat{R}_1\hat{R}_2(\langle\alpha_1\alpha_2\rangle - \langle\alpha_1\rangle\langle\alpha_2\rangle) \\ &= \hat{R}_1\hat{R}_2\hat{K}_{\infty,I} \bullet (G_T + G_A)_1(G_T + G_A)_2 \end{aligned} \quad (9.3.46)$$

and

$$\langle X_1X_2\rangle = \hat{R}_1\hat{R}_2\langle\alpha_1\alpha_2\rangle = \hat{R}_1\hat{R}_2\hat{\mathsf{M}}_{\infty 1}\hat{\mathsf{M}}_{\infty 2}(G_{T1} + G_{A1})(G_{T2} + G_{A2}) \quad (9.3.47)$$

where we have included possible ambient sources. Thus, the mean intensity of the received wave is

$$\left\langle X^2\right\rangle = \hat{R}^{(2)}\left\langle\hat{\mathsf{M}}_\infty^{(2)}(G_T + G_A)^2\right\rangle = \left\langle\left[\hat{R}\hat{\mathsf{M}}_\infty(G_T + G_A)\right]^2\right\rangle. \quad (9.3.47a)$$

Higher order moments are determined in the same fashion:

$$\langle X_1X_2X_3X_4\rangle = \hat{R}_1\hat{R}_2\hat{R}_3\hat{R}_4\langle\alpha_1\ldots\alpha_4\rangle = \hat{R}_1\ldots\hat{R}_4\left\langle\hat{\mathsf{M}}_{\infty,1}\ldots\hat{\mathsf{M}}_{\infty,4}\right\rangle G_1\ldots G_4 \quad (9.3.48a)$$

with

$$\left\langle X_1^2X_2^2\right\rangle = \hat{R}_1^{(2)}\hat{R}_2^{(2)}\left\langle\alpha_1^2\alpha_2^2\right\rangle = \hat{R}_1^{(2)}\hat{R}_2^{(2)}\left\langle\hat{\mathsf{M}}_{\infty,1}^{(2)}\hat{\mathsf{M}}_{\infty,2}^{(2)}\right\rangle G_1^2G_2^2 \quad (9.3.48b)$$

for the second-moment function of the *intensity* X^2, cf. (9.3.47a). From [(9.3.38a) and (9.3.38b)], it is clear that all orders of the operator $\hat{Q}$ (or $\hat{\eta}_\infty = \hat{M}_\infty\hat{Q}$) appear in these moment expressions.

The appropriate diagrams follow from those of the corresponding operators I_∞, or $\hat{M}_\infty$, cf. [(9.3.41a) and (9.3.41b)], [(9.3.43a) and (9.3.43b)] (on preoperating by $\hat{R}$, cf. (9.3.44)). Thus, for $\langle X \rangle$, (9.3.45) we can write directly $\langle X \rangle = \hat{R}\,$⊂⟩, which is developed in a PSS explicitly in (9.3.37), preceded by $\hat{R}$. The diagram for (9.3.47), by direct expansion, is

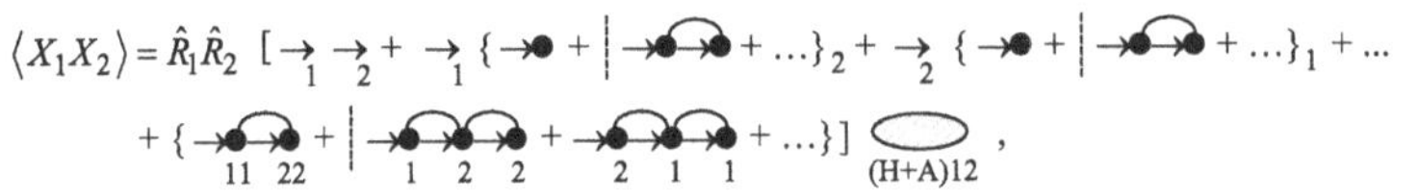

$$\langle X_1 X_2 \rangle = \hat{R}_1 \hat{R}_2 \;\; [\underset{1}{\rightarrow} \, \underset{2}{\rightarrow} + \, \underset{1}{\rightarrow} \{ \rightarrow\!\bullet + | \rightarrow\!\bullet\!\!\bullet + \ldots \}_2 + \, \underset{2}{\rightarrow} \{ \rightarrow\!\bullet + | \rightarrow\!\bullet\!\!\bullet + \ldots \}_1 + \ldots$$
$$+ \{ \underset{11\;22}{\rightarrow\!\bullet\!\!\bullet} + | \underset{1\;2\;2}{\rightarrow\!\bullet\!\!\bullet\!\!\bullet} + \underset{2\;1\;1}{\rightarrow\!\bullet\!\!\bullet\!\!\bullet} + \ldots \}] \underset{(H+A)12}{\text{⟨⟩}} \,, \tag{9.3.49}$$

from which the various higher order moments of $\hat{Q}$ (at points 1, 2) appear clearly. The elements to the left of the vertical dashed lines constitute a form of first-Born approximation to $\langle X_1 X_2 \rangle$, cf. remarks for Eqs. (9.3.37a) et seq.

Finally, the central problem remains of obtaining the PSS solutions, or the equivalent FOS, either by approximation (in the manner of Section 9.3.2), or in the comparatively rare cases where closed form solutions can be generated (cf. 9.1.2). One can examine some new approaches, whereby suitable decompositions of the inhomogeneity operator $\hat{Q}$ into ordered sums of its interaction elements can be achieved, thereby including the primary groups of multiple scatter effects, along with their group probability distributions.

9.4 SUMMARY REMARKS

In the preceding sections we have outlined a "classical" operator formulation for treating scattering in a linear medium. Because of its random nature, exemplified by our quantitative *a priori* ignorance of a typical replica or representation of the scattered field, we cannot best construct the Langevin equation governing propagation and various assumptions about its statistical character. These, however, are (usually) sufficient for us to obtain solutions to the particular Langevin equations for the phenomenon in question. These "solutions", as we have seen, are various statistics of the scattered field, that is, moments and, if possible, probability distributions of the field. As explained at the beginning of this chapter and Chapter 8, this approach constitutes an *aposteriori* or *predictive* theory, since we never know in advance the deterministic nature of the replica which we will actually encounter.

In brief, this chapter has provided a rather elementary and necessarily condensed account of some of the principal results of classical scattering theory (in linear media) from an operational viewpoint. However, we have generalized the treatment to include surfaces as well as volumes, and distributed sources and their accompanying apertures and arrays, as well as the formalism of propagation. The operational form in which our treatment is cast has the advantages of formal exactness and compactness, and as such provides a convenient vehicle for the large-scale computations required to obtain numerical results in most cases. It also allows us to express the detailed and complex structures analytically, from which in turn computations are to be made. As noted earlier, these computations are generally on the sale of those needed for weather predictions and large-scale turbulence. In some cases, however, analytically tractable results can be obtained, often with appropriate approximations.

We may summarize our treatment here with the following brief account of the present chapter.

Section 9.1 describes operational solutions for the first and second moments of the scattered field. In more detail, feedback operational solutions, are obtained for the generic Langevin equation (cf. 9.1.5a). Then Dyson's equation for the first moment is given, a deterministic relation, along with the equivalent mass operator $\hat{Q}_1^{(d)}$, for volumes *and* surfaces, along with its generalization $\hat{\eta}_1^{(d)} \left(= \hat{M}_\infty \hat{Q}^{(d)} \right)$; both $\hat{Q}_1^{(d)}$ and $\hat{\eta}_1^{(d)}$ are global operators. When the medium is Hom-Stat, Dyson's equation can be expressed formally in closed analytical relations, which however must employ a finite approximation of $\hat{Q}_1^{(d)}$, cf. (9.1.26) et seq. for volumes, and (9.1.37) et seq. for surface. Included here is the example of the Langevin equation for the general time-dependent Helmholtz equation with a random index of refraction, cf. Section 9.1.3 Examples.

Section 9.2 presents operational solutions for the higher order moment (i.e. Bethe–Salpeter equations) in feedback form, including the structure of $\hat{Q}_{12}^{(d)}$, $\hat{\eta}_{12}^{(d)}$. (Even in for Hom-Stat media there mth-order moments $(m \geq 2)$are *not* reducible to closed form, unlike the original first-order Dyson equation above.) A brief treatment of the transport equation is also included in Section 9.2.4, along with analysis of the Gaussian cases for $\hat{Q}$ (Section 9.2.5) and an example of very strong scatter — that is, $\|\hat{\eta}\| = 1$, leading to a diffusion medium, for which first and second moments of the field in the volume, that is, $\langle \alpha^{(Q)} \rangle$, $\left\langle \alpha_1^{(Q)} \alpha_2^{(Q)} \right\rangle$, are obtainable explicitly.

In Section 9.3 equivalent representations in terms of simplified Feynman diagrams are discussed for both first- and second-order moments of the field including variety of approximations, such as the *Bourret or bilocal approximation series modification, truncation, Taylor series and Born approximations.* This Section concludes with some lower order statistics of the channel as a whole and in particular for the received field.

REFERENCES

1. W. C. Chew, *Waves and Fields in Inhomogeneous Media*, IEEE Press, Piscataway, NJ, 1995.
2. L. B. Felsen and N. Marcuvitz, *Radiation and Scattering of Waves*, Prentice Hall, Englewood Cliffs, NJ, 1994.
3. U. Frisch, Wave Propagation in Random Media, in A. T. Bharucha-Reid (Ed.), *Probabilistic Methods in Applied Mathematics*, Vol. 1, Academic Press, New York, 1968, pp. 76–198.
4. V. I. Tatarskii, *The Effects of the Turbulent Atmosphere on Wave Propagation*, Vol. TT-68-50464, U. S. Department of Commerce, NTIS, Springfield, VA, 1971.
5. A. Ishimaru, *Wave Propagation and Scattering in Random Media*, Vols. I and II, Academic Press, New York, 1978.
6. F. G. Bass and I. M. Fuks, *Wave Scattering from Statistically Rough Surfaces*; (Translated by C. B. and J. F. Vesecky), International Series in Natural Philosophy, Vol. 93, Pergamon Press, New York, 1979.
7. S. M. Rytov, Yu. A. Kravtsov, and V. I. Tatarskii, *Principles of Statistical Radio Physics*, Vol. 4, (Translated from Russian Edition, in 1989), Springer, New York, 1978. (Chapters 4 and 5 primarily here).
8. V. Twersky, On Multiple Scattering of Waves, *J. Res. Nat. Bureau Standards, Sect. D*, **64**, 715–730 (1960).
9. D. Dence and J. E. Spence, Wave Propagation in Random Anisotropic Media, in A.T. Bharucha-Reid (Ed.), *Probabilistic Methods in Applied Mathematics*, Vol. 3, Academic Press, New York, 1973, pp. 122–182.
10. V. I. Klayatskin, *A Statistical Description of Dynamical Systems with Fluctuating Parameters*, in the series "Contemporary Problems in Physics," Nauka, Moscow, 1975.

11. S. M. Flatté, Wave Propagation through Random Media: Contributions from Ocean Acoustics, *Proc. IEEE*, **71**(11),1267–1294 (1983).
12. K. Furutsu, *Random Media and Boundaries*, Springer-Verlag, New York, 1982.
13. J. A. Oglivy, *Theory of Wave Scattering from Random Surfaces*, Adam-Hilger, New York, 1991.
14. B. J. Uscinski, *The Elements of Wave Propagation in Random Media*, McGraw-Hill, New York, 1977;*Wave Propagation and Scattering*, Clarendon Press, Oxford, UK, 1986.
15. V. I. Tatarskii, A. Ishimaru, and V. U. Zavorotny, *Wave Propagation in Random Media (Scintillation)*, Int'l Conference, August 3–7, 1992, University of Washington, SPIE and Institute of Physics Publishing, Bristol, UK, 1993.
16. A. D. Wheelon, Electromagnetic Scintillation, *Geometrical Optics*, Vol. I. 2001; *Weak Scattering*, Vol. 2, 2003; *Strong Scattering*, Vol. 3, Cambridge University Press, Cambridge, UK.
17. D. Middleton, *An Introduction to Statistical Communication Theory*, Classic Reissue, IEEE Press, Piscataway, NJ, 1996.
18. R. D. Mattuck, *A Guide to Feynman Diagrams in the Many-Body Problems*, 2nd ed., McGraw-Hill, New York, 1976.
19. P. M. Morse and H. Feshbach, *Methods of Theoretical Physics*, Ist ed., Vols. I and II, International Series in Pure and Applied Physics, McGraw-Hill, New York, 1953. [Chapter 8, and in particular, Section 8.4, cf. p. 859 and Section 8.5, and pp. 992–996.]
20. W. A. Lovitt, *Linear Integral Equations*, McGraw-Hill, New York, 1924; reprinted, Dover, New York, 1950.
21. R. B. Lindsey, *Mechanical Radiation*, International Series in Pure and Applied Physics, McGraw-Hill, New York, 1960.
22. J. D. Jackson, *Classical Electrodynamics*, 2nd ed., John Wiley & Sons, Inc., New York, 1975.
23. J. R. Breton and D. Middleton, Stochastic Channels as Generalized Networks, *Int'l Symposium on Information Theory*, Cornell University, Ithaca, NY, Oct. 10–14, 1977, p. 81. Also, *J. Acoustic Soc. Am.*, **69**(5), pp. 1245–1260 (1981); Tech. Rpt. 5871, 16 Jan. 1978, Naval Underwater System Center (NUSC) New London, CT, 16 Jan. 1978; see Breton's Ph.D. Dissertation, University of Rhode Island, Dec. 1977.
24. A. D. Polyanin and A. V. Manzhirov, *Handbook of Integral Equations*, CRC Press, Boca Raton, FL.
25. T. M. Elfouhaily and C. A. Guerin, A Critical Survey of Approximate Scattering Wave Theories from Random Rough Surfaces, *Waves Random Media*, **14**(4),pp. R1–R40 (1) (2004).
26. D. Kaiser, Physics and Feynman's Diagrams, *Am. Scientist*, **93**, 156–165 (2005). For physical application, see Mattuck [18], as well as the references therein.

APPENDIX A1

SELECTED PROBABILITY DISTRIBUTIONS FOR GAUSSIAN FIELDS

In the important cases where the noise (and sometimes the signal) fields are Gaussian, it is well known that it is possible to obtain their probability distributions—in applications the associated probability densities (pdf values)—of such fields, when their space–time variances are specified. Moreover, in such cases, it is also possible to obtain the performance of detection and estimation systems operating in these Gaussian environments, often in exact formulations, or in such forms as to simplify their evaluation greatly. A variety of examples is presented in the early chapters, where obtaining exact generic results under various conditions of signal reception, namely, "coherent," "incoherent," and combinations thereof, is the goal. These in turn involve both linear and quadratic functions and functionals of the received data $x(= [x_j])$ for Bayes optimal detection and estimation algorithms.

In earlier work (Chapter 17, [1]), we have considered similar problems: first-order pdf values of the spectral intensity of Gaussian processes (Section 17.2.3, [1]), pdf values of allowing (quadratic) nonlinear operations and filtering (Section 17.3, [1]), first- and higher-order characteristic functions (cf's), and pdf values under similar nonlinear conditions (cf. Problems 17.15–19, [1]). Our major innovations here are the extensions of this earlier work (1) to *space–time fields*, including the often encountered cases of separable space–time fields, (2) *preformed beams*, where only the temporal portion of the array or aperture outputs are subject to optimal processing, and (3) explicit results for broadband incoherent reception, which involves quadratic functions (and functionals) of the received noise field.

Non-Gaussian Statistical Communication Theory, David Middleton.

Here, we limit our attention to the following generic cases and to the evaluation of their first-order pdf values, mainly under H_0:N and H_1:$S+N$. Specifically, we seek the pdf values of the generic relation $y = \log G(x) = A_J + F(x)$, $x = [x_{j=mn}] = [x(\boldsymbol{R}_m, t_n)]$, when the characteristic functions of the received field x under H_0 and H_1 are given, see Section 2.4, for these Gaussian fields from which the noise field data $\boldsymbol{x}$ are taken. Here $F(x)$ is the Gaussian quadratic form $F = \tilde{x}Lx$ and

$$y = \log G(x) = A_J + \tilde{x}Lx; \quad \text{under} \left. \begin{matrix} H_0{:}N;\ H_1 = S+N \\ H_1{:}N+S_1; H_2 = S_2+N \end{matrix} \right\}, \tag{A1.1}$$

the latter for the binary signal cases $S_1 + N$ versus $S_2 + N$. Generally, field statistic $L = [L_{jj'}]$ is a $(J \times J), J = MN$ square matrix, which may or may not be symmetrical; L itself is often a product of covariance matrices and is at least positive semidefinite, that is, $\tilde{x}Lx \geq 0$. The matrix L is also assumed to have an inverse, that is, $L^{-1}L = I$ and, therefore, $\det L \neq 0$, and to possess distinct eigenvalues, some of which may be negative and all of which are nonzero. Here, the indices (j,j') are double, that is, $j = mn$, where $m = 1, 2, \ldots, M$ and $n = 1, 2, \ldots, N$; $J = MN$; represent respectively points in space and time, for example, $L_{jj'} = L(\boldsymbol{R}_m, t_n; \boldsymbol{R}_{m'}, t_{n'})$. Alternatively, and equivalently, it is often convenient to use a "condensed" or single index number k, where k renumbers $j = mn$. Thus, schematically, we can write the following isomorphisms:

$$\begin{aligned} \begin{pmatrix} j \\ k \end{pmatrix} = mn = \Bigg\{ & \begin{pmatrix} 11 \\ 1 \end{pmatrix}, \begin{pmatrix} 12 \\ 2 \end{pmatrix}, \ldots, \begin{pmatrix} 1N \\ N \end{pmatrix}; \begin{pmatrix} 21 \\ N+1 \end{pmatrix}, \begin{pmatrix} 22 \\ N+2 \end{pmatrix}, \ldots, \begin{pmatrix} 2N \\ \vdots \\ 2N \end{pmatrix}; \\ & \begin{pmatrix} M1 \\ (M-1)N \end{pmatrix}, \begin{pmatrix} M2 \\ (M-1)N+1 \end{pmatrix}, \ldots; \begin{pmatrix} MN \\ MN \end{pmatrix} \Bigg\}. \end{aligned} \tag{A1.2}$$

where notational compactness is useful, particularly in cases for which $\boldsymbol{L}$ does not factor into separable spatial and temporal components.

The pdf values we seek to evaluate, given their associated characteristic functions $F_1(i\xi|H_0, H_1)$, involve integrals of the following type:

$$w_1(y|H_0) = \int_{-\infty}^{\infty} \frac{e^{-i\xi(y-A_J)} d\xi/2\pi}{[\det(\boldsymbol{I} - i\xi a\boldsymbol{L})]^{\gamma}}; \quad w_1(y|H_1) = \int_{-\infty}^{\infty} \frac{e^{-i\xi(y-A_J)+H_1(i\xi)} d\xi/2\pi}{[\det(\boldsymbol{I} - i\xi a\boldsymbol{L})]^{\gamma}}, \tag{A1.3}$$

for the hypothesis states $H_0 : N$ and $H_1 : S+N$, refer to Eq. (A1.1). Here, $\gamma = 1$ or $1/2$ are associated respectively with narrow- and broadband data $\{x\}$ for the most part. In the above $a, (> 0), y, A_J, \gamma,$ and $\boldsymbol{L}$ are generally real quantities, while ξ can be complex, by analytic continuation, and $\boldsymbol{I} = [\delta_{kk'}] = [\delta_{mn,\, m'n'}]$ is the unit matrix. By inspection, we see that the associated cf's here are

$$F_1(i\xi|H_0) = \frac{e^{-i\xi A_J}}{[\det(\boldsymbol{I} - i\xi a\boldsymbol{L})]^{\gamma}}; \quad F_1(i\xi|H_1) = \frac{e^{-i\xi A_J + H_1(i\xi)}}{[\det(\boldsymbol{I} - i\xi a\boldsymbol{L})]^{\gamma}}, \tag{A1.3a}$$

with $H_1(0) = 0$, so that $F_1(0|H_{0,1})$, implying $\int_{\infty} w_1(y|H_0, H_1)dy = 1$. In addition, $F_1(i\xi)$ is assumed to be suitably continuous, so that its Fourier transform is everywhere nonnegative, that is, the resulting pdf values $w_1(y|H_0, H_1)$ are everywhere nonnegative, as required for a proper pdf[1]. The cf's here are obtained for *Gaussian* fields.

This appendix is organized as follows:

Section A1.1 Diagonalization of $\det(\boldsymbol{I}+\gamma\boldsymbol{L})$ for *nonseparable space–time fields*; eigenvalue methods for discrete and continuous sampling; equivalent trace methods for discrete and continuous sampling.

Section A1.2 Extension of the results of Section A1.1 to *separable space–time fields*, including pertinent elements of matrix and Kronecker product algebras, for discrete and continuously sampled fields.

Thus, the initial task for evaluating the expressions (A1.3) is the reduction of $\det(\boldsymbol{I}+\gamma\boldsymbol{L})$ to diagonal form $\det(\boldsymbol{I}+\gamma\boldsymbol{L}) = \prod_{k=1}^{J}(1+\gamma\lambda_k^{(J)})$, where $\lambda_k^{(J)}, k = 1,\ldots,J = MN$ are the eigenvalues, obtained in turn from the determinantal equation $\det(\boldsymbol{L}-\lambda^{(J)}\boldsymbol{I})$. Contour integration of the result in Eq. (A1.3) yields the desired pdf values. (See also the test and Problems in Chapter 17, [1], as well as the footnote on p. 724, *ibid.)*

A1.1 DIAGONALIZATION OF det($I + \gamma L$) FOR NONSEPARABLE FIELDS

We begin here with the nonseparable space–time cases $L(\boldsymbol{R}_m,t_n;\boldsymbol{R}_{m'},t_{n'}) \neq A(\boldsymbol{R}_m,\boldsymbol{R}_{m'})$ $B(t_n,t_{n'})$, where it is convenient not to use the "condensed" indexing (k) for $\boldsymbol{L} = [L_{kk'}]$, k isomorphic to $j = mn$, refer to Eq. (A1.2). In this section, we present two methods for achieving the desired reduction. The first requires the eigenvalues (and associated eigenvectors) of a certain class of homogeneous integral equation and is exact for all γ. The second yields useful approximate expressions that in many instance avoid the calculation of eigenvalues needed in the first method. This second method is well suited to problems of weak signal reception, where the often needed averages over the random signal parameters can then be easily carried out, and the moment of the random variable y, (A1.1) (when γ is set equal to 0 eventually), can be directly calculated.

A1.1.1 The Eigenvalue Method: Discrete Sampling

Using the condensed index k, refer to Eq. (A1.2), we assume initially for the various constituents of $\det(\boldsymbol{I}+\gamma\boldsymbol{L})$ that

$$\left.\begin{array}{ll}(1) & \gamma \text{ is in general a complex quantity } (0 \leq |\gamma| \leq \infty)\\ (2) & \boldsymbol{I} = [\delta_{jj'}] = [\delta_{mn,m'n'}] = [\delta_{kk'}], \text{ by (A1.2)}\\ (3) & \boldsymbol{L} \text{ is a } (J\times J),\ J = MN = K \text{ matrix all of whose eigenvalues are distinct}\\ & \text{and all of whose elements are real quantities. } \boldsymbol{L}\text{, however, is not necessarily}\\ & \text{symmetrical}\end{array}\right\}. \tag{A1.3b}$$

[1] See (2) of Section 3.2 of Ref. [1] and footnote therein.

From (3) in Eq. (A1.3b), we can always find a $(J \times J)$ matrix $\boldsymbol{Q}\left(=\left[Q_{jj'}\right]\right)$, which diagonalizes $\boldsymbol{L}$ by means of the similarity transformation:

$$\mathbf{Q}^{-1}\mathbf{L}\mathbf{Q} = \mathbf{L} = \left[\lambda_k^{(J)}\delta_{kl}\right], \quad or \quad \mathbf{LQ} = \mathbf{QL}, \tag{A1.4}$$

where $\lambda_k^{(J)}(k = 1, 2, \ldots, J)$ are the J real, distinct *eigenvalues*[2] of $\boldsymbol{L}$, and δ_{kl} is the usual Kronecker delta $\delta_{kl} = 1; = 0, k \neq l$. The *eigenvectors* of $\boldsymbol{L}$ are formed from the rows (i) or columns (j) of $\boldsymbol{Q}$ and are found from the J linearly independent relations, expressed by the matrix–vector product

$$\boldsymbol{L}\boldsymbol{f}_l = \lambda_l^{(J)}\boldsymbol{f}_l, \quad l = 1, 2, \ldots, J, \quad \text{with} \quad \boldsymbol{f}_l = \left[Q_j\right]_l \quad \text{or} \quad \left[\boldsymbol{Q}_i\right]_l \tag{A1.5a}$$

subject to the orthonormality conditions

$$\tilde{\boldsymbol{f}}_k\boldsymbol{f}_l = \delta_{kl}, \quad k, l = 1, \ldots, J, \tag{A1.5b}$$

after the eigenvalues have been determined from the secular equation

$$\det\left(\boldsymbol{L} - \lambda^{(J)}\boldsymbol{I}\right) = 0. \tag{A1.5c}$$

For matrices satisfying (3) in Eq. (A1.3b), we can use the fact that $\det \boldsymbol{AB} = \det\boldsymbol{A} \cdot \det\boldsymbol{B}$ (for square matrices) and apply (A1.4) to $\det(\boldsymbol{I} + \gamma\boldsymbol{L})$, to get

$$\begin{aligned}\det(\boldsymbol{I} + \gamma\boldsymbol{L}) &= \det\left(\boldsymbol{Q}_J^{-1}\boldsymbol{I}\boldsymbol{Q} + \boldsymbol{Q}^{-1}\gamma\boldsymbol{L}\boldsymbol{Q}\right) = \det(\boldsymbol{I} + \mathbf{L}_L) \\ &= \prod_{k=1}^{J}\left(1 + \gamma\lambda_k^{(J)}\right).\end{aligned} \tag{A1.6}$$

This is the factored or reduced form of the $J = MN$-th degree polynomial in γ represented by $\det(\boldsymbol{I} + \gamma\boldsymbol{L})$. Note that when $\boldsymbol{L}$ is symmetrical, $L_{jj'} = L_{j'j}$ as well as real,[3] Q can be an orthogonal matrix and we can relax the constraint above on $\boldsymbol{L}$ so that all its eigenvalues be distinct. However, if $\boldsymbol{L}$ is not symmetrical, we must reimpose this constraint.[4] However, for the physical processes and fields considered here, this is not a serious restriction.

[2] The numbering (k) of the eigenvalues is distinct from the numbering of the elements of $\boldsymbol{L}$: the former are usually numbered in decreasing order of magnitude for eventual use in numerical evaluations of integrals like (A1.2).

[3] If $\boldsymbol{L}$ is complex, the argument proceeds as above, except that now $\boldsymbol{Q}$ and $\boldsymbol{f}$ have complex elements and the orthogonality condition (A1.5b) becomes $\tilde{\boldsymbol{f}}_k^*\boldsymbol{f}_l$, and so on. Then, if $\boldsymbol{L}$ is Hermitian, that is, $L_{jj'} = L_{j'j}^*$, $\boldsymbol{Q}$ can be a unitary matrix.

[4] Even if the eigenvalues of $\boldsymbol{L}$ are not distinct and $\boldsymbol{L}$ is not symmetric, it is still possible to write

$$\prod_{k=1}^{J}\left(1 + \gamma\lambda_k^{(J)}\right)$$

for $\det(\boldsymbol{I} + \gamma\boldsymbol{L})$, but now there no longer exists a matrix $\boldsymbol{Q}$, and hence a set of eigenvectors $\boldsymbol{f}^k$ that can be used to diagonalize $\boldsymbol{L}$, since $\boldsymbol{L}$ cannot be then put into completely diagonal form.

A1.1.2 The Eigenvalue Method: Continuous Sampling

In some of our subsequent applications, continuous (or analogue) sampling is ultimately required. In passing from the continuous series to the continuous field, or process, the intervals between sampled values at $(\boldsymbol{R}_m, t_n)$ are allowed to become arbitrarily close, while at the same time the total number (J) of sampled values becomes infinite. We now distinguish *three* situations:

(i) The case where the space–time data or observation interval $\mathbf{D} = [(\boldsymbol{O}, \boldsymbol{R}), (\boldsymbol{O}, \boldsymbol{T})] \equiv [\mathbf{D}_R, \Delta T]$ remains finite and intervals between samples $\mathbf{D}\boldsymbol{R} = \boldsymbol{R}/M; \Delta T = T/N$ go to zero as $M, N \to \infty$, or equivalently, $J \to \infty$, that is, $(\mathbf{D}\boldsymbol{R}, \Delta T) \to 0$, while $\mathbf{D}$ remains finite. Thus, $(m/M)\mathbf{D}\boldsymbol{R} = \boldsymbol{R}_m^{\to \boldsymbol{R}}$, and $(n/N)\Delta t = t_n \to t$ in the respective limits $M \to \infty$, $N \to \infty$, where $R \in \Delta_{wR}$ and $t \in T$ in their respective domains.

(ii) The case where the space–time sample intervals $(\mathbf{D}\boldsymbol{R}, \Delta T) \to 0$ and the data interval itself is then allowed to become infinite, that is, $\mathbf{D} \to \infty$.

(iii) The case where the spatial portion $(O, \boldsymbol{R})$ of the observation interval $\mathbf{D}$ remains finite and discrete, while the temporal part (O, T), though finite, still requires $\mathbf{D}t \to \infty$, that is, $N \to \infty$. (This corresponds to the usual analytic treatment in applications where sensor arrays are included as point, rather than distributed processing elements. Apertures, of course, must be regarded as continuous elements over nonzero spatial regions.)

In all cases here, $\boldsymbol{L} \to L(\boldsymbol{R}_1, t_1; \boldsymbol{R}_2, t_2)$ is postulated to have suitable continuity and convergence properties. For case (iii) specifically, $\boldsymbol{L} \to L(\boldsymbol{R}_1, t_1; \boldsymbol{R}_2, t_2)$. The space–time domains of $L(\boldsymbol{R}_1, t_1; \boldsymbol{R}_2, t_2)$ are $-\infty \leq (t_1, t_2) \leq \infty$; $-\infty \leq (\boldsymbol{R}_1, \boldsymbol{R}_2) \leq \infty$, consistent with the assumed continuity and convergence properties required above to achieve the indicated continuous representation $\boldsymbol{L} \to L$.

We now turn to the determinant (A1.6) and its limiting forms as $(\mathbf{D}\boldsymbol{R}, \Delta T) \to 0$, under case (i). This determinant becomes

$$D_{\mathbf{D}}(\gamma) \equiv \lim_{J \to \infty} \det(\boldsymbol{I} + \gamma \boldsymbol{L}) = \prod_{k=1}^{\infty} \left(1 + \gamma \lambda_R^{(\infty)}\right) \tag{A1.7}$$

where $D_{\mathbf{D}}$ is called a *Fredholm determinant*. Since the eigenvalues of $\boldsymbol{L}$ are all distinct, we may employ the Hilbert theory of integral equations to write λ_k as the appropriate limiting form of the eigenvalues $\lambda_k^{(J)}$, thus symbolically:

$$\lambda_k^{(\infty)} = \lim_{J \to \infty} \left(\frac{|\mathbf{D}|}{J} \lambda_k^{(J)} \right) = \lim_{M,N \to \infty} \left(\frac{|\mathbf{D}_R|}{M} \frac{T}{N} \lambda_k^{(J)} \right), \quad |\boldsymbol{R}|, \quad T < \infty, \tag{A1.8}$$

refer to Eqs. (A1.9a) and (A1.9b). The Fredholm determinant $D_{\mathbf{D}}(\gamma)$ is absolutely convergent for all $0 \leq \mathbf{D} \leq (\boldsymbol{R}, T)$, or equivalently, $\mathbf{0} \leq (\boldsymbol{R}_1, \boldsymbol{R}2) \leq \boldsymbol{R}; 0 \leq (t_1, t_2) \leq T$, provided $\gamma L(\boldsymbol{R}_1, t_1; \boldsymbol{R}_2, t_2) \leq M_L$, where $M_L = \max|\gamma L|$ in $\mathbf{D}$. (Similar remarks apply for the infinite intervals when $\boldsymbol{R} \to \infty$ or $T \to \infty$, or both $\to \infty$. Equation (A1.7) accordingly represents the "factored" or reduced form of $D_{\mathbf{D}}$ for the continuous cases (i) and (ii), analogous to the discrete situation (A1.6).

The integral equation from which $\lambda_k^{(\infty)}$ are found is as expected the limit of the set of J simultaneous equations (A1.5a) in the discrete case. The matrix $\boldsymbol{L}$ becomes the kernel $L(\boldsymbol{R}_1, t_1; \boldsymbol{R}_2, t_2)$; the eigenvectors $\boldsymbol{f}_k$ become the eigenfunctions $\psi_k(\boldsymbol{R}, t)$, and in place of the cases (A1.5a) one gets an integral instead. Accordingly, multiplying both sides of (A1.5a) by $|\mathbf{D}|/J(=|\boldsymbol{R}|T/MN)$ and setting $t_n = n\Delta t$, $\boldsymbol{R}_m = m\mathbf{D}\boldsymbol{R}$, $f_k = \Psi_k(\boldsymbol{R}_m, t_n)$, $k, l = 1, \ldots, J$, with $(\boldsymbol{R}_m, \boldsymbol{R}_{m'}) \in |\boldsymbol{R}|$; $(t_n, t_{n'}) \in (0, T)$, one gets

$$\sum_{m,n}^{M,N} \frac{|\boldsymbol{R}|T}{MN} L(\boldsymbol{R}_m, t_n; \boldsymbol{R}_{m'}, t_{n'})\Psi_k(\boldsymbol{R}_{m'}, t_{n'}) = \lambda_k^{(J)}\Psi_k(\boldsymbol{R}_m, t_n) \cdot \frac{\boldsymbol{R}|T}{MN}, \quad \text{with } MN = J. \tag{A1.9a}$$

With the help of (A1.8), formally in the limit $J \to \infty$, (A1.9a) becomes the homogeneous Fredholm integral equation (of the second kind):

$$\int_{D_R} d\boldsymbol{R} \int_{0(-)}^{T(+)} L(\boldsymbol{R}, t; \boldsymbol{R}', t')\Psi_k(\boldsymbol{R}', t')dt' = \lambda_k^{(\infty)}\Psi_k(\boldsymbol{R}, t);$$

$$\mathbf{0} \le \boldsymbol{R} \le \mathbf{D}_{\boldsymbol{R}}; \quad 0 \le t \le T; \quad (k = 1, \ldots, \infty). \tag{A1.9b}$$

A sufficient condition that the eigenvalues λ_k be discrete and that the eigenvectors $\{\Psi_k\}$ form a complete orthonormal set (A1.6) is that the (real) kernel $L(\boldsymbol{R}, t; \boldsymbol{R}', t')$ be symmetric and positive definite on $(\mathbf{0} \le \boldsymbol{R} \le \mathbf{D}_{\boldsymbol{R}}; 0 \le t \le T)$. Usually, $L(\boldsymbol{R}, t; \boldsymbol{R}', t')$ is positive semi-definite and such that at most there may be only a finite number of negative (real) eigenvalues; all other eigenvalues are (real) and positive. If λ_k remain distinct (as well as discrete), Eq. (A1.9b) holds as well for nonsymmetric kernels. (We remember that symmetry is a sufficient condition, not a necessary one: There are nonsymmetric kernels where $\{\Psi_k\}$ form a complete orthonormal set.) A number of illustrations of this are given in Section 17.3 of Ref. [1]. The orthonormality condition (A1.5b) in either case becomes

$$\int_{\mathbf{D}_R} d\boldsymbol{R} \int_0^T \Psi_k(\boldsymbol{R}, t)\Psi_l(\boldsymbol{R}, t)d\boldsymbol{R}dt = \delta_{kl}; \quad \text{k, l} = 1, \ldots, \infty, \tag{A1.10}$$

where $\{\Psi_k\}$ form a complete orthonormal set (of weight 1) on $\mathbf{D}(=\mathbf{D}_{\boldsymbol{R}}, \Delta T)$. Another useful, sufficient condition that $\{\Psi_k\}$ form a complete set is that the kernel $\boldsymbol{L}$ be the Fourier transform of a spectral density, that is, $L(\boldsymbol{R}, t; \boldsymbol{R}', t') = L(\boldsymbol{R} - \boldsymbol{R}'; t\text{-}t') = \boldsymbol{F}^{-1}\{W(\boldsymbol{k}/2\pi f)\}$. For proofs of these statements, the reader may consult the appropriate references, in particular, Ref. [2], especially Chapters 6–8 for an extensive treatment of eigenvalues, eigenfunctions, and their governing relations and conditions. As is frequently the case in applications, note that unsymmetric kernels can occur, such as *polar kernels*, which have the form $A(\boldsymbol{R}', t')L(\boldsymbol{R} - \boldsymbol{R}'; t\text{-}t')$, $A > 0_m$. These can be handled by the methods discussed in Section 8.1 of Ref. [2]. Additional (temporal) examples are also presented in Chapter 17 of Ref. [1], pp. 727 and 728 therein and Appendix A2.2 of Ref. [2] for rational kernels.

A1.1.3 The Trace Method: Discrete Sampling

Our second method of reducing $\det(\boldsymbol{I} + \gamma\boldsymbol{L})$ to a more manageable form depends on the *trace* of the $(J \times J)$ matrix $\boldsymbol{L}$ and its higher powers. To see how this comes about, let us start with the following development of the determinant as a polynomial in γ:

$$\det(\boldsymbol{I} + \gamma \boldsymbol{L}) = \sum_{k=0}^{J} \frac{\gamma^k}{k!} D_k^{(J)}, \quad D_k^{(J)} \equiv \sum_{l_1 \dots l_k}^{J} \begin{vmatrix} L_{l_1 l_1} & L_{l_1 l_2} & \cdots & L_{l_1 l_k} \\ L_{l_{21} l_1} & L_{l_{21} l_2} & \cdots & \cdots \\ \vdots & \vdots & \vdots & \vdots \\ L_{l_k l_1} & \cdots & \cdots & L_{l_k l_k} \end{vmatrix} \qquad 1 \le k, l \le J. \tag{A1.11}$$

Evaluating the determinants $D_k^{(J)}$ shows that it is a function of the traces of $\boldsymbol{L}, \boldsymbol{L}^2, \dots, \boldsymbol{L}^k$, namely,

$$D_k^{(J)} = D_k^{(J)}\left(\mathrm{tr}\boldsymbol{L}, \dots, \mathrm{tr}\boldsymbol{L}^k\right).$$

One finds that

$$\begin{aligned} &D_0^{(J)} = 1 \quad D_1^{(J)} = \mathrm{tr}\boldsymbol{L} \quad D_2^{(J)} = \mathrm{tr}^2\boldsymbol{L} - \mathrm{tr}\boldsymbol{L}^2 \\ &D_3^{(J)} = \mathrm{tr}^3\boldsymbol{L} - 3\mathrm{tr}\boldsymbol{L} \cdot \mathrm{tr}\boldsymbol{L}^2 + 2\mathrm{tr}\boldsymbol{L}^3 \\ &D_4^{(J)} = \mathrm{tr}^4\boldsymbol{L} - 6\mathrm{tr}^2\boldsymbol{L} \cdot \mathrm{tr}\boldsymbol{L}^2 + 3\mathrm{tr}^2\boldsymbol{L}^2 + 8\mathrm{tr}\boldsymbol{L} \cdot \mathrm{tr}\boldsymbol{L}^3 - 6\mathrm{tr}\boldsymbol{L}^4, \text{ and so on.} \end{aligned} \tag{A1.12}$$

From (A1.11 and A1.12), we can readily establish the following identity, which is basic to the trace method for reducing $\det(\boldsymbol{I} + \gamma \boldsymbol{L})$ to more manageable forms. This identity is specifically

$$\det(\boldsymbol{I} + \gamma \boldsymbol{L}) = \exp\left[\sum_{m=1}^{\infty} \frac{(-1)^{m+1}}{m} \gamma^m \mathrm{tr}(\boldsymbol{L}^m)\right], \tag{A1.13}$$

which holds whenever the exponential series converges (see the footnote following Eq. A1.14). Proof of (A1.13) can be established in several ways: (1) a direct method uses (A1.11): both members of (A1.13) are developed in a power series $(k = m)$ in γ, the coefficients of γ^m are compared with the observation that all coefficients of γ^m are identically zero when $m > J$. For $m \le J$, the result is simply the expansion of the determinant (A1.11), with $D_k^{(J)}$, Eq. (A1.12), the coefficient of $\gamma^{k=m}$. This is much simpler in practice than a direct evaluation of $D_{(k=m)}^{(J)}$ in (A1.11), but does not directly indicate the condition of γ for which the series converges. A more satisfactory approach (2), which yields the conditions on γ for convergence of (A1.13), starts with the eigenvalue (A1.6) and employs the well-known result for any (square) matrix $\boldsymbol{L}$[5]

$$\sum_{k=1}^{J} \lambda_k^{(J)^m} = \mathrm{tr}\boldsymbol{L}^m, \quad m \ge 0. \tag{A1.14}$$

[5] See a matrix algebra text.

The details of the proof are given on pp. 7.29 and 7.30 of Ref. [1].[6] The trace method is particularly useful for obtaining large sample, weak signal versions of the pdf values $w_J(\boldsymbol{x}-\boldsymbol{s})_N, w_J(\boldsymbol{x})_N$ in the Gauss noise cases because here only the first two or three $D_1^{(J)}, \ldots, D_3^{(J)}$ are usually needed; refer to Section 17.2 of Ref. [1], and closing remarks here below.

A1.1.4 The Trace Method: Continuous Sampling [1]

In the continuous cases, we proceed along the lines of (ii) above (cf. (A1.8) et seq.), multiplying both sides of (A1.14) by $(\mathbf{D}_R/M)(T/N)$ and passing to their respective limits, to obtain formally

$$\sum_{k=1}^{\aleph} \lambda_k^{(\infty)m} = \lim_{MN\to\infty} \left[\left(\frac{\mathbf{D}_R T}{MN}\right) \mathrm{tr}\boldsymbol{L}^m\right] = \lim_{J\to\infty}\left[\sum_{l_1,\ldots,l_m}^{J} L_{l_1l_2}L_{l_2l_3}\cdots L_{l_ml_1}(D/J)^m\right]$$

$$= \int_{\mathbf{D}_R} d\boldsymbol{r}\cdots d\boldsymbol{r}_m \int_0^T L(\boldsymbol{R}_1,t_1;\boldsymbol{R}_2,t_2)L(\boldsymbol{R}_2,t_2;\boldsymbol{R}_3,t_3)\cdots L(\boldsymbol{R}_m,t_m;\boldsymbol{R}_1,t_1)dt_1\cdots dt_m$$

$$\equiv B_M^{(\mathbf{D}_R T)} \tag{A1.15a}$$

or with $z \equiv (\boldsymbol{R}, t)$,

$$\sum_{k=1}^{\infty} \lambda_k^{(\infty)m} = \int_{\mathbf{D}} d\boldsymbol{z}_1\cdots d\boldsymbol{z}_m L(z_1,z_2)L(z_2,z_3)\cdots L(z_m,z_1) \equiv B_m^{(\mathbf{D})}, \tag{A1.15b}$$

again where $\boldsymbol{L}$ is $(q_o i_o)$ on $\mathbf{D}$, refer to Eq. (A1.9c). $\mathbf{B}_m^{(\mathbf{D})}$ are the *iterated kernels* of the basic integral equation (A1.9b). Specifically, for $m = 1, 2$, these are

$$\mathbf{B}_1^{(\mathbf{D})} = \int_{\mathbf{D}} L(z_1,z_1)d\boldsymbol{z}_1 = \sum_{k=1}^{\infty}\lambda_k^{(\infty)}; \tag{A1.16a}$$

$$\mathbf{B}_2^{(\mathbf{D})} = \int_{\mathbf{D}}\int_{\mathbf{D}} L(z_1,z_2)L(z_2,z_1)d\boldsymbol{z}_1 d\boldsymbol{z}_2 = \sum_{k=1}^{\infty}\lambda_k^{(\infty)^2}, \text{ and so on,} \tag{A1.16b}$$

(where $\boldsymbol{L}$ need not be symmetrical). The fundamental identity (A1.13) now becomes with the help of (A1.6) into (A1.7) and the above (A1.15a) and (A1.15b)

$$D_{\mathbf{D}}(\gamma) = \prod_{k=1}^{\infty}\left(1+\gamma\lambda_k^{(\infty)}\right) = \exp\left[\sum_{m=1}^{\infty}\frac{(-1)^{m+1}}{m}\gamma^m \mathbf{B}_m^{(\mathbf{D})}\right]. \tag{A1.17}$$

[6] The region of convergence in the complex γ-plane is determined solely by the *largest* eigenvalue of $\boldsymbol{L}$ (in absolute magnitude) and is a circle of radius $m \le \left|\lambda_1^{(J)}\right|^{-1}$. The left-hand member of Eq. (A1.13) is, of course, an entire function of γ, defined for all $|\gamma| < \infty$, and represents, in effect, the analytic continuation of the right-hand member outside the circle of convergence $\left|\lambda_1^{(J)}\right|^{-1}$.

The continuous analogue of Eq. (A1.11) is simply the power series in γ:

$$D_{\boldsymbol{D}}(\gamma) = \sum_{k=0}^{\infty} \frac{\gamma^k}{k!} D_k^{(\infty)}\left(\mathbf{B}_1^{(\boldsymbol{D})}, \ldots, \mathbf{B}_k^{(\boldsymbol{D})}\right) \tag{A1.17a}$$

where $D_k^{(\infty)}$ is given by Eq. (A1.12), with the various traces replaced by the appropriate iterated kernels $B_k^{(\infty)}$(A1.15a) and (A1.15b), namely,

$$D_0^{(\infty)} = 1;\quad D_1^{(\infty)} = \mathbf{B}_1^{(\boldsymbol{D})};\quad D_2^{(\infty)} = \mathbf{B}_1^{(\boldsymbol{D})^2} - \mathbf{B}_2^{(\boldsymbol{D})};\quad D_3^{(\infty)} = \mathbf{B}_1^{(\boldsymbol{D})^3} - 3\mathbf{B}_1^{(\boldsymbol{D})}\mathbf{B}_2^{(\boldsymbol{D})} + 2\mathbf{B}_3^{(\boldsymbol{D})}. \tag{A1.17b}$$

We remark that (A.15a) and (A.15b) and the first equation in the identity (A1.17) are valid for all γ, since the Fredholm determinant is absolutely and permanently convergent for all the kernels postulated here.[7]

The practical importance of the trace method is that it often permits an evaluation of the Fredholm determinant, and integrals depending upon it, without having to solve the associated homogeneous integral equation. In place of the eigenvalues λ_k and eigenfunctions (Ψ_k), we have instead to calculate the iterated kernels of Eq. (A1.15) and then use the fundamental identity (A1.17). For all γ within the circle of convergence,[8] this is exact; for γ outside, we are usually led to asymptotic expressions. The chief utility of this approach occurs in those situations where the principal contributions occur for γ within the integral of convergence and where only the first few iterated kernels are then significant. The method, in any case, is often well suited in these Gaussian regimes to the evaluation of the various integrals that arise in threshold detection theory and in a variety of weak signal estimation problems. (See Sections 17.2.1 and 17.2.2; also see Sections 20.3, 21.3.2 and 21.3.3 of earlier work [1].)

The eigenvalue approach, on the other hand, is exact (as far as the Fredholm determinant is concerned), but frequently requires extensive calculations since the set of eigenvalues $\{\lambda_k\}$ may often form a but slowly converging series.[9] Moreover, subsequent operations (like the integrations (A1.1) in the discrete cases, for example) for the most part cannot be carried out without making use of the approximations inherent in the trace method itself, so that from a practical viewpoint there is then no ultimate advantage of the latter approach over the former. Similar remarks apply as well for the discrete situation: the discrete form (A1.13) of the identity (A1.17) is usually more convenient than the eigenvalue representation.

Finally, it should be noted that the results of Section A1.1 (and Section A1.2 ff.) are not restricted to Gauss processes and fields. They apply equally well to *non-Gaussian* fields and processes (with the same conditions on $\boldsymbol{L}$ and L). However, one cannot expect their use to be similar in the non-Gaussian noise regimes.

[7] When L is symmetrical, these relations can be established alternatively with the aid of Mercer's theorem, refer to Eq. 8.61 of Ref. [1].

[8] See footnote 6.

[9] In the purely temporal cases, the Fredholm determinants, in special situations for specific kernels $L \to G(t, u)$, can be expressed in closed form: see Eq. (17.43a) of Ref. [1].

A1.2 DIAGONALIZATION OF det$(I + \gamma L)$ FOR SEPARABLE SPACE–TIME FIELDS: KRONECKER PRODUCTS

In contrast to the general cases of Section A1.1 where $\boldsymbol{L}$, usually a generic covariance function of the field in question, is not space–time separable, we now consider the separable cases. Physical examples arise in the modeling of noise environments in which reception takes place when the medium supports a homogeneous noise field, at least in the domain of the receiver's sensor elements. For the important case of *preformed beams* $(M = 1)$, separability is clearly not an issue in this regard. On the other hand, an accompanying signal field is usually not separable, mainly because of the field's nonuniform wave fronts vis-à-vis the receiving sensor elements, unless the signal source is sufficiently distant. Inhomogeneity of the input field $\{X(R_m, t_n)\}$, in any case, is the principal enemy of separability.

Space–time separability, denoted in brief by $S \otimes T$, can be compactly and conveniently handled by elements of *Kronecker matrix product algebra*, in conjunction with the usual matrix definitions and techniques. Accordingly, we let $(M \times M)$ matrix $\boldsymbol{A} = [A_{mm'}]$ represent the purely spatial portion of $\boldsymbol{L}$, associated with the field sampled by the receiving array or aperture, at positions $\boldsymbol{R}_m, \boldsymbol{R}_{m'}$ in space. Similarly, we designate $(N \times N)$ matrix $\boldsymbol{B} = [B_{nn'}]$ for the temporally sampled outputs of the sensor elements, at times $t_n, t_{n'}$. We can then express $\boldsymbol{L} = [L_{jj'}] = [L_{mn,m'n'}]$, $(m, n = 1, \ldots, M, N)$ in the separable form

$$S \otimes T: \qquad \boldsymbol{L} = [L_{mn,m'n'}] = \boldsymbol{A} \otimes \boldsymbol{B} = [A_{mm'}\boldsymbol{B}] = [A_{mm'}[B_{nn'}]], \tag{A1.18}$$

where, in more detail,

$$\boldsymbol{A} \otimes \boldsymbol{B} = \begin{bmatrix} a_{11}\begin{bmatrix} b_{11} b_{12} \cdots b_{1n} \\ b_{21} \quad \ddots \quad b_{nn} \end{bmatrix} & a_{12}\,[B_{nn'}] & \cdots & a_{1M}\,[B_{nn'}] \\ & & \ddots & \vdots \\ a_{M1}\,\boldsymbol{B} & \cdots & & a_{NM}\,\boldsymbol{B} \end{bmatrix} \equiv [C_{mm',nn'}], \tag{A1.18a}$$

which defines $\boldsymbol{C}$. Specifically, we have $A_{mm'} = A(\boldsymbol{R}_m, \boldsymbol{R}_{m'})$ and $B^{nn'} = B(t_n, t_{n'})$, so that

$$L(\boldsymbol{R}_m, t_n; \boldsymbol{R}_{m'}, t_{n'}) = A(\boldsymbol{R}_m, \boldsymbol{R}_{m'})B(t_n, t_{n'}). \tag{A.18b}$$

The product matrix, often called the *direct product* of $\boldsymbol{A}$ and $\boldsymbol{B}$, is also known as the *Kronecker product matrix* $\boldsymbol{C}$ of $\boldsymbol{A}$ and $\boldsymbol{B}$. It is a square matrix $(J \times J = MN \times MN)$, which possesses an inverse, provided det$\boldsymbol{A}$, det$\boldsymbol{B} \neq 0$. Note that $[L_{jj'}] = [L_{mn,m'n'}]$ does not generally imply $\boldsymbol{L} = \boldsymbol{A} \otimes \boldsymbol{B} = \boldsymbol{C}$. This is true only if $\boldsymbol{L} \equiv \boldsymbol{A} \otimes \boldsymbol{B}$, that is, *space and time variability are separable*, refer to Eq. (A1.18b).

A1.2.1 Elements of Kronecker Product Algebra

We summarize, mostly without proof,[10] a number of useful results involving Kronecker product matrices. We shall need them in the reduction of the generic determinant

[10] Proof is left to the reader here.

$\det(\boldsymbol{I}+\gamma\boldsymbol{L})$ into separable and diagonalized space–time components. The following direct product relations can be demonstrated,[10] given that $\boldsymbol{A}$ and $\boldsymbol{B}$ are postulated to be respectively $(M\times M)$ and $(N\times N)$ square matrices, with finite inverses, such that $\boldsymbol{C}=\boldsymbol{A}\otimes\boldsymbol{B}=\left[C_{mm',nn'}\right]$. We have the following:

(I) *The Associative Law*: $\boldsymbol{A}\otimes\boldsymbol{B}\otimes\boldsymbol{C}=(\boldsymbol{A}\otimes\boldsymbol{B})\otimes\boldsymbol{C}=\boldsymbol{A}\otimes(\boldsymbol{B}\otimes\boldsymbol{C})$ (A1.19a)

(II) *The Commutative Law*: $(\boldsymbol{A}\otimes\boldsymbol{B})(\boldsymbol{C}\otimes\boldsymbol{D})=(\boldsymbol{AC})\otimes(\boldsymbol{BD})$ (A1.19b)

(III) *The Distributive Law*: $(\boldsymbol{A}+\boldsymbol{B})\otimes(\boldsymbol{C}+\boldsymbol{D})=\boldsymbol{A}\otimes\boldsymbol{C}+\boldsymbol{A}\otimes\boldsymbol{D}+\boldsymbol{B}\otimes\boldsymbol{C}+\boldsymbol{B}\otimes\boldsymbol{D}$ (A1.19c)

(*where* $\boldsymbol{A},\boldsymbol{B}$ *are of the same or smaller order than* $\boldsymbol{C},\boldsymbol{D}$; $\boldsymbol{A},\boldsymbol{B},\boldsymbol{C},\boldsymbol{D}$ *are not necessarily of the same order*).

In addition, one can readily show that

$$\mathrm{tr}\boldsymbol{A}\otimes\boldsymbol{B}=(\mathrm{tr}\boldsymbol{A})(\mathrm{tr}\boldsymbol{B});\quad \det\boldsymbol{A}\otimes\boldsymbol{B}=\det\left(\boldsymbol{A}\boldsymbol{B}^{M}\right)=(\det\boldsymbol{A})(\det\boldsymbol{B})^{M}. \tag{A1.20a}$$

The unit matrices associated with $\boldsymbol{A},\boldsymbol{B}$, and $\boldsymbol{C}\,(\equiv\boldsymbol{A}\otimes\boldsymbol{B})$ here are

$$\left.\begin{aligned} &\boldsymbol{I}_A=\mathbf{d}_A=[\delta_{mm'}];\quad \boldsymbol{I}_B=\mathbf{d}_B=[\delta_{nn'}]\\ \therefore\ &\boldsymbol{I}_C=\boldsymbol{I}_A\times\boldsymbol{I}_B=\mathbf{d}_C=\mathbf{d}_{AB}=\left[\delta_{jj'}\right]=\left[\delta_{mn,m'n'}\right]=\left[\delta_{mm'}\cdot\delta_{nn'}\right]\end{aligned}\right\}, \tag{A1.20b}$$

this last obtained on comparing the direct (simple index) renumeration of $j(=mn)\rightarrow k$, refer to remarks associated with Eq. (A1.2), where in our applications (m) and (n) index "space" and "time," respectively, as used generally throughout the book. From Eqs. (A1.20a) and (A1.20b) we see directly that when $\boldsymbol{C}=\boldsymbol{A}\otimes\boldsymbol{B}$, then

$$\boldsymbol{I}_A\otimes\boldsymbol{B}=\begin{bmatrix}\boldsymbol{B} & \boldsymbol{O}\\ \boldsymbol{O} & \boldsymbol{B}\end{bmatrix}_{(M\times M)};\quad \boldsymbol{I}_B\otimes\boldsymbol{B}=\begin{bmatrix}\boldsymbol{B} & \boldsymbol{O}\\ \boldsymbol{O} & \boldsymbol{B}\end{bmatrix}_{(N\times N)};$$

$$\therefore\ \boldsymbol{I}_C\otimes\boldsymbol{B}=\begin{bmatrix}\boldsymbol{B} & \boldsymbol{O}\\ \boldsymbol{O} & \boldsymbol{B}\end{bmatrix}_{(J\times J)};\quad J=MN. \tag{A1.21}$$

(IV) *Kronecker Product Inverses*:

It is also possible to obtain a Kronecker product inverse $\boldsymbol{C}^{-1}=(\boldsymbol{A}\otimes\boldsymbol{B})^{-1}$ from the above. We start with the definition $\boldsymbol{C}^{-1}\boldsymbol{C}\equiv\boldsymbol{I}_C=\boldsymbol{C}\boldsymbol{C}^{-1}$, which must be satisfied whether or not $\boldsymbol{C}$ is separable, refer to Eq. (A1.18) et seq., as long as $\boldsymbol{C}^{-1}=[\boldsymbol{C}^{j'j}/\det\boldsymbol{C}]$ exists. To obtain $(\boldsymbol{A}\otimes\boldsymbol{B})^{-1}$, let us begin by assuming that

$$(\boldsymbol{A}\otimes\boldsymbol{B})^{-1}=\boldsymbol{A}^{-1}\otimes\boldsymbol{B}^{-1}. \tag{A1.22}$$

This is then used in the definition $\boldsymbol{C}^{-1}\boldsymbol{C}=\boldsymbol{I}_C$ to write with the help of the commutative law (A1.19b)

$$\boldsymbol{C}^{-1}\boldsymbol{C}=\left(\boldsymbol{A}^{-1}\otimes\boldsymbol{B}^{-1}\right)(\boldsymbol{A}\otimes\boldsymbol{B})=\left(\boldsymbol{A}^{-1}\boldsymbol{A}\right)\otimes\left(\boldsymbol{B}^{-1}\boldsymbol{B}\right)=\boldsymbol{I}_A\otimes\boldsymbol{I}_B=\boldsymbol{I}_C, \tag{A1.23}$$

as required, so that (A1.22) is indeed the desired inverse. (This is to be compared with the usual matrix product inverse $(\boldsymbol{AB})^{-1} = \boldsymbol{B}^{-1}\boldsymbol{A}^{-1}$, where, of course, $\boldsymbol{A}$ and $\boldsymbol{B}$ are of the same order and $\det \boldsymbol{A},\ \det \boldsymbol{B} \neq 0$.)

One can also define *Kronecker powers* from:

$$\boldsymbol{A}^{[1]} = \boldsymbol{A}; \boldsymbol{A}^{[2]} = \boldsymbol{A} \otimes \boldsymbol{A}; \quad \boldsymbol{A}^{[k+1]} = \boldsymbol{A} \otimes \boldsymbol{A}^{[k]} = \underbrace{\boldsymbol{A} \otimes \boldsymbol{A} \otimes \cdots \otimes \boldsymbol{A}}_{\leftarrow\ k\ \rightarrow}; \quad \boldsymbol{A}^{[k+l]} = \boldsymbol{A}^{[k]} \otimes \boldsymbol{A}^{[l]}. \tag{A1.24a}$$

When $\boldsymbol{A}$ and $\boldsymbol{B}$ commute, that is, $\boldsymbol{AB} = \boldsymbol{BA}$ (and therefore $\boldsymbol{A}$ and $\boldsymbol{B}$ must be of the same order $(M \times M)$ or $(N \times N)$), one has

$$(\boldsymbol{AB})^{[k]} = \boldsymbol{A}^{[k]}\boldsymbol{B}^{[k]} = (\boldsymbol{BA})^{[k]} = \boldsymbol{B}^{[k]}\boldsymbol{A}^{[k]}. \tag{A1.24b}$$

If $\boldsymbol{A}$ and $\boldsymbol{B}$ do *not* commute, $(\boldsymbol{AB})^{[k]} \neq \boldsymbol{A}^{[k]}\boldsymbol{B}^{[k]}$ generally (never, if $k = 2$), but it is true that $(\boldsymbol{AB})^{[k]} = \boldsymbol{A}^{[k]}\boldsymbol{B}^{[k]}$, all $\boldsymbol{A}$ and $\boldsymbol{B}$.

A1.2.2 Reduction of L to Diagonal Form

The reduction of $\boldsymbol{L} \equiv \boldsymbol{A} \otimes \boldsymbol{B}$ to diagonal form parallels the procedures of *A1.1* above, refer to Eq. (A1.4) et seq., except that now we must use an appropriately separable diagonalizing (i. e., orthogonal for real matrices) or Hermitian (for complex matrices)) matrix $\boldsymbol{Q}_{S\otimes T} = \boldsymbol{Q}_S \otimes \boldsymbol{Q}_T$, itself thus separable into $(M \times M)$ space and $(N \times N)$ time components, respectively. Thus, we have to consider

$$\boldsymbol{Q}_{S\otimes T}^{-1}\boldsymbol{L}\boldsymbol{Q}_{S\otimes T} = (\boldsymbol{Q}_S \otimes \boldsymbol{Q}_T)^{-1}(\boldsymbol{L} = \boldsymbol{A} \otimes \boldsymbol{B})(\boldsymbol{Q}_S \otimes \boldsymbol{Q}_T). \tag{A1.25a}$$

With the result (A1.22) in conjunction with the commutative law (A1.19b) on the bracketed quantities below, we obtain the diagonalized results:

$$\left[\boldsymbol{Q}_S^{-1} \otimes \boldsymbol{Q}_T^{-1}\boldsymbol{A} \otimes \boldsymbol{B}\right]\boldsymbol{Q}_S \otimes \boldsymbol{Q}_T = \boldsymbol{Q}_S^{-1}\left[\boldsymbol{A} \otimes \boldsymbol{Q}_T^{-1}\boldsymbol{B}\right][\boldsymbol{Q}_S \otimes \boldsymbol{Q}_T] = \left(\boldsymbol{Q}_S^{-1}\boldsymbol{A}\boldsymbol{Q}_S\right) \otimes \left(\boldsymbol{Q}_T^{-1}\boldsymbol{B}\boldsymbol{Q}_T\right), \tag{A1.25b}$$

$$\therefore = \boldsymbol{Q}_{S\otimes T}^{-1}\boldsymbol{L}\boldsymbol{Q}_{S\otimes T} = \mathbf{L}_A \otimes \mathbf{L}_B = \mathbf{L}_C = \left[\lambda_j^{(C)}\delta_{jj'}\right] = \left[\lambda_{mn}^{(C)}\delta_{mm'}\delta_{nn'}\right] = \left[\lambda_m^{(A)}\lambda_n^{(B)}\delta_{mm'}\delta_{nn'}\right]. \tag{A.25c}$$

Conversely, we have

$$\boldsymbol{L} = \left(\boldsymbol{Q}_S\mathbf{L}_A\boldsymbol{Q}_S^{-1}\right) \otimes \left(\boldsymbol{Q}_T\mathbf{L}_B\boldsymbol{Q}_T^{-1}\right). \tag{A1.25d}$$

In the above, $\mathbf{L}_A = \mathbf{l}_A = \left[\lambda_m^{(A)}\delta_{mm'}\right]$ and $\mathbf{L}_B = \mathbf{l}_B = \left[\lambda_n^{(B)}\delta_{nn'}\right]$. The matrices determining the respective *eigenvalues* of $\boldsymbol{A}$ and $\boldsymbol{B}$ here are found from

$$\det\left(\boldsymbol{A} - \lambda^{(A)}\boldsymbol{L}_A\right) = 0, \quad \det\left(\boldsymbol{B} - \lambda^{(B)}\boldsymbol{L}_B\right) = 0, \quad \text{and} \quad \therefore \det\left(\boldsymbol{C} - \lambda^{(C)}\boldsymbol{L}_C\right) = 0 \tag{A1.25e}$$

in the usual way, refer to Eq. (A1.5c). Note from (A1.25c) that

$$\lambda_{mn}^{(C)} = \lambda_{mn}^{(A)}\lambda_{mn}^{(B)}, \quad m = 1,\ldots,M; \quad n = 1,\ldots,N. \tag{A1.26}$$

When solving these determinantal equations for the various eigenvalues, we shall index them in descending orders of magnitude, for example, $|\lambda_k| > |\lambda_{k+1}| > |\lambda_{k+2}| > \cdots; k \geq 1$, where $k = mn$, refer to Eq. (A1.2) also.

The associated *eigenvectors* $\boldsymbol{f}_k^{(C)}$ are obtained from relations like (A1.5a), namely, the rows, or columns, of $\boldsymbol{Q}_{S\otimes T} = (\boldsymbol{Q}_S \otimes \boldsymbol{Q}_T)$ and its component matrices, namely,

$$\boldsymbol{A}\boldsymbol{f}_m^{(A)} = \lambda_m^{(A)} f_m^{(A)}, \quad m = 1,\ldots,M; \quad \boldsymbol{B}\boldsymbol{f}_n^{(B)} = \lambda_n^{(B)} f_n^{(B)}, \quad n = 1,\ldots,N. \tag{A1.27a}$$

From the fact that $\lambda_j^{(C)} = \lambda_{mn}^{(C)} = \lambda_m^{(A)}\lambda_n^{(B)}$, Eq. (A1.26), we next obtain the eigenvectors $\boldsymbol{f}_j^{(C)}$, namely,

$$\boldsymbol{L}\boldsymbol{f}_j^{(C)} = \left(\lambda_m^{(A)}\lambda_n^{(B)}\right) f_{j=mn}^{A\otimes B} = \lambda_j^{(C)}\lambda_j^{(C)}, \quad j = mn,\ldots,MN. \tag{A1.27b}$$

Using (A1.27a) and (A1.27b), we see that (A1.27b) may also be written equivalently in detail:

$$\boldsymbol{L}\boldsymbol{f}_{mn}^{(C)} = \boldsymbol{A}\boldsymbol{f}_m^{(A)} \otimes \boldsymbol{B}\boldsymbol{f}_n^{(B)} = \sum_{m'} A_{mm'}\left[f^{(A)}(\boldsymbol{R}_{m'})\right]_m \otimes \sum_{n'} B_{nn'}\left[f^{(\boldsymbol{B})}(t_{n'})\right]_n = \lambda_m^{(A)}\lambda_n^{(B)} f_m^{(A)} f_n^{(B)}, \tag{A.27c}$$

where $\boldsymbol{f}$ are the indicated *eigenvectors*, formed respectively from the rows or columns of the separable diagonalizing matrices $\boldsymbol{Q}_S$ and $\boldsymbol{Q}_T$.

The corresponding orthogonality conditions are

$$f_m^{(A)} f_{m'}^{(A)} = \delta_{mm'}, \; (m, m' = 1,\ldots,M); \quad f_n^{(B)} f_{n'}^{(B)} = \delta_{nn'}, \quad (n, n' = 1,\ldots,N), \tag{A1.28a}$$

so that for $\boldsymbol{L} = \boldsymbol{A} \otimes \boldsymbol{B}$ here, we have

$$f_j^{(C)} f_{j'}^{(C)} = \delta_{jj'} = \delta_{mm}\delta_{nn'}; \quad j,j' = 11, 12,\ldots, MN = J \tag{A1.28b}$$

(see the preceding discussion in Section A1.1).

A1.2.3 The Reduction of $\det(\boldsymbol{I}_C + \gamma\boldsymbol{L})$ to Diagonal Form, $\boldsymbol{L} = \boldsymbol{A} \otimes \boldsymbol{B}$

With the results above, we can obtain the diagonalized expressions for $\det(\boldsymbol{I}_C + \gamma\boldsymbol{L})$ here. Proceeding as in (A1.6), we can write

$$\det(\boldsymbol{I}_C + \gamma\boldsymbol{L}) = \det(\boldsymbol{I}_C + \gamma\boldsymbol{A} \otimes \boldsymbol{B}) = \det\left[\boldsymbol{Q}_{S\otimes T}^{-1}\boldsymbol{L}_C\boldsymbol{Q}_{S\otimes T} + \gamma\boldsymbol{Q}_{S\otimes T}^{-1}(\boldsymbol{A} \otimes \boldsymbol{B})\boldsymbol{Q}_{S\otimes T}\right], \tag{A1.29a}$$

since $\boldsymbol{Q}_{S\otimes T}^{-1}\boldsymbol{I}_C\boldsymbol{Q}_{S\otimes T} = \boldsymbol{I}_C$, and in general $\det(\boldsymbol{F}\boldsymbol{G}) = \det \boldsymbol{F} \det \boldsymbol{G}$, ($\boldsymbol{F}, \boldsymbol{G}$ of same order, $K \times K$). But (A1.25c) applied to (A1.29a) yields the desired result in the following equivalent forms:

Separable L:

$$\det(\boldsymbol{I}_C + \gamma \boldsymbol{L}) = \det\left(\boldsymbol{I}_C + \gamma\lambda_j^{(C)}\boldsymbol{I}_\varepsilon\right) = \prod_{j=1}^{J}\left(1 + \gamma\lambda_j^{(C)}\right)$$
$$= \prod_{m=1}^{M}\prod_{n=1}^{N}\left(1 + \gamma\lambda_m^{(A)}\lambda_n^{(B)}\right), \tag{A1.29b}$$

when $\boldsymbol{L}$ is explicitly separable into space and time components. As expected, this result formally includes the nonseparable cases, Eq. (A1.6), if we omit the last relation of (A1.29b).

A1.2.4 The Trace Method: Discrete Sampling

When $\boldsymbol{L} = \boldsymbol{A} \otimes \boldsymbol{B}$, that is, $\boldsymbol{L}$ is space–time separable, the results of in Section A1.1 are readily extended with the help of (A1.26) applied to (A1.14), namely,

$$\sum_{k=1}^{J}\lambda_k^{(J)P} = \sum_{m,n}^{J=MN}\left[\lambda_m^{(A)}\lambda_n^{(B)}\right] = \text{trace}\,\boldsymbol{L}^p = \text{trace}\{(\boldsymbol{A}\otimes\boldsymbol{B})^p\}. \tag{A1.30}$$

As before, Eq. (A1.11), we have now for (A1.13) and (A1.29b), the following equivalent relations:

$$\left.\begin{aligned}\det(\boldsymbol{I}_C + \gamma\boldsymbol{L}) = \det(\boldsymbol{I}_C + \gamma(\boldsymbol{A}\otimes\boldsymbol{B})) &= \exp\left[\sum_{p=1}^{\infty}\frac{(-1)^{p+1}}{p}\gamma p\,\text{trace}(\boldsymbol{A}\otimes\boldsymbol{B})^p\right] \\ &= \sum_{p=0}^{J}\frac{\gamma^p}{p!}D_p^{(J)}, \quad J = MN \\ &= \prod_{m=1}^{M}\prod_{n=1}^{N}\left(1+\gamma\lambda_m^{(A)}\lambda_n^{(B)}\right)\end{aligned}\right\} \tag{A1.31}$$

with $D_p^{(J=MN)}$ given by (A1.12), where trace $\boldsymbol{L}^p$, $1 \le p \le J$, is provided by (A1.30) and of course $\boldsymbol{L}^p \neq (\text{trace}\,\boldsymbol{L})^p$. (The extension of the proof of Eq. (A1.13) to the exponential relation in Eq. (A1.31) is readily made along the lines of pp. 729 and 730 of Ref. [1].) Again, the trace method, using $D_p^{(J)}$'s, is often the simplest though approximate way to obtain the (first-order) pdf values of noise and signal and noise in the critical limiting cases of weak signal detection (see the discussion in the paragraph following Eq. (A1.17b)).

A1.2.5 Continuous Samplings: Eigenvalue and Trace Methods

Here we can at once use the results (A1.29b) and (A1.31) to the obvious extensions of the continuous cases for both the eigenvalue (1) and trace methods (2) developed in Section A1.1. For the former in the limits (A1.8), we get the Fredholm determinant for these separable cases and data intervals $(0, T)$, $|\boldsymbol{D}_R|$:

Continuous:

$$\mathbf{D_D}(\gamma)_{S\otimes T} = \prod_{m=1}^{\infty}\prod_{n=1}^{\infty}\left[1 + \gamma\lambda_m^{(A,\infty)}\lambda_n^{(B,\infty)}\right], \tag{A1.31a}$$

when a finite continuous aperture in $m|\boldsymbol{D}_R|$ is employed. For most of the examples discussed in the present book, involving a finite number of "point" sensor elements in the spatial interval $|\boldsymbol{D}_R|$, the limit operation $\lim T/N$ (A1.8) is only over T and the following hybrid occurs and the semidiscrete result is as follows:

Semidiscrete:
$$\boldsymbol{D}_D(\gamma)_{S\otimes T} = \prod_{m=1}^{M}\prod_{n=1}^{\infty}\left[1 + \gamma\lambda_m^{(A,\infty)}\lambda_n^{(B,\infty)}\right]. \tag{A1.31b}$$

A1.2.5.1 The Eigenvalue Approach The associated integral equations for the eigenvalues and eigenvectors are given by Eqs. (A1.9b) and (A1.10) for the various combinations of discrete and continuous forms, that is, $(S\otimes T)_{\text{cont.}}, S_{\text{discrete}}\otimes T_{\text{cont.}}$, and $(S\otimes T)_{\text{discrete}}$. Now we have the continuous version of $\boldsymbol{L}$ so that

$$L(\boldsymbol{R}, t; \boldsymbol{R}', t') = L_A(\boldsymbol{R}, \boldsymbol{R}')L_B(t, t'), \qquad \boldsymbol{R}, \boldsymbol{R}' \in \boldsymbol{D}_R;\ (t, t' \in T), \tag{A1.32}$$

so that from Eqs. (A1.9) and (A1.10) we obtain obvious extensions of the nonseparable results of Section A1.1,

(i) $(S\otimes T)_{\text{cont.}}$:
$$\int_{\boldsymbol{D}_R} L_A(\boldsymbol{R}, \boldsymbol{R}')\psi_p(\boldsymbol{R}')\int_0^T L_B(t, t')\phi_q(t')dt' = \lambda_p^{(A,\infty)}\lambda_q^{(B,\infty)}\psi_p(\boldsymbol{R})\phi_q(t), \tag{A1.32}$$

where we have written Eq. (A1.32), namely, $L = L_A L_B$, because of the postulated separability of space and time, according to the procedures of Section A1.1 for this limiting continuous case, and with the help of (A1.26). Clearly, (A1.32) embodies two separate relations:

$$\int_{\boldsymbol{D}_R} L_A(\boldsymbol{R}, \boldsymbol{R}')\psi_p(\boldsymbol{R}')d\boldsymbol{R} = \lambda_p^{(A,\infty)}\psi_p(\boldsymbol{R});$$
$$\int_0^T L_B(t, t')\phi_q(t')dt' = \lambda_q^{(B,\infty)}\phi_q(t); \quad p, q = 1, \ldots, \infty, \tag{A1.33a}$$

where $\{\psi_p\}$ are the respective eigenfunctions, which form a complete orthonormal set, obeying

$$\int_{\boldsymbol{D}_R} \psi_p(\boldsymbol{R})\psi_{p'}(\boldsymbol{R}')d\boldsymbol{R} = \delta_{pp'}; \quad \int_O^T \phi_q(t)\phi_{q'}(t')dt' = \delta_{qq'}. \tag{A1.33b}$$

Again (refer to remarks following Eq. (A1.9b)), a *sufficient* condition that the sets $\{\psi_p\}$ and $\{\phi_q\}$ are complete and orthonormal with discrete eigenvalues is that the (here real) kernels L_A, L_B be symmetric. (For a complete discussion of the various conditions on the kernels for orthonormality, and so on, see Chapters 6 and 7 of Ref. [2].)

(ii) $(S_{\text{disc}}\otimes T_{\text{cont}})$: Semidiscrete. Here the spatial portion of the $S\otimes T$ condition is discrete, representing "point" sensor elements, out of each of which the temporal process flows. Using the discrete form of space (mm'), refer to Eq. (A1.27a), in conjunction with the continuous version of time in (A1.32), we can write the determining relations for the eigenvalues and eigenvectors/eigenfunctions from $L = L_A L_B$, namely,

$$\mathbf{A}\mathbf{f}_m^{(A)} \int_0^T L_B(t, t')\phi_q(t')dt' = \lambda_m^{(A)} \mathbf{f}_m^{(A)} \lambda_q^{(B,\infty)} \phi_q(t); \quad \text{with}$$
$$m = 1, \ldots, M;\ q = 1, \ldots, \infty. \tag{A1.34}$$

This separates into the spatial portion (A1.27a) and temporal part (A1.33a), with their respective orthonormality conditions (A1.5b) and (A1.33b). Here, $\lambda_m^{(A)}$ are found as the solutions of $\det\left(\mathbf{A} - \lambda^{(A)}\mathbf{I}_A\right) = 0$ as before, refer to Eq. (A1.25d).

(iii) $(S \otimes T)_{\text{disc}}$. In this instance, both space and time are handled discretely. The results are given in Eqs. (A1.27) and (A1.28).

A1.2.5.2 The Trace Method $(\boldsymbol{S} \otimes \boldsymbol{T})$ Again using the fact that space and time are separable now, so that $L = L_A L_B$, refer to Eq. (A1.32), we find by direct extension of Section A1.1 that Eqs. (A1.15)–(A1.17) now become

$$\sum_{k=1}^{\infty} \lambda_k^{(\infty)p} = \sum_{mn}^{\infty} \lambda_m^{(A,\infty)p} \lambda_n^{(B,\infty)p} = \sum_{mn}^{\infty} \lambda_m^{(C,\infty)p} \equiv B_p^{(\mathbf{D})} = B_p^{(A,\infty)} B_p^{(B,\infty)}, \tag{A1.35}$$

The Fredholm determinate (A1.7), with (A1.17), now factors (from (A1.31a)).

REFERENCES

1. D. Middleton, *An Introduction to Statistical Communication Theory*, Classic Reissue, IEEE Press, Piscataway, NJ, 1996.
2. P. M. Morse and H. Feshbach, *Methods of Theoretical Physics*, 1st ed., Vols. I and II, McGraw-Hill, NY, 1953.

INDEX

Acoustic communications, 6
Adaptive beam, 167
Adaptive procedures, 308
Additive signals, 242
Ad hoc noise pdf, 9
Ad hoc statistics, 8
All-pass beam pattern, 470
All-pass filter, 469
Alternative hypothesis, 18
Amplitude
  coherent estimation of, 287–293
  estimation of signal amplitudes
      $[P(H_1) \leq 1]$, 350
  incoherent estimation of, 294–300
  random, 70
  spectrum for discrete samples, 102–107
Amplitude estimation, 298, 352–355
  Bayes estimators, 352
    Pdfs of $a^* p = 1(x)$, 352–354
  Bayes risk, minimum average error, 354–355
  coherent estimation, 287–293
    biased/unbiased estimates, 293
    by (real) θ filters, 291–293
    signal amplitude quadratic cost function,
        coherent estimation, 287–290
    simple cost functions, 290–291
  comprehensive theory, 287
Amplitude factor, estimation, 287
Analytic continuation, 61, 602
Aperiodic sampling, 108
Aperture/array, explicit role of, 463–465
Aperture elements, 132
Aperture functions, 134
  beam function, 116
Aperture response, 470
Aperture system function, 117
Aperture weighting function, 125
Array processing (quadratic arrays), 165–169
Arrays and finite duration sources, 527
  scalar fields, 532
  source, field, and aperture dimensions, 534
  vector fields, 533
Associated field and equivalent solutions, for
      volumes and surfaces, 478–479
  simple volume example, 480
Associative law, 611
Asymmetrical intensity spectrum, 187
Autocorrelation, 160
  discrete, 168
  of normalized signal and, 254
  of received data, 160, 167, 168
Average
  ensemble, 597
  error, 359, 372
  intensity constraint, 294
  loss, 32

Non-Gaussian Statistical Communication Theory, David Middleton.

Average (*Continued*)
loss rating, 28–30
risk, 34, 50, 57, 64, 66, 69, 258, 265, 279, 305, 315, 385, 402, 429

Background noise, 141
Bandwidths, 125, 136, 151, 154, 229
Bayes conditional probabilities, 261. *See also* Conditional probability
Bayes criterion, 216
Bayes decision rule, 32, 34, 36, 38, 48, 52, 53
Bayes detection
and estimation with weak coupling, 331
probability, 367
theory, for the cases of overlapping signal, 402
Bayes detection analysis, 199
Bayes detectors, 66, 356
Bayes error, 357–358, 359, 360, 372
actual evaluation, 358
relative, 348
Bayes estimations, 277, 279, 281, 282, 284, 285, 286, 288, 289, 298, 299, 301, 316, 356–357, 358
of amplitude, 295, 356
properties, 281
structure, 379
of waveform, 315
Bayes extraction systems, 282
calculation, 287
coherent estimation of amplitude, 287–293
decision theory formulation, 272–287
with quadratic cost function, 278–281
signal amplitude incoherent estimation, 294–300
signal estimation and analysis, 271–305
with a simple cost function, 274
waveform estimation (random fields), 300–304
Bayesian approach, 310
Bayesian decision methods, 15
Bayesian features of joint detection and estimation under uncertainty, 407
Bayesian formulation, 339
Bayesian framework, 64
Bayesian sequential detectors, 57
Bayesian statistical decision theory (BSDT), 39, 73
Bayes matched filters, 115, 138, 189, 190, 193, 196, 197, 209, 216, 217, 229, 249
bilinear and quadratic forms, 188–219
clutter and reverberation
inverse/matched filters, 227–230
coherent detection, 214–216
coherent reception, causal matched filters, 190–192
detection parameter, separated structure, 204–205
discrete integral equations solutions, 207–214
discrete matched filters in wave number–frequency domain, 223–230
example, 231–233
extensions, 200–202
Fourier transforms of discrete series, 219–222
incoherent reception
causal matched filters, 192–195
realizable matched filters, 195–198
independent beam forming and temporal processing, 230–235
matched filters and separation, 202–207
matrix reduction and evaluation, 214
narrowband incoherent detection, 216–217
narrowband signals, 233–234
remarks, 218–219, 234–235
signal-to-noise intensity ratios, 214–219
space and time separation, 210–212
space–time matched filter as optimum beam former, 225–226
test statistic, separated structure, 203–204
unnormalized covariances, 212–214
wave number as frequency functions, 219–235, 226–227
white noise, 191–192
in space and time, 205–207
Wiener–Kolmogoroff filters, 198–200
Bayes probability of error, 164
Bayes risks, 66, 256, 258, 267, 277, 278, 280, 281, 289, 290, 291, 302, 318, 324, 330, 333, 362, 371
for coherent multiple-alternative detection, 254
Bayes' rule, 392
Bayes sequential detectors, 68
Bayes space–time matched filters, 142
Bayes's theorem, 54, 328, 334
Bayes systems, 32, 35, 49, 272, 273, 282, 311
Bayes tests, 53
Beam pattern, 231
Bessel function, 158, 159, 162, 178, 186
Bethe–Salpeter equation (BSE), 540, 565, 580
feedback operational representation, 567
Betting curves, 67–68
Biased estimators, 293

Binary decisions, 40, 245, 259, 339
Binary detection, 344, 345
  and estimation, 396
  systems, 18, 245–246, 271
    region of decision, 246
Binary hypothesis classes, 41
Binary on–off detection, 396, 408
Binary sources, 543
  in V_T, 543
Binary systems coupling, 414, 416, 417
  joint $D+E$ for, 414
  multiple tracking, 417, 418
  strong coupling, 415
    estimation, 416
  weak coupling, 416, 417
Binary systems with joint uncoupled $D+E$, 407, 408
  consistency and convergence, of estimators, 413, 414
  detection, 408, 409
  uncoupled estimators
    for r intervals, with QCF, 410–412
    simple cost functions, 412–413
Binary theory, 250
Binary (on–off) uncoupled simultaneous detection and estimation sequence, 408
"Black box" approach, 74
Born approximations, 593–594, 596
Branch factor, 185
Broadband systems, 233
Bromwich contour, 121
Bulk modulus, 512

Campbell's theorem, 228
Canonical channel, 8
  propagation, noise, and signal processing in, 2
  scalar channel, 434
  schematic anatomy, 541
Canonical channel II
  "classical" operator solutions, 539–599
  equivalent representations, elementary Feynman diagrams, 580–598
    approximations, 586–594
    elementary statistics of received field, 596–598
    random channel, characterization, 594–596
    vocabulary, 581–586
  Langevin equation, higher order moments operational solutions for, 565–580, 570–572
    integral operator, structure of, 568–570
    remarks, 579–580
    second-order moments, 565–568
    strong scatter, 575–579
    transport equations, 572–574, 574–575
  random media, operational solutions-first/second-order moments, 541–564
    Dyson's equation in statistically homogeneous and stationary media, 551–560
    Gaussian case, 563–564
    mass operator, statistical structure of, 560
    operator forms, moment solutions and Dyson's equation, 543–551
    remarks, 564
    stochastic time-dependent Helmholtz equation, 560–563
  scattering in random media, 539–599
Canonical momentum density, 512
Capacity as space–time filter, 125
Cauchy's theorem, 61
Causal time-invariant filter, 202
Causal time-varying filter, 196
Central limit theorem (CLT), 169, 574
Channel phenomenon, 77
Characteristic functions
  for Bayes risk, 302
  in error probabilities and contour integration, 60
  for Gaussian fields, 602
  in homogeneous—stationary fields—finite and infinite samples, 100, 101
  for incoherent detection, 184, 185
  in probability distributions and detection probabilities, 160
  for suboptimum systems, 252
Cholesky matrices, 194, 214
Classical fixed sample theory, 345
"Classical" statistical-physics (S-P) approach, 540
Coherent detection, 176–180, 214–216
  performance, 180
  problem, 187
Coherent extraction, 21
Coherent reception, 145
  causal matched filters, 190–192
Coherent signals, 20
Collecting coefficients, 242
Column vectors, 146
Communication, 3
  decision theoretic formulation for, 272
  process, 4
  theory, 437

Commutative law, 611
Compact operators, components, 16–17
Complete class theorem, 36, 48
Concomitant probability of detection, 359
Conditional error probabilities, 341
Conditional probabilities, 52, 70
Conditional probability, 25, 68, 250, 260, 275, 367. *See also* Bayes conditional probabilities
Conditional risk, 46
Conditional unbiased estimators, 293
Conditional variance, 281
Confidence intervals, 273
Congruent matrix, 192, 194
Constant cost function, 70, 332
Constant false alarm (CFA) detector, 50
Constant false alarm rate detector (CFAR), 50
Continuous identity, 111
Continuous non-Hom-Stat fields, 95–100
Continuous random field concept, 115
Continuous sampling, 136
Continuous space–time Wiener–Khintchine relations, 91–102
  directly sampled approximation, 93–95
  extended Wiener–Khintchine theorems, 95–100
  homogeneous—stationary fields—finite and infinite samples, special case, 100–102
Convenience, 105
Convergence in mean-square (CMS), 543, 544
Conversion
  factor, 102
  of space–time field into a temporal process, 137
Correction decision, 266
Correlation detector, 192, 229, 400
Correlation function, 589, 597
Correlation receiver, 201
Cost functions, 30, 43, 255, 272, 273, 275, 278, 283, 285, 287, 302, 303, 310, 316, 324, 326, 331, 334, 337, 338, 342, 355, 385, 401, 402, 426. *See also* Simple cost functions (SCFs)
  assigned to original binary on–off problem, 421
  exponential (*See* Exponential cost function)
  quadratic (*See* Quadratic cost function (QCF))
  "rectangular," 283
  for signal, 286
Coupled detection, 326
Coupled joint detection and estimation situation, 309, 327
Coupling, 384–386
Covariances, 78, 89, 90, 100, 111, 193, 208, 212
  function, 78, 595, 596
  matrix, 78, 269, 288, 302, 378, 393
Cross-correlation receiver, 191

Data acquisition interval, 309
Data processing
  elements, 15
  systems, 282
Data space, 263
Decision curves, 66–68
Decision errors, costs, 313
Decision model, formulation, 240–241
Decision processes, 72, 73, 144, 149, 170, 178, 179, 218, 246, 247, 314, 367
  for detection, 340
  procedures, 265
  schematic diagram, 245
Decision regions, 72
Decision rules, 25–26, 240, 241, 256, 258, 266, 273, 274, 315, 328, 340
  of detector's operation, 332
  function, 34
  nonrandomized, 243
Decision space, 242
Decision theory formulation, 17, 24, 272–287
  applications, 19
  Bayes extraction
    with quadratic cost function, 278–281
    with simple cost function, 274–278
  cost functions, 283–287
  framework, 272
  nonrandomized decision rules and average risk, 272–274
    prediction/extrapolation, 274
    simple/coincidental extrapolation, 274
    smoothing/interpolation, 274
  properties, 281–283
  techniques, 272
  theoretical approach, 42
Decision vector, 311
Decoding, 22
Degradation criteria, 342
Degradation factor, 218
Degree of suboptimality, 93
Delay-line filter, 145
Delta functions, 276, 335
Density flux, 514

Detection performance parameter (DPP), 146, 156
Detection process, 21, 315, 344, 370, 372
  algorithm, 143–144, 177–179, 308
    estimation, 308
  decision, 310
  with decisions rejection, nonoverlapping signal classes, 262–267
    optimum $(K+1)$-ary decisions with rejection, 264–266
    remarks, 267
    simple cost assignment, 266
  directed estimation, 315
    with decision rejection, 314
  parameter, 146, 153, 169, 170
    separated structure, 204–205
  probabilities, 160–164, 163, 368
  problems classification, 17
  signal and hypothesis classes in, 16
  system, 65
Detector
  decision, 314, 333
  system, 57
Deterministic inhomogeneous media, 473
  propagation in, 481
Deterministic mass operators (DMO), 545, 546
  estimation, 507–508
Deterministic media, 465
Deterministic medium
  propagation model, 82, 539
Deterministic scattered field in wave number, 494–496
  commutation, and convolution, 498–499
  transform operator solutions, 496
    full spectrum with backscatter, 498
    space-frequency with backscatter, 496–497
    wave number–frequency, 497
Deterministic volume, and surface scatter, 475–478
Diagonal matrix, 193
Differential equation, 540
Diffusion mechanism, 576
Dilation strain D_{0i}, 513
Dimensional error functions, 252
Dimensionless amplitude factor, 169
Dirac delta function, 10, 328
Discrete elements, 126
Discrete filter, 192
Discrete Fourier transforms, 127, 222
Discrete integral equations, 145, 190, 203, 205, 207–214
Discrete matched filters, 168, 215, 219
  in wave number–frequency domain, 223–230
Discrete nonlinear integral equations, 195
Discrete periodic temporal sampling, 220
Discrete point sensors, 225
Discrete sampling, 8, 129, 137
  operator, 102, 136
  procedure, 112
Discrete space–time matched filters, 167
Discrete spatial sampling, 135
Disjoint cases, 239–254
  binary detection, 245–246
  detection, 240–242
  error probabilities, average risk, and system evaluation, 250–252
  geometric interpretation, 244–245
  minimization of average risk, 242–244
  simple $(K+1)$-ary detection, 248–249
  ternary detection, 246–248
Disjoint signal classes, 259
Distortion, 514
Distribution densities, 70, 272
Distributive law, 611
Doppler effects, 472
Doppler scenario, 345
Dyson equations, 540, 552, 555, 556, 557, 558, 560, 564, 565, 566, 570, 571, 572, 580
  equivalent deterministic mass operator, 545
  first-order, 584, 585, 587, 589
  second-order, 585, 586, 587, 588, 589, 590
  third- and higher order, 591
Dyson formulation, 565

Eckart filter, 228, 229
Eigenvectors, 613
Elastic medium, 510
Elastic wave propagation, in viscous media, 522–524
Electromagnetic communications, 6
Electromagnetic energy, 524
  electric displacement vector, 525
  electric field strength, 525
  electromagnetic poynting's vector, 526
  EM field equations, 525
  equations of propagation, 527
  Hertz vector potentials, 526
  total energy flux, 525
  vector Hertzian propagation, solution of, 527
Electromagnetic radiation, 513
Energy densities, 120, 509, 514
  in dissipative media, 522
  fluid, 514

Energy flux, in gases, liquids, and solids
selected models, 528–531
Energy propagation, in gases, liquids, and solids
selected models, 528–531
Engineering approach, 465, 542
inhomogeneous deterministic media, 508–509
Engineering model, for field α_H, 470
Equal nonviscous elastic media, 516–517
acoustic waves, 519–520
elastic wave propagation in ideal solid/elastic fluid medium, 520–522
ideal shearless media, 517–519
Equation of continuity, 514
Equivalent deterministic medium (EDM), 572
Equivalent Fourier transform, 128
Equivalent numbering system, 23
Error probabilities, 61, 205, 251, 260, 262
for incoherent detection, 165
Estimation process, 332, 335
under multiple hypotheses, 383
Estimation theory, 304
Estimator bias, 336
Estimator-dependent cost functions, 342
Estimator rule, 329
Euler equation, 511
Explicit decision systems, 31
Exponential cost function, 283
Extraction process, 20, 273
Extrapolation, 20, 21

Fading mechanism, 176
False alarm probability, 148, 162, 169, 186, 340, 343, 367, 368
False rejection probability, 182
Far-field approximation, 123
Far-field operation, 542
Far-field regimes, 11
Feedback formulations, 437
Feedback operational representation (FOR), 503, 505, 543
Feynman diagram equivalent (FD), 583
FOS representations, 505–507
loop iterations, 506
and solution, 503–507
Feedback operational solution (FOS), 543
FOR representations, 505–507
Feynman diagram, 581
Filter's weighting function, 129
Finite space–time sample interval, 102
Fixed-sample tests, 44
Fluctuating noise intensity, 176
Flux densities, in dissipative media, 522
Formulation, 383
for disjoint signal classes, 384
no coupling, 384
optimum decision rule, 386–389
specific cost functions, 389
strong coupling, 385–386
weak coupling, 384
Fourier representation, 83
Fourier transform (FT), 61, 79, 92, 93, 95–100, 97, 100, 109, 116, 119, 138, 150, 223, 227, 235, 552, 588
definition, 102
of discrete series, 219–222
expressions, 568
space, 220
Fraunhofer approximation, 122
Fraunhofer condition, 122
Fraunhofer constraint, 467
Fraunhofer field region, 121, 130
Fraunhofer regimes. *See* Far-field regimes
Fredholm determinant, 605, 609, 614
Frequency space, 494
Fresnel region, 124

Gaussian case, 574
Gaussian detection, 160
Gaussian distribution, 18
Gaussian environments, 601
Gaussian fields, 142, 175, 603
det$(I+\gamma L)$ for nonseparable fields, diagonalization of, 603–609
continuous samplings, eigenvalue and trace methods, 614–616
eigenvalue approach, 615–616
continuous sampling, 605–606
discrete sampling, 603–604
Kronecker product algebra elements, 610–612
reduction of det$(I_c+\gamma L)$, 614
reduction of L to diagonal form, 612–613
trace method, 616
continuous sampling, 608–609
discrete sampling, 606–608, 614
det$(I+\gamma L)$ for separable space–time fields, diagonalization of, 610–616
probability distributions for, 601–616
Gaussian incoherent narrowband detectors, 165
Gaussian model, 351
Gaussian noise fields, 78, 112, 141–235, 176, 183, 189, 197, 251

Bayes matched filters/bilinear and quadratic forms, 188–219
clutter and reverberation, inverse (Urkowitz) and (Eckart) matched filters, 227–230
coherent detection, 214–216
coherent reception, causal matched filters, 190–192
detection parameter, separated structure, 204–205
discrete integral equations solutions, 207–214
discrete matched filters in wave number–frequency domain, 223–230
extensions, 200–202
Fourier transforms of discrete series, 219–222
incoherent reception
causal matched filters, 192–195
realizable matched filters, 195–198
independent beam forming and temporal processing, 230–235
matched filters and separation in space and time I, 202–207
matrix reduction and evaluation, 214
narrowband incoherent detection, 216–217
narrowband signals, 233–234
remarks, 218–219
signal-to-noise ratios, processing gains, and minimum detectable signals, 214–219
space–time matched filter as optimum beam former, 225–226
space/time separation, 210–212
test statistic, separated structure, 203–204
unnormalized covariances, 212–214
in wave number–frequency domain, 219–235
white noise in space and time, 191–192, 205–207
Wiener–Kolmogoroff filters, 198–200
optimal detection III, slowly fluctuating noise backgrounds, 176–188
broadband signals in normal noise, incoherent detection, 183–188
coherent detection, 176–180, 187
algorithm, 177–179
detector performance, 180
Gaussian noise, optimum threshold detection in, 183–184
incoherent detection, 184–187
detector performance, 181–182
narrowband incoherent detection algorithms, 180–182
with asymmetrical intensity spectrum, 187–188
Neyman–Pearson and ideal observers, 182
optimum detection II, selected Gaussian prototypes—incoherent reception, 154–176
array processing (quadratic arrays), 165–169
incoherent detection II, deterministic narrowband signals with slow Rayleigh fading, 169–171
incoherent detection III, narrowband equivalent envelope inputs—representations, 172–176
incoherent detection, narrowband deterministic signals, 154–169
incoherent detector structures, 155–160
matched filters, 155–160
Neyman–Pearson and ideal observers, 164–165
performance probabilities, 176
distributions and detection probabilities, 160–164
spectral symmetry, 174
statistical tests, 155–160
evaluation, 174–176
optimum detection I, selected gaussian prototypes—coherent reception, 142–154
array processing II, beam forming with linear arrays, 150–154
detection algorithm, 143–144
deterministic signals in Gauss noise, 142–146
Neyman–Pearson and ideal observers, 148–150
performance, 146–150
space–time matched filter, 145–146
optimum threshold detection in, 183–184
Gaussian random processes, 95
Gaussian signals, 300
Gaussian statistics, 154, 563, 575
Gauss noise, 163
deterministic signals in, 142–146
Gauss processes, 609
Generalized cost function for K-signals, 403–406
Generalized Huygens principle, 437, 453
evaluation of surface integrals, 457–459
for field α_H, 453, 455–457

Generalized Huygens principle (*Continued*)
geometry of finite distributed source, 455
homogeneous media, and volume sources only, 461, 463
for inhomogeneous deterministic media including backscatter, 484–487
frequency dispersion, 492–493
integral equations, 487–489
time-dependent Helmholtz equation, 489–492
inhomogeneous media, 459–461
Generalized likelihood ratio (GLR), 47, 138, 141, 242, 265, 312, 313, 317, 331, 332, 388. *See also* Likelihood ratios
Generalized likelihood-ratio test (GLRT), 48
Generalized loss function, 27
Generic channel, components of, 438
Generic detection algorithm, 159
Generic deterministic channel, 437
Global intensity operator (GIO), 565
Global mass operator, 539
Green's function, 8, 86, 116, 117, 126, 465, 466
ideal medium, 441–447
measure of medium's dispersion, 453
reciprocity, 449
regularity at infinity, 448
selected, 449
diffusion equation, 451
diffusion, maximum velocity of heat absorption, 451–453
time-dependent Helmholtz equation, 450
wave equation with radiation absorption, 450, 451
wave equation with relaxation absorption, 451
space–time causality/radiation condition, 448
transient response of, 494
uniqueness, 448–449
Green's functions (GFs), 128, 437, 549, 576
Guarding, advantage, 33

Hamiltonian energy density, 512
Hankel's first exponential integral, 170
Heisenberg's uncertainty principle, 4
Helmholtz equation, 120, 563
in ideal unbounded medium, 447–448
Helmholtz field, 515
Helmholtz medium, 127, 130–132, 134, 137
Hessian matrix, 390, 391
Homogeneous isotropic dispersive unbounded media, 454
Homogeneous–stationary Wiener–Khintchine relations, 97
Homogeneous unbounded media, 437
Hom-Stat assumption, 553
Hom-Stat character, 553
Hom-Stat conditions, 82, 593
Hom-Stat fields, 208, 564
Hom-Stat properties, 552, 557
Hom-Stat situations, 91, 559
Hooke's law, 515
Hypothesis classes, 18, 69
Hypothesis testing, 271

Ideal compressible nonviscous fluid, 515
applying Hooke's law, 515
displacement potential, 516
Ideal incompressible fluid medium, 514
Ideal medium
Green's function for, 441–447
propagation of, 439–441
Ideal observer, 148, 164, 182, 262
approach, 150
detection system, 51–52
system, 51, 52
Ideal observers, 148–150
Identity matrix, 208
Incoherent broadband reception, 187
Incoherent detection, 184–187
broadband signals in normal noise, 183–188
detection II, deterministic narrowband signals with slow Rayleigh fading, 169–171
detection III, narrowband equivalent envelope inputs, 172–176
Incoherent detector, 158
performance, 181–182
structures, 155–160
Incoherent extraction, 21
Incoherent reception, 90, 150
amplitude estimation, 297
causal matched filters, 192–195
realizable matched filters, 195–198
Incoherent signals, 20
Independent beam forming and temporal processing, 230–235
Information theory, 17, 29
Inherent simplification, advantage, 99
Inhomogeneity operator, 434
Inhomogeneous media and channels, 473–475
Inhomogeneous reciprocity, 480
surfaces, 483–484
volume, 482–483

Integration, equivalent contours, 61, 62
Intensity spectrum, 96, 109
Interpolation, 20, 21
Interval estimate concept, 338–339
Invariant filter, 195
Inverse/matched filters, 229
  clutter and reverberation, 227–230
Inverse transform, 96, 103, 220
  Fourier transform, 106
Irreducible risk, 46, 47
Isotropic Helmholtz medium, 116

Joint D + E process
  for binary systems with strong and weak coupling, 414–417
  under multiple hypotheses with strong and weak coupling, 417–419
  with no coupling, 423
Joint detection and estimation $p(H_1) \leq 1^1$, 307–379, 348
  optimal estimation [$p(H_1) \leq 1^1$], 315–325
    quadratic cost function, MMSE and Bayes risk, 316–319
    simple cost functions, UMLE and Bayes Risk, 319–325
  $p \leq 1{:}H_1$ *vs.* H_0, 346–347
  under prior uncertainty, 309–315
    case 2, coupling, 314–315
    case 1, no coupling, 312–314
  signal amplitudes, examples–estimation, 350–372
    acceptance/rejection of estimator, 367–371
    amplitude estimation, 352–355
    Bayes error/relative Bayes error, $p \leq 1$, QCF, 359–365
    Bayes estimators/Bayes error, 355–358
    performance degradation, 358–367
    relative Bayes error, 358–359
    signal intensity $I_o \equiv a^2o$, 371–372
  simultaneous joint detection and estimation, 326–350
    Bayes detection and estimation with weak coupling, 331–333
    detection probabilities, 339–341
    estimator Bias($p \leq 1$), 336–338
    extensions and modifications, 342–345
    $\gamma^*_{p<1QCF}$ for weak/no coupling, 333–336
    interval estimation , $p(H_1) \leq 1$, 338–339
    limited generalization, 335–336
    performance degradation, 348
    Sherman's theorem, 334–335
    strong coupling, 326–331
    waveform estimation, 341–342
  without coupling, 312
Joint D + E under multiple hypotheses with strong and weak coupling, 417–419
  with no coupling, 423–424
  simple cost functions, 421–422
  strong coupling, 419–421
  supervised learning, 425–428
  weak coupling, 422–423
Joint distribution density, 288
Joint hypotheses, 336
Jointly optimum detection, 383
Joint probability densities, 251, 261

K-ary detection, 271
Khintchine (W–Kh) theorem, 77
Kinetic energy density, 514
Kronecker delta, 109, 275
Kronecker powers, 612
Kronecker product, 93, 212
  inverses, 611
  matrix, 610

Lagrange density, 511
Lagrange integral, 511
Lagrange multiplier, 198, 328
LAN detector, 184
Langevin equation, 539, 540, 541, 544, 560, 564, 567, 570, 576, 578, 580
  nth-order moment, 578
Langevin equations, 540, 564, 574
Langevin mass operator, 560
Langevin propagation, 575
Large-sample threshold theory, 299
Least favorable distribution, 33, 282
Lévy continuity theorem, 101
Likelihood detector, 57
Likelihood ratios, 144, 155, 179, 219, 243, 260, 261, 266, 330, 355. *See also* Generalized likelihood ratio (GLR)
Likelihood-ratio test, 52
Linear arrays, 151
  beam forming with, 150–154
Linear equations, 244
Linear time–variable filter, 10
Loève's approach, 91
Lower triangular matrix, 201
Low-frequency components, 176

Marginal probabilities, 275
Markoff assumptions, 9

Mass operators, 541
deterministic nature, 551
Matched filters, 155–160, 170, 202, 204, 205, 224, 227, 228
classes of, 214
factors, 230
response, 157
and separation in space and time I, 202–207
structure, 189
Matrix reduction and evaluation, 214
Maximum aposteriori (MAP) detectors
from Bayesian viewpoint, 53–57
test, 56
Maximum likelihood estimators, 305, 343, 345
Maximum likelihood receiver, 305
Maximum power constraint, 294, 296
Memory (τ), 438
Minimax decision process, 33
Minimax decision rule, 32
Minimax detectors, 52–53
Minimax error probabilities, 53
Minimax extractor, 302
Minimax situation, 34
Minimax systems, 52
Minimization equation, 330, 342
Minimization process, 324
Minimum detectable signal, 66
Minimum mean square error (MMSE), 308, 351, 352
estimators, 316
filter, 198, 291, 365
Monofrequentic signals, 125
Monotonic function, 65
Monotonic transformations, 59
Multimedia, 437
interactions, 500
ocean environments—three media, 502–503
one- and two-media models, 500–502
Multiple alternative detection, 18, 239–269, 267
detection with decisions rejection, nonoverlapping signal classes, 262–267
optimum $(K+1)$-ary decision with rejection, 265–266
remarks, 267
simple cost assignment, 266
disjoint cases, 239–254
binary detection, 245–246
detection, 240–242
error probabilities, average risk, and system evaluation, 250–252
geometric interpretation, 244–245
minimization of average risk, 242–244
simple (K + 1)-ary detection, 248–249
ternary detection, 246–248
overlapping hypothesis classes, 254–262
average risk minimization for overlapping hypothesis classes, 257–259
error probabilities, average and Bayes risk, and system evaluations, 260–262
reformulation, 255–257
simple $(K+1)$-ary detection, 259–260

Narrowband approximations, 156, 157, 173
Narrowband conditions, 173
Narrowband deterministic signals
incoherent detection, 154–169
Narrowband incoherent detection
algorithms, 180–182
with asymmetrical intensity spectrum, 187–188
Narrowband input signal, 135
Narrowband sensor, outputs, 85–88
Narrowband signals, 80, 90–91, 233
Narrowband systems, 81
Navier–Stokes equation, 522–524
Nearest neighbors distributions, 255
Neyman–Pearson (NP) class, 340
Neyman–Pearson detection theory, 50–51
Neyman–Pearson detectors, 50, 144, 148, 149, 164, 182
Neyman–Pearson observers, 148–150, 164–165, 182, 262
Noise covariances, 209
Noise fields, 234
components, 172
Noise processes, 44, 326
Noise theory, 4
Nonadaptive ("one-shot") systems, 309
Non-Gaussian noise, 2, 5, 74, 115, 189
regimes, 609
Non-Gaussian statistical communication theory, 8, 540
elements, 2
Non-Hom-Stat character, 110
Nonlocal propagator, 542
Nonoverlapping hypothesis, 43
Nonrandomized decision rule, 25
Normalization process, 82
Normalized sensor, 81
Normalizing factor, 108
Normal noise process, covariance matrix, 288
Nuisance parameters, 334

Null hypothesis, 18, 185
Null signals, 71, 240
Number–frequency domain, 223

On-Off test, 143
Optimal data processing algorithms, 57
Optimal detection III, slowly fluctuating noise backgrounds, 176–188
Optimum aperture function, 234
Optimum binary on–off detection, 49
Optimum decision process, 31, 48, 250
  rule for detector, 387
Optimum detection I
  selected Gaussian prototypes—coherent reception, 142–154
    performance, 146–150
    space–time matched filter, 145–146
Optimum detection II
  selected Gaussian prototypes—incoherent reception, 154–176
Optimum detection process, 198, 342
Optimum detection rules, 330
Optimum detector analysis, 199
Optimum detector structure incoherent detection, 164
Optimum estimators, 278, 301, 324, 334
  weak-signal forms, 300
Optimum extraction procedure, 274
Optimum multiple-alternative systems, 249
Optimum probability, 148
Optimum systems, 22, 249, 283
  error probabilities for, 253
Orthogonal matrix, 212
Overlapping hypothesis classes, 254–262
  average risk minimization for overlapping hypothesis classes, 257–259
  error probabilities, average and Bayes risk, and system evaluations, 260–262
  reformulation, 255–257
  simple $(K+1)$-ary detection, 259–260

Paley–Wiener criterion, extensions of, 471–473
  causality, 473
  causality condition, 471
  Doppler effects, 472
  Fourier transforms, 472
  inverse relations, for time-variable weighting function, 473
Parameter estimation, 20
Parseval's theorem, 473
Peak-value constraint, 299
Performance probabilities, 160–164, 176, 177
Periodic sampling, 107, 133, 219
Perturbation series solution (PSS), 543, 547, 549
Perturbation techniques (PSS), 544
Physical communication processes, 98
Physically-based non-Gaussian noise models, 7
Planar hypersurfaces, 244
Point estimation, 20, 271
Potential energy, 511
Principal theorems, 37–38
Probability, 340
  density functions, 24, 27, 58, 60, 147, 251, 276, 277, 279, 327, 540
  of detection *vs. a priori* probability, 370
  distributions, 160–164, 434
Propagating velocity potential, 515
Propagation classical theory, 539
Propagation equation, 120

Quadratic cost functions (QCFs), 269, 278, 279, 281, 287, 288, 290, 294–298, 300–301, 308, 338, 361, 389–392. *See also* Cost functions
  amplitude estimator under, 362
  Bayes estimation with, 316–318
  Bayes risk with, 318–319
  estimators, 333
  and Gaussian statistics, 393
  MMSE and Bayes risk, 316–319
  multistage uncoupled detection and estimation with, 412
  sample, 357
Quadratic detector, 188
Quadratic error, 363

Random (scalar) channels, 435
Randomized decision rule, 25
Random operator, 541
Random variables, 252, 261
Rayleigh fading, 235
Rayleigh distribution, 578
Rayleigh fading, 142, 170, 171
Rayleigh pdf, 169
Received-data vector, 241
Receiver operating characteristic (ROC), 49
  diagrams, 180
Receiving array element, 225
Reception process, 129–134
  binary bayes detection, 40–46
    average risk, 43
    cost assignments, 43–45
    error probabilities, 45–46
    on–off signal detection, 42

Reception process (*Continued*)
binary two-signal detection, disjoint and overlapping hypothesis classes, 69–73
disjoint signal classes, 69–70
overlapping hypothesis classes, 70–73
continuous sampling (Helmholtz medium), 130–132
definitions and principal theorems, 35–40
admissible decision rules, 35
complete class theorem, 36
optimum decision rules, properties, 35–36
prior probabilities, cost assignments, and system invariants, 38–40
discrete sampling, 132–134
optimum detection, on–off optimum processing algorithms, 46–50
logarithmic GLRT, 48
remarks on bayes optimality of GLR, 48–50
optimum detection, on–off performance measures and system comparisons, 57–68
betting curves, 67–68
decision curves and system comparisons, 66–68
error probabilities
and contour integration, 60–65
optimum systems, 58–65
suboptimum systems, 65–66
performance measures, 68
performance *vs.* sample size, 68
sufficient statistics and monotonic mapping, 59–60
receiving aperture, geometry, 130
signal detection and estimation, 15–17, 17–22
nature of data processing, 19, 20–21
nature of estimate, 20
nature of hypotheses, 18
number of signal classes to be distinguished, 18
reception problems, 21–22
signal and noise statistics, 19–20, 21
types of extraction, 20–21
situation, 22–27, 23
assumptions, space–time sampling, 22–25
decision problem, 26–27
decision rule, 25–26
generic similarity of detection and extraction, 27
special on-off optimum binary systems, 50–57
Bayesian sequential detectors, 57
ideal observer detection system, 51–52
maximum aposteriori (MAP) detectors from a Bayesian viewpoint, 53–57
minimax detectors, 52–53
Neyman–Pearson detection theory, 50–51
as statistical decision problem, 15–74
system evaluation, 27–35
Bayes systems optimization, 31–32
evaluation functions, 27–30
minimax systems optimization, 32–35
system comparisons and error probabilities, 30–31
Reception situation model, 309
Reciprocal linearity, 104
Reformulation, 255–257
Rejection procedures, 262
Relative Bayes error, 348, 364–366. *See also* Bayes error
Relative normalizing factor, 210
Risk theory, 36
application, 43

Sampling interval, 103
Sampling process, 22, 25, 107, 133, 239
Scale factor, 331
Scale-normalizing factors, 83
Scaling function, 343
Scattering theory, 7
Sensor elements, 86
apertures and arrays, 115
Sensor geometry, 152
Sensor position, 133
Sequential detection, 19
Shape factor, 278
Shear modulus, 512
Sherman's theorem, 334–335
Signal amplitude
coherent Bayes estimator, 292
incoherent estimation, 294–300
quadratic cost function, 294–298
simple cost functions SCF_1, 298–300
Signal processes, 4, 280
binary on–off signals, 396
classes, 241
detection probability, 378
estimation, decision theoretic formulation, 272
extraction system, 271
information-bearing features, 271
null, 383
overlapping signal classes, 404
parameter space Ω, 402
probability distribution, 26

Signal processors, 9
Signal-to-noise intensity ratios, 217–219
Signal-to-noise ratio, 66, 68, 147, 157, 291, 354, 370
Signal uncertainty, 352
Signal waveform, 80
Simple $(K+1)$-ary detection, 248–249, 259–260
Simple cost functions (SCFs), 55, 301–304, 308, 319–325, 336, 393–396, 406
  Bayes error, 357
  Bayes estimators, 322–324
  Bayes risk, 301–303, 319–325
  extensions, 303–304
  minimax estimators, 303
  nonstrict "simple" cost function, 324–325
    Bayes risk and estimators, 324–325
  type one (SCF_1), 285
  types of, 319–320
  UMLE, 319–325
Simple estimation, 20, 21
Single-alternative detection systems theory, 40
Smoothing and prediction theory, 303
Solenoidal vector displacement, 513
Source density function (GT), 438
Source, field, and aperture dimensions, 534
Space–time bandwidth, 136
Space–time Bayes matched filters, 8
Space–time covariance, 80
Space–time covariances, 77–139
  aperture and arrays, 115–138
  continuous space–time Wiener–Khintchine relations, 91–102
    directly sampled approximation, 93–95
    extended Wiener–Khintchine theorems, 95–100
    homogeneous—stationary fields—finite and infinite samples, special case, 100–102
  inhomogeneous and nonstationary signal and noise fields I, 78–91
  Wiener–Khintchine relations for discretely sampled random fields, 108–115
    comments, 112–115
    Hom-Stat Wiener–Khintchine theorem, 110–112
  Wiener–Khintchine relations for discrete samples in non-hom-stat situation, 102–108
    amplitude spectrum for discrete samples, 102–107
    periodic sampling, 107–108
Space–time cross-correlation, 145
Space–time diagram elements
  dictionary of, 582
Space–time discrete matched filters, 141
Space–time fields, 84, 137, 226, 351
  intensity spectra and covariance functions, 113–114
Space–time filter, 117
Space–time formulation, 7, 78
Space–time generalized channel
  temporal channel equivalent, 542
Space–time geometry, 105
Space–time matched filter, 188, 211, 292
Space–time narrowband noise, 84
Space–time processing, 87
Space–time processor, 166
Space–time sampling, 88, 274
  interval, 207
  procedure, 78
Space–time separability, 610
Space–time spectrum, 103
Spatial causality, 120
Spatial Fourier transform, 126
Spatial processing, 166, 183
Spectral symmetry, 174
Square-law rectifier, 196
Statistical communication theory (SCT), 2–5, 540
  elements, 42
  role, 2–5
  scientific method, 2–5
  scope of analysis, 5–6
Statistical decision theory (SDT), 39
  applications, 307
  methods, 40
Statistically equivalent test statistic, 163
Statistical–physical (S-P) approach, 435, 540
Statistical signal processing (SSP)
  role, 4
Statistical tests, 155–160
Steering vector, 89, 123, 131, 133, 152, 181, 227, 231
Stochastic signals, 256
Strain, 510
Stress energy tensor, 511
Strict simple cost function (SCF_2), 285, 301, 321
Strong coupling, 326–331, 338
  alternative approach, 329–331
  nonrandomized decision rules, 328–329
Strong-signals, 295, 296, 299, 300
  estimators, 304

Subjective approach, 38
Suboptimum average risk, 363
Suboptimum detectors, 214
Suboptimum estimator, 358, 361, 365
Suboptimum systems, 251, 261, 292
Superposition principle, 2
Surface elevation, 550
Symmetrical linear arrays, 221
Symmetric channel, 341
Symmetric distance function, 283
Symmetric matrix, 166, 187, 195, 201
Synthetic aperture radar (SAR), 125
Synthetic aperture sonar (SAS), 125
System function, 211

Taylor expansion, 335
Taylor series approximations, 591, 592
Temporal processing, 23
Temporal signals, 19
  waveform, 19
Ternary detection, 246–248
  regions of decision, 247, 248
Test functions, 206
Testing hypotheses, 17
Test statistics, 150, 158, 159, 160, 172, 177, 180, 181, 184, 188
  evaluation, 174–176
Threshold reception, 154, 163, 299
Threshold theory, 9
Time-invariant linear filter $h_{C,}$ 471
  equivalency conditions for, 471
Time processing, 146
Time-variable filters, 195, 470
  matched filters, 221
Time-variable operations, 192
Time-varying linear filters, 465–466
  equivalent temporal filters, 466–471
Total energy density, 514
Trace method, 616
Transmission, geometry of, 121
Transmitted signal, 134

Ultrawide-band (UWB) signals, 7
Unconditional maximum likelihood estimate (UMLE), 55, 276, 277, 278, 284, 285, 290, 298, 304, 308, 322, 333, 352, 393
Unconditional unbiased estimators, 293
Uncoupled detection and estimation, 408
  Bayesian detection, 400
  optimum detection, 400
Unnormalized covariances, 213
Variance matrix, 253
Vector fields, 7
Vector interval, geometry, 106
Velocity potential ψ_v, 514

Wald's assumptions, 37
Wald's complete class theorem, 27
Wald's fundamental theorem, 36
Wald theory of sequential tests, 26
Waveform estimation (random fields), 277, 280, 300–304
  normal noise signals in normal noise fields (*See* Quadratic cost function; Simple cost functions)
Waveform vector, 404
Wavelength–frequency relation, 124
Wave number frequency. *See also* Space-time covariances
Wave number–frequency bounds, 106
Wave number–frequency domain, 219–235
Wave number frequency spectra, 77–139
  aperture and arrays, 115–138
  continuous space–time Wiener–Khintchine relations, 91–102
    directly sampled approximation, 93–95
    extended Wiener–Khintchine theorems, 95–100
    homogeneous—stationary fields—finite and infinite samples, special case, 100–102
  inhomogeneous and nonstationary signal and noise fields I, 78–91
  Wiener–Khintchine relations for discretely sampled random fields, 108–115
    comments, 112–115
    discrete Hom-Stat Wiener–Khintchine Theorem, 110–112
  Wiener–Khintchine relations for discrete samples in non-hom-stat situation, 102–108
    amplitude spectrum for discrete samples, 102–107
    periodic sampling, 107–108
Wave number–frequency spectrum, 221
Wavenumber frequency(WNF)
  amplitude, 554
  transform, 77
  transport equation, 573
Weak coupling, 313, 338
Weak-signals, 295, 296
  estimators, 304

Weighting coefficients, 400
Weighting functions, 118, 128, 190, 199, 229. *See* Green's function
White noise, 225, 232
  cases, 191–192
Wiener-Khintchine filter, 199, 200, 292
Wiener–Khintchine relations, 78, 91, 93, 112, 568
  for discretely sampled random fields, 108–115
    comments, 112–115
    discrete Hom-Stat Wiener–Khintchine Theorem, 110–112
  for discrete samples in non-hom-stat situation, 102–108
    amplitude spectrum for discrete samples, 102–107
    periodic sampling, 107–108
Wiener–Khintchine theorem, 79, 92, 98, 99, 101, 222
Wiener–Kolmogoroff cases, 142
Wiener–Kolmogoroff filter, 214, 291, 292, 297

Zero-memory nonlinear rectifiers, 249

IEEE PRESS SERIES ON DIGITAL AND MOBILE COMMUNICATION

John B. Anderson, *Series Editor*
University of Lund

1. *Wireless Video Communications: Second to Third Generation and Beyond*
   Lajos Hanzo, Peter Cherriman, and Jurgen Streit
2. *Wireless Communications in the 2lst Century*
   Mansoor Sharif, Shigeaki Ogose, and Takeshi Hattori
3. *Introduction to WLLs: Application and Deployment for Fixed and Broadband Services*
   Raj Pandya
4. *Trellis and Turbo Coding*
   Christian Schlegel and Lance Perez
5. *Theory of Code Division Multiple Access Communication*
   Kamil Sh. Zigangirov
6. *Digital Transmission Engineering,* Second Edition
   John B. Anderson
7. *Wireless Broadband: Conflict and Convergence*
   Vern Fotheringham and Shamla Chetan
8. *Wireless LAN Radios: System Definition to Transistor Design*
   Arya Behzad
9. *Millimeter Wave Communication Systems*
   Kao-Cheng Huang and Zhaocheng Wang
10. *Channel Equalization for Wireless Communications: From Concepts to Detailed Mathematics*
    Gregory E. Bottomley
11. *Handbook of Position Location: Theory, Practice, and Advances*
    Edited by Seyed (Reza) Zekavat and R. Michael Buehrer
12. *Digital Filters: Principle and Applications with MATLAB*
    Fred J. Taylor
13. *Resource Allocation in Uplink OFDMA Wireless Systems: Optimal Solutions and Practical Implementations*
    Elias E. Yaacoub and Zaher Dawy
14. *Non-Gaussian Statistical Communication Theory*
    David Middleton

Forthcoming Titles

Frequency Stabilization: Introduction and Applications
Venceslav F. Kroupa
Fundamentals of Convolutional Coding, Second Edition
Rolf Johannesson and Kamil Zigangirov

Printed in the United States
By Bookmasters